Analyzing Spectra

Squeeze as much information as you can from the molecular formula: use chemical arithmetic, deciding where you can how many rings and/or double bonds are present. Combine this with characteristic infrared bands, δ values, proton counts, and splitting of various NMR signals to give you structural units. If the spectrum (or combination of spectra) is unambiguous, you should have only one possible structure left; go back and check this against all the information you have.

For problems on spectra, answers are presented in two stages: names of the unknown compounds are given in their proper sequence along with the other answers; then, at the end of the Study Guide, spectra are reproduced with infrared bands identified and NMR signals assigned. We suggest that you check each of your answers in two stages, too. First, check the name; if your answer is wrong, or if you have not been able to work the problem at all, return to the spectrum in the textbook and, knowing the correct structure, have another go at it: see if you can now identify bands, assign signals, and analyze spin–spin splittings. Then, finally, turn to the back of the Study Guide and check your answer against the analyzed spectrum.

Woodward–Hoffmann Rules for Electrocyclic Reactions

Number of π electrons	Reaction	Motion
$4n$	thermal	conrotatory
$4n$	photochemical	disrotatory
$4n + 2$	thermal	disrotatory
$4n + 2$	photochemical	conrotatory

Woodward–Hoffmann Rules for $[i + j]$ Cycloadditions

$i + j$	Thermal	Photochemical
$4n$	supra,antara antara,supra	supra,supra antara,antara
$4n + 2$	supra,supra antara,antara	supra,antara antara,supra

STUDY GUIDE TO

Organic Chemistry

Sixth Edition

by R. T. Morrison
and R. N. Boyd

Prentice Hall, Englewood Cliffs, New Jersey 07632

Acquisition Editor: ***Diana Farrell***
Production Editor: ***Christine Sharrock, Omega Scientific***
Design Director: ***Florence Dara Silverman***
Cover Designer: ***Bruce Kenselaar***
Prepress Buyer: ***Paula Massenaro***
Manufacturing Buyer: ***Lori Bulwin***
Editorial Assistant: ***Lynne Breitfeller***
Marketing Manager: ***Kelly Albert***

Cover photograph: A computer-generated representation of DNA as viewed looking along the double helix. (Courtesy of the Computer Graphics Laboratory, University of California, San Francisco. © Regents, University of California)

A Simon and Schuster Company
Englewood Cliffs, New Jersey 07632

Printed in the United States of America
10

ISBN 0-13-643677-3

Prentice-Hall International (UK) Limited, *London*
Prentice-Hall of Australia Pty. Limited, *Sydney*
Prentice-Hall Canada Inc., *Toronto*
Prentice-Hall Hispanoamericana, S.A., *Mexico*
Prentice-Hall of India Private Limited, *New Delhi*
Prentice-Hall of Japan, Inc., *Tokyo*
Simon & Schuster Asia Pte. Ltd., *Singapore*
Editora Prentice-Hall do Brasil, Ltda., *Rio de Janeiro*

Contents

Acknowledgments

Our thanks to Sadtler Research Laboratories for the infrared spectra labeled "Sadtler" and to the Infrared Data Committee of Japan for those labeled "IRDC", and to Dr. David Kritchevsky of the Wistar Institute for permission to quote the words of his song, "Farnesol".

To the Student

Right now, confronted with the array of unfamiliar material in your textbook, you must be wondering: *what am I expected to get out of all this?*

The best way to find out is to work problems: first, to see if you understand the facts and principles you have been reading about; second, and more important, to learn how to *use* this chemistry in the same practical ways that an organic chemist does.

Give yourself a fair chance to work each problem. Don't give up too easily. Re-read the pertinent part of the text. *Think* about it. Use paper and pencil and really work at it.

Only after you have done all that, check your answer against the one in this Study Guide. If you were on the right track, fine. If you went off the track, try to see *where*. Follow through the explanation carefully to see how you should approach this kind of problem next time.

You must *learn* how to use your brand-new organic chemistry, and to do this you must push yourself. You must try to work difficult problems, and you will not always succeed. But you can learn from your failures as well as your successes.

With some answers, we have given references to the chemical literature. It is not necessary for you to read all of these papers—or even any of them. But if some topic catches your fancy, follow it up. And, in any case, read one—or two or three—of these papers, so that you can see in down-to-earth detail the kind of experimental work that underlies any science.

Robert Thornton Morrison
Robert Neilson Boyd

Note

Reference to a page in this Study Guide will be given as "page 000 of this Study Guide". Any other reference is understood to be to a page in *Organic Chemistry, Sixth Edition*, by R. T. Morrison and R. N. Boyd.

1

Structure and Properties

1.1 Ionic: a, e, f.

In general, we expect electron transfer—ionic bonding—between atoms that are widely separated, left-to-right, in the Periodic Table: typically, between metals on the far left and non-metals on the far right. We expect electron sharing—covalent bonding—between atoms that are less widely separated: between S and O, for example, or N and C; and between H and almost any atom from groups III–VII. (See, too, the covalent compounds in Problem 1.2.)

(a) K^+ :Br:$^-$ (b) H:S:H (c) :F:N:F: with :F: above N (d) :Cl:C:Cl: with H above C and :Cl: below C

(e) Ca^{2+} :O:S:O:$^{2-}$ with :O: above and below S (f) H:N:H$^+$:Cl:$^-$ with H above and below N (g) H:P:H with H above P (h) H:C:O:H with H above and below C

1.2 (a) H:O:O:H (b) :N:::N: (c) H:O:N::O: with :O: above N (d) :O:N::O:$^-$ with :O: above N

(e) H:C:::N: (f) :O::C::O: (g) H:O:C:O:H with O: below C (h) H:C:C:H with H H above and H H below

1.3 (a)

	1s	2s	2p			3s	3p		
Na	all are like Ne					(·)	( )	( )	( )
Mg						(··)	( )	( )	( )
Al						(··)	(·)	( )	( )
Si	(··)	(··)	(··)	(··)	(··)	(··)	(·)	(·)	( )
P						(··)	(·)	(·)	(·)
S	1	8				(··)	(··)	(·)	(·)
Cl						(··)	(··)	(··)	(·)
Ar						(··)	(··)	(··)	(··)

(b) Elements of the same family have the same electronic configuration for their highest energy level.

(c) Metallic elements on the left of the Periodic Table lose electrons to give a 2,8 configuration; non-metallic elements on the right gain electrons to give a 2,8,8 configuration.

1.4 To arrive at the shape of each molecule, we see how many orbitals the central atom needs to hold not only the atoms attached to it but also any unshared pairs of electrons. If it needs four orbitals, it will use sp^3 orbitals; if three, sp^2; if two, sp. Following is given the shape in each case: first, if unshared pairs are included; and then, in parentheses, if only atomic nuclei are considered.

(a) Tetrahedral, like CH_4 in Figure 1.10, page 16. (*Tetrahedral.*)

(b) Tetrahedral, like NH_3 in Figure 1.12, page 18. (*Pyramidal, with tetrahedral angles.*)

(c) Tetrahedral, like H_2O in Figure 1.14, page 19. (*Flat, with a tetrahedral angle.*)

(d) Tetrahedral, like NH_3. (*Pyramidal, with tetrahedral angles.*)

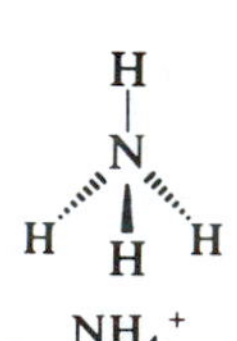
NH_4^+

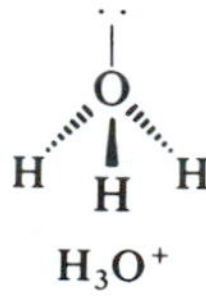
H_3O^+

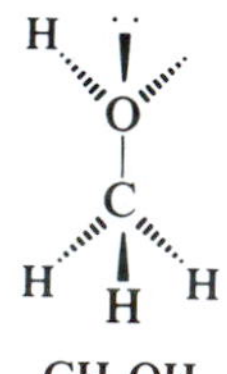
CH_3OH

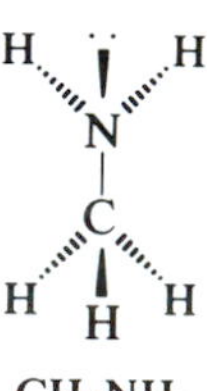
CH_3NH_2

1.5 Structure (a), not (b), since in (a) the dipoles would cancel each other out.

(a)

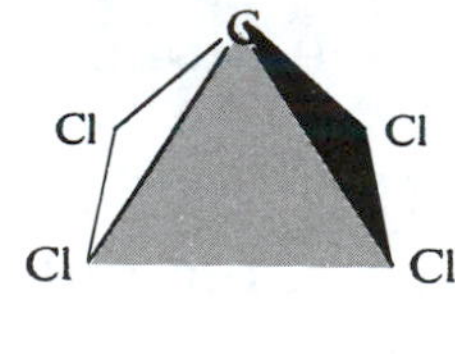
(b)

1.6 Linear. This is the only shape that would permit cancellation of the two carbon–oxygen dipoles.

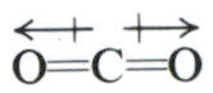

```
 <-+ +->
 O=C=O
```

As with $BeCl_2$ (Figure 1.6, page 13), this linearity stems from *sp* hybridization at the central atom; with CO_2 there is further overlap to give double bonds (Sec. 8.2).

1.7 (a) We would expect a zero dipole moment.

If nitrogen were sp^2-hybridized, the molecule would be flat and symmetrical (like BF_3, Figure 1.8, page 15), and the three bond dipoles would cancel each other out. In fact, NH_3 has $\mu = 1.46$ D. Therefore, the molecule is not flat, and not sp^2-hybridized.

(b) We would expect NF_3 to have a much larger dipole moment than NH_3.

If nitrogen used *p* orbitals for bonding, both NH_3 and NF_3 would be pyramidal: as shown on page 25 except that the bond angles would be 90° rather than tetrahedral. The line of reasoning is the same as that on pages 24–25 up to the point where the unshared pair of electrons is taken into consideration. Now the unshared pair would be in a 2*s* orbital *symmetrical about N*, and would provide no dipole to oppose the bond dipoles. Unopposed, the large N—F dipoles would give NF_3 a much larger dipole moment than NH_3. In fact, NF_3 has a much smaller dipole moment than NH_3. Therefore, the unshared pair is not in a 2*s* orbital, and the molecule is not *p*-hybridized.

1.8 Associated: d, e.

We simply draw the structure of each molecule, and look for H attached to O or N. Only if we find O—H or N—H do we expect the compound to be associated. (We see that the hydrogens of the recurring CH_3 groups cannot be involved in hydrogen bonding.)

```
        H   H                  H               H
        |   |                  |               |
(a)  H—C—O—C—H        (b)  H—C—F      (c)  H—C—Cl
        |   |                  |               |
        H   H                  H               H
```

(a) 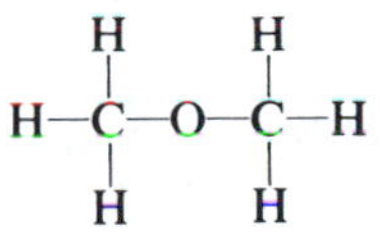 Not associated (b) 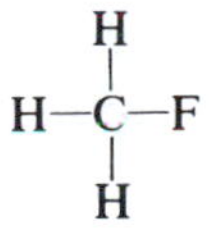 Not associated (c) 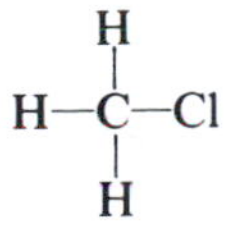 Not associated

```
       H  H     H  H                       H       H
       |  |     |  |                       |       |
(d) H—C—N---H—N—C—H    equivalent to   CH3—N---H—N—CH3
       |  |        |                       |
       H  H        H                       H
```

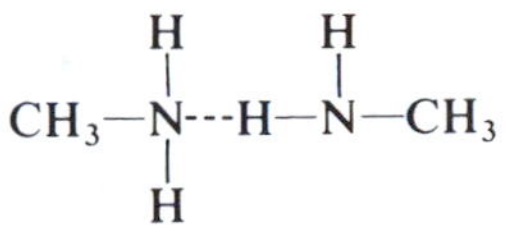

Associated

```
         H        CH3                  CH3
         |        |                    |
(e) CH3—N---H—N             (f) CH3—N
         |        |                    |
         CH3      CH3                  CH3
```

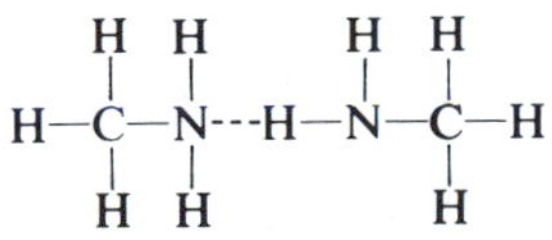

(e) 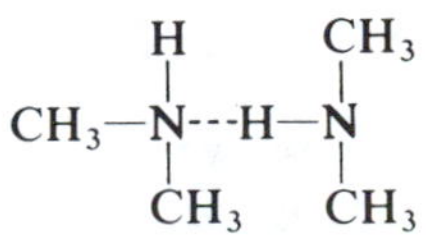 *Associated* (f) 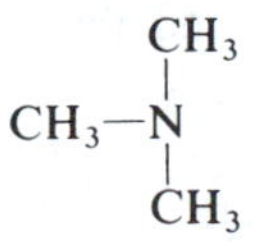 Not associated

1.9 In each case we focus our attention on the atom holding the proton to be lost, that is, the atom that will hold the unshared pair of electrons in the conjugate base being formed. *The better this atom accommodates these electrons*, the greater the extent to which the conjugate base is formed, and hence, by definition, *the stronger the conjugate acid*.

$$\underset{\substack{\text{Acid}\\ \textit{Holds proton}}}{\text{H:Z}} \quad + \quad \text{:B} \rightleftharpoons \underset{\substack{\text{Conjugate base}\\ \textit{Holds unshared}\\ \textit{pair of electrons}}}{\text{:Z}} \quad + \quad \text{H:B}$$

We shall consider two factors that determine an atom's ability to accommodate electrons:

(i) its *electronegativity*, since by definition a more electronegative atom has a greater avidity for electrons. Among atoms of the same row of the Periodic Table, electronegativity increases as we move *to the right*.

(ii) its *size*, since a bigger atom permits greater dispersal of the charge of the electrons, and this, as we shall see (Sec. 5.20), tends to stabilize a charged particle. Among atoms of the same family, size increases as we move *downward* in the Table.

(a) $CH_3NH_2 < CH_3OH$

Oxygen and nitrogen are in the same row of the Periodic Table, and oxygen is the *more electronegative*.

(b) $CH_3OH < CH_3SH$

Sulfur and oxygen are in the same family of the Periodic Table, and sulfur is the *bigger*.

(c) $NH_4^+ < H_3O^+$

Again, oxygen and nitrogen are in the same row of the Periodic Table, and oxygen is the *more electronegative*.

1.10 Drawing upon our general chemical knowledge, we arrive at the following orders of acidity. (We know, for example, that H_3O^+, not H_2O, is the strong acid in aqueous solutions of compounds like sulfuric acid.)

(a) $H_3O^+ > H_2O$

(b) $NH_4^+ > NH_3$

(c) $H_2S > HS^-$

(d) $H_2O > OH^-$

(e) We see that among closely related molecules where the same atom loses the proton,

positive charge $\longrightarrow$ increased acidity

negative charge $\longrightarrow$ decreased acidity

How do we account for this? Following the approach of the preceding problem, we consider the various conjugate bases being formed, and the ability of each to accommodate the electron pair—*negatively charged*, remember—left behind upon loss of the proton.

This accommodation is easiest for the neutral conjugate base formed from a positively charged acid,

$$\underset{Positive}{H_3O^+} + :B \rightleftarrows \underset{Neutral}{H_2O} + H:B$$

harder for the negatively charged conjugate base formed from a neutral acid,

$$\underset{Neutral}{H_2O} + :B \rightleftarrows \underset{Negative}{OH^-} + H:B$$

and still harder for the doubly charged conjugate base formed from a negatively charged acid.

$$\underset{Negative}{OH^-} + :B \rightleftarrows \underset{Doubly\ negative}{O^{2-}} + H:B$$

1.11 Again following the approach of Problem 1.9, we focus our attention on the atom holding the unshared pair of electrons in the base. *The better the atom accommodates these electrons*, the less available they are for sharing, and *the weaker the base*. An order of relative basicity is necessarily opposite to the order of relative acidity of the conjugate acids.

(a) $CH_3^- > NH_2^- > OH^- > F^-$ (b) $NH_3 > H_2O > HF$ (c) $SH^- > Cl^-$

In (a), (b), and (c) the atoms of each set are in the same row of the Periodic Table, and therefore accommodation of the electron pair depends upon electronegativity: the more electronegative the element, the weaker the base.

(d) $F^- > Cl^- > Br^- > I^-$ (e) $OH^- > SH^- > SeH^-$

In (d) and (e) the atoms of each set are in the same family of the Periodic Table, and therefore accommodation depends upon *size*: the bigger the atom, the weaker the base.

1.12 $CH_3NH_2 > CH_3OH > CH_3F$

We apply the same approach as in the preceding problem. In this set basicity varies inversely with *electronegativity*: the more electronegative the atom, the weaker the base.

1.13 We approach this problem as we did Problem 1.10, above. Again we draw upon our general chemical knowledge, and arrive at the following orders of basicity. (We know, for example, that it is OH^-, not H_2O, that makes an aqueous solution of NaOH strongly basic.)

(a) $OH^- > H_2O > H_3O^+$ (b) $NH_2^- > NH_3$ (c) $S^{2-} > HS^- > H_2S$

(d) We see that among closely related molecules where the same atom holds the unshared pair,

negative charge $\longrightarrow$ increased basicity

positive charge $\longrightarrow$ decreased basicity

To account for this, we consider the unshared pair of electrons on each base and how available they are for sharing. For a given atom, availability of electrons is clearly greatest in an electron-rich, negatively charged molecule, and least in an electron-poor, positively charged molecule.

1. Ionic: a, d, e, g.

Following the guidelines of Problem 1.1, we arrive at the following likely structures.

(a) Mg^{2+} 2 :Cl:$^-$ (b) :Cl:C:Cl: with H above and below C (c) :I:Cl: (d) Na^+ :O:Cl:$^-$

(e) K^+ :O:Cl:O:$^-$ with :O: above and below Cl (f) :Cl:Si:Cl: with :Cl: above and below Si (g) Ba^{2+} :O:S:O:$^{2-}$ with :O: above and below S (h) H:C:N:H with H above C and H H below

2. (a) H:N:N:H with H H above (b) H:O:S:O:H with :O: above and below S (c) H:O:S:O:$^-$ with :O: above and below S (d) :Cl:C::O: with :Cl: above C

(e) H:O:N::O: (f) :O:N::O:$^-$ (g) :O:C::O:$^{2-}$ with :O: above C (h) H:C::C:H with H H above

(i) H:C:::C:H (j) H:C::O: with H above C (k) H:C:O:H with O: below C (l) H:C:C:C:H with H H H above and H H H below

3. To arrive at the shape of each molecule, we follow the procedure of Problem 1.4, and see how many orbitals the central atom needs to hold both atoms and unshared pairs of electrons. Following is given the shape in each case: first, if unshared pairs are included; and then, in parentheses, if only atomic nuclei are considered.

(a) Trigonal, like BF_3 in Figure 1.8, page 15. (*Trigonal.*)

(b) Tetrahedral, like NH_3 in Figure 1.12, page 18. (*Pyramidal, with tetrahedral angles.*)

(c) Trigonal, like BF_3. (*Trigonal.*)

(d) Tetrahedral, like H_2O in Figure 1.14, page 19. (*Flat, with a tetrahedral angle.*)

(e) Tetrahedral, like H_2O. (*Flat, with a tetrahedral angle.*)

(f) Tetrahedral, like H_2O. (*Flat, with a tetrahedral angle.*)

(g) Tetrahedral, like CH_4 in Figure 1.10, page 16. (*Tetrahedral.*)

(h) Tetrahedral, like NH_3. (*Pyramidal, with tetrahedral angles.*)

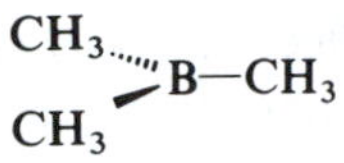

$(CH_3)_3B$ 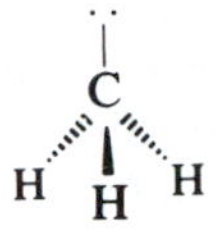 $CH_3{:}^-$ 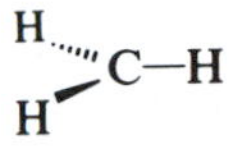 CH_3^+

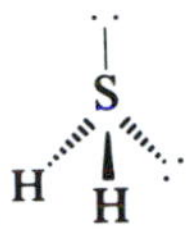

H_2S 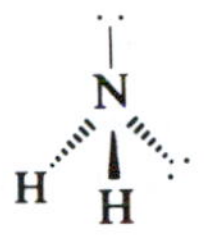 NH_2^- 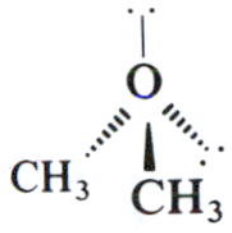 $(CH_3)_2O$

BF_4^- 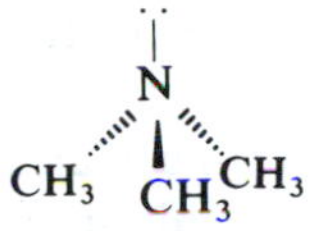 $(CH_3)_3N$

4. Octahedral, that is, two square pyramids base-to-base.

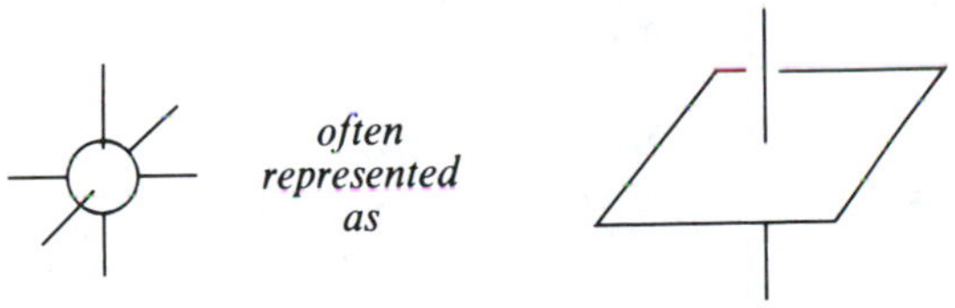

often represented as

5. (a) Toward Br (the more electronegative atom).

(b) Toward Cl (the more electronegative atom).

(c) Non-polar (cancellation).

(d) Away from the H atoms, bisecting the angle between the Cl atoms (vectors).

(e) 180° away from the C—H bond (compare Figure 1.16, page 24).

(f) Similar to the water molecule (Figure 1.17, page 25).

(g) Similar to the water molecule (Figure 1.17, page 25).

(h) Toward the unshared pair on N (Figure 1.17, page 25).

(i) Away from the Cl atoms, bisecting the angle between the F atoms.

6. (a) F is more electronegative than Cl, and attracts electrons more; *e* for HF is bigger than for HCl. This bigger *e* evidently outweighs the smaller *d*, and μ is bigger, too.

(b) Clearly, the direction of the dipole in either case is toward the highly electronegative F. Since *d* is essentially the same for the two molecules, the bigger μ for CD_3F indicates a bigger *e*. Evidently F pulls electrons from D more easily than from H; that is, D is more electron-releasing than H. Relative to the C—H bond, then, the C—D bond has a dipole from D to C.

7. The Li compound is ionic, salt-like; the Be compound is non-ionic, covalent.

8. Boiling point is raised by hydrogen bonding (association) between like molecules of a compound; solubility is increased by hydrogen bonding between solute molecules and solvent molecules. Alcohol molecules form hydrogen bonds to each other, raising the boiling point, and to water molecules, increasing solubility. Ether molecules can form hydrogen bonds only with water, which furnishes H attached to O; this increases solubility, but in the absence of association the low boiling point is "normal" for the molecular weight.

9. (a)

H_3O^+	+	HCO_3^-	$\rightleftharpoons$	H_2CO_3	+	H_2O
Stronger acid		Stronger base		Weaker acid		Weaker base

(b)

OH^-	+	HCO_3^-	$\rightleftharpoons$	CO_3^{2-}	+	H_2O
Stronger base		Stronger acid		Weaker base		Weaker acid

(c)

NH_3	+	H_3O^+	$\rightleftharpoons$	NH_4^+	+	H_2O
Stronger base		Stronger acid		Weaker acid		Weaker base

(d)

CN^-	+	H_2O	$\leftrightharpoons$	HCN	+	OH^-
Weaker base		Weaker acid		Stronger acid		Stronger base

(e)

H^-	+	H_2O	$\longrightarrow$	H_2	+	OH^-
Stronger base		Stronger acid		Weaker acid		Weaker base

(f)

C_2^{2-}	+	H_2O	$\longrightarrow$	$2\ OH^-$	+	C_2H_2
Stronger base		Stronger acid		Weaker base		Weaker acid

10. (a) H_3O^+, since

$$\underset{\text{Stronger acid}}{HCl} + H_2O \rightleftharpoons \underset{\text{Weaker acid}}{H_3O^+} + Cl^-$$

(b) HCl.

(c) The solution in benzene is the more strongly acidic; it contains undissociated HCl, which is stronger than H_3O^+, as shown by the reaction in (a).

11. Reversible protonation of an unshared electron pair on an oxygen atom (page 34) converts the organic compound into an ionic compound—a salt—which is soluble in the highly polar solvent, concentrated sulfuric acid.

$$\underset{\text{Organic compound}}{-\overset{|}{O}:} + H_2SO_4 \rightleftharpoons \underset{\substack{\text{Ionic compound} \\ \textit{Soluble in } H_2SO_4}}{-\overset{|}{O}:H^+} + HSO_4^-$$

12. Within a set of related oxy acids, the greater the number of oxygens, the stronger the acid.

To interpret this, we follow the approach of Problem 1.9, above. We focus our attention on the conjugate base being formed when each acid loses a proton: the better the conjugate base accommodates the electron pair left behind, the weaker the base and hence the stronger the acid. With an uncharged acid, as here, the conjugate base is an anion, and accommodation of the electron pair involves dispersal of the negative charge it gives the base. This negative charge—this extra electron pair—is dispersed over the oxygens of the base; the greater the number of oxygens, the greater the dispersal of charge, the weaker the base, and hence the stronger the acid.

13. What you are being asked to do here is something that you—or any organic chemist—must do a great deal of. It takes practice to draw formulas like these, and to make sure that you have considered all possibilities. At the beginning, it is particularly easy to draw two formulas that seem to be different, and fail to see that they actually represent the same molecule—simply turned end-for-end, perhaps, or bent around a corner. The safest way to avoid this pitfall is to *use molecular models*. If you think two structures are different, make models of them and try to make the models coincide in all their parts. Turn them, twist them, bend them, rotate the atoms about the bonds—do everything except break bonds; then, if the models still do not coincide, you have in your hands two different structures representing two different molecules.

(a)

```
   H  H  H            H  H  H
   |  |  |            |     |
H—C—C—N—H          H—C—N—C—H
   |  |               |     |
   H  H               H     H
```

(b)

```
   H  H  H
   |  |  |
H—C—C—C—H
   |  |  |
   H  H  H
```

```
                                H
                                |
                              H—C—H
        H  H  H  H            H |  H                H  H  H             H  H  H
        |  |  |  |            | |  |                |  |  |             |  |  |
(c)   H—C—C—C—C—H           H—C—C—C—H      (d)  H—C—C—C—Cl          H—C—C—C—H
        |  |  |  |            |  |  |               |  |  |             |  |  |
        H  H  H  H            H  H  H               H  H  H             H  Cl H
```

```
        H  H  H               H  H  H             H       H  H
        |  |  |               |  |  |             |       |  |
(e)   H—C—C—C—O—H           H—C—C—C—H          H—C—O—C—C—H
        |  |  |               |  |  |             |       |  |
        H  H  H               H  O  H             H       H  H
                                 |
                                 H
```

```
        H  H               H  H                H  H
        |  |               |  |                |  |
(f)   H—C—C=O           H—C=C—O—H           H—C—C—H
        |                                       \ /
        H                                        O
```

14. To minimize decomposition of an unstable compound.

Heat breaks bonds and thus causes decomposition. Almost any organic compound undergoes *some* decomposition upon being heated, as can be seen by the gradual darkening of the residue during a distillation. An unstable compound contains relatively weak bonds, and is particularly prone to such decomposition; flash distillation minimizes the heating time and hence the decomposition.

2

Methane

Energy of Activation. Transition State

2.1 We write each equation and identify the bonds being broken (on the left side of the equation) and the bonds being formed (on the right side). Turning to the table of homolytic bond dissociation energies (Table 1.2 on page 21, or inside the front cover), we see how much energy is absorbed or liberated when each of these bonds is broken or formed, and we write this value beneath the bond. Now, following the pattern of calculations shown on page 50, we determine the overall change in the heat content, H, of the system. If more heat is liberated than absorbed, H decreases, ΔH is *negative*, and the reaction is *exothermic*. If more heat is absorbed than liberated, H increases, ΔH is *positive*, and the reaction is *endothermic*.

(a) $CH_3—H + Br—Br \longrightarrow CH_3—Br + H—Br$

$$\underbrace{104 \quad 46}_{150} \qquad \underbrace{70 \quad 88}_{158} \qquad \Delta H = -8\text{ kcal}$$

(b) $CH_3—H + I—I \longrightarrow CH_3—I + H—I$

$$\underbrace{104 \quad 36}_{140} \qquad \underbrace{56 \quad 71}_{127} \qquad \Delta H = +13\text{ kcal}$$

(c) $CH_3—H + F—F \longrightarrow CH_3—F + H—F$

$$\underbrace{104 \quad 38}_{142} \qquad \underbrace{108 \quad 136}_{244} \qquad \Delta H = -102\text{ kcal}$$

2.2 For each halogen, we write equations corresponding to (1), (2), and (3) on page 50. Then we follow the procedure described above for Problem 2.1.

(a)

$$\underset{(46)}{Br—Br} \longrightarrow 2Br\cdot \qquad \Delta H = +46\text{ kcal}$$

$$Br\cdot + \underset{(104)}{CH_3—H} \longrightarrow CH_3\cdot + \underset{(88)}{H—Br} \qquad \Delta H = +16\text{ kcal}$$

$$CH_3\cdot + \underset{(46)}{Br—Br} \longrightarrow \underset{(70)}{CH_3—Br} + Br\cdot \qquad \Delta H = -24\text{ kcal}$$

(b) $\underset{(36)}{I{-}I} \longrightarrow 2I\cdot \qquad \Delta H = +36$ kcal

$I\cdot + \underset{(104)}{CH_3{-}H} \longrightarrow CH_3\cdot + \underset{(71)}{H{-}I} \qquad \Delta H = +33$ kcal

$CH_3\cdot + \underset{(36)}{I{-}I} \longrightarrow \underset{(56)}{CH_3{-}I} + I\cdot \qquad \Delta H = -20$ kcal

(c) $\underset{(38)}{F{-}F} \longrightarrow 2F\cdot \qquad \Delta H = +38$ kcal

$F\cdot + \underset{(104)}{CH_3{-}H} \longrightarrow CH_3\cdot + \underset{(136)}{H{-}F} \qquad \Delta H = -32$ kcal

$CH_3\cdot + \underset{(38)}{F{-}F} \longrightarrow \underset{(108)}{CH_3{-}F} + F\cdot \qquad \Delta H = -70$ kcal

2.3 Since each of these reactions involves only bond-breaking, with *no* bond-making, we expect E_{act} to be equal to ΔH (see page 55).

$$\underset{(58)}{Cl{-}Cl} \longrightarrow Cl\cdot + \cdot Cl \qquad \Delta H = +58 \quad E_{act} = 58$$

$$\underset{(104)}{CH_3{-}H} \longrightarrow CH_3\cdot + \cdot H \qquad \Delta H = +104 \quad E_{act} = 104$$

As we easily calculate, homolysis of the Cl—Cl bond has a *much* lower E_{act} than homolysis of the CH_3—H bond, and occurs so much faster that it is by far the preferred chain-initiating step. As with step (2) of this reaction (Sec. 2.21), the nature of step (1) is ultimately determined by the relative strengths of two bonds. In the first step of (thermal) chlorination, the Cl—Cl bond is the one that breaks because it is the *weaker* bond.

2.4 Our line of reasoning is the same as that applied to chlorination in Sec. 2.21. The mechanism actually followed involves (2a) and (3a).

(2a) $Br\cdot + CH_4 \longrightarrow HBr + CH_3\cdot$

(3a) $CH_3\cdot + Br_2 \longrightarrow CH_3Br + Br\cdot$

The alternative would involve (2b) and (3b).

(2b) $Br\cdot + CH_4 \longrightarrow CH_3Br + H\cdot$

(3b) $H\cdot + Br_2 \longrightarrow HBr + Br\cdot$

Again we focus our attention on step (2), since it is here that the two paths divide; what happens in (2) determines what must happen in (3). Using homolytic bond dissociation energies, we calculate ΔH for the corresponding reactions (2a) and (2b). For (2a) ΔH is +16 kcal; E_{act} must be at least that large and, as we know, is actually 18 kcal.

(2a) $Br\cdot + \underset{(104)}{CH_3{-}H} \longrightarrow \underset{(86)}{H{-}Br} + CH_3\cdot \qquad \Delta H = +16 \quad E_{act} = 18$

For (2b), ΔH is +34 kcal; E_{act} must be *at least* 34 kcal—and is probably considerably larger.

$$\text{(2b)} \quad \text{Br}\cdot + \underset{(104)}{\text{CH}_3\text{—H}} \longrightarrow \underset{(70)}{\text{CH}_3\text{—Br}} + \text{H}\cdot \qquad \Delta H = +34 \quad E_{act} = \textit{at least } 34$$

When a bromine atom collides with a methane molecule, either of two reactions can conceivably occur—but *only if* the collision provides enough energy. The collision is overwhelmingly more likely to provide 18 kcal than 34 kcal. (As it happens, the difference in E_{act} here, 16 kcal, is the same as between the competing steps in chlorination; here, too, we can calculate that (2a) is *2.5 million times more likely* to occur than (2b)!)

Once more, we see, what the molecules do is *what is easiest for them.*

2.5 In the methyl cation, CH_3^+, carbon has the same number of electrons as boron in BF_3 (Sec. 1.10) and, similarly, is bonded to three atoms. Like boron, carbon here is sp^2-hybridized; the molecule is flat, with bond angles of 120°. We could represent it as in Figure 2.10 (page 65), except that the *p* orbital in (*b*) is *empty*.

In the methyl anion, $CH_3:^-$, carbon holds not only three atoms, but also an unshared pair of electrons, and thus resembles nitrogen in NH_3 (Sec. 1.12). Like nitrogen, carbon here is sp^3-hybridized; there is a tetrahedral disposition, with an unshared pair of electrons at one corner of the tetrahedron. (Compare Figure 1.12, page 18.) If only atomic nuclei are considered, we expect $CH_3:^-$ to be pyramidal, with tetrahedral angles.

(We would also expect that, like NH_3, the methyl anion—and its larger relatives—would undergo rapid *inversion* (page 19).)

The shapes of the three methyl particles thus depend upon the number of unshared electrons on carbon: the *cation*, with no unshared electrons, trigonal; the *anion*, with an unshared pair, tetrahedral; and the *free radical*, with a single unshared electron, trigonal or intermediate between trigonal and tetrahedral.

2.6 (a) Forms insoluble silver halide in the presence of nitric acid.

(b) Boiling removes volatile HCN and H_2S which otherwise would interfere (AgCN and Ag_2S) with the halide test.

2.7 (a) Because the %C and %H add up to considerably less than 100%. In the absence of any other elements, the missing portion is assumed to be oxygen.

(b) 100% − (52.1% C + 13.1% H) = 34.8% O

2.8 (a)

$$\text{wt. Cl} = 20.68 \times \frac{\text{Cl}}{\text{AgCl}} = 20.68 + \frac{35.45}{143.32}\,\text{mg}$$

$$\%\text{Cl} = \text{wt. Cl/wt. sample} \times 100 = 20.68 \times \frac{35.45}{143.32} \times \frac{1}{7.36} \times 100$$

$$\%\text{Cl} = 69.5\%$$

(b) $$\%\text{Cl} = \frac{\text{Cl}}{\text{CH}_3\text{Cl}} \times 100 = \frac{35.45}{50.48} \times 100 = 70.2\%$$

(c) $$\text{wt. AgCl} = \text{wt. sample} \times \frac{2\text{AgCl}}{\text{CH}_2\text{Cl}_2} = 7.36 \times \frac{286.64}{84.93} = 24.84 \text{ mg}$$

(d) $$\text{wt. AgCl} = \text{wt. sample} \times \frac{3\text{AgCl}}{\text{CHCl}_3} = 7.36 \times \frac{429.96}{119.37} = 26.51 \text{ mg}$$

(e) $$\text{wt. AgCl} = \text{wt. sample} \times \frac{4\text{AgCl}}{\text{CCl}_4} = 7.36 + \frac{573.28}{153.83} = 27.43 \text{ mg}$$

2.9 (a) $$\text{wt. C} = 8.86 \times \frac{\text{C}}{\text{CO}_2} = 8.86 \times \frac{12.01}{44.01} \text{ mg}$$

$$\%\text{C} = \text{wt. C/wt. sample} \times 100 = 8.86 \times \frac{12.01}{44.01} \times \frac{1}{3.02} \times 100$$

$$\%\text{C} = 80.1\%$$

$$\text{wt. H} = 5.43 \times \frac{2\text{H}}{\text{H}_2\text{O}} = 5.43 \times \frac{2.016}{18.02} \text{ mg}$$

$$\%\text{H} = \text{wt. H/wt. sample} \times 100 = 5.43 \times \frac{2.016}{18.02} \times \frac{1}{3.02} \times 100$$

$$\%\text{H} = 20.1\%$$

$$\text{C: } \frac{80.1}{12.01} = 6.67 \text{ gram-atoms} \quad \text{H: } \frac{20.1}{1.008} = 19.9 \text{ gram-atoms}$$

$$\text{C: } 6.67/6.67 = 1.0 \quad \text{H: } 20.1/6.67 = 3.0$$

$$\text{Empirical formula} = \text{CH}_3$$

(b) As in (a) for %C and %H, using proper weights for sample, CO_2, and H_2O, we get

$$\text{C: } 2.67 \text{ gram-atoms} \quad \text{H: } 5.38 \text{ gram-atoms}$$

$$\text{wt. Cl} = 13.49 \times \frac{\text{Cl}}{\text{AgCl}} = 13.49 \times \frac{35.45}{143.32} \text{ mg}$$

$$\%\text{Cl} = \text{wt. Cl/wt. sample} \times 100 = 13.49 \times \frac{35.45}{143.32} \times \frac{1}{5.32} \times 100$$

$$\%\text{Cl} = 62.7\%$$

$$\text{Cl: } \frac{62.7}{35.45} = 1.77 \text{ gram-atoms}$$

$$\text{C: } \frac{2.67}{1.77} = 1.51 \quad \text{H: } \frac{5.38}{1.77} = 3.04 \quad \text{Cl: } \frac{1.77}{1.77} = 1$$

But we cannot accept $C_{1.5}H_3Cl$, so we multiply by 2 to get a whole number of each kind of atom; this gives us an empirical formula of $C_3H_6Cl_2$.

2.10 CH = 13; 78/13 = 6(CH) units; the molecular formula is C_6H_6.

2.11

$$\text{wt. C} = 10.32 \times \frac{C}{CO_2} = 10.32 \times \frac{12.01}{44.01}\ \text{mg}$$

$$\%C = \text{wt. C/wt. sample} \times 100 = 10.32 \times \frac{12.01}{44.01} \times \frac{1}{5.17} \times 100$$

$$\%C = 54.5\%$$

$$\text{wt. H} = 4.23 \times \frac{2H}{H_2O} = 4.23 \times \frac{2.016}{18.02}\ \text{mg}$$

$$\%H = \text{wt. H/wt. sample} \times 100 = 4.23 \times \frac{2.016}{18.02} \times \frac{1}{5.17} \times 100$$

$$\%H = 9.2\%$$

The deficiency of 36.3% (100 − (54.5 + 9.2)) is due to oxygen.

$$C\colon \frac{54.5}{12.01} = 4.53 \text{ gram-atoms} \quad H\colon \frac{9.2}{1.0} = 9.2 \text{ gram-atoms} \quad O\colon \frac{36.3}{16.0} = 2.27 \text{ gram-atoms}$$

$$C\colon \frac{4.53}{2.27} = 2.0 \qquad H\colon \frac{9.2}{2.27} = 4.0 \qquad O\colon \frac{2.27}{2.27} = 1$$

Empirical formula = C_2H_4O; each unit has weight of 44.

88/44 = 2 C_2H_4O units; thus the molecular formula is $C_4H_8O_2$.

1. Following the procedures of Problems 2.7, 2.8, and 2.9, we obtain

X: 93.9% C, 6.3% H. Y: 64.0% C, 4.5% H, 31.4% Cl. Z: 62.0% C, 10.3% H, 27.7% O.

For compound Z, 27.7% O is calculated from 100 − (62.0% C + 10.3% H).

2. (a)

$$C_3H_7Cl = 78.54\ \text{m.w.}$$

$$\%C = \frac{3C}{C_3H_7Cl} \times 100 = \frac{3 \times 12.01}{78.54} \times 100 = 45.9\%\ C$$

$$\%H = \frac{7H}{C_3H_7Cl} \times 100 = \frac{7 \times 1.008}{78.54} \times 100 = 8.9\%\ H$$

$$\%Cl = \frac{Cl}{C_3H_7Cl} \times 100 = \frac{35.45}{78.54} \times 100 = 45.1\%\ Cl$$

Use of this procedure leads to the following:

(b)	(c)	(d)	(e)	(f)
52.1% C	54.5% C	41.8% C	20.0% C	55.6% C
13.1% H	9.2% H	4.7% H	6.7% H	6.2% H
34.8% O	36.3% O	18.6% O	26.6% O	10.8% N
		16.3% N	46.7% N	27.4% Cl
		18.6% S		

3. We follow the procedure of Problem 2.9. In (c), (d), and (f), oxygen is determined by difference.

(a)

$$\text{C: } \frac{85.6}{12.01} = 7.13 \text{ gram-atoms} \qquad \text{H: } \frac{14.4}{1.008} = 14.29 \text{ gram-atoms}$$

$$\text{C: } 7.13/7.13 = 1.0 \qquad \text{H: } 14.29/7.13 = 2.0$$

Empirical formula CH_2

(b)

$$\text{C: } \frac{92.2}{12.01} = 7.68 \qquad \text{H: } \frac{7.8}{1.008} = 7.74$$

$$\text{C: } 7.68/7.68 = 1.0 \qquad \text{H: } 7.74/7.68 = 1.0$$

Empirical formula CH

(c)

$$\text{C: } \frac{40.0}{12.01} = 3.33 \qquad \text{H: } \frac{6.7}{1.008} = 6.65 \qquad \text{O: } \frac{53.3}{16.00} = 3.33$$

$$\text{C: } 3.33/3.33 = 1.0 \qquad \text{H:} 6.65/3.33 = 2.0 \qquad \text{O: } 3.33/3.33 = 1.0$$

Empirical formula CH_2O

(d)

$$\text{C: } \frac{29.8}{12.01} = 2.48 \qquad \text{H: } \frac{6.3}{1.008} = 6.25 \qquad \text{Cl: } \frac{44.0}{35.45} = 1.24 \qquad \text{O: } \frac{19.9}{16.00} = 1.24$$

$$\text{C: } 2.48/1.24 = 2.0 \qquad \text{H: } 6.25/1.24 = 5.0 \qquad \text{Cl: } 1.24/1.24 = 1.0 \qquad \text{O: } 1.24/1.24 = 1.0$$

Empirical formula C_2H_5OCl

(e)

$$\text{C: } \frac{48.7}{12.01} = 4.05 \qquad \text{H: } \frac{13.6}{1.008} = 13.49 \qquad \text{N: } \frac{37.8}{14.01} = 2.70$$

$$\text{C: } 4.05/2.70 = 1.5 \qquad \text{H: } 13.49/2.70 = 5.0 \qquad \text{N: } 2.70/2.70 = 1.0$$

This gives $C_{1.5}H_5N$, so we multiply by 2 to get whole numbers.

Empirical formula $C_3H_{10}N_2$

(f)

$$\text{C: } \frac{25.2}{12.01} = 2.10 \qquad \text{H: } \frac{2.8}{1.008} = 2.78 \qquad \text{Cl: } \frac{49.6}{35.45} = 1.40 \qquad \text{O: } \frac{22.4}{16.0} = 1.40$$

$$\text{C: } 2.10/1.40 = 1.5 \qquad \text{H: } 2.78/1.40 = 2.0 \qquad \text{Cl: } 1.40/1.40 = 1.0 \qquad \text{O: } 1.40/1.40 = 1.0$$

We multiply $C_{1.5}H_2OCl$ by 2 to get whole numbers.

Empirical formula $C_3H_4O_2Cl_2$

4. 18.9% O is calculated by difference.

$$\text{C: } \frac{70.8}{12.0} = 5.9 \qquad \text{H: } \frac{6.2}{1.0} = 6.2 \qquad \text{O: } \frac{18.9}{16.0} = 1.2 \qquad \text{N: } \frac{4.1}{14.0} = 0.3$$

$$\text{C: } 5.9/0.3 = 20 \qquad \text{H: } 6.2/0.3 = 21 \qquad \text{O: } 1.2/0.3 = 4 \qquad \text{N: } 0.3/0.3 = 1$$

Empirical formula of papaverine = $C_{20}H_{21}O_4N$.

5. In a similar fashion (14.7% O calculated by difference):

$$C = 4.3 \quad H = 4.3 \quad O = 0.9 \quad N = 0.9 \quad S = 0.3 \quad Na = 0.3$$

$$C = \frac{4.3}{0.3} = 14 \quad H = \frac{4.3}{0.3} = 14 \quad O = \frac{0.9}{0.3} = 3 \quad N = \frac{0.9}{0.3} = 3 \quad S = \frac{0.3}{0.3} = 1 \quad Na = \frac{0.3}{0.3} = 1$$

Empirical formula of methyl orange = $C_{14}H_{14}O_3N_3SNa$.

6. (a) As in Problem 2.9: 85.8% C, 14.3% H.

(b)
$$C: \frac{85.8}{12.0} = 7.1 \quad H: \frac{14.3}{1.0} = 14.3$$

Empirical formula = CH_2; unit weight = 14

(c) 84/14 = 6 CH_2 units. Molecular formula = C_6H_{12}

7. 53.3% O is calculated by difference.

C: 40.0% of m.w. of 60 ⟶ 24 g C/mol ⟶ 2 C per molecule
H: 6.7% of m.w. of 60 ⟶ 4.0 g H/mol ⟶ 4 H per molecule
O: 53.3% of m.w. of 60 ⟶ 32 g O/mol ⟶ 2 O per molecule

Molecular formula = $C_2H_4O_2$

8. Half as many atoms as in Problem 7 = CH_2O.

9. As in Problem 7:

12.2% O is calculated by difference.

C: 0.733 × 262 = 192 g C/mol ⟶ 16 C per molecule
H: 0.038 × 262 = 10.0 g H/mol ⟶ 10 H per molecule
O: 0.122 × 262 = 32.0 g O/mol ⟶ 2 O per molecule
N: 0.107 × 262 = 28.0 g N/mol ⟶ 2 N per molecule

Molecular formula of indigo = $C_{16}H_{10}O_2N_2$

10. (a) The smallest possible molecule must contain at least *one* sulfur atom.

$$\%S = \frac{1S}{\text{min. mol. wt.}} \times 100 \quad 3.4 = \frac{32.06}{\text{min. mol. wt.}} \times 100$$

min. mol. wt. = 942

(b) 5734/942 = approx. 6 of the minimum units

Therefore, there are 6 S per molecule.

11. We follow the procedure of Problem 2.1.

(a) $\mathrm{H{-}H + F{-}F \longrightarrow 2H{-}F}$

$$\underbrace{104 \quad 38}_{142} \qquad \underbrace{2 \times 136}_{272} \qquad \Delta H = -130$$

(b) $\mathrm{H{-}H + Cl{-}Cl \longrightarrow 2H{-}Cl}$

$$\underbrace{104 \quad 58}_{162} \qquad \underbrace{2 \times 103}_{206} \qquad \Delta H = -44$$

(c) $\mathrm{H{-}H + Br{-}Br \longrightarrow 2H{-}Br}$

$$\underbrace{104 \quad 46}_{150} \qquad \underbrace{2 \times 88}_{176} \qquad \Delta H = -26$$

(d) $\mathrm{H{-}H + I{-}I \longrightarrow 2H{-}I}$

$$\underbrace{104 \quad 36}_{140} \qquad \underbrace{2 \times 71}_{142} \qquad \Delta H = -2$$

(e) $\mathrm{C_2H_5{-}H + Br{-}Br \longrightarrow C_2H_5{-}Br + H{-}Br}$

$$\underbrace{98 \quad 46}_{144} \qquad \underbrace{69 \quad 88}_{157} \qquad \Delta H = -13$$

(f) $\mathrm{C_6H_5CH_2{-}H + Br{-}Br \longrightarrow C_6H_5CH_2{-}Br + H{-}Br}$

$$\underbrace{85 \quad 46}_{131} \qquad \underbrace{51 \quad 88}_{139} \qquad \Delta H = -8$$

(g) $\mathrm{H_2C{=}CHCH_2{-}H + Br{-}Br \longrightarrow H_2C{=}CHCH_2{-}Br + H{-}Br}$

$$\underbrace{88 \quad 46}_{134} \qquad \underbrace{47 \quad 88}_{135} \qquad \Delta H = -1$$

(h) We follow the procedure of Problem 2.2.

1st step: $\mathrm{Br{-}Br \longrightarrow 2Br\cdot}$ $\quad \Delta H = +46$ kcal

		C_2H_6	$C_6H_5CH_3$	$H_2C{=}CHCH_3$
2nd step:	$\Delta H =$	+10 kcal	−3 kcal	0 kcal
3rd step:	$\Delta H =$	−23 kcal	−5 kcal	−1 kcal

12. (a)

(i) $\mathrm{CH_3\cdot + \underset{(104)}{CH_3{-}H} \longrightarrow \underset{(104)}{CH_3{-}H} + CH_3\cdot}$ $\quad \Delta H = 0 \quad E_{act} = 13$

(ii) $\mathrm{CH_3\cdot + \underset{(104)}{CH_3{-}H} \longrightarrow \underset{(88)}{CH_3{-}CH_3} + H\cdot}$ $\quad \Delta H = +16 \quad E_{act} =$ *at least* 16

For the endothermic reaction (ii), E_{act} must be at least as large as the ΔH, 16 kcal; for reaction (i), E_{act} *could* be zero. (It is actually 13 kcal.) The reaction with the lower E_{act} is the one that takes place.

The same bond (CH_3—H) is being broken in both reactions; what differs is the bond being formed. In the final analysis, reaction takes the course it does because a C—H bond is stronger than a C—C bond.

(b) Highly improbable, since E_{act} for the competing reaction, with Cl_2, is much smaller.

In a 50:50 mixture of CH_4 and Cl_2, collision of $CH_3\cdot$ with either kind of molecule is equally likely. But the highly exothermic reaction with Cl_2 has a very small E_{act}, and is greatly favored over the reaction with an E_{act} of 13 kcal.

$$CH_3\cdot \xrightarrow{Cl_2} CH_3Cl + Cl\cdot \qquad \Delta H = -26 \quad E_{act}\ \textit{very small}$$

$$CH_3\cdot \xrightarrow{CH_4} CH_4 + CH_3\cdot \qquad \Delta H = 0 \quad E_{act} = 13$$

13. The C—Cl bond is weaker than the C—F bond, and easier to break.

For reasons that are not understood, the presence of F instead of H on the carbon makes the C—Cl bond easier to break. (Look at Sec. 9.23).

14. (a) $CH_3\cdot$ can react not only with Br_2 to give product, but also with HBr to regenerate CH_4 and thus slow down reaction. (The latter reaction is the reverse of a chain-propagating step.)

$$CH_3\cdot \xrightarrow{Br_2} CH_3Br + Br\cdot \qquad \Delta H = -24 \quad E_{act}\ \textit{very small}$$

$$CH_3\cdot \xrightarrow{HBr} CH_4 + Br\cdot \qquad \Delta H = -16 \quad E_{act} = 2$$

As we saw on page 54, E_{act} for the reaction with HBr is 2 kcal. This is larger than the very small E_{act} for the reaction with Br_2, but evidently small enough for some reaction to follow this competing route when the concentration of HBr is high and collision likely.

(b) Competition here would involve reaction of $CH_3\cdot$ with HCl, a reaction that has an E_{act} of 3 kcal (page 54). This is a more difficult reaction than the one with HBr ($E_{act} = 2$ kcal), and evidently cannot compete successfully with the reaction between $CH_3\cdot$ and Cl_2.

$$CH_3\cdot \xrightarrow{Cl_2} CH_3Cl + Cl\cdot \qquad \Delta H = -26 \quad E_{act}\ \textit{very small}$$

$$CH_3\cdot \xrightarrow{HCl} CH_4 + Cl\cdot \qquad \Delta H = -1 \quad E_{act} = 3$$

(c) As reaction proceeds, HBr accumulates as one of the reaction products. This begins to compete with Br_2 for $CH_3\cdot$, as in part (a), and thus to reverse the halogenation process.

15. (a) The evidence indicates a free-radical chain reaction analogous to the one for chlorination of methane. The line of reasoning is the same as that given in Sec. 2.12.

$$\text{Cl—Cl} \xrightarrow{\text{light}} 2\text{Cl}\cdot$$

$$\text{Cl}\cdot + \text{H—H} \longrightarrow \text{HCl} + \text{H}\cdot$$

$$\text{H}\cdot + \text{Cl—Cl} \longrightarrow \text{HCl} + \text{Cl}\cdot \; \textit{etc.}$$

(b) An analogous mechanism would contain a chain-propagating step with an E_{act} of at least 33 kcal, too large to compete with the recombination of I· atoms to regenerate I_2.

$$\underset{(104)}{\text{I}\cdot + \text{H—H}} \longrightarrow \underset{(71)}{\text{H—I}} + \text{H}\cdot \qquad \underset{(E_{act} \geq 33\text{ kcal})}{\Delta H = +33\text{ kcal}}$$

16. (a)

$$(\text{CH}_3)_4\text{Pb} \xrightarrow{\text{heat}} \underset{\text{New mirror}}{\text{Pb}} + 4\text{CH}_3\cdot \xrightarrow[\text{Old mirror}]{\text{Pb}} \underset{\text{Effluent}}{(\text{CH}_3)_4\text{Pb}}$$

(b) The farther the CH_3· radicals have to travel, the more time there is for them to combine with each other,

$$\text{CH}_3\cdot + \cdot\text{CH}_3 \longrightarrow \text{CH}_3\text{—CH}_3$$

and the lower their concentration when they reach the old mirror. The lower their concentration, the more slowly they react with the mirror.

Such combination is an *unlikely* process because of the low concentration of these particles; in a chlorination mixture, for example, it happens very seldom because there is a much greater opportunity for them to collide with, and hence react with, Cl_2 molecules (page 48). But under the special conditions of Paneth's experiments—carried along a quartz tube by an inert gas—there is nothing else about for them to react with; and so, they (slowly) react with each other.

(This work is summarized in Wheland, G. W. *Advanced Organic Chemistry*, 3rd ed.; Wiley: New York, 1960; pp 733–737, and, in more detail, in Steacie, E. W. R. *Atom and Free Radical Reactions*, 2nd ed.; Reinhold: New York, 1954; Vol. 1, pp 37–53. For a fascinating outgrowth of this chemistry, see Rice, F. O. "The Chemistry of Jupiter"; *Sci. Am.* **June 1956**.)

This is an example of something that happens quite often in organic chemistry: an advance in theory is made possible only by the invention of a new *experimental technique*—Paneth's mirror method, in this case. As we shall see, carbocations can be studied at leisure because George Olah (page 192) discovered a new medium in which—although, like free radicals, usually highly reactive particles—they can find nothing to react with. Emil Fischer (page 1155) opened up the vast field of carbohydrates by the use of a simple reaction he had discovered, the conversion of sugars into osazones—a reaction which, among other things, changes sugars from hard-to-handle syrups to readily identifiable crystalline compounds. The sequence of the hundreds of amino acid units making up the giant chain of a protein is now routinely determined *automatically* in a computer-controlled analyzer developed by Pehr Edman (page 1219), based upon his discovery of a reaction between phenyl isothiocyanate and one end of a protein chain. And the list could go on and on.

17. $$(C_2H_5)_4Pb \xrightarrow{140\,^\circ C} Pb + 4C_2H_5\cdot$$

We expect $C_2H_5\cdot$ formed in this way to react with Cl_2 in the same manner as $CH_3\cdot$ to give ethyl chloride and $Cl\cdot$, via

$$C_2H_5\cdot + Cl{-}Cl \longrightarrow C_2H_5Cl + Cl\cdot$$

Then $Cl\cdot$ generated in this way can start a halogenation chain involving methane.

3

Alkanes

Free-Radical Substitution

Note: In all problems in this chapter that deal with isomerism, *we shall for the present neglect the existence of stereoisomers*. After you have studied Chapter 4, it would be instructive to return to some of these problems, and see how you might modify your answers.

3.1 Within a few weeks you will be drawing isomeric structures rapidly and easily, and may have trouble remembering that, in the beginning, this is a job that requires work. You are learning a new language, that of organic chemistry, and one of the first steps is learning to *write* it. You had your first practice at doing this in Problem 13 (page 38), and at this point you should re-read the answer to that problem on page 9 of this Study Guide. Take particularly seriously the advice to *use molecular models*: this is the only way really to assure yourself that two formulas actually represent different, isomeric molecules.

At this point it may be best to strip the molecules of hydrogens, and focus our attention on the underlying carbon skeletons and the chlorine atoms attached to these skeletons.

```
                                                                     Cl
                                                                     |
(a) C—C—C—C—Cl   C—C—C—C    C—C—C—C    C—C—C—C    C—C—C—C    C—C—C—C
          |           |  |     |    |   |      |        |       |  |
          Cl          Cl Cl    Cl   Cl  Cl     Cl       Cl      Cl Cl
```

As with all such problems, we shall be *systematic*. We first draw the straight carbon chain of *n*-butane.

```
C—C—C—C
```

(We draw it in a straight line. We *could* bend it around a corner or in a zig-zag, and it would still represent the same compound; but we make things as easy for ourselves as we can.)

We attach a Cl to, say, one end of the chain (to C–1) and, leaving it there, we attach the second Cl successively to each of the carbons: to C–1, C–2, C–3, and C–4.

```
C—C—C—C—Cl   C—C—C—C    C—C—C—C    C—C—C—C
      |           |  |     |    |   |      |
      Cl          Cl Cl    Cl   Cl  Cl     Cl
```

We now have four structures, and have exhausted the possibilities that contain a Cl on a terminal carbon. We can easily assure ourselves that all of these are *different* from each other: they cannot possibly be interconverted without breaking bonds and making new ones.

Next, we place our first Cl on C–2 and repeat the process, remembering that we have already used the terminal positions. This gives us two more isomers, for a total of six.

```
     Cl
     |
C—C—C—C          C—C—C—C
     |              |   |
     Cl             Cl  Cl
```

```
       C           C          C
       |           |          |
(b) C—C—C—Cl    C—C—C      C—C—C
       |           |  |     |    |
       Cl          Cl Cl    Cl   Cl
```

We follow the procedure of (a), this time starting with the branched carbon chain of isobutane. Models will show us a particularly important feature of the isobutane molecule: all three methyl groups are equivalent, regardless of the fact that we may draw two of them horizontally, and the third pointing up (or down).

3.2 No. Each can give rise to two monochloro compounds.

```
                                    C          C
                                    |          |
C—C—C—C     C—C—C—C              C—C—C      C—C—C
      |         |                     |        |
      Cl        Cl                    Cl       Cl
```

3.3 Van der Waals repulsion between "large" methyls.

Crowding between methyls { CH_3, CH_3

3.4 (a)

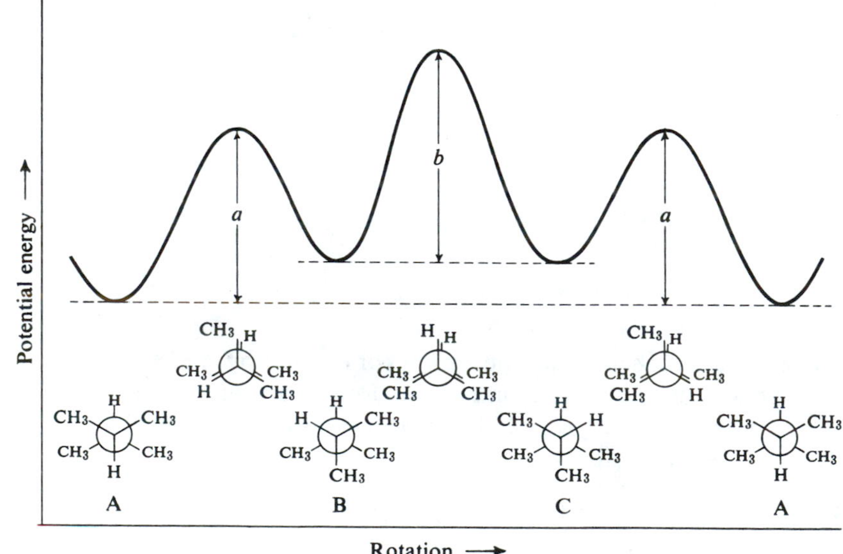

(b)

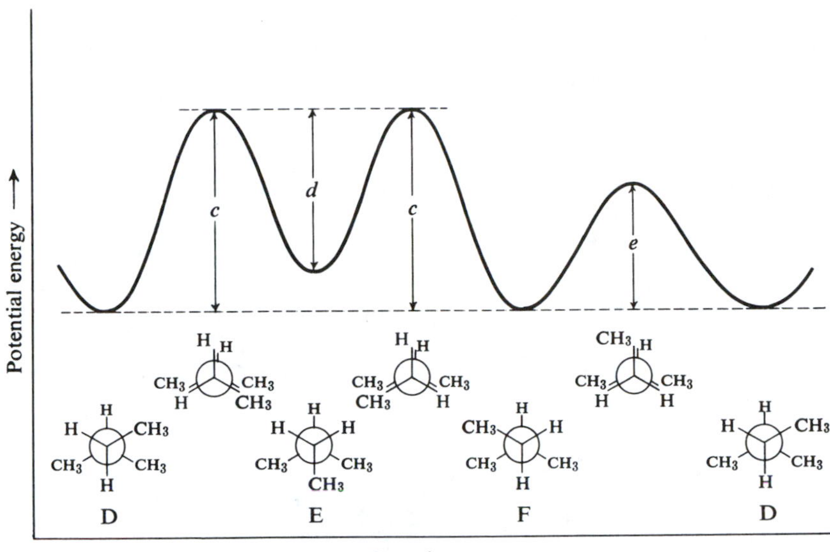

(c)

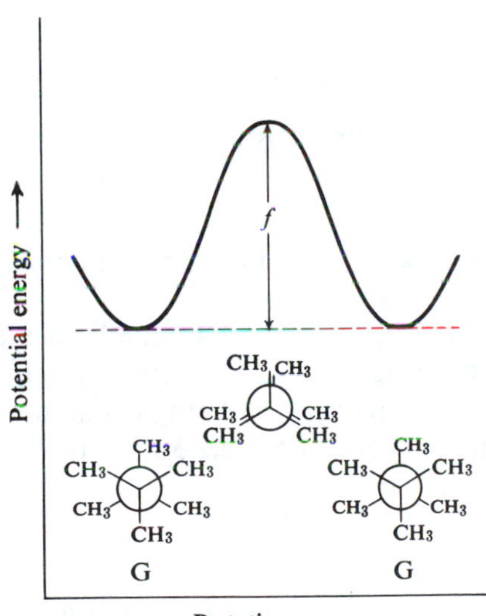

(d) On the assumption of 0.8 kcal per methyl–methyl gauche interaction, and of 3.0 kcal torsional energy plus 0.4 kcal for two methyl–hydrogen eclipsings and 2.2–3.9 kcal per methyl–methyl eclipsing (from Figure 3.8), we arrive at the following tentative predictions:

$$b > 4.4\text{–}6.1 > a > 3.4 \qquad c > 4.4\text{–}6.1 > d > 3.4 > e$$

The size of f depends upon the value for methyl–methyl eclipsing.

3.5 We follow the systematic approach of Problem 3.1.

```
                                                           C
                                                           |
(a) C-C-C-C-C-C-C   C-C-C-C-C-C   C-C-C-C-C-C   C-C-C-C-C
                            |           |              |
                            C           C              C

                                    C                         C
                                    |                         |
    C-C-C-C-C   C-C-C-C-C   C-C-C-C-C   C-C-C-C-C   C-C-C-C
        | |       |   |         |           |         | |
        C C       C   C         C           C         C C
                                            |
                                            C
```

```
                                               C
                                               |
(b) C-C-C-C-C   C-C-C-C-C   C-C-C-C-C   C-C-C-C
            |         |         |             |
            Cl        Cl        Cl            Cl

        C           C           C          C
        |           |           |          |
    C-C-C-C     C-C-C-C     C-C-C-C     C-C-C
        |         |         |            | |
        Cl        Cl        Cl           C Cl
```

```
                                                                  Br
                                                                  |
(c) C-C-C-C-Br   C-C-C-C     C-C-C-C     C-C-C-C       C-C-C-C
          |          | |       |   |     |     |           |
          Br        Br Br     Br   Br    Br    Br          Br

                   C          C          C
                   |          |          |
    C-C-C-C    C-C-C-Br    C-C-C      C-C-C
      | |          |         | |      |   |
     Br Br         Br       Br Br     Br  Br
```

3.6 (a) Order of isomers as on page 87:

- *n*-hexane
- 2-methylpentane
- 3-methylpentane
- 2,2-dimethylbutane
- 2,3-dimethylbutane

(b) Order of isomers as in Problem 3.5(a):

- *n*-heptane
- 2-methylhexane
- 3-methylhexane
- 2,2-dimethylpentane
- 2,3-dimethylpentane
- 2,4-dimethylpentane
- 3,3-dimethylpentane
- 3-ethylpentane
- 2,2,3-trimethylbutane

3.7 (a) Order of isomers as in Problem 3.5(b):

- 1-chloropentane
- 2-chloropentane
- 3-chloropentane
- 1-chloro-2-methylbutane
- 2-chloro-2-methylbutane
- 3-chloro-2-methylbutane
- 1-chloro-3-methylbutane
- 1-chloro-2,2-dimethylpropane

(b) Order of isomers as in Problem 3.5(c):

- 1,1-dibromobutane
- 1,2-dibromobutane
- 1,3-dibromobutane
- 1,4-dibromobutane
- 2,2-dibromobutane
- 2,3-dibromobutane
- 1,1-dibromo-2-methylpropane
- 1,2-dibromo-2-methylpropane
- 1,3-dibromo-2-methylpropane

3.8 All three graphs show a rate of increase falling off with increasing carbon number.

3.9 Hydrogen (H or D) becomes attached to the same carbon that held Mg.

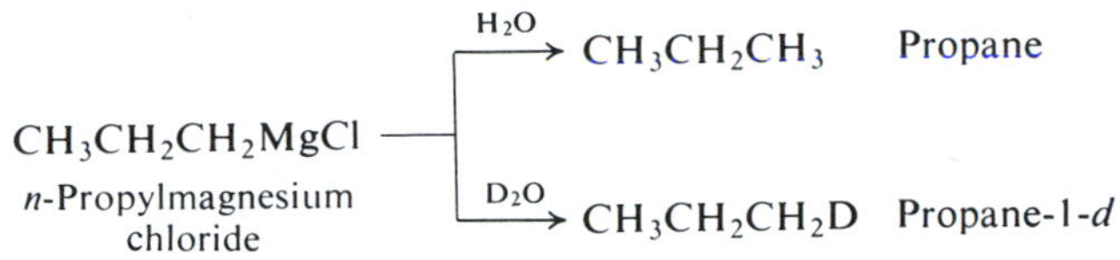

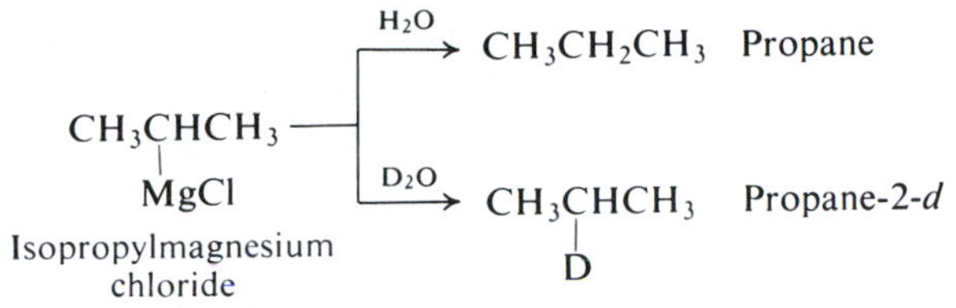

$$CH_3CH_2CH_2MgCl \;(\textit{n}\text{-Propylmagnesium chloride}) \begin{cases} \xrightarrow{H_2O} CH_3CH_2CH_3 & \text{Propane} \\ \xrightarrow{D_2O} CH_3CH_2CH_2D & \text{Propane-1-}d \end{cases}$$

$$\underset{\text{Isopropylmagnesium chloride}}{CH_3\underset{\substack{|\\ MgCl}}{C}HCH_3} \begin{cases} \xrightarrow{H_2O} CH_3CH_2CH_3 & \text{Propane} \\ \xrightarrow{D_2O} CH_3\underset{\substack{|\\ D}}{C}HCH_3 & \text{Propane-2-}d \end{cases}$$

3.10 (a) All that have the carbon skeleton of *n*-pentane:

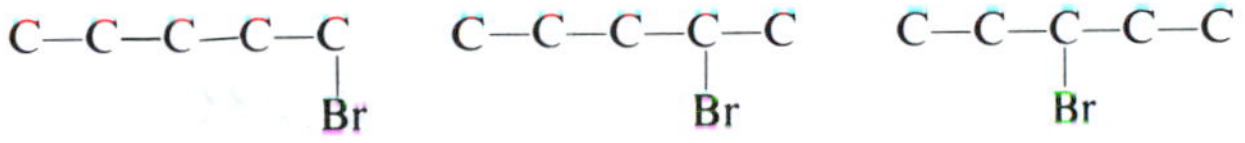

$$C{-}C{-}C{-}C{-}\underset{\substack{|\\ Br}}{C} \qquad C{-}C{-}C{-}\underset{\substack{|\\ Br}}{C}{-}C \qquad C{-}C{-}\underset{\substack{|\\ Br}}{C}{-}C{-}C$$

(b) All that have the carbon skeleton of 2-methylbutane (isopentane):

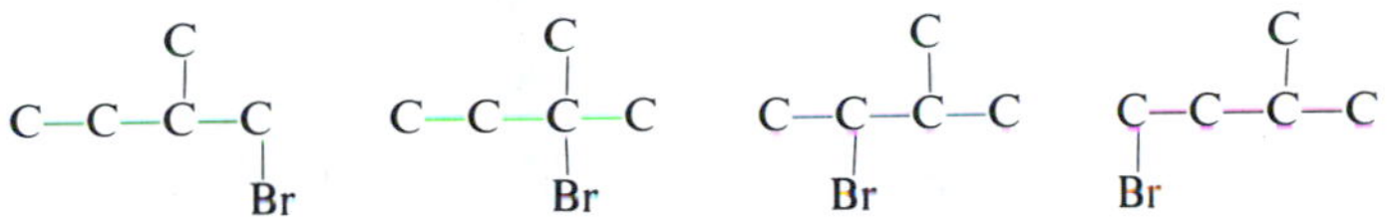

$$C{-}C{-}\overset{\substack{C\\ |}}{C}{-}\underset{\substack{|\\ Br}}{C} \qquad C{-}C{-}\overset{\substack{C\\ |}}{\underset{\substack{|\\ Br}}{C}}{-}C \qquad C{-}\underset{\substack{|\\ Br}}{C}{-}\overset{\substack{C\\ |}}{C}{-}C \qquad \underset{\substack{|\\ Br}}{C}{-}C{-}\overset{\substack{C\\ |}}{C}{-}C$$

(c) All that have the carbon skeleton of 2,3-dimethylbutane:

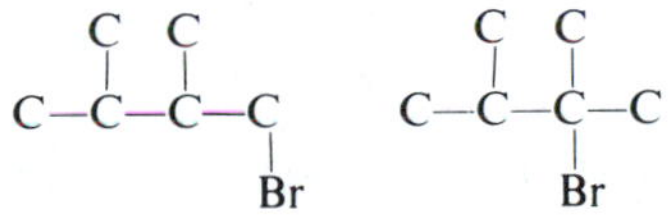

$$C{-}\overset{\substack{C\\ |}}{C}{-}\overset{\substack{C\\ |}}{C}{-}\underset{\substack{|\\ Br}}{C} \qquad C{-}\overset{\substack{C\\ |}}{C}{-}\overset{\substack{C\\ |}}{\underset{\substack{|\\ Br}}{C}}{-}C$$

(d) Only neopentyl bromide has the proper carbon skeleton:

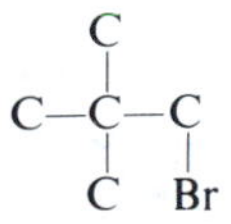

$$C{-}\overset{\substack{C\\ |}}{\underset{\substack{|\\ C}}{C}}{-}\underset{\substack{|\\ Br}}{C}$$

3.11 (a)

$$CH_3{-}\overset{\substack{CH_3\\ |}}{C}H \,\vdots\, CH_2{-}CH_2{-}CH_3 \longleftarrow \begin{cases} (CH_3{-}\overset{\substack{CH_3\\ |}}{C}H)_2CuLi + X{-}CH_2{-}CH_2{-}CH_3 \\ CH_3{-}\overset{\substack{CH_3\\ |}}{C}H{-}X + LiCu(CH_2{-}CH_2{-}CH_3)_2 \end{cases}$$

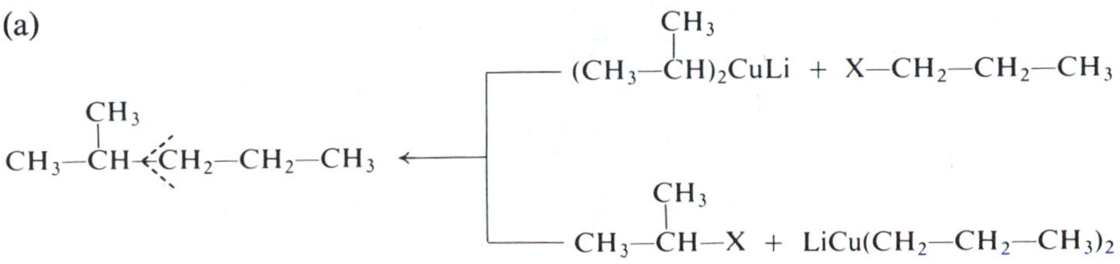

(b) Use the first, since in this synthesis R′X is primary.

(Corey, E. J.; Posner, G. H. "Carbon–Carbon Bond Formation by Selective Coupling of *n*-Alkylcopper Reagents with Organic Halides"; *J. Am. Chem. Soc.* **1968**, *90*, 5615; House, H. O., *et al.* "Reaction of Lithium Dialkyl- and Diarylcuprates with Organic Halides"; *J. Am. Chem. Soc.* **1969**, *91*, 4871.)

3.12

```
(a) C—C—C—C—C      C—C—C—C—C      C—C—C—C—C
            |              |            |
            Cl             Cl           Cl

        C              C              C              C
        |              |              |              |
(b) C—C—C—C        C—C—C—C        C—C—C—C        C—C—C—C
          |            |              |          |
          Cl           Cl             Cl         Cl
```

3.13 The products are:

```
                     C                C  C
                     |                |  |
(a) C—C—C      (b) C—C—C      (c) C—C—C—C      (d) C—C—C—C—C
        |                |                |                  |
        Cl               Cl               Cl                 Cl

                     C                C  C
                     |                |  |
    C—C—C          C—C—C          C—C—C—C          C—C—C—C—C
      |              |                |                  |
      Cl             Cl               Cl                 Cl

                                                   C—C—C—C—C
                                                       |
                                                       Cl

           C                  C  C               C     C
           |                  |  |               |     |
(e) C—C—C—C          (f) C—C—C—C          (g) C—C—C—C—C
             |                |  |                   |  |
             Cl               C  Cl                  C  Cl

           C                  C   C              C     C
           |                  |   |              |     |
    C—C—C—C              C—C—C—C            C—C—C—C—C
           |                  |   |                |   |
           Cl                 Cl  C                Cl  C

           C                  C  C               C     C
           |                  |  |               |     |
    C—C—C—C              C—C—C—C            C—C—C—C—C
      |                  |    |                |     |
      Cl                 Cl   C                Cl    C

           C                                     C     C
           |                                     |     |
    C—C—C—C                                 C—C—C—C—C
    |                                       |        |
    Cl                                      Cl       C
```

The proportions of the products are:

(a) 44% 1-Cl
56% 2-Cl

(b) 64% 1°
36% 3°

(c) 55% 1°
45% 3°

(d) 21% 1-Cl
53% 2-Cl
26% 3-Cl

(e) 28% 1-Cl-2-Me
23% 2-Cl-2-Me
35% 2-Cl-3-Me
14% 1-Cl-3-Me

(f) 45% 1-Cl-2,2,3-triMe
25% 3-Cl-2,2,3-triMe
30% 1-Cl-2,3,3-triMe

(g) 33% 1-Cl-2,2,4-triMe
28% 3-Cl-2,2,4-triMe
18% 4-Cl-2,2,4-triMe
22% 1-Cl-2,4,4-triMe

These predicted proportions of products can be calculated by the method shown on page 109, using the relative reactivities of 5.0:3.8:1.0 given there. For (a), this would be:

$$\frac{n\text{-PrCl}}{i\text{-PrCl}} = \frac{\text{no. of 1° H}}{\text{no. of 2° H}} \times \frac{\text{reactivity of 1° H}}{\text{reactivity of 2° H}} = \frac{6}{2} \times \frac{1.0}{3.8} = \frac{6.0}{7.6}$$

Then,

$$\%1° = \frac{6.0}{6.0 + 7.6} \times 100 = 44\%$$

$$\%2° = \frac{7.6}{6.0 + 7.6} \times 100 = 56\%$$

For (d):

1-Cl: 6H, each with reactivity 1.0 ⟶ 6.0
2-Cl: 4H, each with reactivity 3.8 ⟶ 15.2
3-Cl: 2H, each with reactivity 3.8 ⟶ 7.6

number × reactivity total = 28.8

$$\%\text{1-Cl} = \frac{\text{reactivity 1-Cl}}{\text{total reactivity of molecule}} \times 100 = \frac{6.0}{28.8} \times 100 = 21\%$$

$$\%\text{2-Cl} = \frac{15.2}{28.8} \times 100 = 53\%$$

$$\%\text{3-Cl} = \frac{7.6}{28.8} \times 100 = 26\%$$

The others are calculated similarly.

3.14 The predicted proportions are calculated by the method outlined in the answer to Problem 3.13 above, using the relative reactivities of 1600:82:1 given on page 109.

(a) 4% 1-Br
96% 2-Br

(b) 0.6% 1°
99.4% 3°

(c) 0.3% 1°
99.7% 3°

(d) 1% 1-Br
66% 2-Br
33% 3-Br

(e) 0.3% 1-Br-2-Me
90% 2-Br-2-Me
9% 2-Br-3-Me
0.2% 1-Br-3-Me

(f) 0.6% 1-Br-2,2,3-triMe
99% 3-Br-2,2,3-triMe
0.4% 1-Br-2,3,3-triMe

(g) 0.5% 1-Br-2,2,4-triMe
9% 3-Br-2,2,4-triMe
90% 4-Br-2,2,4-triMe
0.3% 1-Br-2,4,4-triMe

3.15 Instead of 400/1, the ratio of products would be $400/(1 \times 10) = 40:1$, a ratio that would be much easier to measure accurately.

3.16 We allow for the relative numbers of hydrogens in each compound, C_2H_6 and C_5H_{12}, as we did on page 109. Each compound, of course, contains *only* primary hydrogens.

$$\frac{neo\text{-PeCl}}{\text{EtCl}} = \frac{\text{no. of neopentyl H}}{\text{no. of ethyl H}} \times \frac{\text{reactivity of neopentyl H}}{\text{reactivity of ethyl H}}$$

$$2.3 = \frac{12}{6} \times \frac{\text{reactivity of neopentyl H}}{\text{reactivity of ethyl H}}$$

$$\frac{\text{reactivity of neopentyl H}}{\text{reactivity of ethyl H}} = 2.3 \times \frac{6}{12} = 1.15$$

3.17 (a) Some *tert*-butyl chloride would have been formed, in two steps, from initially generated isobutyl radicals. These radicals would have been generated by abstraction of protium from the hydrocarbon, and hence with formation of HCl, not DCl. The *t*-BuCl:*iso*-BuCl ratio would have been larger than the DCl:HCl ratio, contrary to fact.

(b) Same as (a).

(Brown, H. C.; Russell, G. A. "Photochlorination of 2-Methylpropane-2-*d* and α-d_1-Toluene; the Question of Free Radical Rearrangement or Exchange in Substitution Reactions"; *J. Am. Chem. Soc.* **1952**, *74*, 3995.)

3.18 $t\text{-Bu—D} \xleftarrow{D_2O} t\text{-BuMgCl} \xleftarrow[Et_2O]{Mg} t\text{-BuCl}$

3.19 Add DBr, and see if unconsumed methane contains deuterium.

3.20 See if $^{35}Cl^{36}Cl$ (mass 71) and/or $^{36}Cl^{37}Cl$ (mass 73) shows up in the mass spectrum.

$$^{35}Cl\cdot + {}^{36}Cl\text{—}{}^{36}Cl \longrightarrow {}^{35}Cl\text{—}{}^{36}Cl + {}^{36}Cl\cdot$$

Mass = 71

$$^{37}Cl\cdot + {}^{36}Cl\text{—}{}^{36}Cl \longrightarrow {}^{37}Cl\text{—}{}^{36}Cl + {}^{36}Cl\cdot$$

Mass = 73

Other (normal) mass values would be 70 ($^{35}Cl\text{—}^{35}Cl$), 72 ($^{35}Cl\text{—}^{37}Cl$), and 74 ($^{37}Cl\text{—}^{37}Cl$).

3.21 The alkane was 2,2-dimethylhexane. (See Sec. 3.17.)

$$CH_3CH_2CH_2CH_2\text{—Br} + (CH_3\overset{CH_3}{\underset{CH_3}{C}}\text{—})_2CuLi \longrightarrow CH_3CH_2CH_2CH_2\text{—}\overset{CH_3}{\underset{CH_3}{C}}CH_3$$

n-Butyl bromide; Lithium di-*tert*-butylcopper; 2,2-Dimethylhexane

1. (a)
$$\begin{array}{cccc} & \mathrm{H_3C} & \mathrm{CH_3} & \\ \mathrm{CH_3CH_2}- & \mathrm{C}- & \mathrm{C} & -\mathrm{CH_3} \\ & \mathrm{H_3C} & \mathrm{CH_3} & \end{array}$$

(b)
$$\begin{array}{cccc} \mathrm{CH_3}- & \mathrm{CH}- & \mathrm{CH} & -\mathrm{CH_3} \\ & \mathrm{CH_3} & \mathrm{CH_3} & \end{array}$$

(c)
$$\begin{array}{ccccc} & \mathrm{CH_3} & \mathrm{CH_3} & & \\ \mathrm{CH_3CH_2}- & \mathrm{CH}- & \mathrm{C}- & \mathrm{CH} & -\mathrm{CH_2CH_3} \\ & & \mathrm{H_3C} & \mathrm{CH_3} & \end{array}$$

(d)
$$\begin{array}{cccc} & \mathrm{H_3C} & \mathrm{CH_3} & \\ \mathrm{CH_3CH_2CH_2}- & \mathrm{C}- & \mathrm{CH} & -\mathrm{CH_2CH_3} \\ & \mathrm{CH_2CH_3} & & \end{array}$$

(e)
$$\begin{array}{ccccc} & \mathrm{CH_3} & & \mathrm{CH_3} & \\ \mathrm{CH_3CH_2CH_2}- & \mathrm{C}- & \mathrm{CH_2}- & \mathrm{CH} & -\mathrm{CH_3} \\ & \mathrm{CH_2CH_3} & & & \end{array}$$

(f)
$$\begin{array}{ccccc} \mathrm{CH_3}- & \mathrm{CH}- & \mathrm{CH_2CH_2}- & \mathrm{CH} & -\mathrm{CH_3} \\ & \mathrm{CH_3} & & \mathrm{CH_3} & \end{array}$$

(g)
$$\begin{array}{cccc} & & \mathrm{CH_3} & \\ \mathrm{CH_3CH_2}- & \mathrm{CH}- & \mathrm{CH} & -\mathrm{CH_3} \\ & \mathrm{CH_2CH_3} & & \end{array}$$

(h)
$$\begin{array}{ccccc} & & & \mathrm{CH_3} & \\ \mathrm{CH_3}- & \mathrm{CH}- & \mathrm{CH_2}- & \mathrm{C} & -\mathrm{CH_3} \\ & \mathrm{CH_3} & & \mathrm{CH_3} & \end{array}$$

(i)
$$\begin{array}{cccc} & & \mathrm{CH_3} & \\ \mathrm{CH_3CH_2}- & \mathrm{CH}- & \mathrm{CH} & -\mathrm{CH_3} \\ & \mathrm{Cl} & & \end{array}$$

(j)
$$\begin{array}{ccc} & \mathrm{CH_3} & \\ \mathrm{CH_3}- & \mathrm{C}- & \mathrm{CH_2} \\ & \mathrm{Br} & \mathrm{Br} \end{array}$$

2. (a)
$$\begin{array}{ccccc} & \mathrm{CH_3} & & & \\ \mathbf{CH_3}- & \mathbf{CH}- & \mathbf{CH_2}- & \mathbf{CH_2}- & \mathbf{CH_3} \\ 1 & 2 & 3 & 4 & 5 \end{array}$$

. . . pentane	Longest chain: five carbons.
. . . methylpentane	Substituted by a methyl.
2-methylpentane	On C–2. (Numbering from other end would give C–4, a higher number.)

(b)
$$\begin{array}{ccc} & \mathrm{Br} & \\ \mathbf{CH_3}- & \mathbf{C}- & \mathbf{CH_3} \\ & \mathrm{Br} & \\ 3 & 2 & 1 \end{array}$$

. . . propane	Longest chain: three carbons.
. . . dibromopropane	Disubstituted by bromines.
2,2-dibromopropane	Both on C–2. (Could be numbered from other end.)

(c)
$$\begin{array}{ccccccccc} & & & & CH_3 & & & & \\ & & & & | & & & & \\ CH_3 & - & CH_2 & - & C & - & CH_2 & - & CH_3 \\ & & & & | & & & & \\ & & & & CH_3 & & & & \\ 1 & & 2 & & 3 & & 4 & & 5 \end{array}$$

. . . pentane	Longest chain: five carbons.
. . . dimethylpentane	Disubstituted by methyls.
3,3-dimethylpentane	Both on C–3. (Could be numbered from other end.)

(d)
$$\begin{array}{ccccccccc} & & & & CH_3 & & & & \\ & & & & | & & & & \\ CH_3 & - & CH_2 & - & C & - & CH_2 & - & CH_3 \\ & & & & | & & & & \\ & & & & CH_2CH_3 & & & & \\ 1 & & 2 & & 3 & & 4 & & 5 \end{array}$$

. . . pentane	Longest chain: five carbons.
. . . ethyl . . . methylpentane	Substituted by ethyl and methyl.
3-ethyl-3-methylpentane	Both on C–3. (Could be numbered from other end.)

Groups are named *in alphabetical order*.

(e)
$$\begin{array}{ccccccccccc} & & & & CH_3 & & CH_3 & & CH_3 & & \\ & & & & | & & | & & | & & \\ CH_3 & - & CH_2 & - & CH & - & CH & - & CH & - & CH_3 \\ 6 & & 5 & & 4 & & 3 & & 2 & & 1 \end{array}$$

. . . hexane	Longest chain: six carbons.
. . . trimethylhexane	Trisubstituted by methyls.
2,3,4-trimethylhexane	On C–2, C–3, and C–4. (Numbering from other end would give C–3, C–4, and C–5, a higher set of numbers.)

(f)
$$\begin{array}{ccccccccccccccc} & & & & CH_3 & & & & CH_2CH_3 & & & & & & \\ & & & & | & & & & | & & & & & & \\ CH_3 & - & CH_2 & - & CH & - & CH_2 & - & CH & - & CH_2 & - & CH_2 & - & CH_3 \\ 1 & & 2 & & 3 & & 4 & & 5 & & 6 & & 7 & & 8 \end{array}$$

. . . octane	Longest chain: eight carbons.
. . . ethyl . . . methyloctane	Substituted by ethyl and methyl.
5-ethyl-3-methyloctane	On C–5 and C–3. (Numbering from other end would give C–4 and C–6, a higher set of numbers.)

(g)
$$
\begin{array}{c}
\quad CH_3 \qquad\quad CH_3 \\
CH_3-\overset{|}{\underset{|}{C}}-CH_2-\overset{|}{\underset{|}{C}}-CH_3 \\
\quad CH_3 \qquad\quad CH_3 \\
5 \quad 4 \;\; 3 \quad\; 2 \;\; 1
\end{array}
$$

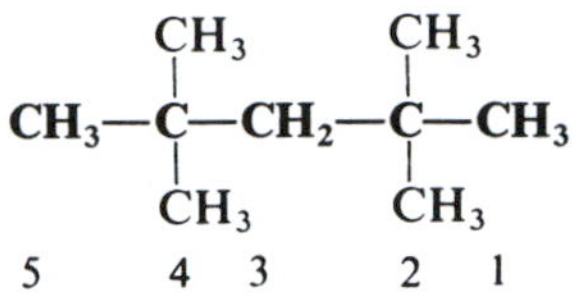

. . . pentane	Longest chain: five carbons.
. . . tetramethylpentane	Tetrasubstituted by methyls.
2,2,4,4-tetramethylpentane	On C–2 and C–4. (Could be numbered from other end.)

(h)
$$
\begin{array}{c}
CH_3 \\
CH_3-\overset{|}{\underset{|}{C}}-\underset{|}{CH}-CH_3 \\
Cl \;\; CH_3 \\
1 \quad 2 \;\; 3 \quad 4
\end{array}
$$

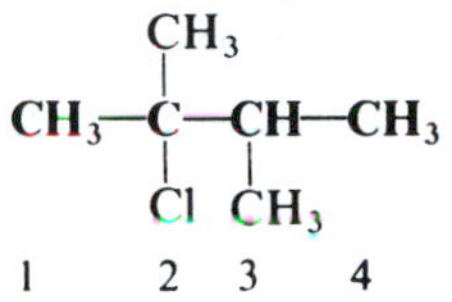

. . . butane	Longest chain: four carbons.
. . . chloro . . . dimethylbutane	Substituted by chlorine and two methyls.
2-chloro-2,3-dimethylbutane	At C–2, and C–2 and C–3. (Numbering from other end would give C–3, and C–2 and C–3, a higher set of numbers.)

(i)
$$
\begin{array}{c}
\quad CH_3 \qquad\qquad\qquad CH_2CH_3 \\
CH_3-\overset{|}{CH}-CH_2-CH_2-\overset{|}{CH}-CH_2-CH_3 \\
1 \quad 2 \quad 3 \quad 4 \quad 5 \quad 6 \quad 7
\end{array}
$$

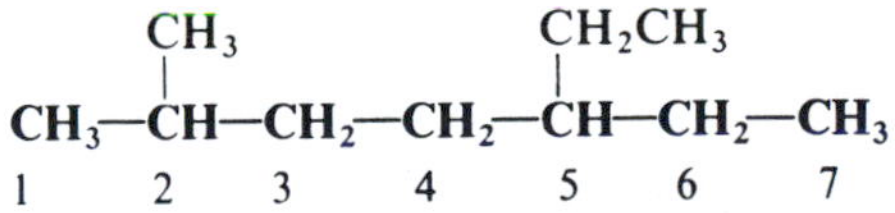

. . . heptane	Longest chain: seven carbons.
. . . ethyl . . . methylheptane	Substituted by ethyl and methyl.
5-ethyl-2-methylheptane	On C–5 and C–2. (Numbering from other end would give C–3 and C–6, a higher set of numbers.)

```
           CH3              CH3
            |                |
(j) CH3—CH—CH—CH2—C—CH2—CH3
                |            |
               CH3         CH2CH3
     1     2    3    4     5  6     7
```

... heptane	Longest chain: seven carbons.
... ethyl ... trimethylheptane	Substituted by ethyl and three methyls.
5-ethyl-2,3,5-trimethylheptane	On C–5, and C–2, C–3, and C–5. (Lowest set of numbers, as judged by first point of difference: C–2 or C–3?)

```
           CH3  CH2CH3
            |    |
(k) CH3—CH—C—CH2—CH2—CH3
                 |
               CH2CH3
     1     2    3  4     5     6
```

... hexane	Longest chain: six carbons.
... diethyl ... methylhexane	Substituted by two ethyls and a methyl.
3,3-diethyl-2-methylhexane	On C–3 and C–3, and C–2.

An alternative six-carbon chain would have only two (bigger) substituents. We pick the more highly substituted of two equal chains, to keep the substituents as simple as possible.

```
                 CH3
                  |
(l) CH3—CH2—CH—CH2—CH—CH2—CH2—CH3
                          |
                          CH
                         /  \
                      CH3    CH3
    1     2     3   4     5    6     7     8
```

... octane	Longest chain: eight carbons.
... isopropyl ... methyloctane	Substituted by isopropyl and methyl.
5-isopropyl-3-methyloctane	On C–5 and C–3. (Lowest set of numbers.)

3. (a) 1a, 2c, 2d, 2g
(b) 1d, 1e, 1h, 2a, 2k
(c) 1b, 1c, 1f, 1g, 2f, 2i, 2j
(d) 1b
(e) 1a, 1h, 2e, 2g
(f) 2d

For example:

```
                     1°   1°
     1°     2°      CH3  CH3     1°
(a) CH3—CH2—C——C—CH3
                     |     |
                    CH3  CH3
                     1°   1°
```
1a: No 3° H

```
                                     1°   3°
                                    CH3   H
     1°     2°     2°              |     |        2°     1°
(b) CH3—CH2—CH2—C——C—CH2—CH3
                                   |     |
                          CH3—CH2  CH3
                           1°    2°    1°
```
1d: One 3° H

```
                   3°      1°    3°
                   H      CH3    H
     1°     2°     |       |     |     2°     1°
(c) CH3—CH2—C——C——C—CH2—CH3
                   |       |     |
                  CH3     CH3   CH3
                   1°      1°    1°
```
1c: Two 3° H

```
            3°  3°
            H   H
     1°     |   |     1°
(d) CH3—C—C—CH3
            |   |
           H3C  CH3
            1°   1°
```
1b: No 2° H

```
          3°              1°
          H               CH3
     1°   |       2°      |        1°
(e) CH3—C—CH2—C——CH3
          |               |
         CH3             CH3
          1°              1°
```
1h: Two 2° H

```
                          1°
                          CH3
     1°     2°            |        2°     1°
(f) CH3—CH2—C——CH2—CH3
                          |
                   CH3—CH2
                    1°    2°
```
2d: Six 2° H *vs.* twelve 1° H

4. (a) 1e, 1g, 1h, 2a, 2e, 2i, 2j, 2k, 2l
(b) 1b, 1f
(c) 1e, 1h, 2a, 2i
(d) 1f
(e) 1d, 2e, 2f, 2l
(f) 1c
(g) 1a, 1h
(h) 2g
(i) 2e, 2l
(j) 1h
(k) 1d, 2f, 2l

For example:

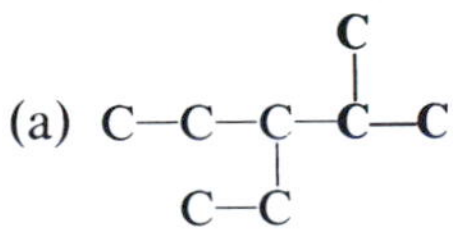
1g: One isopropyl

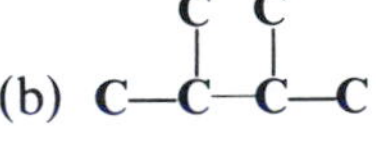
1b: Two isopropyls

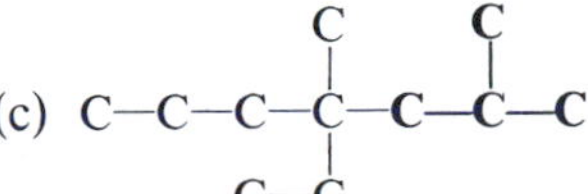
1e: One isobutyl

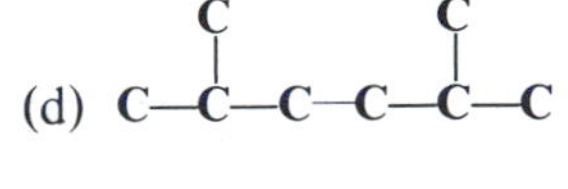
1f: Two isobutyls

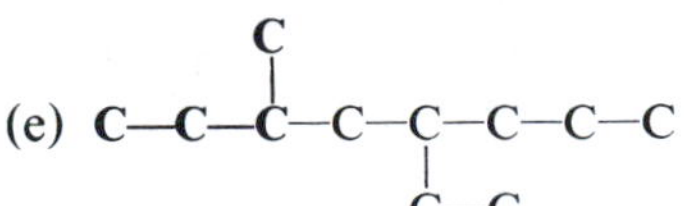
2f: One *sec*-butyl

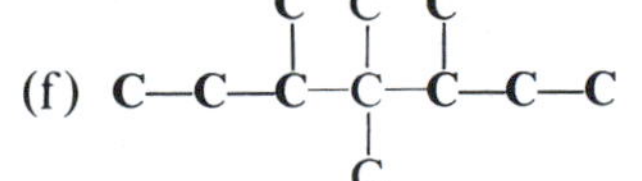
1c: Two *sec*-butyls

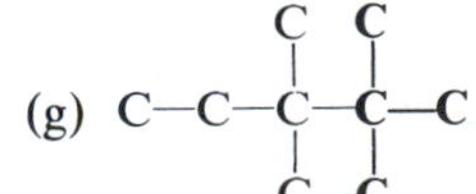
1a: One *tert*-butyl

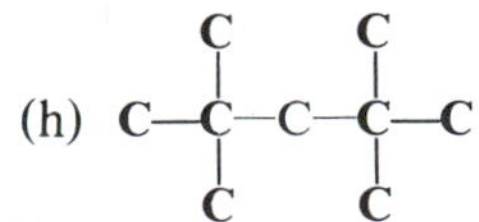
2g: Two *tert*-butyls

```
        C     C
        |     |
(i) C—C—C—C—C—C
          |
          C
```

2e: One isopropyl and one *sec*-butyl

```
      C     C
      |     |
(j) C—C—C—C—C
            |
            C
```

1h: One *tert*-butyl and one isobutyl

```
             C
             |
(k) C—C—C—C—C—C—C
             |  |
           C—C  C
```

1d: A methyl, an ethyl, a *n*-propyl, and a *sec*-butyl

5. We can easily calculate that to have a molecular weight of 86 an alkane must be a *hexane*, C_6H_{14}. As we saw on page 87, there are five of these. Following the approach of Problem 3.1, we arrive at the following answers. (Again we use stripped-down formulas to reveal the skeleton better.)

(a) Two monobromo derivatives: 2,3-dimethylbutane
(b) Three monobromo derivatives: *n*-hexane, 2,2-dimethylbutane
(c) Four monobromo derivatives: 3-methylpentane
(d) Five monobromo derivatives: 2-methylpentane
(e) 2,3-dimethylbutane has six dibromo derivatives
(f) See below.

```
C—C—C—C—C—C
```

n-Hexane

could give

```
C—C—C—C—C—C
          |
          Br

C—C—C—C—C—C
        |
        Br

C—C—C—C—C—C
      |
      Br
```

Three

```
        C
        |
C—C—C—C—C
```

2-Methyl-pentane

could give

```
        C
        |
C—C—C—C—C
          |
          Br

        C
        |
C—C—C—C—C
        |
        Br

        C
        |
C—C—C—C—C
      |
      Br

        C
        |
C—C—C—C—C
  |
  Br

        C
        |
C—C—C—C—C
|
Br
```

Five

```
    C
    |
C—C—C—C—C
```

3-Methyl-pentane

could give

```
    C
    |
C—C—C—C—C
        |
        Br

    C
    |
C—C—C—C—C
      |
      Br

    C
    |
C—C—C—C—C
    |
    Br

    C—Br
    |
C—C—C—C—C
```

Four

```
    C
    |
C—C—C—C
    |
    C
```

2,2-Dimethyl-butane

could give

```
    C
    |
C—C—C—C
    |  |
    C  Br

    C
    |
C—C—C—C
  |  |
  Br C

    C
    |
C—C—C—C
|   |
Br  C
```

Three

```
  C C
  | |
C—C—C—C
```

2,3-Dimethyl-butane

could give

```
  C C
  | |
C—C—C—C
      |
      Br

  C C
  | |
C—C—C—C
    |
    Br
```

Two

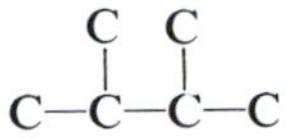

C–C(C)–C(C)–C (2,3-Dimethylbutane) *could give*

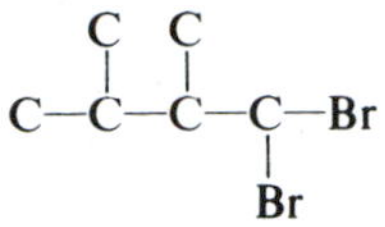

- C–C(C)–C(C)–CH(Br)–Br: 1,1-Dibromo-2,3-dimethylbutane 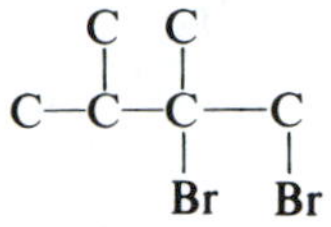
- C–C(C)–C(C)(Br)–C(Br): 1,2-Dibromo-2,3-dimethylbutane 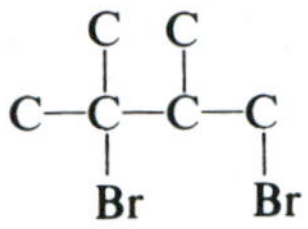
- C–C(C)(Br)–C(C)–C(Br): 1,3-Dibromo-2,3-dimethylbutane 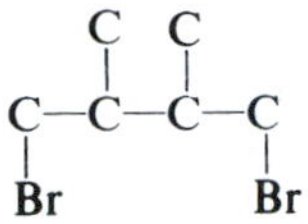
- C(Br)–C(C)–C(C)–C(Br): 1,4-Dibromo-2,3-dimethylbutane 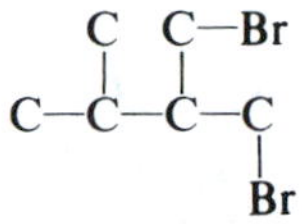
- C–C(C)–C(C–Br)–C(Br): 1-Bromo-1-(bromomethyl)-2,3-dimethylbutane 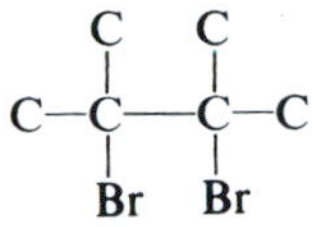
- C–C(C)(Br)–C(C)(Br)–C: 2,3-Dibromo-2,3-dimethylbutane

6. Starting with the structure of cyclopentane on page 444, we arrive at the following answers.

Mono
- Ring: CHCl, CH_2, CH_2, CH_2, CH_2

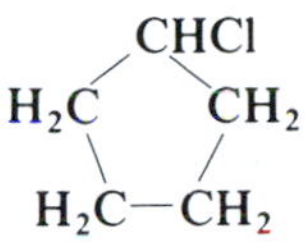

Di
- 1,1-: ring CCl_2, CH_2, CH_2, CH_2, CH_2 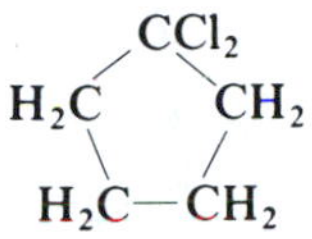
- 1,2-: ring CHCl, CHCl, CH_2, CH_2, CH_2
- 1,3-: ring CHCl, CH_2, CHCl, CH_2, CH_2 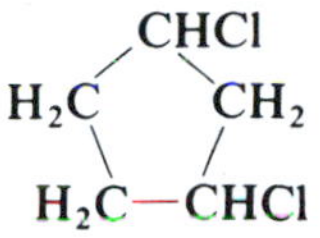

Tri
- 1,1,2-: ring CCl_2, CHCl, CH_2, CH_2, CH_2 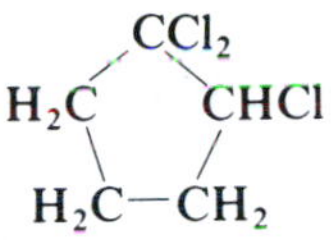
- 1,1,3-: ring CCl_2, CH_2, CHCl, CH_2, CH_2
- 1,2,3-: ring CHCl, CHCl, CHCl, CH_2, CH_2
- 1,2,4-: ring CHCl, CHCl, CH_2, CHCl, CH_2

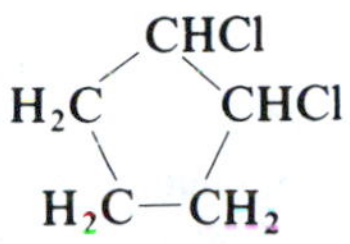

7. c, b, e, a, d.

8. (a) $$CH_3–CH(CH_3)–CH_2–Br + Mg \xrightarrow{\text{ether}} CH_3–CH(CH_3)–CH_2–MgBr$$

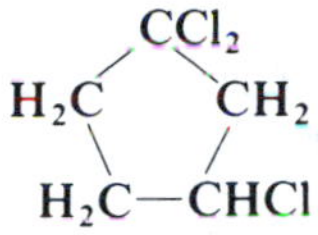

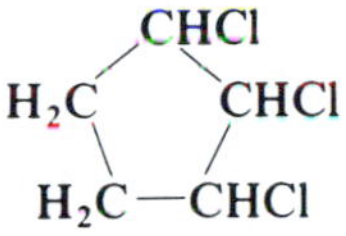

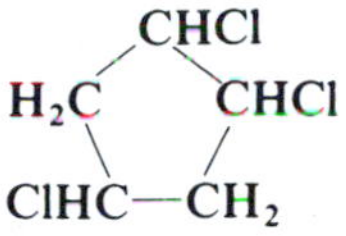

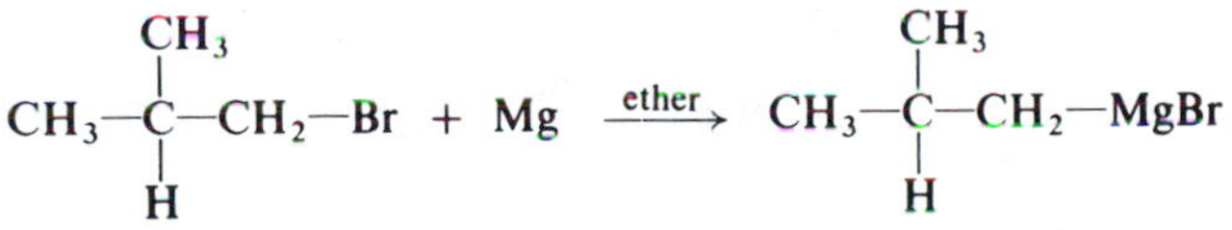

Isobutylmagnesium bromide

(b) $$CH_3–C(CH_3)_2–Br + Mg \xrightarrow{\text{ether}} CH_3–C(CH_3)_2–MgBr$$

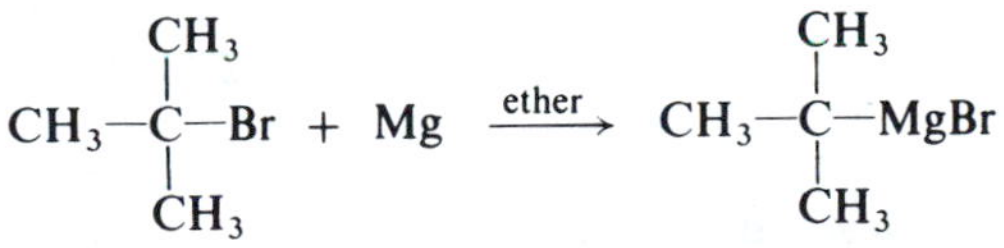

tert-Butylmagnesium bromide

(c) $$CH_3–CH(CH_3)–CH_2–MgBr + H_2O \longrightarrow CH_3–CH(CH_3)–CH_3 + Mg(OH)Br$$

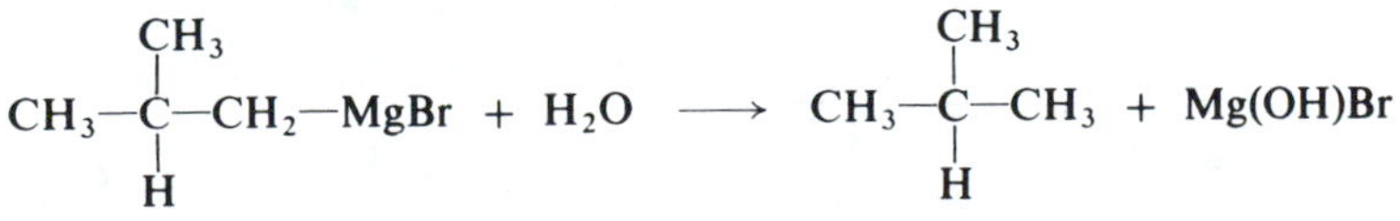

Isobutane

(d) $$CH_3-\underset{CH_3}{\overset{CH_3}{\underset{|}{\overset{|}{C}}}}-MgBr + H_2O \longrightarrow CH_3-\underset{CH_3}{\overset{CH_3}{\underset{|}{\overset{|}{C}}}}-H + Mg(OH)Br$$

Isobutane

(e) $$CH_3-\underset{H}{\overset{CH_3}{\underset{|}{\overset{|}{C}}}}-CH_2-MgBr + D_2O \longrightarrow CH_3-\underset{H}{\overset{CH_3}{\underset{|}{\overset{|}{C}}}}-CH_2D + Mg(OD)Br$$

Isobutane-1-*d*

(f) $$CH_3-CH_2-\overset{CH_3}{\overset{|}{C}}H-Cl + 2Li \longrightarrow CH_3-CH_2-\overset{CH_3}{\overset{|}{C}}HLi + LiCl$$

sec-Butyllithium

$$CH_3-CH_2-\overset{CH_3}{\overset{|}{C}}HLi + CuI \longrightarrow CH_3-CH_2-\overset{CH_3}{\overset{|}{C}}H-)_2CuLi$$

Lithium di-*sec*-butylcopper
Lithium di-*sec*-butylcuprate

(g) $$CH_3-CH_2-\overset{CH_3}{\overset{|}{C}}H-)_2CuLi + 2CH_3-CH_2-Br \longrightarrow 2CH_3-CH_2-\overset{CH_3}{\overset{|}{C}}H-CH_2-CH_3 + CuBr + LiBr$$

3-Methylpentane

9. (a) $CH_3CH_2CH_2CH_2Br \xrightarrow{Mg} CH_3CH_2CH_2CH_2MgBr \xrightarrow{H_2O} CH_3CH_2CH_2CH_3$

(b) $CH_3CH_2\overset{CH_3}{\overset{|}{C}}HBr \xrightarrow{Mg} CH_3CH_2\overset{CH_3}{\overset{|}{C}}HMgBr \xrightarrow{H_2O} CH_3CH_2CH_2CH_3$

(c) $CH_3CH_2Cl \xrightarrow{Li} CH_3CH_2Li \xrightarrow{CuCl} (CH_3CH_2)_2CuCl \xrightarrow{CH_3CH_2Cl} CH_3CH_2CH_2CH_3$

(d) $CH_3CH_2CH{=}CH_2 \xrightarrow{H_2,\ Ni} CH_3CH_2CH_2CH_3$

(e) $CH_3CH{=}CHCH_3 \xrightarrow{H_2,\ Ni} CH_3CH_2CH_2CH_3$

10. Once again we work with carbon skeletons, and follow the systematic approach of Problem 3.1.

(a) $C-C-C-C-C-\underset{Cl}{\underset{|}{C}}$ $\quad C-C-C-C-\underset{Cl}{\underset{|}{C}}-C$ $\quad C-C-C-\underset{Cl}{\underset{|}{C}}-C-C$

(b) $C-\overset{C}{\overset{|}{C}}-C-C-\underset{Cl}{\underset{|}{C}}$ $\quad C-\overset{C}{\overset{|}{C}}-C-\underset{Cl}{\underset{|}{C}}-C$ $\quad C-\overset{C}{\overset{|}{C}}-\underset{Cl}{\underset{|}{C}}-C-C$ $\quad C-\underset{Cl}{\underset{|}{\overset{C}{\overset{|}{C}}}}-C-C-C$ $\quad \underset{Cl}{\underset{|}{C}}-\overset{C}{\overset{|}{C}}-C-C-C$

```
        C     C             C     C             C     C             C     C
        |     |             |     |             |     |             |     |
(c) C—C—C—C—C       C—C—C—C—C       C—C—C—C—C       C—C—C—C—C
        |         |         |     |             |   |             |   |
        C         Cl        C     Cl            C   Cl            Cl  C

        C                 C               C
        |                 |               |
(d) C—C—C—C       C—C—C—C       C—C—C—C
        |   |             |   |         |   |
        C   Cl            C   Cl        Cl  C
```

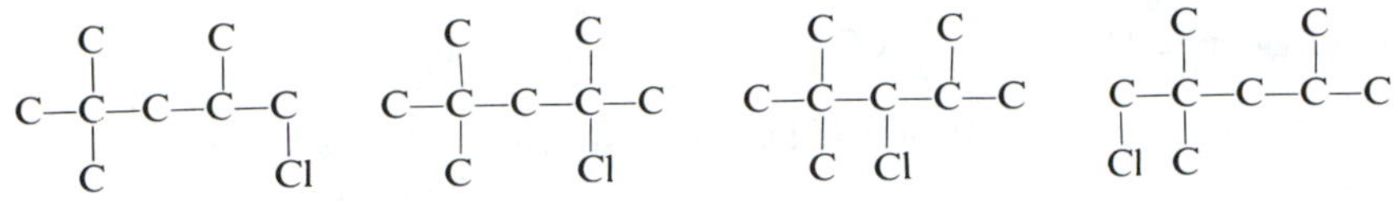

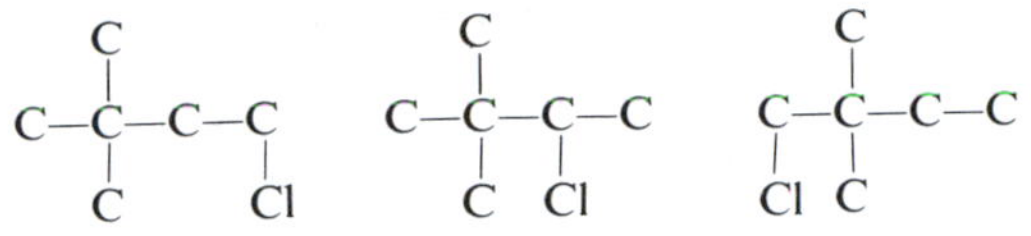

11. Order of isomers as in Problem 10:

(a) 16, 42, 42%
(b) 21, 17, 26, 26, 10%
(c) 33, 28, 18, 22%
(d) 46, 39, 15%

The calculations are made as described for Problem 3.13, page 28 of this Study Guide.

12. Allyl, benzyl > 3° > 2° > 1° > methyl, vinyl

13. Allylic, benzylic > 3° > 2° > 1° > CH_4, vinylic

14. (a) Water rapidly destroys a Grignard reagent.

(b) The —OH group is acidic enough to destroy a Grignard reagent, if indeed one could even be formed.

15. Rearrangement, *by migration of Br*, of initially formed 1° radical into more stable 2° or 3° radical:

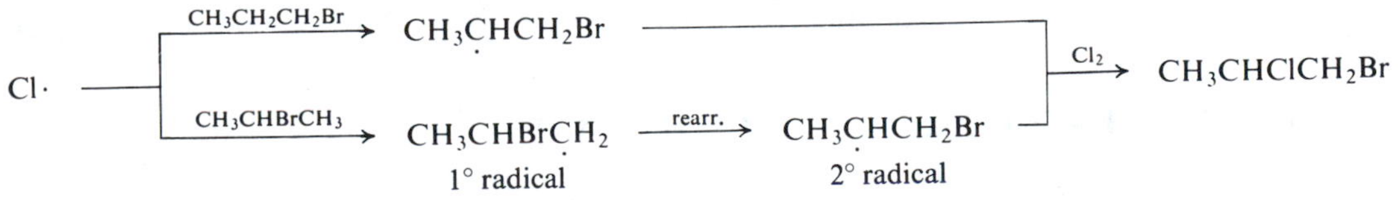

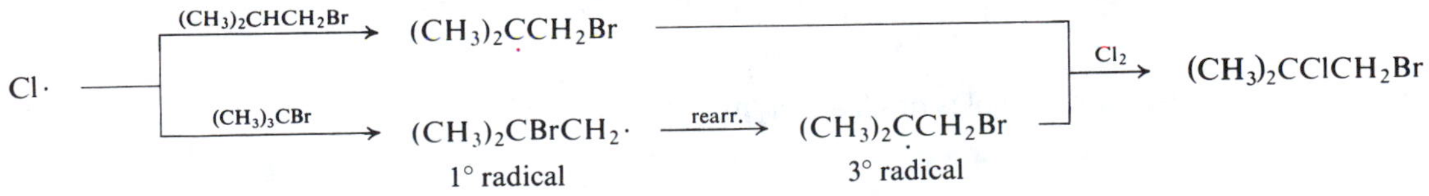

$Cl\cdot \xrightarrow{CH_3CH_2CH_2Br} CH_3\dot{C}HCH_2Br \xrightarrow{Cl_2} CH_3CHClCH_2Br$

$Cl\cdot \xrightarrow{CH_3CHBrCH_3} CH_3CHBr\dot{C}H_2$ (1° radical) $\xrightarrow{rearr.} CH_3\dot{C}HCH_2Br$ (2° radical) $\xrightarrow{Cl_2} CH_3CHClCH_2Br$

$Cl\cdot \xrightarrow{(CH_3)_2CHCH_2Br} (CH_3)_2\dot{C}CH_2Br \xrightarrow{Cl_2} (CH_3)_2CClCH_2Br$

$Cl\cdot \xrightarrow{(CH_3)_3CBr} (CH_3)_2CBrCH_2\cdot$ (1° radical) $\xrightarrow{rearr.} (CH_3)_2\dot{C}CH_2Br$ (3° radical) $\xrightarrow{Cl_2} (CH_3)_2CClCH_2Br$

Although alkyl free radicals seldom if ever rearrange by migration of hydrogen or alkyl, it seems clear that they can rearrange by migration of halogen.

16. (a)

$$n\text{-}C_{14}H_{30} + \frac{43}{2}O_2 \longrightarrow 14CO_2 + 15H_2O$$

From Table 3.3, n-$C_{14}H_{30}$ (m.w. 198) has a density of 0.764 g/mL. One liter of kerosine thus weighs 764 g, and is $\frac{764 \text{ g}}{198 \text{ g/mol}}$ moles. This amount of kerosine requires $\frac{764}{198} \times \frac{43}{2}$ moles of oxygen, which weighs $\frac{764}{198} \times \frac{43}{2} \times 32$ g, or 2650 g.

(b)

$$\begin{array}{llll} CH_3(CH_2)_{12}CH_3: & 12\text{—}CH_2\text{—} & \longrightarrow & 12 \times 157 \text{ kcal} = 1884 \text{ kcal} \\ & 2\ \ CH_3\text{—} & \longrightarrow & 2 \times 186 \text{ kcal} = \underline{\ 327} \text{ kcal} \\ & & & \qquad 2256 \text{ kcal/mol} \end{array}$$

One liter of kerosine:

$$\frac{764}{198} \text{ mol} \times 2256 \text{ kcal/mol} \longrightarrow 8710 \text{ kcal}$$

(c)

$$H\cdot + \cdot H \longrightarrow H_2 \qquad \Delta H = -104 \text{ kcal/mol}$$

To produce 8710 kcal requires

$$\frac{8710 \text{ kcal}}{104 \text{ kcal/mol}} = 84 \text{ mol } H_2 = 169 \text{ g}$$

17. Carius: mono, 45.3% Cl; di, 62.8% Cl.
Mol. wt. detn.: mono, 78.5; di, 113.

18. Try to synthesize it by the Corey–House method from isopentyl bromide.

$$CH_3\underset{\overset{|}{CH_3}}{C}HCH_2CH_2 \vdots CH_2CH_2\underset{\overset{|}{CH_3}}{C}HCH_3 \longleftarrow \left[\begin{array}{l} (CH_3\underset{\overset{|}{CH_3}}{C}HCH_2CH_2)_2CuLi \xleftarrow{\text{CuI}} \xleftarrow{\text{Li}} CH_3\underset{\overset{|}{CH_3}}{C}HCH_2CH_2Br \\ BrCH_2CH_2\underset{\overset{|}{CH_3}}{C}HCH_3 \end{array}\right.$$

(In each formula the CH_3 group is attached above the marked carbon.)

19. (a) Methane, formed by $CH_3OH + CH_3MgI \longrightarrow CH_4 + CH_3OMgI$.

$$1.04 \text{ mL gas is } \frac{1.04 \text{ mL}}{22.4 \text{ mL/mmol}} \text{ or } \frac{1.04}{22.4} \text{ mmol } CH_4,$$

evolved from the same number of millimoles of CH_3OH, m.w. 32.

$$\text{wt. } CH_3OH = \frac{1.04}{22.4} \text{ mmol} \times 32 \text{ mg/mmol} = 1.49 \text{ mg } CH_3OH$$

(no. of moles) × (wt. per mole)

(b)

$$\text{no. mmol alcohol} = \frac{1.57\text{ mL}}{22.4\text{ mL/mmol}}$$

$$\text{m.w. alc. (mg/mmol)} = \frac{\text{wt. alc. (mg)}}{\text{no. mmol alc.}} = \frac{4.12\text{ mg}}{1.57/22.4} = 59$$

$$59 = C_nH_{2n+1}OH = 12n + (2n+1)(1) + 16 + 1 = 14n + 18$$

$$n = 3 \longrightarrow C_3H_7OH\text{, which is } n\text{-PrOH or } iso\text{-PrOH.}$$

(c)

$$\text{mmol of alcohol} = \frac{1.79\text{ mg}}{90\text{ mg/mmol}} = \text{approx. 0.02 mmol alcohol}$$

$$0.02\text{ mmol alcohol} \longrightarrow \frac{1.34\text{ mL}}{22.4\text{ mL/mmol}} = 0.06\text{ mmol } H_2$$

$$\text{each mole alcohol} \longrightarrow 3\text{ moles } H_2$$

alcohol contains 3 —OH groups

$$\text{m.w. } 90 = C_nH_{2n-1}(OH)_3 = 12n + (2n-1)(1) + (3 \times 16) + (3 \times 1)$$

$$90 = 14n + 50$$

$n = 3$ gives $CH_2(OH)—CH(OH)—CH_2(OH)$, glycerol (page 214).

(We might draw alternative structures containing more than one —OH per carbon, but such compounds are, in general, highly unstable.)

20. (a) (1) $(CH_3)_3CO—OC(CH_3)_3 \xrightarrow{130\,°C} 2(CH_3)_3CO\cdot$

(2) $(CH_3)_3CO\cdot + (CH_3)_3CH \longrightarrow (CH_3)_3COH + (CH_3)_3C\cdot$

(3) $(CH_3)_3C\cdot + CCl_4 \longrightarrow (CH_3)_3CCl + Cl_3C\cdot$

(4) $Cl_3C\cdot + (CH_3)_3CH \longrightarrow Cl_3CH + (CH_3)_3C\cdot$

then (3), (4), (3), (4), etc.

(b) (1) $(CH_3)_3C—O—Cl \xrightarrow{\text{light}} (CH_3)_3CO\cdot + Cl\cdot$

(2) $(CH_3)_3CO\cdot + RH \longrightarrow (CH_3)_3COH + R\cdot$

(3) $R\cdot + (CH_3)_3C—O—Cl \longrightarrow RCl + (CH_3)_3CO\cdot$

then (2), (3), (2), (3), etc.

4

Stereochemistry I. Stereoisomers

4.1 If methane were a pyramid with a rectangular base: two stereoisomers.

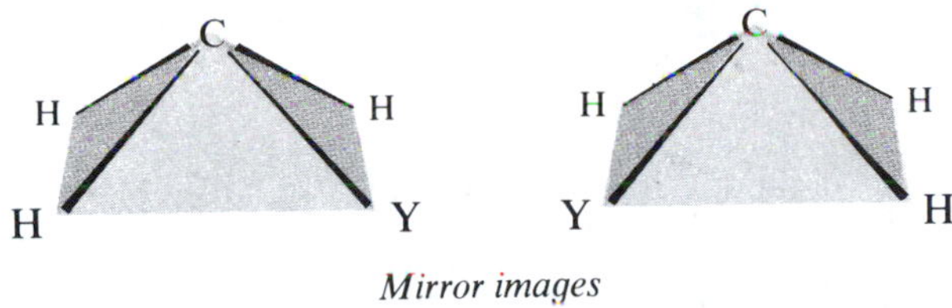

Mirror images

These are mirror images but, as we can see, they are *not* superimposable.

4.2 (a) If methane were rectangular: three stereoisomers.

(b) If methane were square: two stereoisomers.

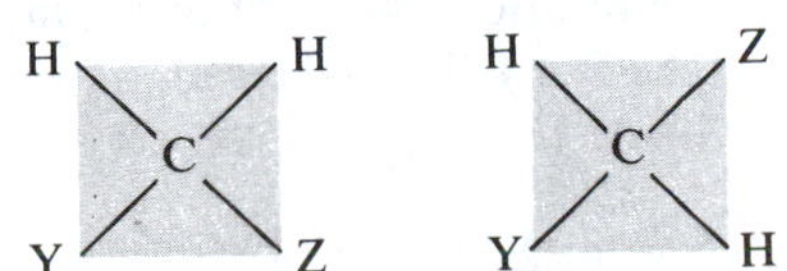

(c) If methane were a pyramid with a square base: three stereoisomers.

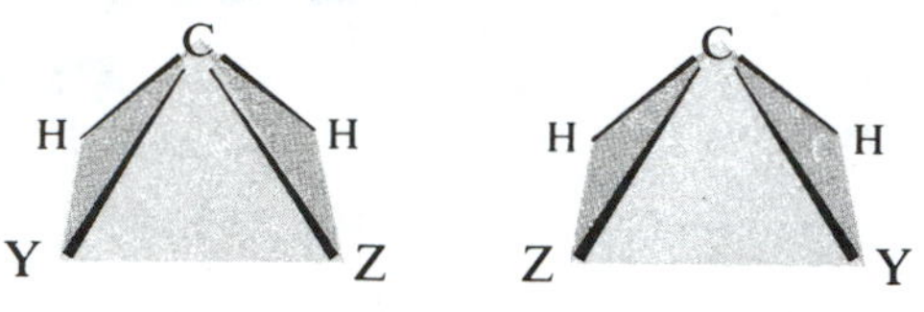

Mirror images

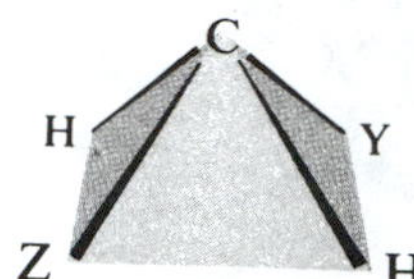

Two of these are mirror images of each other, but not superimposable.

(d) If methane were tetrahedral: only one structure is possible.

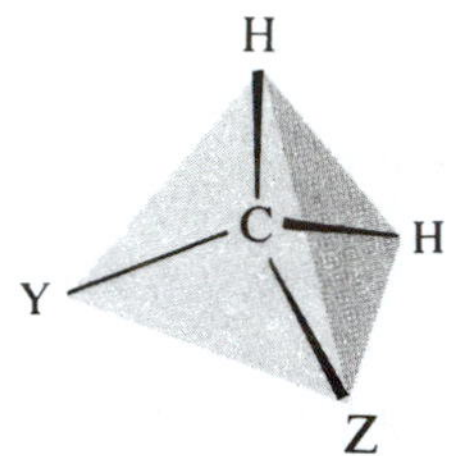

(The mirror image of the structure shown is superimposable on the original.)

4.3 (a)

$$[\alpha] = \frac{\alpha}{l \times d} = \frac{-1.2^\circ}{0.5 \text{ dm} \times \frac{6.15}{100}} = -39.0^\circ$$

(b) $-39.0^\circ = \frac{\alpha}{1.0 \times \frac{6.15}{100}}$ A tube twice as long $\longrightarrow$ double observed rotation

$$\alpha = -2.4^\circ$$

(c) $-39.0^\circ = \frac{\alpha}{0.5 \times \frac{6.15}{200}}$ Halved conc. $\longrightarrow$ halved no. of molecules $\longrightarrow$ halved rotn.

$$\alpha = -0.6^\circ$$

4.4 Use a shorter or longer tube, and measure the rotation again. For example, if $[\alpha]$ were actually $+45^\circ$, a 2.5-cm tube $\longrightarrow$ α_{obs} of $+11.25^\circ$. Other values of $[\alpha]$ would give α_{obs} as follows:

$$-315^\circ \longrightarrow -78.75^\circ; \quad +405^\circ \longrightarrow +101.25^\circ; \quad +765^\circ \longrightarrow +191.25^\circ.$$

4.5 Chiral: b, d, f, g, h.

We first look for chiral centers:

(b) C—C—C—C*H(Cl)—C (d) C—C—C—C*H(C)—CH_2Cl (f) C—C—C*H(Cl)—C(C)—C

(g) C—C*H(Cl)—C—C(C)—C (h) C—C—C*H(Br)—CH_2Cl

If we find a chiral center—and *if* we are certain that it is the *only* one—we can be certain that the molecule is chiral. (If there is more than one chiral center, the molecule may or may not be chiral, as we shall see in Sec. 4.17.) At this stage, it is safest to check our conclusion by drawing cross formulas or, better, wedge formulas or, best of all, by making models. For example, for compound (h):

mirror | mirror

CH_2Cl / H—+—Br / C_2H_5 ⋮ CH_2Cl / Br—+—H / C_2H_5 *or* CH_2Cl / H—C—Br / C_2H_5 ⋮ CH_2Cl / Br—C—H / C_2H_5

Non-superimposable: molecule is chiral

4.6 (a) $CH_3CH_2\overset{*}{C}HDCl$ $CH_3\overset{*}{C}HDCH_2Cl$ $DCH_2CH_2CH_2Cl$ $CH_3\overset{*}{C}HClCH_2D$ $(CH_3)_2CDCl$

(b) All those containing a chiral center ($\overset{*}{C}$) are chiral.

After identifying chiral centers, we follow the procedure of Problem 4.5. For example, for $CH_3CH_2\overset{*}{C}HDCl$:

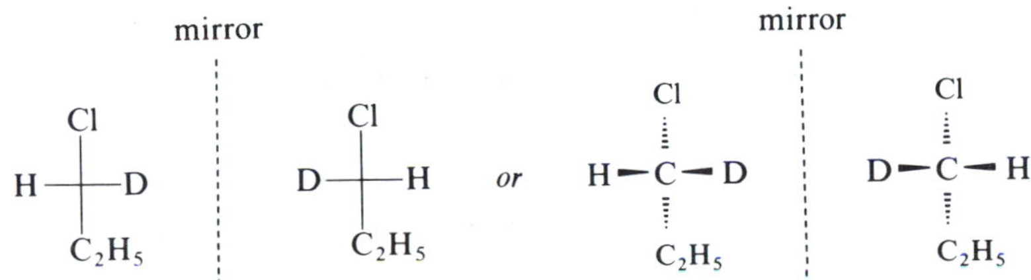

Non-superimposable: molecule is chiral

4.7 (a)–(c) Work with models; (d) mirror images: a, b.

4.8 (a)

Cl	Cl
I—┼—SO_3H	HO_3S—┼—I
H	H
R	*S*

I > Cl > S > H

(b)

D	D
CH_3—┼—Br	Br—┼—CH_3
H	H
R	*S*

Br > C > D > H

4.9 3° > 2° > 1° > CH_3

C	C	H	H
—C—C	—C—C	—C—C	—C—H
C	H	H	H
C,C,C >	C,C,H >	C,H,H >	H,H,H
3°	2°	1°	CH_3

4.10 (a)

H	H	CH_3	CH_3
n-C_3H_7—┼—CH_3	CH_3—┼—C_3H_7-*n*	*n*-C_3H_7—┼—CH_2Cl	$ClCH_2$—┼—C_3H_7-*n*
Cl	Cl	H	H
R	*S*	*R*	*S*

Cl > 1° > CH_3 > H

C—Cl > C—C > C—H > H

H	H	H	H
iso-C_3H_7—┼—C_2H_5	C_2H_5—┼—C_3H_7-*iso*	*iso*-C_4H_9—┼—CH_3	CH_3—┼—C_4H_9-*iso*
Cl	Cl	Cl	Cl
R	*S*	*R*	*S*

Cl > 2° > 1° > H

Cl > 1° > CH_3 > H

H	H
$ClCH_2$—┼—C_2H_5	C_2H_5—┼—CH_2Cl
Br	Br
R	*S*

Br > C—Cl > C—C > H

(b)

H	H	H	H
C_2H_5—┼—D	D—┼—C_2H_5	DCH_2—┼—CH_3	CH_3—┼—CH_2D
Cl	Cl	Cl	Cl
R	*S*	*R*	*S*

Cl > C > D > H

Cl > C—D > C—H > H

H	H
$ClCH_2$—┼—CH_3	CH_3—┼—CH_2Cl
D	D
R	*S*

C—Cl > C—H > D > H

4.11

(a)

$$\begin{array}{c} Cl \\ C_2H_5 - \!\!+\!\! - CH{=}CH_2 \\ H \\ \textit{R} \end{array} \qquad \begin{array}{c} Cl \\ CH_2{=}CH - \!\!+\!\! - C_2H_5 \\ H \\ \textit{S} \end{array}$$

Cl > C═C > C—C > H

(b)

$$\begin{array}{c} Cl \\ (CH_3)_2CH - \!\!+\!\! - CH{=}CH_2 \\ H \\ \textit{R} \end{array} \qquad \begin{array}{c} Cl \\ CH_2{=}CH - \!\!+\!\! - CH(CH_3)_2 \\ H \\ \textit{S} \end{array}$$

Cl > C═C > C—C > H

(c)

$$\begin{array}{c} H \\ HOOC - \!\!+\!\! - CH_2COOH \\ OH \\ \textit{R} \end{array} \qquad \begin{array}{c} H \\ HOOCCH_2 - \!\!+\!\! - COOH \\ OH \\ \textit{S} \end{array}$$

O > C—O > C—C > H

(d)

$$\begin{array}{c} H \\ C_6H_5 - \!\!+\!\! - CH_3 \\ NH_2 \\ \textit{R} \end{array} \qquad \begin{array}{c} H \\ CH_3 - \!\!+\!\! - C_6H_5 \\ NH_2 \\ \textit{S} \end{array}$$

N > C_{aryl} > C_{aliph} > H

(e)

$$\begin{array}{c} CH_3 \\ (CH_3)_2CH - \!\!+\!\! - C_3H_7\text{-}n \\ C_2H_5 \\ \textit{R} \end{array} \qquad \begin{array}{c} CH_3 \\ n\text{-}C_3H_7 - \!\!+\!\! - CH(CH_3)_2 \\ C_2H_5 \\ \textit{S} \end{array}$$

2° > 1° C—C—C > 1° C—C—H > CH_3

(f)

$$\begin{array}{c} H \\ HOOC - \!\!+\!\! - C_6H_5 \\ OH \\ \textit{R} \end{array} \qquad \begin{array}{c} H \\ C_6H_5 - \!\!+\!\! - COOH \\ OH \\ \textit{S} \end{array}$$

O > C—O > C_{aryl} > H

(g)

$$\begin{array}{c} H \\ HOOC - \!\!+\!\! - CH_3 \\ NH_2 \\ \textit{R} \end{array} \qquad \begin{array}{c} H \\ CH_3 - \!\!+\!\! - COOH \\ NH_2 \\ \textit{S} \end{array}$$

N > C—O > C—H > H

4.12 (a), (g): two enantiomers, both active
(b), (c), (e): two enantiomers, both active; one inactive *meso* compound
(f), (h): two pairs of enantiomers, all active
(d): four pairs of enantiomers, all active

Draw cross formulas as below, and check your conclusions by use of models.

Every enantiomer, if separated from all other stereoisomers, will be optically active. Every *meso* compound will be optically inactive. Any stereoisomers that are not enantiomers are diastereomers.

(a)

$$\begin{array}{c} CH_2Br \\ H - \!\!+\!\! - Br \\ CH_3 \end{array} \qquad \begin{array}{c} CH_2Br \\ Br - \!\!+\!\! - H \\ CH_3 \end{array}$$

Enantiomers

(b)

$$\begin{array}{c} C_2H_5 \\ CH_3 - \!\!+\!\! - Br \\ Br - \!\!+\!\! - CH_3 \\ C_2H_5 \end{array} \qquad \begin{array}{c} C_2H_5 \\ Br - \!\!+\!\! - CH_3 \\ CH_3 - \!\!+\!\! - Br \\ C_2H_5 \end{array} \qquad \begin{array}{c} C_2H_5 \\ CH_3 - \!\!+\!\! - Br \\ CH_3 - \!\!+\!\! - Br \\ C_2H_5 \end{array}$$

Enantiomers *Meso*

(c)

CH_3	CH_3	CH_3
H—Br	Br—H	H—Br
CH_2	CH_2	CH_2
Br—H	H—Br	H—Br
CH_3	CH_3	CH_3
Enantiomers		*Meso*

(d)

CH_3	CH_3	CH_3	CH_3
H—Br	Br—H	Br—H	H—Br
H—Br	Br—H	H—Br	Br—H
H—Br	Br—H	H—Br	Br—H
C_2H_5	C_2H_5	C_2H_5	C_2H_5
Enantiomers		*Enantiomers*	

CH_3	CH_3	CH_3	CH_3
H—Br	Br—H	H—Br	Br—H
Br—H	H—Br	H—Br	Br—H
H—Br	Br—H	Br—H	H—Br
C_2H_5	C_2H_5	C_2H_5	C_2H_5
Enantiomers		*Enantiomers*	

(e)

CH_2Br	CH_2Br	CH_2Br
H—Br	Br—H	H—Br
Br—H	H—Br	H—Br
CH_2Br	CH_2Br	CH_2Br
Enantiomers		*Meso*

(f)

CH_3	CH_3	CH_3	CH_3
H—Br	Br—H	H—Br	Br—H
Cl—H	H—Cl	H—Cl	Cl—H
CH_3	CH_3	CH_3	CH_3
Enantiomers		*Enantiomers*	

(g)

CH_2Cl	CH_2Cl
CH_3—H	H—CH_3
C_2H_5	C_2H_5
Enantiomers	

(h)

CH_2Cl	CH_2Cl	CH_2Cl	CH_2Cl
CH_3—H	H—CH_3	CH_3—H	H—CH_3
H—Cl	Cl—H	Cl—H	H—Cl
CH_3	CH_3	CH_3	CH_3

Enantiomers *Enantiomers*

4.13 The order of isomers is the same as that given above in Problem 4.12.

(a) (R); (S). (b) (S,S); (R,R); (S,R) or (R,S). (c) (S,S); (R,R); (S,R) or (R, S).

(d) (2S,3S,4R); (2R,3R,4S); (2R,3S,4R); (2S,3R,4S); (2S,3R,4R); (2R,3S,4S); (2S,3S,4S); (2R,3R,4R).

(e) (S,S); (R,R); (S,R) or (R,S). (f) (2S,3S); (2R,3R); (2S,3R); (2R,3S).

(g) (S); (R). (h) (2R,3R); (2S,3S); (2R,3S); (2S,3R).

4.14 Refer to the curves on pages 24–25 of this Study Guide.

(a) three conformers: A, B, C; B and C are enantiomers
(b) three conformers: D, E, F; D and F are enantiomers
(c) one conformer: G

In (a), B and C are less abundant (more Me–Me *gauche* interactions).

In (b), D and F are more abundant (fewer Me–Me *gauche* interactions).

4.15 (a)

I: front H (top), Br, Br; back Br, Br, H (bottom)

II: front H (top), H, Br; back Br, Br, Br (bottom)

III: front H (top), Br, H; back Br, Br, Br (bottom)

The two forms are made up of I, and II plus III.

(b) Neither form is active: I is achiral, and II plus III is a racemic modification.

4.16 (a) Abstraction of either 2° hydrogen is equally likely, giving equal amounts of enantiomeric pyramidal radicals, each of which reacts equally readily with chlorine, thus giving equal amounts of enantiomeric *sec*-BuCl molecules (i.e., the racemic modification); in any case, a pyramidal radical would probably undergo rapid inversion and lose configuration.

(b) Displacement of either 2° hydrogen is equally likely, giving equal amounts of enantiomeric *sec*-BuCl molecules (racemic modification).

4.17 (a) Four fractions.

$$\underset{\text{Isopentane}}{CH_3CH_2\overset{CH_3}{\overset{|}{C}}HCH_3} \xrightarrow{Cl_2} \underset{\text{1-Chloro-2-methylbutane}}{CH_3CH_2\overset{CH_3}{\overset{|}{C}}HCH_2Cl} + \underset{\text{2-Chloro-2-methylbutane}}{CH_3CH_2\overset{CH_3}{\overset{|}{C}}CH_3 \atop \quad\quad Cl} + \underset{\text{2-Chloro-3-methylbutane}}{CH_3\underset{Cl}{C}H\overset{CH_3}{\overset{|}{C}}HCH_3} + \underset{\text{1-Chloro-3-methylbutane}}{ClCH_2CH_2\overset{CH_3}{\overset{|}{C}}HCH_3}$$

(b)

CH_2Cl	CH_2Cl	$CH(CH_3)_2$	$CH(CH_3)_2$
CH_3—+—H	H—+—CH_3	H—+—Cl	Cl—+—H
C_2H_5	C_2H_5	CH_3	CH_3
S	*R*	*R*	*S*

C—Cl > C—C > C—H > H (1-Chloro-2-methylbutane, *Racemic*)

Cl > C—C > C—H > H (2-Chloro-3-methylbutane, *Racemic*)

$$CH_3CH_2\text{—}\overset{CH_3}{\underset{Cl}{C}}\text{—}CH_3 \qquad ClCH_2CH_2\text{—}\overset{CH_3}{\underset{H}{C}}\text{—}CH_3$$

2-Chloro-2-methylbutane *Achiral* 1-Chloro-3-methylbutane *Achiral*

(c) No fraction is optically active: each fraction is either a single achiral compound or a racemic modification. (Inactive reactants give inactive products.)

(d) Replacement of either H on —CH_2— to give 1-chloro-2-methylbutane is equally likely, and replacement of any H on either —CH_3 to give 2-chloro-3-methylbutane is equally likely, in each case giving rise to the racemic modification.

4.18 In answer to the question in the middle of page 153: further (mono)chlorination of *sec*-butyl chloride can occur at any one of four carbons:

$$CH_3CH_2\underset{Cl}{C}HCH_3 \xrightarrow{Cl_2} \underset{\text{1,2-Dichlorobutane}}{CH_3CH_2\underset{Cl}{C}H\text{—}\underset{Cl}{C}H_2} + \underset{\text{2,2-Dichlorobutane}}{CH_3CH_2\text{—}\overset{Cl}{\underset{Cl}{C}}\text{—}CH_3} + \underset{\text{2,3-Dichlorobutane}}{CH_3\underset{Cl}{C}H\text{—}\underset{Cl}{C}HCH_3} + \underset{\text{1,3-Dichlorobutane}}{\underset{Cl}{C}H_2CH_2\underset{Cl}{C}HCH_3}$$

We shall discuss the stereochemical nature of the 2,3-dichlorobutane in Sec. 4.26. At this point we are asked about the stereochemical composition of the other products if the starting material is the optically active compound, (*S*)-*sec*-butyl chloride.

(a)

CH_3		CH_2Cl		CH_3		CH_3
H—C—Cl	⟶	H—C—Cl	+	Cl—C—Cl	+	H—C—Cl
C_2H_5		C_2H_5		C_2H_5		CH_2CH_2Cl
(*S*)-*sec*-Butyl chloride		(*R*)-1,2-Dichlorobutane		2,2-Dichlorobutane		(*S*)-1,3-Dichlorobutane
Cl > C—C > C—H > H		Cl > C—Cl > C—C > H		*Achiral*		Cl > C—C > C—H > H

In the formation of the 1,3-dichlorobutane, as in the formation of the 1,2 compound (Sec. 4.23), no bond to the chiral center is broken, and hence configuration is retained about that center. (The product has the same specification, *S*, as the starting material, but that is just accidental: $—CH_2CH_2Cl$ happens to be of the same relative priority as the $—C_2H_5$. Indeed, we notice that the 1,2-dichlorobutane obtained has the *opposite* specification—yet it, too, is formed with retention of configuration.)

In the formation of the 2,2-dichlorobutane, the chiral center is destroyed—the carbon now holds two identical ligands—and the compound is achiral.

(b) The 1,2- and 1,3-dichlorobutanes will be optically active; the achiral 2,2-dichlorobutane will be optically inactive.

4.19 c, d, e, g. In the others, a bond to chiral carbon is broken.

To answer this, we need to draw out the structures of the reactants and the products, and identify any chiral centers.

4.20 (a) The C—O bond is broken, and re-formed.

Chemically, the starting material and the product are the same: *sec*-butyl alcohol. Stereochemically, however, they are different: the starting material is optically active, and must consist predominantly of one enantiomer; the product is optically inactive, and must consist of equal parts of both enantiomers. Some of the molecules must, then, change their configuration; they can do this only through the breaking of a bond to the chiral center. But the product, like the starting material, is *sec*-butyl alcohol; and therefore, for every bond broken, a bond to the chiral center must be made.

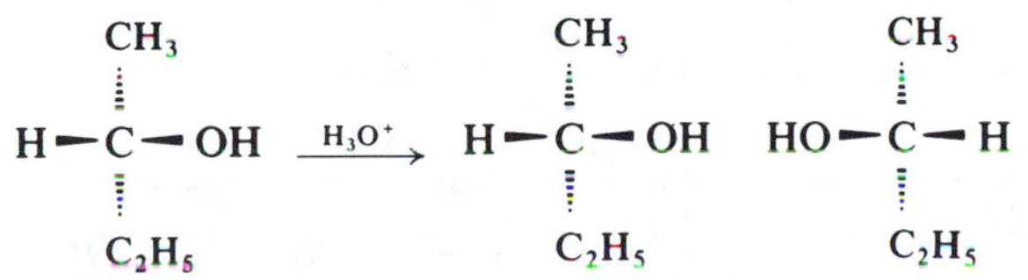

$$\begin{array}{c} CH_3 \\ | \\ H—C—OH \\ | \\ C_2H_5 \end{array} \xrightarrow{H_3O^+} \begin{array}{c} CH_3 \\ | \\ H—C—OH \\ | \\ C_2H_5 \end{array} \quad \begin{array}{c} CH_3 \\ | \\ HO—C—H \\ | \\ C_2H_5 \end{array}$$

(or enantiomer) — Racemic modification

Optically active — *Optically inactive*

Drawing the structure, we find that the chiral center is the carbon holding the —OH. This carbon has three other bonds; but we know that alkanes are inert to aqueous acids (Sec. 3.18), and it seems highly unlikely that a C—H or C—C bond is the one being broken—and re-formed—under the experimental conditions.

By contrast, we can visualize, in a general way, how aqueous acid might bring about the breaking and re-making of the C—O bond. The —OH of an alcohol is basic, and accepts a proton to form the protonated alcohol (Sec. 1.22). Now, if the C—O bond of this

$$\underset{\text{Alcohol}}{R—OH} + H^+ \rightleftarrows \underset{\text{Protonated alcohol}}{R—OH_2^+} \rightleftarrows \underset{\text{Water}}{H_2O} + \,?$$

were to break, one of the products would be H_2O. Since H_2O is the solvent, there is plenty of it about, and the C—O bond could be re-formed—either subsequently to the bond-breaking, or simultaneously with it—by the reaction of one of these abundant molecules.

We conclude that, most likely, it is the C—O bond that is being broken and re-formed under the influence of acid.

(As we shall see later (Sec. 6.13), this is exactly what we would expect of an alcohol under these conditions.)

(b) The C—I bond is broken, and re-formed.

Our line of reasoning here is similar to that in part (a).

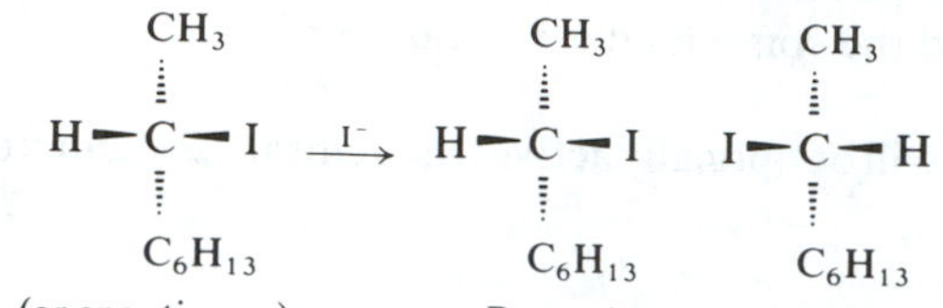

$$\begin{array}{c} CH_3 \\ \vdots \\ H - C - I \\ \vdots \\ C_6H_{13} \end{array} \xrightarrow{I^-} \begin{array}{c} CH_3 \\ \vdots \\ H - C - I \\ \vdots \\ C_6H_{13} \end{array} \quad \begin{array}{c} CH_3 \\ \vdots \\ I - C - H \\ \vdots \\ C_6H_{13} \end{array}$$

(or enantiomer) Racemic modification

Optically active *Optically active*

Of the four bonds to the chiral center, it is the C—I that is most likely to break, and probably in such a way as to form that familiar species, the iodide ion, I^-. Although we cannot at this point suggest a specific way (like the protonation in part (a)) in which I^- actually *brings about* reaction, it is clear that its presence gives opportunity for the remaking of the C—I bond.

$$R{-}I \overset{I^-}{\rightleftharpoons} I^- + ?$$

Alkyl iodide Iodide ion

(c) We could carry out each of these experiments with $H_2{}^{18}O$ or radioactive I^-. If our interpretation is correct, we would find ^{18}O or radioactive I^- in the products.

Let us begin with part (a). The C—O bond breaks, and the —OH group leaves the chiral center. It probably leaves as H_2O; but, even if it does not, it would certainly be *converted into* H_2O very rapidly by the medium. (An OH^- does not last long in an acidic solution.) And so, in one way or another, the —OH group ends up in the solution as H_2O.

Next, the other half of the process. A C—O bond forms, and an —OH group becomes attached to the chiral carbon: either after the C—O bond breaks, or at the same time. This —OH must come, again in one way or another, from H_2O. But the solvent is water and, in the experiment we are proposing, it is $H_2{}^{18}O$. The —OH that becomes attached to carbon need not be the same —OH that left—and, indeed, it is highly unlikely that it would be. It is overwhelmingly more likely to come from one of the abundant solvent molecules, and thus be —^{18}OH.

If we are right, then, and the loss of optical activity involves the breaking and making of C—O bonds, then this loss of activity should go hand-in-hand with the exchange of one oxygen isotope for another.

For the experiment with the alkyl iodide and iodide ion, our line of reasoning is exactly analogous. The iodine that becomes attached to carbon is more likely to be one of the abundant radioactive I^- ions of the reagent than one of the relatively few ordinary I^- ions provided by the organic compound. Again, we expect loss of optical activity to go hand-in-hand with isotopic exchange.

Both of these experiments have actually been carried out, and with the expected results. As we shall see later (Problem 5.4, page 183), the work with the alkyl iodide, done in 1935 in a particularly elegant way, is considered to have established one of the mechanisms for what is perhaps the most important class of organic reactions, nucleophilic aliphatic substitution.

4.21

$$(+3.54°/+5.90°) \times (-1.67°) = 1.00°$$

Purity of product same as purity of starting material

4.22 As always in analyzing the stereochemistry of such reactions, we follow the guidelines given at the bottom of page 157: (1) unless a bond to an original chiral center is broken, the configuration is retained at that center; and (2) if a new chiral center is generated, *both* possible configurations about the new center result.

(a) There will be five diastereomeric fractions: two optically inactive and three optically active.

CH_2Cl / Cl—C—H / CH_2CH_3	CH_3 / Cl—C—Cl / CH_2CH_3	CH_3 / Cl—C—H / Cl—C—H / CH_3	CH_3 / Cl—C—H / H—C—Cl / CH_3	CH_3 / Cl—C—H / CH_2CH_2Cl
S *Active*	*Achiral* *Inactive*	*R,S* *Meso* *Achiral* *Inactive*	*R,R* *Active*	*R* *Active*

(b) There will be five diastereomeric fractions: all optically inactive. They will be the same as those in (a), except that each chiral structure is now accompanied by an equal amount of its enantiomer.

(c) There will be six diastereomeric fractions: all optically inactive.

$CHCl_2$ / H—C—CH_3 / CH_2CH_3	CH_2Cl / CH_3—C—Cl / CH_2CH_3	CH_2Cl / CH_3—C—H / H—C—Cl / CH_3	CH_2Cl / CH_3—C—H / Cl—C—H / CH_3	CH_2Cl / H—C—CH_3 / CH_2CH_2Cl	CH_2Cl / H—C—CH_2Cl / CH_2CH_3
R +enantiomer (*S*) *Inactive*	*R* +enantiomer (*S*) *Inactive*	2*R*,3*R* +enantiomer (2*S*,3*S*) *Inactive*	2*R*,3*S* +enantiomer (2*S*,3*R*) *Inactive*	*R* +enantiomer (*S*) *Inactive*	*Achiral* *Inactive*

Formation of enantiomers and diastereomers: a closer look

If you have worked very carefully with models in following Secs. 4.22 and 4.26 and in working Problem 4.22, you may have made observations that raised questions in your mind. To understand better how formation of diastereomers differs from formation of enantiomers, let us contrast the reaction of the chiral 3-chloro-2-butyl radical shown in Fig. 4.4 with the reaction of the achiral *sec*-butyl radical.

In Sec. 4.22, we said that attachment of chlorine to either face of the *sec*-butyl radical is equally likely. This is in effect true, but deserves closer examination. Consider any conformation of the free radical: I, for example. It is clear that attack by chlorine from the top of I and attack from the bottom are *not* equally likely. But a rotation of 180° about

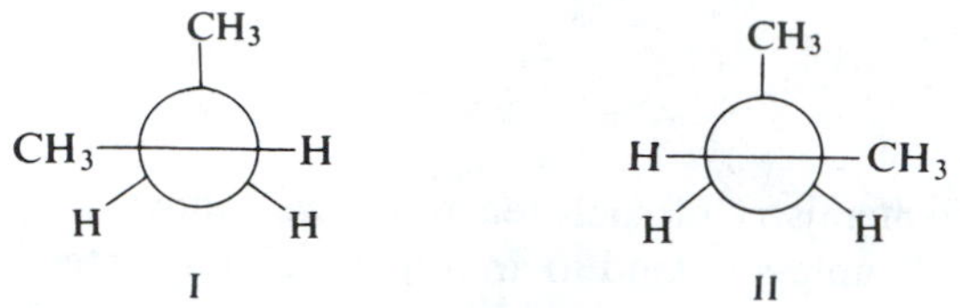

I II

sec-Butyl radical
Achiral

the single bond converts I into II; these are two conformations of the same free radical, and are, of course, in equilibrium with each other. They are mirror images, and hence of equal energy and equal abundance; any preferred attack from, say, the bottom of I to give the *R* product will be exactly counterbalanced by attack from the bottom of II to give the *S* product.

The "randomness of attack" that yields the racemic modification from achiral reactants is not necessarily due to the symmetry of any individual reactant molecule, but rather to the random distribution of such molecules between mirror-image conformations (or to random selection between mirror-image transition states).

Now, let us turn to reaction of the chiral 3-chloro-2-butyl radical (Fig. 4.4). Here, the free radical we are concerned with already contains a chiral center, about which it has the *S* configuration; attack is *not* random on such a radical because mirror-image conformations are not present—they could only come from *R* free radicals, and there are none of those radicals present.

Preferred attack from, say, the bottom of conformation III—a likely preference since this would keep the two chlorine atoms as far apart as possible in the transition state—would yield *meso*-2,3-dichlorobutane. A rotation of 180° about the single bond would convert III into IV. Attack from the bottom of IV would yield the *S,S* isomer. But III

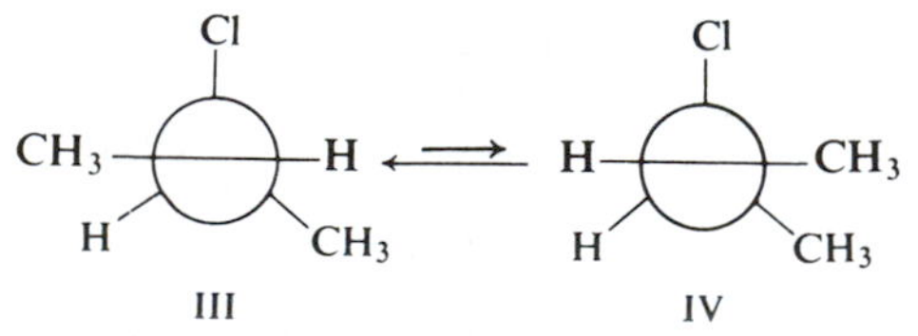

III IV

3-Chloro-2-butyl radical
Chiral

and IV are not mirror images, are not of equal energy, and are not of equal abundance. In particular, because of lesser crowding between the methyl groups, we would expect III to be more stable and hence more abundant than IV, and the *meso* product to predominate over the *S,S* isomer (as it actually does).

We might have made a different guess about the preferred direction of attack, and even a different estimate about relative stabilities of conformations, but we would still arrive at the same basic conclusion: except by sheer coincidence, the two diastereomers would not be formed in equal amounts.

(To test your understanding of what has just been said, try Problem 4.25, below. You can check your answer on page 66 of this Study Guide.)

In this discussion, we have assumed that the relative rates of competing reactions depend on relative populations of the conformations of the reactants. This assumption is correct here if, as seems likely, reaction of the free radicals with chlorine is easier and faster than the rotation that interconverts conformations.

If, on the other hand, reaction with chlorine were a relatively difficult reaction and much slower than interconversion of conformations, then relative rates would be determined by relative stabilities of the transition states. We would still draw the same general conclusions. In the reaction of the achiral *sec*-butyl radical, the transition states are mirror images and therefore of the same stability, and the rates of formation of the two products would be exactly the same. In the reaction of the chiral 3-chloro-2-butyl radical, the transition states are not mirror images and therefore not of the same stability, and rates of formation of the two products would be different. (In the latter case, we would even make the same prediction, that the *meso* product would predominate, since the same relationship between methyl groups that would make conformation III more stable would also make the transition state resembling conformation III more stable.)

Problem 4.25 Answer the following questions about the formation of 2,3-dichlorobutane from (*R*)-*sec*-butyl chloride. (a) Draw conformations (V and VI) of the intermediate radicals that correspond to III and IV above. (b) What is the relationship between V and VI? (c) How will the V : VI ratio compare with the III : IV ratio? (d) Assuming the same preferred direction of attack by chlorine as on III and IV, which stereoisomeric product would be formed from V? From VI? (e) Which product would you expect to predominate? (f) In view of the ratio of products actually obtained from (*S*)-*sec*-butyl chloride, what ratio of products must be obtained from (*R*)-*sec*-butyl chloride?

4.23 Racemization is also consistent with a pyramidal free radical that undergoes inversion (like ammonia, page 19) faster than it reacts with chlorine, and thus gives equal amounts of the enantiomers.

The important thing about the evidence is that it is consistent with the mechanism involving an intermediate free radical—whether flat or an inverting pyramid—and is inconsistent with the alternative mechanism.

4.24 Once more we follow the guidelines summarized above for Problem 4.22.

In each active fraction there is only one compound, which is chiral.

$CHCl_2$ (top); CH_3—C—H; CH_2CH_3 (bottom)

S

CH_2Cl (top); CH_3—C—H; H—C—Cl; CH_3 (bottom)

2*R*,3*R*

CH_2Cl (top); CH_3—C—H; Cl—C—H; CH_3 (bottom)

2*R*,3*S*

CH_2Cl (top); CH_3—C—H; CH_2CH_2Cl (bottom)

S

In one of the inactive fractions there is only one compound; it is achiral (only three different groups on C–2).

$$\begin{array}{c} CH_2Cl \\ | \\ ClCH_2{-}C{-}H \\ | \\ CH_2CH_3 \end{array}$$

Achiral

The other inactive fraction is a racemic modification.

$$\begin{array}{ccc} CH_2Cl & & CH_2Cl \\ CH_3{-}\!\!+\!\!{-}Cl & & Cl{-}\!\!+\!\!{-}CH_3 \\ CH_2CH_3 & & CH_2CH_3 \\ R & & S \end{array}$$

Enantiomers

When C–2 is attacked, the resultant planar free radical gives equal numbers of enantiomeric product molecules.

(Brown, H. C.; Kharasch, M. S.; Chao, T. H. *J. Am. Chem. Soc.* **1940**, *62*, 3435.)

1. (a) Secs. 4.3 and 4.4 (b) Sec. 4.4 (c) Sec. 4.4 (d) Sec. 4.5
(e) Sec. 4.9 (f) Sec. 4.9 (g) Sec. 4.10 (h) Sec. 4.2
(i) Secs. 4.7 and 4.11 (j) Secs. 4.7 and 4.17 (k) Sec. 4.18 (l) Sec. 4.12
(m) Sec. 4.14 (n) Sec. 3.3 (o) Sec. 4.15 (p) Sec. 4.15
(q) Sec. 4.4 (r) Sec. 4.4 (s) Sec. 4.20 (t) Sec. 4.20

2. (a) Chirality. (b) Chirality. (c) Usually an excess of one enantiomer that persists long enough to permit measurement. (d) It has a mirror image that is not superimposable on the original. (e) Restrictions on planar formulas are discussed in Sec. 4.10. In models, no bond to a chiral carbon should be broken. (f) We draw a wedge formula or build a model of the molecule, and then follow the steps in Sec. 4.15 and the Sequence Rules in Sec. 4.16.

3. Equal but opposite specific rotations; opposite *R*/*S* specifications; all other properties are the same.

4. (a) Screw, scissors, spool of thread.
(b) Glove, shoe, coat sweater, tied scarf.
(c) Helix, double helix.
(d) Football (laced), tennis racket (looped trim), golf club, rifle barrel.
(e) Hand, foot, ear, nose, yourself.

5. (a) Sawing. (b) Opening a soft-drink can. (c) Throwing a ball.

6. (a)

H above, n-C_3H_7 left, C_2H_5 right, Br below — *R*

H above, C_2H_5 left, C_3H_7-n right, Br below — *S*

Br > C,C,C > C,C,H > H

(b) Achiral.

(c)

CH_3 above, $BrCH_2$ left, C_2H_5 right, Br below — *R*

CH_3 above, C_2H_5 left, CH_2Br right, Br below — *S*

Br > C,Br > C,C > C,H

(d)

H above, $ClCH_2CH_2$ left, CH_2CH_3 right, Cl below — *R*

H above, CH_3CH_2 left, CH_2CH_2Cl right, Cl below — *S*

Cl > C,C,Cl > C,C,H > H

(e)

H above, t-C_4H_9 left, C_4H_9-*iso* right, Cl below — *R*

H above, *iso*-C_4H_9 left, C_4H_9-t right, Cl below — *S*

Cl > 3° (C,C,C,C) > 1° (C,H,H,C) > H

(f)

H above, n-C_3H_7 left, D right, Cl below — *R*

H above, D left, C_3H_7-n right, Cl below — *S*

Cl > C > D > H

7. (a) and (b) 3-Methylhexane and 2,3-dimethylpentane.

8. (a)

CH_3	CH_3	CH_3	CH_3
H—2—OH	HO—H	H—OH	HO—H
H—3—Br	Br—H	Br—H	H—Br
CH_3	CH_3	CH_3	CH_3
2*S*,3*R*	2*R*,3*S*	2*S*,3*S*	2*R*,3*R*

Enantiomers (2*S*,3*R* and 2*R*,3*S*); *Enantiomers* (2*S*,3*S* and 2*R*,3*R*)

(b)

CH_2Br	CH_2Br	CH_2Br	CH_2Br
H—2—Br	Br—H	H—Br	Br—H
H—3—Br	Br—H	Br—H	H—Br
CH_3	CH_3	CH_3	CH_3
2*S*,3*R*	2*R*,3*S*	2*S*,3*S*	2*R*,3*R*

Enantiomers (2*S*,3*R* and 2*R*,3*S*); *Enantiomers* (2*S*,3*S* and 2*R*,3*R*)

(c)

C_6H_5	C_6H_5	C_6H_5
H—CH_3	H—CH_3	CH_3—H
H—CH_3	CH_3—H	H—CH_3
C_6H_5	C_6H_5	C_6H_5
R,*S* (or *S*,*R*)	*R*,*R*	*S*,*S*
Meso	*Enantiomers*	

(d)

C_2H_5	C_2H_5	C_2H_5
H—CH_3	H—CH_3	CH_3—H
CH_2	CH_2	CH_2
CH_2	CH_2	CH_2
H—CH_3	CH_3—H	H—CH_3
C_2H_5	C_2H_5	C_2H_5
S,*R* (or *R*,*S*)	*S*,*S*	*R*,*R*
Meso	*Enantiomers*	

(e)

CH_3	CH_3	CH_3	CH_3
H—2—OH	HO—H	H—OH	HO—H
H—3—C_6H_5	C_6H_5—H	C_6H_5—H	H—C_6H_5
CH_3	CH_3	CH_3	CH_3
2*S*,3*S*	2*R*,3*R*	2*S*,3*R*	2*R*,3*S*
Enantiomers		*Enantiomers*	

(f)

CH_2OH	CH_2OH	CH_2OH	CH_2OH
H—2—OH	HO—H	H—OH	HO—H
H—3—OH	HO—H	H—OH	HO—H
H—4—OH	HO—H	HO—H	H—OH
CH_3	CH_3	CH_3	CH_3
2*S*,3*R*,4*R*	2*R*,3*S*,4*S*	2*S*,3*R*,4*S*	2*R*,3*S*,4*R*
Enantiomers		*Enantiomers*	

CH_2OH	CH_2OH	CH_2OH	CH_2OH
H—OH	HO—H	H—OH	HO—H
HO—H	H—OH	HO—H	H—OH
H—OH	HO—H	HO—H	H—OH
CH_3	CH_3	CH_3	CH_3
2*S*,3*S*,4*R*	2*R*,3*R*,4*S*	2*S*,3*S*,4*S*	2*R*,3*R*,4*R*
Enantiomers		*Enantiomers*	

(g) In the first and fourth isomers, C–3 is a *pseudoasymmetric carbon*: it holds four different groups, but two of these differ only in configuration. We specify configuration about this carbon as *r* or *s*, arbitrarily giving the *R* substituent (C–4) higher priority than the *S* substituent (C–2).

CH_2OH	CH_2OH	CH_2OH	CH_2OH
H—(2)—OH	H—OH	HO—H	H—OH
H—(3)—OH	H—OH	HO—H	HO—H
H—(4)—OH	HO—H	H—OH	H—OH
CH_2OH	CH_2OH	CH_2OH	CH_2OH
2*S*,3*s*,4*R*	2*S*,4*S*	2*R*,4*R*	2*S*,3*r*,4*R*
Meso	*Enantiomers*		*Meso*

(h)

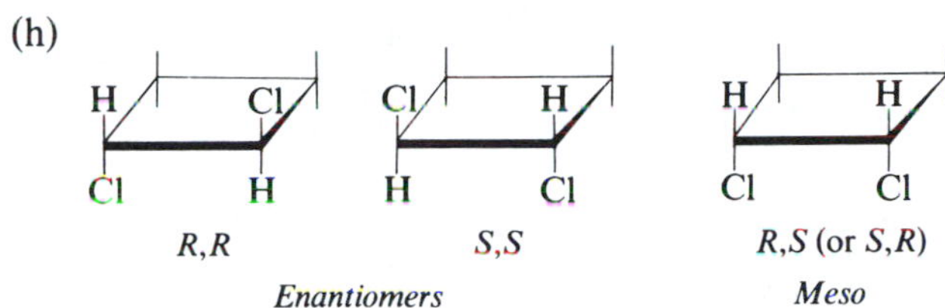

R,*R* *S*,*S* *R*,*S* (or *S*,*R*)

Enantiomers *Meso*

(i) In each structure, we note the vertical plane of symmetry cutting through the molecule from the left-front to the right-rear corner. Alternatively, we simply flip the mirror image over (*trans* isomer) or rotate it through 180° (*cis* isomer).

trans (Sec. 13.13) *cis* (Sec. 13.13)

Achiral *Achiral*

(j)

$n\text{-}C_3H_7\text{—}N^{\oplus}(CH_3)(C_2H_5)\text{—}C_4H_9\text{-}n$ Cl^- $n\text{-}C_4H_9\text{—}N^{\oplus}(CH_3)(C_2H_5)\text{—}C_3H_7\text{-}n$ Cl^-

S *R*

Enantiomers

(k)

C_2H_5 / CH_3—C—H / C_2H_5—$\overset{\oplus}{N}$—CH_3 Cl^- / C_3H_7-*n*	C_2H_5 / H—C—CH_3 / CH_3—$\overset{\oplus}{N}$—C_2H_5 Cl^- / C_3H_7-*n*	C_2H_5 / CH_3—C—H / CH_3—$\overset{\oplus}{N}$—C_2H_5 Cl^- / C_3H_7-*n*	C_2H_5 / H—C—CH_3 / C_2H_5—$\overset{\oplus}{N}$—CH_3 Cl^- / C_3H_7-*n*
C*R*,N*S*	C*S*,N*R*	C*R*,N*R*	C*S*,N*S*

Enantiomers *Enantiomers*

9. A, $CH_3CCl_2CH_3$; B, $ClCH_2CH_2CH_2Cl$; C, $CH_3CHClCH_2Cl$, chiral; D, $CH_3CH_2CHCl_2$.

Active C gives: $CH_3CHClCHCl_2$, chiral; $CH_3CCl_2CH_2Cl$ and $ClCH_2CHClCH_2Cl$, both achiral.

(a) and (b) The possible dichloro products are 1,1-, 1,2-, 1,3-, and 2,2-dichloropropane.

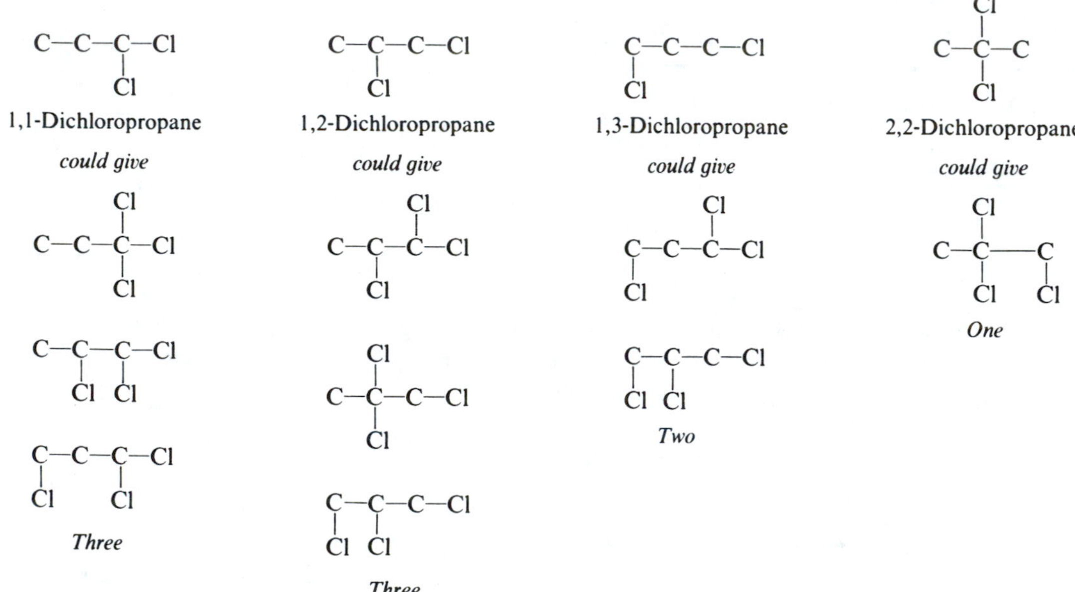

From this information, we conclude that A is 2,2-dichloropropane and B is 1,3-dichloropropane. Compounds C and D are the 1,1- and 1,2-isomers, but the data so far do not permit us to say which is which.

(c) C must be 1,2-dichloropropane, since, of the two unassigned compounds, this is the only one capable of optical activity. D, therefore, is 1,1-dichloropropane.

CH_3—$\overset{*}{C}H(Cl)$—CH_2Cl
1,2-Dichloropropane
Chiral

CH_3—CH_2—$CHCl_2$
1,1-Dichloropropane
Achiral

(d) 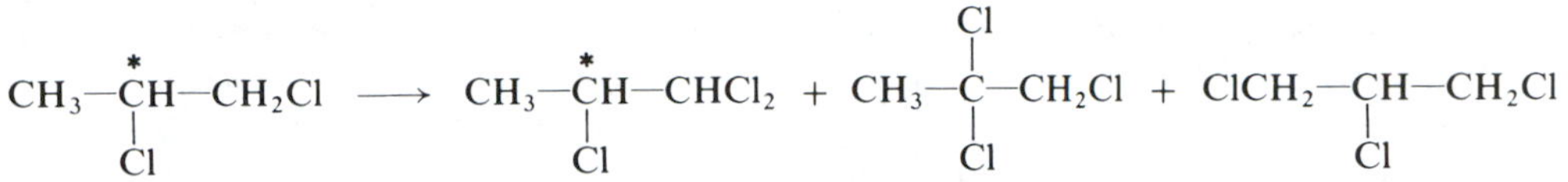

Active — 1,1,2-Trichloropropane *Chiral* *Active* — 1,2,2-Trichloropropane *Achiral* *Inactive* — 1,2,3-Trichloropropane *Achiral* *Inactive*

10. (a) None. (b) One pair of configurational enantiomers.

1-Bromo-2-chloroethane *Achiral* — 2-Bromo-1-chloropropane *Chiral*

(c) I and II are conformational enantiomers. IV, V, and VI are conformers of one configurational enantiomer; VII, VIII, and IX are conformers of the other.

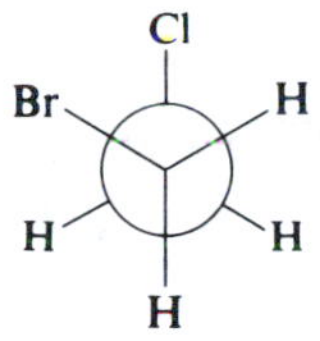

I

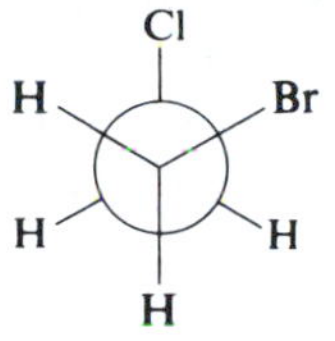

II

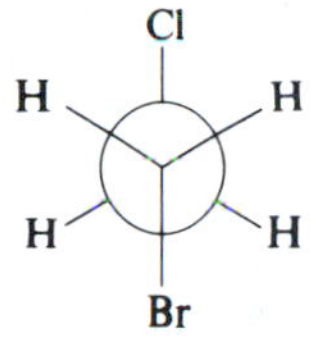

III

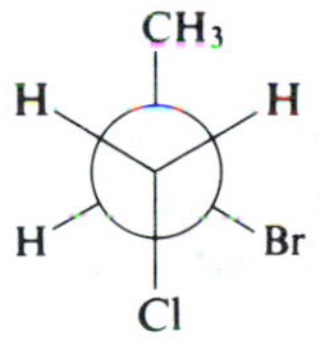

IV

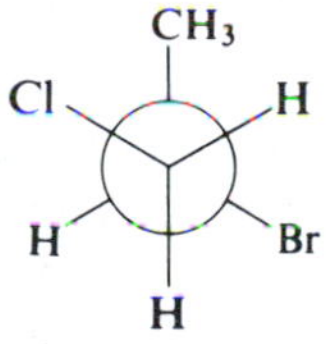

V

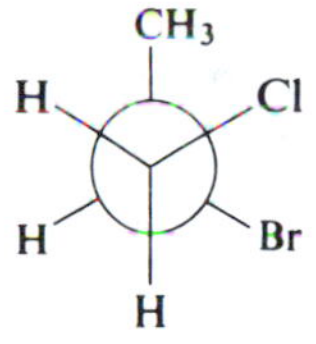

VI

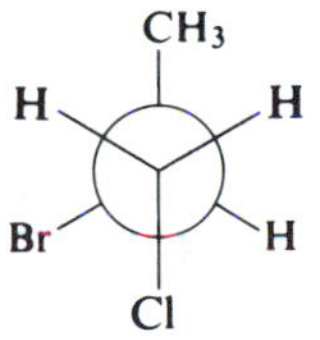

VII

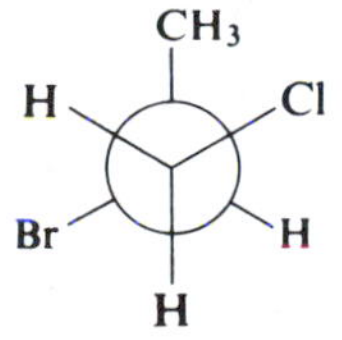

VIII

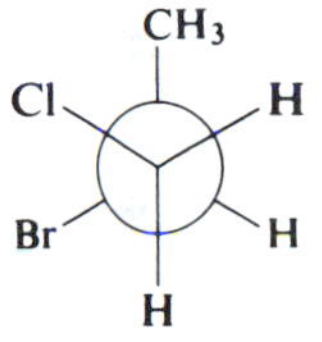

IX

11. Attractive dipole–dipole interaction between —Cl and —CH_3. The groups can get far enough apart to avoid van der Waals repulsion, and settle down at the distance of maximum stability, where attraction exists.

12. You can place the solution in a polarimeter, and see whether the muscarine is optically active or inactive. If it is optically active, it can be—and very probably is—the natural substance, and has come from a fly agaric; accidental death is quite possible and, indeed, most likely. But if it is optically inactive, it must be synthetic material, and cannot have been produced by a mushroom; death cannot have been an accident, and almost certainly was . . . *murder*.

Like nearly every alkaloid (page 159), muscarine is complex enough to be chiral, and a few minutes' work with models will show you that the structure given on page 164 is not superimposable on its mirror image. Like nearly every chiral substance, when produced in a biological system—that is, by the action of optically active enzymes—it is produced in optically active form; one enantiomer is produced selectively (*stereoselectively*, Sec. 10.2) in preference to the other. But when synthesized in the laboratory by use of ordinary, optically inactive reagents, muscarine is, of course, obtained in the optically inactive, racemic form. (See Sec. 4.11.)

The structure of muscarine has been even more deeply shrouded in mystery than the death of George Harrison; only after 150 years of research was the structure finally elucidated, in 1957. The stereochemistry of this compound illustrates a great deal that we have learned in this chapter. The molecule contains three chiral centers (*Can you find them?*), and exists in 2^3 or eight stereoisomeric forms: four pairs of enantiomers. All eight have been prepared and their configurations have been assigned. Only one, the 2*S*,3*R*,5*R* isomer, is the naturally occurring muscarine. (The numbering of positions begins at O and proceeds clockwise.) In the form of the chloride ($X^- = Cl^-$), this isomer has an $[\alpha]_D$ of +8.1. Its absolute configuration has been established by a synthesis that, in effect, goes back to the (+)-tartaric acid studied by Bijvoet (page 140). Just as the formation of (+)-muscarine in a biological system is stereoselective, so its biological action is *stereospecific* (Sec. 10.5): the enantiomeric (−) isomer is physiologically inactive.

(Wilkinson, S. "The History and Chemistry of Muscarine"; *Q. Rev., Chem. Soc.* **1961**, *15*, 153.)

The case of George Harrison has been chronicled in detail by Dorothy L. Sayers and Robert Eustace in their novel *The Documents in the Case* (Gollancz, London, 1930; Avon Books, New York, 1971). A year after the publication of *The Documents*, Miss Sayers confessed to having made "a first-class howler" in the book: natural muscarine, she now thought, would, like the synthetic substance, be optically inactive. Actually, she should have stuck by her guns; her confession was inaccurate, and based upon an incorrect structure that had been proposed for muscarine. The basis of the book is sound: as became clear some years later, natural muscarine *is* optically active, and *can* be distinguished from synthetic material by use of the polarimeter.

Oh, yes: was the muscarine in George Harrison's stew optically active or inactive? Who—if anyone—dunnit? Ah, that would be telling!

13. We follow the guidelines laid down above for Problem 4.22.

(a) Three fractions: all inactive (achiral or racemic).

$CH_3CH_2CH_2CH_2$—CH_2—Cl (H, H on C)
Achiral

Cl—C—H with CH_3 above and $CH_2CH_2CH_3$ below
R
+enantiomer (*S*)

CH_3CH_2—CHCl—CH_2CH_3
Achiral

(b) Five fractions: all inactive (achiral or racemic).

$CH_3CH_2CH_2CH_2$—CHCl—Cl
Achiral

H—C—Cl with CH_2Cl above and $CH_2CH_2CH_3$ below
R
+enantiomer (*S*)

H—C—Cl with CH_2CH_2Cl above and CH_2CH_3 below
R
+enantiomer (*S*)

H—C—Cl with $CH_2CH_2CH_2Cl$ above and CH_3 below
R
+enantiomer (*S*)

Cl—CH_2—$CH_2CH_2CH_2$—CH_2—Cl
Achiral

(c) Seven fractions: five active, two inactive (both achiral).

We note that *configuration* about the original chiral center (the lower one in these formulas) is maintained—although the *specification* may change (as it does in the last formula).

Cl—C—H with $CH_2CH_2CH_3$ above and CH_3 below
(*S*)-2-Chloropentane

would give

$CH_2CH_2CH_2Cl$ / Cl—C—H / CH_3	CH_3 / Cl—C—H / CH_2 / Cl—C—H / CH_3	CH_3 / H—C—Cl / CH_2 / Cl—C—H / CH_3	CH_2CH_3 / Cl—C—H / Cl—C—H / CH_3	CH_2CH_3 / H—C—Cl / Cl—C—H / CH_3	$CH_2CH_2CH_3$ / Cl—C—Cl / CH_3	$CH_2CH_2CH_3$ / Cl—C—H / CH_2Cl
S	*S,R*	*S,S*	2*S*,3*R*	2*S*,3*S*	*Achiral*	*R*
Active	*Achiral* (*meso*) *Inactive*	*Active*	*Active*	*Active*	*Inactive*	*Active*

(d) Seven fractions: six active, one inactive (racemic).

Here again, we note that configuration about the original chiral center (C–3) is maintained—except, of course, when a bond to it is broken. In this particular case, bond-breaking gives an intermediate free radical, which loses configuration, and a racemic modification is formed.

$$\begin{array}{c} CCl(CH_3)_2 \\ H\!-\!\overset{3}{C}\!-\!CH_3 \\ CH_2CH_3 \end{array}$$

(*R*)-2-Chloro-2,3-dimethylpentane

would give

$$\begin{array}{c} CH_2Cl \\ Cl\!-\!\overset{2}{C}\!-\!CH_3 \\ H\!-\!\overset{3}{C}\!-\!CH_3 \\ CH_2CH_3 \end{array}$$

2*S*,3*R*
Active

$$\begin{array}{c} CH_3 \\ Cl\!-\!\overset{2}{C}\!-\!CH_2Cl \\ H\!-\!\overset{3}{C}\!-\!CH_3 \\ CH_2CH_3 \end{array} \equiv \begin{array}{c} CH_2Cl \\ CH_3\!-\!\overset{2}{C}\!-\!Cl \\ H\!-\!\overset{3}{C}\!-\!CH_3 \\ CH_2CH_3 \end{array}$$

2*R*,3*R*
Active

$$\begin{array}{c} CH_3 \\ CH_3\!-\!C\!-\!Cl \\ Cl\!-\!C\!-\!CH_3 \\ CH_2CH_3 \end{array} \qquad \begin{array}{c} CH_3 \\ CH_3\!-\!C\!-\!Cl \\ CH_3\!-\!C\!-\!Cl \\ CH_2CH_3 \end{array}$$

S *R*
Racemic modification
Inactive fraction

$$\begin{array}{c} CH_3 \\ CH_3\!-\!C\!-\!Cl \\ H\!-\!C\!-\!CH_2Cl \\ CH_2CH_3 \end{array}$$

S
Active

$$\begin{array}{c} CH_3 \\ CH_3\!-\!C\!-\!Cl \\ H\!-\!\overset{3}{C}\!-\!CH_3 \\ H\!-\!\overset{4}{C}\!-\!Cl \\ CH_3 \end{array}$$

3*R*,4*R*
Active

$$\begin{array}{c} CH_3 \\ CH_3\!-\!C\!-\!Cl \\ H\!-\!\overset{3}{C}\!-\!CH_3 \\ Cl\!-\!\overset{4}{C}\!-\!H \\ CH_3 \end{array}$$

3*R*,4*S*
Active

$$\begin{array}{c} CH_3 \\ CH_3\!-\!C\!-\!Cl \\ H\!-\!C\!-\!CH_3 \\ CH_2CH_2Cl \end{array}$$

R
Active

(e) One fraction: inactive (racemic). (Oxidation of either $—CH_2OH$.)

$$\begin{array}{c} CH_2OH \\ H\!-\!\overset{2}{C}\!-\!OH \\ H\!-\!\overset{3}{C}\!-\!OH \\ CH_2OH \end{array} \quad \textit{would give} \quad \begin{array}{c} COOH \\ H\!-\!\overset{2}{C}\!-\!OH \\ H\!-\!\overset{3}{C}\!-\!OH \\ CH_2OH \end{array} \qquad \begin{array}{c} CH_2OH \\ H\!-\!\overset{3}{C}\!-\!OH \\ H\!-\!\overset{2}{C}\!-\!OH \\ COOH \end{array} \equiv \begin{array}{c} COOH \\ HO\!-\!\overset{2}{C}\!-\!H \\ HO\!-\!\overset{3}{C}\!-\!H \\ CH_2OH \end{array}$$

meso-1,2,3,4-Butanetetraol

2*R*,3*R* 2*S*,3*S*
Racemic modification

(f) Two fractions: one active, one inactive (*meso* compound).

$CH{=}CH_2$

Cl—+—H

CH_3

(*S*)-3-Chloro-1-butene

would give

CH_3

H—+—Cl

Cl—+—H

CH_3

S,*S*

Active

CH_3

Cl—+—H

Cl—+—H

CH_3

S,*R*

Achiral (*meso*)

Inactive

(g) Two fractions: both inactive (one racemic, one *meso*).

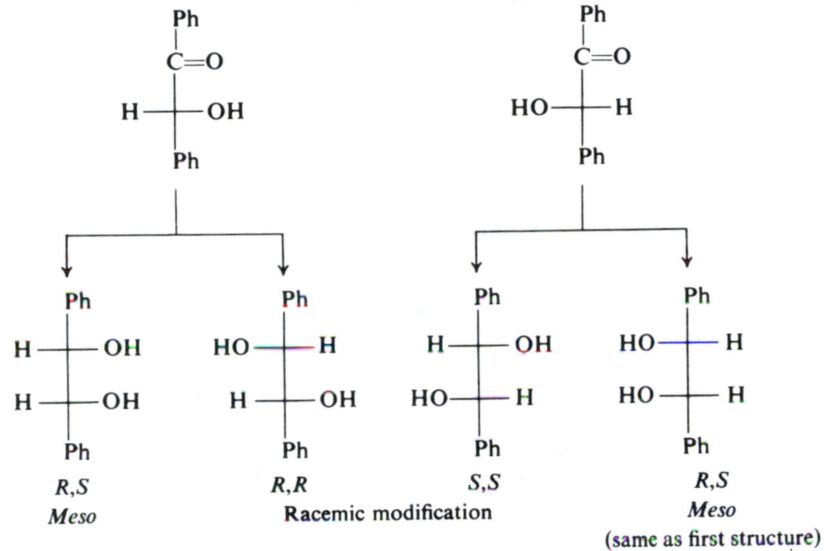

14. (a)

$CH{=}CH_2$

HO—+—H

CH_2OH

R

$\longrightarrow$

CH_2OH

H—+—OH

HO—+—H

CH_2OH

S,*S*

Active

E

CH_2OH

HO—+—H

HO—+—H

CH_2OH

R,*S*

Achiral (*meso*)

F

(b)

CH_2Cl

CH_3—+—H

C_2H_5

(*S*)-1-Chloro-2-methylbutane

$\longrightarrow$

C_2H_5

H—+—CH_3

CH_2

Cu

CH_2

CH_3—+—H

C_2H_5

S,*S*

G

(c)

$$\mathrm{G} + \begin{array}{c} CH_2Cl \\ CH_3\text{—}\!\!+\!\!\text{—}H \\ C_2H_5 \end{array} \longrightarrow \begin{array}{c} C_2H_5 \\ H\text{—}\!\!+\!\!\text{—}CH_3 \\ CH_2 \\ CH_2 \\ CH_3\text{—}\!\!+\!\!\text{—}H \\ C_2H_5 \end{array}$$

(*S*)-1-Chloro-2-methylbutane → *S,S* H

(d) Whichever $—CH_2OH$ reacts with HBr, the product is the same.

$$\begin{array}{c} CH_2OH \\ HO\text{—}\!\!+\!\!\text{—}H \\ H\text{—}\!\!+\!\!\text{—}OH \\ CH_2OH \end{array} \longrightarrow \begin{array}{c} CH_2Br \\ HO\text{—}\!\!+^3\!\!\text{—}H \\ H\text{—}\!\!+^2\!\!\text{—}OH \\ CH_2OH \end{array} \quad \left[\begin{array}{c} CH_2OH \\ HO\text{—}\!\!+^2\!\!\text{—}H \\ H\text{—}\!\!+^3\!\!\text{—}OH \\ CH_2Br \end{array} \equiv \begin{array}{c} CH_2Br \\ HO\text{—}\!\!+^3\!\!\text{—}H \\ H\text{—}\!\!+^2\!\!\text{—}OH \\ CH_2OH \end{array} \right]$$

R,R *Active* — 2*R*,3*S* *Active* I — *Identical: turned end-for-end*

(e)

$$\begin{array}{c} C_2H_5 \\ C{=}CH_2 \\ H\text{—}\!\!+\!\!\text{—}CH_3 \\ C_2H_5 \end{array} \longrightarrow \begin{array}{c} C_2H_5 \\ CH_3\text{—}\!\!+\!\!\text{—}H \\ H\text{—}\!\!+\!\!\text{—}CH_3 \\ C_2H_5 \end{array} \quad \begin{array}{c} C_2H_5 \\ H\text{—}\!\!+\!\!\text{—}CH_3 \\ H\text{—}\!\!+\!\!\text{—}CH_3 \\ C_2H_5 \end{array}$$

(*R*)-2-Ethyl-3-methyl-1-pentene — *R,R* *Active* J — *R,S* *Achiral* (*meso*) *Inactive* K

15. The enantiomeric acids react at different rates with the optically active reagent, the alcohol, to form diastereomeric esters. Unequal amounts of the acids are consumed, to give unequal amounts of the esters, and to leave unconsumed unequal amounts of the enantiomeric acids. The three products are, then: the two diastereomeric esters, and unconsumed acid that now contains an excess of one enantiomer and hence is optically active.

4.25 (On page 55 of this Study Guide.)

(a) V is the mirror image of III; VI is the mirror image of IV.

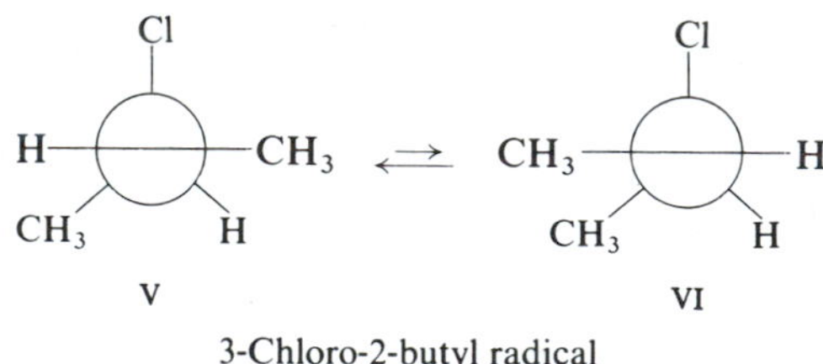

3-Chloro-2-butyl radical from (*R*)-*sec*-butyl chloride

(b) V and VI are diastereomers.

(c) The V:VI ratio will be the same as the III:IV ratio, since the same methyl–methyl interaction exists.

(d) V ⟶ *meso*; VI ⟶ *R*,*R* isomer.

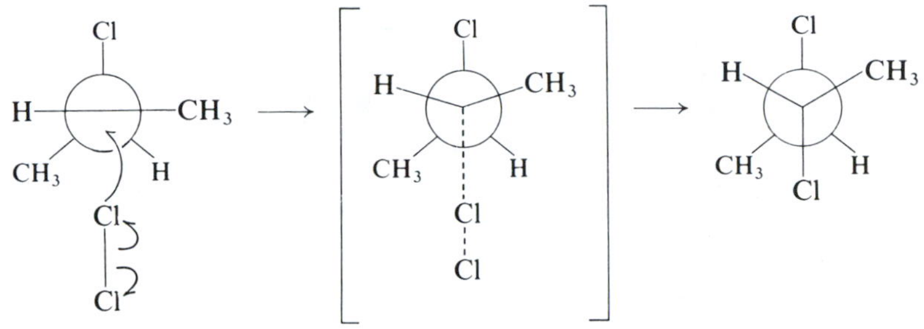

Cl
H — CH$_3$ ⟶
CH$_3$ H
Cl
Cl
V

[Cl
H CH$_3$
CH$_3$ H
Cl
Cl]
Transition state

⟶ Cl
H CH$_3$
CH$_3$ H
Cl

(e) The *meso* product will predominate, for the same reasons that it predominates in the reaction of (*S*)-*sec*-butyl chloride.

(f) *R*,*R*:*meso* = 29:71.

5

Alkyl Halides

Nucleophilic Aliphatic Substitution

5.1

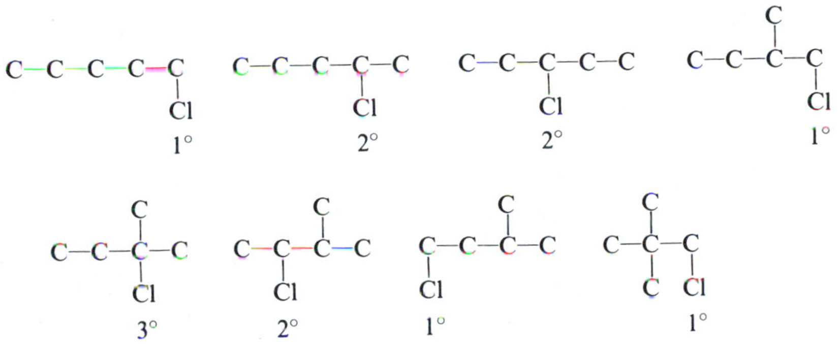

```
C—C—C—C—C      C—C—C—C—C      C—C—C—C—C      C—C—C—C
        |            |            |            |  |
        Cl           Cl           Cl           C  Cl
        1°           2°           2°              1°

     C              C              C            C
     |              |              |            |
C—C—C—C        C—C—C—C        C—C—C—C        C—C—C
     |            |            |                |  |
     Cl           Cl           Cl               C  Cl
     3°           2°           1°                  1°
```

5.2

$$CH_3CH_2CH_2{-}X + CH_3OH \longrightarrow CH_3CH_2CH_2{-}\overset{\overset{\displaystyle H}{|}}{O}{}^{\oplus}{-}CH_3 + X^-$$

Alkyl halide (*Substrate*) — Alcohol (*Nucleophile*) — Protonated ether — Halide ion (*Leaving group*)

$$CH_3CH_2CH_2{-}\overset{\overset{\displaystyle H}{|}}{O}{}^{\oplus}{-}CH_3 + CH_3OH \rightleftharpoons CH_3CH_2CH_2{-}O{-}CH_3 + CH_3OH_2^{\ +}$$

Protonated ether — Alcohol — Ether — Protonated alcohol

5.3 (a) Clearly, the two ethers have opposite configurations, and therefore both cannot have the same configuration as the 1-phenyl-2-propanol. One of them must have the configuration opposite to that of the alcohol from which it was made, and hence one of the reaction routes must have brought about an inversion of configuration.

In which reaction did the inversion take place? To find the answer, let us examine the individual steps of each route. (The absolute configuration of the (+)-1-phenyl-2-propanol shown below is the correct one, but that does not really matter: it was not known, of course, in 1923, and you could have worked the problem equally well assuming either configuration for the starting material.)

There are four reactions in all. In three of these, as the nature of the products shows, no bond to the chiral carbon is broken; these reactions must have proceeded with retention of configuration.

$$\begin{array}{c} CH_3 \\ \vdots \\ H-C-O\!\mid\!H \\ \vdots \\ CH_2C_6H_5 \end{array} + K \longrightarrow \begin{array}{c} CH_3 \\ \vdots \\ H-C-O^-K^+ \\ \vdots \\ CH_2C_6H_5 \end{array} + \tfrac{1}{2}H_2$$

(+)-1-Phenyl-2-propanol

*No breaking of bonds to C**

Retention

$$\begin{array}{c} CH_3 \\ \vdots \\ H-C-O^-K^+ \\ \vdots \\ CH_2C_6H_5 \end{array} + C_2H_5\!\mid\!OTs \longrightarrow \begin{array}{c} CH_3 \\ \vdots \\ H-C-O-C_2H_5 \\ \vdots \\ CH_2C_6H_5 \end{array} + K^+OTs^-$$

*No breaking of bonds to C**

Retention

(+)-Ethyl 1-phenyl-2-propyl ether

$$\begin{array}{c} CH_3 \\ \vdots \\ H-C-O\!\mid\!H \\ \vdots \\ CH_2C_6H_5 \end{array} + Cl\!\mid\!Ts \longrightarrow \begin{array}{c} CH_3 \\ \vdots \\ H-C-O-Ts \\ \vdots \\ CH_2C_6H_5 \end{array} + HCl$$

(+)-1-Phenyl-2-propanol

*No breaking of bonds to C**

Retention

Inversion must have occurred, then, in the remaining reaction, the one between 1-phenyl-2-propyl tosylate and potassium ethoxide. Here, breaking of a bond to the chiral center is quite consistent with the nature of the product: one anion displaces another.

$$C_2H_5O^-K^+ + \begin{array}{c} CH_3 \\ \vdots \\ H-C\!\mid\!O-Ts \\ \vdots \\ CH_2C_6H_5 \end{array} \longrightarrow \begin{array}{c} CH_3 \\ \vdots \\ C_2H_5O-C-H \\ \vdots \\ CH_2C_6H_5 \end{array} + K^+OTs^-$$

(−)-Ethyl 1-phenyl-2-propyl ether

Bond to C broken*

(b) It does not matter what the optical purity of the starting alcohol is. The important thing is that the optical purity of the ethers formed by the two routes is the *same*, as shown by their equal (though opposite) rotations. Since the (+) ether was formed with *complete* retention, the (−) ether must have been formed with *complete* inversion.

(c) Here is a clear-cut example of complete inversion of configuration taking place *in a readily identifiable step*: a step in which one anion displaces another in a reaction of a kind soon to become known as nucleophilic substitution. The recognition that inversion took place, and was complete, did not depend upon knowing the optical purity either of the starting material or of the products; it did not depend upon (prior) knowledge of the relative configurations of starting material and either product.

The axiom that configuration cannot change in a reaction involving no breaking of bonds to the chiral center was not so firmly established then as now, and Phillips himself simply says of such a reaction that "change of configuration . . . is unlikely".

Phillips, Henry. "A New Type of Walden Inversion"; *J. Chem. Soc.* **1923**, 44.

5.4 When *one* molecule of halide suffers exchange and inversion, the optical activity of *two* molecules is lost, since the optical rotation of the inverted molecule cancels the optical rotation of one unreacted molecule.

The reaction here involves an alkyl halide, the classical substrate for nucleophilic substitution. It was not possible to be certain about the relative configuration of an alkyl halide, or its optical purity, because its preparation had to involve the breaking of a bond to the carbon receiving the halogen. Yet here we have an experiment from which we can draw a conclusion that does not depend upon knowing the relative configuration of an alkyl halide, or its optical purity: every molecule undergoing substitution suffers inversion of configuration. Furthermore, this was observed for a reaction following second-order kinetics, and hence clearly of the kind named earlier in that same year as S_N2 (pages 179–180).

Hughes, E. D.; Juliusberger, F.; Masterman, S.; Topley, B.; Weiss, J. "Aliphatic Substitution and the Walden Inversion. Part I"; *J. Chem. Soc.* **1935**, 1525. For the paper first laying out the broad theory of nucleophilic aliphatic substitution, see Hughes, E. D.; Ingold, C. K. "Mechanism of Substitution at a Saturated Carbon Atom"; *J. Chem. Soc.* **1935**, 244.

5.5 We expect to obtain (*R*)-2-octanol of [α] −6.5°.

$$\begin{array}{c} C_6H_{13} \\ \vdots \\ Br-C-H \\ \vdots \\ CH_3 \end{array} \xrightarrow[S_N2]{NaOH} \begin{array}{c} C_6H_{13} \\ \vdots \\ H-C-OH \\ \vdots \\ CH_3 \end{array}$$

S-(+)-2-Bromooctane	*R*-(−)-2-Octanol
[α] = +24.9°	[α] = −6.5°
Optical purity 63%	*Optical purity 63%*

We refer to the assigned configurations and rotations given on page 182. Since our reactant is (+)-2-bromooctane, it has the opposite configuration to that shown for the (−) bromide. If optically pure 2-bromooctane has a specific rotation of 39.6°, our reactant is 63% optically pure.

$$\text{Optical purity} = \frac{24.9°}{39.6°} \times 100 = 63\%$$

Reaction is S_N2 and hence should take place with *complete inversion*: the 2-octanol will have the opposite configuration from the starting bromide, and will be of the same optical purity, 63%. (It *happens to have* the opposite specification, *R*.)

$$\frac{63}{100} \times -10.3^\circ = -6.5^\circ$$

5.6 (a) We are given the configurations and specific rotations of optically pure chloride and alcohol:

$$\begin{array}{c} C_6H_5 \\ \vdots \\ H-C-Cl \\ \vdots \\ CH_3 \end{array} \qquad \qquad \begin{array}{c} C_6H_5 \\ \vdots \\ H-C-OH \\ \vdots \\ CH_3 \end{array}$$

(*R*)-(−)-α-Phenylethyl chloride

[α] = −109°

(*R*)-(−)-α-Phenylethyl alcohol

[α] = −42.3°

Since our reactant has [α] −34°, it is

$$\frac{34^\circ}{109^\circ} \times 100 = 31\% \text{ optically pure}$$

and has the same configuration as that shown above.

Since our product has [α] +1.7°, it is

$$\frac{1.7^\circ}{42.3^\circ} \times 100 = 4.0\% \text{ optically pure}$$

and has (an excess of) the configuration opposite to that shown above.

$$\begin{array}{c} C_6H_5 \\ \vdots \\ H-C-Cl \\ \vdots \\ CH_3 \end{array} \xrightarrow{\text{NaOH}} \begin{array}{c} C_6H_5 \\ \vdots \\ HO-C-H \\ \vdots \\ CH_3 \end{array}$$

[α] = −34°

Optical purity 31%

[α] = +1.7°

Optical purity 4%

(b) and (c) Since the product has the opposite configuration from the starting material and is of lower optical purity,

$$\frac{\text{opt. pur. product}}{\text{opt. pur. reactant}} = \frac{4.0}{31.0} = 13\%$$

reaction has taken place with inversion (13%) plus racemization (87%).

For each molecule that undergoes back-side attack, a racemic pair is formed, and the optical activity of *two* molecules is thus lost. In this reaction, then, 43.5% of the molecules undergo front-side attack (retention), and 56.5% undergo back-side attack (inversion). The

43.5% with retained configuration cancel the activity of another 43.5% with inverted configuration to give 87% racemic product, and leave the optical purity at 13% of the original value.

5.7 The S_N1 reaction is slow because the neopentyl cation is primary, and therefore slow to form. The S_N2 reaction is slow because of the steric factor: although there is only *one* substituent attached to $-CH_2X$, it is a very large one, *t*-Bu. (See page 188.)

5.8 Rearrangement of a 2° cation into a 3° cation appears likely. See page 335 for equations.

5.9 The rate is made up of two parts, S_N2 and S_N1:

$$\text{rate} = \overbrace{4.7 \times 10^{-5}[RX][OH^-]}^{S_N2} + \overbrace{0.24 \times 10^{-5}[RX]}^{S_N1}$$

From this we can establish the percentage due to S_N2:

$$\%S_N2 = \frac{S_N2}{S_N2 + S_N1} \times 100 = \frac{4.7 \times 10^{-5}[RX][OH^-]}{4.7 \times 10^{-5}[RX][OH^-] + 0.24 \times 10^{-5}[RX]} \times 100$$

$$\%S_N2 = \frac{4.7[OH^-]}{4.7[OH^-] + 0.24} \times 100$$

Finally, we can calculate the percentage due to S_N2 at various $[OH^-]$ values:

(a) When $[OH^-] = 0.001$, $\%S_N2 = \dfrac{4.7 \times 0.001}{4.7 \times 0.001 + 0.24} \times 100 = 1.9\%$

Similarly:

(b) $[OH^-] = 0.01$, $\%S_N2 = 16.4$

(c) $[OH^-] = 0.1$, $\%S_N2 = 66.2$

(d) $[OH^-] = 1.0$, $\%S_N2 = 95.1$

(e) $[OH^-] = 5.0$, $\%S_N2 = 99.0$

1. (a) $CH_3-C(CH_3)_2-CH_2Br$

(b) $CH_3-CH_2-CH(CH_3)-C(CH_3)(Cl)-CH_3$

(c) $CH_3-CH_2-CH_2-CH(CH_3)-CH_2-CH_2Br$

(d) $CH_3-C(CH_3)(Cl)-CH_3$

(e) $CH_3-CH(CH_3)-CH(Cl)-CH_2Cl$

(f) $CH_3-CH(CH_3)-CH(Br)-CH(CH_3)-CH_3$

2. (a)

$$\begin{array}{c} CH_3 \\ | \\ CH_3-CH-CH_2 \\ \quad\quad\quad | \\ \quad\quad\quad I \end{array}$$

1-Iodo-2-methylpropane

(b)

$$\begin{array}{c} CH_3 \\ | \\ CH_3-CH-CH-CH_3 \\ | \\ Cl \end{array}$$

2-Chloro-3-methylbutane

(c)

$$\begin{array}{c} CH_3 \\ | \\ CH_3-CH-C-CH_2-CH_3 \\ | \quad\;\; | \\ Br \quad CH_3 \end{array}$$

2-Bromo-3,3-dimethylpentane

(d)

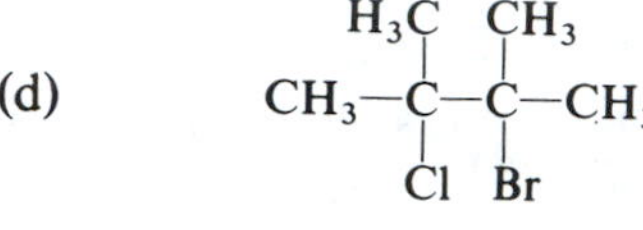

$$\begin{array}{c} H_3C \;\; CH_3 \\ | \quad\;\; | \\ CH_3-C-C-CH_3 \\ | \quad\;\; | \\ Cl \;\; Br \end{array}$$

2-Bromo-3-chloro-2,3-dimethylbutane

3. (a) $CH_3CH_2CH_2CH_2OH$ *n*-Butyl alcohol

(b) no reaction

(c) $CH_3CH_2CH_2CH_3$ *n*-Butane

(d) $CH_3CH_2CH_2CH_2-CH_2CH_3$ *n*-Hexane

(e) $CH_3CH_2CH_2CH_2MgBr$ *n*-Butylmagnesium bromide

(f) $CH_3CH_2CH_2CH_2D$ *n*-Butane-1-*d*

(g) no reaction

(h) $CH_3CH_2CH_2CH_2I$ *n*-Butyl iodide

(i) no reaction

4. (a) *n*-$BuNH_2$

(b) *n*-BuNH—Ph (Ph on N)

$$\begin{array}{c} n\text{-}BuNH \\ | \\ Ph \end{array}$$

(c) *n*-BuCN

(d) *n*-BuOEt

(e)

$$\begin{array}{c} n\text{-}BuO-C-CH_3 \\ \| \\ O \end{array}$$

(f) *n*-$BuSCH_3$

5. (a) NaI, acetone (b) NaCl (c) CH_3ONa
(d) NaOH(aq) (e) KCN (f) NH_3 (excess)
(g) Mg, ether (h) Li; then CuI

6. Steric factors are controlling. The usual S_N2 order is 1° > 2° > 3°, and slow for large G in $G-CH_2X$.

(a)

$$\underset{1^\circ}{\begin{array}{c} CH_3CH_2CH_2CH_2-CH_2 \\ | \\ Br \end{array}} > \underset{2^\circ}{\begin{array}{c} CH_3CH_2CH_2-CH-CH_3 \\ | \\ Br \end{array}} > \underset{3^\circ}{\begin{array}{c} CH_3 \\ | \\ CH_3CH_2-C-CH_3 \\ | \\ Br \end{array}}$$

(b)

$$\underset{1^\circ}{\begin{array}{c} CH_3 \\ | \\ CH_3CHCH_2-CH_2 \\ | \\ Br \end{array}} > \underset{2^\circ}{\begin{array}{c} CH_3 \\ | \\ CH_3CH-CH-CH_3 \\ | \\ Br \end{array}} > \underset{3^\circ}{\begin{array}{c} CH_3 \\ | \\ CH_3-C-CH_2CH_3 \\ | \\ Br \end{array}}$$

(c)

$$\underset{G = n\text{-}Pr-}{CH_3CH_2CH_2-CH_2Br} > \underset{iso\text{-}Bu-}{\begin{array}{c} CH_3 \\ | \\ CH_3CHCH_2-CH_2Br \end{array}} > \underset{sec\text{-}Bu-}{\begin{array}{c} CH_3 \\ | \\ CH_3CH_2CH-CH_2Br \end{array}} > \underset{tert\text{-}Bu-}{\begin{array}{c} CH_3 \\ | \\ CH_3-C-CH_2Br \\ | \\ CH_3 \end{array}}$$

7. Polar factors are controlling: the more stable the cation being formed in the initial ionization, the faster the reaction. The usual order is

$$3° > 2° > 1°$$

(a) $CH_3CH_2C(CH_3)(Br)CH_3$ (3°) > $CH_3CH_2CH_2CH(Br)CH_3$ (2°) > $CH_3CH_2CH_2CH_2CH_2Br$ (1°)

(b) $CH_3C(CH_3)(Br)CH_2CH_3$ (3°) > $CH_3CH(CH_3)CH(Br)CH_3$ (2°) > $CH_3CH(CH_3)CH_2CH_2Br$ (1°)

8.

	S_N2	S_N1
(a) Stereochemistry	Inversion	Racemization
(b) Kinetic order	2nd order	1st order
(c) Rearrangements	No	Yes
(d) Relative rates	Me > Et > *i*-Pr > *t*-Bu	*t*-Bu > *i*-Pr > Et > Me
(e) Relative rates	RI > RBr > RCl	RI > RBr > RCl
(f) Temperature inc.	Faster	Faster
(g) Doubling [RX]	Rate doubled	Rate doubled
(h) Doubling $[OH^-]$	Rate doubled	Rate unaffected

9. (a) See Sec. 6.13, especially page 230. The electrical conductivity of the solution of HX in ROH indicates ionization to form ROH_2^+ and X^-.
(b) See the bottom of page 231 and the top of page 232.

10. (a) CH_3I, CH_2Cl_2, and possibly C_2H_5Br. (b) CH_3I (violet color due to I_2).
(c) $Br^- \longrightarrow Br_2$, which is red-brown in CCl_4; Cl^- is unaffected.

6

Alcohols and Ethers

6.1 There are many —OH groups capable of hydrogen bonding with water molecules (see page 1192).

6.2 See the answer to Problem 8, Chapter 1.

6.3 (a) $(CH_3)_2CHCH_2\overset{*}{C}H(NH_3^+)COO^-$ ⟶ $(CH_3)_2CHCH_2CH_2OH$

Leucine — *Chiral*; Isopentyl alcohol — *Achiral*

$CH_3CH_2\overset{*}{C}H(CH_3)\overset{*}{C}H(NH_3^+)COO^-$ ⟶ $CH_3CH_2\overset{*}{C}H(CH_3)CH_2OH$

Isoleucine — *Chiral*; Active amyl alcohol — *Chiral*

(b) The chirality of one of the two original chiral carbons of isoleucine is retained.

6.4 The first material to distill is the ternary azeotrope; 100 g of it will distill, carrying over 7.5 g of water, 18.5 g of alcohol, and 74 g of benzene. This leaves 124 g of pure anhydrous alcohol. In actual practice, a slight excess of benzene is added; this is removed, after the distillation of the ternary mixture, as a binary azeotrope with alcohol (b.p. 68.3 °C).

6.5 Electron-withdrawing substituents (—Cl, —F, —OH, —NO_2) increase acid strength by stabilizing the anions.

(a) $ClCH_2CH_2OH > CH_3CH_2OH$

(b) $F_3CCHOHCF_3 > CH_3CHOHCH_3$

(c) $HOCH_2CHOHCH_2OH > CH_3CH_2CH_2OH$

(d) In each case, the weaker acid is the stronger nucleophile.

6.6 (a)

$$t\text{-BuOH} + Na \longrightarrow \tfrac{1}{2}H_2 + t\text{-BuO}^-Na^+$$

$$t\text{-BuO}^-Na^+ + EtBr \longrightarrow t\text{-BuOEt} + NaBr$$

(b) Nucleophilic substitution, with t-BuO$^-$ acting as the nucleophile. In the second case, elimination, with EtO$^-$ acting as the base.

6.7 Free-radical chlorination of neopentane. Simple alkyl free radicals show little tendency to rearrange. (*Note:* This is one of the very rare instances where halogenation of an alkane is a feasible laboratory preparation.)

6.8 (a) and (b) Reaction involves nucleophilic substitution with the protonated alcohol as substrate and a second molecule of alcohol as nucleophile.

First,

$$\underset{\text{Alcohol}}{ROH} + H^+ \rightleftharpoons \underset{\text{Protonated alcohol}}{ROH_2^+}$$

then, either S_N2

$$\underset{\substack{\text{Alcohol}\\ \textit{Nucleophile}}}{ROH} + \underset{\substack{\text{Protonated alcohol}\\ \textit{Substrate}}}{R{-}OH_2^+} \longrightarrow \left[\overset{\overset{H}{|\delta_+}}{RO}\cdots R\cdots\overset{\delta_+}{OH_2}\right] \longrightarrow \underset{\text{Protonated ether}}{R{-}\overset{\overset{H}{|}}{O^{\oplus}}{-}R} + \underset{\substack{\text{Water}\\ \textit{Leaving group}}}{H_2O}$$

or S_N1

$$(1)\quad \underset{\text{Protonated alcohol}}{R{-}OH_2^+} \longrightarrow \underset{\text{Carbocation}}{R^+} + H_2O$$

$$(2)\quad \underset{\text{Alcohol}}{ROH} + \underset{\text{Carbocation}}{R^+} \longrightarrow \underset{\text{Protonated ether}}{R{-}\overset{\overset{H}{|}}{O^{\oplus}}{-}R}$$

and, finally,

$$\underset{\text{Protonated ether}}{R{-}\overset{\overset{H}{|}}{O^{\oplus}}{-}R} \rightleftharpoons \underset{\text{Ether}}{R{-}O{-}R} + H^+$$

6.9 There is clearly overall inversion of configuration of the alcohol. There are only two steps in the reaction sequence. The first, preparation of the *sec*-butyl tosylate, must involve cleavage of the O—H bond of the alcohol (see page 235), and hence cannot change the configuration about the chiral carbon. Inversion must, therefore, have occurred in the only other step, hydrolysis of the tosylate.

6.10 (a) Diethyl ether, ethyl *n*-propyl ether, di-*n*-propyl ether.

(b) *tert*-Butyl ethyl ether in good yield, because *t*-BuOH gives a carbocation so much faster than EtOH.

$$t\text{-BuOH} \xrightarrow[-H_2O]{H^+} t\text{-Bu}^{\oplus} \xrightarrow{\text{EtOH}} t\text{-Bu—}\overset{\text{H}}{\underset{\oplus}{\text{O}}}\text{—Et} \xrightarrow{-H^+} t\text{-Bu—O—Et}$$

6.11 (a) and (b) No bond to the chiral carbon of 2-octanol is broken in the reaction sequence.

$$\underset{\substack{(-)\text{-2-Octanol} \\ \textit{Optical purity 80\%}}}{n\text{-Hex—}\overset{\text{CH}_3}{\text{CH}}\text{—O}\vdots\text{H}} \xrightarrow{\text{Na}} n\text{-Hex—}\overset{\text{CH}_3}{\text{CH}}\text{—O}^-\text{Na}^+ \xrightarrow{\text{EtBr}} \underset{\substack{(-)\text{-2-Ethoxyoctane} \\ \textit{Optical purity 80\%}}}{n\text{-Hex—}\overset{\text{CH}_3}{\text{CH}}\text{—O—Et}}$$

As a result, the (−)-ether has the same configuration as the (−)-alcohol, and is of the same optical purity.

The starting alcohol is

$$\frac{8.24}{10.3} \times 100 = 80\% \text{ optically pure}$$

The product ether (−15.6°) is also 80% optically pure, which means that its rotation of −15.6° is 80% of the maximum rotation (for the pure ether). That is,

$$-15.6° = 0.80[\alpha] \quad \text{and} \quad [\alpha] = -19.5°$$

6.12 (a) The starting bromide is

$$\frac{30.3°}{39.6°} \times 100 = 76.5\% \text{ optically pure}$$

The product ether is

$$\frac{15.3°}{19.5°} \times 100 = 78.5\% \text{ optically pure}$$

and of opposite configuration. Since configuration is changed without significant change of optical purity, the reaction proceeds with complete inversion.

$$\underset{\substack{(-)\text{-2-Bromooctane} \\ \textit{Optical purity 76.5\%}}}{n\text{-Hex—}\overset{\text{CH}_3}{\text{CH}}\text{—Br}} \xrightarrow{\text{EtO}^-\text{Na}^+} \underset{\substack{(+)\text{-2-Ethoxyoctane} \\ \textit{Optical purity 78.5\%}}}{n\text{-Hex—}\overset{\text{CH}_3}{\text{CH}}\text{—OEt}}$$

(b), (c) S_N2, as might be expected in a solvent of low polarity, and with a reactant of high basicity.

(d) S_N2 displacement of —Br from EtBr.

(e) In one there is attack (with inversion) on the chiral carbon; in the other there is retention of configuration since there is no attack on chiral carbon.

6.13 Reaction of the protonated ether with bromide ion would be expected to take either of two courses: (a) if S_N2, by attack at the least hindered group, methyl, to yield MeBr and *sec*-BuOH, or (b) if S_N1, by formation of the more stable carbocation, *sec*-Bu^+, to yield *sec*-BuBr and MeOH. The evidence indicates that reaction is actually S_N2. (Since the *sec*-Bu—O bond is not broken, we observe retention of configuration and no loss of optical purity.)

6.14 (a) solubility of ether in cold conc. H_2SO_4 (b) same as (a)
(c) CrO_3/H_2SO_4 (d) warm acid followed by $KMnO_4$

1. (a)–(c)

$CH_3CH_2CH_2CH_2CH_2OH$
1-Pentanol
1°

$CH_3CH_2CH_2CHOHCH_3$
2-Pentanol
2°

$CH_3CH_2CHOHCH_2CH_3$
3-Pentanol
2°

$CH_3CH_2CH(CH_3)CH_2OH$
2-Methyl-1-butanol
1°

$CH_3CH_2C(OH)(CH_3)_2$
2-Methyl-2-butanol
3°

$CH_3CHOHCH(CH_3)_2$
3-Methyl-2-butanol
2°

$(CH_3)_2CHCH_2CH_2OH$
3-Methyl-1-butanol
1°

$(CH_3)_3CCH_2OH$
2,2-Dimethylpropanol
1°

(d) $(CH_3)_2CHCH_2CH_2OH$
Isopentyl alcohol

$CH_3CH_2CH_2CH_2CH_2OH$
n-Pentyl alcohol

$CH_3CH_2C(OH)(CH_3)_2$
tert-Pentyl alcohol

(e) Examples are:

n-$C_5H_{11}CH_2OH$
1°

n-$C_4H_9CHOHCH_3$
2°

n-$C_3H_7C(OH)(CH_3)_2$
3°

2. (a) CH_3OCH_3 (b) $(CH_3)_2CHOCH(CH_3)_2$
(c) *n*-$C_4H_9OCH_3$ (d) $(CH_3)_2CHCH_2OC(CH_3)_3$
(e) $CH_3CH_2CH_2CH(OCH_3)CH_3$ (f) $CH_3CH(OH)CH_2OCH_3$

3. (a) diisobutyl ether (b) isopropyl methyl ether
(c) *tert*-butyl ethyl ether (d) 4-methoxyheptane

4. d (highest b.p.), e, a, c, b.

5. (a) alcohols (Chapter 6)
acids (Chapter 19)
amides (in Chapter 20)
amines (Chapters 22 and 23)
phenols (Chapter 24)
carbohydrates (Chapters 34 and 35)
heterocyclic compounds with —N—H (Chapter 30)
amino acids (Chapter 36)
proteins (Chapter 36)
(b) All, except hydrocarbons and halides.

6. (a) $(CH_3)_2CHOH_2{}^+HSO_4{}^-$, isopropoxonium hydrogen sulfate; $(CH_3)_2CHOSO_2OH$, isopropyl hydrogen sulfate
(b) no reaction
(c) $(CH_3)_2C{=}O$, acetone
(d) no reaction
(e) $(CH_3)_2CHBr$, isopropyl bromide
(f) $(CH_3)_2CHMgBr$, isopropylmagnesium bromide
(g) $(CH_3)_2CHI$, isopropyl iodide
(h) $(CH_3)_2CHONa$, sodium isopropoxide
(i) no reaction
(j) CH_4, methane; $(CH_3)_2CHOMgBr$, magnesium isopropoxide bromide
(k) no reaction
(l) $(CH_3)_2CHOTs$, isopropyl tosylate

7. The products are:

(a) *t*-BuOEt
(b) isobutylene
(c) Et_2O
(d) no reaction
(e) MeI + EtI
(f) no reaction
(g) $(Et_2OH)^+HSO_4{}^-$ (see page 34)
(h) $2EtOSO_3H$

8. In general, the order of reactivity is

$$3° > 2° > 1° < \text{MeOH}$$

Electronic factors are important except for 1° alcohols, for which steric factors are controlling.

(a) $\underset{3°}{CH_3CH_2{-}C(CH_3)(OH){-}CH_3} > \underset{2°}{CH_3CH(CH_3){-}CH(OH){-}CH_3},\ \underset{2°}{CH_3CH_2{-}CH(OH){-}CH_2CH_3},\ \underset{2°}{CH_3CH_2CH_2{-}CH(OH){-}CH_3} >$

$CH_3CH_2CH_2CH_2{-}CH_2OH > CH_3CH(CH_3)CH_2{-}CH_2OH > CH_3CH_2CH(CH_3){-}CH_2OH > CH_3{-}C(CH_3)_2{-}CH_2OH$

1° (rate falling off with increasing size of G in $G{-}CH_2OH$)

(b) $\underset{3°}{CH_3{-}C(CH_3)(OH){-}CH_3} > \underset{2°}{CH_3CH_2{-}CH(OH){-}CH_3} > \underset{1°}{CH_3{-}CH(CH_3){-}CH_2OH}$

(c) $CH_3CH_2{-}CH(OH){-}CH_2CH_3 > CH_3CH_2{-}CH(OH){-}CH_2{-}CH_2F > CH_3CH_2{-}CH(OH){-}CHF{-}CH_3 > CH_3CH_2{-}CH(OH){-}CF_2{-}CH_3$

All are 2° (rate falling off with nearness and number of electron-withdrawing fluorines).

9. 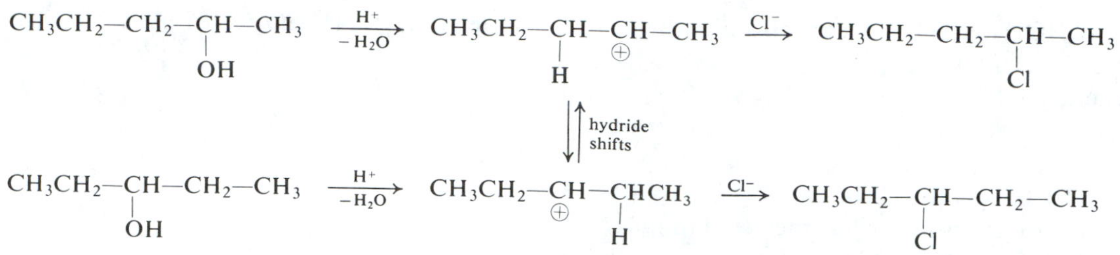 $CH_3CH_2{-}CH_2{-}\underset{\large OH}{\underset{|}{CH}}{-}CH_3 \xrightarrow[-H_2O]{H^+} CH_3CH_2{-}\underset{\large H}{\underset{|}{CH}}{-}\overset{}{\underset{\oplus}{CH}}{-}CH_3 \xrightarrow{Cl^-} CH_3CH_2{-}CH_2{-}\underset{\large Cl}{\underset{|}{CH}}{-}CH_3$

$\updownarrow$ hydride shifts

$CH_3CH_2{-}\underset{\large OH}{\underset{|}{CH}}{-}CH_2{-}CH_3 \xrightarrow[-H_2O]{H^+} CH_3CH_2{-}\underset{\oplus}{CH}{-}\underset{\large H}{\underset{|}{C}}HCH_3 \xrightarrow{Cl^-} CH_3CH_2{-}\underset{\large Cl}{\underset{|}{CH}}{-}CH_2{-}CH_3$

10. Yellow dichromate is reduced to green Cr(III) by ethyl alcohol; the motorist is reduced to tears or rage and, often, to walking.

11. See the bottom of page 408 and the top of page 409.

12. (a) CrO_3/H_2SO_4, or conc. H_2SO_4: alcohol positive.
(b) CrO_3/H_2SO_4, or conc. H_2SO_4: alcohol positive.
(c) CrO_3/H_2SO_4: alcohol positive.
(d) CrO_3/H_2SO_4: primary alcohol positive.
(e) Conc. H_2SO_4: ether positive.
(f) CrO_3/H_2SO_4: primary alcohol positive.
(g) Conc. H_2SO_4: ether positive.
(h) Only bromo compound gives halide test.

13. (a)

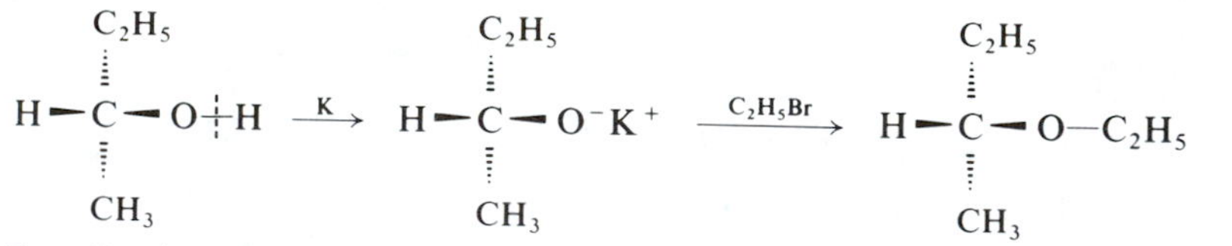

$H{-}\underset{CH_3}{\overset{C_2H_5}{C}}{-}O{\vdots}H \xrightarrow{K} H{-}\underset{CH_3}{\overset{C_2H_5}{C}}{-}O^-K^+ \xrightarrow{C_2H_5Br} H{-}\underset{CH_3}{\overset{C_2H_5}{C}}{-}O{-}C_2H_5$

(*R*)-*sec*-Butyl alcohol → (*R*)-*sec*-Butyl ethyl ether

(b)

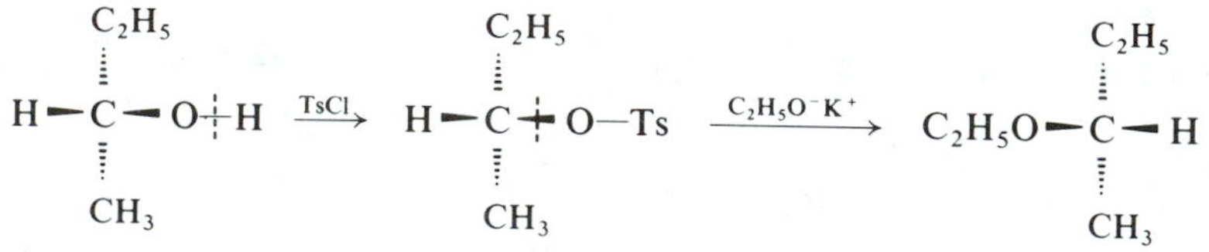

$H{-}\underset{CH_3}{\overset{C_2H_5}{C}}{-}O{\vdots}H \xrightarrow{TsCl} H{-}\underset{CH_3}{\overset{C_2H_5}{C}}{\vdots}O{-}Ts \xrightarrow{C_2H_5O^-K^+} C_2H_5O{-}\underset{CH_3}{\overset{C_2H_5}{C}}{-}H$

(*R*)-*sec*-Butyl alcohol → (*S*)-*sec*-Butyl ethyl ether

Retention of configuration in (a); inversion of configuration in the last step of (b). (Compare the answer to Problem 5.3.)

14. There are two extreme possibilities. Either (1) the protonated alcohol dissociates to a planar carbocation, hydration of which can occur at either face to give racemic alcohol,

(1a) $ROH + H^+ \rightleftharpoons ROH_2^+ \longrightarrow R^+ + H_2O$

(1b) $R^+ + H_2O \longrightarrow ROH_2^+ \rightleftharpoons ROH + H^+$

or (2) the protonated alcohol suffers S_N2 attack by water, with inversion and loss of optical activity. (See the answer to Problem 5.4, above.)

(2a) $$ROH + H^+ \rightleftharpoons ROH_2^+$$

(2b) $$H_2O + ROH_2^+ \longrightarrow \left[H_2\overset{\delta_+}{O}\text{----}R\text{----}\overset{\delta_+}{O}H_2\right] \longrightarrow H_2\overset{+}{O}R + H_2O$$

(2c) $$H_2\overset{+}{O}R \rightleftharpoons HOR + H^+$$

Actually, studies with $H_2{}^{18}O$ gave results similar to those in Problem 5.4: every replacement of oxygen is accompanied by inversion. Evidently the reaction is S_N2. (Or has a great deal of S_N2 *character*. If a cationic intermediate—an encumbered carbocation (Sec. 7.9)—*is* involved, it does not last long enough to exchange its front-side (departing) H_2O for solvent H_2O; for, in that case, re-attachment of front-side H_2O would give exchange without inversion—contrary to fact.)

15. (a) Rearrangement of an intermediate carbocation. See the E1 mechanism on page 303; for the rearrangement, see pages 305–306.

(b) See the E2 mechanism on page 294. The only alkene that can be formed by simultaneous loss of H and Br is III.

(c) What we see here for elimination is competition between alternative mechanisms, just as we saw for nucleophilic substitution in Sec. 5.23. Again, one (the E1) is unimolecular, and the other (the E2) is bimolecular; indeed, the first—and rate-controlling—step of the E1 is the *same* as the first step of S_N1.

We follow exactly the same line of reasoning here as in the third paragraph of page 210, except that now we are concerned with a reagent that acts as a *base* rather than as a nucleophile. In (a), the base is the weakly basic solvent, ethanol, which waits until the carbocation is formed before abstracting a proton; formation of a carbocation, of course, gives the opportunity for rearrangement. In (b), the base is the strongly basic ethoxide ion, which attacks the substrate itself in a one-step reaction that simply does not offer the opportunity for rearrangement to occur.

(The situation here is strictly analogous to the one described on page 204 for substitution in neopentyl bromide; indeed, the reagents are even the same—ethanol, on the one hand, and ethoxide ion—but acting as nucleophiles in that case, not bases.)

(d) In (a), the S_N1 products: IV (and possibly a little V).

```
   H3C  CH3                          CH3
    |    |                            |
CH3—C——CH—CH3                    CH3—C——CH—CH3
    |                                 |    |
   OC2H5                             H3C  OC2H5
```

2-Ethoxy-2,3-dimethylbutane
IV

3-Ethoxy-2,2-dimethylbutane
V

In (b), the S_N2 product: only V.

16. An intramolecular hydrogen bond between —OH and —G stabilizes the *gauche* conformation.

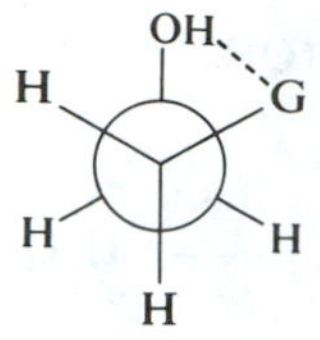

17. Let us begin our work with synthesis as we mean to go on: by **working backwards.** There are relatively few ways to make almost any compound we might want; there are relatively few ways to make its immediate precursors; and so on back to our primary starting materials. On the other hand, most starting materials—alcohols, say—can undergo so many different reactions that, if we go at the problem the other way around, we find a bewildering number of paths, few of which take us where we want to go. And so, to begin a synthesis we set down the structure of our **target molecule**, and see what we need to make it.

We do not know very much yet, and our ignorance will actually help: there is not a great deal in our minds for us to sort through. But this will change—and rapidly. Very soon we shall be designing syntheses of many steps, and drawing upon a considerable store of organic chemistry. Let us therefore approach these first, simple syntheses systematically.

(a) $CH_3CH_2CH_2CH_2Br \xleftarrow{PBr_3} CH_3CH_2CH_2CH_2OH$

(b) $CH_3CH_2CH_2CH_2I \xleftarrow{HI} CH_3CH_2CH_2CH_2OH$

(c) $CH_3CH_2CH_2CH_2OSO_3H \xleftarrow{H_2SO_4} CH_3CH_2CH_2CH_2OH$

(d) $CH_3CH_2CH_2CH_2ONa \xleftarrow{Na} CH_3CH_2CH_2CH_2OH$

(e) We know of only one way to make *n*-BuCN (and this *is* the only way): from the halide. And only one way to make the (pure) bromide.

$$CH_3CH_2CH_2CH_2CN \xleftarrow{NaCN} CH_3CH_2CH_2CH_2Br \longleftarrow \text{(a)}$$

(f) We must select our oxidizing agent carefully to avoid oxidation to the carboxylic acid.

$$CH_3CH_2CH_2CHO \xleftarrow{C_5H_5NHCrO_3Cl} CH_3CH_2CH_2CH_2OH$$

(g) $CH_3CH_2CH_2COOH \xleftarrow{KMnO_4} CH_3CH_2CH_2CH_2OH$

(h) At this point, we know only one way—a very good one—to make a pure alkane from a compound of the same carbon skeleton: hydrolysis of a Grignard reagent. A Grignard reagent is (nearly) always made from an alkyl halide, and this, in turn, (nearly) always from an alcohol. Which, not surprisingly, is what we have on hand.

$$CH_3CH_2CH_2CH_3 \xleftarrow{H_2O} CH_3CH_2CH_2CH_2MgBr \xleftarrow{Mg} CH_3CH_2CH_2CH_2Br \longleftarrow \text{(a)}$$

(i) A slight variation on (h), this time making use of deuterium oxide.

$$CH_3CH_2CH_2CH_2D \xleftarrow{D_2O} CH_3CH_2CH_2CH_2MgBr \xleftarrow{Mg} CH_3CH_2CH_2CH_2Br \longleftarrow \text{(a)}$$

(j) Although we work backwards, we must keep our eye on where we are starting from. Here, an eight-carbon product must involve the joining together of two four-carbon units.

$$CH_3CH_2CH_2CH_2 \vdots CH_2CH_2CH_2CH_3 \longleftarrow \begin{cases} CH_3CH_2CH_2CH_2Br \longleftarrow \text{(a)} \\ \downarrow Li \\ (CH_3CH_2CH_2CH_2)_2CuLi \xleftarrow{CuI} CH_3CH_2CH_2CH_2Li \end{cases}$$

18. Following the same system as for Problem 17, we arrive at the following syntheses.

(a) $CH_3CH(Cl)CH_3 \xleftarrow{HCl} CH_3CH(OH)CH_3$

(b) $CH_3CH_2OTs \xleftarrow{TsCl} CH_3CH_2OH$

(c) $(CH_3)_3C{-}OK \xleftarrow{K} (CH_3)_3C{-}OH$

(d) $CH_3CH_2CN \xleftarrow{CN^-} CH_3CH_2Br \xleftarrow{HBr} CH_3CH_2OH$

(e) $(CH_3)_3C{-}H \xleftarrow{H_2O} (CH_3)_3C{-}MgBr \xleftarrow{Mg} (CH_3)_3C{-}Br \xleftarrow{HBr} (CH_3)_3C{-}OH$

(f) We see two possible ways to put together the ether molecule: one from a two-carbon halide and a three-carbon alkoxide; the other from a three-carbon halide and a two-carbon alkoxide. Since both halides are primary and hence unlikely to undergo elimination (page 242), either route seems acceptable.

$$CH_3CH_2CH_2{-}O \vdots C_2H_5 \longleftarrow \begin{cases} CH_3CH_2Br \xleftarrow{HBr} CH_3CH_2OH \\ CH_3CH_2CH_2ONa \xleftarrow{Na} CH_3CH_2CH_2OH \end{cases}$$

or

$$CH_3CH_2CH_2 \vdots O{-}CH_2CH_3 \longleftarrow \begin{cases} CH_3CH_2CH_2Br \xleftarrow{PBr_3} CH_3CH_2CH_2OH \\ CH_3CH_2ONa \xleftarrow{Na} CH_3CH_2OH \end{cases}$$

(g) $\underset{\displaystyle |\atop \displaystyle D}{CH_3CH_2CHCH_3} \xleftarrow{D_2O} \underset{\displaystyle |\atop \displaystyle MgBr}{CH_3CH_2CHCH_3} \xleftarrow{Mg} \underset{\displaystyle |\atop \displaystyle Br}{CH_3CH_2CHCH_3} \xleftarrow{HBr} \underset{\displaystyle |\atop \displaystyle OH}{CH_3CH_2CHCH_3}$

(h) As in (f), we consider two possible combinations of reagents for the final step: reaction of a lithium dialkylcopper with either a primary alkyl halide or a secondary alkyl halide. Since only primary halides give good yields in this reaction (page 102), we select the route shown.

$$\underset{\displaystyle |\atop \displaystyle CH_3}{CH_3CH_2CH}\!\vdots\!CH_2CH_2CH_3 \longleftarrow \begin{cases} CH_3CH_2CH_2Br \xleftarrow{PBr_3} CH_3CH_2CH_2OH \\ (\underset{\displaystyle |\atop \displaystyle CH_3}{CH_3CH_2CH}-)_2CuLi \xleftarrow{CuI} \xleftarrow{Li} \underset{\displaystyle |\atop \displaystyle Br}{CH_3CH_2CHCH_3} \xleftarrow{HBr} \underset{\displaystyle |\atop \displaystyle CH_3}{CH_3CH_2CHOH} \end{cases}$$

(i) $\overset{\displaystyle CH_3\atop \displaystyle |}{CH_3CHCOOH} \xleftarrow{KMnO_4} \overset{\displaystyle CH_3\atop \displaystyle |}{CH_3CHCH_2OH}$

(j) $\overset{\displaystyle H\atop \displaystyle |}{CH_3C{=}O} \xleftarrow{C_5H_5NHCrO_3Cl} CH_3CH_2OH$

(k) $\underset{\displaystyle \|\atop \displaystyle O}{CH_3CH_2CCH_3} \xleftarrow{CrO_3} \underset{\displaystyle |\atop \displaystyle OH}{CH_3CH_2CHCH_3}$

7

Role of the Solvent

Secondary Bonding

7.1 Protic: a, d, f, h, i.

We draw the structure of each molecule and look for H attached to O or N. If there is an O—H or N—H grouping, the compound is protic; if not, it is aprotic.

(a) 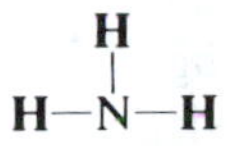 H—N(—H)—H
Ammonia
Protic

(b) 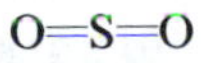 O=S=O
Sulfur dioxide
Aprotic

(c) 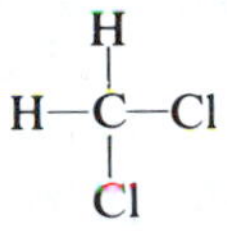 H—C(H)(Cl)—Cl
Methylene chloride
Aprotic

(d) 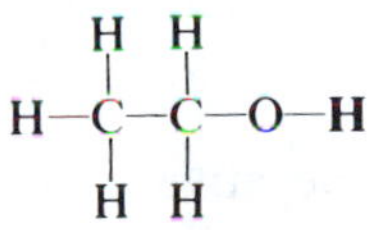 H—C(H)(H)—C(H)(H)—O—H
Ethanol
Protic

(e) 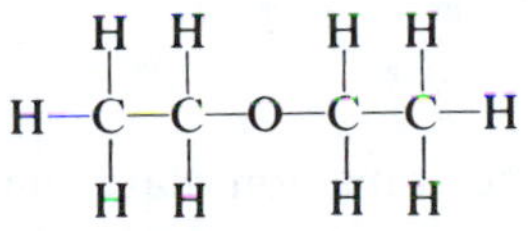 H—C(H)(H)—C(H)(H)—O—C(H)(H)—C(H)(H)—H
Diethyl ether
Aprotic

(f) 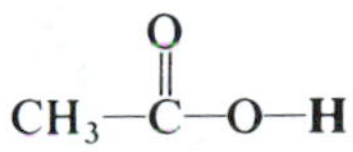 CH_3—C(=O)—O—H
Acetic acid
Protic

(g) 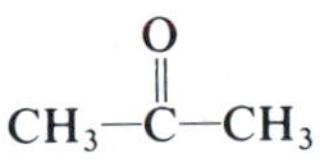 CH_3—C(=O)—CH_3
Acetone
Aprotic

(h) H—C(=O)—N(H)—H 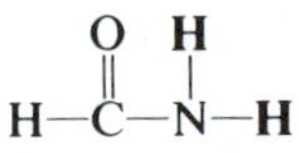
Formamide
Protic

(i) 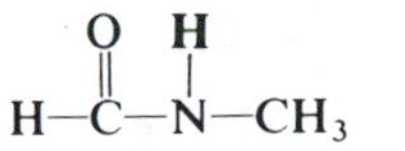 H—C(=O)—N(H)—CH_3
N-Methylformamide
Protic

(j) CH_3—C≡N
Acetonitrile
Aprotic

(k) 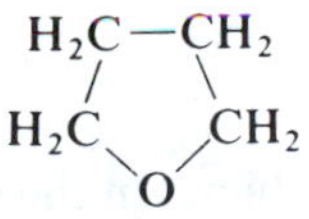 H_2C—CH_2, H_2C CH_2, O (ring)
Tetrahydrofuran
Aprotic

(l) 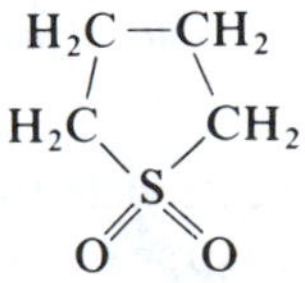 H_2C—CH_2, H_2C CH_2, S(=O)(=O) (ring)
Sulfolane
Aprotic

7.2 (a) CH_3 groups are hydrocarbon, and lipophilic. Attached to the ammonium nitrogen, they make this cation lipophilic, too, and hence better solvated by non-polar solvents; when the cation enters such a solvent, it pulls the chloride ion along to balance its charge, and the salt dissolves.

(b) By increasing the size of each hydrocarbon group—to $CH_3CH_2CH_2CH_2$, for example—we could increase the lipophilic character of the cation and the solubility of the salt in non-polar solvents. (As we shall see in our discussion of *phase transfer*, Sec. 7.7, such solubility is of great practical importance.)

7.3 In the reactant the charge is essentially concentrated on the sulfur atom; in the transition state it is dispersed over sulfur and carbon. A polar solvent forms stronger ion–dipole bonds to the reactant than to the transition state and thus stabilizes the reactant more.

$$\underset{\text{Reactant}}{R{-}\overset{+}{S}(CH_3)_2} \longrightarrow \underset{\text{Transition state}}{\left[\overset{\delta_+}{R}{\cdots}\overset{\delta_+}{S}(CH_3)_2\right]} \longrightarrow \underset{\text{Products}}{R^+ + S(CH_3)_2}$$

Concentrated charge: *stabilized more than transition state by solvation* (Reactant)

Dispersed charge (Transition state)

The deactivating effect here is much weaker than the powerful activation described for heterolysis of an alkyl halide (Sec. 7.5). In that case, a neutral molecule was being converted into two ions, and considerable charge was being generated on going from the reactant to the transition state. In the present case, one ion (a sulfonium ion) is being converted into another (a carbocation); on going from reactant to transition state, charge is simply being dispersed. In general, we can expect solvent polarity to exert powerful effects where charge is being created or destroyed, and weak effects where charge is being concentrated or dispersed.

7.4 (a) The reactants are neutral molecules; in the transition state, positive and negative charges begin to develop. A polar solvent forms much stronger dipole–dipole bonds to the polar transition state than to the weakly polar reactants, and thus stabilizes the transition state much more.

$$\underset{\text{Reactants}}{H_3N + RX} \longrightarrow \underset{\text{Transition state}}{\left[H_3\overset{\delta_+}{N}{\cdots}R{\cdots}\overset{\delta_-}{X}\right]} \longrightarrow \underset{\text{Products}}{H_3\overset{+}{N}{-}R + X^-}$$

More polar than reactants: stabilized more by solvation

(b) Here, two charged reactants yield two uncharged products; in the transition state, both charges are beginning to disappear. A polar solvent forms much stronger (ion–dipole) bonds to the full-fledged charges on the reactants than to the diminished charges on the transition state, and thus stabilizes the reactants much more.

$$HO^- + (CH_3)_3S^+ \longrightarrow \left[\overset{\delta_-}{HO}\text{----}CH_3\text{----}\overset{\delta_+}{S}(CH_3)_2\right] \longrightarrow HO{-}CH_3 + S(CH_3)_2$$

Reactants — *Full-fledged charges: stabilized more than transition state by solvation*

Transition state — *Diminished charges*

Products

(c) In one of the reactants the charge is essentially concentrated on the sulfur atom; in the transition state it is dispersed over sulfur and nitrogen. A polar solvent forms stronger ion–dipole bonds to the reactants than to the transition state, and thus stabilizes the reactants more.

$$(CH_3)_3N + CH_3{-}\overset{+}{S}(CH_3)_2 \longrightarrow \left[(CH_3)_3\overset{\delta_+}{N}\text{----}CH_3\text{----}\overset{\delta_+}{S}(CH_3)_2\right] \longrightarrow (CH_3)_3\overset{+}{N}{-}CH_3 + S(CH_3)_2$$

Reactants — *Concentrated charge: stabilized more than transition state by solvation*

Transition state — *Dispersed charge*

Products

7.5 The very weakly basic bisulfate ion has virtually no nucleophilic power and, unlike hydroxide ion, cannot compete with the nucleophile that we want to attack the substrate.

1. In interpreting experimental evidence, we typically proceed by stages. First, we summarize what the evidence *shows*: interpret what is observed in terms of what is probably happening, point out a pattern, perhaps, or make a (tentative) generalization. Then, if possible, we take further steps, and try to *account for* what is happening: to relate our first, low-level interpretation to chemical behavior in other areas and, if we can, to explain what is happening on the basis of fundamental principles. Our answer thus starts with statements about which there can be little or no argument, and moves toward ideas that are increasingly speculative.

Bonding between a cation and a molecule of water is *ion–dipole*: attraction between the positive charge of the ion and the negative pole (the unshared electrons, essentially) of water. The strength of this bond is measured by the heat liberated when the bond is formed.

(a) This evidence shows that the strength of bonding depends upon the *size* of the cation: it is strongest for the smallest cation, H^+, and weakest for the biggest, Rb^+.

Our interpretation is the following. The amount of the charge is the same (plus one) on all the cations. What varies with the size of the ion, however, is the degree to which this charge is dispersed. Electrostatic attraction is strongest for the concentrated charge on a small ion, and weakens as the ion becomes bigger and the charge more diffuse.

(b) The evidence shows that the strength of bonding depends upon how many water molecules the ion holds: the greater the number already attached, the weaker the bond to the next one.

Again we are dealing with dispersal of charge, this time over the entire hydrated ion. As each water molecule is attached, the ion becomes bigger and the charge more dispersed (page 257), and the electrostatic attraction for the next water molecule is weaker.

2. The alkyl substituents in a bulky carbocation stabilize the cation just as a cluster of solvent molecules does: through dispersal of charge (page 257). The only difference is that the substituents are held by covalent bonds and the solvent molecules are held by ion–dipole bonds.

3. The predominant species is the one held together by the strongest hydrogen bonds: those in which the more acidic molecule, water, is the hydrogen-bond donor, and the more basic atom, nitrogen in ammonia, is the hydrogen-bond acceptor (page 251).

4. Add these rows to your table.

	S_N2	S_N1
(i) Increase H_2O	Little effect	Faster in more polar solvent
(j) Increase EtOH	Little effect	Slower
(k) HMPT as solvent	Much faster	Slower

5. In the polar solvent DMSO the halide ions are little solvated, and the order of their reactivity is the same as that shown by the unsolvated ions in the gas phase, reflecting the strength of the C—X bond being formed (page 263). This is opposite to the order of their reactivity in protic solvents like methanol, where the smaller, harder ions are more strongly solvated, through hydrogen bonding, and hence more strongly deactivated.

6. (a) Since these compounds have about the same molecular weight, and roughly the same shape (straight-chain), we look for other causes of the differences in boiling point.

n-Pentane is non-polar and hence its molecules are held to one another only by van der Waals forces. Diethyl ether has a small dipole moment, which could give rise to only weak dipole–dipole bonds. These two compounds have the lowest boiling points.

With *n*-propyl chloride, the dipole moment rises; the stronger dipole–dipole bonds cause a modest rise in boiling point. With *n*-butyraldehyde, this effect is more marked: a larger dipole moment, still stronger dipole–dipole bonds, and a still higher boiling point.

With *n*-butyl alcohol, we see a large jump in boiling point, despite a marked drop in dipole moment. The explanation is that the alcohol is an *associated liquid* (Sec. 6.5); its "abnormally" high boiling point is due to the greater energy needed to break the hydrogen bonds that hold its molecules to one another. (Although the ether and the aldehyde contain oxygen, they contain hydrogen that is bonded only to carbon; these hydrogens are not positive enough to form hydrogen bonds.)

(b) Solubility in water depends chiefly on the ability of a solute to form hydrogen bonds to the solvent.

n-Pentane and *n*-propyl chloride do not contain F, O, or N, cannot form hydrogen bonds, and are insoluble in water.

n-Butyl alcohol contains an —OH group through which, acting as either an acceptor or a donor, the alcohol can form hydrogen bonds to water. Diethyl ether and *n*-butyraldehyde cannot serve as hydrogen-bond donors (see (a), above), but through their oxygens they can accept hydrogen bonds from the water. The degree of solubility of each of these three compounds (7–8 g per 100 g H_2O) reflects the balance between its hydrocarbon portion and its polar, hydrogen-bonding portion.

7. See pages 478–479.

8. The second-order kinetics indicates that all these reactions are S_N2. We follow the approach of Sec. 7.6 and Problems 7.3 and 7.4, above.

(a) Here we have reaction between a neutral substrate and a negatively charged reagent, just as in the attack by OH^- on RX discussed on pages 261–262. A polar protic solvent stabilizes the iodide ion more than the transition state, and tends to slow down reaction. As the polarity and hydrogen-bonding ability of the solvent decreases ($H_2O > CH_3OH > C_2H_5OH$), so does deactivation of the nucleophile, and there is a moderate increase in rate.

$$ {}^*I^- + CH_3{-}I \longrightarrow \left[\overset{\delta_-}{{}^*I}{\cdots}CH_3{\cdots}\overset{\delta_-}{I} \right] \longrightarrow {}^*I{-}CH_3 + I^- $$

Reactants	Transition state	Products
Concentrated charge: stabilized more than transition state by solvation	*Dispersed charge*	

(b) As in Problem 7.4(a), above, the reactants are neutral, and the transition state is polar. *n*-Hexane is non-polar; for chloroform we would expect a large dipole moment (Sec. 1.16). With this large increase in polarity of the solvent, there is a large increase in rate.

$$ R_3N + CH_3{-}I \longrightarrow \left[R_3\overset{\delta_+}{N}{\cdots}CH_3{\cdots}\overset{\delta_-}{I} \right] \longrightarrow R_3\overset{+}{N}{-}CH_3 + I^- $$

Reactants	Transition state	Products
	More polar than reactants: stabilized more by solvation	

(c) The system is similar to the one in part (a), with powerful stabilization—and thus deactivation—of the bromide ion by the protic solvent, methanol. The aprotic solvent HMPT bonds chiefly to the cation (page 263) and leaves the anion relatively free and highly reactive; the result is an enormous increase in rate.

$$ Br^- + CH_3{-}OTs \longrightarrow \left[\overset{\delta_-}{Br}{\cdots}CH_3{\cdots}\overset{\delta_-}{OTs} \right] \longrightarrow Br{-}OCH_3 + {}^-OTs $$

Reactants	Transition state	Products
Concentrated charge: stabilized more than transition state by solvation	*Dispersed charge*	

9. (a) Ion–dipole bonds involving the unshared pairs of electrons on the oxygen atoms.

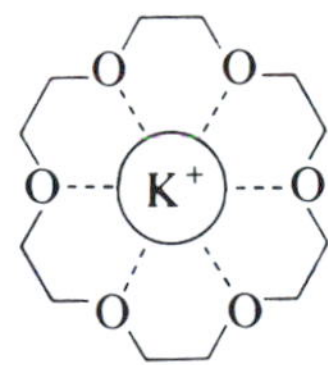

(b) See the last paragraph on page 479.

10. In this weakly ionizing solvent, we expect considerable ion pairing, which stabilizes and thus deactivates the halide ions (pages 265–266). This electrostatic attraction should be strongest between the concentrated charges on smaller ions. The small Li^+ forms tight ion pairs; the large quat ion, with its charge shielded by the bulky alkyl groups, forms only very loose ion pairs.

In the presence of the quat ion, then, the halide ions are relatively free, and show the same order of reactivity as they do in the gas phase (compare Problem 5, above). In the presence of Li^+, the smaller the halide, the stronger the ion pairing and the less reactive the nucleophiles; the order of reactivity is the same as that observed in protic solvents—and for much the same reason.

11. The molecules in concentrated sulfuric acid are held together by many very strong hydrogen bonds. There are two OH groups, very acidic and hence powerful hydrogen-bond donors (page 251), and there are four oxygens to serve as hydrogen-bond acceptors. The result is a three-dimensional network within which sulfuric acid molecules can move about only sluggishly: a thick, sirupy liquid.

For glycerol, with its three OH groups, the situation is much the same. (A contributing factor: each OH is more acidic than in an ordinary alcohol because of the electron-withdrawing effect of the two other OH groups.)

The best known sirup-formers—notorious among organic chemists—are, of course, the sugars. Look, for example, at D-(+)-glucose on page 1145.

12. What we are seeing is an example of phase-transfer catalysis (Sec. 7.7). In the aqueous phase, some of the alcohol is converted into alkoxide ion by the action of hydroxide ion:

$$n\text{-OctOH} + OH^- \rightleftharpoons n\text{-OctO}^- + H_2O$$

Alkoxide ion is then carried into the organic phase (the alkyl halide layer) by the lipophilic quat ion. In the organic phase, the alkoxide ion—virtually liberated from both solvation and ion-pairing forces—acts as a powerful nucleophile and reacts with the alkyl halide.

$$n\text{-OctO}^- + n\text{-BuCl} \longrightarrow n\text{-Oct—O—Bu-}n + Cl^-$$

If hydroxide ion were escorted into the organic layer, it would be expected to convert the alkyl halide into alcohol. But the high yield of ether shows that alkoxide is transferred in preference to hydroxide, evidently because the *n*-octyl group confers lipophilicity on this anion.

This method avoids a separate operation of the classical synthesis: reaction of the (dry) alcohol with Na or K.

8

Alkenes I. Structure and Preparation

Elimination

8.1 Let us first draw all the isomers we can, without taking into consideration geometric isomerism. We must be especially careful to see that no carbon has more than four bonds. As always, let us be systematic.

(a)

$CH_3CH_2CH_2{-}CH{=}CH{-}H$ $CH_3CH_2{-}CH{=}CH{-}CH_3$ (*Z* and *E*) $CH_3CH_2{-}C(CH_3){=}CH{-}H$

$CH_3{-}CH{=}C(CH_3){-}CH_3$ $H{-}CH{=}CH{-}CH(CH_3){-}CH_3$

(b)

$CH_3{-}CH{=}CH{-}Cl$ (*Z* and *E*) $CH_3{-}CCl{=}CH{-}H$ $CH_2Cl{-}CH{=}CH{-}H$

(c)

$CH_3CH_2{-}CH{=}CH{-}Cl$ (*Z* and *E*) $CH_3CH_2{-}CCl{=}CH{-}H$ $CH_3CHCl{-}CH{=}CH{-}H$ $CH_2ClCH_2{-}CH{=}CH{-}H$

$CH_3{-}CH{=}C{-}CH_2Cl$ (*Z* and *E*) $CH_3{-}CH{=}CCl{-}CH_3$ (*Z* and *E*) $CH_3{-}C(CH_3){=}CHCl$ $CH_2Cl{-}C(CH_3){=}CH_2$

Now let us see which of these can exist as a pair of geometric isomers. As we saw on page 280, geometric isomers can exist only if *each* of the doubly bonded carbons holds two different groups: if either one holds two identical groups, there are no geometric isomers.

We can check our conclusions by drawing possible isomers as symmetrically as we can (as on pages 280–281) and trying to superimpose them. *Remember*: These formulas represent flat molecules, and can be (mentally) handled as actual models would be—rotated or flipped over. If we cannot superimpose them, they represent a pair of geometric isomers.

(a) 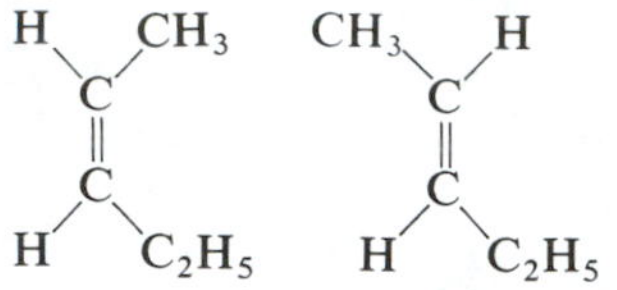 H CH_3 CH_3 H
C C
‖ ‖
C C
H C_2H_5 H C_2H_5
Z *E*

$CH_3 > H$
$C_2H_5 > H$

(b) 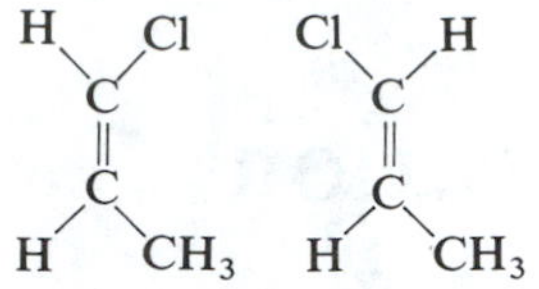 H Cl Cl H
C C
‖ ‖
C C
H CH_3 H CH_3
Z *E*

$Cl > H$
$CH_3 > H$

(c) 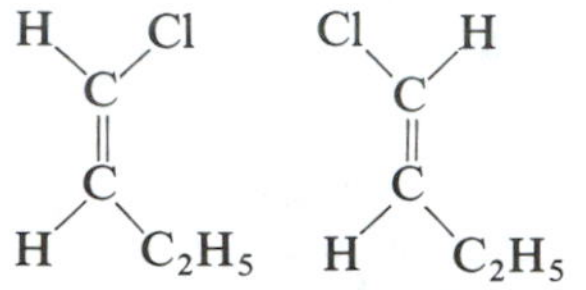 H Cl Cl H
C C
‖ ‖
C C
H C_2H_5 H C_2H_5
Z *E*

$Cl > H$
$C_2H_5 > H$

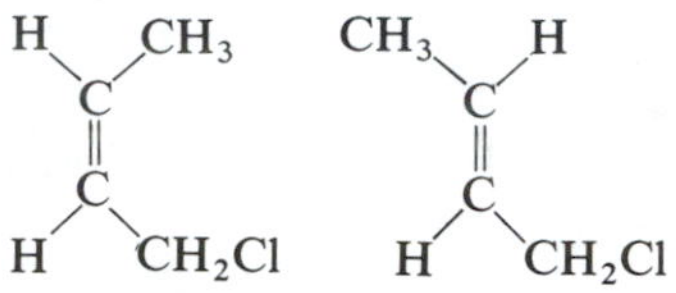

H CH_3 CH_3 H
C C
‖ ‖
C C
H CH_2Cl H CH_2Cl
Z *E*

$CH_3 > H$
$CH_2Cl > H$

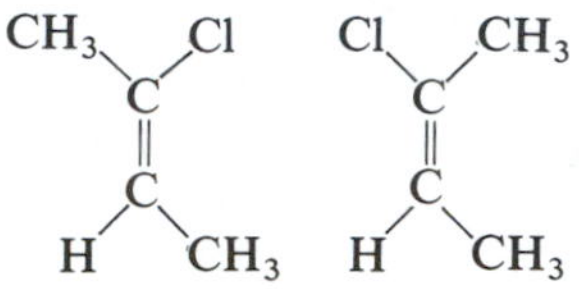

CH_3 Cl Cl CH_3
C C
‖ ‖
C C
H CH_3 H CH_3
Z *E*

$Cl > CH_3$
$CH_3 > H$

8.2 (a) $(CH_3)_2C{=}C(CH_3)_2$

(b) $H_2C{=}C(CH_3)CH_2Br$

(c) 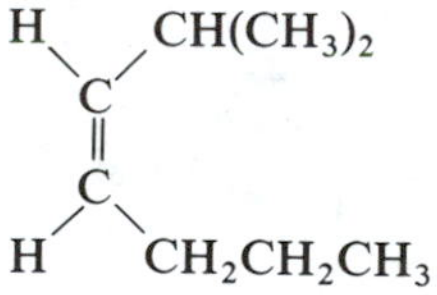 H $CH(CH_3)_2$
C
‖
C
H $CH_2CH_2CH_3$

(d) 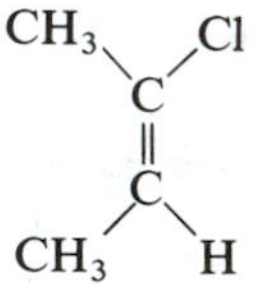 CH_3 Cl
C
‖
C
CH_3 H

8.3 Order of isomers as in answer to Problem 8.1. Geometric isomers are specified as *Z* or *E* in the second part of Problem 8.1.

(a) 1-Pentene
(*Z*)-2-pentene
(*E*)-2-pentene
2-methyl-1-butene
2-methyl-2-butene
3-methyl-1-butene

(b) (*Z*)-1-chloropropene
(*E*)-1-chloropropene
2-chloropropene
3-chloropropene

8.4 (a) *trans*-1,2-Dichloroethene is non-polar; in 1,1- and *cis*-1,2-dichloroethene, the dipole lies in the plane of the molecule along the bisector of the angle between the Cl atoms.

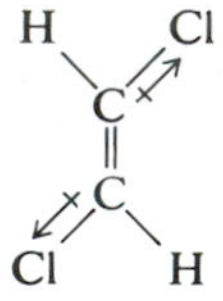

H Cl
C
‖
C
Cl H

$\mu = 0$

H Cl
C
‖
C
H Cl

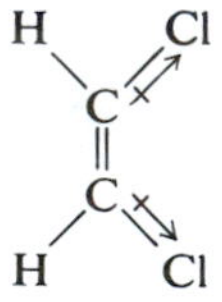

net dipole
⟼

(b) The C_4 compound has the larger dipole moment because of electron release by the two $—CH_3$ groups in the same direction as the C—Cl dipole.

H Cl / C=C / H Cl CH_3 Cl / C=C / CH_3 Cl

(c) The net dipole is in the direction of the Cl atoms, but is smaller because the C—Br dipoles oppose the C—Cl dipoles.

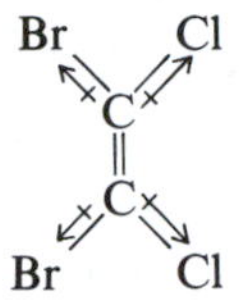

Br Cl / C=C / Br Cl

net dipole
↔

8.5 This is 1,2-elimination: for the double bond to form, a hydrogen must be lost from the carbon *next to* the carbon losing the halogen; that is, a β-hydrogen must be lost.

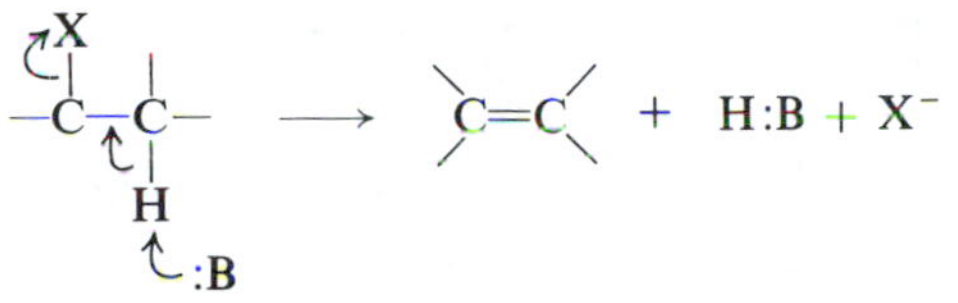

$$-\overset{X}{\underset{|}{C}}-\underset{H}{\overset{|}{C}}- \quad \longrightarrow \quad \rangle C{=}C\langle \quad + \quad H{:}B + X^-$$

:B

To predict which products can be formed in each of these cases, we follow the procedure of page 291. We simply draw the structure of the alkyl halide and identify all β-hydrogens. We can expect an alkene corresponding to the loss of *any* one of these β-hydrogens—*but no other alkenes.*

(a) 1-Pentene.

1-Chloropentane can lose hydrogen only from C–2, and hence can yield only 1-pentene.

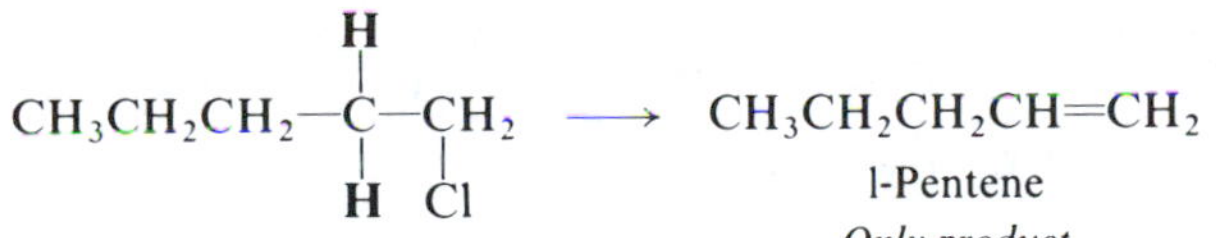

$$CH_3CH_2CH_2-\overset{H}{\underset{H}{C}}-\underset{Cl}{CH_2} \longrightarrow CH_3CH_2CH_2CH{=}CH_2$$

1-Chloropentane   1-Pentene *Only product*

(b) 1-Pentene and 2-pentene (*Z* and *E*).

2-Chloropentane can lose hydrogen either from C–1 and yield 1-pentene

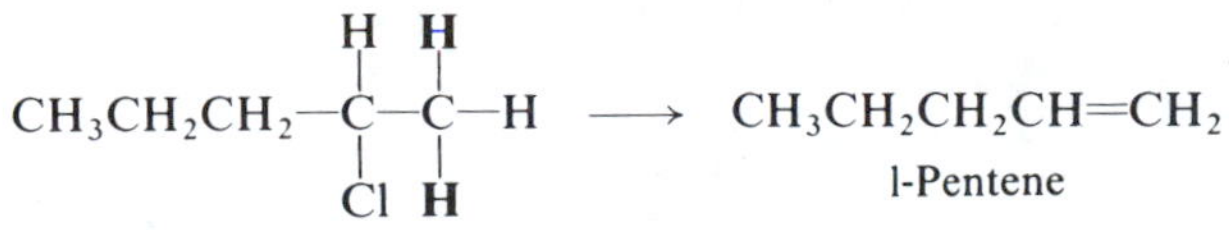

$$CH_3CH_2CH_2-\overset{H}{\underset{Cl}{C}}-\overset{H}{\underset{H}{C}}-H \longrightarrow CH_3CH_2CH_2CH{=}CH_2$$

2-Chloropentane   1-Pentene

or from C–2 and yield 2-pentene.

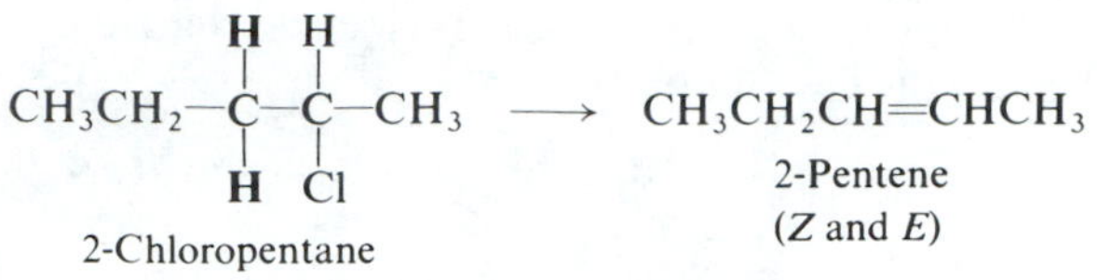

$$CH_3CH_2-\underset{H}{\overset{H}{C}}-\underset{Cl}{\overset{H}{C}}-CH_3 \longrightarrow CH_3CH_2CH{=}CHCH_3$$

2-Chloropentane 2-Pentene (*Z* and *E*)

2-Pentene exists as geometric isomers, and either of these can be formed.

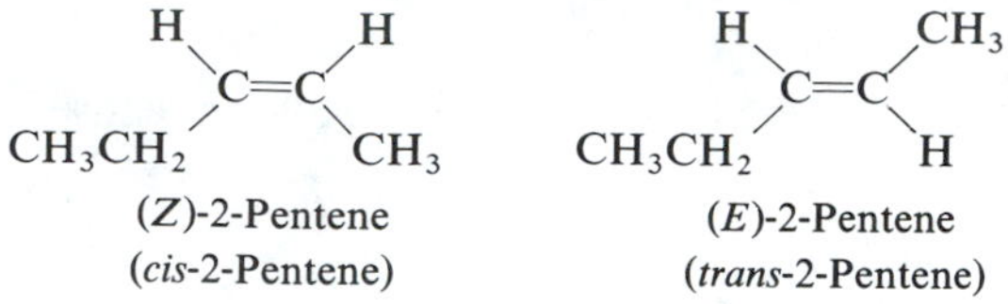

$$\underset{(\textit{cis}\text{-2-Pentene})}{\underset{(Z)\text{-2-Pentene}}{(H)(CH_3CH_2)C{=}C(H)(CH_3)}} \qquad \underset{(\textit{trans}\text{-2-Pentene})}{\underset{(E)\text{-2-Pentene}}{(H)(CH_3CH_2)C{=}C(CH_3)(H)}}$$

(c) 2-Pentene (*Z* and *E*).

3-Chloropentane can lose hydrogen from the carbon on either side of C–3, but the product is the same: 2-pentene.

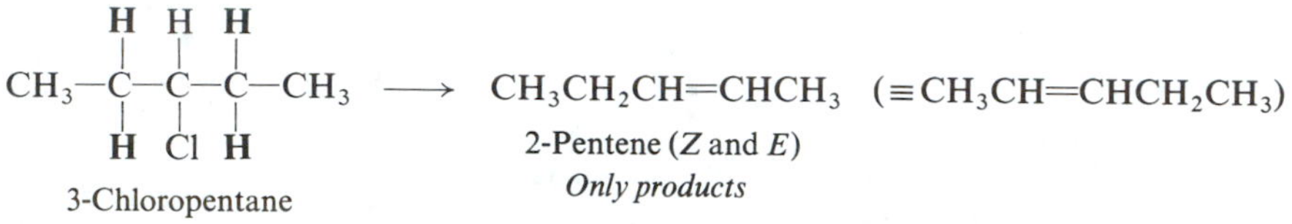

$$CH_3-\underset{H}{\overset{H}{C}}-\underset{Cl}{\overset{H}{C}}-\underset{H}{\overset{H}{C}}-CH_3 \longrightarrow CH_3CH_2CH{=}CHCH_3 \quad (\equiv CH_3CH{=}CHCH_2CH_3)$$

3-Chloropentane 2-Pentene (*Z* and *E*) *Only products*

(d) 2-Methyl-2-butene and 2-methyl-1-butene.

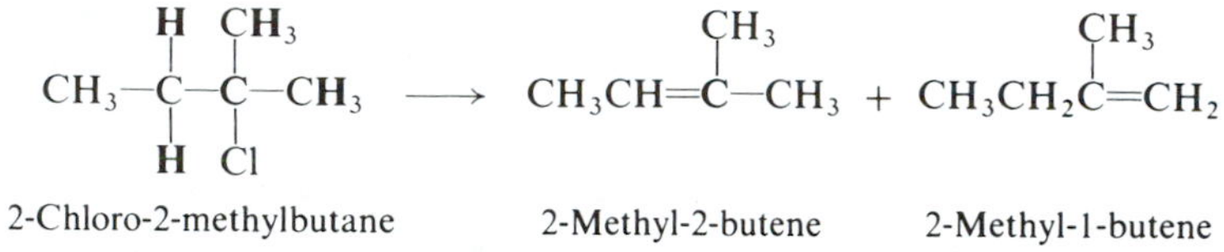

$$CH_3-\underset{H}{\overset{H}{C}}-\underset{Cl}{\overset{CH_3}{C}}-CH_3 \longrightarrow CH_3CH{=}\overset{CH_3}{C}-CH_3 + CH_3CH_2\overset{CH_3}{C}{=}CH_2$$

2-Chloro-2-methylbutane 2-Methyl-2-butene 2-Methyl-1-butene

(e) 2-Methyl-2-butene and 3-methyl-1-butene.

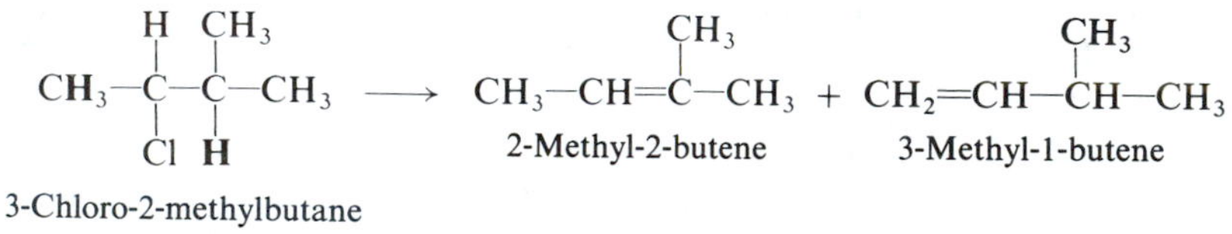

$$CH_3-\underset{Cl}{\overset{H}{C}}-\underset{H}{\overset{CH_3}{C}}-CH_3 \longrightarrow CH_3-CH{=}\overset{CH_3}{C}-CH_3 + CH_2{=}CH-\overset{CH_3}{CH}-CH_3$$

3-Chloro-2-methylbutane 2-Methyl-2-butene 3-Methyl-1-butene

(f) 2,3-Dimethyl-2-butene and 2,3-dimethyl-1-butene.

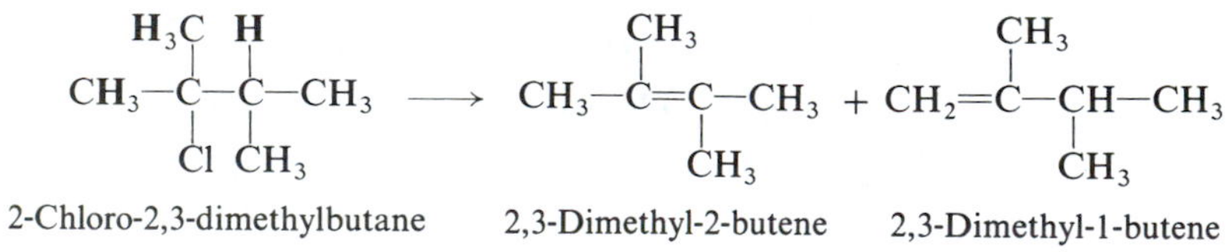

$$CH_3-\underset{Cl}{\overset{H_3C}{C}}-\underset{CH_3}{\overset{H}{C}}-CH_3 \longrightarrow CH_3-\overset{CH_3}{C}{=}\underset{CH_3}{C}-CH_3 + CH_2{=}\overset{CH_3}{C}-\underset{CH_3}{CH}-CH_3$$

2-Chloro-2,3-dimethylbutane 2,3-Dimethyl-2-butene 2,3-Dimethyl-1-butene

(g) None.

There is no β-hydrogen, and hence no 1,2-elimination.

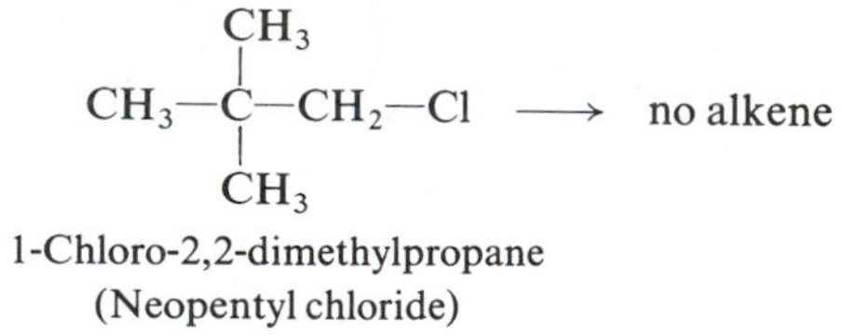

$$CH_3-\underset{CH_3}{\overset{CH_3}{C}}-CH_2-Cl \longrightarrow \text{no alkene}$$

1-Chloro-2,2-dimethylpropane
(Neopentyl chloride)

8.6 We apply the approach of Problem 8.5.

(a)

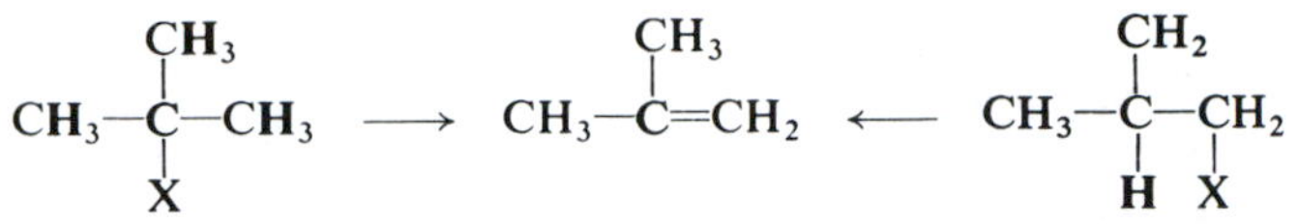

$$\underset{\textit{tert}\text{-Butyl halide}}{\mathrm{CH_3{-}\overset{\displaystyle CH_3 \atop |}{\underset{| \atop \displaystyle X}{C}}{-}CH_3}} \longrightarrow \mathrm{CH_3{-}\overset{\displaystyle CH_3 \atop |}{C}{=}CH_2} \longleftarrow \underset{\text{Isobutyl halide}}{\mathrm{CH_3{-}\overset{\displaystyle CH_2 \atop |}{\underset{| \atop \displaystyle H}{C}}{-}\underset{| \atop \displaystyle X}{CH_2}}}$$

(b)

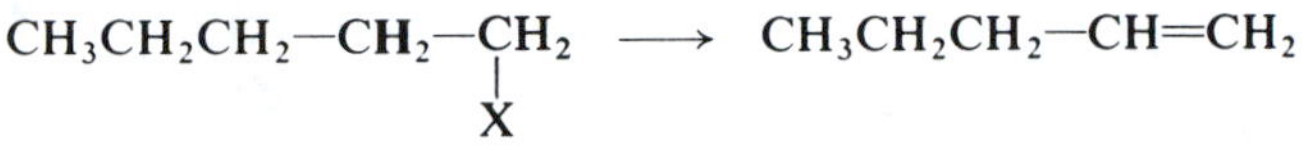

$$\underset{\text{1-Halopentane}}{\mathrm{CH_3CH_2CH_2{-}CH_2{-}\underset{| \atop \displaystyle X}{CH_2}}} \longrightarrow \mathrm{CH_3CH_2CH_2{-}CH{=}CH_2}$$

(2-Halopentane gives mixture.)

(c)

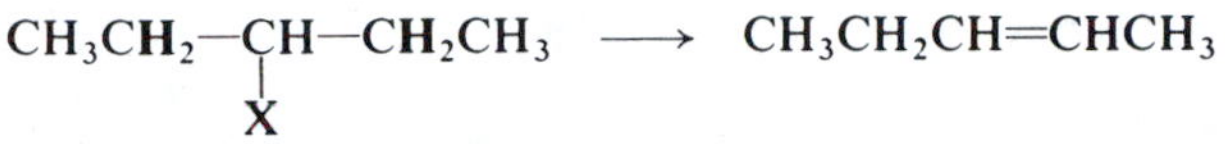

$$\underset{\text{3-Halopentane}}{\mathrm{CH_3CH_2{-}\underset{| \atop \displaystyle X}{CH}{-}CH_2CH_3}} \longrightarrow \mathrm{CH_3CH_2CH{=}CHCH_3}$$

(2-Halopentane gives mixture.)

(d)

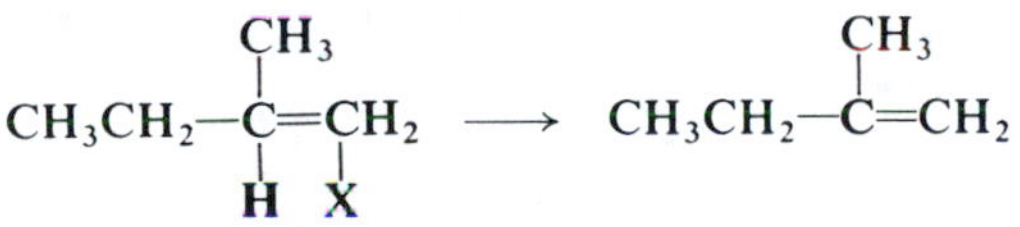

$$\underset{\text{1-Halo-2-methylbutane}}{\mathrm{CH_3CH_2{-}\overset{\displaystyle CH_3 \atop |}{\underset{| \atop \displaystyle H}{C}}{=}\underset{| \atop \displaystyle X}{CH_2}}} \longrightarrow \mathrm{CH_3CH_2{-}\overset{\displaystyle CH_3 \atop |}{C}{=}CH_2}$$

(2-Halo-2-methylbutane gives mixture.)

(e) none

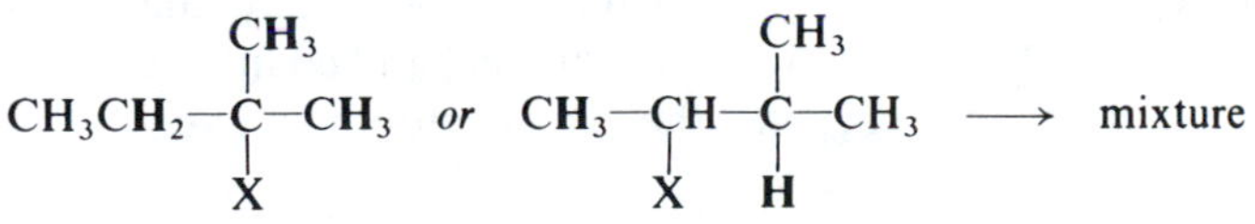

$$\underset{\text{2-Halo-2-methylbutane}}{\mathrm{CH_3CH_2{-}\overset{\displaystyle CH_3 \atop |}{\underset{| \atop \displaystyle X}{C}}{-}CH_3}} \quad \textit{or} \quad \underset{\text{2-Halo-3-methylbutane}}{\mathrm{CH_3{-}\underset{| \atop \displaystyle X}{CH}{-}\overset{\displaystyle CH_3 \atop |}{\underset{| \atop \displaystyle H}{C}}{-}CH_3}} \longrightarrow \text{mixture}$$

(f)

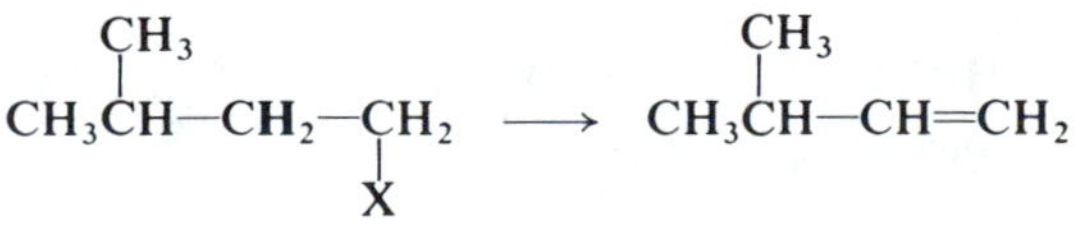

$$\underset{\text{Isobutyl halide}}{\mathrm{CH_3\overset{\displaystyle CH_3 \atop |}{C}H{-}CH_2{-}\underset{| \atop \displaystyle X}{CH_2}}} \longrightarrow \mathrm{CH_3\overset{\displaystyle CH_3 \atop |}{C}H{-}CH{=}CH_2}$$

(2-Halo-3-methylbutane gives mixture.)

8.7

(Scheme: H^+, heat converts *tert*-butyl alcohol directly to the alkene.)

$$CH_3-\underset{|}{\overset{CH_3}{\overset{|}{C}}}=CH_2 \xleftarrow{\text{KOH(alc)}} CH_3-\underset{\underset{Cl}{|}}{\overset{CH_3}{\overset{|}{C}}}-CH_3 \xleftarrow{\text{HCl}} CH_3-\underset{\underset{OH}{|}}{\overset{CH_3}{\overset{|}{C}}}-CH_3$$

tert-Butyl chloride *tert*-Butyl alcohol

$$\xleftarrow[t\text{-BuOH}]{t\text{-BuOK}} CH_3-\underset{\underset{OTs}{|}}{\overset{CH_3}{\overset{|}{C}}}-CH_3 \xleftarrow{\text{TsCl}}$$

tert-Butyl tosylate

8.8 We must allow for the fact that there are two H and only one D to be abstracted from the substrate.

(a) $k^H/k^D = 2.05$, calculated as follows:

$$k^H/k^D = \text{rate per H/rate per D}$$

$$k^H/k^D = \frac{\text{moles HCl formed/H atoms available}}{\text{moles DCl formed/D atoms available}} = \frac{0.0868/2}{0.0212/1} = 2.05$$

(b)

$$k^H/k^D = 2.05 = \frac{\text{moles HCl/1}}{\text{moles DCl/2}}$$

$$\text{moles HCl/moles DCl} = 2.05/2 = 1.02$$

8.9 Orientation in dehydrohalogenation follows Saytzeff's rule: the preferred product is the more stable alkene. For these simple alkenes, stability depends upon how many alkyl groups are attached to the doubly bonded carbons: the more highly substituted, the more stable.

We have already decided (Problem 8.5, above) which alkenes *can* be formed from each of these alkyl halides. Now, to predict the preferred product in each case, we follow the procedure of pages 300–301. We draw the structure of each possible alkene, and count the number of alkyl groups attached to the doubly bonded carbons; the alkene with the greater number is the preferred product.

(a) 1-Pentene, the only product.

(b) 2-Pentene (predominantly *E*).

$$CH_3CH_2-\overset{H}{\overset{|}{C}}=\overset{H}{\overset{|}{C}}-CH_3 \quad \textit{preferred over} \quad CH_3CH_2CH_2-\overset{H}{\overset{|}{C}}=\overset{H}{\overset{|}{C}}-H$$

Disubstituted alkene Monosubstituted alkene

(Of the stereoisomeric 2-pentenes, *E* exceeds *Z* for conformational reasons. Problem 7, page 385.)

(c) 2-Pentene (predominantly *E*).

Only the stereoisomeric 2-pentenes can be formed by 1,2-elimination. (Of the two, *E* exceeds *Z*.)

(d) 2-Methyl-2-butene.

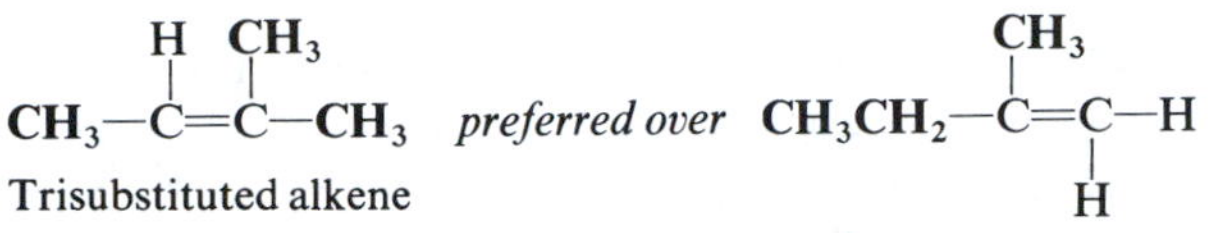

$$\underset{\text{Trisubstituted alkene}}{CH_3-\overset{H}{\overset{|}{C}}=\overset{CH_3}{\overset{|}{C}}-CH_3} \quad \textit{preferred over} \quad \underset{\text{Disubstituted alkene}}{CH_3CH_2-\overset{CH_3}{\overset{|}{C}}=\underset{H}{\underset{|}{C}}-H}$$

(e) 2-Methyl-2-butene.

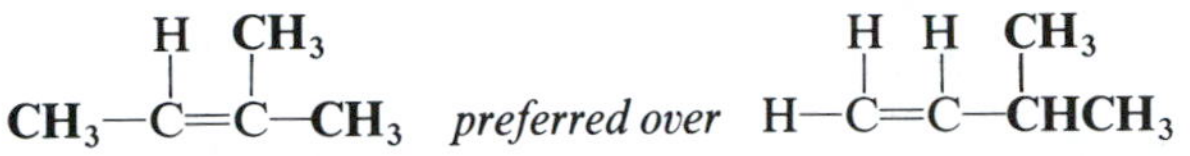

$$\underset{\text{Trisubstituted alkene}}{CH_3-\overset{H}{\overset{|}{C}}=\overset{CH_3}{\overset{|}{C}}-CH_3} \quad \textit{preferred over} \quad \underset{\text{Monosubstituted alkene}}{H-\overset{H}{\overset{|}{C}}=\overset{H}{\overset{|}{C}}-\overset{CH_3}{\overset{|}{C}}HCH_3}$$

(f) 2,3-Dimethyl-2-butene.

$$\underset{\text{Tetrasubstituted alkene}}{CH_3-\overset{CH_3}{\overset{|}{C}}=\underset{CH_3}{\underset{|}{C}}-CH_3} \quad \textit{preferred over} \quad \underset{\text{Disubstituted alkene}}{H-\overset{H}{\overset{|}{C}}=\overset{CH_3}{\overset{|}{C}}-\underset{CH_3}{\underset{|}{C}}HCH_3}$$

(g) No alkene.

8.10 isobutyl > *n*-propyl > ethyl (≫ neopentyl)

Reactivity in E2 dehydrohalogenation depends chiefly upon the stability of the alkenes being formed. We draw the structure of the alkene (or alkenes) expected from each halide, and estimate relative stabilities, as before, on the basis of the number of substituents on the doubly bonded carbons.

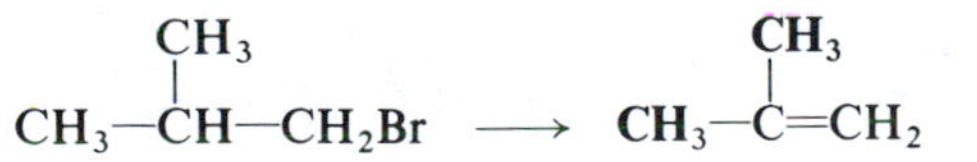

$$\underset{\text{Isobutyl bromide}}{CH_3-\overset{CH_3}{\overset{|}{C}}H-CH_2Br} \longrightarrow \underset{\text{Disubstituted alkene}}{CH_3-\overset{CH_3}{\overset{|}{C}}=CH_2}$$

$$\underset{n\text{-Propyl bromide}}{CH_3CH_2CH_2Br} \longrightarrow \underset{\text{Monosubstituted alkene}}{CH_3-CH=CH_2}$$

$$\underset{\text{Ethyl bromide}}{CH_3CH_2Br} \longrightarrow \underset{\text{Unsubstituted alkene}}{CH_2=CH_2}$$

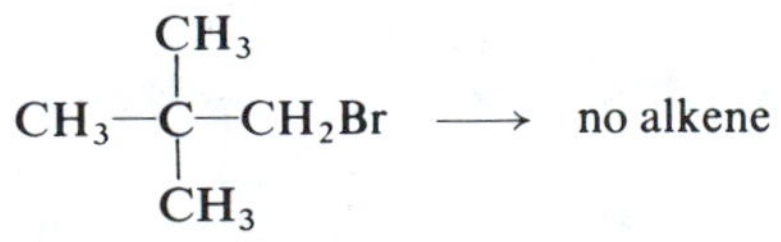

$$\underset{\text{Neopentyl bromide}}{CH_3-\underset{CH_3}{\underset{|}{\overset{CH_3}{\overset{|}{C}}}}-CH_2Br} \longrightarrow \text{no alkene}$$

8.11 The order of reactivity, 3° > 2° > 1°, suggests a carbocation mechanism, and this idea is strongly supported by the rearrangement of the carbon skeleton. *If* 3,3-dimethyl-2-butanol were to form a carbocation, it would be the 3,3-dimethyl-2-butyl cation; this is the same cation as that formed from 2-bromo-3,3-dimethylbutane in E1 elimination, and would be expected to yield the same rearranged alkenes (see pages 305–306).

We have already (Sec. 6.13) seen how acid catalyzes the formation of carbocations from alcohols through protonation of the —OH group.

Putting together the first steps in the S_N1 conversion of alcohols into alkyl halides (page 230) and the second step of the E1 reaction of alkyl halides (page 303), we arrive at the following likely mechanism for the dehydration of alcohols: a kind of E1 elimination from the protonated alcohol. (See Sec. 8.26.)

(1) $-\underset{\text{H}}{\overset{|}{\underset{|}{\text{C}}}}-\underset{\text{OH}}{\overset{|}{\underset{|}{\text{C}}}}- + \text{H:B} \rightleftharpoons -\underset{\text{H}}{\overset{|}{\underset{|}{\text{C}}}}-\underset{\text{OH}_2^+}{\overset{|}{\underset{|}{\text{C}}}}- + \text{B:}$

(2) $-\underset{\text{H}}{\overset{|}{\underset{|}{\text{C}}}}-\underset{\text{OH}_2^+}{\overset{|}{\underset{|}{\text{C}}}}- \rightleftharpoons -\underset{\text{H}}{\overset{|}{\underset{|}{\text{C}}}}-\underset{\oplus}{\overset{|}{\text{C}}}- + \text{H}_2\text{O}$

(3) $\text{B:} + -\underset{\text{H}}{\overset{|}{\underset{|}{\text{C}}}}-\underset{\oplus}{\overset{|}{\text{C}}}- \rightleftharpoons \rangle\text{C}=\text{C}\langle + \text{H:B}$

8.12 (a) and (b)

$\text{CH}_3-\overset{\text{CH}_3}{\overset{|}{\text{CH}}}-\underset{\text{OTs}}{\underset{|}{\text{CH}}}-\text{CH}_2\text{CH}_3$

$\downarrow -\text{OTs}^-$

$\text{CH}_3-\underset{\text{H}}{\overset{\text{CH}_3}{\overset{|}{\underset{|}{\text{C}}}}}-\underset{\oplus}{\text{CH}}-\text{CH}_2\text{CH}_3$ (2°) $\xrightarrow{\text{rearr}}$ $\text{CH}_3-\underset{\oplus}{\overset{\text{CH}_3}{\overset{|}{\text{C}}}}-\text{CH}_2\text{CH}_2\text{CH}_3$ (3°)

$-\text{H}^+$ (from the 2° cation) gives:

$\text{CH}_3\overset{\text{CH}_3}{\overset{|}{\text{CH}}}-\text{CH}=\text{CH}-\text{CH}_3$
4-Methyl-2-pentene
Disubstituted alkene
11%

$\text{CH}_3-\overset{\text{CH}_3}{\overset{|}{\text{C}}}=\text{CH}-\text{CH}_2\text{CH}_3$
2-Methyl-2-pentene
Trisubstituted alkene
80%

$-\text{H}^+$ (from the 3° cation) gives 2-methyl-2-pentene (above) and:

$\text{CH}_2=\overset{\text{CH}_3}{\overset{|}{\text{C}}}-\text{CH}_2\text{CH}_2\text{CH}_3$
2-Methyl-1-pentene
Disubstituted alkene
9%

One product (4-methyl-2-pentene) must be the product of rearrangement, and a second product—the major one (2-methyl-2-pentene)—*could* be. The nature of the substrate (2°) and the reaction conditions (no added strong base) are those that favor such a unimolecular reaction.

(c) As the more stable geometric isomer (Sec. 8.4), the *trans* alkene is the favored product.

8.13 *t*-BuO^- in DMSO is not solvated via hydrogen bonding, and hence is a much stronger base than highly solvated OH^- in alcohol.

8.14 As we discussed in Sec. 6.13, it is not the classification as 1°, 2°, or 3°—*as such*—that is important. It is the factors actually at work: here, *steric hindrance*, which determines reactivity by S_N2; and *alkene stability*, which chiefly determines reactivity by E2. These factors give rise to the relationship between 1°, 2°, 3° and the substitution–elimination competition. But they do more than that, as the present examples show. Branching at the β-carbon does not change the classification of the substrate. Yet it (a) increases steric hindrance and thus slows down S_N2 (Sec. 5.14), and (b) increases the branching of the alkene being formed and thus speeds up E2. The net result is a dramatic change in the course of reaction. Of the three substrates, ethyl bromide is the least hindered and gives the least stable alkene, and isobutyl bromide is the most hindered and gives the most stable alkene.

8.15 We are dealing here with solvolysis of 2° and 3° substrates—that is, reactions in the presence of no added strong nucleophile or strong base—and hence most probably with competition between S_N1 and E1 reactions. Once formed, does a carbocation combine with a nucleophile to form the substitution product, or lose a proton to form the alkene?

The carbocations formed from each set of substrates are of the same class—both secondary in (a) and (b), and both tertiary in (c)—and hence should probably not differ greatly in their rate of combining with a nucleophile. The rate at which they lose a proton, however, depends upon the stability of the alkene being formed, and it is here that we should look for a difference.

(a) $CH_3-CH(Br)-CH_3 \longrightarrow$ $\mathbf{CH_3-CH{=}CH_2}$ (Monosubstituted alkene) — *5% alkene*

$CH_3CH_2-CH(Br)-CH_3 \longrightarrow$ $\mathbf{CH_3-CH{=}CH-CH_3}$ (Disubstituted alkene, *Major product*) + $\mathbf{CH_3CH_2-CH{=}CH_2}$ (Monosubstituted alkene) — *9% alkene*

(b) $CH_3CH_2CH_2-CH(Br)-CH_3 \longrightarrow$ $\mathbf{CH_3CH_2-CH{=}CH-CH_3}$ (Disubstituted alkene) + $\mathbf{CH_3CH_2CH_2-CH{=}CH_2}$ (Monosubstituted alkene) — *7% alkene*

$CH_3CH_2-CH(Br)-CH_2CH_3 \longrightarrow$ $\mathbf{CH_3CH_2-CH{=}CH-CH_3}$ (Disubstituted alkene) — *15% alkene*

In 3-bromopentane any of four H's can be lost to give the more stable 2-pentene; in 2-bromopentane there are only two such H's.

(c) 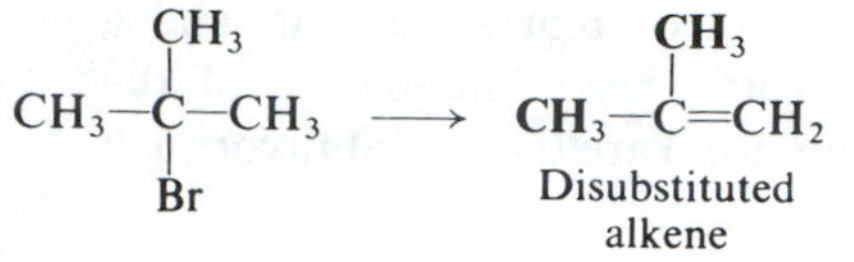 $CH_3-C(CH_3)(Br)-CH_3 \longrightarrow CH_3-C(CH_3){=}CH_2$ (Disubstituted alkene) *19% alkene*

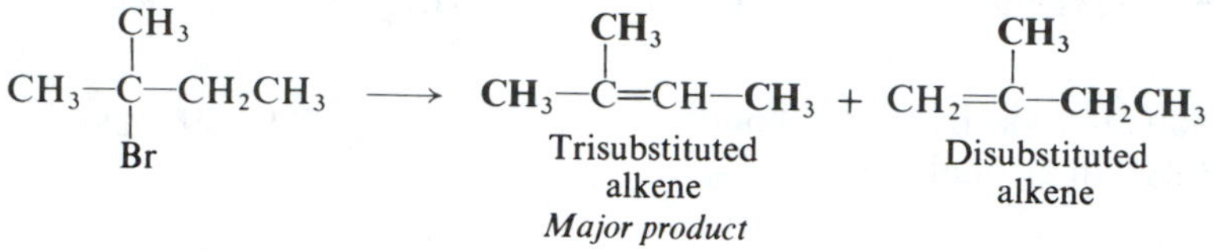

$CH_3-C(CH_3)(Br)-CH_2CH_3 \longrightarrow CH_3-C(CH_3){=}CH-CH_3 + CH_2{=}C(CH_3)-CH_2CH_3$ *36% alkene*

Trisubstituted alkene, *Major product*; Disubstituted alkene

8.16 F^- in DMSO is not solvated via hydrogen bonding, and is a strong base.

8.17 The principal base present is the solvent, *t*-BuOH; the principal acid is the protonated alcohol, *t*-$BuOH_2^+$.

When H_2SO_4 dissolves in *t*-BuOH, it is completely dissociated:

(1) 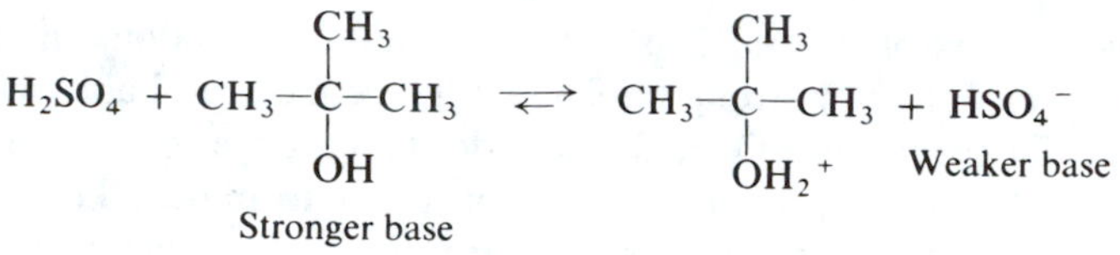 $H_2SO_4 + CH_3-C(CH_3)(OH)-CH_3 \rightleftharpoons CH_3-C(CH_3)(OH_2^+)-CH_3 + HSO_4^-$

Stronger base (*t*-BuOH); Weaker base (HSO_4^-)

For this step then, H:B is H_2SO_4, and :B is HSO_4^-.

Now the rest of the mechanism follows:

(2) 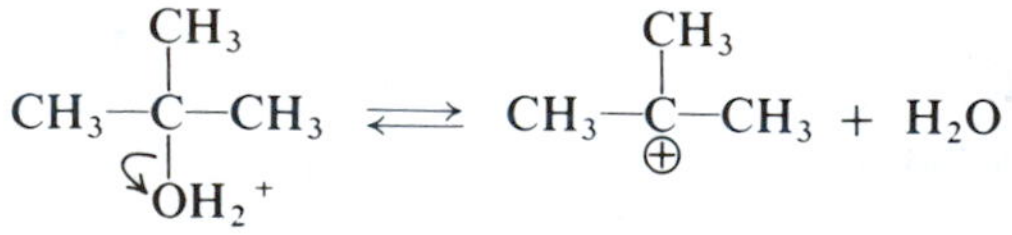 $CH_3-C(CH_3)(OH_2^+)-CH_3 \rightleftharpoons CH_3-\overset{\oplus}{C}(CH_3)-CH_3 + H_2O$

(3) $H-CH_2-\overset{\oplus}{C}(CH_3)-CH_3 + (CH_3)_3C-\ddot{O}H \;(:B) \rightleftharpoons H_2C{=}C(CH_3)-CH_3 + CH_3-C(CH_3)(OH_2^+)-CH_3 \;(H:B)$

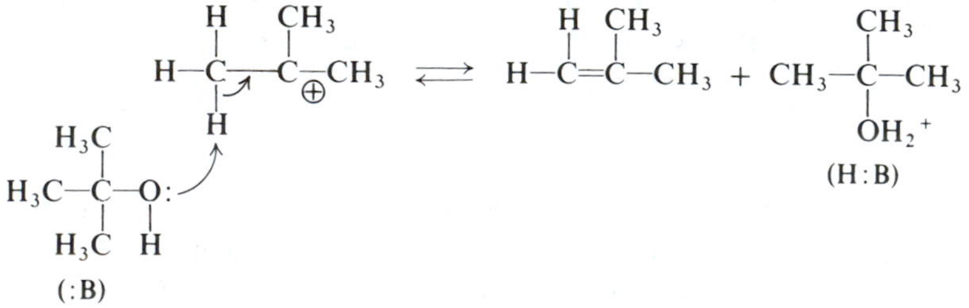

By far the most abundant—and by far the strongest—base present is the solvent, *t*-BuOH; and it is this base that abstracts the proton in step (3). In step (3), then, :B is *t*-BuOH, and H:B is *t*-$BuOH_2^+$.

But the *t*-$BuOH_2^+$ formed in step (3) as H:B is the substrate for step (2); we have generated a new molecule of protonated alcohol directly, and without the recurrence of step (1). Under these conditions, step (1) is a chain-initiating step and, after this has taken place, the chain-propagating steps (2) and (3) occur over and over.

(As the product H_2O collects, of course, it begins to compete with *t*-BuOH as the base in step (3), forming H_3O^+. Now step (1) begins to be important, with H_3O^+ as H:B transferring a proton to *t*-BuOH as the base.)

8.18 Following the principle of microscopic reversibility (page 313), we simply write the three steps of page 312 *in reverse*, with H_2O as the base :B and H_3O^+ as the acid H:B.

(1) $$H_3O^+ + \text{>C=C<} \rightleftharpoons H_2O + \text{—C(H)—C}^{\oplus}\text{—}$$

(2) $$\text{—C(H)—C}^{\oplus}\text{—} + H_2O \rightleftharpoons \text{—C(H)—C(OH}_2^{\,+}\text{)—}$$

(3) $$\text{—C(H)—C(OH}_2^{\,+}\text{)—} + H_2O \rightleftharpoons \text{—C(H)—C(OH)—} + H_3O^+$$

We note that, as in dehydration, the steps are all reversible.

8.19 (1) $$CH_3\text{—C(CH}_3\text{)(OH)—}CH_3 + H_3{}^{18}O^+ \underset{k_{-1}}{\overset{k_1}{\rightleftharpoons}} CH_3\text{—C(CH}_3\text{)(OH}_2^{\,+}\text{)—}CH_3 + H_2{}^{18}O$$

(2) $$CH_3\text{—C(CH}_3\text{)(OH}_2^{\,+}\text{)—}CH_3 \underset{k_{-2}}{\overset{k_2}{\rightleftharpoons}} CH_3\text{—C}^{\oplus}\text{(CH}_3\text{)—}CH_3 + H_2O$$

(3) $$H_2{}^{18}O + CH_3\text{—C}^{\oplus}\text{(CH}_3\text{)—}CH_3 \underset{k_{-3}}{\overset{k_3}{\rightleftharpoons}} CH_2\text{=C(CH}_3\text{)—}CH_3 + H_3{}^{18}O^+$$

$$k_{-2} > k_3$$

The intermediate carbocation recombines with water to regenerate the alcohol (the reverse of steps (2) and (1)) considerably faster than it loses a proton (step (3)) to form the alkene. That is to say, k_{-2} is considerably larger than k_3. But most of the water around it is not the ordinary H_2O that it lost, but the labeled solvent $H_2{}^{18}O$; and so the regenerated alcohol is mostly the labeled alcohol, *t*-Bu^{18}OH.

From this it is clear that here (in contrast to a true E1 reaction) the formation of the carbocation (step (2)) is not the rate-determining step in the elimination; for that to be true, k_3 would have to be much larger than k_{-2}, so that every carbocation formed would go on rapidly to form alkene. Step (2) *contributes to* the overall rate, but so does step (3).

8.20 Here again we see a familiar pattern: rearrangement of an initially formed carbocation into a more stable one, and then loss of a proton from this new cation—and possibly, in some cases, from the original cation as well—to give alkenes, with the more stable ones the preferred products.

$$CH_3CH_2CH(CH_3)CH_2OH_2^+ \underset{}{\overset{H^+}{\rightleftharpoons}} CH_3CH_2CH(CH_3)CH_2OH$$

2-Methyl-1-butanol

$\updownarrow -H_2O$

$$\underset{(1°)}{CH_3CH_2-C(CH_3)(H)-CH_2^{\oplus}} \xrightarrow{\text{rearr.}} \underset{(3°)}{CH_3CH_2-\overset{\oplus}{C}(CH_3)-CH_3}$$

$-H^+$ (from the 1° cation) and $-H^+$ (from the 3° cation):

$$CH_3CH_2-C(CH_3)=CH_2$$

2-Methyl-1-butene
Disubstituted alkene

$$CH_3-CH=C(CH_3)-CH_3$$

2-Methyl-2-butene
Trisubstituted alkene
Chief product

$$CH_3-C(CH_3)_2-CH(OH_2^+)-CH_3 \overset{H^+}{\rightleftharpoons} CH_3-C(CH_3)(H_3C)-CH(OH)-CH_3$$

3,3-Dimethyl-2-butanol

$\updownarrow -H_2O$

$$\underset{(2°)}{CH_3-C(CH_3)_2-\overset{\oplus}{CH}-CH_3} \xrightarrow{\text{rearr}} \underset{(3°)}{CH_3-\overset{\oplus}{C}(CH_3)-CH(CH_3)-CH_3}$$

$-H^+$

$$CH_3-C(CH_3)=C(CH_3)-CH_3$$

2,3-Dimethyl-2-butene
Tetrasubstituted alkene
Chief product

$$CH_2=C(CH_3)-CH(CH_3)CH_3$$

2,3-Dimethyl-1-butene
Disubstituted alkene

1. (a) $CH_3CH_2CH(CH_3)CH_2CH_2CH(CH_3)CH{=}CH_2$ (b) $ClCH_2CH{=}CH_2$ (c) $(CH_3)_3CCH{=}C(CH_3)_2$

(d) $(C_2H_5)(CH_3)C{=}C(CH_3)(C_2H_5)$ — C_2H_5 and CH_3 on the upper C; CH_3 and C_2H_5 on the lower C

(e) $(C_2H_5)(Cl)C{=}C(CH_3)(C_2H_5)$ — C_2H_5 and Cl on the upper C; CH_3 and C_2H_5 on the lower C

(f) $(CH_3)(Cl)C{=}C(D)(H)$ — CH_3 and Cl on the upper C; D and H on the lower C

(g) $CH_2{=}CH{-}CH(H)(Br){-}CH_3$ (H above and Br below the central carbon)

(h) $CH_3CH_2{-}C(H)(CH_3){-}CH{=}CH{-}CH_3$ (H above and CH_3 below the carbon bonded to CH_3CH_2)

2. (a) $CH_3{-}C(CH_3){=}CH_2$
2-Methylpropene

(b) $(H)(CH_2CH_3)C{=}C(H)(CH_2CH_3)$ — H and CH_2CH_3 on the upper C; H and CH_2CH_3 on the lower C
(*Z*)-3-Hexene

(c) $CH_3{-}C(CH_3)_2{-}CH{=}CH_2$
3,3-Dimethyl-1-butene

(d) $(CH_3)_2CH{-}CH{=}CH{-}CH(CH_3){-}CH_3$ — $CH_3{-}CH(CH_3)$ and H on the upper C; H and $CH(CH_3){-}CH_3$ on the lower C
(*E*)-2,5-Dimethyl-3-hexene

(e) $CH_3{-}CH(CH_3){-}CH_2{-}CH{=}C(CH_3){-}CH_3$
2,5-Dimethyl-2-hexene

(f) $CH_3CH_2{-}C(CH_2CH_3){=}CH_2$
2-Ethyl-1-butene

3. b, d, g, h, i, k (three isomers: *Z,Z*-, *E,E*-, *Z,E*-).

(b) *Z*: $(H)(CH_3)C{=}C(H)(CH_3)$ — H and CH_3 on the upper C; H and CH_3 on the lower C
E: $(CH_3)(H)C{=}C(H)(CH_3)$ — CH_3 and H on the upper C; H and CH_3 on the lower C

$CH_3 > H$

(d) *Z*: $(H)(Cl)C{=}C(H)(Cl)$ — H and Cl on the upper C; H and Cl on the lower C
E: $(Cl)(H)C{=}C(H)(Cl)$ — Cl and H on the upper C; H and Cl on the lower C

$Cl > H$

(g) *Z*: $(H)(CH_3)C{=}C(H)(C_2H_5)$ — H and CH_3 on the upper C; H and C_2H_5 on the lower C
E: $(CH_3)(H)C{=}C(H)(C_2H_5)$ — CH_3 and H on the upper C; H and C_2H_5 on the lower C

$CH_3 > H$
$C_2H_5 > H$

(h) *Z*: $(H)(Cl)C{=}C(H)(CH_3)$ — H and Cl on the upper C; H and CH_3 on the lower C
E: $(Cl)(H)C{=}C(H)(CH_3)$ — Cl and H on the upper C; H and CH_3 on the lower C

$Cl > H$
$CH_3 > H$

(i)

H, CH_3 / C=C / CH_3, CH_2Cl	CH_3, H / C=C / CH_3, CH_2Cl
Z	*E*

$CH_3 > H$
$CH_2Cl > CH_3$

(k)

H, CH_3 / C=C / H, (C=C: H / CH_3, H)	CH_3, H / C=C / H, (C=C: H / CH_3, H)	CH_3, H / C=C / H, (C=C: H / H, CH_3)
Z,Z	*E,Z*	*E,E*

C > H

4. Differ in all except (h); dipole moment would tell.

5. (a) $CH_3CH_2CH_2OH \xrightarrow{H^+,\ heat} CH_3CH{=}CH_2 + H_2O$

(b) $CH_3CHOHCH_3 \xrightarrow{H^+,\ heat} CH_3CH{=}CH_2 + H_2O$

(c) $CH_3CHClCH_3 + KOH \longrightarrow CH_3CH{=}CH_2 + KCl + H_2O$

(d) $CH_3CH(OTs)CH_3 + t\text{-BuOK} \xrightarrow{t\text{-BuOH}} CH_3CH{=}CH_2 + KOTs + t\text{-BuOH}$

(e) $CH_3CHBrCH_2Br + Zn \longrightarrow CH_3CH{=}CH_2 + ZnBr_2$

(f) $CH_3C{\equiv}CH + H_2 \xrightarrow[\text{catalyst}]{\text{Lindlar}} CH_3CH{=}CH_2$ (See Sec. 12.8)

6. We follow the procedure of Problem 8.5, above.

(a) 1-hexene

(b) 1-hexene
(*Z*)-2-hexene
(*E*)-2-hexene

(c) 2-methyl-1-pentene

(d) 2-methyl-2-pentene
2-methyl-1-pentene

(e) 2-methyl-2-pentene
(*Z*)-4-methyl-2-pentene
(*E*)-4-methyl-2-pentene

(f) 4-methyl-1-pentene
(*Z*)-4-methyl-2-pentene
(*E*)-4-methyl-2-pentene

(g) 4-methyl-1-pentene

(h) 2,3-dimethyl-2-pentene
(*Z*)-3,4-dimethyl-2-pentene
(*E*)-3,4-dimethyl-2-pentene
2-ethyl-3-methyl-1-butene

7. We follow the procedure of Problem 8.9, above. For these dehydrohalogenations, we expect orientation to follow Saytzeff's rule, and yield predominantly the more stable—more highly substituted—alkene.

(b) 2-hexene (predominantly *E*)
(d) 2-methyl-2-pentene
(e) 2-methyl-2-pentene
(f) 4-methyl-2-pentene (predominantly *E*)
(h) 2,3-dimethyl-2-pentene

8. As we saw in Sec. 8.26, the order of reactivity is generally 3° > 2° > 1°.

(a) $CH_3CH_2CH_2CHOHCH_3$ (b) $(CH_3)_2C(OH)CH_2CH_3$ (c) $(CH_3)_2CHC(OH)(CH_3)_2$

9. We follow the procedure of Problem 8.10, above, and expect reactivity to depend primarily upon the stability of the alkene being formed.

(a) $CH_3CH_2C(CH_3)(Br)CH_3 > CH_3CH_2CH(Br)CH_2CH_3 > CH_3CH_2CH_2CH(Br)CH_3 > CH_3CH_2CH_2CH_2CH_2Br$

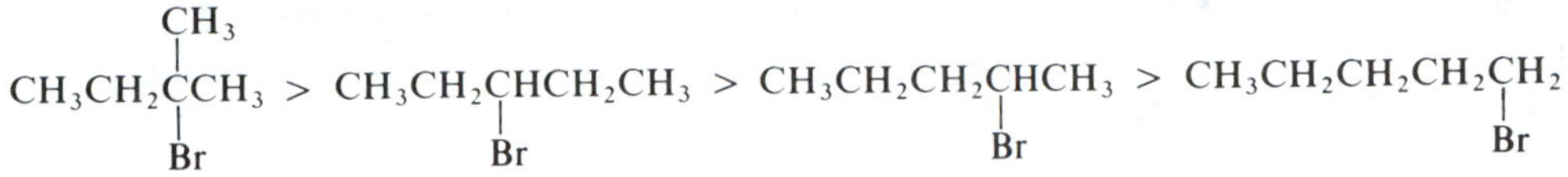

(b) $CH_3C(CH_3)(Br)CH_2CH_3 > CH_3CH(CH_3)CH(Br)CH_3 > CH_3CH(CH_3)CH_2CH_2Br$

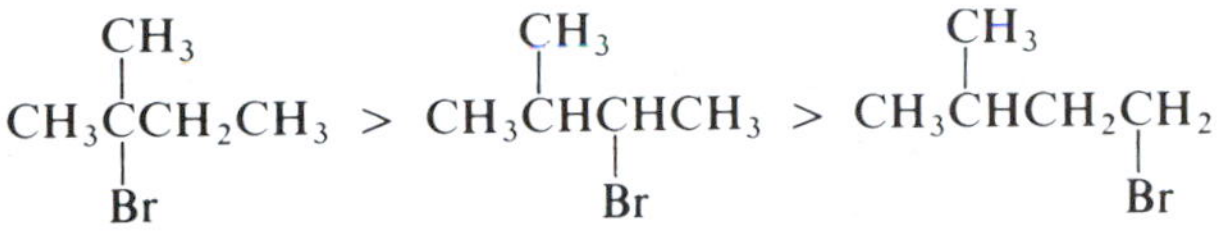

(c) $CH_3CH_2CH(CH_3)CH_2Br > CH_3CH(CH_3)CH_2CH_2Br, CH_3CH_2CH_2CH_2CH_2Br \gg CH_3-C(CH_3)_2-CH_2Br$

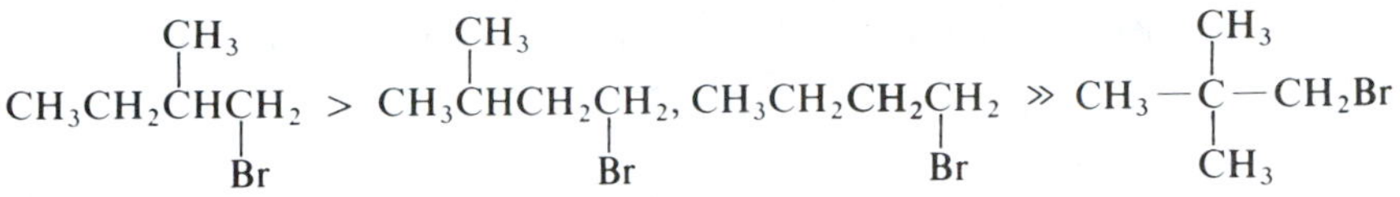

No reaction

10. (a) The first step is the formation of a 1° carbocation. This rearranges by a methyl shift to the more stable 3° cation, from which the alkenes are obtained. As usual, the more highly branched alkene predominates.

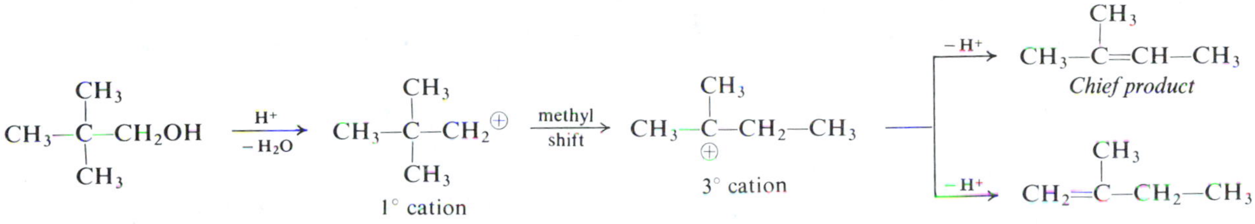

$$CH_3-C(CH_3)_2-CH_2OH \xrightarrow[-H_2O]{H^+} \underset{1^\circ \text{ cation}}{CH_3-C(CH_3)_2-CH_2^{\oplus}} \xrightarrow[\text{shift}]{\text{methyl}} \underset{3^\circ \text{ cation}}{CH_3-\overset{\oplus}{C}(CH_3)-CH_2-CH_3}$$

$$\xrightarrow{-H^+} \underset{\textit{Chief product}}{CH_3-C(CH_3)=CH-CH_3}$$

$$\xrightarrow{-H^+} CH_2=C(CH_3)-CH_2-CH_3$$

11. The reaction of the quaternary ammonium ions, R_4N^+, would seem to be an example of base-promoted 1,2-elimination (which it is, Sec. 23.5), with the only moderately basic amine (R_3N) as the leaving group, and the hydroxide ion abstracting a β-hydrogen.

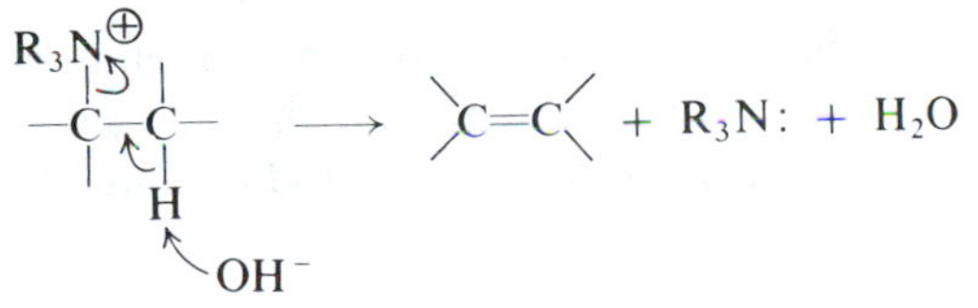

$$R_3N^{\oplus}-\overset{|}{\underset{|}{C}}-\overset{|}{\underset{H}{C}}- \;(\text{H abstracted by } OH^-) \longrightarrow \;>C=C< \; + \; R_3N: \; + \; H_2O$$

Unlike the quaternary ion, the corresponding ammonium ion, RNH_3^+, bears protons on nitrogen. These protons are enormously more acidic than a β-hydrogen attached to carbon (they are about as acidic as the protons in NH_4^+) and hence one of them is abstracted by the strong base, OH^-. Once this proton has been lost, we no longer have an ammonium

$$\underset{\text{An ammonium ion}}{RNH_3^+} + OH^- \rightleftarrows \underset{\text{An amine}}{RNH_2} + H_2O$$

ion at all, but an *amine*, and amines do not undergo this kind of elimination. (For an amine to undergo base-promoted 1,2-elimination, the leaving group would have to be the RNH^- ion, very strongly basic and an extremely poor leaving group—poorer even than OH^-.)

12. Formation of the second product clearly involves rearrangement, and of exactly the kind observed in the S_N1 reaction of 3,3-dimethyl-2-butyl substrates (page 204). This strongly suggests a carbocation mechanism, with rearrangement of the secondary 3,3-dimethyl-2-butyl cation into the tertiary 2,3-dimethyl-2-butyl cation, as shown on pages 206–207. For the most likely mechanism we are led to the sequence of reactions shown in Sec. 9.9, which leads specifically to the rearrangement shown on page 335.

13. Following the approach of Problem 17 on page 84 of this Study Guide, we set down the formula of our target molecule and see what we need to make it.

(a) A single step, dehydration of an alcohol of the correct carbon skeleton. Either *tert*-butyl alcohol or isobutyl alcohol could be used; we would probably use the tertiary alcohol because of its ease of dehydration.

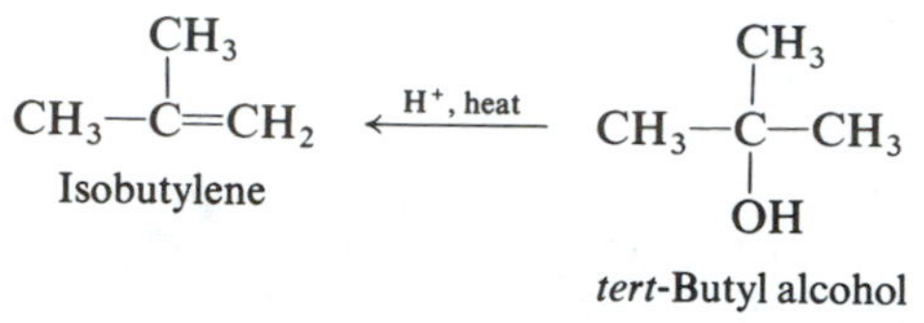

$$\underset{\text{Isobutylene}}{CH_3-\overset{CH_3}{\overset{|}{C}}=CH_2} \xleftarrow{H^+,\ \text{heat}} \underset{tert\text{-Butyl alcohol}}{CH_3-\underset{OH}{\underset{|}{\overset{CH_3}{\overset{|}{C}}}}-CH_3}$$

or

$$\underset{\text{Isobutylene}}{CH_3-\overset{CH_3}{\overset{|}{C}}=CH_2} \xleftarrow{H^+,\ \text{heat}} \underset{\text{Isobutyl alcohol}}{CH_3-\underset{H}{\underset{|}{\overset{CH_3}{\overset{|}{C}}}}-CH_2OH}$$

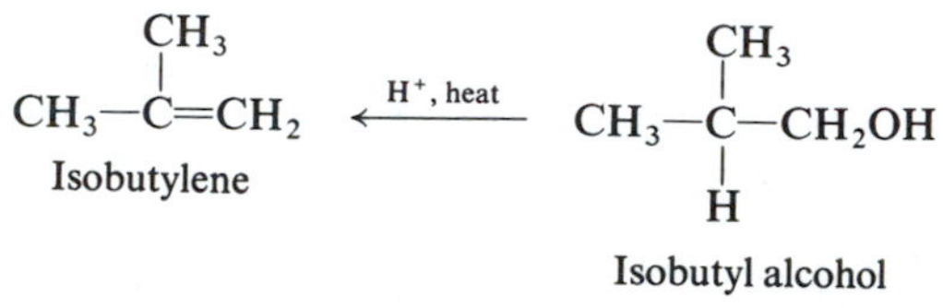

(b) Designing this synthesis may appear simple, but it actually requires considerable thinking. Let us set down our target molecule

$$\underset{\text{1-Butene}}{CH_3CH_2CH{=}CH_2}$$

and examine its structure. It could be prepared by dehydrohalogenation of an alkyl halide of the same carbon skeleton, or by dehydration of an alcohol. If the halogen or hydroxyl group were attached to C–2, we would obtain some of the desired product, but much more of its isomer, 2-butene:

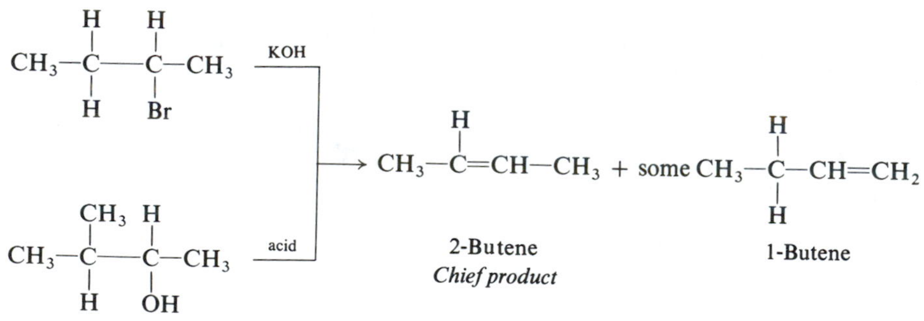

$$CH_3-\underset{H}{\underset{|}{\overset{H}{\overset{|}{C}}}}-\underset{Br}{\underset{|}{\overset{H}{\overset{|}{C}}}}-CH_3 \xrightarrow{KOH}$$

$$CH_3-\underset{H}{\underset{|}{\overset{CH_3}{\overset{|}{C}}}}-\underset{OH}{\underset{|}{\overset{H}{\overset{|}{C}}}}-CH_3 \xrightarrow{acid}$$

$$\longrightarrow \underset{\substack{\text{2-Butene} \\ \textit{Chief product}}}{CH_3-\overset{H}{\overset{|}{C}}=CH-CH_3} + \text{some } \underset{\text{1-Butene}}{CH_3-\underset{H}{\underset{|}{\overset{H}{\overset{|}{C}}}}-CH=CH_2}$$

We would select, then, a compound with the functional group attached to C–1. Even so, if we were to use the alcohol, there would be extensive rearrangement to yield, again, the more stable 2-butene:

$$\underset{\text{1-Butanol}}{CH_3-CH_2-CH_2-CH_2OH} \xrightarrow{\text{acid}} \underset{\text{1-Butene}}{CH_3-CH_2-CH{=}CH_2} \text{ and mostly } \underset{\text{2-Butene}}{CH_3-CH{=}CH-CH_3}$$

Only dehydrohalogenation of 1-bromobutane would yield the desired product in pure form:

$$\underset{\text{1-Bromobutane}}{CH_3-CH_2-CH_2-CH_2Br} \xrightarrow{\text{alcoholic KOH}} \underset{\text{1-Butene}}{CH_3-CH_2-CH{=}CH_2}$$

How do we prepare the necessary alkyl halide? Certainly not by bromination of an alkane, since even if we could make the proper alkane in some way, bromination would occur mostly at a secondary position to give the wrong product. As usual, then, we would prepare the halide from the corresponding alcohol, in this case *n*-butyl alcohol. Since this is a primary alcohol (without branching near the —OH group), and hence does not form the halide via the carbocation, rearrangement is not likely; we might use, then, either hydrogen bromide or PBr_3.

$$CH_3-CH_2-CH_2-CH_2-Br \xleftarrow{PBr_3} \underset{\text{1-Butanol}}{CH_3-CH_2-CH_2-CH_2-OH}$$

(c) To make an ether by the Williamson synthesis, we use an alkoxide and alkyl halide. As usual (Problem 18(f), Chapter 6), there are two choices: we may consider the molecule to be put together either (*a*) between *n*-propyl and oxygen or (*b*) between *tert*-butyl and oxygen.

$$CH_3CH_2CH_2 \overset{(a)}{\vdots} O \overset{(b)}{\vdots} C(CH_3)_3$$

Of the two possibilities we reject (*b*), since this would involve the reaction of a tertiary alkyl halide with a strong base, the alkoxide; this we know (Sec. 8.25) would proceed, not by substitution to give the ether, but by elimination to give the alkene. The alternative (*a*), on the other hand, involves the reaction of a primary alkyl halide, and is our choice. The entire synthesis, then, is as follows:

$$t\text{-Bu}-O \vdots \text{Pr-}n \longleftarrow \begin{cases} t\text{-BuO}^-Na^+ \xleftarrow{Na} t\text{-BuOH} \\ n\text{-PrBr} \xleftarrow{PBr_3} n\text{-PrOH} \end{cases}$$

(d) As in part (c) above, we consider the two possible ways of putting the molecule together. We select the reaction of the primary halide rather than the reaction of the secondary.

sec-Bu—O ⁞ Bu-*iso* ← [*sec*-BuO$^-$Na$^+$ $\xleftarrow{\text{Na}}$ *sec*-BuOH] and [*iso*-BuBr $\xleftarrow{\text{PBr}_3}$ *iso*-BuOH]

9

Alkenes II. Reactions of the Carbon–Carbon Double Bond

Electrophilic and Free-Radical Addition

9.1 (a) $C_4H_8 + 6O_2 \longrightarrow 4CO_2 + 4H_2O$ (b) same as (a).

(c)

1-butene	649.8
cis-2-butene	648.1
trans-2-butene	647.1

(d)

1-pentene	806.9
cis-2-pentene	805.3
trans-2-pentene	804.3

The most highly branched alkene is usually the most stable (lowest energy content); a *trans* isomer is usually more stable (lower energy content) than a *cis* isomer.

9.2 (a) H_3O^+; HBr. (b) HBr. (c) HBr.

In aqueous solution, HBr reacts to yield the weaker acid, H_3O^+. (Compare Problem 10, Chapter 1.)

9.3 Removal of hydrogen occurs from the carbon carrying the fewer hydrogens. ("From him that hath not, it shall be taken, even that which he hath.")

9.4 (a) Nucleophilic aliphatic substitution, with water as the nucleophile and the weakly basic HSO_4^- ion as the leaving group.

$$\underset{\text{Substrate}}{R\!\!-\!\!\!\mid\!\!-OSO_3H} + \underset{\text{Nucleophile}}{H_2O} \longrightarrow \underset{\text{Product}}{R\!-\!OH_2^+} + \underset{\text{Leaving group}}{HSO_4^-}$$

$$R\!-\!OH_2^+ + H_2O \rightleftarrows R\!-\!OH + H_3O^+$$

(b) We would expect S_N2.

$$H_2O + C_2H_5\!-\!OSO_3H \longrightarrow \left[\overset{\delta+}{H_2O}\text{----}C_2H_5\ \ \overset{\delta-}{OSO_3H}\right] \longrightarrow H_2\overset{+}{O}\!-\!C_2H_5$$

$$C_2H_5OH_2^+ + H_2O \rightleftarrows C_2H_5OH + H_3O^+$$

(c) We would expect S_N1.

$$(CH_3)_3C{-}OSO_3H \longrightarrow (CH_3)_3C^+ + HSO_4^- \qquad \textit{Slow}$$

$$(CH_3)_3C^+ + H_2O \longrightarrow (CH_3)_3C{-}OH_2^+ \qquad \textit{Fast}$$

$$(CH_3)_3C{-}OH_2^+ + H_2O \rightleftharpoons (CH_3)_3C{-}OH + H_3O^+$$

9.5 What is the significance of this finding? (We have already encountered this technique (Sec. 8.18), and should have a suspicion of where we are heading.) Consider what would happen if carbocations were formed rapidly and reversibly in step (1), and then—every so often—slowly combined with the base to complete the addition. In that case most carbocations

$$CH_3{-}\overset{CH_3}{\underset{}{C}}{=}\underset{H}{C}{-}CH_3 \underset{D_2O}{\overset{(1)\ D^+}{\rightleftharpoons}} CH_3{-}\overset{H_3C}{\underset{\oplus}{C}}{-}\overset{H}{\underset{D}{C}}{-}CH_3 \underset{D_2O}{\rightleftharpoons} CH_3{-}\overset{CH_3}{C}{=}\underset{D}{C}{-}CH_3$$

Unlabeled alkene
Starting material

Labeled alkene
Exchange product: not obtained

(2) D_2O ↓

$$CH_3{-}\overset{CH_3}{\underset{+OD_2}{C}}{-}CHD{-}CH_3 \xrightarrow{-D^+} CH_3{-}\overset{CH_3}{\underset{OD}{C}}{-}CHD{-}CH_3$$

would lose hydrogen many times to regenerate the alkene before eventually going on to product. But in the carbocation there are *two* hydrogens on C-3: the protium (H) that has been there all along, and the newly acquired deuterium (D). In reverting to alkene, the carbocation would be just as likely—*more* likely, actually (*Why?*)—to lose protium as deuterium, and thus leave deuterium in the alkene. While stewing in the solution until half-reaction time, unconsumed alkene would exchange much of its protium for deuterium and, on recovery, would be found to be heavily deuterated—*contrary to fact*.

What this evidence shows is that *if* carbocations are formed—and other evidence shows that they *are*—they combine with base much faster than they revert to alkene. That is to say, as the mechanism on page 332 shows, step (1) is the slow, rate-determining step. How fast addition takes place depends chiefly on how fast the carbocation is formed.

9.6

$$CH_3{-}\overset{CH_3}{\underset{H}{C}}{-}CH{=}CH_2 \xrightarrow{HCl} CH_3{-}\overset{CH_3}{\underset{H}{C}}{-}\underset{\oplus}{C}H{-}CH_3 \xrightarrow{Cl^-} CH_3{-}\overset{CH_3}{\underset{H}{C}}{-}\underset{Cl}{C}H{-}CH_3$$

3-Methyl-1-butene — *A 2° cation* — 3-Chloro-2-methylbutane

↓ hydride shift

$$CH_3{-}\overset{CH_3}{\underset{\oplus}{C}}{-}\underset{H}{C}H{-}CH_3 \xrightarrow{Cl^-} CH_3{-}\overset{CH_3}{\underset{Cl}{C}}{-}\underset{H}{C}H{-}CH_3$$

A 3° cation — 2-Chloro-2-methylbutane

Since a 1,2-shift of hydrogen can convert the initially formed secondary cation into a more stable tertiary cation, such a rearrangement does occur, and much of the product is derived from this new cation.

9.7 (a) Stability of carbocations *t*-Bu$^+$ > *i*-Pr$^+$ > Et$^+$

See Figure 9.5, below.

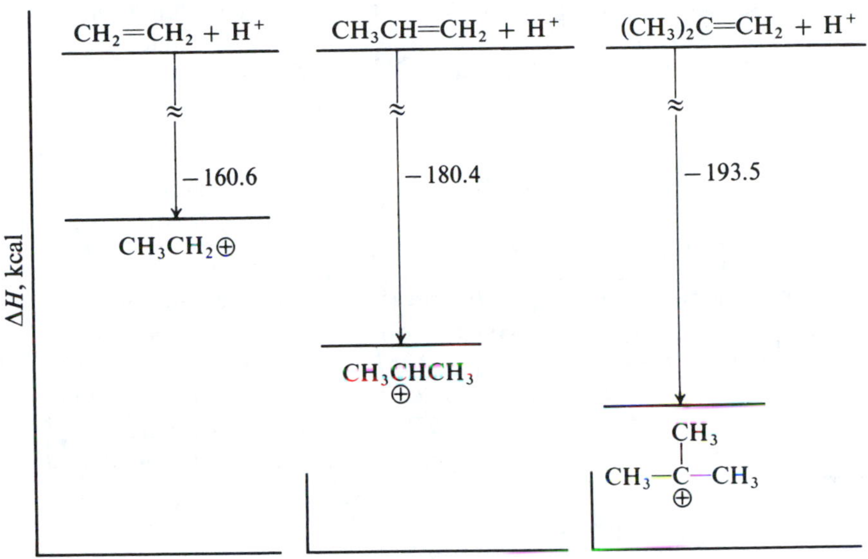

Figure 9.5 Stabilities of carbocations relative to alkenes. (The plots are aligned with each other for easy comparison.)

(b) and (c) The differences are essentially the same whatever the standard, alkene or alkyl halide.

Standard	Energy difference, kcal/mol	
	Et$^+$ − *i*-Pr$^+$	Et$^+$ − *t*-Bu$^+$
alkene	19.8	32.9
RBr	20	35
RCl	21	34
RI	20	36

9.8

$$CH_3CH_2CH_2CH_2OH \xrightarrow[E2]{H^+} CH_3CH_2CH{=}CH_2 \xrightarrow[H^+]{H_2O} \underset{\textit{Markovnikov orientation}}{CH_3CH_2\underset{OH}{\underset{|}{C}}HCH_3} \xrightarrow[E2\ or\ E1]{H^+} \underset{\textit{Saytzeff orientation}}{CH_3CH{=}CHCH_3}$$

9.9 (a) Isopropyl methyl ether, $(CH_3)_2CH{-}O{-}CH_3$.

$$CH_3{-}CH{=}CH_2 \xrightarrow{Hg^{2+}} CH_3{-}\underset{\diagdown\;\diagup\atop Hg_{2+}}{CH{-}CH_2} \xrightarrow{CH_3OH} \left[CH_3{-}\overset{HOCH_3}{\underset{\diagdown\;\diagup\atop Hg_{2+}}{CH{-}CH_2}} \right] \xrightarrow{-H^+} CH_3{-}\overset{OCH_3}{\overset{|}{C}H}{-}CH_2Hg^+ \xrightarrow{NaBH_4} CH_3{-}\overset{OCH_3}{\overset{|}{C}H}{-}CH_3$$

Reaction is exactly analogous to oxymercuration, except that the mercurinium ion suffers nucleophilic attack by methanol instead of by water. Attack has much S_N1 character (Secs. 9.14 and 13.24), and results in Markovnikov orientation.

(b) Ethers.

(c) *Solvomercuration* (with an alcohol as the solvent) can be used to make an ether in a case where the Williamson synthesis is ruled out by competing elimination, or where the appropriate alkene is the more readily available precursor.

(Brown, H. C.; Geoghegan, P., Jr. "The Oxymercuration–Demercuration of Representative Olefins. A Convenient, Mild Procedure for the Markovnikov Hydration of the Carbon–Carbon Double Bond"; *J. Am. Chem. Soc.* **1967**, *89*, 1522.)

(Olah, G. A.; Clifford, P. R. "Organometallic Chemistry. I. The Ethylene- and Norbornylenemercurinium Ions"; *J. Am. Chem. Soc.* **1971**, *93*, 1261; "Organometallic Chemistry. II. Direct Mercuration of Olefins to Stable Mercurinium Ions"; *J. Am. Chem. Soc.* **1971**, *93*, 2320.)

9.10 The same mechanism as on page 355, with radicals in steps (2) and (4) abstracting: (a) H, (b) Br, (c) Br, (d) H from S, (e) H. For example, in (e):

(1) $$\text{peroxide} \longrightarrow \text{Rad}\cdot$$

(2) $$\text{Rad}\cdot + R{-}\underset{H}{\underset{\cdot\cdot}{C}}{=}O \longrightarrow \text{Rad}{:}H + R{-}\dot{C}{=}O$$

(3) $$R{-}\dot{C}{=}O + n\text{-}C_6H_{13}CH{=}CH_2 \longrightarrow n\text{-}C_6H_{13}\dot{C}H{-}CH_2{-}\underset{O}{\underset{\|}{C}}{-}R$$

(4) $$n\text{-}C_6H_{13}\dot{C}H{-}CH_2{-}\underset{O}{\underset{\|}{C}}{-}R + R{-}\underset{H}{\underset{\cdot\cdot}{C}}{=}O \longrightarrow n\text{-}C_6H_{13}CH_2{-}CH_2{-}\underset{O}{\underset{\|}{C}}{-}R + R{-}\dot{C}{=}O$$

then (3), (4), (3), (4), etc.

(Key steps in the development of the theory of free-radical addition are described in: Kharasch, M. S.; Mayo, F. R.; *J. Am. Chem. Soc.* **1933**, *55*, 2468; Kharasch, M. S.; Engelmann, H.; Mayo, F. R. *J. Org. Chem.* **1937**, *2*, 288; Kharasch, M. S.; Jensen, E. V.; Urry, W. H. *Science (Washington, D.C.)* **1945**, *102*, 128. For a personal account of the discovery of the peroxide effect, see Mayo, F. R. In *Vistas in Free Radical Chemistry*; Waters, W. A., Ed.; Pergamon: New York, 1959; pp 139–142.)

9.11 The radical produced in step (3) of the CCl_4 sequence on page 355 adds to $RCH{=}CH_2$, and the new radical so formed then attacks CCl_4 in the manner of step (4).

$$R\dot{C}H{-}CH_2{-}CCl_3 + RCH{=}CH_2 \longrightarrow R\dot{C}H{-}CH_2{-}CH(R){-}CH_2{-}CCl_3$$

$$R\dot{C}H{-}CH_2{-}CH(R){-}CH_2{-}CCl_3 + Cl{:}CCl_3 \longrightarrow RCH(Cl){-}CH_2{-}CH(R){-}CH_2{-}CCl_3 + \cdot CCl_3$$

There is thus *competition* between two familiar reactions of free radicals (in this case, of the $R\dot{C}HCH_2CCl_3$ radical): addition to a double bond, or abstraction of an atom. This competition exists in all reactions of this kind; the relative importance of the two paths depends on how reactive the alkene is, how reactive CX_4 is, and how selective $R\dot{C}HCH_2CCl_3$ is. (See Problem 7, page 1097.)

9.12 (a) No free radicals are formed in the dark, and the electrophilic reaction is too slow with deactivated tetrachloroethylene.

(b) (1) $$Cl_2 \xrightarrow{\text{light}} 2Cl\cdot$$

(2) $$Cl\cdot + Cl_2C{=}CCl_2 \longrightarrow Cl_3C{-}\dot{C}Cl_2$$

(3) $$Cl_3C{-}\dot{C}Cl_2 + Cl_2 \longrightarrow Cl_3C{-}CCl_3 + Cl\cdot$$

then (2), (3), (2), (3), etc.

Oxygen stops the chain by reacting with the free radical or $Cl\cdot$.

9.13 (a) Orlon, $CH_2{=}CH{-}CN$, acrylonitrile
(b) Saran, $CH_2{=}CCl_2$, 1,1-dichloroethene (vinylidene chloride)
(c) Teflon, $CF_2{=}CF_2$, tetrafluoroethylene

9.14 If the more stable radical is formed faster in every step—a 2° radical, say, rather than a 1° radical—orientation will always be the same.

9.15 (a) and (b)

$$CH_2{=}C(CH_3){-}CH_2{-}C(CH_3)_2{-}CH_3 \xrightarrow{O_3} \xrightarrow{H_2O,\ Zn} H_2C{=}O + O{=}C(CH_3){-}CH_2{-}C(CH_3)_2{-}CH_3$$

$$CH_2{=}C(CH_3){-}CH_2{-}C(CH_3)_2{-}CH_3 \xrightarrow{KMnO_4/NaIO_4} CO_2 + O{=}C(CH_3){-}CH_2{-}C(CH_3)_2{-}CH_3$$

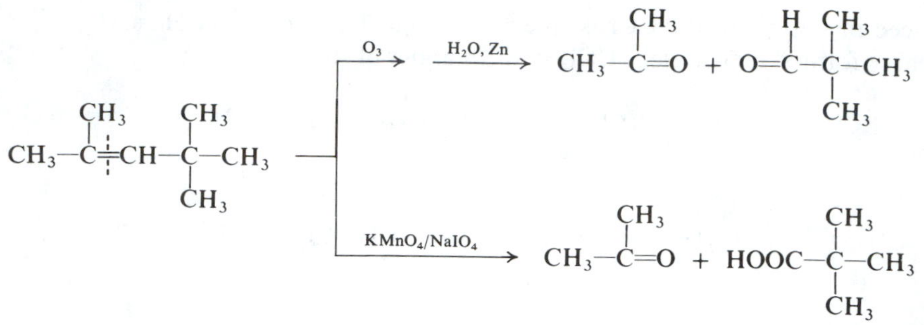

$$CH_3-\overset{CH_3}{\overset{|}{C}}=CH-\overset{CH_3}{\overset{|}{\underset{CH_3}{\underset{|}{C}}}}-CH_3 \xrightarrow{O_3} \xrightarrow{H_2O,\ Zn} CH_3-\overset{CH_3}{\overset{|}{C}}=O + O=\overset{H}{\overset{|}{C}}-\overset{CH_3}{\overset{|}{\underset{CH_3}{\underset{|}{C}}}}-CH_3$$

$$CH_3-\overset{CH_3}{\overset{|}{C}}=CH-\overset{CH_3}{\overset{|}{\underset{CH_3}{\underset{|}{C}}}}-CH_3 \xrightarrow{KMnO_4/NaIO_4} CH_3-\overset{CH_3}{\overset{|}{C}}=O + HOOC-\overset{CH_3}{\overset{|}{\underset{CH_3}{\underset{|}{C}}}}-CH_3$$

Ozonolysis or cleavage by $KMnO_4/NaIO_4$ effectively replaces a double bond to carbon by a double bond to oxygen. $KMnO_4$ oxidizes aldehydes further to carboxylic acids,

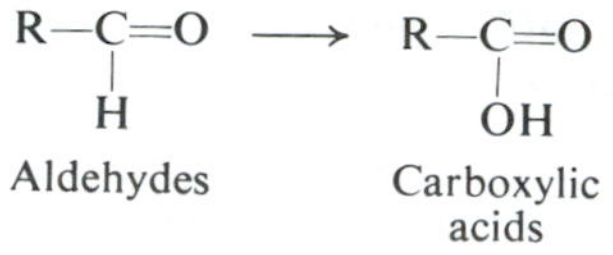

$$\underset{\text{Aldehydes}}{R-\underset{H}{\underset{|}{C}}=O} \longrightarrow \underset{\text{Carboxylic acids}}{R-\underset{OH}{\underset{|}{C}}=O}$$

and specifically formaldehyde is oxidized to carbon dioxide.

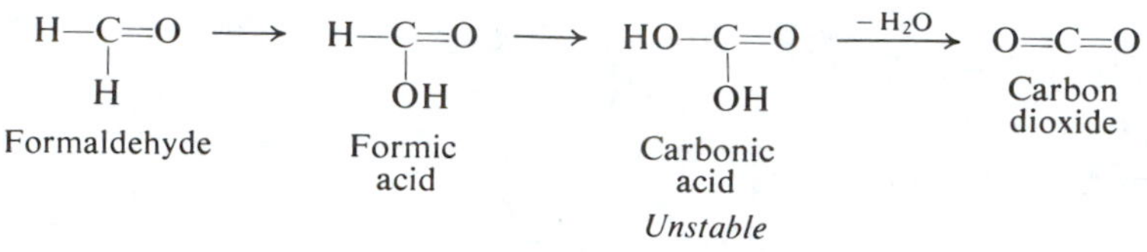

$$\underset{\text{Formaldehyde}}{H-\underset{H}{\underset{|}{C}}=O} \longrightarrow \underset{\text{Formic acid}}{H-\underset{OH}{\underset{|}{C}}=O} \longrightarrow \underset{\substack{\text{Carbonic acid}\\ \textit{Unstable}}}{HO-\underset{OH}{\underset{|}{C}}=O} \xrightarrow{-H_2O} \underset{\text{Carbon dioxide}}{O=C=O}$$

9.16 (a) Br_2/CCl_4, or $KMnO_4$, or conc. H_2SO_4: alkene positive.
(b) Br_2/CCl_4, or $KMnO_4$, or conc. H_2SO_4: alkene positive. Or, $AgNO_3$: alkyl halide positive.
(c) Br_2/CCl_4, or $KMnO_4$: alkene positive. Or, CrO_3/H_2SO_4: 2° alcohol positive.
(d) Br_2/CCl_4, or $KMnO_4$: alkene positive.
(e) Only alkene gives positive Br_2/CCl_4 test; only alkyl halide gives halogen test; only 2° alcohol gives positive CrO_3/H_2SO_4 test; only ether is negative to these tests, yet sol. in conc. H_2SO_4.

9.17 A, alkane B, 2° alcohol C, alkyl halide D, alkene E, 3° alcohol

9.18 (a) CrO_3/H_2SO_4 (b) Br_2/CCl_4, or $KMnO_4$
(c) Br_2/CCl_4, or $KMnO_4$, or CrO_3/H_2SO_4 (d) conc. H_2SO_4

1. (a) isobutane (b) 1,2-dichloro-2-methylpropane
(c) 1,2-dibromo-2-methylpropane (d) no reaction
(e) *tert*-butyl bromide (f) isobutyl bromide
(g) *tert*-butyl iodide (h) *tert*-butyl iodide
(i) *tert*-butyl hydrogen sulfate (j) *tert*-butyl alcohol
(k) 1-bromo-2-methyl-2-propanol (l) 1-bromo-2-chloro-2-methylpropane and products (c) and (k)
(m) 2,4,4-trimethyl-1-pentene and 2,4,4-trimethyl-2-pentene (n) 2,2,4-trimethylpentane
(o) 2-methyl-1,2-propanediol (p) acetone and carbon dioxide
(q) same as (o) (r) acetone and formaldehyde
(s) *tert*-butyl alcohol (t) isobutyl alcohol

2. (a) propylene (b) ethylene (c) 2-butene (d) isobutylene
(e) vinyl chloride (f) 2-methyl-1-butene (g) ethylene (h) propylene

Relative reactivities follow the sequence given on page 338. The electronegative oxygens of —COOH make this group electron-withdrawing, like the halogens (see page 338).

3. (a) 2-iodobutane (b) 3-iodopentane and 2-iodopentane
(c) *tert*-pentyl iodide (d) *tert*-pentyl iodide
(e) 2-iodo-3-methylbutane and 2-iodo-2-methylbutane (f) 1-bromo-1-iodoethane
(g) 2-iodo-2,3-dimethylbutane (h) 4-iodo-2,2,4-trimethylpentane

In each case orientation follows Markovnikov's rule (Secs. 9.5 and 9.11). In (b), the rule indicates no preference—a 2° cation is formed either way—and both products are obtained. In (e), some of the intermediate 2° cation rearranges to the more stable 3° cation.

4.

	Oxymercuration–demercuration	Hydroboration–oxidation
$CH_3CH_2CH_2CH_2CH_2OH$		$CH_3CH_2CH_2CH{=}CH_2$
$CH_3CH_2CH_2CH(OH)CH_3$	$CH_3CH_2CH_2CH{=}CH_2$	
$CH_3CH_2CH(OH)CH_2CH_3$	None	
$CH_3CH_2CH(CH_3)CH_2OH$		$CH_3CH_2C(CH_3){=}CH_2$
$CH_3CH_2C(CH_3)(OH)CH_3$	$CH_3CH{=}C(CH_3)CH_3$ *or* $CH_3CH_2C(CH_3){=}CH_2$	
$CH_3CH(OH)CH(CH_3)CH_3$	$CH_2{=}CHCH(CH_3)CH_3$	$CH_3CH{=}C(CH_3)CH_3$
$CH_3CH(CH_3)CH_2CH_2OH$		$CH_3CH(CH_3)CH{=}CH_2$
$CH_3C(CH_3)_2CH_2OH$	None	

5. The 3° radical is more stable than the 2° radical, and forms faster.

$$\underset{\text{2-Ethyl-1-hexene}}{CH_3CH_2CH_2CH_2\overset{\overset{C_2H_5}{|}}{C}{=}CH_2} \xrightarrow{Rad\cdot} \underset{\substack{\text{A 3° radical}\\ \textit{More stable: forms faster}}}{CH_3CH_2CH_2CH_2\overset{\overset{C_2H_5}{|}}{\dot{C}}{-}CH_2{-}Rad}$$

$$\underset{\text{1-Octene}}{CH_3(CH_2)_4CH_2CH{=}CH_2} \xrightarrow{Rad\cdot} \underset{\text{A 2° radical}}{CH_3(CH_2)_4CH_2\dot{C}H{-}CH_2{-}Rad}$$

6. (a) Methyl alcohol (rather than water) reacts with the intermediate bromonium ion, to form the protonated ether; subsequent loss of a proton gives the ether.

$$CH_2{=}CH_2 \xrightarrow{Br_2} \overset{Br\oplus}{CH_2{-}CH_2} \xrightarrow{CH_3OH} \underset{\text{A protonated ether}}{\underset{Br\quad \oplus\underset{H}{O}{-}CH_3}{CH_2{-}CH_2}} \xrightarrow{-H^+} \underset{\text{An ether}}{\underset{Br\quad OCH_3}{CH_2{-}CH_2}}$$

(b) $CH_3{-}\underset{OCH_3}{CH}{-}CH_2Br$, 1-bromo-2-methoxypropane.

The ring-opening step—attack on the bromonium ion by methyl alcohol—takes place at the 2° position (rather than the 1°) because of S_N1 character of the transition state. (See Sec. 13.24.)

$$CH_3{-}\overset{Br\oplus}{CH{-}CH_2} + CH_3{-}\underset{H}{O}{:} \longrightarrow CH_3{-}\underset{\oplus\underset{H}{O}{-}CH_3}{CH}{-}CH_2Br \xrightarrow{-H^+} \underset{\text{1-Bromo-2-methoxypropane}}{CH_3{-}\underset{OCH_3}{CH}{-}CH_2Br}$$

7. If an alkene were an intermediate, it should undergo reaction with D^+ as well as with H^+, and the product (alcohol) should contain D attached to carbon.

$$-\underset{H}{\overset{|}{C}}-\overset{|}{\underset{\oplus}{C}}- \xrightarrow{-H^+} -\overset{|}{C}{=}\overset{|}{C}- \xrightarrow{D^+} -\underset{D}{\overset{|}{C}}-\overset{|}{\underset{\oplus}{C}}- \xrightarrow{\text{water}} \text{C-deuterated alcohol}$$

8.

	(a)	(b)	(c)	(d)
Acid:	H_2O	Et_3B	$(BH_3)_2$	$(BH_3)_2$
Base:	H^-	NH_3	Me_3N	H^-

9. (a) The alternative mechanism is much less likely because one chain-propagating step (4a) is highly endothermic and hence must have a high E_{act}, in contrast to step (4), which is exothermic and can have a low E_{act}.

Let us first compare alternative steps (3) and (3a), using ethylene as our alkene. (For want of other values, we use the Br—CH_2CH_3 and H—CH_2CH_3 bond dissociation energies for Br—$CH_2CH_2\cdot$ and H—$CH_2CH_2\cdot$; whatever the true values, the C—H bond is stronger than the C—Br, and our conclusions will be unaffected.

$$\text{(3)}\quad \text{Br}\cdot + \underset{(51)}{CH_2{=}CH_2} \longrightarrow \underset{(69)}{\text{Br—}CH_2CH_2\cdot} \qquad \Delta H = -18$$

$$\text{(3a)}\quad \text{H}\cdot + \underset{(51)}{CH_2{=}CH_2} \longrightarrow \underset{(98)}{\text{H—}CH_2CH_2\cdot} \qquad \Delta H = -47$$

As we see, step (3a) is even more exothermic than (3), and could have a low E_{act}. *If* H· were present, we would expect it to add to an alkene. From energy considerations alone, then, we would expect (3a) to be at least as likely as (3).

Now let us compare steps (4) and (4a).

$$\text{(4)}\quad BrCH_2CH_2\cdot + \underset{(88)}{\text{H—Br}} \longrightarrow \underset{(98)}{BrCH_2CH_2\text{—H}} + \text{Br}\cdot \qquad \Delta H = -10 \qquad E_{act}\ \textit{could be small}$$

$$\text{(4a)}\quad CH_3CH_2\cdot + \underset{(88)}{\text{H—Br}} \longrightarrow \underset{(69)}{CH_3CH_2\text{—Br}} + \text{H}\cdot \qquad \Delta H = +19 \qquad E_{act}\ \textit{at least}\ 19$$

Here we see a marked contrast. Step (4) is exothermic and could have a small E_{act}, but (4a) is highly endothermic and must have an E_{act} of at least 19 kcal. In an attack on HBr, any free radical will abstract H in strong preference to Br. (What we have just said applies equally well to steps (2) and (2a), and the reaction of the radical, Rad·, generated from the peroxide.)

Ultimately, then, the reaction takes the course it does because C—H bonds are stronger than C—Br bonds.

(b) The intermediate radical (step 3, page 351) must be

$$\begin{array}{cc} | & | \\ \text{—C—} & \dot{\text{C}}\text{—} \\ | & \\ \text{Br} & \end{array}$$

since otherwise (step 3a, page 364) it would be

$$\begin{array}{cc} | & | \\ \text{—C—} & \dot{\text{C}}\text{—} \\ | & \\ \text{H} & \end{array}$$

or

$$\begin{array}{cc} | & | \\ \text{—C—} & \dot{\text{C}}\text{—} \\ | & \\ \text{D} & \end{array}$$

depending on whether HBr or DBr was used.

10. (a) (3) $\underset{(51)}{CH_3CH{=}CH_2} + Br\cdot \longrightarrow \underset{(69)}{CH_3\dot{C}HCH_2{-}Br}$ $\quad \Delta H = -18$

(4) $CH_3\dot{C}HCH_2Br + \underset{(88)}{H{-}Br} \longrightarrow CH_3\underset{\underset{(95)}{H}}{C}HCH_2Br + Br\cdot$ $\quad \Delta H = -7$ $\quad$ E_{act} *very small*

(b) (3) $\underset{(51)}{CH_3CH{=}CH_2} + Cl\cdot \longrightarrow \underset{(82)}{CH_3\dot{C}HCH_2{-}Cl}$ $\quad \Delta H = -31$

(4) $CH_3\dot{C}HCH_2Cl + \underset{(103)}{H{-}Cl} \longrightarrow CH_3\underset{\underset{(95)}{H}}{C}HCH_2Cl + Cl\cdot$ $\quad \Delta H = +8$ $\quad$ E_{act} *at least* 8

(c) See the values given above for each equation. (Step (2) would be similar to step (4).)

(d) The difference between HBr and HCl does not lie in step (3), which is exothermic in both cases. The difference lies in step (4): while this is exothermic for HBr, it is endothermic for HCl, with a minimum E_{act} of 8 kcal. Now, an E_{act} of 8 kcal is not too high for a chain-carrying step—witness the free-radical bromination of methane (Sec. 2.17). But it makes the free-radical reaction slower for HCl than for HBr: so slow, evidently, that it cannot compete with heterolytic addition—which is, after all, steaming along at its usual rate, peroxides or no peroxides.

11. (a) $\underset{\text{Chloroacetone}}{Cl^{14}CH_2{-}\overset{\overset{CH_3}{|}}{C}{=}O} + \underset{\text{Formaldehyde}}{O{=}CH_2} \xleftarrow{O_3} \underset{\text{Methallyl chloride}}{Cl^{14}CH_2{-}\overset{\overset{CH_3}{|}}{C}{=}CH_2}$

(b) The reaction is heterolytic (ionic). There is no light or peroxides to generate free radicals; if free-radical chains *were* started, oxygen would break many chains and thus lower the yield of methallyl chloride—contrary to fact.

(c) Reaction of Cl_2 with the alkene to form a cation, which then loses a proton to give the product: the first step of electrophilic addition followed by the second step of E1 elimination.

$$CH_3{-}\overset{\overset{CH_3}{|}}{C}{=}CH_2 + Cl_2 \longrightarrow CH_3{-}\underset{\oplus}{\overset{\overset{CH_3}{|}}{C}}{-}CH_2Cl + Cl^-$$

$$CH_3{-}\underset{\oplus}{\overset{\overset{CH_3}{|}}{C}}{-}CH_2Cl \longrightarrow CH_2{=}\overset{\overset{CH_3}{|}}{C}{-}CH_2Cl + H^+$$

For simplicity, we have shown the intermediate cation as an open carbocation rather than a cyclic chloronium ion. It may well *be* an open cation, or something very close to it. The carbocation here would be 3° and relatively stable; the electron-deficient carbon may have little need of electrons from chlorine. Chlorine may be joined chiefly to C–1, with only very weak bonding to C–2; most of the charge would then be on carbon. Whatever bonding there is between C–2 and chlorine is broken as the proton is lost and the π bond forms; the reaction then takes on aspects of E2, with $ClCH_2$〰 as the leaving group.

$$\text{B:} \curvearrowright \overset{\text{H}}{\underset{|}{}}\!\!CH_2\!-\!\overset{CH_3}{\overset{|}{C}}\!-\!CH_2 \;(\text{bridged by } Cl^{\oplus}) \longrightarrow \text{B:H} + CH_2\!=\!\overset{CH_3}{\overset{|}{C}}\!-\!\underset{Cl}{\underset{|}{CH_2}}$$

(d) The more stable 3° cation is less reactive toward Cl^-, and more prone to lose one of six hydrogens to form the branched alkene.

(e) To a small extent, rearrangement of the intermediate cation to a 3° cation, which, like the 3° cation in (c), loses a proton (one of six) to form the branched alkene.

$$CH_3\!-\!\overset{CH_3}{\underset{CH_3}{\overset{|}{\underset{|}{C}}}}\!-\!CH\!=\!CH_2 + Cl_2 \longrightarrow CH_3\!-\!\overset{CH_3}{\underset{CH_3}{\overset{|}{\underset{|}{C}}}}\!-\!\underset{\oplus}{CH}\!-\!CH_2Cl + Cl^-$$

$$CH_3\!-\!\overset{CH_3}{\underset{CH_3}{\overset{|}{\underset{|}{C}}}}\!-\!\underset{\oplus}{CH}\!-\!CH_2Cl \xrightarrow{\text{rearr.}} CH_3\!-\!\overset{CH_3}{\underset{\oplus}{\overset{|}{C}}}\!-\!\underset{CH_3}{\underset{|}{CH}}\!-\!CH_2Cl$$

$$CH_3\!-\!\overset{CH_3}{\underset{\oplus}{\overset{|}{C}}}\!-\!\underset{CH_3}{\underset{|}{CH}}\!-\!CH_2Cl \longrightarrow CH_2\!=\!\overset{CH_3}{\overset{|}{C}}\!-\!\underset{CH_3}{\underset{|}{CH}}\!-\!CH_2Cl + H^+$$

12. The usual addition of HOX to simple alkenes, we saw in Sec. 9.14, indicates preferential attack by X^- at the more substituted carbon of the cyclic halonium ion: by path (*b*), for example, in the propylene reaction (page 342). We attributed this preference to the fact that bond-breaking exceeds bond-making in the transition state (pages 486–487), and the carbon under attack has acquired considerable positive charge. Attack occurs, then, at the carbon best able to accommodate this charge.

Now, in the propylene intermediate, C–2 is the carbon best able to accommodate the positive charge because of electron release by the CH_3— group. But the allyl bromide intermediate contains, not CH_3—, but $BrCH_2$— which, because of the electronegative Br, is electron-withdrawing instead of electron-releasing. It is C–1 that can best accommodate the positive charge of the transition state, and it is here that attack (path *a*) preferentially takes place.

$$Br\!\leftarrow\!CH_2\!-\!\overset{Br^{\oplus}}{CH\!-\!CH_2} \;(H_2O \text{ attack: } b \text{ at CH}, a \text{ at } CH_2)$$

$$\xrightarrow{a} BrCH_2\!-\!\overset{Br}{\overset{|}{CH}}\!-\!CH_2OH_2^{\,+} \xrightarrow{-H^+} BrCH_2\!-\!\overset{CH_3}{\overset{|}{CH}}\!-\!CH_2OH \quad \textit{Chief product}$$

$$\xrightarrow{b} BrCH_2\!-\!\underset{OH_2^{\,+}}{\underset{|}{CH}}\!-\!CH_2Br \xrightarrow{-H^+} BrCH_2\!-\!\underset{OH}{\underset{|}{CH}}\!-\!CH_2Br$$

13. In view of chemistry we have learned, combination of isobutane with ethylene would seem to require generation of a *tert*-butyl cation,

$$CH_2{=}CH_2 \xrightarrow{H^+} CH_3CH_2^{\oplus} \xrightarrow{t\text{-BuH}} CH_3CH_3 + CH_3{-}\underset{CH_3}{\overset{CH_3}{C}}{}^{\oplus}$$

followed by its addition to ethylene.

$$CH_3{-}\underset{CH_3}{\overset{CH_3}{C}}{}^{\oplus} + CH_2{=}CH_2 \longrightarrow CH_3{-}\underset{CH_3}{\overset{CH_3}{C}}{-}CH_2{-}CH_2^{\oplus}$$

The skeleton of the alkylation product requires rearrangement of this 1° cation: by first a hydride shift to give a 2° cation, and then an alkyl shift to give a 3° cation, which has the required carbon skeleton.

$$\underset{\text{1° cation}}{CH_3{-}\underset{H_3C}{\overset{CH_3}{C}}{-}\underset{H}{CH}{-}CH_2^{\oplus}} \longrightarrow \underset{\text{2° cation}}{CH_3{-}\underset{CH_3}{\overset{CH_3}{C}}{-}\overset{\oplus}{C}H{-}CH_3} \longrightarrow \underset{\text{3° cation}}{CH_3{-}\underset{\oplus}{\overset{CH_3}{C}}{-}\underset{CH_3}{CH}{-}CH_3}$$

Finally, the 3° cation abstracts hydride from isobutane to give the alkylation product and a new *tert*-butyl cation, which continues the chain.

$$CH_3{-}\underset{\oplus}{\overset{CH_3}{C}}{-}\underset{CH_3}{CH}{-}CH_3 + H{-}\underset{CH_3}{\overset{CH_3}{C}}{-}CH_3 \longrightarrow CH_3{-}\underset{H}{\overset{CH_3}{C}}{-}\underset{CH_3}{CH}{-}CH_3 + {}^{\oplus}\underset{CH_3}{\overset{CH_3}{C}}{-}CH_3$$

14. The ether is cleaved to isobutylene, which undergoes *acid-catalyzed* polymerization, an extension of dimerization (Sec. 9.15).

$$n\text{-Bu}{-}O{-}\text{Bu-}t \xrightarrow{H^+} n\text{-Bu}{-}\underset{\oplus}{\overset{H}{O}}{-}\text{Bu-}t \longrightarrow n\text{-BuOH} + (CH_3)_3C^{\oplus} \xrightarrow{-H^+} (CH_3)_2C{=}CH_2$$

$$H^+\ CH_2{=}\underset{CH_3}{\overset{CH_3}{C}}\quad CH_2{=}\underset{CH_3}{\overset{CH_3}{C}}\quad CH_2{=}\underset{CH_3}{\overset{CH_3}{C}} \longrightarrow \underset{\text{Polyisobutylene}}{CH_3{-}\underset{CH_3}{\overset{CH_3}{C}}{-}CH_2{-}\underset{CH_3}{\overset{CH_3}{C}}{-}CH_2{-}\underset{CH_3}{\overset{CH_3}{C}}{\sim}}$$

15. (a) Both alkenes form the same 3° carbocation, and hence give the same alcohol.

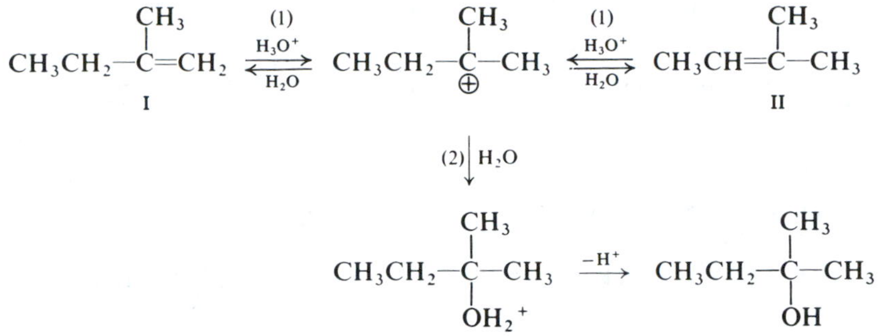

$$\underset{\text{I}}{CH_3CH_2\text{—}\overset{\overset{\displaystyle CH_3}{|}}{C}{=}CH_2} \underset{H_2O}{\overset{(1)\ H_3O^+}{\rightleftarrows}} CH_3CH_2\text{—}\underset{\oplus}{\overset{\overset{\displaystyle CH_3}{|}}{C}}\text{—}CH_3 \underset{H_2O}{\overset{(1)\ H_3O^+}{\leftrightarrows}} \underset{\text{II}}{CH_3CH{=}\overset{\overset{\displaystyle CH_3}{|}}{C}\text{—}CH_3}$$

$$\Big\downarrow (2)\ H_2O$$

$$CH_3CH_2\text{—}\underset{\underset{\displaystyle OH_2^{\ +}}{|}}{\overset{\overset{\displaystyle CH_3}{|}}{C}}\text{—}CH_3 \xrightarrow{-H^+} CH_3CH_2\text{—}\underset{\underset{\displaystyle OH}{|}}{\overset{\overset{\displaystyle CH_3}{|}}{C}}\text{—}CH_3$$

(b) These experiments show that (at least for hydration) the first step of electrophilic addition, the formation of the carbocation, is slow and rate-determining, as shown on page 332.

Starting with alkene I, let us consider the relative rates of the two chief steps, (1) and (2). First, what would we expect if carbocations were formed rapidly and reversibly in step (1), and then—every so often—slowly combined with water (step 2) to give the protonated alcohol? If that were so, most carbocations would lose a hydrogen many times to regenerate the alkene before eventually going on to product. But in the carbocation there are hydrogens on C–3 as well as on C–1. In "reverting" to alkene, the carbocation is just as likely—*more* likely, actually, because of the greater stability of the alkene being formed—to lose a hydrogen to form alkene II as to form alkene I. Exposed to the acidic medium until half-reaction time, much I would be converted into II, and recovered unconsumed alkene would be a mixture of I and II—*contrary to fact*. (We could apply exactly the same line of reasoning to the experiments starting with alkene II, and draw exactly the same conclusions.)

Like the work on hydrogen exchange (Problem 9.5), these experiments show that any carbocations formed combine much more rapidly with the nucleophile, water, than they revert to alkene. (That is, k_2 is much larger than k_{-1}.) Step (1) is the slow, rate-determining step. How fast addition takes place depends upon how fast the carbocation is formed.

16. (a) $n\text{-Pr—}\overset{\overset{\displaystyle H}{|}}{C}{=}O + O{=}\overset{\overset{\displaystyle H}{|}}{C}H \xleftarrow{O_3} \underset{\text{1-Pentene}}{n\text{-Pr—}\overset{\overset{\displaystyle H}{|}}{C}{=}\overset{\overset{\displaystyle H}{|}}{C}H}$

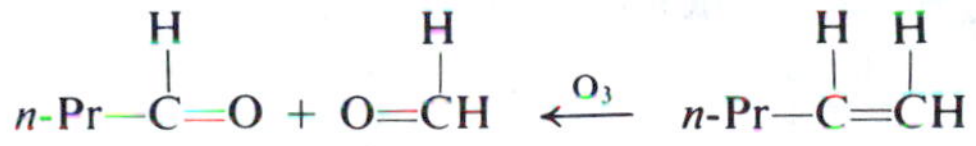

(b) $i\text{-Pr—}\overset{\overset{\displaystyle H}{|}}{C}{=}O + O{=}\overset{\overset{\displaystyle H}{|}}{C}\text{—}CH_3 \xleftarrow{O_3} \underset{\text{4-Methyl-2-pentene}}{CH_3\text{—}\underset{\underset{\displaystyle CH_3}{|}}{CH}\text{—}\overset{\overset{\displaystyle H}{|}}{C}{=}\overset{\overset{\displaystyle H}{|}}{C}\text{—}CH_3}$

(c) $CH_3\text{—}\overset{\overset{\displaystyle CH_3}{|}}{C}{=}O + O{=}\overset{\overset{\displaystyle CH_3}{|}}{C}\text{—}CH_3 \xleftarrow{O_3} \underset{\text{2,3-Dimethyl-2-butene}}{CH_3\text{—}\overset{\overset{\displaystyle H_3C}{|}}{C}{=}\overset{\overset{\displaystyle CH_3}{|}}{C}\text{—}CH_3}$

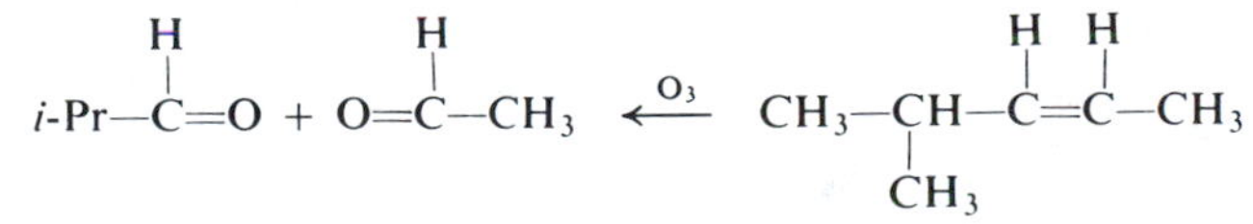

(d) $CH_3\text{—}\overset{\overset{\displaystyle H}{|}}{C}{=}O + O{=}\overset{\overset{\displaystyle H}{|}}{C}\text{—}CH_2\text{—}\overset{\overset{\displaystyle H}{|}}{C}{=}O + O{=}\overset{\overset{\displaystyle H}{|}}{C}H \xleftarrow{O_3} \underset{\text{1,4-Hexadiene}}{CH_3\text{—}\overset{\overset{\displaystyle H}{|}}{C}{=}\overset{\overset{\displaystyle H}{|}}{C}\text{—}CH_2\text{—}\overset{\overset{\displaystyle H}{|}}{C}{=}\overset{\overset{\displaystyle H}{|}}{C}H}$

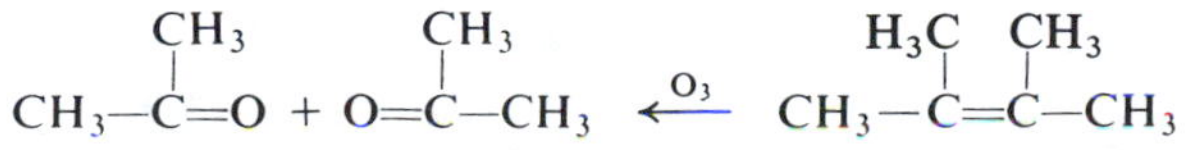

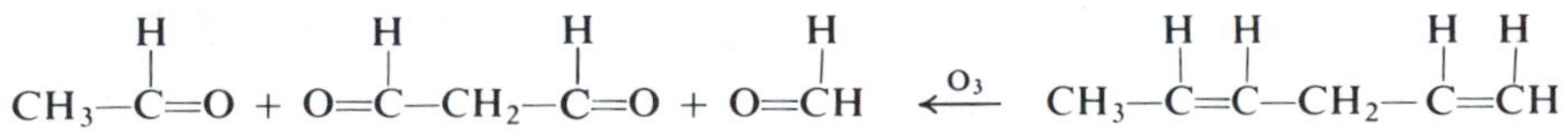

(e) 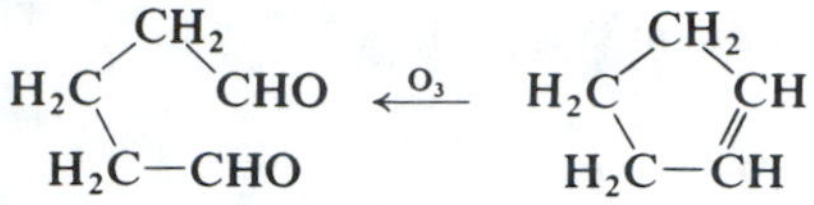 $H_2C(CH_2)$–CHO / H_2C–CHO $\xleftarrow{O_3}$ cyclopentene (H_2C–CH_2–H_2C, H_2C–CH=CH)

```
     CH2                    CH2
    /   \                  /   \
(e) H2C   CHO  <--O3--  H2C     CH
    \                     \     //
    H2C—CHO               H2C—CH
```

(f) Carboxylic acids (RCOOH) instead of aldehydes (RCHO); CO_2 instead of formaldehyde (HCHO); ketones (RCOR′) in (c).

17. (a) $KMnO_4$, or Br_2/CCl_4, or conc. H_2SO_4: alkene positive.

(b) $KMnO_4$, or Br_2/CCl_4, or conc. H_2SO_4: alkene positive. Or, $AgNO_3$: halide positive.

(c) $AgNO_3$: halide positive.

(d) Conc. H_2SO_4: alcohol positive. (Actually, alcohol is water-soluble.)

(e) $KMnO_4$, or Br_2/CCl_4: unsaturated ether positive.

(f) Br_2/CCl_4: unsaturated alcohol positive.

(g) CrO_3/H_2SO_4, or conc. H_2SO_4: alcohol positive. (Actually, alcohol is water-soluble.)

(h) Br_2/CCl_4: alkene positive. Or, CrO_3/H_2SO_4: alcohol positive.

(i) $AgNO_3$: halide positive.

(j) $KMnO_4$, or Br_2/CCl_4: unsaturated ether positive.

(k) $AgNO_3$: halide positive. Br_2/CCl_4: alkene positive. (This alcohol is water-soluble.)

(l) CrO_3/H_2SO_4: dihalide negative, diol and halohydrin positive. $AgNO_3$: halohydrin positive. (Dihalide positive, but eliminated by first test.)

(m) Br_2/CCl_4, or $KMnO_4$: alkene positive. CrO_3/H_2SO_4: alcohol positive. $AgNO_3$: halide positive. Alkane negative to all the tests.

18. (a) 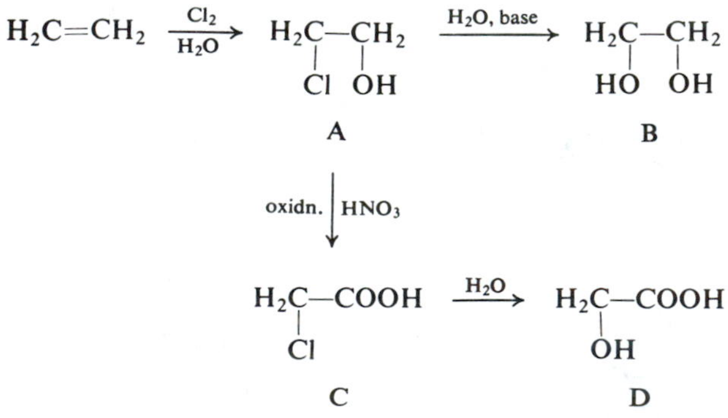 $H_2C{=}CH_2 \xrightarrow[H_2O]{Cl_2} \underset{\text{A}}{\underset{Cl\quad OH}{H_2C{-}CH_2}} \xrightarrow{H_2O,\ base} \underset{\text{B}}{\underset{HO\quad OH}{H_2C{-}CH_2}}$

A $\xrightarrow{\text{oxidn.}\ HNO_3}$ C

(b) $\underset{\text{C}}{\underset{Cl}{H_2C{-}COOH}} \xrightarrow{H_2O} \underset{\text{D}}{\underset{OH}{H_2C{-}COOH}}$

(c) $CH_2{=}CHCH_2OH \xrightarrow{Br_2} \underset{\text{E}}{\underset{Br\quad Br}{CH_2CHCH_2OH}} \xrightarrow[\text{oxidn.}]{HNO_3} \underset{\text{F}}{\underset{Br\quad Br}{CH_2CHCOOH}} \xrightarrow{Zn} \underset{\text{G}}{CH_2{=}CHCOOH}$

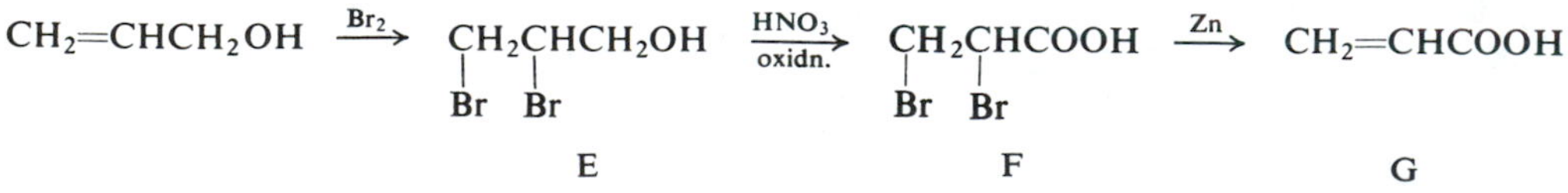

(d) $CH_2{=}CH_2 \xrightarrow[(+Cl,\ +O,\ +H)]{Cl_2,\ H_2O} \underset{Cl}{CH_2}-\underset{OH}{CH_2}$ (A)

The transformation of A (two carbons) into H (four carbons) obviously requires two molecules of A, and an atom count shows the loss of H_2O in the reaction. An ether is formed.

$$\underset{\text{A (two moles)}}{ClCH_2CH_2OH + HOCH_2CH_2Cl} \xrightarrow{-H_2O} \underset{\text{H}}{ClCH_2CH_2{-}O{-}CH_2CH_2Cl} \xrightarrow[-2HCl]{KOH} \underset{\text{I, Divinyl ether}}{CH_2{=}CH{-}O{-}CH{=}CH_2}$$

(e) Intramolecular alkoxymercuration–demercuration ⟶ cyclic ether.

$CH_2{=}CH{-}CH_2CH_2CH_2OH \xrightarrow{Hg(OAc)_2}$ (mercurinium ion, Hg^{2+} bridging $H_2C{-}CH$, attacked by OH) $\xrightarrow{-H^+}$ ring $H_2C(Hg^+){-}CH{-}O$ (five-membered ring with CH_2, H_2C, CH_2) $\xrightarrow{NaBH_4}$ 2-methyltetrahydrofuran (CH_3 on ring carbon next to O), J

19. 3-Hexene. The carbon skeleton must be the same as that of *n*-hexane; there is one double bond, which must be located in the center of the chain (three carbons on each side).

20. (a) $CH_3{-}C(CH_3)(OH){-}CH_3 \xleftarrow{H_2O,\ H^+} CH_3{-}C(CH_3){=}CH_2$

tert-Butyl alcohol ← Isobutylene

(b) $CH_3{-}CH(I){-}CH_3 \xleftarrow{HI} CH_3{-}CH{=}CH_2$

Isopropyl iodide ← Propylene

(c) $CH_3{-}CH(CH_3){-}CH_2Br \xleftarrow{\text{HBr, peroxides}} CH_3{-}C(CH_3){=}CH_2$

Isobutyl bromide ← Isobutylene

(d) $CH_3CH_2C(CH_3)(HO){-}CH_2Cl \xleftarrow{Cl_2,\ H_2O} CH_3CH_2C(CH_3){=}CH_2$

1-Chloro-2-methyl-2-butanol ← 2-Methyl-1-butene

(e) $CH_3{-}CH_2{-}CH_2{-}CH(CH_3){-}CH_3 \xleftarrow{H_2,\ Ni} CH_3{-}CH_2{-}CH{=}C(CH_3){-}CH_3$

2-Methylpentane ← 2-Methyl-2-butene

(or from any other alkene with the same carbon skeleton)

(f) $CH_3-C(CH_3)(OH)-C(CH_3)(OH)-CH_3 \xleftarrow[\text{or } HCO_2OH]{\text{cold alkaline } KMnO_4} CH_3-C(CH_3)=C(CH_3)-CH_3$

2,3-Dimethyl-2,3-butanediol ← 2,3-Dimethyl-2-butene

21. As we did for Problem 17 in Chapter 6, let us draw the structure of our target molecule and work backwards to it. In most of these we see two substituents *on adjacent carbons*, and naturally first consider an alkene as a likely precursor.

(a) $CH_3-CH(Cl)-CH_2Cl \xleftarrow{Cl_2} CH_3-CH=CH_2 \xleftarrow{H^+,\ heat} CH_3-CH(OH)-CH_3$

1,2-Dichloropropane ... Isopropyl alcohol

(b) Here, it is clear which alkene we want as the immediate precursor. But, as in Problem 13(b) in Chapter 8, we must consider carefully just how to make it—*pure*.

$CH_3CH_2CH(Cl)-CH_2Cl \xleftarrow{Cl_2} CH_3CH_2CH=CH_2 \xleftarrow{KOH(alc)} CH_3CH_2CH_2CH_2Br \xleftarrow{HBr} CH_3CH_2CH_2CH_2OH$

1,2-Dichlorobutane ... *n*-Butyl alcohol

(c) $CH_3-CH(OH)-CH_2OH \xleftarrow{KMnO_4} CH_3-CH=CH_2 \xleftarrow{H^+,\ heat} CH_3-CH(OH)-CH_3$

1,2-Propanediol ... Isopropyl alcohol

(d) $CH_3-C(CH_3)(OH)-CH_2Br \xleftarrow{Br_2,\ H_2O} CH_3-C(CH_3)=CH_2 \xleftarrow{H^+,\ heat} CH_3-C(CH_3)(OH)-CH_3$

1-Bromo-2-methyl-2-propanol ... *tert*-Butyl alcohol

(e) $CH_3-C(CH_3)(H)-CH_2H \xleftarrow{H_2,\ Pt} CH_3-C(CH_3)=CH_2 \xleftarrow{H^+,\ heat} CH_3-C(CH_3)(OH)-CH_3$ *or* $CH_3-C(CH_3)(H)-CH_2OH$

Isobutane ... *tert*-Butyl alcohol ... Isobutyl alcohol

(f) Here, following the approach of Problem 13(c) in Chapter 8, we consider the two ways to put together the ether molecule. With the danger of competing elimination in mind, we choose the reaction of the primary halide, and not of the secondary halide.

$CH_3CH_2 \vdots O-CH(CH_3)CH_3 \leftarrow$ [$CH_3CH_2Br \xleftarrow{HBr} CH_3CH_2OH$ (Ethyl alcohol)] + [$CH_3CH(CH_3)ONa \xleftarrow{Na} CH_3CH(CH_3)OH$ (Isopropyl alcohol)]

Ethyl isopropyl ether

(g) As an alternative method we could use *solvomercuration* of propylene (Problem 9.9). This, we know, follows Markovnikov orientation, and would yield the product we want.

$$\underset{\displaystyle\;\;\;|\;\;\;\;\;\;\atop \displaystyle OC_2H_5}{CH_3-CH-CH_3} \xleftarrow{NaBH_4} \xleftarrow{Hg(OOCCF_3)_2} \underset{\text{Ethyl alcohol}}{C_2H_5OH} + CH_3-CH=CH_2$$

$$CH_3-CH=CH_2 \xleftarrow{H^+,\ heat} \underset{\text{Isopropyl alcohol}}{\underset{\displaystyle OH}{CH_3CHCH_3}}$$

10

Stereochemistry II. Stereoselective and Stereospecific Reactions

10.1 (a) The diol of m.p. 19 °C, inactive but resolvable, is a racemic modification that consists of equal amounts of (2*R*,3*R*)- and (2*S*,3*S*)-2,3-butanediol (II and I, below). The non-resolvable diol of m.p. 34 °C is *meso*-2,3-butanediol (below).

(b) Permanganate gives *syn*-hydroxylation. Top-side and bottom-side attachments are equally likely.

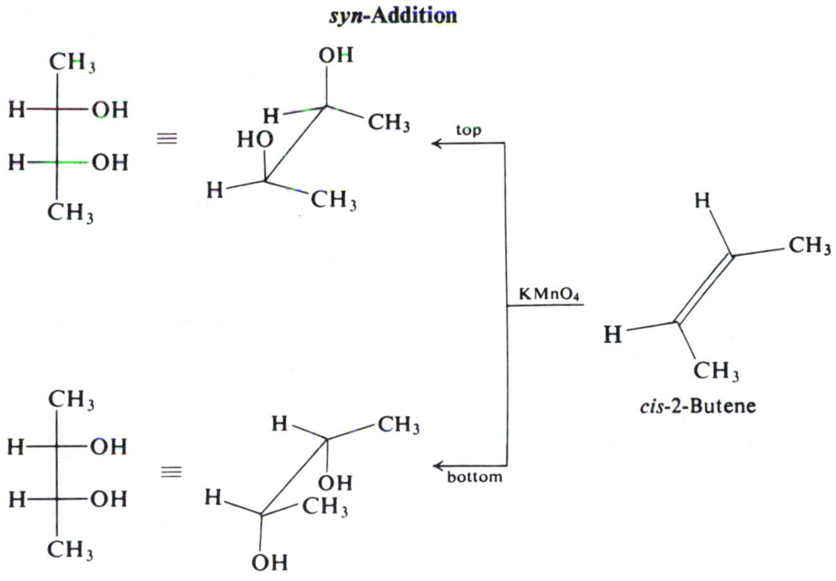

***syn*-Addition**

CH₃ / H—OH / H—OH / CH₃ ≡ (top) ← KMnO₄ — *cis*-2-Butene → (bottom) CH₃ / H—OH / H—OH / CH₃

meso-2,3-Butanediol
M.p. 34 °C

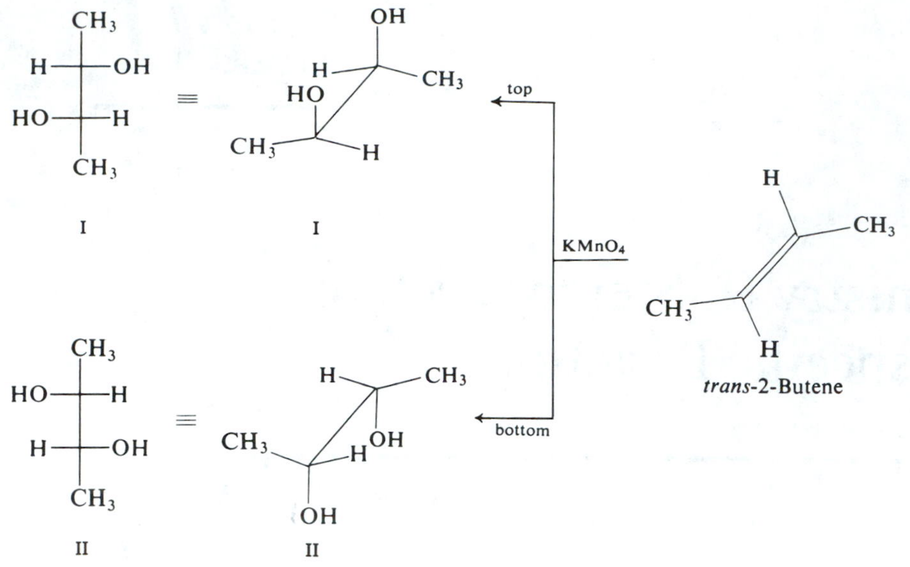

top
KMnO4
bottom
trans-2-Butene
I
I
II
II

I and II *are enantiomers*
Racemic 2,3-butanediol
M.p. 19 °C

(c) Peroxy acids give *anti*-hydroxylation. Attachments as in *a* and *b* (or *c* and *d*) are equally likely.

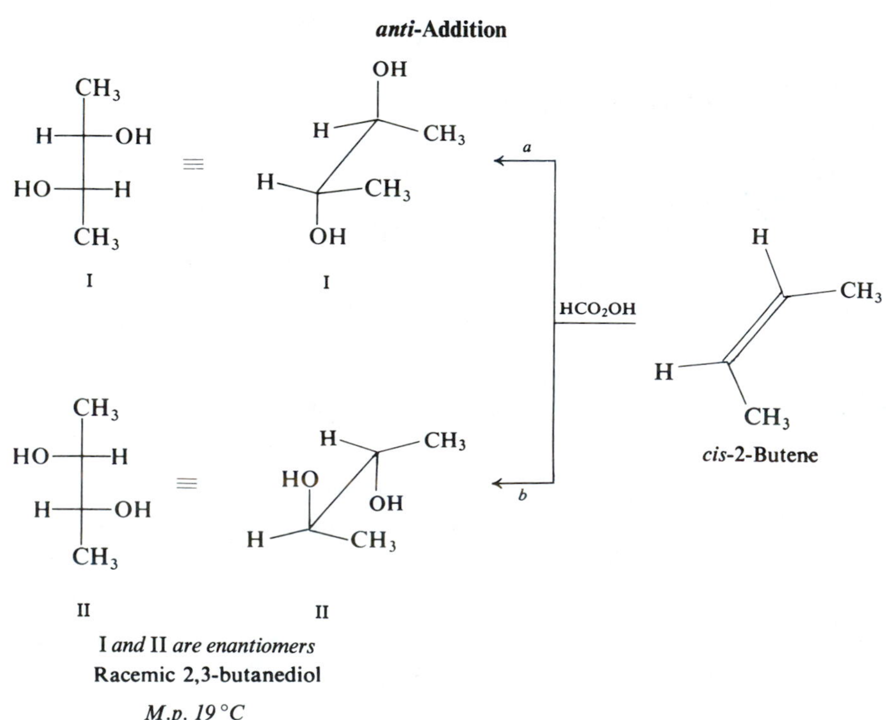

***anti*-Addition**

a
HCO2OH
b
cis-2-Butene
I
I
II
II

I and II *are enantiomers*
Racemic 2,3-butanediol
M.p. 19 °C

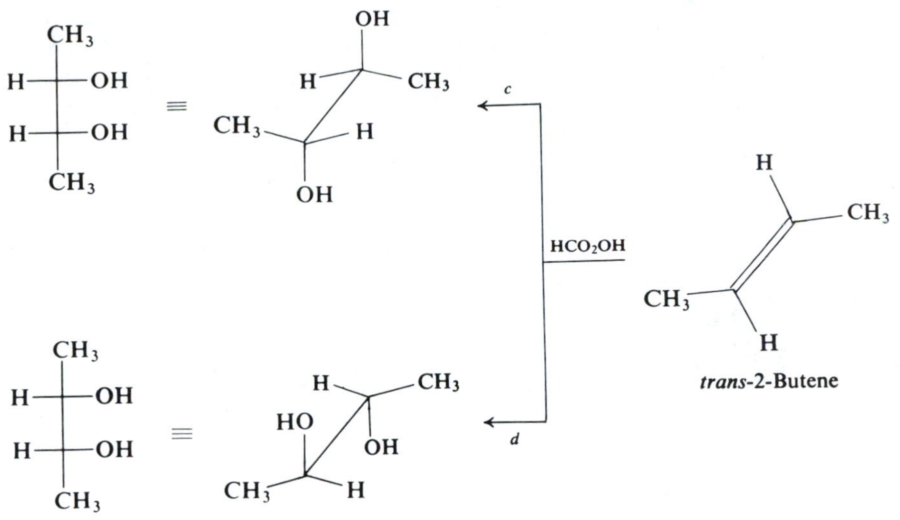

CH_3, H, OH, H, OH, CH_3 ≡ OH, H, CH_3, CH_3, H, OH

c

HCO_2OH

H, CH_3, CH_3, H

trans-2-Butene

d

CH_3, H, OH, H, OH, CH_3 ≡ H, CH_3, HO, OH, CH_3, H

meso-2,3-Butanediol

M.p. 34 °C

10.2 (a) Enantiomers, formed in equal amounts.

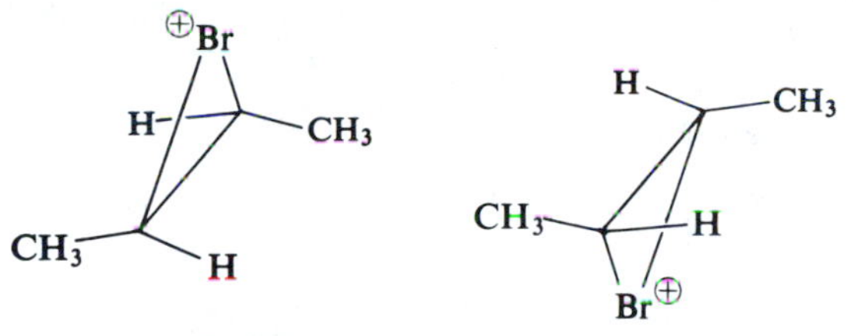

$\oplus$Br, H, CH_3, CH_3, H   H, CH_3, CH_3, H, Br$\oplus$

(b) Identical, achiral.

$\oplus$Br, CH_3, H, CH_3, H CH_3, H, CH_3, H, Br$\oplus$

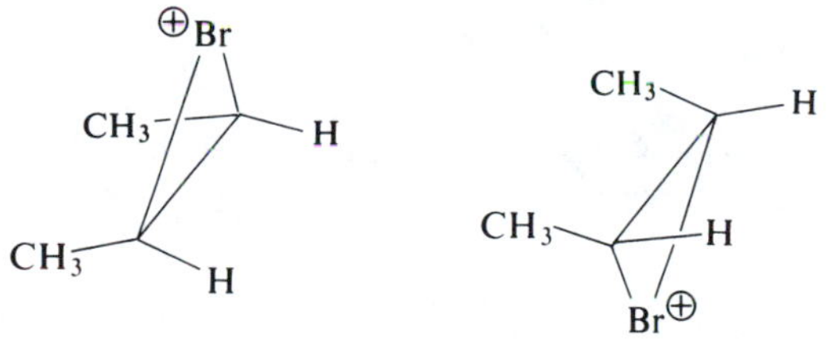

(c) Enantiomers, formed in equal amounts.

$\oplus$Br, H, C_2H_5, CH_3, H   H, C_2H_5, CH_3, H, Br$\oplus$

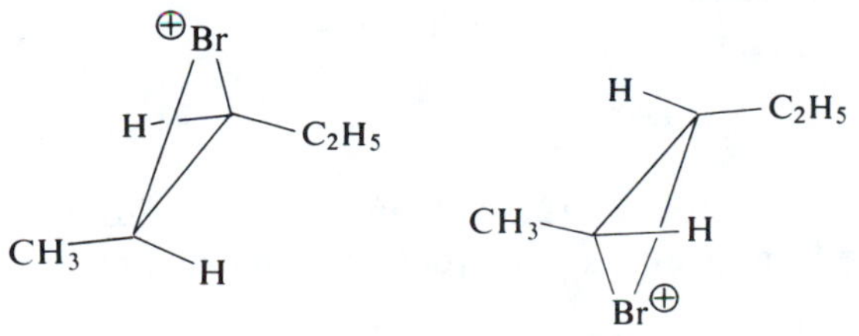

(d) Enantiomers, formed in equal amounts.

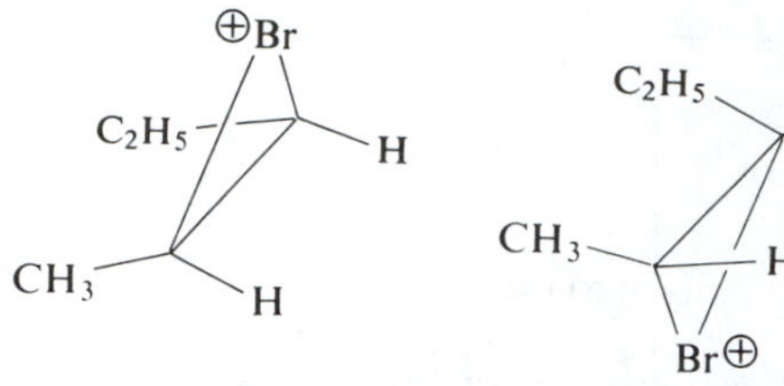

(Roberts, I.; Kimball, G. E. "The Halogenation of Ethylenes"; *J. Am. Chem. Soc.* **1937**, *59*, 947.)

10.3 (a) Even though attacks by the two paths—at the methyl end and at the ethyl end—are not equally likely, the product is racemic: (2*R*,3*S*)- and (2*S*,3*R*)-2,3-dibromopentane. There are equal amounts of the enantiomeric cyclic bromonium ions (see Problem 10.2c) undergoing attack. The product from one bromonium ion undoubtedly consists of unequal amounts of the two possible enantiomeric dibromides; if, say, attack at the methyl end were preferred, then *R*,*S* > *S*,*R*.

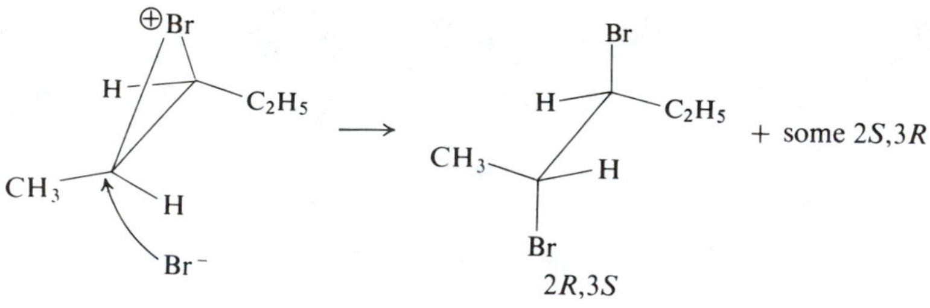

→ + some 2*S*,3*R*

2*R*,3*S*

But this would be exactly balanced by the same preference for attack at the methyl end of the other (enantiomeric) bromonium ion, to give *S*,*R* > *R*,*S*.

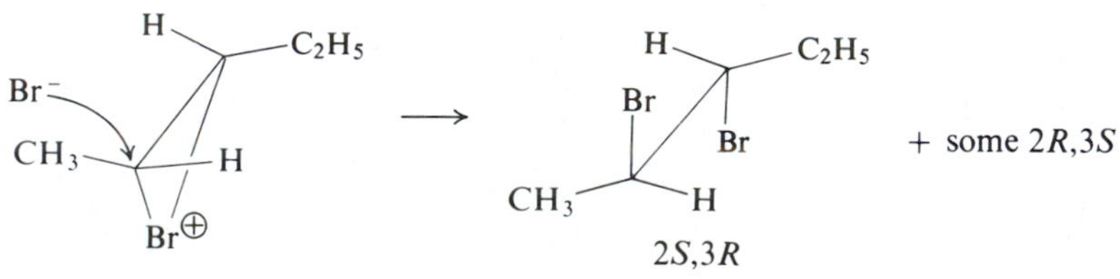

→ + some 2*R*,3*S*

2*S*,3*R*

(b) Similar to (a), with the enantiomeric products (2*R*,3*R*)- and (2*S*,3*S*)-2,3-dibromopentane. Here, if *R*,*R* > *S*,*S* from one bromonium ion, it would be balanced by *S*,*S* > *R*,*R* from the other.

10.4

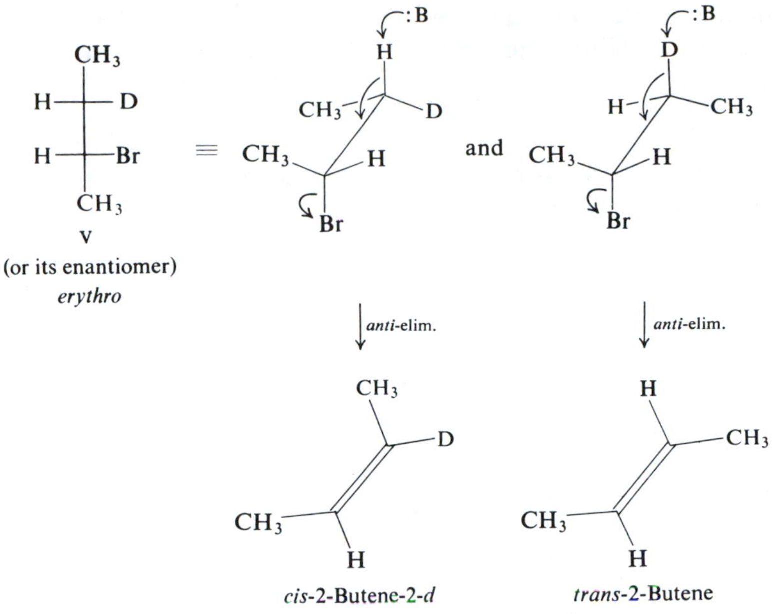

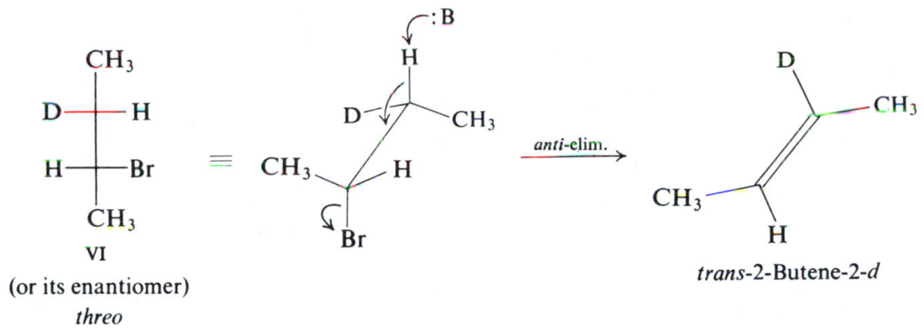

CH_3 H—D H—Br CH_3 ≡ (:B, H, CH_3, D, CH_3, H, Br) and (:B, D, H, CH_3, CH_3, H, Br)

V
(or its enantiomer)
erythro

↓ *anti*-elim. ↓ *anti*-elim.

CH_3, D, CH_3, H
cis-2-Butene-2-*d*

H, CH_3, CH_3, H
trans-2-Butene

CH_3 D—H H—Br CH_3 ≡ (:B, H, D, CH_3, CH_3, H, Br) —*anti*-elim.→ D, CH_3, CH_3, H

VI
(or its enantiomer)
threo

trans-2-Butene-2-*d*

The reactions must proceed as shown above, via *anti*-elimination of —Br and —H (or —D). *syn*-Elimination from V, say, to give *cis*-2-butene would have to result in loss of deuterium, contrary to fact.

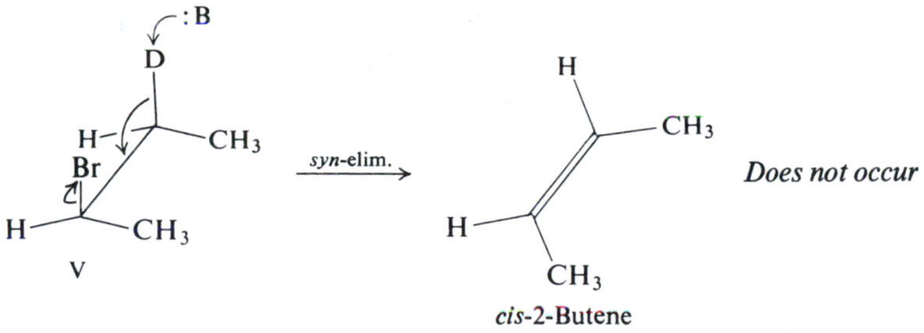

(:B, D, H, CH_3, Br, H, CH_3) —*syn*-elim.→ H, CH_3, H, CH_3 *Does not occur*

V

cis-2-Butene

1. Homogeneous hydrogenation involves *syn*-addition. See Figures 29.4 and 29.5 on pages 1047–1048. (*cis*-Butenedioic acid is called *maleic acid*, and *trans*-butenedioic acid is called *fumaric acid*.)

2. We write a mechanism analogous to the one for the addition of bromine shown in Figures 10.2 and 10.3 on pages 374–375, but with additional protonation and deprotonation steps as shown on page 483.

3. (a) Addition is *anti*, as shown by examination (disregarding the intermediate chloronium ion) of the structures in Figures 10.5 and 10.6, below.
(b) See Figure 10.5, below.

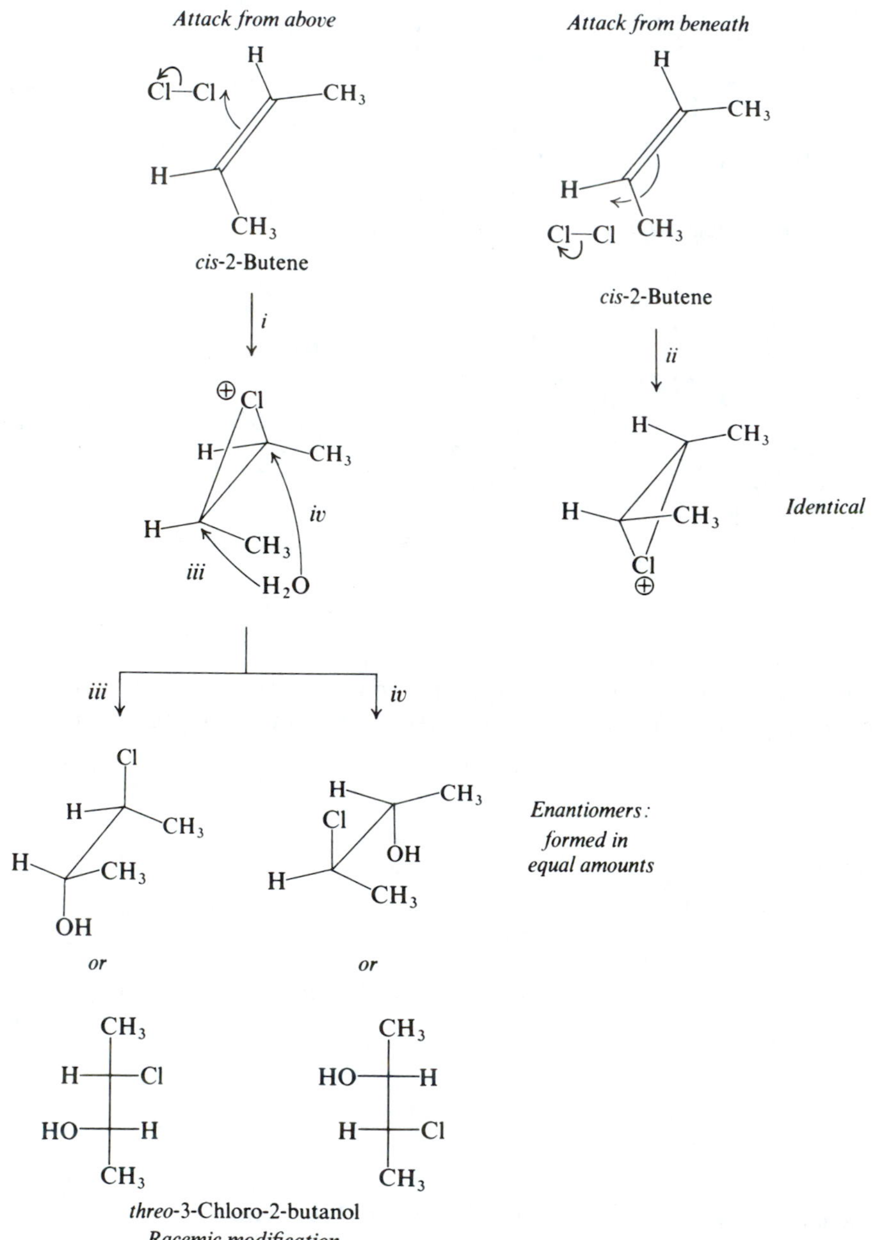

Figure 10.5 Reaction of *cis*-2-butene in Problem 3(b).

(c) See Figure 10.6, below.

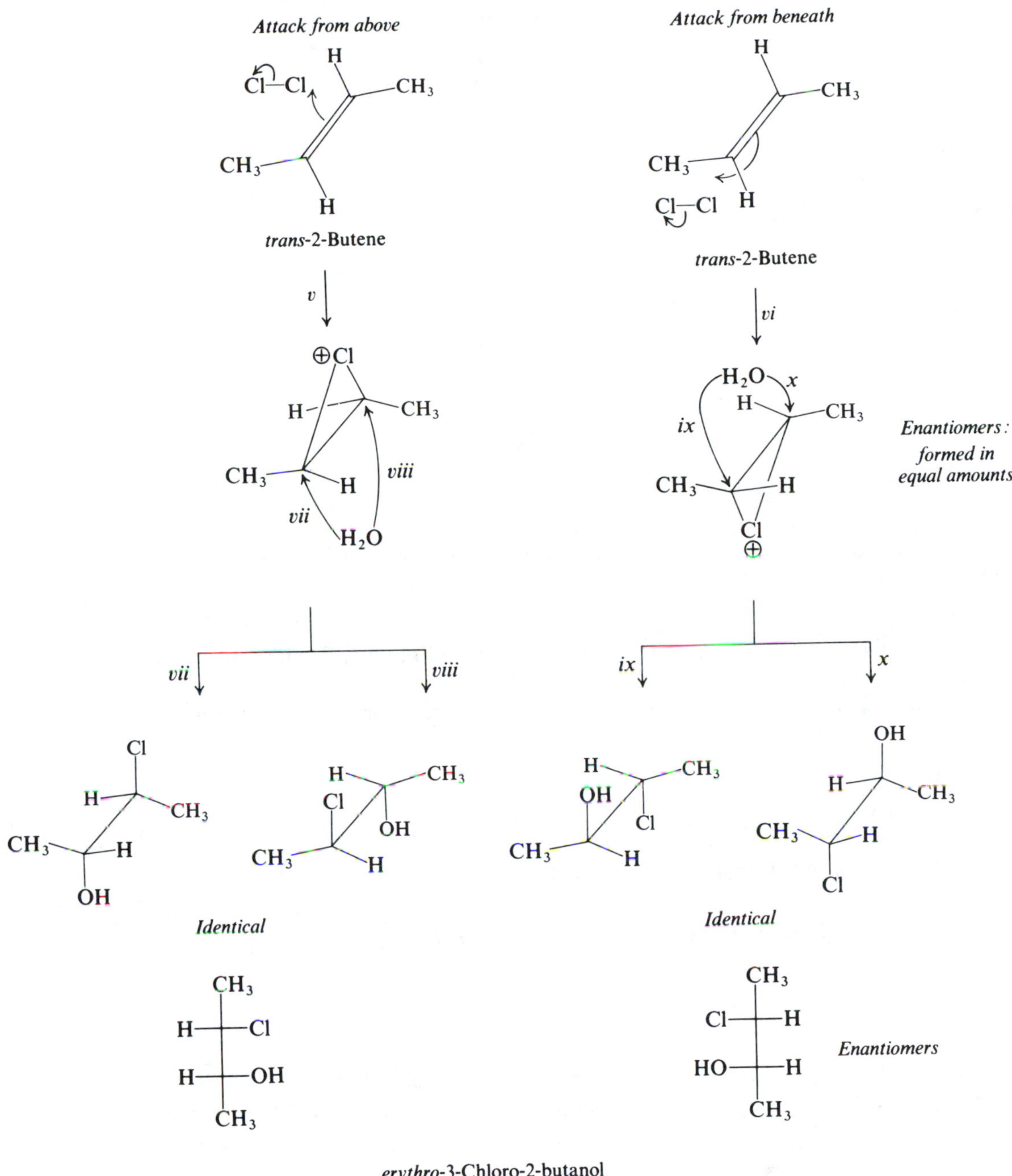

Figure 10.6 Reaction of *trans*-2-butene in Problem 3(c).

(d) In (a), it is cleavage of the cyclic chloronium ion (via *iii* or *iv*, equally likely) that leads to a racemic product. In (b), it is formation of the cyclic chloronium ion (via *v* or *vi*, equally likely) that leads to a racemic product.

4. We would expect E1 elimination to be essentially non-stereoselective and non-stereospecific, since the leaving groups are lost *in different steps*. The carbocation formed in the first step could exist in various conformations—either as initially formed or through subsequent rotation about a carbon–carbon bond (page 373)—from which a β-proton could be lost.

5. (a) Heterolytic addition via a cyclic iodonium ion.

$$R{-}CH{=}CH_2 + IN_3 \longrightarrow R{-}\overset{I^{\oplus}}{CH{-}CH_2} + N_3^-$$

$$N_3^- + R{-}\overset{I^{\oplus}}{CH{-}CH_2} \longrightarrow R{-}\underset{N_3}{CH}{-}CH_2I$$

The mechanism is analogous to that for the addition of X_2 and HOX, and takes place with the same orientation (Sec. 9.12).

(b) Polar solvents favor a heterolytic reaction, with formation of intermediate ions, like that of IN_3 in part (a).

Light or peroxides promote a free-radical addition, a chain reaction analogous to that of HBr (Sec. 9.21).

$$Rad\cdot + BrN_3 \longrightarrow RadBr + N_3\cdot$$

$$N_3\cdot + RCH{=}CH_2 \longrightarrow R\dot{C}HCH_2N_3$$

$$R\dot{C}HCH_2N_3 + BrN_3 \longrightarrow R\underset{Br}{CH}CH_2N_3 + N_3\cdot$$

and so on

Oxygen breaks the chain by combining with the organic free radicals (Sec. 2.14). In non-polar solvents (where the heterolytic reaction is not fast), slow spontaneous free-radical formation

$$BrN_3 \longrightarrow Br\cdot + N_3\cdot$$

makes itself evident.

$$BrN_3 + RCH{=}CH_2 \longrightarrow R\dot{C}HCH_2Br + N_3\cdot$$

(Hassner, Alfred. "Regiospecific and Stereospecific Introduction of Azide Function into Organic Molecules"; *Acc. Chem. Res.* **1971**, *4*, 9.)

6. (a) *syn*-Hydrogenation (Problem 1, above). One fraction: inactive (racemic).

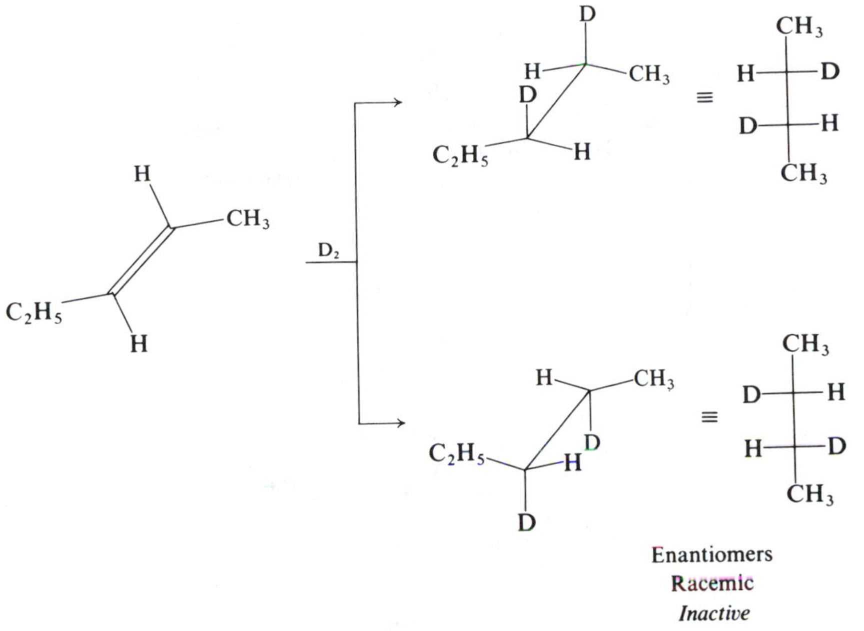

Enantiomers
Racemic
Inactive

(b) *syn*-Hydroxylation (Problem 10.1(b), above). Two fractions: one active, one inactive (*meso*).

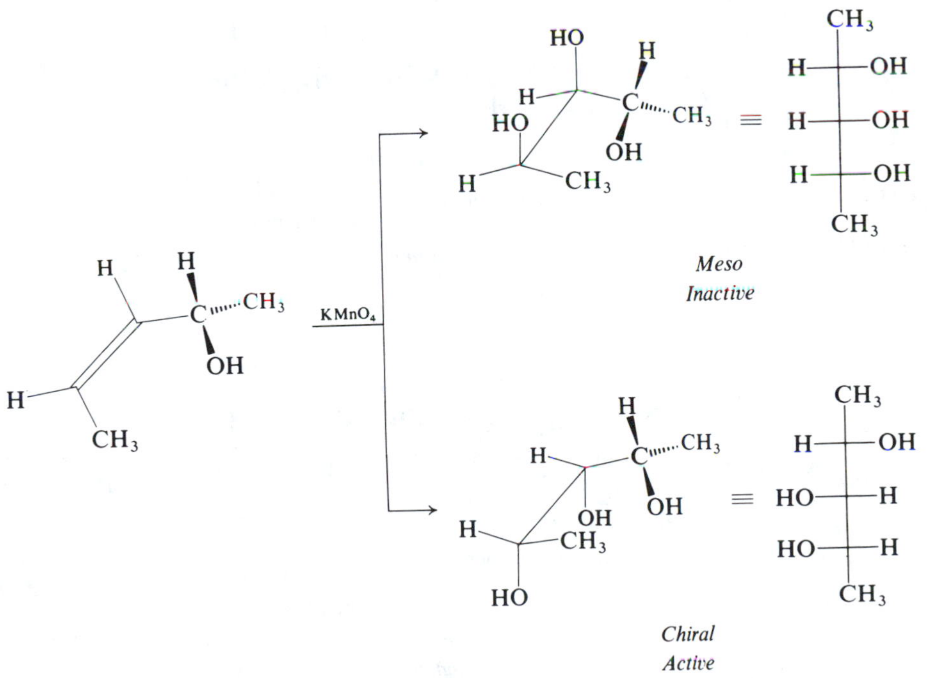

Meso
Inactive

Chiral
Active

(c) *anti*-Hydroxylation (Problem 10.1(c), above). Two fractions: one active, one inactive (*meso*).

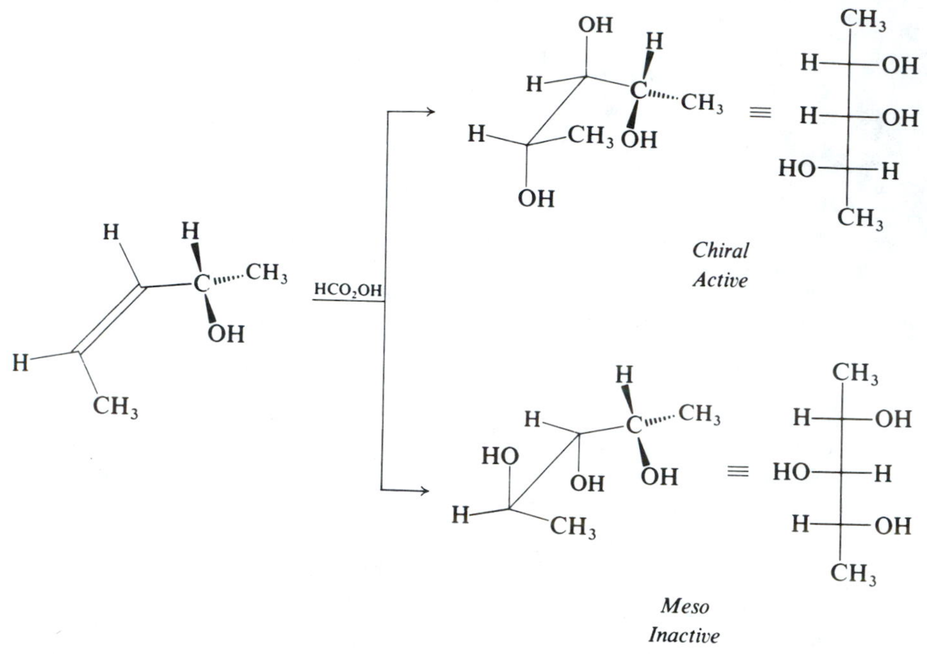

H H CH$_3$ OH H CH$_3$ HCO_2OH

OH H H CH$_3$ H CH$_3$ OH OH ≡ CH$_3$ H—OH H—OH HO—H CH$_3$

Chiral
Active

H H CH$_3$ HO OH OH H CH$_3$ ≡ CH$_3$ H—OH HO—H H—OH CH$_3$

Meso
Inactive

(d) *anti*-Addition of bromine. Two fractions: both inactive (both racemic).

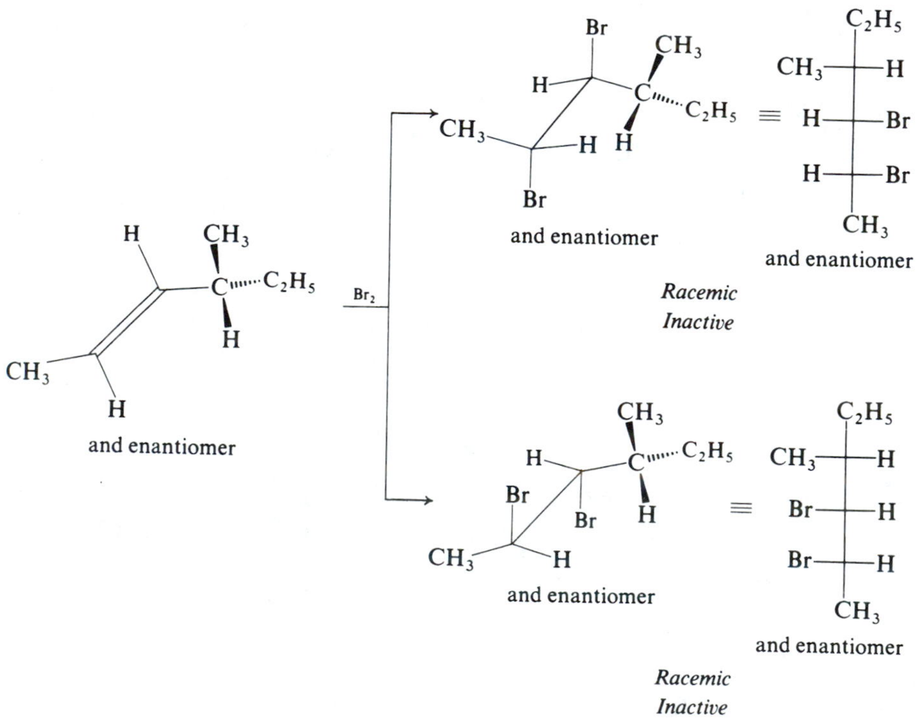

H CH$_3$ C$_2$H$_5$ H CH$_3$ H and enantiomer Br_2

Br CH$_3$ H C$_2$H$_5$ CH$_3$ H H Br and enantiomer ≡ C$_2$H$_5$ CH$_3$—H H—Br H—Br CH$_3$ and enantiomer

Racemic
Inactive

CH$_3$ H C$_2$H$_5$ Br Br H CH$_3$ H and enantiomer ≡ C$_2$H$_5$ CH$_3$—H Br—H Br—H CH$_3$ and enantiomer

Racemic
Inactive

(e) Hydroxylation. Two fractions: one inactive, one active. (Whether the hydroxylation is *syn* or *anti* has no bearing here, since only one new chiral center is being generated; what is important is that, when this chiral center is generated, both possible configurations result.)

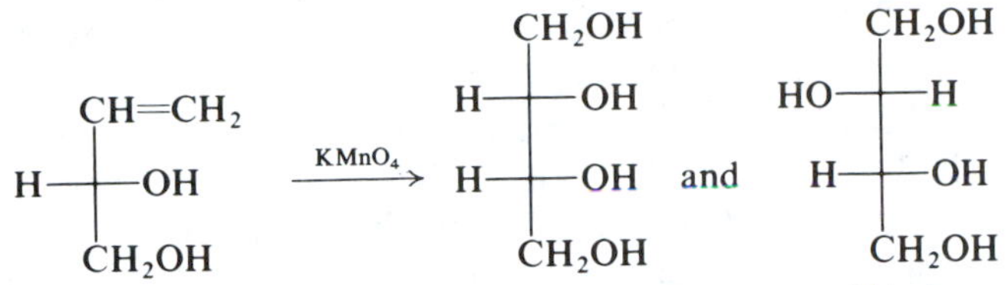

$$\begin{array}{c} CH{=}CH_2 \\ H-\!\!\!+\!\!\!-OH \\ CH_2OH \end{array} \xrightarrow{KMnO_4} \begin{array}{c} CH_2OH \\ H-\!\!\!+\!\!\!-OH \\ H-\!\!\!+\!\!\!-OH \\ CH_2OH \end{array} \text{ and } \begin{array}{c} CH_2OH \\ HO-\!\!\!+\!\!\!-H \\ H-\!\!\!+\!\!\!-OH \\ CH_2OH \end{array}$$

Meso *Inactive* — *Chiral* *Active*

(f) Two fractions: one inactive, one active. (Again a new chiral center is being generated, and in both possible configurations.)

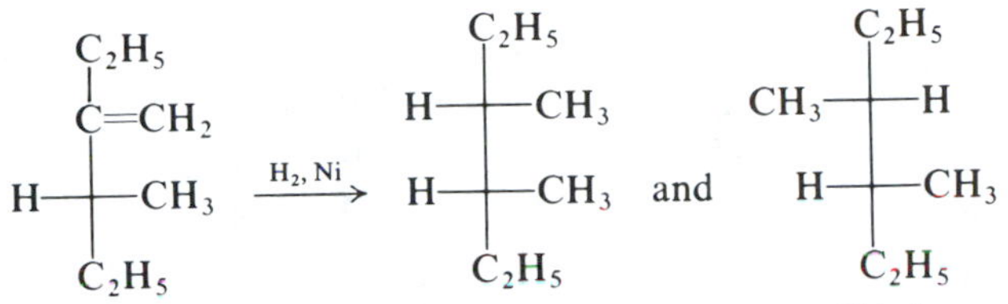

$$\begin{array}{c} C_2H_5 \\ C{=}CH_2 \\ H-\!\!\!+\!\!\!-CH_3 \\ C_2H_5 \end{array} \xrightarrow{H_2,\ Ni} \begin{array}{c} C_2H_5 \\ H-\!\!\!+\!\!\!-CH_3 \\ H-\!\!\!+\!\!\!-CH_3 \\ C_2H_5 \end{array} \text{ and } \begin{array}{c} C_2H_5 \\ CH_3-\!\!\!+\!\!\!-H \\ H-\!\!\!+\!\!\!-CH_3 \\ C_2H_5 \end{array}$$

Meso *Inactive* — *Chiral* *Active*

7. (a) What we are seeing here, superimposed on the basic pattern of *anti*-elimination for the E2 reaction, is another stereochemical factor: *conformational effects*. Let us examine these effects closely, since they are typical of the kind of thing we may expect to encounter in a wide variety of reactions.

The *trans*-2-butene, we saw in Secs. 8.6 and 9.4, is more stable than the *cis*. This difference in stability we attributed to a difference in van der Waals strain: in the *cis* isomer the two bulky methyl groups are crowded together on the same side of the molecule; in the *trans* isomer they lie on opposite sides (see Figure 8.7, page 279). So here, as in orientation,

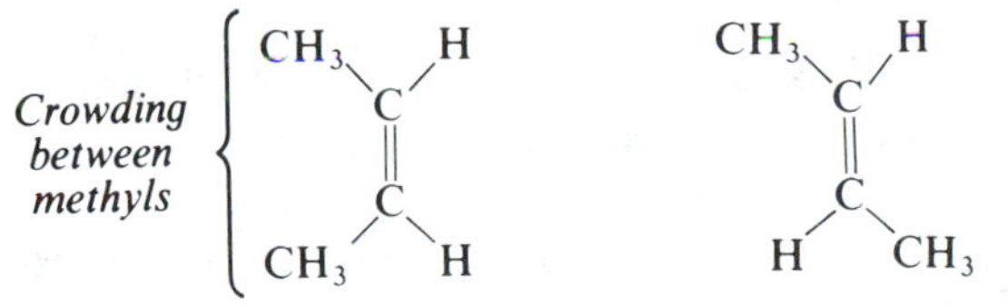

Crowding between methyls

$$\begin{array}{ccc} CH_3 & & H \\ & C & \\ & \| & \\ & C & \\ CH_3 & & H \end{array} \qquad \begin{array}{ccc} CH_3 & & H \\ & C & \\ & \| & \\ & C & \\ H & & CH_3 \end{array}$$

cis-2-Butene — *trans*-2-Butene

Less van der Waals strain: more stable

the preferred product of elimination is the more stable alkene: more stable this time, not because of the position of the double bond in the chain, but because of the geometry of the molecule.

Now, the predominance of *trans* alkene in the product means that it is formed *faster* than the *cis* alkene. We are dealing once more with relative rates of reaction, and must again compare transition states for the competing reactions.

Elimination is *anti*. To yield the *cis* alkene, reaction must pass through transition state I‡, derived from conformation I of the substrate; in I‡, as in conformation I, the methyl groups are on the same side of the molecule—as they will be in the product (Figure 10.7, below). To yield the *trans* alkene, reaction must pass through transition state II‡, derived from conformation II of the substrate; in II‡, as in conformation II, the methyl groups are on opposite sides of the molecule—again, as they will be in the product.

We have, then, the following situation. The reactant exists as conformers, two of which can give rise to different products via different transition states. The E_{act} for the reaction, elimination, is very large (20–25 kcal) compared with the E_{act} for interconversion of the conformers, which requires only rotation about a single bond. In a situation like this, it can be shown mathematically, the relative rates of formation of the two products are independent of the relative populations of the two conformers, and *depend only on the relative stabilities of the two transition states.*

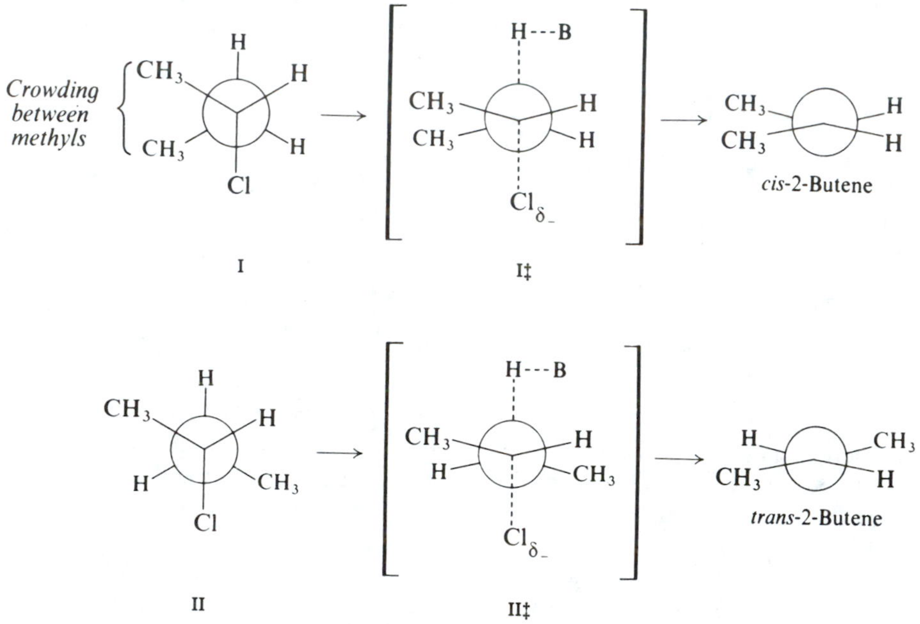

Figure 10.7 Conformational effects on *anti*-elimination: E2 dehydrohalogenation of *sec*-butyl chloride. Crowding between methyl groups on the same side of the molecule makes transition state I‡ less stable than transition state II‡.

Let us, then, compare transition states I‡ and II‡. In I‡ the bulky methyl groups are crowded together on the same side of the molecule: at least as close together as in conformer I and, to the extent that the double bond has formed and the molecule started to flatten, even closer together, as in the *cis* alkene itself. In II‡, by contrast, the methyls lie in roomier positions on opposite sides of the molecule. Because of greater van der Waals strain, I‡ is less stable than II‡. E_{act} for formation of the *cis* isomer is greater than for formation of the *trans* isomer; the *cis* isomer is formed more slowly and hence in lesser amount. The same factor, van der Waals strain, that determines the relative stabilities of the products also determines the relative stabilities of the transition states leading to their formation.

We have already discussed (page 55 of this Study Guide) the alternative kind of situation, where reaction is much easier and faster than interconversion of conformers. There, we saw, the relative rates of competing reactions—and hence the ratio of products—were determined only by the relative populations of the conformers. If that *were* the case in E2 elimination, we would still expect preferential formation of the *trans* isomer, since II is more stable than I for the same basic reason that II‡ is more stable than I‡. But the preference for *trans* should be less marked, since the crowding in I is less than in I‡, where flattening of the molecule has begun.

(b) Because of the weaker carbon–halogen bond (Sec. 8.19), the alkyl bromide is more reactive than the chloride. The transition state is therefore reached earlier with the bromide (Sec. 2.24), and has less product character; that is, it is less like the alkene it will become. The methyl groups in the transition state leading to the *cis* alkene are not as close together as in the case of the alkyl chloride (Figure 10.7, above), and there is less crowding between them. The difference in stability between I‡ and II‡ is less for the bromide than for the chloride, and the preference for the *trans* alkene is less marked.

8. (a)

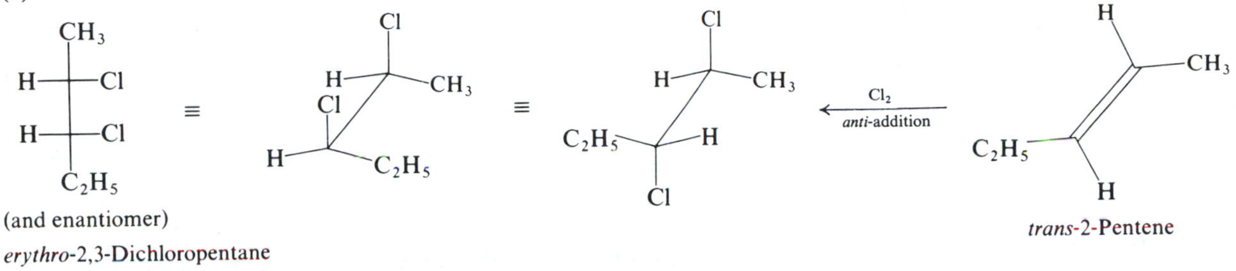

(and enantiomer)
erythro-2,3-Dichloropentane ≡ ≡ ← Cl_2 *anti*-addition ← *trans*-2-Pentene

(b)

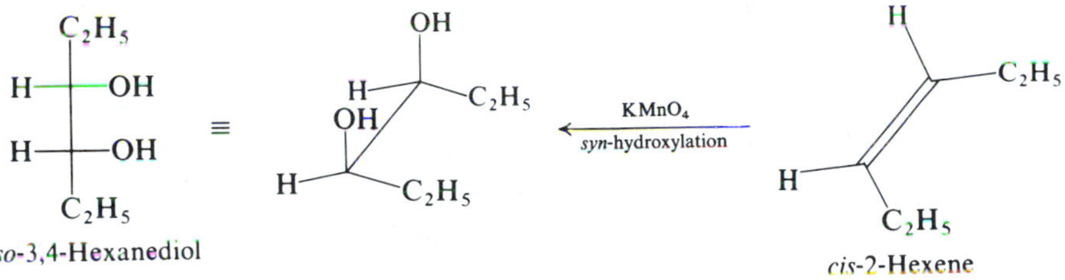

meso-3,4-Hexanediol ≡ ← $KMnO_4$ *syn*-hydroxylation ← *cis*-2-Hexene

(c)

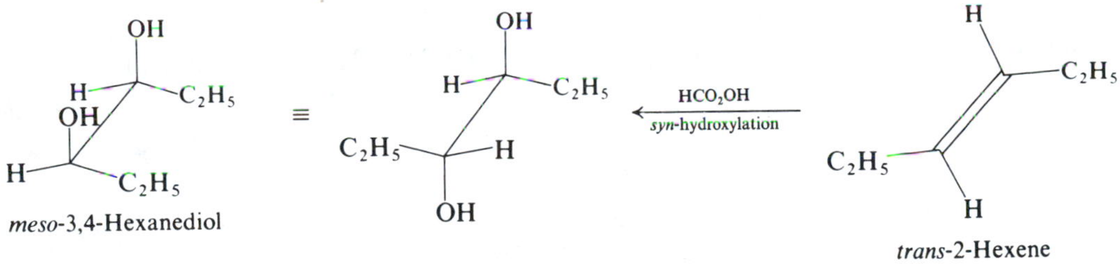

meso-3,4-Hexanediol ≡ ← HCO_2OH *syn*-hydroxylation ← *trans*-2-Hexene

(d)

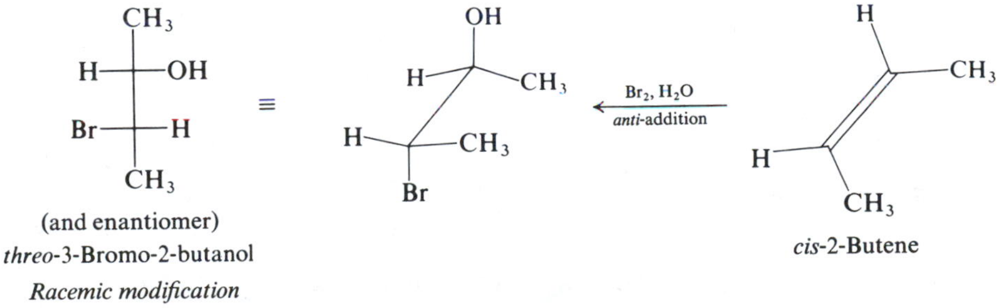

(and enantiomer)
threo-3-Bromo-2-butanol
Racemic modification ≡ ← Br_2, H_2O *anti*-addition ← *cis*-2-Butene

(e)

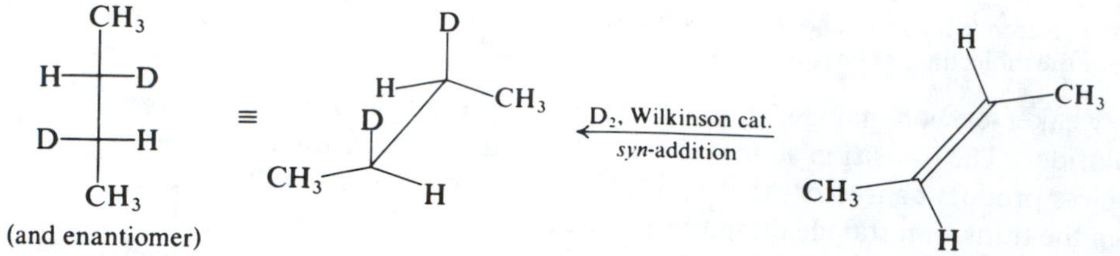

(and enantiomer)
rac-Butane-2,3-d_2

D_2, Wilkinson cat.
syn-addition

trans-2-Butene

9. (a) (i) *anti*-Elimination, with I^- as base.

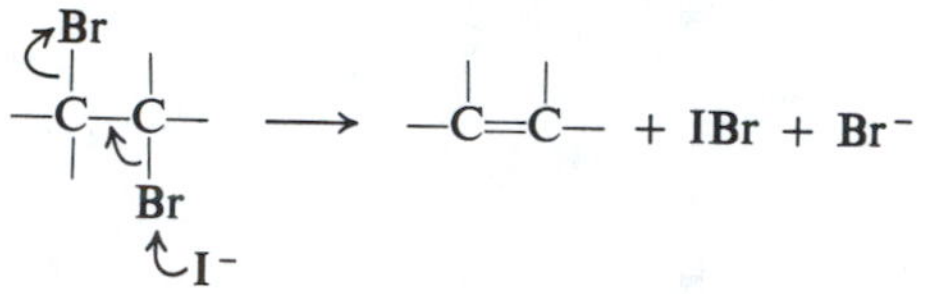

$$-C-C- \longrightarrow -C{=}C- + IBr + Br^-$$

(ii) Intermolecular nucleophilic attack, with inversion; followed by intramolecular nucleophilic attack, also with inversion; and finally elimination, which is necessarily *syn*. Net result: overall *syn*-elimination.

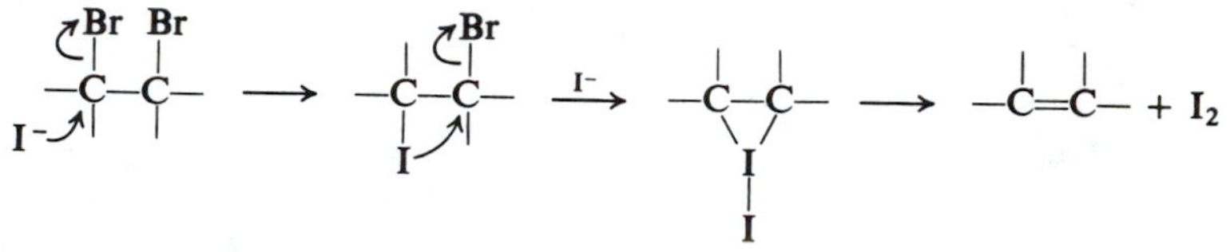

$$\xrightarrow{I^-} \quad -C{=}C- + I_2$$

(b) Only net *syn*-elimination (mechanism ii) can give the *cis*-CHD=CHD actually obtained from *meso*-CHDBr—CHDBr.

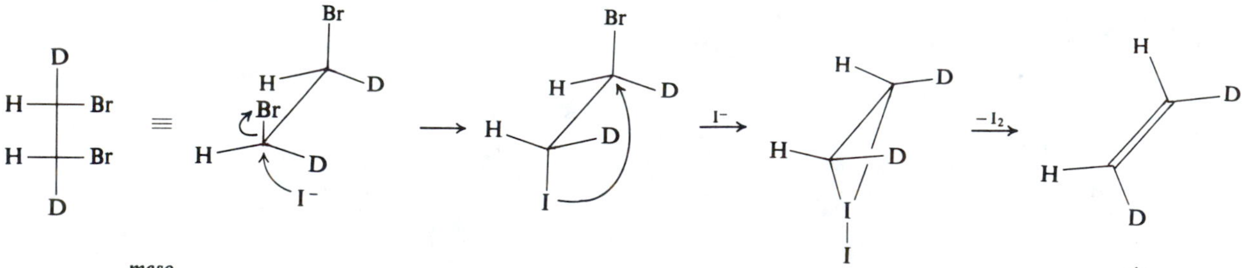

meso $\xrightarrow{I^-}$ $\xrightarrow{-I_2}$ *cis*

Only net *anti*-elimination (mechanism i) can give the observed results for the dibromobutanes.

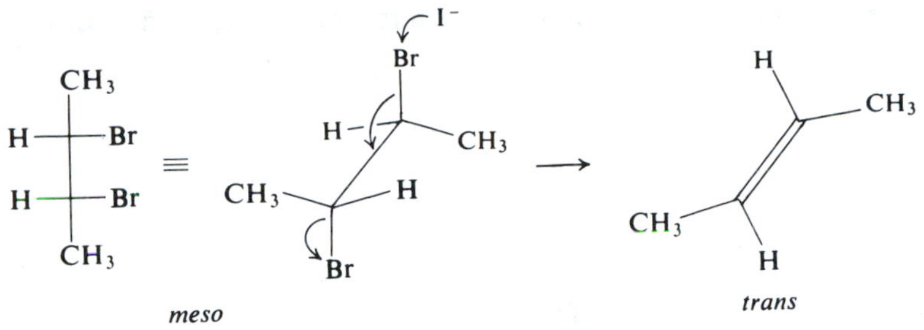

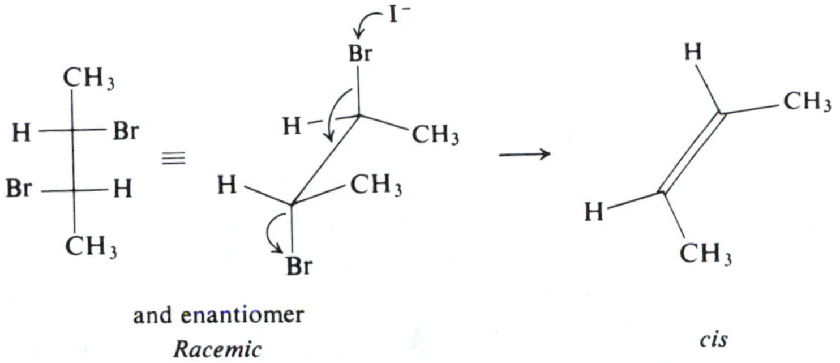

meso

trans

and enantiomer
Racemic

cis

Both nucleophilic attacks of mechanism (ii) are more difficult at the 2° carbons of the dibromobutanes than at the 1° carbon of the dibromoethane. At the same time, the methyl groups of the dibromobutanes help to stabilize the incipient double bond of mechanism (i). The result is a shift in mechanism from (ii) for the dibromoethane to (i) for the dibromobutanes.

10. (a) The racemic dibromide undergoes the usual *anti*-dehydrohalogenation, with pyridine attacking H.

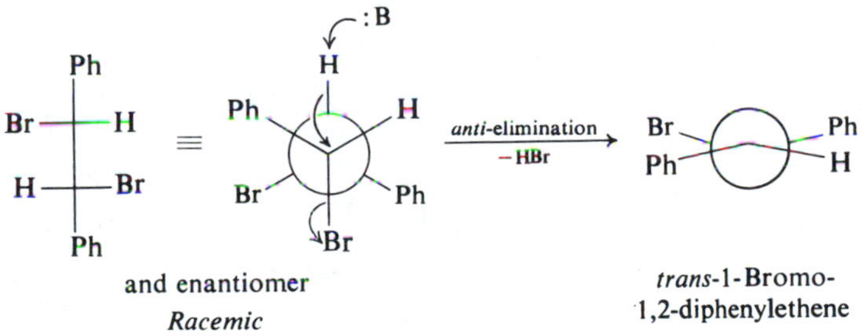

anti-elimination
$-HBr$

and enantiomer
Racemic

trans-1-Bromo-1,2-diphenylethene

The *meso* dibromide undergoes *anti*-elimination by mechanism (i) of the preceding problem, with pyridine (instead of iodide ion) attacking Br.

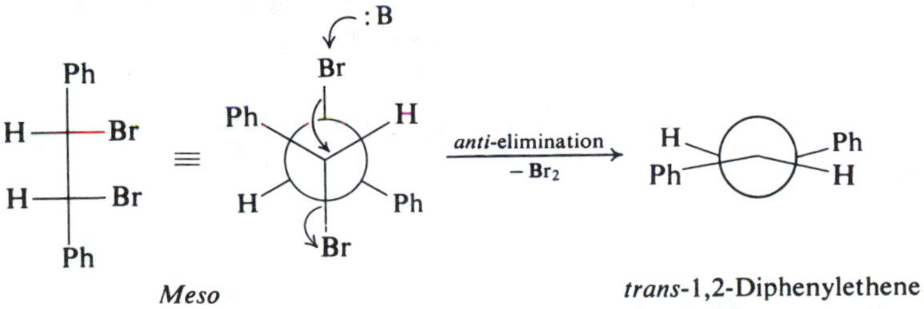

anti-elimination
$-Br_2$

Meso

trans-1,2-Diphenylethene

(b) Each reaction proceeds via the transition state in which the leaving groups are *anti*-periplanar and the bulky phenyl groups are as far apart as possible.

11. Highly electronegative F is less able than Cl, Br, or I to share electrons to form a three-membered ring.

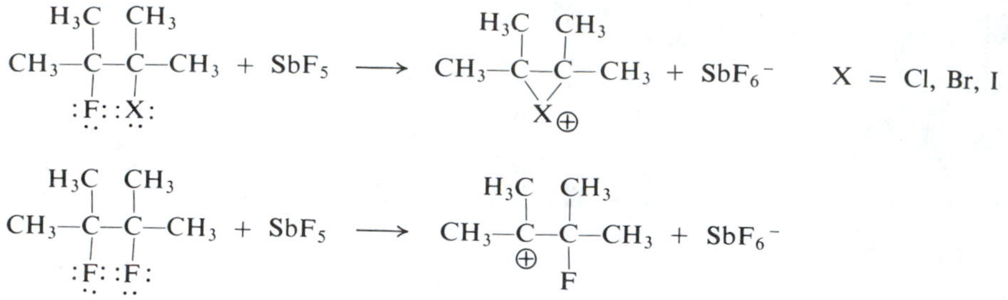

$$CH_3-C(CH_3)(F)-C(CH_3)(X)-CH_3 + SbF_5 \longrightarrow CH_3-C(CH_3)-C(CH_3)-CH_3 \text{ (bridged by } X^{\oplus}) + SbF_6^- \qquad X = Cl, Br, I$$

$$CH_3-C(CH_3)(F)-C(CH_3)(F)-CH_3 + SbF_5 \longrightarrow CH_3-\overset{\oplus}{C}(CH_3)-C(CH_3)(F)-CH_3 + SbF_6^-$$

12. (a)–(c) See Sec. 29.2.

(Winstein, S.; Lucas, H. J. "Retention of Configuration in the Reaction of the 3-Bromo-2-butanols with Hydrogen Bromide"; *J. Am. Chem. Soc.* **1939**, *61,* 1576.)

13. Two stereochemical combinations could give the observed results: (i) *anti*-addition of bromine, followed by *anti*-elimination; (ii) *syn*-addition followed by *syn*-elimination. Since (stereoselective) addition of bromine is known to be *anti* (Sec. 10.2), elimination must be *anti*, too.

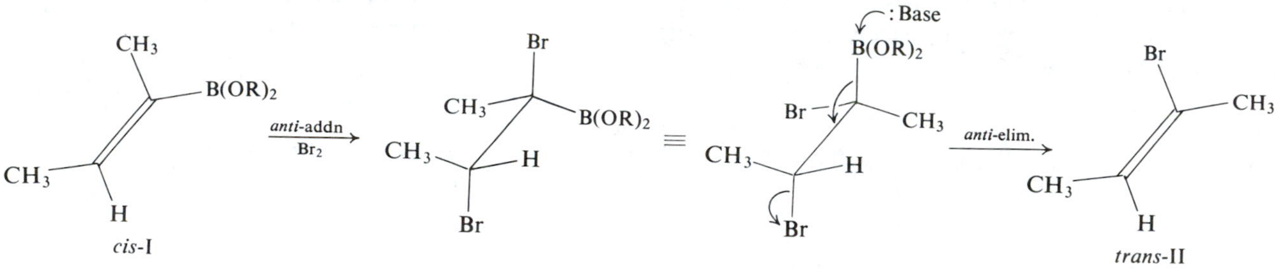

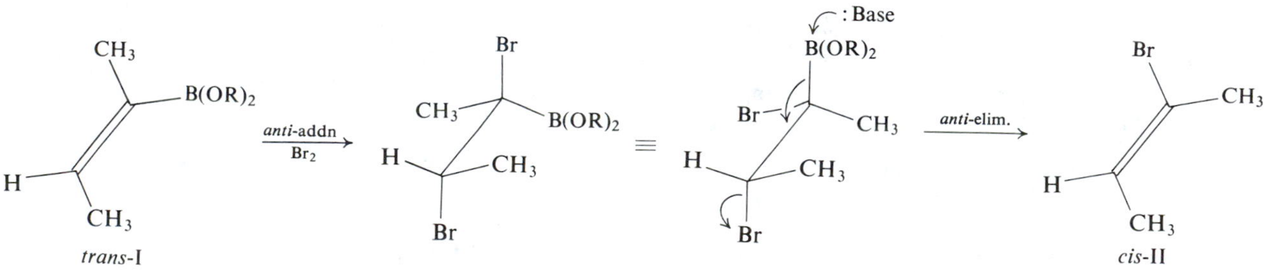

cis-I $\xrightarrow[Br_2]{anti\text{-addn}}$ $CH_3CH(Br)-C(Br)(CH_3)B(OR)_2$ $\equiv$ (:Base attacks B(OR)$_2$; Br leaves) $\xrightarrow{anti\text{-elim.}}$ *trans*-II

trans-I $\xrightarrow[Br_2]{anti\text{-addn}}$ $CH_3CH(Br)-C(Br)(CH_3)B(OR)_2$ $\equiv$ (:Base attacks B(OR)$_2$; Br leaves) $\xrightarrow{anti\text{-elim.}}$ *cis*-II

Water (or some other base) attacks electron-poor (Lewis acidic) boron, much as OH^-

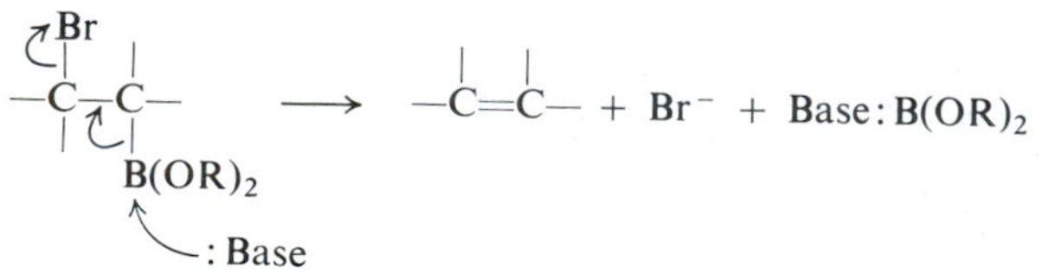

$$-\underset{|}{\overset{Br}{C}}-\underset{B(OR)_2}{\overset{|}{C}}- \;(\leftarrow :Base) \longrightarrow -\overset{|}{C}=\overset{|}{C}- + Br^- + Base{:}B(OR)_2$$

attacks —H in dehydrohalogenation, and brings about E2 elimination.

$$(RO)_2BCH_2CH_2Br + 3H_2O \longrightarrow CH_2{=}CH_2 + HBr + 2ROH + H_3BO_3$$

11

Conjugation and Resonance

Dienes

11.1 (a) The allyl radical would have to react with HCl to abstract a hydrogen atom.

$$\underset{(103)}{\text{H—Cl}} + \cdot\text{CH}_2\text{CH}{=}\text{CH} \longrightarrow \text{Cl}\cdot + \underset{(88)}{\text{H—CH}_2\text{CH}{=}\text{CH}_2} \qquad \Delta H = +15\text{ kcal} \qquad E_{act}\ \textit{at least}\ 15$$

(b) As shown, the E_{act} would have to be at least 15 kcal, and is probably higher. This is in marked contrast to the exothermic reaction with the halogen molecule to complete the addition.

11.2 From each alkene, the most stable free radical that can be formed is an allylic one, and this is the preferred intermediate. Through allylic rearrangement, this intermediate yields two products (neglecting stereoisomerism), the same ones in both cases: 3-chloro-1-butene and 1-chloro-2-butene.

$$\underset{\text{1-Butene}}{\text{CH}_3\text{CH}_2\text{CH}{=}\text{CH}_2} + t\text{-BuO}\cdot \longrightarrow \underset{\substack{\text{An allylic radical}\\ \textit{Preferred product}}}{\text{CH}_3\dot{\text{C}}\text{HCH}{=}\text{CH}_2} + t\text{-BuOH}$$

$$\underset{\text{2-Butene}}{\text{CH}_3\text{CH}{=}\text{CHCH}_3} + t\text{-BuO}\cdot \longrightarrow \underset{\substack{\text{An allylic radical}\\ \textit{Preferred product}}}{\text{CH}_3\text{CH}{=}\text{CHCH}_2\cdot} + t\text{-BuOH}$$

$$\text{CH}_3\dot{\text{C}}\text{HCH}{=}\text{CH}_2 \xrightarrow{t\text{-BuOCl}} \underset{\text{3-Chloro-1-butene}}{\text{CH}_3\underset{\text{Cl}}{\underset{|}{\text{C}}}\text{HCH}{=}\text{CH}_2} + \underset{\text{1-Chloro-2-butene}}{\text{CH}_3\text{CH}{=}\text{CH}\underset{\text{Cl}}{\underset{|}{\text{C}}}\text{H}_2} \xleftarrow{t\text{-BuOH}} \text{CH}_3\text{CH}{=}\text{CHCH}_2\cdot$$

As we shall soon see, the two alkenes give the same products because they form the *same* allylic intermediate, a hybrid of the same two structures.

$$[CH_3-\dot{C}H-CH=CH_2 \quad CH_3-CH=CH-\dot{C}H_2] \text{ equivalent to } \underbrace{CH_3-CH\!\stackrel{...}{-}\!CH\!\stackrel{...}{-}\!CH_2}_{\cdot}$$

This hybrid radical reacts at either of two positions, to give either of two products.

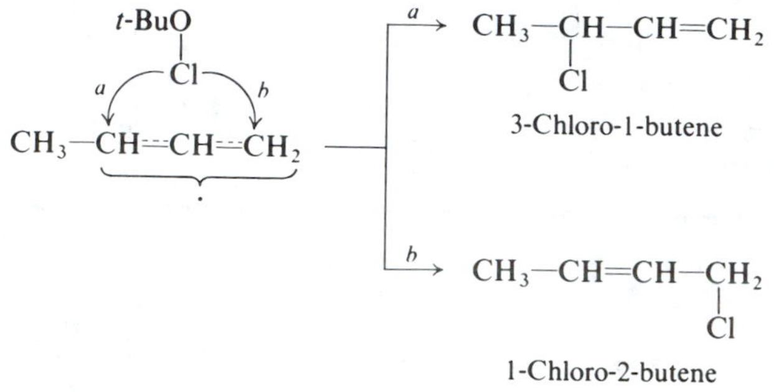

$$t\text{-BuO}-Cl + \underbrace{CH_3-CH\!\stackrel{...}{-}\!CH\!\stackrel{...}{-}\!CH_2}_{\cdot} \xrightarrow{a} CH_3-CH(Cl)-CH=CH_2$$

3-Chloro-1-butene

$$\xrightarrow{b} CH_3-CH=CH-CH_2Cl$$

1-Chloro-2-butene

11.3 Use isotopically labeled propylene, $^{14}CH_3CH=CH_2$, say, and see if the label appears at two different positions in the product.

$$^{14}CH_3-CH=CH_2 \xrightarrow{NBS} Br^{14}CH_2-CH=CH_2 + {}^{14}CH_2=CH-CH_2Br$$

On ozonolysis, we would expect the isotopic carbon to appear, not only in the bromoacetaldehyde,

$$Br^{14}CH_2CH=CH_2 \xrightarrow{O_3} \underset{\text{Bromoacetaldehyde}}{Br^{14}CH_2CH=O} + \underset{\text{Formaldehyde}}{O=CH_2}$$

but also in the formaldehyde.

$$^{14}CH_2=CHCH_2Br \xrightarrow{O_3} \underset{\text{Formaldehyde}}{H_2{}^{14}C=O} + \underset{\text{Bromoacetaldehyde}}{O=CHCH_2Br}$$

11.4 $\left[-\overset{+}{N}(=O)(O^-) \quad -\overset{+}{N}(O^-)(=O) \right]$ *equivalent to* $\left\{ -\overset{+}{N}(\stackrel{...}{-}O)_2 \right\}^-$

11.5 $\left[O=C(O^-)_2 \quad {}^-O-C(=O)(O^-) \quad {}^-O-C(O^-)(=O) \right]$ *equivalent to* $\left\{ O\stackrel{...}{-}C(\stackrel{...}{-}O)_2 \right\}^{2-}$

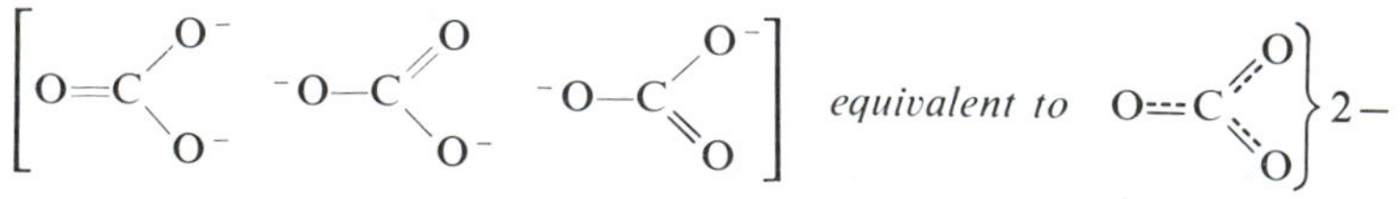

11.6 See Sec. 19.13.

11.7 The one-and-a-half bonds in the intermediate allylic radicals effectively prevent the rotation needed to convert one stereoisomeric radical into the other.

cis-2-Butene $\xrightarrow{t\text{-BuO}\cdot}$ *cis* Radical $\xrightarrow{t\text{-BuOCl}}$ *cis*-1-Chloro-2-butene + 3-Chloro-1-butene

trans-2-Butene $\xrightarrow{t\text{-BuO}\cdot}$ *trans*-Radical $\xrightarrow{t\text{-BuOCl}}$ *trans*-1-Chloro-2-butene + 3-Chloro-1-butene

11.8 (a) No. The measured heat of hydrogenation (49.8 kcal/mol) is 36.0 kcal/mol less than that calculated for three double bonds ($3 \times 28.6 = 85.8$ kcal/mol). From the Kekulé structure we would expect three long single bonds and three short double bonds—contrary to fact.

(b) A better representation is:

equivalent to

As a hybrid of two equivalent structures, all bonds would be the same and intermediate in length between single and double bonds. Resonance would be important and would lead to considerable stabilization of the molecule—as it does.

11.9 Stabilization of a carbocation depends upon the electron-deficient carbon getting electrons from elsewhere in the molecule (Sec. 5.20). Hyperconjugation would involve getting electrons from carbon–hydrogen bonds. It has been proposed that, like the p orbital of a free radical, the empty p orbital of a carbocation can overlap σ orbitals of alkyl groups to which it is attached (Fig. 11.7).

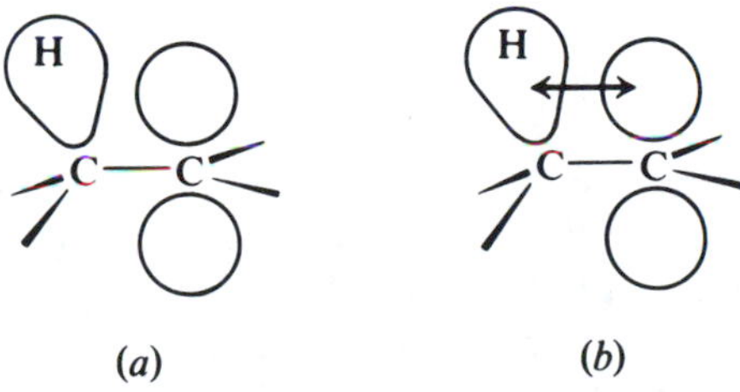

Figure 11.7 Hyperconjugation in a carbocation. (*a*) Separate σ and p orbitals. (*b*) Overlapping orbitals.

As we discussed for free radicals, this kind of overlap permits individual electrons to help bind together three nuclei, two carbons and one hydrogen.

In resonance language, the ethyl cation, for example, would be described as a hybrid of not only structure I, but also the three structures II, III, and IV, in which a double bond joins the two carbons, and the positive charge is carried by a hydrogen. To the extent

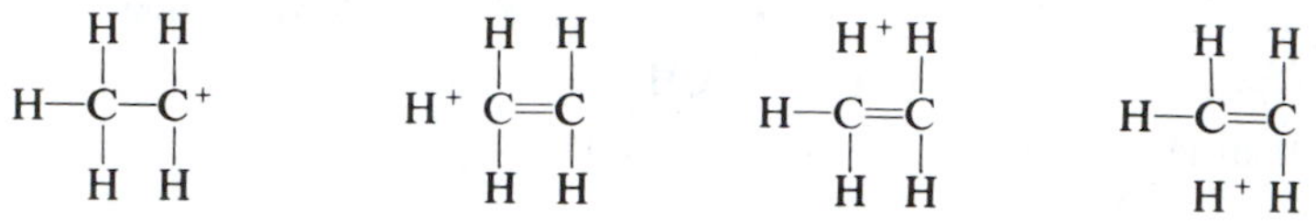

```
   H  H            H  H          H⁺H            H  H
   |  |            |  |            |            |  |
 H—C—C⁺        H⁺  C=C        H—C=C         H—C=C
   |  |            |  |          |  |             |
   H  H            H  H          H  H           H⁺ H

    I               II            III            IV
```

that the carbon–carbon bond has acquired double-bond character, the electron-deficient carbon has gained electrons, and the positive charge is dispersed over the three hydrogens. With increased branching of the carbocation, dispersal of charge and the resulting stabilization is greater: dispersal over six hydrogens for the isopropyl cation and nine for the *tert*-butyl cation.

Earlier (Sec. 5.20), we described electron release by alkyl groups as an inductive effect; here, we see it as a resonance effect. Despite a great deal of work on the problem, the relative importance of these two factors is not clear. One frequently finds the two lumped together as the "inductive–hyperconjugative" effect of alkyl groups. In this book we may often refer to the inductive effect of alkyl groups, but it should be understood that this may well include a contribution from hyperconjugation.

11.10 We expect S_N1 reactions proceeding via the following allylic cations:

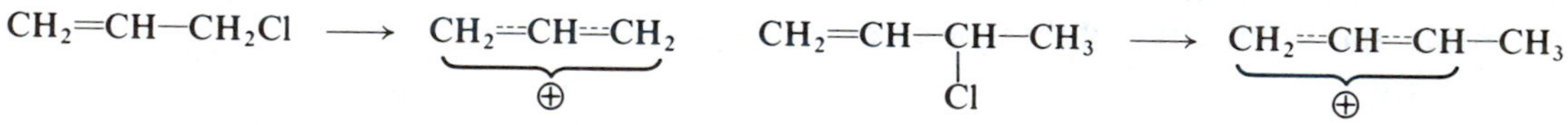

$$\underset{\text{I}}{CH_2{=}CH{-}CH_2Cl} \longrightarrow \underset{\text{I}^+}{\underbrace{CH_2{\overset{...}{=}}CH{\overset{...}{=}}CH_2}_{\oplus}} \qquad \underset{\text{II}}{CH_2{=}CH{-}\underset{Cl}{\underset{|}{C}H}{-}CH_3} \longrightarrow \underset{\text{II}^+}{\underbrace{CH_2{\overset{...}{=}}CH{\overset{...}{=}}CH}_{\oplus}{-}CH_3}$$

$$\underset{\text{III}}{CH_2{=}\overset{CH_3}{\overset{|}{C}}{-}CH_2Cl} \longrightarrow \underset{\text{III}^+}{\underbrace{CH_2{\overset{...}{=}}\overset{CH_3}{\overset{|}{C}}{\overset{...}{=}}CH_2}_{\oplus}} \qquad \underset{\text{IV}}{CH_3{-}CH{=}CH{-}CH_2Cl} \longrightarrow \underset{\text{IV}^+}{CH_3{-}\underbrace{CH{\overset{...}{=}}CH{\overset{...}{=}}CH_2}_{\oplus}}$$

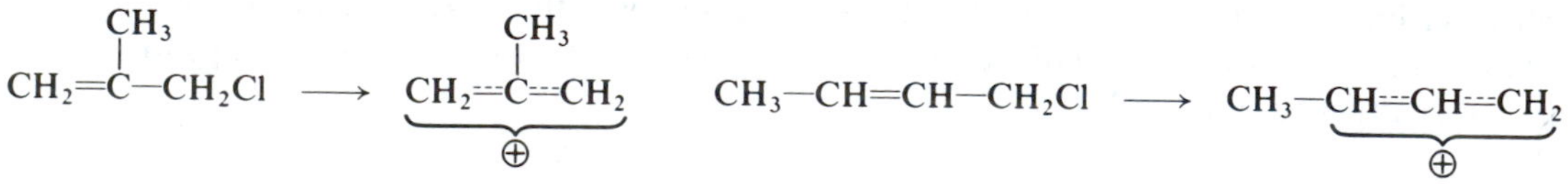

In cations II^+ and IV^+, methyl is attached to one of the terminal carbons of the allylic system, the carbons that carry most of the positive charge; through electron release, methyl helps to disperse the charge and thus powerfully to stabilize the cation and the transition state leading to its formation. In cation III^+, methyl is attached to the central carbon of the allylic system, which carries little of the positive charge; electron release by methyl is of little help in stabilizing the cation or transition state, and may actually interfere sterically with solvent assistance.

11.11

$$CH_2{=}CH{-}CH{=}CH_2 \xrightarrow{HCl} \underset{|}{CH_2}{-}\overset{\oplus}{\overbrace{CH{\cdots}CH{\cdots}CH_2}} \quad (H \text{ on } CH_2;\ Cl^- \text{ attacks at } a \text{ or } b)$$

$$\xrightarrow{a} CH_2{-}CH(Cl){-}CH{=}CH_2 \text{ (H on first } CH_2\text{)}$$

3-Chloro-1-butene

$$\xrightarrow{b} CH_2{-}CH{=}CH{-}CH_2Cl \text{ (H on first } CH_2\text{)}$$

1-Chloro-2-butene

In the first step of the electrophilic addition, the proton becomes attached to one of the terminal carbons since this gives the most stable intermediate carbocation, an allylic cation. The allylic cation, like those formed via heterolysis, combines with the nucleophile at either of two positions to give either of two products. (This is discussed in detail in Sec. 11.22.)

11.12 As always with S_N1 reactions, let us look at the structure of the intermediate carbocation. To the electron-deficient carbon is attached an oxygen atom. Oxygen is electronegative and, like halogen, should exert an electron-withdrawing inductive effect—an effect that, as we have seen (Sec. 6.13), tends to destabilize a carbocation. Yet here, if the parallel between rate of formation and stability holds, we have evidence of powerful *stabilization*.

The parallel *does* hold: measurements have shown the methoxymethyl cation to be 76 kcal/mol more stable than the methyl cation—more stable, even, than *tert*-butyl. Can oxygen, then, release electrons? The answer is *yes, by its resonance effect*.

Although electronegative, the oxygen of the $CH_3O{-}$ group is basic; it has unshared pairs of electrons that it tends to share, thus acquiring a positive charge. Just as water accepts a proton to form the hydronium (oxonium) ion,

$$H_2\ddot{O} + H^+ \longrightarrow H_3O^+$$

so alcohols and ethers, as we have found, accept protons to form substituted oxonium ions.

$$R\ddot{O}H + H^+ \longrightarrow ROH_2^+ \qquad R\ddot{O}R + H^+ \longrightarrow R\overset{\overset{H}{|}}{O}R^+$$

The effects of properly placed oxygen on the stability of carbocations—here, and in other kinds of reactions—can be accounted for by assuming that oxygen can share more than one pair of electrons with electron-deficient carbon and can accommodate a positive charge. Fundamentally, it is the *basicity* of oxygen that is involved.

With that background, let us return to the structure of the methoxymethyl cation. We have written its structure as I, but we could just as well have written it as II. Again the conditions for resonance have been met: two structures that differ only in the arrangement of electrons.

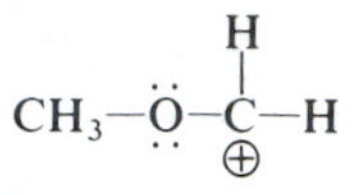

$$CH_3-\ddot{\underset{..}{O}}-\overset{\oplus}{C}H_2 \qquad\qquad CH_3-\overset{\oplus}{\ddot{O}}=CH_2$$

I 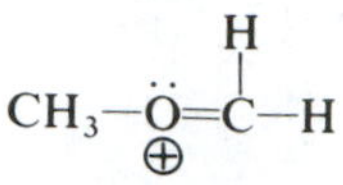 II

Especially stable: every atom has octet

Of the two structures, we would expect II to be by far the more stable, since in it *every atom* (except hydrogen, of course) *has a complete octet of electrons*. By sharing *two* pairs of electrons with carbon—and thus acquiring the positive charge itself—oxygen has provided carbon with the electrons needed to complete its octet. Structure II is so much more stable than I that, by itself, it must pretty well represent the structure of the cation. The cation is thus hardly a *carbo*cation at all, but an *oxonium ion*. This oxonium ion formed from methoxymethyl chloride is enormously more stable than the carbocation that would be formed from a simple alkyl chloride, and it is formed at a vastly faster rate.

(Compare, for example, the structures of H_3O^+ and $CH_3{}^+$. Here it is not a matter of which atom, oxygen or carbon, can better accommodate a positive charge; it is a matter of complete *vs.* incomplete octets.)

This is a conjugated system: the double bond is formed by overlap of the empty *p* orbital of carbon with a filled *p* orbital of oxygen. The electron-deficient carbon of a carbocation can be conjugated with an unshared pair on atoms other than oxygen: nitrogen, sulfur, even halogen. There, too, we shall find, the resulting stabilization

$$-\overset{\oplus}{\underset{|}{C}}-\ddot{A} \qquad\qquad -\underset{|}{C}=\overset{\oplus}{A}$$

Conjugation with an unshared pair

of the carbocation can have spectacular effects on the rate, not just of heterolysis, but of reactions of many other types.

In the case of the oxonium ion, then, electron release through the resonance effect is clearly much more powerful than electron withdrawal through the inductive effect, and controls reactivity. In Sec. 15.19 we shall find that in some cases the two factors can be more closely balanced.

11.13 In the S_N1 solvolysis, methyl speeds up reaction by releasing electrons and thus stabilizing the carbocation being formed (Problem 11.10, above). In the presence of the strongly nucleophilic ethoxide ion, reaction shifts to S_N2—witness the second-order kinetics—and methyl slows reaction down through steric hindrance.

11.14 By loss of triflate ion, I gives a vinylic cation which:

(i) combines with the solvent to give the ether II;
(ii) loses a proton from C–1 to give the alkyne III;
(iii) loses a proton from C–3 to give the allene IV; and
(iv) rearranges by a hydride shift to a more stable, allylic cation that then gives the ether V.

$$CH_3-\underset{\underset{CH_3}{|}}{CH}-\underset{\underset{OTf}{|}}{C}=CH_2 \text{ (I)} \xrightarrow{-OTf^-} CH_3-\overset{CH_3}{\overset{|}{CH}}-\underset{\oplus}{C}=CH_2 \text{ (A vinylic cation)}$$

$$\xrightarrow{CF_3CH_2OH} CH_3-\overset{CH_3}{\overset{|}{CH}}-\underset{\underset{OCH_2CF_3}{|}}{C}=CH_2 \text{ (II)}$$

$$\xrightarrow{-H^+} CH_3-\overset{CH_3}{\overset{|}{CH}}-C\equiv CH \text{ (III)}$$

$$\xrightarrow{-H^+} CH_3-\overset{CH_3}{\overset{|}{C}}=C=CH_2 \text{ (IV)}$$

$$\xrightarrow[\text{shift}]{\text{hydride}} CH_3-\overset{CH_3}{\overset{|}{\underset{\oplus}{C}}}-\underset{\underset{H}{|}}{C}=CH_2 \text{ (An allylic cation)} \xrightarrow{CF_3CH_2OH} CH_3-\overset{CH_3}{\overset{|}{\underset{\underset{OCH_2CF_3}{|}}{C}}}-CH=CH_2 \text{ (V)}$$

11.15 (a) Assuming a value of 28–30 kcal for each double bond, we calculate a value of 56 to 60 kcal.

$$CH_2=C=CH_2 + 2H_2 \longrightarrow CH_3CH_2CH_3 \qquad \Delta H = \underset{(2\times 28)}{-56} \text{ to } \underset{(2\times 30)}{-60} \text{ kcal}$$

(b) The actual value (71 kcal) is considerably higher than what we estimated, indicating that cumulated double bonds are unstable relative to conjugated or isolated double bonds.

11.16

$$CH_3-\overset{H}{\overset{|}{\underset{\underset{H}{|}}{C}}}-\underset{\underset{Br}{|}}{CH}-\overset{H}{\overset{|}{\underset{\underset{H}{|}}{C}}}-CH=CH_2 \xrightarrow{-HBr} \underset{\text{1,3-Hexadiene}}{CH_3CH_2-CH=CH-CH=CH_2} + \underset{\text{1,4-Hexadiene}}{CH_3CH=CH-CH_2-CH=CH_2}$$

Conjugated: more stable
Chief product

11.17 (a) Initial attachment of the proton must be to C–1, since only in this way can the observed products be formed, by 1,2- and 1,4-addition.

Attachment of the proton to either end of the conjugated system gives an allylic cation. But attachment to the C–1 end is favored, since this allylic cation is also a 3° cation. That is to say, part of the positive charge develops on the carbon (C–2) to which the methyl group is attached; electron release by methyl helps to stabilize the incipient carbocation.

$$CH_2=\overset{CH_3}{\overset{|}{C}}-CH=CH_2 \xrightarrow{H^+} \left[\underset{\underset{H}{|}}{CH_2}-\overset{CH_3}{\overset{\downarrow}{\underset{\oplus}{C}}}-CH=CH_2 \quad \underset{\underset{H}{|}}{CH_2}-\overset{CH_3}{\overset{\downarrow}{C}}=CH-\underset{\oplus}{CH_2} \right] \textit{ equivalent to } \underset{\underset{H}{|}}{CH_2}-\overset{CH_3}{\overset{\downarrow}{C}}\text{=-=}CH\text{=-=}CH_2 \; (\oplus)$$

(b) Similar to (a), with initial attachment of Br^+ to the C–1 end of the conjugated system.

11.18 Evidently—and not surprisingly—conjugated dienes can react with free radicals as they do with electrophilic reagents: by 1,4-addition as well as 1,2-addition. To understand this, we have only to examine the structure of the free radical that is the intermediate in this reaction.

As we have seen (Sec. 9.23), the peroxide decomposes (step 1) to yield a free radical, which abstracts bromine from $BrCCl_3$ (step 2) to generate a $\cdot CCl_3$ radical.

(1) $$\text{Peroxide} \longrightarrow \text{Rad}\cdot$$

(2) $$\text{Rad}\cdot + BrCCl_3 \longrightarrow \text{Rad—Br} + \cdot CCl_3$$

The $\cdot CCl_3$ radical thus formed adds to the butadiene (step 3). The products obtained show that this addition is to one of the *ends* of the conjugated system. Why is this? In free-radical addition to conjugated dienes, it seems clear that orientation is controlled by the stability of the radical being formed (Sec. 9.22). Thus $\cdot CCl_3$ adds where it does because in this way a resonance-stabilized allylic free radical is formed.

(3) $$\overset{\cdot CCl_3}{\underset{1\quad\ \ 2\quad\ \ 3\quad\ \ 4}{CH_2{=}CH{-}CH{=}CH_2}} \longrightarrow \left[\begin{array}{l} Cl_3C{-}CH_2{-}\dot{C}H{-}CH{=}CH_2 \\ Cl_3C{-}CH_2{-}CH{=}CH{-}\dot{C}H_2 \end{array}\right]$$

Addition to end of conjugated system

equivalent to

$$Cl_3C{-}CH_2{-}\underbrace{CH\text{---}CH\text{---}CH_2}_{\cdot}$$

Allylic free radical

The allylic free radical then abstracts bromine from a molecule of $BrCCl_3$ (step 4) to complete the addition, and in doing so forms a new $\cdot CCl_3$ radical which can carry on the chain. In step (4) bromine can become attached to either C–2 or C–4 to yield either the 1,2- or 1,4-product.

(4) $$Cl_3C{-}CH_2{-}\underbrace{CH\text{---}CH\text{---}CH_2}_{\cdot} \xrightarrow{BrCCl_3} Cl_3C{-}CH_2{-}\underset{Br}{\underset{|}{CH}}{-}CH{=}CH_2$$

Allylic free radical

1,2-Addition product

$$+\ Cl_3C{-}CH_2{-}CH{=}CH{-}CH_2{-}Br$$

1,4-Addition product

11.19 (a) $CH_3CH{=}CH{-}CH{=}CH_2 \longrightarrow CH_3CHO + OHC{-}CHO + HCHO$

(b) $CH_2{=}CH{-}CH_2{-}CH{=}CH_2 \longrightarrow HCHO + OHC{-}CH_2{-}CHO + HCHO$

(c) $CH_2{=}\underset{CH_3}{\underset{|}{C}}{-}CH{=}CH_2 \longrightarrow HCHO + O{=}\underset{CH_3}{\underset{|}{C}}{-}CHO + HCHO$

11.20 (a) $n\,CH_2{=}\underset{CH=CH_2}{\underset{|}{CH}} \xrightarrow{1,2} ({-}CH_2{-}\underset{CH=CH_2}{\underset{|}{CH}}{-})_n \xrightarrow{O_3} ({-}CH_2{-}\underset{CHO}{\underset{|}{CH}}{-})_n + n\,HCHO$

(b) $n\text{-}CH_2{=}CH{-}CH{=}CH_2 \xrightarrow{1,4} (-CH_2{-}CH{=}CH{-}CH_2{-}CH_2{-}CH{=}CH{-}CH_2-)_{n/2}$

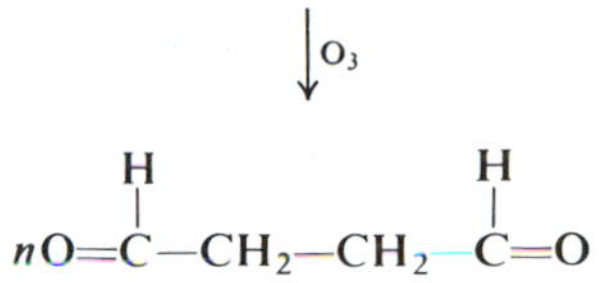

$$\downarrow O_3$$

$$nO{=}\overset{\displaystyle H}{\overset{|}{C}}{-}CH_2{-}CH_2{-}\overset{\displaystyle H}{\overset{|}{C}}{=}O$$

11.21 In natural rubber, the isoprene units are mostly combined as though by head-to-tail 1,4-addition.

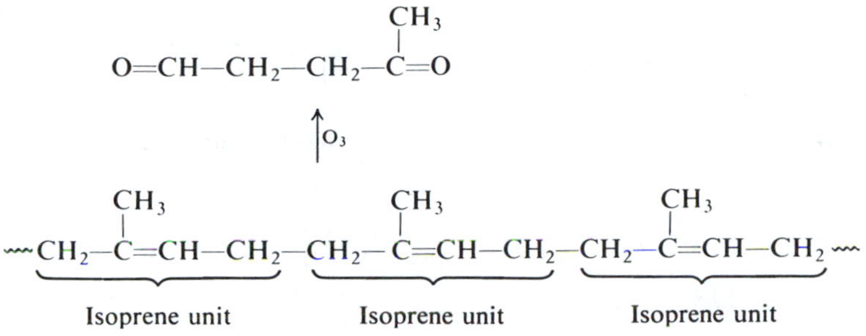

$$O{=}CH{-}CH_2{-}CH_2{-}\overset{\displaystyle CH_3}{\overset{|}{C}}{=}O$$

$$\uparrow O_3$$

$$\sim\underbrace{CH_2{-}\overset{\displaystyle CH_3}{\overset{|}{C}}{=}CH{-}CH_2}_{\text{Isoprene unit}}{-}\underbrace{CH_2{-}\overset{\displaystyle CH_3}{\overset{|}{C}}{=}CH{-}CH_2}_{\text{Isoprene unit}}{-}\underbrace{CH_2{-}\overset{\displaystyle CH_3}{\overset{|}{C}}{=}CH{-}CH_2}_{\text{Isoprene unit}}\sim$$

1.

$$CH_3{-}\underset{\underset{\displaystyle H}{|}}{\overset{\overset{\displaystyle H_3C}{|}}{C}}{-}\underset{\underset{\displaystyle H}{|}}{\overset{\overset{\displaystyle H}{|}}{C}}{-}\underset{\underset{\displaystyle H}{|}}{\overset{\overset{\displaystyle H}{|}}{C}}{-}\overset{\overset{\displaystyle H}{|}}{C}{=}\overset{\overset{\displaystyle H}{|}}{C}{-}CH_3$$

5 3 4 1 6 6 2 *order of reactivity*

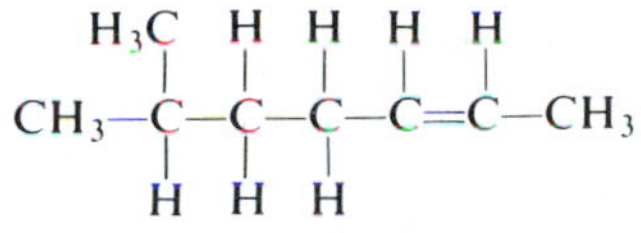

H that is both 2° and allylic is more reactive than H that is 1° and allylic.

2. (a)–(e)

C—C—C=C—C=C
1,3-Hexadiene
Conjugated
Geom. isom.

C—C=C—C—C=C
1,4-Hexadiene
Geom. isom.

Gives: C—C—CHO + OHC—CHO + HCHO   C—CHO + OHC—C—CHO + HCHO

C=C—C—C—C=C
1,5-Hexadiene

C—C=C—C=C—C
2,4-Hexadiene
Conjugated
Geom. isom.

Gives: HCHO + OHC—C—C—CHO + HCHO   C—CHO + OHC—CHO + OHC—C

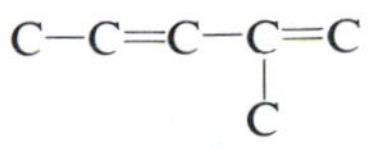

$$C{-}C{=}C{-}\underset{\underset{\displaystyle C}{|}}{C}{=}C$$
2-Methyl-1,3-pentadiene
Conjugated
Geom. isom.

$$C{=}C{-}C{-}\underset{\underset{\displaystyle C}{|}}{C}{=}C$$
2-Methyl-1,4-pentadiene

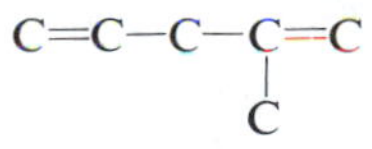

Gives: $C{-}CHO + OHC{-}\underset{\underset{\displaystyle C}{|}}{C}{=}O + HCHO$ $HCHO + OHC{-}C{-}\underset{\underset{\displaystyle C}{|}}{C}{=}O + HCHO$

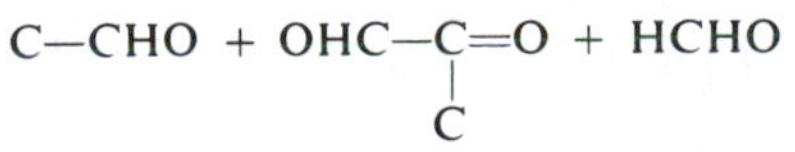

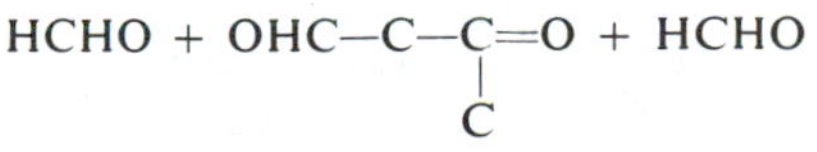

C=C—C=C—C (with C on C-4)
4-Methyl-1,3-pentadiene
Conjugated

Gives: HCHO + OHC—CHO + O=C—C (with C on the carbonyl carbon)

C—C=C—C=C (with C on C-3)
3-Methyl-1,3-pentadiene
Conjugated
Geom. isom.

Gives: C—CHO + O=C—CHO (with C on the carbonyl carbon) + HCHO

C=C—C—C=C (with C on C-3)
3-Methyl-1,4-pentadiene

Gives: HCHO + OHC—C—CHO (with C on the central carbon) + HCHO

C=C—C=C (with C on C-2 and C-3)
2,3-Dimethyl-1,3-butadiene
Conjugated

Gives: HCHO + O=C—C=O (with C on each carbonyl carbon) + HCHO

C=C—C=C (with C—C on C-2)
2-Ethyl-1,3-butadiene
Conjugated

Gives: HCHO + OHC—C=O (with C—C on the carbonyl carbon) + HCHO

(f) 2-Methyl-1,3-pentadiene and 3-methyl-1,3-pentadiene cannot be distinguished by their ozonolysis products.

3. We expect the products of both 1,2- and 1,4-addition. (We indicate the existence of geometric isomers, but not other diastereomers or enantiomers.)

(a) $CH_2CHCH{=}CH_2$ (H on C-1 and C-2) + $CH_2CH{=}CHCH_2$ (H on C-1 and C-4)

1-Butene 2-Butene (*cis* and *trans*)

(b) $CH_2CHCHCH_2$ (H on each carbon)

n-Butane

(c) $CH_2CHCH{=}CH_2$ (Br on C-1 and C-2) + $CH_2CH{=}CHCH_2$ (Br on C-1 and C-4)

3,4-Dibromo-1-butene 1,4-Dibromo-2-butene (*cis* and *trans*)

(d) $BrCH_2CH—CHCH_2Br$ (Br on C-2 and C-3)

1,2,3,4-Tetrabromobutane

(e) $CH_2(H)CH(Cl)CH{=}CH_2$ + $CH_2(H)CH{=}CHCH_2Cl$
3-Chloro-1-butene; 1-Chloro-2-butene (*Z* and *E*)

(f) $CH_3CH(Cl){-}CH(Cl)CH_3$ + $CH_3CH(Cl)CH_2CH_2Cl$
2,3-Dichlorobutane; 1,3-Dichlorobutane

(g) 2HCHO + OHC—CHO

(h) $2CO_2$ + HOOC—COOH

Part (f) is worth a closer look. Each of the products of (e) reacts with another mole of HCl via the more stable carbocation: 2° rather than 1°,

$$CH_3CH(Cl)CH{=}CH_2 \xrightarrow{HCl} CH_3CH(Cl){-}\overset{\oplus}{C}HCH_2(H) \xrightarrow{Cl^-} CH_3CH(Cl){-}CH(Cl){-}CH_2(H)$$

rather than

$$CH_3CH(Cl){-}CH(H)CH_2^{\oplus}$$

and, of two possible 2° carbocations, the one with the positive charge farther from the electron-withdrawing —Cl.

$$CH_3CH{=}CHCH_2Cl \xrightarrow{HCl} CH_3\overset{\oplus}{C}HCH(H)CH_2Cl \xrightarrow{Cl^-} CH_3CH(Cl){-}CH(H)CH_2Cl$$

rather than

$$CH_3CH(H)\overset{\oplus}{C}HCH_2Cl$$

4. We treat these two double bonds as though they were in different molecules; either one or both react, depending upon the quantity of reagent.

(a) $CH_2{=}CHCH_2CH(H)CH_2(H)$ — 1-Pentene

(b) $CH_2(H)CH(H)CH_2CH(H)CH_2(H)$ — *n*-Pentane

(c) $CH_2{=}CHCH_2CH(Br){-}CH_2Br$ — 4,5-Dibromo-1-pentene

(d) $CH_2(Br)CH(Br)CH_2CH(Br){-}CH_2(Br)$ — 1,2,4,5-Tetrabromopentane

(e) $CH_2{=}CHCH_2CH(Cl){-}CH_2(H)$ — 4-Chloro-1-pentene

(f) $CH_2(H)CH(Cl)CH_2CH(Cl){-}CH_2(H)$ — 2,4-Dichlorobutane

(g) 2HCHO + $OHCCH_2CHO$

(h) $2CO_2$ + $HOOCCH_2COOH$

5. In each case we expect the major product to be the alkene that is more stable: the one that is more highly substituted, or is conjugated.

(a) $CH_3CH_2CH_2CH_2Cl \longrightarrow CH_3CH_2CH{=}CH_2$
1-Butene

$CH_3CH_2CH(Cl)CH_3 \longrightarrow CH_3CH{=}CHCH_3$ + $CH_3CH_2CH{=}CH_2$
2-Butene (*cis* and *trans*) *Chief product*; 1-Butene

(b) $$\mathrm{CH_3CH_2CH_2CH_2Cl \longrightarrow \underset{\text{1-Butene}}{CH_3CH_2CH{=}CH_2}}$$

$$\mathrm{ClCH_2CHCH{=}CH_2 \longrightarrow \underset{\text{1,3-Butadiene}}{CH_2{=}CHCH{=}CH_2}}$$

(c) $$\mathrm{CH_3CH_2\overset{CH_3}{\underset{Br}{C}}CH_3 \longrightarrow \underset{\substack{\text{2-Methyl-2-butene}\\ \textit{Chief product}}}{CH_3CH{=}\overset{CH_3}{C}CH_3} + \underset{\text{2-Methyl-1-butene}}{CH_3CH_2\overset{CH_3}{C}{=}CH_2}}$$

$$\mathrm{CH_3\underset{Br}{C}H\overset{CH_3}{C}HCH_3 \longrightarrow \underset{\substack{\text{2-Methyl-2-butene}\\ \textit{Chief product}}}{CH_3CH{=}\overset{CH_3}{C}{-}CH_3} + \underset{\text{3-Methyl-1-butene}}{CH_2{=}CH\overset{CH_3}{C}HCH_3}}$$

(d) $$\mathrm{CH_3CH_2\overset{CH_3}{C}HCH_2Br \longrightarrow \underset{\text{2-Methyl-1-butene}}{CH_3CH_2\overset{CH_3}{C}{=}CH_2}}$$

$$\mathrm{BrCH_2CH_2\overset{CH_3}{C}HCH_3 \longrightarrow \underset{\text{3-Methyl-1-butene}}{CH_2{=}CH\overset{CH_3}{C}HCH_3}}$$

(e) $$\mathrm{CH_3\overset{CH_3}{C}H{-}\overset{CH_3}{C}HCH_2Cl \longrightarrow \underset{\text{2,3-Dimethyl-1-butene}}{CH_3\overset{CH_3}{C}H{-}\overset{CH_3}{C}{=}CH_2}}$$

$$\mathrm{CH_3\overset{CH_3}{C}H{-}\overset{CH_3}{\underset{Cl}{C}}CH_3 \longrightarrow \underset{\substack{\text{2,3-Dimethyl-2-butene}\\ \textit{Chief product}}}{CH_3\overset{CH_3}{C}{=}\underset{CH_3}{C}CH_3} + \underset{\text{2,3-Dimethyl-1-butene}}{CH_3CH{-}\underset{CH_3}{\overset{CH_3}{C}}{=}CH_2}}$$

(f) $$\mathrm{ClCH_2CH_2CH{=}CH_2 \longrightarrow \underset{\text{1,3-Butadiene}}{CH_2{=}CHCH{=}CH_2}}$$

$$\mathrm{ClCH_2CH_2CH_2CH{=}CH_2 \longrightarrow \underset{\text{1,4-Pentadiene}}{CH_2{=}CHCH_2CH{=}CH_2}}$$

6. We compare the structures of the alkenes being formed. The more reactive alkyl halide is the one that gives the more stable alkene or alkenes: more highly substituted, or conjugated.

(a) 2-chlorobutane
(b) 4-chloro-1-butene
(c) 2-bromo-2-methylbutane
(d) 1-bromo-2-methylbutane
(e) 2-chloro-2,3-dimethylbutane
(f) 4-chloro-1-butene

7. In each case the initial attack is by H^+ to give the most stable intermediate carbocation (Markovnikov's rule). In the case of the conjugated dienes, an allylic cation is preferred over a non-allylic, a 3° allylic over a 2° allylic, and a 2° allylic over a 1° allylic. (Look again at the answer to Problem 11.17, above.)

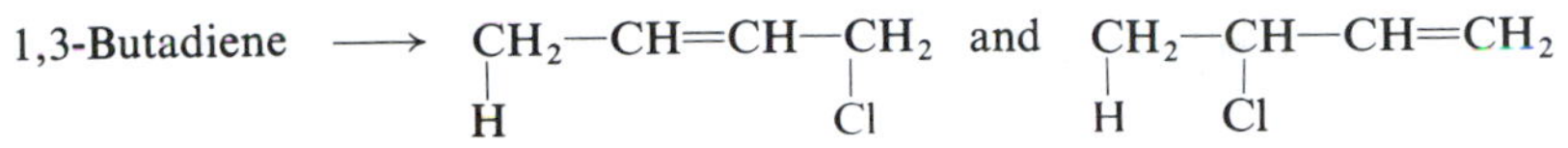

$$\text{1,3-Butadiene} \longrightarrow \underset{H}{CH_2}-CH=CH-\underset{Cl}{CH_2} \quad \text{and} \quad \underset{H}{CH_2}-\underset{Cl}{CH}-CH=CH_2$$

The other products are:

(a) $\underset{H}{CH_2}-\underset{Cl}{CH}-CH_2CH_3$

(b) $\underset{H}{CH_2}-\underset{Cl}{CH}-CH_2-CH=CH_2$

(c) $\underset{H}{CH_2}-\underset{Cl}{\overset{CH_3}{C}}-CH=CH_2$ and $\underset{H}{CH_2}-\overset{CH_3}{C}=CH-\underset{Cl}{CH_2}$

(d) $\underset{H}{CH_2}CH=CH-\underset{Cl}{CH}-CH_3$

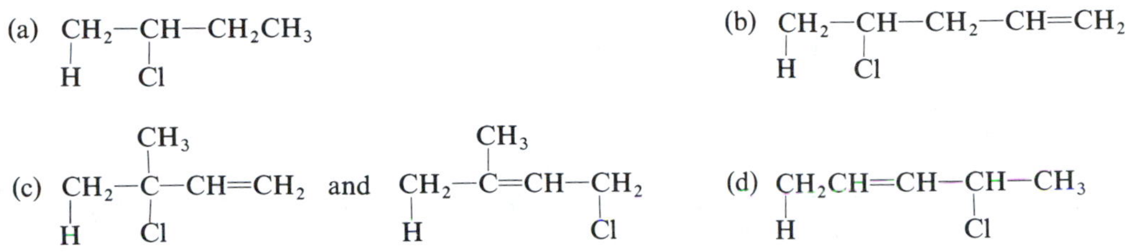

8. In each case the initial attack is by $\cdot CCl_3$ to give the most stable intermediate free radical. We follow an approach analogous to that in the preceding problem.

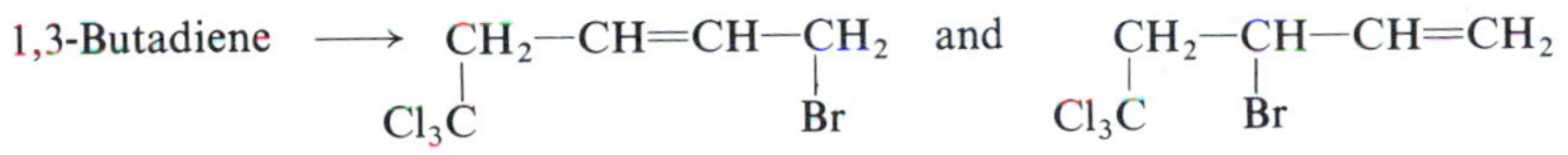

$$\text{1,3-Butadiene} \longrightarrow \underset{Cl_3C}{CH_2}-CH=CH-\underset{Br}{CH_2} \quad \text{and} \quad \underset{Cl_3C}{CH_2}-\underset{Br}{CH}-CH=CH_2$$

The other products are:

(a) $\underset{Cl_3C}{CH_2}-\underset{Br}{CH}-CH_2-CH_3$

(b) $\underset{Cl_3C}{CH_2}-\underset{Br}{CH}-CH_2-CH=CH_2$

(c) $\underset{Cl_3C}{CH_2}-\underset{Br}{\overset{CH_3}{C}}-CH=CH_2$ and $\underset{Cl_3C}{CH_2}-\overset{CH_3}{C}=CH-\underset{Br}{CH}$

(d) $\underset{Cl_3C}{CH_2}-CH=CH-\underset{Br}{CH}-CH_3$

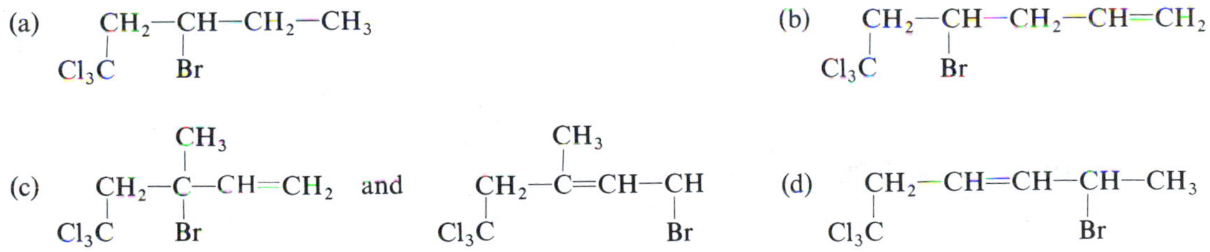

9. We expect reactivity to depend upon the stability of the carbocation being formed in (a)–(d), and on leaving-group ability in (e). We draw out and examine the structure of each cation; for allylic cations, we draw the contributing structures to remind us just where the positive charge is strongest. (Study again the answer to Problem 11.10, above.)

(a) 3-chloropropene > *n*-propyl chloride > 1-chloropropene
(b) 3-bromo-1-butene > 2-bromobutane > 2-bromo-1-butene
(c) 4-bromo-2-methyl-2-pentene > 4-bromo-2-pentene, 4-bromo-3-methyl-2-pentene
(d) 4-methyl-2-penten-4-yl tosylate > 2-penten-4-yl tosylate > 2-buten-1-yl tosylate
(e) triflate > tosylate > bromide > chloride

10. (a) The CH_2 planes are perpendicular to each other.

(b) Yes; even though no chiral carbon is present, the molecules are chiral.

R(H)C=C=C(H)R   R(H)C=C=C(H)R

(c) Central carbon *sp*, linear; terminal carbons sp^2, trigonal; planes of the two double bonds perpendicular. This leads to the same shape as in (a) and (b).

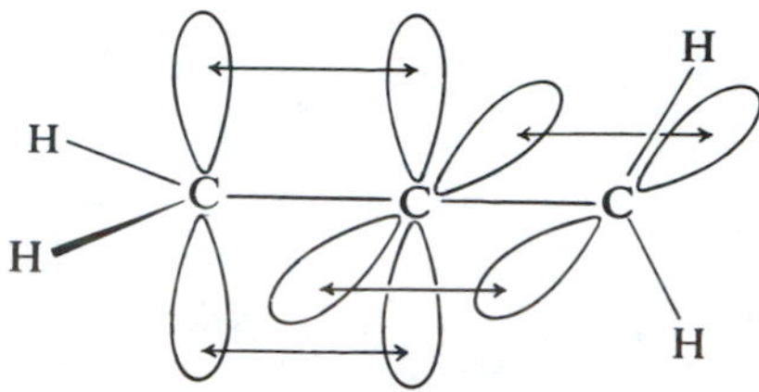

Allene

(See Eliel, E. L. *Stereochemistry of Carbon Compounds*; McGraw-Hill: New York, 1962; pp 307–311.)

11. (a) The allylic alcohol gives a hybrid allylic cation, which yields two products:

$$CH_3CHOHCH{=}CH_2 \xrightarrow[-H_2O]{H^+} \underbrace{CH_3CH{\overset{\dots}{-}}CH{\overset{\dots}{-}}CH_2}_{\oplus} \longrightarrow \begin{cases} CH_3CHBrCH{=}CH_2 \\ CH_3CH{=}CHCH_2Br \end{cases}$$

(b) This alcohol gives the same allylic cation, and hence the same products, as in (a):

$$CH_3CH{=}CHCH_2OH \xrightarrow[-H_2O]{H^+} CH_3{-}\underbrace{CH{\overset{\dots}{-}}CH{\overset{\dots}{-}}CH_2}_{\oplus} \longrightarrow \text{same products as in (a)}$$

12.

$$(F_3C)(Ph)C{=}CF_2 + OEt^- \longrightarrow F_3C{-}\overset{\ominus}{C}(Ph){-}C(OEt)F{-}F \longrightarrow (F_3C)(Ph)C{=}C(OEt)(F) + F^-$$

$$(ClF_2C)(Ph)C{=}CF_2 + OEt^- \longrightarrow Cl{-}F_2C{-}\overset{\ominus}{C}(Ph){-}C(OEt)F{-}F \longrightarrow F_2C{=}C(Ph){-}C(OEt)(F){-}F + Cl^-$$

Formation of the carbanion is the rate-determining step, nearly identical for the two similar alkenes. Then the weakest C—X bond is broken, but the rate of this step does not affect the overall rate; that is, there is no "element effect" (see Sec. 8.19).

The overall reaction is thus a kind of nucleophilic substitution, but by what we might call an *addition–elimination* mechanism.

(Koch, H. F.; Kielbanian, A. J., Jr. "Nucleophilic Reactions of Fluoroolefins. Evidence for a Carbanion Intermediate in Vinyl and Allyl Displacement Reactions"; *J. Am. Chem. Soc.* **1970**, *92*, 729.)

13.

Diene $\xrightarrow[\text{to C-3}]{H^+}$ 3° cation $\xrightarrow[\text{C-2 to C-6}]{\text{cyclize}}$ 3° cation $\xrightarrow[\text{from C-8}]{-H^+}$ Alkene

This reaction is simply an intramolecular version of the acid-catalyzed dimerization of alkenes (Sec. 9.15). Since both double bonds—both "alkenes"—are part of the same molecule, reaction leads to *cyclization*, that is, formation of a ring (Sec. 13.4).

14. (a) $KMnO_4$, or Br_2/CCl_4: diene positive.
(b) $AgNO_3$: allyl bromide positive.
(c) $AgNO_3$: 1-chloro-2-butene (allylic) positive; 2-chloro-2-butene (vinylic) negative.

15.

$CH_3{-}CH{=}CH{-}CH(OH){-}CH{=}CH_2$
1,4-Hexadien-3-ol
Starting material

$\Big\uparrow\!\!\downarrow$ $-H^+$ / $+H^+$

$CH_3{-}CH{=}CH{-}CH(OH_2^+){-}CH{=}CH_2$ $\underset{+H_2O}{\overset{-H_2O}{\rightleftharpoons}}$ $CH_3{-}\underbrace{CH{\cdots}CH{\cdots}CH{\cdots}CH{\cdots}CH_2}_{\oplus}$ $\underset{+H_2O}{\overset{-H_2O}{\rightleftharpoons}}$ $CH_3{-}CH{=}CH{-}CH{=}CH{-}CH_2(OH_2^+)$

$CH_3{-}CH{=}CH{-}CH{=}CH{-}CH_2OH$
2,4-Hexadien-1-ol
($-H^+$ / $+H^+$ to the oxonium ion above)

$-H_2O$ / $+H_2O$ (from the cation)

$CH_3{-}CH(OH_2^+){-}CH{=}CH{-}CH{=}CH_2$ $\underset{+H^+}{\overset{-H^+}{\rightleftharpoons}}$ $CH_3{-}CH(OH){-}CH{=}CH{-}CH{=}CH_2$
3,5-Hexadien-2-ol
Product

In the presence of acid, the starting alcohol undergoes reversible heterolysis to form a hybrid, doubly allylic cation. Upon recombination with water, this cation can form any of three isomeric alcohols. The system is in equilibrium, and the equilibrium favors the most stable of the alcohols: 3,5-hexadien-2-ol, in which the two double bonds are conjugated. (We could not have predicted that this alcohol would be more stable than 2,4-hexadien-1-ol, but we are not surprised that the product is one of two conjugated alcohols.)

16. (a) Yes, a doubly allylic cation.

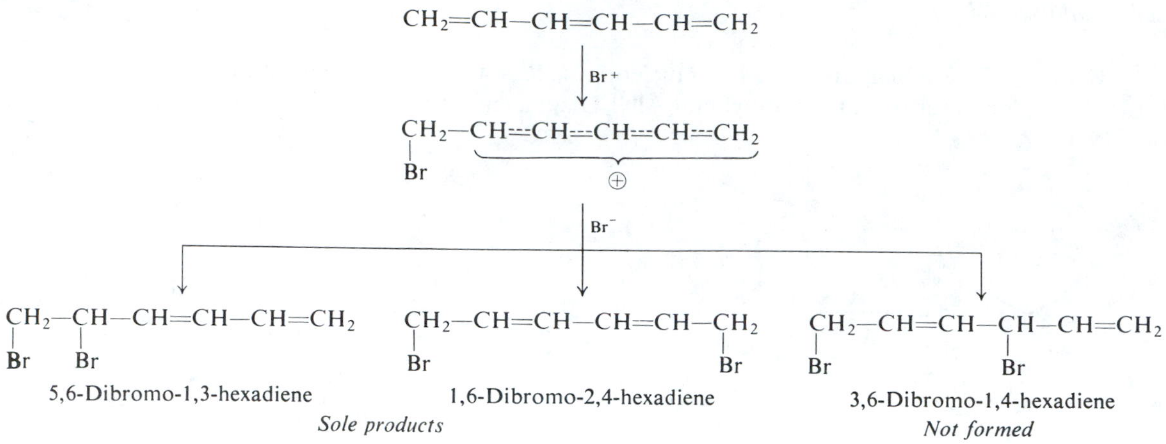

$$CH_2{=}CH{-}CH{=}CH{-}CH{=}CH_2 \xrightarrow{Br^+} \underset{\displaystyle Br}{CH_2}{-}\underbrace{CH{\cdots}CH{\cdots}CH{\cdots}CH{\cdots}CH_2}_{\oplus} \xrightarrow{Br^-}$$

$$\underset{Br}{CH_2}{-}\underset{Br}{CH}{-}CH{=}CH{-}CH{=}CH_2 \qquad \underset{Br}{CH_2}{-}CH{=}CH{-}CH{=}CH{-}\underset{Br}{CH_2} \qquad \underset{Br}{CH_2}{-}CH{=}CH{-}\underset{Br}{CH}{-}CH{=}CH_2$$

5,6-Dibromo-1,3-hexadiene; 1,6-Dibromo-2,4-hexadiene — *Sole products*

3,6-Dibromo-1,4-hexadiene — *Not formed*

(b) 3,6-Dibromo-1,4-hexadiene.

(c) Equilibrium control, which favors the more stable products, the two conjugated dienes.

17.

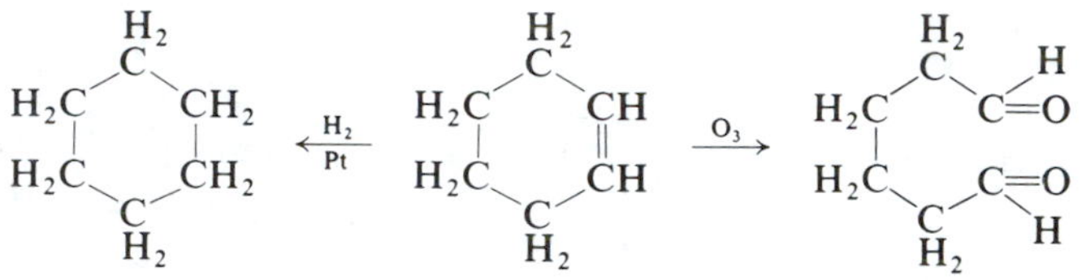

Cyclohexane $\xleftarrow{H_2/Pt}$ cyclohexene $\xrightarrow{O_3}$ OHC$-CH_2CH_2CH_2CH_2-$CHO

18. We can attack the problem as follows:

$$\text{mmoles } H_2 \text{ taken up} = \frac{8.40 \text{ mL}}{22.4 \text{ mL/mmole}}$$

$$\text{mmoles compound} \sim \frac{10.02 \text{ mg}}{80 \text{ mg/mmole}} \quad \text{(uncertainty in m.w.)}$$

$$\frac{\text{mmoles } H_2 \text{ taken up}}{\text{mmole compound}} = \frac{8.40/22.4}{10.02/80} = 3$$

The hydrocarbon therefore contains three double bonds, or one double bond and one triple bond. The ozonolysis results show only three carbons, but a m.w. in the range of 80–85 indicates six carbon atoms ($6 \times 12 = 72$). Evidently two moles of each ozonolysis product are obtained per mole of hydrocarbon. Since HCHO can only come from a terminal unsaturation, and the di-aldehyde from an inner segment of a chain, there is only one way of putting the pieces together: the hydrocarbon is 1,3,5-hexatriene.

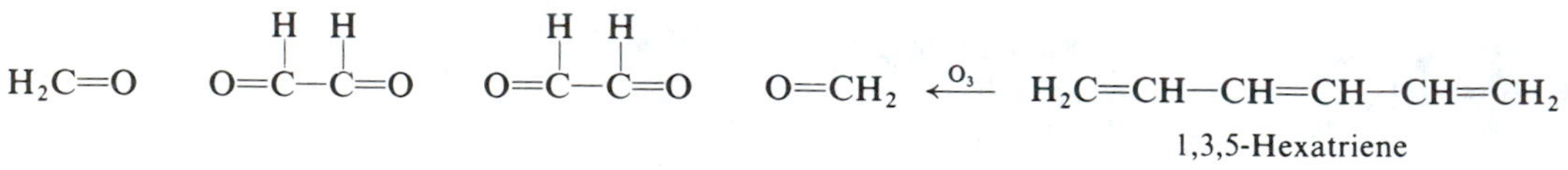

$$H_2C{=}O \quad O{=}\underset{}{\overset{H}{C}}{-}\overset{H}{C}{=}O \quad O{=}\overset{H}{C}{-}\overset{H}{C}{=}O \quad O{=}CH_2 \xleftarrow{O_3} H_2C{=}CH{-}CH{=}CH{-}CH{=}CH_2$$

1,3,5-Hexatriene

On checking, we find that the proposed triene structure accommodates all the analytical data. (Any alkyne or cyclic structure will be found unsuitable.)

19. Myrcene contains ten carbons, but the ozonolysis products show only nine; the missing carbon must be found in a second molecule of HCHO. This gives four molecules of ozonolysis products, which agrees with the three double bonds shown by hydrogenation. (Cleavage at three places gives four products.)

(a) The fragments can be put back together in three ways:

```
   C          H           H                               C
   |          |           |                               |
C—C=O    O=C—C—C—C—C=O    O=CH2  ←—   C—C=C—C—C—C—C=C
                     ||                                   ||
                     O                                    C

                   O=CH2

   C          H           H                        C
   |          |           |                        |
C—C=O    O=C—C—C—C—C=O    O=CH2  ←—   C—C=C—C—C—C—C=C
               ||                                  ||
               O                                   C

              O=CH2

   C                 H                       C
   |                 |                       |
C—C=O    O=C—C—C—C=O     O=CH2  ←—   C—C=C—C—C—C=C
            |                                |
           HC=O  O=CH2                       C=C
```

(b) Myrcene is made up of two head-to-tail isoprene units.

```
        CH3
        |                :
CH3—C=CH—CH2—┼—CH2—C—CH=CH2
                         :        ||
                                  CH2
                  Myrcene
```

20. (a) Starting with the carbon skeleton of myrcene (Problem 19, above), we find that the ten carbons of the oxidation fragments must be fitted together as follows:

```
   C                                                    C
   |                                                    |
C—C=O    O=C—C—C—C=O    O=C—C  ←—   C—C=C—C—C—C=C—C
            |         |       |                               |
           HO        C      HO                             C
                                                 Dihydromyrcene
```

(b) Dihydromyrcene is therefore the result of 1,4-addition of hydrogen to the conjugated system of myrcene.

```
    CH3                    CH2                           CH3
    |                      ||          H2                |
CH3C=CH—CH2—CH2—C—CH=CH2  ———————→  CH3—C=CH—CH2—CH2—C=CH—CH3
            Myrcene                 1,4-addn.                          |
                                                                      CH3
                                                          Dihydromyrcene
```

12

Alkynes

12.1 CO_2: O=C=O, linear; C uses two *sp* orbitals, and two *p* orbitals at right angles (compare allene, Problem 11, page 423). H_2O: sp^3, tetrahedral, with unshared pairs in two lobes (Sec. 1.12).

12.2 (a) $CH_3C{\equiv}CCH_3 \xrightarrow{Br_2} CH_3\underset{Br}{C}{=}\underset{Br}{C}CH_3 \xrightarrow{Br_2} CH_3\underset{Br}{\overset{Br}{C}}{-}\underset{Br}{\overset{Br}{C}}CH_3$

(b) Through electron withdrawal, —Br will deactivate the 2,3-dibromo-2-butene.

(c) This will favor the addition to 2-butyne.

(d) Use excess 2-butyne.

(e) Drip Br_2 into a solution of 2-butyne, to avoid a temporary excess of Br_2.

12.3 $\underset{H}{\overset{CH_3}{}}C{=}C\underset{H}{\overset{C_2H_5}{}}$ (*cis*-2-Pentene) $\xleftarrow[\text{Lindlar}]{H_2} CH_3C{\equiv}CC_2H_5 \xleftarrow{NaNH_2} \xleftarrow{KOH(alc)} CH_3\underset{Br}{C}H{-}\underset{Br}{C}HC_2H_5 \xleftarrow{Br_2}$ 2-Pentene (*cis–trans* mixture)

12.4

$$CH_3-C(CH_3)_2-C\equiv CH \xrightarrow{HCl} CH_3-C(CH_3)_2-\overset{\oplus}{C}=CH_2 \xrightarrow{Cl^-} CH_3-C(CH_3)_2-CCl=CH_2 \xrightarrow{HCl} CH_3-C(CH_3)_2-CCl_2-CH_3$$

Vinylic cation → (Cl⁻) ... 2,2-Dichloro-3,3-dimethylbutane

↓ methyl shift

$$CH_3-C(CH_3)\overset{\oplus}{=\!=}C(CH_3)\!=\!=CH_2 \xrightarrow{Cl^-} CH_3-CCl(CH_3)-C(CH_3)=CH_2 \xrightarrow{HCl} CH_3-CCl(CH_3)-CCl(CH_3)-CH_3$$

Allylic cation ... 2,3-Dichloro-2,3-dimethylbutane

↓ Cl^-

$$CH_3-C(CH_3)=C(CH_3)-CH_2Cl \xrightarrow{HCl} CH_3-CCl(CH_3)-CH(CH_3)-CH_2Cl$$

1,3-Dichloro-2,3-dimethylbutane

(In the formation of the last compound, 1,3-dichloro-2,3-dimethylbutane, the orientation of addition of HCl is controlled by the inductive effect of the —Cl already in the molecule.)

12.5 H goes to the terminal C; that is, there is Markovnikov orientation, via a 2° vinylic cation rather than a 1° vinylic cation.

$$\underset{\substack{\text{Acetone}\\ \textit{Keto}}}{CH_3-\underset{\underset{O}{\|}}{C}-CH_3} \longleftarrow \underset{\textit{Enol}}{CH_3-\underset{OH}{C}=CH_2} \xleftarrow{-H^+} CH_3-\underset{OH_2^+}{C}=CH_2 \xleftarrow{H_2O} CH_3\overset{\oplus}{C}=\underset{H}{C}-H \xleftarrow{H^+} \underset{\text{Propyne}}{CH_3C\equiv CH}$$

12.6 (a) It is the calcium salt of acetylene, $Ca^{2+}C\equiv C^{2-}$, and could be called *calcium acetylide*. (b) Its reaction with water is an acid–base reaction in which the weaker acid (acetylene) is displaced from its salt by a stronger acid (water).

12.7 (a) $n\text{-}PrC\equiv CH \longrightarrow n\text{-}PrCOOH + HOOCH$

(b) $EtC\equiv CCH_3 \longrightarrow EtCOOH + HOOCCH_3$

(c) $i\text{-}PrC\equiv CH \longrightarrow i\text{-}PrCOOH + HOOCH$

(d) $CH_3CH=CH-CH=CH_2 \longrightarrow CH_3CHO + OHC-CHO + HCHO$

(e) $CH_2=CH-CH_2-CH=CH_2 \longrightarrow HCHO + OHC-CH_2-CHO + HCHO$

1.

C—C—C—C—C≡CH
1-Hexyne
(*n*-Butylacetylene)
Forms salts
Gives: *n*-BuCOOH + HOOCH

C—C—C—C≡C—C
2-Hexyne
(*n*-Propylmethylacetylene)
Gives: *n*-PrCOOH + $HOOCCH_3$

C—C—C≡C—C—C
3-Hexyne
(Diethylacetylene)
Gives: 2EtCOOH

```
    C
    |
C—C—C—C≡CH
```
4-Methyl-1-pentyne
(Isobutylacetylene)
Forms salts
Gives: *iso*-BuCOOH + HOOCH

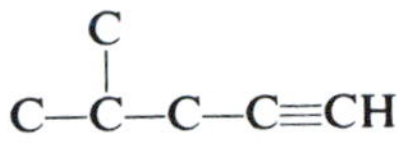

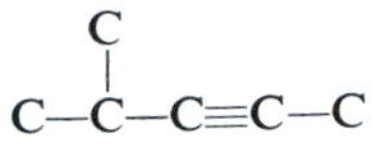

```
  C
  |
C—C—C≡C—C
```
4-Methyl-2-pentyne
(Isopropylmethylacetylene)
Gives: *i*-PrCOOH + $HOOCCH_3$

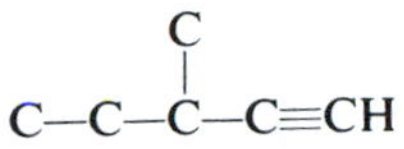

```
      C
      |
C—C—C—C≡CH
```
3-Methyl-1-pentyne
(*sec*-Butylacetylene)
Forms salts
Gives: *sec*-BuCOOH + HOOCH

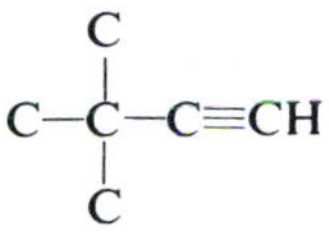

```
  C
  |
C—C—C≡CH
  |
  C
```
3,3-Dimethyl-1-butyne
(*tert*-Butylacetylene)
Forms salts
Gives: *t*-BuCOOH + HOOCH

2. (a) Remove 2HBr: KOH(alc), heat

(b) Br_2/CCl_4; KOH(alc), heat

(c) KOH(alc), heat; then as in (b)

(d) Acid, heat; then as in (b)

(e) Remove 2HCl: KOH(alc), heat; or KOH(alc), heat, followed by $NaNH_2$

(f) Attach methyl group

$$CH_3C{\equiv}CH \xleftarrow{CH_3I} NaC{\equiv}CH \xleftarrow{NaNH_2} HC{\equiv}CH$$

3. (a) $1H_2$, cat.

(b) As in (a); then excess H_2, Ni, heat, pressure

(c) $$CH_3{-}CHBr_2 \xleftarrow{HBr} CH_2{=}CHBr \xleftarrow{HBr} HC{\equiv}CH$$

(d) HCl, CuCl

(e) $$CH_2Cl{-}CH_2Cl \xleftarrow{Cl_2} CH_2{=}CH_2 \longleftarrow \text{(a)}$$

(f) H_2O, H^+, Hg^{2+}

(g) $$CH_3{-}C{\equiv}CH \xleftarrow{CH_3I} NaC{\equiv}CH \xleftarrow{NaNH_2} HC{\equiv}CH$$

(h) $$C_2H_5{-}C{\equiv}CH \xleftarrow{C_2H_2Br} NaC{\equiv}CH \xleftarrow{NaNH_2} HC{\equiv}CH$$

(i) $$CH_3{-}C{\equiv}C{-}CH_3 \xleftarrow{CH_3I} NaC{\equiv}C{-}CH_3 \xleftarrow{NaNH_2} HC{\equiv}C{-}CH_3 \longleftarrow \text{(g)}$$

(j) $$\textit{cis-}CH_3CH{=}CHCH_3 \xleftarrow[\text{Lindlar cat.}]{H_2} CH_3C{\equiv}CCH_3 \longleftarrow \text{(i)}$$

(k) *trans*-$CH_3CH{=}CHCH_3 \xleftarrow{\text{Na, NH}_3} CH_3C{\equiv}CCH_3 \longleftarrow$ (i)

(l) $n\text{-Pr—C}{\equiv}\text{CH} \xleftarrow{n\text{-PrBr}} \text{NaC}{\equiv}\text{CH} \xleftarrow{\text{NaNH}_2} \text{HC}{\equiv}\text{CH}$

(m) $CH_3\text{—C}{\equiv}\text{C—Et} \xleftarrow{CH_3I} \text{NaC}{\equiv}\text{C—Et} \xleftarrow{\text{NaNH}_2} \text{HC}{\equiv}\text{C—Et} \longleftarrow$ (h)

(n) $\text{Et—C}{\equiv}\text{C—Et} \xleftarrow{\text{EtBr}} \text{NaC}{\equiv}\text{C—Et} \xleftarrow{\text{NaNH}_2} \text{HC}{\equiv}\text{C—Et} \longleftarrow$ (h)

4. (a) 1-butene (b) *n*-butane (c) 1,2-dibromo-1-butene
(d) 1,1,2,2-tetrabromobutane (e) 2-chloro-1-butene (f) 2,2-dichlorobutane
(g) methyl ethyl ketone (2-butanone) (h) $AgC{\equiv}CC_2H_5$ (i) 1-butyne
(j) $LiC{\equiv}CC_2H_5$ (k) 3-hexyne
(l) 1-butyne and isobutylene (m) $C_2H_5C{\equiv}CMgBr$ and ethane (n) 1-butyne
(o) $CH_3CH_2COOH + HCOOH$ (p) $CH_3CH_2COOH + CO_2$

5. (a) *syn*-Reduction.

cis-2-Butene $\xleftarrow{H_2,\ \text{Lindlar cat.}}$ $CH_3\text{—C}{\equiv}\text{C—}CH_3$ (2-Butyne)

(b) *anti*-Reduction.

trans-2-Butene $\xleftarrow{\text{Na, NH}_3}$ $CH_3\text{—C}{\equiv}\text{C—}CH_3$ (2-Butyne)

(c) Since addition of bromine is *anti*, we must use *anti*-reduction in the first step.

meso-2,3-Dibromobutane $\xleftarrow{Br_2}$ *trans*-2-Butene $\longleftarrow$ (b)

(d) Since halohydrin formation is *anti*, we must use *syn*-reduction in the first step.

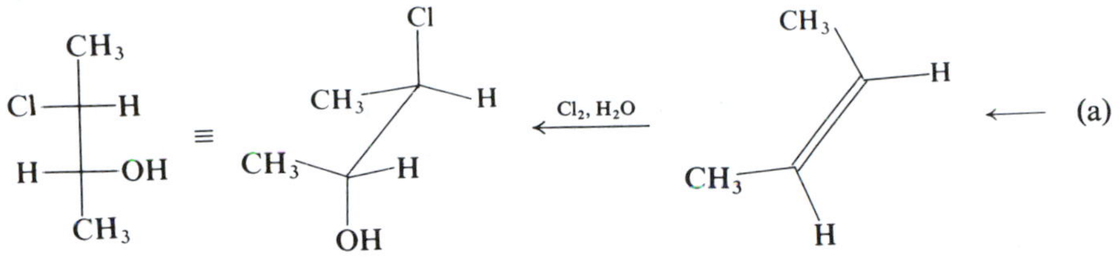

(and enantiomer)
rac-3-Chloro-2-butanol $\xleftarrow{Cl_2, H_2O}$ *cis*-2-Butene ← (a)

(e) Either *anti*-reduction, *anti*-hydroxylation; or *syn*-reduction, *syn*-hydroxylation.

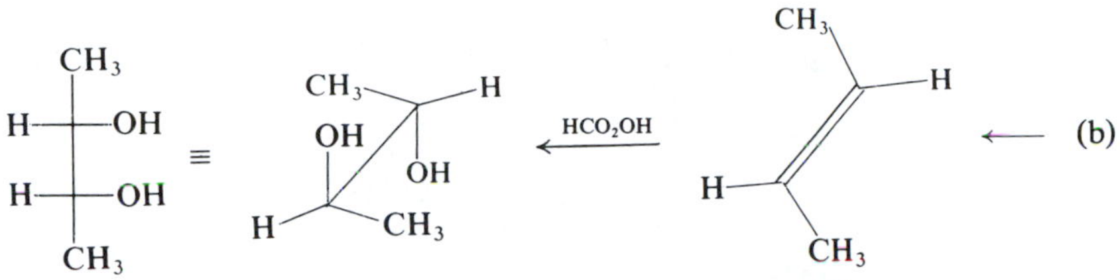

$\xleftarrow{HCO_2OH}$ *trans*-2-Butene ← (b)

or

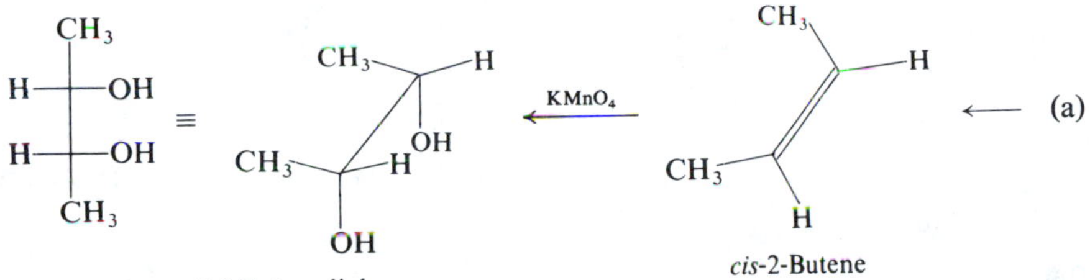

meso-2,3-Butanediol $\xleftarrow{KMnO_4}$ *cis*-2-Butene ← (a)

(f) Either *syn*-reduction, *anti*-hydroxylation; or *anti*-reduction, *syn*-hydroxylation.

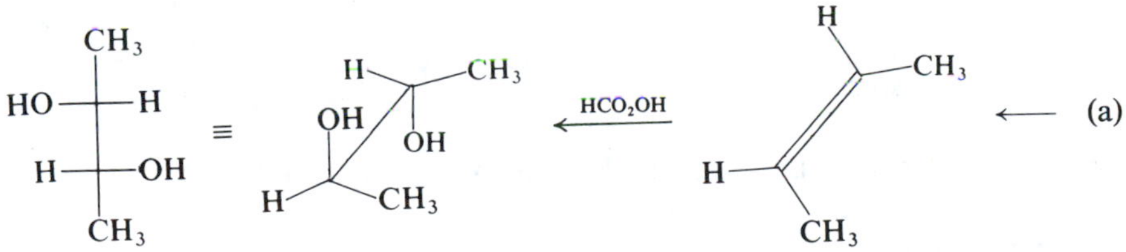

$\xleftarrow{HCO_2OH}$ *cis*-2-Butene ← (a)

or

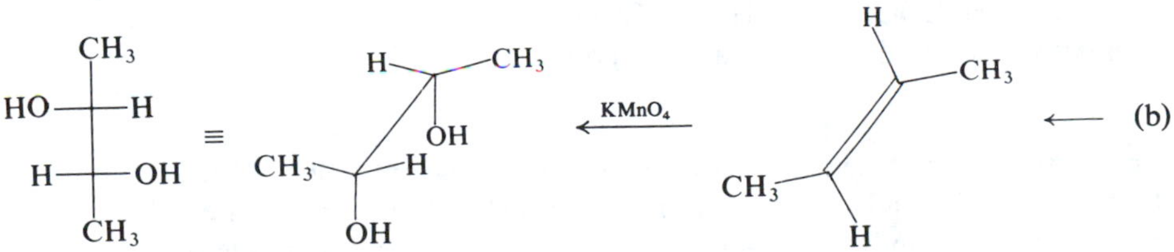

(and enantiomer)
rac-2,3-Butanediol $\xleftarrow{KMnO_4}$ *trans*-2-Butene ← (b)

(g) Certain ketones can be formed by Markovnikov addition of H_2O to a triple bond (Problem 12.5, above), followed by ketonization of the enol.

$$\underset{\substack{\text{2-Butanone}\\ \textit{Keto}}}{CH_3-CH_2-\underset{\underset{O}{\|}}{C}-CH_3} \longleftarrow \underset{\textit{Enol}}{CH_3-\overset{\overset{H}{|}}{C}=\underset{\underset{OH}{|}}{C}-CH_3} \xleftarrow[H^+,\ Hg^{2+}]{H_2O} \underset{\text{2-Butyne}}{CH_3-C\equiv C-CH_3}$$

6. (a) *anti*-Reduction, then *anti*-addition of bromine.

Meso (Fischer projection: Et top; H—Br; H—Br; Et bottom) $\xleftarrow{Br_2}$ *trans* (H, Et on upper C; Et, H on lower C) $\xleftarrow{\text{Li, } NH_3}$ EtC≡CEt ⟵ HC≡CH, 2EtBr, $NaNH_2$ ↑ EtOH

Meso

trans

(b) Either *anti*-reduction, *syn*-hydroxylation; or *syn*-reduction, *anti*-hydroxylation.

Racemic product (Fischer projection: CH_3 top; HO—H; H—OH; C_4H_9 bottom) and enantiomer ⟵ [$KMnO_4$ from *trans* (H, Me on upper C; *n*-Bu, H on lower C); HCO_2OH from *cis* (Me, H on upper C; *n*-Bu, H on lower C)]

trans $\xleftarrow{\text{Li, } NH_3}$ *n*-BuC≡CMe

cis $\xleftarrow[\text{Lindlar cat.}]{H_2}$ *n*-BuC≡CMe

n-BuC≡CMe ⟵ HC≡CH, *n*-BuBr, MeI; *n*-BuBr ↑ *n*-BuOH; MeI ↑ MeOH

CH$_3$
HO—H
H—OH
C$_4$H$_9$
and enantiomer
Racemic

trans

cis

7. (a) $KMnO_4$, or Br_2/CCl_4: alkyne positive.
(b) $Ag(NH_3)_2OH$: alkyne positive.
(c) Quantitative hydrogenation: alkyne takes up two moles H_2; alkene, only one mole.
(d) $Ag(NH_3)_2OH$: terminal alkyne positive.
(e) $KMnO_4$, or Br_2/CCl_4: diene positive.
(f) $Ag(NH_3)_2OH$: alkyne positive.
(g) Ozonolysis, followed by identification of the products: diene gives HCHO and $OHCCH_2CHO$; alkyne gives CH_3COOH and CH_3CH_2COOH.
(h) Br_2/CCl_4: alkyne positive. Or CrO_3/H_2SO_4: alcohol positive.

8. Only diethyl ether, *n*-pentane, and methylene chloride give a negative $KMnO_4$ test; of these, only the ether is soluble in H_2SO_4, and the other two can be distinguished by elemental analysis. Of the unsaturated compounds, only 1-chloropropene gives a positive halide test, and only 1-pentyne gives a precipitate with $Ag(NH_3)_2OH$. Distinguish among the others by ozonolysis and identification of the cleavage products.

9. $n\text{-}C_{13}H_{27}C{\equiv}CH \xrightarrow{n\text{-BuLi}} n\text{-}C_{13}H_{27}C{\equiv}CLi \xrightarrow{n\text{-}C_8H_{17}Br} n\text{-}C_{13}H_{27}C{\equiv}CC_8H_{17}\text{-}n$

A B

$B \xrightarrow[\text{Lindlar}]{H_2}$ $(n\text{-}C_{13}H_{27})(H)C{=}C(H)(C_8H_{17}\text{-}n)$ (cis)

(*Z*)-9-Tricosene
Sex attractant of common house fly
Muscalure

13

Cyclic Aliphatic Compounds

13.1 (a) (b) (c)

Cyclohexanol $\xrightarrow{H^+, \text{ heat}}$ cyclohexene $\xrightarrow{NBS}$ 3-bromocyclohexene $\xrightarrow{KOH(alc)}$ 1,3-cyclohexadiene

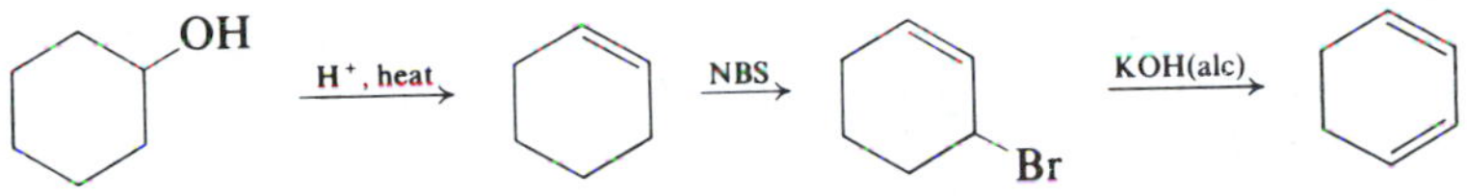

NBS = *N*-bromosuccinimide (p. 390)

13.2 Using a model of the axial conformation, we consider rotation about the bond between the alkyl group and C–1 of the ring. Ethyl and isopropyl can be rotated so that a hydrogen, —CH_2CH_3 or —$CH(CH_3)_2$, is nearest the axial hydrogens on C–3 and C–5; but no matter how *tert*-butyl is rotated, a large methyl group interferes with the axial hydrogens.

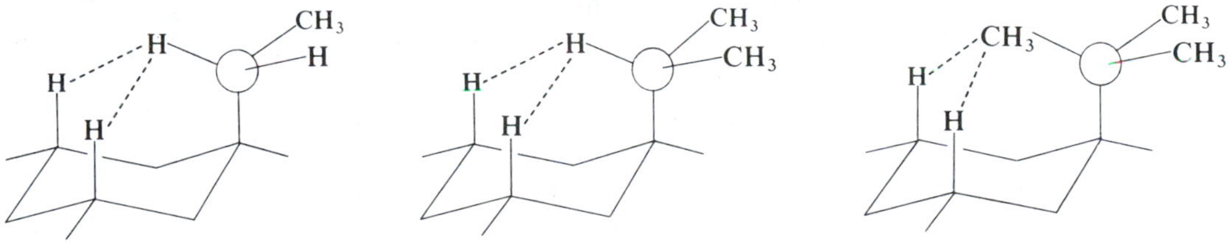

Ethylcyclohexane Isopropylcyclohexane *tert*-Butylcyclohexane

13.3 Only the *trans* diol is resolvable into enantiomers; the *cis* diol is a non-resolvable *meso* compound. (See the discussion on pages 465–466.)

13.4 *syn*-Hydroxylation with $KMnO_4$; *anti*-hydroxylation with HCO_2OH.

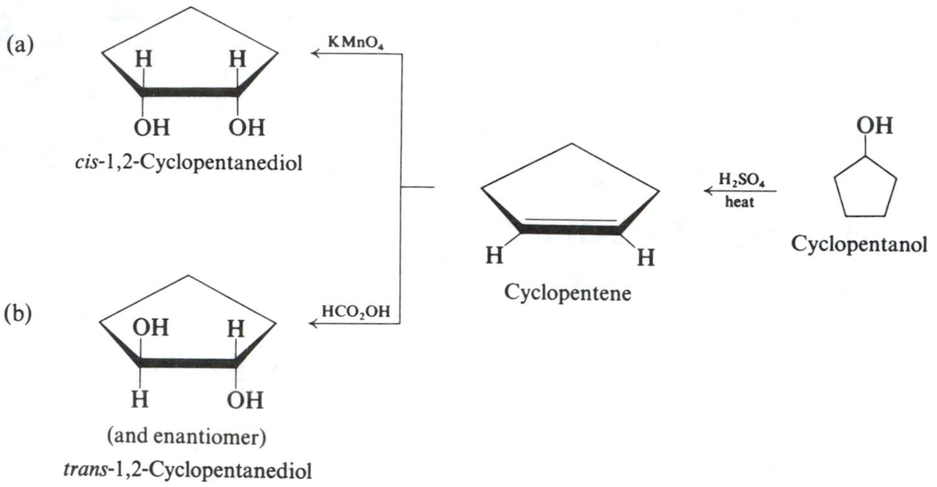

(a) *cis*-1,2-Cyclopentanediol $\xleftarrow{KMnO_4}$ Cyclopentene $\xleftarrow[\text{heat}]{H_2SO_4}$ Cyclopentanol

(b) *trans*-1,2-Cyclopentanediol (and enantiomer) $\xleftarrow{HCO_2OH}$ Cyclopentene

13.5 All the *cis* compounds, and *trans*-1,3-cyclobutanedicarboxylic acid.

For each of these, we can make a model. We then make a model of its mirror image, and try to superimpose this mirror image on the original. If it can be superimposed, the two are identical, and the molecule is achiral; if it cannot be superimposed, the molecule is chiral.

But we must learn to work with structural formulas, too. We draw a formula of the molecule, and then a formula of its mirror image—just as is shown for the 1,2-cyclopentanediols on pages 465–466. We next try mentally to superimpose the mirror image on the original. We are not dealing here with cross formulas, and are not bound by the rules given on pages 135–136. The kind of formula we use for cyclic compounds represents, in a stylized way, a model, and we can—in our mind's eye—do anything with it that we could do with an actual model: we can flip it over, for example, or rotate it about—so long as we keep in mind which bonds are pointing up or down, and which part of the ring is projecting toward us and which is pointing away.

It helps to redraw a formula as it will look after it has been flipped over or rotated, and then compare this newly drawn formula with that of the original. With *trans*-1,2-cyclobutanedicarboxylic acid, for example, we draw the molecule, III, and its mirror image, IV;

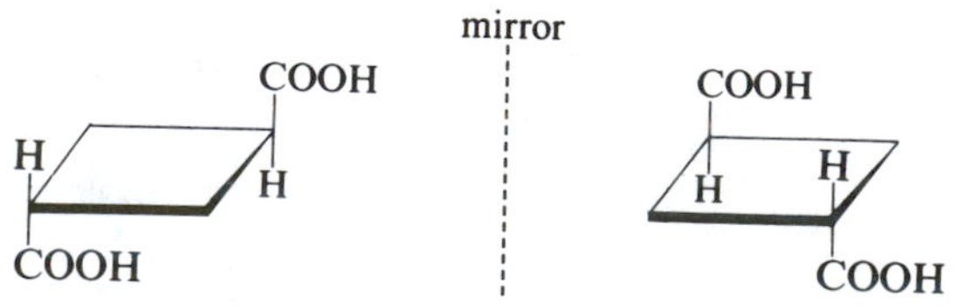

mirror

III

IV
Mirror image of III

we then rotate the mirror image from its original position, IV*a*, to make IV*b*, which we find to be identical with III.

COOH H H COOH — *by rotation gives* — COOH H H COOH

IV*a* IV*b*

Identical to III

The molecule is identical with its mirror image, and is achiral.

It is particularly helpful to examine these formulas for a plane of symmetry, as shown on page 466: if we see such a plane, we know the compound cannot be chiral. (If we do *not* see one, we had better go through the molecule-and-mirror-image procedure.) Notice that both the *cis*- and *trans*-1,3-cyclobutanedicarboxylic acids have a plane of symmetry cutting vertically through the molecule from the left-front to the right-rear corner (when oriented as on page 465).

13.6 (a), (d), (e): axial–equatorial = equatorial–axial.

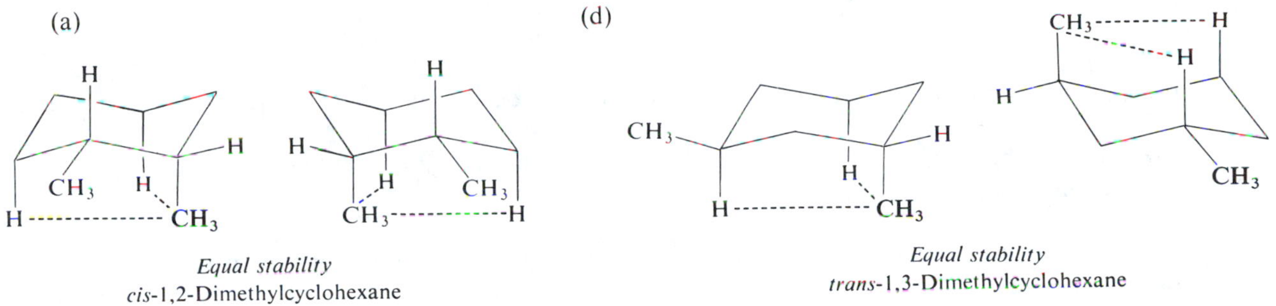

Equal stability
cis-1,2-Dimethylcyclohexane

Equal stability
trans-1,3-Dimethylcyclohexane

(e)

H CH3 H H CH3 H — CH3 H H H CH3 H

Equal stability
cis-1,4-Dimethylcyclohexane

(b), (c), (f): equatorial–equatorial more stable than axial–axial.

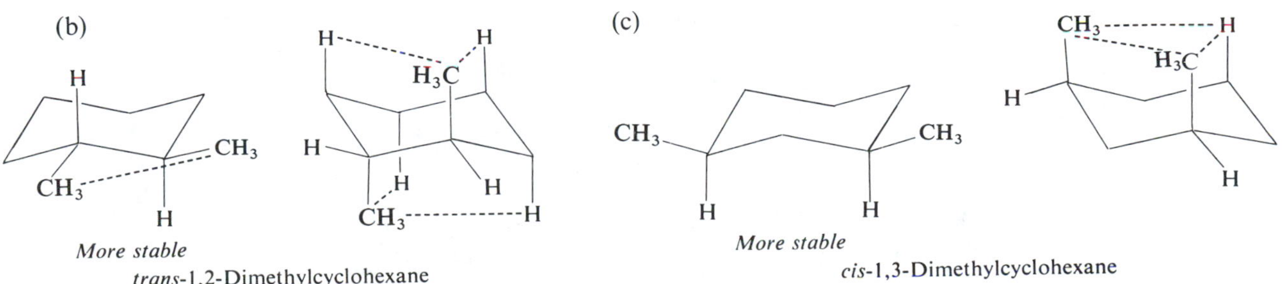

More stable
trans-1,2-Dimethylcyclohexane

More stable
cis-1,3-Dimethylcyclohexane

(f)

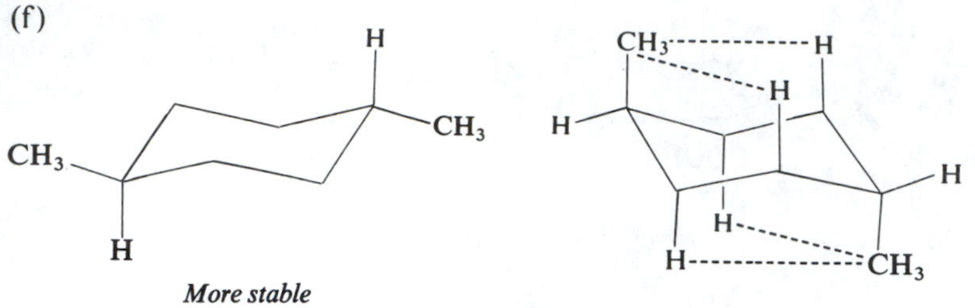

More stable

trans-1,4-Dimethylcyclohexane

(g) 0 kcal difference for (a), (d), (e): same number (two) of methyl–hydrogen interactions in each.

2.7 kcal difference for (b): one butane-*gauche* (0.9 kcal) *vs.* four methyl–hydrogen interactions (4×0.9 kcal).

5.4 kcal difference for (c): no interactions *vs.* two methyl–hydrogen interactions (2×0.9 kcal) + methyl–methyl interaction (3.6 kcal, see Problem 13.7).

3.6 kcal difference for (f): no interactions *vs.* four methyl–hydrogen interactions (4×0.9 kcal).

13.7 (a) 1,3-Diaxial interaction of two —CH_3 groups (see 13.6(c), above).

(b) 3.6 kcal, the difference between the total of 5.4 kcal and two methyl–hydrogen interactions (2×0.9 kcal).

(c) The *trans* isomer exists as either of two equivalent chair conformations, with two methyl–hydrogen interactions and one methyl–methyl interaction. The *cis* isomer exists (almost) exclusively in the chair conformation with only one axial methyl group and hence two methyl–hydrogen interactions.

trans:

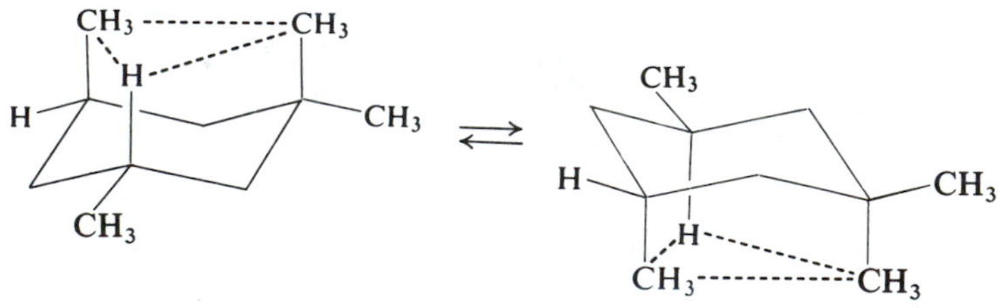

Equivalent

One methyl–methyl and two methyl–hydrogen interactions in each

cis:

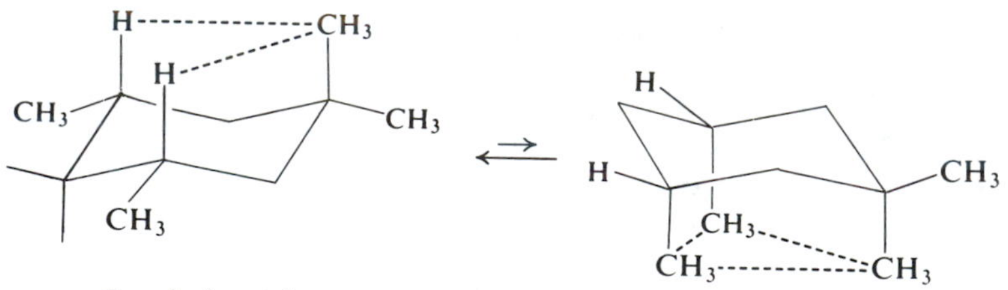

Greatly favored

Only two methyl–hydrogen interactions

Can be neglected in equilibrium

Three methyl–methyl interactions

Thus the difference in stability between the two isomers (3.7 kcal) is due to one methyl–methyl interaction, in excellent agreement with Pitzer's calculations.

(Allinger, N. L.; Miller, M. A. "The 1,3-Diaxial Methyl–Methyl Interaction"; *J. Am. Chem. Soc.* **1961**, *83*, 2145.)

13.8 (a) The *cis* isomer is more stable than the *trans*.

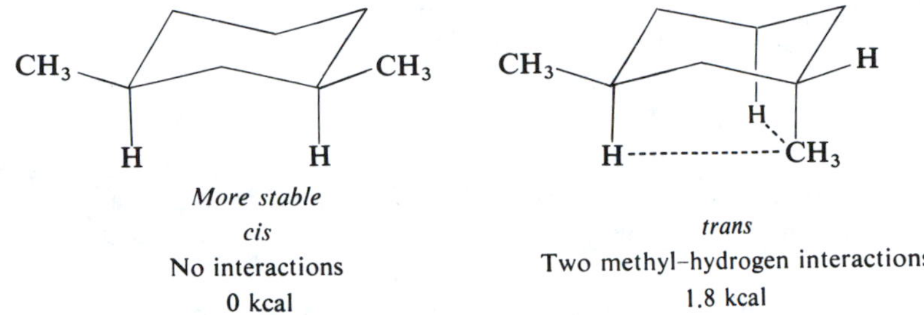

More stable
cis
No interactions
0 kcal

trans
Two methyl–hydrogen interactions
1.8 kcal

(b) The *trans* isomer is more stable than the *cis*.

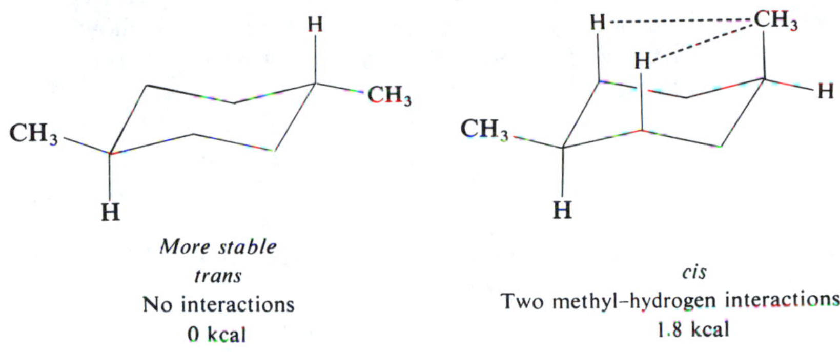

More stable
trans
No interactions
0 kcal

cis
Two methyl–hydrogen interactions
1.8 kcal

(c) A difference of 1.8 kcal/mol in each case.

13.9 (a)

(b)

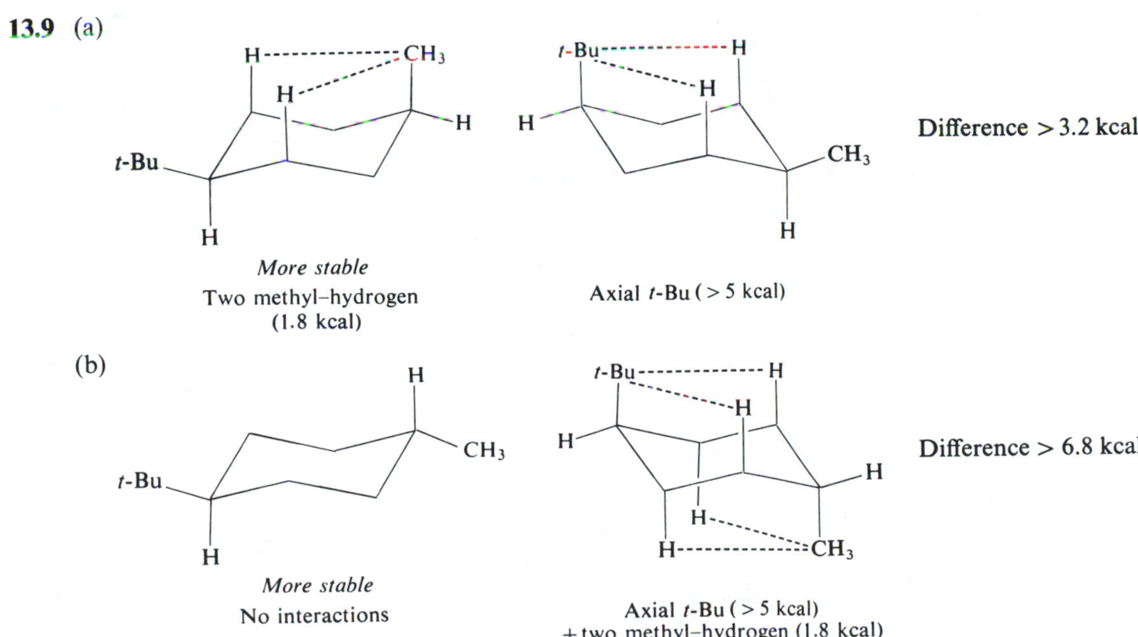

(c)

More stable
Four methyl–hydrogen (3.6 kcal)

Axial *t*-Bu (> 5 kcal) + one butane-*gauche* (0.9 kcal)

Difference > 2.3 kcal

13.10 Resolvable: b, d. *Meso*: c. (Neither e nor f contains chiral carbons, and hence they are not *meso*; each one is simply an achiral compound.)

We can find the answer to each part most simply by drawing a planar formula and looking for a plane of symmetry, and then confirming this by comparing the original formula with a formula of the mirror image. To see what is really involved, however, we must make models of molecule and mirror image and compare these, being sure to allow the models to do *what the actual molecules can do*: flip readily from one chair conformation into the other. (Compare Problem 13.11.)

All molecules not superimposable on their mirror images are chiral, and hence resolvable. All molecules superimposable on their mirror images are achiral; *if they contain chiral centers*, they belong to the special class of achiral compounds called *meso*.

(a) See page 470. There is a vertical plane of symmetry, passing between C–1 and C–2, and C–4 and C–5.

(b) See page 470.

(c) There is a vertical plane of symmetry passing through C–2 and C–5.

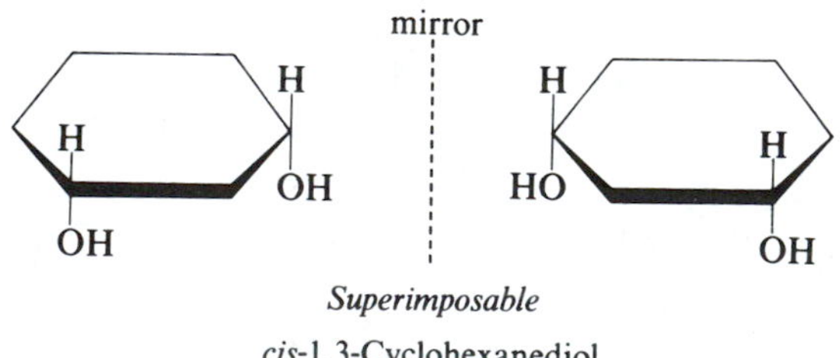

Superimposable
cis-1,3-Cyclohexanediol

(d)

mirror

Not superimposable
trans-1,3-Cyclohexanediol

(e) There is a vertical plane of symmetry passing through C–1 and C–4. (There are no chiral centers, and hence the molecule is not *meso*.)

mirror

H H H H
HO OH HO OH

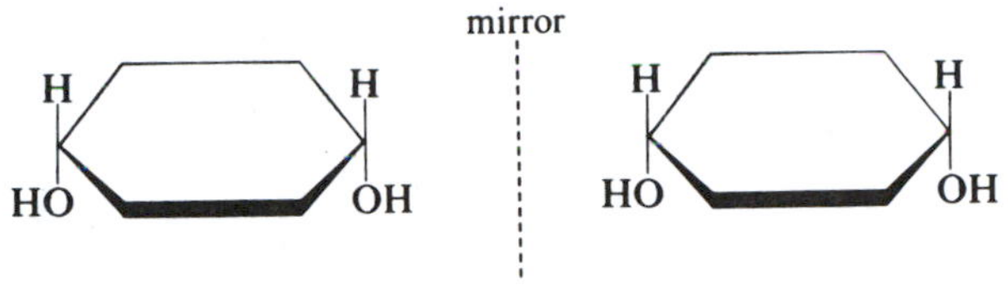

Superimposable
cis-1,4-Cyclohexanediol

(f) There is a vertical plane of symmetry passing through C–1 and C–4. (There are no chiral centers, and hence the molecule is not *meso*.)

mirror

HO H H OH
H OH HO H

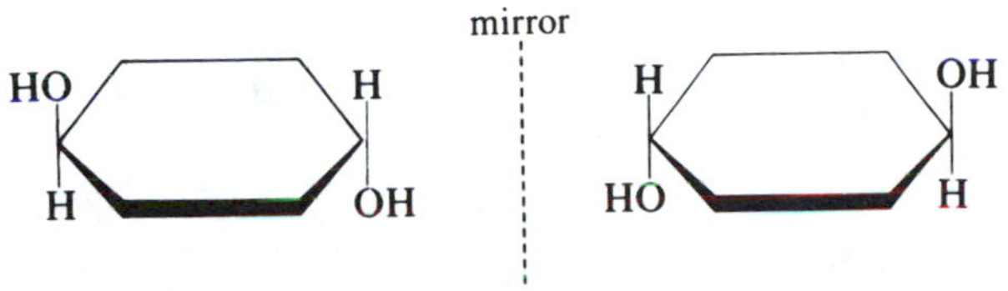

Superimposable
trans-1,4-Cyclohexanediol

13.11 *cis*-1,2-Cyclohexanediol: a pair of conformational enantiomers.

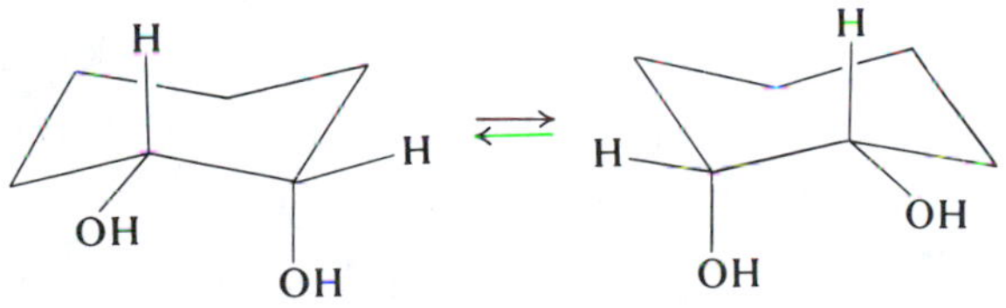

H H
H ⇄ H
OH OH
OH OH

trans-1,2-Cyclohexanediol: a pair of configurational enantiomers, each of which exists as a pair of conformational diastereomers.

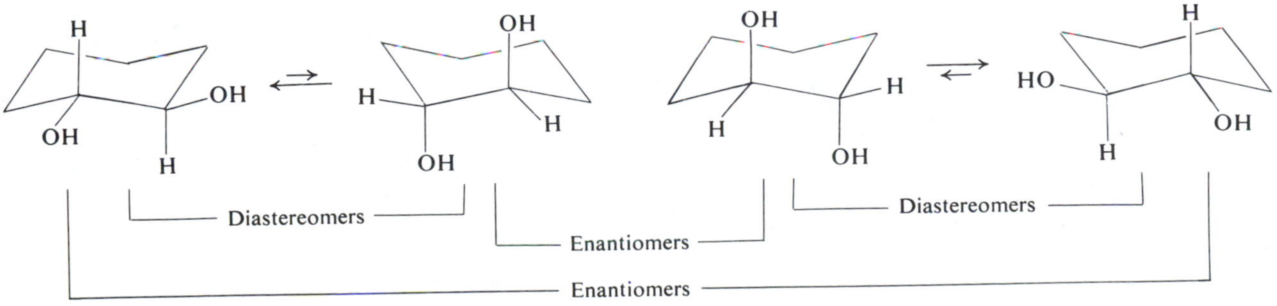

H OH OH H
OH ⇄ H H ⇄ HO
OH H H OH H OH
H OH OH H

Diastereomers Diastereomers
Enantiomers
Enantiomers

cis-1,3-Cyclohexanediol: a pair of conformational diastereomers.

HO OH H H ⇄ OH OH H H

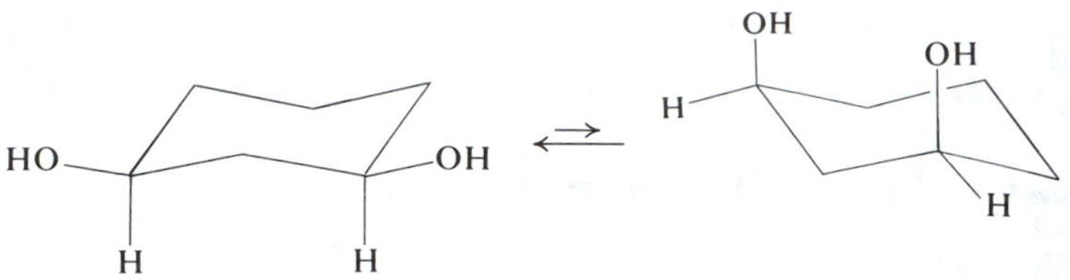

trans-1,3-Cyclohexanediol: a pair of configurational enantiomers, each of which exists as a single conformation. (Use models to convince yourself that certain structures are identical.)

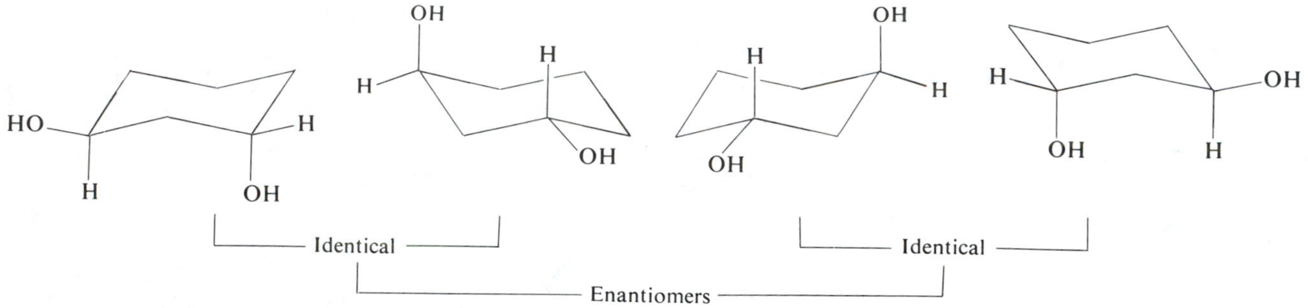

Identical — Identical — Enantiomers

cis-1,4-Cyclohexanediol: exists as a single conformation.

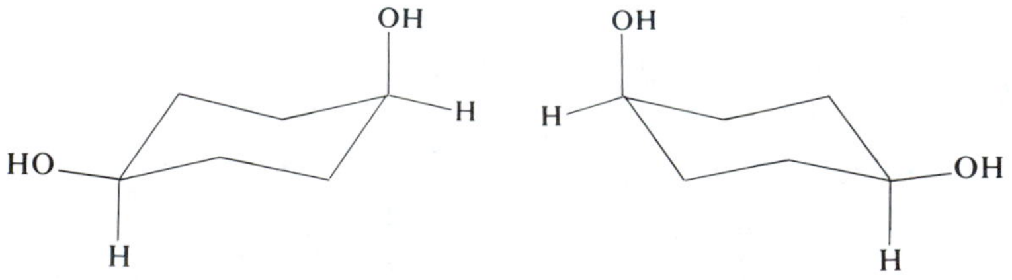

Identical

trans-1,4-Cyclohexanediol: a pair of conformational diastereomers.

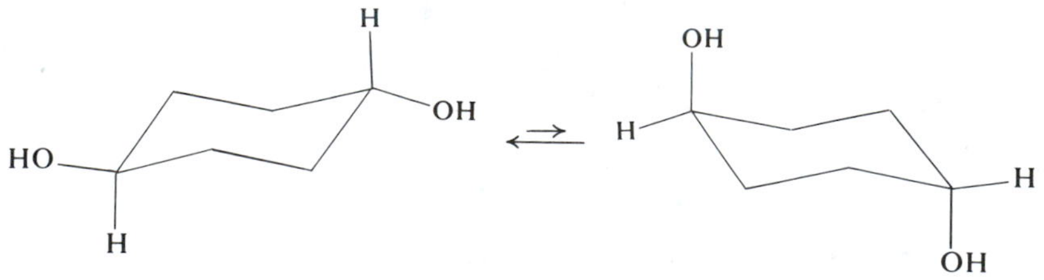

13.12 Pairs of enantiomers: a, b, c, d.
Achiral: e, f.
No *meso* compounds.
None are non-resolvable racemic modifications.

(a) *cis*-2-Chlorocyclohexanol: a pair of configurational enantiomers. (Each of these is a pair of conformational diastereomers.)

mirror

Non-superimposable
Enantiomers

Conformational diastereomers

Enantiomers

Conformational diastereomers

Enantiomers

(b) *trans*-2-Chlorocyclohexanol: a pair of configurational enantiomers. (Each of these is a pair of conformational diastereomers.)

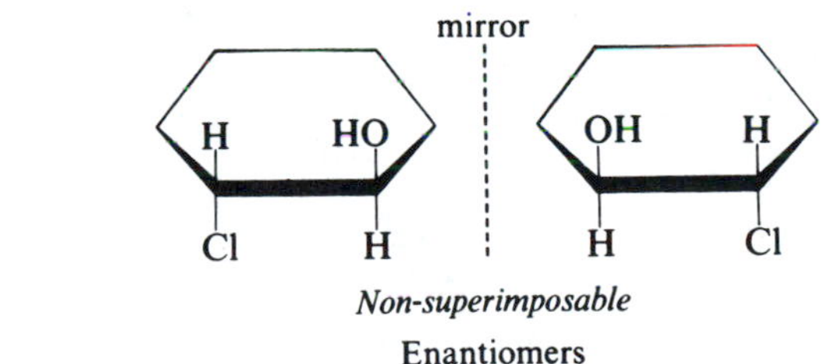

Non-superimposable
Enantiomers

Conformational diastereomers

Enantiomers

Conformational diastereomers

Enantiomers

(c) *cis*-3-Chlorocyclohexanol: a pair of configurational enantiomers. (Each of these exists as a single conformation.)

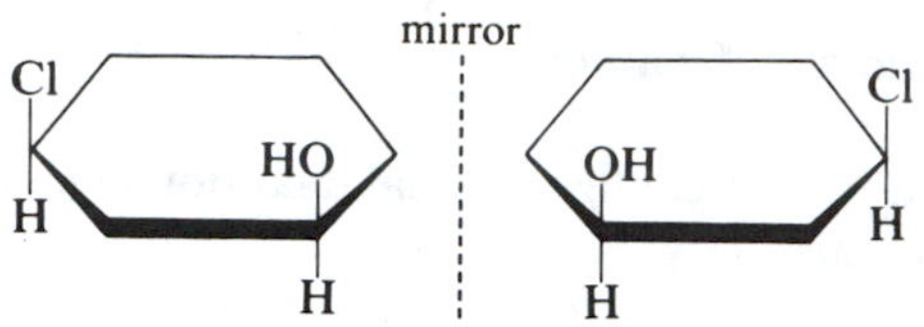

mirror

Cl HO H H | OH Cl H H

Non-superimposable

Enantiomers

Cl OH H Cl H H ⇄ Cl OH H H | OH Cl H H HO Cl ⇄ H H

Conformational diastereomers — Conformational diastereomers

Enantiomers

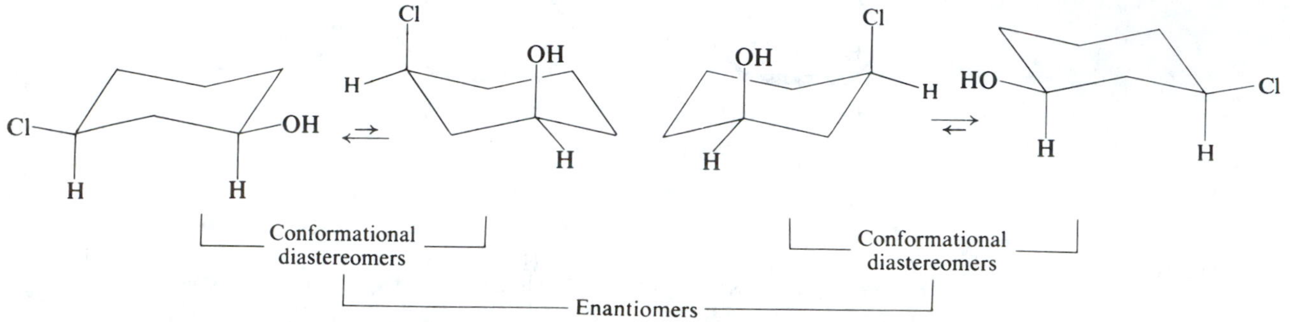

(d) *trans*-3-Chlorocyclohexanol: a pair of configurational enantiomers. (Each of these exists as a single conformation.)

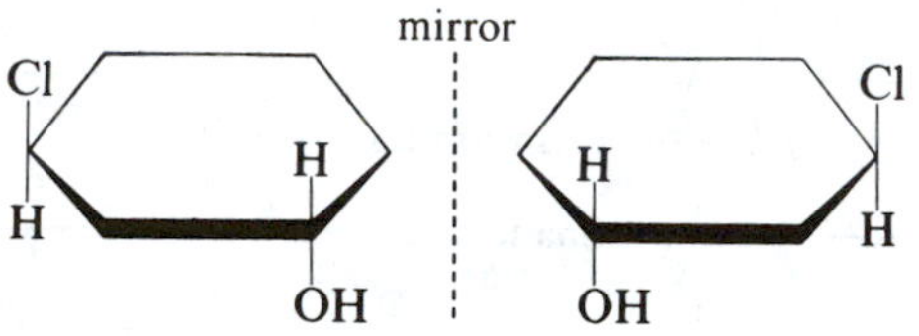

mirror

Cl H H OH | H Cl H OH

Non-superimposable

Enantiomers

Cl H OH H ... OH Cl H H | H Cl OH H H Cl OH H

Conformational diastereomers — Conformational diastereomers

Enantiomers

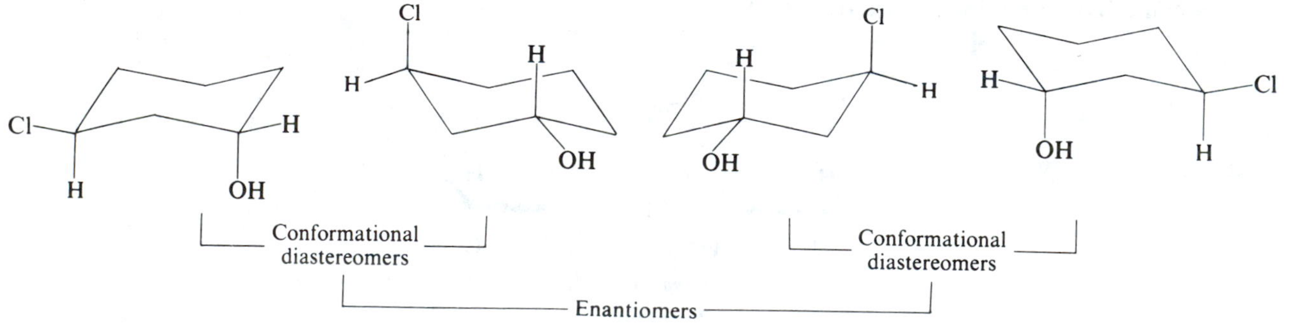

(e) *cis*-4-Chlorocyclohexanol: achiral, and contains no chiral carbons. (It exists as a pair of conformational diastereomers.)

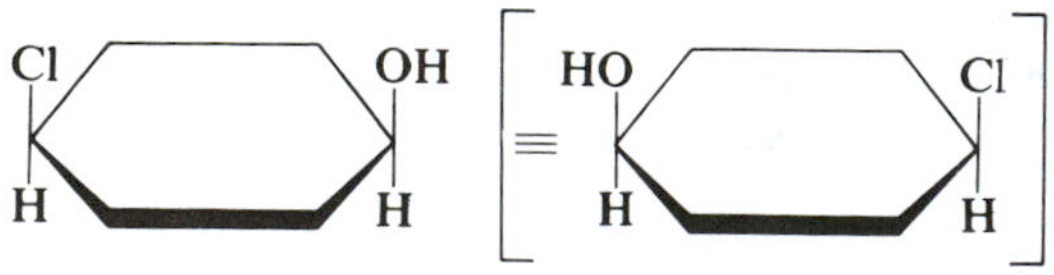

Cl OH H H [≡ HO Cl H H]

Achiral

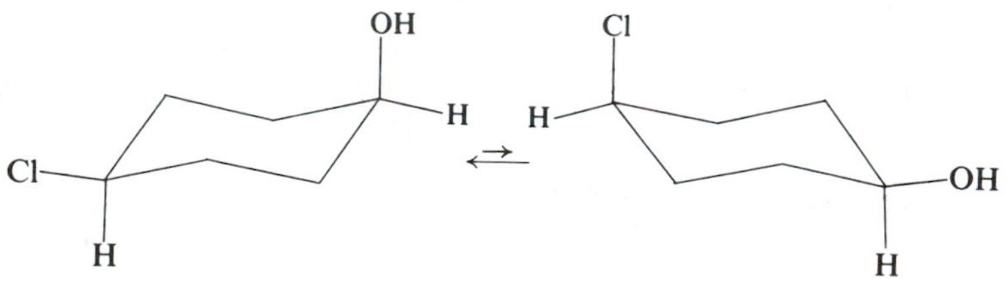

OH Cl H H ⇄ Cl H OH H

Conformational diastereomers

(f) *trans*-4-Chlorocyclohexanol: achiral, and contains no chiral carbons. (It exists as a pair of conformational diastereomers.)

Achiral

Conformational diastereomers

13.13 The low rate of elimination raises the suspicion that in that one isomer it is impossible for —H and —Cl to get into an *anti*-periplanar relationship. Only one isomer,

All Cl atoms equatorial; all —Cl *anti* to —H

fills the bill. (Draw some of the other structures to check the uniqueness of this isomer.)

13.14 In the absence of added base the E2 reaction of menthyl chloride is slowed down to such an extent that it is outpaced by the E1 reaction. There are no stereochemical restrictions on this reaction, since the leaving groups are lost in different steps. The carbocation loses a proton from either of two positions to give both alkenes, with the more stable one predominating.

13.15 (a) Orientation of elimination is powerfully controlled by the phenyl group, which stabilizes the incipient double bond in the transition state.

With the nature of the product thus ordained, each tosylate reacts by the only periplanar elimination open to it: *anti* from the *cis* isomer, and *syn* from the *trans* isomer.

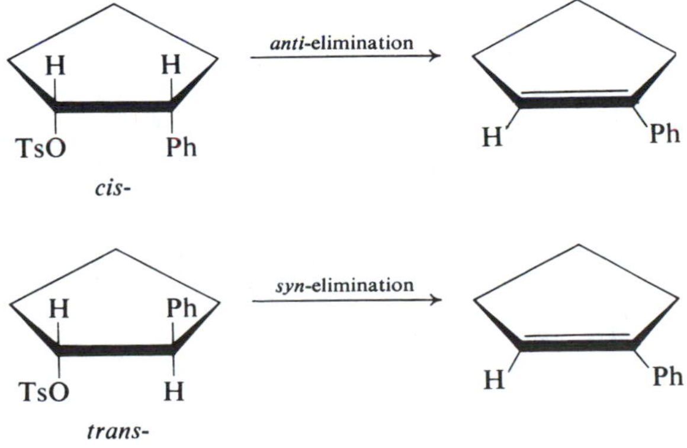

Although the *anti*-elimination requires some twisting of the ring, it is still faster than the *syn*-elimination. The significant thing is that *syn*-elimination occurs as fast as it does; it requires eclipsing of groups, but in cyclopentane compounds, groups are already badly eclipsed. (The corresponding cyclohexane compounds give analogous results, but there the *cis* isomer reacts over 10 000 times as fast as the *trans*.)

(b) I must react by *anti*-elimination, and II by *syn*-elimination. Here, *syn*-elimination is actually faster. In this rigid bicyclic system, the twisting of I required for *anti*-periplanar elimination is more difficult than in a simple cyclopentane derivative, whereas in II the leaving groups are already held *syn*-periplanar. (In addition, in I both Cl's are tucked in a fold of the molecule, making solvation of a leaving Cl^- difficult.)

13.16 Oxygen reacts with triplet methylene or with a diradical intermediate, leaving only singlet methylene to add. (Compare the action of oxygen as an inhibitor of free-radical chlorination, Sec. 2.14.)

13.17 (a) (b)

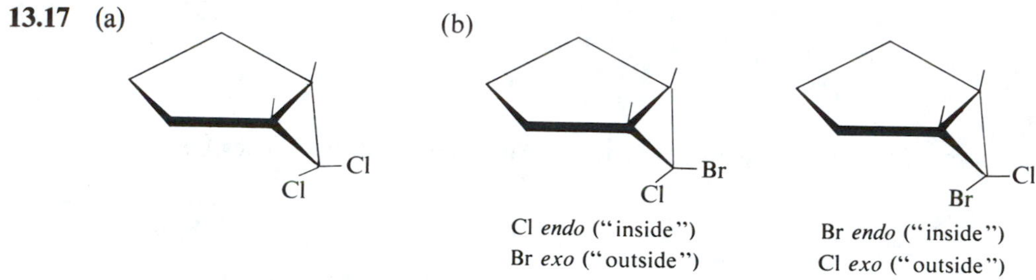

Cl *endo* ("inside")
Br *exo* ("outside")

Br *endo* ("inside")
Cl *exo* ("outside")

13.18 (a) $CHCl_3$ has no β-carbon. (b) Electron withdrawal by chlorines stabilizes the anion, speeds up its formation and/or shifts equilibrium (1) (on page 477) to the right.

(Moss, Robert A. "Carbene Chemistry"; *Chem. Eng. News* **June 16, 1969**, pp 60–68; **June 30, 1969**, pp 50–58.)

13.19

1,4-Dioxane (ring: O, H_2C, CH_2, H_2C, CH_2, O) $\xleftarrow[-2H_2O]{H^+}$ 1,2-Ethanediol ($HOCH_2CH_2OH$ + $HOCH_2CH_2OH$)

1,4-Dioxane

1,2-Ethanediol
Two moles

13.20

Furan $\xrightarrow{H_2,\ Ni}$ Tetrahydrofuran $\xrightarrow[heat]{HCl}$ $HOCH_2CH_2CH_2CH_2Cl$ $\xrightarrow[heat]{HCl}$ $ClCH_2CH_2CH_2CH_2Cl$

Furan Tetrahydrofuran 4-Chloro-1-butanol 1,4-Dichlorobutane

13.21 (a) The conversion of halohydrins into epoxides by the action of base is simply an adaptation of the Williamson synthesis (Sec. 6.20); a cyclic compound is obtained because both alcohol and halide happen to be part of the same molecule. In the presence of hydroxide ion a small proportion of the alcohol exists as alkoxide; this alkoxide displaces halide ion from another portion of the same molecule to yield the cyclic ether.

(1) $BrCH_2{-}CH_2OH + OH^- \rightleftharpoons H_2O + BrCH_2{-}CH_2O^{\ominus}$

(2) $BrCH_2{-}CH_2{-}\ddot{O}{:}^{\ominus} \longrightarrow \left[\overset{\delta-}{Br}\cdots CH_2{-}CH_2\cdots\overset{\delta-}{O} \right] \longrightarrow CH_2{-}CH_2\ (\text{epoxide, ring O}) + Br^-$

Since halohydrins are nearly always prepared from alkenes by addition of halogen and water to the carbon–carbon double bond (Sec. 9.14), this method amounts to the conversion of an alkene into an epoxide. (Alternatively, the carbon–carbon double bond may be oxidized directly to the epoxide group by peroxy compounds, such as peroxybenzoic acid.)

(b) $CH_3{-}CH{=}CH_2 \xrightarrow{Cl_2,\ H_2O} CH_3{-}CH(OH){-}CH_2Cl \xrightarrow{\text{conc. aq. } OH^-} CH_3{-}CH{-}CH_2\ (\text{ring O})$

Propylene chlorohydrin; Propylene oxide

13.22 (a) $Et{-}O{-}CH_2CH_2{-}O{\vdots}CH_2CH_2{-}OH \xleftarrow[H^+]{H_2C{-}CH_2\ (\text{O})} Et{-}O{\vdots}CH_2CH_2{-}OH \xleftarrow[H^+]{H_2C{-}CH_2\ (\text{O})} EtOH$

(b) As in (a), replacing EtOH by PhOH.

(c) As in (a), replacing EtOH by HOH.

(d) $HO{-}CH_2CH_2{-}O{-}CH_2CH_2{-}O{\vdots}CH_2CH_2{-}OH \xleftarrow[H^+]{H_2C{-}CH_2\ (\text{O})} HO{-}CH_2CH_2{-}O{-}CH_2CH_2{-}OH$ (From c)

13.23 S_N1 $H^+ + CH_2{-}CH_2\ (\text{ring O}) \rightleftharpoons CH_2{-}CH_2\ (\text{ring } \overset{\oplus}{O}H)$

$CH_2{-}CH_2\ (\text{ring } \overset{\oplus}{O}H) \longrightarrow HOCH_2{-}CH_2^{\oplus}$ *Slow*

$H_2O + HOCH_2{-}CH_2^{\oplus} \longrightarrow HOCH_2{-}CH_2OH_2^+ \xrightarrow{-H^+} HOCH_2{-}CH_2OH$

S_N2 as shown on page 486.

13.24 (i) Attachment of oxygen can occur with equal ease to either face of the alkene. In (a), the two (epoxide) structures are identical and achiral; the same is true in (b). In (c), (d), and (e), equal amounts of enantiomeric epoxides are formed. (Compare the answer to Problem 10.2, page 375.)

(ii) Back-side attack on the epoxide ring can occur with equal ease (is equally likely) at either carbon in (a), in (b), and in (c), giving in (a) and (b) equal amounts of enantiomers, in (c) the same inactive, *meso* compound, whichever enantiomer or whichever carbon is attacked. (See Figure 10.2 on page 374, and Figure 10.3 on page 375.)

(iii) Even though attacks by the two paths—at the methyl end and at the ethyl end—are not equally likely in (d) and (e), the product is racemic in each case. There are equal amounts of enantiomeric epoxides undergoing attack. The product from one epoxide undoubtedly consists of unequal amounts of the two possible enantiomeric diols; attack at the methyl end, for example, may be preferred. But any excess of, say, (*S*,*S*)-diol from one epoxide stereoisomer would be exactly balanced by an excess of (*R*,*R*)-diol from the enantiomeric epoxide. (Compare the answer to Problem 10.3, page 375.)

(a) H H O → H HO OH H and OH H H OH

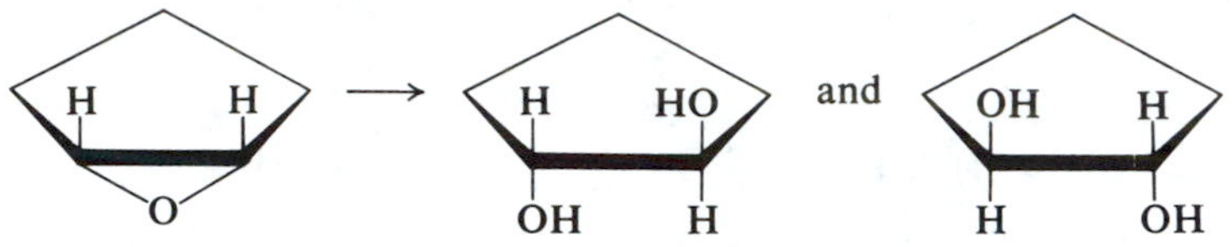

Achiral

trans-1,2-Cyclopentanediol
Racemic modification

(b) O H CH_3 H CH_3 → OH H CH_3 H CH_3 OH and H CH_3 HO OH H CH_3

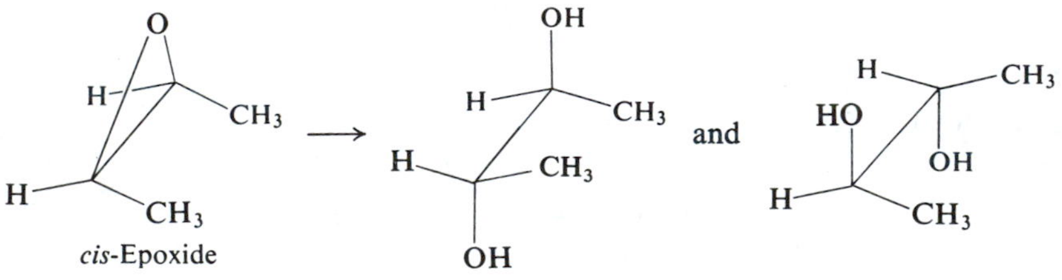

cis-Epoxide
Achiral

2,3-Butanediol
Racemic

(c) O H CH_3 CH_3 H + H CH_3 CH_3 H O → OH H CH_3 CH_3 H OH

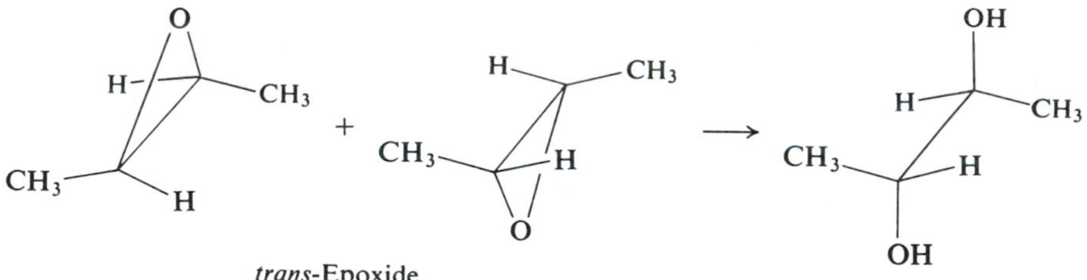

trans-Epoxide
Enantiomers

2,3-Butanediol
Meso

(d) and

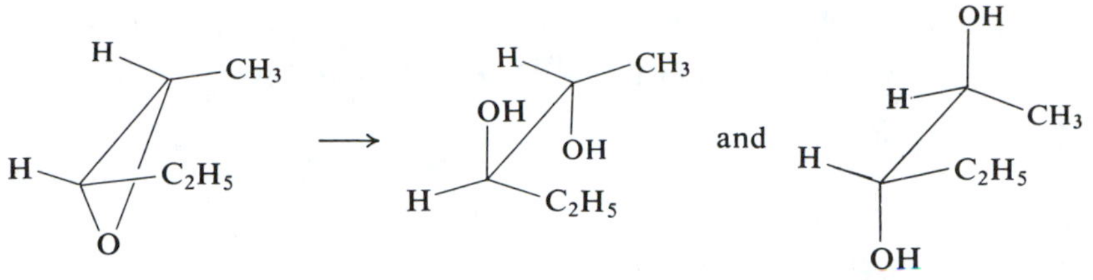

R,*R*- *S*,*S*-

and

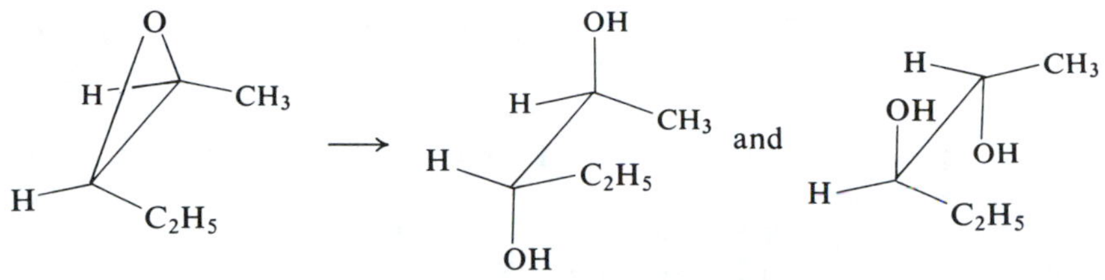

cis-Epoxide
Enantiomers

S,*S*- *R*,*R*-

2,3-Pentanediol
Racemic modification

(e) and

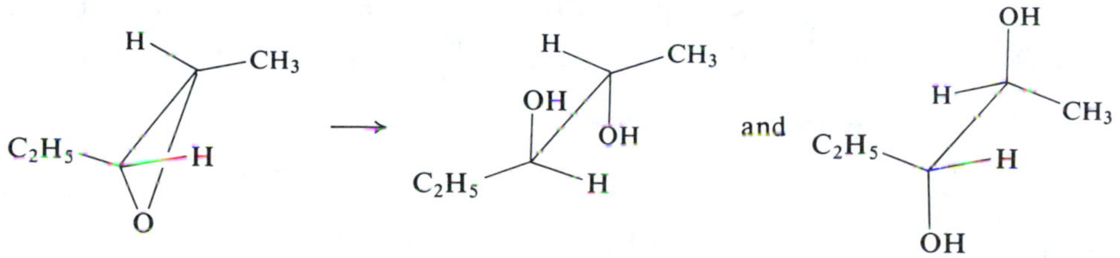

2*R*,3*S*- 2*S*,3*R*-

and

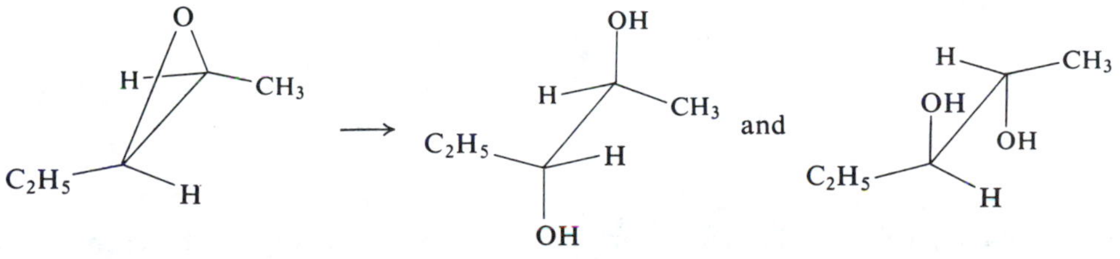

trans-Epoxide
Enantiomers

2*S*,3*R*- 2*R*,3*S*-

2,3-Pentanediol
Racemic modification

(f) None of the products, as obtained, would be optically active.

13.25

(a) $CH_3OH + H_2C{-}CH_2$ (epoxide, $\overset{\oplus}{O}$–H) $\longrightarrow$ $CH_3\overset{H}{\underset{\oplus}{O}}CH_2CH_2OH \xrightarrow{-H^+} CH_3OCH_2CH_2OH$

(b) $CH_3O^- + H_2C{-}CH_2$ (epoxide, O) $\longrightarrow CH_3OCH_2CH_2O^- \xrightarrow{+H^+} CH_3OCH_2CH_2OH$

(c) $CH_3NH_2 + H_2C{-}CH_2$ (epoxide, O) $\longrightarrow CH_3\overset{H}{\underset{H}{N^{\oplus}}}{-}CH_2CH_2O^- \longrightarrow CH_3\overset{H}{N}CH_2CH_2OH$

13.26 $CH_3CHOHCH_2OH + OH^- \rightleftharpoons CH_3CHOHCH_2O^- + H_2O$

$CH_3CHOHCH_2O^- + H_2C{-}\overset{CH_3}{CH}$ (epoxide, O) $\longrightarrow CH_3CHOHCH_2O{-}CH_2{-}\overset{CH_3}{CH}{-}O^-$

$CH_3CHOHCH_2OCH_2\overset{CH_3}{CH}O^- + H_2C{-}\overset{CH_3}{CH}$ (epoxide, O) $\longrightarrow CH_3CHOHCH_2OCH_2\overset{CH_3}{CH}O{-}CH_2{-}\overset{CH_3}{CH}O^-$

and so on, $\longrightarrow CH_3CHOHCH_2O\left[CH_2\overset{CH_3}{CH}O\right]_n CH_2\overset{CH_3}{CH}O^- \xrightarrow{H_2O} CH_3CHOHCH_2O\left[CH_2\overset{CH_3}{CH}O\right]_n CH_2\overset{CH_3}{CH}OH$

13.27 (a) $CH_3HC{-}CH_2$ (epoxide, O) $+ NH_3 \longrightarrow CH_3\underset{O^-}{CH}{-}\underset{\overset{+}{N}H_3}{CH_2} \longrightarrow CH_3\underset{OH}{CH}{-}\underset{NH_2}{CH_2}$

(b) $Cl^- + (CH_3)_2C{-}CHCH_3$ (epoxide, $\overset{\oplus}{O}$–H) $\longrightarrow CH_3{-}\underset{Cl}{\overset{CH_3}{C}}{-}\underset{OH}{CH}{-}CH_3$

13.28 (a) C_6H_{14}, C_6H_{12} (b) C_5H_{12}, C_5H_{10} (c) C_6H_{12}, C_6H_{10} (d) $C_{12}H_{26}$, $C_{12}H_{24}$, $C_{12}H_{22}$. (e) For the same degree of unsaturation, there are two fewer hydrogens for each ring that is present.

13.29 All are C_6H_{12}; no information about ring size.

13.30 α-Carotene:

$$\begin{array}{l} C_{40}H_{78} \text{ sat'd} \\ -C_{40}H_{56} \text{ polyene} \\ \hline 22H \text{ taken up to saturate 11 double bonds} \end{array} \qquad \begin{array}{l} C_{40}H_{82} \text{ alkane} \\ -C_{40}H_{78} \\ \hline 4H \text{ still missing} \end{array}$$

The missing 4H means *two rings*.

β-Carotene: similarly, 11 double bonds and *two rings*.

γ-Carotene:

$$\begin{array}{l} C_{40}H_{80} \text{ sat'd} \\ -C_{40}H_{56} \text{ polyene} \\ \hline 24H \text{ taken up to saturate 12 double bonds} \end{array} \qquad \begin{array}{l} C_{40}H_{82} \text{ alkane} \\ -C_{40}H_{80} \\ \hline 2H \text{ still missing} \end{array}$$

The missing 2H means *one ring*.

Lycopene:

$$\begin{array}{l} C_{40}H_{82} \text{ sat'd} \\ -C_{40}H_{56} \text{ polyene} \\ \hline 26H \text{ taken up to saturate 13 double bonds} \end{array} \qquad \begin{array}{l} C_{40}H_{82} \text{ alkane} \\ -C_{40}H_{82} \\ \hline \text{no H missing} \end{array}$$

No missing H means *no ring*.

13.31 (a) cyclopentene $\xrightarrow{O_3}$ $H_2C(-CH_2-CHO)(-H_2C-CHO)$, i.e. $OHC-CH_2-CH_2-CH_2-CHO$

(b) 1-methylcyclopentene (CH_3) $\xrightarrow{O_3}$ $H_2C(-CH_2-C(=O)CH_3)(-H_2C-CHO)$

(c) 3-methylcyclopentene (CH_3) $\xrightarrow{O_3}$ $H_2C(-CH_2-CHO)(-HC(CH_3)-CHO)$

(d) 1,3-cyclohexadiene $\xrightarrow{O_3}$ $H_2C(CHO)-H_2C(CHO)$ + $OHC-OHC$

(e) 1,4-cyclohexadiene $\xrightarrow{O_3}$ $H_2C(CHO)(CHO)$ + $OHC-CH_2-CHO$

13.32 The diene takes up two moles H_2, cyclohexene only one. The diene yields HCHO, cyclohexene does not.

Stereochemical Formulas of Cyclic Compounds

For convenience, organic chemists use a variety of ways to show the stereochemistry of cyclic compounds.

A solid line (often thickened) indicates a bond coming *out of* the plane of the paper; a dashed line indicates a bond going *behind* the plane.

Alternatively, a round dot represents a hydrogen atom coming out of the plane of the paper; the other bond to that carbon is then understood to be going behind the plane. Where a dot is absent, hydrogen lies behind the plane, and the bond shown is understood to be coming out of the plane.

Thus, we may encounter *trans*-1,2-dibromocyclopentane represented as

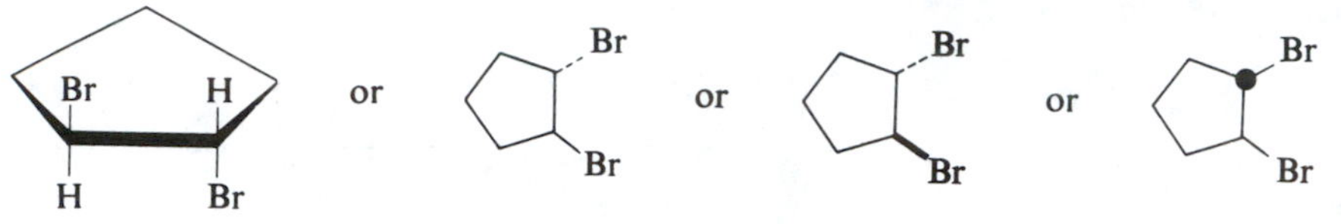

Br H H Br or Br Br or Br Br or Br Br

and the *cis* isomer represented as

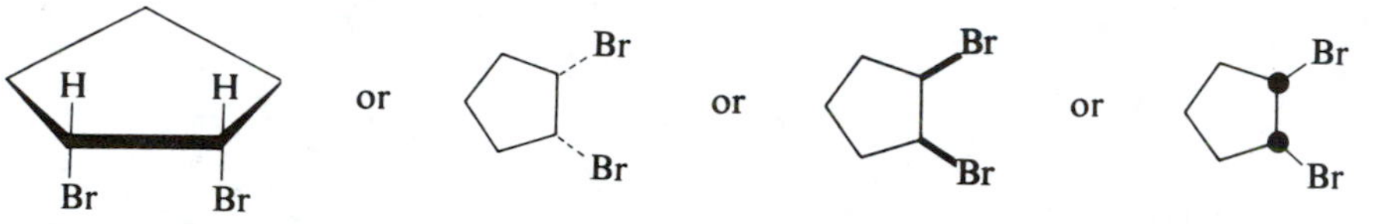

H H Br Br or Br Br or Br Br or Br Br

1. (a) 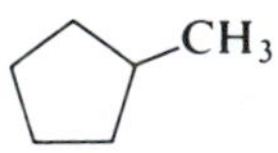 CH_3 (b) 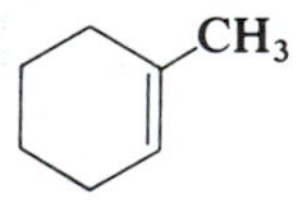 CH_3 (c) 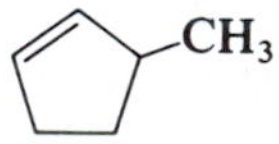 CH_3

(d) 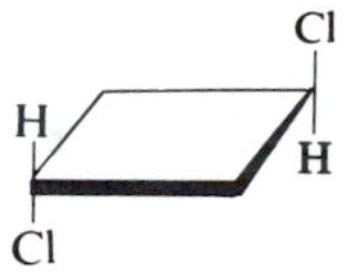 Cl H H Cl (e) 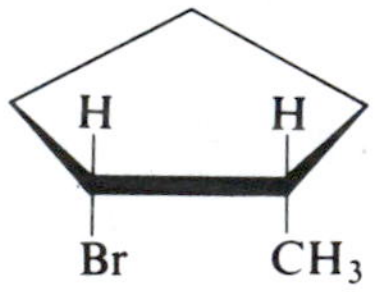 H H Br CH_3

(f) 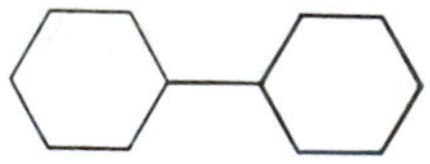 (g) 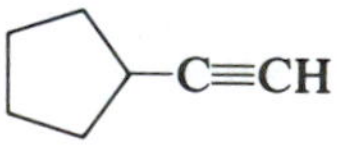 $C{\equiv}CH$

(h) 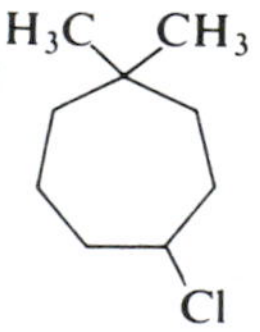 H_3C CH_3 Cl (i) 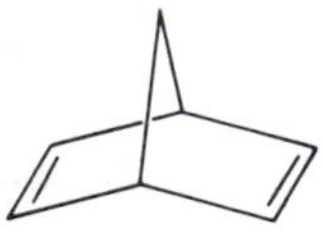 (j) 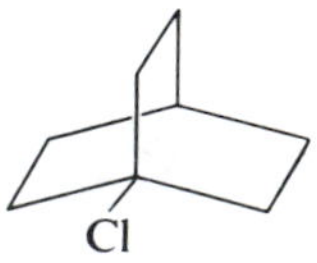 Cl

2. (a) $ClCH_2CH_2CH_2Cl$ *(ring opens)* (b) Cl (c) $CH_3CH_2CH_2OSO_3H$ *(ring opens)*

(d) no reaction (e) Cl (f) no reaction

(g) *anti*-Addition

H Br Br H + enantiomer

(h) Allylic substitution

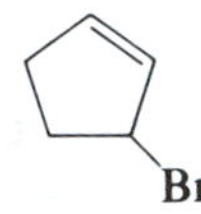

(i) 

(j) *anti*-Addition via cyclic bromonium ion

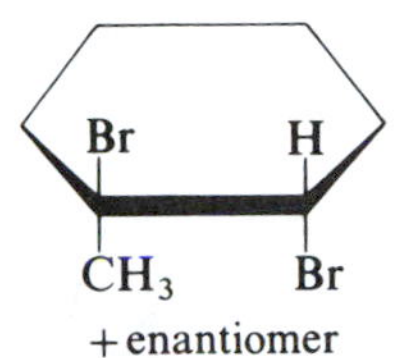

+enantiomer

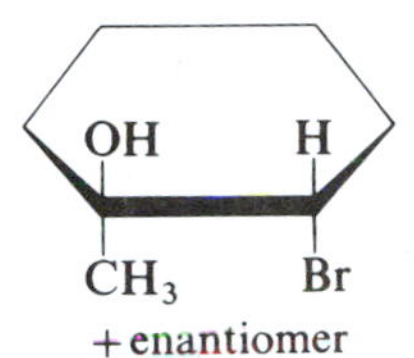

+enantiomer

(k) Free-radical addition of HBr is often stereoselective (*anti*) as it is here.

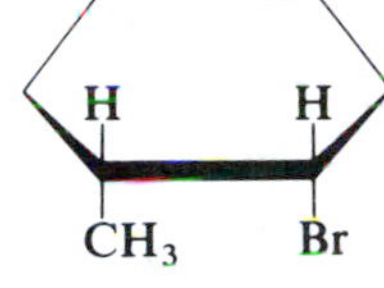

(l) Allylic substitution

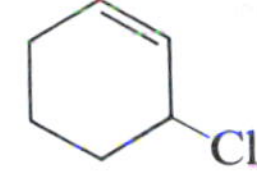

(m) 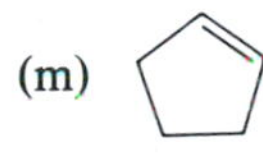

(n) 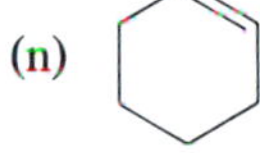

(o) *syn*-Hydroxylation

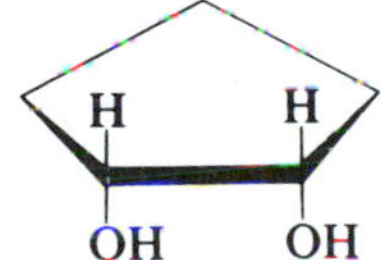

(p) *anti*-Hydroxylation

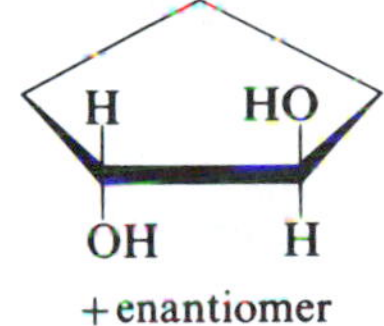

+enantiomer

(q) 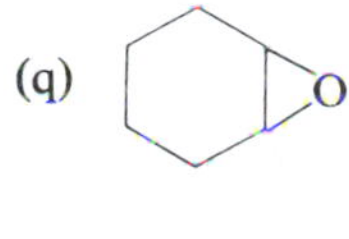

(r) $HOOC(CH_2)_3COOH$

(s) Allylic substitution

Br

(t)

(u)

(v) Dimerization of alkene

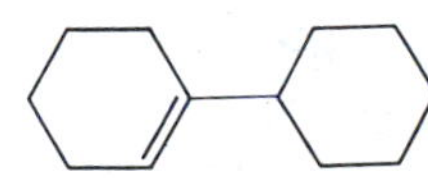

(w) 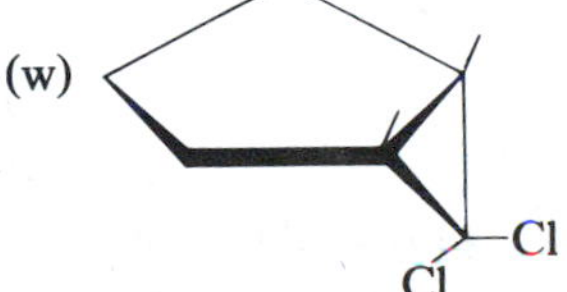

6,6-Dichlorobicyclo[3.1.0]hexane

(x) 
Bicyclo[3.1.0]hexane

(y) 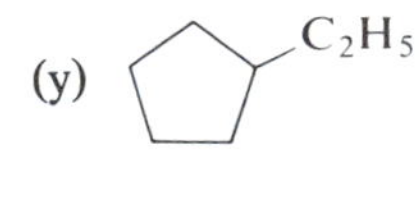

(z) 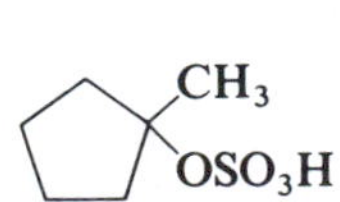

(aa) 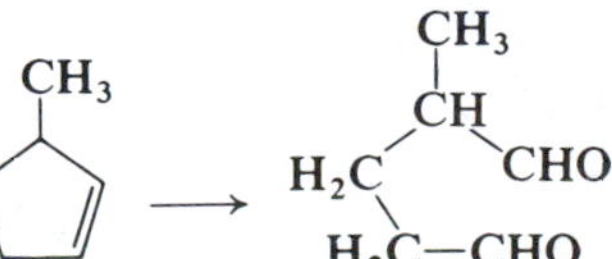

(bb) 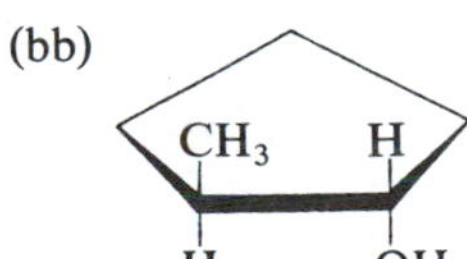

(cc) 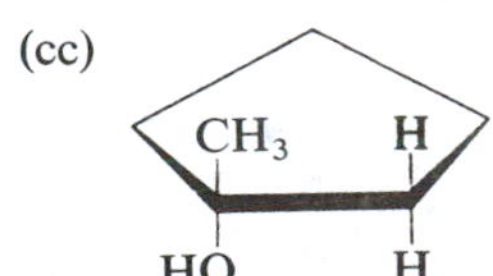

3.

(a) H_2SO_4, heat

(b) Product (a), cat. H_2

(c) 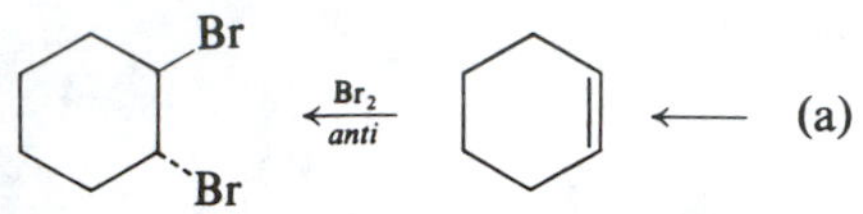 trans-1,2-dibromocyclohexane $\xleftarrow[anti]{Br_2}$ cyclohexene $\longleftarrow$ (a)

(d) 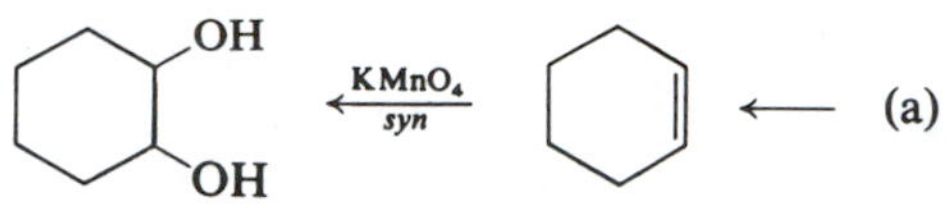 cis-1,2-cyclohexanediol $\xleftarrow[syn]{KMnO_4}$ cyclohexene $\longleftarrow$ (a)

(e) 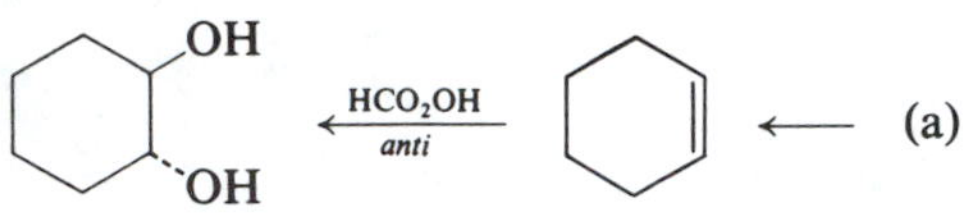 trans-1,2-cyclohexanediol $\xleftarrow[anti]{HCO_2OH}$ cyclohexene $\longleftarrow$ (a)

(f) 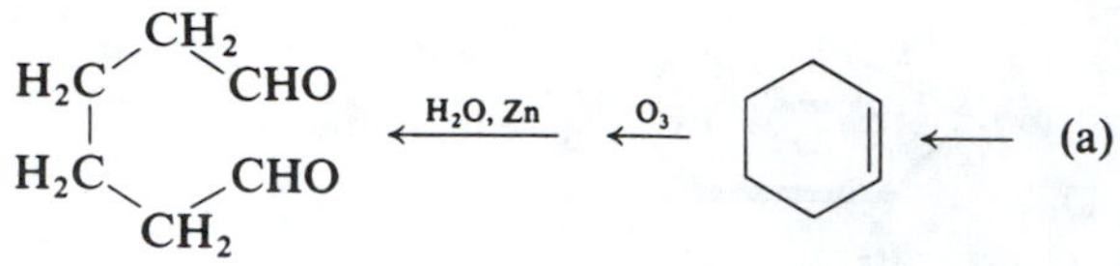 $OHC-CH_2-CH_2-CH_2-CH_2-CHO$ $\xleftarrow{H_2O, Zn}$ $\xleftarrow{O_3}$ cyclohexene $\longleftarrow$ (a)

(g) 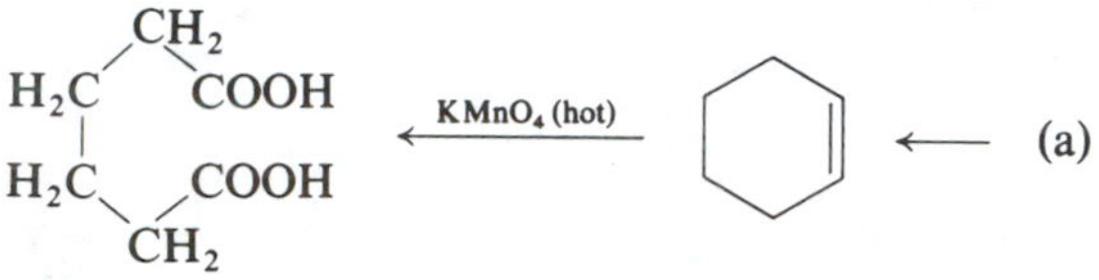 $HOOC-CH_2-CH_2-CH_2-CH_2-COOH$ $\xleftarrow{KMnO_4\ (hot)}$ cyclohexene $\longleftarrow$ (a)

(h) bromocyclohexane (Br, H) $\xleftarrow{HBr}$ cyclohexene $\longleftarrow$ (a)

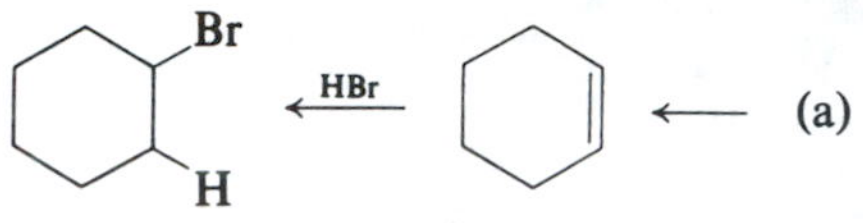

(i) 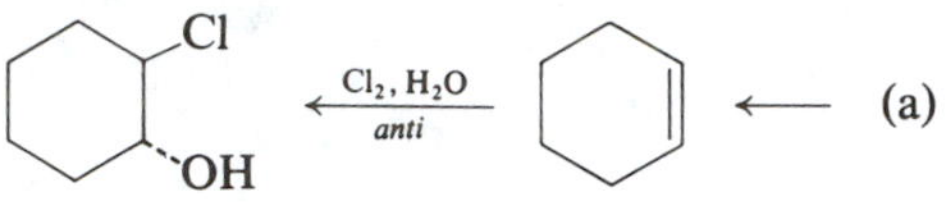 trans-2-chlorocyclohexanol (Cl, OH) $\xleftarrow[anti]{Cl_2, H_2O}$ cyclohexene $\longleftarrow$ (a)

(j) 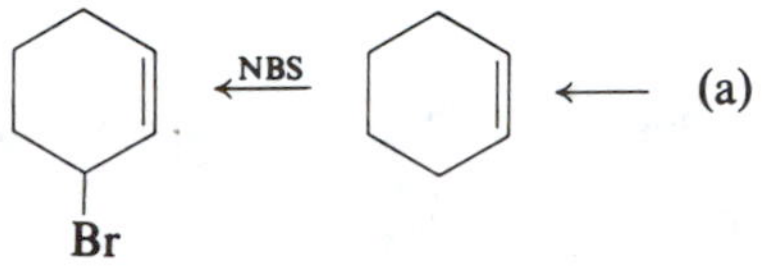 3-bromocyclohexene (Br) $\xleftarrow{NBS}$ cyclohexene $\longleftarrow$ (a)

(k) 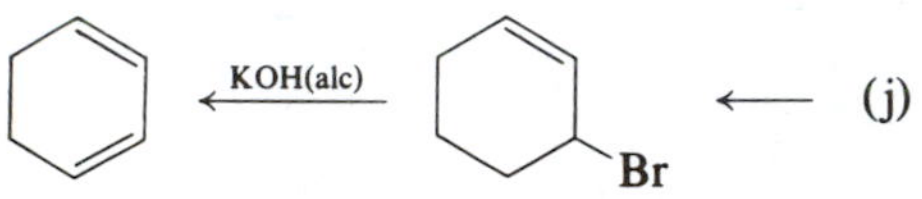 1,3-cyclohexadiene $\xleftarrow{KOH(alc)}$ 3-bromocyclohexene (Br) $\longleftarrow$ (j)

(l) 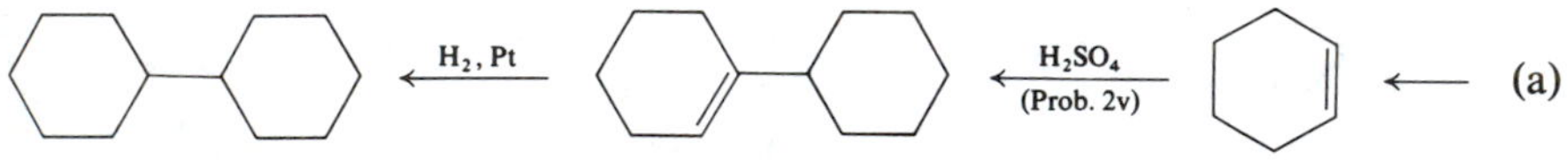 bicyclohexyl $\xleftarrow{H_2, Pt}$ 1-cyclohexylcyclohexene $\xleftarrow[(Prob. 2v)]{H_2SO_4}$ cyclohexene $\longleftarrow$ (a)

(m)

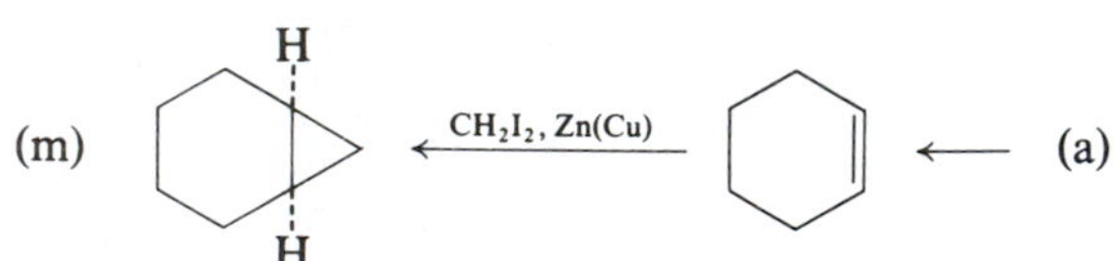

4. We follow the approach outlined in Problems 13.5 and 13.10, above: we draw planar formulas of molecules, look for planes of symmetry, and draw formulas of mirror images; we check superimposability by rotating formulas and by flipping them over; if we feel uncertain, we make models. As always, we try to be *systematic*.

(a)

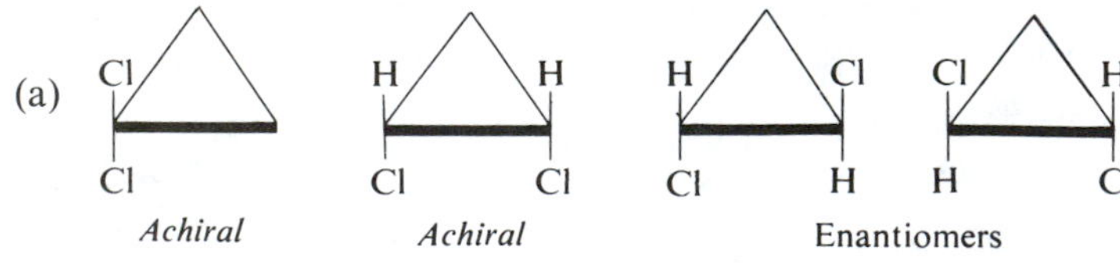

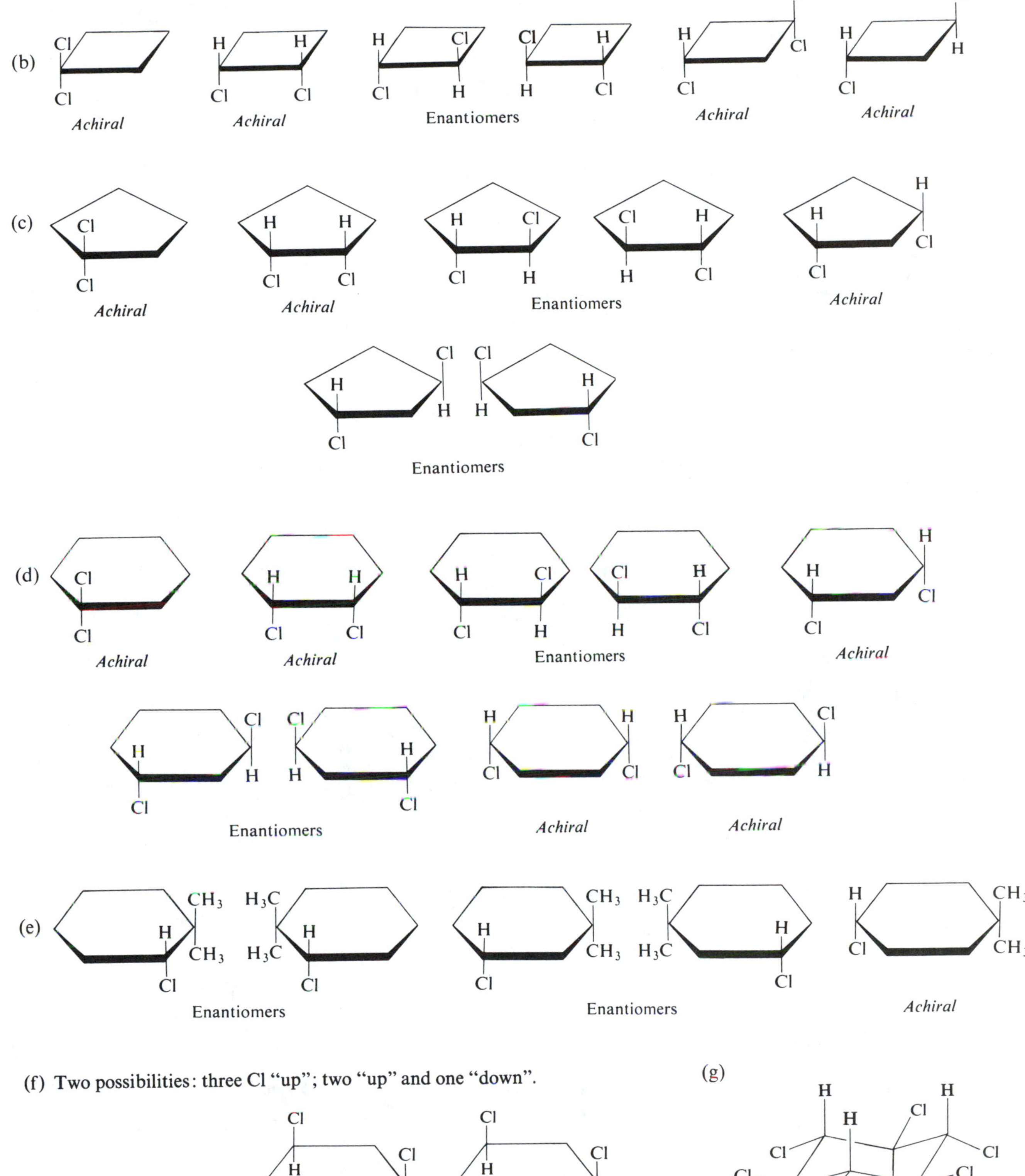

(f) Two possibilities: three Cl "up"; two "up" and one "down".

Achiral Achiral

(g)

All-equatorial

5. (a)

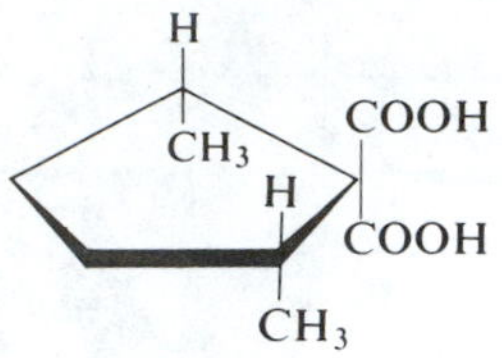

Methyls *cis* 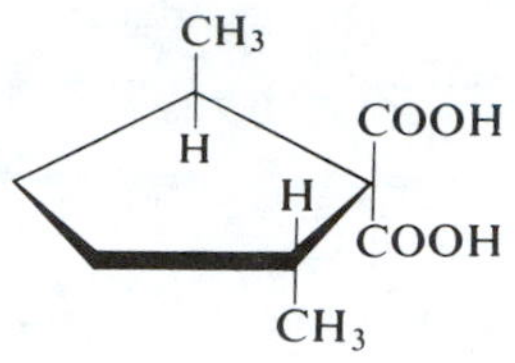 Methyls *trans*

(b) $\xrightarrow[-CO_2]{heat}$ and

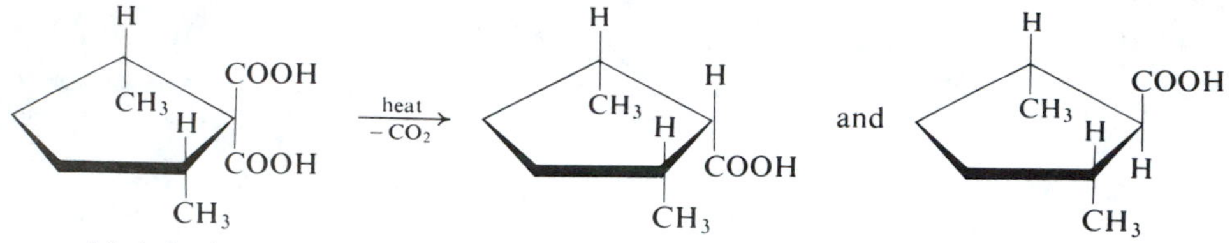

Methyls *cis* *Diastereomers*

Therefore, this must be A.

6. (a)

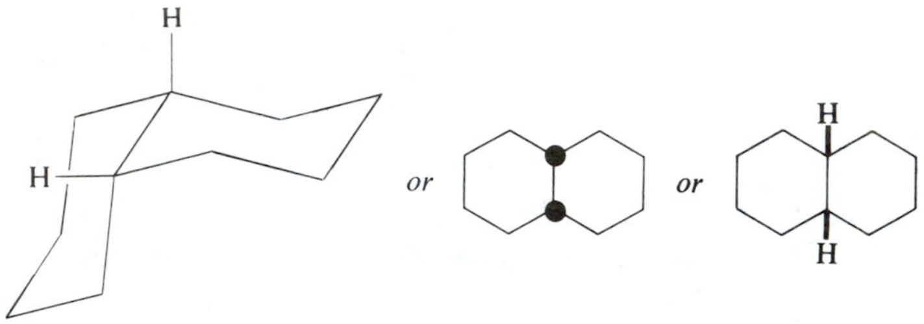

or *or*

cis-Decalin

or *or*

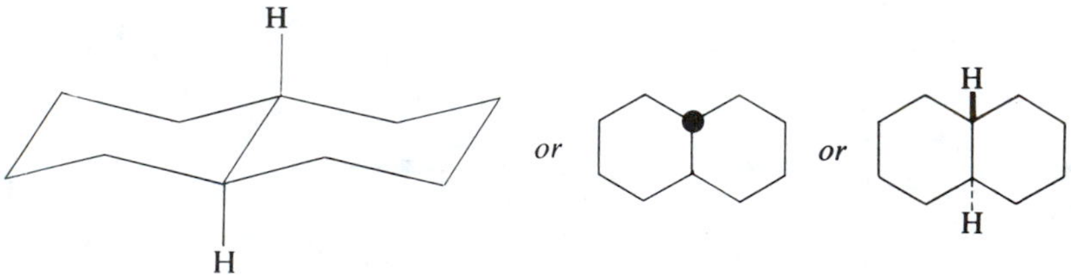

trans-Decalin

(b)

cis:

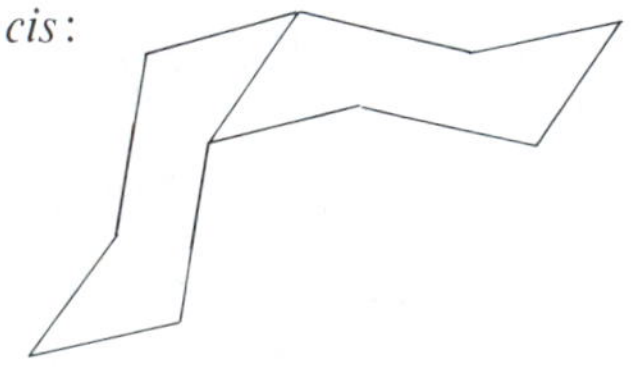

All chair 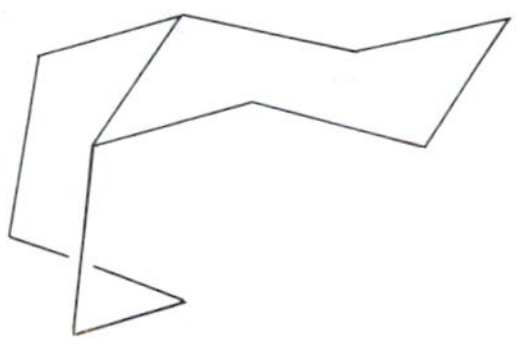 Twist-boat chair 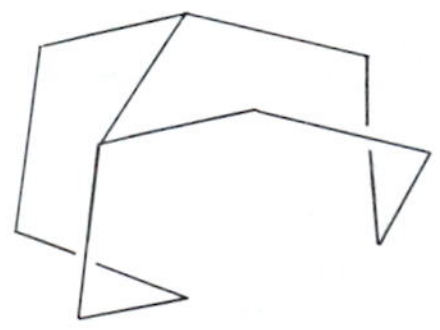 All twist-boat

trans:

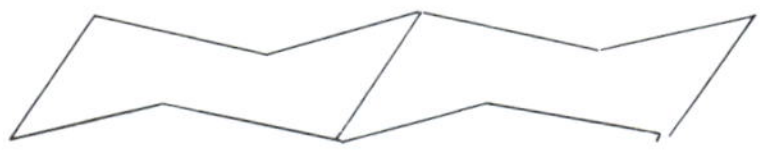

All chair 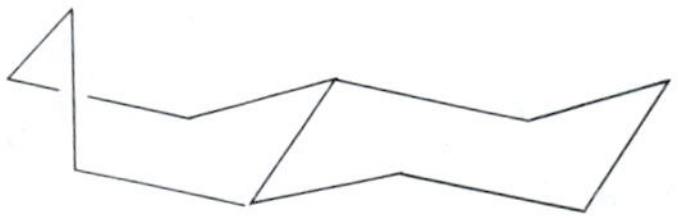 Twist-boat chair

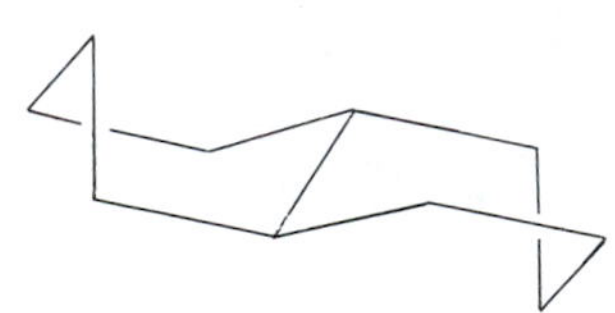

All twist-boat

(c) The all-chair conformation is the most stable in each case.

(d) In the *trans*-decalin, both large substituents (the other ring) are equatorial.

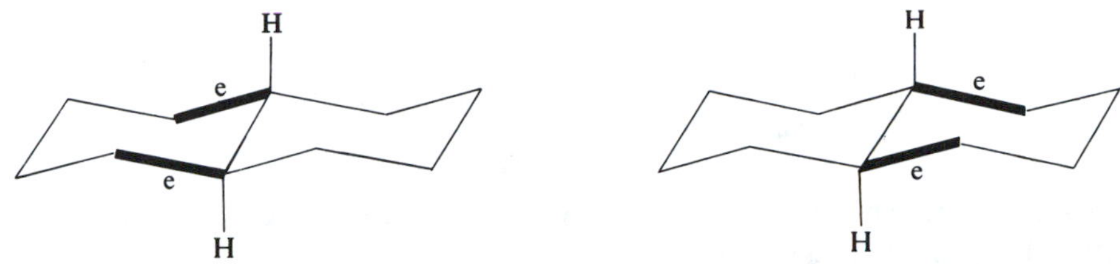

In the *cis* isomer, on the other hand, one of the two large substituents is axial.

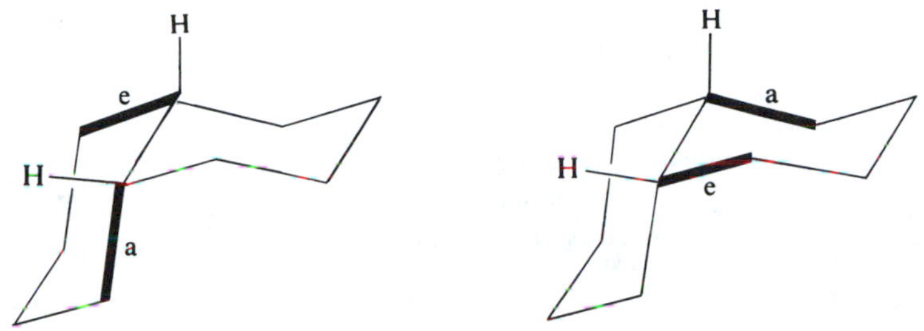

(e) The ease of interconversion does not depend on the energy difference between forms, but on the height of the energy barrier. There is a high energy barrier between the decalins, since a carbon–carbon bond must be broken in the process of interconversion. There is a low energy barrier between the chair and twist-boat forms of cyclohexane, since there interconversion requires only rotation about single bonds (with some increase in angle and torsional strain in the transition state).

7. The *cis* isomer exists in a chair conformation with both *tert*-butyl groups equatorial. The *trans* isomer exists in a twist-boat conformation that accommodates both *tert*-butyl groups in *quasi*-equatorial positions. The difference in energy (5.9 kcal) is thus due essentially to the interactions normally present in the twist conformation.

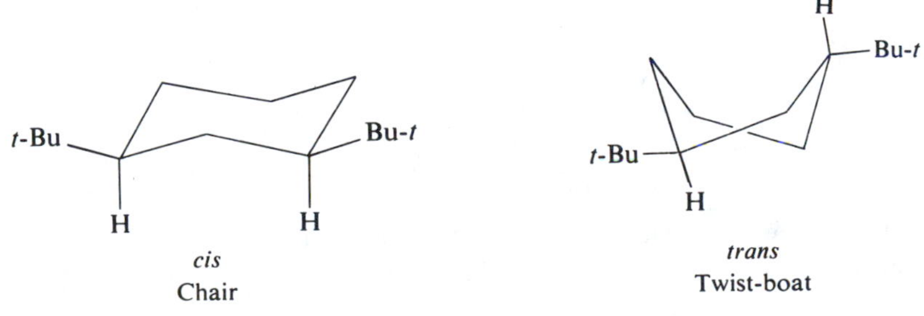

cis
Chair

trans
Twist-boat

(Allinger, N. L.; Freiberg, L. A. "The Energy of the Boat Form of the Cyclohexane Ring"; *J. Am. Chem. Soc.* **1960**, *82*, 2393; or Eliel, E. L.; Allinger, N. L.; Angyal, S. J.; Morrison, G. A. *Conformational Analysis*; Wiley-Interscience: New York, 1965; pp 38–39.)

8. (a) This is essentially an equatorial —CH_3.

CH_3

H

Methylcyclopentane

(b) The situation is very much as for the corresponding cyclohexane derivatives with regard to equatorial–axial relationships and diaxial interactions.

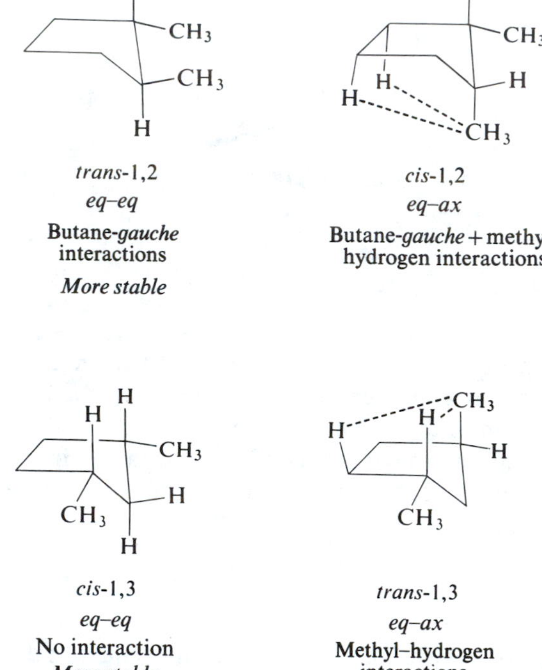

trans-1,2
eq–eq
Butane-*gauche* interactions
More stable

cis-1,2
eq–ax
Butane-*gauche* + methyl–hydrogen interactions

cis-1,3
eq–eq
No interaction
More stable

trans-1,3
eq–ax
Methyl–hydrogen interactions

(Eliel, E. L., *et al.*, reference to Problem 7 above.)

9. (a) Steric factors are controlling. The usual S_N2 order is 1° > 2° > 3°.

Cyclohexyl–CH_2Br (1°) > Cyclohexyl–Br (2°) > 1-CH_3-1-Br-cyclohexane (3°)

(b) Polar factors are controlling: the more stable the cation being formed in the initial ionization, the faster the reaction. The usual order is 3° > 2° > 1°.

1-CH_3-1-Br-cyclohexane (3°) > Cyclohexyl–Br (2°) > Cyclohexyl–CH_2Br (1°)

(c) Under E2 conditions, the more stable the alkene being formed, the faster the reaction. Here, the stability of the products is benzene > conjugated diene > alkene.

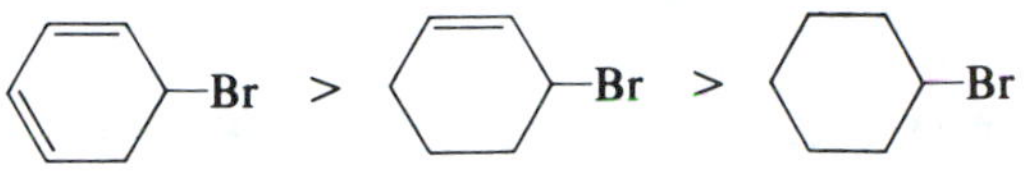

—Br > —Br > —Br

(d) *Anti*-periplanar elimination to the more stable, branched alkene is possible from the *cis* isomer, but not from the *trans* isomer; we therefore expect the *cis* isomer to react more rapidly. (Compare the reactions of neomenthyl and menthyl chlorides, pages 472–473.)

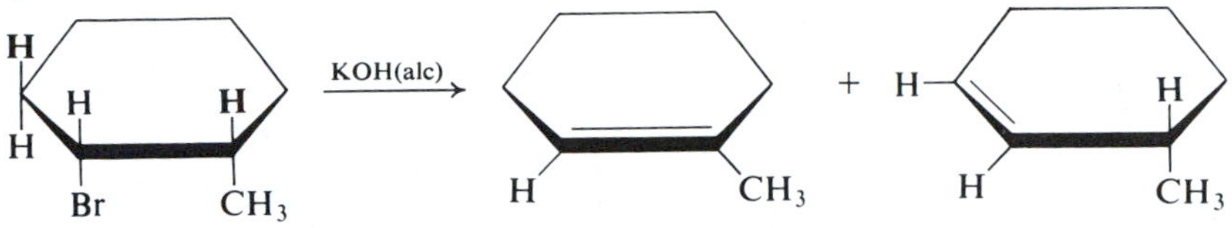

KOH(alc)

cis-2-Bromo-1-methylcyclohexane

More stable alkene
Major product

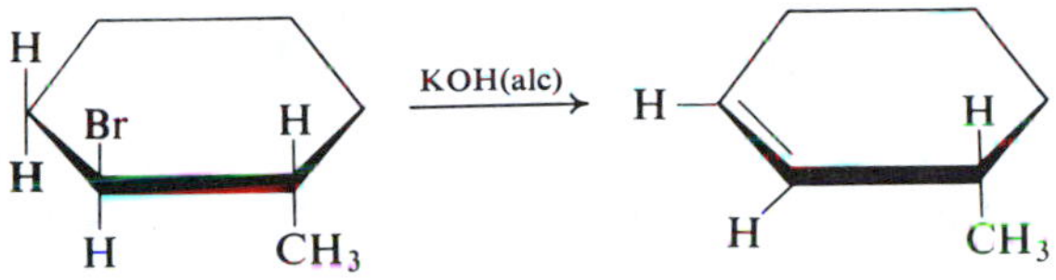

KOH(alc)

trans-2-Bromo-1-methylcyclohexane

Less stable alkene
Only product

10. 1,3,5,5-Tetramethyl-1,3-cyclohexadiene $+ H_2SO_4 \rightleftarrows$ Carbocation bisulfate HSO_4^-

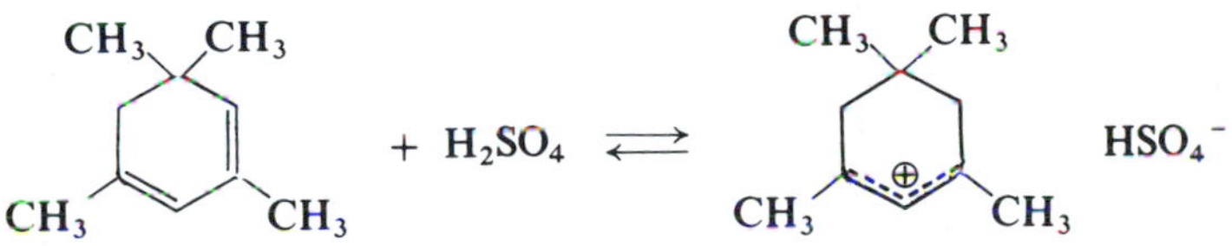

Carbocation bisulfate
A salt

The two particles are the ions: the carbocation and HSO_4^-. In the strongly acidic medium, the diene is converted completely into the allylic cation, which is stabilized by electron release from two methyl groups, one at each end of the allylic system. Upon addition of water, this base abstracts a proton to regenerate the conjugated diene.

(*Problem*: To which end of the conjugated system does the proton add? Why here and not at the other end?)

11. The reaction of 1,2-dimethylcyclopentene illustrates the stereochemistry of the synthesis: *hydroboration–oxidation involves overall* ***syn-addition.***

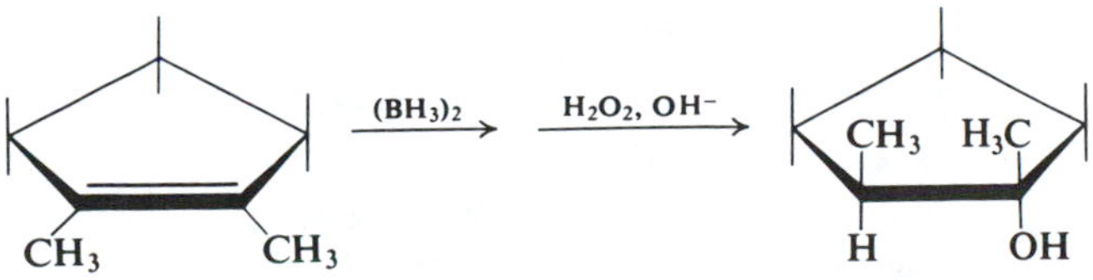

$(BH_3)_2$ → H_2O_2, OH^- →

1,2-Dimethylcyclopentene

cis-1,2-Dimethylcyclopentanol

(To see how this stereochemistry can be applied to open-chain alkenes, try Problem 13.33, below. You can check your answer on page 207 of this Study Guide.)

Problem 13.33 Predict the products of hydroboration–oxidation of:
(a) (*E*)-2-phenyl-2-butene; (b) (*Z*)-2-phenyl-2-butene; (c) 1-methylcyclohexene. (Phenyl is C_6H_5.)

12. (a) *syn*-Hydroxylation. Two fractions: one active, one inactive.

H HO H H —$KMnO_4$→ H HO H H OH OH and H HO OH HO H H

Meso *Inactive* *Chiral* *Active*

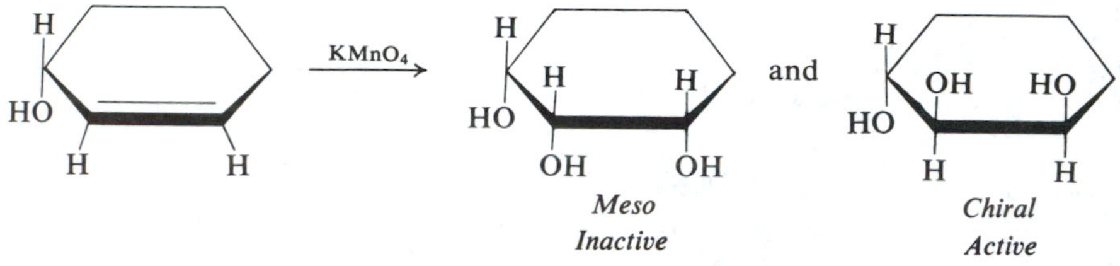

(b) *anti*-Hydroxylation. Two fractions: one active, one inactive.

H HO H H —HCO_2OH→ H HO OH H H OH and H HO H HO OH H

Meso *Inactive* *Chiral* *Active*

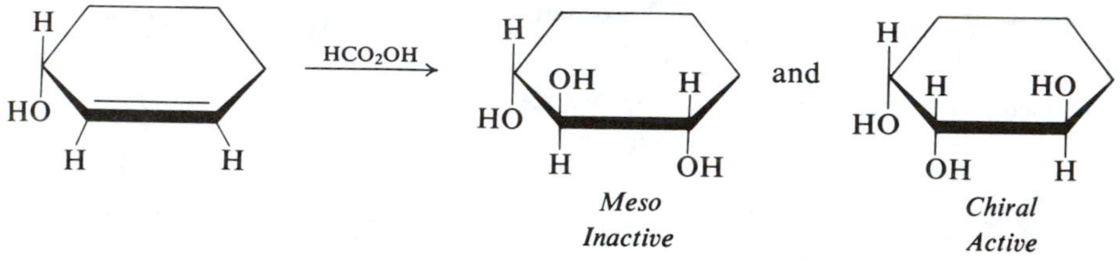

(c) Free-radical substitution. Two fractions: one active, one inactive.

Cl H H H Cl H —Cl_2, 300 °C→ Cl H H Cl Cl H *or* Cl H H H Cl Cl and Cl Cl H H Cl H *or* Cl H Cl H Cl H

Identical *Meso* *Inactive* Identical *Chiral* *Active*

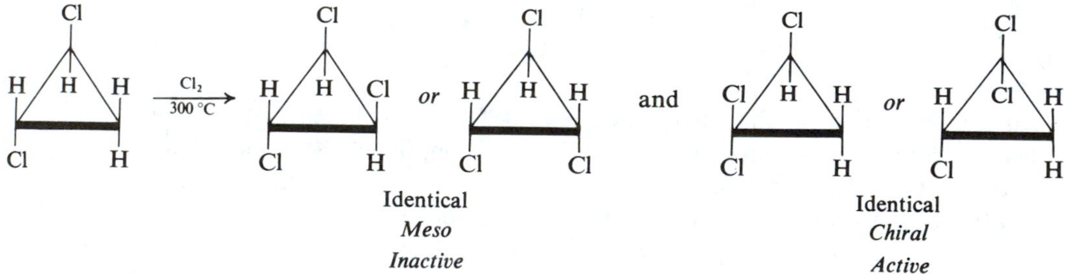

(d) *anti*-Addition. Two fractions: both inactive (racemic).

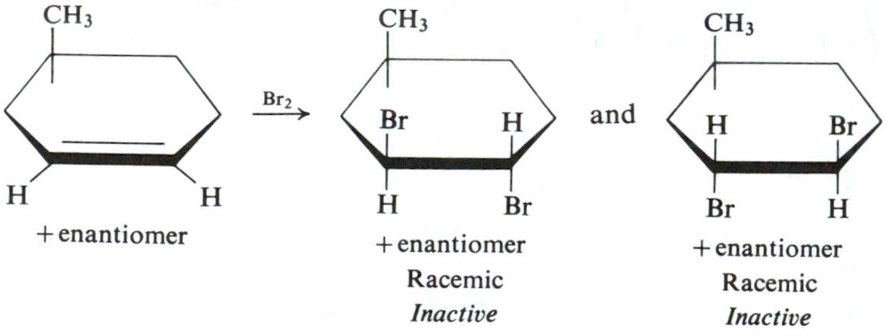

CH_3 H H +enantiomer —Br_2→ CH_3 Br H H Br +enantiomer Racemic *Inactive* and CH_3 H Br Br H +enantiomer Racemic *Inactive*

(e) S_N2 with complete inversion. E2 with loss of either of two equivalent protons. Two fractions: both inactive.

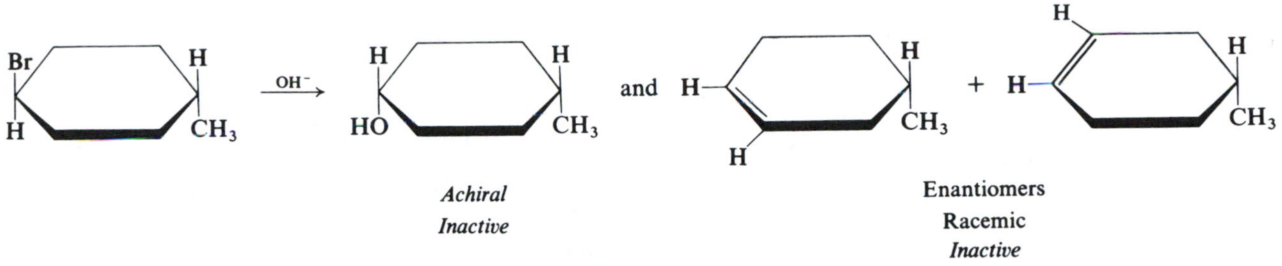

Achiral
Inactive

Enantiomers
Racemic
Inactive

13. Let us look first at the TsCl/*t*-BuOK route. Elimination is evidently *anti*, which can take place only with the orientation observed:

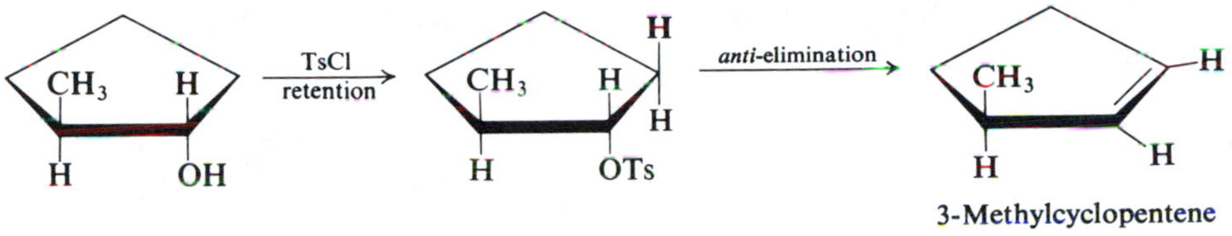

3-Methylcyclopentene

If it were *syn*, elimination could take place with either orientation, and almost certainly would give chiefly the more stable alkene, 1-methylcyclopentene:

+ some 3-methylcyclopentene

1-Methylcyclopentene
Not obtained

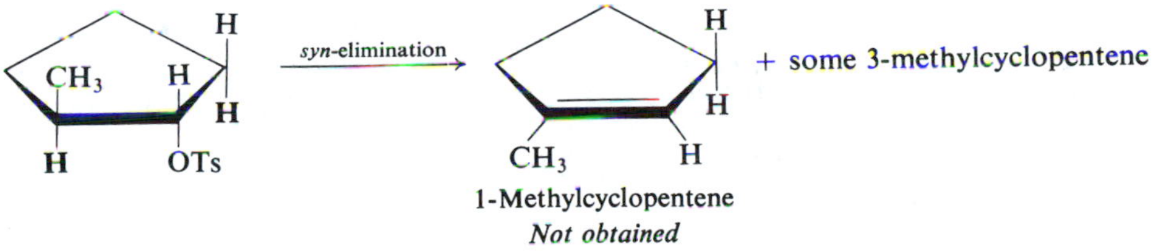

(Note the contrast to the behavior of *trans*-2-phenylcyclopentyl tosylate, in Problem 13.15 above. There, the more powerful effect of the phenyl group forced Saytzeff orientation even though it had to be *syn*.)

The dehydration route involves elimination via a planar carbocation. There are no stereochemical restrictions on this reaction, since the leaving groups are lost in different steps. (Compare Problem 13.14, above.) Orientation is straightforward Saytzeff, with the more stable alkene, 1-methylcyclopentene, predominating.

14. In the transition state for *anti*-elimination from menthyl chloride, for —Cl to be in the required axial position, the bulky methyl and isopropyl groups must also be axial; in either transition state for *anti*-elimination from neomenthyl chloride, the alkyl groups are equatorial.

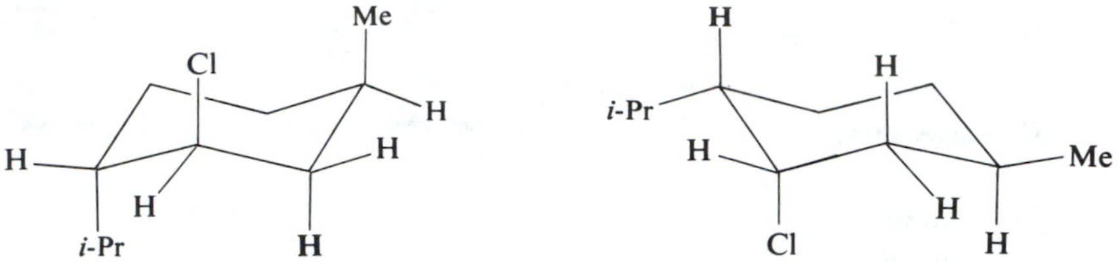

Menthyl chloride

Neomenthyl chloride

15. The *cis* isomer reacts via E2 *anti*-elimination, with —H and —OTs in axial positions. In the *trans* isomer, however, the leaving groups can be axial only if the bulky *tert*-butyl group is also put in an axial position; this is so difficult that the molecule reacts via a carbocation (E1, with some S_N1) mechanism instead.

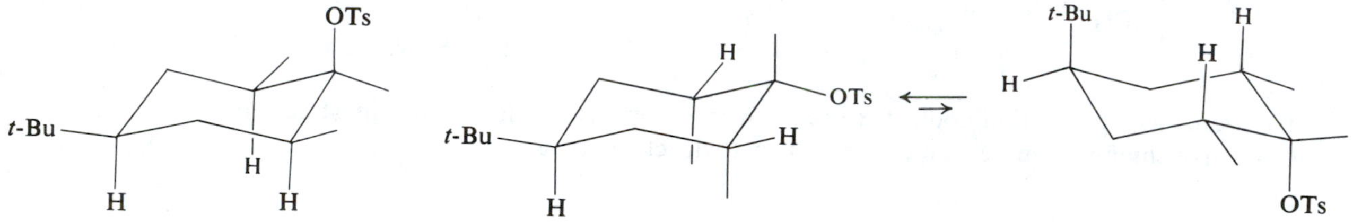

cis Isomer

t-Bu equatorial
—OTs and —H axial
anti E2 *possible*

trans Isomer

t-Bu equatorial
—OTs equatorial
anti E2 *not possible*

t-Bu axial
Highly unstable

16. (a) $HOCH_2CH_2OH$, 1,2-ethanediol

(b) same as (a)

(c) $HOCH_2CH_2OEt$, 2-ethoxyethanol

(d) $HOCH_2CH_2OCH_2CH_2OEt$, ethyl ether of diethylene glycol (see e)

(e) $HOCH_2CH_2OCH_2CH_2OH$, diethylene glycol

(f) $HO(CH_2CH_2O)_3H$, triethylene glycol

(g) $BrCH_2CH_2OH$, 2-bromoethanol

(h) $N{\equiv}CCH_2CH_2OH$, 2-cyanoethanol

(i) $HCOOCH_2CH_2OH$, 1,2-ethanediol monoformate

(j) $H_2NCH_2CH_2OH$, 2-aminoethanol (ethanolamine)

(k) $Et_2NCH_2CH_2OH$, 2-(*N*,*N*-diethylamino)ethanol

(l) $HC{\equiv}COCH_2CH_2OH$, 2-ethynoxyethanol

17. Alkaline hydrolysis takes place with retention of configuration because attack by OH^- occurs at the 1° carbon; no bond to the chiral carbon is broken. Acidic hydrolysis takes place with inversion of configuration because attack by H_2O on the protonated epoxide occurs at the 2° (in this case, chiral) carbon atom.

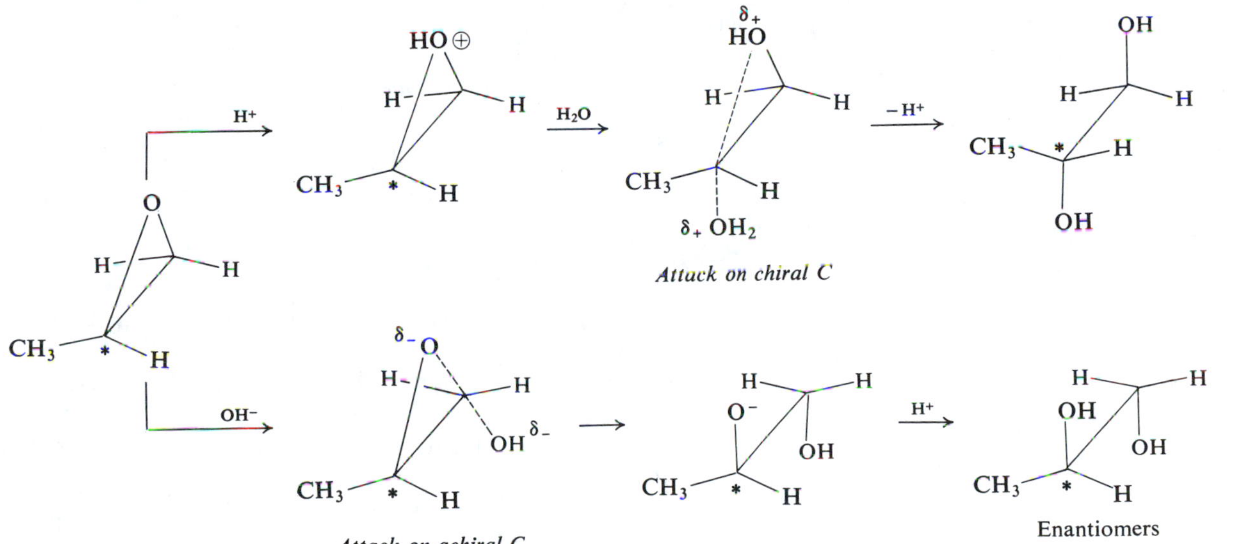

Enantiomers
Opposite rotations

18. (a)

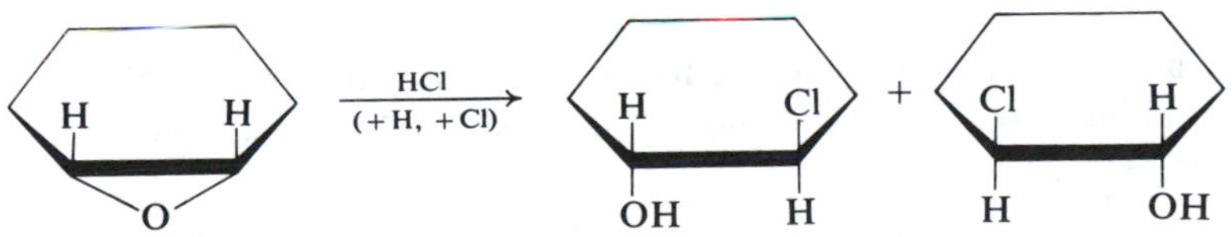

I
trans-2-Chlorocyclohexanol
Racemic

(b) CH$_3$ H $\xrightarrow[\textit{anti}\text{-hydroxylation}]{HCO_2OH}$ OH H CH$_3$ OH + CH$_3$ HO HO H

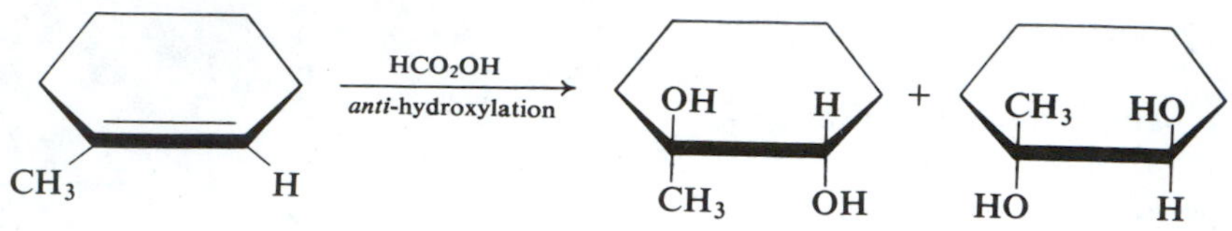

J
Racemic

(c) *syn*-Hydroxylation of the double bond, producing one new chiral carbon (in both configurations), followed by cleavage of the epoxide.

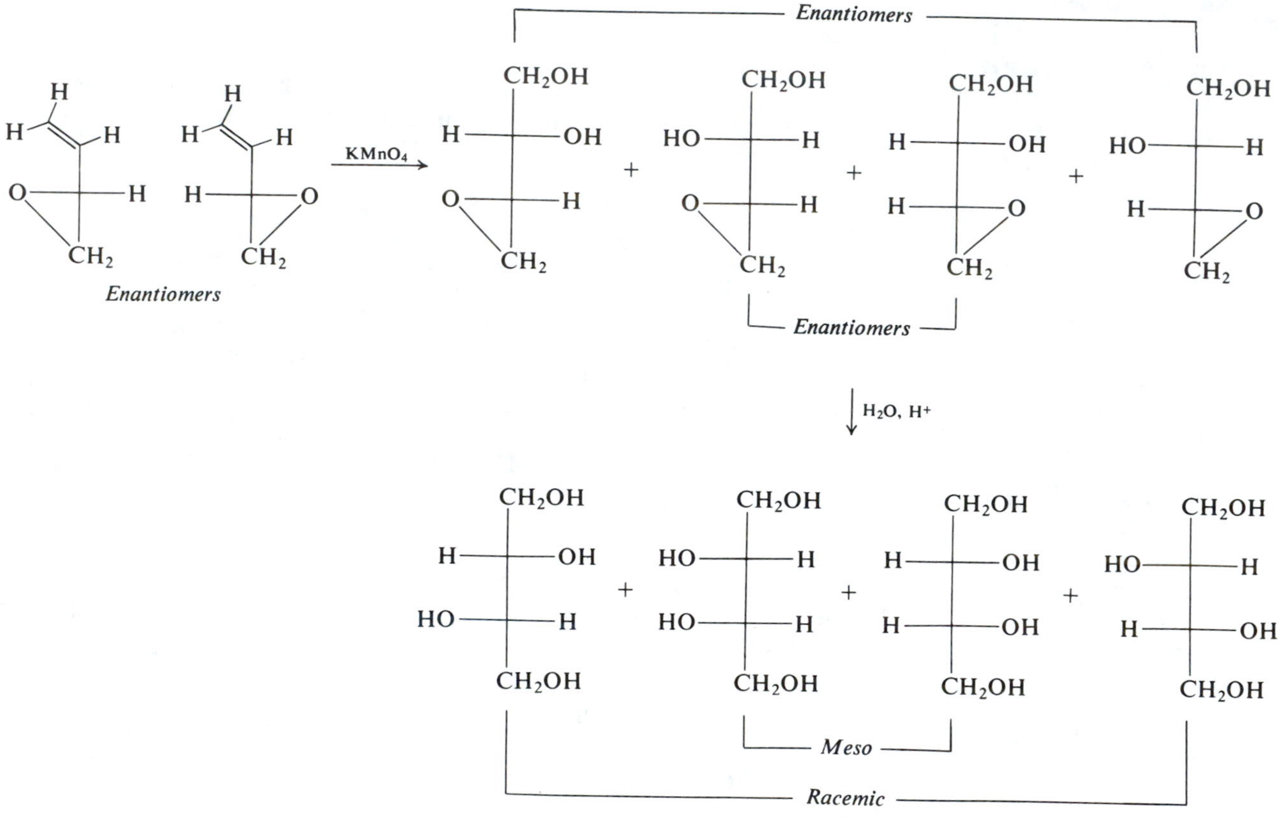

Enantiomers $\xrightarrow{KMnO_4}$ Enantiomers; Enantiomers

$\downarrow$ H_2O, H^+

CH$_2$OH CH$_2$OH CH$_2$OH CH$_2$OH

Meso

Racemic
K

Whatever the mechanism of cleavage, a mixture (K) of *meso* and racemic products will be obtained.

(d) Chlorohydrin formation via a bridged chloronium ion, followed by an intramolecular Williamson synthesis to give an epoxide, and finally cleavage of the epoxide to a diol. (See Figure 13.23, page 201 of this Study Guide.)

(e) Similar to (d), but starting with *trans*-2-butene, and ending up with *meso*-2,3-butanediol. (See Figure 13.24, page 202 of this Study Guide.)

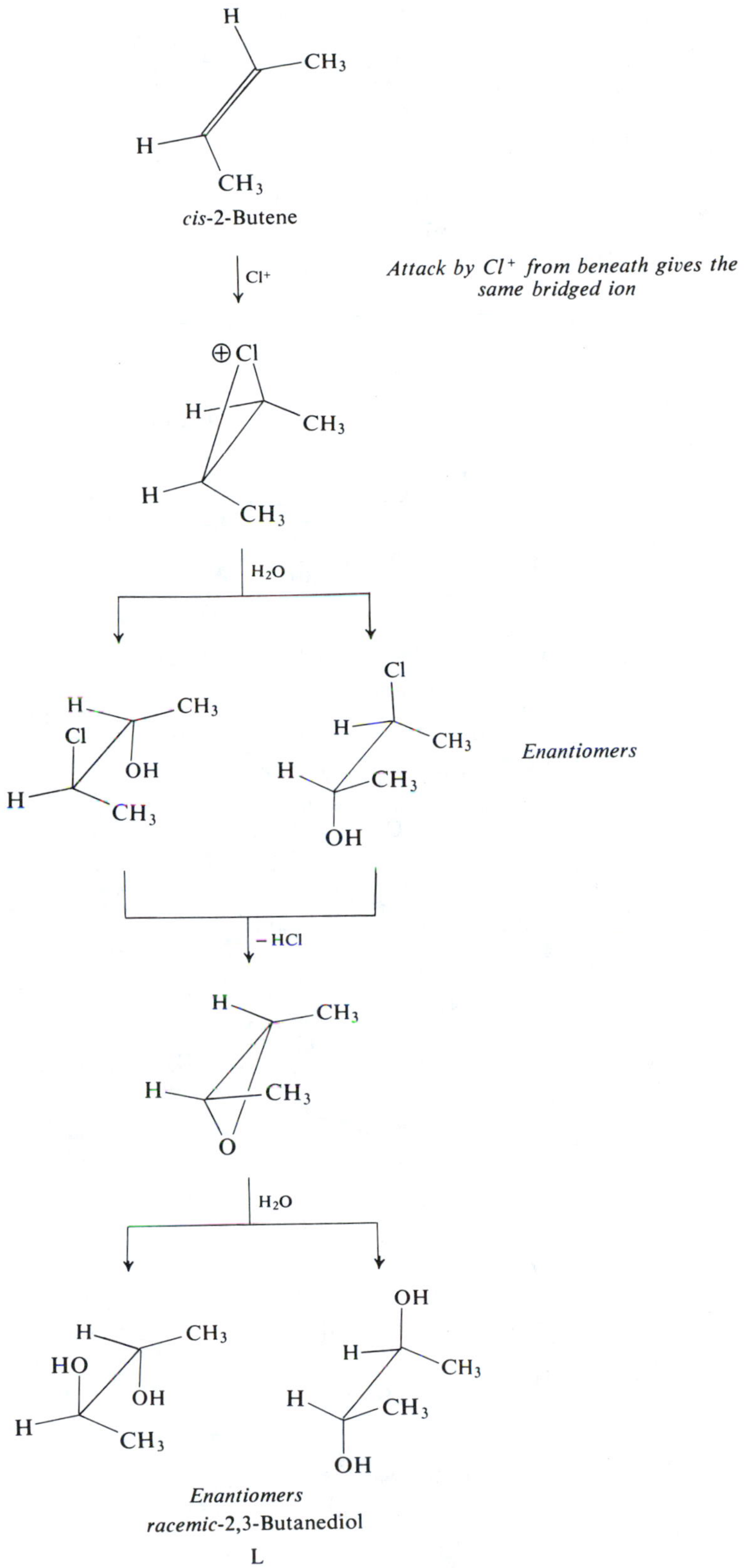

H
CH_3
H
CH_3
cis-2-Butene

Cl^+

Attack by Cl^+ from beneath gives the same bridged ion

⊕Cl
H
CH_3
H
CH_3

H_2O

H CH_3 Cl OH H CH_3

Cl H CH_3 H CH_3 OH

Enantiomers

− HCl

H CH_3 H CH_3 O

H_2O

H CH_3 HO OH H CH_3

OH H CH_3 H CH_3 OH

Enantiomers
racemic-2,3-Butanediol
L

Figure 13.23 Reaction of *cis*-2-butene in Problem 18(d).

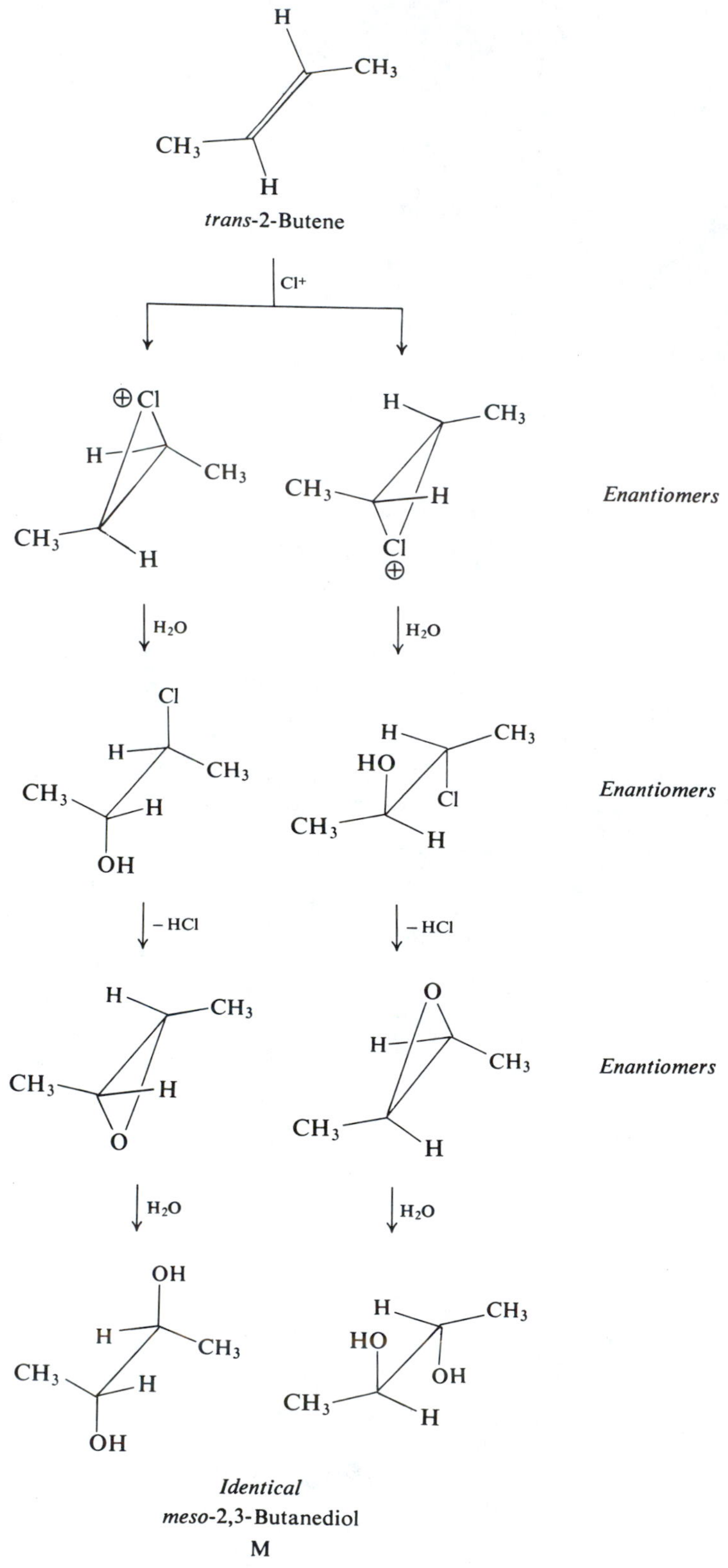

H
CH_3
CH_3
H
trans-2-Butene

Cl^+

⊕Cl
H
CH_3
CH_3
H

H
CH_3
CH_3
H
Cl
⊕

Enantiomers

H_2O H_2O

Cl
H
CH_3
CH_3
H
OH

H
CH_3
HO
Cl
CH_3
H

Enantiomers

– HCl – HCl

H
CH_3
CH_3
H
O

O
H
CH_3
CH_3
H

Enantiomers

H_2O H_2O

OH
H
CH_3
CH_3
H
OH

H
CH_3
HO
OH
CH_3
H

Identical
meso-2,3-Butanediol
M

Figure 13.24 Reaction of *trans*-2-butene in Problem 18(e).

Answers to Synthesis Problems

To save space in this Study Guide, we have sometimes had to omit certain parts of a complete synthesis scheme. When we say "from ROH" or "from RCH_2OH" or "from R_2CHOH", for example, this is understood to mean

$$ROH \xrightarrow{PBr_3} RBr$$

or

$$ROH \xrightarrow{HBr} RBr$$

or

$$RCH_2OH \xrightarrow{C_5H_5NHCrO_3Cl} RCHO$$

or

$$R_2CHOH \xrightarrow{CrO_3} R_2C{=}O$$

or whatever specific process is the correct one. We omit the treatment with water that is an essential step in the work-up of a Grignard reaction mixture:

$$RMgBr + {>}C{=}O \longrightarrow -\underset{R}{\overset{|}{C}}-O^-MgBr^+ \xrightarrow{H_2O} -\underset{R}{\overset{|}{C}}-OH$$

Unless the ground rules for your particular course permit otherwise, you should show all these omitted steps in *your* synthesis schemes.

19. (a) *syn*-Addition of methylene; *syn*-hydrogenation.

$$\text{cis-1,2-Dimethylcyclopropane} \xleftarrow[Zn(Cu)]{CH_2I_2} \text{cis-2-Butene} \xleftarrow{H_2,\ \text{Lindlar cat.}} \underset{\text{2-Butyne}}{CH_3-C{\equiv}C-CH_3}$$

cis-1,2-Dimethylcyclopropane *cis*-2-Butene

$$CH_3-C{\equiv}C-CH_3 \xleftarrow{CH_3I} NaC{\equiv}C-CH_3 \xleftarrow{NaNH_2} HC{\equiv}C-CH_3 \xleftarrow{CH_3I} HC{\equiv}CNa \xleftarrow{NaNH_2} HC{\equiv}CH$$

$$CH_3I \xleftarrow{P + I_2} CH_3OH$$

(b) *syn*-Addition of methylene; *anti*-hydroxylation.

$$\text{trans-1,2-Dimethylcyclopropane (and enantiomer), Racemic} \xleftarrow[\text{Zn(Cu)}]{CH_2I_2} \text{trans-2-Butene} \xleftarrow{\text{Na, NH}_3} CH_3{-}C{\equiv}C{-}CH_3 \text{ (2-Butyne)} \longleftarrow \text{(a)}$$

(and enantiomer)
trans-1,2-Dimethylcyclopropane
Racemic

trans-2-Butene

2-Butyne

(c)

$$\textit{cis}\text{-1,2-di-}n\text{-propylcyclopropane} \xleftarrow{a} \textit{cis}\text{-}(n\text{-Pr})HC{=}CH(n\text{-Pr}) \xleftarrow{b} n\text{-Pr}{-}C{\equiv}C{-}\text{Pr-}n \xleftarrow{c} HC{\equiv}CH + 2\ n\text{-PrBr}$$

$$n\text{-PrBr} \xleftarrow{PBr_3} n\text{-PrOH}$$

cis *cis*

a: CH_2I_2, Zn. (*syn*-Addition of methylene.)

b: H_2, Lindlar's catalyst. (*syn*-Hydrogenation.)

c: Two stages, each involving $NaNH_2$ and then *n*-PrBr (from *n*-PrOH).

(d)

$$\text{1,1-dichloro-2-ethyl-3-methylcyclopropane (+ enantiomer, Racemic)} \xleftarrow{a} \textit{trans}\text{-MeHC}{=}\text{CHEt} \xleftarrow{b} \text{MeC}{\equiv}\text{CEt} \xleftarrow{c} \text{HC}{\equiv}\text{CEt} \xleftarrow{d} \text{HC}{\equiv}\text{CH}$$

+enantiomer
Racemic

trans

a: $CHCl_3$, *t*-BuO^- (from *t*-BuOH). (*syn*-Addition of CCl_2.)

b: Li, NH_3. (*anti*-Hydrogenation.)

c: $NaNH_2$, then MeI (from MeOH).

d: $NaNH_2$, then EtBr (from EtOH).

20. (a) Conc. H_2SO_4: cyclopropane positive.
(b) $KMnO_4$, or Br_2/CCl_4: alkene positive.
(c) Conc. H_2SO_4: cyclopropane positive.
(d) $KMnO_4$, or Br_2/CCl_4: alkene positive.
(e) $KMnO_4$, or Br_2/CCl_4: alkene positive.

(f) $KMnO_4$, or Br_2/CCl_4: alkene positive.
(g) CrO_3/H_2SO_4: alcohol positive.
(h) Br_2/CCl_4: alkene positive. Or, CrO_3/H_2SO_4: alcohol positive.
(i) Br_2/CCl_4, or $KMnO_4$: alkene positive. CrO_3/H_2SO_4: alcohol positive. $AgNO_3$: halide positive. Alkane negative to all the tests.

21. This elimination (like addition) is *anti* (Problem 9, page 385), and removes the same (radioactive) Br's that were added.

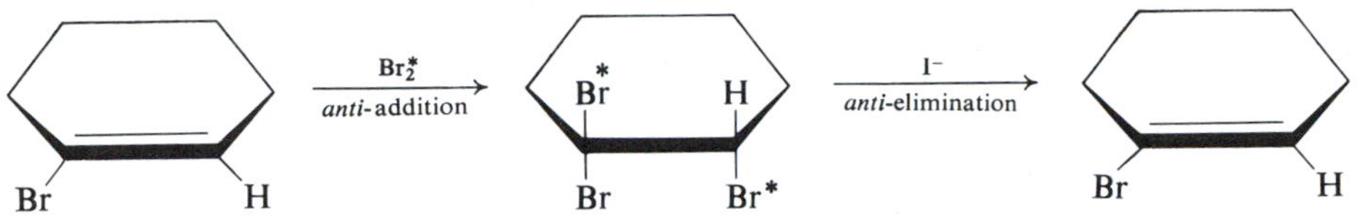

22. (a) $C_{10}H_{22}$ alkane
$-C_{10}H_{18}$
4H *Two rings*

(b)

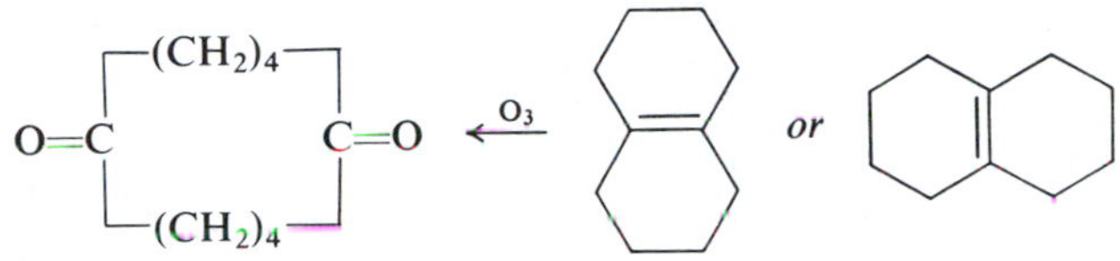

III

23. (a) $C_{10}H_{20}$ *p*-menthane
$-C_{10}H_{16}$ limonene
4H *Two double bonds* in limonene

$C_{10}H_{22}$ alkane
$-C_{10}H_{20}$ *p*-menthane
2H *One ring* in limonene

(b) One of the original 10 carbons is missing from the oxidation product IV; presumably it was lost as CO_2. On this assumption, and knowing that there are two double bonds and one ring, we arrive at the following possible structures for limonene:

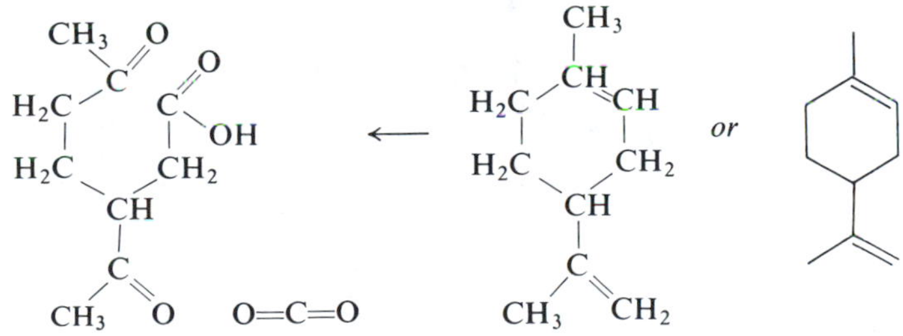

IV + CO_2

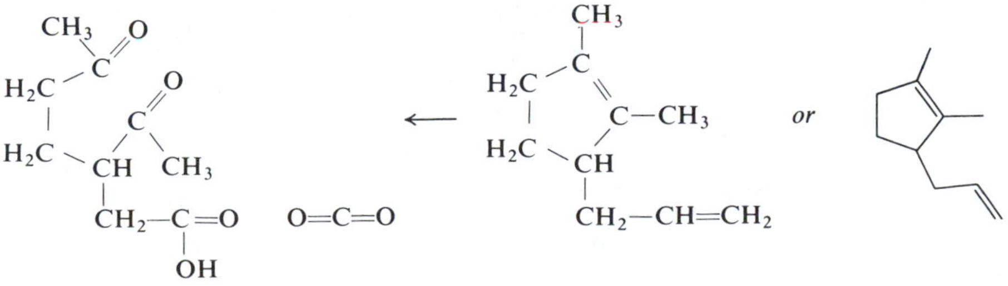

IV + CO_2

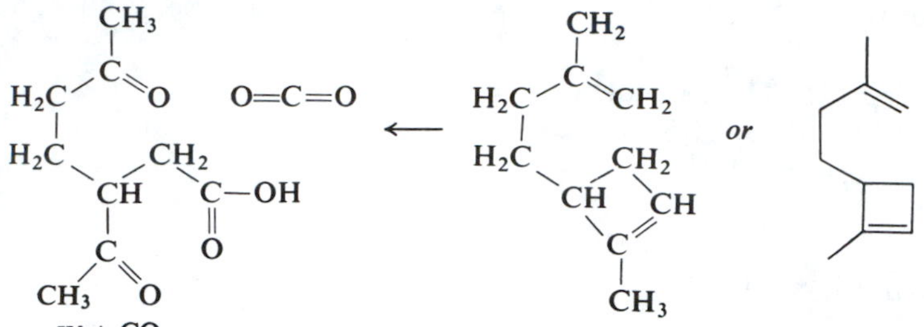

CH_3 C=O, H_2C, H_2C, CH, CH_2, C—OH (=O), C(=O) CH_3 $O=C=O$ ← CH_2 C, H_2C, CH_2, H_2C, CH, CH, C, CH_3 *or*

IV + CO_2

(c) The most likely structure is

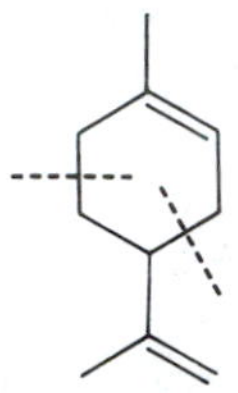

Limonene
Two isoprene units

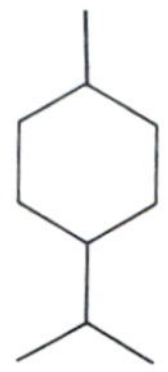

p-Menthane

(d) The alcohol could be either of the tertiary alcohols shown; actually, it is the one labeled α-terpineol.

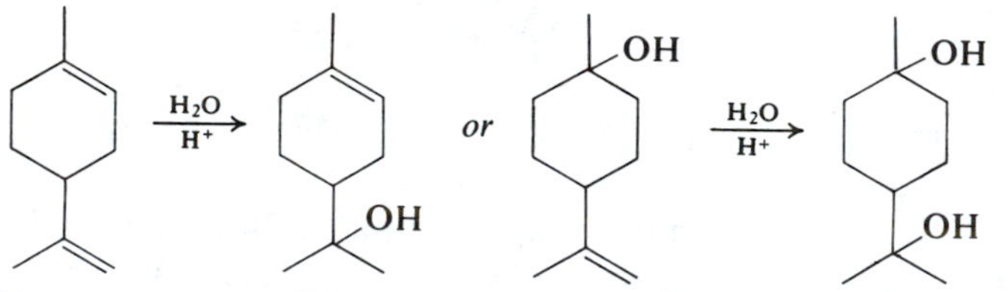

Limonene $\xrightarrow[H^+]{H_2O}$ α-Terpineol *or* (OH) $\xrightarrow[H^+]{H_2O}$ Terpin hydrate

(e) Terpin hydrate is the di-*tert*-alcohol.

24. (a)

$C_{10}H_{20}$ *p*-menthane
$-C_{10}H_{16}$ α-terpinene

4H *Two double bonds* in α-terpinene

$C_{10}H_{22}$ alkane
$-C_{10}H_{20}$ *p*-menthane

2H *One ring* in α-terpinene

(b) Using the skeleton of the reduction product, *p*-menthane (Problem 23, above), as a clue, we can account for the eight-carbon oxidation product V and the missing two carbons as follows:

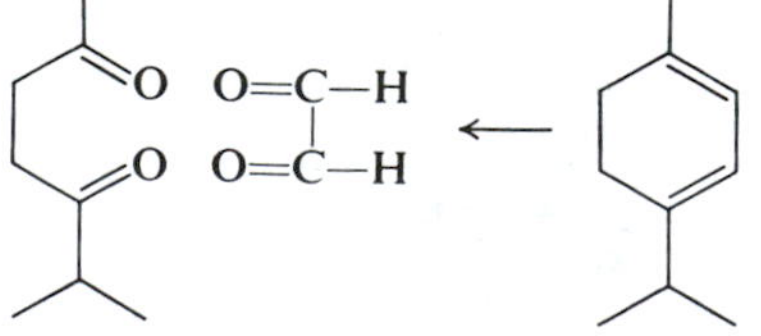

V (O, O) O=C—H, O=C—H ← α-Terpinene

(c) VI is most likely formed as follows:

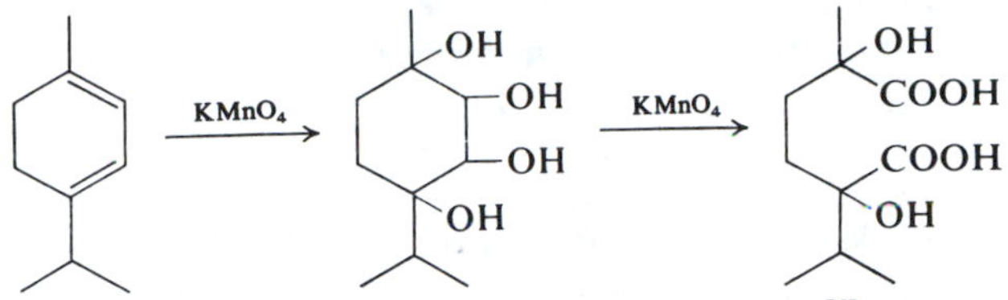

α-Terpinene $\xrightarrow{KMnO_4}$ Tetrahydroxy derivative $\xrightarrow{KMnO_4}$ VI

OH OH OH OH — OH COOH COOH OH

VI
3° alcohols unoxidized

25. The key step, ring closure, involves the familiar addition of a carbocation to an alkene (Sec. 9.15). (Compare the answer to Problem 13, page 423.)

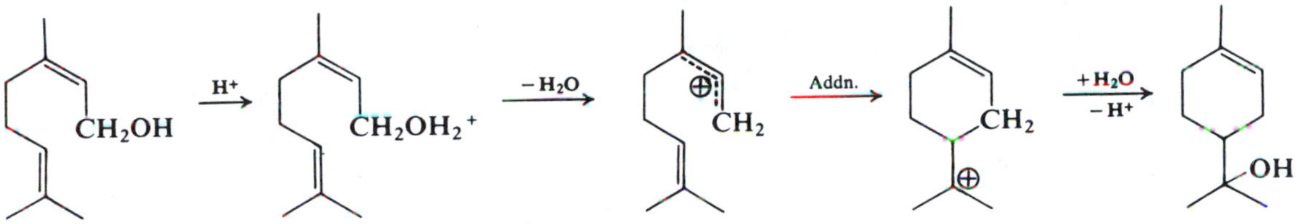

Nerol (CH_2OH) $\xrightarrow{H^+}$ Oxonium ion ($CH_2OH_2^+$) $\xrightarrow{-H_2O}$ Allylic cation *Open-chain* ($\oplus$ CH_2) $\xrightarrow{\text{Addn.}}$ 3° cation *Cyclic* (CH_2, $\oplus$) $\xrightarrow[-H^+]{+H_2O}$ α-Terpineol (OH)

13.33 (On page 196 of this Study Guide.) *syn*-Hydration, with anti-Markovnikov orientation in each case.

(a) *cis* (*E*) alkene: H, CH_3, C_6H_5, CH_3 $\xrightarrow{HB/O}$ product (OH, H, CH_3, H, C_6H_5, CH_3) and (H, CH_3, OH, C_6H_5, CH_3, H) ≡

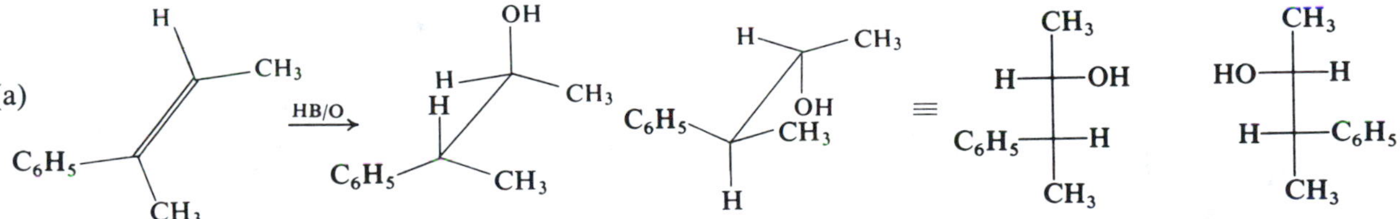

$$\begin{array}{c} CH_3 \\ H-\!\!\!-\!\!\!-OH \\ C_6H_5-\!\!\!-\!\!\!-H \\ CH_3 \end{array} \qquad \begin{array}{c} CH_3 \\ HO-\!\!\!-\!\!\!-H \\ H-\!\!\!-\!\!\!-C_6H_5 \\ CH_3 \end{array}$$

Enantiomers

(b) (*Z*) alkene: H, CH_3, CH_3, C_6H_5 $\xrightarrow{HB/O}$ product (OH, H, CH_3, H, CH_3, C_6H_5) and (H, CH_3, OH, CH_3, C_6H_5, H) ≡

$$\begin{array}{c} CH_3 \\ H-\!\!\!-\!\!\!-OH \\ H-\!\!\!-\!\!\!-C_6H_5 \\ CH_3 \end{array} \qquad \begin{array}{c} CH_3 \\ HO-\!\!\!-\!\!\!-H \\ C_6H_5-\!\!\!-\!\!\!-H \\ CH_3 \end{array}$$

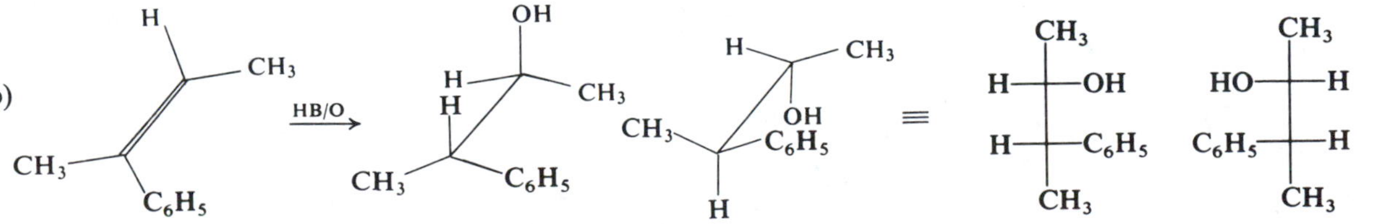

Enantiomers

(c)

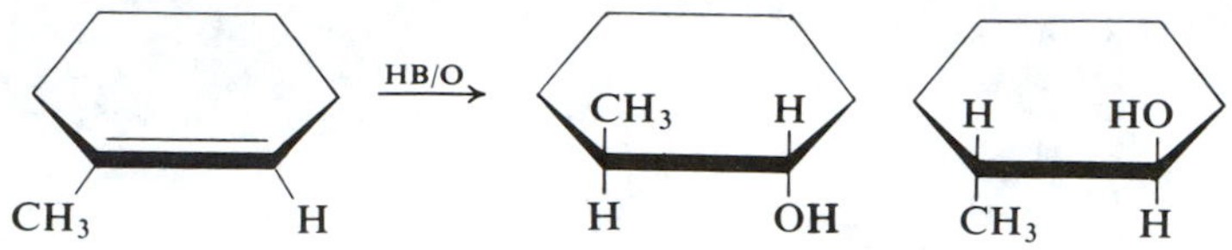

CH_3 H $\xrightarrow{HB/O}$ CH_3 H H OH H HO CH_3 H

Enantiomers

(Brown, H. C. *Hydroboration*; W. A. Benjamin: New York, 1962; especially Chapters 6–8. Brown, H. C. *Boranes in Organic Chemistry*; Cornell University Press: Ithaca, NY, 1972; pp 255–280. The latter book contains a fascinating personal account not only of the discovery of organoboranes and their chemistry, but of the career of one of the most productive of all organic chemists.)

14

Aromaticity

Benzene

14.1 (a)

$$\begin{array}{lcll} \text{benzene} + 3H_2 = \text{cyclohexane} & \Delta H & -49.8 \text{ kcal} \\ -(\text{cyclohexadiene} + 2H_2 = \text{cyclohexane} & \Delta H & -55.4 \text{ kcal}) \\ \hline \text{benzene} + H_2 = \text{cyclohexadiene} & \Delta H & +5.6 \text{ kcal} \end{array}$$

(b)

$$\begin{array}{lcll} \text{cyclohexadiene} + 2H_2 = \text{cyclohexane} & \Delta H & -55.4 \text{ kcal} \\ -(\text{cyclohexene} + H_2 = \text{cyclohexane} & \Delta H & -28.6 \text{ kcal}) \\ \hline \text{cyclohexadiene} + H_2 = \text{cyclohexene} & \Delta H & -26.8 \text{ kcal} \end{array}$$

14.2 (a)

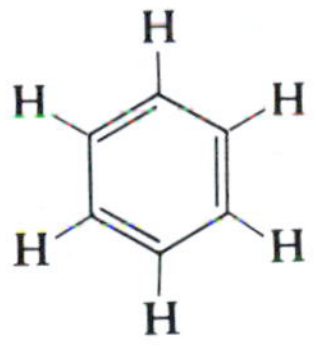

$$\begin{array}{lcl} 6\ \text{C—H bonds} = 6 \times 54.0 & = & 324.0 \text{ kcal} \\ 3\ \text{C=C bonds} = 3 \times 117.4 & = & 352.2 \\ 3\ \text{C—C bonds} = 3 \times 49.3 & = & \underline{147.9} \\ \text{calcd.} & = & 824.1 \text{ kcal} \end{array}$$

(b) 824.1 – 789.1 = 35.0 kcal greater than observed value.
(calc) (obs)

14.3 The sp^2–s character of the C—H bond in benzene. It should be (and is) shorter, and thus stronger, than the sp^3–s C—H bond in cyclohexane.

14.4 (a) No; (b) see Sec. 30.6.

14.5 (a)

(b) Similar to Figs. 14.2 and 14.3.

(c) Six, two "from" each N.

14.6 (a) Compound I is the salt, cyclopropenyl hexachloroantimonate:

$SbCl_6^-$

The same cation is formed by the $AgBF_4$ treatment. It is symmetrical, and all three protons are equivalent.

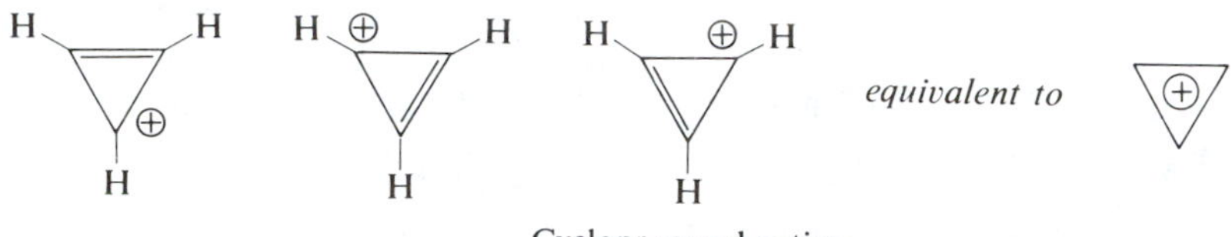

Cyclopropenyl cation

The cyclopropenyl cation is even more stable than the allyl cation: 20 kcal more stable, relative to the parent chloride. The unusual stability suggests not just resonance stabilization, but aromaticity.

(b) The cyclopropenyl cation contains two π electrons, which fits the Hückel $4n + 2$ rule for $n = 0$.

(Breslow, R.; Groves, J. T. "Cyclopropenyl Cation. Synthesis and Characterization"; *J. Am. Chem. Soc.* **1970**, *92*, 984.)

14.7

Six dibromonitrobenzenes

M.p. +6 °C
ortho

M.p. −7 °C
meta

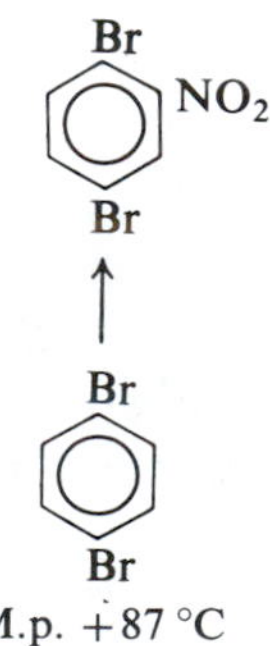

Dibromobenzenes

14.8

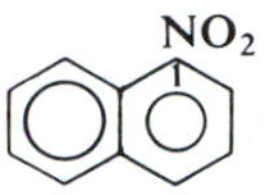

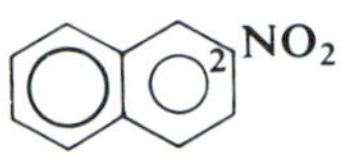

NO_2 (1-nitronaphthalene) NO_2 (2-nitronaphthalene)

Mononitro: two

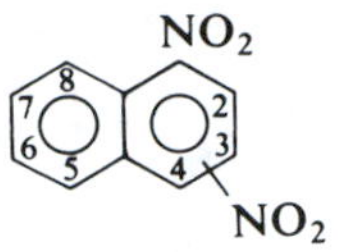

Seven 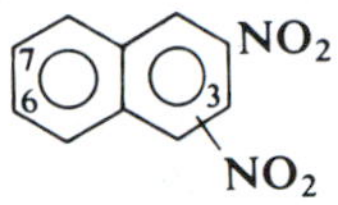 *Three*

Dinitro: ten

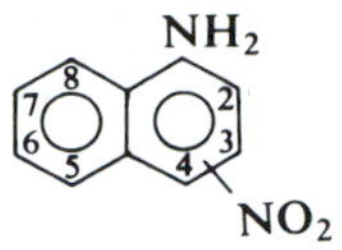

Seven 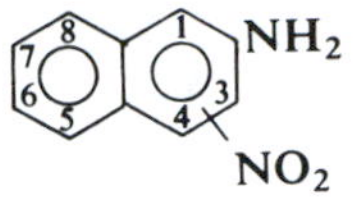 *Seven*

Nitronaphthylamines: fourteen

14.9 Examination of structures I, II, and III on page 511 shows us that this difference in bond lengths is to be expected. The C(1)–C(2) bond is double in two structures and single in only one; the C(2)–C(3) bond is single in two structures and double in only one. We would therefore expect the C(1)–C(2) bond to have more double-bond character than single, and the C(2)–C(3) bond to have more single-bond character than double. The greater the double-bond character, the shorter the bond.

14.10 KOH takes out CO_2 and H_2O; the other liquids would not do this.

14.11

$$\underset{(740\text{ mm})}{\text{partial pressure } N_2} = \underset{(746\text{ mm})}{\text{total pressure}} - \underset{(6\text{ mm})}{\text{partial pressure } H_2O}$$

$$\text{vol. } N_2 \text{ at S.T.P.} = 1.31 \times \frac{273}{293} \times \frac{740}{760} \text{ mL}$$

$$\text{mmoles } N_2 = \frac{\text{mL gas (at S.T.P.)}}{22.4 \text{ mL/mmol}}$$

$$\text{wt. N} = \text{mmoles } N_2 \times 28.0 \text{ mg/mmol}$$

$$\%\text{N} = \frac{\text{wt. N}}{\text{wt. sample}} \times 100 = 1.31 \times \frac{273}{293} \times \frac{740}{760} \times \frac{1}{22.4} \times 28.0 \times \frac{1}{5.72} \times 100$$

$$\%\text{N} = 26.0\% \text{ N}$$

14.12

$$\text{mmoles } NH_3 = \text{mL HCl} \times \text{conc. HCl} = 5.73 \text{ mL} \times 0.0110 \text{ mmol/mL}$$

$$\text{wt. N} = \text{mmoles } NH_3 \times 17.0 \text{ mg/mmol} \times \frac{N}{NH_3} = 5.73 \times 0.0110 \times 17.0 \times \frac{14.0}{17.0} \text{ mg}$$

$$\%\text{N} = \frac{\text{wt. N}}{\text{wt. sample}} \times 100 = 5.73 \times 0.0110 \times 14.0 \times \frac{1}{3.88} \times 100$$

$$\%\text{N} = 22.8\% \text{ N}$$

14.13

$$\text{wt. S} = \text{wt. } BaSO_4 \times \frac{S}{BaSO_4} = 6.48 \times \frac{32.0}{233.4} \text{ mg}$$

$$\%\text{S} = \frac{\text{wt. S}}{\text{wt. sample}} \times 100 = 6.48 \times \frac{32.0}{233.4} \times \frac{1}{4.81} \times 100 = 18.5\% \text{ S}$$

14.14

p-$H_2NC_6H_4NH_2$ $\quad \frac{2N}{C_6H_8N_2} \times 100 = \frac{28.0}{108} \times 100 = 25.9\%$ N (calcd.)

$HOCH_2CH_2NH_2$ $\quad \frac{N}{C_2H_7ON} \times 100 = \frac{14.0}{61.0} \times 100 = 22.9\%$ N (calcd.)

p-$CH_3C_6H_4SO_3H$ $\quad \frac{S}{C_7H_8O_3S} \times 100 = \frac{32.0}{172} \times 100 = 18.6\%$ S (calcd.)

1. (h) $-SO_3H$ on C–1 (i) $-CH_3$ on C–1 (j) $-COOH$ on C–1 (l) $-OH$ on C–1

2. (a) Three (o, m, p) (b) Three (o, m, p)

(c) Three: 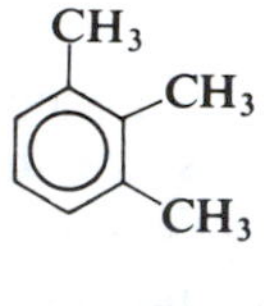 1,2,3 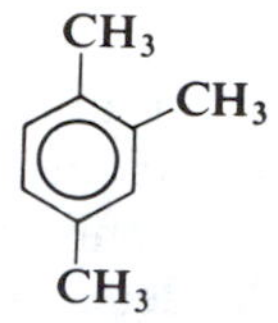 1,2,4 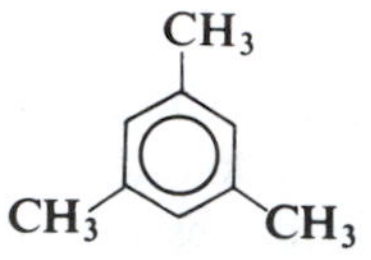 1,3,5

(d) Six: see answer to Problem 14.7

(e) Ten: 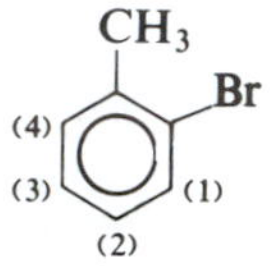 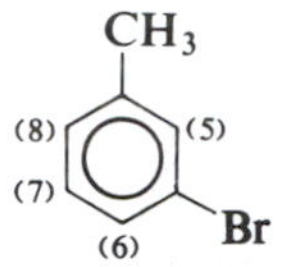 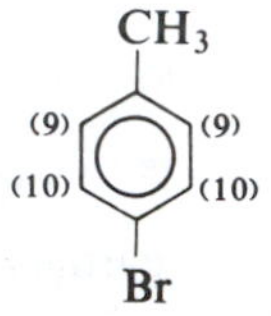 (*Each number in parentheses indicates a different isomer.*)

(f) Six: 2,3,4-, 2,3,5-, 2,3,6-, 2,4,5-, 2,4,6-, 3,4,5-

3. (a) Two, three, three, one, two.
(b) Five, five, five, two, four (neglecting stereoisomers).
(c) None.

4. (a) 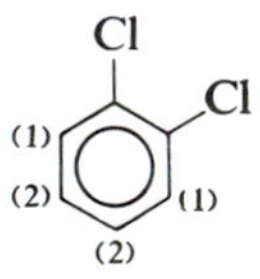 (b) 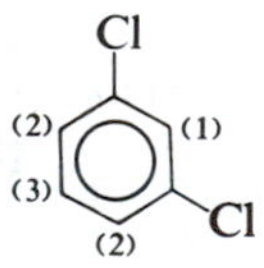 (c) 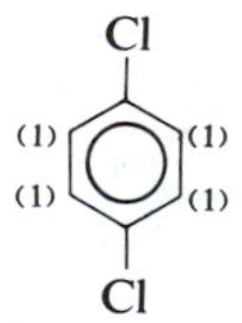 (d)

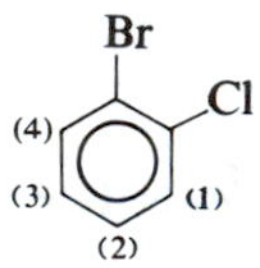

(e) 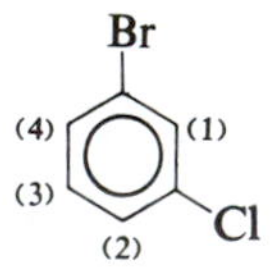 (f) 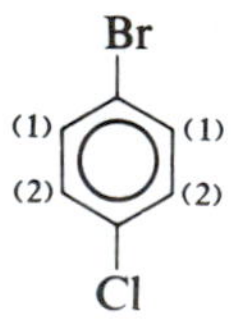 (g) 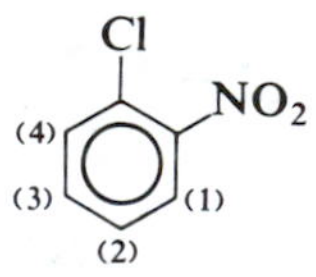 (h)

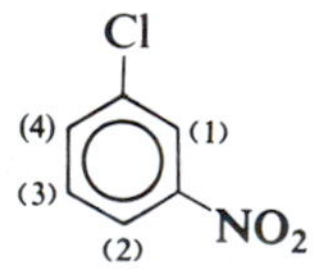

(i) 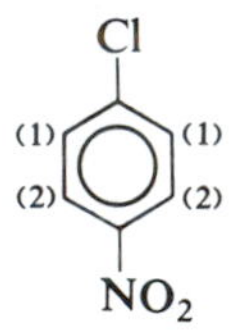 (j) 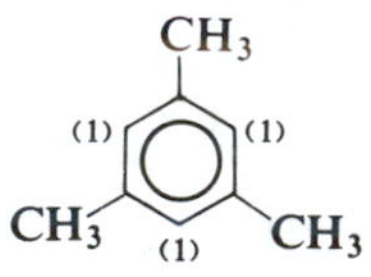 (k) 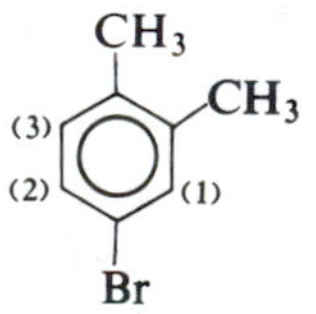 (l)

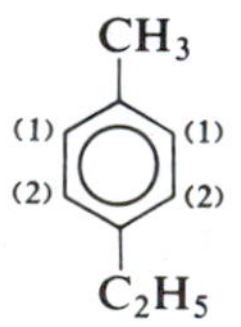

5. (a) Two substituents can be attached to VI in three ways: (i) one above the other on any vertical edge; (ii) at opposite corners of any square; (iii) at two corners of one triangle.

(b) (i) (ii) (iii)

One Two Three

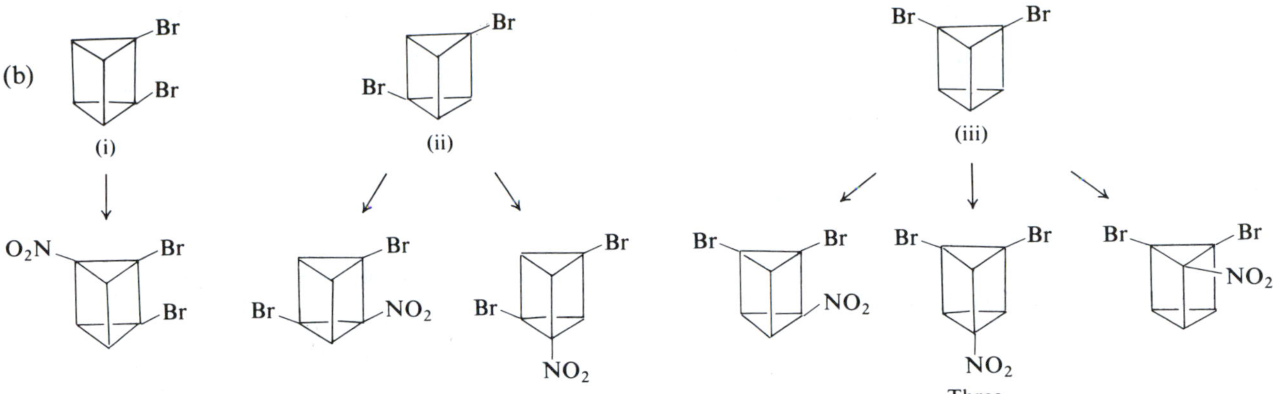

Therefore, (i) is *para*, (ii) is *ortho*, and (iii) is *meta*.

(c) No, the *ortho* isomer is chiral; enantiomeric structures are possible:

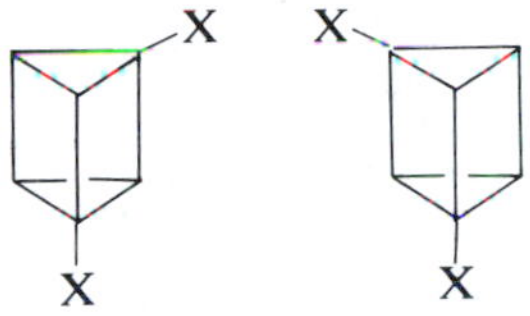

Enantiomers

6. C_8H_{10}. Two carbons attached to the ring: one Et, or two Me. The possible isomeric hydrocarbons—with the *different* substitution sites marked by numbers—are:

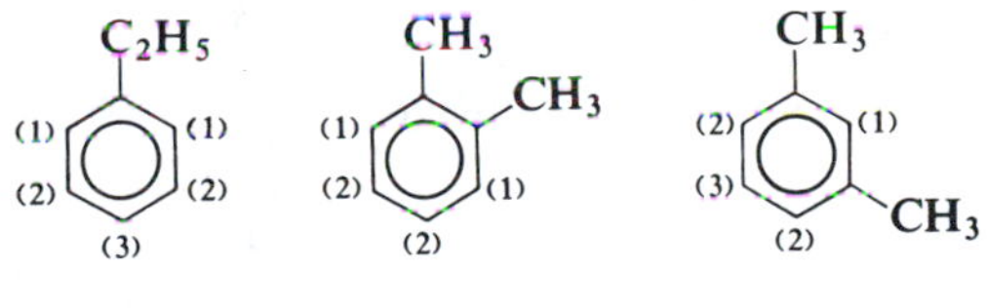

Three (c)	Two (b)	Three (c)	One (a)

C_9H_{12}. Three carbons attached to the ring: one *n*-Pr or *i*-Pr; one Et and one Me; or three Me. The possible isomeric hydrocarbons—with the *different* substitution sites marked by numbers—are:

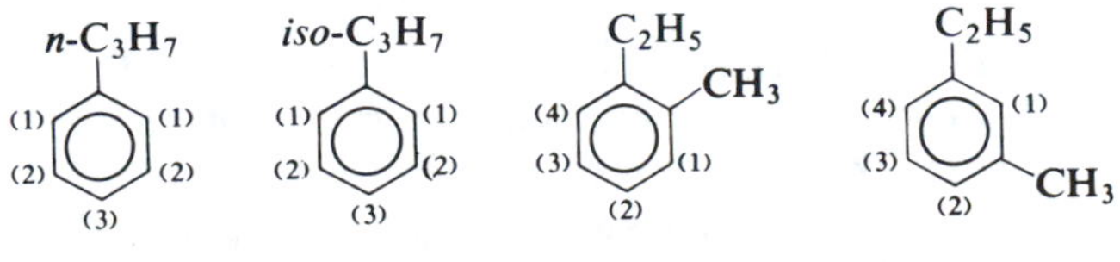

n-C_3H_7: Three (f)	*iso*-C_3H_7: Three (f)	Four (g)	Four (g)	Two (e)

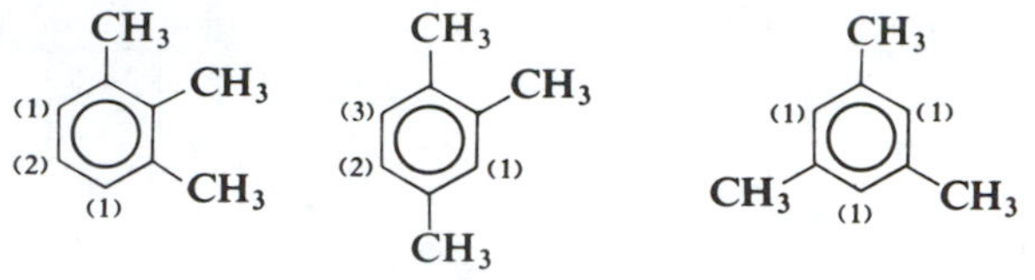

CH_3 CH_3 CH_3 (1) (2) (1)
Two
(e)

CH_3 CH_3 CH_3 (3) (2) (1)
Three
(f)

CH_3 CH_3 CH_3 (1) (1) (1)
One
(d)

7. Yes. Each isomer has a different number of mononitro compounds that can be related to (derived from or convertible into) it.

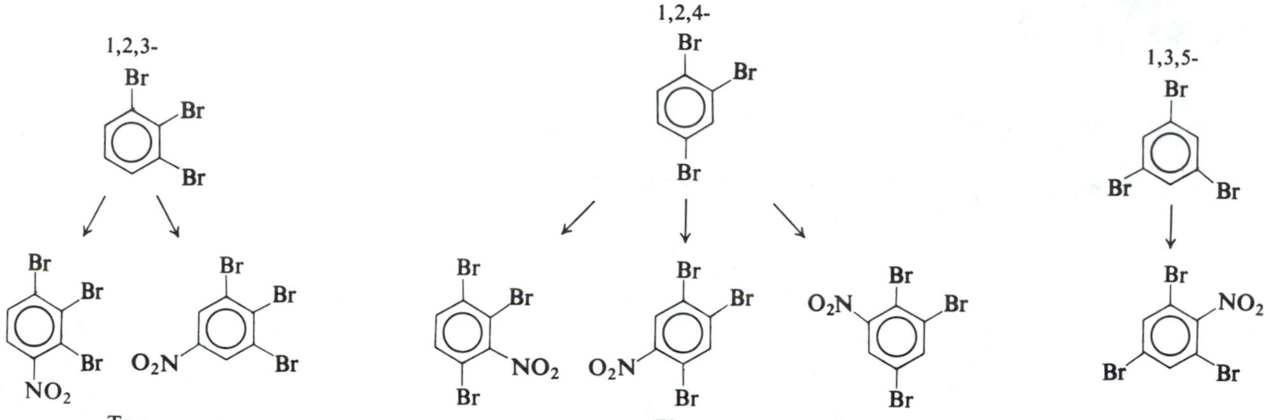

1,2,3- Br Br Br → Two

1,2,4- Br Br Br → Three

1,3,5- Br Br Br → One

8. The six possible diaminobenzoic acids and the structures of the diamines that could be formed from them are:

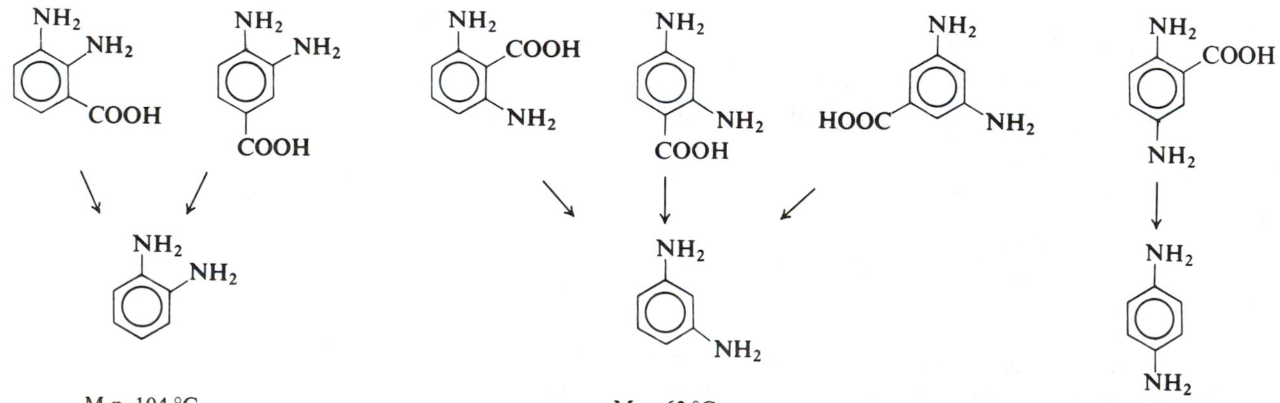

M.p. 104 °C
Ortho

M.p. 63 °C
Meta
Diaminobenzenes

M.p. 142 °C
Para

9. (a) To be aromatic, annulenes should follow the Hückel $4n + 2$ rule for the π electrons: 6 for $n = 1$, 10 for $n = 2$, 14 for $n = 3$, 18 for $n = 4$. The annulenes (with the number of π electrons indicated within the brackets in each name) that are expected to be aromatic are: [6]annulene (benzene); [10]annulene, [14]annulene, [18]annulene. (Actually, for [10] and [14]annulenes, the geometry is unfavorable: crowding of hydrogens inside the ring prevents planarity and hence interferes with π overlap.)

(b)

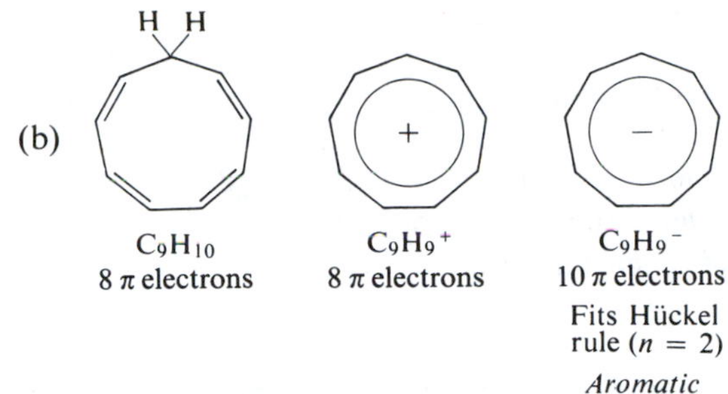

10. Six π electrons (see Sec. 30.2), four from carbons, two from N.

11. (a)

$$\text{unit wt. CHCl} = 12 + 1 + 35.5 = 48.5$$

$$\frac{\text{wt./molecule}}{\text{wt./unit}} = \frac{291}{48.5} = 6 \text{ CHCl units per molecule}$$

$$\text{molecular formula} = C_6H_6Cl_6$$

(b) A possible compound is 1,2,3,4,5,6-hexachlorocyclohexane.

(c) Formed by an addition reaction:

$$C_6H_6 + 3Cl_2 \xrightarrow{\text{light}} C_6H_6Cl_6$$

1,2,3,4,5,6-Hexachlorocyclohexane

(d) Not aromatic: no π electrons available.

(e)–(f) These are stereoisomers, with various combinations of *cis* and *trans* Cl's. Actually, a total of nine such stereoisomers is possible. Two are a pair of enantiomers:

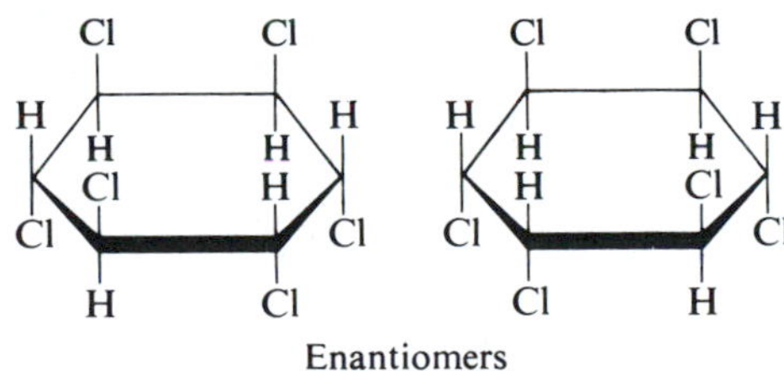

Enantiomers

The other seven are achiral.

12. The argument is an extension of that in Sec. 12.11. In the phenyl anion, $C_6H_5^-$, the unshared pair occupies an sp^2 orbital, which is intermediate in *s* character between the sp^3 orbital of the pentyl anion and the *sp* orbital of the acetylide anion. Electrons in the sp^2 orbital are held more tightly than those in an sp^3 orbital but not so tightly as those in an *sp* orbital; in basicity, therefore, phenyl anion lies between pentyl anion and acetylide anion.

13. (a) Cyclic structure with alternating double and single bonds.

(b) The number of π electrons is eight, which is not a Hückel number.

(c) The number of π electrons in the anion is 10, fitting the Hückel $4n + 2$ rule for $n = 2$. Evidently stabilization due to aromaticity is enough to outweigh the double negative charge and angle strain (the ring must be flat for π overlap, and hence must have C—C—C angles of 135°).

(d) Cyclooctatetraene: puckered rings maintaining the geometry of the carbon–carbon double bonds (x-ray diffraction shows it to be a "tub"). $C_8H_8^{2-}$: flat, regular octagon.

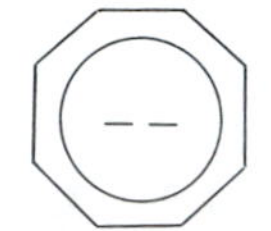

Cyclooctatetraene   Cyclooctatetraenyl dianion

15

Electrophilic Aromatic Substitution

15.1

$$HONO_2 + HONO_2 \rightleftharpoons H_2\overset{+}{O}NO_2 + ONO_2^-$$

$$H_2\overset{+}{O}NO_2 \rightleftharpoons H_2O + {}^+NO_2$$

$$H_2O + HONO_2 \rightleftharpoons H_3O^+ + ONO_2^-$$

$$3HONO_2 \rightleftharpoons {}^+NO_2 + 2NO_3^- + H_3O^+$$

The first step is a Lowry–Brønsted acid–base equilibrium in which one molecule of nitric acid serves as acid, and another as base. This produces protonated nitric acid, which can then form NO_2^+ through loss of the good leaving group H_2O; unprotonated nitric acid would have to lose the very poor leaving group, the strongly basic OH^- ion.

We see here the function of *protonation* (Sec. 6.13), this time in the reaction of inorganic compounds.

15.2

$$HO\overset{O}{\underset{O}{S}}OH + HO\overset{O}{\underset{O}{S}}OH \rightleftharpoons HO\overset{O}{\underset{O}{S}}OH_2^+ + O\overset{O}{\underset{O}{S}}OH^-$$

$$HO\overset{O}{\underset{O}{S}}OH_2^+ \rightleftharpoons HO\overset{O}{\underset{O}{S}}{}^+ + H_2O$$

$$HO\overset{O}{\underset{O}{S}}{}^+ + O\overset{O}{\underset{O}{S}}OH^- \rightleftharpoons HO\overset{O}{\underset{O}{S}}O\overset{O}{\underset{O}{S}}OH$$

15.3 (a) $n\text{-Pr}^+ \longrightarrow iso\text{-Pr}^+$ (b) $iso\text{-Bu}^+ \longrightarrow t\text{-Bu}^+$

(c) $(CH_3)_3CCH_2^+ \longrightarrow$ *tert*-pentyl cation (d) carbocation mechanism

15.4 (a) $t\text{-BuOH} \xrightarrow{H^+} t\text{-BuOH}_2^+ \xrightarrow{-H_2O} t\text{-Bu}^{\oplus} \xrightarrow{PhH} \text{Ph—Bu-}t + H^+$

(b) $CH_3CH{=}CH_2 \xrightarrow{H^+} CH_3\overset{}{\underset{\oplus}{C}}HCH_3 \xrightarrow{PhH} \text{Ph—CH}(CH_3)_2 + H^+$

15.5 The chlorinating agent is $H_2\overset{+}{O}\text{—Cl}$ or possibly Cl^+, generated in the following manner:

$$HOCl + H^+ \rightleftarrows H_2\overset{+}{O}\text{—Cl} \rightleftarrows H_2O + Cl^+$$

15.6 An acid–base complex is formed, which, like II on page 529, transfers halogen without its electrons. Because of its bulk, this reagent attacks a substituted aromatic ring most readily at the least hindered position: *para* to a substituent.

$$(AcO)_3Tl + Br_2 \rightleftarrows (AcO)_3\overset{\ominus}{Tl}\text{—}\overset{\oplus}{Br}\text{—Br}$$

15.7 (a) 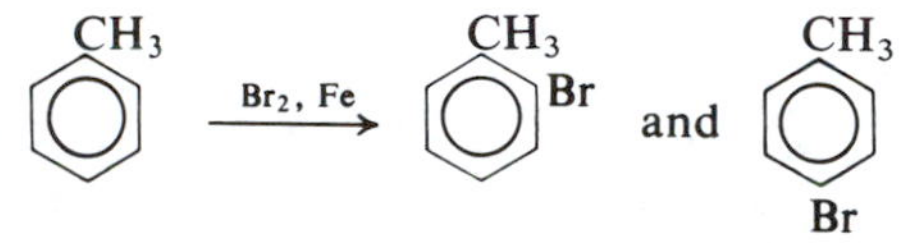 CH_3 $\xrightarrow{Br_2,\ Fe}$ CH_3, Br and CH_3, Br (*Ortho* and *para*)

(b) 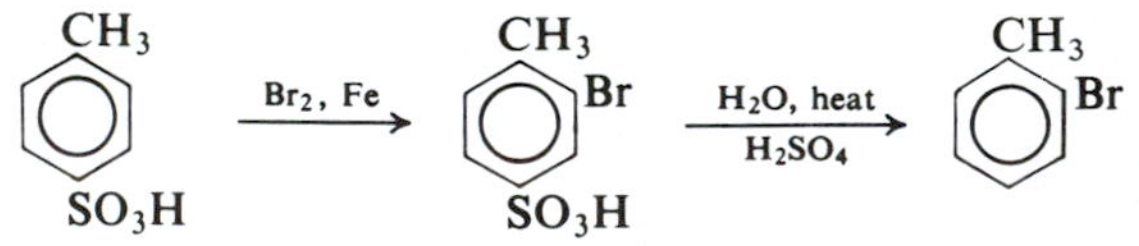 CH_3, SO_3H $\xrightarrow{Br_2,\ Fe}$ CH_3, Br, SO_3H $\xrightarrow[H_2SO_4]{H_2O,\ heat}$ CH_3, Br (*Ortho* only)

(c) 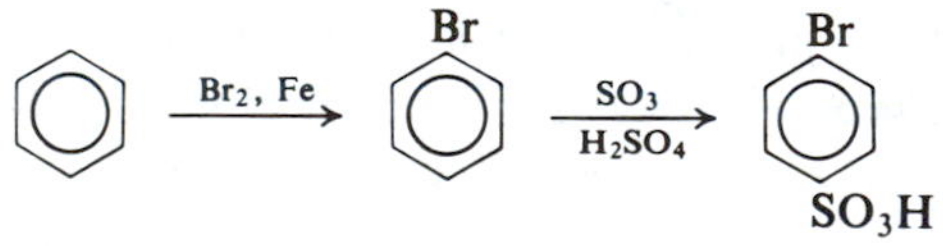 $\xrightarrow{Br_2,\ Fe}$ Br $\xrightarrow[H_2SO_4]{SO_3}$ Br, SO_3H

Br, SO_3H — $\xrightarrow{Br_2,\ Fe}$ Br, Br, SO_3H $\xrightarrow[H_2SO_4]{H_2O,\ heat}$ Br, Br

Br, SO_3H — $\xrightarrow{Cl_2,\ Fe}$ Br, Cl, SO_3H $\xrightarrow[H_2SO_4]{H_2O,\ heat}$ Br, Cl

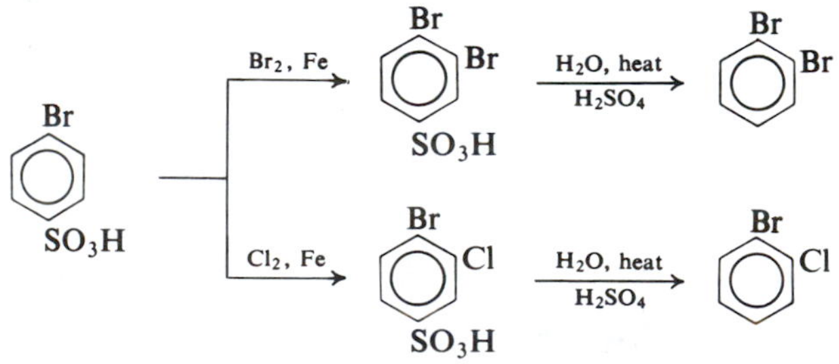

15.8 (a) $R\text{—}\underset{\displaystyle Cl}{\underset{|}{C}}{=}\ddot{O}{:} + AlCl_3 \longrightarrow (R\text{—}C{\equiv}O{:})^+ AlCl_4^-$

(b) $(Ar'\text{—}N{\equiv}N{:})^+Cl^-$

(c) $H\text{—}\ddot{\underset{..}{O}}\text{—}\ddot{N}{=}\ddot{O}{:} + H^+ \longrightarrow H\text{—}\underset{\displaystyle H}{\underset{|}{\overset{+}{\ddot{O}}}}\text{—}\ddot{N}{=}\ddot{O}{:} \longrightarrow ({:}N{\equiv}O{:})^+ + H_2O$

In each case, the cation plays the role of Y^+ in a typical electrophilic aromatic substitution.

$$\text{ArH} + (\text{R—C}\equiv\text{O:})^+ \longrightarrow \overset{\oplus}{\text{Ar}}\begin{matrix}\text{H} \\ \text{C=}\ddot{\text{O}}\text{:} \\ | \\ \text{R}\end{matrix} \xrightarrow{-\text{H}^+} \text{Ar—}\underset{\text{R}}{\underset{|}{\text{C}}}\text{=}\ddot{\text{O}}\text{:}$$

Acylium ion — Aryl ketone — *Friedel–Crafts acylation*

$$\text{ArH} + (\text{Ar}'\text{—N}\equiv\text{N:})^+ \longrightarrow \overset{\oplus}{\text{Ar}}\begin{matrix}\text{H} \\ \ddot{\text{N}}\text{=}\ddot{\text{N}}\text{—Ar}'\end{matrix} \xrightarrow{-\text{H}^+} \text{Ar—}\ddot{\text{N}}\text{=}\ddot{\text{N}}\text{—Ar}'$$

Diazonium ion — Azo compound — *Coupling*

$$\text{ArH} + (\text{:N}\equiv\text{O:})^+ \longrightarrow \overset{\oplus}{\text{Ar}}\begin{matrix}\text{H} \\ \ddot{\text{N}}\text{=}\ddot{\text{O}}\text{:}\end{matrix} \xrightarrow{-\text{H}^+} \text{Ar—}\ddot{\text{N}}\text{=}\ddot{\text{O}}\text{:}$$

Nitrosonium ion — Nitroso compound — *Nitrosation*

15.9 (a) The intermediate benzenonium ion is $(ArHD)^+$, which can lose either H or D.

(b) Hydrogen ion is displaced from the ring by another hydrogen (deuterium) ion in a typical electrophilic aromatic substitution (*deuterodeprotonation*) that is fast with (activated) phenol (PhOH), slow with benzene (PhH), and negligible with (deactivated) benzenesulfonic acid ($PhSO_3H$).

15.10 [2,4,6-trimethylbenzenonium ion: ring (+) bearing CH_3 groups and a CH_2 (H, H) carbon] BF_4^- Shows that $\overset{\oplus}{\text{Ar}}\begin{matrix}\text{H}\\\text{H}\end{matrix}$ and presumably $\overset{\oplus}{\text{Ar}}\begin{matrix}\text{H}\\\text{Y}\end{matrix}$ intermediates can actually exist.

15.11 (a) CH_3CHCl^+. (b) $CH_3CH_2^+$. (c) Carbocation (b). (d) Through its inductive effect (page 542) chlorine tends to intensify the positive charge and thus to destabilize carbocation (a).

(e) $^+CH_2CH_2Cl$. (f) Carbocation (a). (g) Through its resonance effect (page 542) chlorine tends to stabilize carbocation (a).

(h), (i) As in electrophilic substitution (Sec. 15.19), reactivity is controlled by the stronger inductive effect, and orientation is controlled by the weaker but more selective resonance effect.

15.12 These facts become understandable when we recall that sulfonation is readily reversible (Sec. 15.12).

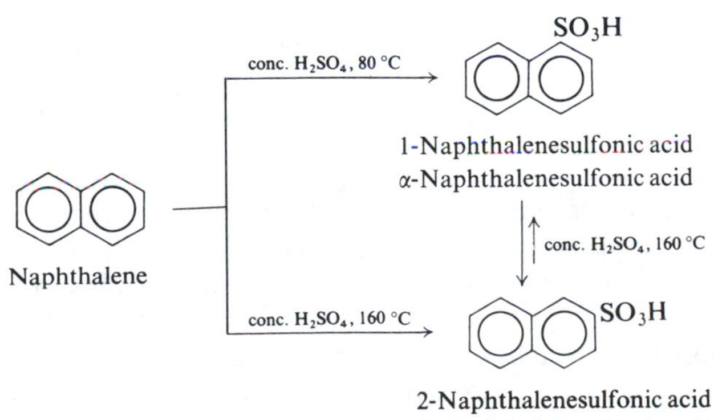

Sulfonation, like nitration and halogenation, occurs more rapidly at the α-position, since this involves the more stable intermediate carbocation. But, for the same reason, attack by a proton, with subsequent desulfonation, also occurs more readily at the α-position. Sulfonation at the β-position occurs more slowly but, once formed, the β-sulfonic acid test tends to resist desulfonation. At low temperatures desulfonation is slow and we isolate the product that is formed faster, the *alpha* naphthalenesulfonic acid. At higher temperatures, desulfonation becomes important, equilibrium is more readily established, and we isolate the product that is more stable, the *beta* naphthalenesulfonic acid. (See Figure 15.4, below.)

SO_3H ⇌ (SO_3 / H^+) naphthalene ⇌ (SO_3 / H^+) SO_3H

α-Isomer
Formed rapidly; desulfonated rapidly

β-Isomer
Formed slowly; desulfonated slowly

We see here a situation exactly analogous to one we have encountered several times before: in 1,2- and 1,4-addition to conjugated dienes (Sec. 11.23) and in Friedel–Crafts alkylation of toluene (Sec. 16.12). At low temperatures the controlling factor is *rate of reaction*, at high temperatures, *position of equilibrium*.

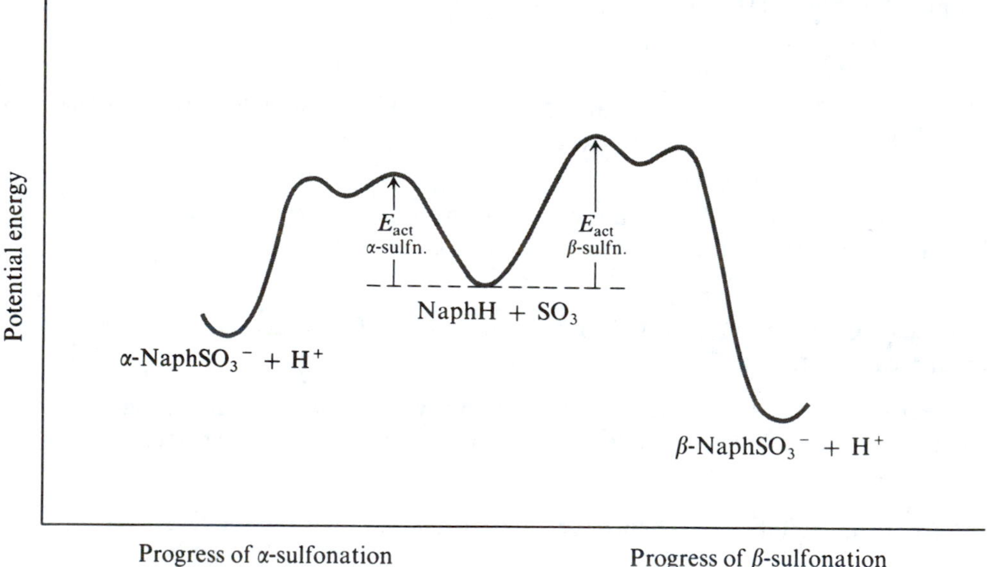

Figure 15.4 Potential energy changes during the course of reaction: α- *vs.* β-sulfonation of naphthalene (Problem 15.12).

15.13 Orientation in naphthalene can be accounted for on the same basis as orientation in substituted benzenes: formation of the more stable intermediate carbocation. In judging the relative stabilities of these naphthalene carbocations, we consider that those in which an aromatic sextet is preserved are by far the more stable and hence the more important. Let us see if we can account for orientation in substituted naphthalenes in the same way.

The structures preserving an aromatic sextet are those in which the positive charge is carried by the ring under attack; it is in this ring, therefore, that the charge chiefly develops. Consequently, attack occurs most readily on whichever ring can best accommodate the positive charge: the ring that carries an electron-releasing (activating) group or the ring that does *not* carry an electron-withdrawing (deactivating) group. (We have arrived at the quite reasonable conclusion that a substituent exerts its greatest effect—activating or deactivating—on the ring to which it is attached.)

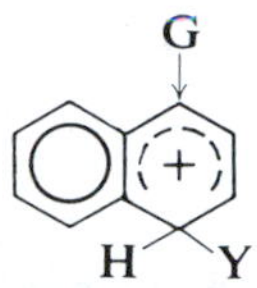

G *is electron-releasing*: *activating, attack in same ring*

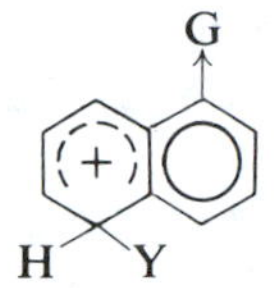

G *is electron-withdrawing*: *deactivating, attack in other ring*

An electron-releasing group located at position 1 can best help accommodate the positive charge if attack occurs at position 4 (or position 2), through the contribution of structures like I and II.

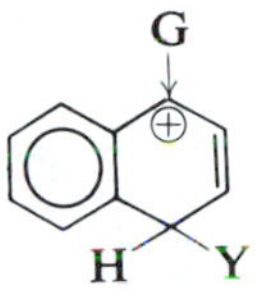

I

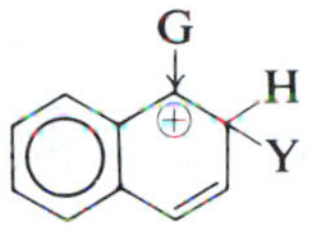

II

G *is electron-releasing*: *when on position 1, it directs attack to positions 4 or 2*

This is true whether the group releases electrons by an inductive effect or by a resonance effect. For example:

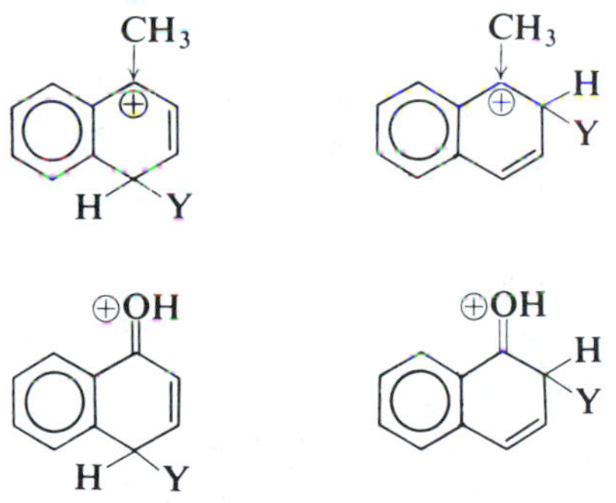

An electron-releasing group located at position 2 could help accommodate the positive charge if attack occurred at position 1 (through structures like III), or if attack occurred at position 3 (through structures like IV).

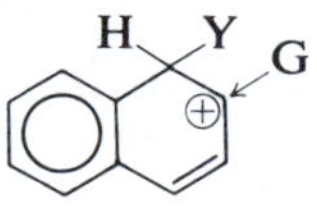

III

More stable: *aromatic sextet preserved*

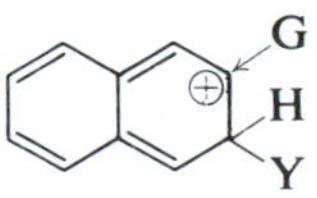

IV

Less stable: *aromatic sextet disrupted*

G *is electron-releasing*: *when on position 2, it directs attack to position 1*

However, we can see that only the structures like III preserve an aromatic sextet; these are much more stable than the structures like IV, and are the important ones. It is not surprising, therefore, that substitution occurs almost entirely at position 1.

The major products of further substitution in a monosubstituted naphthalene can, then, usually be predicted by the following rules.

(a) An activating group (electron-releasing group) tends to direct further substitution into the same ring. An activating group in position 1 directs further substitution to position 4 (and, to a lesser extent, to position 2). An activating group in position 2 directs further substitution to position 1.

(b) A deactivating group (electron-withdrawing group) tends to direct further substitution into the other ring: at an α-position in nitration or halogenation, or at an α- or β-position (depending upon temperature) in sulfonation.

We thus find that the strongly activating —OH in 1-naphthol directs to its own ring and to *both* position 1 and position 2:

$$\text{1-Naphthol} + HNO_3 \xrightarrow{H_2SO_4,\ 20\ ^\circ C} \text{2,4-Dinitro-1-naphthol}$$

The deactivating $—NO_2$ in 1-nitronaphthalene directs to the other ring and to either α-position there.

$$\text{1-Nitronaphthalene} + HNO_3 \xrightarrow{H_2SO_4,\ 0\ ^\circ C} \text{1,5-Dinitronaphthalene} + \underset{\textit{Chief product}}{\text{1,8-Dinitronaphthalene}}$$

1. Faster (activated): a, c, d, g, h, k. Slower (deactivated): b, e, f, i, j.

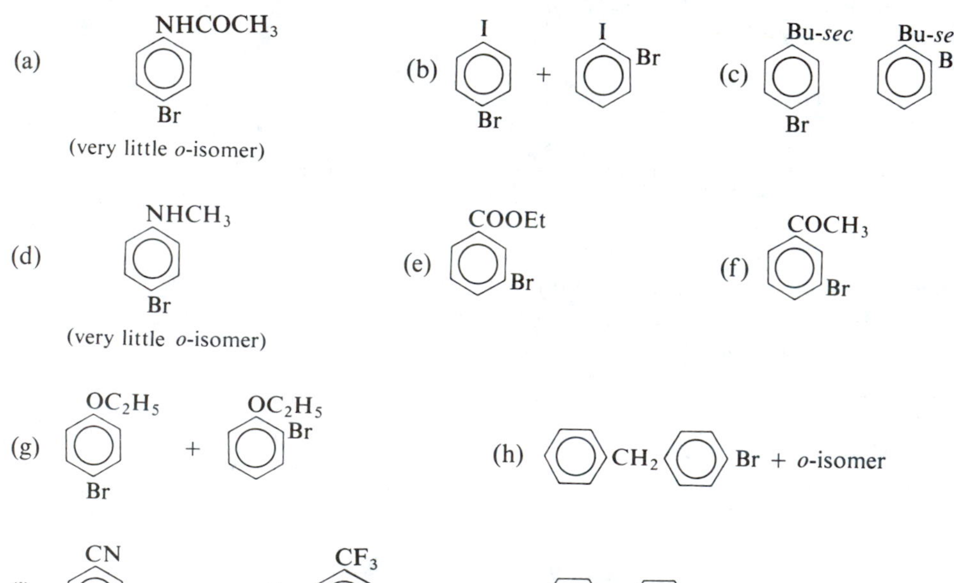

2. (a) 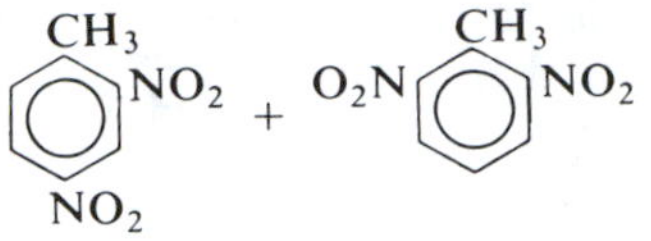 CH_3, NO_2, NO_2 + CH_3, O_2N, NO_2 (b) 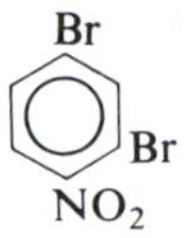 Br, Br, NO_2 (c) 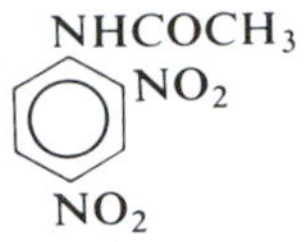 $NHCOCH_3$, NO_2, NO_2

(d) 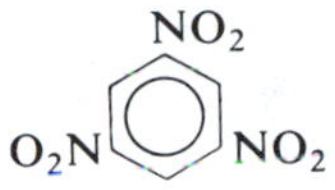 NO_2, O_2N, NO_2 (e) 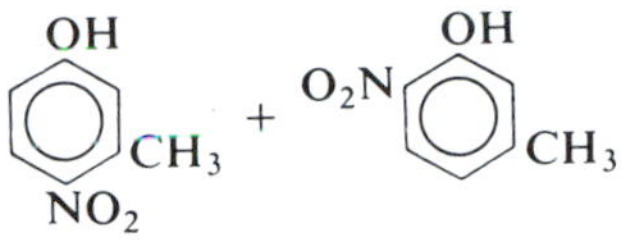 OH, CH_3, NO_2 + OH, O_2N, CH_3 (—OH >> —CH_3) (f) 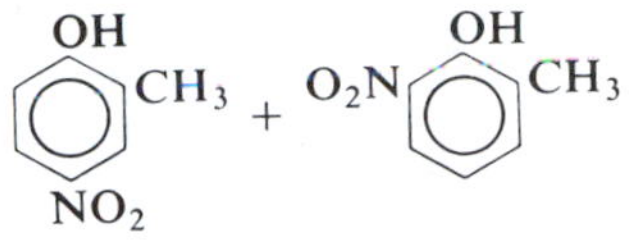 OH, CH_3, NO_2 + OH, O_2N, CH_3 (—OH >> —CH_3)

(g) 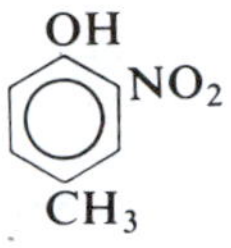 OH, NO_2, CH_3 (—OH >> —CH_3) (h) 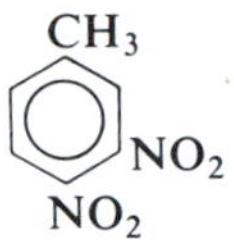 CH_3, NO_2, NO_2 (i) 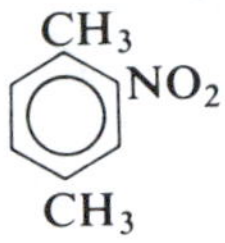 CH_3, NO_2, CH_3 (All positions in substrate equal)

(j) 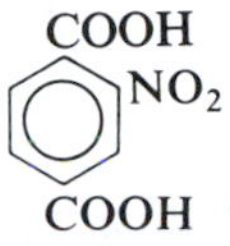 COOH, NO_2, COOH (All positions in substrate equal) (k) 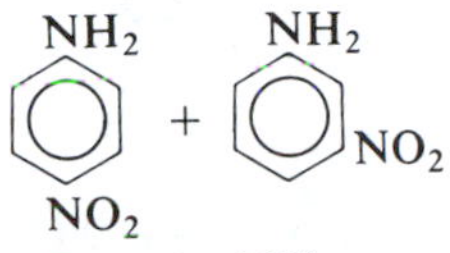 NH_2, NO_2 + NH_2, NO_2 (See page 861)

3. When an *o,p*-directing group is present (as is true in every case except b and d), sulfonation *para* (and *ortho*) to that group predominates. All products are named as sulfonic acids, with —SO_3H considered to be on C-1; (d) is *m*-benzenedisulfonic acid.

4. *m*-Directing groups are deactivating; the more of them there are present, the lower the reactivity. Most *o,p*-directing groups (except halogen) are activating; the more activating groups there are present, and the more of them that activate the same position, the more reactive the compound. For example, in (a): the three —CH_3 groups in mesitylene (triply) activate the same positions, the two —CH_3 groups in *m*-xylene (doubly) activate the same positions, but the two —CH_3 groups in *p*-xylene activate different positions.

(a) mesitylene > *m*-xylene > *p*-xylene > toluene > benzene

(b) toluene > benzene > bromobenzene > nitrobenzene

(c) aniline > acetanilide > benzene > acetophenone

(d) *p*-xylene > toluene > *p*-toluic acid > terephthalic acid

(e) C_6H_5Cl > *p*-$O_2NC_6H_4Cl$ > 2,4-$(O_2N)_2C_6H_3Cl$

(f) 2,4-dinitrophenol > 2,4-dinitrochlorobenzene

(g) 2,4-dinitrotoluene > *m*-dinitrobenzene

5. (a) Substitution is faster in the ring that is not deactivated by —NO_2. Orientation is *o,p* to the other ring (see Table 15.3, page 522).

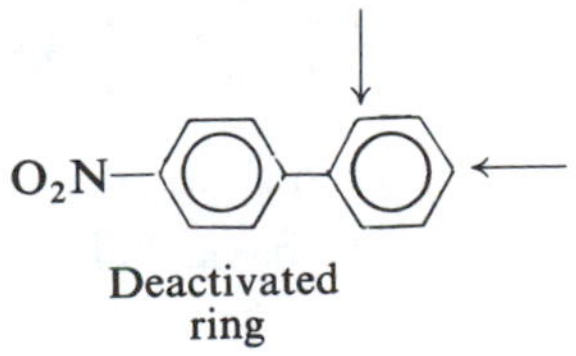

O_2N—

Deactivated ring

(b) Substitution is faster in the ring that is not deactivated by $—NO_2$. Orientation is *o,p* to the substituent, $—CH_2Ar$.

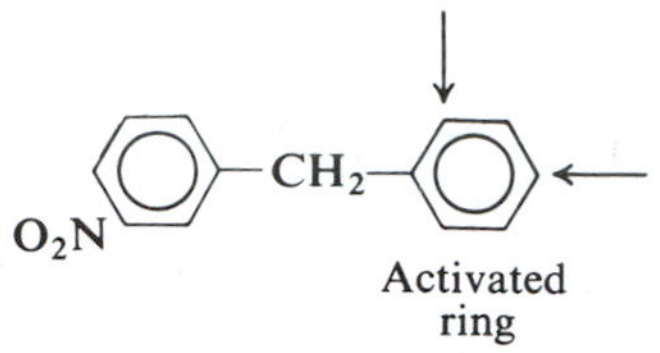

O_2N CH_2

Activated ring

(c) Substitution is faster in the ring that is activated by phenolic oxygen (similar to —OR). The other ring is actually deactivated by —COOAr.

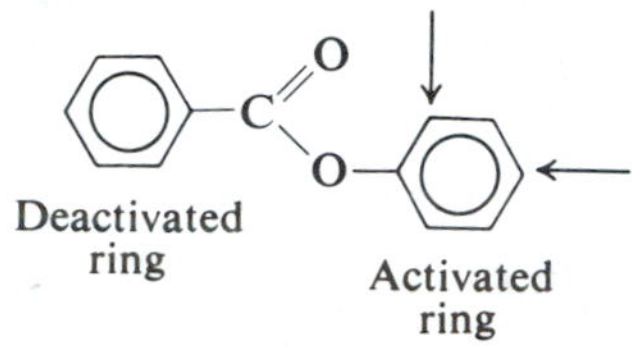

Deactivated ring

Activated ring

6. The farther away a deactivating $—NR_3^+$ or $—NO_2$ group is from the ring, the less effective it is. In (c), the more electron-withdrawing groups there are, the greater the deactivation.

(a) $Ph(CH_2)_3NMe_3^+ > Ph(CH_2)_2NMe_3^+ > PhCH_2NMe_3^+ > PhNMe_3^+$

(b) $Ph(CH_2)_2NO_2 > PhCH_2NO_2 > PhNO_2$

(c) $PhCH_3 > PhCH_2COOEt > PhCH(COOEt)_2 > PhC(COOEt)_3$

The most active compound in each set gives the lowest percentage of *meta* isomer, the least active the highest percentage.

7. The positive charge of the intermediate can be dispersed (through three extra structures) by the second phenyl group when attack is *ortho* or when it is *para*, but not when it is *meta*. For example:

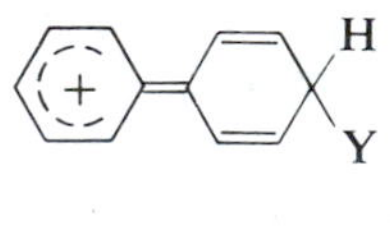

H Y

8. $HONO_2 + H_2SO_4 \rightleftarrows H_2\overset{+}{O}NO_2 + HSO_4^-$

$H_2\overset{+}{O}NO_2 \longrightarrow H_2O + {}^+NO_2$ *Slow*

$H_2O + H_2SO_4 \rightleftarrows H_3O^+ + HSO_4^-$

Protonated nitric acid forms NO_2^+ through loss of the good leaving group H_2O; unprotonated nitric acid would have to lose the more strongly basic OH^- ion. (See Sec. 6.13 and the answer to Problem 15.3.)

9. The reactions are electrophilic substitutions, with activation as usual by $-NH_2$ and $-OH$. Attacking electrophiles are the familiar Br^+ (or its equivalent) and NO_2^+. In each case, however, the displaced group is not the usual H^+ (deprotonation), but SO_3, as in desulfonation. Thus,

$$ArSO_3^- \xrightarrow{Br^+} \overset{\oplus}{Ar}\genfrac{}{}{0pt}{}{Br}{SO_3^-} \xrightarrow{-SO_3} ArBr \qquad \textit{Bromodesulfonation}$$

$$ArSO_3^- \xrightarrow{NO_2^+} \overset{\oplus}{Ar}\genfrac{}{}{0pt}{}{NO_2}{SO_3^-} \xrightarrow{-SO_3} ArNO_2 \qquad \textit{Nitrodesulfonation}$$

as compared with, say,

$$ArH \xrightarrow{Br^+} \overset{\oplus}{Ar}\genfrac{}{}{0pt}{}{Br}{H} \xrightarrow{-H^+} ArBr \qquad \textit{Bromodeprotonation}$$

or

$$ArSO_3^- \xrightarrow{H^+} \overset{\oplus}{Ar}\genfrac{}{}{0pt}{}{H}{SO_3^-} \xrightarrow{-SO_3} ArH \qquad \textit{Protodesulfonation}$$

10. This is an example of *bromodealkylation.*

$$Br_2 + AlBr_3 \rightleftarrows Br_3\overset{\ominus}{Al}-\overset{\oplus}{Br}-Br$$

$$Br_3\overset{\ominus}{Al}-\overset{\oplus}{Br}-Br + C_6H_5C(CH_3)_3 \longrightarrow \overset{\oplus}{C_6H_5}\genfrac{}{}{0pt}{}{Br}{C(CH_3)_3} + AlBr_4^-$$

$$\overset{\oplus}{C_6H_5}\genfrac{}{}{0pt}{}{Br}{C(CH_3)_3} \longrightarrow C_6H_5Br + (CH_3)_3C^+$$

$$(CH_3)_3C^+ + AlBr_4^- \longrightarrow (CH_3)_2C{=}CH_2 + HBr + AlBr_3$$

11. The substituent in each case, $-N(CH_3)_3^+$ or $-CF_3$, is powerfully electron-withdrawing and (in the absence of an opposing resonance effect) favors formation of the cation with the charge on the more remote carbon.

(a) $Me_3\overset{+}{N} \leftarrow \overset{H}{\overset{|}{C}}{=}CH_2 \xrightarrow{H^+}$

$$\rightarrow Me_3\overset{+}{N} \leftarrow CH_2-CH_2^{\oplus} \xrightarrow{I^-} Me_3\overset{+}{N}-CH_2CH_2I \qquad \textit{Actual product}$$

More stable cation

$$\rightarrow Me_3\overset{+}{N} \leftarrow \overset{\oplus}{C}H-CH_3$$

(b) $CH_2{=}\overset{H}{C}{\rightarrow}CF_3 \xrightarrow{H^+}$

$\rightarrow CH_2{-}\overset{H}{\underset{H}{C}}{\rightarrow}CF_3 \xrightarrow{Br^-} BrCH_2CH_2CF_3$ *Actual product*

More stable cation

$\rightarrow \underset{H}{C}H_2{-}\overset{H}{\underset{\oplus}{C}}{\rightarrow}CF_3$

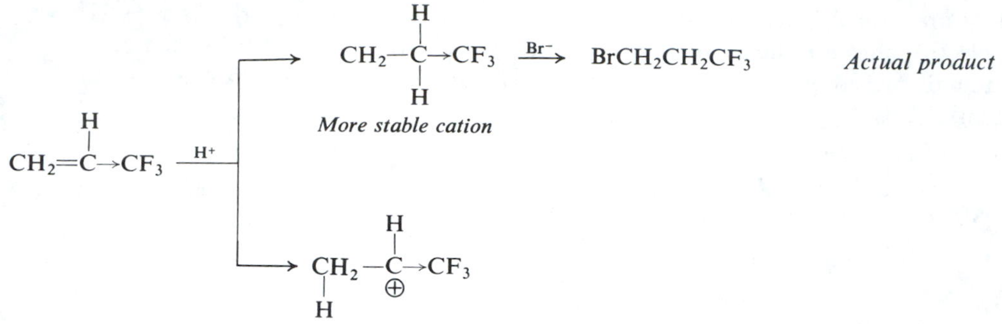

(c) $AlBr_3$ is used to provide the stronger acid ($HAlBr_4$) that is needed for attack on the highly deactivated alkene.

12. (a) Measure the relative rates of reaction; you would expect a lower rate with C_6D_6.

(b) Measure the relative amounts of C_6H_5Y and C_6D_5Y by mass spectrometry; you would expect more C_6H_5Y.

(c) Carry out a substitution reaction with ordinary anisole,

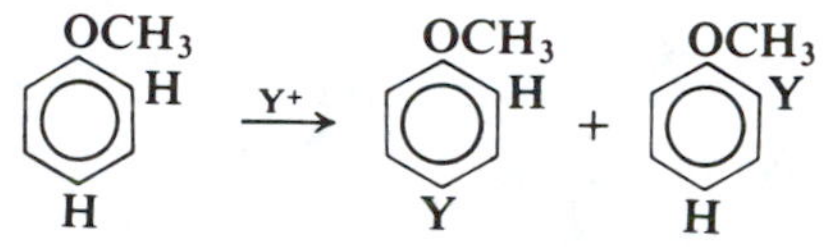

OCH₃ (ring, H ortho, H para) $\xrightarrow{Y^+}$ OCH₃ (ring, H ortho, Y para) + OCH₃ (ring, Y ortho, H para)

and determine (by gas chromatography, say) the ratio of *o*-product to *p*-product (*o*/*p* ratio). Now carry out the same reaction under the same conditions, but start with anisole-4-*d*.

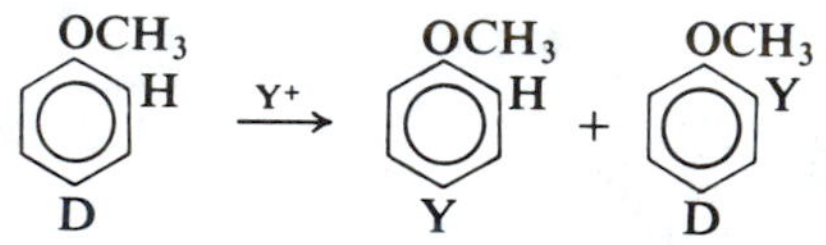

OCH_3 (ring, H ortho, D para) $\xrightarrow{Y^+}$ OCH_3 (ring, H ortho, Y para) + OCH_3 (ring, Y ortho, D para)

If there is an isotope effect, less of the *p*-product will be formed (C—D bond has to be broken) and the *o*/*p* ratio will be larger than before.

(d) Carry out a substitution reaction, and analyze the product for D/H ratio. If there is an isotope effect, fewer C—D bonds will be broken than C—H bonds, $C_6H_2D_3Y$ will exceed $C_6H_3D_2Y$, and the D/H ratio will be higher in the product than in the reactant.

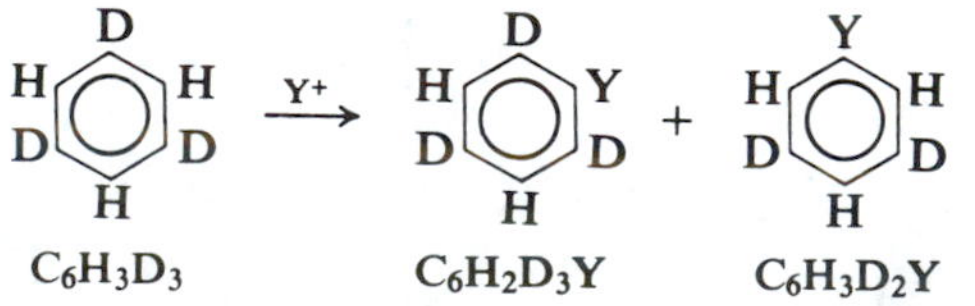

$C_6H_3D_3$ $\xrightarrow{Y^+}$ $C_6H_2D_3Y$ + $C_6H_3D_2Y$

13. See Sec. 30.4.

14. Naphthalene is stabilized by resonance to the extent of 61 kcal/mol; benzene is stabilized to the extent of 36 kcal/mol. When the aromatic character of one ring of naphthalene is destroyed, only 25 kcal of resonance energy is sacrificed; in the next stage, 36 kcal has to be sacrificed.

Oxidation of naphthalene by oxygen in the presence of vanadium pentoxide destroys one ring and yields phthalic anhydride. Because of the availability of naphthalene from coal tar, and the large demand for phthalic anhydride (for example, see Sec. 31.7), this is an important industrial process.

Naphthalene $\xrightarrow{CrO_3,\ HOAc,\ 25\ ^\circ C}$ 1,4-Naphthoquinone α-Naphthoquinone (*40% yield*)

Naphthalene $\xrightarrow{O_2,\ V_2O_5,\ 460–480\ ^\circ C}$ Phthalic anhydride (*76% yield*)

Oxidation of certain naphthalene derivatives destroys the aromatic character of one ring in a somewhat different way, and yields diketo compounds known as *quinones* (Sec. 27.9). Because of this tendency to form quinones, it is not always feasible to prepare naphthalenecarboxylic acids as we do benzoic acids, by oxidation of methyl side chains.

In contrast to benzene, naphthalene can be reduced by chemical reducing agents. It is converted by sodium and ethanol into 1,4-dihydronaphthalene, and by sodium and isopentyl alcohol into 1,2,3,4-tetrahydronaphthalene (*tetralin*). The temperature at which each of these sodium reductions is carried out is the boiling point of the alcohol used; at the higher temperature permitted by isopentyl alcohol (b.p. 132 °C), reduction proceeds further than with the lower-boiling ethyl alcohol (b.p. 78 °C). The tetrahydronaphthalene is simply a dialkyl derivative of benzene. As with other benzene derivatives, the aromatic ring that remains is reduced only by vigorous catalytic hydrogenation.

15. (a) $p\text{-}CH_3C_6H_4NO_2 \xleftarrow{HNO_3,\ H_2SO_4} C_6H_5CH_3$

(b) $p\text{-}BrC_6H_4NO_2 \xleftarrow{HNO_3,\ H_2SO_4} BrC_6H_5 \xleftarrow{Br_2,\ Fe} C_6H_6$

(c) $p\text{-}ClC_6H_4Cl \xleftarrow{Cl_2,\ Fe} ClC_6H_5 \xleftarrow{Cl_2,\ Fe} C_6H_6$

(d) $m\text{-}BrC_6H_4SO_3H \xleftarrow{Br_2,\ Fe} C_6H_5SO_3H \xleftarrow{SO_3,\ H_2SO_4} C_6H_6$

(e) $p\text{-}BrC_6H_4SO_3H \xleftarrow{SO_3,\ H_2SO_4} BrC_6H_5 \xleftarrow{Br_2,\ Fe} C_6H_6$

(f) $p\text{-}BrC_6H_4COOH \xleftarrow{KMnO_4} p\text{-}BrC_6H_4CH_3 \xleftarrow{Br_2,\ Fe} C_6H_5CH_3$

(g) m-BrC_6H_4COOH $\xleftarrow{Br_2,\ Fe}$ C_6H_5COOH $\xleftarrow{KMnO_4}$ $C_6H_5CH_3$

(h) 1,3,5-trinitrobenzene (O_2N, NO_2, NO_2) $\xleftarrow{a}$ m-dinitrobenzene (NO_2, NO_2) $\xleftarrow{b}$ nitrobenzene (NO_2) $\xleftarrow{c}$ benzene

a: fuming HNO_3, fuming H_2SO_4, 100–110 °C, 5 days, 45% yield.

b: fuming HNO_3, conc. H_2SO_4, 95 °C, 84% yield.

c: conc. HNO_3, conc. H_2SO_4, 55–60 °C, 99% yield.

(i) (CH_3, Br, NO_2) $\xleftarrow{Br_2,\ Fe}$ (CH_3, NO_2) $\longleftarrow$ as in (a)

(j) (COOH, Br, NO_2) $\xleftarrow{K_2Cr_2O_7}$ (CH_3, Br, NO_2) $\longleftarrow$ as in (i)

(k) (COOH, NO_2, Br) $\xleftarrow{HNO_3,\ H_2SO_4}$ (COOH, Br) $\longleftarrow$ as in (f)

(l) (COOH, O_2N, NO_2) $\xleftarrow[\text{high heat}]{HNO_3,\ H_2SO_4}$ (COOH, NO_2) $\xleftarrow[\text{heat}]{HNO_3,\ H_2SO_4}$ (COOH) $\xleftarrow{KMnO_4}$ (CH_3)

(m) (Br, Br, NO_2) $\xleftarrow{Br_2,\ Fe}$ (Br, NO_2) $\longleftarrow$ as in (b)

(n) (Cl, NO_2, Cl) $\xleftarrow{HNO_3,\ H_2SO_4}$ (Cl, Cl) $\longleftarrow$ as in (c)

16. (a) (Br, OCH_3, Br, NO_2) $\xleftarrow{2Br_2,\ Fe}$ (OCH_3, NO_2) $\xleftarrow{HNO_3,\ H_2SO_4}$ (OCH_3)

(b) (COOH, NO_2, Br) $\xleftarrow{K_2Cr_2O_7}$ (CH_3, NO_2, Br) $\xleftarrow{Br_2,\ Fe}$ (CH_3, NO_2)

(c) $2,4,6\text{-}Br_3C_6H_2NH_2 \xleftarrow{3Br_2(aq)} C_6H_5NH_2$

(d) $2,4\text{-}(O_2N)_2C_6H_3NHCOCH_3 \xleftarrow{HNO_3,\ H_2SO_4}$ 2- and 4-$O_2NC_6H_4NHCOCH_3 \xleftarrow{HNO_3,\ H_2SO_4} PhNHCOCH_3$

(e) 5-nitro-1,3-benzenedicarboxylic acid (O_2N, COOH, COOH) $\xleftarrow[\text{high heat}]{HNO_3,\ H_2SO_4}$ 1,3-benzenedicarboxylic acid (COOH, COOH) $\xleftarrow{KMnO_4}$ m-xylene (CH_3, CH_3)

(f) 3-nitro-1,2-benzenedicarboxylic acid (COOH, COOH, NO_2) $\xleftarrow{K_2Cr_2O_7}$ 3-nitro-o-xylene (CH_3, CH_3, NO_2) $\xleftarrow{HNO_3,\ H_2SO_4}$ o-xylene (CH_3, CH_3)

(g) Nitro-1,4-benzenedicarboxylic acid (COOH, NO_2, COOH) $\leftarrow$

- upper route: $\xleftarrow{K_2Cr_2O_7}$ 2-nitro-p-xylene (CH_3, NO_2, CH_3) $\xleftarrow{HNO_3,\ H_2SO_4}$ p-xylene (CH_3, CH_3)
- lower route: $\xleftarrow[H_2SO_4]{HNO_3}$ 1,4-benzenedicarboxylic acid (COOH, COOH) $\xleftarrow{KMnO_4}$ p-xylene (CH_3, CH_3)

(h) Nitration of *p*-xylene (upper route) takes advantage of the activating effect of the two $-CH_3$ groups. Nitration of a ring deactivated by two $-COOH$ groups would be very difficult.

16

Aromatic–Aliphatic Compounds

Arenes and Their Derivatives

16.1 (a) H_2SO_4, heat (b) Zn(Hg), HCl
(c) H_2, Pt, heat, pressure (d) H_2SO_4, heat; then as in (c)
(e) KOH(alc); then as in (c). Or: Mg, anhyd. Et_2O; H_2O

16.2 *tert*-Pentylbenzene results in each case from attack on benzene by the *tert*-pentyl cation:

$$CH_3CH_2-\overset{\overset{CH_3}{|}}{\underset{\underset{CH_3}{|}}{C}}\oplus + C_6H_6 \longrightarrow C_6H_5\overset{\overset{CH_3}{|}}{\underset{\underset{CH_3}{|}}{C}}-CH_2CH_3 + H^+$$

This cation is formed as follows:

$$\underset{\text{2-Methyl-1-butanol}}{CH_3CH_2-\overset{\overset{CH_3}{|}}{\underset{\underset{H}{|}}{C}}-CH_2OH} \xrightarrow[-H_2O]{H^+} \underset{1^\circ\text{ cation}}{CH_3CH_2-\overset{\overset{CH_3}{|}}{\underset{\underset{H}{|}}{C}}-CH_2^{\oplus}}$$

$$\underset{\text{3-Methyl-2-butanol}}{CH_3-\underset{\underset{OH}{|}}{CH}-\overset{\overset{CH_3}{|}}{\underset{\underset{H}{|}}{C}}-CH_3} \xrightarrow[-H_2O]{H^+} \underset{2^\circ\text{ cation}}{CH_3-\overset{\oplus}{CH}-\overset{\overset{CH_3}{|}}{\underset{\underset{H}{|}}{C}}-CH_3}$$

$$\underset{\text{3-Methyl-1-butanol}}{HOCH_2-\underset{\underset{H}{|}}{CH}-\overset{\overset{CH_3}{|}}{\underset{\underset{H}{|}}{C}}-CH_3} \xrightarrow[-H_2O]{H^+} \underset{1^\circ\text{ cation}}{{}^{\oplus}CH_2-\underset{\underset{H}{|}}{CH}-\overset{\overset{CH_3}{|}}{\underset{\underset{H}{|}}{C}}-CH_3} \longrightarrow \underset{2^\circ\text{ cation}}{CH_3-\overset{\oplus}{CH}-\overset{\overset{CH_3}{|}}{\underset{\underset{H}{|}}{C}}-CH_3}$$

$$\underset{\text{Neopentyl alcohol}}{HO-CH_2-\overset{\overset{CH_3}{|}}{\underset{\underset{CH_3}{|}}{C}}-CH_3} \xrightarrow[-H_2O]{H^+} \underset{1^\circ\text{ cation}}{{}^{\oplus}CH_2-\overset{\overset{CH_3}{|}}{\underset{\underset{CH_3}{|}}{C}}-CH_3}$$

All four cations $\longrightarrow$

$$\underset{3^\circ\text{ cation}}{CH_3-CH_2-\overset{\overset{CH_3}{|}}{\underset{\oplus}{C}}-CH_3}$$

16.3 Because of its positive charge, the substituent in I should be strongly deactivating, comparable to $—NR_3^+$.

16.4 The situation reminds us of the competition between 1,2- and 1,4-addition (Sec. 11.23),

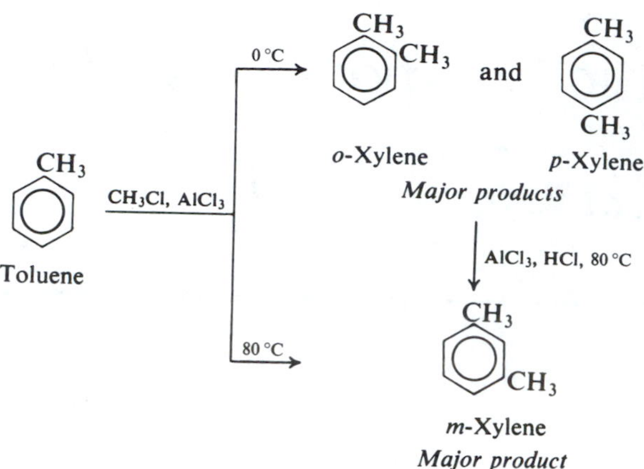

and we make the following hypothesis.

At 0 °C, we are observing rate control: *o*- and *p*-xylene are formed faster. At 80 °C, we are observing equilibrium control: *m*-xylene is the more stable product.

Methyl activates most strongly at the *ortho* and *para* positions. This favors alkylation at these positions, but it also favors *de*alkylation—via electrophilic attack by a proton—at these same positions. The *ortho* and *para* isomers are formed more rapidly, but are also dealkylated more rapidly; the *meta* isomer is formed more slowly, but, once formed, tends to persist.

Hydrogen chloride provides H^+ for reversal of alkylation:

$$HCl + AlCl_3 \rightleftarrows H^+AlCl_4^-$$

Experiment has shown that the above hypothesis is correct in broad outline, but needs modification. Let us consider conversion of *p*-xylene as a rearrangement involving migration of an alkyl group. As we have pictured it, the alkyl group leaves (1) the *para* position—as the cation—and then attaches itself (2) to a *meta* position, most likely in another molecule of toluene. Reaction would thus be *intermolecular* (between molecules).

With alkyl groups bigger than methyl, this does seem to happen. But with methyl, there is evidence that rearrangement is *intramolecular* (within a molecule). A free methyl cation does not separate. Instead, in a 1,2-shift of exactly the kind we have already encountered (Sec. 5.22), methyl migrates from one position to the next in the intermediate carbocation.

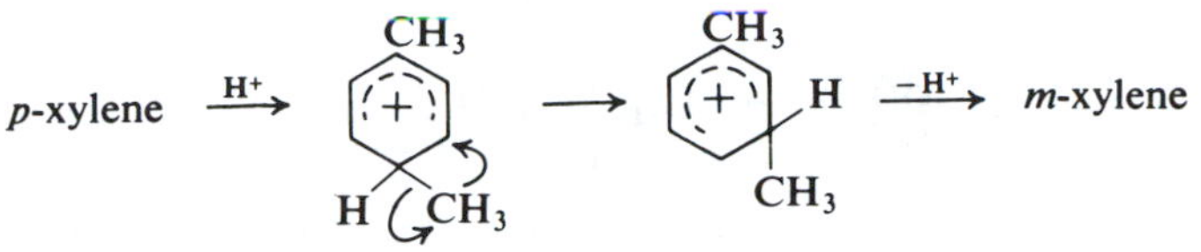

p-xylene $\xrightarrow{H^+}$ [cation: CH_3, (+), H, CH_3] $\longrightarrow$ [cation: CH_3, (+), H, CH_3] $\xrightarrow{-H^+}$ *m*-xylene

(For a general discussion, see Roberts, R. M. "Friedel–Crafts Chemistry"; *Chem. Eng. News* **Jan. 25, 1965,** 96, or Norman, R. O. C.; Taylor, R.; *Electrophilic Substitution in Benzenoid Compounds*; Elsevier: New York, 1965; pp 160–168.)

16.5 The newly introduced alkyl group activates the aromatic ring toward further substitution; in the other reactions, the newly introduced group deactivates the ring.

16.6 To permit overlap of π cloud and p orbital. (See Figure 16.4, page 569.)

16.7 Hyperconjugation stabilizes *o*- or *p*-$CH_3C_6H_4CH_2\cdot$, through contribution from structures like these:

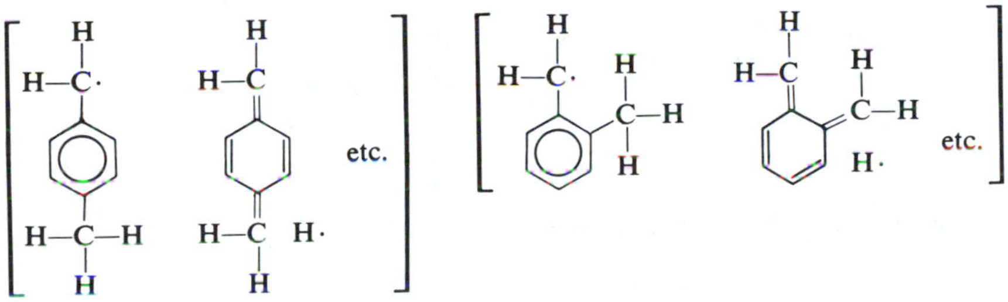

etc. etc.

Such structures are not possible for *m*-$CH_3C_6H_4CH_2\cdot$.

16.8 (a) Similar to Figure 2.4, page 53, with E_{act} = 19 kcal, and ΔH = +11 kcal.

(b) 8 kcal (the difference between E_{act} of 19 kcal and ΔH of +11 kcal).

(c) Steric hindrance to combination of the radicals.

16.9 The freezing point of a 1-molal solution (one mole of solute in 1000 g of solvent, or one millimole per gram of solvent) in benzene is 5 °C lower than that of pure benzene. If the freezing point of a solution is depressed by only one-tenth (0.5/5.0) of that amount, its concentration is only one-tenth as great, or 0.1 molal.

$$\text{molality} = \frac{\text{mmoles solute}}{\text{g solvent}} = \frac{\text{mg solute/m.w.}}{\text{g solvent}}$$

$$0.1 = \frac{1500 \text{ mg/m.w.}}{50} \qquad \text{m.w.} = 300$$

An apparent m.w. of 300 compared with an expected m.w. of 542 for $C_{42}H_{38}$ indicates considerable dissociation into free radicals.

16.10 (a) $Ph_3C{:}\ddot{\underset{..}{Cl}}{:} \longrightarrow Ph_3C^{\oplus} + {:}\ddot{\underset{..}{Cl}}{:}^-$

The Ph_3C^+ ion is stabilized by dispersal of charge over three rings, and ionization is promoted by polar solvents.

(b) The same ion is formed from triphenylmethanol.

$$Ph_3C\text{—}OH \xrightarrow{H^+} Ph_3C\text{—}\overset{\oplus}{O}H \longrightarrow \underset{\textit{Yellow}}{Ph_3C^{\oplus}} + H_2O$$
H

16.11 CCl_4

$\downarrow AlCl_3$

$^{\oplus}CCl_3 + AlCl_4^- \xrightarrow{PhH} PhCCl_3$

$\downarrow AlCl_3$

$Ph\overset{}{\underset{\oplus}{C}}Cl_2 + AlCl_4^- \xrightarrow{PhH} Ph_2CCl_2$

$\downarrow AlCl_3$

$Ph_2\underset{\oplus}{C}Cl + AlCl_4 \xrightarrow{PhH} Ph_3CCl$

$\downarrow AlCl_3$

$Ph_3C^{\oplus} + AlCl_4^- \xrightarrow{PhH}$ no reaction

Ph_3C^+ is too stable to react with C_6H_6 in the final stage.

16.12 Resonance stabilization of the anion, with dispersal of negative charge, is greatest for Ph_3C^-, least for $C_5H_{11}^-$.

16.13 The first reaction proceeds by S_N1, and the second by S_N2, for a combination of reasons.

An S_N1 reaction of a neutral substrate, we have seen (Sec. 7.5), involves a transition state that is more polar than the reactants and contains an anionic leaving group. Such a transition state is stabilized, relative to the reactants, by a solvent that is polar and, most important, that can hydrogen bond to the leaving group.

Here, the S_N1 reaction is favored by formic acid, which not only is more polar than ethanol but also is more acidic and forms much stronger hydrogen bonds (page 251). At the same time, the less polar ethanol favors the competing S_N2 reaction, which has a transition state less polar than the reactants (Sec. 7.6). Reinforcing the solvent effect is the nature of the nucleophile: the weak nucleophile, water; and the strong nucleophile, ethoxide ion (Sec. 5.23).

As we would expect, the S_N1 reaction is speeded up strongly by the presence of the methyl group, whose electron-releasing inductive effect helps to disperse the developing positive charge in the transition state.

Since a methyl group in the *para* position exerts no steric hindrance, and since no particular charge develops on the central carbon in the S_N2 transition state, we expect at most a weak effect. This is, of course, what is observed. The fact that it is weak activation would suggest that carbon is slightly positive in the transition state of this particular S_N2 reaction, but we could not have predicted this.

16.14 For these alcohols, reactivity is largely determined by the stability of the carbocations being formed. This is affected by, among other things, electron release or withdrawal by substituents on the aromatic ring.

(a) $Ph—CH(OH)—CH_2CH_3$ > $Ph—CH_2—CH(OH)—CH_3$ > $Ph—CH_2CH_2—CH_2OH$

benzylic — 2° — 1°

(b)  $HO—C_6H_4—CH_2OH$ > $C_6H_5—CH_2OH$ > $N{\equiv}C—C_6H_4—CH_2OH$

(c) Ph_3COH > Ph_2CHOH > $PhCH_2OH$

16.15 As usual, reactivity by S_N1 is controlled by the stability of the carbocation being formed, and reactivity by S_N2 is controlled chiefly by steric hindrance. The effects of the methyl groups are their customary ones, superimposed on the effect of C_6H_5.

Reactivity by S_N1 $C_6H_5CH_2Cl$ < $C_6H_5CHClCH_3$ < $C_6H_5CCl(CH_3)_2$

Reactivity by S_N2 $C_6H_5CH_2Cl$ > $C_6H_5CHClCH_3$ > $C_6H_5CCl(CH_3)_2$

16.16 (a) 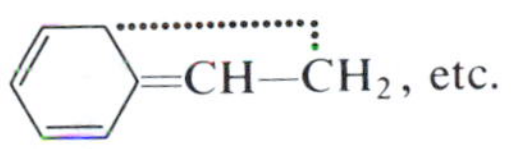 (cyclohexadienylidene)$=CH—CH_2$, etc.

(b) The same argument as for conjugated dienes (Sec. 11.19).

16.17 The base abstracts a proton to generate an allylic–benzylic *carbanion*. This, as a hybrid, can regain the proton at either of two positions, and either regenerate the starting material or form 1-phenylpropene. Equilibrium is reached, and favors the more conjugated, more stable hydrocarbon.

$$C_6H_5—CH_2—CH{=}CH_2 + OH^- \rightleftarrows \underbrace{C_6H_5—CH{\cdots}CH{\cdots}CH_2}_{\ominus} + H_2O \rightleftarrows C_6H_5—CH{=}CH—CH_3 + OH^-$$

A hybrid carbanion

More stable: favored by equilibrium

16.18 See Figure 16.6, below. An analogous figure would apply to the formation of a benzylic cation.

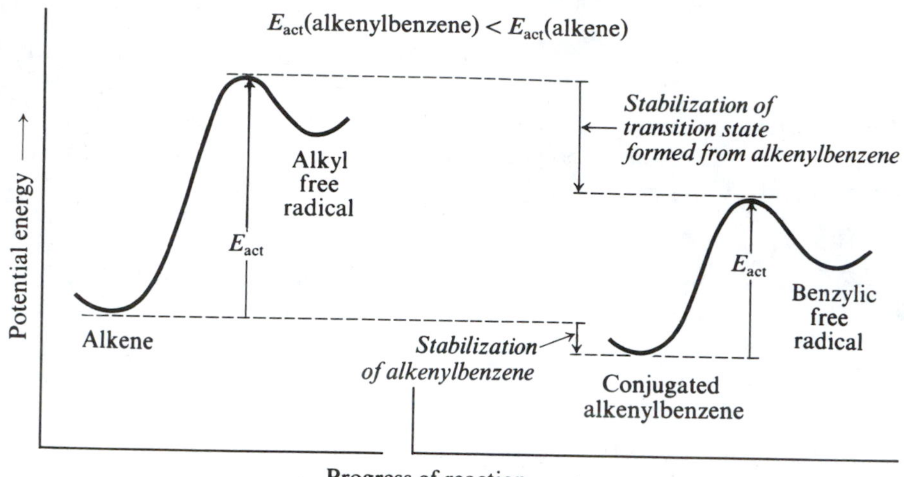

Figure 16.6 Molecular structure and rate of reaction. The transition state from the conjugated alkenylbenzene is stabilized more than the alkenylbenzene itself: E_{act} is lowered. (The plots are aligned with each other for easy comparison.)

16.19 Because of conjugation of both rings with the double bond, and perhaps between rings through the double bond, the reactant may be stabilized more than the transition state.

16.20 (a) $PhC{\equiv}CH \xleftarrow{NaNH_2} \xleftarrow{KOH(alc)} PhCHBrCH_2Br \xleftarrow{Br_2} PhCH{=}CH_2 \xleftarrow[-H_2]{cat., heat} PhCH_2CH_3$

(b) cis-PhCH=CHCH₃ (Ph and CH_3 on the same side; H, H) $\xleftarrow[Lindlar]{H_2} PhC{\equiv}CCH_3 \xleftarrow{NaNH_2} \xleftarrow{KOH(alc)} PhCHBrCHBrCH_3 \xleftarrow{Br_2}$ trans-PhCH=CHCH₃ (Ph and CH_3 on opposite sides)

16.21 (a) elemental analysis (b) hot $KMnO_4$ (c) hot $KMnO_4$
(d) fuming sulfuric acid (e) Br_2/CCl_4, or $KMnO_4$ (f) CrO_3/H_2SO_4

16.22 Upon oxidation by $KMnO_4$, *n*-butylbenzene gives C_6H_5COOH (m.p. 122 °C), and *m*-diethylbenzene gives *m*-$C_6H_4(COOH)_2$ (m.p. 348 °C).

16.23 (a) soluble (or polymerizes) (b) discharge of color
(c) discharge of color, brown MnO_2 (d) orange-red color
(e) negative test (no color change)

16.24 (a) Br_2/CCl_4, or $KMnO_4$ (b) $Ag(NH_3)_2OH$ or $Cu(NH_3)_2OH$
(c) oxidation of allylbenzene gives PhCOOH (m.p. 122 °C); 1-nonene gives $C_7H_{15}COOH$ (m.p. 16 °C, b.p. 239 °C)
(d) CrO_3/H_2SO_4

16.25

	conc. H_2SO_4[a]	*cold* $KMnO_4$[b]	Br_2[b]	CrO_3[c]	*fum.* H_2SO_4[a]	$CHCl_3$, $AlCl_3$[c]	Na[d]
Alkanes	–	–	–	–	–	–	–
Alkenes	+	+	+	–	+	–	–
Alkynes	+	+	+	–	+	–	–[e]
Alkyl halides	–	–	–	–	–	–	–
Alkylbenzenes	–	–[f]	–	–	+	+	–
1° alcohols	+	–	–	+	+	–	+
2° alcohols	+	–	–	+	+	–	+
3° alcohols	+	–	–	–	+	–	+

[a] Dissolves. [b] Decolorizes. [c] Changes color. [d] Hydrogen bubbles. [e] 1-Alkynes give test.
[f] Decolorizes hot $KMnO_4$.

1. (a) 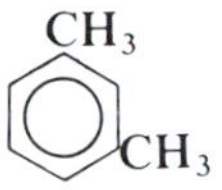 CH_3, CH_3 (b) 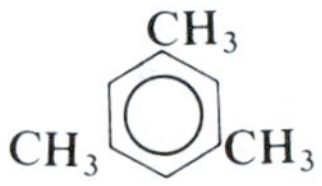 CH_3, CH_3, CH_3 (c) 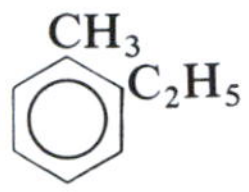 CH_3, C_2H_5

(d)  *t*-Bu, *t*-Bu (e) 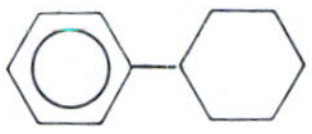 (f) 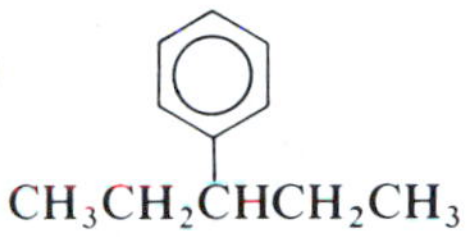 $CH_3CH_2CHCH_2CH_3$

(g) 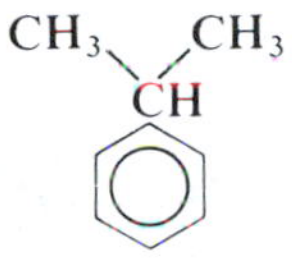 CH_3, CH_3, CH (h) C_6H_5, C_6H_5, C=C, H, H 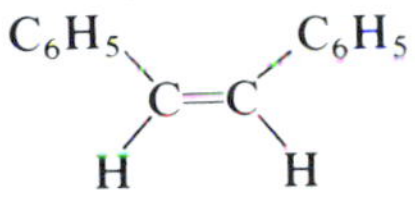 (i) $C_6H_5CH{=}CH{-}CH{=}CHC_6H_5$

(j) 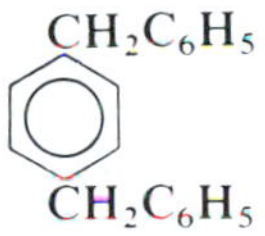 $CH_2C_6H_5$, $CH_2C_6H_5$ (k) 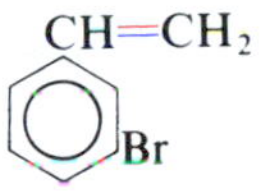 $CH{=}CH_2$, Br (l) $C_6H_5C{\equiv}CC_6H_5$

2. (a) ethylene, HF
(b) cat. H_2
(c) $2H_2$, Pt
(d) H_2SO_4, heat; H_2, Pt
(e) H_2SO_4, heat; H_2, Pt. Or: PBr_3; Mg, Et_2O; H_2O
(f) KOH(alc); H_2, Pt
(g) KOH(alc); H_2, Pt. Or: Mg, Et_2O; H_2O
(h) Mg, Et_2O; H_2O
(i) Zn(Hg), HCl

3. No reaction: a, c, f, g, l, p.

(b) *n*-propylcyclohexane
(d) benzoic acid (or salt)
(e) benzoic acid
(h) *o*- and *p*-*n*-$PrC_6H_4NO_2$
(i) *o*- and *p*-*n*-$PrC_6H_4SO_3H$
(j) *o*- and *p*-*n*-PrC_6H_4Cl
(k) *o*- and *p*-*n*-PrC_6H_4Br
(m) $PhCHBrCH_2CH_3$
(n) *o*- and *p*-*n*-$PrC_6H_4CH_3$
(o) *o*- and *p*-*n*-$PrC_6H_4CH_2C_6H_5$
(q) *p*-*t*-BuC_6H_4Pr-*n*
(r) *p*-*t*-BuC_6H_4Pr-*n*
(s) *p*-cyclohexyl-*n*-propylbenzene

4. (a) *n*-PrPh
(b) *n*-propylcyclohexane (alkyl group equatorial)
(c) $PhCHBrCHBrCH_3$ (for stereochemistry, wait for Problem 16, this chapter)
(d) *p*-$BrC_6H_4CHBrCHBrCH_3$
(e) $PhCHClCH_2CH_3$
(f) $PhCHBrCH_2CH_3$
(g) $PhCH_2CHBrCH_3$
(h) $PhCH(OSO_3H)CH_2CH_3$
(i) $PhCH(OH){-}CH(Br)CH_3$

(j) *syn*-hydroxylation

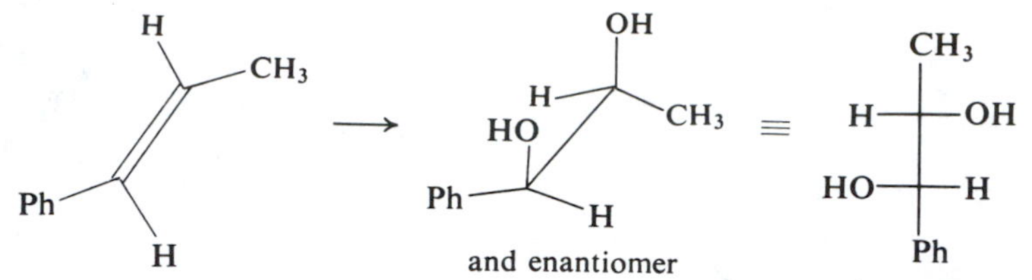

H, CH_3 / Ph, H → OH, H, HO, CH_3, Ph, H ≡ CH_3; H—OH; HO—H; Ph

and enantiomer

and enantiomer
Threo
(1*S*,2*S*) and (1*R*,2*R*)

(k) $PhCOOH + CH_3COOH$

(l) *anti*-hydroxylation

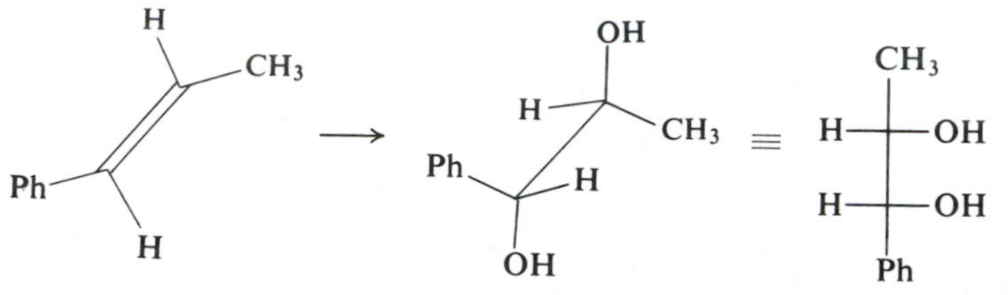

H, CH_3 / Ph, H → OH, H, CH_3, Ph, H, OH ≡ CH_3; H—OH; H—OH; Ph

and enantiomer

and enantiomer
Erythro
(1*R*,2*S*) and (1*S*,2*R*)

(m) $PhCHO + OHCCH_3$

(n) $PhCH{=}CHCH_2Br$ (allylic substitution)

(o) *syn*-addition

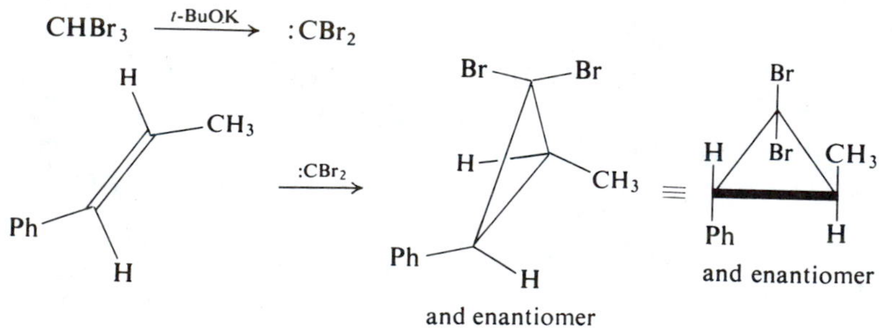

$$CHBr_3 \xrightarrow{t\text{-BuOK}} :CBr_2$$

H, CH_3 / Ph, H $\xrightarrow{:CBr_2}$ Br, Br, H, CH_3, Ph, H ≡ Br, H, Br, CH_3, Ph, H

and enantiomer

and enantiomer

(p) $PhC{\equiv}CCH_3$

5. (a) cyclohexylbenzene
(b) $PhC{\equiv}CAg$
(c) m-$O_2NC_6H_4COOH$ (side-chain oxidation)
(d) $PhCH_2CHClCH_3$
(e) p-ClC_6H_4COOH
(f) isomerization to more stable, conjugated alkene (see Sec. 16.19)

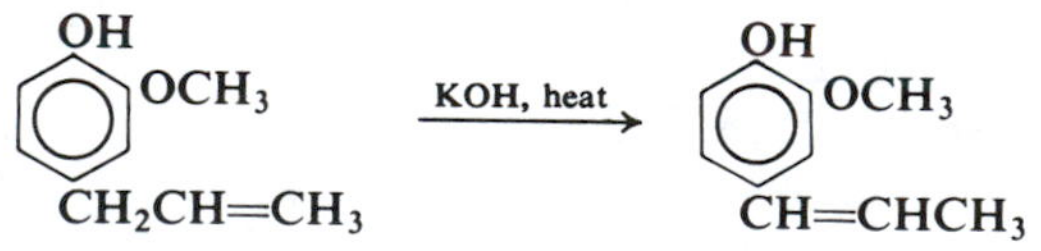

OH, OCH_3, $CH_2CH{=}CH_3$ $\xrightarrow{\text{KOH, heat}}$ OH, OCH_3, $CH{=}CHCH_3$

Eugenol → Isoeugenol

(g) $PhCH_2MgCl$
(h) $PhCH_3$
(i) 2-bromo-1,4-dimethylbenzene

(j) $PhCH{=}CH{-}CH(H){-}CH_2(H)$ $PhCH(H){-}CH{=}CH{-}CH_2(H)$ $PhCH(H){-}CH(H){-}CH{=}CH_2$

Only product reported   *Not formed*

Actually, only the most stable of the possible products is obtained—the one in which the double bond is conjugated with the ring—suggesting that reaction is equilibrium-controlled. (Compare Problem 16.17, page 424.)

(k) PhCHO + PhCHO

(l) *syn*-hydrogenation ⟶ (*Z*)-$PhCH{=}CHCH_2Ph$ (m) *anti*-hydrogenation ⟶ (*E*)-$PhCH{=}CHCH_2Ph$

(n) $CH_3O{-}C_6H_4{-}CH{=}CH{-}C_6H_5 \xrightarrow{H^+} CH_3\overset{\oplus}{O}{=}C_6H_4{=}CH{-}CH(H){-}C_6H_5 \xrightarrow{Br^-} CH_3O{-}C_6H_4{-}CH(Br){-}CH(H){-}C_6H_5$

Initial attack gives the more stable cation

6. (a) The most reactive hydrogen is both benzylic and allylic.

$Ph{-}C(H)(H){-}C(H){=}C(H){-}C(H)(H){-}C(H)(H){-}CH_3$

1 5 5 2 3 4

(b) The most reactive hydrogen is doubly benzylic.

$CH_3{-}C_6H_4{-}CH_2{-}C_6H_4{-}CH_2{-}CH_2{-}CH_3$

3 1 2 4 5

(c) Removal of hydrogen from C–1 gives a benzylic free radical stabilized by hyperconjugation (see Problem 16.7, page 570).

$1\text{-}CH_3 > 2\text{-}CH_3, 4\text{-}CH_3$

(d) Removal of allylic hydrogen leads to two products via a delocalized allylic radical.

(1) $PhCH(Br)CH{=}CHCH_2CH_2CH_3$ and $PhCH{=}CHCH(Br)CH_2CH_2CH_3$

(2) $PhCH_2CH{=}CHCH(Br)CH_2CH_3$ and $PhCH_2CH(Br)CH{=}CHCH_2CH_3$

(3) $PhCH_2CH{=}CHCH_2CH(Br)CH_3$

(4) $PhCH_2CH{=}CHCH_2CH_2CH_2Br$

(5) $PhCH_2C(Br){=}CHCH_2CH_2CH_3$ or $PhCH_2CH{=}C(Br)CH_2CH_2CH_3$

7. We expect the preferred product (i) to have the double bond conjugated with the ring, and (ii) to be the less crowded of a pair of geometric isomers.

(a) $PhCH{=}CHCH_2CH_3$ (chiefly *E*) (b) $PhC(Et){=}CH_2$ (c) major, $PhC(CH_3){=}CHCH_3$ (chiefly *E*); $PhC(Et){=}CH_2$

(d) major, $PhCH{=}CHCH_2CH_3$ (chiefly *E*); $PhCH_2CH{=}CHCH_3$ (chiefly *E*)

(e) major, $PhC(CH_3){=}CHCH_3$ (chiefly *E*); $PhCH(CH_3)CH{=}CH_2$

8. We expect the more highly branched alkenes to predominate; these may be formed through rearrangement of the initially formed carbocations into more stable ones.

(a) major, $PhCH{=}CHCH_2CH_3$ (chiefly *E*) (b) major, $PhC(CH_3){=}CHCH_3$ (chiefly *E*); $PhC(Et){=}CH_2$

(c) as in (b) (d) major, $PhCH{=}CHCH_2CH_3$ (chiefly *E*) (e) as in (b)

9. We expect the alcohol that forms the most stable carbocation to be dehydrated fastest.

(a) c > a > e, d > b (b) $Ph_2C(OH)CH_3 > PhCHOHCH_3 > PhCH_2CH_2OH$

(c) p-$CH_3C_6H_4CHOHCH_3 > C_6H_5CHOHCH_3 > p$-$BrC_6H_4CHOHCH_3$

10. In (a)–(d), reactivity is determined chiefly by the stability of the carbocation being formed. This is affected by, among other things, electron release or withdrawal by substituents on the ring. In (e), the more stable, conjugated alkene is formed faster.

(a) p-$CH_3C_6H_4CH{=}CH_2 > C_6H_5CH{=}CH > p$-$ClC_6H_4CH{=}CH_2$

(b) p-$H_2NC_6H_4CH(OH)CH_3 > C_6H_5CH(OH)CH_3 > p$-$O_2NC_6H_4CH(OH)CH_3$

(c) p-$MeOC_6H_4CH_2Cl > p$-$MeC_6H_4CH_2Cl > C_6H_5CH_2Cl > p$-$ClC_6H_4CH_2Cl > p$-$O_2NC_6H_4CH_2Cl$

(d) $PhCH(Br)CH_3 > PhCH_2Br > PhCH_2CH_2Br$

2° benzylic benzylic 1°

(e) $C_6H_5CH_2CH(Br)CH_3 > C_6H_5CH_2CH_2CH_2Br$

11. (a) PhCH(Br)CH=CHCH$_2$Br   PhCH(Br)—CH(Br)CH=CH$_2$ PhCH=CHCH(Br)—CH$_2$Br

(b) The first and third structures, formed via initial addition to C–4.

(c) The most stable product (third structure) is formed, suggesting equilibrium control. (Compare Problem 16.17, page 424.)

12. (a) The *trans* isomer is more stable by 5.7 kcal (26.3 – 20.6).

(b) *cis*-Stilbene $\xrightarrow{Br\cdot}$ [radical] $\xrightarrow{\text{rotation}}$ [radical] $\xrightarrow{-Br\cdot}$ *trans*-Stilbene

The agent is Br·, formed in either of two familiar ways. Br· adds to *cis*-stilbene to give a free radical that rotates to another conformation (actually a more stable one, with Ph's farther apart). Before the second step of addition (reaction with *scarce* HBr or Br_2) can occur, Br· is lost, to yield *trans*-stilbene.

(c) Equilibrium favors the more stable stereoisomer.

13. Ionization of Ph_3COH in H_2SO_4 (to give Ph_3C^+) produces twice as many ions per mole as does MeOH (which gives $MeOH_2^+$), hence twice the lowering of the freezing point.

$$MeOH + H_2SO_4 \rightleftharpoons MeOH_2^+ + HSO_4^- \quad \textit{two ions}$$

$$Ph_3COH + H_2SO_4 \rightleftharpoons Ph_3COH_2^+ + HSO_4^-$$

$$Ph_3COH_2^+ \rightleftharpoons Ph_3C^+ + H_2O$$

$$H_2O + H_2SO_4 \rightleftharpoons H_3O^+ + HSO_4^-$$

$$Ph_3COH + 2H_2SO_4 \rightleftharpoons Ph_3C^+ + H_3O^+ + 2HSO_4^- \quad \textit{four ions}$$

14. (a) (1) $CBrCl_3 \xrightarrow{\text{light}} Br\cdot + \cdot CCl_3$

(2) $\cdot CCl_3 + PhCH_3 \longrightarrow PhCH_2\cdot + HCCl_3$

(3) $PhCH_2\cdot + CBrCl_3 \longrightarrow PhCH_2Br + \cdot CCl_3$

then (2), (3), (2), (3), etc.

(b) (4) $\cdot Br + PhCH_3 \longrightarrow PhCH_2\cdot + HBr$

(5) $Cl_3C\cdot + \cdot CCl_3 \longrightarrow C_2Cl_6$

C_2Cl_6 is formed in the chain-terminating step (5).

Every time (1) occurs, there is formed not only $\cdot CCl_3$ but also $Br\cdot$. Like $\cdot CCl_3$, $Br\cdot$ abstracts hydrogen from toluene (4) to give $PhCH_2\cdot$ and thus starts a chain: (4), (3), (2), (3), (2), etc. For every (1), there are therefore two similar parallel chains, one started by the sequence (1), (2), and the other by the sequence (1), (4).

$CHCl_3$ is formed in each (2), (3) combination. HBr is formed only in (4), and is thus a measure of how many times (1) occurs (one HBr for each $Br\cdot$, and one $Br\cdot$ for each photon of light absorbed). The 20:1 ratio shows that the average chain length (*two* chains, remember) is 10.

15. 2-, 3-, 4-, 5-, and 6-phenyldodecanes, from rearrangement of an initial secondary cation having the charge on C–2 to other secondary cations with the charge on C–3, C–4, C–5, or C–6.

16. Addition is predominantly, but not exclusively, *anti*. This lack of (complete) stereoselectivity indicates that much of the reaction proceeds via the open benzylic cation, which is subject to attack at either face, either before or after rotation. For example, from a *trans*-alkene:

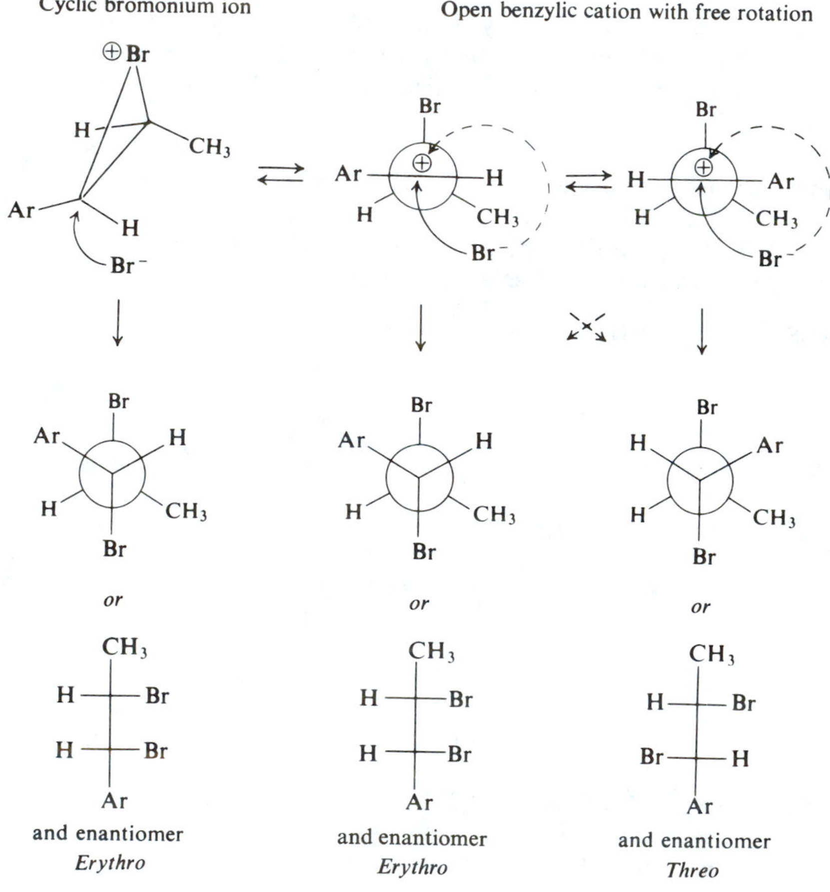

Through resonance, the electron-deficient carbon of the open benzylic cation gets electrons from the ring, and has less need of sharing an extra pair from bromine; bridging is weak, and easily broken.

The electron-releasing —OCH_3 group helps further to stabilize the benzylic cation, and hence increases the importance of the open cation in the reaction mechanism.

(Fahey, R. C.; Schneider, H.-J.; *J. Am. Chem. Soc.* **1968**, *90*, 4429.)

17. (a) *m*-Xylene. Both —CH_3 groups in *m*-xylene activate the same positions toward electrophilic aromatic substitution. *m*-Xylene is thus preferentially sulfonated; the resulting sulfonic acid dissolves in the sulfuric acid, while unreacted *o*- and *p*-xylene remain insoluble.

(b) *m*-Xylene. Desulfonation is electrophilic aromatic substitution and, as in (a), the *meta* isomer is the most reactive. Preferential desulfonation thus frees insoluble (and volatile) *m*-xylene. The non-volatile *o*- and *p*-xylenesulfonic acids remain in the aqueous acid.

(c) *m*-Xylene. Reaction with $H^+BF_4^-$ involves the first step of electrophilic aromatic substitution, formation of the benzenonium ion, or *sigma complex* (compare Problem 15.10).

CH_3, H, H, BF_4^-, CH_3

For the reasons given in (a), the sigma complex from *m*-xylene is the most stable, and hence the one favored by equilibrium. The ionic sigma complex dissolves in the polar solvent BF_3/HF; *o*- and *p*-xylenes remain insoluble.

(d) *m*-Xylene. X is *m*-$CH_3C_6H_4CH_2^-Na^+$.

We are dealing with equilibria involving *carbanions*: here, stability is decreased by electron-releasing groups, which tend to intensify the negative charge. Equilibrium favors anions from the xylenes, rather than the one from isopropylbenzene with the two methyls attached to the negative site.

$$C_6H_5\overset{CH_3\downarrow}{\underset{CH_3\uparrow}{C}}:^{\ominus}Na^+ + CH_3C_6H_4CH_3 \rightleftharpoons C_6H_5Pr\text{-}iso + \underset{\textit{More stable}}{CH_3C_6H_4CH_2:^{\ominus}Na^+}$$

Of the xylenes, *m*-xylene gives the most stable anion, *m*-$CH_3C_6H_4CH_2^-$. Like all benzylic anions, it is stabilized by dispersal of the negative charge over the ring, particularly to the positions *ortho* and *para* to the —CH_2^-. When we draw the contributing structures for *m*-$CH_3C_6H_4CH_2^-$ we find that in none of them is the negative charge located on the

$CH_2^{\ominus}$, CH_3 — CH_2, $\ominus$, CH_3 — CH_2, $\ominus$, CH_3 — CH_2, $\ominus$, CH_3

carbon to which the (destabilizing) methyl group is attached. This is in contrast to what we find for *o*- or *p*-$CH_3C_6H_4CH_2^-$. (*Draw these contributing structures.*) Methyl thus destabilizes the *meta* isomer *least*.

Separation depends upon the relative non-volatility of organosodium compounds.

(See *Chem. Eng. News* **June 14, 1971,** 30–32.)

18. (a) $PhCH_2CH_3 \xleftarrow{HF} CH_2{=}CH_2 + C_6H_6$

(b) $PhCH{=}CH_2 \xleftarrow{KOH(alc)} PhCHBrCH_3 \xleftarrow[heat]{Br_2} PhCH_2CH_3 \longleftarrow$ (a)

(c) $PhC{\equiv}CH \xleftarrow{NaNH_2} \xleftarrow{KOH(alc)} PhCH(Br){-}CH_2Br \xleftarrow{Br_2, CCl_4} PhCH{=}CH_2 \longleftarrow$ (b)

(d) $PhCH(CH_3)_2 \xleftarrow{HF} CH_3CH{=}CH_2 + C_6H_6$

(e) $Ph{-}C(CH_3){=}CH_2 \xleftarrow{KOH(alc)} Ph{-}C(CH_3)(Br){-}CH_3 \xleftarrow[heat]{Br_2} PhCH(CH_3)_2 \longleftarrow$ (d)

(f) $PhCH_2CH{=}CH_2 \xleftarrow{HF} ClCH_2CH{=}CH_2 + C_6H_6$

(g) $PhC{\equiv}CCH_3 \longleftarrow$
- $\xleftarrow{CH_3I} PhC{\equiv}CNa \xleftarrow{NaNH_2} PhC{\equiv}CH \longleftarrow$ (c)
- $\xleftarrow{NaNH_2} \xleftarrow{KOH(alc)} PhCH(Br){-}CH(Br)Ph \xleftarrow[CCl_4]{Br_2} PhCH{=}CHCH_3 \xleftarrow[isom.]{KOH,\ heat} PhCH_2CH{=}CH_2 \longleftarrow$ (f)

(h) (H)(Ph)C=C(CH$_3$)(H) [trans] $\xleftarrow{Li,\ NH_3} PhC{\equiv}CCH_3 \longleftarrow$ (g)

(i) (H)(Ph)C=C(H)(CH$_3$) [cis] $\xleftarrow[Lindlar\ cat.]{H_2} PhC{\equiv}CCH_3 \longleftarrow$ (g)

(j) $p\text{-}(CH_3)_3CC_6H_4CH_3 \xleftarrow{HF} (CH_3)_2C{=}CH_2 + C_6H_5CH_3$

(k) $p\text{-}O_2NC_6H_4CH{=}CH_2 \xleftarrow{KOH} p\text{-}O_2NC_6H_4CHBrCH_3 \xleftarrow[heat]{Br_2} p\text{-}O_2NC_6H_4CH_2CH_3 \xleftarrow[H_2SO_4]{HNO_3} PhEt \longleftarrow$ (a)

(l) $p\text{-}BrC_6H_4CH_2Br \xleftarrow[heat]{Br_2} p\text{-}BrC_6H_4CH_3 \xleftarrow{Br_2,\ Fe} C_6H_5CH_3$

(m) $p\text{-}O_2NC_6H_4CHBr_2 \xleftarrow[heat]{2Br_2} p\text{-}O_2NC_6H_4CH_3 \xleftarrow{HNO_3,\ H_2SO_4} C_6H_5CH_3$

(n) $p\text{-}BrC_6H_4COOH \xleftarrow{KMnO_4} p\text{-}BrC_6H_4CH_3 \xleftarrow{Br_2,\ Fe} C_6H_5CH_3$

(o) $m\text{-}BrC_6H_4COOH \xleftarrow{Br_2,\ Fe} C_6H_5COOH \xleftarrow{KMnO_4} C_6H_5CH_3$

(p)

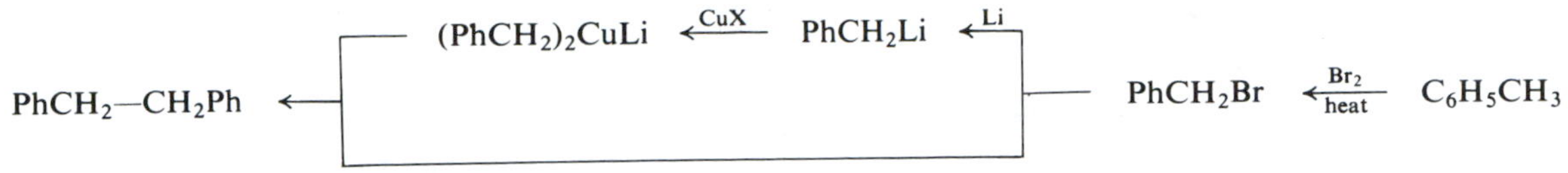

$$PhCH_2{-}CH_2Ph \leftarrow [(PhCH_2)_2CuLi \xleftarrow{CuX} PhCH_2Li \xleftarrow{Li}] \text{ and } PhCH_2Br \xleftarrow[\text{heat}]{Br_2} C_6H_5CH_3$$

(q)

$$O_2N\text{-}C_6H_4\text{-}CH_2\text{-}C_6H_5 \xleftarrow[AlCl_3]{C_6H_6} O_2N\text{-}C_6H_4\text{-}CH_2Cl \xleftarrow[\text{heat}]{Cl_2} O_2N\text{-}C_6H_4\text{-}CH_3 \xleftarrow[H_2SO_4]{HNO_3} C_6H_5CH_3$$

19. (a) Ozonolysis and identification of products.

(b) Oxidation and determination of the m.p.s of the resulting acids for the isomeric trimethylbenzenes and ethyltoluenes; side-chain chlorination followed by dehydrohalogenation and then ozonolysis of the resulting alkenes will distinguish between *n*- and isopropylbenzene.

(c), (d), (e): Oxidation and determination of the m.p.s of the resulting acids.

20. Bromobenzene is the only one that will give a Br test.

The three that give a Cl test can be distinguished from each other by oxidation to the acids and determination of m.p.s.

The two unsaturated compounds (positive $KMnO_4$ test) can be distinguished by ozonolysis and identification of the products.

The five arenes can be oxidized to carboxylic acids, which can be distinguished by their m.p.s. The very high-melting acids from mesitylene and *m*-ethyltoluene can be further distinguished by their neutralization equivalents (Sec. 19.21).

21. The empirical formulas of A, B, and C indicate that one phenyl group is present:

$$\begin{array}{r} C_8H_9 \\ -C_6H_5 \\ \hline C_2H_4 \end{array} \qquad \begin{array}{r} C_9H_{11} \\ -C_6H_5 \\ \hline C_3H_6 \end{array}$$

which suggests *which suggests*

$$Ph{-}CH_2{-}CH_2{-}H \text{ (i.e. } Ph{-}CH{-}CH_3\text{, one open bond) or } Ph{-}CH_2{-}CH_2{-} \qquad Ph{-}C(CH_3)(CH_3){-} \text{ or } Ph{-}CH(CH_3){-}CH_2{-}$$

The aliphatic residues, C_2H_4 and C_3H_6, are impossible for monosubstituted units. This predicament is eliminated by the m.w. determinations which show doubled empirical formulas, enabling us to write:

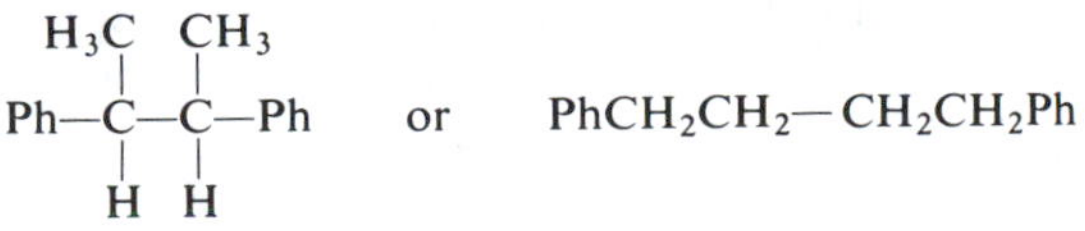

$$Ph{-}CH(CH_3){-}CH(CH_3){-}Ph \quad \text{or} \quad PhCH_2CH_2{-}CH_2CH_2Ph$$

I
2,3-Diphenylbutane

II
1,4-Diphenylbutane

and

$$\begin{array}{ccc} & H_3C\quad CH_3 & \\ & |\qquad | & \\ Ph- & C-C & -Ph \\ & |\qquad | & \\ & H_3C\quad CH_3 & \end{array} \qquad \text{or} \qquad \begin{array}{ccccc} & CH_3 & & CH_3 & \\ & | & & | & \\ Ph- & C & -CH_2-CH_2- & C & -Ph \\ & | & & | & \\ & H & & H & \end{array}$$

III
2,3-Diphenyl-2,3-dimethylbutane

IV
3,4-Diphenylhexane

We now can tackle the chemical reactions leading to A, B, and C. Free-radical attack seems likely under the conditions (see Problem 20, page 124). Knowing that benzylic hydrogen is most easily abstracted, we are led to write the following equations. For ethylbenzene:

$$t\text{-Bu}-O{:}O-\text{Bu-}t \longrightarrow 2t\text{-Bu}-O\cdot$$

$$2Ph-\underset{}{\overset{CH_3}{\overset{|}{C}}}H_2 + 2t\text{-Bu}-O\cdot \longrightarrow 2Ph-\overset{CH_3}{\overset{|}{\underset{\cdot}{C}}}H + 2t\text{-BuOH}$$

$$Ph-\underset{H}{\underset{|}{\overset{CH_3}{\overset{|}{C}}}}\cdot + \cdot\underset{H}{\underset{|}{\overset{CH_3}{\overset{|}{C}}}}-Ph \longrightarrow Ph-\underset{H}{\underset{|}{\overset{H_3C}{\overset{|}{C}}}}-\underset{H}{\underset{|}{\overset{CH_3}{\overset{|}{C}}}}-Ph$$

The overall reaction, then, is:

$$2Ph-\overset{CH_3}{\overset{|}{C}}H_2 + (t\text{-Bu})_2O_2 \longrightarrow \underset{\text{A and B}}{Ph-\overset{H_3C}{\overset{|}{C}}H-\overset{CH_3}{\overset{|}{C}}H-Ph} + 2t\text{-BuOH}$$

The observed yields fit this equation: there were obtained 0.02 mole of *tert*-butyl alcohol, and (1 + 1)/210 or about 0.01 mole of A plus B.

For isopropylbenzene, in a similar manner:

$$2Ph-\underset{CH_3}{\underset{|}{\overset{CH_3}{\overset{|}{C}}}}H + (t\text{-Bu})_2O_2 \longrightarrow \underset{\text{C}}{Ph-\underset{H_3C}{\underset{|}{\overset{H_3C}{\overset{|}{C}}}}-\underset{CH_3}{\underset{|}{\overset{CH_3}{\overset{|}{C}}}}-Ph} + 2t\text{-BuOH}$$

with the observed yield again fitting the equation.

Now, why are there two products, A and B? We might at first consider that they are I and II. But this would require formation of equal numbers of the benzylic free radicals $Ph(CH_3)CH\cdot$ and the primary free radicals $PhCH_2CH_2\cdot$, which seems highly unlikely. Besides, in that case we would expect a *third* isomer, formed by combination of unlike free radicals, and only two isomers were actually obtained.

$$Ph-\overset{CH_3}{\overset{|}{C}}H-CH_2CH_2Ph$$

Examination of the formula for I shows that three stereoisomers are possible: a pair of enantiomers and a *meso* structure. It seems most likely, then, that A and B are racemic and *meso*-2,3-diphenylbutane.

Ph	Ph	Ph
CH_3—+—H	H—+—CH_3	CH_3—+—H
H—+—CH_3	CH_3—+—H	CH_3—+—H
Ph	Ph	Ph

Enantiomers
One fraction

Meso

This conclusion is supported by the evidence from the isopropylbenzene experiment. The structure of III does not permit stereoisomerism; only one product is predicted, and only one was obtained.

22. The tricyclopropylmethyl free radical is stabilized—much as triphenylmethyl is—through delocalization of the odd electron over the rings. This is believed to involve overlap of the

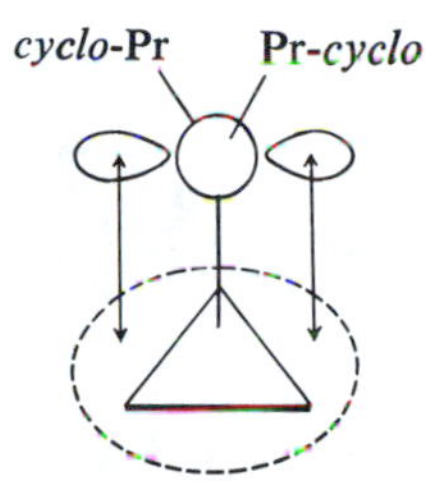

Tricyclopropylmethyl free radical

p orbital with the C—C bonds of the cyclopropane rings, which (Sec. 13.9) have considerable π character. (We shall encounter evidence of this kind of overlap in Problem 18, page 646.)

17

Spectroscopy and Structure

17.1 (a) $(CH_3)_3C^+$ $CH_2{=}CH{-}CH_2^+$ $CH_3CH_2^+$ $CH_2{=}CH^+$

(b) $C_5H_{12}^{+\cdot} \longrightarrow C_4H_9^+ + CH_3\cdot$

17.2 There is no unsaturation apparent (no absorption in the 1650 cm^{-1} range), so there must be one ring ($C_6H_{14} - C_6H_{12} = 2H$ missing, hence one ring). There is no indication of $-CH_3$ (2960 cm^{-1} and 2870 cm^{-1}), so the ring must have no side chain. The one compound that fits the data is cyclohexane.

17.3 (a) isobutylbenzene (b) *tert*-butylbenzene
(c) *p*-isopropyltoluene

(See the labeled spectra on page 678 of this Study Guide.)

17.4 $H_2C{=}C(CH_3){-}CH_2OH$ $CH_3{-}CH(CH_3){-}CH_2OH$

2-Methyl-2-propen-1-ol (Methallyl alcohol) **A**

Isobutyl alcohol **B**

(See the labeled spectra on page 679 of this Study Guide.)

17.5 *m*-Methylanisole, *m*-$CH_3C_6H_4OCH_3$.

(See the labeled spectrum on page 679 of this Study Guide.)

17.6 [structure: β-Carotene, $[\ldots CH{=}]_2$]

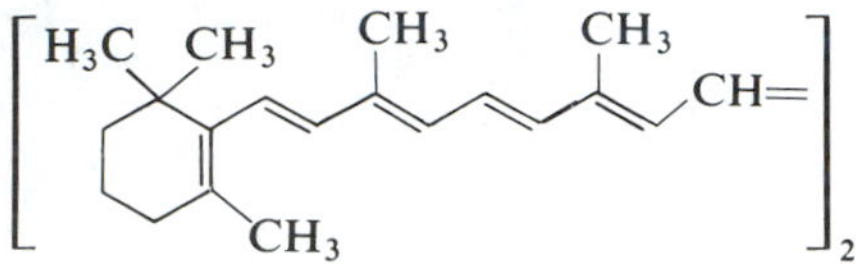

β-Carotene

17.7 (a) C, 1,4-pentadiene; D and E, (*Z*)- and (*E*)-1,3-pentadiene. (b) Heats of hydrogenation, infrared spectra.

17.8 (a) $\underset{a}{CH_3}\underset{b}{CHCl_2}$, two signals; $\underset{a}{CH_2Cl}\underset{a}{CH_2Cl}$, one signal.

(b) $\underset{a}{CH_3}CBr_2\underset{a}{CH_3}$, one signal; $\underset{b}{CH_2Br}\underset{a}{CH_2}\underset{b}{CH_2Br}$, two signals; $\underset{a}{CH_3}\underset{b}{CH_2}\underset{c}{CHBR_2}$, three signals;

$\underset{a}{CH_3}\underset{d}{CHBr}{-}C(H^{b\,(\text{or }c)})(H^{c\,(\text{or }b)}){-}Br$, four signals.

(c) $\underset{c}{C_6H_5}\underset{b}{CH_2}\underset{a}{CH_3}$, three signals; *p*-$\underset{a}{CH_3}\underset{b}{C_6H_4}\underset{a}{CH_3}$, two signals.

(d) 1,3,5-$\underset{b}{C_6H_3}\underset{a}{(CH_3)_3}$, two signals; *p*-$\underset{b}{CH_3}\underset{d}{C_6H_4}\underset{c}{CH_2}\underset{a}{CH_3}$, four signals; $\underset{c}{C_6H_5}\underset{b}{CH}\underset{a}{(CH_3)_2}$, three signals.

(e) $\underset{a}{CH_3}\underset{b}{CH_2}\underset{c}{OH}$, three signals; $\underset{a}{CH_3}O\underset{a}{CH_3}$, one signal.

(f) $\underset{a}{CH_3}\underset{b}{CH_2}O\underset{b}{CH_2}\underset{a}{CH_3}$, two signals; $\underset{c}{CH_3}O\underset{d}{CH_2}\underset{b}{CH_2}\underset{a}{CH_3}$, four signals; $\underset{b}{CH_3}O\underset{c}{CH}\underset{a}{(CH_3)_2}$, three signals;

$\underset{a}{CH_3}\underset{b}{CH_2}\underset{c}{CH_2}\underset{d}{CH_2}\underset{e}{OH}$, five signals.

(g) Oxetane ($\overset{a}{H_2C}$—$\overset{b}{CH_2}$, $\underset{b}{H_2C}$—O ring), two signals; $\underset{a}{CH_3}$—$\overset{d}{HC}$—C(H b (or c))(H c (or b)), epoxide ring through O, four signals.

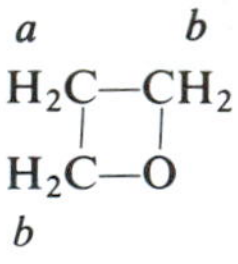

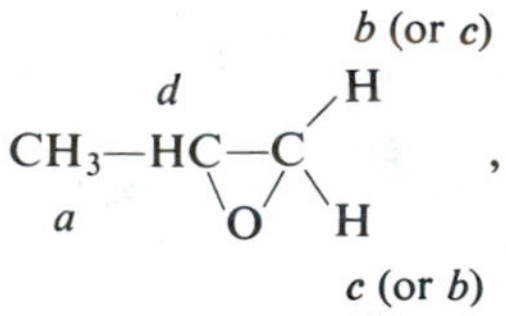

(h) $\underset{a}{CH_3}\underset{b}{CH_2}\underset{c}{CHO}$, three signals; $\underset{a}{CH_3}CO\underset{a}{CH_3}$, one signal; $(\overset{c}{H})(\underset{d}{H})C{=}C(\overset{e}{H})(\underset{a}{CH_2}\underset{b}{OH})$, five signals.

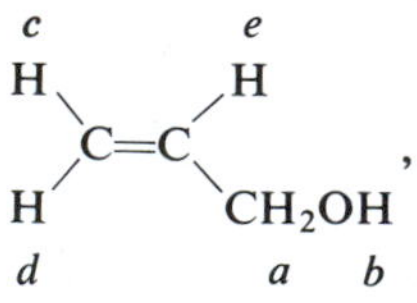

17.9

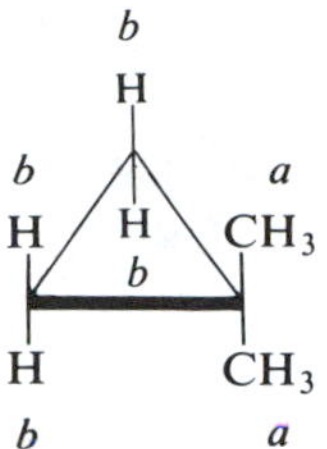

1,1-Dimethylcyclopropane
Two signals

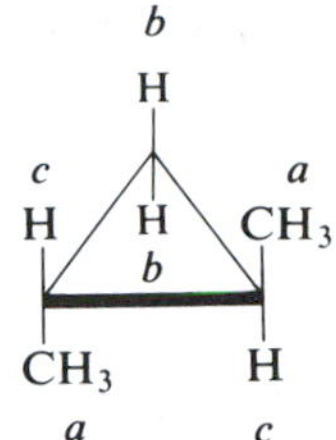

trans-1,2-Dimethylcyclopropane
Three signals

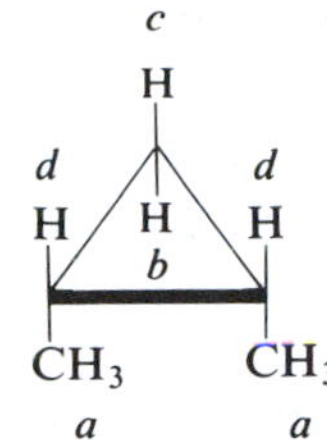

cis-1,2-Dimethylcyclopropane
Four signals

17.10 One signal, because of rapid interconversion of equatorial and axial protons.

17.11 The relative positions of protons are indicated by the sequence of letters in the answer to Problem 17.8; that is, *a* is farthest upfield, *b* is next, and so on. (Shift of —OH varies, Sec. 17.22.)

(a) *a* 3H, *b* 1H; *a*.
(b) *a*; *a* 2H, *b* 4H; *a* 3H, *b* 2H, *c* 1H; *a* 3H, *b* 1H, *c* 1H, *d* 1H.
(c) *a* 3H, *b* 2H, *c* 5H; *a* 6H, *b* 4H.
(d) *a* 9H, *b* 3H; *a* 3H, *b* 3H, *c* 2H, *d* 4H; *a* 6H, *b* 1H, *c* 5H.
(e) *a* 3H, *b* 2H, *c* 1H; *a*.
(f) *a* 6H, *b* 4H; *a* 3H, *b* 2H, *c* 3H, *d* 2H; *a* 6H, *b* 3H, *c* 1H; *a* 3H, *b* 2H, *c* 2H, *d* 2H, *e* 1H.
(g) *a* 2H, *b* 4H; *a* 3H, *b* 1H, *c* 1H, *d* 1H.
(h) *a* 3H, *b* 2H, *c* 1H; *a*; *a* 2H, *b* 1H, *c, d, e* 1H each.

Analyzing Spectra

Squeeze as much information as you can from the molecular formula: use chemical arithmetic, deciding where you can how many rings and/or double bonds are present. Combine this with characteristic infrared bands, δ values, proton counts, and splitting of various NMR signals to give you structural units. If the spectrum (or combination of spectra) is unambiguous, you should have only one possible structure left; go back and check this against all the information you have.

For problems on spectra, answers are presented in two stages: names of the unknown compounds are given in their proper sequence along with the other answers; then, at the end of the Study Guide, spectra are reproduced with infrared bands identified and NMR signals assigned. We suggest that you check each of your answers in two stages, too. First, check the name; if your answer is wrong, or if you have not been able to work the problem at all, return to the spectrum in the textbook and, knowing the correct structure, have another go at it: see if you can now identify bands, assign signals, and analyze spin–spin splittings. Then, finally, turn to the back of the Study Guide and check your answer against the analyzed spectrum.

17.12 (a) neopentylbenzene (b) 1,2-dibromo-2-methylpropane (c) benzyl alcohol

Try to fit these answers to the spectra, following the general approach outlined above. Next, see the labeled spectra on page 680 of this Study Guide, and try again. Finally, study the following.

(a) The relative peak heights are: *a*, 9H; *b*, 2H; *c*, 5H. Downfield peak *c* is clearly due to five aromatic protons: C_6H_5—. Peak *a* is clearly nine equivalent 1° aliphatic protons: $3CH_3$—. We do a little chemical arithmetic at this point,

$$\begin{array}{r}C_{11}H_{16}\\ -\,C_6H_5\text{—}\\ \hline C_5H_{11}\end{array}\qquad \begin{array}{r}C_5H_{11}\\ -\,3CH_3\text{—}\\ \hline C_2H_2\end{array}\qquad \begin{array}{r}C_2H_2\\ -\;\text{—}\overset{|}{\underset{|}{C}}\text{—}\\ \hline \text{—}CH_2\text{—}\end{array}\quad \text{(to hold the 3Me's)}$$

and end up with a residue of —CH_2—. This corresponds to peak *b*, which has a δ value of about that of benzylic H. There is only one way to put the pieces together:

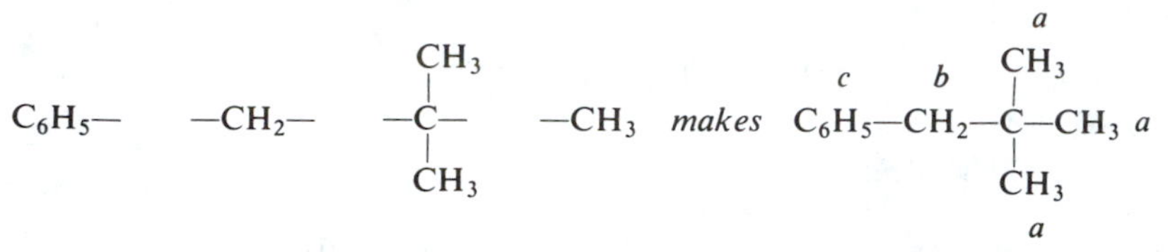

$$C_6H_5\text{—}\qquad \text{—}CH_2\text{—}\qquad \text{—}\overset{CH_3}{\underset{CH_3}{\overset{|}{\underset{|}{C}}}}\text{—}\qquad \text{—}CH_3\quad \textit{makes}\quad \overset{c}{C_6H_5}\text{—}\overset{b}{CH_2}\text{—}\overset{\overset{a}{CH_3}}{\underset{\underset{a}{CH_3}}{\overset{|}{\underset{|}{C}}}}\text{—}CH_3\ a$$

Neopentylbenzene

(b) The relative peak heights are $a{:}b = 3{:}1$. In view of the molecular formula, this means: *a* 6H; *b* 2H. The molecule is saturated, open-chain ($C_4H_8Br_2$ corresponds to C_4H_{10}), and must have the carbon skeleton of either *n*-butane or isobutane. The six protons of signal *a* are equivalent, and are probably in two —CH_3 groups, shifted

$$\begin{array}{r}C_4H_8Br_2\\ -\,2CH_3\\ \hline C_2H_2Br_2\end{array}\qquad \begin{array}{r}C_2H_2Br_2\\ -\,CH_2\\ \hline CBr_2\end{array}$$

downfield by —Br. Signal *b* is due, then, to —CH_2Br. On this assumption (supported by absence of any splitting in the signals due to protons on adjacent carbons, Sec. 17.13), we arrive at the following structure:

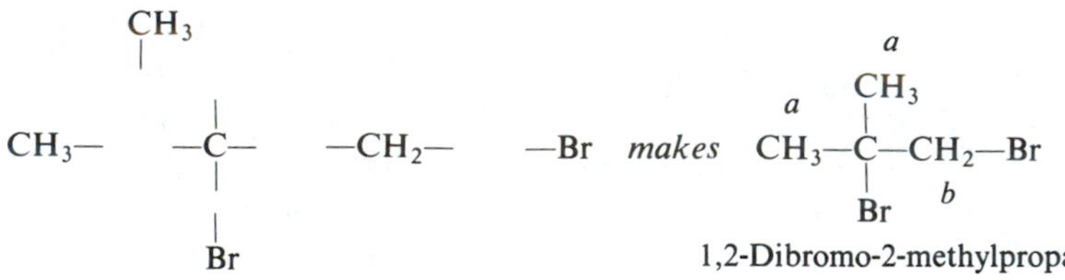

$$CH_3\text{—}\qquad \overset{CH_3}{\overset{|}{}}\qquad \text{—}\overset{|}{\underset{\underset{Br}{|}}{C}}\text{—}\qquad \text{—}CH_2\text{—}\qquad \text{—}Br\quad \textit{makes}\quad \overset{a}{CH_3}\text{—}\overset{\overset{a}{CH_3}}{\underset{Br}{\overset{|}{\underset{|}{C}}}}\text{—}\underset{b}{CH_2}\text{—}Br$$

1,2-Dibromo-2-methylpropane

(c) The relative peak heights are: *a*, 1H; *b*, 2H; *c*, 5H. The broad signal *a* indicates acidic hydrogen; in view of the molecular formula, it must be attached to oxygen: —OH. Signal *c* clearly indicates aromatic protons: C_6H_5—. Some more chemical arithmetic,

$$\begin{array}{r} C_7H_8O \\ -\,C_6H_5\text{—} \\ \hline CH_3O \end{array} \qquad \begin{array}{r} CH_3O \\ -\,\text{—}OH \\ \hline \text{—}CH_2\text{—} \end{array}$$

leaves us with a residue of —CH_2—. The pieces go together in only one way:

$$C_6H_5\text{—} \qquad \text{—}CH_2\text{—} \qquad \text{—}OH \quad \textit{makes} \quad \overset{c}{C_6H_5}\text{—}\overset{b}{CH_2}\text{—}\overset{a}{OH}$$

Benzyl alcohol

17.13 The order of the compounds is the same as in the answer to Problem 17.8.

(a) *a*, doublet, 3H
b, quartet, 1H

a, singlet

(b) *a*, singlet

a, quintet, 2H
b, triplet, 4H

a, triplet, 3H
b, multiplet, 2H
c, triplet, 1H

a, doublet, 3H
b, pair of doublets, 1H
c, pair of doublets, 1H
d, complex, 1H

(c) *a*, triplet, 3H
b, quartet, 2H
c, complex, 5H

a, singlet, 6H
b, singlet, 4H

(d) *a*, singlet, 9H
b, singlet, 3H

a, triplet, 3H
b, singlet, 3H
c, quartet, 2H
d, complex, 4H

a, doublet, 6H
b, heptet, 1H
c, complex, 5H

(e) *a*, triplet, 3H
b, quartet, 2H
c, singlet, 1H

a, singlet

(f) *a*, triplet, 6H
b, quartet, 4H

a, triplet, 3H
b, multiplet, 2H
c, singlet, 3H
d, triplet, 2H

a, doublet, 6H
b, singlet, 3H
c, heptet, 1H

a, triplet, 3H
b, multiplet, 2H
c, multiplet, 2H
d, triplet, 2H
e, singlet, 1H

(g) *a*, quintet, 2H
b, triplet, 4H

a, doublet, 3H
b, pair of doublets, 1H
c, pair of doublets, 1H
d, complex, 1H

(h) *a*, triplet, 3H
b, multiplet, 2H
c, triplet, 1H

a, singlet

a, multiplet, 2H
b, singlet, 1H
c, *d*, and *e*, multiplets, 1H each

17.14 No; the same compounds as in the answer to Problem 17.12.

17.15 (a) ethylbenzene (b) 1,3-dibromopropane (c) *n*-propyl bromide

Try to fit these answers to the spectra, following the general approach outlined on page 251 of this Study Guide. Next, see the labeled spectra on page 681 of this Study Guide, and try again. Finally, study the following.

(a) The quartet–triplet combination of signals *a* and *b* is characteristic of a CH_3CH_2— group, as in Figure 17.13(c), page 612: upfield triplet, splitting of —CH_3 by —CH_2—; downfield quartet, splitting of —CH_2— by —CH_3; *J* values identical. Subtracting C_2H_5— from the molecular formula leaves C_6H_5—,

$$\begin{array}{r} C_8H_{10} \\ -\,C_2H_5— \\ \hline C_6H_5— \end{array}$$

which gives signal *c* in the aromatic region. The compound is

$$\overset{c}{C_6H_5}\overset{b}{CH_2}\overset{a}{CH_3}$$

Ethylbenzene

The relative peak heights corroborate this: *a*, 3H; *b*, 2H; *c*, 5H.

(b) From its formula, it is clearly open-chain, saturated, and one of four possible isomeric dibromopropanes. The triplet–quintet combination indicates 4H's split by —CH_2— (signal *b*), and 2H's split by four protons (signal *a*). Only one of the possible isomers fits this pattern,

$$—\overset{b}{CH_2}—\overset{a}{CH_2}—\overset{b}{CH_2}—$$

and it takes no great imagination to attach 2Br atoms to give

$$Br\overset{b}{CH_2}—\overset{a}{CH_2}—\overset{b}{CH_2}Br$$

1,3-Dibromopropane

in which the signal for the terminal protons will be shifted far downfield by —Br.

(c) The formula requires a simple choice between *n*-propyl bromide and isopropyl bromide:

$$\overset{a}{CH_3}—\overset{b}{CH_2}—\overset{c}{CH_2}—Br \qquad \overset{a}{CH_3}—\underset{\large Br}{\overset{b}{CH}}—\overset{a}{CH_3}$$

Expect three signals	*Expect two signals*
Triplet–multiplet–triplet	Septet–doublet
3H 2H 3H	6H 1H

Clearly, the compound is *n*-propyl bromide,

$$\overset{a}{CH_3}—\overset{b}{CH_2}—\overset{c}{CH_2}—Br$$

n-Propyl bromide

17.16 (a) $(CH_3)_3C{-}O{-}CH_2CH_3$ *tert*-Butyl ethyl ether (b) $(CH_3CH_2CH_2)_2O$ Di-*n*-propyl ether (c) $(CH_3)_2CH{-}O{-}CH(CH_3)_2$ Diisopropyl ether

(See the labeled spectra on page 682 of this Study Guide.)

17.17 At room temperature, interconversion of the three possible conformers is so rapid that a single average signal is given; at −120 °C, interconversion is so slow that separate signals are given by the achiral structure I and the (racemic) chiral structures II and III.

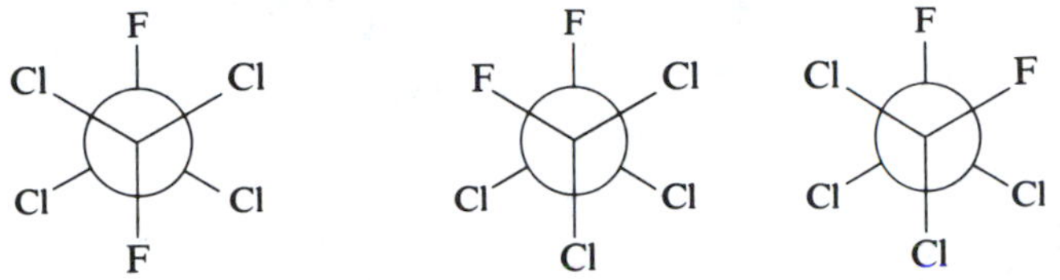

I: front F (top), Cl, Cl; back Cl, Cl, F (bottom)
II: front F (top), F, Cl; back Cl, Cl, Cl (bottom)
III: front F (top), Cl, F; back Cl, Cl, Cl (bottom)

The unequal peak areas indicate different amounts of the two components; there is no splitting of signals because in any conformation the fluorines are equivalent.

17.18 At room temperature, interconversion of the three possible conformers is so rapid that a single average signal is given; at −98 °C, interconversion is so slow that separate signals are given by the achiral structure IV and the (racemic) chiral structures V and VI.

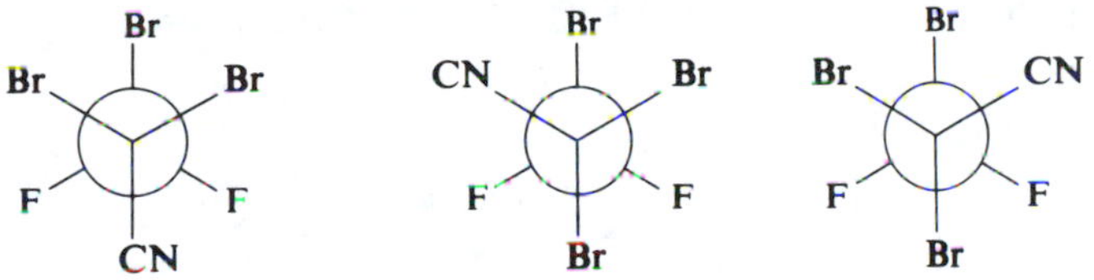

IV: front Br (top), Br, Br; back F, F, CN (bottom)
V: front Br (top), CN, Br; back F, F, Br (bottom)
VI: front Br (top), Br, CN; back F, F, Br (bottom)

The pair of doublets is given by V and VI: in each, the fluorines are not equivalent—one is "between" two —Br's, the other "between" —Br and —CN. The relative peak areas indicate that V and VI are favored: there is less crowding of Br atoms than in IV.

17.19 (i) (a) $\overset{a}{CH_3}\overset{b}{CH_2}\overset{c}{CH_2}Cl$ *Three signals*

$\overset{a}{CH_3}\overset{b}{CH}(Cl)\overset{a}{CH_3}$ *Two signals*

(b) $\overset{a}{CH_3}\overset{b}{CH_2}\overset{c}{CH_2}\overset{b}{CH_2}\overset{a}{CH_3}$ *Three signals*

$\overset{a}{CH_3}\overset{d}{CH_2}\overset{c}{CH}(\overset{b}{CH_3})\overset{b}{CH_3}$ *Four signals*

$\overset{b}{CH_3}{-}\overset{a}{C}(\overset{b}{CH_3})_2{-}\overset{b}{CH_3}$ *Two signals*

(c) $\overset{a}{CH_3}\overset{b}{CH_2}\overset{c}{CH_2}\overset{c}{CH_2}\overset{b}{CH_2}\overset{a}{CH_3}$

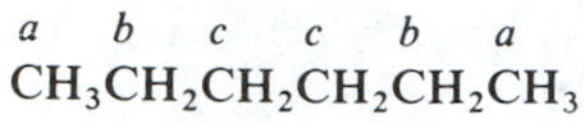

Three signals

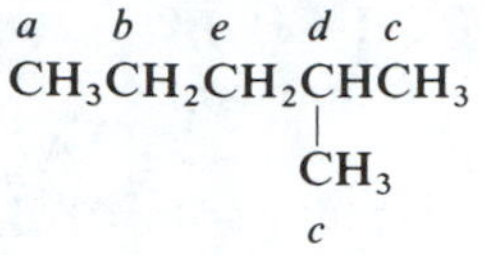

$\overset{a}{CH_3}\overset{b}{CH_2}\overset{e}{CH_2}\overset{d}{C}H\overset{c}{CH_3}$, with $\underset{c}{CH_3}$ on the CH

Five signals

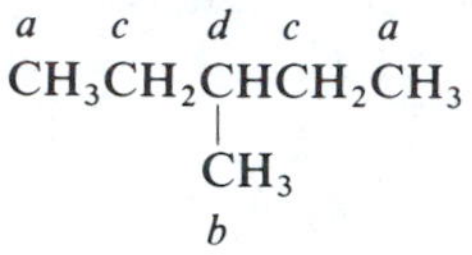

$\overset{a}{CH_3}\overset{c}{CH_2}\overset{d}{C}H\overset{c}{CH_2}\overset{a}{CH_3}$, with $\underset{b}{CH_3}$ on the CH

Four signals

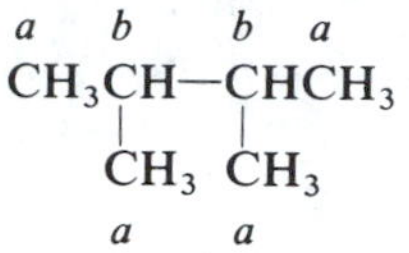

$\overset{a}{CH_3}\overset{b}{CH}-\overset{b}{CH}\overset{a}{CH_3}$, with $\underset{a}{CH_3}$ on each CH

Two signals

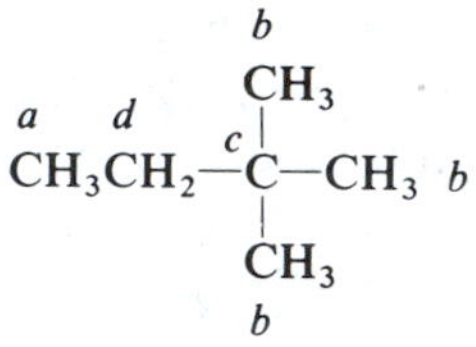

$\overset{a}{CH_3}\overset{d}{CH_2}-\overset{c}{C}(\overset{b}{CH_3})_2-CH_3\ b$

Four signals

(d) o-Xylene: CH_3 (a), CH_3 (a); ring carbons d, d, c, c, b, b

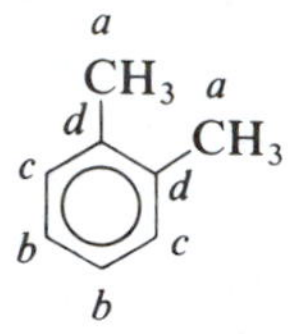

Four signals

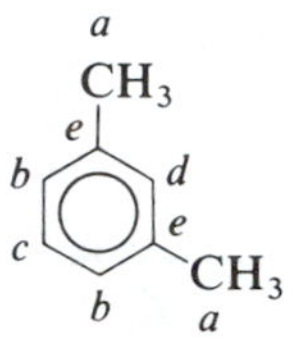

m-Xylene: CH_3 (a), CH_3 (a); ring carbons e, e, d, c, b, b

Five signals

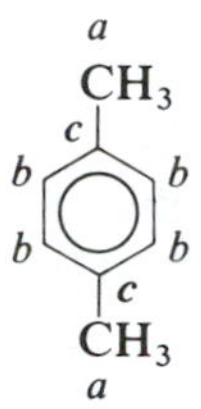

p-Xylene: CH_3 (a), CH_3 (a); ring carbons c, c, b, b, b, b

Three signals

(e) $\overset{a}{CH_3}\overset{b}{CH_2}\overset{d}{CH_2}\overset{c}{CH_2}\overset{c}{CH_2}\overset{d}{CH_2}\overset{b}{CH_2}\overset{a}{CH_3}$

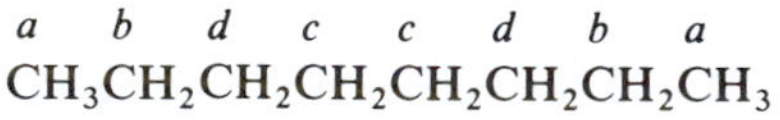

Four signals

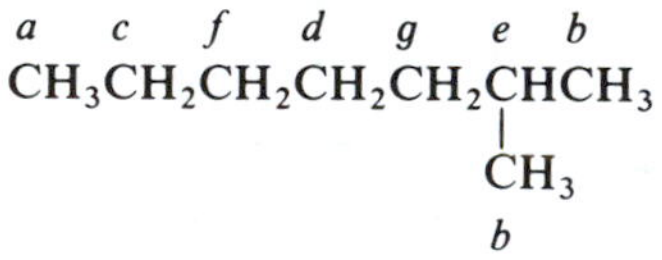

$\overset{a}{CH_3}\overset{c}{CH_2}\overset{f}{CH_2}\overset{d}{CH_2}\overset{g}{CH_2}\overset{e}{C}H\overset{b}{CH_3}$, with $\underset{b}{CH_3}$ on the CH

Seven signals

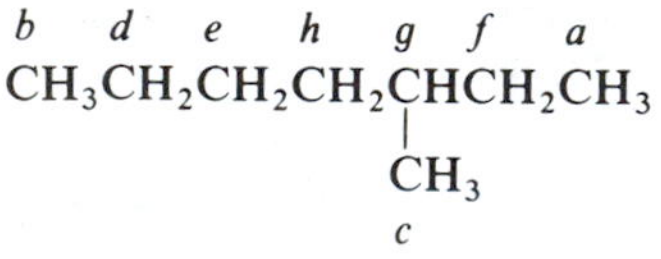

$\overset{b}{CH_3}\overset{d}{CH_2}\overset{e}{CH_2}\overset{h}{CH_2}\overset{g}{C}H\overset{f}{CH_2}\overset{a}{CH_3}$, with $\underset{c}{CH_3}$ on the CH

Eight signals

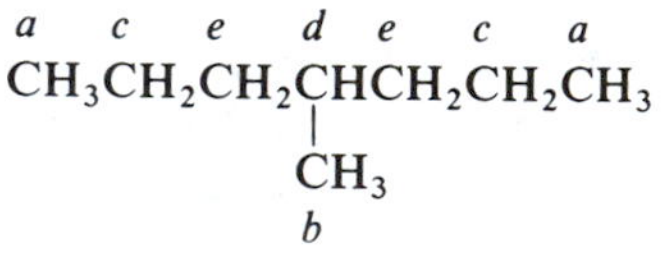

$\overset{a}{CH_3}\overset{c}{CH_2}\overset{e}{CH_2}\overset{d}{C}H\overset{e}{CH_2}\overset{c}{CH_2}\overset{a}{CH_3}$, with $\underset{b}{CH_3}$ on the CH

Five signals

(f) Mesitylene: three CH_3 (a); ring carbons c, c, c, b, b, b

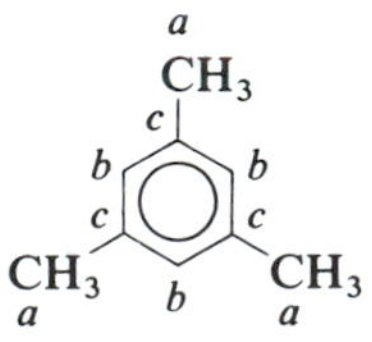

Three signals

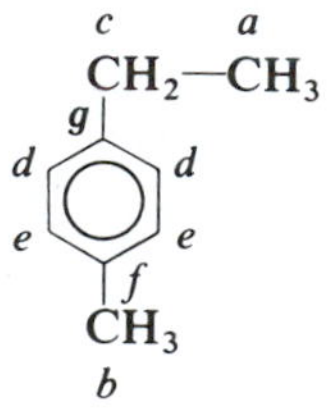

p-Ethyltoluene: $\overset{c}{CH_2}-\overset{a}{CH_3}$ on ring carbon g; CH_3 (b) on ring carbon f; ring carbons d, d, e, e

Seven signals

Isopropylbenzene: $\overset{a}{CH_3}$, $\overset{a}{CH_3}$ on $\overset{b}{CH}$; ring carbons f, d, d, e, e, c

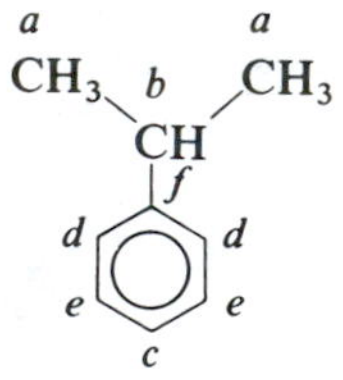

Six signals

(g)

$\overset{a}{CH_3}\overset{b}{CH_2}\overset{c}{CH_2}\overset{d}{CH_2}\overset{f}{CH}{=}\overset{e}{CH_2}$
Six signals

$\overset{a}{CH_3}\overset{b}{CH_2}\overset{c}{C}H{=}\overset{c}{C}H\overset{b}{CH_2}\overset{a}{CH_3}$ (both H on the same side)
Three signals

$\overset{a}{CH_3}\overset{b}{CH_2}\overset{c}{C}H{=}\overset{c}{C}H\overset{b}{CH_2}\overset{a}{CH_3}$ (H on opposite sides)
Three signals

$\overset{b}{CH_3}\overset{c}{CH_2}\overset{e}{C}H{=}\overset{f}{C}(\overset{a}{CH_3})\overset{d}{CH_3}$
Six signals

In 2-methyl-2-hexene we would see different signals for the diastereotopic methyl carbons: *a* is *cis* to ethyl, and *d* is *trans*.

(ii) Each of the above compounds can be distinguished from all others of its set on the basis of the numbers of signals *except*: (c) 3-methylpentane and 2,2-dimethylbutane; (g) 1-hexene and 2-methyl-2-pentene, *cis*- and *trans*-3-hexene.

17.20 Since, at this stage, we are considering only the numbers of signals, our approach is more laborious than it will be later; it will, however, give us necessary practice in recognizing equivalent and non-equivalent carbons. We draw structural formulas of all possible isomers; for each one, we label all sets of equivalent carbons, and compare the numbers of signals expected with the number in the actual spectrum.

(a) The formula shows an open-chain saturated structure ($C_5H_{11}Cl$ is equivalent to C_5H_{12}). We follow the systematic approach of Problem 3.1.

I: $\overset{a}{C}-\overset{b}{C}-\overset{c}{C}-\overset{d}{C}-\overset{e}{C}-Cl$ — *Five signals*

II: $\overset{a}{C}-\overset{b}{C}-\overset{d}{C}-\overset{e}{C}(Cl)-\overset{c}{C}$ — *Five signals*

III: $\overset{a}{C}-\overset{b}{C}-\overset{c}{C}(Cl)-\overset{b}{C}-\overset{a}{C}$ — *Three signals*

IV: $\overset{a}{C}-\overset{c}{C}-\overset{d}{C}(\overset{b}{C})-\overset{e}{C}-Cl$ — *Five signals*

V: $\overset{a}{C}-\overset{c}{C}-\overset{d}{C}(Cl)(\overset{b}{C})-\overset{b}{C}$ — *Four signals*

VI: $\overset{c}{C}-\overset{e}{C}(Cl)-\overset{d}{C}(\overset{a}{C})-\overset{b}{C}$ — *Five signals*

VII: $Cl-\overset{d}{C}-\overset{c}{C}-\overset{b}{C}(\overset{a}{C})-\overset{a}{C}$ — *Four signals*

VIII: $\overset{a}{C}-\overset{b}{C}(\overset{a}{C})(\overset{a}{C})-\overset{c}{C}-Cl$ — *Three signals*

Structure VI deserves a special look. We have marked the methyl groups *a* and *b* as non-equivalent; they are *diastereotopic*, and we would expect them to give rise to different signals. To see that this is so, we follow the same procedure as for determining the equivalence or non-equivalence of protons (Sec. 17.10). We imagine each $-CH_3$ in turn to be replaced by some other group, such as $-CH_2Z$. As we see, such replacement here

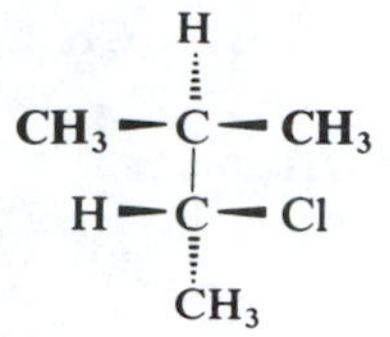

```
      H                       H                        H
      ⋮                       ⋮                        ⋮
CH3—C—CH3          [  CH3—C—CH2Z          ZCH2—C—CH3  ]
      |                       |                        |
   H—C—Cl                  H—C—Cl                   H—C—Cl
      ⋮                       ⋮                        ⋮
     CH3                     CH3                      CH3
```

Diastereotopic methyl groups

1-Chloro-3-methylbutane

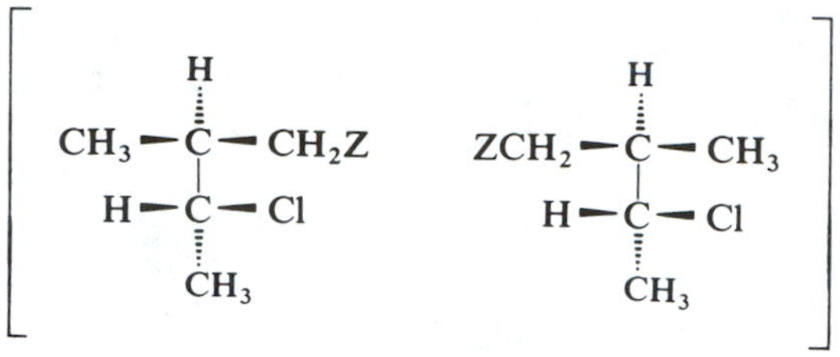

gives one or the other of a pair of diastereomers. The environments of these methyls are neither identical nor mirror images of each other; these methyls are diastereotopic, non-equivalent, and give separate CMR signals.

Since the unknown gives four signals, it can have either structure V or VII.

(b) We consider the same possible structures as for part (a). Since this unknown also gives four signals, it can have either structure V or VII.

(c) We consider the same possible structures as for part (a), with Br taking the place of Cl. Since the unknown gives five signals, it can be either

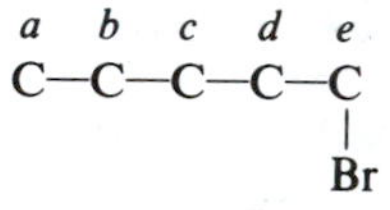

```
                                                        a
                                                        C
a b c d e               a b e d c                 c e   |  b
C—C—C—C—C      or       C—C—C—C—C       or        C—C—C—C
        |                     |                     |   | d
        Br                    Br                    Br
   IX                      X                        XI
```

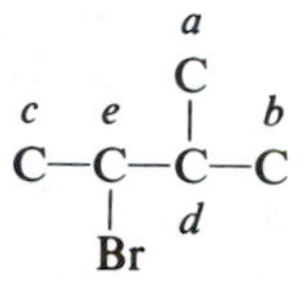

17.21 We simply look at the structural formula and count the number of hydrogens attached to each carbon; the number of peaks in each signal will equal the number of hydrogens *plus one* (page 631). The order of the compounds is the same as in the answer to Problem 17.19.

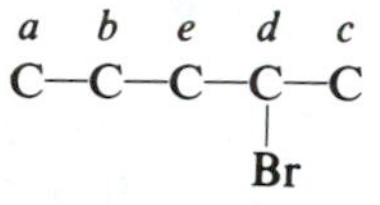

(a) $\begin{matrix}a&b&c\\ q&t&t\end{matrix}$; $\begin{matrix}a&b\\ q&d\end{matrix}$. (b) $\begin{matrix}a&b&c\\ q&t&t\end{matrix}$; $\begin{matrix}a&b&c&d\\ q&q&d&t\end{matrix}$; $\begin{matrix}a&b\\ s&q\end{matrix}$.

(c) $\begin{matrix}a&b&c\\ q&t&t\end{matrix}$; $\begin{matrix}a&b&c&d&e\\ q&t&q&d&t\end{matrix}$; $\begin{matrix}a&b&c&d\\ q&q&t&d\end{matrix}$; $\begin{matrix}a&b\\ q&d\end{matrix}$; $\begin{matrix}a&b&c&d\\ q&q&s&t\end{matrix}$.

(d) $\begin{matrix}a&b&c&d\\ q&d&d&s\end{matrix}$; $\begin{matrix}a&b&c&d&e\\ q&d&d&d&s\end{matrix}$; $\begin{matrix}a&b&c\\ q&d&s\end{matrix}$.

(e) $\begin{matrix}a&b&c&d\\ q&t&t&t\end{matrix}$; $\begin{matrix}a&b&c&d&e&f&g\\ q&q&t&t&d&t&t\end{matrix}$; $\begin{matrix}a&b&c&d&e&f&g&h\\ q&q&q&t&t&t&d&t\end{matrix}$; $\begin{matrix}a&b&c&d&e\\ q&q&t&d&t\end{matrix}$.

(f) $\begin{matrix}a&b&c\\ q&d&s\end{matrix}$; $\begin{matrix}a&b&c&d&e&f&g\\ q&q&t&d&d&s&s\end{matrix}$; $\begin{matrix}a&b&c&d&e&f\\ q&d&d&d&d&s\end{matrix}$.

(g) *a b c d e f* / q t t t t d ; *a b c* / q t d ; *a b c* / q t d ; *a b c d e f* / q q t q d s .

We could now readily distinguish between the alternatives in part (c) and in part (g).

17.22 (a) 2-chloro-2-methylbutane (b) 1-chloro-3-methylbutane (c) 1-bromopentane

From the number of signals alone (four), we concluded that (a) and (b) had formulas V and VII, but we could not tell which was which. Now we count the hydrogens on each

$$\overset{a}{CH_3}\overset{c}{CH_2}-\overset{d}{C}(\overset{b}{CH_3})(Cl)-CH_3\ b \qquad \overset{d}{CH_2}(Cl)\overset{c}{CH_2}\underset{b}{CH}(\overset{a}{CH_3})CH_3\ a$$

V | VII

a b c d / q q t s ; *a b c d* / q d t t

carbon and predict the multiplicities shown under each formula. When we compare these with the multiplicities listed for the spectra, we see that they match exactly: (a) is V, and (b) is VII.

From the number of signals alone (five), we concluded that (c) was IX, X, or XI.

$$\overset{a}{CH_3}\overset{b}{CH_2}\overset{c}{CH_2}\overset{e}{CH_2}\overset{d}{CH_2}Br \qquad \overset{a}{CH_3}\overset{b}{CH_2}\overset{e}{CH_2}\overset{d}{CH}(Br)\overset{c}{CH_3} \qquad \overset{c}{CH_3}\overset{e}{CH}(Br)\underset{d}{CH}(\overset{a}{CH_3})CH_3\ b$$

IX | X | XI

a b c d e / q t t t t ; *a b c d e* / q t q t d ; *a b c d e* / q q q d d

When we consider the splitting of signals, we see that (c) can only be IX.

An alternative approach The way we have analyzed the spectra in Figure 17.21 is feasible only for rather simple compounds, for which there are relatively few isomers; even here, as we have found, it can be a laborious process. Let us examine these same spectra afresh, this time as we will ordinarily approach such a problem: with realization of what the splitting of signals signifies.

(a) For each signal we write down the structural unit that must give rise to it,

$$CH_3- \qquad CH_3- \qquad -CH_2- \qquad -\overset{|}{\underset{|}{C}}-$$

and to this, as given by the molecular formula, we add

$$-Cl$$

We add these units up, and find that they total C_4H_8Cl. A little arithmetic shows us

$$\begin{array}{r} C_5H_{11}Cl \\ -\,C_4H_8Cl \\ \hline CH_3 \end{array}$$

that we are lacking one $—CH_3$. One of the quartets, then, must be due to *two equivalent* CH_3 groups. Keeping in mind the equivalence of two of the methyl groups, we find that we can put together the pieces in only one way:

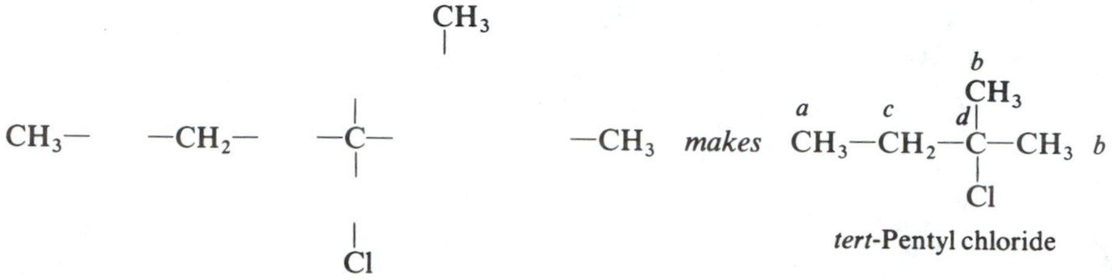

$$CH_3— \quad —CH_2— \quad —\overset{\overset{CH_3}{|}}{\underset{\underset{Cl}{|}}{C}}— \quad —CH_3 \quad \textit{makes} \quad \overset{a}{CH_3}—\overset{c}{CH_2}—\overset{\overset{b}{CH_3}}{\underset{Cl}{\overset{d|}{\underset{|}{C}}}}—CH_3\ b$$

tert-Pentyl chloride

(b) As before, we write down the units indicated by the multiplicities, and include the Cl:

$$CH_3— \quad —CH_2— \quad —CH_2— \quad —\overset{|}{C}H— \quad —Cl$$

On addition, these total C_4H_8Cl. Arithmetic shows us

$$\begin{array}{r} C_5H_{11}Cl \\ -\,C_4H_8Cl \\ \hline CH_3 \end{array}$$

that we are lacking one $—CH_3$, and that the quartet must be due to two equivalent CH_3 groups. These pieces go together in only one way:

$$CH_3— \quad —\overset{\overset{CH_3}{|}}{C}H— \quad —CH_2— \quad —CH_2— \quad —Cl \quad \textit{makes} \quad \overset{a}{CH_3}—\underset{b}{\overset{\overset{\overset{a}{CH_3}}{|}}{C}H}—\overset{c}{CH_2}—\overset{d}{CH_2}—Cl$$

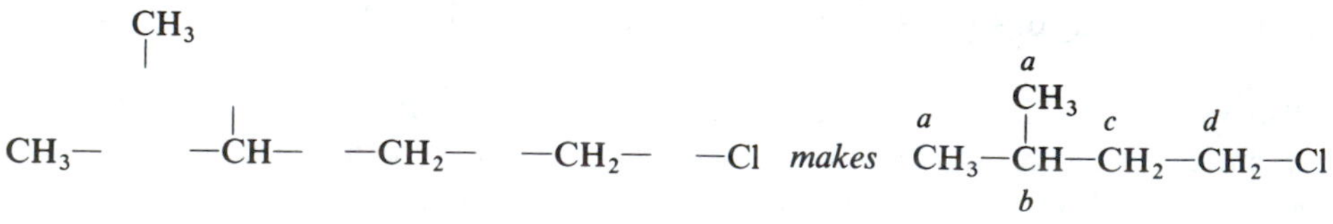

Isopentyl chloride

(c) We write down the structural units indicated by the multiplicities, and include the Br:

$$CH_3— \quad —CH_2— \quad —CH_2— \quad —CH_2— \quad —CH_2— \quad —Br$$

On addition, these total $C_5H_{11}Br$. Since this equals the molecular formula, we know that no units are repeated—that is, that there are no sets of equivalent carbons. The pieces can be put together in only one way:

$CH_3—$ $—CH_2—$ $—CH_2—$ $—CH_2—$ $—CH_2—$ $—Br$ *makes* $\overset{a}{CH_3}—\overset{b}{CH_2}—\overset{c}{CH_2}—\overset{d}{CH_2}—\overset{e}{CH_2}—Br$

n-Pentyl bromide

Assignment of peaks Having arrived at the structures of these three compounds, we find that we can now assign certain signals in the spectra to specific carbons in the compounds simply on the basis of the splitting. If there is only one singlet or doublet or triplet or quartet in a spectrum, we can with confidence assign such a group to the proper carbon in the molecule.

(a) $CH_3—C(CH_3)(Cl)—CH_2—CH_3$ (labels *c*, *d*)

(b) *a* $CH_3—CH(\overset{a}{CH_3})—CH_2—CH_2—Cl$ (label *b* on CH)

(c) $\overset{a}{CH_3}—CH_2—CH_2—CH_2—CH_2—Br$

(We shall return to the assignment of peaks in these molecules in Problem 17.25.)

17.23

$\overset{13.7}{CH_3}—\overset{22.6}{CH_2}—\overset{34.5}{CH_2}—\overset{22.6}{CH}(H)—\overset{13.7}{CH_3}$

$\overset{5}{CH_3}—\overset{4}{CH_2}—\overset{3}{CH_2}—\overset{2}{CH}(\underset{1}{CH_3})—\overset{1}{CH_3}$

	Calc.	*Actual*
C-1	13.7 + 9.4 = 23.1	*22.5*
C-2	22.6 + 9.2 = 31.8	*27.8*
C-3	34.5 + 9.4 = 43.9	*41.8*
C-4	22.6 − 2.5 = 20.1	*20.7*
C-5	13.7 = 13.7	*14.1*

There is rough agreement between the calculated and actual δ values. In this case, the agreement is good enough to give the correct sequence of peaks in the spectra; this is not always true, however, and such calculations must be used with caution.

17.24 (a) Working with these values,

$CH_3—\overset{136}{CH}=\overset{115}{CH_2}$ $CH_3—\overset{124}{CH}=\overset{124}{CH}—CH_3$ (*cis*) $CH_3—\overset{125}{CH}=\overset{125}{CH}—CH_3$ (*trans*)

we calculate the following:

$CH_3—\overset{\beta}{\overset{-12}{CH}}=\overset{\alpha}{\overset{+9}{CH}}—\mathbf{CH_3}$ (*cis*) $CH_3—\overset{\beta}{\overset{-11}{CH}}=\overset{\alpha}{\overset{+10}{CH}}—\mathbf{CH_3}$ (*trans*)

α-effects +9 and +10

β-effects −12 and −11

(b) Working with these values,

$$\overset{136}{CH_3-CH}\overset{115}{=CH_2} \qquad CH_3-CH_2-\overset{140}{CH}\overset{113}{=CH_2}$$

we calculate the following:

$$CH_3-\overset{\overset{\beta}{+4}}{CH_2}-CH\overset{\overset{\gamma}{-2}}{=}CH_2$$

β-effect $+4$
γ-effect -2

Here, the β-effect is much smaller than in part (a), and is opposite in sign.

(c) No, the effects are not really comparable. In part (a) the β-effect is exerted *across the double bond*; in part (b) it is not. (This contrast in β-effects is generally observed in the CMR spectra of alkenes.)

17.25 On the basis of the numbers of signals and of the splitting (Problem 17.22, above), we have already been able to assign a single structure to each compound. As we shall see below, the size of the chemical shifts may serve, roughly, to confirm our earlier conclusions.

Using only the splitting, we have also been able to assign certain peaks to particular carbons in each molecule: specifically, where there is only one signal of a particular multiplicity. Now let us see if the values of chemical shifts will let us assign still more peaks.

(a) For this compound we have the following data:

a	9.5	q
b	32.0	q
c	39.0	t
d	70.9	s

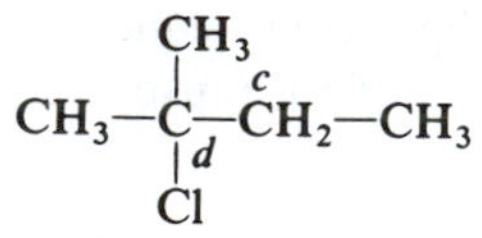

$$CH_3-\overset{\overset{\large CH_3}{|}}{\underset{\underset{\large Cl}{|}}{C}}{}_d-\overset{c}{CH_2}-CH_3$$

We have assigned peaks *c* and *d* on the basis of their multiplicities. The remaining question is: to which methyl groups should we assign peaks *a* and *b*? As in *i* or in *ii*?

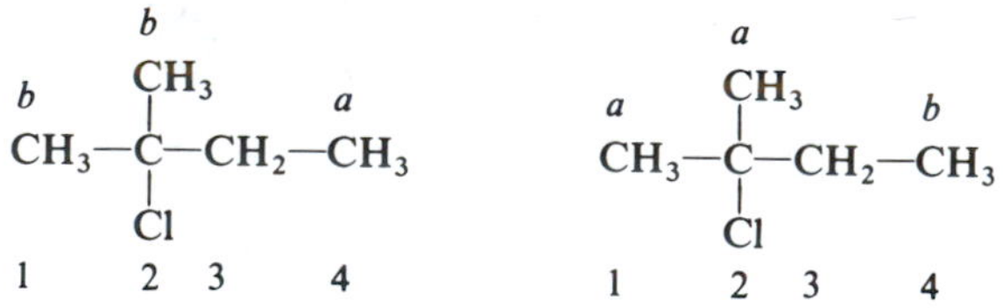

$$\underset{1}{\overset{b}{CH_3}}-\underset{2}{\overset{\overset{b}{CH_3}}{\underset{Cl}{C}}}-\underset{3}{CH_2}-\underset{4}{\overset{a}{CH_3}} \qquad \underset{1}{\overset{a}{CH_3}}-\underset{2}{\overset{\overset{a}{CH_3}}{\underset{Cl}{C}}}-\underset{3}{CH_2}-\underset{4}{\overset{b}{CH_3}}$$

i *ii*

In *n*-butane,

$$\underset{1}{CH_3}-\underset{2}{CH_2}-\underset{3}{CH_2}-\underset{4}{CH_3}$$

C–1 and C–4 are, of course, equivalent, and would have the same δ value. In mentally converting *n*-butane into 2-chloro-2-methylbutane, what changes do we make to each of these carbons? To C–1 we add a β-methyl and a β-chloro; to C–4 we add a γ-methyl and a γ-chloro. Using the values for β- and γ-effects given on pages 636–637,

C–1		C–4	
β-CH_3	+9.4	γ-CH_3	−2.5
β-Cl	+10.1	γ-Cl	−5.3
	+19.5		−7.8

we predict that δ will increase by 19.5 for C–1 and decrease by 7.8 for C–4; that is to say, C–1 should absorb 27.3 ppm downfield from C–4. This is a very large difference, and we have no hesitation in assigning the downfield quartet to C–1 and the upfield quartet to C–4, as in *i*. (The actual difference between the peaks, we see, is 22.5 ppm.)

As we can see from the labeled spectrum on page 683 of this Study Guide, we have now correctly assigned all the peaks for this compound.

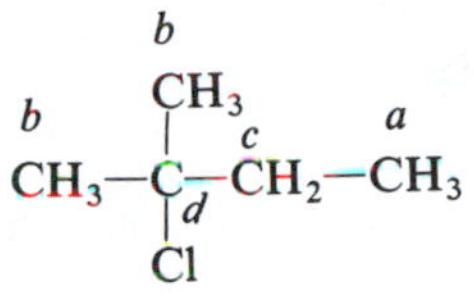

$$\overset{b}{CH_3}\text{—}\underset{\underset{Cl}{|}\,d}{\overset{\overset{b}{CH_3}}{\overset{|}{C}}}\text{—}\overset{c}{CH_2}\text{—}\overset{a}{CH_3}$$

tert-Pentyl chloride

(b) For this compound we have the following data:

a	22.1	q
b	25.8	d
c	41.8	t
d	43.2	t

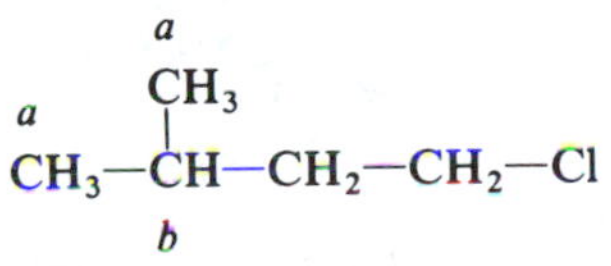

$$\overset{a}{CH_3}\text{—}\underset{b}{\overset{\overset{a}{CH_3}}{\overset{|}{CH}}}\text{—}CH_2\text{—}CH_2\text{—}Cl$$

We have already assigned peaks *a* and *b*, and are left with the question: which methylene group gives peak *c* and which gives peak *d*? As in *iii* or in *iv*?

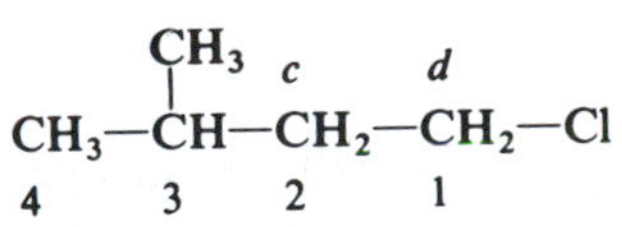

$$\underset{4}{CH_3}\text{—}\underset{3}{\overset{CH_3}{\overset{|}{CH}}}\text{—}\underset{2}{\overset{c}{CH_2}}\text{—}\underset{1}{\overset{d}{CH_2}}\text{—}Cl \qquad \underset{4}{CH_3}\text{—}\underset{3}{\overset{CH_3}{\overset{|}{CH}}}\text{—}\underset{2}{\overset{d}{CH_2}}\text{—}\underset{1}{\overset{c}{CH_2}}\text{—}Cl$$

iii 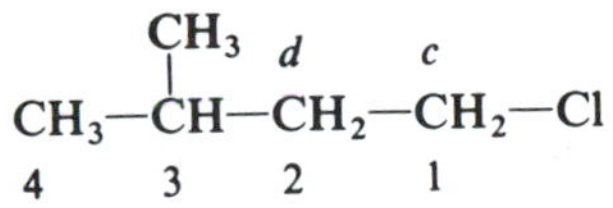 *iv*

Let us compare this molecule with the 2-chloro-2-methylbutane of part (a), using the actual δ values for that compound. (We renumber this for ease of comparison.) In

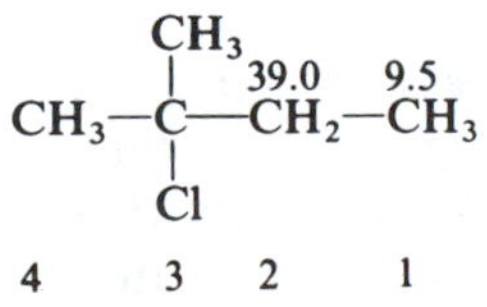

$$\underset{4}{CH_3}\text{—}\underset{3}{\underset{\underset{Cl}{|}}{\overset{\overset{CH_3}{|}}{C}}}\text{—}\underset{2}{\overset{39.0}{CH_2}}\text{—}\underset{1}{\overset{9.5}{CH_3}}$$

mentally converting *tert*-pentyl chloride into 2-chloro-2-methylbutane, we move Cl along the skeleton. C–1 gains an α-chloro, and loses a γ-chloro. C–2 is unchanged: it keeps a β-chloro. Using the values for α- and γ-effects given on page 636

	C–1
α-Cl	+30.6
γ-Cl	−(−5.3)
	+35.9

we predict for C–1 a δ of 45.4 (9.5 + 35.9) and for C–2 a δ of 39.0. On that basis, we would assign peaks *c* and *d* as in *iii*.

As we can see in the labeled spectrum on page 683 of this Study Guide, this is the correct assignment. However, the predicted difference in δ values is very much larger

$$\overset{a}{CH_3}-\underset{b}{\overset{\overset{a}{CH_3}|}{CH}}-\overset{c}{CH_2}-\overset{d}{CH_2}-Cl$$

Isopentyl chloride

than the actual one, demonstrating how crude these calculations are, and suggesting that the correctness of our prediction may be fortuitous.

(c) For this compound we have the following data:

a	13.9	q
b	22.0	t
c	30.5	t
d	32.8	t
e	33.4	t

$$\underset{5}{\overset{a}{CH_3}}-\underset{4}{CH_2}-\underset{3}{CH_2}-\underset{2}{CH_2}-\underset{1}{CH_2}-Br$$

We have already assigned peak *a*, but are left with four unassigned triplets.

Let us compare the δ values here with those for *n*-pentane shown on page 636.

$$\underset{5}{\overset{13.7}{CH_3}}-\underset{4}{\overset{22.6}{CH_2}}-\underset{3}{\overset{34.5}{CH_2}}-\underset{2}{\overset{22.6}{CH_2}}-\underset{1}{\overset{13.7}{CH_3}}$$

We see that δ for C–5, to which we have already assigned peak *a*, is virtually unchanged from δ for C–5 in *n*-pentane; this is what we would expect because of the distance of the carbon from the added Br. We also see that peak *b* for *n*-pentyl bromide has a δ very near to that for C–4 in *n*-pentane. Since C–4 in *n*-pentyl bromide is also far from the Br (in the δ position), we can fairly confidently assign peak *b* to C–4.

We have left unassigned three triplets with quite similar δ values. In view of the roughness of the calculations we can make (in part (b), for example), we hesitate to assign these peaks on that basis.

We might, however, venture a tentative assignment: peak *c* to C–3. The δ here (30.5 ppm) is the only one of the three unassigned peaks that is *smaller* than δ for a possibly corresponding peak in *n*-pentane (δ 34.5 for C–3); whatever the exact value for the γ-effect due to Br, we would expect it to be small and negative.

On that basis we are left with peaks *d* and *e*, both considerably downfield from the peaks for C–4 and C–5—as they would be as the result of sizable positive α- and β-effects.

In the labeled spectrum on page 683 of this Study Guide, we see that we have correctly assigned peaks *a*, *b*, and *c*.

$$\overset{a}{CH_3}-\overset{b}{CH_2}-\overset{c}{CH_2}-\overset{d}{CH_2}-\overset{e}{CH_2}-Br$$

n-Pentyl bromide

17.26 The size of a γ-effect across a carbon–carbon double bond depends upon the stereochemical relationship between the interacting groups: it is stronger between carbons that are *cis* to each other than between carbons that are *trans*. In a pair of geometric isomers, we saw on page 638, the γ-effect is more negative by about 5.4 ppm in the *cis* isomer than in the *trans* isomer; that is to say, the effect moves a peak farther upfield in the *cis* isomer.

Let us draw formulas for our stereoisomeric 3-hexenes, and focus our attention on the δ values for C–3 and C–4, which exert γ-effects on each other.

$$CH_3-\overset{25.8}{\mathbf{CH_2}}-CH{=}CH-\overset{25.8}{\mathbf{CH_2}}-CH_3$$

Isomer A
trans

$$CH_3-\overset{20.6}{\mathbf{CH_2}}-CH{=}CH-\overset{20.6}{\mathbf{CH_2}}-CH_3$$

Isomer B
cis

As we see, absorption by these carbons is farther upfield—and by 5.2 ppm—in isomer B, indicating clearly that this is the *cis* isomer.

17.27 Change the concentration.

17.28 (a) 1°, $-\overset{H}{\underset{H}{C}}-OH$ gives triplet (splitting by 2H on adjacent C);

2°, $-\underset{H}{\overset{|}{C}}-OH$ gives doublet (splitting by 1H on adjacent C);

3°, $-\underset{|}{\overset{|}{C}}-OH$ gives singlet (no splitting, no H on adjacent C).

(b) Acid:

$$R^*OH^* + H{:}B \rightleftarrows \begin{bmatrix} R^*\underset{H}{O}H^* \end{bmatrix}^+ + {:}B^-$$

$$\begin{bmatrix} R^*\underset{H}{O}H^* \end{bmatrix}^+ + ROH \rightleftarrows R^*OH + \begin{bmatrix} R\underset{H^*}{O}H \end{bmatrix}^+$$

$$\begin{bmatrix} R\underset{H^*}{O}H \end{bmatrix}^+ + {:}B^- \rightleftarrows ROH^* + H{:}B$$

Base: $R^*OH^* + :B \rightleftharpoons R^*O^- + [H^*:B]^+$

$$R^*O^- + ROH \rightleftharpoons R^*OH + RO^-$$

$$RO^- + [H^*:B]^+ \rightleftharpoons ROH^* + :B$$

17.29 (a) $PhCHOHCH_3$ α-Phenylethyl alcohol (b) $PhCH_2CH_2OH$ β-Phenylethyl alcohol (c) $PhCH_2OCH_3$ Benzyl methyl ether

(See the labeled spectra on page 684 of this Study Guide.)

17.30 (a) It is impossible to reverse the spin of only one of a pair: it would be a violation of the Pauli principle.

(b) If both spins of a pair are reversed, there is no net change in energy, and no signal.

17.31 (a) $CH_3\cdot$ (b) $CH_3\dot{C}HCH_3$ $CH_3CH_2\dot{C}HCH_3$ (c) $Ph_3C\cdot$

1. (a) The formula shows an open-chain, saturated structure ($C_3H_3Cl_5$ is equivalent to C_3H_8). One hydrogen split by two hydrogens (to give a triplet) and two hydrogens split by one hydrogen (to give a doublet) leave few choices for the structure:

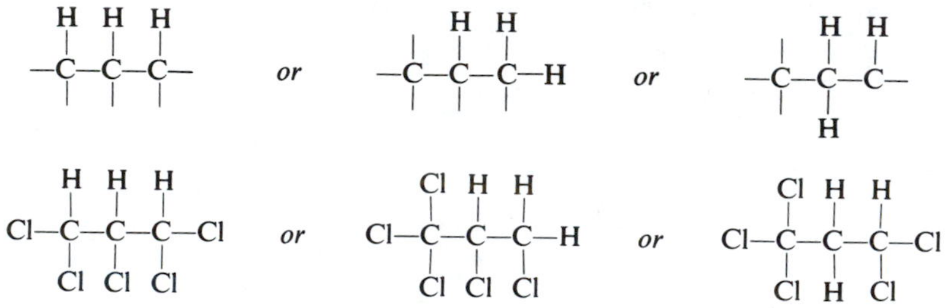

```
 H  H  H                 H  H                    H  H
 |  |  |              |  |  |                    |  |
—C—C—C—       or    —C—C—C—H        or      —C—C—C—
 |  |  |              |  |  |                 |  |
                                                 H

   H  H  H              Cl H  H               Cl H  H
   |  |  |              |  |  |               |  |  |
Cl—C—C—C—Cl    or    Cl—C—C—C—H       or   Cl—C—C—C—Cl
   |  |  |              |  |  |               |  |  |
   Cl Cl Cl             Cl Cl Cl              Cl H  Cl
      i                    ii                    iii
```

Since the 2H signal is farther downfield, *ii* can be ruled out on the basis that a 1° H (in —CH_2Cl) would be expected upfield from a 2° H, and *iii* can be ruled out on the basis that —CH_2— would be expected upfield from —$CHCl_2$. This leaves only *i*:

b *a* *b*

$CHCl_2CHClCHCl_2$

1,1,2,3,3-Pentachloropropane

A 1° H is shifted farther downfield by two —Cl's than is a 2° H by one —Cl.

(b) The formula shows an open-chain, saturated structure. An unsplit singlet for 3H indicates —CH_3 attached to carbon carrying no hydrogen: C—C—CH_3. This leaves us with no other choice but to write

b *a*

$CH_2Cl—CCl_2—CH_3$

1,2,2-Trichloropropane

(c) From the formula, we can see that we are dealing with one of four isomeric bromobutanes:

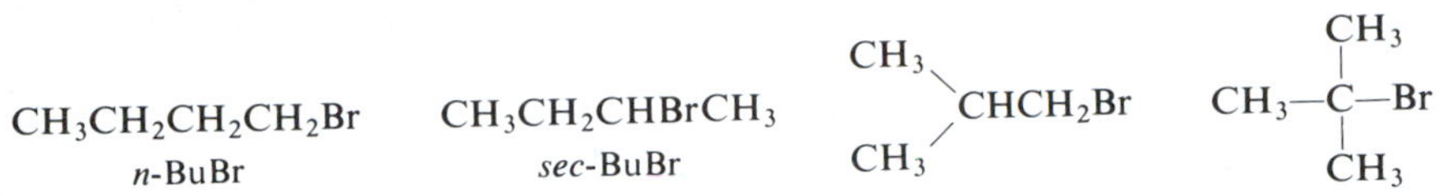

$CH_3CH_2CH_2CH_2Br$ *n*-BuBr $CH_3CH_2CHBrCH_3$ *sec*-BuBr $(CH_3)_2CHCH_2Br$ *iso*-BuBr $CH_3—C(CH_3)_2—Br$ *tert*-BuBr

Only one of these, isobutyl bromide, can give rise to a 6H signal split into a doublet. Therefore, the compound must be

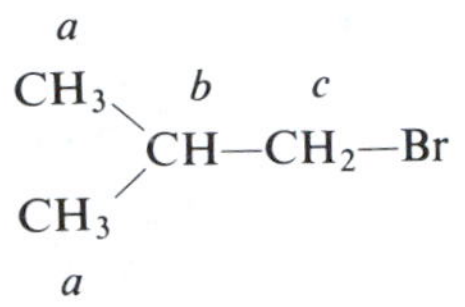

$$\overset{a}{(CH_3)_2}\overset{b}{CH}—\overset{c}{CH_2}—Br$$

Isobutyl bromide

The $—CH_2—$ signal, split into a doublet by the 3° H, is shifted downfield by —Br.

(d) The formula, in conjunction with the 5H signal *b* at δ 7.28, indicates immediately the presence of a $C_6H_5—$ group, and consequently ($C_{10}H_{14} - C_6H_5 = C_4H_9$) 4C and 9H attached to it. A single, unsplit 9H signal at δ 1.30 clearly indicates $3CH_3—$ groups, forcing us to conclude that C_4H_9 is *tert*-Bu, and that the compound is

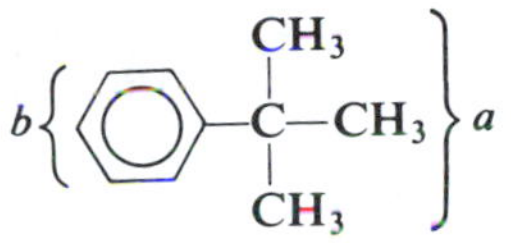

$$b\{C_6H_5\}—C(CH_3)_3\ \}a$$

tert-Butylbenzene

(e) This is a bit more formidable than (d), but a C_6H_5 group is present, and we need to work out only the C_4H_9 side chain, for which there are only three possibilities (the fourth was used in (d), above).

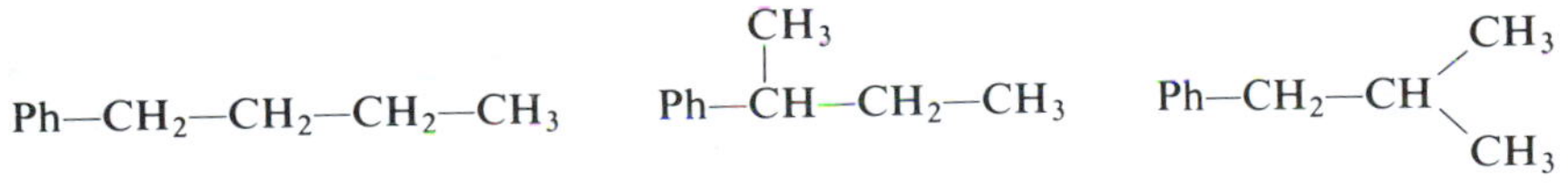

$Ph—CH_2—CH_2—CH_2—CH_3$ *n*-butyl $Ph—CH(CH_3)—CH_2—CH_3$ *sec*-butyl $Ph—CH_2—CH(CH_3)_2$ isobutyl

Again (compare (c), above), the only structure that can give a 6H signal split into a doublet is the isobutyl side chain, and the compound is isobutylbenzene:

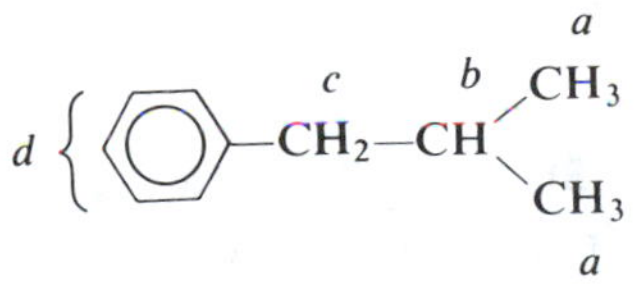

$$d\{C_6H_5\}—\overset{c}{CH_2}—\overset{b}{CH}(\overset{a}{CH_3})_2$$

The $—CH_2—$ doublet (signal *c*) is shifted downfield by the phenyl group.

(f) The presence of a disubstituted ($—C_6H_4—$) benzene ring is shown by the downfield signal *c* (of 4H). This leaves ($C_9H_{10} - C_6H_4 = C_3H_6$) 3C and 6H to be accounted for. The quintet–triplet combination of 2H (split into a quintet by 4H's) and 4H (split into a triplet by 2H's) leads to $\underset{b}{-CH_2}-\underset{a}{CH_2}-\underset{b}{CH_2}-$

This can be fitted to $C_6H_4\langle$ in only one way:

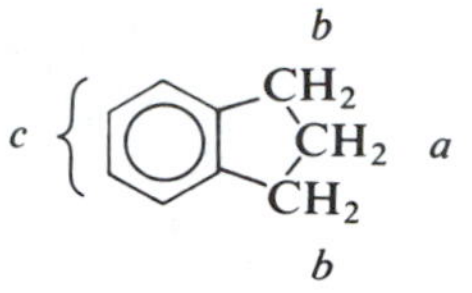

c { benzene ring fused to $\underset{b}{CH_2}$–$\underset{a}{CH_2}$–$\underset{b}{CH_2}$ }

Indane

As a benzylic H, signal *b* is shifted downfield to δ 2.91.

(g) Once again, a phenyl group (C_6H_5-) shows up (signal *c*). Subtraction ($C_{10}H_{13}Cl - C_6H_5$) leaves C_4H_8Cl, which indicates a single (only one group can be attached to C_6H_5-) saturated C_4 side chain. Signal *a* shows two unsplit CH_3- groups; signal *b* shows an unsplit $-CH_2-$. This leaves one carbon and one $-Cl$. The pieces can be put together in two ways (of which the first is the actual compound).

C_6H_5- CH_3-, CH_3- $-CH_2-$ $-Cl$ $-\overset{|}{\underset{|}{C}}-$ *makes*

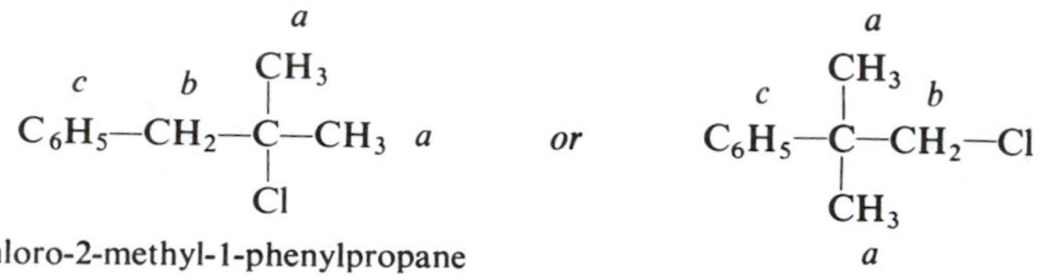

$\underset{c}{C_6H_5}-\underset{b}{CH_2}-C(\underset{a}{CH_3})(Cl)-\underset{a}{CH_3}$

2-Chloro-2-methyl-1-phenylpropane

or

$\underset{c}{C_6H_5}-C(\underset{a}{CH_3})(\underset{a}{CH_3})-\underset{b}{CH_2}-Cl$

1-Chloro-2-methyl-2-phenylpropane

(h) The two signals, *a* and *b*, are unusually far upfield, and indicate the presence of a cyclopropane ring (Table 17.4, page 607, or inside front cover) carrying two groups of 2H each. Signal *c* shows an unsplit CH_3-, and far downfield signal *d* shows a C_6H_5- group. This accounts for all 10C and 12H, and the only reasonable combination of the fragments gives

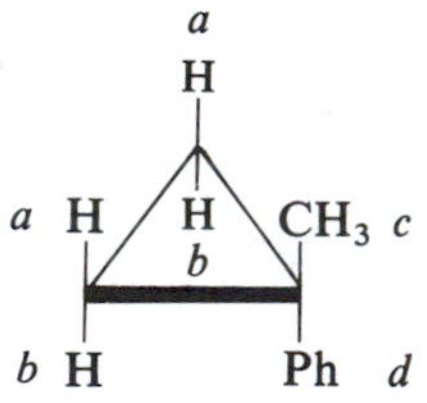

Cyclopropane ring bearing H (*a*), H (*a*), H (*b*), H (*b*), CH_3 (*c*), Ph (*d*)

a and b may be reversed

1-Methyl-1-phenylcyclopropane

(i) The presence of a C_6H_5- group (signal *d*, 5H) is easily seen, leaving a side chain of $-C_3H_6Br$ to be figured out. The absence of any signal for 3H means the absence of any CH_3- group, and since any branched side chain,

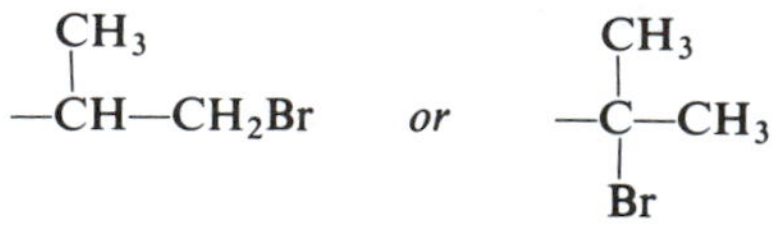

$-\overset{CH_3}{\overset{|}{CH}}-CH_2Br$ *or* $-\overset{CH_3}{\overset{|}{\underset{Br}{\underset{|}{C}}}}-CH_3$

would contain a CH_3— group, we can eliminate those isomers. Of those remaining,

$$-CH_2CH_2CH_2Br \qquad -CH_2\underset{\displaystyle Br}{\underset{|}{C}}HCH_3 \qquad -\underset{\displaystyle Br}{\underset{|}{C}}HCH_2CH_3$$

only the first has no CH_3— group. The compound must be, then,

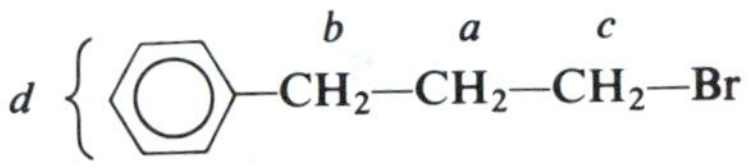

$$d\ \{C_6H_5\overset{b}{-CH_2}-\overset{a}{CH_2}-\overset{c}{CH_2}-Br$$

3-Bromo-1-phenylpropane

(j) Signal *a* indicates that a CH_3— is present in this saturated, open-chain molecule ($C_3H_5ClF_2$ is equivalent to C_3H_8). That signal *b* is a triplet and not a quartet shows us that an ethyl group (the triplet–quartet is characteristic) cannot be present. Hence, the only possible distribution of C's and H's is

$$-CH_2-\overset{|}{\underset{|}{C}}-CH_3$$

The question still remains: what splits signal *a* into a triplet, and signal *b* also into a triplet? The answer to this is, of course, that the two ^{19}F atoms are doing it (page 617, and Problems 17.17 and 17.18). The compound, therefore, must be

$$Cl-\overset{b}{CH_2}-\overset{\displaystyle F}{\overset{|}{\underset{\displaystyle F}{\underset{|}{C}}}}-\overset{a}{CH_3}$$

1-Chloro-2,2-difluoropropane

(Remember, *absorption* by fluorine does not appear in this proton NMR spectrum—only the *splitting* by fluorine.)

2. We follow the alternative approach that we followed in Problem 17.22 on pages 259–261 of this Study Guide.

(a) The compound is open-chain and saturated. ($C_3H_5Cl_3$ is equivalent to C_3H_8.) For each signal we write down the structural unit that must give rise to it, and include the three Cl's given by the formula.

$$-CH_2- \qquad -\underset{|}{C}H- \qquad -Cl \qquad -Cl \qquad -Cl$$

We add these units up, and find that they total $C_2H_3Cl_3$. Arithmetic shows us

$$\begin{array}{r} C_3H_5Cl_3 \\ -C_2H_3Cl_3 \\ \hline CH_2 \end{array}$$

that we are lacking one —CH_2—. The triplet, then, must be due to *two equivalent* CH_2 groups. We find that we can put these units together in only one way:

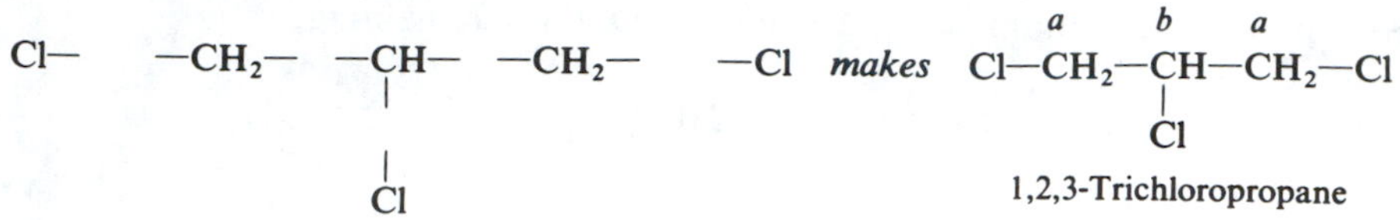

$$\text{Cl—} \quad \text{—CH}_2\text{—} \quad \text{—}\underset{|}{\overset{}{\text{CH}}}\text{—} \quad \text{—CH}_2\text{—} \quad \text{—Cl} \quad \textit{makes} \quad \overset{a}{\text{Cl—CH}_2}\text{—}\overset{b}{\underset{\text{Cl}}{\text{CH}}}\text{—}\overset{a}{\text{CH}_2}\text{—Cl}$$

$$\underset{\text{Cl}}{|}$$

1,2,3-Trichloropropane

(b) Again we have an open-chain, saturated compound. We write down the units indicated,

$$CH_3\text{—} \quad \text{—}CH_2\text{—} \quad \text{—}\overset{|}{C}H\text{—} \quad \text{—Br}$$

and find that they total C_3H_6Br. Arithmetic shows us

$$\begin{array}{r} C_4H_9Br \\ -\,C_3H_6Br \\ \hline CH_3 \end{array}$$

that we are lacking one $\text{—}CH_3$, and that the quartet is due to two equivalent CH_3 groups. These pieces go together in only one way:

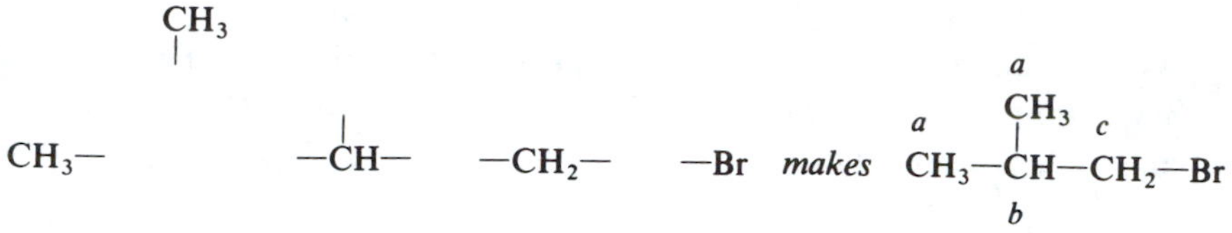

$$\underset{|}{CH_3} \qquad CH_3\text{—} \quad \text{—}\overset{|}{C}H\text{—} \quad \text{—}CH_2\text{—} \quad \text{—Br} \quad \textit{makes} \quad \overset{a}{CH_3}\text{—}\underset{b}{\overset{\overset{a}{CH_3}}{\overset{|}{C}H}}\text{—}\overset{c}{CH_2}\text{—Br}$$

Isobutyl bromide

(c) Once more an open-chain, saturated compound. We write down the units,

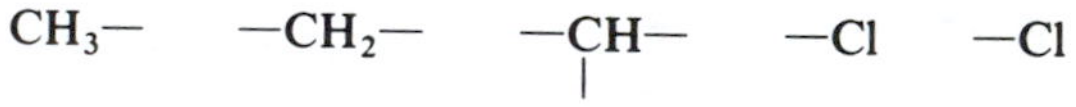

$$CH_3\text{—} \quad \text{—}CH_2\text{—} \quad \text{—}\underset{|}{C}H\text{—} \quad \text{—Cl} \quad \text{—Cl}$$

and find that they total $C_3H_6Cl_2$, the molecular formula. There are no equivalent sets of carbons. The pieces can be put together in two ways:

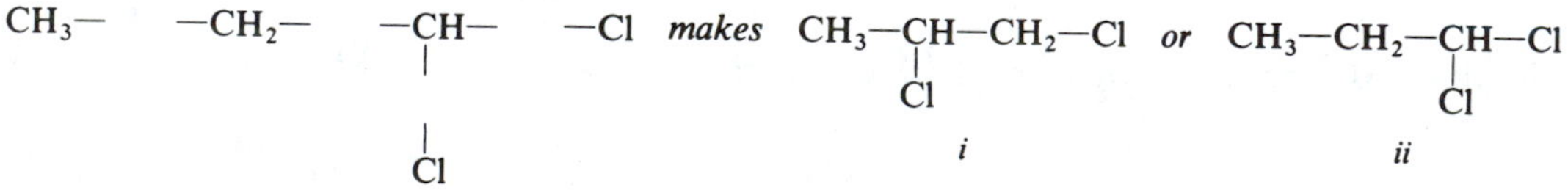

$$CH_3\text{—} \quad \text{—}CH_2\text{—} \quad \text{—}\underset{|}{C}H\text{—} \quad \text{—Cl} \quad \underset{|}{Cl} \quad \textit{makes} \quad \underset{i}{CH_3\text{—}\underset{Cl}{\underset{|}{C}H}\text{—}CH_2\text{—Cl}} \quad \textit{or} \quad \underset{ii}{CH_3\text{—}CH_2\text{—}\underset{Cl}{\underset{|}{C}H}\text{—Cl}}$$

We can decide fairly confidently that the compound is actually *i* rather than *ii* on the basis of the chemical shifts. To begin with, the quartet has δ 22.4, consistent with a downfield effect due to a β-Cl as in *i*, but inconsistent with an upfield effect due to two γ-Cl's in *ii*. Next, the shifts for the doublet (δ 55.8) and the triplet (δ 49.5) are consistent with each having an α-Cl and a β-Cl, with the doublet (C–2) downfield from the triplet (C–1). In *ii*, we would expect a very large downfield shift for the doublet (two α-Cl's), certainly larger than for the triplet.

We conclude that the compound is

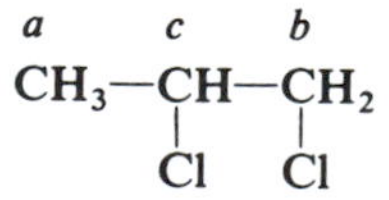

$$\overset{a}{CH_3}\text{—}\underset{Cl}{\underset{|}{\overset{c}{C}H}}\text{—}\underset{Cl}{\underset{|}{\overset{b}{C}H_2}}$$

1,2-Dichloropropane

(d) The downfield triplet and doublet indicate a carbon–carbon double bond, and this is confirmed by the formula (C_3H_5Br is equivalent to C_3H_6). The units indicated can be put together in just one way:

$$CH_2{=} \quad {=}CH{-} \quad {-}CH_2{-} \quad {-}Br \quad \textit{makes} \quad \overset{b}{C}H_2{=}\overset{c}{C}H{-}\overset{a}{C}H_2Br$$

Allyl bromide

(e) The formula is consistent with an alkyne, a diene, or a cycloalkene. The downfield doublet (δ 127.2) indicates a carbon–carbon double bond rather than a triple bond. We write down the units indicated,

$$-CH{=}CH- \quad -CH_2- \quad -CH_2-$$

add them up (C_4H_6), and find that we are lacking C_2H_4, or two CH_2 groups.

$$\begin{array}{r} C_6H_{10} \\ -C_4H_6 \\ \hline C_2H_4 \end{array}$$

These units fit together in only one way:

$$-CH{=}CH- \quad -CH_2- \quad -CH_2- \quad -CH_2- \quad -CH_2- \quad \textit{makes}$$

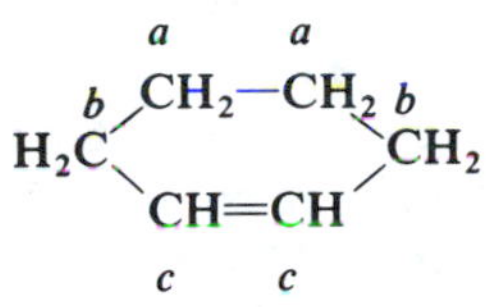

```
        a       a
     b CH2—CH2 b
  H2C          CH2
     CH=CH
     c   c
```

Cyclohexene

In this structure, we see, the doubly bonded carbons are equivalent, as they must be to give one doublet; and there are two equivalent pairs of methylene carbons, consistent with the spectrum.

(f) This is an open-chain, saturated compound that contains these units:

$$CH_3- \quad -CH_2- \quad -CH_2- \quad -\underset{|}{C}H- \quad -Br \quad -Br$$

These total $C_4H_8Br_2$, the molecular formula. We can put the pieces together in three ways:

$$\underset{i}{CH_3{-}CH_2{-}\underset{Br}{\underset{|}{C}H}{-}\underset{Br}{\underset{|}{C}H_2}} \quad or \quad \underset{ii}{CH_3{-}CH_2{-}CH_2{-}\underset{Br}{\underset{|}{C}H}{-}Br} \quad or \quad \underset{iii}{CH_3{-}\underset{Br}{\underset{|}{C}H}{-}CH_2{-}\underset{Br}{\underset{|}{C}H_2}}$$

Let us examine these possibilities in light of the chemical shifts.

In *iii* C–4 has a β-Br, which should have a downfield effect. But the quartet has δ 10.9, which is, if anything, upfield from where we would expect a methyl in the parent *n*-butane to absorb. We reject *iii*.

In *ii* C–3 has two γ-Br's, which should exert an upfield effect. Yet even the most upfield of the triplets (δ 29.0) is downfield from where we would expect a methylene in *n*-butane to be (compare C–2 in *n*-pentane, on page 636).

In *i*, on the other hand, one methylene (C–1) has an α-Br and a β-Br, and the other (C–3) has a β-Br. We would expect both triplets to be shifted downfield, as they actually are.

There is another point in favor of *i*. In *ii* the Br's should have negligible effects on C–4; in *i* one Br should exert an upfield γ-effect. As we saw above, the quartet is actually upfield from where we would expect it to be in the unsubstituted molecule.

With some trepidation, we pick *i* as the structure of our unknown—and we are right!

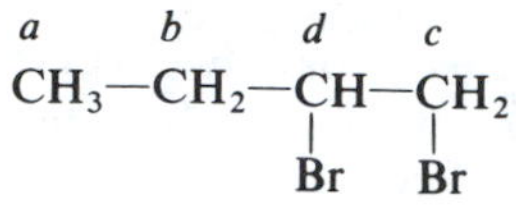

$$\overset{a}{CH_3}-\overset{b}{CH_2}-\underset{Br}{\overset{d}{CH}}-\underset{Br}{\overset{c}{CH_2}}$$

1,2-Dibromobutane

3. The two possible structures (both achiral) are:

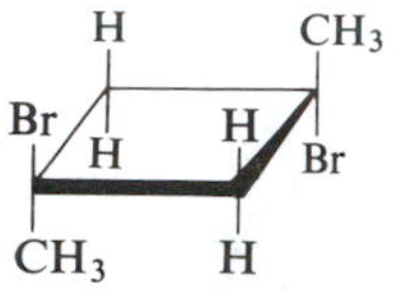

trans-1,3-Dibromo-1,3-dimethylcyclobutane
X

and

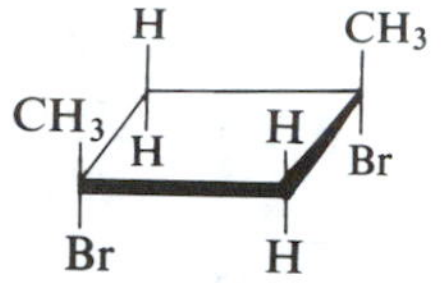

cis-1,3-Dibromo-1,3-dimethylcyclobutane
Y

The ring protons in the *trans* isomer are all equivalent (each is *cis* to a —Br and to a $—CH_3$) and will give rise to one unsplit 4H signal; the other singlet (6H) is of course due to the two $—CH_3$'s. Obviously, X is the *trans* isomer. The ring protons in the *cis* isomer fall into two groups: two (equivalent to each other) are *cis* to $—CH_3$ groups, two (equivalent to each other) are *cis* to —Br's. Thus there will be two signals for ring protons, each split into a doublet by H's on the opposite face of the ring.

4. (structure: ring cation with CH_3 (*b*), two CH_3 (*a*), ring H's (*d*), CH_2 H's (*c*)) SbF_6^-

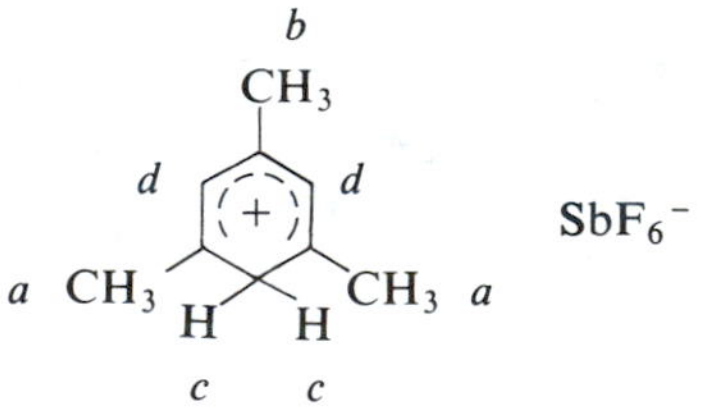

It shows that $Ar^{\oplus}(H)(H)$ and presumably $Ar^{\oplus}(H)(Y)$ intermediates can exist.

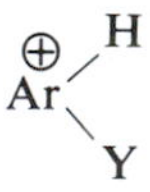

(See also Problem 15.10, page 535.)

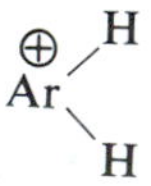

5. Electron release by the methyl groups lowers the deshielding of the ring protons.

6. (a) (three methylcyclopropene/methylenecyclopropane structures)

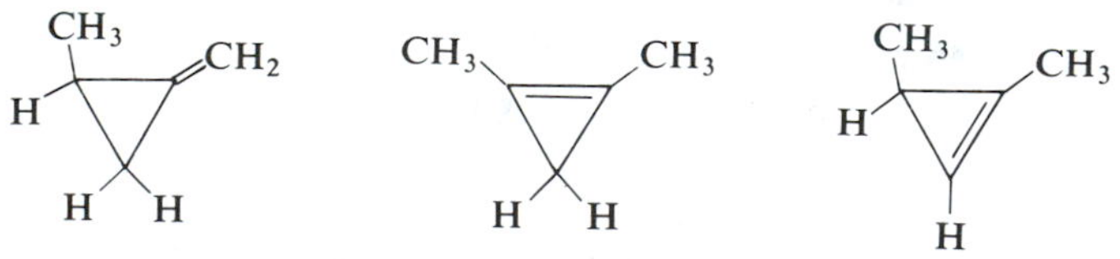

(b) The one with the $=CH_2$ group.

(c) A, 1,2-dimethylcyclopropene, the only structure with but two sets of protons.

(d) The aromatic (two π electrons) cyclopropenyl cation formed by loss of a ring hydrogen from A (compare Problem 14.6, page 507).

(e) $CH_3C{\equiv}CCH_3$ + CH_2N_2 + light.

7. See page 1000.

8. (a) This situation is exactly analogous to the one in Problem 17.17 on page 629, except that here we are dealing with proton NMR instead of fluorine NMR. The compound exists as three conformers: achiral structure I and (racemic) chiral structures II and III.

At room temperature, interconversion is so fast that a single average signal is given. At −45 °C, interconversion is so slow that separate signals are given: one by I, and the other by II and III.

(b) The unequal intensities of the signals indicate different proportions of the two components. As we would expect, these diastereomeric structures are of different stabilities.

(c) Each methyl group is *gauche* to two chlorines in I, and to only one chlorine in II and III. The methyl protons in I should be more strongly deshielded by the halogen, and should absorb downfield from those in II and III. Conformer I is evidently more stable: this could be due to less repulsive van der Waals methyl–methyl interaction and/or more attractive methyl–chlorine interaction (see Problem 11, Chapter 4).

Abraham, R. J.; Loftus, P. *Proton and Carbon-13 NMR Spectroscopy*; Heyden and Son: Philadelphia, 1978; pp 178–179.

9. (a) Two rings in C. (b) Two rings, two double bonds in B.

(c) B is bicyclo[2.2.0]hexa-2,5-diene ("Dewar benzene", pages 495 and 497). C is bicyclo-[2.2.0]hexane.

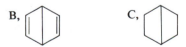

(d) The 3° H's are split by all the vinylic protons; in turn, these are split by both 3° H's.

10. (a)–(c)

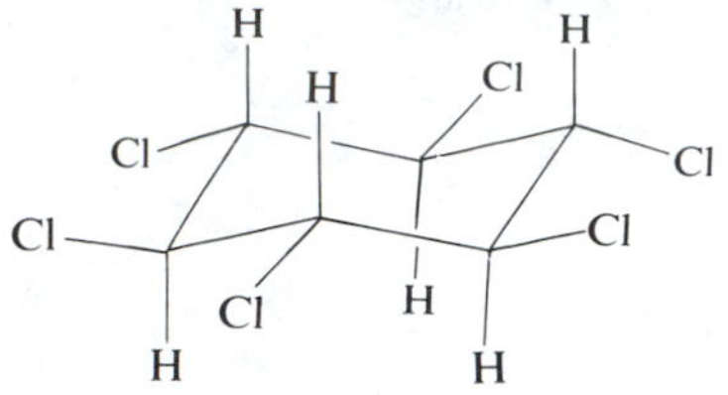

eeeeee
All H's equivalent (all axial)
One signal

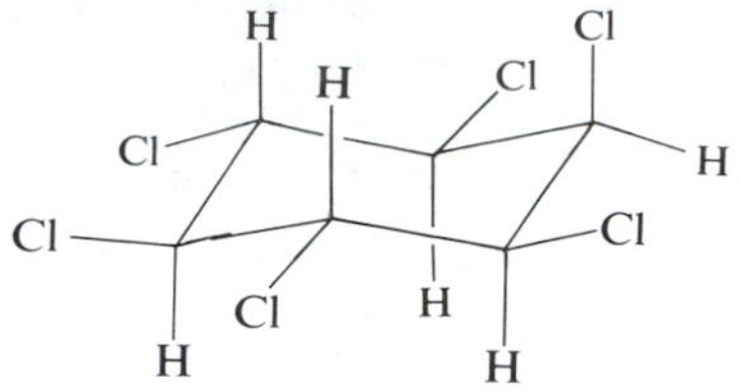

eeeeea
5 axial H's, 1 equatorial H
Two signals (5:1)

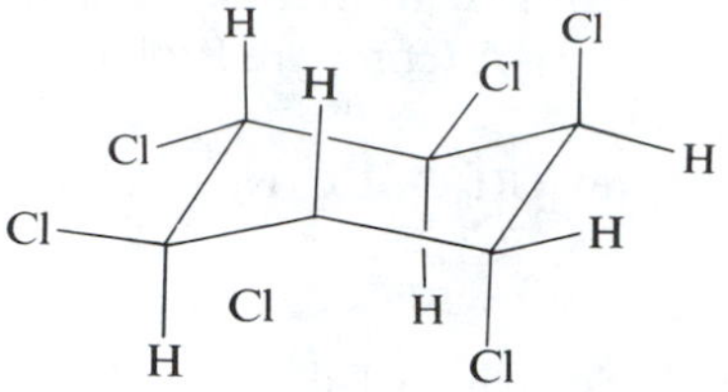

eeeeaa
4 axial H's, 2 equatorial H's
Two signals (4:2)

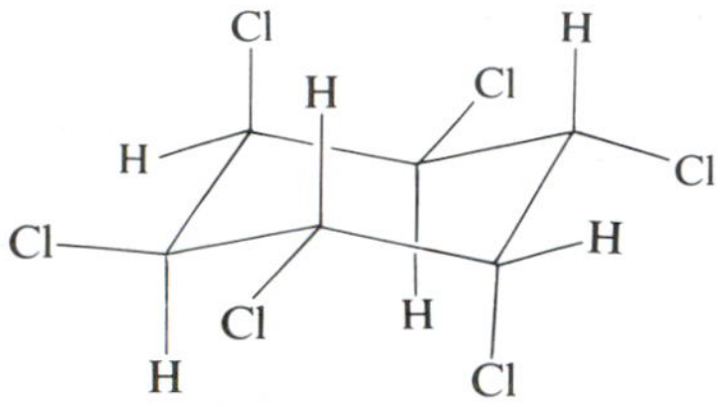

eeaeea
4 axial H's, 2 equatorial H's
Two signals (4:2)

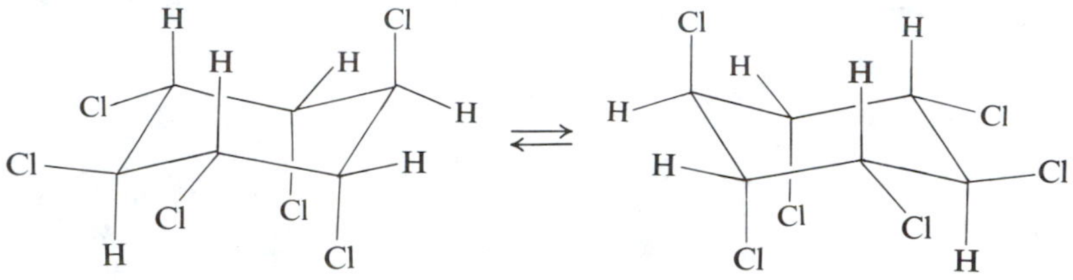

eeeaaa ⇄ aaaeee

Conformations of equal stability
All H's equivalent by rapid flip-flop between axial and equatorial positions
One signal

(d) eeeeee: no change.
eeeaaa: split into two signals of equal area (one for three axial H's, one for three equatorial H's).

11. (a)

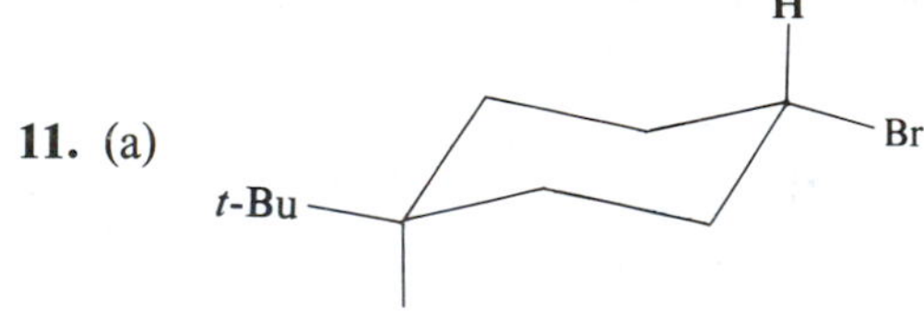

trans

Signal for axial —H on C–1 shifted downfield (clear of other proton signals) by adjacent —Br.

(b)

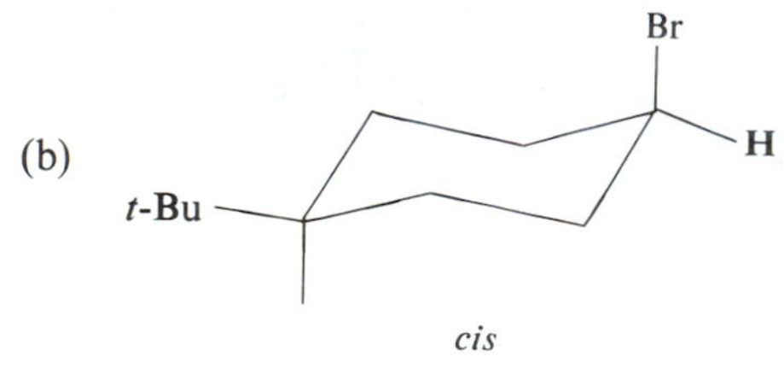

cis

Signal for equatorial —H on C–1 even farther downfield than signal for axial —H.

Generally, a signal for an axial —H will appear upfield from a signal for an equatorial —H.

12. At -75 °C, interconversion of chair conformers becomes so slow that signals for both axial and equatorial —H are seen. The conformation with equatorial —Br (axial —H, upfield signal) predominates, accounting for

$$\frac{4.6}{4.6 + 1.0} \times 100 = 82\%$$

of the molecules.

13. Since *cis*-decalin is made up of two cyclohexane rings, we suspect that the temperature effect arises as it does in the proton NMR spectrum of cyclohexane itself (pages 628–629): at a low temperature the interconversion of equivalent chair conformations is slowed down so that we are "seeing" the molecule in one or another conformation, and not as a blur with its atoms in some average positions.

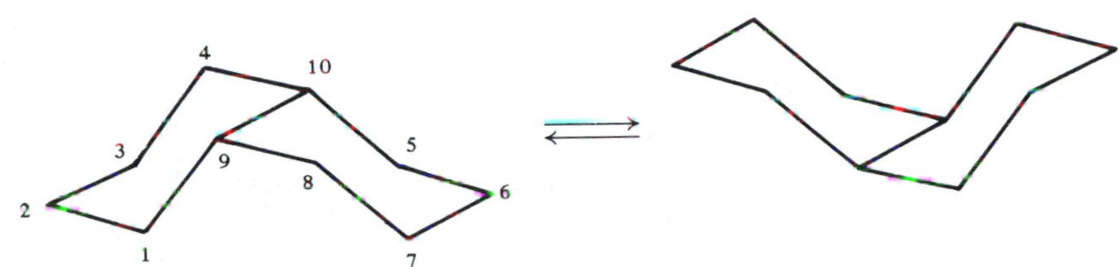

Let us analyze the structure of *cis*-decalin in terms, first, of the low-temperature spectrum. There are ten carbons in the molecule, and five peaks of equal intensity suggest five pairs of equivalent carbons. Which are the paired carbons? Examination of the formula—or, better, a molecular model—shows that these pairs are:

C–1 and C–5
C–2 and C–6
C–3 and C–7
C–4 and C–8
C–9 and C–10

Some of the differences between one pair and another are easy to see: C–1 or C–5, for example, is attached to a fused carbon, but C–2 or C–6 is once removed. Other differences are conformational and harder to see: C–1 or C–5 projects into the fold of the molecule, but C–4 or C–8 projects outward. (With models we can see the distinction more clearly: C–1 or C–5 is *anti* to H on a fused carbon, but C–4 or C–6 is *gauche*.)

At the low temperature, then, each of the five peaks is given by one of these pairs.

At the high temperature, interconversion of the chair conformations occurs so rapidly that, in so far as CMR is concerned, the purely conformational differences disappear. C–1 and C–5 rapidly exchange their conformational locations with C–4 and C–8. We see a single peak for these four carbons; this peak reflects the average environment of the carbons and is at a different frequency from the two peaks it replaces. In a similar way the peaks for C–2 and C–6 and for C–3 and C–7 coalesce into a single peak at a different frequency.

The smaller peak is, of course, due to C–9 and C–10. It is smaller because it is due to only two carbons. It is not shifted because the environment of these carbons—which are the pivots about which the two rings flip—remains unchanged.

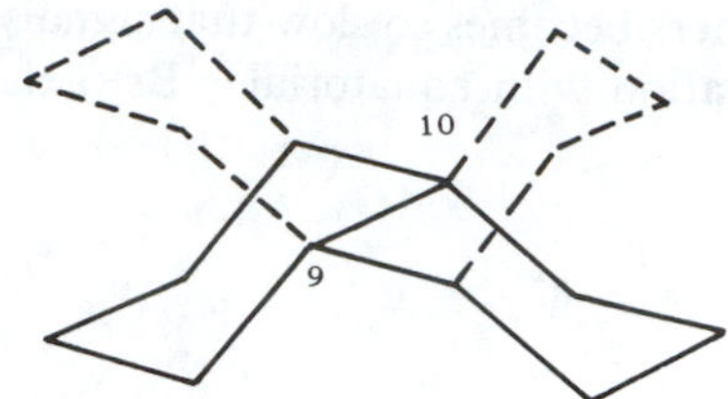

Abraham, R. J.; Loftus, P. *Proton and Carbon-13 NMR Spectroscopy*; Heyden and Son: Philadelphia, 1978; pp 180–182.

14. (a) *t*-BuF: doublet at δ 1.30 is the —CH_3 signal split by —F ($J = 20$).

i-PrF: two doublets centered on δ 1.23 are the —CH_3 signal split by —H ($J = 4$) and split again by —F ($J = 23$);

two multiplets centered on δ 4.64 are the —H signal split by —CH_3 ($J = 4$) and split again by —F ($J = 48$).

(b) $t\text{-BuF} + SbF_5 \longrightarrow t\text{-Bu}^+SbF_6^-$

$i\text{-PrF} + SbF_5 \longrightarrow i\text{-Pr}^+SbF_6^-$

Removed from the molecules, —F no longer splits the proton signals. At the same time, deshielding by the positive charge of the cation causes a very strong downfield shift. (See the spectrum in Figure 5.6, page 192. This was the first *direct observation* of simple alkyl cations, as was discussed in Sec. 5.16.)

15. Let us look first at the structure of this hydrocarbon formed by the action of $NaNH_2$ on neopentyl chloride. The fact that it does not decolorize permanganate, yet has the formula C_5H_{10}, shows that it contains one ring. Solubility in concentrated H_2SO_4 suggests that it is a cyclopropane; this is confirmed by the far upfield NMR peak at δ 0.20, characteristic of cyclopropane ring protons. The molecular formula requires it to be ethylcyclopropane or a dimethylcyclopropane. Of these, only one is consistent with the NMR spectrum: 1,1-dimethylcyclopropane. (See the answer to Problem 17.9, page 604.)

CH_2—CH_2 / C / CH_3 CH_3
1,1-Dimethylcyclopropane

CH_2—CHD / C / CH_3 CH_3
1,1-Dimethylcyclopropane-2-*d*

The reaction with the labeled alkyl halide must give an analogous hydrocarbon. The molecular weight of 71 shows that it contains *only one* deuterium atom per molecule, and is C_5H_9D.

These hydrocarbons are evidently formed through intramolecular insertion of a methylene generated by 1,1-elimination (Sec. 13.17).

$(CH_3)_3C$—CD_2Cl $\xrightarrow{\text{base}}$ $(CH_3)_2C(CH_2$—H$)$—CD: $\xrightarrow{\text{insertion}}$ $(CH_3)_2C$ with ring CH_2—CHD

The labeling experiment rules out an alternative mechanism (involving "γ-elimination") proposed in 1942 by Frank Whitmore—before methylenes were conceived of, even by D. Duck. This mechanism would require *two* deuteriums per molecule of product, contrary to fact:

$(CH_3)_2C(CH_3)CD_2Cl \xrightarrow{\text{base}} (CH_3)_2C(\ominus CH_2{:})CD_2{-}Cl \longrightarrow (CH_3)_2C(CH_2{-}CD_2)$

Not obtained

(Friedman, L.; Berger, J. *J. Am. Chem. Soc.* **1961**, *83*, 500.)

16. (a) The hydrocarbon formed by the action of $NaNH_2$ on methallyl chloride is 1-methylcyclopropene, formed by intramolecular insertion of a methylene produced by 1,1-elimination.

$H_2C{=}C(CH_3)CH_2Cl \xrightarrow{\text{base}} H_2C{=}C(CH_3)CH{:} \xrightarrow{\text{insertion}}$ 1-Methylcyclopropene (H *c*, CH_3 *b*, H H *a a*)

(Fisher, F.; Applequist, D. E. *J. Org. Chem.* **1965**, *30*, 2089.)

(b) We would expect an analogous reaction to convert allyl chloride into cyclopropene.

$H_2C{=}CHCH_2Cl \xrightarrow{\text{base}} H_2C{=}CHCH{:} \xrightarrow{\text{insertion}}$ Cyclopropene

(A low yield of cyclopropene has actually been obtained in this manner by Closs, G. L.; Krantz, K. D. *J. Org. Chem.* **1966**, *31*, 638.)

17. (a) Hydrocarbon D is C_4H_6, and has the structure

[diamond] *or* (H *a*, H *a*, H *c*, H *c*, H *b*, H *b*)

Bicyclo[1.1.0]butane

We can see three sets of protons; NMR signals have been assigned as shown.

(b) Synthesis (i) is an intramolecular coupling (compare pages 101 and 102), and involves nucleophilic displacement of halide by carbanionoid carbon:

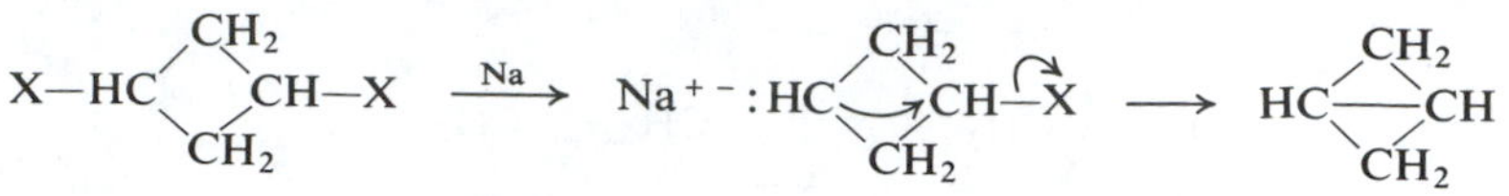

$$X{-}HC\langle{}^{CH_2}_{CH_2}\rangle CH{-}X \xrightarrow{Na} Na^{+}\,{}^{-}{:}HC\langle{}^{CH_2}_{CH_2}\rangle CH{-}X \longrightarrow HC\langle{}^{CH_2}_{CH_2}\rangle CH$$

Bicyclobutane

Synthesis (ii) involves intramolecular addition of a methylene:

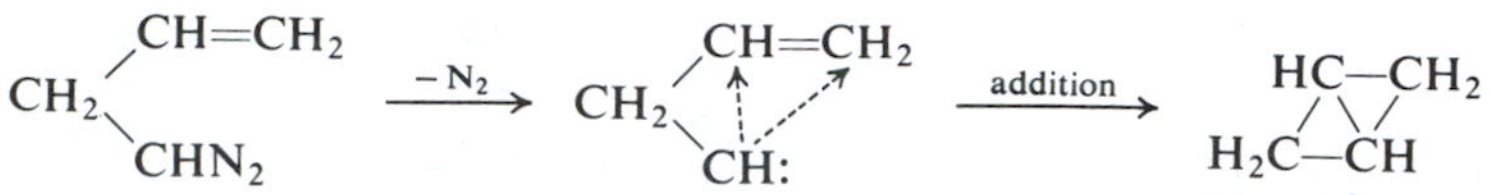

$$CH_2\langle{}^{CH=CH_2}_{CHN_2} \xrightarrow{-N_2} CH_2\langle{}^{CH=CH_2}_{CH:} \xrightarrow{\text{addition}} \begin{matrix} HC{-}CH_2 \\ H_2C{-}CH \end{matrix}$$

Bicyclobutane

We have now seen, in Problems 15 and 16 and the present problem, *intramolecular* (within-a-molecule) examples of the two principal reactions of methylenes: insertion and addition.

18. (a) R_3C^+ is formed in acid.

(i) $R_3COH + 2H_2SO_4 \longrightarrow R_3C^+ + H_3O^+ + 2HSO_4^-$

(ii) Conjugation of positive charge with cyclopropyl rings.

(iii) Downfield shift due to deshielding by positive charge.

The tricyclopropylmethanol cation, R_3C^+, is stabilized—much as the triphenylmethyl cation is (Problem 16.10, page 575)—through dispersal of the positive charge over the rings. This is believed to involve overlap of the empty p orbital with the C—C bonds of the cyclopropane rings, which (Sec. 13.9) have considerable π character. (See the answer to Problem 22, Chapter 16.)

(b)

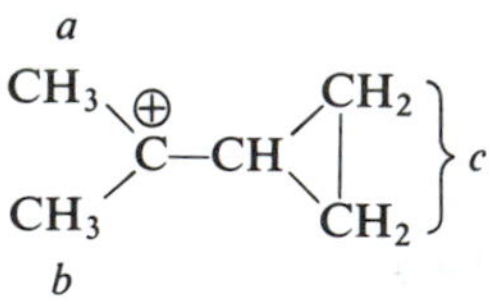

$$\begin{matrix} a \\ CH_3 \\ CH_3 \\ b \end{matrix} \rangle \overset{\oplus}{C}{-}CH\langle \begin{matrix} CH_2 \\ | \\ CH_2 \end{matrix} \rbrace c$$

The two methyls are clearly not equivalent, since they give different NMR signals, and must therefore be located unsymmetrically. The plane of the methyls and trigonal carbon is perpendicular to, and bisects, the ring; this geometry permits the overlap described in (a):

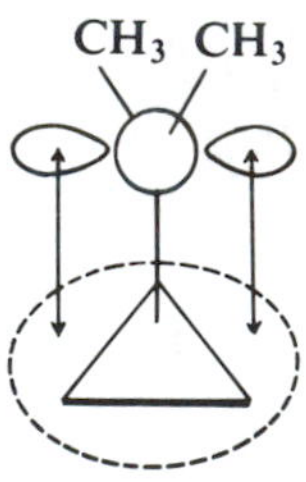

CH_3 CH_3

2-Cyclopropyl-2-propyl cation

The ring is *cis* to one methyl, *trans* to the other:

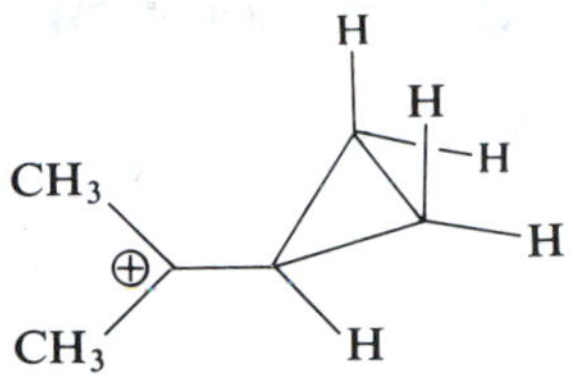

Either methyl can, of course, be *cis* or *trans* to the ring; when $—CD_3$, which gives no NMR signal, is present, only half as many molecules as before have $—CH_3$ *cis* and half as many have $—CH_3$ *trans*.

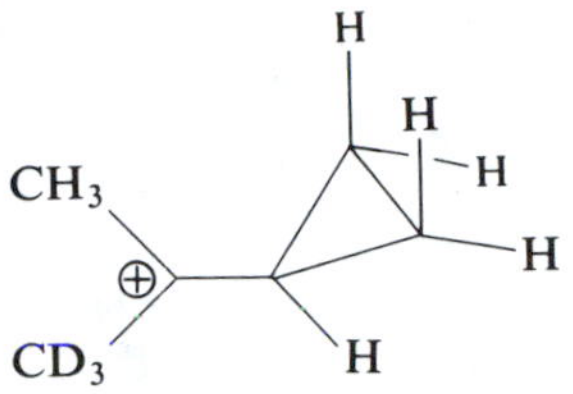

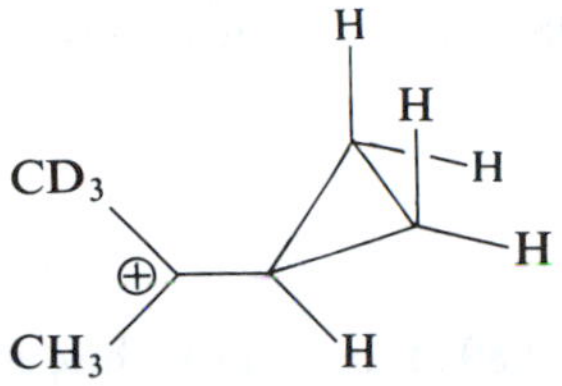

Each NMR signal is reduced to half its previous area.

(Deno, N. C., *et al.* "Carbonium Ions. XIX. The Intense Conjugation in Cyclopropyl Carbonium Ions"; *J. Am. Chem. Soc.* **1965**, *87*, 4533; Pittman, C. U., Jr; Olah, G. A. "Stable Carbonium Ions. XVII. Cyclopropyl Carbonium Ions and Protonated Cyclopropyl Ketones"; *J. Am. Chem. Soc.* **1965**, *87*, 5123.)

19. E, $Ph—CH(Ph)—CH(OH)—Ph$ F, $Ph—CH_2—C(Ph)(OH)—Ph$

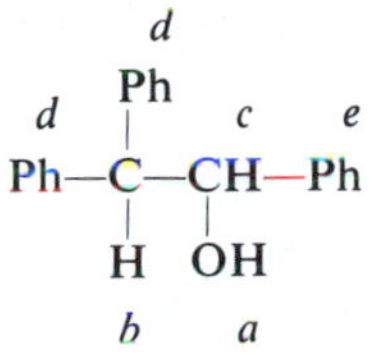

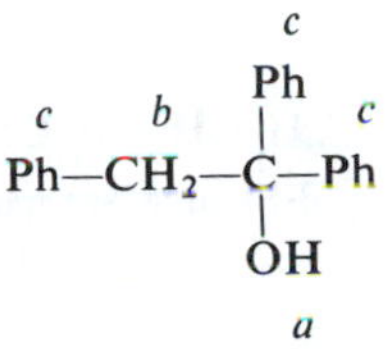

Distinguish between the 2° and 3° alcohols by CrO_3/H_2SO_4 test.

20. There is an intramolecular hydrogen bond in the *cis* isomer (see Sec. 24.2).

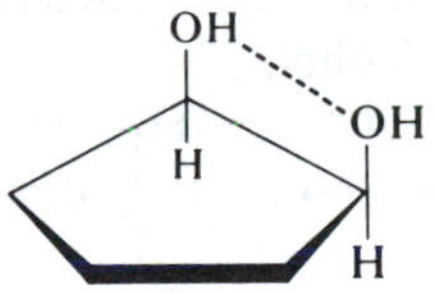

21. (a) α-phenylethyl bromide, $C_6H_5CHBrCH_3$ (b) *tert*-pentylbenzene (c) *sec*-butyl bromide

(See the labeled spectra on page 685 of this Study Guide).

22. (a) 3,3-dimethyl-1-butene (b) methylcyclopentane (c) *trans*-4-octene

(See the labeled spectra on page 686 of this Study Guide.)

23. (a) $CH_3CH_2CHCH_3$ with OH on the CH (sec-Butyl alcohol) (b) $CH_3-C(CH_3)(H)-CH_2OH$ (Isobutyl alcohol) (c) $CH_3CH_2OCH_2CH_3$ (Diethyl ether)

$$\underset{\textit{sec-}\text{Butyl alcohol}}{CH_3CH_2\underset{\displaystyle OH}{\underset{|}{C}}HCH_3} \qquad \underset{\text{Isobutyl alcohol}}{CH_3-\overset{\displaystyle CH_3}{\overset{|}{\underset{\displaystyle H}{\underset{|}{C}}}}-CH_2OH} \qquad \underset{\text{Diethyl ether}}{CH_3CH_2OCH_2CH_3}$$

(See the labeled spectra on page 687 of this Study Guide.)

24. (a) $CH_3CH_2OCH_2CH_2OH$
2-Ethoxyethanol

(b) $CH_3\underset{\displaystyle OCH_3}{\underset{|}{C}}HCH_2CH_2OH$
3-Methoxy-1-butanol

(c)
2,5-Dihydrofuran

(See the labeled spectra on page 688 of this Study Guide.)

25. 2,4,4-trimethyl-2-pentene

(See the labeled spectrum on page 689 of this Study Guide.)

26. $CH_3C{\equiv}CCH_2OH$
2-Butyn-1-ol
G

(See the labeled spectra on page 689 of this Study Guide.)

27. p-$CH_3C_6H_4OCH_2CH_3$ — Ethyl p-tolyl ether — H

$C_6H_5CH_2OC_2H_5$ — Benzyl ethyl ether — I

$C_6H_5CH_2CH_2CH_2OH$ — 3-Phenyl-1-propanol — J

(See the labeled spectra on pages 690 and 691 of this Study Guide.)

28. (a) The absence of the strong absorption below 900 cm^{-1} that is characteristic of aromatic compounds indicates that geraniol is aliphatic.

The strong, broad band centered at about 3300 cm^{-1} is characteristic of an OH group; the band centered at 999 cm^{-1} indicates a primary alcohol.

The molecular formula $C_{10}H_{17}OH$ is equivalent to $C_{10}H_{18}$, indicating unsaturation and/or rings. The presence of at least one carbon–carbon double bond is confirmed by the bands at 1680 and 840 cm^{-1}.

(b) There are ten signals in the CMR spectrum, one for every carbon of the molecular formula. No two carbons in geraniol are equivalent.

(c) From the multiplicities in the CMR spectrum, we have three CH_3 (9H), three CH_2 (6H), two CH (2H), and two C without H.

(d) The CMR spectrum accounts for 17 hydrogens. The missing hydrogen plus the oxygen gives an OH group, as was indicated by the infrared spectrum in part (a). The OH is almost certainly on the carbon that gives peak f (δ 59.0); this carbon has two hydrogens, making $-CH_2OH$ and confirming our conclusion of part (a).

(e) The four downfield peaks in the CMR spectrum indicate two carbon–carbon double bonds. These, as we saw in part (a), are consistent with the molecular formula and the infrared spectrum.

(f) Moving downfield, we see: nine protons (methyl, allylic?), four protons (methylene, allylic?), one proton (—**OH**), two protons (—$\mathbf{CH_2}$OH), two protons (vinylic).

The nine protons suggest three methyl groups; the four protons suggest two methylene groups. The δ values are what we might expect if all five of these groups were allylic.

(g) There is one easily exchanged hydrogen, clearly the one in —OH.

(h) The only strong splitting we see in the NMR spectrum is that between the two-proton doublet at δ 4.06 (—$\mathbf{CH_2}$OH) and the one-proton triplet at δ 5.4 (one of the vinylic protons). This is consistent with the unit

$$\mathrm{C{=}C(H)CH_2OH}$$

The absence of (strong) splitting of the peaks for the other two methylene groups and for the three methyl groups, taken with the δ values for these, strongly suggests that they are all allylic. In particular, there is no triplet–quartet and hence no attachment of CH_3 to CH_2.

(i) At this point, then, we have the following units (disregarding stereochemistry):

$$\mathrm{{>}C{=}C(H)CH_2OH \quad (H)}\qquad \mathrm{{>}C{=}C{<}} \quad \textit{or} \quad \mathrm{{>}C{-}C(H)CH_2OH} \qquad \mathrm{{>}C{=}C(H){-}}$$

To these we attach

$$CH_3— \quad CH_3— \quad CH_3— \quad —CH_2— \quad —CH_2—$$

in such a way that each of these groups is connected to a doubly bonded carbon and no CH_3 is attached to a CH_2. We arrive at three possible structures (disregarding geometric isomerism).

$$\underset{i}{\mathrm{CH_3{-}C(CH_3){=}CH{-}CH_2{-}CH_2{-}C(CH_3){=}CH{-}CH_2OH}}$$

$$\underset{ii}{\mathrm{CH_3{-}C(CH_3){=}C(CH_3){-}CH_2{-}CH_2{-}CH{=}CH{-}CH_2OH}}$$

$$\underset{iii}{\mathrm{CH_3{-}CH{=}C(CH_3){-}CH_2{-}CH_2{-}C(CH_3){=}CH{-}CH_2OH}}$$

In all of these, to meet our requirements, we have had to attach two methylene groups to each other. In the four-proton signal at δ 2.05 we see indications of splitting, such as might well occur from coupling between two not-very-different methylene groups.

In the CMR spectrum there is a large difference in shift between the two methylene groups (δ 26.7 and δ 39.7). This is consistent with *i* and *ii*, in which one methylene has a β-methyl that the other has not; it is not consistent with *iii*, in which each methylene has a β-methyl.

This leaves *i* and *ii* as our most likely candidates. But geraniol is a ten-carbon natural product, and to make our decision we turn to the isoprene rule (Sec. 11.25). In *i* we see two head-to-tail isoprene units; in *ii* (or *iii*) we do not. We select *i* as the structure of geraniol. (We cannot specify the stereochemistry about the double bond in the 2-position.)

The actual structure of geraniol is given with the labeled spectra on page 692 of this Study Guide. Before looking there, however, try the following problem, to see what the (much older) chemical evidence tells you about geraniol.

29. Oxidation to RCHO and RCOOH without loss of carbon shows RCH_2OH.

(a) $CH_3-C(CH_3)=O$ $\quad$ $O=C(HO)-CH_2-CH_2-C(CH_3)=O$ $\quad$ $O=C(HO)-C(OH)=O$ $\longleftarrow$ $CH_3-C(CH_3)=CH-CH_2-CH_2-C(CH_3)=CH-CH_2OH$

Geraniol

(b) Geometric isomers.

(c) Nerol must be the *Z* isomer, with $-CH_2OH$ in the more favorable position for ring closure.

$(CH_3)_2C=CH-CH_2-CH_2-C(CH_3)=CH-CH_2OH$ (CH_2OH cis to CH_2)

Nerol

Z

$(CH_3)_2C=CH-CH_2-CH_2-C(CH_3)=CH-CH_2OH$ (CH_2OH trans to CH_2)

Geraniol

E

30. Both alcohols give the same allylic cation, and hence the same bromide.

$$R-C(CH_3)(OH)-CH=CH_2 \xrightarrow[-H_2O]{H^+} \underbrace{R-\overset{CH_3}{C}{\cdots}CH{\cdots}CH_2}_{\oplus} \xleftarrow[-H_2O]{H^+} R-C(CH_3)=CH-CH_2OH$$

$$\downarrow Br^-$$

$$C_{10}H_{17}Br$$

If you would like further practice at analyzing spectra, try your hand at the following problems. The answers to these are given on page 675 of this Study Guide; labeled spectra begin on page 693 of this Study Guide.

Problem 17.32 Give a structure or structures consistent with each of the proton NMR spectra in Figure 17.40, page 284 of this Study Guide.

Problem 17.33 Give a structure or structures consistent with each of the CMR spectra in Figure 17.41, page 285 of this Study Guide.

Problem 17.34 Give a structure or structures consistent with each of the CMR spectra in Figure 17.42, page 286 of this Study Guide.

Problem 17.35 Give a structure or structures consistent with each of the CMR spectra in Figure 17.43, page 287 of this Study Guide.

Problem 17.36 Give a structure or structures for compound K, whose infrared and proton NMR spectra are shown in Figure 17.44, page 288 of this Study Guide.

Problem 17.37 Give a structure or structures for the compound L, whose infrared and proton NMR spectra are shown in Figure 17.45, page 289 of this Study Guide.

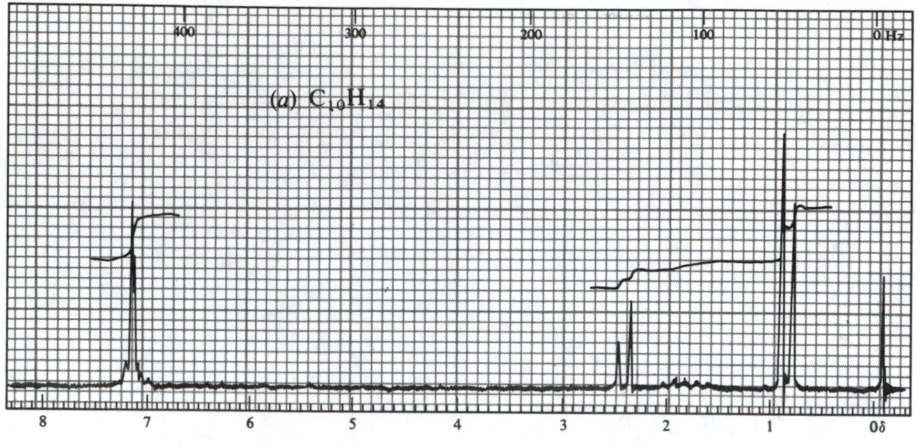

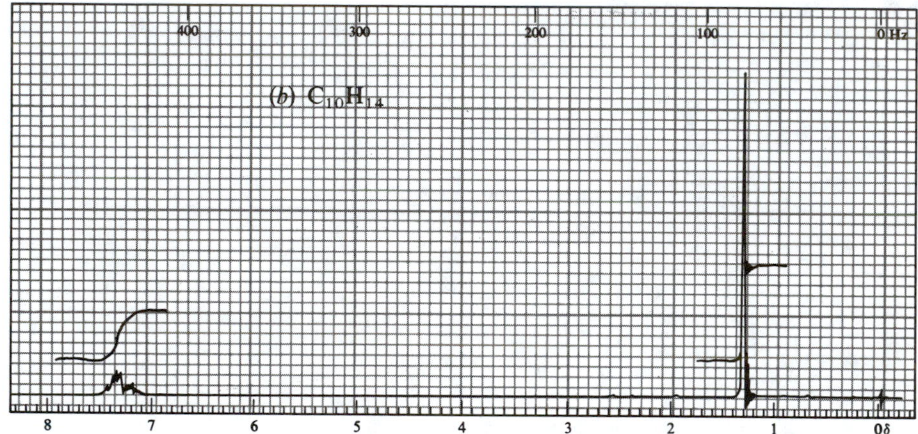

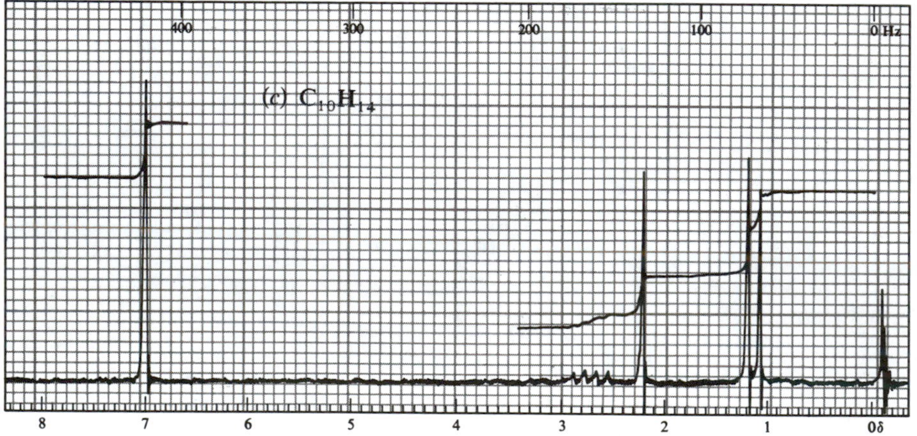

Figure 17.40 Proton NMR spectra for Problem 17.32 on page 283 of this Study Guide.

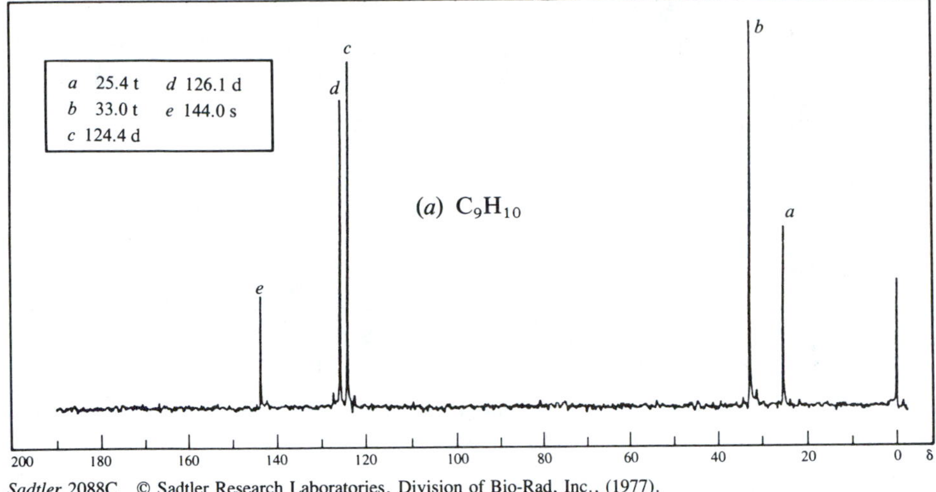

Sadtler 2088C © Sadtler Research Laboratories, Division of Bio-Rad, Inc., (1977).

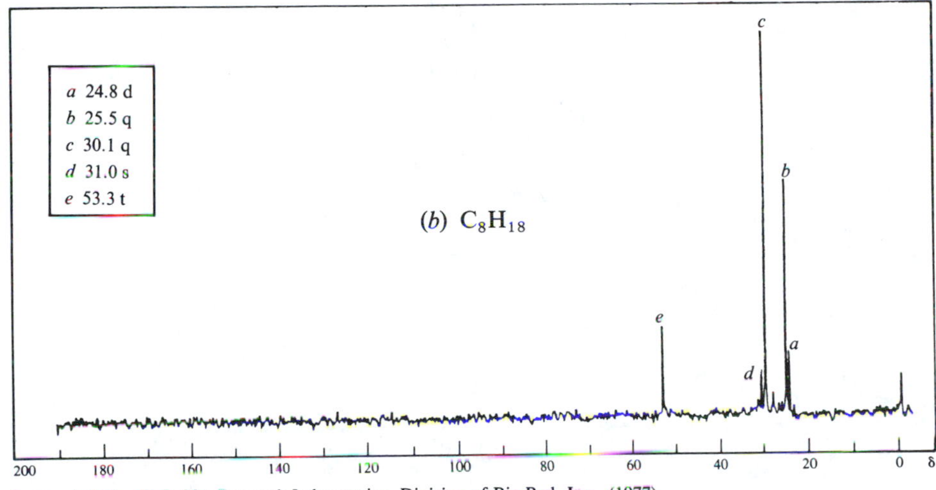

Sadtler 1613C © Sadtler Research Laboratories, Division of Bio-Rad, Inc., (1977).

a 20.8 q e 128.8 d
b 31.4 q f 134.4 s
c 34.2 s g 147.9 s
d 125.0 d

(c) $C_{11}H_{16}$

200 180 160 140 120 100 80 60 40 20 0 δ

Sadtler 1477C © Sadtler Research Laboratories, Division of Bio-Rad, Inc., (1977).

Figure 17.41 CMR spectra for Problem 17.33 on page 283 of this Study Guide.

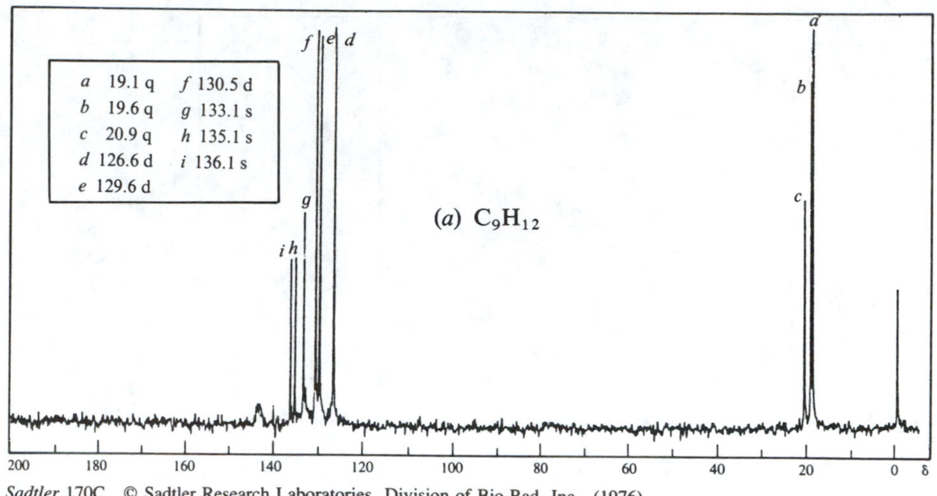

Sadtler 170C © Sadtler Research Laboratories, Division of Bio-Rad, Inc., (1976).

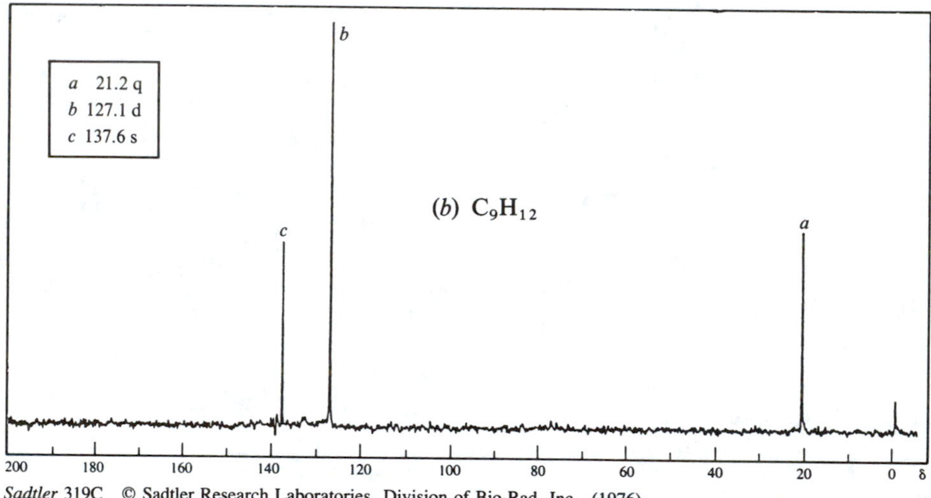

Sadtler 319C © Sadtler Research Laboratories, Division of Bio-Rad, Inc., (1976).

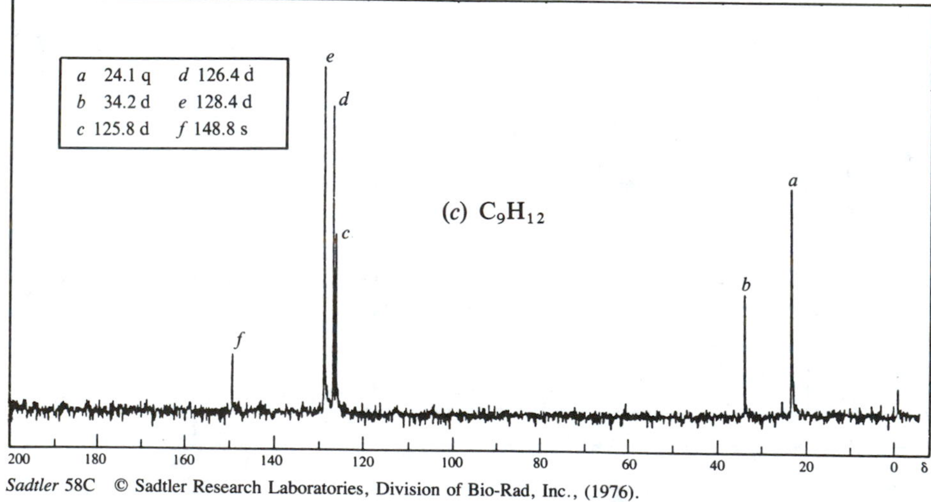

Sadtler 58C © Sadtler Research Laboratories, Division of Bio-Rad, Inc., (1976).

Figure 17.42 CMR spectra for Problem 17.34 on page 283 of this Study Guide.

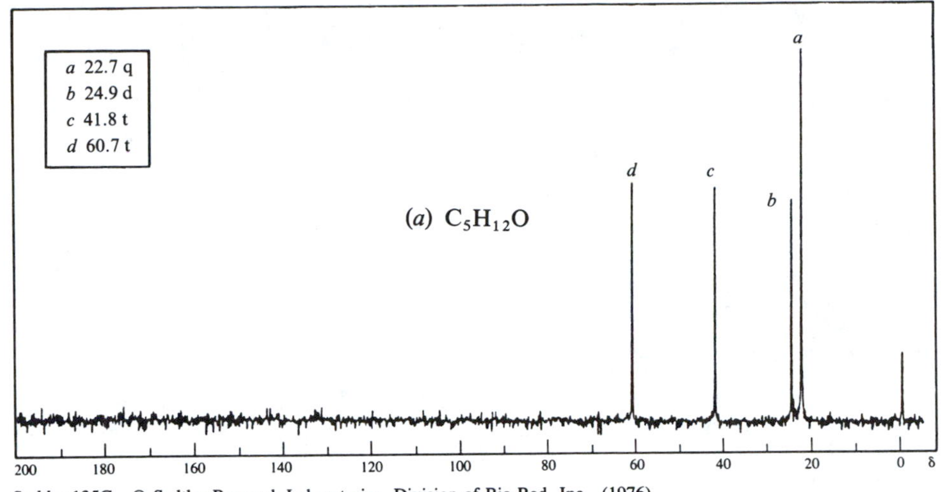

Sadtler 135C © Sadtler Research Laboratories, Division of Bio-Rad, Inc., (1976).

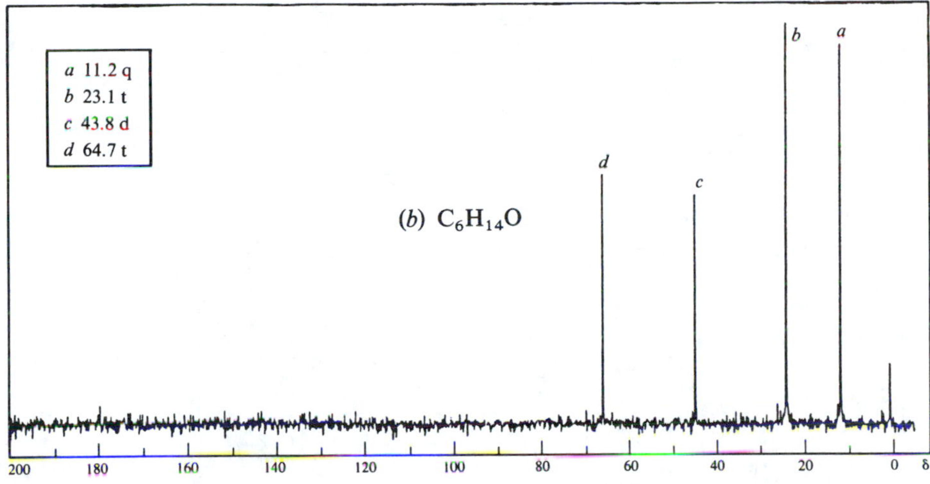

Sadtler 30C © Sadtler Research Laboratories, Division of Bio-Rad, Inc., (1976).

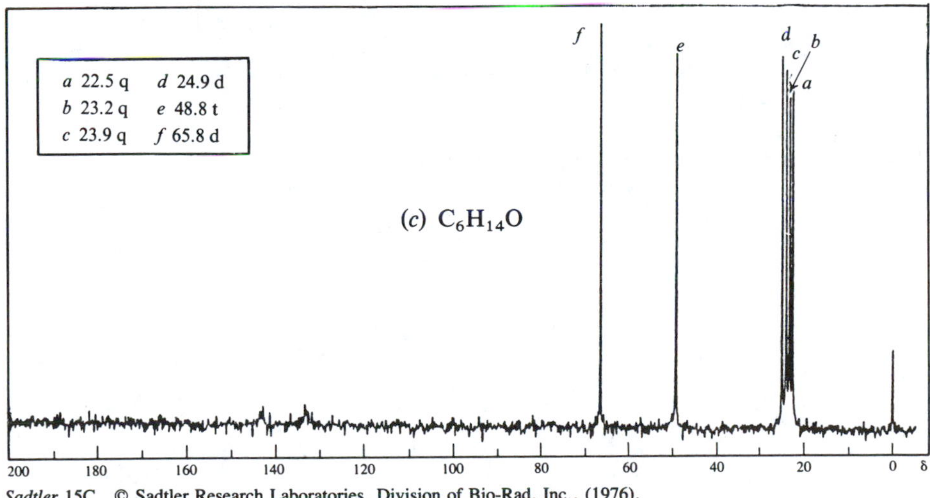

Sadtler 15C © Sadtler Research Laboratories, Division of Bio-Rad, Inc., (1976).

Figure 17.43 CMR spectra for Problem 17.35 on page 283 of this Study Guide.

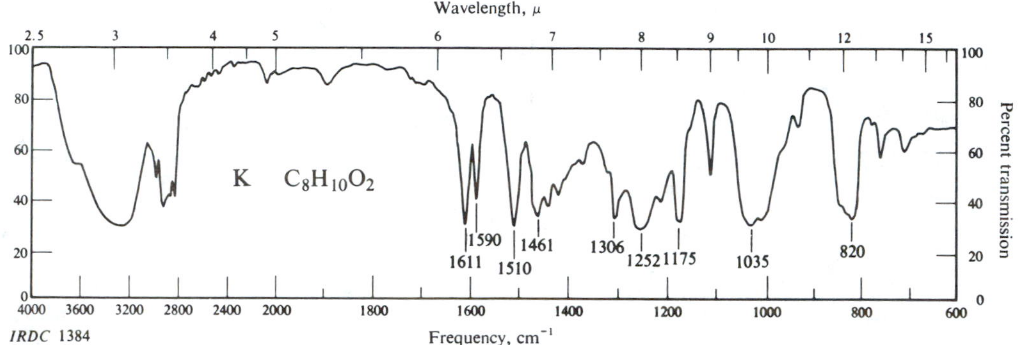

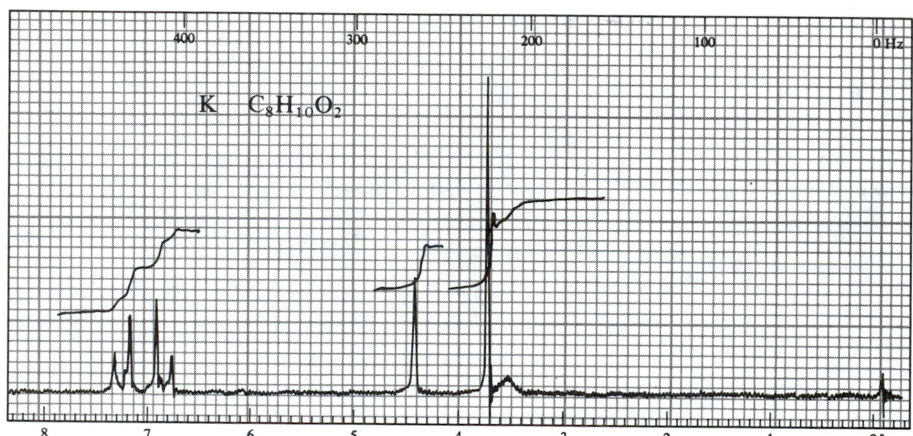

Figure 17.44 Infrared and proton NMR spectra for Problem 17.36 on page 283 of this Study Guide.

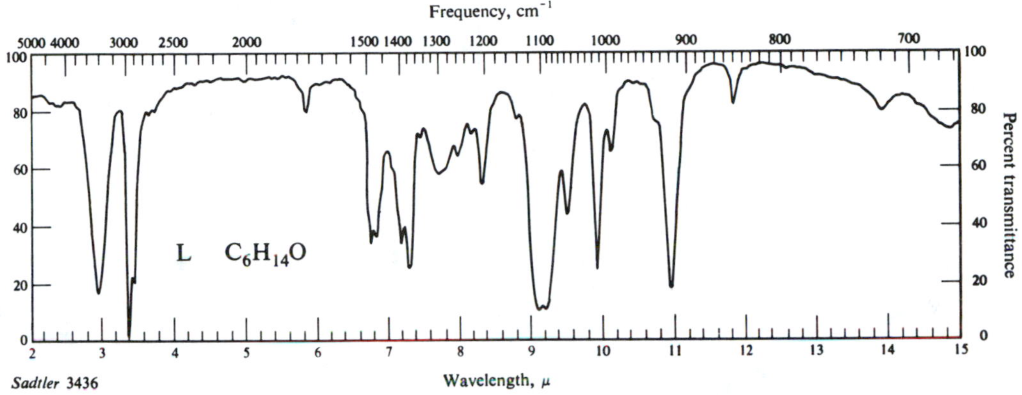

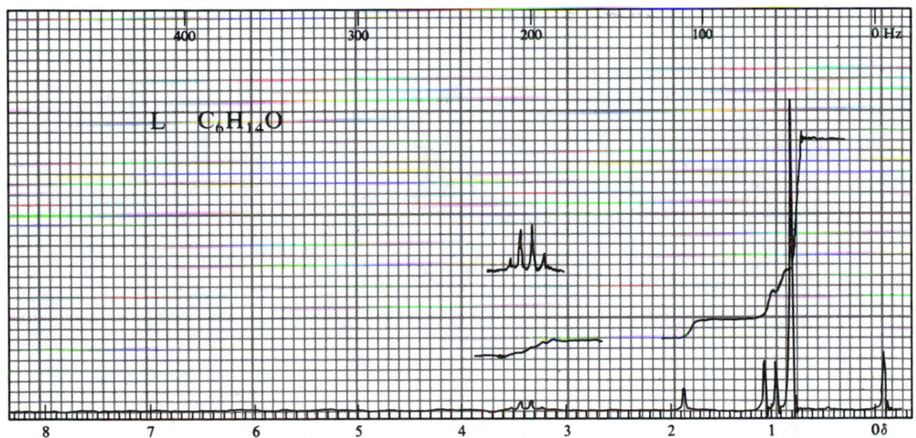

Figure 17.45 Infrared and proton NMR spectra for Problem 17.37 on page 283 of this Study Guide.

18

Aldehydes and Ketones

Nucleophilic Addition

18.1 No, because one has to make the organocopper compound from the organolithium reagent, and the nitro group would interfere with any attempts to form that reagent.

18.2 (a) 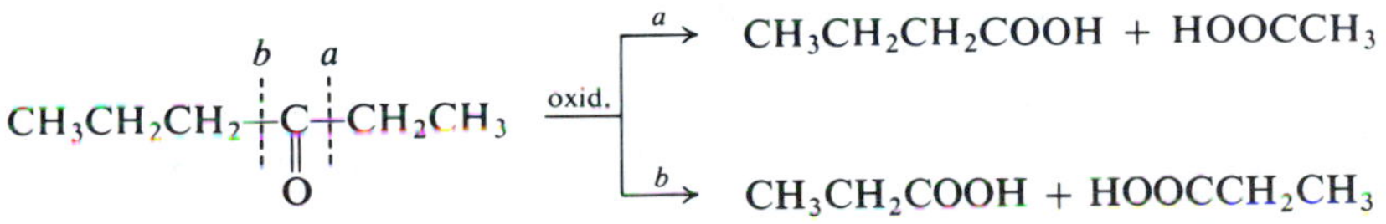 $CH_3CH_2CH_2\overset{b}{\vdots}\underset{\underset{O}{\|}}{C}\overset{a}{\vdots}CH_2CH_3 \xrightarrow{\text{oxid.}}$

$\xrightarrow{a} CH_3CH_2CH_2COOH + HOOCCH_3$

$\xrightarrow{b} CH_3CH_2COOH + HOOCCH_2CH_3$

(b) cyclohexanone $\xrightarrow{\text{oxid.}}$ $HO-\underset{\underset{O}{\|}}{C}(CH_2)_4\underset{\underset{O}{\|}}{C}-OH$, adipic acid

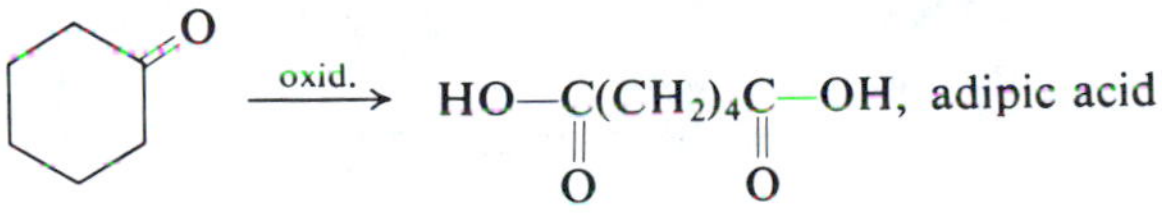

18.3 (a) 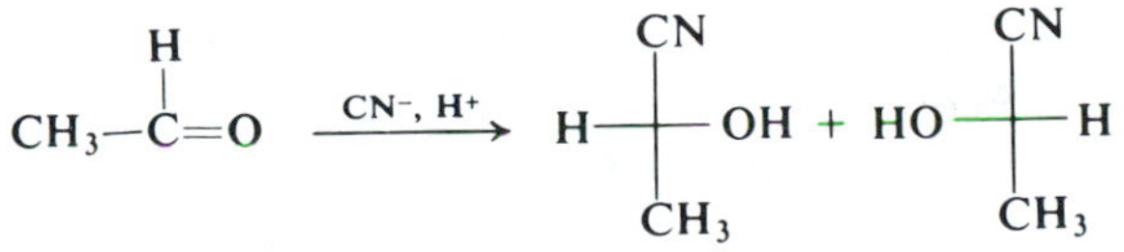 $CH_3-\overset{H}{\overset{|}{C}}=O \xrightarrow{CN^-,\ H^+}$ $H-\underset{CH_3}{\overset{CN}{C}}-OH + HO-\underset{CH_3}{\overset{CN}{C}}-H$

Racemic modification
One fraction
Inactive, but resolvable

(b) 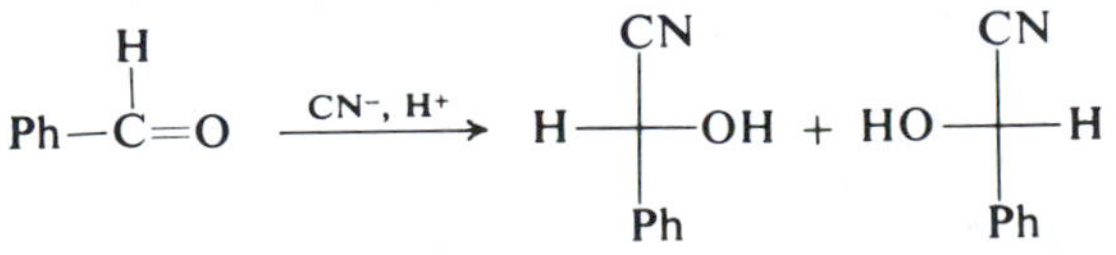 $Ph-\overset{H}{\overset{|}{C}}=O \xrightarrow{CN^-,\ H^+}$ $H-\underset{Ph}{\overset{CN}{C}}-OH + HO-\underset{Ph}{\overset{CN}{C}}-H$

Racemic modification
One fraction
Inactive, but resolvable

(c) 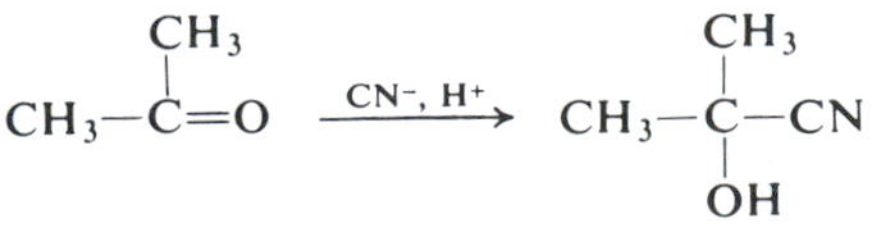 $CH_3-\overset{CH_3}{\overset{|}{C}}=O \xrightarrow{CN^-,\ H^+}$ $CH_3-\underset{OH}{\overset{CH_3}{C}}-CN$

Single compound
One fraction
Inactive; not resolvable

(d)

CHO
H—|—OH
CH_2OH

$\xrightarrow{CN^-,\ H^+}$

CN
H—|—OH
H—|—OH
CH_2OH
Active

\+

CN
HO—|—H
H—|—OH
CH_2OH
Active

Diastereomers
Two fractions

(e)

CHO
H—|—OH
CH_2OH

CHO
HO—|—H
CH_2OH
Racemic

$\xrightarrow{CN^-,\ H^+}$

CN
H—|—OH
H—|—OH
CH_2OH

\+

CN
HO—|—H
H—|—OH
CH_2OH

CN
HO—|—H
HO—|—H
CH_2OH
Racemic

\+

CN
H—|—OH
HO—|—H
CH_2OH
Racemic

Diastereomeric
Two fractions
Each inactive, but resolvable

(f) No change, since no bond to chiral carbon is broken.

18.4 Semicarbazide formation is reversible. Cyclohexanone reacts more rapidly, but benzaldehyde gives the more stable product. Initially, one isolates the product of rate control; later, after equilibrium is established, one isolates the product of equilibrium control.

18.5 High alcohol, low water concentrations shift the hemiacetal–acetal and aldehyde–hemiacetal equilibria (page 681) in the direction of the acetal. Low alcohol, high water concentrations shift the equilibria in the direction of the aldehyde.

18.6 (a) Williamson synthesis of ethers. (b) Symphoria. (c) Acetals (cyclic).

(d) Treatment with acid gives HCHO and catechol, *o*-$C_6H_4(OH)_2$; there is no reaction with base, typical of acetals (and other ethers).

18.7 An attempt to hydroxylate acrolein itself would result in oxidation of the —CHO group. The acetal grouping, however, is not affected by permanganate or the alkaline reaction medium, and is converted into —CHO after hydroxylation is complete.

$$CH_2{=}CH{-}CH(OEt)_2 \xrightarrow{KMnO_4} \underset{OH}{CH_2}{-}\underset{OH}{CH}{-}CH(OEt)_2 \xrightarrow{H_2O,\ H^+} \underset{OH}{CH_2}{-}\underset{OH}{CH}{-}CHO$$

Acrolein diethyl acetal — Glyceraldehyde

18.8 The rate-determining step is formation of a cation analogous to I (page 681); the more stable the cation, the faster it is formed.

(a) $R{-}C\{{:}OR'\}_2^{\oplus} > R{-}C(H){=}\overset{+}{O}R' > R{-}CH_2^+$

From: ortho ester — acetal — ether

(b) $R_2C{=}\overset{+}{O}R' > RCH{=}\overset{+}{O}R' > H_2C{=}\overset{+}{O}R'$

From: ketal — acetal — formal

In (b), R stabilizes the cation (relative to the reactant) more than H does because (*i*) R releases electrons, and (*ii*) R is bigger and hence favors change from the tetrahedral (sp^3) reactant to the trigonal (sp^2) product.

18.9

$$R{-}CH{=}O \underset{}{\overset{H^+}{\rightleftharpoons}} R{-}CH{=}OH^+ \overset{H_2{}^{18}O}{\rightleftharpoons} R{-}CH(OH){-}{}^{18}OH_2^+$$

Unlabeled aldehyde
Starting material

$$R{-}CH(OH){-}{}^{18}OH_2^+ \underset{H^+}{\overset{-H^+}{\rightleftharpoons}} R{-}CH(OH){-}{}^{18}OH \underset{-H^+}{\overset{H^+}{\rightleftharpoons}} R{-}CH(OH_2^+){-}{}^{18}OH$$

$$R{-}CH(OH_2^+){-}{}^{18}OH \overset{H_2O}{\rightleftharpoons} R{-}CH{=}{}^{18}OH^+ \overset{H^+}{\rightleftharpoons} R{-}CH{=}{}^{18}O$$

Labeled aldehyde
Exchange product

18.10 (a)

$$\underset{\text{I}}{R{-}CH(OH){-}O^-} + OH^- \rightleftharpoons \underset{\text{II}}{R{-}CH(O^-){-}O^-} + H_2O$$

(b) Loss of hydride ion yields the resonance-stabilized $RCOO^-$ (III) directly.

$$R-\overset{H}{C}=O + R-\underset{O^-}{\overset{H}{C}}-O^- \longrightarrow R-\underset{H}{\overset{H}{C}}-O^- + R-C\left\{\begin{matrix}O\\O\end{matrix}\right\}^{\ominus}$$

II III

18.11 Use $R-\overset{D}{C}=O$, and see if a second D shows up in the alcohol, $R-\underset{D}{\overset{D}{C}}-OH$.

$$R-\overset{D}{C}=O + R-\underset{OH}{\overset{D}{C}}-O^- \longrightarrow R-\underset{D}{\overset{D}{C}}-O^- + RCOOH$$

$$\downarrow +H^+ \qquad \downarrow -H^+$$

$$R-\underset{D}{\overset{D}{C}}-OH \qquad RCOO^-$$

18.12 On both electronic and steric grounds (Sec. 18.7), one would expect reaction (1), page 684, to be *faster* for HCHO than for another aldehyde, and the position of *equilibrium* to lie farther to the right. If reaction (1) is rate-determining, HCHO is the chief hydride donor because it forms I faster than the other aldehyde does. If reaction (2) is rate-determining, HCHO is the chief hydride donor because equilibrium (1) provides more I derived from HCHO than from the other aldehyde. (Kinetics studies indicate that reaction (2) is rate-determining.)

18.13 Internal crossed Cannizzaro reaction:

$$Ph-\underset{O}{\overset{\|}{C}}-\overset{H}{C}=O \xrightleftharpoons{OH^-} Ph-\underset{O}{\overset{\|}{C}}-\underset{OH}{\overset{H}{C}}-O^- \longrightarrow Ph-\underset{^-O}{\overset{H}{C}}-\underset{OH}{C}=O \longrightarrow Ph-\underset{OH}{\overset{H}{C}}-COO^-$$

18.14 $$Ph-\underset{O}{\overset{\|}{C}}-\underset{O}{\overset{\|}{C}}-Ph \xrightleftharpoons{OH^-} Ph-\underset{O}{\overset{\|}{C}}-\underset{O^-}{\overset{Ph}{C}}-OH \longrightarrow Ph-\underset{^-O}{\overset{Ph}{C}}-\underset{O}{\overset{\|}{C}}-OH \longrightarrow Ph-\underset{OH}{\overset{Ph}{C}}-COO^-$$

Use of OCH_3^- instead of OH^- gives the ester by the same route.

18.15 (a) $CH_3CH_2CH_2CH_2Li + H_2O \longrightarrow CH_3CH_2CH_2CH_3 + Li^+OH^-$

(b) $CH_3CH_2CH_2CH_2Li + D_2O \longrightarrow CH_3CH_2CH_2CH_2D + Li^+OD^-$

(c) $CH_3CH_2CH_2CH_2Li + C_2H_5OH \longrightarrow CH_3CH_2CH_2CH_3 + C_2H_5O^-Li^+$

(d) $CH_3CH_2CH_2CH_2Li + CH_3NH_2 \longrightarrow CH_3CH_2CH_2CH_3 + CH_3NH^-Li^+$

(e) $CH_3CH_2CH_2CH_2Li + C_2H_5C{\equiv}CH \longrightarrow CH_3CH_2CH_2CH_3 + C_2H_5C{\equiv}C^-Li^+$

(f) $CH_3CH_2CH_2CH_2Li + CH_3\underset{\underset{O}{\|}}{C}CH_3 \longrightarrow CH_3\overset{\overset{CH_3}{|}}{\underset{\underset{O—Li}{|}}{C}}CH_2CH_2CH_2CH_3$

18.16 Hydrolysis of the intermediate Mg salt is not shown, but is part of each synthesis.

$n\text{-Bu} \vdots CH_2OH \longleftarrow n\text{-BuMgBr} + HCHO$

$n\text{-Pr} \overset{a}{\vdots} CHOH \underset{b}{\vdots} CH_3 \longleftarrow$
- a: $n\text{-PrMgBr} + OHCCH_3$
- b: $n\text{-PrCHO} + CH_3MgBr$

$EtCHOH \vdots Et \longleftarrow EtCHO + EtMgBr$

$sec\text{-Bu} \vdots CH_2OH \longleftarrow sec\text{-BuMgBr} + HCHO$

$CH_3CH_2 \overset{a}{\vdots} \overset{\overset{OH}{|}}{\underset{\underset{CH_3}{|}}{C}} \overset{b}{\vdots} CH_3 \longleftarrow$
- a: $CH_3CH_2MgBr + O{=}C(CH_3)_2$
- b: $CH_3CH_2\underset{\underset{CH_3}{|}}{C}{=}O + CH_3MgBr$

$i\text{-Pr} \overset{a}{\vdots} CHOH \underset{b}{\vdots} CH_3 \longleftarrow$
- a: $i\text{-PrMgBr} + OHC—CH_3$
- b: $i\text{-Pr—CHO} + CH_3MgBr$

$iso\text{-Bu} \vdots CH_2OH \longleftarrow iso\text{-BuMgBr} + HCHO$

$t\text{-Bu} \vdots CH_2OH \longleftarrow t\text{-BuMgBr} + HCHO$

18.17 (a) Electrophilic addition of the elements of ROH to the double bond.

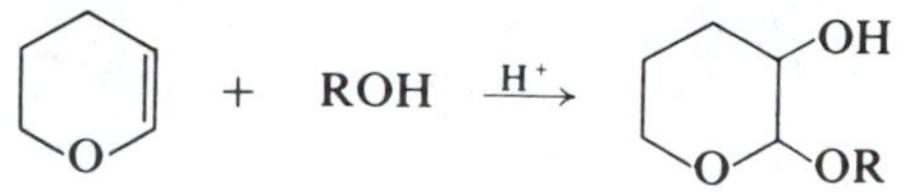

DHP + ROH $\xrightarrow{H^+}$ A THP ether

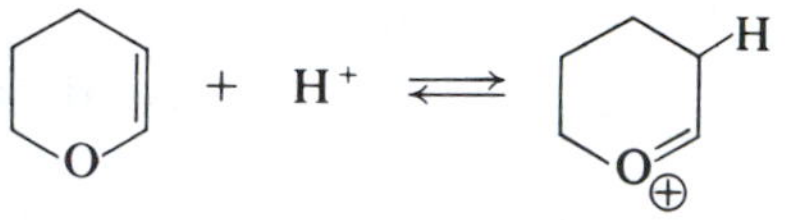

(b)

(1) DHP + H^+ $\rightleftharpoons$ I

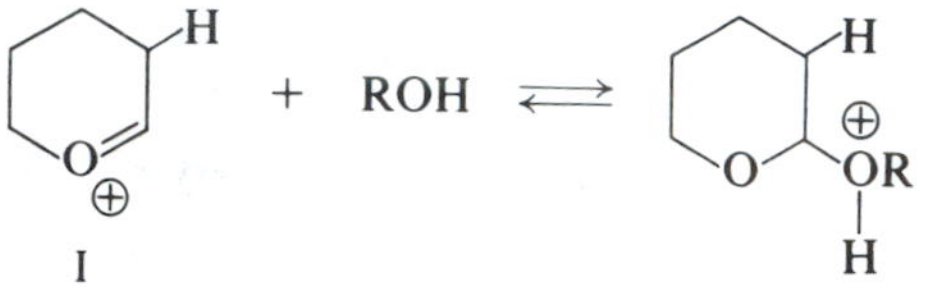

(2) I + ROH $\rightleftharpoons$

(3) $\rightleftharpoons$ A THP ether

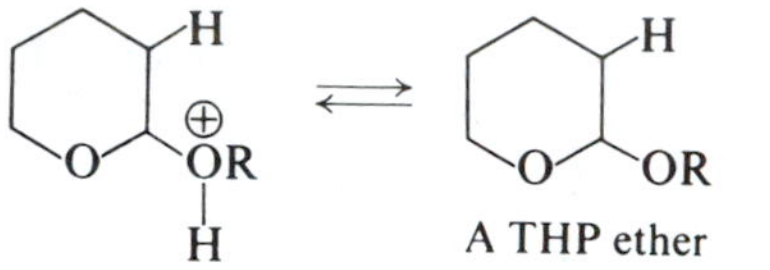

(c) Like other examples of electrophilic addition that we have encountered (Sec. 9.9), the first—and rate-controlling—step is the attachment of a proton to the alkene to form a carbocation (step 1). In doing this the proton could add to either of the two doubly bonded carbons; it actually adds in the way that gives the more stable carbocation (Sec. 9.11).

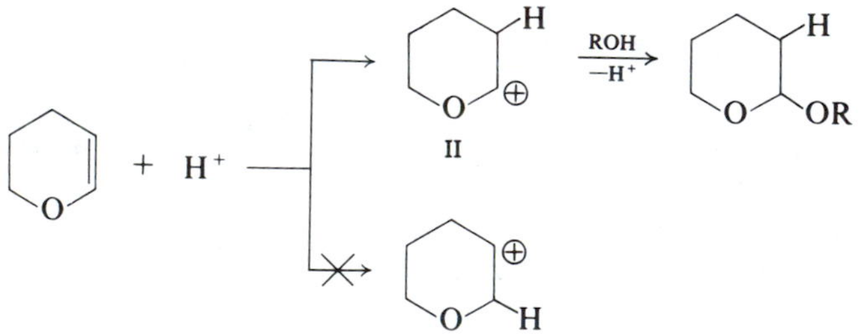

+ H^+ → II $\xrightarrow[-H^+]{ROH}$; ✗→ III

The observed product shows that cation II is formed in preference to cation III, and if we examine the structure of II we can see why this is so. We have written its structure as II, but we could just as well have written it as I.

II I

Especially stable: every atom has octet

As we saw in Sec. 15.18, a structure like I should be especially stable, since in it every atom has an octet of electrons. Indeed, I must by itself pretty well represent the structure of the cation; it is an *oxonium ion*, and is enormously more stable than an ordinary carbocation.

Both orientation and reactivity are determined by the unusual stability of cation I: I is formed in preference to III, and ultimately yields the observed product; I is formed much faster than a carbocation from a simple alkene, and leads to the unusually high reactivity of DHP.

But the stability of I has still another effect: it results in the extreme ease with which the THP ether is cleaved by dilute aqueous acid—a property on which the usefulness of DHP depends. Cleavage involves the exact reversal of steps (1)–(3), and it is the reversal of step (2)—the formation of the cation, once again—that determines the rate at which this cleavage takes place.

(d)

$$CH_3\underset{\underset{OH}{|}}{CH}\vdots CH_2CH_2OH \xleftarrow[H^+]{H_2O} CH_3\underset{\underset{OMgBr}{|}}{CH}—CH_2CH_2OTHP \longleftarrow \begin{cases} CH_3CHO \\ BrMgCH_2CH_2OTHP \end{cases}$$

$$CH_3CHO \xleftarrow{C_5H_5NHCrO_3Cl} CH_3CH_2OH$$

$$BrMgCH_2CH_2OTHP \xleftarrow{Mg} BrCH_2CH_2OTHP \xleftarrow[H^+]{DHP} BrCH_2CH_2OH \xleftarrow[H_2O]{Br_2} CH_2{=}CH_2 \xleftarrow[heat]{H^+} CH_3CH_2OH$$

18.18 Add the following to the table in the answer to Problem 16.25.

	conc. H_2SO_4[a]	*cold* $KMnO_4$[b]	Br_2[b]	CrO_3[c]	*fum.* H_2SO_4[a]	$CHCl_3$, $AlCl_3$[c]	*Na*[d]
Aldehydes	+	+	–[e]	+	+	–	–
Ketones	+	–	–[e]	–	+	–	–

[a] Dissolves. [b] Decolorizes. [c] Color changes. [d] Hydrogen bubbles. [e] Slow, liberates HBr.

18.19 At the site of cleavage:

(1) Replace the bond to carbon by a bond to —OH. Each such cleavage requires one HIO_4.

(2) If a resulting fragment is now a *gem*-diol (unstable), remove H_2O to give a stable carbonyl compound.

$$—\overset{|}{\underset{\underset{OH}{|}}{C}}—OH \longrightarrow H_2O + —\overset{|}{C}{=}O$$

A *gem*-diol *Unstable* — A carbonyl compound

For example:

$$\underset{\text{A glycol}}{\underset{HO\quad OH\quad OH}{H_2C \vdots CH \vdots CH_2}} \xrightarrow{2HIO_4} \underset{OH}{H_2C-OH} + HO-\underset{OH}{\overset{H}{C}}-OH + HO-\underset{OH}{CH_2}$$

$$\downarrow -H_2O \qquad \downarrow -H_2O \qquad \downarrow -H_2O$$

$$H_2C{=}O \qquad H-C(=O)OH \qquad O{=}CH_2$$

Cleavage products

(a) $CH_3CHOH \vdots CH_2OH \xrightarrow{1HIO_4} CH_3CHO + HCHO$

(b) $CH_3CHOH \vdots CHO \xrightarrow{1HIO_4} CH_3CHO + HCOOH$

(c) $CH_2OH \vdots CHOHCH_2OCH_3 \xrightarrow{1HIO_4} HCHO + OHCCH_2OCH_3$

(d) $CH_2OH\underset{OCH_3}{C}HCH_2OH \longrightarrow$ no reaction with HIO_4

(e) cyclopentane-1,2-diol (OH, OH; cleavage at the C–C bond between them) $\xrightarrow{1HIO_4} OHC(CH_2)_3CHO$

(f) $CH_2OH \vdots CHOH \vdots CHOH \vdots CHOH \vdots CHO \xrightarrow{4HIO_4} HCHO + HCOOH + HCOOH + HCOOH + HCOOH$

(g) $CH_2OH \vdots CHOH \vdots CHOH \vdots CHOH \vdots CH_2OH \xrightarrow{4HIO_4} HCHO + HCOOH + HCOOH + HCOOH + HCHO$

18.20 To reconstruct (mentally) the compound from the cleavage fragments, reverse the procedure of the preceding problem. (1) Replace a carbonyl group in a cleavage product by two —OH's (unstable *gem*-diol). (2) Now delete —OH's in pairs, one from each of two fragments, and join the carbons together. For example:

$$H_2C{=}O \qquad H-C(=O)OH \qquad O{=}CH_2 \quad \text{(Cleavage products)}$$

$$\uparrow \qquad \uparrow \qquad \uparrow$$

$$\underset{OH}{H_2C-OH} \quad HO-\underset{OH}{\overset{H}{C}}-OH \quad HO-\underset{HO}{CH_2} \longleftarrow \underset{HO\quad OH\quad OH}{H_2C-CH-CH_2}$$

$$[HIO_4] \qquad [HIO_4]$$

$$CH_3-\overset{CH_3}{\overset{|}{C}}=O + O=CH_2 \xleftarrow{1HIO_4} CH_3-\underset{HO}{\overset{CH_3}{\overset{|}{\underset{|}{C}}}}-\underset{OH}{\underset{|}{CH_2}} \quad (A)$$

$$(CH_2)_4\begin{cases} CH=O \\ O=CH \end{cases} \xleftarrow{1HIO_4} \text{cyclohexane-1,2-diol} \quad (B)$$

$$(CH_2)_4\begin{cases} C(=O)-OH \\ O=CH \end{cases} \xleftarrow{1HIO_4} \text{2-hydroxycyclohexanone} \quad (C)$$

$$HOOC-\overset{H}{\overset{|}{C}}=O + O=\overset{H}{\overset{|}{C}}-COOH \xleftarrow{1HIO_4} HOOC\underset{OH}{\underset{|}{C}}H-\underset{OH}{\underset{|}{C}}HCOOH \quad (D)$$

$$H_2C=O + \underset{O \quad OH}{\overset{H}{\overset{|}{C}}} + \underset{O \quad OH}{\overset{H}{\overset{|}{C}}} + O=CH_2 \xleftarrow{3HIO_4} \underset{HO}{\underset{|}{H_2C}}-\underset{OH}{\underset{|}{CH}}-\underset{OH}{\underset{|}{CH}}-\underset{OH}{\underset{|}{CH_2}} \quad (E)$$

$$H_2C=O + HCOOH + CO_2 + HCOOH$$

$$\uparrow$$

$$(\underset{OH}{\underset{|}{H_2C}}-OH + HO-\underset{OH}{\overset{H}{\overset{|}{\underset{|}{C}}}}-OH + HO-\underset{O}{\underset{\|}{C}}-OH + HO-\underset{OH}{\overset{H}{\overset{|}{\underset{|}{C}}}}-OH)$$

$$\uparrow 3HIO_4$$

$$\underset{HO}{\underset{|}{H_2C}}-\underset{OH}{\underset{|}{CH}}-\underset{O}{\underset{\|}{C}}-\underset{H}{\underset{|}{C}}=O \quad (F)$$

$$H_2C=O + 4HCOOH + HCOOH \xleftarrow{5HIO_4} H_2C-\underset{OH}{\overset{H}{\overset{|}{\underset{|}{C}}}}-\underset{OH}{\overset{H}{\overset{|}{\underset{|}{C}}}}-\underset{OH}{\overset{H}{\overset{|}{\underset{|}{C}}}}-\underset{OH}{\overset{H}{\overset{|}{\underset{|}{C}}}}-\underset{OH}{\overset{H}{\overset{|}{\underset{|}{C}}}}=O \quad (G)$$

1. (a) $CH_3CH_2CH_2CH_2CHO$
n-Valeraldehyde
Pentanal

$CH_3CH_2CH_2COCH_3$
Methyl *n*-propyl ketone
2-Pentanone

$CH_3CH_2COCH_2CH_3$
Ethyl ketone
3-Pentanone

$CH_3CH_2CH(CH_3)CHO$
α-Methylbutyraldehyde
2-Methylbutanal

$(CH_3)_2CHCH_2CHO$
Isovaleraldehyde
3-Methylbutanal

$(CH_3)_2CHCOCH_3$
Isopropyl methyl ketone
3-Methyl-2-butanone

$(CH_3)_3CCHO$
Trimethylacetaldehyde
2,2-Dimethylpropanal

(b) $PhCH_2CHO$ Phenylacetaldehyde Phenylethanal; $PhCOCH_3$ Acetophenone; o-$CH_3C_6H_4CHO$ o-Tolualdehyde

m-$CH_3C_6H_4CHO$ m-Tolualdehyde; p-$CH_3C_6H_4CHO$ p-Tolualdehyde

2. (a) CH_3COCH_3
(b) $PhCHO$
(c) $CH_3COCH_2CH(CH_3)_2$
(d) $(CH_3)_3CCHO$
(e) $PhCOCH_3$
(f) $(CH_3)_2CHCH_2CH_2CHO$
(g) $PhCH_2CHO$
(h) $PhCOPh$
(i) $C_2H_5CH(CH_3)CH_2CH(CH_3)CHO$
(j) $C_2H_5CH(CH_3)COCH_3$
(k) $CH_3CH{=}CHCHO$
(l) $(CH_3)_2C{=}CHCOCH_3$
(m) $PhCH{=}CHCOPh$
(n) $CH_3CH_2CH(OH)CH_2CHO$
(o) $PhCH_2COPh$
(p) $(p$-$HOC_6H_4)_2C{=}O$
(q) m-$CH_3C_6H_4CHO$

3. (a) $PhCH_2COO^- + Ag$
(b) $PhCH_2COOH$
(c) $PhCH_2COOH$
(d) $PhCOOH$
(e) $PhCH_2CH_2OH$
(f) $PhCH_2CH_2OH$
(g) $PhCH_2CH_2OH$
(h) $PhCH_2CHOHPh$
(i) $PhCH_2CHOHCH(CH_3)_2$
(j) $PhCH_2CHOHC{\equiv}CH$
(k) $PhCH_2CHOHCN$
(l) $PhCH_2CH{=}NOH$
(m) $PhCH_2CH{=}NNHPh$
(n) $PhCH_2CH{=}NNHAr$
(o) $PhCH_2CH{=}NNHCONH_2$
(p) $PhCH_2CH(OEt)_2$

4. (a) no reaction
(b) no reaction
(c) no reaction
(d) $HOOC(CH_2)_4COOH$
(e) cyclohexanol
(f) cyclohexanol
(g) cyclohexanol
(h) cyclohexane ring with Ph and OH on same carbon
(i) cyclohexane ring with $CH(CH_3)_2$ and OH on same carbon
(j) cyclohexane ring with $C{\equiv}CH$ and OH on same carbon
(k) cyclohexane ring with CN and OH on same carbon
(l) cyclohexane ring $=NOH$

(m) 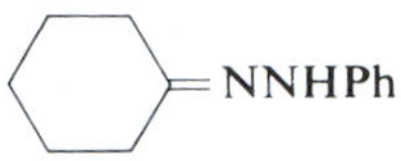 Cyclohexylidene $=NNHPh$ (n) 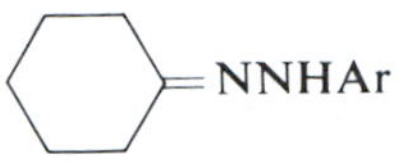 Cyclohexylidene $=NNHAr$ (o) 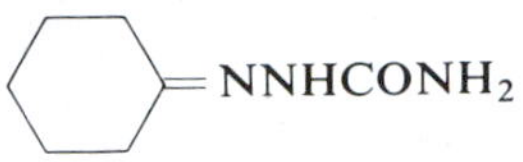 Cyclohexylidene $=NNHCONH_2$

(p) Might predict no reaction; actually get the ketal (cyclohexane with OEt, OEt on one carbon)

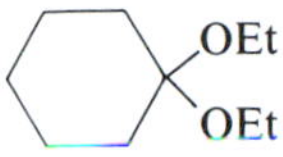

5. (a) Cannizzaro reaction:

$$2PhCHO \xrightarrow{OH^-} PhCH_2OH + PhCOO^-Na^+$$

(b) Crossed Cannizzaro reaction:

$$PhCHO + HCHO \xrightarrow{OH^-} PhCH_2OH + HCOO^-Na^+$$

(c) $PhCHOHCN$, mandelonitrile

(d) $PhCHOHCOOH$, mandelic acid

(e) $PhCHOHCH_3$, α-phenylethyl alcohol

(f) $PhCH{=}CH_2$, styrene

(g) $PhCHOH^{14}CH(CH_3)_2$

(h) $PhCH^{18}O$

6.

(a) 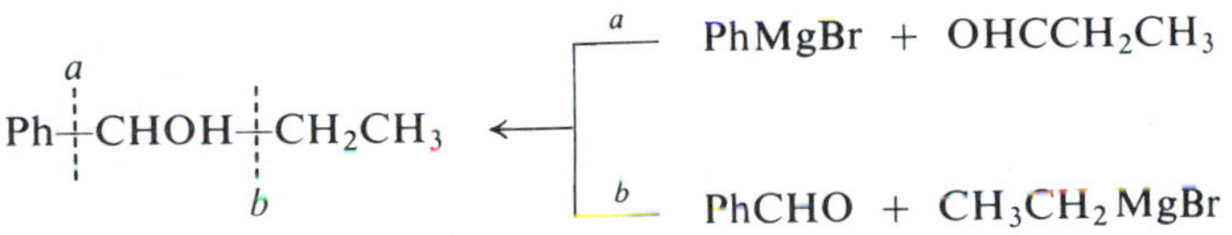 $Ph \overset{a}{|} CHOH \overset{b}{|} CH_2CH_3 \leftarrow$ (a) $PhMgBr + OHCCH_2CH_3$; (b) $PhCHO + CH_3CH_2MgBr$

(b) $Ph \overset{a}{|} C(OH)(CH_3) \overset{b}{|} CH_3 \leftarrow$ (a) $PhMgBr + O{=}C(CH_3)_2$; (b) $PhC(CH_3){=}O + CH_3MgBr$

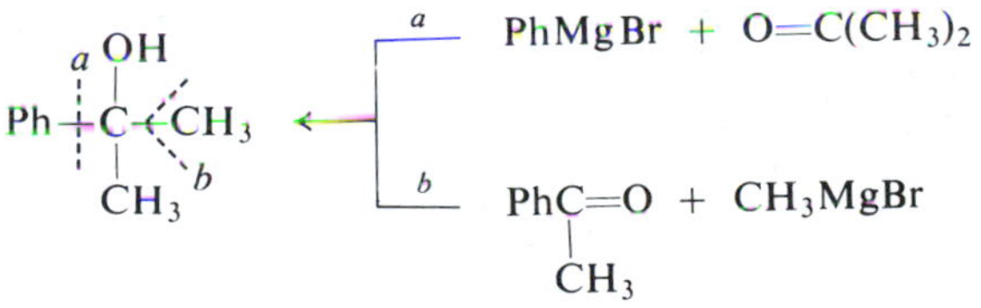

(c) $PhCH_2CHOH | CH_3 \longleftarrow PhCH_2CHO + CH_3MgBr$

($PhCH_2MgX$ reacts abnormally with CH_3CHO and many other aldehydes.)

(d) $PhCH_2CH_2 | CH_2OH \longleftarrow PhCH_2CH_2MgBr + HCHO$

(e) 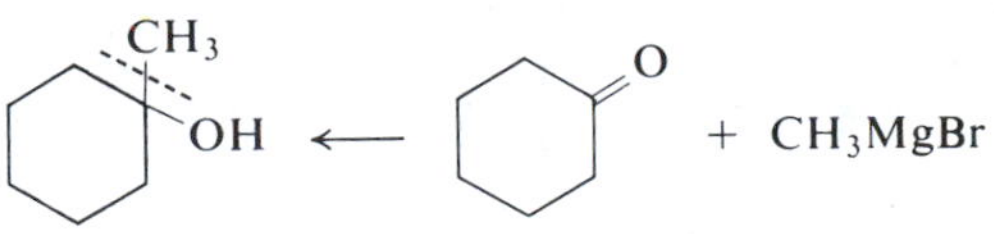 1-methylcyclohexanol (CH_3, OH) $\longleftarrow$ cyclohexanone (O) $+ CH_3MgBr$

(f) 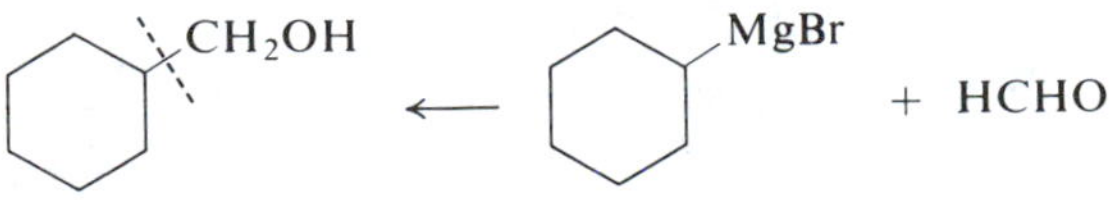 Cyclohexyl-CH_2OH $\longleftarrow$ cyclohexyl-MgBr $+ HCHO$

(g) Cyclohexyl–CH(OH)–CH$_3$ (bonds *a* and *b* marked) ⟵ *a*: Cyclohexyl–MgX + OHCCH$_3$; *b*: Cyclohexyl–CHO + CH$_3$MgBr

(h) *i*-Pr—CHOH⫶Pr-*i* ⟵ *i*-PrCHO + *i*-PrMgBr

(i) *p*-CH$_3$C$_6$H$_4$⫶(*a*)CHOH⫶(*b*)CH$_3$ ⟵ *a*: *p*-CH$_3$C$_6$H$_4$MgBr + OHCCH$_3$; *b*: *p*-CH$_3$C$_6$H$_4$CHO + CH$_3$MgBr

(j) CH$_3$CH$_2$CH(OH)⫶C≡CH ⟵ CH$_3$CH$_2$CHO + HC≡CLi

(k) CH$_3$CH(OH)⫶C≡CCH$_3$ ⟵ CH$_3$CHO + CH$_3$C≡CLi

7. (a) CH$_3$CH$_2$CH$_2$OH $\xleftarrow[\text{or } H_2,\ Ni]{LiAlH_4}$ CH$_3$CH$_2$CHO

(b) CH$_3$CH$_2$COOH $\xleftarrow{KMnO_4}$ CH$_3$CH$_2$CHO

(c) CH$_3$CH$_2$CH(OH)COOH $\xleftarrow{H_2O,\ H^+}$ CH$_3$CH$_2$CH(OH)C≡N $\xleftarrow{CN^-,\ H^+}$ CH$_3$CH$_2$CHO

(d) CH$_3$CH$_2$CH(OH)⫶CH$_3$ ⟵ CH$_3$CH$_2$CHO + CH$_3$MgBr

(e) CH$_3$CH$_2$CH(OH)⫶Ph ⟵ CH$_3$CH$_2$CHO + PhMgBr

(f) CH$_3$CH$_2$C(=O)CH$_3$ $\xleftarrow{CrO_3}$ CH$_3$CH$_2$CH(OH)CH$_3$ ⟵ (d)

(g) CH$_3$CH$_2$CH(OH)⫶CH(CH$_3$)$_2$ ⟵ CH$_3$CH$_2$CHO + BrMgCH(CH$_3$)$_2$

8. (a) $PhCH_2CH_3 \xleftarrow{Zn(Hg),\ H^+} PhCOCH_3$

(b) $PhCOOH \xleftarrow{H^+} \xleftarrow{NaOI} PhCOCH_3$

(c) $PhCH(OH)CH_3 \xleftarrow[\text{or } H_2,\ Ni]{LiAlH_4} PhC(=O)CH_3$

(d) $Ph-C(CH_3)(OH)-CH_2CH_3 \longleftarrow PhC(=O)CH_3 + C_2H_5MgBr$

(e) $Ph-C(CH_3)(OH)-Ph \longleftarrow PhC(=O)CH_3 + PhMgBr$

(f) $Ph-C(CH_3)(OH)-COOH \xleftarrow{H_2O,\ H^+} Ph-C(CH_3)(OH)-C{\equiv}N \xleftarrow{CN^-,\ H^+} PhC(=O)CH_3$

Synthesis: Working Backwards

Granting that we know the chemistry of the individual steps, how do we go about planning a route to more complicated compounds—alcohols, say? In almost every organic synthesis it is best to begin with the molecule we want—the *target* molecule—and *work backwards* from it. There are relatively few ways to make a complicated alcohol, for example; there are relatively few ways to make a Grignard reagent or an aldehyde or ketone; and so on back to our primary starting materials. On the other hand, our starting materials can undergo so many different reactions that, if we go at the problem the other way round, we find a bewildering number of paths, few of which take us where we want to go.

We try to limit a synthesis to as few steps as possible, but nevertheless do not sacrifice purity for time. To avoid a rearrangement in the preparation of an alkene, for example, we take two steps via the halide rather than the single step of dehydration.

9. (a) $(CH_3)_2CHCHO \xleftarrow{C_5H_5NHCrO_3Cl}$ *iso*-BuOH

(b) $PhCH_2CHO \xleftarrow{C_5H_5NHCrO_3Cl} Ph{\vdots}CH_2CH_2OH \longleftarrow$ PhMgBr (from C_6H_6) + $H_2C{-}CH_2$ (epoxide, O bridging) (Sec. 13.20)

(c) $Br\text{-}C_6H_4\text{-}CHO \xleftarrow[H^+]{H_2O} \xleftarrow[Ac_2O]{CrO_3} Br\text{-}C_6H_4\text{-}CH_3 \xleftarrow[Fe]{Br_2} C_6H_5CH_3$

(d) $CH_3CH_2COCH_3 \xleftarrow{CrO_3}$ *sec*-BuOH

(e) 2,4-$(NO_2)_2C_6H_3CHO \xleftarrow{H_2O}$ 2,4-$(NO_2)_2C_6H_3CHCl_2 \xleftarrow[heat]{2Cl_2}$ 2,4-$(NO_2)_2C_6H_3CH_3 \xleftarrow[\text{in stages}]{\text{nitration}} C_6H_5CH_3$

(f) $O_2N\text{-}C_6H_4\text{-}C(=O)Ph \xleftarrow[AlCl_3]{C_6H_6} O_2N\text{-}C_6H_4\text{-}C(=O)Cl \xleftarrow{SOCl_2} O_2N\text{-}C_6H_4\text{-}C(=O)OH$ (page 525)

(g) $Et \vdots C(=O)\text{—}CH(CH_3)_2 \longleftarrow$
- Et_2CuLi (from EtBr)
- $ClC(=O)\text{—}CH(CH_3)_2 \xleftarrow{SOCl_2} HOC(=O)\text{—}CH(CH_3)_2 \xleftarrow{KMnO_4}$ *iso*-BuOH

(h) $PhCH_2C(=O)CH_3 \xleftarrow{CrO_3} PhCH_2CH(OH) \vdots CH_3 \longleftarrow PhCH_2CHO$ (from b) + CH_3MgBr

(i) $m\text{-}O_2N\text{-}C_6H_4\text{-}C(=O)Ph \xleftarrow[AlCl_3]{C_6H_6} m\text{-}O_2N\text{-}C_6H_4\text{-}C(=O)Cl \xleftarrow{SOCl_2} m\text{-}O_2N\text{-}C_6H_4\text{-}C(=O)OH$ (page 525)

(j) $n\text{-}PrC(=O)\text{-}C_6H_4\text{-}CH_3 \xleftarrow[AlCl_3]{PhCH_3} n\text{-}PrC(=O)Cl \xleftarrow{SOCl_2} n\text{-}C_3H_7C(=O)OH \xleftarrow{KMnO_4}$ *n*-BuOH

(k) $CH_3CH_2CH(CH_3)CHO \xleftarrow{K_2Cr_2O_7} CH_3CH_2CH(CH_3) \vdots CH_2OH \longleftarrow$
- $HCHO \xleftarrow{Cu,\ heat}$ MeOH
- *sec*-BuMgBr

(l) $i\text{-}Bu\text{—}C(=O) \vdots Bu\text{-}n \longleftarrow$
- *n*-Bu_2CuLi (from *n*-BuBr)
- $i\text{-}Bu\text{—}C(=O)\text{—}Cl \xleftarrow{SOCl_2} i\text{-}Bu\text{—}C(=O)\text{—}OH \xleftarrow{H^+} \xleftarrow{CO_2} i\text{-}BuMgBr$

(m) $O_2N\text{-}C_6H_4\text{-}C(=O) \vdots CH_3 \longleftarrow$
- Me_2CuLi (from MeBr)
- $O_2N\text{-}C_6H_4\text{-}C(=O)Cl \xleftarrow{SOCl_2} O_2N\text{-}C_6H_4\text{-}C(=O)OH$ (page 525)

(n) $CH_3C_6H_4$—$C(=O)$—$C_6H_4NO_2$ $\xleftarrow[AlCl_3]{PhCH_3}$ $ClC(=O)C_6H_4NO_2$ $\xleftarrow{SOCl_2}$ $HOC(=O)C_6H_4NO_2$ (page 525)

(o) $O_2NC_6H_4$—$C(=O)$—Et ⟵ Et_2CuLi (from EtBr) + $O_2NC_6H_4CCl(=O)$ $\xleftarrow{SOCl_2}$ $O_2NC_6H_4COH(=O)$ (page 525)

10. (a) $CH_3CH(CH_3)$—$C(CH_3)(OH)$—CH_3 ⟵ $CH_3CH(CH_3)MgBr$ + $O{=}C(CH_3)$—CH_3 (both from *iso*-PrOH)

(b) CH_3—$C(Ph)(OH)$—CH_3 ⟵ PhMgBr (from C_6H_6) + CH_3—$C(=O)$—CH_3 (from *iso*-PrOH)

(c) CH_3—$C(Ph){=}CH_2$ $\xleftarrow{H^+,\ heat}$ CH_3—$C(Ph)(OH)$—CH_3 ⟵ (b)

(d) CH_3CH_2—$C(CH_3){=}CH_2$ $\xleftarrow[(alc)]{KOH}$ $\xleftarrow{PBr_3}$ $CH_3CH_2C(H_3C)(H)$—CH_2OH ⟵ *sec*-BuMgBr + HCHO (from MeOH)

(e) $CH_3CH_2CH(CH_3)CH_3$ $\xleftarrow{H_2,\ Pt}$ $CH_3CH_2C(CH_3){=}CH_2$ ⟵ (d)

(f) $CH_3CH_2C(CH_3)(Br)$—CH_2Br $\xleftarrow{Br_2,\ CCl_4}$ $CH_3CH_2C(CH_3){=}CH_2$ ⟵ (d)

(g) *n*-Pr—CH(OH)—Et ⟵
- *a*: *n*-PrMgBr + OHC—Et (both from *n*-PrOH)
- *b*: *n*-PrCHO (from *n*-BuOH) + EtMgBr (from EtOH)

(h) *n*-Pr—C(=O)—Et $\xleftarrow{K_2Cr_2O_7,\ H^+}$ *n*-Pr—CH(OH)—Et ⟵ (g)

(i) *n*-Pr—C(Et)(OH)—*n*-Pr ⟵ *n*-Pr—C(Et)(=O) (from f) + *n*-PrMgBr (from *n*-PrOH)

(j) $n\text{-Bu}-\underset{\displaystyle Br}{\overset{\displaystyle CH_3}{C}}-CH_3 \xleftarrow{HBr} n\text{-Bu}\vdots\underset{\displaystyle OH}{\overset{\displaystyle CH_3}{C}}-CH_3 \longleftarrow n\text{-BuMgBr (from } n\text{-BuOH)} + O{=}\overset{\displaystyle CH_3}{C}-CH_3 \text{ (from } i\text{-PrOH)}$

(k) $Ph\underset{\displaystyle Cl}{C}HCH_3 \xleftarrow{HCl} Ph\vdots\underset{\displaystyle OH}{C}HCH_3 \longleftarrow PhMgBr + OHCCH_3 \text{ (from EtOH)}$

(l) $PhCH_2CH_2CH_2CH_3 \xleftarrow[HCl]{Zn(Hg)} Ph-\underset{\displaystyle O}{\overset{\|}{C}}CH_2CH_2CH_3 \xleftarrow[AlCl_3]{C_6H_6} Cl\underset{\displaystyle O}{\overset{\|}{C}}CH_2CH_2CH_3$

(m) $n\text{-Pr}\underset{\displaystyle OH}{C}HCOOH \xleftarrow[H^+]{H_2O} n\text{-Pr}\underset{\displaystyle OH}{C}HC{\equiv}N \xleftarrow[H^+]{CN^-} n\text{-PrCHO} \xleftarrow{K_2Cr_2O_7} n\text{-PrCH}_2OH$

(n) $CH_3CH_2CH_2CH_2\vdots CH_2\underset{\displaystyle CH_3}{C}HCH_3 \xleftarrow[HCl]{Zn(Hg)} CH_3CH_2CH_2CH_2\vdots\underset{\displaystyle O}{\overset{\|}{C}}\overset{\displaystyle CH_3}{C}HCH_3 \longleftarrow$
- $(CH_3CH_2CH_2CH_2)_2CuLi$ (from n-BuOH)
- $CH_3\overset{\displaystyle CH_3}{C}HCOCl$ (from i-BuOH as in 9g)

(o) $CH_3\underset{\displaystyle CH_3}{C}HCH_2-\underset{\displaystyle HO}{\overset{\displaystyle CH_3}{C}}-\underset{\displaystyle CH_3}{C}HCH_3 \longleftarrow$
- CH_3MgBr (from MeOH)
- $CH_3\underset{\displaystyle CH_3}{C}HCH_2\vdots\underset{\displaystyle O}{\overset{\|}{C}}-\underset{\displaystyle CH_3}{C}HCH_3 \longleftarrow$
  - $i\text{-Bu}_2CuLi$ (from i-BuOH)
  - $CH_3\underset{\displaystyle CH_3}{C}HCOCl$ (as in 9g)

(p) $O_2N-C_6H_4-\underset{\displaystyle OH}{C}HCOOH \xleftarrow[H^+]{H_2O} O_2N-C_6H_4-\underset{\displaystyle OH}{C}HC{\equiv}N \xleftarrow[H^+]{CN^-} O_2N-C_6H_4-CHO$ (page 662)

(q) $PhCH_2\vdots\underset{\displaystyle OH}{\overset{\displaystyle Ph}{C}}-CH_3 \longleftarrow PhCH_2MgCl + PhCOCH_3$ (page 664)

(r) $Br-C_6H_4-\underset{\displaystyle OH}{\overset{\displaystyle Ph}{C}}\vdots Et \longleftarrow$
- EtMgBr
- $Br-C_6H_4-\underset{\displaystyle O}{\overset{\|}{C}}Ph \xleftarrow{AlCl_3}$
  - $PhBr \xleftarrow{Br_2,\ Fe} C_6H_6$
  - $Ph\underset{\displaystyle O}{\overset{\|}{C}}Cl \xleftarrow{PCl_5} Ph\underset{\displaystyle O}{\overset{\|}{C}}OH$

(s) $CH_3C(CH_3){=}CHCOOH \xleftarrow[-H_2O]{heat} CH_3CH(CH_3)CH(OH)COOH \xleftarrow[H^+]{H_2O} CH_3CH(CH_3)CH(OH)CN \xleftarrow[H^+]{CN^-} CH_3CH(CH_3)CHO$

$CH_3CH(CH_3)CHO \xleftarrow{C_5H_5NHCrO_3Cl} i\text{-BuOH}$

11. (a) $CH_3{-}CH(CH_3){-}^{14}CH_2OH \xleftarrow{i\text{-PrMgBr}} H^{14}CHO \xleftarrow{C_5H_5NHCrO_3Cl} {}^{14}CH_3OH$

(b) $CH_3{-}^{14}CH(CH_3){-}CH_2OH \xleftarrow{HCHO} CH_3{-}^{14}CH(CH_3)MgBr \xleftarrow{Mg} \xleftarrow{PBr_3} CH_3{-}^{14}CH(CH_3)OH$

$CH_3{-}^{14}CH(CH_3)OH \xleftarrow{CH_3MgBr} CH_3{-}^{14}CHO \xleftarrow{K_2Cr_2O_7} CH_3{-}^{14}CH_2OH$

$CH_3{-}^{14}CH_2OH \xleftarrow{CH_3MgBr} H^{14}CHO \xleftarrow{C_5H_5NHCrO_3Cl} {}^{14}CH_3OH$

(c) ${}^{14}CH_3{-}CH(CH_3){-}CH_2OH \xleftarrow{HCHO} {}^{14}CH_3{-}CH(CH_3)MgBr \xleftarrow{Mg} \xleftarrow{PBr_3} {}^{14}CH_3{-}CH(CH_3)OH$

${}^{14}CH_3{-}CH(CH_3)OH \xleftarrow{CH_3CHO} {}^{14}CH_3MgBr \xleftarrow{Mg} \xleftarrow{HBr} {}^{14}CH_3OH$

(d) $CH_3{-}CH{=}^{14}CH_2 \xleftarrow[(alc)]{KOH} CH_3{-}CH_2{-}^{14}CH_2Br \xleftarrow{HBr} CH_3CH_2{-}^{14}CH_2OH$

$CH_3CH_2{-}^{14}CH_2OH \xleftarrow{CH_3CH_2MgBr} H^{14}CHO \xleftarrow{C_5H_5NHCrO_3Cl} {}^{14}CH_3OH$

(e) $CH_3{-}^{14}CH{=}CH_2 \xleftarrow{H^+,\ heat} CH_3{-}^{14}CH(OH){-}CH_3$ (made in b)

(f) ${}^{14}CH_3{-}CH{=}CH_2 \xleftarrow{KOH(alc)} {}^{14}CH_3CH_2CH_2Br \xleftarrow{PBr_3} {}^{14}CH_3CH_2CH_2OH$

${}^{14}CH_3CH_2CH_2OH \xleftarrow{H_2C{-}CH_2\ (\text{epoxide, O})} {}^{14}CH_3MgBr$ (made in c)

(g) $PhD \xleftarrow{D_2O} PhMgBr \xleftarrow{Mg} PhBr \xleftarrow{Br_2,\ Fe}$ benzene

(h) $CH_3CH_2CH(D)^{14}CH_3 \xleftarrow{D_2O} CH_3CH_2CH(MgBr)^{14}CH_3 \xleftarrow{Mg} CH_3CH_2CH(Br)^{14}CH_3$

$CH_3CH_2CH(Br)^{14}CH_3 \xleftarrow{PBr_3} CH_3CH_2CH(OH)^{14}CH_3 \xleftarrow{CH_3CH_2CHO} {}^{14}CH_3MgBr$ (made in c)

12. (a) *Anti*-elimination.

(b)

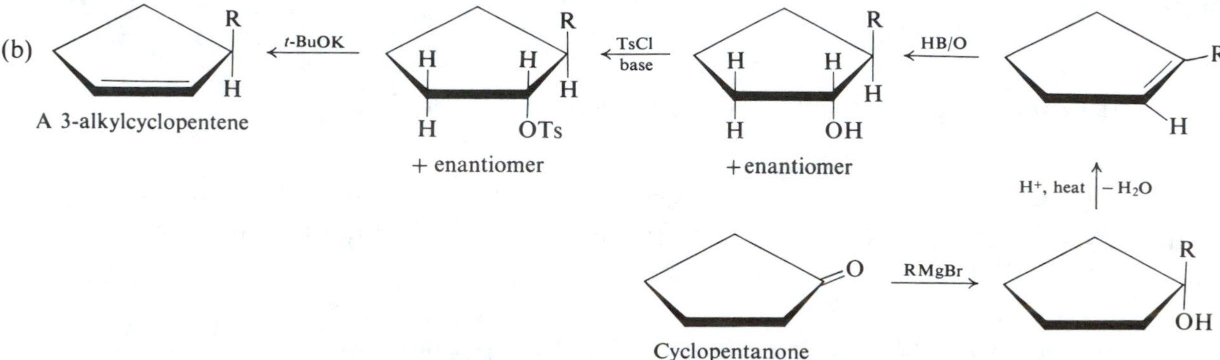

(c) Inversion at C-2 during replacement of —OH by —X would give the *cis*-2-halo-1-alkylcyclopentane, which would undergo *anti*-elimination to yield chiefly 1-alkylcyclopentene, the more stable alkene. But this would simply bring us back to the same alkene that we had subjected to hydroboration–oxidation (HB/O).

1-alkylcyclopentene $\xrightarrow{HB/O}$ *trans*-2-alkylcyclopentanol $\xrightarrow[\text{or HX}]{PX_3}$ *cis*-1-alkyl-2-halocyclopentane (*Chief product*)

Chief product $\xrightarrow[-HX]{\text{base}}$ 1-alkylcyclopentene

13. (a) $C_{11}H_{14}O_2 \xrightarrow{PCl_3} Ph{-}C(CH_3)_2{-}CH_2{-}C(=O)Cl$

A

Loss of HCl but no change in carbon number in going from A to B indicates that the Friedel–Crafts acylation is intramolecular:

A $\xrightarrow{AlCl_3}$ B (3,3-dimethyl-1-indanone) $\xrightarrow{\text{Wolff–Kishner reduction}}$ C (1,1-dimethylindane; protons labelled *a*, *a*, *b*, *c*, *d*)

(b) C was also produced by Friedel–Crafts alkylation.

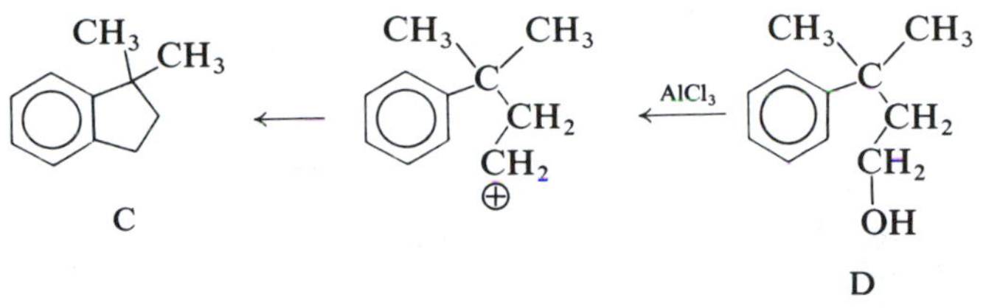

C ⟵ $C_6H_5C(CH_3)_2CH_2CH_2^{\oplus}$ $\xleftarrow{AlCl_3}$ $C_6H_5C(CH_3)_2CH_2CH_2OH$ (D)

14. Compare Sec. 34.14.

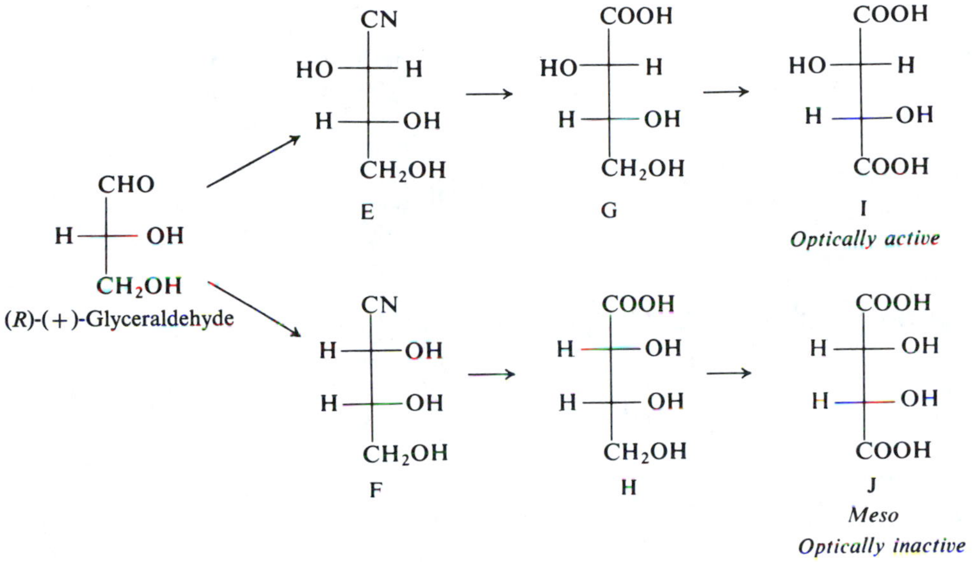

(R)-(+)-Glyceraldehyde (CHO, H–C–OH, CH_2OH) → E (CN; HO–C–H; H–C–OH; CH_2OH) → G (COOH; HO–C–H; H–C–OH; CH_2OH) → I (COOH; HO–C–H; H–C–OH; COOH)

I: *Optically active*

(R)-(+)-Glyceraldehyde → F (CN; H–C–OH; H–C–OH; CH_2OH) → H (COOH; H–C–OH; H–C–OH; CH_2OH) → J (COOH; H–C–OH; H–C–OH; COOH)

J: *Meso*
Optically inactive

15. (a) K (cis-cyclopentane-1,2-diol acetonide: ring C–H, C–H bonded through O, O to $C(CH_3)_2$)

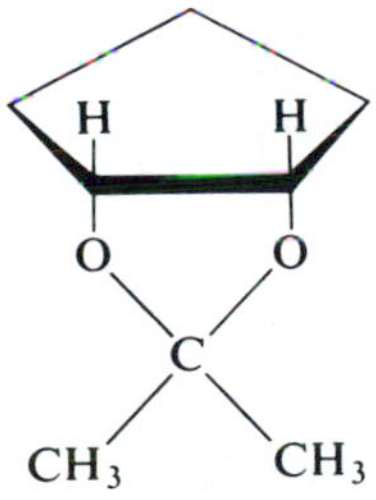

K
A cyclic ketal

(b) The —OH groups in the *trans* isomer are too far apart for the cyclic structure to form.

16.

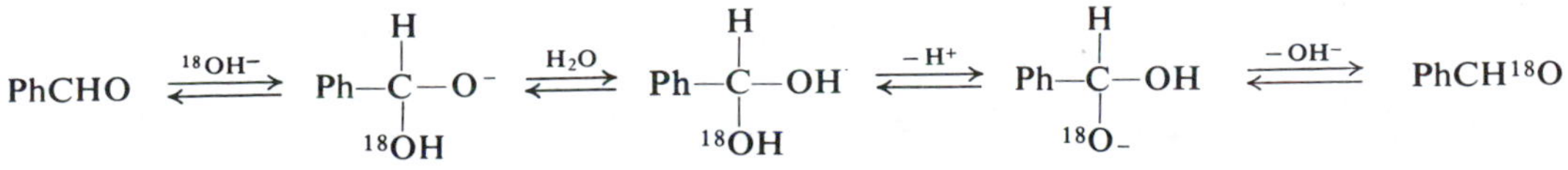

$$PhCHO \underset{}{\overset{^{18}OH^-}{\rightleftharpoons}} Ph{-}\underset{^{18}OH}{\overset{H}{C}}{-}O^- \overset{H_2O}{\rightleftharpoons} Ph{-}\underset{^{18}OH}{\overset{H}{C}}{-}OH \overset{-H^+}{\rightleftharpoons} Ph{-}\underset{^{18}O^-}{\overset{H}{C}}{-}OH \overset{-OH^-}{\rightleftharpoons} PhCH^{18}O$$

17. A proton adds very rapidly to the double bond since this addition gives the comparatively stable oxonium ion VII. Completion of addition then yields the hemiacetal, which is, of course, rapidly hydrolyzed.

$$R-CH=CH-OR' \overset{H^+}{\rightleftharpoons} \underset{\text{VII}}{R-CH_2-CH=\overset{\oplus}{O}R'} \overset{H_2{}^{18}O}{\rightleftharpoons} RCH_2-CH(\overset{\oplus}{{}^{18}OH_2})-OR' \overset{-H^+}{\rightleftharpoons} \underset{\text{Hemiacetal}}{RCH_2-CH({}^{18}OH)-OR'}$$

$$RCH_2-CH({}^{18}OH)-OR' \overset{H^+}{\rightleftharpoons} RCH_2-CH({}^{18}OH)-\overset{\oplus}{O}HR' \overset{-R'OH}{\rightleftharpoons} RCH_2-CH=\overset{\oplus}{{}^{18}O}H \overset{-H^+}{\rightleftharpoons} RCH_2CH^{18}O$$

18. This is an example of electrophilic aromatic substitution, with the leaving group IX a protonated aldehyde. The intermediate *sigma* complex, VIII, is an oxonium ion with most

(1) $CH_3O-C_6H_4-CHOH-C_6H_4-G + Br_2 \rightleftharpoons \underset{\text{VIII}}{CH_3\overset{+}{O}=C_6H_4(Br)(CHOH-C_6H_4-G)} + Br^-$

(2) $CH_3\overset{+}{O}=C_6H_4(Br)(CHOH-C_6H_4-G) \longrightarrow CH_3O-C_6H_4-Br + \underset{\text{IX}}{H\overset{+}{O}=CH-C_6H_4-G}$

$$\xrightarrow{-H^+} OHC-C_6H_4-G$$

of the charge located on the *p*-methoxy group. Formation (1) of VIII is reversible; addition of Br^- speeds up reversal of (1) and raises the fraction of VIII that reverts to starting material. (That is, $k_{-1}[Br^-]$ is increased relative to k_2.) Electron release by G stabilizes the leaving group IX and the transition state leading to its formation; k_2 is thus increased relative to $k_{-1}[Br^-]$, and the rate of formation of products is speeded up.

19. The initial product of the Grignard reaction is, of course, the magnesium salt of the alcohol, benzhydrol. This is oxidized to the ketone benzophenone, presumably by a hydride transfer—to the excess benzaldehyde—which is aided by the negative charge on alkoxide oxygen.

$$PhMgBr + PhCHO \longrightarrow Ph-CH(Ph)-O^-(MgBr)^+$$

$$\underset{\substack{\text{Anion of}\\\text{benzhydrol}}}{Ph-CH(Ph)-O^-} + \underset{\substack{\text{Benzaldehyde}\\\textit{Excess}}}{Ph-CH=O} \longrightarrow \underset{\text{Benzophenone}}{Ph-C(Ph)=O} + Ph-CH_2-O^- \xrightarrow{H_2O} \underset{\text{Benzyl alcohol}}{PhCH_2OH}$$

Although an unwanted side reaction here, hydride transfer from an alkoxide to a carbonyl group is the basis of the *Oppenauer oxidation* (Djerassi, C. *Organic Reactions*; Wiley: New York; Vol. VI, Chapter 5),

$$R_2CHOH + CH_3COCH_3 \xrightleftharpoons{t\text{-BuOK}} R_2C{=}O + CH_3CHOHCH_3$$

and of its reverse, the *Meerwein–Ponndorf–Verley* reduction (Wilds, A. L. *Organic Reactions*; Wiley: New York; Vol. II, Chapter 5),

$$R_2C{=}O + CH_3CHOHCH_3 \xrightleftharpoons{Al(OPr\text{-}i)_3} R_2CHOH + CH_3COCH_3 \uparrow$$

both very mild, selective synthetic reactions.

20. (a) A THP ether is an acetal: a cyclic acetal in which one of the parent alcohol molecules is part of the ring. Like any acetal, in the presence of aqueous acid it rapidly undergoes the reverse of reactions (1) and (2) on page 681.

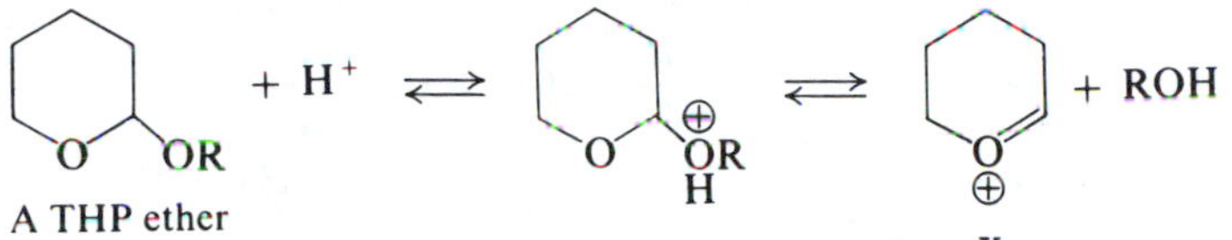

A THP ether (OR) + H^+ ⇄ (protonated, $\overset{\oplus}{O}R$ H) ⇄ X + ROH

A THP ether
An acetal

X
An oxonium ion

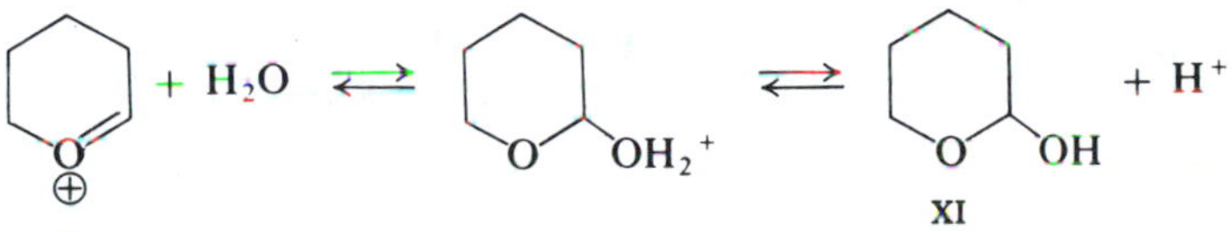

X + H_2O ⇄ (OH_2^+) ⇄ XI (OH) + H^+

X

XI
A hemiacetal

This gives the hemiacetal XI which, like other hemiacetals, rapidly undergoes hydrolysis to yield the aldehyde and the alcohol—which happen to be part of the same molecule.

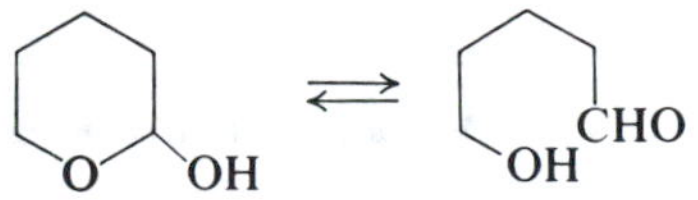

XI ⇄ (CHO, OH)

XI
A hemiacetal

Aldehyde and alcohol

(We note that the key intermediate—the one whose stability is responsible for the facility with which reaction in both directions takes place—is oxonium ion X, which corresponds to oxonium ion I on page 681.)

(b) 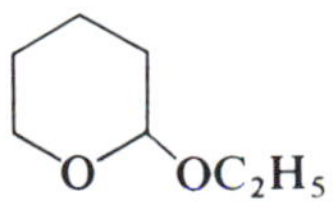 (ring, O, OC_2H_5) $\xrightarrow[H^+]{H_2O}$ $HOCH_2CH_2CH_2CH_2CHO + C_2H_5OH$

21. The protonated aldehyde is an electrophile, the double bond is a nucleophile; reaction between the two functions closes the ring; hydration of the resulting carbocation produces a diol.

H^+ + [aldehyde] ⇄ [protonated aldehyde] $\xrightarrow{\text{ring closure}}$ [3° cation] $\xrightarrow{+H_2O,\ -H^+}$ [diol]

Protonated aldehyde 3° cation *A six-membered ring* Diol

22. We note application of our general principles of page 53 of this Study Guide: (1) since no bond is broken to the original chiral center, configuration is maintained about that center; and (2) each time a new chiral center is generated, *both* possible configurations about that center result, and mixtures of diastereomers are obtained.

CH_2Br, H—C—CH_3, *i*-Bu ((*R*)-) $\xrightarrow{Mg}$ CH_2MgBr, H—C—CH_3, *i*-Bu (L) $\xrightarrow{i\text{-BuCHO}}$ *i*-Bu, H—C—OH, CH_2, H—C—CH_3, *i*-Bu and *i*-Bu, HO—C—H, CH_2, H—C—CH_3, *i*-Bu

Mixture of diastereomers

M

M $\xrightarrow{CrO_3}$ *i*-Bu, C=O, CH_2, H—C—CH_3, *i*-Bu (N) $\xrightarrow{CH_3MgBr}$ $\xrightarrow{H_2O}$ *i*-Bu, CH_3—C—OH, CH_2, H—C—CH_3, *i*-Bu and *i*-Bu, HO—C—CH_3, CH_2, H—C—CH_3, *i*-Bu

Mixture of diastereomers

O

O $\xrightarrow[-H_2O]{I_2,\ \text{heat}}$ *i*-Bu—C—CH_3, ‖ CH, H—C—CH_3, *i*-Bu (*Mixture of diastereomers*) and *i*-Pr—C—H, ‖ CCH_3, CH_2, H—C—CH_3, *i*-Bu (*Mixture of diastereomers*)

P

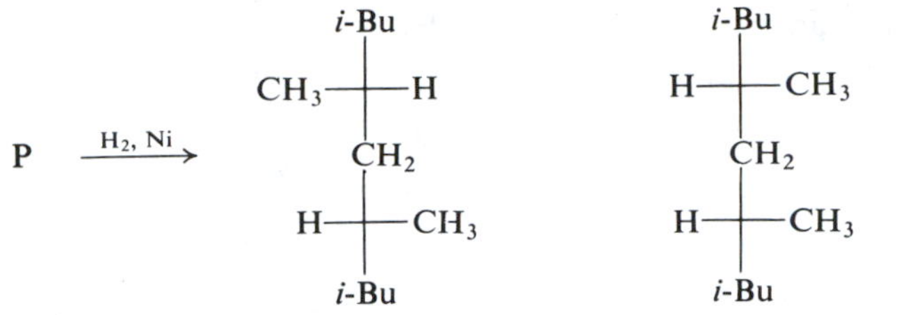

P $\xrightarrow{H_2,\ Ni}$

Q: *i*-Bu / CH_3—C—H / CH_2 / H—C—CH_3 / *i*-Bu
Optically active
Q
(4*R*,6*R*)-

R: *i*-Bu / H—C—CH_3 / CH_2 / H—C—CH_3 / *i*-Bu
Optically inactive (*meso*)
R
(4*R*,6*S*)- or (4*S*,6*R*)-

2,4,6,8-Tetramethylnonane

23. (a) Tollens' reagent
(b) Tollens' reagent
(c) 2,4-dinitrophenylhydrazine
(d) iodoform test
(e) Schiff or Tollens' test
(f) Tollens' test
(g) acid, then Schiff or Tollens' test
(h) iodoform test
(i) 2,4-dinitrophenylhydrazine, or CrO_3/H_2SO_4
(j) acid, then Schiff or Tollens' test
(k) acid and heat, then Schiff or Tollens' test

24. The Tollens' test shows citral to be aldehyde; cleavage into three fragments by oxidation indicates two double bonds. Putting the fragments together according to the isoprene rule,

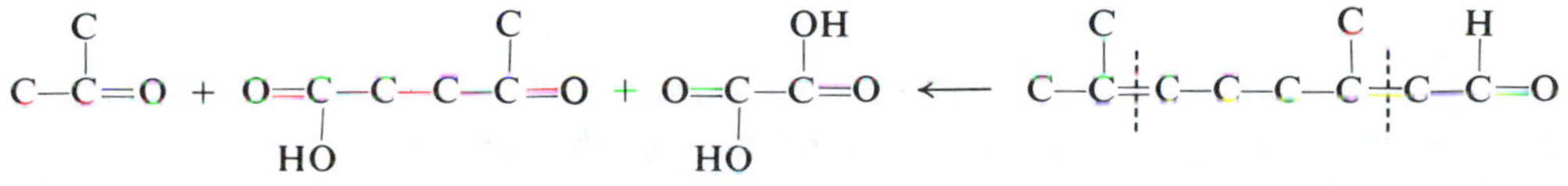

$$(C)_2C{=}O + O{=}C(HO)—C—C—C(C){=}O + O{=}C(HO)—C(OH){=}O \longleftarrow C—C(C)\,\vdots\,C—C—C—C(C)\,\vdots\,C—C(H){=}O$$

we arrive at:

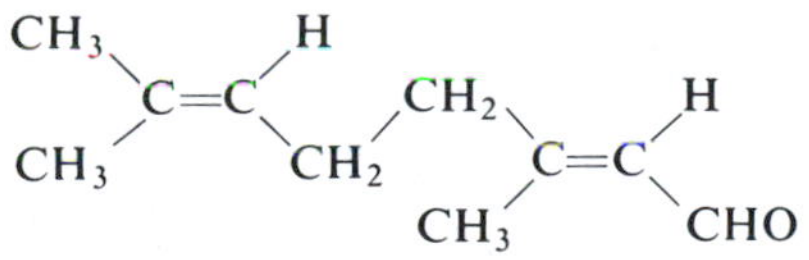

$(CH_3)_2C{=}CH—CH_2—CH_2—C(CH_3){=}CH—CHO$ (H and CHO on second double bond: CHO *trans* to CH_2)
Citral *a*
E

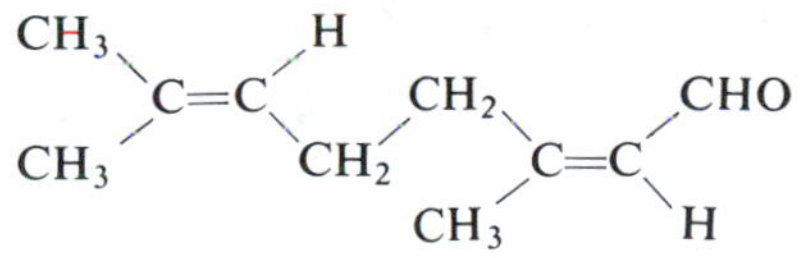

$(CH_3)_2C{=}CH—CH_2—CH_2—C(CH_3){=}CH—CHO$ (CHO *cis* to CH_2)
Citral *b*
Z

Citral *a*, like geraniol, has H and CH_3 *trans*; citral *b*, like nerol, has H and CH_3 *cis*.

25. The compound is evidently unsaturated ($KMnO_4$ positive) and a ketone (NH_2OH positive, Tollens' negative). The first hydrogenation gives a saturated ketone ($KMnO_4$ negative, NH_2OH positive), the second leads to an alcohol ($KMnO_4$ negative, NH_2OH negative, CrO_3/H_2SO_4 positive). The formula of the alcohol shows the presence of one ring.

	$C_{10}H_{22}O$	satd. open-chain alcohol
−	$C_{10}H_{20}O$	carvomenthol
	2H	missing means 1 ring

Fitting together the oxidation fragments, with an eye on the isoprene rule,

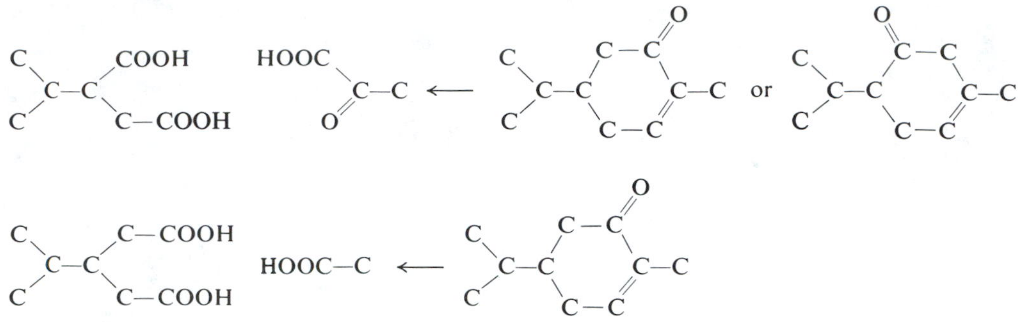

we conclude that carvotanacetone is

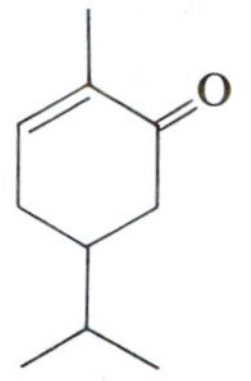

Carvotanacetone

26. The pinacol rearrangement is believed to involve two important steps: (1) loss of water from the protonated diol to form a carbocation; and (2) rearrangement of the carbocation by a 1,2-shift to yield the protonated ketone. Each step involves two reactions, one of which is a reversible protonation–deprotonation.

(1) 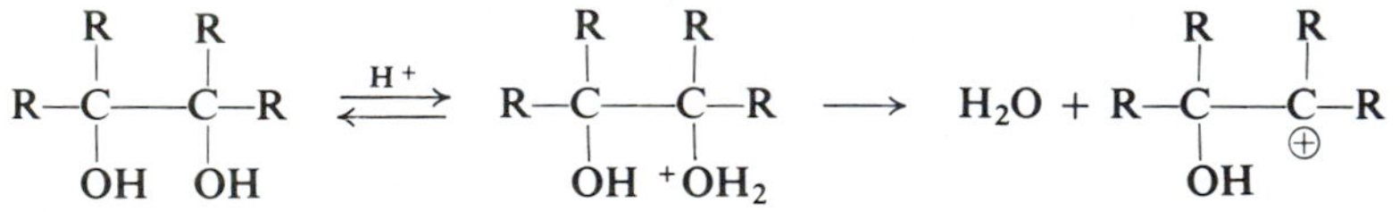 $R_2C(OH)-C(OH)R_2 \underset{}{\overset{H^+}{\rightleftharpoons}} R_2C(OH)-C(^+OH_2)R_2 \longrightarrow H_2O + R_2C(OH)-\overset{\oplus}{C}R_2$

(2) $R_2C(OH)-\overset{\oplus}{C}R_2 \longrightarrow R-C(=\overset{\oplus}{O}H)-CR_3 \rightleftharpoons H^+ + R-C(=O)-CR_3$

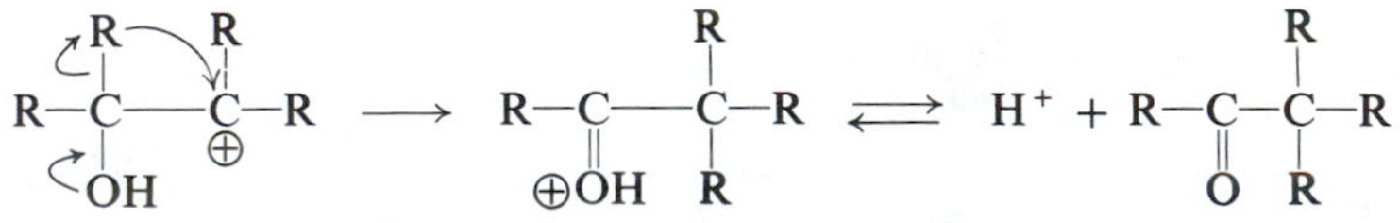

Both steps in this reaction are already familiar to us: formation of a carbocation from an alcohol under the influence of acid, followed by a 1,2-shift to the electron-deficient atom. The pattern is also familiar: rearrangement of a cation to a more stable cation, in this case to the protonated ketone. The driving force is the usual one behind carbocation reactions: the need to provide the electron-deficient carbon with electrons. The special feature of the pinacol rearrangement is the presence in the molecule of the second oxygen atom; it is this oxygen atom, with its unshared pairs, that ultimately provides the needed electrons.

27. $n\text{-}C_5H_{11}C{\equiv}CH \xrightarrow{LiNH_2} n\text{-}C_5H_{11}C{\equiv}CLi$ (S) $\xrightarrow{BrCH_2CH_2CH_2Cl} n\text{-}C_5H_{11}C{\equiv}CCH_2CH_2CH_2Cl$ (T)

T $\xrightarrow{Mg} n\text{-}C_5H_{11}C{\equiv}CCH_2CH_2CH_2MgCl \xrightarrow{n\text{-}C_{10}H_{21}CHO} \xrightarrow{H^+} n\text{-}C_5H_{11}C{\equiv}CCH_2CH_2CH_2CH(OH)C_{10}H_{21}\text{-}n$ (U)

U $\xrightarrow[\text{Lindlar}]{H_2}$ (*n*-C_5H_{11})(H)C=C(H)($CH_2CH_2CH_2CH(OH)C_{10}H_{21}$-*n*) (V) $\xrightarrow{CrO_3}$ (*n*-C_5H_{11})(H)C=C(H)($CH_2CH_2CH_2C(=O)C_{10}H_{21}$-*n*)

(*Z*)-6-Henicosen-11-one

Sex attractant of Douglas-fir tussock moth

28. (a) $HO(CH_2)_8OH \xrightarrow{HBr} Br(CH_2)_8OH$ (W) $\xrightarrow[H^+]{DHP} Br(CH_2)_8OTHP$ (X) $\xrightarrow{HC{\equiv}CLi} HC{\equiv}C(CH_2)_8OTHP$ (Y)

Y $\xrightarrow{LiNH_2} LiC{\equiv}C(CH_2)_8OTHP \xrightarrow{C_2H_5Br} C_2H_5C{\equiv}C(CH_2)_8OTHP$ (Z)

Z $\xrightarrow{H_2O,\ H^+} C_2H_5C{\equiv}C(CH_2)_8OH$ (AA) $\xrightarrow{CH_3COCl} C_2H_5C{\equiv}C(CH_2)_8OCOCH_3$ (BB) $\xrightarrow[\text{Lindlar}]{H_2}$ (C_2H_5)(H)C=C(H)($(CH_2)_8OCOCH_3$)

(*Z*)-9-Dodecen-1-yl acetate

Sex attractant of grape berry moth (page 382)

(b) At the hydrogenation stage, treat BB with Na in liquid NH_3, and thus obtain the (*E*) isomer.

29. (a) $n\text{-}C_4H_9C{\equiv}CH \xrightarrow{n\text{-BuLi}} n\text{-}C_4H_9C{\equiv}CLi \xrightarrow{CH_2\text{–}CH_2\ (\text{epoxide, }O)} \xrightarrow{H^+} n\text{-}C_4H_9C{\equiv}CCH_2CH_2OH$ (CC)

CC $\xrightarrow[\text{Lindlar}]{H_2}$ (*n*-C_4H_9)(H)C=C(H)(CH_2CH_2OH) (DD) $\xrightarrow{PBr_3}$ (*n*-C_4H_9)(H)C=C(H)(CH_2CH_2Br) (EE)

$Br(CH_2)_6OH \xrightarrow[H^+]{DHP} Br(CH_2)_6OTHP$ (FF) $\xrightarrow{HC{\equiv}CLi} HC{\equiv}C(CH_2)_6OTHP$ (GG)

GG $\xrightarrow{n\text{-BuLi}} LiC{\equiv}C(CH_2)_6OTHP \xrightarrow{EE}$ (*n*-C_4H_9)(H)C=C(H)($CH_2CH_2C{\equiv}C(CH_2)_6OTHP$) (HH)

HH $\xrightarrow[\text{Lindlar}]{H_2}$ (*n*-C_4H_9)(H)C=C(H)–CH_2CH(H)=C(H)($(CH_2)_6OTHP$) (II)

II $\xrightarrow[H^+]{H_2O}$ (Z,Z)-n-C_4H_9CH=CHCH_2CH_2CH=CH$(CH_2)_6OH$ (JJ) $\xrightarrow{CH_3COCl}$ (Z,Z)-n-C_4H_9CH=CHCH_2CH_2CH=CH$(CH_2)_6OCOCH_3$

JJ

(7Z, 11Z)-Hexadecadien-1-yl acetate

Sex attractant of pink bollworm moth

Gossyplure

(b) Use Na and liquid NH_3 instead of the Lindlar catalyst for either or both of the hydrogenations.

30. 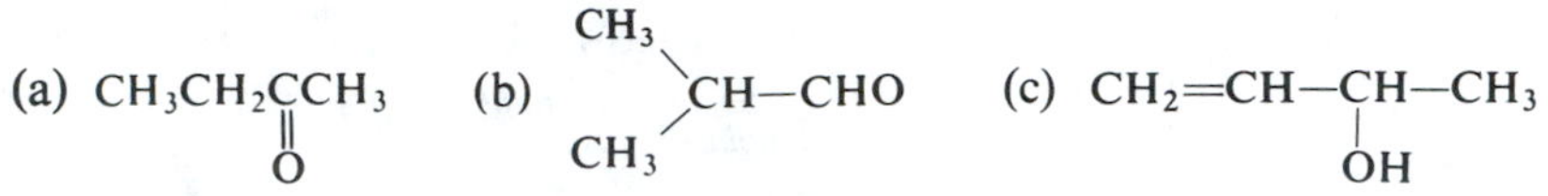 (a) $CH_3CH_2CCH_3$ (C=O) — 2-Butanone (b) $(CH_3)_2CH{-}CHO$ — Isobutyraldehyde (c) $CH_2{=}CH{-}CH(OH){-}CH_3$ — 3-Buten-2-ol

(See the labeled spectra on page 699 of this Study Guide.)

31. 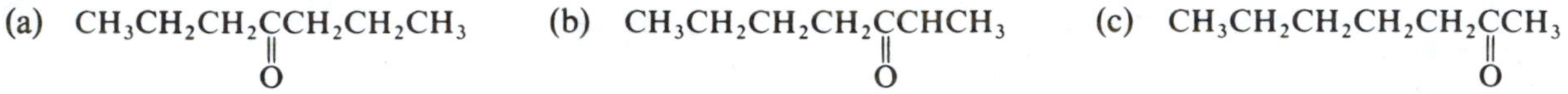 (a) $CH_3CH_2CH_2CCH_2CH_2CH_3$ (C=O) — 4-Heptanone (b) $CH_3CH_2CH_2CH_2CCHCH_3$ (C=O) — 3-Heptanone (c) $CH_3CH_2CH_2CH_2CH_2CCH_3$ (C=O) — 2-Heptanone

(See the labeled spectra on page 700 of this Study Guide.)

32. 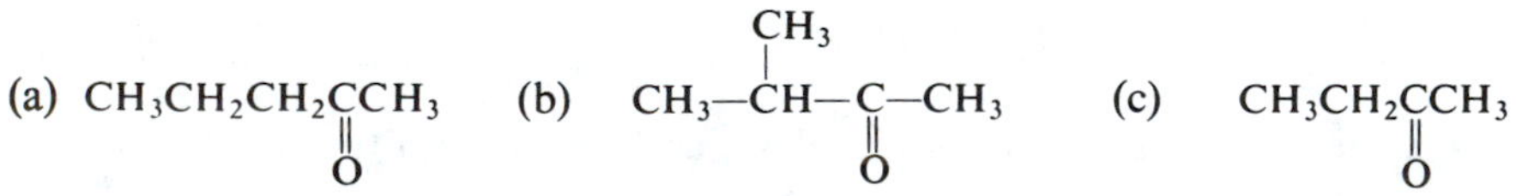 (a) $CH_3CH_2CH_2CCH_3$ (C=O) — 2-Pentanone (b) $CH_3{-}CH(CH_3){-}C(=O){-}CH_3$ — Isopropyl methyl ketone (c) $CH_3CH_2CCH_3$ (C=O) — Ethyl methyl ketone

(See the labeled spectra on page 701 of this Study Guide.)

33. 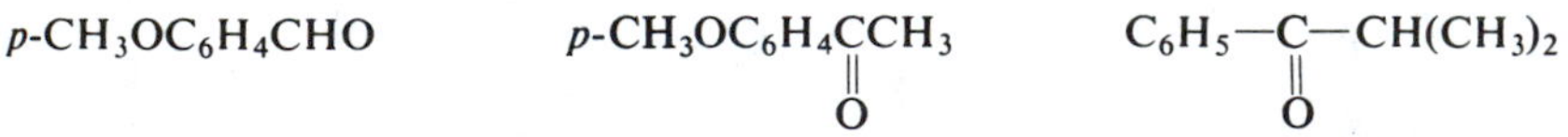 p-$CH_3OC_6H_4CHO$ — p-Methoxybenzaldehyde, KK

p-$CH_3OC_6H_4CCH_3$ (C=O) — p-Methoxyacetophenone, LL

$C_6H_5{-}C(=O){-}CH(CH_3)_2$ — Isobutyrophenone (Isopropyl phenyl ketone), MM

(See the labeled spectra on pages 702 and 703 of this Study Guide.)

34. 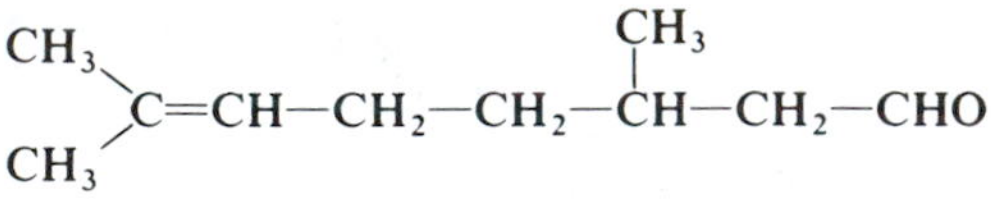 $(CH_3)_2C{=}CH{-}CH_2{-}CH_2{-}CH(CH_3){-}CH_2{-}CHO$

3,7-Dimethyl-6-octenal

Citronellal

NN

(See the labeled spectra on page 704 of this Study Guide.)

19

Carboxylic Acids

19.1 At 110 °C: $$\frac{0.11\ \text{g}}{\text{m.w. (g/mol)}} = \frac{\text{vol. at STP}}{22400\ \text{mL/mol}} = \frac{63.7 \times \frac{273}{383} \times \frac{454}{760}}{22400}$$

$$\text{m.w.} = \frac{0.11 \times 22400}{63.7} \times \frac{383}{273} \times \frac{760}{454} = 91 \text{ (at 110 °C)}$$

At 156 °C: $$\frac{0.081}{\text{m.w.}} = \frac{66.4 \times \frac{273}{429} \times \frac{458}{760}}{22400}$$

$$\text{m.w.} = \frac{0.081 \times 22400}{66.4} \times \frac{429}{273} \times \frac{760}{458} = 71 \text{ (at 156 °C)}$$

The monomer, CH_3COOH, would have m.w. 60. Association occurs even in the vapor phase, decreasing as the temperature increases.

19.2 (a) (1) $$\text{peroxide} \longrightarrow \text{Rad}\cdot$$

(2) $$\text{Rad}\cdot + CH_3CH_2CH_2COOH \longrightarrow \text{RadH} + CH_3CH_2\underset{\cdot}{C}HCOOH$$

(3) $$n\text{-BuCH}{=}CH_2 + CH_3CH_2\underset{\cdot}{C}HCOOH \longrightarrow n\text{-Bu}\underset{\cdot}{C}H{-}CH_2{-}\underset{\displaystyle C_2H_5}{\underset{|}{C}H}{-}COOH$$

(4) $$n\text{-Bu}\underset{\cdot}{C}HCH_2\underset{\displaystyle C_2H_5}{\underset{|}{C}H}COOH + CH_3CH_2CH_2COOH \longrightarrow n\text{-BuCH}_2CH_2\underset{\displaystyle C_2H_5}{\underset{|}{C}H}COOH + CH_3CH_2\underset{\cdot}{C}HCOOH$$

then (3), (4), (3), (4), etc.

Free-radical attack favors abstraction of an *alpha*-hydrogen, presumably because of resonance stabilization of the resulting free radical:

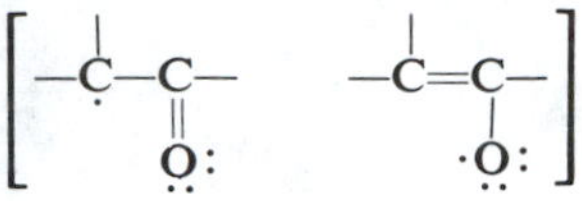

$$\left[-\overset{|}{\dot{C}}-\underset{\underset{\ddot{O}:}{\|}}{C}- \quad -\overset{|}{C}=\underset{\underset{\cdot\ddot{O}:}{|}}{C}- \right]$$

(b) 2-Methyldecanoic acid, $CH_3(CH_2)_7\underset{\underset{CH_3}{|}}{C}HCOOH$

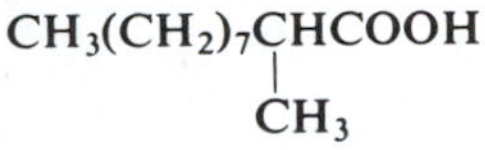

(c) 2,2-Dimethyldodecanoic acid, $CH_3(CH_2)_9\overset{\overset{CH_3}{|}}{\underset{\underset{CH_3}{|}}{C}}COOH$

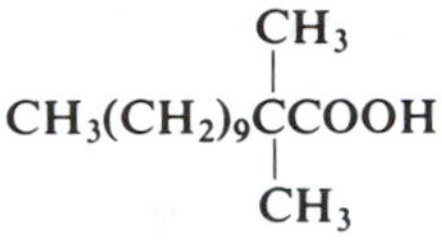

(d) Ethyl *n*-octylmalonate, $CH_3(CH_2)_7\overset{\overset{COOEt}{|}}{\underset{\underset{COOEt}{|}}{C}}H$

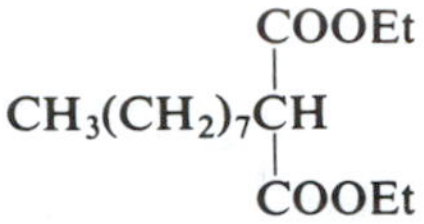

(Petrow, A. D.; Nikischin, G. I.; Ogibin, J. N. "Freiradikale Anlagerung von Säuren und Alkoholen zu den α-Olefinen"; *Third International Congress on Surface-Active Agents, Cologne, September 1960*; Vol. 1, Sec. A, pp 78–83.)

19.3 (a) Clearly, the *tert*-pentyl cation is an intermediate:

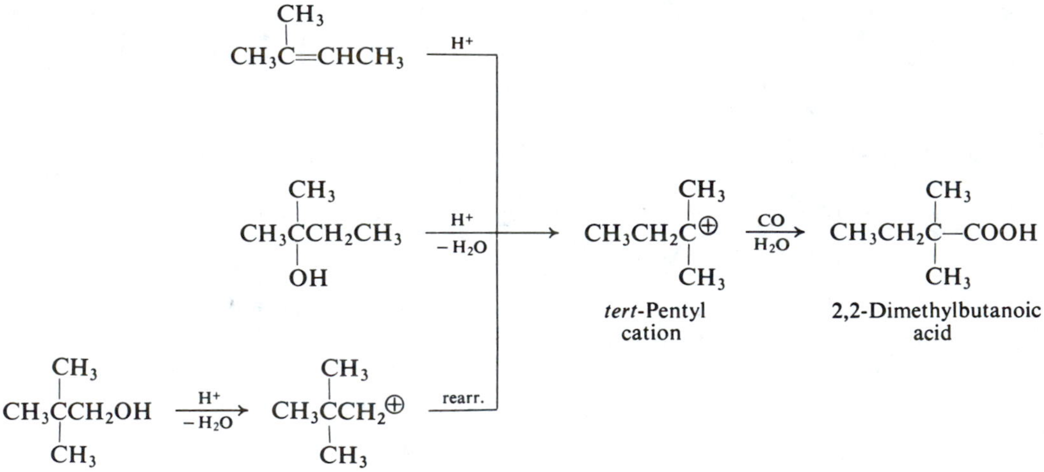

$$CH_3\overset{\overset{CH_3}{|}}{C}=CHCH_3 \xrightarrow{H^+}$$

$$CH_3\overset{\overset{CH_3}{|}}{\underset{\underset{OH}{|}}{C}}CH_2CH_3 \xrightarrow[-H_2O]{H^+} CH_3CH_2\overset{\overset{CH_3}{|}}{\underset{\underset{CH_3}{|}}{C}}\oplus \xrightarrow[H_2O]{CO} CH_3CH_2\overset{\overset{CH_3}{|}}{\underset{\underset{CH_3}{|}}{C}}-COOH$$

tert-Pentyl cation 2,2-Dimethylbutanoic acid

$$CH_3\overset{\overset{CH_3}{|}}{\underset{\underset{CH_3}{|}}{C}}CH_2OH \xrightarrow[-H_2O]{H^+} CH_3\overset{\overset{CH_3}{|}}{\underset{\underset{CH_3}{|}}{C}}CH_2\oplus \xrightarrow{\text{rearr.}}$$

The most likely mechanism of its reaction with carbon monoxide involves the formation and hydration of an *acylium ion*, a cation that owes its considerable stability to the fact

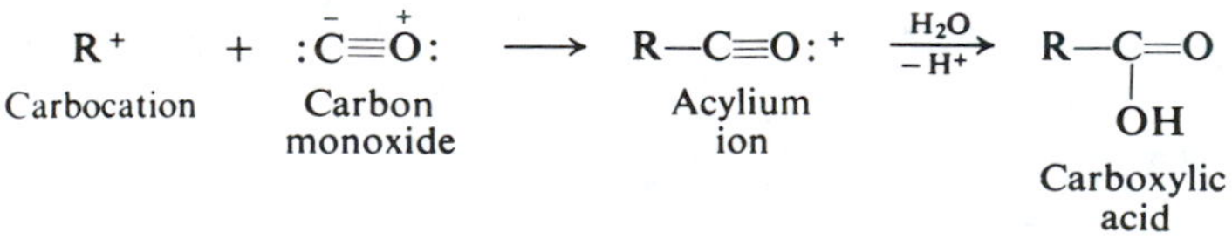

$$\underset{\text{Carbocation}}{R^+} + \underset{\text{Carbon monoxide}}{:\overset{-}{C}\equiv\overset{+}{O}:} \longrightarrow \underset{\text{Acylium ion}}{R-C\equiv O:^+} \xrightarrow[-H^+]{H_2O} \underset{\text{Carboxylic acid}}{R-\underset{\underset{OH}{|}}{C}=O}$$

that in it every atom has a complete octet (compare Sec. 15.18).

(b) 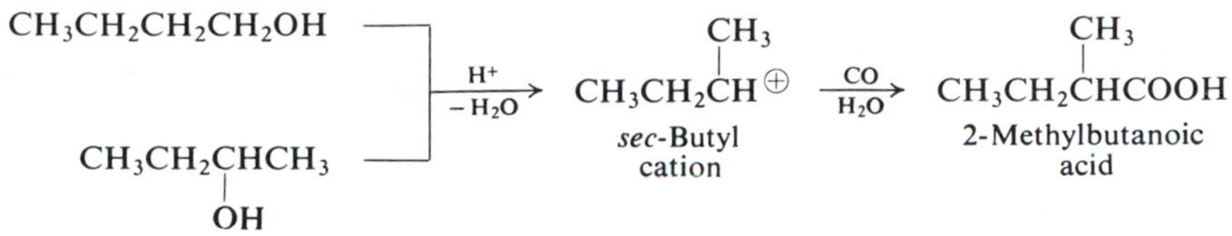 $CH_3CH_2CH_2CH_2OH$ and $CH_3CH_2CH(OH)CH_3$ $\xrightarrow[-H_2O]{H^+}$ $CH_3CH_2\overset{CH_3}{\overset{|}{C}}H^{\oplus}$ (*sec*-Butyl cation) $\xrightarrow[H_2O]{CO}$ $CH_3CH_2\overset{CH_3}{\overset{|}{C}}HCOOH$ (2-Methylbutanoic acid)

(Koch, H.; Haaf, W. "Über die Synthese verzweigter Carbonsäuren nach der Ameisensäure-Methode"; *Ann. Chem.* **1958**, *618*, 251.)

19.4 (a) 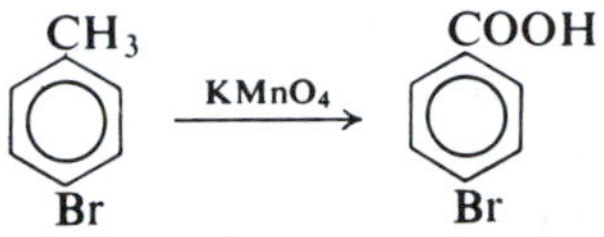 p-$BrC_6H_4CH_3$ $\xrightarrow{KMnO_4}$ p-BrC_6H_4COOH

p-Bromobenzoic acid

(b) 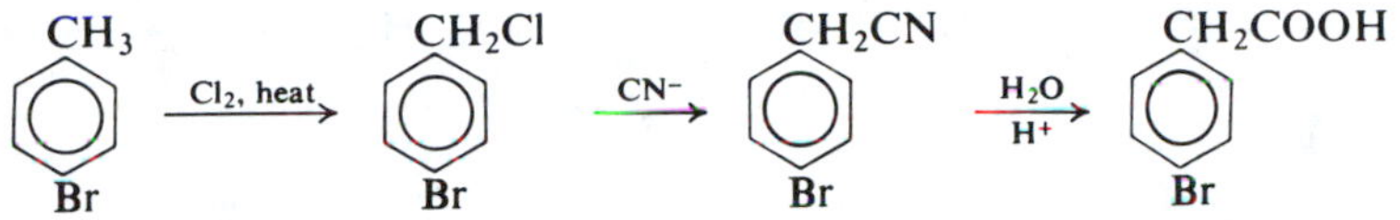 p-$BrC_6H_4CH_3$ $\xrightarrow{Cl_2,\ heat}$ p-$BrC_6H_4CH_2Cl$ $\xrightarrow{CN^-}$ p-$BrC_6H_4CH_2CN$ $\xrightarrow[H^+]{H_2O}$ p-$BrC_6H_4CH_2COOH$

p-Bromophenylacetic acid

19.5 The carbonate ion is a hybrid of three structures of equal stability. Each C—O bond is double in one structure and single in two; it has less double-bond character, and hence is longer, than the bonds in formate ion.

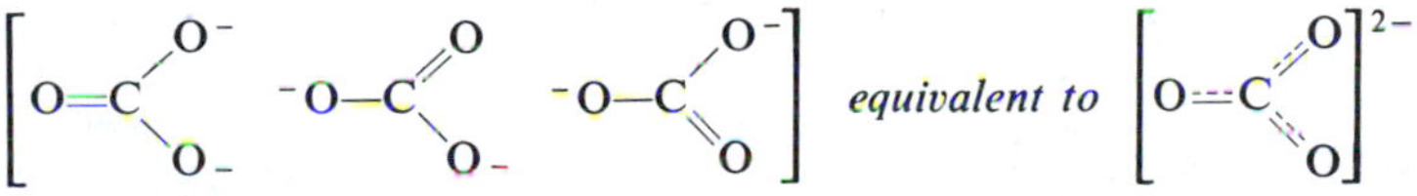

$$\left[O{=}C(O^-)O^- \quad {}^-O{-}C({=}O)O^- \quad {}^-O{-}C(O^-){=}O \right] \quad \textit{equivalent to} \quad \left[O\text{=:=}C(\text{=:=}O)\text{=:=}O \right]^{2-}$$

19.6 See Sec. 21.1.

19.7 (a) F > Cl > Br > I.

(b) Electron-withdrawing.

19.8 $^+OCH_3{=}C_6H_4{=}C(O^-)OH$ (quinoid structure)

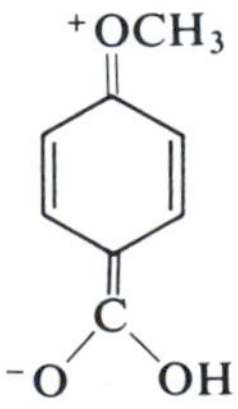

For the anion, such a structure would involve separation of positive and *double* negative charges.

19.9 (a) The C—OH bond is broken:

$$R-C(=O)\text{-}\!\!\!\!\times\!\!\!\!\text{-}OH \longrightarrow R-C(=O)-Cl$$

(b) The C—OH bond of the acid and the CO—H bond of the alcohol are broken:

$$Ph-C(=O)\text{-}\!\!\!\!\times\!\!\!\!\text{-}OH + H\vdots{}^{18}O-CH_3 \longrightarrow Ph-C(=O)-{}^{18}OCH_3 + HOH$$

19.10 (a) $n\text{-}C_{11}H_{23}CH_2Br \xleftarrow{PBr_3} n\text{-}C_{11}H_{23}CH_2OH \xleftarrow{LiAlH_4} n\text{-}C_{11}H_{23}COOH$

(b) $n\text{-}C_{11}H_{23}CH_2COOH \xleftarrow{H^+} \xleftarrow{CO_2} n\text{-}C_{11}H_{23}CH_2MgBr \xleftarrow{Mg} n\text{-}C_{11}H_{23}CH_2Br \longleftarrow$ (a)

Or via nitrile: $RBr \longrightarrow RCN \longrightarrow RCOOH$

(c) $n\text{-}C_{11}H_{23}CH_2CH_2CH_2OH \xleftarrow{H_2C\!-\!CH_2\ (\text{epoxide, }O)} n\text{-}C_{11}H_{23}CH_2MgBr \longleftarrow$ (b)

(d) $n\text{-}C_{10}H_{21}CH{=}CH_2 \xleftarrow{KOH(alc)} n\text{-}C_{11}H_{23}CH_2Br \longleftarrow$ (a)

(e) $n\text{-}C_{10}H_{21}CH_2CH_3 \xleftarrow{H_2,\ Pt} n\text{-}C_{10}H_{21}CH{=}CH_2 \longleftarrow$ (d)

(f) $n\text{-}C_{10}H_{21}C{\equiv}CH \xleftarrow{NaNH_2} \xleftarrow{KOH} n\text{-}C_{10}H_{21}CH(Br)-CH_2(Br) \xleftarrow{Br_2,\ CCl_4} n\text{-}C_{10}H_{21}CH{=}CH_2 \longleftarrow$ (d)

(g) $n\text{-}C_{10}H_{21}C(=O)-CH_3 \xleftarrow[\text{(See Sec. 12.10)}]{H_2O,\ H_2SO_4,\ Hg^{2+}} n\text{-}C_{10}H_{21}C{\equiv}CH \longleftarrow$ (f)

(h) $n\text{-}C_{10}H_{21}CHOH-CH_3 \xleftarrow{H_2,\ Pt} n\text{-}C_{10}H_{21}C(=O)-CH_3 \longleftarrow$ (g)

(i) $n\text{-}C_{10}H_{21}COOH \xleftarrow{H^+} \xleftarrow{OI^-} n\text{-}C_{10}H_{21}C(=O)-CH_3 \longleftarrow$ (g) (Sec. 18.21)

(j) $n\text{-}C_{12}H_{25}CHOH-CH_3 \xleftarrow{H^+} \xleftarrow{CH_3CHO} n\text{-}C_{12}H_{25}MgBr \xleftarrow{Mg} n\text{-}C_{12}H_{25}Br \longleftarrow$ (a)

(k) $n\text{-}C_{12}H_{25}C(CH_3)(OH)-CH_3 \xleftarrow{H^+} \xleftarrow{CH_3COCH_3} n\text{-}C_{12}H_{25}MgBr \xleftarrow{Mg} n\text{-}C_{12}H_{25}Br \longleftarrow$ (a)

19.11 (a) CH_3CH_2COOH.

(b)

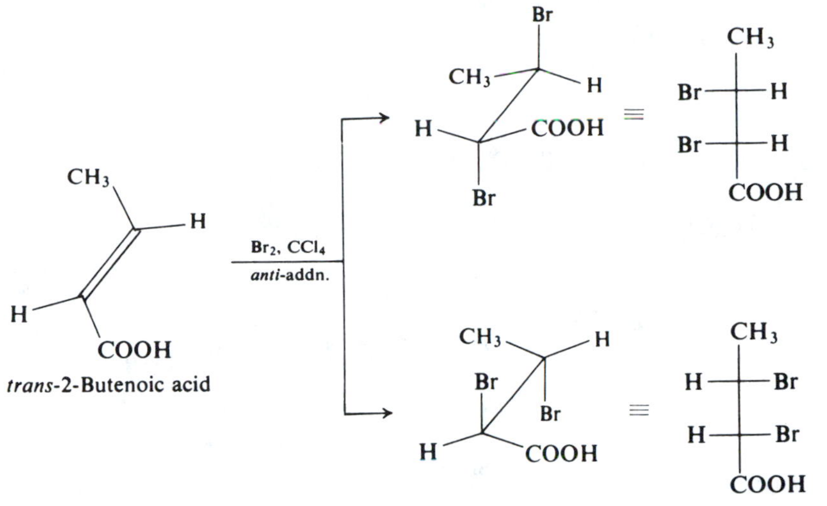

Enantiomers
racemic *erythro*-2,3-Dibromobutanoic acid

(c) $PhCHOHCH_2COOH \xrightarrow[-H_2O]{H^+}$ 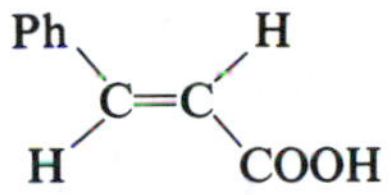 (chiefly *trans*)

(d) The only way H_2O can be lost ($C_8H_8O_3 - H_2O = C_8H_6O_2$) within a single molecule (no increase in carbon number) is through intramolecular esterification to give a cyclic ester (a *lactone*, Sec. 20.15).

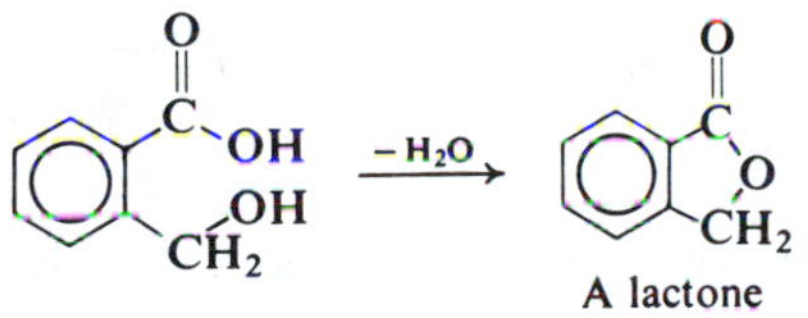

A lactone
A cyclic ester

19.12 To prevent generation of HCN by the action of the acid on sodium cyanide.

19.13 (a) We need to add two carbons:

$$HOOC(CH_2)_3COOH \xleftarrow[\text{heat}]{H_2O,\ H^+} N{\equiv}C(CH_2)_3C{\equiv}N \xleftarrow{2CN^-} Br(CH_2)_3Br \xleftarrow[\text{excess}]{2HBr} HOCH_2CH_2CH_2OH$$

(b) We need to break a C_{18}-acid into a C_9-fragment:

$$CH_3(CH_2)_7CH{=}CH(CH_2)_7COOH \xrightarrow{O_3} \xrightarrow{H_2O,\ H^+} HOOC(CH_2)_7COOH$$

(c) $HOCH_2C{\equiv}CCH_2OH \xrightarrow{H_2,\ Pt} HOCH_2CH_2CH_2CH_2OH \xrightarrow{KMnO_4} HOOCCH_2CH_2COOH$

19.14 (a) Diester, $EtOOC(CH_2)_4COOEt$.

(b) Monoester, $EtOOC(CH_2)_4COOH$.

(c) Diol, $HOCH_2CH_2CH_2CH_2OH$.

(d) Monobromo acid, $HOOCCH_2CH_2CHBrCOOH$.

(e) Diacyl chloride, p-$C_6H_4(COCl)_2$.

(f)

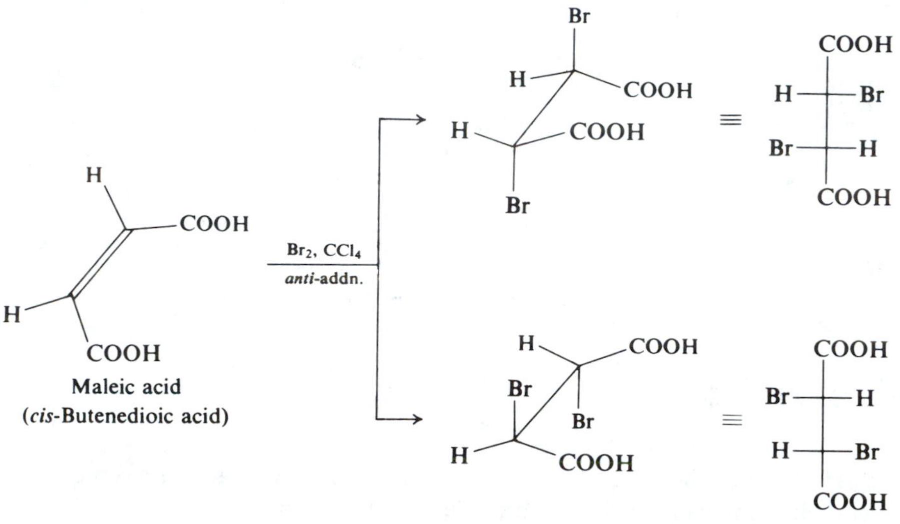

Maleic acid
(*cis*-Butenedioic acid)

$\xrightarrow[\textit{anti}\text{-addn.}]{Br_2,\ CCl_4}$

Enantiomers
racemic-2,3-Dibromobutanedioic acid

19.15 The —COOH group is electron-withdrawing, acid-strengthening compared to —H or —CH_3.

	HOOC—COOH		H—COOH	HOOC—CH_2—COOH		H—CH_2—COOH
K_a:	5400×10^{-5}	>	17.7×10^{-5}	140×10^{-5}	>	1.75×10^{-5}

19.16 As the distance between —COOH groups is increased, the inductive effect of one on the other weakens.

	HOOC—COOH	>	HOOC—CH_2—COOH	>	HOOC—CH_2CH_2—COOH	>	HOOC—$CH_2CH_2CH_2$—COOH
K_a:	5400×10^{-5}		140×10^{-5}		6.4×10^{-5}		4.5×10^{-5}

19.17 (a)

$C_2H_2O_4$	oxalic acid
$+C_2H_6O_2$	1,2-ethanediol
$C_4H_8O_6$	
$-C_4H_4O_4$	product
$2H_2O$	lost: there must be two esterifications

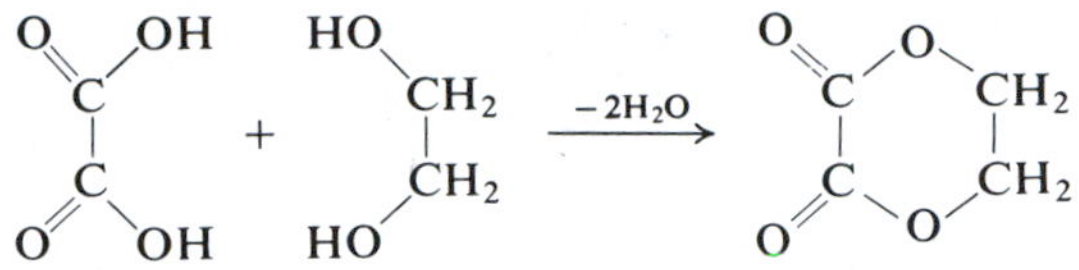

$HOOC{-}COOH$ + $HOCH_2CH_2OH$ $\xrightarrow{-2H_2O}$ cyclic $(O{=}C{-}O{-}CH_2{-}CH_2{-}O{-}C{=}O)$

Oxalic acid 1,2-Ethanediol *A cyclic diester*

(b) $C_4H_6O_4$ succinic acid
$-C_4H_4O_3$ product
H_2O lost: there must be intramolecular anhydride formation

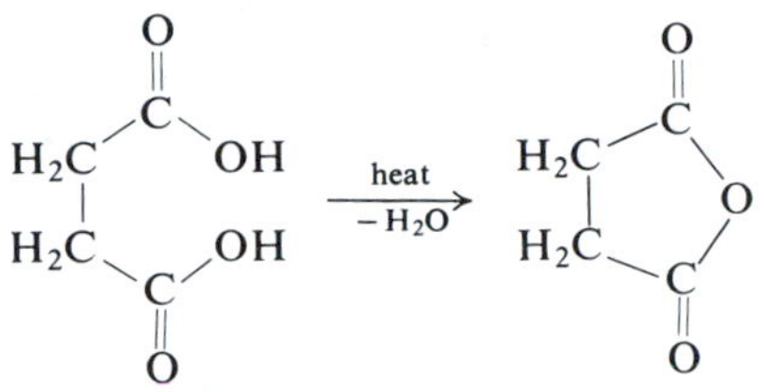

$H_2C(COOH){-}H_2C(COOH)$ $\xrightarrow[-H_2O]{heat}$ cyclic $H_2C{-}C(=O){-}O{-}C(=O){-}CH_2$

Succinic acid Succinic anhydride *A cyclic anhydride*

(c) $C_8H_6O_4$ terephthalic acid
$+C_2H_6O_2$ 1,2-ethanediol
$C_{10}H_{12}O_6$
$-C_{10}H_8O_4$ Dacron unit
$2H_2O$: double esterification

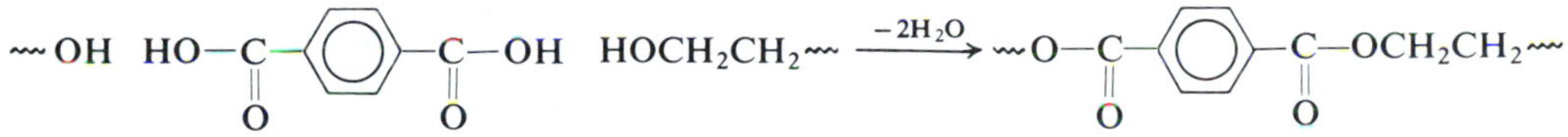

~OH HO—C(=O)—C_6H_4—C(=O)—OH HOCH$_2$CH$_2$~ $\xrightarrow{-2H_2O}$ ~O—C(=O)—C_6H_4—C(=O)—OCH$_2$CH$_2$~

Dacron
A polyester

19.18

	conc. H_2SO_4[a]	*cold* $KMnO_4$[b]	Br_2[b]	CrO_3[c]	*fum.* H_2SO_4[a]	$CHCl_3$, $AlCl_3$[c]	Na[d]	$NaOH$[a]	$NaHCO_3$[e]
Alkanes	−	−	−	−	−	−	−	−	−
Alkenes	+	+	+	−	+	−	−	−	−
Alkynes	+	+	+	−	+	−	−[f]	−	−
Alkyl halides	−	−	−	−	−	−	−	−	−
Alkylbenzenes	−	−[g]	−	−	+	+	−	−	−
1° alcohols	+	−	−	+	+	−	+	−	−
2° alcohols	+	−	−	+	+	−	+	−	−
3° alcohols	+	−	−	−	+	−	+	−	−
Ethers	+	−	−	−	+	−	−	−	−
Aldehydes	+	+	−[i]	+	+	−	−	−	−
Ketones	+	−	−[i]	−	+	−	−	−	−
Carboxylic acids	+	−	−	−	+	−	+[h]	+	+
Phenols	+	−	+[i]	−	+	−	+	+	−[j]

[a] Dissolve. [b] Decolorize. [c] Change color. [d] Hydrogen bubbles. [e] Dissolve, with CO_2 bubbles. [f] 1-Alkynes give test. [g] Decolorize hot $KMnO_4$. [h] Explosive reaction. [i] Evolve HBr. [j] Phenols with several electron-withdrawing substituents give test.

19.19 *o*-Chlorobenzoic acid, $C_7H_5O_2Cl$, m.w. (and equiv. weight) 156.5. ($2,6\text{-}Cl_2C_6H_3COOH$ has m.w. 191.)

19.20 (a)

$$\frac{18.7}{1000} \times 0.0972 \text{ equiv. of acid} = 0.187 \text{ g}$$

$$1 \text{ equiv. of acid} = 0.187 \times \frac{1000}{18.7} \times \frac{1}{0.0972} = 103 \text{ g}$$

(b) Ethoxyacetic acid, $C_2H_5OCH_2COOH$, m.w. (and equiv. weight) 104. The others are not possible: *n*-caproic acid, *n*-$C_5H_{11}COOH$, m.w. (and equiv. wt.) 116; methoxyacetic acid, CH_3OCH_2COOH, m.w. (and equiv. wt.) 90.

19.21 (a) Two; equiv. wt. = m.w./2 = 166/2 = 83.

(b) N.E. = mol.wt./number of acidic H per molecule.

(c) 70 (m.w./3), 57 (m.w./6).

19.22 Sodium carbonate, Na_2CO_3.

1. C–1: formic, methanoic
C–2: acetic, ethanoic
C–3: propionic, propanoic
C–4: *n*-butyric, butanoic
C–5: *n*-valeric, pentanoic
C–6: *n*-caproic, hexanoic
C–8: *n*-caprylic, octanoic
C–10: *n*-capric, decanoic
C–12: lauric, dodecanoic
C–16: palmitic, hexadecanoic
C–18: stearic, octadecanoic

2. (a) $(CH_3)_2CHCH_2COOH$
3-Methylbutanoic acid

(b) $(CH_3)_3CCOOH$
2,2-Dimethylpropanoic acid

(c) $CH_3CH_2CH_2CH(CH_3)CH(CH_3)COOH$
2,3-Dimethylhexanoic acid

(d) *n*-BuCH(Et)$CH_2CH(CH_3)COOH$
α-Methyl-γ-ethylcaprylic acid

(e) $PhCH_2COOH$
Phenylethanoic acid

(f) $PhCH_2CH_2CH_2COOH$
4-Phenylbutanoic acid

(g) $HOOCCH_2CH_2CH_2CH_2COOH$
Hexanedioic acid

(h) *p*-$CH_3C_6H_4COOH$
4-Methylbenzoic acid

(i) *o*-$C_6H_4(COOH)_2$
1,2-Benzenedicarboxylic acid

(j) *m*-$C_6H_4(COOH)_2$
1,3-Benzenedicarboxylic acid

(k) *p*-$C_6H_4(COOH)_2$
1,4-Benzenedicarboxylic acid

(l) *p*-HOC_6H_4COOH
4-Hydroxybenzoic acid

(m) $CH_3CH_2CH(CH_3)COOK$
Potassium 2-methylbutanoate

(n) $(CH_3CHClCOO)_2Mg$
Magnesium α-chloropropionate

(o) (*Z*)-HOOC—CH=CH—COOH
(*Z*)-Butenedioic acid

(p) HOOCCHBrCHBrCOOH
2,3-Dibromobutanedioic acid

(q) $(CH_3)_2CHC{\equiv}N$
2-Methylethanenitrile

(r) 2,4-$(O_2N)_2C_6H_3C{\equiv}N$
2,4-Dinitrobenzonitrile

3. (a) Same number of carbons:

$$\text{PhCH}_3 \xrightarrow[\text{heat}]{\text{KMnO}_4,\ \text{OH}^-} \text{PhCOOH}$$

(b) Add one carbon:

$$\text{PhBr} \xrightarrow{\text{Mg}} \text{PhMgBr} \xrightarrow{\text{CO}_2} \text{PhCOOMgBr} \xrightarrow{\text{H}^+} \text{PhCOOH}$$

(c) Same number of carbons:

$$\text{PhC}{\equiv}\text{N} \xrightarrow[\text{H}^+]{\text{H}_2\text{O}} \text{PhCOOH} + \text{NH}_4^+$$

(d) Same number of carbons:

$$\text{PhCH}_2\text{OH} \xrightarrow[\text{heat}]{\text{KMnO}_4,\ \text{H}^+} \text{PhCOOH}$$

(e) Same number of carbons: hydrolyze trihalide.

$$\text{PhCCl}_3 \xrightarrow{\text{OH}^-} \left[\text{Ph—C(OH)}_3\right] \xrightarrow{-\text{H}_2\text{O}} \text{Ph—C(=O)OH}$$

(f) Lose one carbon (via the haloform reaction, Sec. 18.21):

$$\text{Ph—C(=O)}\vdots\text{CH}_3 \xrightarrow{\text{OI}^-} \underset{\text{Iodoform}}{\text{HCI}_3} + \text{PhCOO}^- \xrightarrow{\text{H}^+} \text{PhCOOH}$$

4. (a) Same number of carbons: 1° alcohol $\xrightarrow{\text{oxidn}}$ acid.

$$n\text{-PrCH}_2\text{OH} \xrightarrow[\text{heat}]{\text{KMnO}_4,\ \text{H}^+} n\text{-PrCOOH}$$

(b) Add one carbon: via cyanide ion.

$$n\text{-PrCOOH} \xleftarrow[\text{H}^+]{\text{H}_2\text{O}} n\text{-PrC}{\equiv}\text{N} \xleftarrow{\text{CN}^-} n\text{-PrBr} \xleftarrow{\text{PBr}_3} n\text{-PrOH}$$

It cannot be used to prepare $(CH_3)_3$—COOH from *t*-BuBr. (See elimination *vs.* substitution, Sec. 8.25.)

(c) Add one carbon: via Grignard and CO_2.

$$n\text{-PrCOOH} \xleftarrow{\text{H}^+} \xleftarrow{\text{CO}_2} n\text{-PrMgBr} \xleftarrow{\text{Mg}} n\text{-PrBr} \xleftarrow{\text{PBr}_3} n\text{-PrOH}$$

(d) Remove one carbon: via the haloform reaction, Sec. 18.21.

$$n\text{-Pr—C(=O)}\vdots\text{CH}_3 \xrightarrow{\text{OI}^-} \underset{\text{Iodoform}}{\text{HCI}_3} + n\text{-PrCOO}^- \xrightarrow{\text{H}^+} n\text{-PrCOOH}$$

5. In every case, the ether has to be cleaved (ring opened) to yield the halide and/or alcohol.

(a) Same number of carbons:

$$\text{Ether (4C)} \xrightarrow[\text{heat}]{H_2O,\ H^+} \underset{\substack{(Di)alcohol\\ 4C}}{HOCH_2CH_2CH_2CH_2OH} \xrightarrow[\text{heat}]{KMnO_4,\ H^+} \underset{\substack{(Di)acid\\ 4C}}{HOOCCH_2CH_2COOH}$$

(b) We need to add one extra carbon, probably via CN^-:

$$\text{(4C)} \xrightarrow[\text{heat}]{HCl(aq)} \underset{\substack{Monohalide\\ 4C}}{HOCH_2(CH_2)_2CH_2Cl} \xrightarrow{CN^-} \underset{\substack{1^\circ\ alc.\quad Nitrile\\ 5C}}{HOCH_2(CH_2)_3CN} \xrightarrow{KMnO_4} \underset{\substack{Acid\quad Nitrile\\ 5C}}{HOOC(CH_2)_3CN}$$

$$\xrightarrow{H_2O,\ H^+} \underset{\substack{Diacid\\ 5C}}{HOOC(CH_2)_3COOH}$$

(c) We need to add two extra carbons, probably via CN^-:

$$\text{(4C)} \xrightarrow[\text{heat}]{HCl(aq)} \underset{\substack{Dihalide\\ 4C}}{Cl(CH_2)_4Cl} \xrightarrow{2CN^-} \underset{\substack{Dinitrile\\ 6C}}{NC(CH_2)_4CN} \xrightarrow[H^+]{H_2O} \underset{\substack{Diacid\\ 6C}}{HOOC(CH_2)_4COOH}$$

6. (a) PhCOOK (b) $(PhCOO)_3Al$ (c) $(PhCOO)_2Ca$
(d) PhCOONa (e) $PhCOONH_4$ (g) $PhCH_2OH$
(i) PhCOCl (j) PhCOCl (k) PhCOCl

(l) COOH-substituted benzene with Br (n) COOH-substituted benzene with NO_2 (o) COOH-substituted benzene with SO_3H

(q) PhCOO-*n*-Pr

No reaction: f, h, m, p.

7. (a) C_4H_9COOK (b) $(C_4H_9COO)_3Al$ (c) $(C_4H_9COO)_2Ca$
(d) C_4H_9COONa (e) $C_4H_9COONH_4$ (g) $C_4H_9CH_2OH$
(i) C_4H_9COCl (j) C_4H_9COCl (k) C_4H_9COCl
(m) $C_3H_7CH(Br)COOH$ (q) $C_4H_9COOC_3H_7$-*n*

No reaction: f, h, l, n, o, p.

8. (a) $i\text{-Pr—C}(=O)\text{OEt} \xleftarrow{H^+} i\text{-PrCOOH} + \text{HOEt}$
Ester

(b) $i\text{-PrC}(=O)\text{Cl} \xleftarrow{SOCl_2} i\text{-PrCOOH}$
Acid chloride

(c) $i\text{-Pr—C}(=O)\text{NH}_2 \xleftarrow{NH_3} i\text{-Pr—C}(=O)\text{Cl} \longleftarrow$ (b)
Amide

(d) $[i\text{-Pr—C}(=O)\text{O}]^-_2\ \text{Mg}^{2+} \xleftarrow{Mg} i\text{-Pr—C}(=O)\text{OH}$
Salt

(e) $i\text{-Pr—CH}_2\text{OH} \xleftarrow{H^+} \xleftarrow{LiAlH_4} i\text{-Pr—C}(=O)\text{OH}$
1° alcohol

9. (a) $\text{PhCOO}^-\text{Na}^+ \xleftarrow{NaOH} \text{PhCOOH}$

(b) $\text{PhC}(=O)\text{Cl} \xleftarrow{PCl_5} \text{PhCOOH}$

(c) $\text{Ph—C}(=O)\text{NH}_2 \xleftarrow{NH_3} \text{Ph—C}(=O)\text{Cl} \longleftarrow$ (b)

(d) $\text{Ph—C}(=O)\text{O—Pr-}n \longleftarrow \text{Ph—C}(=O)\text{Cl}$ (from b) + n-PrOH

(e) $\text{Ph—C}(=O)\text{O—C}_6\text{H}_4\text{—CH}_3 \longleftarrow \text{PhC}(=O)\text{Cl}$ (from b) + $\text{HO—C}_6\text{H}_4\text{—CH}_3$
p-Cresol

(f) $\text{Ph—C}(=O)\text{O—C}_6\text{H}_4\text{—Br} \xleftarrow{OH^-} \text{Ph—C}(=O)\text{Cl}$ (from b) + $\text{HO—C}_6\text{H}_4\text{—Br}$
m-Bromophenol

(g) $\text{PhCH}_2\text{OH} \xleftarrow{H^+} \xleftarrow{LiAlH_4} \text{PhCOOH}$

10. (a) Neutralize acid with NaOH(aq).

(b) Esterify acid with EtOH.

(c) Heat acid with $SOCl_2$.

(d) Allow acid chloride (from c) to react with NH_3.

(e) Brominate ring with Br_2, Fe.

(f) Nitrate ring with HNO_3/H_2SO_4.

(g) Reduce —COOH to —CH_2OH with $LiAlH_4$.

(h) Brominate side chain with Br_2/P (Hell–Volhard–Zelinsky reaction).

(i) Treat bromoacid (from h) with NH_3.

(j) Hydrolyze bromoacid (from h) with NaOH(aq), warm.

(k) Add an extra carbon: via CN^-, probably.

Ph—CH(COOH)—COOH $\xleftarrow[\text{warm}]{H_2O,\ H^+}$ Ph—CH(C≡N)—COOH $\xleftarrow{CN^-}$ Ph—CH(Br)—COO^- ⟵ (h)

An alternative way will be found in Chapter 21: a crossed Claisen condensation between ethyl phenylacetate and ethyl carbonate,

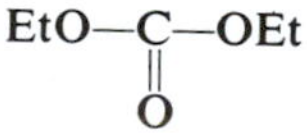

EtO—C(=O)—OEt

11. (a) PhCOOK

(b) 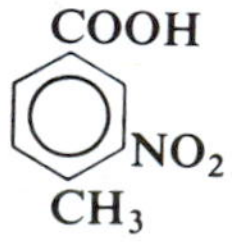 benzene ring with COOH, NO_2, CH_3

(c) $HOCH_2CH_2CH_2CH_2OH$

(d) $PhCOOCH_2Ph$, benzyl benzoate

(e) 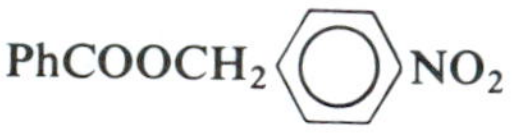 $PhCOOCH_2$–C_6H_4–NO_2 (and *o*-isomer)

(f) $CH_3CH_2CHBrCOOH$

(g) 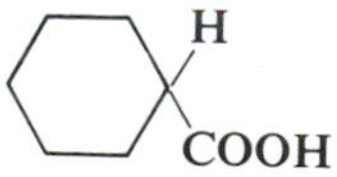 cyclohexane ring with H, COOH

Cyclohexanecarboxylic acid

(h) 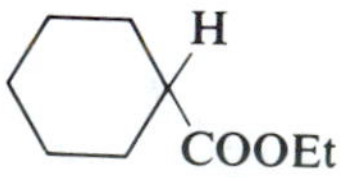 cyclohexane ring with H, COOEt

(i)  cyclohexane ring with H, COCl

(j) 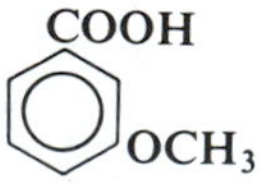 benzene ring with COOH, OCH_3

(k) 1,3,5-$C_6H_3(COOH)_3$

(l) $(CH_3)_2CHCOOCH_2CH(CH_3)_2$, isobutyl isobutyrate

(m) COOH OH Br (and COOH OH Br)

(n) $O_2N-C_6H_4-CH_2O-C(=O)-CH_3$
p-Nitrobenzyl acetate

(o) n-$C_{17}H_{35}COOH$, stearic acid

(p) n-$C_8H_{17}COOH + HOOC(CH_2)_7COOH$

(q) $CH_3(CH_2)_4CH{=}O + O{=}CHCH_2CH{=}O + O{=}CH(CH_2)_7COOH$

(r) same as product (g)

(s) $PhCOOCH_2CH_2OOCPh$, ethylene glycol dibenzoate

(t) o-$C_6H_4(COOEt)_2$, ethyl phthalate

(u) racemic *threo*-9,10-dibromooctadecanoic acid (*anti*-addition) (for *threo*, see page 380)

(v) $CH_3(CH_2)_7C{\equiv}C(CH_2)_7COOH$, 9-octadecynoic acid

(w) racemic *threo*-9,10-dihydroxyoctadecanoic acid (*anti*-hydroxylation) (for *threo*, see page 380)

12. (a) $CH_3CH_2CH_2{}^{14}COOH \longleftarrow CH_3CH_2CH_2MgBr + {}^{14}CO_2$

(b) $CH_3CH_2{}^{14}CH_2COOH \xleftarrow{CO_2} Et^{14}CH_2MgBr \xleftarrow{Mg} \xleftarrow{PBr_3} Et^{14}CH_2OH$

$Et^{14}CH_2OH \xleftarrow{LiAlH_4} Et^{14}COOH \longleftarrow EtMgBr + {}^{14}CO_2$

(c) $CH_3{}^{14}CH_2CH_2COOH \xleftarrow{KMnO_4} CH_3{}^{14}CH_2CH_2CH_2OH \longleftarrow CH_3{}^{14}CH_2MgBr + H_2C\underset{O}{-}CH_2$

$CH_3{}^{14}CH_2MgBr \xleftarrow{Mg} \xleftarrow{PBr_3} CH_3{}^{14}CH_2OH \xleftarrow{LiAlH_4} CH_3{}^{14}COOH \longleftarrow CH_3MgBr + {}^{14}CO_2$

(d) ${}^{14}CH_3CH_2CH_2COOH \xleftarrow{KMnO_4} {}^{14}CH_3CH_2CH_2CH_2OH \longleftarrow {}^{14}CH_3CH_2MgBr + H_2C\underset{O}{-}CH_2$

${}^{14}CH_3CH_2MgBr \xleftarrow{Mg} \xleftarrow{PBr_3} {}^{14}CH_3CH_2OH \xleftarrow{HCHO} {}^{14}CH_3MgBr \xleftarrow{Mg} \xleftarrow{HBr} {}^{14}CH_3OH$

13. (a) $PhCOOH \xleftarrow{KMnO_4} PhCH_3$

(b) $PhCH_2COOH \xleftarrow[H^+]{H_2O} PhCH_2C{\equiv}N \xleftarrow{CN^-} PhCH_2Cl \xleftarrow[heat]{Cl_2} PhCH_3$

(c) $CH_3-C_6H_4-COOH \xleftarrow{CO_2} CH_3-C_6H_4-MgBr \xleftarrow{Mg} CH_3-C_6H_4-Br \xleftarrow{Br_2,\ Fe} PhCH_3$

(d) m-$ClC_6H_4COOH \xleftarrow{Cl_2,\ Fe} C_6H_5COOH \longleftarrow$ (a)

(e) $p\text{-}ClC_6H_4COOH \xleftarrow{KMnO_4} p\text{-}ClC_6H_4CH_3 \xleftarrow{Cl_2,\ Fe} PhCH_3$

(f) Similar to (b), except brominate toluene first in the *para* position.

(g) $PhCH(Br)COOH \xleftarrow{Br_2,\ P} PhCH_2COOH \longleftarrow$ (b)

14. (a) $CH_3CH_2CH(CH_3)COOEt \xleftarrow{EtOH,\ H^+} CH_3CH_2CH(CH_3)\text{—}COOH \xleftarrow{CO_2}$ *sec*-BuMgBr (from *sec*-BuOH)

(b) $3,5\text{-}(O_2N)_2C_6H_3COCl \xleftarrow{PCl_5} 3,5\text{-}(O_2N)_2C_6H_3COOH \xleftarrow[\text{twice}]{HNO_3,\ H_2SO_4} C_6H_5COOH \xleftarrow{KMnO_4} PhCH_3$

(c) $p\text{-}BrC_6H_4CH(NH_2)COOH \xleftarrow{NH_3} p\text{-}BrC_6H_4CH(Br)COOH \xleftarrow{Br_2,\ P} p\text{-}BrC_6H_4CH_2COOH \longleftarrow$ 13(f)

(d) $CH_3CH(OH)COOH \xleftarrow{OH^-} CH_3CH(Br)COOH \xleftarrow[P]{Br_2} CH_3CH_2COOH \xleftarrow{KMnO_4} CH_3CH_2CH_2OH$

(e) $p\text{-}HO_3SC_6H_4COOH \xleftarrow{KMnO_4} p\text{-}HO_3SC_6H_4CH_3 \xleftarrow{H_2SO_4,\ SO_3} PhCH_3$

(f) $CH_3CH_2CH{=}CHCOOH \xleftarrow{KOH} CH_3CH_2CH_2CH(Br)COOH \xleftarrow[P]{Br_2} CH_3CH_2CH_2CH_2COOH$

$n\text{-}BuCOOH \xleftarrow{CO_2} n\text{-}BuMgBr \xleftarrow{Mg} \xleftarrow{PBr_3} n\text{-}BuOH$

(g) $p\text{-}CH_3C_6H_4C({=}O)NH_2 \xleftarrow{NH_3} p\text{-}CH_3C_6H_4C({=}O)Cl \xleftarrow{SOCl_2} p\text{-}CH_3C_6H_4COOH \longleftarrow$ 13(c)

(h) $n\text{-}BuCH_2CH_2O\text{—}C({=}O)\text{—}Ph \xleftarrow{H^+} n\text{-}BuCH_2CH_2OH + PhCOOH$ (from $PhCH_3$)
n-Hexyl alcohol

$n\text{-}Bu \vdots CH_2CH_2OH \longleftarrow n\text{-}BuMgBr$ (from *n*-BuOH) $+ H_2C\text{—}CH_2$ (ethylene oxide, bridged by O) (Sec. 13.20)

(i) 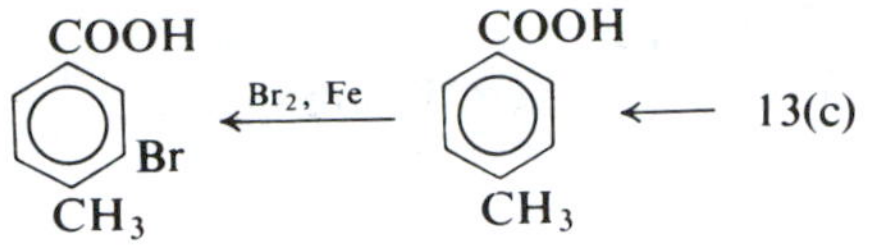 COOH, Br, CH_3 (2-bromo-4-methylbenzoic acid) $\xleftarrow{Br_2,\ Fe}$ COOH, CH_3 (p-toluic acid) ← 13(c)

(j) $PhCH(CH_3)$—$COOH$ $\xleftarrow{CO_2}$ $PhCH(CH_3)MgCl$ $\xleftarrow{Mg}$ $\xleftarrow{HCl}$ Ph—$CH(CH_3)OH$ ⟵ PhMgBr (from C_6H_6) + CH_3CHO (from EtOH)

(k) 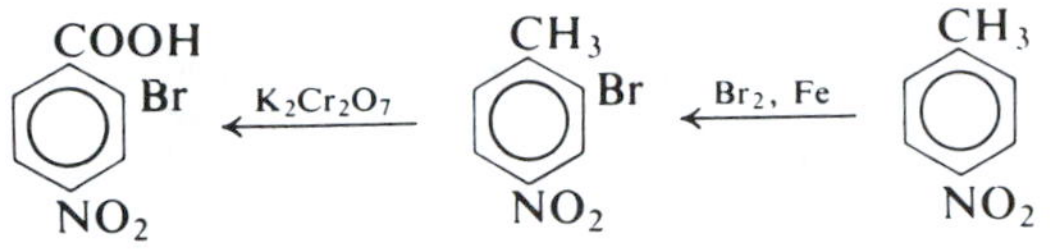 COOH, Br, NO_2 $\xleftarrow{K_2Cr_2O_7}$ CH_3, Br, NO_2 $\xleftarrow{Br_2,\ Fe}$ CH_3, NO_2 (page 525)

(l) 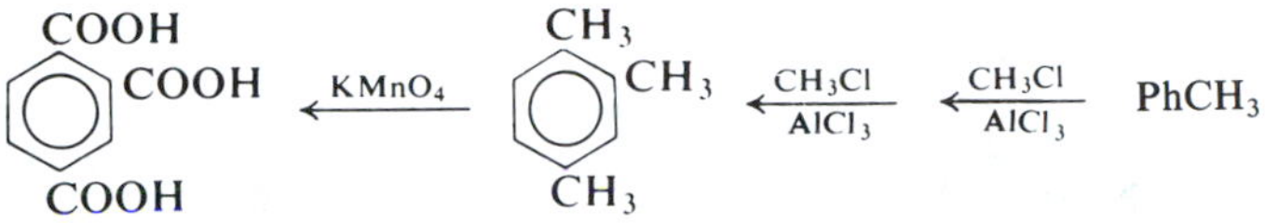 COOH, COOH, COOH $\xleftarrow{KMnO_4}$ CH_3, CH_3, CH_3 $\xleftarrow[AlCl_3]{CH_3Cl}$ $\xleftarrow[AlCl_3]{CH_3Cl}$ $PhCH_3$

15. Electron-withdrawing substituents increase acidity; electron-releasing substituents decrease acidity. The more substituents there are, the greater the effect. In saturated compounds, the closer the substituent is to —COOH, the greater its effect.

(a) $CH_3CH_2CHBrCOOH > CH_3CHBrCH_2COOH > CH_2BrCH_2CH_2COOH > CH_3CH_2CH_2COOH$

(b) tri-Cl > di-Cl > mono-Cl > unsubstituted acid

(c)  O_2N—⬡—COOH > ⬡—COOH > CH_3—⬡—COOH

(d) 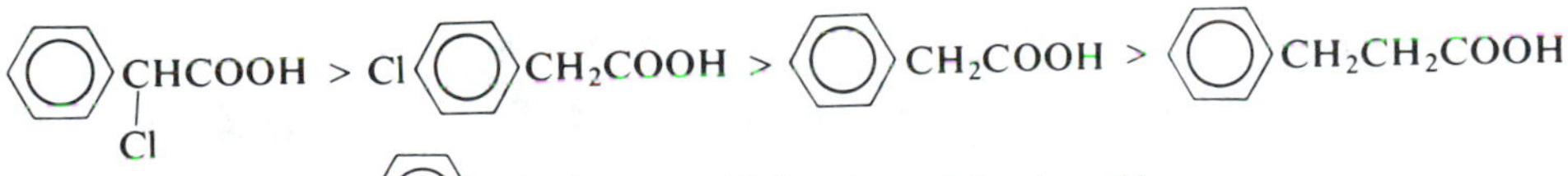 ⬡—CHClCOOH > Cl—⬡—CH_2COOH > ⬡—CH_2COOH > ⬡—CH_2CH_2COOH

⬡— is electron-withdrawing relative to —H

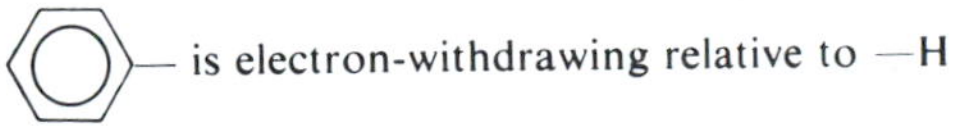

(e)  O_2N—⬡—COOH > O_2N—⬡—CH_2COOH > O_2N—⬡—CH_2CH_2COOH

(f) $H_2SO_4 > CH_3COOH > H_2O >$ EtOH $> HC{\equiv}CH > NH_3 > C_2H_6$

(g) $HOOCCH_2COOH > HOOCCH_2CH_2COOH > CH_3COOH$

16. The weaker the acid, the more powerfully basic will be its anion.

$NaC_2H_5 > NaNH_2 > NaC{\equiv}CH >$ NaOEt $>$ NaOH $> NaOOCCH_3 > NaHSO_4$

17. The separation here is exactly as described in Sec. 19.4, with sodium formate taking the place of sodium bicarbonate. Sodium formate is chosen because formic acid is stronger than benzoic acid and weaker than *o*-chlorobenzoic acid.

$$o\text{-}ClC_6H_4COOH + HCOO^- \rightleftarrows o\text{-}ClC_6H_4COO^- + HCOOH$$

o-ClC$_6$H$_4$COOH	HCOO$^-$	*o*-ClC$_6$H$_4$COO$^-$	HCOOH
Stronger acid	*Stronger base*	*Weaker base*	*Weaker acid*
K_a 120 × 10^{-5}			K_a 17.7 × 10^{-5}

$$C_6H_5COOH + HCOO^- \rightleftarrows C_6H_5COO^- + HCOOH$$

C$_6$H$_5$COOH	HCOO$^-$	C$_6$H$_5$COO$^-$	HCOOH
Weaker acid	*Weaker base*	*Stronger base*	*Stronger acid*
K_a 6.3 × 10^{-5}			K_a 17.7 × 10^{-5}

18. Steric factors are controlling in esterification (or in its reverse, hydrolysis): the more hindered the acid or alcohol, the slower the reaction. In particular, the usual order for alcohols is MeOH > 1° > 2° > 3°.

(a) MeOH > *n*-PrOH > *sec*-BuOH > *tert*-pentyl alcohol

(b) C_6H_5COOH > *o*-$CH_3C_6H_4COOH$ > 2,6-$(CH_3)_2C_6H_3COOH$

(c) $HCOOH$ > CH_3COOH > CH_3CH_2COOH > $(CH_3)_2CHCOOH$ > $(CH_3)_3CCOOH$

19. (a) See Figure 19.8, below.

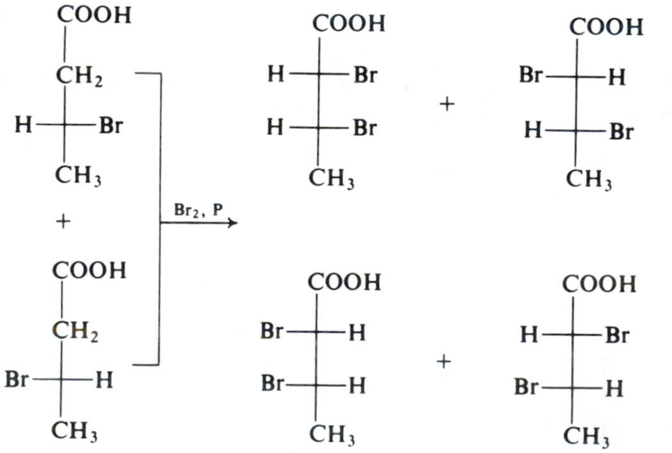

COOH / CH$_2$ / H—C—Br / CH$_3$ + COOH / CH$_2$ / Br—C—H / CH$_3$ —Br$_2$, P→

COOH / H—C—Br / H—C—Br / CH$_3$ + COOH / Br—C—H / H—C—Br / CH$_3$

COOH / Br—C—H / Br—C—H / CH$_3$ + COOH / H—C—Br / Br—C—H / CH$_3$

erythro-2,3-Dibromobutanoic acid
Racemic
A or B

threo-2,3-Dibromobutanoic acid
Racemic
B or A

Figure 19.8 α-Bromination of β-bromobutyric acid (Problem 19a).

(b) See Figure 19.9, below.

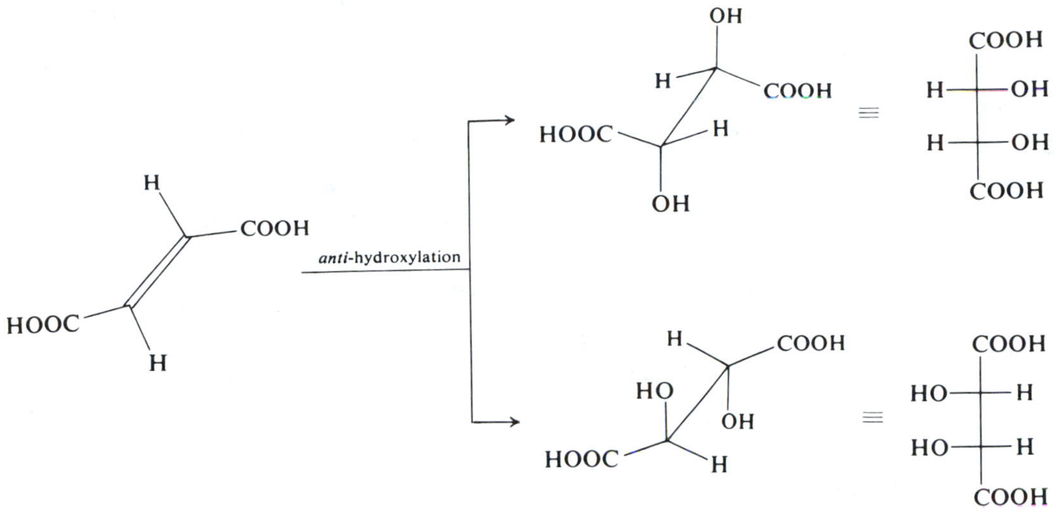

H COOH HOOC H

anti-hydroxylation

OH H COOH HOOC H OH ≡ COOH H—OH H—OH COOH

H COOH HO OH HOOC H ≡ COOH HO—H HO—H COOH

Identical
Mesotartaric acid
Achiral
C

Figure 19.9 Hydroxylation of fumaric acid with HCO_2OH (Problem 19b).

(c) $C_7H_8Br_2 - C_6H_8 = CBr_2$, showing that the methylene CBr_2 adds to the diene. The *cis* structure of D, E, and F follows from the stereochemistry (*syn*) of the addition of CBr_2.

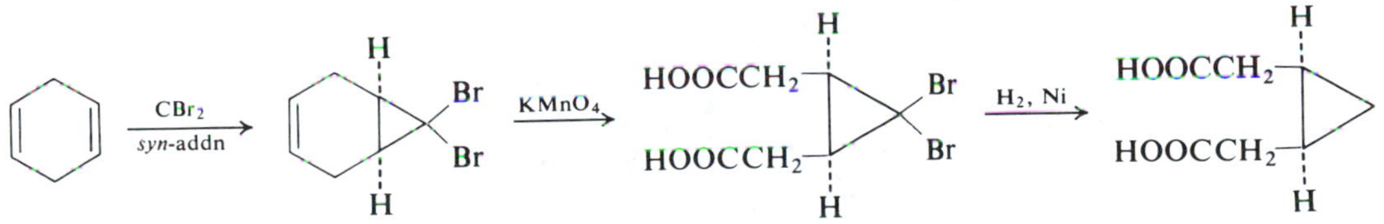

CBr_2 *syn*-addn → D (H, Br, Br, H) $\xrightarrow{KMnO_4}$ E ($HOOCCH_2$, $HOOCCH_2$, H, Br, Br, H) $\xrightarrow{H_2,\ Ni}$ F ($HOOCCH_2$, $HOOCCH_2$, H, H)

D

E
A *meso* compound

F
A *meso* compound

20. $HC{\equiv}CH + CH_3MgBr \longrightarrow CH_4 + HC{\equiv}CMgBr$

Stronger acid — *Weaker acid* — G

$HC{\equiv}CMgBr + CO_2 \longrightarrow HC{\equiv}CCOOMgBr \xrightarrow{H^+} HC{\equiv}CCOOH$

G — H — I

Compound I undergoes hydration, which could take place with either of two orientations, *a* or *b*:

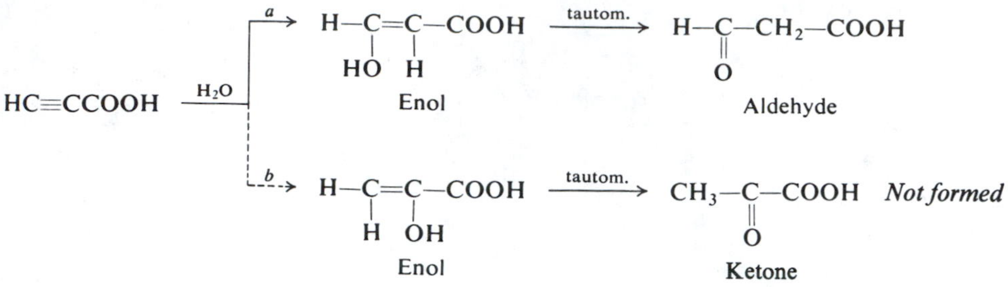

$$HC{\equiv}CCOOH \xrightarrow{H_2O} \begin{cases} \xrightarrow{a} \underset{\text{Enol}}{H{-}\underset{HO}{C}{=}\underset{H}{C}{-}COOH} \xrightarrow{\text{tautom.}} \underset{\text{Aldehyde}}{H{-}\underset{O}{\overset{}{C}}{-}CH_2{-}COOH} \\ \xrightarrow{b} \underset{\text{Enol}}{H{-}\underset{H}{C}{=}\underset{OH}{C}{-}COOH} \xrightarrow{\text{tautom.}} \underset{\text{Ketone}}{CH_3{-}\underset{O}{C}{-}COOH} \quad \textit{Not formed} \end{cases}$$

The fact that J is oxidized to an acid without loss of carbon gives us its structure, and shows that orientation is as in *a*.

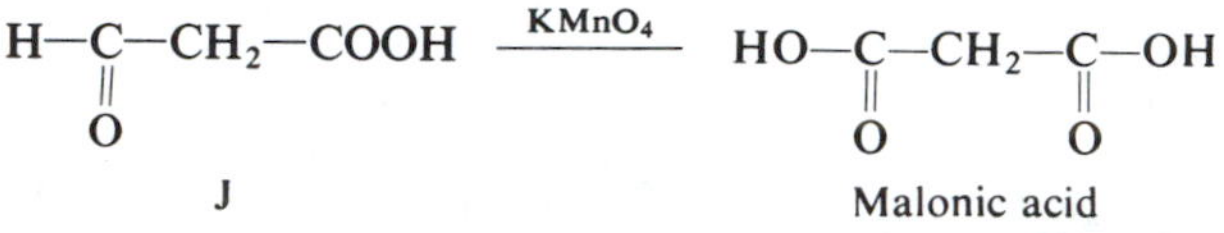

$$\underset{\text{J}}{H{-}\underset{O}{C}{-}CH_2{-}COOH} \xrightarrow{KMnO_4} \underset{\text{Malonic acid}}{HO{-}\underset{O}{C}{-}CH_2{-}\underset{O}{C}{-}OH}$$

21. (a) $NaHCO_3(aq)$ (b) $NaHCO_3(aq)$ (c) $NaHCO_3(aq)$
(d) $AgNO_3$ (e) $NaHCO_3(aq)$ (f) $NaHCO_3(aq)$

Ppt. of AgCl in (d); evolution of CO_2 in all the others.

22. (a)

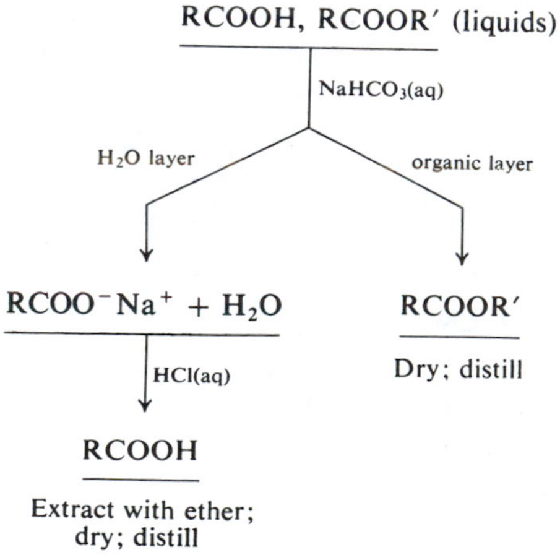

RCOOH, RCOOR′ (liquids)

↓ $NaHCO_3(aq)$

- H_2O layer → $RCOO^-Na^+ + H_2O$ → $HCl(aq)$ → RCOOH — Extract with ether; dry; distill
- organic layer → RCOOR′ — Dry; distill

(b) *n*-Butyric acid is too soluble in water for us to use Procedure (a).

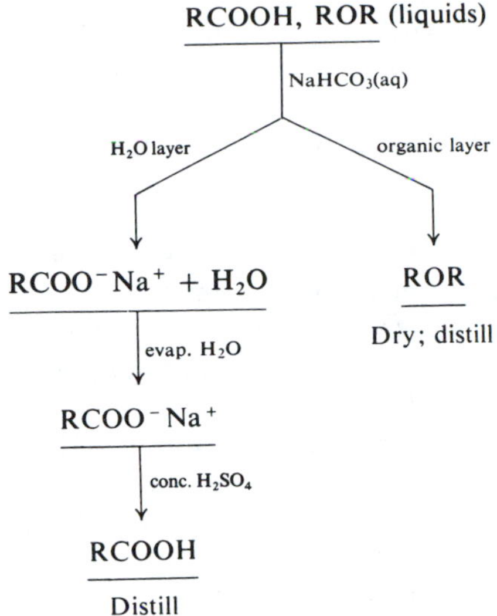

(c) As in (b), with 1-hexanol being in the organic layer.

(d)

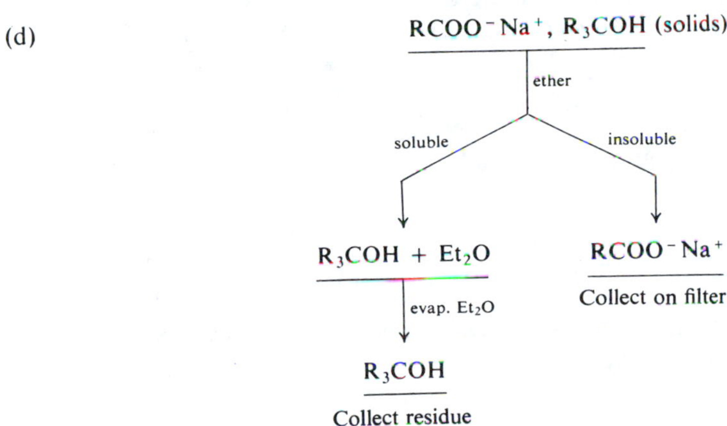

23. (a) $KMnO_4$. (b) CrO_3/H_2SO_4.

(c) Mesotartaric acid reacts with hot $KMnO_4$; elemental analysis for the other three acids.

(d) Valeric acid gives negative halogen test; use neutralization equivalents for the others.

(e) Neutralization equivalents; or 3-nitrophthalic acid gives an anhydride (m.p. 162 °C, neutral compound) when heated.

(f) The cinnamic acid reacts with $KMnO_4$; elemental analysis for the others.

(g) Try the easiest tests first. If a test is negative, go on to the next.

(1) Only 4-methylpentanoic acid will react with $NaHCO_3$(aq) with evolution of CO_2, and go into solution.

(2) Only α-phenylethyl chloride will react with $AgNO_3$ to ppt. AgCl.

(3) Only *o*-toluidine is basic enough to dissolve in dilute HCl(aq).

(4) Linalool is the only alcohol and the only compound that will respond quickly to CrO_3/H_2SO_4 reagent (orange-red ⟶ green).

(5) *β*-Chlorostyrene is the only alkene; try cold $KMnO_4$ (red-violet) ⟶ MnO_2 (brown).

(6) Try to dissolve the unknown in warm conc. H_2SO_4. Of the possibilities remaining at this point, only *cis*-decalin and 2,4-dichlorotoluene are insoluble in warm conc. H_2SO_4; and of these, only 2,4-dichlorotoluene will give a Cl test in an elemental analysis when subjected to sodium fusion. (Isodurene is easily sulfonated by conc. H_2SO_4; the others form soluble oxonium salts, $R_2OH^+HSO_4^-$.)

If the unknown dissolves in warm conc. H_2SO_4, go on to (7). (Three possibilities left.)

(7) Subject the unknown to sodium fusion and elemental analysis for halogen. If a positive test for Cl is observed, the unknown is *o*-chloroanisole.

(8) If test (7) is negative, the unknown (now down to two possibilities) must be an ether or a hydrocarbon, *p*-cresyl ethyl ether or isodurene. The ether would be cleaved by hot concentrated HI to give EtI, b.p. 72 °C; if the test is negative, the unknown must be isodurene.

24. (a) Tropic acid contains —COOH, and a 1° or 2° alcohol group; and because tropic acid yields benzoic acid, all this must be on only one side chain on a benzene ring.

$$\begin{array}{r} C_9H_{10}O_3 \\ -\ COOH \\ \hline C_8H_9O \end{array} \qquad \begin{array}{r} C_8H_9O \\ -\ OH \\ \hline C_8H_8 \end{array} \qquad \begin{array}{r} C_8H_8 \\ -\ C_6H_5 \\ \hline C_2H_3 \end{array} \text{ left for side chain}$$

Hydratropic acid has the same carbon skeleton; but, since —OH has been replaced by —H, the side chain carries no functional group other than —COOH.

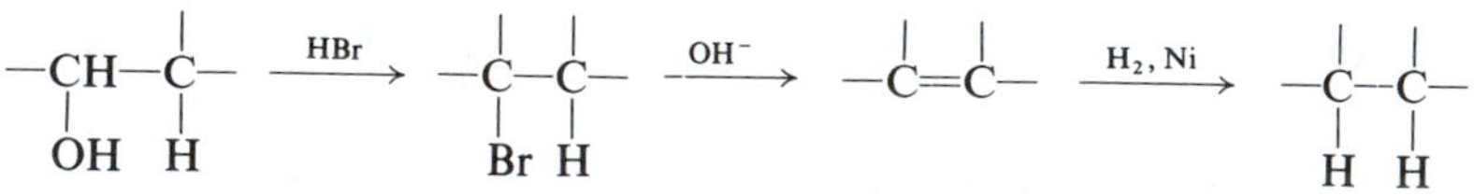

$$\underset{\textit{In tropic acid}}{—\underset{OH}{CH}—\underset{H}{C}—} \xrightarrow{HBr} —\underset{Br}{C}—\underset{H}{C}— \xrightarrow{OH^-} \underset{\textit{In atropic acid}}{—C{=}C—} \xrightarrow{H_2,\ Ni} \underset{\textit{In hydratropic acid}}{—\underset{H}{C}—\underset{H}{C}—}$$

At this point, the acids could be

$$Ph—\underset{CH_2OH}{CH}—COOH \longrightarrow Ph—\underset{CH_3}{CH}—COOH$$

or

$$Ph—\underset{OH}{CH}—CH_2—COOH$$

or

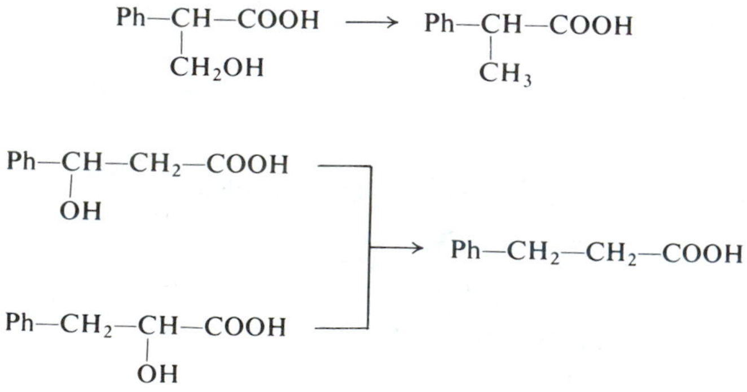

$$Ph—CH_2—\underset{OH}{CH}—COOH$$

(both) ⟶ $Ph—CH_2—CH_2—COOH$

Tropic acid: *possible structures*

Hydratropic acid: *possible structures*

(b) On the basis of the following synthesis,

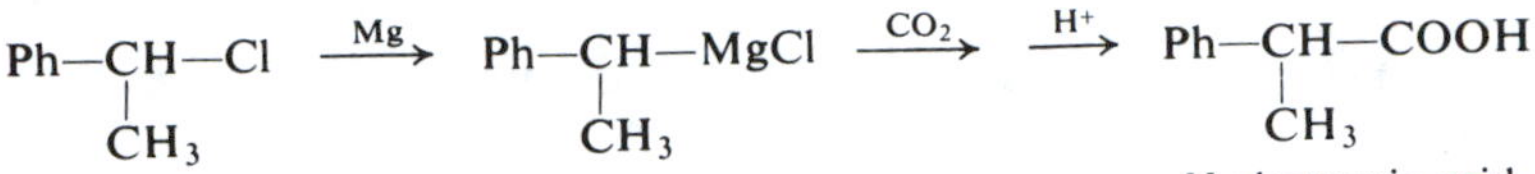

$$\underset{\alpha\text{-Phenylethyl chloride}}{\text{Ph—}\underset{\text{CH}_3}{\underset{|}{\text{CH}}}\text{—Cl}} \xrightarrow{\text{Mg}} \text{Ph—}\underset{\text{CH}_3}{\underset{|}{\text{CH}}}\text{—MgCl} \xrightarrow{\text{CO}_2} \xrightarrow{\text{H}^+} \underset{\text{Hydratropic acid}}{\text{Ph—}\underset{\text{CH}_3}{\underset{|}{\text{CH}}}\text{—COOH}}$$

we can assign the following structures:

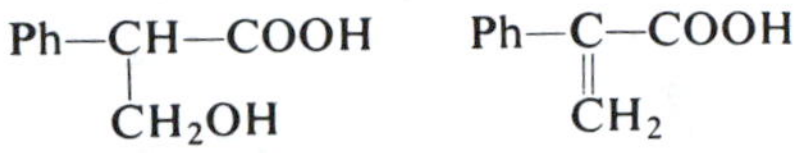

$$\underset{\text{Tropic acid}}{\text{Ph—}\underset{\text{CH}_2\text{OH}}{\underset{|}{\text{CH}}}\text{—COOH}} \qquad \underset{\text{Atropic acid}}{\text{Ph—}\underset{\text{CH}_2}{\underset{\|}{\text{C}}}\text{—COOH}}$$

25. Following the general procedure described in the box on page 251 of this Study Guide, we do some chemical arithmetic, draw tentative structures, eliminate the impossible isomers, and try to decide among the possible isomers.

(a) $C_3H_5ClO_2$
$-$ COOH (tentative for δ 11.22 signal)
C_2H_4Cl means a saturated group equivalent to C_2H_5

Two structures can be drawn:

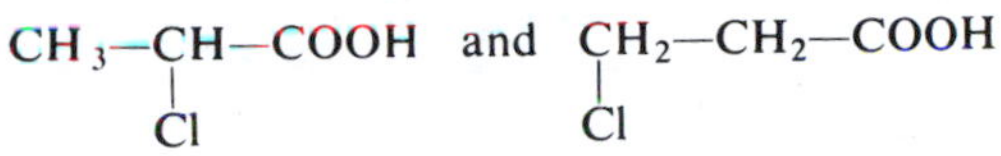

$$\text{CH}_3\text{—}\underset{\text{Cl}}{\underset{|}{\text{CH}}}\text{—COOH} \quad \text{and} \quad \underset{\text{Cl}}{\underset{|}{\text{CH}_2}}\text{—CH}_2\text{—COOH}$$

Of these, only the first can give a 3H signal. The compound is

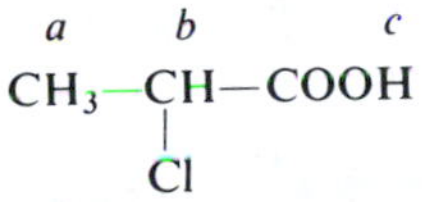

$$\overset{a}{\text{CH}_3}\text{—}\underset{\text{Cl}}{\underset{|}{\overset{b}{\text{CH}}}}\text{—}\overset{c}{\text{COOH}}$$

α-Chloropropionic acid

(b) The absence of any signal for acidic H leads us to consider esters (two oxygens), RCOOR′, for this compound.

$C_3H_5ClO_2$
$-$ COO from RCOOR′
C_2H_5Cl, to be divided between R and R′

Without looking further at the NMR data, we see four possibilities:

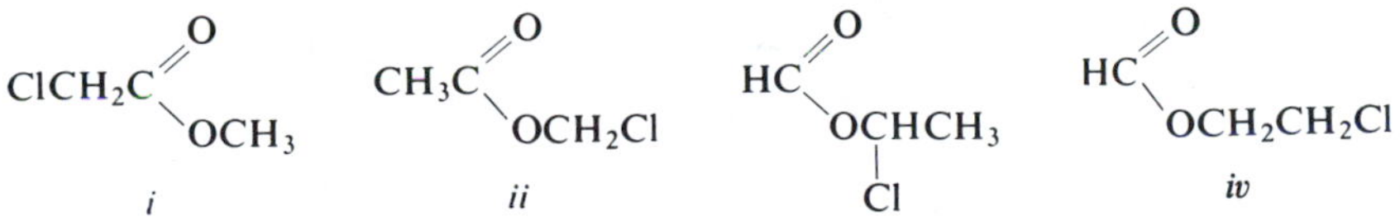

$$\underset{i}{\text{ClCH}_2\text{C}(\text{=O})\text{OCH}_3} \qquad \underset{ii}{\text{CH}_3\text{C}(\text{=O})\text{OCH}_2\text{Cl}} \qquad \underset{iii}{\text{HC}(\text{=O})\text{O}\underset{\text{Cl}}{\underset{|}{\text{CH}}}\text{CH}_3} \qquad \underset{iv}{\text{HC}(\text{=O})\text{OCH}_2\text{CH}_2\text{Cl}}$$

The appearance of two unsplit 3H and 2H signals in the NMR spectrum rules out all except *i* and *ii*. From *ii* we would expect 3H at about δ 2.0, and 2H considerably farther downfield than δ 4.08. The compound must be

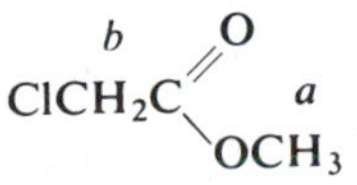

$$\text{Cl}\overset{b}{\text{CH}_2}\text{C}(\text{=O})\overset{a}{\text{OCH}_3}$$

Methyl chloroacetate

(c) As in (b), an ester rather than an acid seems likely.

$$\begin{array}{ll} C_4H_7BrO_2 & \\ \underline{-\ COO} & \text{from RCOOR}' \\ C_3H_7Br, & \text{to be divided between R and R}' \end{array}$$

The 3H + 2H triplet–quartet points to a $—CH_2CH_3$, as usual,

$$\begin{array}{r} C_3H_7Br \\ \underline{-\ C_2H_5} \\ CH_2Br \end{array}$$

leaving only $—CH_2Br$ to be fitted in. The 2H singlet at δ 3.77 is $—CH_2—$ shifted downfield by —Br and —COOR′ (compare signal *b* from compound (b), above). The compound has to be

$$\overset{b}{BrCH_2}C(=O)\underset{c\quad a}{OCH_2CH_3}$$

Ethyl bromoacetate

(d) This looks like an acid (signal *d* at δ 10.97).

$$\begin{array}{r} C_4H_7BrO_2 \\ \underline{-\ COOH} \\ C_3H_6Br \end{array}$$

Signal *a* is $—CH_3$ split by two protons; signal *b*, a quintet, is $—CH_2—$ split by $—CH_3$ and by $—\underset{|}{C}H—$; signal *c* is $—\underset{|}{C}H$ split by $—CH_2—$ and shifted far downfield. The pieces thus are:

$$CH_3— \qquad —CH_2— \qquad —\underset{|}{C}H— \qquad —Br \qquad —COOH$$

On the basis of the splittings of the signals and their δ values, only one structure is possible:

$$\overset{a}{CH_3}\overset{b}{CH_2}\underset{Br}{\overset{c}{\underset{|}{C}H}}\overset{d}{COOH}$$

α-Bromobutyric acid

(e) The singlet at δ 10.95 indicates —COOH; the triplet–quartet system indicates $—CH_2CH_3$.

$$\begin{array}{rl} C_4H_8O_3 & \\ \underline{-\ COOH} & \text{signal } d \\ C_3H_7O & \end{array} \qquad \begin{array}{rl} C_3H_7O & \\ \underline{-\ C_2H_5} & \text{signals } a \text{ and } b \\ CH_2O & \end{array} \qquad \begin{array}{rl} CH_2O & \\ \underline{-\ CH_2} & \text{signal } c \\ O & \text{oxygen} \end{array}$$

This leaves only oxygen. In the absence of a signal for —OH, it must be in an ether linkage; furthermore it must separate Et— from $-CH_2-$ since signal *b* is unaffected by the protons of the $-CH_2-$ group. The compound can only be:

$$\overset{a}{CH_3}\overset{b}{CH_2}O\overset{c}{CH_2}CO\overset{d}{OH}$$

Ethoxyacetic acid

26.

	Benzoic acid	*Sodium benzoate*
(a)	low b.p.	high b.p. (*dec*)
(b)	low m.p.	high m.p. (*dec*)
(c)	insol. H_2O	sol. H_2O
(d)	sol. Et_2O	insol. Et_2O
(e)	complete ionization	complete ionization
(f)	partial ionization	complete ionization
(g)	stronger acid, weaker base	weaker acid, stronger base

This comparison generally holds for carboxylic acids and their salts. (Generally, only alkali metal salts are soluble in water.)

27. (a) $CH_3CH{=}CHCOOH$ Crotonic acid (b) $C_6H_5CHOHCOOH$ Mandelic acid (c) *p*-$O_2NC_6H_4COOH$ *p*-Nitrobenzoic acid

(See the labeled spectra on page 705 of this Study Guide.)

20

Functional Derivatives of Carboxylic Acids

Nucleophilic Acyl Substitution

20.1 (a) $HO^- + R{-}\overset{O}{\overset{\|}{C}}{-}CX_3 \longrightarrow R{-}\overset{O^-}{\overset{|}{C}}{-}CX_3 \longrightarrow R{-}C\overset{O}{\underset{OH}{\lessgtr}} + {}^-CX_3 \xrightarrow{+H^+} HCX_3$
(the OH group shown below the central C of the intermediate; the acid $RCOOH$ loses H^+ ($-H^+$) to give $RCOO^-$)

This is an exception to our generalization of Sec. 20.4; here a ketone undergoes nucleophilic *substitution*—but only because the three halogen atoms make CX_3^- a weaker base and hence better leaving group than most R^-.

(b) $C_6H_5{-}CO{-}C_6H_4F(o) + H_2N^- \longrightarrow C_6H_5{-}C(O^-)(NH_2){-}C_6H_4F \longrightarrow C_6H_5{-}C(=O)NH_2 + {}^-C_6H_4F \xrightarrow{NH_3} C_6H_5F$

Here is another exception, but again one we can understand: the powerfully electron-withdrawing *o*-F stabilizes the leaving anion, making it a weaker base and better leaving group than most Ar^-.

20.2 The —COOH groups are (very nearly) the same distance apart in the *cis*- and *trans*-1,2-cyclohexanedicarboxylic acids, and the anhydride bridge can form easily from either. See the answer to Problem 20.3(b).

In *trans*-1,2-cyclopentanedicarboxylic acid, formation of the anhydride bridge would place prohibitive strain on the molecule.

20.3 (a) The *cis* acid is the only one capable of forming a cyclic anhydride.

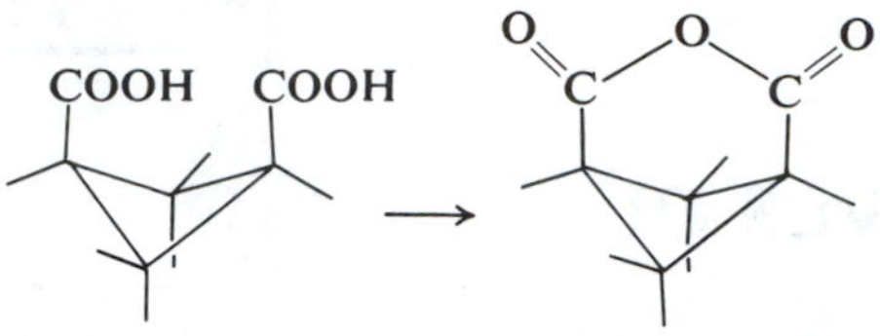

COOH COOH →

cis Acid Cyclic anhydride

(b) In both acids, the two —COOH groups are the same distance apart.

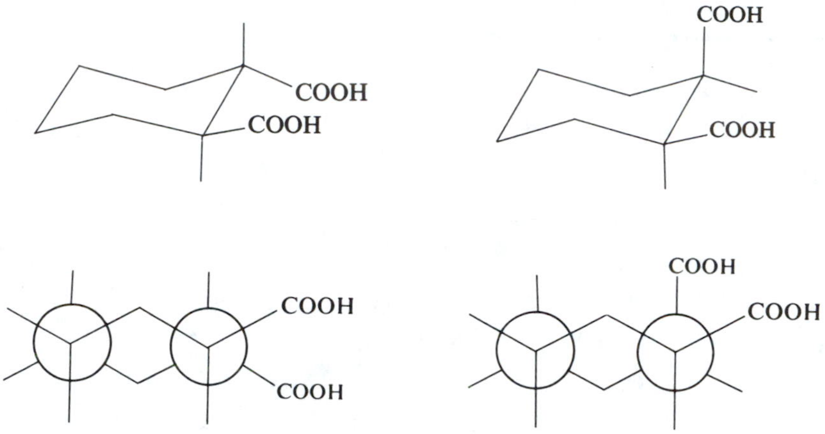

trans-1,2-Diacid *cis*-1,2-Diacid

(c) Yes; only the *cis* acid can form the cyclic (diaxial) anhydride.

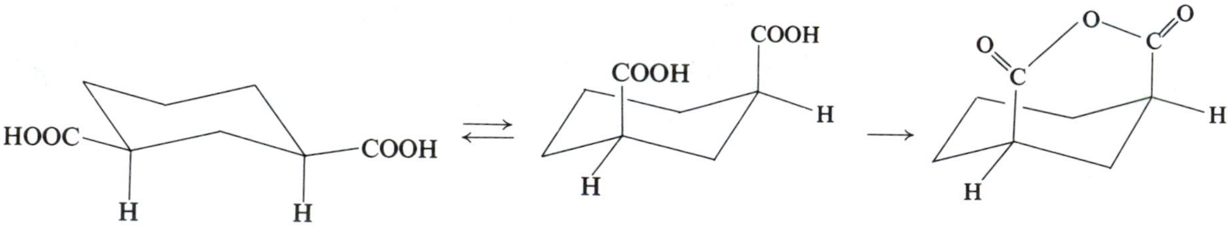

cis-1,3-Diacid
Diequatorial conformation

cis-1,3-Diacid
Diaxial conformation

Cyclic anhydride

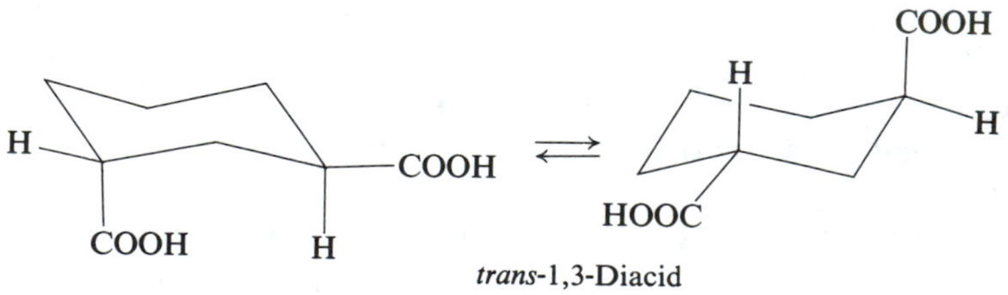

trans-1,3-Diacid

No cyclic anhydride possible

20.4 We recognize a succession of familiar reaction types: Friedel–Crafts acylation; Clemmensen reduction; acid chloride formation; Friedel–Crafts acylation with ring closure; hydrogenation; dehydration.

The final step is *aromatization*. A hydrogenation catalyst like platinum lowers the energy barrier between hydrogenated and dehydrogenated compounds, and thus speeds up reaction in *both* directions. The position of the equilibrium is determined by other factors: hydrogenation is favored by an excess of hydrogen under pressure; dehydrogenation is favored—as here—by sweeping the hydrogen away in a stream of inert gas.

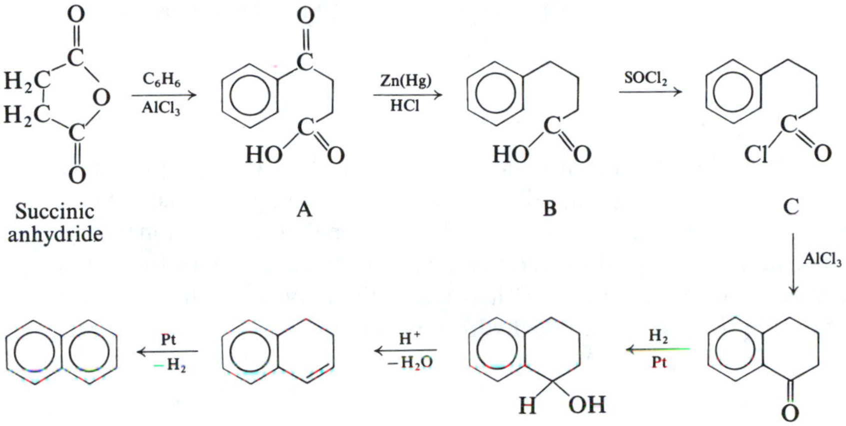

Succinic anhydride $\xrightarrow[AlCl_3]{C_6H_6}$ A $\xrightarrow[HCl]{Zn(Hg)}$ B $\xrightarrow{SOCl_2}$ C $\xrightarrow{AlCl_3}$ D $\xrightarrow[Pt]{H_2}$ E $\xrightarrow[-H_2O]{H^+}$ F $\xrightarrow[-H_2]{Pt}$ G

G Naphthalene

20.5 (a) D (Problem 20.4) $\xrightarrow{PhMgBr}$ (Ph, OMgBr) $\xrightarrow{H_2O}$ (Ph, OH)

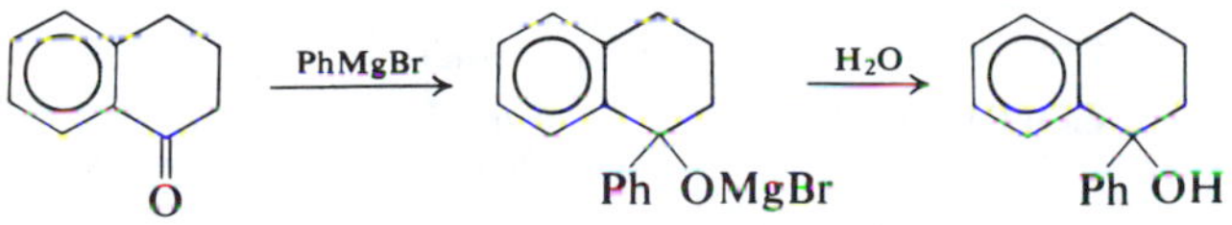

(b) (Ph, OH) $\xrightarrow[-H_2O]{H^+}$ (Ph) $\xrightarrow[-H_2]{Pt}$ (Ph)

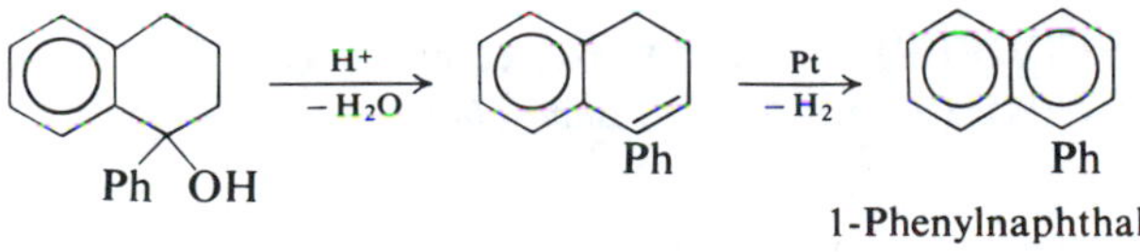

1-Phenylnaphthalene

20.6 Friedel–Crafts acylation.

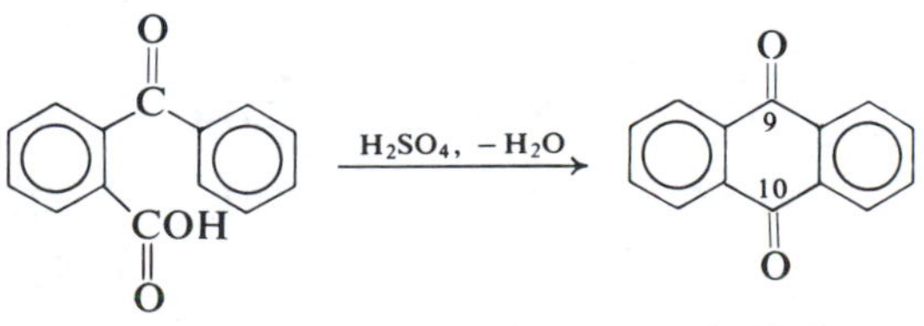

$C_{14}H_{10}O_3$ *o*-Benzoylbenzoic acid $\xrightarrow{H_2SO_4,\ -H_2O}$ $C_{14}H_8O_2$ 9,10-Anthraquinone

20.7 Two successive Friedel–Crafts reactions are involved. In the first one, acylation takes place chiefly *para* to the methyl group.

Phthalic anhydride + Toluene $\xrightarrow{AlCl_3}$ *o*-(*p*-Toluyl)benzoic acid $\xrightarrow{H_2SO_4,\ heat}$ 2-Methyl-9,10-anthraquinone

20.8 Reaction of phthalic anhydride with a racemic alcohol yields an alkyl hydrogen phthalate, which has an acidic "handle". This handle allows reaction with an optically active base to form diastereomeric salts, which can be separated by fractional crystallization. Once separated, each ester can be hydrolyzed to an optically active alcohol. Since hydrolysis of a carboxylic ester does not usually involve cleavage of the alkyl–oxygen bond (Sec. 20.17), there is no loss of optical activity in the hydrolysis step.

Phthalic anhydride + (+)-ROH, (−)-ROH — Enantiomers *racemic modification*

↓

C_6H_4(COOR(+))(COOH) and C_6H_4(COOR(−))(COOH) — Enantiomers *racemic modification*

↓ (+)-B

C_6H_4(COOR(+))(COO$^-$(+)-BH$^+$) and C_6H_4(COOR(−))(COO$^-$(+)-BH$^+$) — Diastereomers *separable*

↓ OH$^-$ ↓ OH$^-$

(+)-ROH (−)-ROH — Resolved enantiomers

20.9 Succinic acid ($H_2C(COOH)$–$H_2C(COOH)$) $\xrightarrow[-H_2O]{heat}$ Succinic anhydride $\xrightarrow{2NH_3}$ $H_2C(CONH_2)$–$H_2C(COO^-NH_4^+)$ $\xrightarrow{H^+}$ Succinamic acid ($H_2C(CONH_2)$–$H_2C(COOH)$) $\xrightarrow{heat}$ Succinimide

20.10 In the anion from ammonia, the negative charge is localized on nitrogen. In the anion from benzamide, it is accommodated by nitrogen and one oxygen, and in the anion from phthalimide by nitrogen and *two* oxygens.

Most stable NH_2^- *Least stable*

20.11 The mixture is shaken with benzene, which extracts the ester and benzoic acid. Treatment of the benzene extract with Na_2CO_3(aq) removes benzoic acid as the water-soluble sodium salt. Distillation of the wet benzene layer removes water and benzene first, and the dry ester is finally collected at 200 °C.

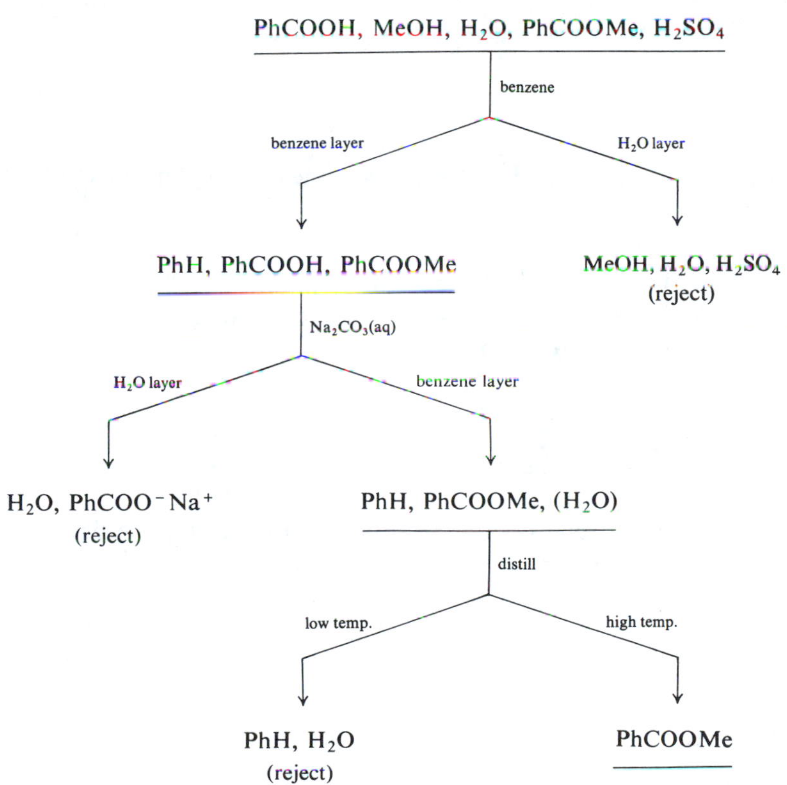

20.12 (a) A six-carbon product requires combination of two lactic acid residues:

$$\begin{array}{rl} C_6H_{12}O_6 & \text{2 lactic acid} \\ -\ C_6H_8O_4 & \text{product} \\ \hline 4H + 2O & \text{means } 2H_2O \text{ removed} \end{array}$$

Water is lost through a double esterification involving opposite ends of two reactant molecules.

$$2\ CH_3CH(OH)COOH \xrightarrow{-2H_2O} \text{Lactide (cyclic: } CH_3CH{-}C(=O){-}O{-}CH(CH_3){-}C(=O){-}O{-})$$

Lactic acid
Two moles

Lactide
A cyclic double ester

(b) Linkage of polyfunctional molecules through esterification of —OH on one molecule by —COOH from another to give a linear polymer. (Such a process is called *step-reaction polymerization,* Sec. 31.7.)

$$\sim O(CH_2)_9\overset{O}{\overset{\|}{C}}{-}O(CH_2)_9\overset{O}{\overset{\|}{C}}{-}O(CH_2)_9\overset{O}{\overset{\|}{C}}\sim$$

Polymer: 5 to 55 units

20.13 (a) Electron withdrawal by G speeds up reaction by speeding up the first step: it helps to disperse the negative charge developing on oxygen.

$$\underset{W}{\overset{R}{}}{>}C{=}O + :Z \rightleftarrows \left[R{-}\overset{Z}{\overset{\vdots}{C}}(W){\cdots}O^{\delta-} \right] \rightleftarrows R{-}\overset{Z}{\overset{|}{C}}(W){-}O^- \longrightarrow \underset{Z}{\overset{R}{}}{>}C{=}O + :W^-$$

$R = p\text{-}GC_6H_4$

(b) Br, activating; NH_2, deactivating; *t*-Bu, deactivating.

(c) $NO_2 > H > CH_3 > NH_2$. Here, electron withdrawal speeds up not only the first step as in (a), but also the second step: it lowers the basicity of W^-, and makes it a better leaving group.

20.14 (a) Steric effect, and electron release by alkyl groups (compare Sec. 18.7).

(b) formate > acetate > propionate > isobutyrate > trimethylacetate

$HCOOMe > CH_3COOMe > CH_3CH_2COOMe > (CH_3)_2CHCOOMe > (CH_3)_3CCOOMe$

20.15 The basicity of the leaving group is one factor: the weaker bases leave the intermediate faster.

Basicity: $Cl^- < RCOO^- < OR^- < NH_2^-$

20.16

$$R{-}\overset{^{18}O}{\overset{\|}{C}}{-}OR' + H^+ \rightleftharpoons R{-}\overset{^{18}OH}{\overset{\vdots}{C}}{\cdots}OR'^{\oplus} + H_2O \rightleftharpoons R{-}\underset{OH_2^+}{\overset{^{18}OH}{C}}{-}OR' \rightleftharpoons \text{hydrolysis}$$

$$\downarrow\uparrow$$

$$H^+ + R{-}\underset{OH}{\overset{^{18}OH}{C}}{-}OR'$$ Compare with II on page 775

$$\downarrow\uparrow$$

$$R{-}\underset{O}{\underset{\|}{C}}{-}OR' + H^+ \rightleftharpoons R{-}\underset{OH}{\underset{\vdots}{C}}{\cdots}OR'^{\oplus} + H_2{}^{18}O \rightleftharpoons R{-}\underset{OH}{\overset{^{18}OH_2^+}{C}}{-}OR' \rightleftharpoons \text{hydrolysis}$$

20.17 Crowding in the tetrahedral intermediates by bulky groups makes them less stable, and the corresponding transition states more difficult to achieve, thus decreasing reaction rates.

20.18 (a) Hydrolysis of these esters of 3° alcohols evidently proceeds via the 3° cations, in an S_N1 reaction of the protonated esters.

$$CH_3{-}\overset{O}{\overset{\|}{C}}{-}O{-}R + H^+ \rightleftharpoons CH_3{-}\overset{OH}{\overset{\vdots}{C}}{\cdots}O{-}R^{\oplus} \text{ and/or } CH_3{-}\overset{O}{\overset{\|}{C}}{-}\underset{H}{\overset{\oplus}{O}}{-}R \longrightarrow CH_3COOH + R^{\oplus}$$

$$R^{\oplus} \xrightarrow{H_2O} ROH_2^+ \xrightarrow{-H^+} ROH$$

(b) Here, as in S_N2 *vs.* S_N1, we have competition between a bimolecular reaction and a unimolecular reaction. But in ester hydrolysis we have many possible reactions, and competition is between the *easiest* of these. The easiest unimolecular reaction is S_N1. The easiest bimolecular reaction is attack by water, not at alkyl carbon, but at acyl carbon (see Sec. 20.4). The change in mechanism results in, among other things, a change in point of cleavage.

tert-Alkyl esters undergo S_N1 fastest for electronic reasons, and undergo acyl substitutions slowest for steric reasons. If there is to be a shift in mechanism of ester hydrolysis, this is where we would certainly expect to observe it.

20.19

$$n\text{-Bu}-\underset{\underset{\text{OH}}{|}}{\overset{\overset{\text{Et}}{|}}{\text{C}}}-\text{Et} \xleftarrow{\ a:\ \text{EtMgBr}\ } n\text{-Bu}-\underset{\underset{\text{O}}{\|}}{\text{C}}-\text{Et} \xleftarrow{\text{Et}_2\text{CuLi}} n\text{-BuCOCl} \xleftarrow{\text{SOCl}_2} n\text{-BuCOOH}$$

$$n\text{-Bu}-\text{C(Et)(OH)}-\text{Et} \xleftarrow{\ b:\ 2\text{EtMgBr}\ } n\text{-BuCOOMe} \xleftarrow{\text{MeOH, H}^+} n\text{-BuCOOH}$$

3-Ethyl-3-heptanol

20.20 (a) Esters of formic acid.

$$\text{H}-\underset{\underset{\text{OH}}{|}}{\overset{\overset{\text{R}}{|}}{\text{C}}}-\text{R} \longleftarrow \text{H}-\text{C}\overset{\text{O}}{\underset{\text{OR}'}{\lessgtr}} + 2\text{RMgX}$$

A 2° alcohol A formate

(b)

$$n\text{-Pr}-\underset{\underset{\text{OH}}{|}}{\overset{\overset{\text{H}}{|}}{\text{C}}}-\text{Pr-}n \longleftarrow 2n\text{-PrMgBr (from } n\text{-PrOH)} + \text{HCOOEt (from MeOH and EtOH)}$$

4-Heptanol

20.21 1-Octadecanol, $CH_3(CH_2)_{16}CH_2OH$, and 1-butanol. The conditions of hydrogenolysis lead also to hydrogenation of the double bond.

20.22 (a)

$$\text{H}_2\text{N}-\underset{\underset{\text{O}}{\|}}{\text{C}}-\text{O}-\overset{\overset{\text{CH}_3}{|}}{\text{C}}\text{HCH}_2\text{CH}_2\text{CH}_2 \xleftarrow{\text{NH}_3} \text{Cl}-\underset{\underset{\text{O}}{\|}}{\text{C}}-\text{OPe-2} \longleftarrow \text{Cl}-\underset{\underset{\text{O}}{\|}}{\text{C}}-\text{Cl} + \text{2-PeOH}$$

Amide–ester *Ester–acid chloride* Phosgene

(b)

$$\text{PhCH}_2\text{O}-\underset{\underset{\text{O}}{\|}}{\text{C}}-\text{Cl} \longleftarrow \text{Cl}-\underset{\underset{\text{O}}{\|}}{\text{C}}-\text{Cl} + \text{PhCH}_2\text{OH}$$

Benzyl chloroformate Phosgene Benzyl alcohol
Ester–acid chloride *One mole*

20.23 Great resonance stabilization of the cation, $[C(NH_2)_3]^+$, with contribution from three equivalent structures, and accommodation of positive charge by three nitrogens.

$$\left[(\text{H}_2\text{N})_2\text{C}=\overset{\oplus}{\text{N}}\text{H}_2 \quad \text{H}_2\overset{\oplus}{\text{N}}=\text{C}(\text{NH}_2)_2 \quad (\text{H}_2\text{N})_2\text{C}=\overset{\oplus}{\text{N}}\text{H}_2 \right] \quad \textit{equivalent to} \quad \left\{ \text{C}(\text{NH}_2)_3 \right\}^+$$

20.24

$$\left[{}^{2-}\!:\ddot{\text{N}}-\text{C}\equiv\text{N}: \quad :\text{N}\equiv\text{C}-\ddot{\text{N}}:^{2-} \quad {}^{-}\!:\ddot{\text{N}}=\text{C}=\ddot{\text{N}}:^{-} \right]$$

Linear: *sp* hybridization of carbon. Bonds of equal length (longer than triple, shorter than single); charge equally distributed between nitrogens.

20.25 Hydrolysis of nitrile and amide (or of diimide). Products are urea (intermediate), calcium carbonate, ammonia.

20.26 (a) Carbon–nitrogen triple bond (nitrile group). For example:

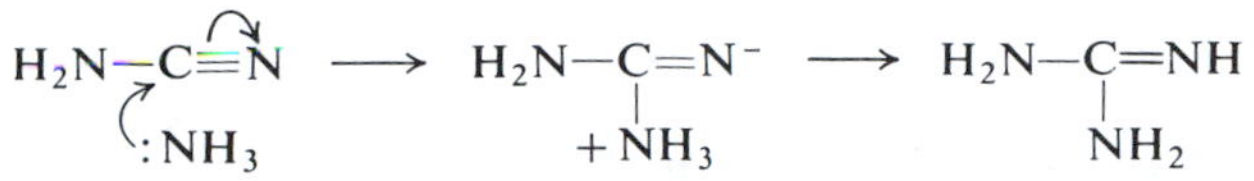

$$H_2N{-}C{\equiv}N + :NH_3 \longrightarrow H_2N{-}\underset{\displaystyle {}^{+}NH_3}{\underset{|}{C}}{=}N^- \longrightarrow H_2N{-}\underset{\displaystyle NH_2}{\underset{|}{C}}{=}NH$$

Guanidine

(b) Nucleophilic addition (followed by proton shifts).

(c) H_2O, MeOH, H_2S, NH_3 are nucleophilic reagents that add to the carbon–nitrogen triple bond of cyanamide. In acid, protonation of $\equiv$N occurs first; in aqueous base, OH^- is the nucleophile.

20.27 (a) RCOCl

(b) $RCOO^-NH_4^+$, $RCONH_2$, RCN, amides of low mol. wt. amines

(c) $RCOO^-NH_4^+$ (d) $(RCO)_2O$ (e) RCOOR′

20.28 (a)

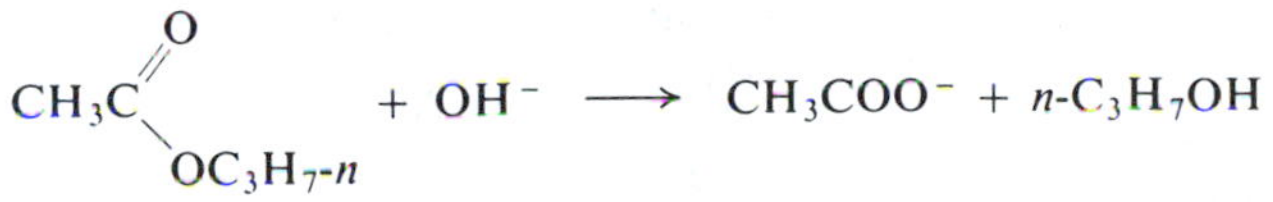

$$CH_3C(=O)OC_3H_7\text{-}n + OH^- \longrightarrow CH_3COO^- + n\text{-}C_3H_7OH$$

n-Propyl acetate

One mole of the ester requires one mole of OH^- for hydrolysis (saponification): the saponification equivalent (S.E.) = m.w. of the ester = 102 ($C_5H_{10}O_2$).

(b) All eight have the same formula, $C_5H_{10}O_2$, with four carbons divided between acid and alcohol. In RCOOR′, therefore, R can be H, Me, Et, *n*-Pr, *iso*-Pr; R′ can be Me, Et, *n*-Pr, *iso*-Pr, *n*-Bu, *iso*-Bu, *sec*-Bu, *tert*-Bu (but not H). The nine allowable combinations are:

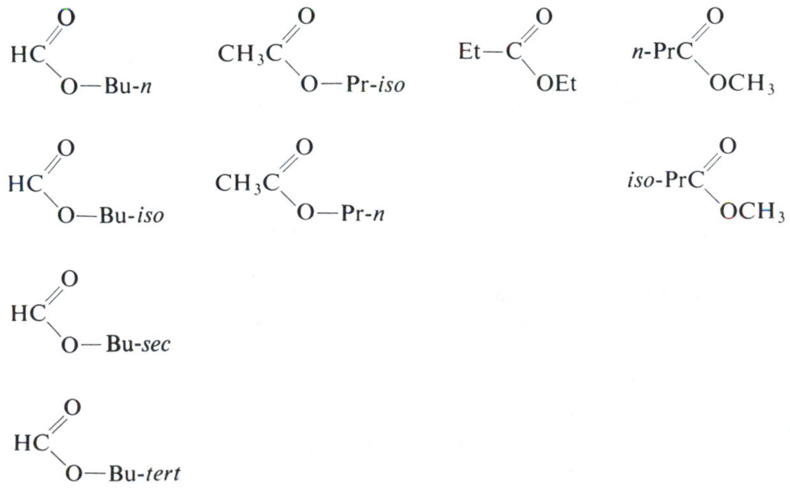

$HC(=O){-}O{-}Bu\text{-}n$ $CH_3C(=O){-}O{-}Pr\text{-}iso$ $Et{-}C(=O)OEt$ $n\text{-}PrC(=O)OCH_3$

$HC(=O){-}O{-}Bu\text{-}iso$ $CH_3C(=O){-}O{-}Pr\text{-}n$ $iso\text{-}PrC(=O)OCH_3$

$HC(=O){-}O{-}Bu\text{-}sec$

$HC(=O){-}O{-}Bu\text{-}tert$

(c) In RCOOH, $C_5H_{10}O_2$, R must be a four-carbon group: *n*-Bu, *iso*-Bu, *sec*-Bu, *tert*-Bu are possible, giving rise to only four acids of equiv. wt. 102.

(d) Less useful, because of the great number of possible combinations of R and R′ in RCOOR′.

20.29 (a) $C_6H_4(COOMe)_2 + 2OH^- \longrightarrow C_6H_4(COO^-)_2 + 2MeOH$

Methyl phthalate
M.w. 194

Uses 2 moles OH^-; S.E. = m.w./2 = 194/2 = 97.

(b) S.E. = m.w./number of ester groups per molecule.

(c) S.E. = m.w./3 = $C_{57}H_{110}O_6$/3 = 890/3 = 297.

1. (a) In the order in which their structures appear in the answer to Problem 20.28(b) above:

n-butyl formate	isopropyl acetate	ethyl propionate	methyl *n*-butyrate
isobutyl formate	*n*-propyl acetate		methyl isobutyrate
sec-butyl formate			
tert-butyl formate			

(b) These obviously contain an aromatic ring. For example, $C_8H_8O_2 - COOCH_3 = C_6H_5$.

$C_6H_5-C(=O)-OCH_3$
Methyl benzoate

$C_6H_5CH_2O-CH(=O)$
Benzyl formate

$C_6H_5O-C(=O)-CH_3$
Phenyl acetate

$o\text{-}CH_3C_6H_4O-CH(=O)$
o-Cresyl formate

$m\text{-}CH_3C_6H_4O-CH(=O)$
m-Cresyl formate

$p\text{-}CH_3C_6H_4O-CH(=O)$
p-Cresyl formate

(c) These are diesters (accounting for the four oxygens):

$$\begin{array}{l} C_7H_{12}O_4 \\ \underline{-\ 2\ COOCH_3} \\ C_3H_6 \end{array}$$

C_3H_6 means satd. open-chain: $-C-C-C-$, $-C-C(C)-$, $-C(C)(C)-$

The three possibilities are:

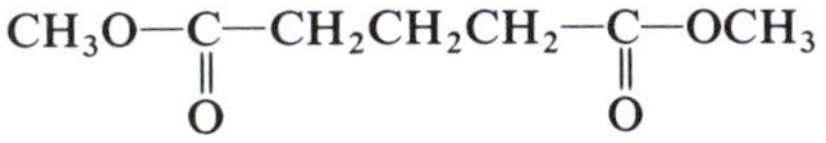

$$CH_3O-\underset{\underset{O}{\|}}{C}-CH_2CH_2CH_2-\underset{\underset{O}{\|}}{C}-OCH_3$$

Methyl glutarate

$$CH_3O-\underset{\underset{O}{\|}}{C}-CH_2-\overset{\overset{CH_3}{|}}{CH}-\underset{\underset{O}{\|}}{C}-OCH_3$$

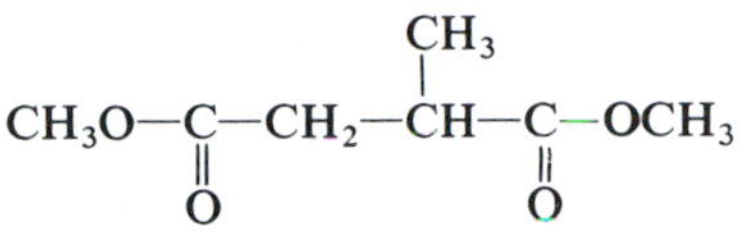

Methyl α-methylsuccinate

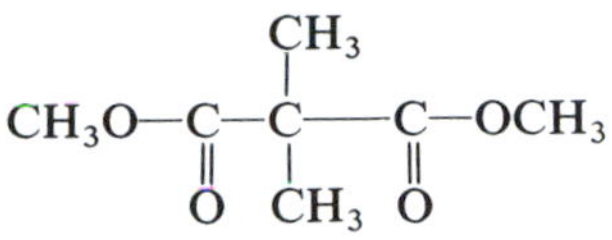

$$CH_3O-\underset{\underset{O}{\|}}{C}-\overset{\overset{CH_3}{|}}{\underset{\underset{CH_3}{|}}{C}}-\underset{\underset{O}{\|}}{C}-OCH_3$$

Methyl dimethylmalonate

2. (a) $n\text{-}C_3H_7COOH$
n-Butyric acid

(b) $n\text{-}C_3H_7COOCH(CH_3)_2$
Isopropyl *n*-butyrate

(c) 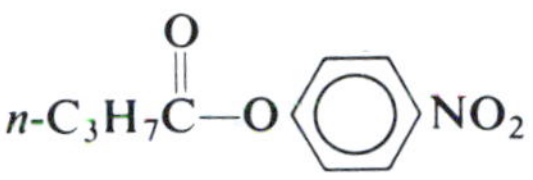 $n\text{-}C_3H_7\overset{\overset{O}{\|}}{C}-O-C_6H_4-NO_2$
p-Nitrophenyl *n*-butyrate

(d) $n\text{-}C_3H_7CONH_2$
n-Butyramide

(e) 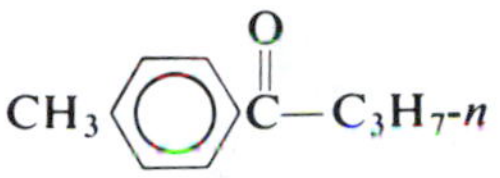 $CH_3-C_6H_4-\overset{\overset{O}{\|}}{C}-C_3H_7\text{-}n$
n-Propyl *p*-tolyl ketone
(*p*-Methyl-*n*-butyrophenone)

(f) no reaction

(g) $n\text{-}C_3H_7COONa$
Sodium *n*-butyrate

(h) $n\text{-}C_3H_7COOEt$ and AgCl
Ethyl *n*-butyrate

(i) 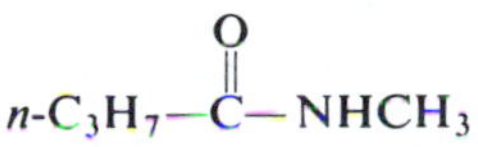 $n\text{-}C_3H_7-\overset{\overset{O}{\|}}{C}-NHCH_3$
N-Methyl-*n*-butyramide
+
$CH_3NH_3^+Cl$
Methylammonium chloride

(j) 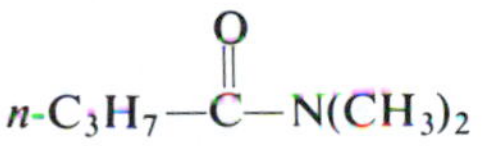 $n\text{-}C_3H_7-\overset{\overset{O}{\|}}{C}-N(CH_3)_2$
N,*N*-Dimethyl-*n*-butyramide
+
$(CH_3)_2NH_2^+Cl^-$
Dimethylammonium chloride

(k) no reaction

(l) 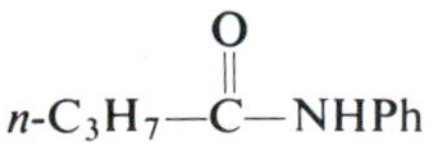 $n\text{-}C_3H_7-\overset{\overset{O}{\|}}{C}-NHPh$
n-Butyranilide
+
$PhNH_3^+Cl^-$
Anilinium chloride

(m) 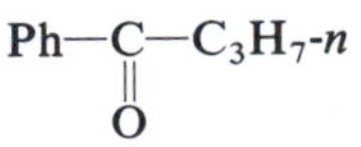 $Ph-\underset{\underset{O}{\|}}{C}-C_3H_7\text{-}n$
n-Butyrophenone
(Phenyl *n*-propyl ketone)

(n) 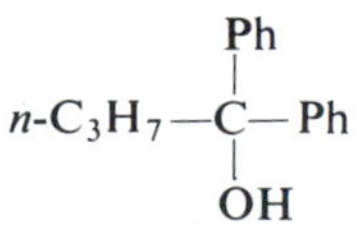 $n\text{-}C_3H_7-\overset{\overset{Ph}{|}}{\underset{\underset{OH}{|}}{C}}-Ph$
1,1-Diphenyl-1-butanol

3. In (b), (c), (e), and (l), an equimolar amount of acetic acid is formed along with the indicated product.

(a) $2CH_3COOH$
Acetic acid
(*Two moles*)

(b) $CH_3COOCH(CH_3)_2$
Isopropyl acetate

(c) $CH_3-\overset{O}{\overset{\|}{C}}-O-C_6H_4-NO_2$
p-Nitrophenyl acetate

(d) $CH_3CONH_2 + CH_3COO^-NH_4^+$
Acetamide Ammonium acetate

(e) $CH_3-C_6H_4-\underset{\underset{O}{\|}}{C}-CH_3$
Methyl *p*-tolyl ketone
(*p*-Methylacetophenone)

(f) no reaction

(g) $CH_3COO^-Na^+$
Sodium acetate

(h) $CH_3COOEt + CH_3COOAg$
Ethyl acetate Silver acetate

(i) $CH_3-\overset{O}{\overset{\|}{C}}-NHCH_3$
N-Methylacetamide
+
$CH_3COO^-\,{}^+H_3NCH_3$
Methylammonium acetate

(j) $CH_3-\overset{O}{\overset{\|}{C}}-N(CH_3)_2$
N,*N*-Dimethylacetamide
+
$CH_3COO^-\,{}^+H_2N(CH_3)_2$
Dimethylammonium acetate

(k) no reaction

(l) $CH_3CONHPh$
Acetanilide

4. (a) $Na^+\,{}^-OOCCH_2CH_2COO^-Na^+$
Sodium succinate

(b) $NH_4^+\,{}^-OOCCH_2CH_2CONH_2$
Ammonium succinamate

(c) $HOOC(CH_2)_2CONH_2$
Succinamic acid

(d) Cyclic imide: $H_2C-C(=O)-NH-C(=O)-CH_2$ (ring)
Succinimide

(e) $HOOC(CH_2)_2COOCH_2Ph$
Benzyl hydrogen succinate

(f) $CH_3-C_6H_4-\underset{\underset{O}{\|}}{C}(CH_2)_2COOH$
β-(*p*-Toluyl)propionic acid

5. (a) $PhCH_2COOH + NH_4Cl$ (b) $PhCH_2COO^-Na^+ + NH_3$

6. (a) $PhCH_2COOH + NH_4Cl$ (b) $PhCH_2COO^-Na^+ + NH_3$

7. In each case an equimolar amount of methanol is formed along with the indicated product.

(a) n-PrCOOH
n-Butyric acid

(b) n-PrCOO$^-$K$^+$
Potassium *n*-butyrate

(c) n-PrCOOCH$(CH_3)_3$
Isopropyl *n*-butyrate

(d) n-PrCOOCH$_2$Ph
Benzyl *n*-butyrate

(e) n-PrCONH$_2$
n-Butyramide

(f) $Ph_2C(OH)CH_2CH_2CH_3$
1,1-Diphenyl-1-butanol

(g) $(iso\text{-}Bu)_2C(OH)CH_2CH_2CH_3$
2,6-Dimethyl-4-*n*-propyl-4-heptanol

(h) n-PrCH$_2$OH
n-Butyl alcohol

8. (a) $Ph-\overset{O}{\overset{\|}{C}}-{}^{18}OCH_3 \xleftarrow{Ph-\overset{O}{\overset{\|}{C}}-Cl} CH_3{}^{18}OH \longleftarrow CH_3Br + H_2{}^{18}O$

(b) $Ph-\overset{^{18}O}{\overset{\|}{C}}-OCH_3 \xleftarrow{MeOH} Ph-\overset{^{18}O}{\overset{\|}{C}}-Cl \xleftarrow{SOCl_2} Ph-\overset{^{18}O}{\overset{\|}{C}}-{}^{18}OH \xleftarrow{H^+} PhC{\equiv}N + H_2{}^{18}O$

$Ph-\overset{^{18}O}{\overset{\|}{C}}-Cl \longleftarrow PhCCl_3 + H_2{}^{18}O$ (one mole)

(c) $Ph-\overset{^{18}O}{\overset{\|}{C}}-{}^{18}OCH_3 \longleftarrow Ph-\overset{^{18}O}{\overset{\|}{C}}-Cl$ (from b) + $CH_3{}^{18}OH$ (from a)

Hydrolysis of $PhCO^{18}OCH_3$ or $PhC^{18}O^{18}OCH_3$ gives $CH_3{}^{18}OH$; hydrolysis of $PhC^{18}OOCH_3$ gives CH_3OH.

9. (a) $Et \vdots {}^{14}\underset{\underset{O}{\|}}{C}-Me \xleftarrow{Et_2CuLi} Cl-{}^{14}\underset{\underset{O}{\|}}{C}-Me \xleftarrow{PCl_3} HOO^{14}C-Me \longleftarrow {}^{14}CO_2 + MeMgBr$

(b) $Et-\underset{\underset{O}{\|}}{C} \vdots {}^{14}CH_3 \xleftarrow{EtCOCl} ({}^{14}CH_3)_2CuLi \xleftarrow{CuI} {}^{14}CH_3Li \xleftarrow{Li} \xleftarrow{PBr_3} {}^{14}CH_3OH$

(c) $CH_3-{}^{14}CH_2 \vdots \underset{\underset{O}{\|}}{C}-CH_3 \xleftarrow{CH_3COCl} (CH_3{}^{14}CH_2)_2CuLi \xleftarrow{CuI} CH_3{}^{14}CH_2Li$

$CH_3{}^{14}CH_2Li \xleftarrow{Li} \xleftarrow{PBr_3} CH_3{}^{14}CH_2OH \xleftarrow{LiAlH_4} CH_3{}^{14}COOH \xleftarrow{CH_3MgBr} {}^{14}CO_2$

(d) $^{14}CH_3CH_2 \vdots COCH_3 \xleftarrow{CH_3COCl} (^{14}CH_3CH_2)_2CuLi \xleftarrow{\text{as in (c)}} {}^{14}CH_3CH_2OH$

$^{14}CH_3 \vdots CH_2OH \xleftarrow{HCHO} {}^{14}CH_3MgBr \longleftarrow \text{as in (b)}$

(e) $Ph^{14}CH_2CH_3 \xleftarrow[HCl]{Zn(Hg)} Ph^{14}C(=O) \vdots CH_3 \xleftarrow{Me_2CuLi} Ph^{14}COCl \xleftarrow{PCl_5} Ph^{14}COOH \xleftarrow{PhMgBr} {}^{14}CO_2$

(f) $PhCH_2{}^{14}CH_3 \xleftarrow[HCl]{Zn(Hg)} Ph{-}C(=O) \vdots {}^{14}CH_3 \xleftarrow{PhCOCl} (^{14}CH_3)_2CuLi \longleftarrow \text{as in (b)}$

(g) $Et{-}C(={}^{18}O){-}Me \xleftarrow{Et_2CuLi} Cl{-}C(={}^{18}O){-}Me \xleftarrow{PCl_3} CH_3C^{18}O_2H \xleftarrow{CH_3C{\equiv}N} H_2{}^{18}O$

10. (a) An ester reacts with ammonia to give an amide and an alcohol.

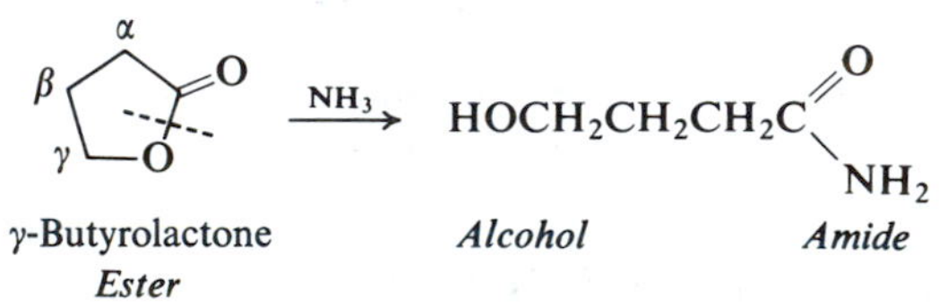

(b) An ester is reduced by $LiAlH_4$ to two alcohols.

(c) An ester undergoes transesterification.

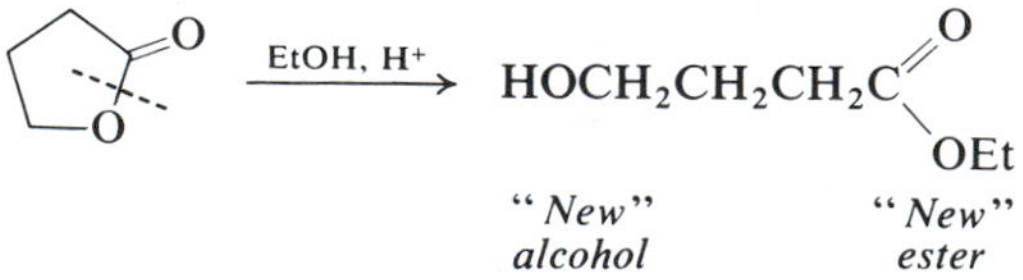

11. There are just three steps. A bond to chiral carbon cannot be broken in the first step,

$$\text{Ts}\vdots\text{Cl} \quad \text{H}\vdots\text{O}-\underset{\text{Me}}{\overset{\text{Et}}{\text{C}}}-\text{H} \longrightarrow \text{Ts}-\text{O}\vdots\underset{\text{Me}}{\overset{\text{Et}}{\text{C}}}-\text{H} \xrightarrow[{^-}\text{OOCPh}]{\text{inversion}} \text{H}-\underset{\text{Me}}{\overset{\text{Et}}{\text{C}}}-\text{O}\vdots\overset{\text{O}}{\overset{\|}{\text{C}}}\text{Ph} \xrightarrow{\text{HO}^-} \text{H}-\underset{\text{Me}}{\overset{\text{Et}}{\text{C}}}-\text{O}-\text{H}$$

(+)-Alcohol (+)-Tosylate (−)-Benzoate (−)-Alcohol

because of the very nature of the reactants and products, and is not broken in the third step (Sec. 6.14). Inversion *must*, then, take place in the attack by benzoate ion on *sec*-butyl tosylate, evidently in an S_N2 reaction.

12. Here, as in Problem 20.18, there is competition between a bimolecular reaction and a unimolecular reaction. With a high concentration of nucleophile (5 M NaOH) there is (bimolecular) nucleophilic acyl substitution, with cleavage of the bond between oxygen and the acyl group:

$$\text{RCO}\vdots\text{OR}'$$

The bond to the chiral carbon is not broken, and there is retention of configuration in the product.

With a low concentration of nucleophile (dil. NaOH), the bimolecular reaction is outpaced by the unimolecular reaction: an S_N1 reaction of the neutral ester, made possible by the stability of the benzylic–allylic cation generated. Both esters give the same hybrid

$$\left.\begin{array}{l}\text{PhCH(OCOR)}-\text{CH}=\text{CH}-\text{CH}_3 \\ \text{PhCH}=\text{CH}-\text{CH(OCOR)}-\text{CH}_3\end{array}\right] \xrightarrow{-\text{RCOO}^-} \underbrace{\text{PhCH}\text{:::}\text{CH}\text{:::}\text{CH}}_{\oplus}-\text{CH}_3 \xrightarrow[\text{H}_2\text{O}]{\text{OH}^-} \text{PhCH}=\text{CH}-\underset{\text{OH}}{\underset{|}{\text{CH}}}-\text{CH}_3$$

cation and hence yield the same alcohol—the one with the double bond in conjugation with the ring. Configuration is lost in the achiral cation, and hence the alcohol is optically inactive.

13. (a) $NaHCO_3$(aq) gives CO_2 with acid.

(b) EtOH gives a pleasant-smelling ester from RCOCl; RCOCl also gives CO_2 from $NaHCO_3$(aq).

(c) NaOH, heat; test for NH_3 by litmus.

(d) Br_2/CCl_4, or $KMnO_4$.

(e) NaOH, heat; test for NH_3 by litmus. Also see slow dissolution of nitrile.

(f) $NaHCO_3$(aq), warm.

(g) CrO_3/H_2SO_4.

(h) NH_4^+ salt gives immediate evolution of NH_3 by the action of cold NaOH(aq).

(i) Alcoholic $AgNO_3$.

14. (a) (b) (c)

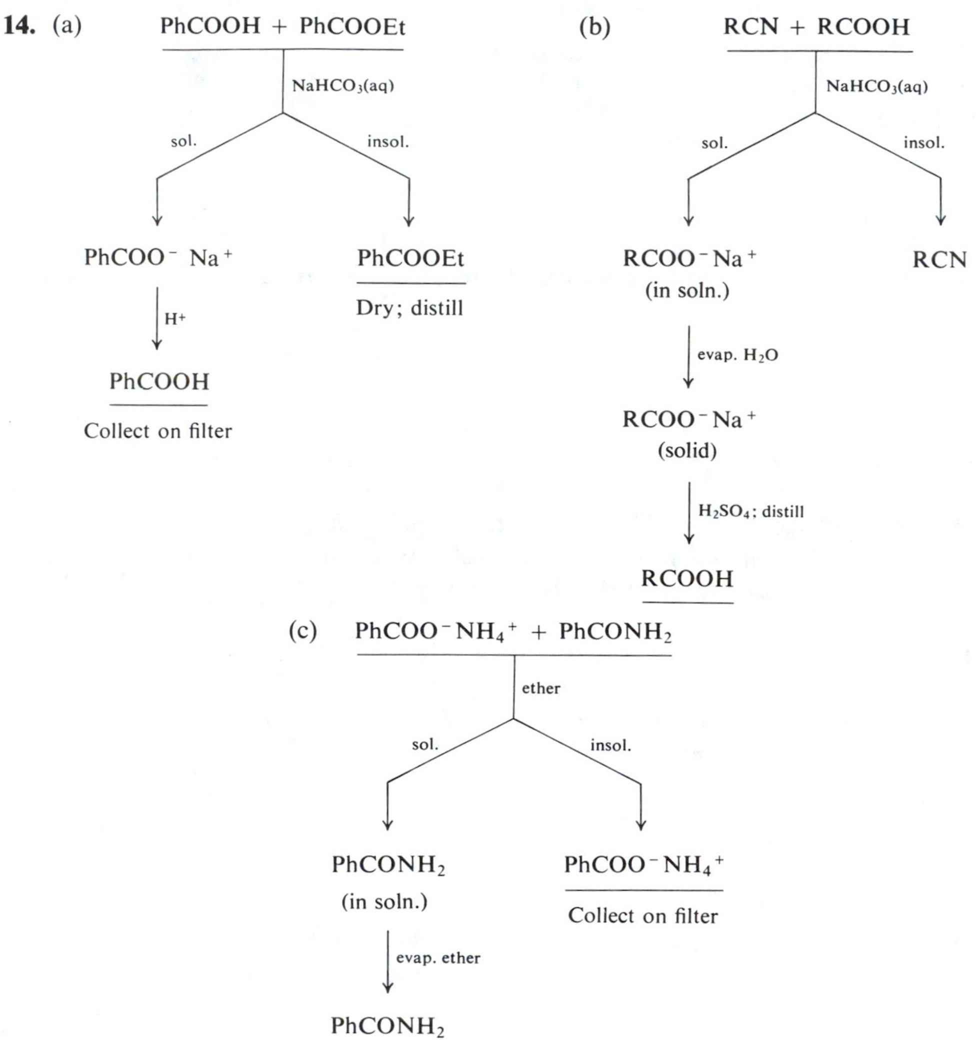

15. The THP ester is formed by acid-catalyzed addition of RCOOH to the double bond, made easy by the stability of the intermediate cation I, which is actually an oxonium ion:

Dihydropyran (DHP) $+ H^+ \rightleftharpoons$ I $\xrightleftharpoons{RCOOH}$ R—C(OH)=O⁺(THP) $\rightleftharpoons$ R—C(=O)—O—THP $+ H^+$

Tetrahydropyranyl ester (RCOOTHP)

Acid-catalyzed hydrolysis of the THP ester is also easy, since it involves formation of the same cation I. A cyclic hemiacetal is formed and, like other hemiacetals, regenerates the aldehyde and alcohol, in this case parts of the same molecule. (Compare Problem 20, Chapter 18, page 704.)

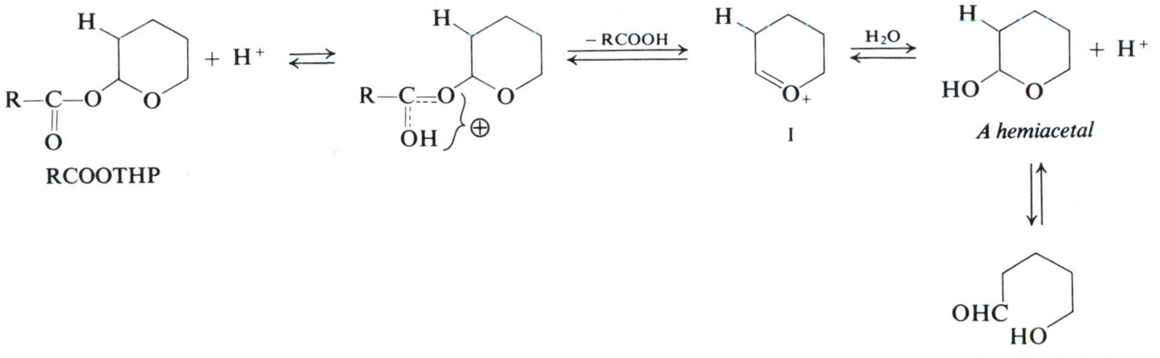

RCOOTHP + H^+ ⇌ [protonated ester] $\xrightarrow{-RCOOH}$ I $\xrightarrow{H_2O}$ A hemiacetal + H^+ ⇌ Aldehyde–alcohol

RCOOTHP

I

A hemiacetal

OHC HO

Aldehyde–alcohol

16. $CH_3COCH_2COCH_3$ $\xrightarrow[H^+]{CN^-}$ $\xrightarrow[H^+]{H_2O}$

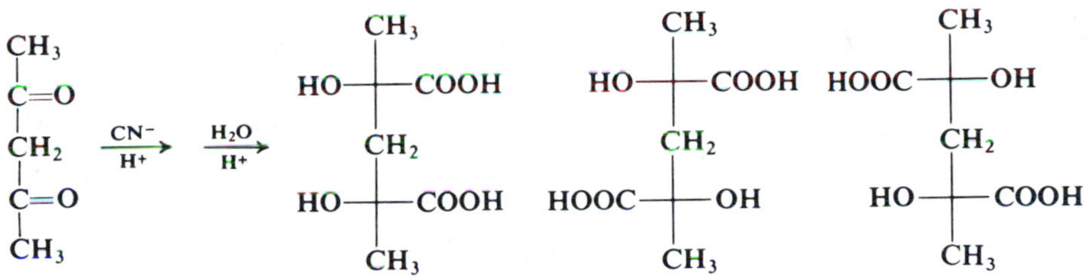

meso
A

racemic
B

B is the racemic modification. It gives monolactone II; but since the remaining —OH and —COOH are *cis*, II can react further to form the dilactone.

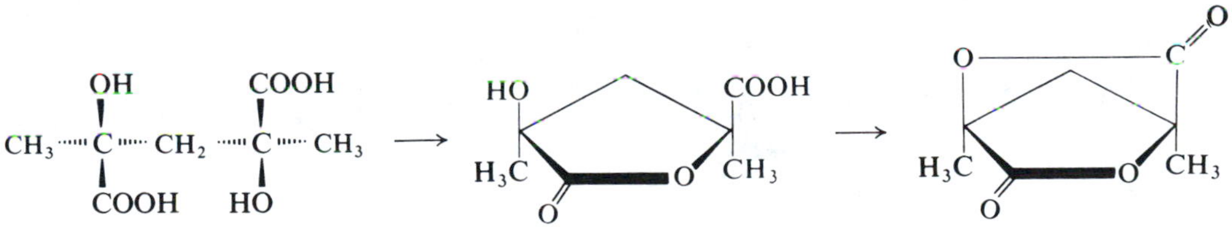

B → II (*Monolactone*) → *Dilactone*

A is the *meso* compound. It gives monolactone III; here, the remaining —OH and —COOH are *trans*, and further reaction is not possible.

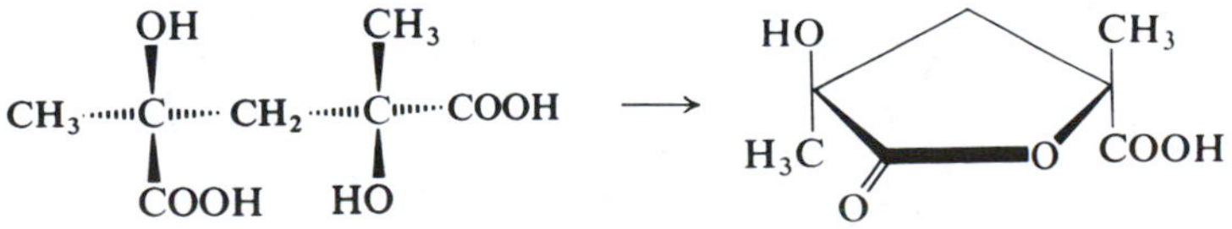

A → III (*Monolactone*)

17. (a) $H_2N{-}\overset{\overset{O}{\|}}{C}{-}NH_2 \xrightarrow[H_2O]{NaOH} 2NH_3 + Na_2CO_3$

Di-amide C

(b) $Cl{-}C({=}O){-}Cl \xrightarrow{EtOH} EtO{-}C({=}O){-}Cl \xrightarrow{NH_3} EtO{-}C({=}O){-}NH_2$

(Di)acid chloride — *Ester–acid chloride* — *Ester–amide* D

(c) $PhBr \xrightarrow{Mg} PhMgBr$ (E) $\xrightarrow{CH_2{-}CH_2\ (O)} \xrightarrow{H^+} PhCH_2CH_2OH$ (F) $\xrightarrow{PBr_3} PhCH_2CH_2Br$ (G)

$PhCH_2CH_2Br$ (G) $\xrightarrow{CN^-} PhCH_2CH_2C{\equiv}N$ (H) $\xrightarrow[H^+]{H_2O} PhCH_2CH_2COOH + NH_4^+$ (I)

$C_6H_5CH_2CH_2COOH$ (I) $\xrightarrow{SOCl_2} C_6H_5CH_2CH_2C({=}O)Cl$ (J) $\xrightarrow{HF}$ K *Friedel–Crafts acylation*

K

1-Indanone

(See Problem 13a, page 703)

K $\xrightarrow{H_2,\ cat.}$ L (H, OH) $\xrightarrow[-H_2O]{H^+,\ heat}$ M

K — L, 1-Indanol — M, Indene

(d) *trans* (ring with CH_3, H / H, OH) $\xrightarrow{CH_3COCl}$ N (ring with CH_3, H / H, $O{-}CCH_3$, C=O) $\xrightarrow{H_2O,\ OH^-}$ O (*trans*) (ring with CH_3, H / H, OH)

trans — N — O (*trans*)

18. (a) Erythrose is an aldehyde and contains three —OH's. Cleavage by $3HIO_4$ shows that it has the following structure, containing two chiral carbons:

$$H{-}CH{=}O + HC({=}O)OH + HC({=}O)OH + HC({=}O)OH \longleftarrow H{-}\underset{OH}{\overset{H}{C}}{-}\overset{*}{\underset{OH}{\overset{H}{C}}}{-}\overset{*}{\underset{OH}{\overset{H}{C}}}{-}\overset{H}{C}{=}O$$

(HIO_4) (HIO_4) (HIO_4) — Erythrose

Oxidation of (−)-erythrose yields a dicarboxylic acid, $HOO\overset{*}{C}CHOH\overset{*}{C}HOHCOOH$. Since the acid is optically inactive, it must have the *meso* configuration, showing that (−)-erythrose is the 2*R*,3*R* compound or its enantiomer.

CHO; H—OH; H—OH; CH_2OH (2*R*,3*R*) $\xrightarrow{HNO_3}$ COOH; H—OH; H—OH; COOH (*meso*-acid)

or

CHO; HO—H; HO—H; COOH (2*S*,3*S*) $\xrightarrow{HNO_3}$ COOH; HO—H; HO—H; COOH (same *meso*-acid)

(−)-Threose must be a diastereomer of erythrose, either the 2*S*,3*R* or 2*R*,3*S* isomer.

CHO; HO—H; H—OH; CH_2OH (2*S*,3*R*) $\xrightarrow{HNO_3}$ COOH; HO—H; H—OH; COOH (Active acid)

CHO; H—OH; HO—H; CH_2OH (2*R*,3*S*) $\xrightarrow{HNO_3}$ COOH; H—OH; HO—H; COOH (Active acid)

(b) CHO; H—OH; CH_2OH (*R*-Glyceraldehyde) $\xrightarrow{CN^-,\ H^+}$ CN; H—OH; H—OH; CH_2OH + CN; HO—H; H—OH; CH_2OH

each $\downarrow H^+,\ H_2O$

CHO; H—OH; H—OH; CH_2OH ((−)-Erythrose) $\xrightarrow{Br_2,\ H_2O}$ COOH; H—OH; H—OH; CH_2OH

COOH; HO—H; H—OH; CH_2OH $\xleftarrow{Br_2,\ H_2O}$ CHO; HO—H; H—OH; CH_2OH ((−)-Threose)

(Compare with Problem 14, Chapter 18, page 703.)

The names of these compounds are the basis for the designations *erythro* and *threo* (see page 380) used to specify certain configurations of compounds containing two chiral carbons. The *erythro* isomer is the one that is convertible (in principle, at least) into a *meso* structure, whereas the *threo* isomer is convertible into a racemic modification.

19. (a) The three signals of equal area in the room temperature NMR spectrum must, of course, come from the three methyl groups of the amide. The two methyl groups on nitrogen are

$$CH_3-C(=O)-N(CH_3)_2$$

not equivalent, showing hindered rotation about the C–N bond. Even at room temperature, rotation is slow enough that the NMR spectrometer sees these groups separately: one *cis* and the other *trans* to oxygen. As the temperature is raised, rotation speeds up, and at 110 °C the two groups are seen in a single average environment.

This hindered rotation suggests double-bond character for the C–N bond, which we attribute to resonance involving these structures:

$$CH_3-C(=O)-N(CH_3)_2 \qquad CH_3-C(-O^-)=N^+(CH_3)_2$$

As we shall see later (Sec. 36.8), there is other evidence supporting partial double-bond character for the C–N bond in amides: the bond length and the coplanarity of the six atoms involved.

(b) We would expect the CMR spectrum to show the same temperature effect as the proton NMR spectrum: three signals at room temperature, with two of them coalescing into one at 110 °C. This has actually been observed.

Abraham, R. J.; Loftus, P. *Proton and Carbon-13 NMR Spectroscopy*; Heyden: Philadelphia, 1978; pp 171–174.

20. (a) $CH_3C(=O)OCH_2CH_3$
Ethyl acetate

(b) $CH_2{=}C(CH_3){-}COOH$
Methacrylic acid

(c) $C_6H_5CH_2CONH_2$
Phenylacetamide

(See the labeled spectra on page 706 of this Study Guide.)

21. (a) $HC(=O)OCH_2CH_2CH_3$
n-Propyl formate

(b) $CH_3CH_2C(=O)OCH_3$
Methyl propionate

(c) $CH_3C(=O)OCH_2CH_3$
Ethyl acetate

(See the labeled spectra on page 707 of this Study Guide.)

22. $CH_3O-C_6H_4-C(=O)OCH_2CH_3$
Ethyl *p*-methoxybenzoate

(See the labeled spectra on page 708 of this Study Guide.)

23. (a) 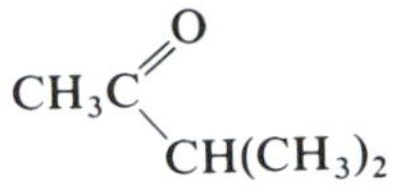 $CH_3C(=O)OCH(CH_3)_2$

Isopropyl acetate

(b) 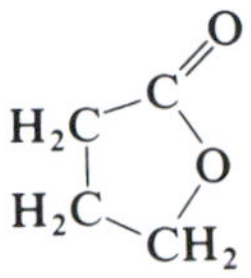 H_2C–$C(=O)$–O–CH_2–H_2C

γ-Butyrolactone

(See the labeled spectra on page 708 of this Study Guide.)

24. (a) 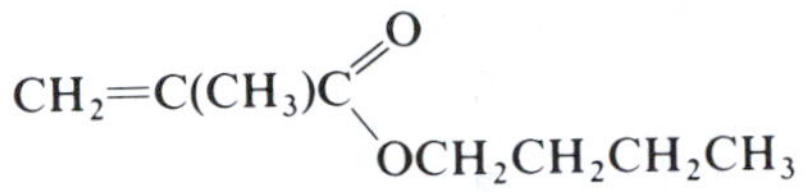 $CH_2{=}C(CH_3)C(=O)OCH_2CH_2CH_2CH_3$

n-Butyl methacrylate

(b) 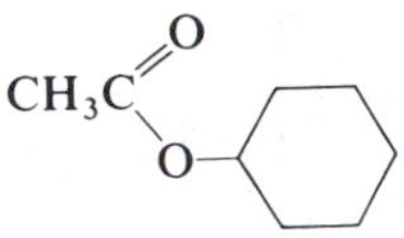 $CH_3C(=O)O$–cyclohexyl

Cyclohexyl acetate

(c) 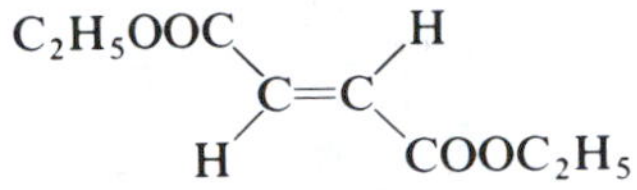 $C_2H_5OOC(H)C{=}C(H)COOC_2H_5$

Diethyl fumarate

(See the labeled spectra on page 709 of this Study Guide.)

25. 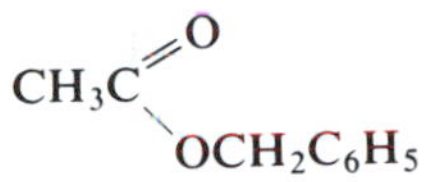 $CH_3C(=O)OCH_2C_6H_5$

Benzyl acetate

P

$C_6H_5CH_2C(=O)OCH_3$

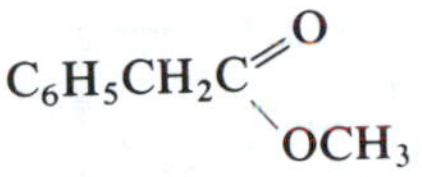

Methyl phenylacetate

Q

$C_6H_5CH_2CH_2COOH$

Hydrocinnamic acid

R

(See the labeled spectra on pages 710 and 711 of this Study Guide.)

26. 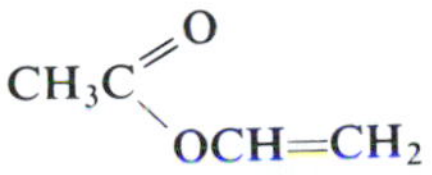 $CH_3C(=O)OCH{=}CH_2$

Vinyl acetate

S

(See the labeled spectra on page 712 of this Study Guide.)

Now try your hand at this problem. The answer is on page 675 of this Study Guide.

Problem 20.30 Give a structure or structures consistent with each of the proton NMR spectra shown in Figure 20.11 on page 362 of this Study Guide.

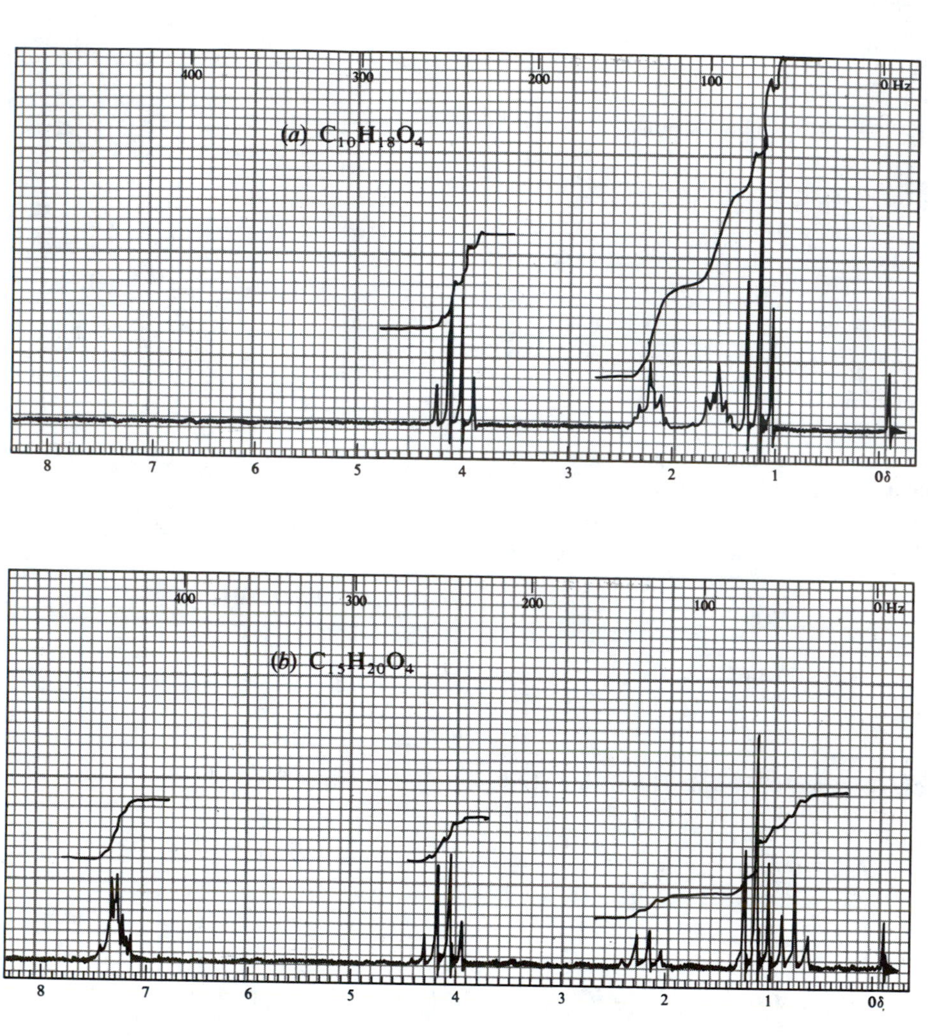

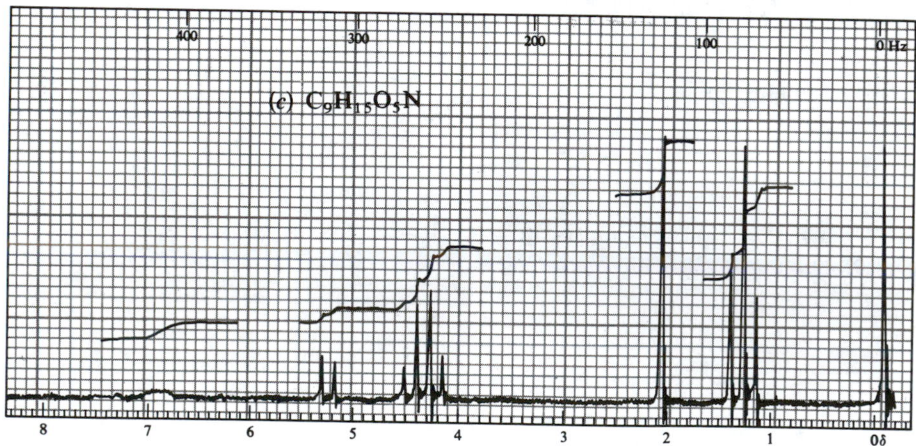

Figure 20.11 Proton NMR spectra for Problem 20.30, p. 361 of this Study Guide.

21

Carbanions I

Aldol and Claisen Condensations

21.1 III, in which the negative charge resides on oxygen, the atom that can best accommodate it.

21.2 Hydrogen on the carbon lying between the two carbonyl groups is the most acidic. Removal of a proton yields an anion that is highly stabilized by accommodation of the negative charge by two oxygens (rather than by only one).

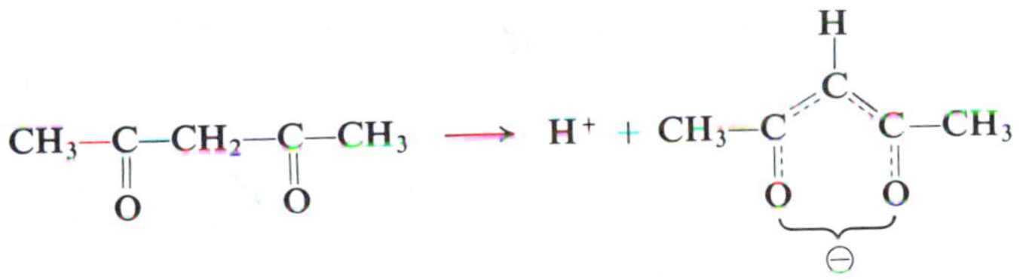

$$CH_3-\underset{\underset{O}{\|}}{C}-CH_2-\underset{\underset{O}{\|}}{C}-CH_3 \longrightarrow H^+ + CH_3-\underset{\underset{O}{\vdots}}{C}\overset{\overset{H}{|}}{\cdots C \cdots}\underset{\underset{O}{\vdots}}{C}-CH_3 \quad (\ominus \text{ shared by both O})$$

21.3 Through resonance like that in free radicals (Sec. 16.15) and carbocations (Sec. 16.17), a phenyl group stabilizes a benzylic anion; the more phenyl groups there are, the greater the stabilization.

21.4 (a) CN^- adds to the conjugated system in the way that gives the more stable anion, the one in which the charge is partly accommodated by oxygen, the atom best able to accommodate it.

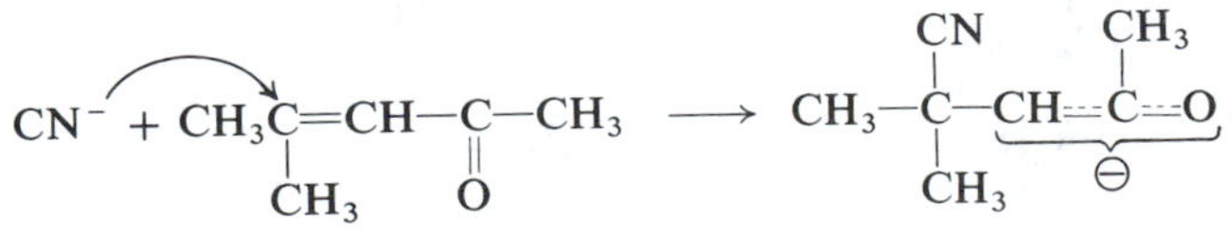

$$CN^- + CH_3\underset{\underset{CH_3}{|}}{C}{=}CH-\underset{\underset{O}{\|}}{C}-CH_3 \longrightarrow CH_3-\underset{\underset{CH_3}{|}}{\overset{\overset{CN}{|}}{C}}-\underbrace{CH{\cdots}\overset{\overset{CH_3}{|}}{C}{\cdots}O}_{\ominus}$$

(b) This anion accepts a proton at either the α-carbon or oxygen to yield either the keto or enol form of the product. In either case, the same product, chiefly keto, is finally obtained.

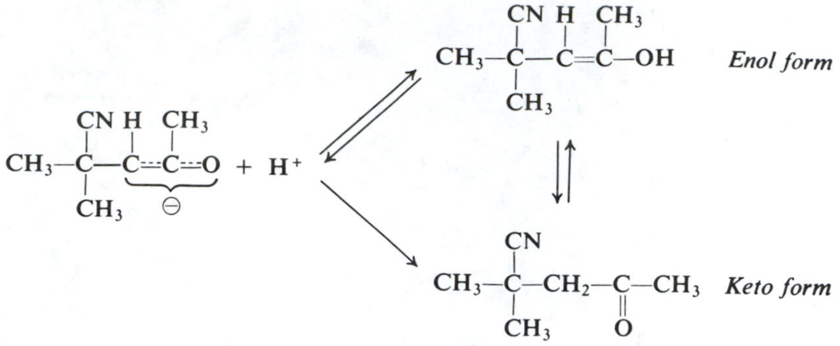

$$CH_3-\underset{CH_3}{\overset{CN}{C}}-\overset{H}{C}\overset{---}{=}\overset{CH_3}{C}\overset{---}{=}O^{\ominus} + H^+ \rightleftharpoons CH_3-\underset{CH_3}{\overset{CN}{C}}-\overset{H}{C}=\overset{CH_3}{C}-OH \quad \textit{Enol form}$$

$$\downarrow\uparrow$$

$$\longrightarrow CH_3-\underset{CH_3}{\overset{CN}{C}}-CH_2-\underset{O}{\overset{\|}{C}}-CH_3 \quad \textit{Keto form}$$

21.5 (a) (1) $(+)\text{-}Ph-\underset{O}{\overset{\|}{C}}-\overset{CH_3}{CH}-C_2H_5 + :B \longrightarrow Ph-\underset{O^{\ominus}}{C}\overset{---}{=}\overset{CH_3}{C}-C_2H_5 + B:H$ *Slow*

Optically active — *Achiral*

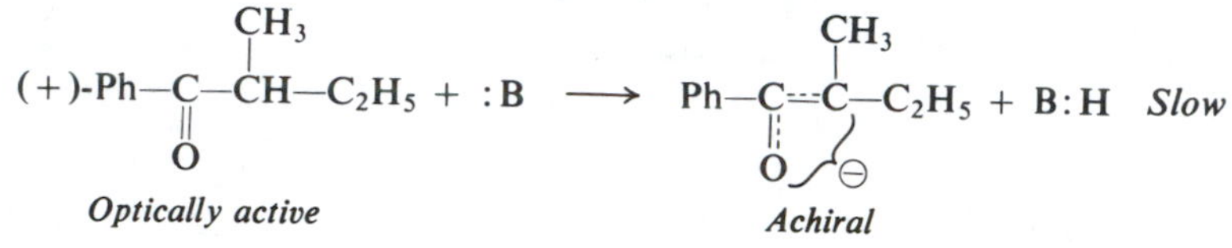

(2) $Ph-\underset{O^{\ominus}}{C}\overset{---}{=}\overset{CH_3}{C}-C_2H_5 + B:H \longrightarrow (\pm)\text{-}Ph-\underset{O}{\overset{\|}{C}}-\overset{CH_3}{CH}-C_2H_5 + :B$

Optically inactive: racemic modification

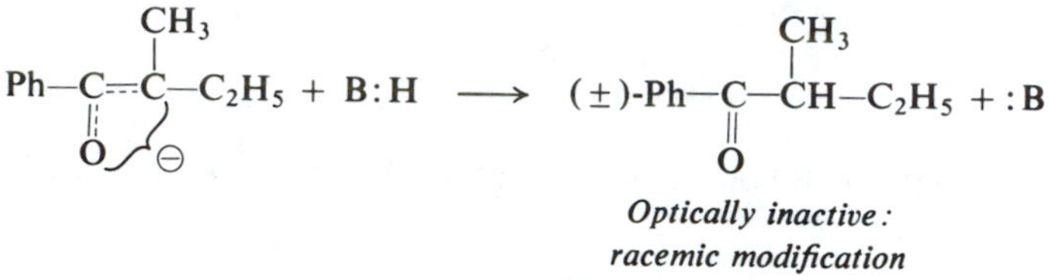

The rate-determining step in racemization is the same as in bromination: formation of the carbanion. The carbanion is flat (and hence *achiral*) to permit accommodation of the negative charge by oxygen, which requires overlap of the p orbital of negative

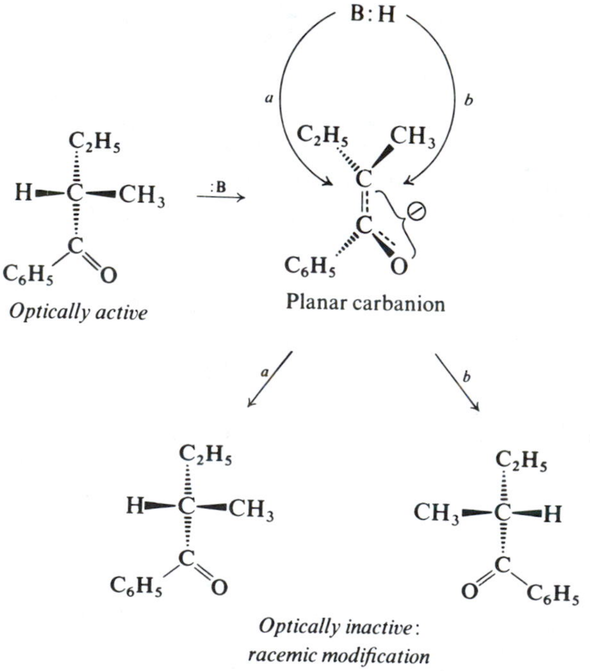

C_2H_5, H, CH_3, C_6H_5, C=O — *Optically active* $\xrightarrow{:B}$ Planar carbanion (C_2H_5, CH_3, C_6H_5, O, $\ominus$), protonated by B:H via path *a* or *b*

a → H—C(C_2H_5)—CH_3, C_6H_5C=O; *b* → CH_3—C(C_2H_5)—H, O=CC_6H_5

Optically inactive: racemic modification

carbon with the π cloud of the C=O group. When the carbanion takes back a proton from H:B the proton may become attached to either face of the carbanion and, depending upon which face, yields one or the other enantiomer. Since attachment to either face is equally likely, the enantiomers are formed in equal amounts.

(b) II can give a carbanion which, as in (a), loses configuration readily. III has no α-hydrogen, and cannot form a carbanion.

(c) The rate-determining step in hydrogen exchange is the same as in racemization, and hence in bromination: formation of the carbanion.

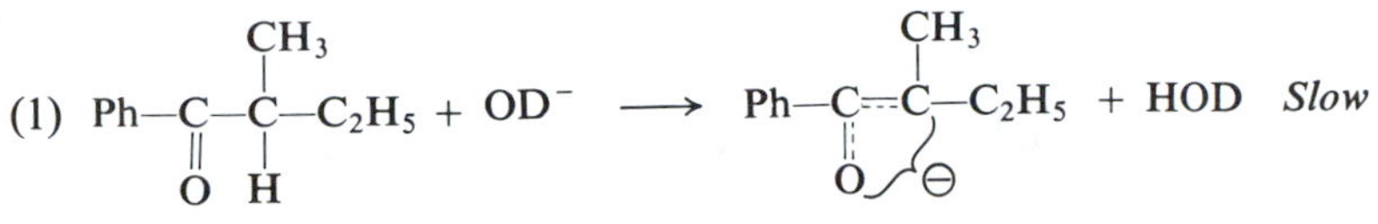

(1) Ph—C(=O)—C(CH$_3$)(H)—C_2H_5 + OD^- ⟶ Ph—C(⋯O$^\ominus$)⋯C(CH$_3$)—C_2H_5 + HOD *Slow*

Unlabeled / *Optically active* — *Hydrogen lost* / *Achiral*

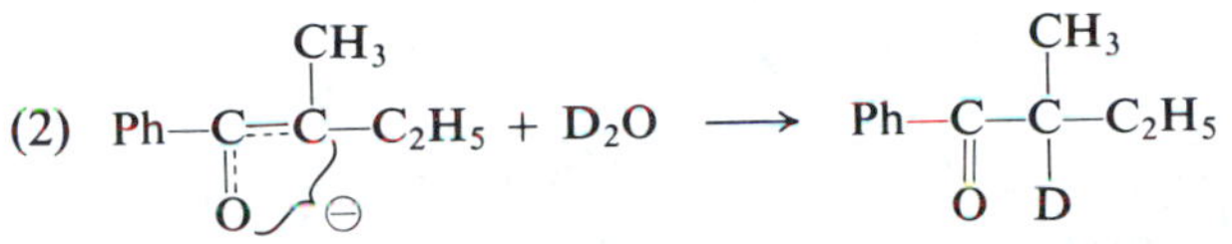

(2) Ph—C(⋯O$^\ominus$)⋯C(CH$_3$)—C_2H_5 + D_2O ⟶ Ph—C(=O)—C(CH$_3$)(D)—C_2H_5

Labeled with D / *Optically inactive*

This is the final link in the chain of evidence showing that the loss of α-hydrogen is the rate-determining step in halogenation. When the carbanion regains a hydrogen ion, (i) this hydrogen ion, by statistical chance, is equally likely to attack either face of the carbanion and thus gives the racemic modification; and (ii) this hydrogen ion, also by statistical chance, is almost certain to be deuterium, since the acid (the solvent) is almost entirely D_2O. With the carbanion thus established as the intermediate in racemization and hydrogen exchange, and in view of the relationship between racemization and halogenation, we can quite reasonably conclude that the carbanion is the intermediate in halogenation, too.

21.6 (a) As in Problem 21.5(a), racemization takes place via an achiral carbanion, with accommodation of the negative charge by acyl oxygen.

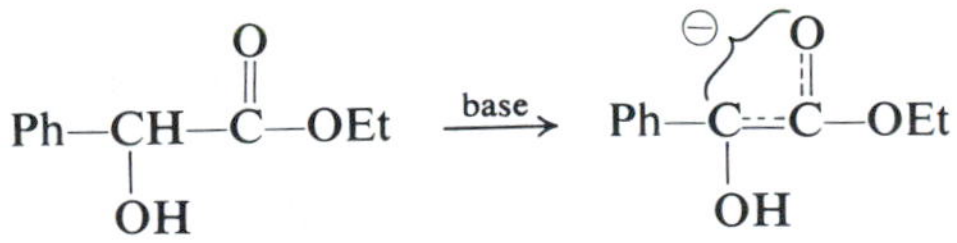

Ph—CH(OH)—C(=O)—OEt $\xrightarrow{\text{base}}$ Ph—C(OH)⋯C(⋯O$^\ominus$)—OEt

Achiral

(b) The doubly charged anion (carbanion and carboxylate ion) is more difficult to produce than a singly charged anion.

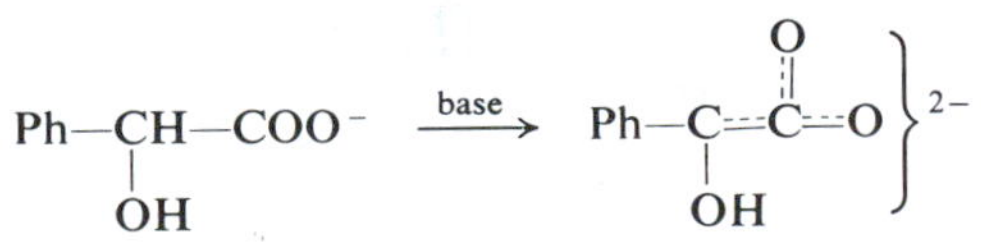

Ph—CH(OH)—COO^- $\xrightarrow{\text{base}}$ $\{$Ph—C(OH)⋯C(⋯O)⋯O$\}^{2-}$

(c) Very slow (if it occurs at all), since there is no α-hydrogen available for removal to form a carbanion.

21.7 We would expect the rate of racemization to be *twice* the rate of exchange, since the gain of one deuterium would bring about the loss of optical activity of two molecules. (See the answer to Problem 5.4, page 183, in which the rate of racemization *is* twice the rate of isotopic exchange.)

21.8 (a) Both reactions go through the same slow step (2), on page 804, formation of the enol.
(b) Same as (a).

21.9 (a) HSO_4^- (b) H_2O or D_2O

21.10 (a) Reaction could proceed either via the carbanion (as on page 803) or via the enol (as on pages 804–805). In either case, halogenation could occur only at the α-position.

(b) Formation of the carbanion from the carboxylic acid would be difficult because of the double negative charge on the anion. (See the answer to Problem 21.6(b).)

21.11 (a) I, III

$$CH_3CH_2C(H){=}O + [CH_3CHCHO]^- \longrightarrow CH_3CH_2CH(OH){-}CH(CH_3)CHO$$

(b)

$$CH_3C(H_3C){=}O + [CH_3COCH_2]^- \longrightarrow CH_3{-}C(CH_3)(OH){-}CH_2COCH_3$$

(c)

$$Ph{-}C(H_3C){=}O + [PhCOCH_2]^- \longrightarrow Ph{-}C(CH_3)(OH){-}CH_2COPh \xrightarrow{-H_2O} Ph{-}C(CH_3){=}CH{-}C(Ph){=}O$$

(d) cyclohexanone + cyclohexanone enolate ⟶ 2-(1-hydroxycyclohexyl)cyclohexanone

O, OH

(e)

$$PhCH_2C(H){=}O + [PhCHCHO]^- \longrightarrow PhCH_2CH(OH){-}CH(Ph)CHO$$

(For a review of the aldol condensation, see Nielsen, A. J.; Houlihan, W. J. "The Aldol Condensation"; *Organic Reactions*; Wiley: New York; Vol. XVI.)

21.12 (a) Step (2) is much faster than the reverse of step (1); as soon as a carbanion is formed (step 1), it reacts with a second CH_3CHO molecule (step 2) before it can react with water (reverse of step 1) to form an acetaldehyde molecule which, in D_2O, would contain deuterium.

More exactly,

$$\text{rate}_2 = k_2[\text{carbanion}][\text{aldehyde}]$$
$$\text{rate}_{-1} = k_{-1}[\text{carbanion}]$$

Evidently k_2[aldehyde] is much bigger than k_{-1}, and rate_2 is much bigger than rate_{-1}.

(b) Second-order.

$$\text{rate} = k_1[\text{aldehyde}][\text{base}]$$

In the more exact terms of the fine print on pages 189–190,

$$\text{rate} = \frac{k_1[\text{aldehyde}][\text{base}]}{1 + \dfrac{k_{-1}}{k_2[\text{aldehyde}]}}$$

When [aldehyde] is high, k_2[aldehyde] is much bigger than k_{-1}, and the term k_{-1}/k_2[aldehyde] is very small relative to 1 and drops out.

(c) When [aldehyde] is low, k_2[aldehyde] is no longer much bigger than k_{-1}, and rate_2 is no longer much bigger than rate_{-1}. As a result, a significant number of carbanions undergo the reverse of step (1) instead of undergoing step (2)—and pick up deuterium to form CH_2DCHO. When these labeled acetaldehyde molecules eventually undergo aldol condensation, they give labeled aldol.

(d) Step (2), carbonyl addition, is slower for a ketone, and is no longer much faster than the reverse of step (1).

21.13 [—C=C—C=O ⟷ —C—C=C—O] (with single-bond substituent lines on each carbon; dotted line connecting the terminal C and O in the second structure)

Overlap of π orbitals of carbon–carbon and carbon–oxygen double bonds, much as in conjugated dienes (see Figure 11.5, page 412) and in alkenylbenzenes.

21.14 (a) $CH_3CH_2CH_2\text{—}\underset{}{\overset{CH_3}{|}}{CH}\text{—}CH_2OH \xleftarrow{H_2,\ Ni} CH_3CH_2CH{=}\overset{CH_3}{\overset{|}{C}}\text{—}CHO \xleftarrow{-H_2O} CH_3CH_2\underset{OH}{\underset{|}{CH}}\,\vdots\,\overset{CH_3}{\overset{|}{C}}HCHO$

$\uparrow$ OH^-

$2CH_3CH_2CHO$
(from *n*-PrOH)

(b) $CH_3-CH(CH_3)-CH_2-CH(OH)-CH_3 \xleftarrow{H_2,\ Ni} CH_3-C(CH_3)=CH-C(=O)-CH_3 \xleftarrow{-H_2O} CH_3C(CH_3)(OH)\text{-}CH_2CCH_3(=O)$

$\uparrow OH^-$

$2CH_3CCH_3$ (C=O)

(from *i*-PrOH)

(c) 2-cyclohexylcyclohexanol (OH) $\xleftarrow{H_2,\ Ni}$ 2-cyclohexylidenecyclohexanone (O) $\xleftarrow{-H_2O}$ 2-(1-hydroxycyclohexyl)cyclohexanone (HO, O) $\xleftarrow{OH^-}$ 2 cyclohexanone (=O)

(from cyclohexanol)

(d) $PhCH_2CH_2CH(Ph)-CH_2OH \xleftarrow{H_2,\ Ni} PhCH_2CH=C(Ph)-CHO \xleftarrow{-H_2O} PhCH_2CH(OH)\text{-}CH(Ph)CHO$

$\uparrow OH^-$

$2PhCH_2CHO$

(from $PhCH_2CH_2OH$)

(e) $CH_3-C(Ph)=CH-CH(OH)-Ph \xleftarrow[HOCH_2CH_2NH_2]{9\text{-BBN}} CH_3-C(Ph)=CH-C(=O)-Ph \xleftarrow{-H_2O} CH_3C(Ph)(OH)\text{-}CH_2CPh(=O)$

$\uparrow OH^-$

$2PhCCH_3$ (C=O)

(from $PhCHOHCH_3$)

21.15 (a) $PhCH_2CH_2CH(OH)CH_3 \xleftarrow{H_2,\ Ni} PhCH=CHC(=O)CH_3 \xleftarrow[-H_2O]{OH^-}$ $PhCHO$ (from $PhCH_3$) + CH_3COCH_3 (from *i*-PrOH)

(b) $PhCH_2CH_2CH(OH)Ph \xleftarrow{H_2,\ Ni} PhCH=CHC(=O)Ph \xleftarrow[-H_2O]{OH^-}$ $PhCHO \longleftarrow PhCH_3$ + $CH_3C(=O)Ph$ (page 664)

(c) $PhCH_2CH_2CH_2Ph \xleftarrow{H_2,\ Ni} PhCH_2CH=CHPh \xleftarrow{-H_2O} PhCH_2CH_2CH(OH)Ph \longleftarrow$ (b)

(d) $PhCH_2CH(Ph)CH_2OH \xleftarrow{H_2,\ Ni} PhCH=C(Ph)CHO \xleftarrow[-H_2O]{OH^-}$ $PhCHO$ (from $PhCH_3$) + $PhCH_2CHO$ (from $PhCH_2COCl$)

(e) $PhCH{=}CHC(=O)CH{=}CHPh \xleftarrow[-2H_2O]{OH^-}$ [$2PhCHO$ (from $PhCH_3$) ; CH_3COCH_3 (from *i*-PrOH)]

21.16 (a) The γ-hydrogen will be acidic, since the negative charge of the anion is accommodated by the electronegative oxygen through the conjugated system.

$$\underset{\gamma}{-\overset{|}{\underset{H}{\underset{|}{C}}}}-\underset{\beta}{\overset{|}{C}}=\underset{\alpha}{\overset{|}{C}}-\overset{|}{C}=O + :B \rightleftarrows \underbrace{-\overset{|}{C}\text{:::}\overset{|}{C}\text{:::}\overset{|}{C}\text{:::}\overset{|}{C}\text{:::}O}_{\ominus} + H:B$$

(b) $PhCH{=}CH{-}CH{=}CH{-}CHO \xleftarrow[-H_2O]{OH^-} PhCHO + CH_3CH{=}CHCHO \xleftarrow[-H_2O]{OH^-} 2CH_3CHO$

21.17 $CH_3COOEt \xrightarrow{OEt^-} [CH_2COOEt]^- \xrightarrow{PhCHO} PhCH(O^-){-}CH_2COOEt \xrightarrow{+H^+} PhCH(OH){-}CH_2COOEt \xrightarrow{-H_2O} PhCH{=}CHCOOEt$

Ethyl 3-phenylpropenoate

21.18 In each case (except c) base abstracts a proton that is *alpha* to one or more electron-withdrawing groups ($-NO_2$, $-C{\equiv}N$, $-C{=}O$) to give one of the following anions:

(a) $[CH_2NO_2]^-$ (b) $[PhCHCN]^-$ (c) $HC{\equiv}C^-$ (d) $\left[CH_3\overset{O}{\overset{\|}{C}}{-}O{-}\overset{O}{\overset{\vdots}{C}}\text{:::}CH_2\right]^-$

The anion then adds to the aldehyde or ketone to give an aldol-like product, the alkoxy group is protonated, and water is lost. In (d), hydrolysis of an intermediate anhydride is needed as well.

(For a review of these reactions, see Johnson, J. R. "The Perkin Reaction and Related Reactions"; *Organic Reactions*; Wiley: New York; Vol. I, Chapter 8; Jones, G. "The Knoevenagel Condensation"; *Organic Reactions*; Wiley: New York; Vol. XV, Chapter 2.)

21.19 Elimination $\longrightarrow$ 1- and 2-butene.

(For a general discussion of the Wittig reaction, see Trippett, S. "The Wittig Reaction"; *Q. Rev., Chem. Soc.* **1963**, *17*, 406.)

21.20 In each case except (c), there are two combinations of reagents that, on paper, would give the desired product. In (a), (b), and (d), the better combination is shown; in each alternative, preparation of the ylide would be accompanied by much elimination. Compare Problem 21.19.

(a) $CH_3CH_2CH_2CH{=}C(CH_3)CH_2CH_3 \longleftarrow CH_3CH_2CH_2CH{=}PPh_3 + O{=}C(CH_3)CH_2CH_3$ (2-Butanone)

(b) $Ph{-}C(CH_3){=}CHCH_2Ph \longleftarrow Ph{-}C(CH_3){=}O + Ph_3P{=}CHCH_2Ph$

(c) $Ph{-}CH{=}CH{-}Ph \longleftarrow Ph{-}C(H){=}O + Ph_3P{=}CHPh$

(d) $C_5H_8{=}CHCH_3$ (cyclopentylidene) $\longleftarrow C_5H_8{=}O$ (cyclopentanone) $+ Ph_3P{=}CHCH_3$

(e) $PhCH{=}CH{-}CH{=}CHPh \longleftarrow PhCH{=}CH{-}C(H){=}O + Ph_3P{=}CHPh$

(f) $CH_2{=}CHCH{=}C(CH_3)COOMe \longleftarrow CH_2{=}CHC(H){=}O + Ph_3P{=}C(CH_3)COOMe$

21.21 (a) $Ph_3P{=}CHC_3H_7\text{-}n \xleftarrow[\text{base}]{Ph_3P} BrCH_2C_3H_7\text{-}n \xleftarrow{PBr_3} n\text{-BuOH}$

$CH_3CH_2COCH_3 \xleftarrow{CrO_3} sec\text{-BuOH}$

(b) $PhCOCH_3 \xleftarrow{AlCl_3}$ benzene + acetic anhydride

$Ph_3P{=}CHCH_2Ph \xleftarrow[\text{base}]{Ph_3P} BrCH_2CH_2Ph \xleftarrow{PBr_3} HOCH_2CH_2Ph$ (page 688)

(c) $PhCHO \xleftarrow{OH^-} PhCHCl_2 \xleftarrow{2Cl_2,\ \text{heat}} PhCH_3$

$Ph_3P{=}CHPh \xleftarrow[\text{base}]{Ph_3P} ClCH_2Ph \xleftarrow{Cl_2,\ \text{heat}} PhCH_3$

(d) $C_5H_8{=}O$ (cyclopentanone) $\xleftarrow{CrO_3} C_5H_9{-}OH$ (cyclopentanol)

$Ph_3P{=}CHCH_3 \xleftarrow[\text{base}]{Ph_3P} BrCH_2CH_3 \xleftarrow{PBr_3} EtOH$

(e) $PhCH{=}CHCHO \xleftarrow[-H_2O]{OH^-} PhCHO$ (as in c) $+ CH_3CHO \xleftarrow{K_2Cr_2O_7} EtOH$

$Ph_3P{=}CHPh \longleftarrow$ as in (c)

(f) $CH_2{=}CHCHO \xleftarrow{heat}$ glycerol + $NaHSO_4$ (see Problem 27.5, page 974)

$Ph_3P{=}C(CH_3)COOMe \xleftarrow[base]{Ph_3P} BrCH(CH_3)COOMe \xleftarrow[P]{Br_2} CH_3CH_2COOMe \xleftarrow[H^+]{MeOH} CH_3CH_2COOH \xleftarrow{KMnO_4} n\text{-}PrOH$

21.22 $PhOCH_2Cl \xrightarrow{Ph_3P} PhOCH_2\overset{+}{P}Ph_3\,Cl^- \xrightarrow{t\text{-}BuOK} PhOCH{=}PPh_3$ (A)

$PhOCH{=}PPh_3$ (A) $+ O{=}C(CH_3)CH_2CH_3 \longrightarrow Ph_3PO + PhOCH{=}C(CH_3)CH_2CH_3$ (B)

$PhOCH{=}C(CH_3)CH_2CH_3$ (B) $\xrightarrow{H^+} Ph\overset{+}{O}{=}CH{-}CH(CH_3)CH_2CH_3 \xrightarrow{H_2O} PhO{-}CH(OH){-}CH(CH_3)CH_2CH_3$ (Hemiacetal) $\longrightarrow O{=}CH{-}CH(CH_3)CH_2CH_3$ (C)

A general rule: cleavage of vinyl ethers is particularly easy (compare Problem 17, Chapter 18, page 704).

The sequence is a general route to aldehydes.

21.23 (a) $PhC({=}O)CH_2CH_2CH_2CH_2Br \xrightarrow{Ph_3P} PhC({=}O)CH_2CH_2CH_2CH_2\overset{+}{P}Ph_3\,Br^- \xrightarrow{NaOEt} [PhC({=}O)CH_2CH_2CH_2CH{=}PPh_3]$

Not isolated

$\downarrow$ $-Ph_3PO$, internal Wittig

D: 1-phenylcyclopentene (Ph—C=CH—CH$_2$—CH$_2$—CH$_2$ ring)

(b) $Ph_3P + Br(CH_2)_3Br + PPh_3 \longrightarrow Ph_3\overset{+}{P}(CH_2)_3\overset{+}{P}Ph_3 \; Br^- \; Br^- \xrightarrow{\text{base}} Ph_3P{=}\overset{H}{C}CH_2\overset{H}{C}{=}PPh_3$ (E)

Phthalaldehyde + E ($Ph_3P{=}CH{-}CH_2{-}CH{=}PPh_3$) $\xrightarrow[-2Ph_3PO]{\text{double Wittig}}$ F (benzocycloheptadiene)

21.24 *trans*-2-Octene (H, C_5H_{11}-*n* / CH_3, H) $\xrightarrow{PhCO_2OH}$ epoxide G (and enantiomer) $\xrightarrow{Ph_2PLi}$ O^-Li^+ / $Ph_2P:$ adduct $\xrightarrow{CH_3I}$ betaine H (O^-, $Ph_2\overset{+}{P}CH_3$; and enantiomer) $\equiv$ (conformer with n-C_5H_{11}, H, O^-, CH_3, H, $Ph_2\overset{+}{P}CH_3$) $\xrightarrow{syn\text{-elim.}}$ *cis*-2-Octene + $Ph_2\underset{CH_3}{PO}$

Here, a betaine is formed by nucleophilic attack on the epoxide by the basic and nucleophilic $Ph_2P—$ group.

21.25 (a) *Dieckmann condensation*: intramolecular Claisen condensation leading to cyclization.

Ethyl adipate (EtO–C(=O)–(CH₂)₃–CH_2COOEt) $\xrightarrow{OEt^-}$ [EtO–C(=O)… $CHCOOEt]^-$ $\xrightarrow{-OEt}$ II (2-carbethoxycyclopentanone: O, H, COOEt)

(b) EtO–C(=O)–(CH₂)₄–CH_2COOEt $\xrightarrow{OEt^-}$ [EtO–C(=O)… $CHCOOEt]^-$ $\xrightarrow{-OEt}$ 2-carbethoxycyclohexanone (O, H, COOEt)

(c) No, because three- and four-membered rings are difficult to make.

The appearance of 12 carbons in the product shows that two molecules of ethyl succinate are involved; loss of $2C_2H_5OH$ suggests a double Claisen condensation.

$C_{16}H_{28}O_8$	2 ethyl succinate
$-C_{12}H_{16}O_6$	product
$C_4H_{12}O_2$	equivalent to $2C_2H_5OH$

EtO O / C / H_2C CH_2COOEt / H_2C CH_2 / EtOOC COOEt $\xrightarrow{OEt^-}$ EtO O / C / H_2C $CHCOOEt]^-$ / H_2C CH_2 / EtOOC COOEt $\xrightarrow{\text{Claisen}}$ O / C / H_2C CHCOOEt / H_2C CH_2 / EtOOC COOEt

$\downarrow OEt^-$

O / C / H_2C CHCOOEt / $^-$[EtOOCCH CH_2 / C / O OEt $\xrightarrow{\text{Claisen}}$ (cyclohexane-1,4-dione ring with COOEt and EtOOC substituents)

(For a review, see Schaefer, J. P.; Bloomfield, J. J. "The Dieckmann Condensation"; *Organic Reactions*; Wiley: New York; Vol. XV, Chapter 1.)

21.26 PhCOOEt is mixed with NaOEt, and then CH_3COOEt is added slowly.
HCOOEt and NaOEt, then CH_3COOEt slowly.
Ethyl oxalate and NaOEt, then CH_3COOEt slowly.
Ethyl carbonate and NaOEt, then $PhCH_2COOEt$ slowly.

In every case, an ester without an α-H is mixed with the base, and then an ester carrying an α-H is added. The first ester cannot undergo self-condensation, and so is stable toward the base. When the second ester is added slowly, it is converted into an anion in the presence of a great deal of the first ester and very little of the second ester; for statistical reasons, then, it adds chiefly to the first ester to give a good yield of the crossed product.

21.27 (a) Replace anion I of page 814 by $^-CH_2COCH_3$, formed from acetone by the action of OEt^-.

(b) $CH_3CH_2C(=O)-OEt + {}^-CH_2COCH_3 \longrightarrow CH_3CH_2-C(=O)-CH_2-C(=O)-CH_3$

2,4-Hexanedione

(c) $PhC(=O)-OEt + {}^-CH_2COPh \longrightarrow Ph-C(=O)-CH_2-C(=O)-Ph$

Dibenzoylmethane

(d) 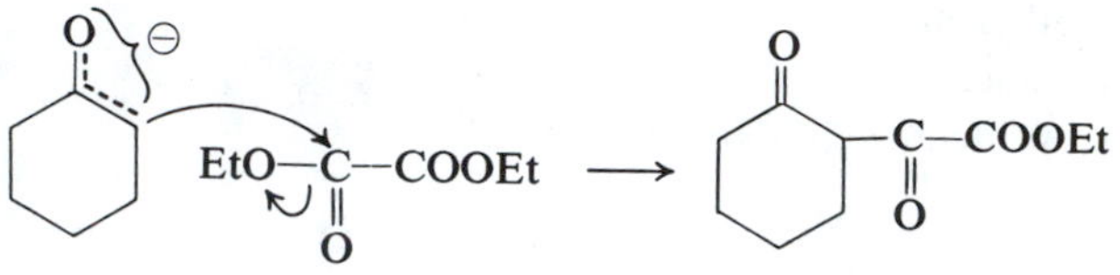 cyclohexanone enolate $+ EtO-C(=O)-COOEt \longrightarrow$ 2-(ethoxalyl)cyclohexanone ($C_6H_9O-C(=O)-COOEt$)

21.28 (a) $PhC(=O)-CH(Ph)-C(=O)-OEt \xleftarrow{OEt^-} PhCOOEt + PhCH_2COOEt$

(b) 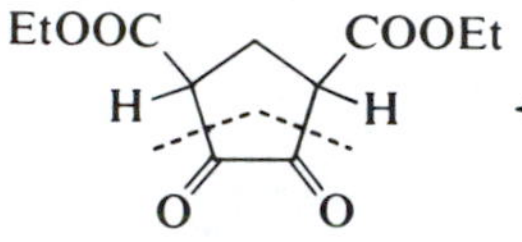 diethyl 4,5-dioxocyclopentane-1,3-dicarboxylate (EtOOC, H; COOEt, H; O, O) $\xleftarrow{OEt^-}$ EtOOC—COOEt + $EtOOCCH_2CH_2CH_2COOEt$

Ethyl oxalate Ethyl glutarate

(c) 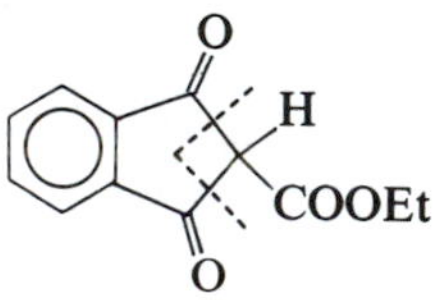 2-carbethoxy-1,3-indandione (O, H, COOEt, O) $\xleftarrow{OEt^-}$ $C_6H_4(COOEt)_2$ $+ CH_3COOEt$

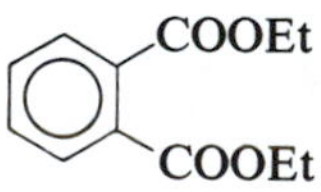

Ethyl phthalate

1. (a) $PhCH_2C(H)=O + [PhCHCHO]^- \xrightarrow{\text{aldol cond.}} PhCH_2CH(OH)-CH(Ph)CHO$

(b) As in (a), then $PhCH_2CH(OH)-C(Ph)(H)-CHO \xrightarrow[-H_2O]{H^+} PhCH_2CH=C(Ph)-C(H)=O$

(c) Same as (a)

(d) $PhCH_2CHO \xrightarrow{Br_2,\ CCl_4} PhCH(Br)CHO$

(e) $PhCH_2\overset{H}{C}{=}O + Ph_3P{=}CH_2 \xrightarrow{-Ph_3PO} PhCH_2CH{=}CH_2$

2. (a) 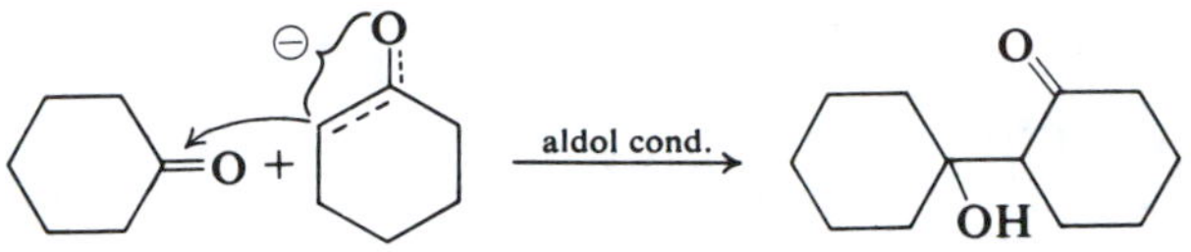 $=O$ + $\ominus$ O $\xrightarrow{\text{aldol cond.}}$ O OH

(b) As in (a), then O OH $\xrightarrow[-H_2O]{H^+}$ O

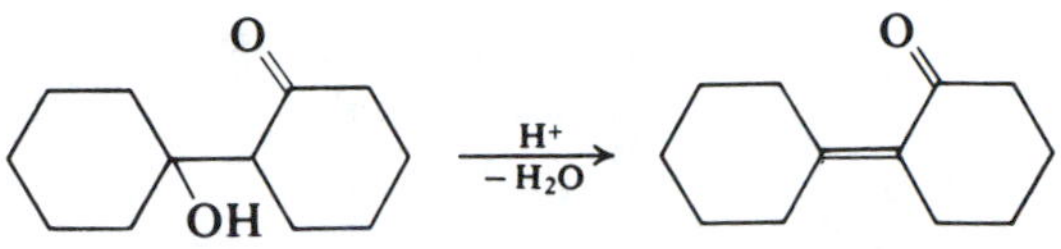

(c) Same as (a)

(d) 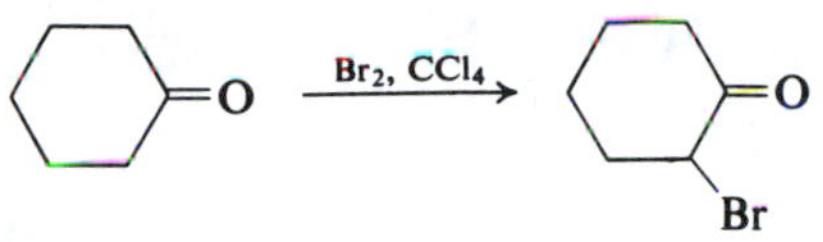 $=O \xrightarrow{Br_2,\ CCl_4} =O$ Br

(e) 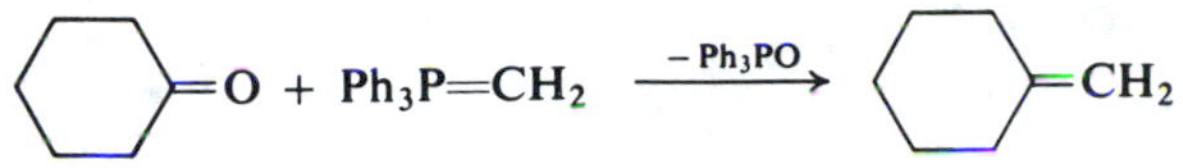 $=O + Ph_3P{=}CH_2 \xrightarrow{-Ph_3PO}$ $=CH_2$ (see page 801)

3. (a) No reaction

(b) Cannizzaro reaction: $2PhCHO \xrightarrow{OH^-} PhCH_2OH + PhCOO^-Na^+$

(c) $Ph\overset{H}{C}{=}O + CH_3{-}CHO \xrightarrow{OH^-} Ph\overset{H}{C}{=}\overset{H}{C}{-}\overset{H}{C}{=}O$

Cinnamaldehyde

(d) $Ph\overset{H}{C}{=}O + \underset{CH_3}{CH_2}{-}CHO \xrightarrow{OH^-} Ph\overset{H}{C}{=}\underset{CH_3}{C}{-}\overset{H}{C}{=}O$

(e) $Ph\overset{H}{C}{=}O + CH_3{-}\underset{O}{\underset{\|}{C}}{-}CH_3 \xrightarrow{OH^-} Ph\overset{H}{C}{=}\overset{H}{C}{-}\underset{O}{\underset{\|}{C}}{-}CH_3$

Benzalacetone (*Benzal* is PhCH=)

(f) $PhCH{=}CH{-}C({=}O){-}CH_3 + O{=}CHPh \xrightarrow{OH^-} PhCH{=}CH{-}C({=}O){-}CH{=}CHPh$

Product (e) Dibenzalacetone

(g) $PhCH{=}O + CH_3{-}C({=}O){-}Ph \xrightarrow{OH^-} PhCH{=}CH{-}C({=}O){-}Ph$

Benzalacetophenone (Chalcone)

(h) Perkin reaction: $PhCH{=}O + CH_3{-}C({=}O){-}O{-}C({=}O)CH_3 \xrightarrow{OAc^-} \xrightarrow{\text{hydrolysis}} PhCH{=}CH{-}C({=}O)OH$

3-Phenylpropenoic acid

(i) $PhCH{=}O + CH_3{-}C({=}O)OEt \xrightarrow{OEt^-} PhCH{=}CH{-}C({=}O)OEt$

Ethyl cinnamate

(j) $PhCH{=}O + CH_2(Ph){-}C({=}O)OEt \xrightarrow{OEt^-} PhCH{=}C(Ph){-}C({=}O)OEt$

(k) Crossed Cannizzaro: $PhCHO + HCHO \xrightarrow{OH^-} PhCH_2OH + HCOO^-Na^+$

(l) $PhCH{=}O + CH_3CH{=}CHCHO \xrightarrow{OH^-} PhCH{=}CH{-}CH{=}CH{-}CHO$

(Compare Problem 21.16, page 810)

(m) $PhCH{=}O + Ph_3P{=}CHCH{=}CH_2 \xrightarrow{-Ph_3PO} PhCH{=}CH{-}CH{=}CH_2$

1-Phenyl-1,3-butadiene

(n) $PhCH{=}O + Ph_3P{=}CH{-}OPh \xrightarrow{-Ph_3PO} PhCH{=}CH{-}OPh$

(o) $PhCH{=}CH{-}OPh \xrightarrow[H_2O]{H^+} PhCH_2CHO$

(from n) Phenylacetaldehyde

Vinyl ethers are easily cleaved; compare Problem 17, Chapter 18, page 704.

4. (a) $\underset{\text{OH}}{\text{CH}_3\text{CH}_2\text{CH}}\text{—}\underset{\text{CH}_3}{\text{CHCHO}} \xleftarrow{\text{OH}^-} 2\text{CH}_3\text{CH}_2\text{CHO}$

(b) $\text{CH}_3\text{CH}_2\text{CH}_2\underset{\text{CH}_3}{\text{CH}}\text{CH}_2\text{OH} \xleftarrow{\text{H}_2,\ \text{Ni}} \text{CH}_3\text{CH}_2\text{CH}{=}\underset{\text{CH}_3}{\text{C}}\text{CHO} \xleftarrow{-\text{H}_2\text{O}}$ product (a)

(c) $\text{CH}_3\text{CH}_2\text{CH}{=}\underset{\text{CH}_3}{\text{C}}\text{CHO} \xleftarrow{-\text{H}_2\text{O}}$ product (a)

(d) $\text{CH}_3\text{CH}_2\text{CH}{=}\underset{\text{CH}_3}{\text{C}}\text{CH}_2\text{OH} \xleftarrow[\text{HOCH}_2\text{CH}_2\text{NH}_2]{\text{9-BBN}}$ product (c)

(e) $\underset{\text{OH}}{\text{CH}_3\text{CH}_2\text{CH}}\text{—}\underset{\text{CH}_3}{\text{CH}}\text{CH}_2\text{OH} \xleftarrow{\text{H}_2,\ \text{Ni}}$ product (a)

(f) $\text{CH}_3\text{CH}_2\text{CH}_2\underset{\text{CH}_3}{\text{CH}}\text{COOH} \xleftarrow[\text{heat}]{\text{KMnO}_4}$ product (b)

(g) $\text{PhCH}{=}\underset{\text{CH}_3}{\text{C}}\text{CHO} \xleftarrow[-\text{H}_2\text{O}]{\text{OH}^-} \text{PhCHO} + \text{CH}_3\text{CH}_2\text{CHO}$

(h) $\text{CH}_3\text{CD}_2\text{CHO} \longleftarrow \text{CH}_3\text{CH}_2\text{CHO} + \text{D}_2\text{O} + \text{OD}^-$

Acidic α-hydrogen easily exchanged with deuterium.

(i) $\text{CH}_3\text{CH}_2\text{CH}^{18}\text{O} \longleftarrow \text{CH}_3\text{CH}_2\text{CHO} + \text{H}_2{}^{18}\text{O} + \text{H}^+$

Compare Problem 18.9, page 683.

(j) $\text{CH}_3\text{CH}_2\text{CH}{=}\text{CHCH}(\text{CH}_3)_2 \xleftarrow{\text{Wittig}} \text{CH}_3\text{CH}_2\overset{\text{H}}{\text{C}}{=}\text{O} + \text{Ph}_3\text{P}{=}\text{CHCH}(\text{CH}_3)_2$ (from *iso*-BuBr)

5. (a) Need to remove 1C: try the haloform reaction (Sec. 18.21).

$\text{PhCOOH} + \text{CHI}_3 \xleftarrow{\text{H}^+} \xleftarrow{\text{I}_2,\ \text{OH}} \text{Ph—}\underset{\text{O}}{\overset{\|}{\text{C}}}\text{—CH}_3$

(b) $\text{CH}_3\text{—}\overset{\text{Ph}}{\text{C}}{=}\text{CH—}\underset{\text{O}}{\overset{\|}{\text{C}}}\text{—Ph} \xleftarrow[-\text{H}_2\text{O}]{\text{OH}^-} 2\text{PhCOCH}_3$

(c) $\text{CH}_3\text{—}\overset{\text{Ph}}{\text{CH}}\text{—CH}_2\text{—}\underset{\text{OH}}{\text{CH}}\text{—Ph} \xleftarrow{\text{H}_2,\ \text{Ni}}$ product (b)

(d) $CH_3-\underset{}{\overset{Ph}{C}}=CH-\underset{OH}{CH}-Ph \xleftarrow[HOCH_2CH_2NH_2]{9\text{-BBN}}$ product (b)

(e) $PhCH=CHCOPh \xleftarrow[-H_2O]{OH^-} PhCHO + CH_3COPh$

(f) $CH_3\underset{Ph}{CH}-CHO \xleftarrow{H_2O,\ H^+} CH_3\underset{Ph}{C}=CHOPh \xleftarrow{\text{Wittig}} CH_3\underset{Ph}{C}=O + ClCH_2OPh$

6. These are all Claisen condensations, and give β-dicarbonyl products.

(a) $EtCH_2\underset{O}{\underset{\|}{C}}OEt + \underset{Et}{CH_2}COOEt \xrightarrow{OEt^-} EtCH_2\underset{O}{\underset{\|}{C}}-\underset{Et}{CH}COOEt$

(b) $PhCH_2\underset{O}{\underset{\|}{C}}OEt + \underset{Ph}{CH_2}COOEt \xrightarrow{OEt^-} PhCH_2\underset{O}{\underset{\|}{C}}-\underset{Ph}{CH_2}COOEt$

(c) $i\text{-}PrCH_2\underset{O}{\underset{\|}{C}}OEt + \underset{Pr\text{-}i}{CH_2}COOEt \xrightarrow{OEt^-} i\text{-}PrCH_2\underset{O}{\underset{\|}{C}}-\underset{Pr\text{-}i}{CH}COOEt$

(d) $H\underset{O}{\underset{\|}{C}}OEt + \underset{CH_3}{CH_2}COOEt \xrightarrow{OEt^-} H\underset{O}{\underset{\|}{C}}-\underset{CH_3}{CH}COOEt$

(known only as the sodio derivative)

(e) Ethyl oxalate is a diester, and can take part in two successive Claisen condensations:

$EtOOC-\underset{O}{\underset{\|}{C}}OEt + \underset{CH_2COOEt}{CH_2}COOEt \xrightarrow{OEt^-} EtOOC-\underset{O}{\underset{\|}{C}}-\underset{CH_2COOEt}{CH}COOEt$

Ethyl oxalate — Ethyl succinate

$\underset{EtOOCCH_2}{EtOOCCH_2} + EtO\underset{O}{\underset{\|}{C}}\ \underset{O}{\underset{\|}{C}}-\underset{CH_2COOEt}{CH}COOEt \xrightarrow{OEt^-} \underset{EtOOCCH_2}{EtOOCCH}-\underset{O}{\underset{\|}{C}}-\underset{O}{\underset{\|}{C}}-\underset{CH_2COOEt}{CH}COOEt$

Ethyl succinate *Second mole* — Product of first step — An α-diketone

The initial product does not undergo an intramolecular Claisen condensation (Dieckmann condensation), evidently because of a reluctance to form a four-membered ring.

The crossed Claisen condensation of ethyl oxalate is the first step in a useful route to α-keto acids (see Problem 25.9, page 930) and thence to α-amino acids (see Problem 36.14, page 1214).

(f) $Ph\underset{O}{\underset{\|}{C}}OEt + \underset{Ph}{CH_2}COOEt \xrightarrow{OEt^-} Ph\underset{O}{\underset{\|}{C}}-\underset{Ph}{CH}COOEt$

(g) 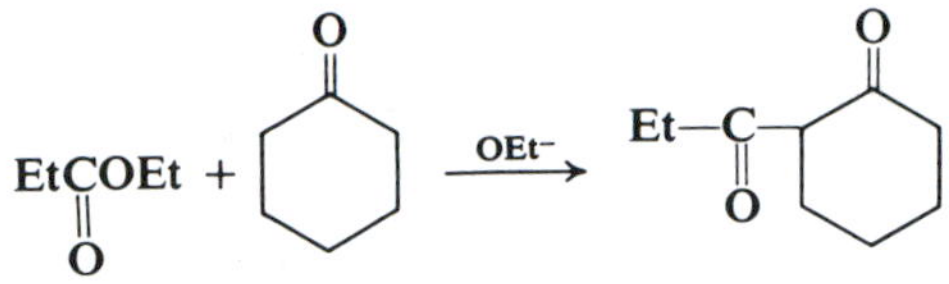 $\mathrm{Et\underset{\underset{O}{\|}}{C}OEt}$ + cyclohexanone $\xrightarrow{\mathrm{OEt^-}}$ $\mathrm{Et{-}\underset{\underset{O}{\|}}{C}{-}}$(2-oxocyclohexyl)

(h) 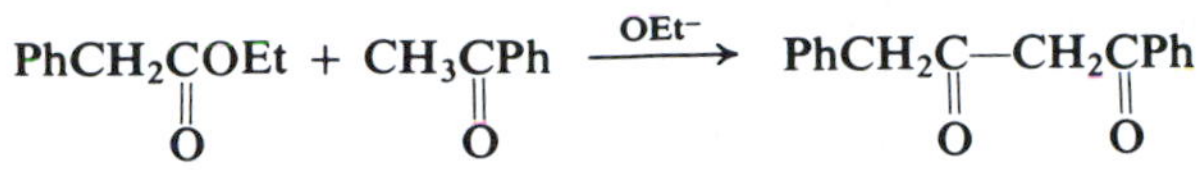 $\mathrm{PhCH_2\underset{\underset{O}{\|}}{C}OEt + CH_3\underset{\underset{O}{\|}}{C}Ph \xrightarrow{OEt^-} PhCH_2\underset{\underset{O}{\|}}{C}{-}CH_2\underset{\underset{O}{\|}}{C}Ph}$

(i) Ethyl carbonate, too, is a diester, and can take part in two successive Claisen condensations:

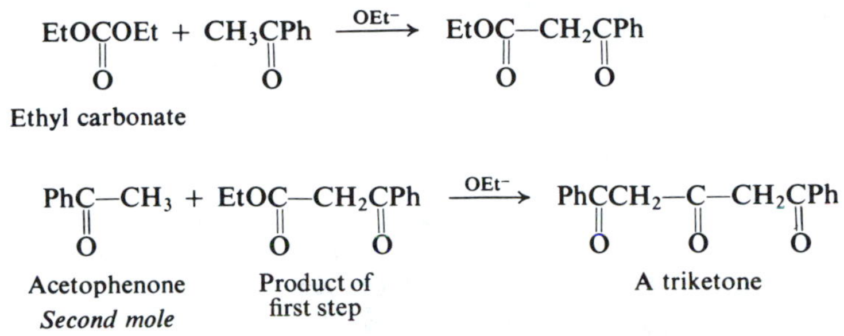

$$\underset{\text{Ethyl carbonate}}{\mathrm{EtO\underset{\underset{O}{\|}}{C}OEt}} + \mathrm{CH_3\underset{\underset{O}{\|}}{C}Ph} \xrightarrow{\mathrm{OEt^-}} \mathrm{EtO\underset{\underset{O}{\|}}{C}{-}CH_2\underset{\underset{O}{\|}}{C}Ph}$$

$$\underset{\substack{\text{Acetophenone}\\ \textit{Second mole}}}{\mathrm{Ph\underset{\underset{O}{\|}}{C}{-}CH_3}} + \underset{\substack{\text{Product of}\\ \text{first step}}}{\mathrm{EtO\underset{\underset{O}{\|}}{C}{-}CH_2\underset{\underset{O}{\|}}{C}Ph}} \xrightarrow{\mathrm{OEt^-}} \underset{\text{A triketone}}{\mathrm{Ph\underset{\underset{O}{\|}}{C}CH_2{-}\underset{\underset{O}{\|}}{C}{-}CH_2\underset{\underset{O}{\|}}{C}Ph}}$$

Ethyl carbonate is useful in synthetic work for the introduction of a carbethoxy group, —COOEt, as in the first step here. (See another example on page 816.)

7. (a) 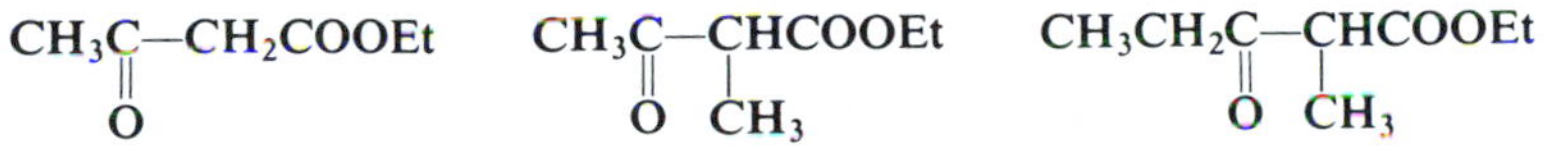 $\mathrm{CH_3\underset{\underset{O}{\|}}{C}{-}CH_2COOEt}$ $\quad$ $\mathrm{CH_3\underset{\underset{O}{\|}}{C}{-}\underset{\underset{CH_3}{|}}{C}HCOOEt}$ $\quad$ $\mathrm{CH_3CH_2\underset{\underset{O}{\|}}{C}{-}\underset{\underset{CH_3}{|}}{C}HCOOEt}$

(b) No. It would give a poor yield of any one of them, contaminated by the other two.

8. (a) 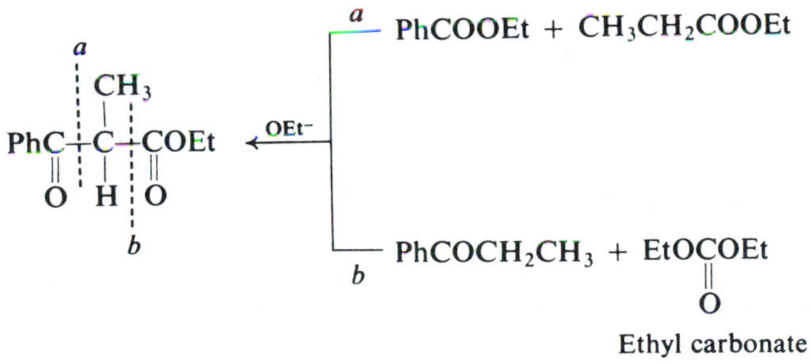 $\mathrm{Ph\underset{\underset{O}{\|}}{C}\overset{a}{\vdots}\underset{\underset{H}{|}}{\overset{\overset{CH_3}{|}}{C}}\overset{}{\underset{b}{\vdots}}\underset{\underset{O}{\|}}{C}OEt}$ $\xleftarrow{\mathrm{OEt^-}}$
- a: $\mathrm{PhCOOEt + CH_3CH_2COOEt}$
- b: $\mathrm{PhCOCH_2CH_3 + EtO\underset{\underset{O}{\|}}{C}OEt}$ (Ethyl carbonate)

(b) 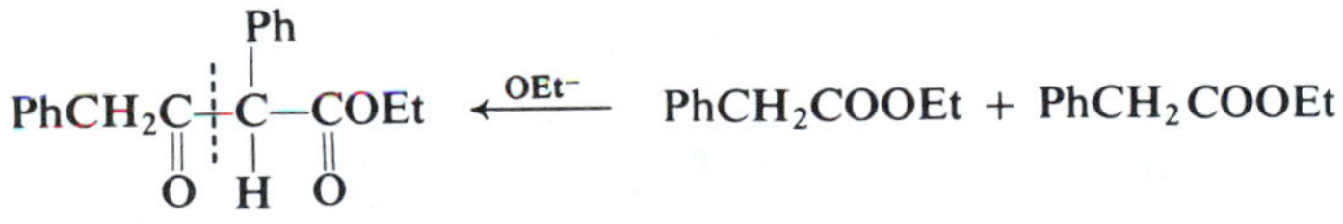 $\mathrm{PhCH_2\underset{\underset{O}{\|}}{C}\vdots\underset{\underset{H}{|}}{\overset{\overset{Ph}{|}}{C}}{-}\underset{\underset{O}{\|}}{C}OEt \xleftarrow{OEt^-} PhCH_2COOEt + PhCH_2COOEt}$

(c) $\mathrm{EtOOC{-}\underset{\underset{O}{\|}}{C}\vdots\underset{\underset{H}{|}}{\overset{\overset{CH_3}{|}}{C}}{-}\underset{\underset{O}{\|}}{C}OEt \xleftarrow{OEt^-}}$ $\underset{\text{Ethyl oxalate}}{\mathrm{EtOOC{-}COOEt}}$ + $\underset{\textit{One mole}}{\mathrm{CH_3CH_2COOEt}}$

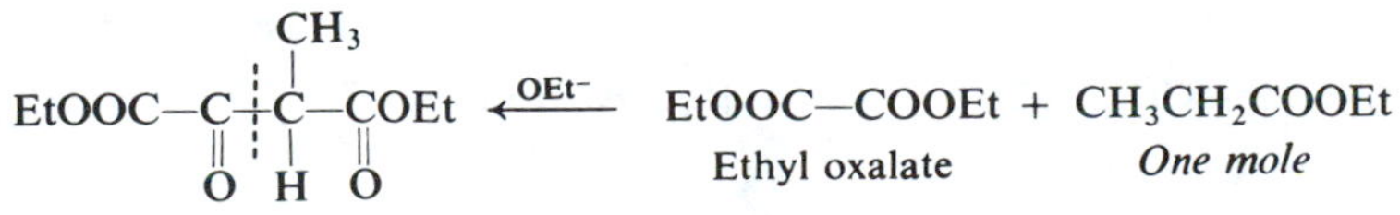

(d) $Ph\!-\!\underset{HC=O}{\overset{H}{C}}\!-\!\overset{O}{\overset{\|}{C}}OEt \xleftarrow{OEt^-} HCOOEt + PhCH_2COOEt$

(e) $(CH_3)_2CH\underset{O}{\overset{}{\underset{\|}{C}}}\!-\!CH_2\underset{O}{\underset{\|}{C}}CH_3 \xleftarrow{OEt^-} (CH_3)_2CHCOOEt + CH_3COCH_3$

(f) $Ph\underset{O}{\underset{\|}{C}}\!-\!CH_2\underset{O}{\underset{\|}{C}}CH_3 \xleftarrow{OEt^-} PhCOOEt + CH_3COCH_3$

(g) 2-Benzoylcyclohexanone $\xleftarrow{OEt^-}$ PhCOOEt + cyclohexanone

(h) $EtO\underset{O}{\underset{\|}{C}}\!-\!\underset{HC=O}{CH}CH_2\underset{O}{\underset{\|}{C}}OEt \xleftarrow{OEt^-} HCOOEt + EtOOCCH_2CH_2COOEt$

Ethyl formate *One mole* Ethyl succinate

9. Remember that a *cis* isomer can often be made by hydrogenation of a triple bond over Lindlar's catalyst (Sec. 12.8).

cis-3-Phenylpropenoic acid (Ph and COOH on the same side of C=C) $\xleftarrow[\text{Lindlar}]{H_2}$ $Ph\!-\!C\equiv C\!-\!COOH \xleftarrow{NaNH_2} \xleftarrow{KOH(alc)} PhCH(Br)\!-\!CH(Br)COOH$

$PhCH(Br)\!-\!CH(Br)COOH \xleftarrow{Br_2,\ CCl_4}$ *trans*-3-Phenylpropenoic acid (Ph and COOH on opposite sides of C=C)

10. (a) $CH_3\!-\!\underset{OH}{\overset{CH_3}{C}}\!-\!CH_2\!-\!\underset{O}{\underset{\|}{C}}\!-\!CH_3 \xleftarrow{OH^-} 2CH_3\underset{O}{\underset{\|}{C}}CH_3 \xleftarrow{CrO_3} i\text{-PrOH}$

(b) $CH_3\overset{CH_3}{CH}\!-\!CH_2\underset{OH}{CH}CH_3 \xleftarrow{H_2,\ Ni} CH_3\overset{CH_3}{C}\!=\!CH\underset{O}{\underset{\|}{C}}CH_3 \xleftarrow{-H_2O} CH_3\underset{OH}{\overset{CH_3}{C}}CH_2\underset{O}{\underset{\|}{C}}CH_3 \longleftarrow \text{(a)}$

(c) $CH_3CH\!=\!CHCHO \xleftarrow{-H_2O} CH_3\underset{OH}{CH}\!-\!CH_2CHO \xleftarrow{OH^-} 2CH_3CHO$ (from EtOH)

(d) $PhCH{=}CHCH_2OH \xleftarrow{NaBH_4} PhCH{=}CHCHO \xleftarrow{-H_2O} [PhCH(OH){-}CH_2CHO] \longleftarrow PhCHO + CH_3CHO$
(PhCHO from $PhCH_3$; CH_3CHO from EtOH)

(e) $NO_2C_6H_4CH{=}CHCHO \xleftarrow{-H_2O} [NO_2C_6H_4CH(OH){-}CH_2CHO] \xleftarrow{OH^-} NO_2C_6H_4CHO + CH_3CHO$
($NO_2C_6H_4CHO$: page 662; CH_3CHO: from EtOH)

(f) $CH_3C(CH_3){=}CHCOOH \xleftarrow{OX^-} CH_3C(CH_3){=}CH{-}C(=O){-}CH_3 \longleftarrow$ as in (b)

(g) $CH_3CH(OH){-}CH_2CH_2(OH) \xleftarrow{H_2,\ Ni} CH_3CH(OH){-}CH_2CHO \xleftarrow{OH^-} 2CH_3CHO$ (from EtOH)

(h) $CH_3CH_2C(CH_3)(OH){-}C{\equiv}CH \longleftarrow CH_3CH_2C(CH_3){=}O + NaC{\equiv}CH$
($CH_3CH_2C(CH_3){=}O$ from *sec*-BuOH; $NaC{\equiv}CH$ from acetylene)

(i) $PhCH{=}CHCH{=}CHCH{=}CH_2 \xleftarrow{-Ph_3PO}$
- $PhCH{=}CHCHO$ (page 810)
- $Ph_3P{=}CH{-}CH{=}CH_2 \xleftarrow[base]{Ph_3P} ClCH_2CH{=}CH_2 \xleftarrow{Cl_2,\ heat} CH_3CH{=}CH_2$ (from *i*-PrOH)

(j) $PhCH{=}CHCH{=}CHCH{=}CHPh \xleftarrow{-Ph_3PO}$
- $Ph_3P{=}CHPh \xleftarrow[base]{Ph_3P} ClCH_2Ph$ (from $PhCH_3$)
- $PhCH{=}CHCH{=}CHCHO \xleftarrow{OH^-} PhCH{=}CHCHO$ (page 810) + CH_3CHO (from EtOH)

(k) $C_2H_5C(CH_3){=}C(CH_3)COOH \xleftarrow{hydrol.} C_2H_5C(CH_3){=}C(CH_3)COOEt \longleftarrow C_2H_5{-}C(CH_3){=}O + Ph_3P{=}C(CH_3)COOEt$
($C_2H_5C(CH_3){=}O$ from *sec*-BuOH; $Ph_3P{=}C(CH_3)COOEt$ from *n*-PrOH, below)

$Ph_3P{=}C(CH_3)COOEt \xleftarrow[base]{Ph_3P} BrCH(CH_3)COOEt \xleftarrow{EtOH,\ H^+} BrCH(CH_3)COOH \xleftarrow{Br_2,\ P} CH_3CH_2COOH$ (from *n*-PrOH)

(l) [indanone] $\xleftarrow{HF}$ $C_6H_5CH_2CH_2COCl$ $\xleftarrow{SOCl_2}$ $\xleftarrow{H_2,\ Ni}$ $C_6H_5CH{=}CHCOOH$ (page 811)

3-Phenylpropenoic acid

(m) [Fischer projection: COOH; H—C—OH; H—C—OH; Ph] ≡ [Ph(H)(OH)C—C(H)(OH)COOH] $\xleftarrow[anti\text{-addn.}]{HCO_2OH}$ [H(Ph)C=C(H)COOH]

trans-3-Phenylpropenoic acid
(Problem 21.18(d), page 811)

11. An aldol-like condensation leads to a γ-hydroxy acid, which subsequently forms the γ-lactone (cyclic ester).

$$\text{A }\gamma\text{-lactone (A cyclic ester)} \longleftarrow \underset{OH}{\overset{\gamma}{MeCH}}-\underset{CH_2COOH\ (\alpha)}{\overset{\beta}{CHCOOH}} \longleftarrow CH_3CHO + \begin{matrix} H_2CCOOH \\ | \\ H_2CCOOH \end{matrix}$$

A γ-lactone
A cyclic ester

A γ-hydroxy acid

12. It gives a mixture of aldol products, since either of two carbanions can form,

$$CH_3CH_2COCH_3 + :B \longrightarrow [CH_3CHCOCH_3]^- \quad or \quad [CH_3CH_2COCH_2]^-$$

and then add to the ketone.

13. A crossed aldol converts HCHO into a *methylol* group, $—CH_2OH$.

$$O{=}\overset{H}{\overset{|}{C}}—CH_3 + O{=}\overset{H}{\overset{|}{C}}—H \xrightarrow{OH^-} O{=}\overset{H}{\overset{|}{C}}—CH_2—CH_2OH$$

Two more of these conversions (there are two more α-hydrogens) gives a tri-methylol compound.

$$O{=}\overset{H}{\overset{|}{C}}—CH_2—CH_2OH \xrightarrow[OH^-]{HCHO} O{=}\overset{H}{\overset{|}{C}}—\underset{CH_2OH}{\underset{|}{CH}}—CH_2OH \xrightarrow[OH^-]{HCHO} O{=}\overset{H}{\overset{|}{C}}—\underset{CH_2OH}{\underset{|}{\overset{CH_2OH}{\overset{|}{C}}}}—CH_2OH$$

At this point, we have run out of α-hydrogens, but now the Cannizzaro reaction can take over.

$$\mathrm{HCHO} + \mathrm{O{=}C(H){-}C(CH_2OH)_3} \xrightarrow{\text{base}} \mathrm{HCOO^-} + \mathrm{HOCH_2{-}C(CH_2OH)_2{-}CH_2OH}$$

No α-hydrogens — Pentaerythritol

14. We exchange protium for deuterium conveniently and regioselectively by treatment of the ketone with D_2O in the presence of base or acid (see Problem 21.5(c), page 804).

Cyclohexanone $\xrightarrow{OD^-/D_2O}$ 2,2,6,6-tetradeuteriocyclohexanone $\xrightarrow{LiAlH_4}$ 2,2,6,6-tetradeuteriocyclohexanol $\xrightarrow{PBr_3}$ 1-bromo-2,2,6,6-tetradeuteriocyclohexane $\xrightarrow{KOH(alc)}$ 1,3,3-trideuteriocyclohexene (D, D, D)

15. See page 934.

16. $$\mathrm{CH_3{-}C({=}O){-}H} + \mathrm{H^+} \rightleftarrows \mathrm{CH_3{-}C({=}\overset{\oplus}{O}H){-}H} \rightleftarrows \mathrm{CH_2{=}C(OH){-}H} + \mathrm{H^+}$$

$$\underset{\text{Electrophile}}{\mathrm{CH_3{-}C({=}\overset{\oplus}{O}H){-}H}} + \underset{\text{Nucleophile}}{\mathrm{CH_2{=}C(OH){-}H}} \rightleftarrows \underset{\text{Protonated adduct}}{\mathrm{CH_3{-}CH(OH){-}CH_2{-}C({=}\overset{\oplus}{O}H){-}H}} \rightleftarrows \underset{\text{Aldol}}{\mathrm{CH_3{-}CH(OH){-}CH_2{-}C({=}O){-}H}} + \mathrm{H^+}$$

17. Reverse of the aldol condensation of acetone (*retrograde aldol* or *retro aldol*). According to the principle of microscopic reversibility (page 313), all steps in the aldol condensation are involved, in reverse order:

$$\mathrm{CH_3{-}C(CH_3)(OH){-}CH_2{-}C({=}O){-}CH_3} + \mathrm{OH^-} \rightleftarrows \mathrm{CH_3{-}C(CH_3)(O^-){-}CH_2{-}C({=}O){-}CH_3} + \mathrm{H_2O}$$

$$\mathrm{CH_3{-}C(CH_3)(O^-){-}CH_2{-}C({=}O){-}CH_3} \rightleftarrows \mathrm{CH_3{-}C(CH_3){=}O} + [\mathrm{CH_2COCH_3}]^-$$

$$[\mathrm{CH_2COCH_3}]^- + \mathrm{H_2O} \rightleftarrows \mathrm{CH_3{-}C(CH_3){=}O} + \mathrm{OH^-}$$

18. (a)

$$\mathrm{H{-}\overset{H}{\underset{H}{C}}{-}\underset{O}{\overset{\|}{C}}{-}\overset{X}{\underset{H}{C}}{-}H} \xrightarrow{OH^-} \mathrm{H{-}\overset{H}{\underset{H}{C}}{-}C(\dot{=}O){=}\overset{X}{C}{-}H}^{\ominus} \quad \textit{More stable anion}$$

$$\dashrightarrow \mathrm{H{-}\overset{H}{C}{=}C(\dot{=}O){-}\overset{X}{\underset{H}{C}}{-}H}^{\ominus} \quad \textit{Not formed}$$

Electron withdrawal by halogen makes hydrogens on the carbon to which halogen has already become attached more acidic and hence more readily removed by base to give further substitution.

(b) $$\mathrm{R{-}\underset{O}{\overset{\|}{C}}{-}CX_3} \xrightarrow{OH^-} \mathrm{R{-}\overset{OH}{\underset{O^-}{C}}{-}CX_3} \longrightarrow \mathrm{R{-}\overset{OH}{\underset{O}{\overset{\|}{C}}}} + \mathrm{CX_3^-} \longrightarrow \mathrm{RCOO^-} + \mathrm{HCX_3}$$

Electron withdrawal by three halogens makes CX_3^- comparatively weakly basic (for a carbanion) and hence a good leaving group.

Thus both essential aspects of the haloform reaction—regiospecificity of halogenation, and cleavage—are controlled by the same factor: stabilization of a carbanion through electron withdrawal by halogen.

19. Recalling the acidity of γ-hydrogens of α,β-unsaturated aldehydes (Problem 21.16, page 810) and looking at the structure for citral (Problem 24, Chapter 18), we propose:

$$\mathrm{CH_3{-}\overset{CH_3}{C}{=}CH{-}\overset{H}{C}{=}O} + \mathrm{CH_3{-}\overset{CH_3}{C}{=}CH{-}\overset{H}{C}{=}O} \xrightarrow[OH^-]{\text{aldol. cond.}} \mathrm{CH_3{-}\overset{CH_3}{C}{=}CH{-}CH{=}CH{-}\overset{CH_3}{C}{=}CH{-}\overset{H}{C}{=}O}$$

3-Methyl-2-butenal *Two moles* — Dehydrocitral

20. Acetoacetic ester is acidic enough to displace CH_4 from CH_3MgI.

$$\mathrm{CH_3COCH_2COOEt + CH_3MgI \longrightarrow \underset{Methane}{CH_4}\uparrow + CH_3COCHCOOEt\ {}^-(MgI)^+}$$

21. (a) $BrCH_2CH_2(CH_2)_6CH_2OH \xrightarrow[H^+]{DHP} BrCH_2CH_2(CH_2)_6CH_2OTHP$ (A) $\xrightarrow{Ph_3P} \xrightarrow{base} Ph_3P{=}CHCH_2(CH_2)_6CH_2OTHP$ (B)

$(C_2H_5)(H)C{=}C(H)(CHO)$ + B ⟶ $C_2H_5CH{=}CH{-}CH{=}CH{-}CH_2(CH_2)_6CH_2OTHP$ (C) + $C_2H_5CH{=}CH{-}CH{=}CH{-}CH_2(CH_2)_6CH_2OTHP$

C $\xrightarrow{HCl}$ $C_2H_5CH{=}CH{-}CH{=}CH{-}CH_2(CH_2)_6CH_2OH$ (D)

(second isomer) $\xrightarrow{HCl}$ $C_6H_5CH{=}CH{-}CH{=}CH{-}CH_2(CH_2)_6CH_2OH$

D $\xrightarrow{AcO}$ $C_2H_5CH{=}CH{-}CH{=}CH{-}CH_2(CH_2)_6CH_2OCOCH_3$ (E)

E
(9*Z*, 11*E*)-Tetradecadien-1-yl acetate
Sex attractant of Egyptian cotton leafworm

(second isomer) $\xrightarrow{Ac_2O}$ $C_2H_5CH{=}CH{-}CH{=}CH{-}CH_2(CH_2)_6CH_2OCOCH_3$

(9*E*, 11*E*)-Tetradecadien-1-yl acetate

(b) In the formation of C, the new carbon–carbon double bond is generated in the two possible configurations, to give a mixture of diastereomers: (9*Z*,11*E*) and (9*E*,11*E*). Although we could not have told this from the evidence presented, the actual pheromone has the (9*Z*,11*E*) configuration.

22. $n\text{-}C_3H_7C{\equiv}CH \xrightarrow{n\text{-}BuMgBr} n\text{-}C_3H_7C{\equiv}CMgBr$ (F) $\xrightarrow{HCHO} \xrightarrow{H^+} n\text{-}C_3H_7C{\equiv}CCH_2OH$ (G)

G $\xrightarrow{PBr_3} n\text{-}C_3H_7C{\equiv}CCH_2Br$ (H) $\xrightarrow{Ph_3P} \xrightarrow{base} n\text{-}C_3H_7C{\equiv}CCH{=}PPh_3$ (I)

I + $OHCCH_2(CH_2)_7COOEt \longrightarrow n\text{-}C_3H_7C{\equiv}CCH{=}CHCH_2(CH_2)_7COOEt$ (J)

J $\xrightarrow[Lindlar]{H_2}$ $(n\text{-}C_3H_7)(H)C{=}C(H)(CH{=}CH(CH_2)_7COOEt)$ (K) $\xrightarrow{LiAlH_4}$ $(n\text{-}C_3H_7)(H)C{=}C(H)(CH{=}CHCH_2(CH_2)_7CH_2OH)$

(10*E*,12*Z*)-Hexadecadien-1-ol

From this evidence we cannot tell the stereochemistry at the 10-position, the double bond generated in the formation of J. The actual pheromone is the (10*E*,12*Z*) isomer.

n-C_3H_7–CH=CH–CH=CH–$CH_2(CH_2)_7CH_2OH$

(10*E*,12*Z*)-Hexadecadien-1-ol
Bombykol
Sex attractant of silkworm moth

23. (a) Base-catalyzed tautomerization:

$$-\underset{\underset{O}{\|}}{C}-\overset{H}{\overset{|}{\underset{|}{C}}}- + :B \rightleftarrows H:B + -\underset{\underset{O}{\vdots}}{C}\overset{..}{=}\overset{|}{C}-\;(\ominus) \rightleftarrows :B + -\underset{\underset{OH}{|}}{C}=\overset{|}{C}-$$

Keto form — Base — Hybrid anion — Base — Enol form

Acid-catalyzed tautomerization:

$$-\underset{\underset{O}{\|}}{C}-\overset{H}{\overset{|}{\underset{|}{C}}}- + H:B \rightleftarrows :B + -\underset{\underset{\oplus OH}{\|}}{C}-\overset{H}{\overset{|}{\underset{|}{C}}}- \rightleftarrows H:B + -\underset{\underset{OH}{|}}{C}=\overset{|}{C}-$$

Keto form — Acid — Cation — Acid — Enol form

(b) and (c) The enol is stabilized by (i) conjugation of the remaining C=O with C=C, and (ii) intramolecular hydrogen bonding:

$$-\underset{\underset{O}{\|}}{C}-\overset{|}{\underset{\underset{H}{|}}{C}}-\underset{\underset{O}{\|}}{C}- \rightleftarrows -\underset{\underset{OH}{|}}{C}=\overset{|}{C}-\underset{\underset{O}{\|}}{C}- \quad (OH\cdots O)$$

Keto form — Enol form
Conjugation
Chelation

1,3- or β-Dicarbonyl compounds

24. (a) *a*, enol $-CH_3$ *b*, keto $-CH_3$ *c*, keto $-CH_2-$ *d*, enol $-CH=$ *e*, enol $-OH$

$$\overset{b}{CH_3}-\underset{\underset{O}{\|}}{C}-\overset{c}{CH_2}-\underset{\underset{O}{\|}}{C}-\overset{b}{CH_3} \qquad \overset{a}{CH_3}-\underset{\underset{\underset{e}{OH}}{|}}{C}=\overset{d}{CH}-\underset{\underset{O}{\|}}{C}-\overset{a}{CH_3} \quad (OH\text{------}O)$$

Ratios $a{:}b$ and $2d{:}c$ are equal; the value of the ratios (5.6:1) shows that approximately 85% enol is present.

(b) *a*, enol —CH_3 *b*, enol —CH= *c*, C_6H_5— *d*, enol —OH

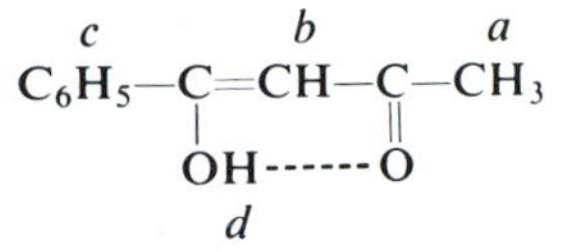

$$\begin{array}{c} \quad\ c \qquad\qquad b \qquad\qquad a \\ C_6H_5{-}\underset{\underset{OH\cdots\cdots O}{|}}{C}{=}CH{-}\underset{\|}{C}{-}CH_3 \\ d \end{array}$$

All enol. Conjugation of the carbon–carbon double bond with the ring stabilizes the enol to the exclusion of the keto form.

(See the labeled spectra on page 714 of this Study Guide.)

22

Amines I. Preparation and Physical Properties

22.1

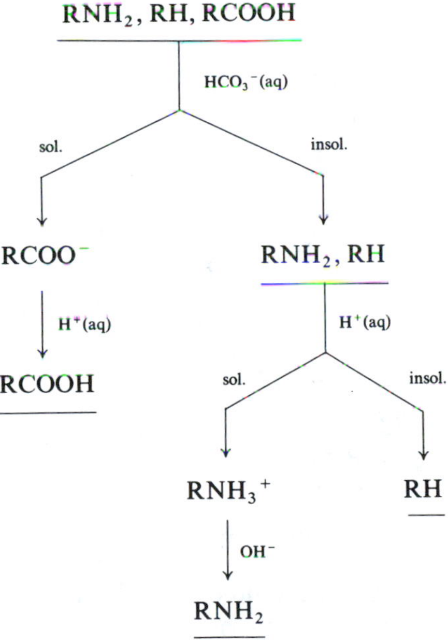

The solid mixture can be treated directly, with each insoluble compound being collected on a filter. The liquids are easier to handle if they are dissolved in ether first; each separation is carried out in the separatory funnel, with the water-soluble salts being ether-insoluble, and the water-insoluble compounds being ether-soluble.

22.2 A carbanion, like ammonia, has a low, easily surmountable energy barrier between mirror-image pyramidal arrangements. Attachment of a fourth group to the carbanion can occur to either arrangement, and thus give the racemic modification.

22.3 At room temperature, inversion about nitrogen is slow enough that NMR "sees" two kinds of ring protons: two protons *cis* and two protons *trans* to ethyl. At 120 °C, inversion is so fast that all four protons are seen in average positions: they are equivalent.

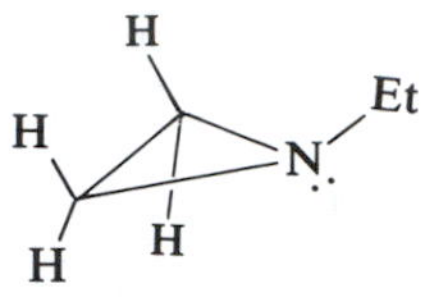

22.4 (a) IV exists as two pairs of enantiomers, each pair diastereomeric to the other pair. In one pair $—CH_3$ and $—Cl$ are *cis*, in the other pair *trans*.

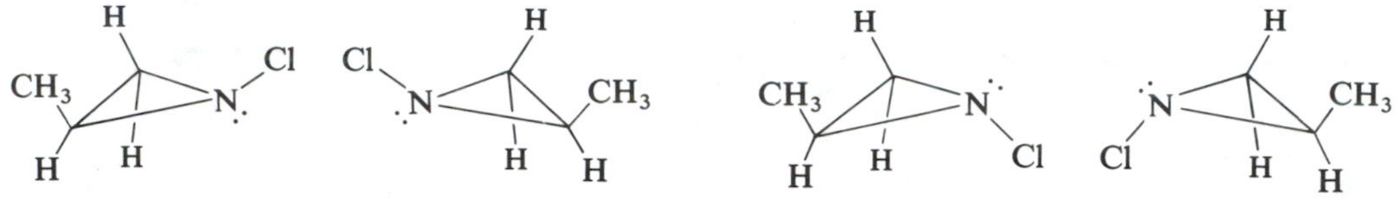

cis *trans*

Interconversion of diastereomers requires inversion about nitrogen, which—presumably because of the rigidity of the three-membered ring—is so slow at 25 °C that the diastereomers can be separated.

(b) Peroxidation of a carbon–nitrogen double bond, like that of a carbon–carbon double bond, produces a rigid three-membered ring containing nitrogen and oxygen.

$$Ph_2C{=}NCH_3 + R^*CO_2OH \longrightarrow Ph_2C\text{—}NCH_3 \text{ (ring with O)} + R^*COOH$$

This exists in enantiomeric forms:

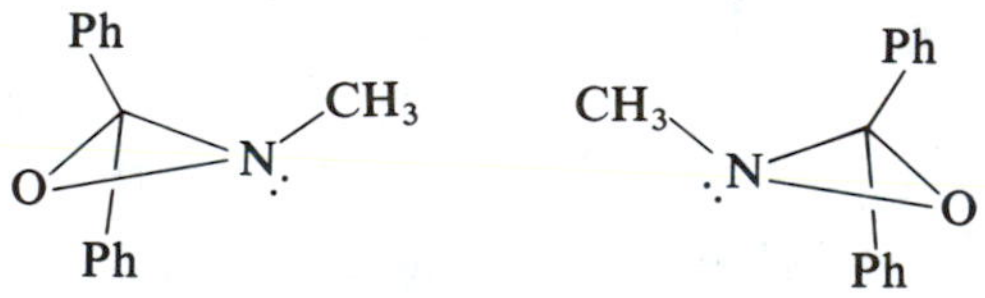

Enantiomers

The transition state for the epoxidation contains both the chiral center in the peroxy acid and the chiral center developing in the epoxide; diastereomeric configurations—of different energy—therefore exist. Since the peroxy acid is optically active, one diastereomer predominates in the transition state, and yields a predominance of one enantiomeric epoxide—and hence an optically active product. (This process is called *asymmetric induction.*) Interconversion of the enantiomeric epoxides requires inversion about nitrogen which, as in (a), is evidently very slow.

(Brois, S. J. "Aziridines. XII. Isolation of a Stable Nitrogen Pyramid"; *J. Am. Chem. Soc.* **1968**, *90*, 508.)

22.5 (i) Hofmann degradation:

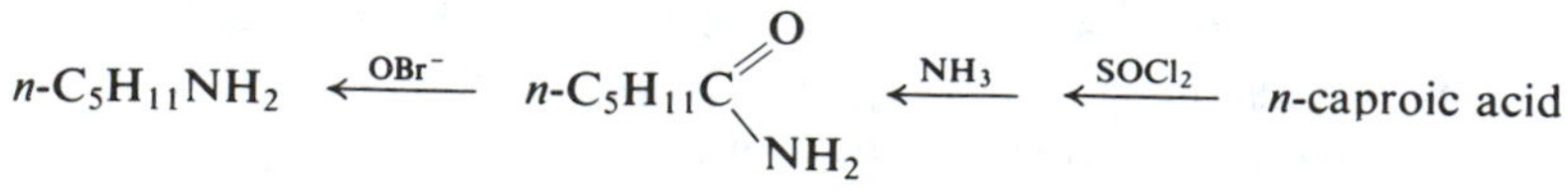

$$n\text{-}C_5H_{11}NH_2 \xleftarrow{OBr^-} n\text{-}C_5H_{11}CONH_2 \xleftarrow{NH_3} \xleftarrow{SOCl_2} n\text{-caproic acid}$$

(ii) Via nitrile:

$$n\text{-}C_5H_{11}NH_2 \xleftarrow{H_2,\ Ni} n\text{-}C_4H_9C{\equiv}N \xleftarrow{CN^-} n\text{-BuBr} \xleftarrow{PBr_3} n\text{-BuOH}$$

(iii) Via halide:

$$n\text{-}C_5H_{11}NH_2 \xleftarrow{NH_3} n\text{-}C_5H_{11}Br \xleftarrow{PBr_3} n\text{-pentyl alcohol}$$

(iv) Reductive amination:

$$n\text{-}C_5H_{11}NH_2 \xleftarrow{NH_3,\ H_2,\ Ni} n\text{-}C_4H_9CHO \xleftarrow{C_5H_5NHCrO_3Cl} n\text{-pentyl alcohol}$$

22.6 (a) $PhCH_2NH_2 \xleftarrow{NH_3} PhCH_2Cl \xleftarrow{Cl_2,\ heat} PhCH_3$

(b) $CH_3C_6H_4CH(NH_2)CH_3 \xleftarrow{NH_3,\ H_2,\ Ni} CH_3C_6H_4C(=O)CH_3 \xleftarrow{Ac_2O,\ AlCl_3} PhCH_3$

(c) $PhCH_2CH_2NH_2 \xleftarrow{H_2,\ Ni} PhCH_2C{\equiv}N \xleftarrow{CN^-} PhCH_2Cl$ (as in a)

(d) $CH_3C_6H_4NH_2 \xleftarrow{Fe,\ H^+} CH_3C_6H_4NO_2 \xleftarrow{HNO_3,\ H_2SO_4} PhCH_3$

(e) $PhNH_2 \xleftarrow{OBr^-} PhCONH_2 \xleftarrow{NH_3} \xleftarrow{PCl_5} PhCOOH \xleftarrow{KMnO_4} PhCH_3$

22.7 See page 854.

22.8 The Hofmann rearrangement is *intra*molecular; that is, the migrating group moves from one atom to another atom within the *same* molecule:

$C_6H_4D\text{-}CONH_2 + C_6H_5CO^{15}NH_2 \xrightarrow{OCl^-} C_6H_4D\text{-}NH_2 + C_6H_5{}^{15}NH_2$

$C_6H_5NH_2 + C_6H_4D\text{-}{}^{15}NH_2$ Cross-products: *not formed*

If, instead, the rearrangement had been *inter*molecular, a phenyl group from one molecule could have become attached to the nitrogen of another. In that case, there would have been obtained, along with the observed products, a mixture of cross-products—contrary to fact.

This elegant double-labeling experiment was carried out by T. J. Prosser and E. L. Eliel at the University of Notre Dame.

22.9 The Curtius rearrangement involves a 1,2-shift to electron-deficient nitrogen, with N_2—that best of all leaving groups—being lost.

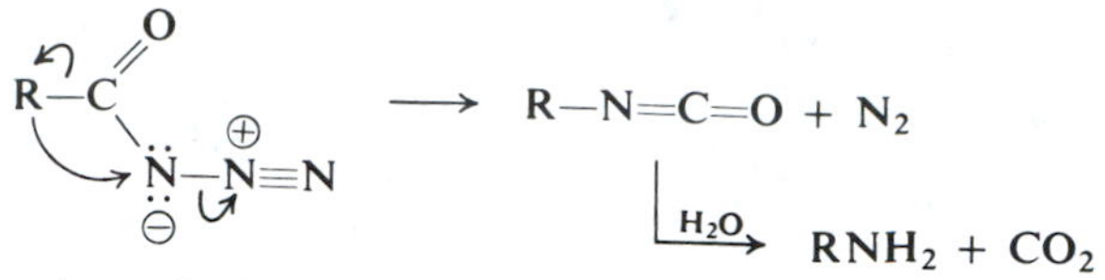

$$R{-}C(=O){-}\ddot{N}^{\ominus}{-}N^{\oplus}{\equiv}N \longrightarrow R{-}N{=}C{=}O + N_2 \xrightarrow{H_2O} RNH_2 + CO_2$$

An acyl azide

(Smith, P. A. S. "The Curtius Rearrangement"; *Organic Reactions*; Wiley: New York; Vol. III, Chapter 8.)

22.10 The lactam IV can form only if III has the *cis* configuration. Since II is also *cis*, the reaction must have taken place so that $-NH_2$ occupies the same position in III that

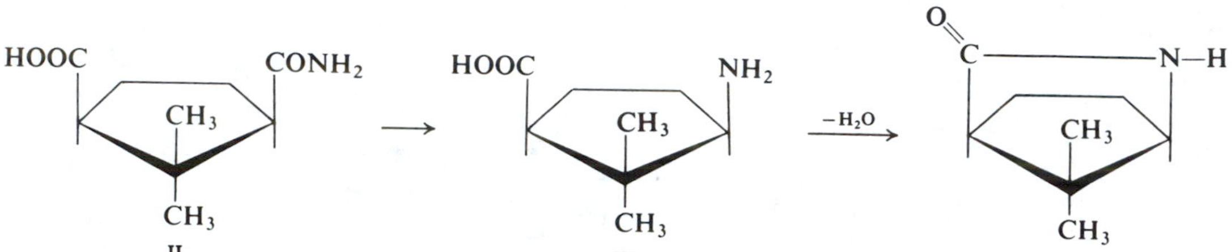

HOOC, CONH$_2$, CH$_3$, CH$_3$ — II, Amide, *cis* $\longrightarrow$ HOOC, NH$_2$, CH$_3$, CH$_3$ — III, Amine, *cis* $\xrightarrow{-H_2O}$ O=C—N—H, CH$_3$, CH$_3$ — IV, Lactam, *cis*

$-CONH_2$ occupied in II; that is, reaction must have taken place with retention of configuration at the migrating group.

1. (a) 1° $CH_3CH_2CH_2CH_2{-}NH_2$ *n*-Butylamine; $CH_3CH_2CH(NH_2)CH_3$ *sec*-Butylamine; $CH_3CH(CH_3)CH_2{-}NH_2$ Isobutylamine; $CH_3{-}C(CH_3)(NH_2){-}CH_3$ *tert*-Butylamine

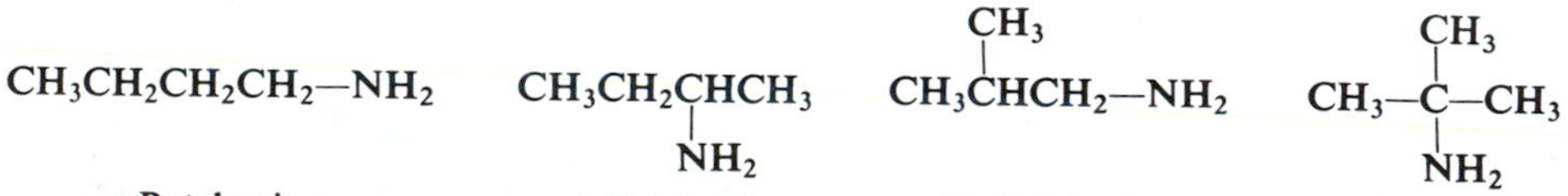

2° $CH_3CH_2CH_2{-}NH{-}CH_3$ Methyl-*n*-propylamine; $CH_3CH(CH_3){-}NH{-}CH_3$ Methylisopropylamine; $CH_3CH_2{-}NH{-}CH_2CH_3$ Diethylamine

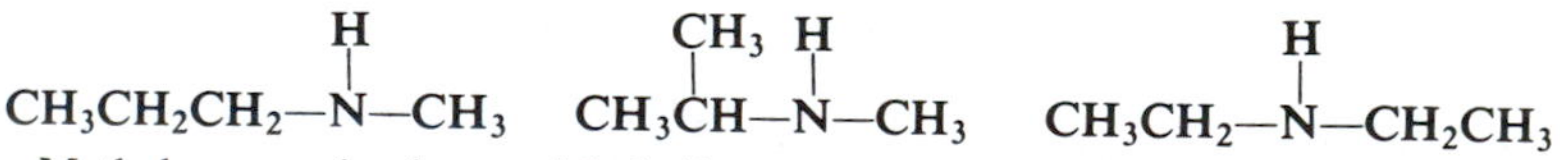

3° $CH_3CH_2{-}N(CH_3){-}CH_3$ Ethyldimethylamine

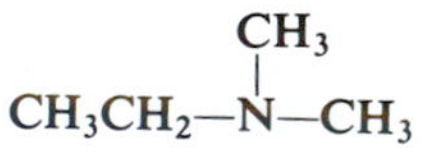

(b) 1° $PhCH_2NH_2$, benzylamine; *o*-, *m*-, and *p*-$CH_3C_6H_4NH_2$ (the toluidines)

2° $Ph{-}NH{-}CH_3$, *N*-methylaniline

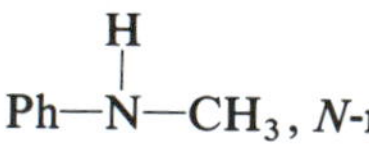

3° none

2. (a) $CH_3CH_2CH(NH_2)CH_3$ (b) $o-CH_3C_6H_4NH_2$ (c) $C_6H_5NH_3^+Cl^-$ (d) $(C_2H_5)_2NH$

(e) $p-H_2NC_6H_4COOH$ (f) $C_6H_5CH_2NH_2$ (g) $(CH_3)_2CHNH_3^{+}\,{}^{-}OOCPh$ (h) $C_6H_5N(CH_3)_2$

(i) $HOCH_2CH_2NH_2$ (j) $C_6H_5CH_2CH_2NH_2$ (k) $C_6H_{11}N(CH_3)_2$ (cyclohexyl) (l) $(C_6H_5)_2NH$

(m) 2,4-$(CH_3)_2C_6H_3NH_2$ (n) $(n\text{-}C_4H_9)_4N^+I^-$

3. (a) Excess NH_3
(b) $C_5H_5NHCrO_3Cl$; NH_3, H_2, Ni
(c) NH_3, H_2, Ni
(d) Fe, H^+, heat
(e) H_2, Ni
(f) Br_2, OH^-
(g) $K_2Cr_2O_7$; $SOCl_2$; NH_3; OBr^-
(h) HBr, NaCN; H_2, Ni

Aniline: d (and, under very vigorous conditions, a)
Benzylamine: all (except the first stage of h)

4. (a) $CH_3-CH(CH_3)-NH_2 \xleftarrow{NH_3,\ H_2,\ Ni} CH_3-C(CH_3)=O \xleftarrow{CrO_3} CH_3CHOHCH_3$

(b) $n\text{-}BuCH_2NH_2 \xleftarrow{H_2,\ Ni} n\text{-}Bu-C\equiv N \xleftarrow{CN^-} n\text{-}BuBr \xleftarrow{HBr} n\text{-}BuOH$

(c) $p\text{-}CH_3C_6H_4NH_2 \xleftarrow{Fe,\ H^+} p\text{-}CH_3C_6H_4NO_2 \xleftarrow{HNO_3,\ H_2SO_4} PhCH_3$

(d) $CH_3CH(CH_3)-NH-CH_2CH_3 \xleftarrow{H_2,\ Ni}$ [$EtNH_2 \xleftarrow{NH_3} EtBr \xleftarrow{HBr} EtOH$] and [$CH_3COCH_3 \xleftarrow{CrO_3} CH_3CHOHCH_3$]

(e) $PhCH(NH_2)CH_3 \xleftarrow{NH_3,\ H_2,\ Ni} Ph-C(=O)CH_3 \xleftarrow{AlCl_3}$ [acetic anhydride (page 763)] and [C_6H_6]

(f) $PhCH_2CH_2NH_2 \xleftarrow{H_2,\ Ni} PhCH_2-C\equiv N \xleftarrow{CN^-} PhCH_2Cl \xleftarrow{Cl_2,\ heat} PhCH_3$

(g) $m\text{-}ClC_6H_4NH_2 \xleftarrow{Fe,\ H^+} m\text{-}ClC_6H_4NO_2 \xleftarrow[FeCl_3]{Cl_2} PhNO_2 \xleftarrow{HNO_3,\ H_2SO_4} C_6H_6$

(h) $p\text{-}H_2NC_6H_4COOH \xleftarrow{Fe,\ H^+} p\text{-}O_2NC_6H_4COOH$ (page 525)

(i) $n\text{-Bu—CH(NH}_2\text{)—Et} \xleftarrow{NH_3,\ H_2,\ Ni} n\text{-Bu—C(=O)—Et} \leftarrow$
- $EtCOCl \xleftarrow{SOCl_2} EtCOOH$ (from n-PrOH)
- $n\text{-}Bu_2CuLi \xleftarrow{CuI} n\text{-BuLi}$ (from n-BuOH)

(j) PhNH—Et $\leftarrow$
- EtBr (from EtOH)
- $PhNH_2 \xleftarrow{Fe,\ H^+} PhNO_2 \xleftarrow{HNO_3,\ H_2SO_4} C_6H_6$

(k) 2,4-dinitroaniline (NH_2, NO_2, NO_2 on benzene ring) $\xleftarrow{NH_3}$ 2,4-dinitrochlorobenzene (Cl, NO_2, NO_2 on benzene ring) $\xleftarrow{HNO_3,\ H_2SO_4} PhCl \xleftarrow{Cl_2,\ Fe} C_6H_6$

(l) $PhCH_2CH(NH_2)CH_3 \xleftarrow{NH_3,\ H_2,\ Ni} PhCH_2C(=O)—CH_3 \leftarrow$
- $Me_2CuLi \xleftarrow{CuI}$ MeLi (from MeOH)
- $PhCH_2COCl \xleftarrow{SOCl_2} PhCH_2COOH$ (page 722)

(m) $O_2N\text{—}C_6H_4\text{—}CH_2NH_2 \xleftarrow{NH_3} O_2N\text{—}C_6H_4\text{—}CH_2Cl \xleftarrow[heat]{Cl_2} O_2N\text{—}C_6H_4\text{—}CH_3$ (from $PhCH_3$)

(n) $PhCH(OH)—CH_2NH_2 \xleftarrow{H_2,\ Ni} PhCH(OH)—C{\equiv}N \xleftarrow[H^+]{CN^-} PhCHO \xleftarrow{H_2O,\ OH^-} PhCHCl_2$ (from $PhCH_3$)

5. (a) Same number of carbons:

$n\text{-}C_{15}H_{31}CH_2NH_2 \xleftarrow{NH_3} n\text{-}C_{15}H_{31}CH_2Br \xleftarrow{HBr} n\text{-}C_{15}H_{31}CH_2OH \xleftarrow{LiAlH_4} n\text{-}C_{15}H_{31}COOH$

(b) One more carbon:

$n\text{-}C_{15}H_{31}CH_2CH_2NH_2 \xleftarrow{H_2,\ Ni} n\text{-}C_{15}H_{31}CH_2C{\equiv}N \xleftarrow{CN^-} n\text{-}C_{15}H_{31}CH_2Br \longleftarrow$ as in (a)

(c) One less carbon:

$n\text{-}C_{15}H_{31}NH_2 \xleftarrow{OBr^-} n\text{-}C_{15}H_{31}CONH_2 \xleftarrow{NH_3} n\text{-}C_{15}H_{31}COCl \xleftarrow{SOCl_2} n\text{-}C_{15}H_{31}COOH$

(d) Two C_{16} units:

$n\text{-}C_{15}H_{31}—CH(NH_2)—CH_2—C_{15}H_{31}\text{-}n \xleftarrow{NH_3,\ H_2,\ Ni} n\text{-}C_{15}H_{31}—C(=O)—CH_2—C_{15}H_{31}\text{-}n$

$n\text{-}C_{15}H_{31}—C(=O)—CH_2—C_{15}H_{31}\text{-}n \leftarrow$
- $n\text{-}C_{15}H_{31}COCl \longleftarrow$ as in (c)
- $(n\text{-}C_{15}H_{31}CH_2)_2CuLi \longleftarrow n\text{-}C_{15}H_{31}CH_2Li \xleftarrow{Li} n\text{-}C_{15}H_{31}CH_2Br \uparrow$ (a)

6. (a) $BrCH_2CH_2Br \xrightarrow[(+2CN,\ -2Br)]{2CN^-} NCC_2H_4CN \xrightarrow[(+8H)]{\text{redn.}} H_2N(CH_2)_4NH_2$ Putrescine

(b) $Br(CH_2)_5Br \xrightarrow[(+2NH_2,\ -2Br)]{\text{excess } NH_3} H_2N(CH_2)_5NH_2$ Cadaverine

7. $H_2N(CH_2)_6NH_2 \xleftarrow{H_2,\ Ni} NC(CH_2)_4CN \xleftarrow{2CN^-} Cl(CH_2)_4Cl \xleftarrow{H_2,\ Ni} ClCH_2CH{=}CHCH_2Cl$

$ClCH_2CH{=}CHCH_2Cl$ ← (1,4-addn., Cl_2) ← $CH_2{=}CH{-}CH{=}CH_2$

8. 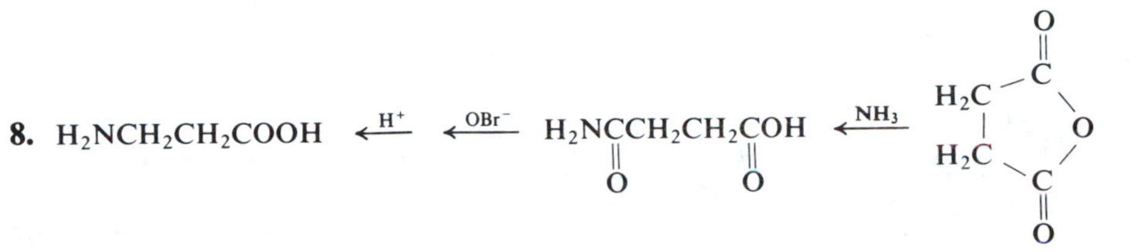 $H_2NCH_2CH_2COOH \xleftarrow{H^+} \xleftarrow{OBr^-} H_2NC(=O)CH_2CH_2C(=O)OH \xleftarrow{NH_3}$ Succinic anhydride

9. (a) Pair of enantiomers: chiral carbon.

(b) One inactive compound: achiral.

(c) Pair of enantiomers: chiral nitrogen.

(d) Pair of enantiomers: chiral molecule. The planes of the rings are perpendicular to each other.

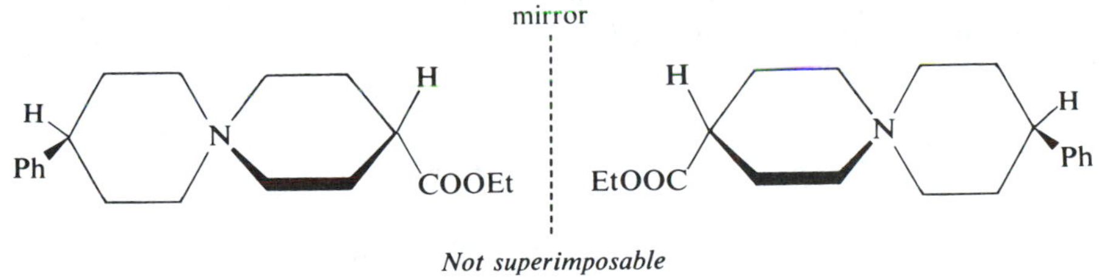

mirror

H, Ph, N, H, COOEt | EtOOC, H, N, H, Ph

Not superimposable

(e) Pair of enantiomers; chiral nitrogen.

10. (a)

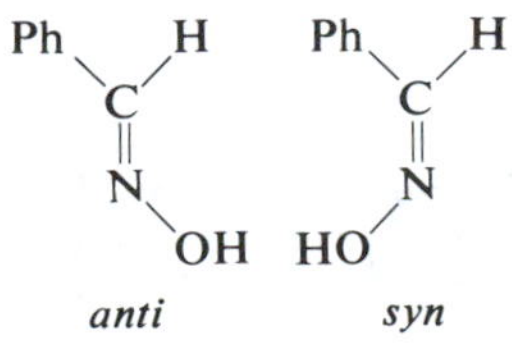

Ph(H)C=N–OH *anti*   Ph(H)C=N–HO *syn*

(b) The electronic configuration about C=N is similar to that about C=C (Secs. 8.2, 8.5, and 8.6). Where both sp^2 orbitals of C= hold substituents, one sp^2 orbital of N= contains an unshared pair. Hindered rotation about the carbon–nitrogen double bond permits stereoisomers to be isolated.

(c) $Ph_2C{=}NOH$: no (the two groups on carbon must be different, Sec. 8.6). The other two: yes.

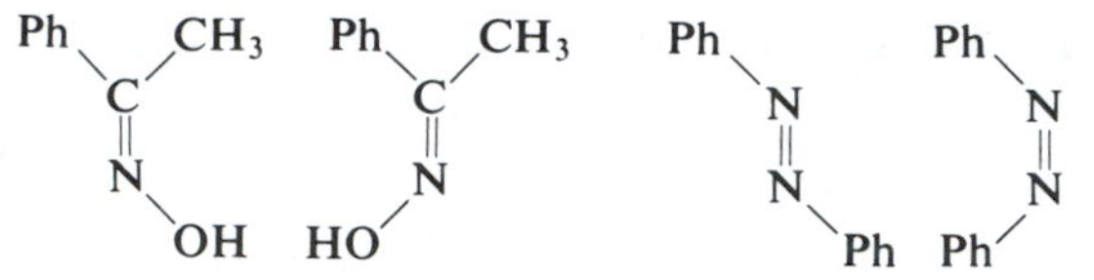

Ph(CH₃)C=N–OH and Ph(CH₃)C=N–OH (two stereoisomers)

Acetophenoneoxime

Ph–N=N–Ph *trans*, Ph–N=N–Ph *cis*

Azobenzene

11. (a) 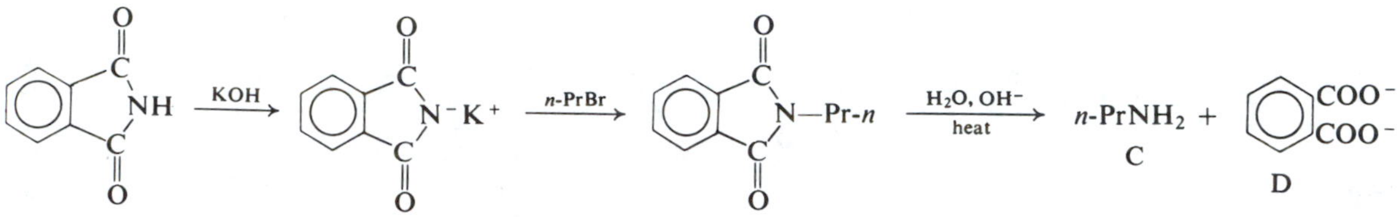 Phthalimide (*Acidic*) $\xrightarrow{KOH}$ A (*Salt of imide*, N^-K^+) $\xrightarrow{n\text{-PrBr}}$ B (N–Pr-*n*) $\xrightarrow[\text{heat}]{H_2O,\ OH^-}$ $n\text{-PrNH}_2$ (C) + $C_6H_4(COO^-)_2$ (D)

(b) Because the N can be alkylated only once, the Gabriel synthesis produces *pure* primary amines, free from contaminating secondary or tertiary amines. The synthesis depends upon the acidity of the N—H bond of the imide (there are two acid-strengthening acyl groups present).

12. (a) The Lossen rearrangement involves migration to electron-deficient nitrogen, with $^-OOCR'$ playing the same part that Br^- does in the Hofmann rearrangement (Sec. 22.15) or N_2 in the Curtius rearrangement (Problem 22.9, page 839).

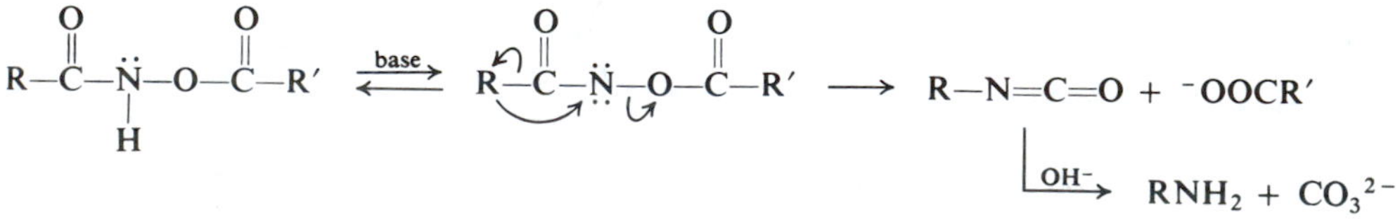

$$R{-}\overset{O}{\overset{\|}{C}}{-}\underset{H}{\ddot{N}}{-}O{-}\overset{O}{\overset{\|}{C}}{-}R' \underset{}{\overset{\text{base}}{\rightleftharpoons}} R{-}\overset{O}{\overset{\|}{C}}{-}\ddot{N}^-{-}O{-}\overset{O}{\overset{\|}{C}}{-}R' \longrightarrow R{-}N{=}C{=}O + {}^-OOCR'$$

$$R{-}N{=}C{=}O \xrightarrow{OH^-} RNH_2 + CO_3^{2-}$$

(b) Electron-releasing substituents in R speed up migration, as they do in the Hofmann rearrangement (Sec. 22.17). Electron-withdrawing substituents in R′ make $^-OOCR'$ less basic and hence a better leaving group. The overall rate evidently depends on both the rate of migration of R and the rate of departure of $^-OOCR'$: as it must if the two processes are concerted and occur in the same step.

(For a review of the Lossen and related rearrangements, see Smith, P. A. S. In *Molecular Rearrangements*, Part 1; de Mayo, P., Ed.; Wiley-Interscience: New York, 1963; pp 528–558.)

23

Amines II. Reactions

23.1 (a) A nitro group *ortho* or *para* to $—NH_2$ stabilizes the amine through structures like

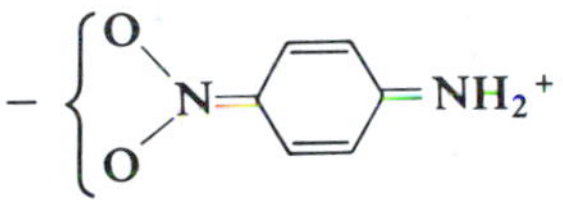

$$^{-}\left\{\begin{matrix} O \\ O \end{matrix}\right\rangle N{=}\langle\bigcirc\rangle{=}NH_2^+$$

(b) No such structures are possible for the *meta* isomer. (Try to draw them, just to prove it to yourself.)

23.2 $(CH_3)_3\overset{+}{N}:\overset{-}{B}F_3$

23.3 E2 elimination from 'onium ions (ammonium, sulfonium) proceeds via a carbanion-like transition state. Orientation is Hofmann: there is preferential abstraction of a proton from the carbon that can best accommodate the partial negative charge, that is, preferential abstraction of the most acidic proton. For example:

Acidity of protons $1° > 2° > 3°$ and $—CH_2Cl > —CH_3$

(a) Here, elimination can occur only in the 2-methyl-3-pentyl group. The question is: from which *branch* of that group is the proton lost?

$$CH_3—\overset{\overset{H}{|}}{C}H—\underset{\underset{\oplus NMe_3}{|}}{C}H—\overset{\overset{CH_3}{|}}{C}H—CH_3 \xrightarrow{E2} CH_3—CH{=}CH—\overset{\overset{CH_3}{|}}{C}H—CH_3 + Me_3N$$

Major product: proton from 2°C

$$CH_3{-}CH_2{-}\underset{\underset{\oplus}{Me_3N}}{CH}{-}\underset{H}{\overset{CH_3}{C}}{-}CH_3 \xrightarrow{E2} CH_3{-}CH_2{-}CH{=}\overset{CH_3}{C}{-}CH_3 + Me_3N$$

Minor product: proton from 3°C

(b) Here, elimination can occur in either of two groups, an ethyl or a *n*-propyl. The question is: from which group is the proton lost?

$$CH_3{-}CH_2{-}CH_2{-}\underset{CH_2-\underset{H}{CH_2}}{\overset{CH_2CH_3}{\overset{\oplus}{N}}}{-}CH_2{-}CH_2{-}CH_3 \xrightarrow{E2} CH_2{=}CH_2 + n\text{-}Pr_2NEt$$

Major product: proton from 1° C

$$CH_3{-}\underset{H}{CH}{-}CH_2{-}\underset{CH_2-CH_3}{\overset{CH_2CH_3}{\overset{\oplus}{N}}}{-}CH_2{-}CH_2{-}CH_3 \xrightarrow{E2} CH_3CH{=}CH_2 + n\text{-}PrNEt_2$$

Minor product: proton from 2° C

(c) As in (b), with the two groups being ethyl and 2-chloroethyl.

$$CH_3{-}CH_2{-}\underset{CH_3}{\overset{CH_3}{\overset{\oplus}{N}}}{-}CH_2{-}\underset{H}{CHCl} \xrightarrow{E2} CH_2{=}CHCl + EtNMe_2$$

Major product: proton from —CH_2Cl

$$\underset{H}{CH_2}{-}CH_2{-}\underset{CH_3}{\overset{CH_3}{\overset{\oplus}{N}}}{-}CH_2{-}CH_2Cl \xrightarrow{E2} CH_2{=}CH_2 + ClCH_2CH_2NMe_2$$

Minor product: proton from —CH_3

(d) As in (b), with the two groups being ethyl and *n*-propyl.

$$\underset{H}{CH_2}{-}CH_2{-}\underset{CH_3}{\overset{CH_3}{\overset{\oplus}{N}}}{-}CH_2{-}CH_2{-}CH_3 \xrightarrow{E2} CH_2{=}CH_2 + n\text{-}PrNMe_2$$

Major product: proton from 1°C

$$CH_3{-}CH_2{-}\underset{CH_3}{\overset{CH_3}{\overset{\oplus}{N}}}{-}CH_2{-}\underset{H}{CH}{-}CH_3 \xrightarrow{E2} CH_2{=}CH{-}CH_3 + EtNMe_2$$

Minor product: proton from 2°C

23.4 (a) Ethoxide ion is a strong base, and gives E2 elimination with the expected Hofmann orientation. Iodide ion and the solvent ethanol are weak bases, and E1 elimination is the principal reaction; the *tert*-pentyl cation is the one formed, and it loses a proton with the expected Saytzeff orientation.

(b) The ether is formed by the competing substitution reaction, which in this case is S_N1.

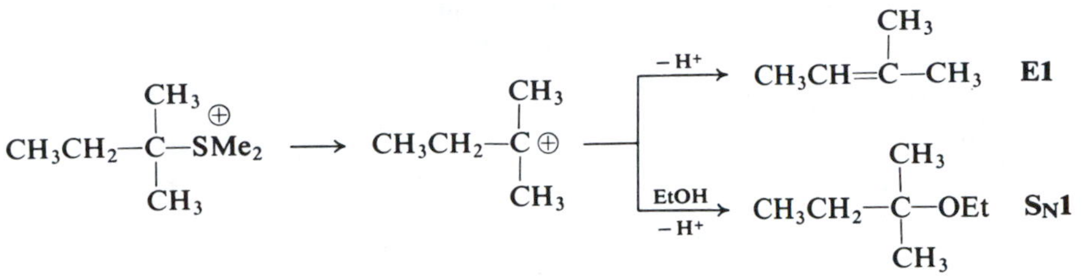

$$CH_3CH_2-\overset{\displaystyle CH_3}{\underset{\displaystyle CH_3}{C}}-\overset{\oplus}{S}Me_2 \longrightarrow CH_3CH_2-\overset{\displaystyle CH_3}{\underset{\displaystyle CH_3}{C}}\oplus \begin{cases} \xrightarrow{-H^+} CH_3CH=\overset{\displaystyle CH_3}{C}-CH_3 & \textbf{E1} \\ \xrightarrow[-H^+]{EtOH} CH_3CH_2-\overset{\displaystyle CH_3}{\underset{\displaystyle CH_3}{C}}-OEt & \textbf{S}_\textbf{N}\textbf{1} \end{cases}$$

(c) Here, the substitution competing with elimination is S_N2, with preferential attack at the less hindered methyl group.

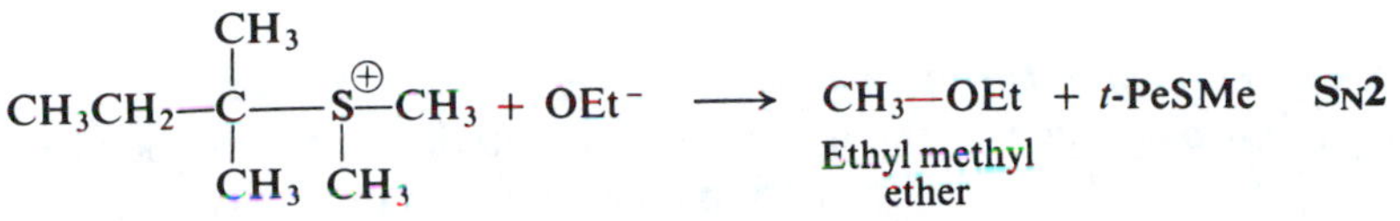

$$CH_3CH_2-\overset{\displaystyle CH_3}{\underset{\displaystyle CH_3}{C}}-\underset{\displaystyle CH_3}{\overset{\oplus}{S}}-CH_3 + OEt^- \longrightarrow \underset{\text{Ethyl methyl ether}}{CH_3-OEt} + t\text{-PeSMe} \quad \textbf{S}_\textbf{N}\textbf{2}$$

23.5 In 2-phenylethyl bromide, the electron-withdrawing phenyl group is attached to the carbon from which H is lost; by dispersing a partial negative charge, phenyl stabilizes a carbanion-like transition state and speeds up reaction.

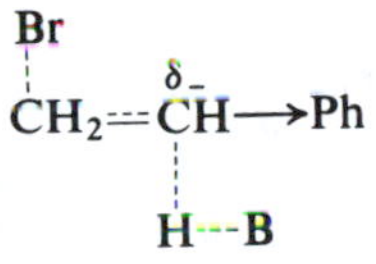

$$\overset{\displaystyle Br}{\overset{\vdots}{CH_2}}{=\!=}\underset{\displaystyle \vdots \atop \displaystyle H\cdots B}{\overset{\delta-}{CH}}\longrightarrow Ph$$

(This effect is *in addition to* the one exerted in both 2- and 1-phenylethyl bromides: through conjugation, phenyl stabilizes the incipient double bond of the transition state.)

23.6 (a) and (b) In the anion from ammonia, the negative charge is localized on nitrogen. In the anion from an amide, it is accommodated by nitrogen and one oxygen. In the anion from diacetamide, the charge is accommodated by nitrogen and *two* oxygens, as it is in the

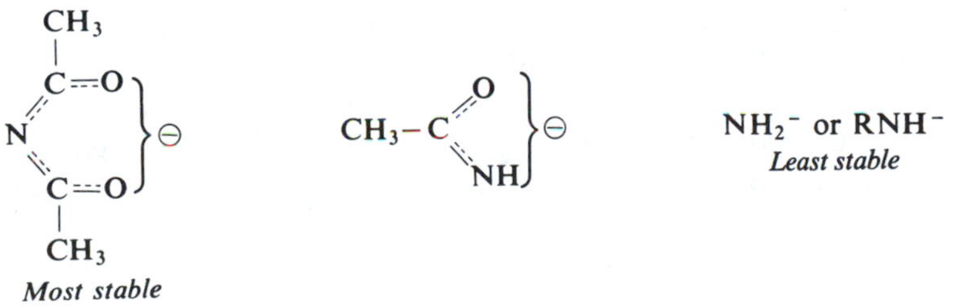

$$\left\{\begin{matrix} CH_3-C{=\!=}O \\ N \\ CH_3-C{=\!=}O \end{matrix}\right\}^{\ominus} \qquad CH_3-C\left\{\begin{matrix} {=\!=}O \\ {=\!=}NH \end{matrix}\right\}^{\ominus} \qquad NH_2^- \text{ or } RNH^-$$

Most stable ... *Least stable*

anion from phthalimide (Problem 20.10, page 768) or the anion from a sulfonamide, and with much the same effect on acidity.

23.7 (i) *Nucleophilic attack at sulfonyl sulfur is more difficult than attack at acyl carbon.* Acyl carbon is trigonal; nucleophilic attack is unhindered, and involves formation of a tetrahedral intermediate. Sulfonyl sulfur is tetrahedral; nucleophilic attack is relatively hindered, and involves the temporary attachment of a fifth group.

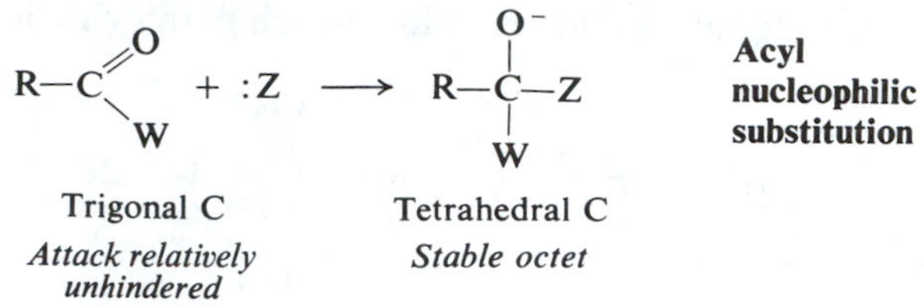

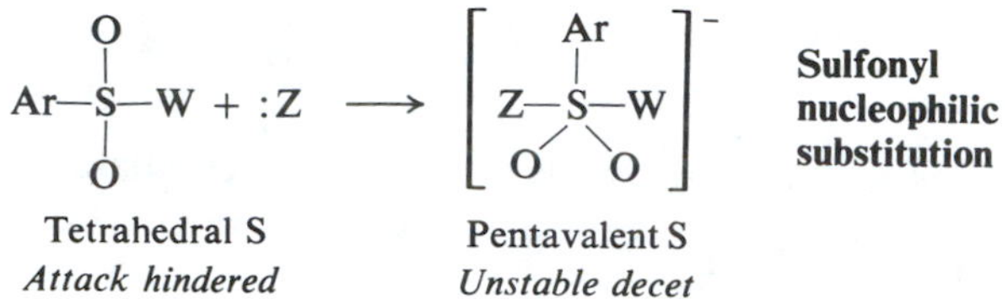

R—C(=O)—W + :Z ⟶ R—C(O⁻)(W)—Z **Acyl nucleophilic substitution**

Trigonal C — *Attack relatively unhindered*

Tetrahedral C — *Stable octet*

Ar—S(=O)(=O)—W + :Z ⟶ [Z—S(Ar)(O)(O)—W]⁻ **Sulfonyl nucleophilic substitution**

Tetrahedral S — *Attack hindered*

Pentavalent S — *Unstable decet*

(ii) *Nucleophilic attack at the alkyl carbon of a sulfonic ester is easier than attack at alkyl carbon of a carboxylic ester.* Displacement of the less basic sulfonate ion is easier than displacement of the carboxylate ion. Just as sulfonate separates with a pair of electrons from hydrogen more readily than does carboxylate (as shown by the relative acidities of the two kinds of acid), so sulfonate separates with a pair of electrons from an alkyl group more readily than does carboxylate.

Ar—S(=O)(=O)—O—R + Z: ⟶ R—Z + $ArSO_3^-$

Weak base: good leaving group

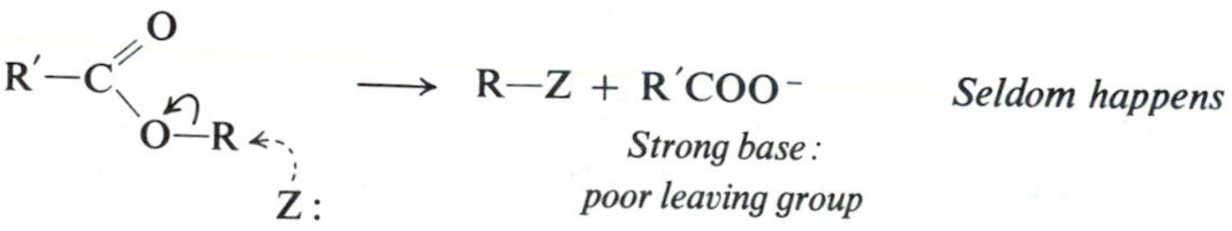

R′—C(=O)—O—R + Z: ⟶ R—Z + R′COO⁻ *Seldom happens*

Strong base: poor leaving group

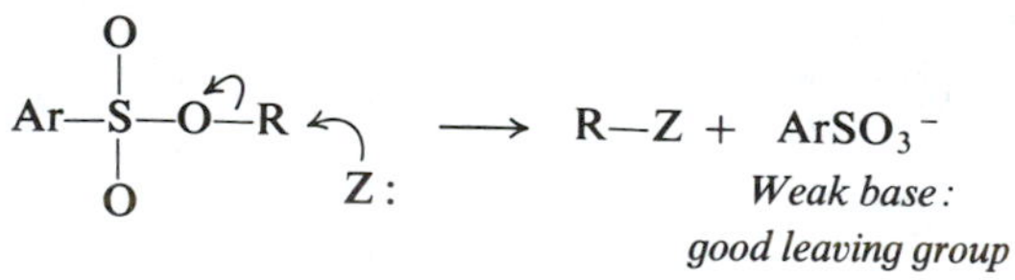

23.8 A logical explanation would be the following. Although there is very little free aniline present, what there is is *very* much more reactive than the anilinium ion and gives the *para* product. (Activation by amino groups is enormous; in halogenation, where protonation of the amine is not important, dimethylaniline is 10^{19} times as reactive as benzene!)

However, from work of J. H. Ridd (University College, London), it appears that the anilinium ion itself undergoes considerable *para* substitution. He proposed that, in comparison with $—N(CH_3)_3^+$, the positive charge on the $—NH_3^+$ is much dispersed by hydrogen bonding of its very acidic protons with the solvent.

23.9 An aromatic $—NH_2$ group ($K_a \approx 10^{-10}$) is considerably weaker than an aliphatic $—NH_2$ group ($K_a \approx 10^{-5}$), and cannot appreciably neutralize the —COOH group ($K_a \approx 10^{-5}$). It *can*, of course, neutralize the strongly acidic $—SO_3H$ group, as in sulfanilic acid.

23.10 (a) $ONO^- \overset{H^+}{\rightleftharpoons} HONO \overset{H^+}{\rightleftharpoons} H_2\overset{+}{O}{-}NO$

(b) $ArH + HONO \longrightarrow \overset{\oplus}{Ar}(H)(NO) + OH^-$

$ArH + H_2\overset{+}{O}{-}NO \longrightarrow \overset{\oplus}{Ar}(H)(NO) + H_2O$ *Less basic: better leaving group*

(c) $H_2\overset{+}{O}{-}NO + Cl^- \longrightarrow Cl{-}NO + H_2O$ SN1-*like* or SN2-*like*

(d) $ArH + Cl{-}NO \longrightarrow \overset{\oplus}{Ar}(H)(NO) + Cl^-$ *Weak base: good leaving group*

23.11 (a) Neither is likely. (i) The two groups, $-N(CH_3)_2$ and $-NHCH_3$, release electrons to the ring to about the same degree. (ii) Aside from a (probably modest) steric effect, nitrogen in the two groups should be about equally susceptible to attack.

(b) There is, of course, one way in which $-NHCH_3$ differs completely from $-N(CH_3)_2$: it contains a proton attached to nitrogen.

In both cases the initial electrophilic attack probably occurs at the same place, the place where unshared electrons are most available: at *nitrogen*, to give an unstable quaternary ammonium ion.

In the case of the secondary amine, this intermediate can lose a proton from nitrogen to give the *N*-nitroso product. This *N-nitrosation* is two-step electrophilic substitution, analogous to electrophilic aromatic substitution.

$$NO^+ + Ph{-}\ddot{N}(CH_3)(H) \longrightarrow Ph{-}\overset{\oplus}{N}(CH_3)(H){-}N{=}O \xrightarrow{-H} Ph{-}N(CH_3){-}N{=}O$$

N-nitroso compound

In the case of the tertiary amine, the intermediate quaternary ion has no proton to lose. Instead, it loses NO^+ and regenerates the amine, which eventually undergoes

$$NO^+ + Ph{-}\ddot{N}(CH_3)_2 \rightleftharpoons Ph{-}\overset{\oplus}{N}(CH_3)_2{-}N{=}O$$ *Cannot lose a proton*

$$NO^+ + Ph{-}\ddot{N}(CH_3)_2 \longrightarrow (H)(O{=}N)C_6H_4{=}\overset{\oplus}{N}(CH_3)_2 \xrightarrow{-H^+} O{=}N{-}C_6H_4{-}N(CH_3)_2$$

C-nitroso compound

substitution in the ring (*C-nitrosation*). *C*-nitrosation is thus a second choice to which the reaction turns when *N*-nitrosation is unsuccessful.

23.12 See page 408 of this Study Guide for procedures in solving aromatic synthesis problems.

CH_3, NO_2 (m-Nitrotoluene) $\xleftarrow{H_3PO_2}$ CH_3, NO_2, N_2^+ $\xleftarrow{HONO}$ CH_3, NO_2, NH_2 $\xleftarrow[H^+]{H_2O}$ CH_3, NO_2, $NHCOCH_3$ $\xleftarrow[H_2SO_4]{HNO_3}$ CH_3, $NHCOCH_3$ (page 872)

m-Nitrotoluene

CH_3, I (m-Iodotoluene) $\xleftarrow{KI}$ CH_3, N_2^+ $\xleftarrow{HONO}$ CH_3, NO_2 (as above)

m-Iodotoluene

CH_3, Br, Br (3,5-Dibromotoluene) $\xleftarrow{H_3PO_2}$ CH_3, Br, Br, N_2^+ $\xleftarrow{HONO}$ CH_3, Br, Br, NH_2 $\xleftarrow{2Br_2(aq)}$ CH_3, NH_2 (page 871)

3,5-Dibromotoluene

Br, Br, Br (1,3,5-Tribromobenzene) $\xleftarrow{H_3PO_2}$ Br, Br, Br, N_2^+ $\xleftarrow{HONO}$ Br, Br, Br, NH_2 $\xleftarrow{3Br_2(aq)}$ NH_2 (page 827)

1,3,5-Tribromobenzene

COOH, CH_3 (o-Toluic acid) $\xleftarrow[heat]{H_2O,\ H^+}$ CN, CH_3 $\xleftarrow{CuCN}$ N_2^+, CH_3 $\xleftarrow{HONO}$ NH_2, CH_3 (page 871)

o-Toluic acid

COOH, CH_3 (m-Toluic acid) $\xleftarrow[heat]{H_2O,\ H^+}$ CN, CH_3 $\xleftarrow{CuCN}$ N_2^+, CH_3 $\xleftarrow{HONO}$ NH_2, CH_3 $\xleftarrow[H^+]{Fe}$ NO_2, CH_3 (as above)

m-Toluic acid

COOH, CH_3 (p-Toluic acid) $\xleftarrow[heat]{H_2O,\ H^+}$ CN, CH_3 $\xleftarrow{CuCN}$ N_2^+, CH_3 $\xleftarrow{HONO}$ NH_2, CH_3 (page 871)

p-Toluic acid

OH, CH_3 (o-Cresol) $\xleftarrow[warm]{H_2O,\ H^+}$ N_2^+, CH_3 $\xleftarrow{HONO}$ NH_2, CH_3 (page 871)

o-Cresol

OH, CH_3 (m-Cresol) $\xleftarrow[warm]{H_2O,\ H^+}$ N_2^+, CH_3 $\xleftarrow{HONO}$ NH_2, CH_3 $\xleftarrow{Fe,\ H^+}$ NO_2, CH_3 (as above)

m-Cresol

OH, CH_3 (p-Cresol) $\xleftarrow[warm]{H_2O,\ H^+}$ N_2^+, CH_3 $\xleftarrow{HONO}$ NH_2, CH_3 (page 871)

p-Cresol

23.13

m-Dibromobenzene ←(CuBr)— m-bromobenzenediazonium ion (N_2^+, Br) ←(HONO)— m-bromoaniline (NH_2, Br) ←(Fe, H^+)— m-bromonitrobenzene (NO_2, Br) (page 872)

m-Bromoiodobenzene ←(KI)— m-bromobenzenediazonium ion (N_2^+, Br) (as above)

23.14 (a) Electron withdrawal by nitro groups makes diazonium ions more electrophilic.

(b) Less reactive due to electron release by $—CH_3$.

23.15 To prevent coupling of diazonium ion with unconsumed amine. In excess acid, most of the amine exists in the protonated form, which does not couple. The concentration of free amine is low, and its rate of coupling is slow.

23.16 (a) In an aromatic amine there are two electron-rich sites open to electrophilic attack: nitrogen and the ring. Like NO^+ (Problem 23.11), ArN_2^+ finds attack at nitrogen easier, and forms the intermediate I. In the case of a primary or secondary amine, I can lose a proton to yield II, a *diazoamino* compound, isomeric with the expected azo compound IV.

$$C_6H_5NHR + ArN_2^+ \rightleftarrows C_6H_5\overset{\oplus}{N}(R)(H)—N{=}N—Ar \rightleftarrows C_6H_5N(R)—N{=}N—Ar + H^+$$

Substitution on nitrogen

I

II
A diazoamino compound

$$C_6H_5NHR + ArN_2^+ \longrightarrow \text{ArN=N(H)C}_6\text{H}_4\text{=}\overset{+}{N}\text{HR} \longrightarrow ArN{=}N\text{—}C_6H_4\text{—}NHR + H^+$$

Substitution on carbon

III

IV
An aminoazo compound

(b) Formation of II is easy, but reversible; electrophilic attack by H^+ regenerates the amine and ArN_2^+. These react again and again, and eventually with attack on the ring to form the *sigma* complex III and from it the azo compound IV. Once formed, IV persists; its formation is *not* reversible.

23.17 (a) $CH_3C_6H_3(Br)NH_2 + H_2NC_6H_3(CH_3)OH \xleftarrow{SnCl_2} CH_3C_6H_3(Br)\text{—N=N—}C_6H_3(CH_3)OH$

(b) $CH_3C_6H_3(Br)\text{—N=N—}C_6H_3(CH_3)OH \longleftarrow CH_3C_6H_3(Br)N_2^+ + C_6H_4(CH_3)OH$ (page 868)

$\uparrow$ HONO

$CH_3C_6H_3(Br)NH_2$ (page 861)

23.18 (a) Nucleophilic substitution.

$$Ph\text{—}SO_2\text{—}\overset{+}{N}(CH_3)_2\text{—}CH_3 \;\; :N(CH_3)_3 \longrightarrow Ph\text{—}SO_2\text{—}N(CH_3)_2 + CH_3:\overset{+}{N}(CH_3)_3$$

(b) A tertiary amine can react as in (a), to yield the derivative of a *secondary* amine. Incorrect conclusion: that the amine is secondary.

23.19 (a) Incorrect conclusion: that the amine is tertiary.

(b) Filter (or separate) and acidify the aqueous solution. A precipitate will form if the amine was primary.

1. (a) $n\text{-BuNH}_3{}^+Cl^-$

(b) $n\text{-BuNH}_3{}^+HSO_4{}^-$

(c) $n\text{-BuNH}_3{}^+\,{}^-OOCCH_3$

(d) no reaction

(e) $CH_3C(=O)\text{—NH—Bu-}n$ and product (c)

(f) $(CH_3)_2CHC(=O)\text{—NH—Bu-}n$ and $n\text{-BuNH}_3{}^+\,{}^-OOCCH(CH_3)_2$

(g) $O_2NC_6H_4C(=O)\text{—NH—Bu-}n$ and $C_5H_5NH^+Cl^-$

(h) $PhSO_2\overset{\ominus}{N}\text{—Bu-}n\ K^+$

(i) $n\text{-BuNH}_2(Et)^+Br^-$, $n\text{-BuNH}(Et)_2{}^+Br^-$, $n\text{-BuN}(Et)_2\text{—Et}^+Br^-$

(j) $n\text{-BuNH}_2(CH_2Ph)^+Br^-$, etc.

(k) no reaction

(l) $n\text{-BuNMe}_3{}^+OH^-$

(m) $CH_3CH_2CH\text{=}CH_2 + Me_3N$

(n) $n\text{-BuNHCH(CH}_3)_2$

(o) see Problem 27, page 883

(p) 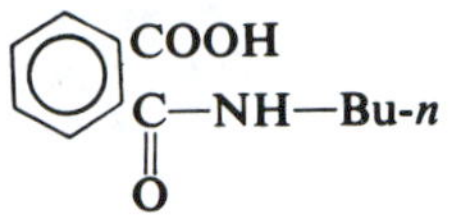 $C_6H_4(COOH)$—C(=O)—NH—Bu-*n*

(q) $n\text{-BuN}^{\oplus}H_2\text{—CH}_2\text{COO}^-$, *N*-(*n*-butyl)glycine

(r) 2,4,6-$(O_2N)_3C_6H_2NH$—Bu-*n*

2. Aliphatic amines are more basic than aromatic amines (Sec. 23.3). Electron-releasing groups raise basicity; electron-withdrawing groups lower basicity (Sec. 23.4).

(a) *cyclo*-$C_6H_{11}NH_2$ > NH_3 > $C_6H_5NH_2$

(b) $CH_3CH_2NH_2$ > $HOCH_2CH_2CH_2NH_2$ > $HOCH_2CH_2NH_2$

(c) *p*-$CH_3OC_6H_4NH_2$ > $C_6H_5NH_2$ > *p*-$O_2NC_6H_4NH_2$

(d) *m*-$EtC_6H_4CH_2NH_2$ > $C_6H_5CH_2NH_2$ > *m*-$ClC_6H_4CH_2NH_2$

(e) *p*-ClC_6H_4NHMe > 2,4-$Cl_2C_6H_3NHMe$ > 2,4,6-$Cl_3C_6H_2NHMe$

3. In an aqueous solution of $Me_4N^+OH^-$ the base is OH^-, which is much stronger than Me_3N.

4. (a) All three form soluble ammonium salts.

(b) $PhNH_2 \xrightarrow{HONO} PhN_2^+$ *Colorless solution*

$PhN(CH_3)\text{—H} \xrightarrow{HONO} PhN(CH_3)\text{—N=O}$ *Neutral, yellow compound*

$PhN(CH_3)\text{—CH}_3 \xrightarrow{HONO} O\text{=N}C_6H_4N(CH_3)\text{—CH}_3$ *Green solid*

(c) All three form quaternary ammonium salts.

(d) $PhNH_2 \longrightarrow ArSO_2N(H)\text{—Ph}$ **Acidic:** *soluble in aqueous base*

$PhN(CH_3)\text{—H} \longrightarrow ArSO_2N(CH_3)\text{—Ph}$ **Neutral:** *insoluble in acid or base*

$PhN(CH_3)\text{—CH}_3 \longrightarrow$ no reaction *If careful* **Remains basic:** *insoluble in base, soluble in acid*

(e) $PhNH_2 \longrightarrow PhNHCOCH_3$ **Neutral:** *insoluble in dilute acid or base*

$PhN(CH_3){-}H \longrightarrow PhN(CH_3)COCH_3$ **Neutral:** *insoluble in dilute acid or base*

$PhN(CH_3){-}CH_3 \longrightarrow$ no reaction **Remains basic:** *soluble in acid, insoluble in base*

(f) As in (e), with $-COPh$ in place of $-COCH_3$.

(g) All three undergo very fast ring bromination to give tribromo products.

5. (a) All three form ammonium salts (same as 4a).

(b) $EtNH_2$ yields N_2; Et_2NH yields a neutral (yellow) *N*-nitroso compound; Et_3N yields the same *N*-nitroso compound, and cleavage products.

(c) All three form quaternary ammonium salts.

(d) $EtNH_2$ gives a solution, Et_2NH gives a solid, Et_3N does not react (as in 4d).

(e) Neutral amides from 1° and 2° amines, no reaction with 3° amine (as in 4e).

(f) Same as (e) (and as in 4f).

(g) We could not have known this, but 1° and 2° amines give *N*-bromoamines (compare page 838). No reaction with 3° amines.

6. (a) *p*-$CH_3C_6H_4N_2^+Cl^-$

(b) *p*-$O{=}NC_6H_4NEt_2$

(c) *n*-PrOH, *iso*-PrOH, $CH_3CH{=}CH_2$ (compare Problem 27, page 883)

(d) *p*-$^-O_3SC_6H_4N_2^+$

(e) $C_6H_5N(CH_3){-}N{=}O$

(f) 2-methyl-2-butene, *tert*-PeOH

(g) 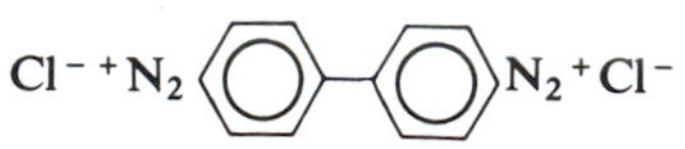 $Cl^-\,{}^+N_2{-}C_6H_4{-}C_6H_4{-}N_2^+Cl^-$

(h) $C_6H_5CH_2OH$

7. (a) 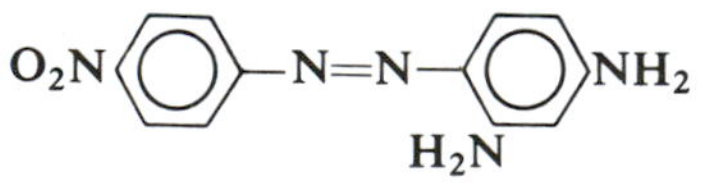 $O_2N{-}C_6H_4{-}N{=}N{-}C_6H_3(NH_2){-}NH_2$ (2,4-diamino)

(b) $O_2N{-}C_6H_4{-}OH$

(c) $O_2N{-}C_6H_4{-}Br$

(d) 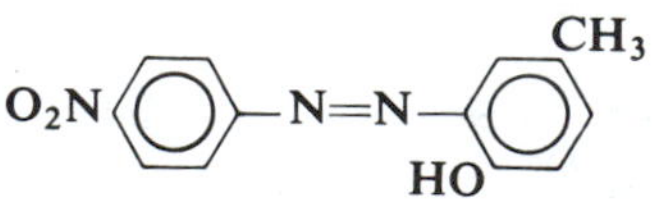 $O_2N{-}C_6H_4{-}N{=}N{-}C_6H_3(HO)(CH_3)$

(e)  $O_2N-C_6H_4-I$ (f) $O_2N-C_6H_4-Cl$

(g) $O_2N-C_6H_4-CN$ (h) $O_2N-C_6H_4-F$

(i) $O_2N-C_6H_4-H$

8. (a) H_3PO_2, H_2O.

(b) H_2O, H^+, heat, removing the cresol immediately by steam distillation.

(c) CuCl. Heat solution of $ArN_2^+CuCl_2^-$ salt.

(d) CuBr. Heat solution of $ArN_2^+CuBr_2^-$ salt.

(e) KI.

(f) HBF_4; isolate $ArN_2^+BF_4^-$; heat dry salt.

(g) CuCN. Heat solution of $ArN_2^+Cu(CN)_2^-$ salt.

(h) 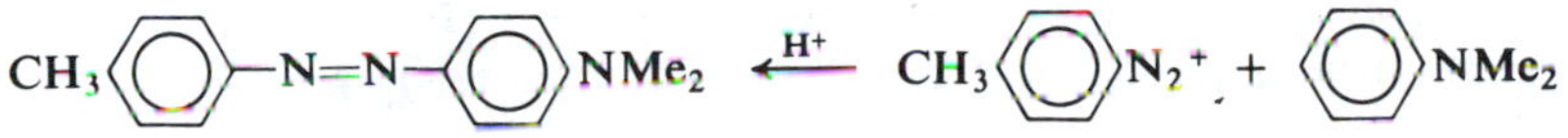 $CH_3-C_6H_4-N{=}N-C_6H_4-NMe_2 \xleftarrow{H^+} CH_3-C_6H_4-N_2^+ + C_6H_5-NMe_2$

(i) 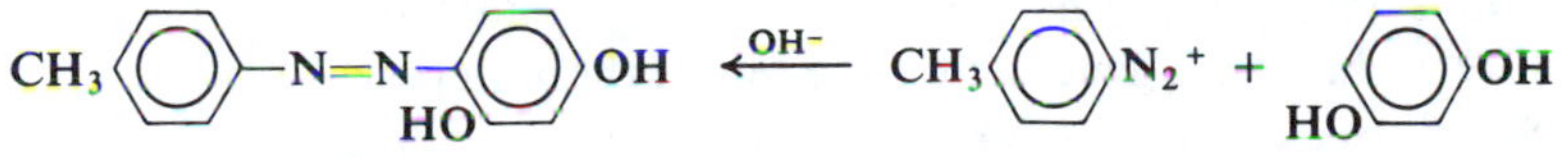 $CH_3-C_6H_4-N{=}N-C_6H_3(HO)-OH \xleftarrow{OH^-} CH_3-C_6H_4-N_2^+ + HO-C_6H_4-OH$

Resorcinol

9. (a) n-Pr$C(=O)$—$NHCH_3$ + $MeNH_3^+Cl^-$

(b) PhN(Me)—$C(=O)CH_3$ + CH_3COOH

(c) n-Pr_3N + $CH_3CH{=}CH_2$

(d) *iso*-Bu$C(=O)$—NEt_2 + $Et_2NH_2^+Cl^-$

(e) MeOH + Me_3N

(f) $Me_3NH^{+}\,^{-}OOCCH_3$

(g) $Me_2NH_2^+Cl^-$ + CH_3COOH

(h) $PhNH_2$ + $PhCOO^-Na^+$

(i) MeOH + PhNH—$CH(=O)$

(j) MeNH$C(=O)$NHMe (*N,N′*-dimethylurea) + $MeNH_3^+Cl^-$

Synthesis: A Systematic Approach

As the syntheses we plan become more and more complicated, we need to follow a systematic approach: something we have probably already been doing without fully realizing it. Let us outline formally the mental steps of such an approach. We shall use synthesis of aromatic compounds as our example here, but the approach is a general one and, with modifications, can be applied to synthesis of all kinds of complicated molecules.

We start, as usual, with the molecule we want—the target molecule—and work backwards. We ask ourselves these questions in this order:

Question (1). Can any of the substituents in the target molecule be introduced by direct electrophilic aromatic substitution, and with the proper orientation? If not, then:

Question (2). Is there any substituent in the target molecule that can be formed by the oxidation or reduction of some related group? For example, $—COOH \longleftarrow —CH_3$ or $—NH_2 \longleftarrow —NO_2$. If not, then:

Question (3). Is there any substituent in the target molecule that can be introduced by displacement of some other group? For example, $—CN \longleftarrow —N_2^+$ or $—OH \longleftarrow —Cl$ (in activated rings). If not, then finally:

Question (4). Is there a hydrogen in the target molecule that got there by displacement of some substituent? For example, $—H \longleftarrow —N_2^+$ or $—H \longleftarrow —SO_3H$.

Whenever the answer to a question is "Yes": we take off that group (Question 1); or transform that group (Question 2) into its precursor ($—COOH \longleftarrow —CH_3$, say); or replace that group (Question 3) by its precursor(s) ($ArCN \longleftarrow ArN_2^+ \longleftarrow ArNH_2$, say); or (Question 4) replace that —H (or perhaps —D or —T, in labeled compounds) by the previous occupant of the site.

Having done the transformations implied by the "Yes" answer, we start the questions again with the *new* structure we have just generated. We repeat these steps as many times as we have to until we ultimately work back to a starting material like benzene or toluene.

We now have a set of backward steps called *transforms*. This uncovers the various needed reactions (nitration, reduction of a diazonium ion, Sandmeyer reaction, etc.), but not necessarily in the proper order. Next, then, we check our *anti*thetic procedure by arranging the separate *transforms* into a workable *syn*thetic sequence of forward steps (*reactions*).

In doing this, we should be aware of the frequent need for *control elements*: groups that we introduce to insure selectivity (regioselectivity or stereoselectivity) in a later step. Control elements may protect a group or lower its activating effect (acetylation of an amine, say); they may block a position or, conversely, increase reactivity at a particular spot. The need for these control elements may not be obvious as we work out the antithetic transforms, but will usually appear when we try to put everything together in a series of synthetic reactions.

(The particular way these questions are asked may have a familiar ring—and so it should. This approach is derived from the one that has been worked out—most notably by Nobel laureate E. J. Corey of Harvard University—for the use of computers to design organic syntheses. See Corey, E. J.; Wipke, W. T. *Science (Washington, D.C.)* **1969**, *166*, 178; Corey, E. J., *et al. J. Am. Chem. Soc.* **1972**, *94*, 421, 431, 440, 460.)

(k) CH_3, N—NO, NO_2

(l) NH_2, Br, Br, Br

(m) NH_2, Br, Br, CH_3, Br

(n) NH_2, Br, Br, CH_3

(o) $N_2{}^+Cl$, CH_3

(p) $NHCOCH_3$, NO_2 (and *o*-isomer)

(q) $NHCOCH_3$, NO_2, CH_3

(r) $NMe_3{}^+I^-$, Et

(s) Br—C₆H₄—N(H)—C(=O)—C₆H₅ (and *o*-isomer)

10. (a) CH_3, Br, NH_2

Let us apply our general procedure to the synthesis of this compound, and see how it works.

(i) Question (1): none of the groups can be introduced (with, of course, the correct orientation) by direct substitution, so we proceed to the next question.

Question (2): —NH_2 can be formed by reduction of —NO_2, so we write out this step in the synthesis.

CH_3, Br, NH_2 $\xleftarrow{\text{Fe, H}^+}$ CH_3, Br, NO_2

(ii) Having done this, we begin again, and ask our questions about the new (nitro) compound.

Question (1): —NO_2 could be put in by direct substitution. But so can —Br, and regioselectively. So let us try a bromination step, working back to a simpler compound.

CH_3, Br, NO_2 $\xleftarrow{\text{Br}_2\text{, Fe}}$ CH_3, NO_2

(iii) Having done this, we begin again, and ask questions about this new compound.

Question (1): —NO_2 can be put in by nitration—and, incidentally, the *para* isomer is easily separated from the *ortho* isomer. Thus, we have worked back to toluene, an acceptable starting material.

CH_3, NO_2 $\xleftarrow{\text{HNO}_3\text{, H}_2\text{SO}_4}$ CH_3

Putting all this together, we get:

$$CH_3C_6H_3(Br)NH_2 \xleftarrow{\text{Fe, H}^+} CH_3C_6H_3(Br)NO_2 \xleftarrow{\text{Br}_2\text{, Fe}} CH_3C_6H_4NO_2 \xleftarrow{\text{HNO}_3\text{, H}_2\text{SO}_4} CH_3C_6H_5$$

When we check this sequence of transforms, we find that it corresponds to a workable sequence of reactions, and constitutes an acceptable synthesis of the product we want.

(b) $CH_3C_6H_3(Br)NH_2$ (CH₃ top, Br, NH₂ bottom)

Applying our general procedure here, we arrive at the following sequence of transforms.

$$CH_3C_6H_3(Br)NH_2 \xleftarrow[\text{(Br}_2)]{\text{Question 1}} CH_3C_6H_4NH_2 \xleftarrow[\text{(reduction)}]{\text{Question 2}} CH_3C_6H_4NO_2 \xleftarrow[\text{(nitration)}]{\text{Question 1}} \underset{\text{Toluene}}{CH_3C_6H_5}$$

Now we check this to see if it corresponds to a workable synthetic route. We realize that, while we can indeed brominate *p*-toluidine, in doing this we would introduce two —Br. If, however, we acetylate the $—NH_2$ first, we slow down substitution enough to get monobromination; and, after this is done, we can easily remove the acetyl group.

We arrive, then, at the following acceptable synthesis.

$$CH_3C_6H_3(Br)NH_2 \xleftarrow[\text{H}^+]{\text{H}_2\text{O}} CH_3C_6H_3(Br)NHCOCH_3 \xleftarrow[\text{Fe}]{\text{Br}_2} CH_3C_6H_4NHCOCH_3 \xleftarrow{\text{Ac}_2\text{O}} CH_3C_6H_4NH_2$$ (page 871)

We have added two steps, acetylation and subsequent hydrolysis, but by these steps we have gained control over the most critical part of the whole synthesis, introduction of bromine.

(c) $NH_2C_6H_4SO_2NHPh$

Here, our general procedure leads us to this sequence of transforms.

$$NH_2C_6H_4SO_2NHPh \xleftarrow[\text{(PhNH}_2)]{\text{Question 2}} NH_2C_6H_4SO_2Cl \xleftarrow[\text{(ClSO}_3\text{H)}]{\text{Question 2}} C_6H_5NH_2 \longleftarrow PhNO_2 \longleftarrow C_6H_6$$

This we find to be defective as a synthetic route. We cannot have a free $—NH_2$ and $—SO_2Cl$ in the same compound: they would react with each other. As in (b), we temporarily protect $—NH_2$ by acetylation. In the final step we remove the acetyl group by hydrolysis under conditions that do not also hydrolyze the less reactive sulfonamide group (see page 864).

p-$NH_2C_6H_4SO_2NHPh$ $\xleftarrow{H_2O,\ H^+}$ p-$CH_3CONHC_6H_4SO_2NHPh$ $\longleftarrow$ $PhNH_2$ (from C_6H_6) + p-$CH_3CONHC_6H_4SO_2Cl$ $\xleftarrow{ClSO_3H}$ $C_6H_5NHCOCH_3$ (page 848)

(d) p-$CH_3CONHC_6H_4NH_2$ $\xleftarrow{Fe,\ H^+}$ p-$CH_3CONHC_6H_4NO_2$ (page 861)

(e) p-$ON{-}C_6H_4NEt_2$ $\xleftarrow{HONO}$ $PhNEt_2$ $\xleftarrow{2EtI}$ $PhNH_2$ (page 827)

(f) $HOOC(NO_2)C_6H_3NH_2$ $\xleftarrow{H_2O,\ H^+}$ $HOOC(NO_2)C_6H_3NHCOCH_3$ $\xleftarrow{HNO_3,\ H_2SO_4}$ p-$HOOCC_6H_4NHCOCH_3$ $\xleftarrow{KMnO_4,\ warm}$ p-$CH_3C_6H_4NHCOCH_3$ (page 872)

(g) 2,6-dibromo-4-isopropylaniline (Br, Br, NH_2, $CH(CH_3)_2$) $\xleftarrow{2Br_2(aq)}$ p-$NH_2C_6H_4CH(CH_3)_2$ $\xleftarrow{Fe,\ H^+}$ p-$NO_2C_6H_4CH(CH_3)_2$ $\xleftarrow{HNO_3,\ H_2SO_4}$ $C_6H_5CH(CH_3)_2$ (from C_6H_6)

(h) p-$H_2NCH_2C_6H_4NH_2$ $\xleftarrow{Fe,\ H^+}$ p-$H_2NCH_2C_6H_4NO_2$ $\xleftarrow{NH_3}$ p-$ClCH_2C_6H_4NO_2$ $\xleftarrow{Cl_2,\ heat}$ p-$CH_3C_6H_4NO_2$ $\xleftarrow{HNO_3,\ H_2SO_4}$ $C_6H_5CH_3$

(i) $Ph{-}N(NO){-}Pr\text{-}i$ $\xleftarrow{HONO}$ $Ph{-}NH{-}Pr\text{-}i$ $\longleftarrow$
- $PhNH_2$ (page 827)
- i-PrBr $\xleftarrow{HBr}$ i-PrOH

(j) n-$BuC(=O){-}N(Et)(Me)$ $\longleftarrow$
- n-BuCOCl $\xleftarrow{SOCl_2}$ n-BuCOOH $\longleftarrow$ n-BuCN $\longleftarrow$ n-BuBr
- EtNHMe $\xleftarrow{MeI}$ $EtNH_2$ $\xleftarrow{NH_3}$ EtBr

(k) n-$BuCH_2CH_2NH_2$ $\xleftarrow{NH_3}$ n-$BuCH_2CH_2Br$ $\longleftarrow$ n-$BuCH_2CH_2OH$ $\xleftarrow{H_2C{-}CH_2\ (O)}$ n-BuMgBr (from n-BuOH)

(l) $PhCH(NH_2)Pr\text{-}n$ $\xleftarrow{H_2,\ NH_3,\ Ni}$ $PhC(=O)Pr\text{-}n$ $\xleftarrow{PhH,\ AlCl_3}$ n-PrCOCl $\xleftarrow{SOCl_2}$ n-PrCOOH $\xleftarrow{KMnO_4}$ n-BuOH

(m) $H_2NCH_2C(=O)NH_2$ $\xleftarrow{NH_3}$ $ClCH_2COCl$ $\xleftarrow{SOCl_2}$ $ClCH_2COOH$ $\xleftarrow{Cl_2,\ P}$ CH_3COOH (from EtOH)

(n) $PhC(=O)NHCH_2COOH$ $\longleftarrow$
- PhCOCl $\xleftarrow{SOCl_2}$ PhCOOH $\xleftarrow{KMnO_4}$ $PhCH_3$
- H_2NCH_2COOH $\xleftarrow{NH_3\ excess}$ $ClCH_2COOH$ (as in m, above)

11. (a) Of all the six isomers, 2,3-dibromotoluene is the hardest to make. Use the antithetic approach outlined in the box above, and see how neatly the question-and-answer procedure gets you back to a readily available starting compound.

(i) 2,3-dibromotoluene $\xleftarrow{H_3PO_2}$ diazonium salt (N_2^+) $\xleftarrow{HONO}$ amine (H_2N) $\xleftarrow{Fe, H^+}$ O_2N compound $\xleftarrow{CuBr}$ O_2N, N_2^+, Br compound $\xleftarrow{HONO}$ O_2N, NH_2, Br compound

Isomer most difficult to make

CH_3/$NHCOCH_3$ $\xrightarrow{HNO_3, H_2SO_4}$ O_2N/$NHCOCH_3$ $\xrightarrow{H_2O, H^+}$ O_2N/NH_2 $\xrightarrow{Br_2}$ O_2N/NH_2/Br

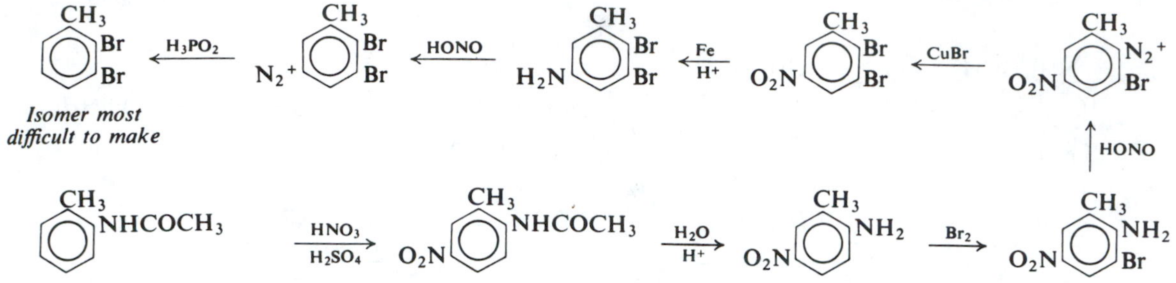

(from o-$CH_3C_6H_4NH_2$, page 871)

About half-way back, you may face the following choice: which —Br should be introduced last, the one next to —CH_3, or the other one?

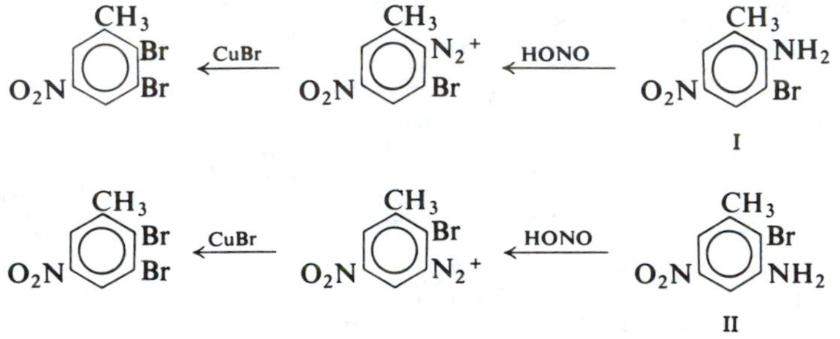

O_2N/Br/Br $\xleftarrow{CuBr}$ O_2N/N_2^+/Br $\xleftarrow{HONO}$ O_2N/NH_2/Br

I

O_2N/Br/Br $\xleftarrow{CuBr}$ O_2N/Br/N_2^+ $\xleftarrow{HONO}$ O_2N/Br/NH_2

II

The answer lies in another question: which amine of the two shown above is the easier to make? The answer to this question is amine I, as shown in the complete answer outlined above, but you should try to work out a route to amine II just to prove to yourself which is easier to make.

A key role is played by —NO_2, which must be put on the ring temporarily to block one of the positions activated by —NH_2.

(ii) CH_3/Br/Br $\xleftarrow{CuBr}$ CH_3/Br/N_2^+ $\xleftarrow{HONO}$ CH_3/Br/NH_2 (as in Problem 10a)

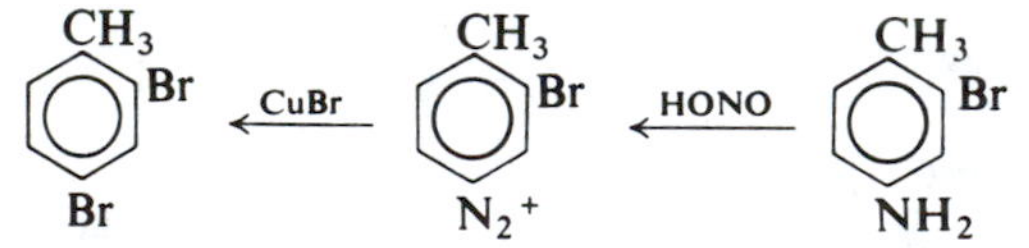

(iii) CH_3/Br/Br $\xleftarrow{CuBr}$ CH_3/N_2^+/Br $\xleftarrow{HONO}$ CH_3/NH_2/Br $\xleftarrow{H_2O, H^+}$ CH_3/$NHCOCH_3$/Br $\xleftarrow{Br_2, Fe}$ CH_3/$NHCOCH_3$

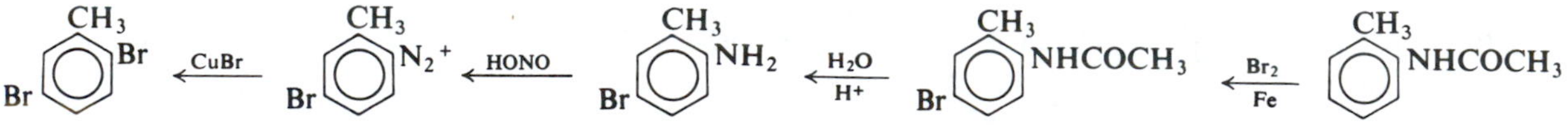

(from o-$CH_3C_6H_4NH_2$, page 871)

(iv) Br/CH_3/Br $\xleftarrow{H_3PO_2}$ Br/CH_3/Br/N_2^+ $\xleftarrow{HONO}$ Br/CH_3/Br/NH_2 $\xleftarrow{Fe, H^+}$ Br/CH_3/Br/NO_2 $\xleftarrow{2Br_2, Fe}$ CH_3/NO_2 (page 525)

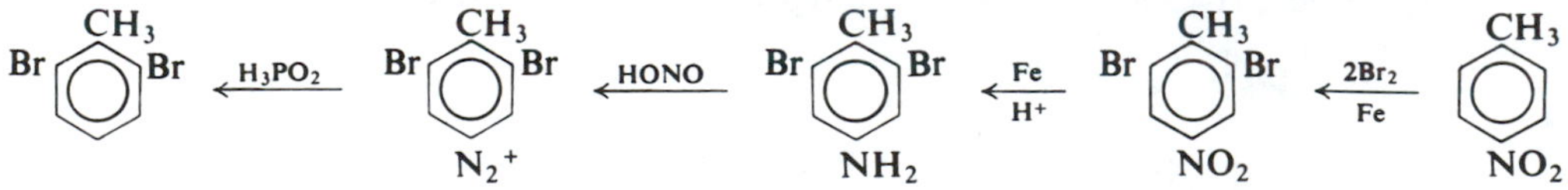

In making isomer (iv) we get all the way to Question 4 (see the box above) before we answer, "Yes". The sequence is —H $\longleftarrow$ —N_2^+ $\longleftarrow$ —NH_2 $\longleftarrow$ —NO_2.

(v) CH_3, Br, Br $\xleftarrow{CuBr}$ CH_3, Br, N_2^+ $\xleftarrow{HONO}$ CH_3, Br, NH_2 (page 871)

(vi) CH_3, Br, Br $\xleftarrow{H_3PO_2}$ CH_3, Br, Br, N_2^+ $\xleftarrow{HONO}$ CH_3, Br, Br, NH_2 $\xleftarrow{2Br_2(aq)}$ CH_3, NH_2 (page 871)

In looking at isomer (vi) we may recognize that the two —Br's must have got onto the ring under the influence of a now-vanished group much more powerfully directing than —CH_3. The —NH_2 group is the logical choice, a powerful *ortho,para*-director easily removed by reductive deamination (diazotization and reduction).

(b) (i) COOH, Cl $\xleftarrow{KMnO_4}$ CH_3, Cl $\xleftarrow{CuCl}$ CH_3, N_2^+ $\xleftarrow{HONO}$ CH_3, NH_2 (page 871)

(ii) COOH, Cl $\xleftarrow[H^+]{H_2O}$ CN, Cl $\xleftarrow{CuCN}$ N_2^+, Cl $\xleftarrow{HONO}$ NH_2, Cl $\xleftarrow[H^+]{Fe}$ NO_2, Cl $\xleftarrow[AlCl_3]{Cl_2}$ NO_2

(*Or*: start with *m*-nitrotoluene, from Problem 23.12, page 872.)

(iii) COOH, Cl $\longleftarrow$ as for *o*-isomer, except start with *p*-nitrotoluene (page 525)

(c) (i) Br, F $\xleftarrow[heat]{HBF_4}$ Br, N_2^+ $\xleftarrow{HONO}$ Br, NH_2 $\xleftarrow[H^+]{Fe}$ Br, NO_2 $\xleftarrow{HNO_3, H_2SO_4}$ Br

(*Separate from p-isomer*)

(ii) Br, F $\xleftarrow[heat]{HBF_4}$ Br, N_2^+ $\xleftarrow{HONO}$ Br, NH_2 $\xleftarrow[H^+]{Fe}$ Br, NO_2 $\xleftarrow[AlCl_3]{Br_2}$ $PhNO_2$

(iii) Br, F $\longleftarrow$ as for *o*-isomer, above, except start with *p*-bromonitrobenzene

12. (a) CH_3, F $\xleftarrow[heat]{HBF_4}$ CH_3, N_2^+ $\xleftarrow{HONO}$ CH_3, NH_2 (page 871)

(b) $CH_3C_6H_4F$ $\xleftarrow[\text{heat}]{HBF_4}$ $CH_3C_6H_4N_2^+$ $\xleftarrow{HONO}$ $CH_3C_6H_4NH_2$ $\xleftarrow[H^+]{Fe}$ $CH_3C_6H_4NO_2$ (as in Problem 23.12, page 872)

(c) p-IC_6H_4COOH $\xleftarrow{KMnO_4}$ p-$CH_3C_6H_4I$ $\xleftarrow{KI}$ p-$CH_3C_6H_4N_2^+$ $\xleftarrow{HONO}$ p-$CH_3C_6H_4NH_2$ (page 871)

(d) m-$BrC_6H_4NH_2$ $\xleftarrow[H^+]{Fe}$ m-$BrC_6H_4NO_2$ $\xleftarrow{Br_2,\ Fe}$ $C_6H_5NO_2$ $\xleftarrow{HNO_3,\ H_2SO_4}$ C_6H_6

(e) 3-Br-4-$CH_3C_6H_3COOH$ $\xleftarrow{Br_2,\ Fe}$ p-$CH_3C_6H_4COOH$ (page 870)

(f) 2-Br-4-$CH_3C_6H_3COOH$ $\xleftarrow[\text{heat}]{H_2O,\ H^+}$ 2-Br-4-$CH_3C_6H_3CN$ $\xleftarrow{CuCN}$ 2-Br-4-$CH_3C_6H_3N_2^+$ $\xleftarrow{HONO}$ 2-Br-4-$CH_3C_6H_3NH_2$ (page 872)

(g) m-EtC_6H_4OH $\xleftarrow[\text{warm}]{H_2O,\ H^+}$ m-$EtC_6H_4N_2^+$ $\xleftarrow{HONO}$ m-$EtC_6H_4NH_2$ $\xleftarrow[H^+]{Fe}$ m-$EtC_6H_4NO_2$

The required *m*-nitroethylbenzene is made in the same manner as *m*-nitrotoluene (in Problem 23.12), starting with ethylbenzene (page 577).

(h) 3,5-$Br_2C_6H_3NH_2$ $\xleftarrow[H^+]{Fe}$ 3,5-$Br_2C_6H_3NO_2$ $\longleftarrow$ as in Problem 12(p), below

(i) 3-Br-4-I-$C_6H_3CH_3$ $\xleftarrow{KI}$ 3-Br-4-N_2^+-$C_6H_3CH_3$ $\xleftarrow{HONO}$ 3-Br-4-NH_2-$C_6H_3CH_3$ (page 872)

(j) 2-NH_2-4-$CH_3C_6H_3OH$ $\xleftarrow[H^+]{Fe}$ 2-NO_2-4-$CH_3C_6H_3OH$ $\xleftarrow[\text{warm}]{H_2O,\ H^+}$ 2-NO_2-4-$CH_3C_6H_3N_2^+$ $\xleftarrow{HONO}$ 2-NO_2-4-$CH_3C_6H_3NH_2$ $\xleftarrow{H_2O,\ H^+}$ 2-NO_2-4-$CH_3C_6H_3NHCOCH_3$ $\xleftarrow{HNO_3,\ H_2SO_4}$ p-$CH_3C_6H_4NHCOCH_3$ (page 872)

(k) 3,5-$Br_2C_6H_3I$ (Br, Br, I) $\xleftarrow{H_3PO_2}$ $\xleftarrow{HONO}$ 2,6-Br_2-4-$NH_2C_6H_2I$ $\xleftarrow[H^+]{Fe}$ 2,6-Br_2-4-$NO_2C_6H_2I$ $\xleftarrow{KI}$ $\xleftarrow{HONO}$ 2,6-Br_2-4-$NO_2C_6H_2NH_2$ (page 873)

(l) 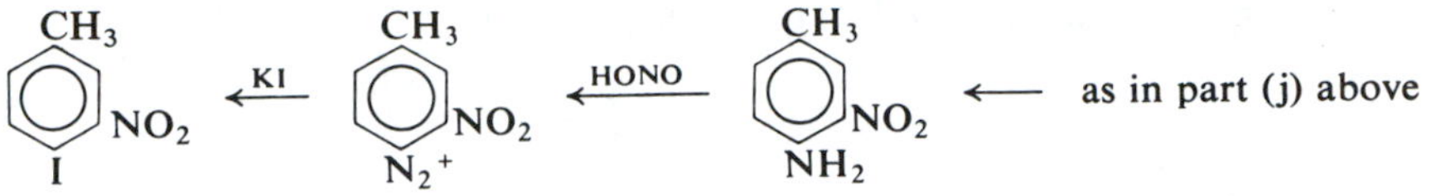 CH_3, NO_2, I benzene $\xleftarrow{KI}$ CH_3, NO_2, N_2^+ benzene $\xleftarrow{HONO}$ CH_3, NO_2, NH_2 benzene $\longleftarrow$ as in part (j) above

(m) 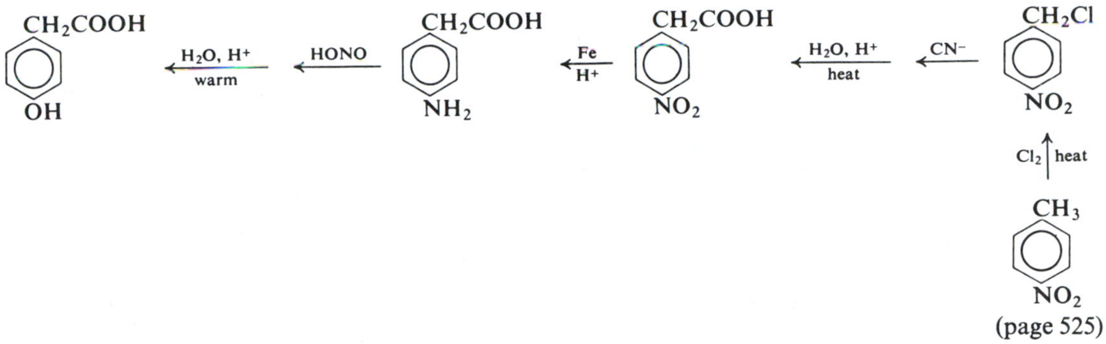 CH_2COOH, OH benzene $\xleftarrow[\text{warm}]{H_2O,\ H^+}$ $\xleftarrow{HONO}$ CH_2COOH, NH_2 benzene $\xleftarrow[H^+]{Fe}$ CH_2COOH, NO_2 benzene $\xleftarrow[\text{heat}]{H_2O,\ H^+}$ $\xleftarrow{CN^-}$ CH_2Cl, NO_2 benzene $\xleftarrow{Cl_2,\ \text{heat}}$ CH_3, NO_2 benzene (page 525)

(n) CH_3, Br, Cl benzene $\xleftarrow{CuCl}$ $\xleftarrow{HONO}$ CH_3, Br, NH_2 benzene $\xleftarrow[H^+]{Fe}$ CH_3, Br, NO_2 benzene $\xleftarrow[Fe]{Br_2}$ CH_3, NO_2 benzene (page 525)

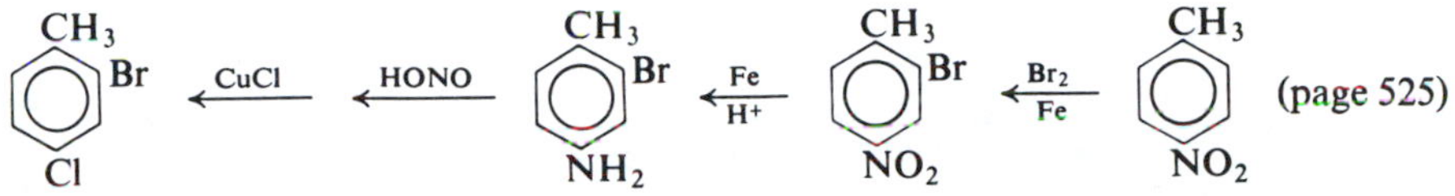

(o) 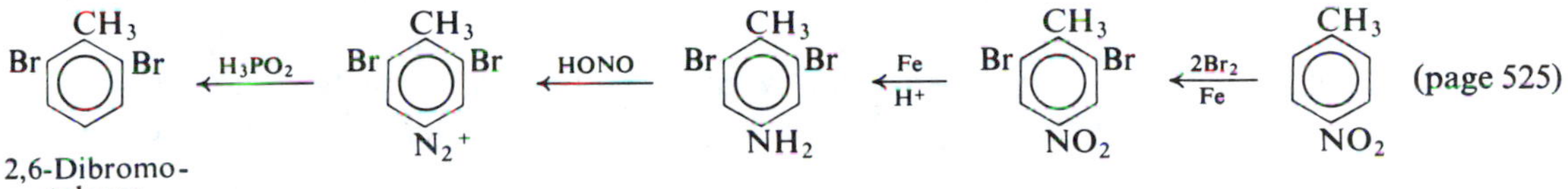 Br, CH_3, Br benzene $\xleftarrow{H_3PO_2}$ Br, CH_3, Br, N_2^+ benzene $\xleftarrow{HONO}$ Br, CH_3, Br, NH_2 benzene $\xleftarrow[H^+]{Fe}$ Br, CH_3, Br, NO_2 benzene $\xleftarrow[Fe]{2Br_2}$ CH_3, NO_2 benzene (page 525)

2,6-Dibromo-toluene

(p) 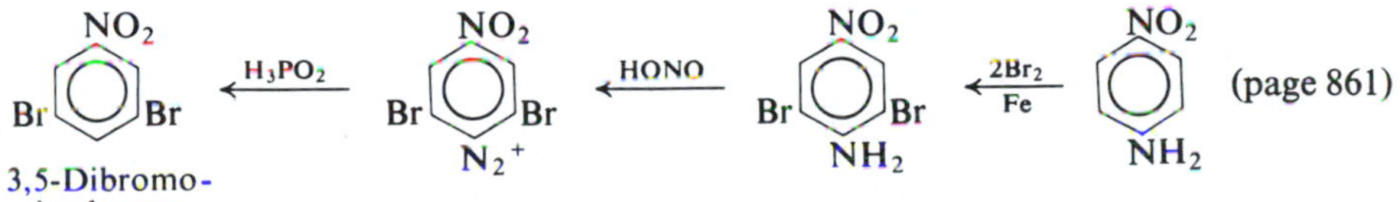 NO_2, Br, Br benzene $\xleftarrow{H_3PO_2}$ NO_2, Br, Br, N_2^+ benzene $\xleftarrow{HONO}$ NO_2, Br, Br, NH_2 benzene $\xleftarrow[Fe]{2Br_2}$ NO_2, NH_2 benzene (page 861)

3,5-Dibromo-nitrobenzene

13. (a) $HOOC(CH_2)_4COOH + H_2N(CH_2)_6NH_2 \longrightarrow$ salt

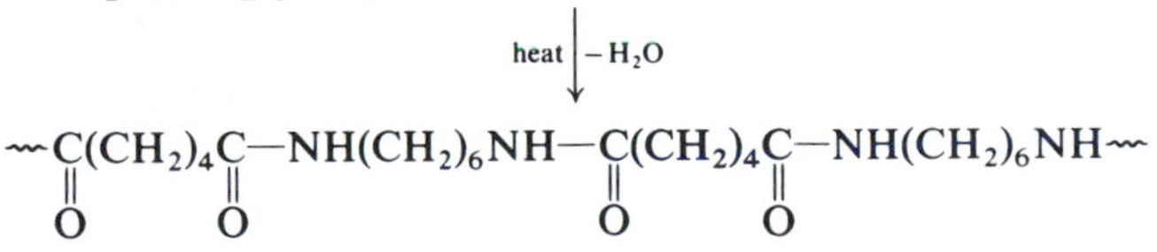

$$\downarrow \text{heat},\ -H_2O$$

$$\sim C(=O)(CH_2)_4C(=O)-NH(CH_2)_6NH-C(=O)(CH_2)_4C(=O)-NH(CH_2)_6NH\sim$$

A polyamide

(b) Acidic hydrolysis of the amide linkages.

14. (a) See page 1060. (b) See page 1065.

15. A poor leaving group (OH^-) is converted into a good leaving group (OTs^-).

[Structure: biphenyl with $-CH_2-N$(pyrrolidine) and $-CH_2-OTs$ → cyclic quaternary ammonium salt, N^+ spiro-pyrrolidinium, OTs^-]

16. (a) Halide ion competes with water as the nucleophile:

$$Ar{-}N_2^+ \xrightarrow{H_2O} Ar{-}OH \qquad Ar{-}N_2^+ \xrightarrow{X^-} Ar{-}X$$

Reaction of $Ph{-}N_2^+$ is S_N1-like:

(1) $Ph{-}N_2^+ \longrightarrow Ph^+ + N_2^+$ *Slow: rate-determining*

(2) $Ph^+ + :Z \longrightarrow Ph{-}Z$

The concentration and nature of the nucleophiles do not affect the *rate* of reaction (step 1), but they do affect *product composition*, controlled in step (2): the more halide ion present, the more halobenzene formed.

Electron withdrawal by $-NO_2$ slows down formation of the aryl cation from p-$NO_2C_6H_4N_2^+$, and the reaction takes place by an S_N2-like substitution:

$$p\text{-}O_2NPh{-}N_2^+ \longrightarrow \left[p\text{-}O_2NPh\begin{smallmatrix} Z \\ N_2^{\delta+} \end{smallmatrix} \right] \longrightarrow p\text{-}O_2NPh{-}Z + N_2$$

In this single-step reaction, the concentration and nature of the nucleophiles affect both rate and product composition.

(b) N_2, with virtually no basicity, is an extraordinarily good leaving group.

17. (a) Hinsberg test

(b) HONO, then β-naphthol

(c) Hinsberg test

(d) Hinsberg test

(e) 3° salt $\xrightarrow{OH^-}$ odor or basic vapors of amine

(f) $AgNO_3$ ($\longrightarrow$ AgCl from $PhNH_3^+Cl^-$)

(g) CrO_3/H_2SO_4

(h) HCl(aq)

(i) $BaCl_2$ ($\longrightarrow$ $BaSO_4$ from sulfate)

(j) $AgNO_3$ ($\longrightarrow$ AgCl from $EtNH_3^+Cl$)

(k) HCl(aq) dissolves $PhNH_2$

(l) $BaCl_2$ ($\longrightarrow$ $BaSO_4$ from sulfate), or H_2O (dissolves salt), or NaOH (sets free $PhNH_2$)

18. Compare the answer to Problem 22.1, above.

(a)–(d) Dissolve the basic amine in aqueous acid. The neutral compound (alkane, ether, amide, dipolar ion) is unaffected and can be collected by distillation or filtration. The amine is then regenerated from the acidic solution of its salt by the action of strong base.

(e) Transform the 2° amine into a neutral amide by acetic anhydride, the 3° amine remaining unacetylated. Then proceed with the separation of the basic amine and the neutral amide as in (c). Finally, hydrolyze the amide back to the 2° amine by basic hydrolysis, and collect the amine by distillation.

(f) Separate the acid first by dissolving it in aqueous NaOH; then proceed to separate the insoluble amine and hydrocarbon as in (a). Regenerate the acid from the alkaline solution of its salt by the action of strong acid.

(g)–(h) Proceed as in (c).

19. (a) HONO, then β-naphthol for aniline; then the Hinsberg test (2° *vs.* 3°).

(b) HCl(aq): base *vs.* neutral amide.

(c) Hinsberg to distinguish the 3° and 2° amines; only one of the 1° amines (*o*-toluidine) will give a diazotization and coupling test.

(d) Hot NaOH(aq) gives NH_3 from ethyl oxamate ($RCONH_2$).

(e) $HCONH_2$ is soluble in H_2O; the basic 3° amine is soluble in HCl(aq), but the neutral nitrile is insoluble.

(f) Only the basic amine is soluble in HCl(aq); hot NaOH(aq) on the nitrile gives NH_3; $PhNO_2$ unreactive.

(g) Elemental analysis distinguishes tosyl chloride (S, Cl) and the two sulfonamides (S, N), which can be further distinguished by NaOH(aq) in the manner of the Hinsberg test.

Elemental analysis distinguishes *p*-chloroaniline (Cl, N) and *p*-nitrobenzyl chloride (Cl, N) which can be further distinguished by $AgNO_3$ (side-chain Cl $\longrightarrow$ AgCl).

Of the compounds remaining: only the 3° amine is colorless; *o*-nitroaniline (K_b 6×10^{-14}) is more soluble than 2,4-dinitroaniline (K_b about 10^{-19}) in HCl(aq); and 2,4-dinitroaniline gives NH_3 by nucleophilic aromatic substitution (see Sec. 26.7) when warmed with NaOH(aq).

20. (a) A, PhCONHPh; B, $PhNH_2$; C, PhCOOH.

(b) Clearly, a rearrangement has occurred: both Ph's are bonded to carbon in the oxime, and one is bonded to nitrogen in the product.

(c) $Ph_2C{=}N{-}OH$ (*Oxime*) $\overset{H^+}{\rightleftharpoons}$ $Ph_2C{=}N{-}OH_2^+$ $\longrightarrow$ $Ph{-}\overset{\oplus}{C}{=}N{-}Ph$ $\xrightarrow[-H^+]{H_2O}$ $Ph{-}C(OH){=}N{-}Ph$ $\xrightarrow{\text{taut.}}$ $Ph{-}C(=O){-}NH{-}Ph$

Amide

N-Phenylbenzamide

(d) By conversion of —OH into a better (less basic) leaving group.

$$\mathrm{Ph_2C{=}N{-}OH} \xrightarrow{PCl_5} \mathrm{Ph_2C{=}N{-}Cl} \longrightarrow \mathrm{Ph{-}\overset{\oplus}{C}{=}N{-}Ph} + Cl^-$$

$$\xrightarrow{H_2O,\ -H^+} \mathrm{PhCONHPh}$$

$$\mathrm{Ph_2C{=}N{-}OH} \xrightarrow{TsCl} \mathrm{Ph_2C{=}N{-}OTs} \longrightarrow \mathrm{Ph\overset{\oplus}{C}{=}N{-}Ph} + OTs^-$$

$$\xrightarrow{H_2O,\ -H^+} \mathrm{PhCONHPh}$$

(Actually, the exact role of acids in the Beckmann rearrangement is not understood. In (c), we have shown sulfuric acid as protonating —OH, but it may instead convert it into $-OSO_3H$. PCl_5 may convert —OH into $-OPCl_4$. See the references given below.)

(e) acetone oxime $\overset{H^+}{\rightleftharpoons}$ $(CH_3)_2C{=}N{-}OH_2^+$ $\longrightarrow$ $CH_3{-}\overset{O}{\overset{\|}{C}}{-}\overset{H}{\overset{|}{N}}{-}CH_3$ *N*-Methylacetamide

(*Z*)-acetophenone oxime $\overset{H^+}{\rightleftharpoons}$ $(CH_3)(Ph)C{=}N{-}OH_2^+$ $\longrightarrow$ $Ph{-}\overset{O}{\overset{\|}{C}}{-}\overset{H}{\overset{|}{N}}{-}CH_3$ *N*-Methylbenzamide

(*E*)-acetophenone oxime $\overset{H^+}{\rightleftharpoons}$ $(Ph)(CH_3)C{=}N{-}OH_2^+$ $\longrightarrow$ $CH_3{-}\overset{O}{\overset{\|}{C}}{-}\overset{H}{\overset{|}{N}}{-}Ph$ *N*-Phenylacetamide

(*Z*)-*p*-nitrobenzophenone oxime $\overset{H^+}{\rightleftharpoons}$ $(Ph)(p\text{-}NO_2C_6H_4)C{=}N{-}OH_2^+$ $\longrightarrow$ $p\text{-}NO_2C_6H_4{-}\overset{O}{\overset{\|}{C}}{-}\overset{H}{\overset{|}{N}}{-}Ph$ *N*-Phenyl-*p*-nitrobenzamide

(*E*)-*p*-nitrobenzophenone oxime $\overset{H^+}{\rightleftharpoons}$ $(p\text{-}NO_2C_6H_4)(Ph)C{=}N{-}OH_2^+$ $\longrightarrow$ $Ph{-}\overset{O}{\overset{\|}{C}}{-}\overset{H}{\overset{|}{N}}{-}C_6H_4NO_2\text{-}p$ *N*-(*p*-Nitrophenyl)benzamide

(*Z*)-methyl-*n*-propyl ketoxime $\overset{H^+}{\rightleftharpoons}$ $(CH_3)(n\text{-}Pr)C{=}N{-}OH_2^+$ $\longrightarrow$ $n\text{-}Pr{-}\overset{O}{\overset{\|}{C}}{-}\overset{H}{\overset{|}{N}}{-}CH_3$ *N*-Methyl-*n*-butyramide

(*E*)-methyl-*n*-propyl ketoxime $\overset{H^+}{\rightleftharpoons}$ $(n\text{-}Pr)(CH_3)C{=}N{-}OH_2^+$ $\longrightarrow$ $CH_3{-}\overset{O}{\overset{\|}{C}}{-}N{-}Pr\text{-}n$ *N*-*n*-Propylacetamide

(f) Hydrolyze the amides and identify the resulting amines and acids.

(g) Cyclohexanone.

Suitably substituted oximes exist as geometric isomers (Problem 10, Chapter 22), and the product of rearrangement depends upon which isomer is involved: as shown in (e), the group *trans* (*anti*) to the leaving group is the one that migrates.

(For a review of the Beckmann rearrangement, see: de la Mare, P. B. D. In *Molecular Rearrangements*, Part 1; de Mayo, P., Ed.; Wiley-Interscience: New York, 1963; pp 483–507; or Heldt, W. Z.; Donaruma, L. G. *Organic Reactions*; Wiley: New York; Vol. XI, Chapter 1.)

21. (a)

C_3H_9N	trimethylamine	$C_5H_{15}O_2N$	choline
$+\ C_2H_4O$	ethylene oxide	$-\ C_5H_{13}ON$	
$C_5H_{13}ON$		H_2O	has been added

$$H_2C{-}CH_2(O) + NMe_3 \longrightarrow {}^{-}OCH_2CH_2\overset{+}{N}Me_3 \xrightarrow{H_2O} HOCH_2CH_2NMe_3{}^{+}OH^{-}$$

$(C_5H_{13}ON)$ Choline $(C_5H_{15}O_2N)$

The quaternary ammonium hydroxide is completely ionized, and gives a strongly basic solution.

(b) The formula is increased by C_2H_2O, indicating replacement of H— by CH_3CO—.

$$CH_3\underset{\underset{O}{\|}}{C}OCH_2CH_2NMe_3{}^{+}OH^{-}$$

Acetylcholine

22. (a) Paragraph 1: a basic 1° aromatic amine.

Paragraph 2: hydrolysis of an acid derivative gives the salt of the acid and a water-soluble alcohol or amine.

Paragraph 3: pptn. of RCOOH. Subsequent dissolution indicates that the acid is amphoteric, with an amino group also present.

Chemical arithmetic:

$C_7H_7O_2N$	C_6H_6N
$-\ COOH$	$-\ C_6H_4$
C_6H_6N	NH_2

shows $C_6H_4(NH_2)(COOH)$

M.p. leaves no choice but *p*-aminobenzoic acid for D (Table 19.1, page 715).

Let us go back to the original Novocaine. Since it is an acid derivative, we can subtract the acyl group of D.

$C_{13}H_{20}O_2N_2$	Novocaine
$-\ H_2NC_6H_4CO$	acyl group
$C_6H_{14}ON$	

We have one nitrogen still unassigned; we cannot say yet whether it is in an amide or amino group, since in either case Novocaine would be acid-soluble by virtue of the $-NH_2$ group of D. So we must investigate the rest of the hydrolysis mixture, the ether solution of an amine (or possibly an aminoalcohol).

Paragraph 4: E is an amine. Reaction with acetic anhydride produces a material, F, which is still basic and therefore still an amine. E thus can only be a tertiary amine. (Tertiary amines do not form amides.) This means that Novocaine is an ester, not an amide. F must be an acetate ester of an alcohol.

$$\begin{array}{rl} C_8H_{17}O_2N & \text{F} \\ -\ C_2H_2O & \text{one ester group adds} \\ \hline C_6H_{15}ON = \text{E} & \end{array}$$

Paragraph 5: $H_2C{-}CH_2$ (epoxide, O bridging) $+ NHEt_2 \longrightarrow HOCH_2CH_2NEt_2$ (E)

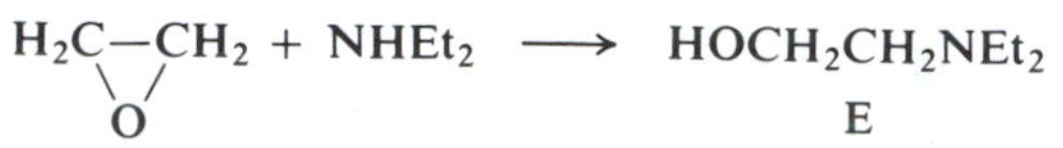

Novocaine, therefore, is the *p*-aminobenzoic ester of alcohol E;

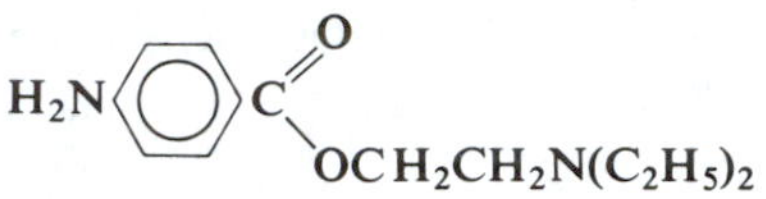

$$H_2N{-}C_6H_4{-}C(=O){-}OCH_2CH_2N(C_2H_5)_2$$

Novocaine

(b)

p-$O_2NC_6H_4COOH \xrightarrow{SOCl_2} p$-$O_2NC_6H_4COCl$ (page 525)

$H_2C{-}CH_2$ (epoxide) $+ Et_2NH \longrightarrow HOCH_2CH_2NEt_2$

p-$O_2NC_6H_4COCl + HOCH_2CH_2NEt_2 \longrightarrow p$-$O_2NC_6H_4COOCH_2CH_2NEt_2 \xrightarrow{H_2,\ Ni} p$-$H_2NC_6H_4COOCH_2CH_2NEt_2$

Novocaine

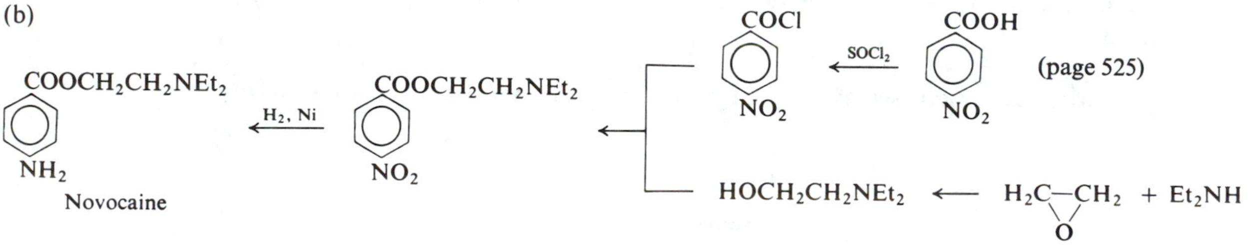

23. G is a neutral compound. Its slow dissolution in hot NaOH points to its being an acid derivative. Formation of H, a base and thus an amine, points to G's being an amide. H itself is a 2° amine, giving J, $PhSO_2NRR'$, a neutral sulfonamide. The nitrogen-free acid I, m.p. 180 °C, is *p*-toluic acid (Table 19.1, page 715).

As usual, chemical arithmetic is helpful:

$$\begin{array}{rl} C_{15}H_{15}ON & \text{cpd. G} \\ -\ CH_3C_6H_4CO & \text{acyl group of I} \\ \hline C_7H_8N & \text{amino portion of amide} \end{array}$$

H begins to look like an aromatic amine,

$$\begin{array}{r} C_7H_8N \\ -\ C_6H_5 \\ \hline CH_3N \end{array}$$

and is undoubtedly *N*-methylaniline, $C_6H_5NHCH_3$. J, then, is $PhSO_2N(CH_3)Ph$.

Putting the pieces together gives us G:

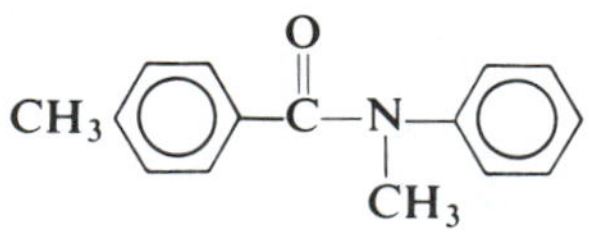

$$CH_3{-}C_6H_4{-}C(=O){-}N(CH_3){-}C_6H_5$$

G

24. CH_3N (bicyclic ketone) $\xrightarrow{\text{reduction}}$ K $\xrightarrow{-H_2O}$ L $\xrightarrow[\text{methylation}]{\text{exhaustive}}$ M ($CH_3\overset{+}{N}CH_3$ OH^-)

M $\xrightarrow{-H_2O}$ N ($N(CH_3)_2$) $\xrightarrow[\text{methylation}]{\text{exhaustive}}$ O ($NMe_3{}^+OH^-$) $\xrightarrow[-NMe_3]{-H_2O}$ P (1,3,5-Cyclooctatriene) $\xrightarrow[\text{1,6-addition}]{Br_2}$ Q (Br, Br)

Q $\xrightarrow{2Me_2NH}$ R (NMe_2, NMe_2) $\xrightarrow[\text{methylation}]{\text{exhaustive}}$ S (HO^- $^+NMe_3$, HO^- $^+NMe_3$) $\xrightarrow[-2NMe_3]{-2H_2O}$ T

1,3,5,7-Cyclooctatetraene

(see Problem 13, page 516)

This synthesis of cyclooctatetraene was carried out in 1911–1913 by the great German chemist and Nobel Prize winner Richard Willstätter. As we saw earlier (Problem 13, page 516), the product was found to have the normal properties of an alkene, and hence is not aromatic. Today, a beginning student finds this not surprising: ten π electrons does not fit Hückel's $4n + 2$ rule for aromaticity. But fifty years ago this finding was one of the principal pieces of evidence that the benzene structure was very special—but in just what way was not realized until the Hückel rule in 1931. (On various grounds, many chemists doubted that Willstätter had actually made cyclooctatetraene; the synthesis was laboriously repeated—and his findings verified—by other workers in 1948. If Willstätter had observed the properties then *expected* of cyclooctatetraene, one wonders, would these doubts have been quite so strong?)

25. Pantothenic acid is a monocarboxylic acid: it yields a monosalt and a monoester. It appears to be an amide as well: it contains non-basic nitrogen, and is hydrolyzed to β-aminopropionic acid and Y. Since Y contains no nitrogen, β-aminopropionic acid must provide the nitrogen of the amide:

$$R{-}\underset{\underset{O}{\|}}{C}{-}NHCH_2CH_2COOH$$

Chemical arithmetic,

$C_9H_{17}O_5N$	pantothenic acid
$-$ $CONHCH_2CH_2COOH$	amide
$C_5H_{11}O_2$	*indicates* that Y is $(C_5H_{11}O_2)$—COONa

shows that Y is saturated and open-chain, and contains oxygen in alcohol or ether linkages.

The synthetic route leads us to compound W.

$$\underset{\text{Isobutyraldehyde}}{H-\underset{CH_3}{\overset{CH_3}{C}}-CHO} \xrightarrow[\text{base}]{HCHO} \underset{\text{U}}{HOCH_2-\underset{CH_3}{\overset{CH_3}{C}}-CHO} \xrightarrow[H^+]{CN^-} \underset{\text{V}}{HOCH_2-\underset{H_3C}{\overset{H_3C}{C}}-\underset{OH}{CH}-CN} \xrightarrow[H^+]{H_2O} \left[\underset{\text{W}}{HOCH_2-\underset{H_3C}{\overset{H_3C}{C}}-\underset{OH}{CH}-COOH}\right]$$

W: A γ-hydroxy acid → X: A γ-lactone (five-membered ring: HO–CH–C(=O)–O–CH₂–C(CH₃)₂)

This is a γ-hydroxy acid and spontaneously forms the γ-lactone X. Treatment of X with base gives Y, the sodium salt of W.

$$\text{X} \xrightarrow{NaOH} \underset{\text{Y}}{HOCH_2-\underset{H_3C}{\overset{H_3C}{C}}-\underset{OH}{CH}-COO^-Na^+}$$

A lactone is an ester. Esters react with ammonia to give amides (Sec. 20.19) and, as we would have expected, with amines to give substituted amides. X reacts with the amino group of β-aminopropionic acid to give the substituted amide that is pantothenic acid.

$$\text{X} + \underset{\beta\text{-Aminopropionic acid}}{NH_2CH_2CH_2COOH} \longrightarrow HOCH_2-\underset{H_3C}{\overset{H_3C}{C}}-\underset{OH}{CH}-\overset{O}{\overset{\|}{C}}-NHCH_2CH_2COOH$$

Pantothenic acid

The S-(+)-form is biologically active

26. Evidently, Z is the hydrochloric acid salt (water-soluble, reacts with NaOH to lose Cl) of a 1° aromatic amine (diazotization and coupling with β-naphthol). Assuming it is the salt of a monoamine, $ArNH_3^+Cl^-$, we can use the neutralization equivalent to determine the structure.

$$\begin{array}{rl} 131 \pm 2 & ArNH_3^+Cl^- \\ -\ 52.5 & NH_3^+Cl^- \\ \hline 78.5 \pm 2 & \textit{equivalent to } C_6H_5- \end{array}$$

Z is anilinium chloride, $C_6H_5NH_3^+Cl^-$, and the liquid is aniline, $C_6H_5NH_2$.

27. (a) *n*-Butyl cation.

(b)

$$CH_3CH_2CH_2CH_2NH_2 \xrightarrow[-N_2]{HONO} CH_3CH_2CH_2CH_2^{\oplus} \begin{cases} \xrightarrow{Cl^-} n\text{-BuCl} & 5\% \\ \xrightarrow{H_2O} n\text{-BuOH} & 25\% \\ \xrightarrow{-H^+} CH_3CH_2CH{=}CH_2 & \\ \end{cases}$$

$$CH_3CH_2CH_2CH_2^{\oplus} \xrightarrow{\text{rearr.}} CH_3CH_2\overset{\oplus}{C}HCH_3 \begin{cases} \xrightarrow{-H^+} CH_3CH{=}CHCH_3 & \\ \xrightarrow{H_2O} sec\text{-BuOH} & 13\% \\ \xrightarrow{Cl^-} sec\text{-BuCl} & 3\% \end{cases}$$

($CH_3CH_2CH{=}CH_2$ and $CH_3CH{=}CHCH_3$ together: 37%)

28. (a)

$$iso\text{-BuNH}_2 \xrightarrow[-N_2]{HONO} iso\text{-Bu}^+ \begin{cases} \xrightarrow{Cl^-} iso\text{-BuCl} \\ \xrightarrow{H_2O} iso\text{-BuOH} \\ \xrightarrow{-H^+} (CH_3)_2C{=}CH_2 \end{cases}$$

$$iso\text{-Bu}^+ \xrightarrow{\text{rearr.}} tert\text{-Bu}^+ \begin{cases} \xrightarrow{-H^+} (CH_3)_2C{=}CH_2 \\ \xrightarrow{H_2O} tert\text{-BuOH} \\ \xrightarrow{Cl^-} tert\text{-BuCl} \end{cases}$$

(b)

$$CH_3\text{—}\underset{CH_3}{\overset{CH_3}{C}}\text{—}CH_2NH_2 \xrightarrow[-N_2]{HONO} \underset{1^\circ\ \text{cation}}{CH_3\text{—}\underset{CH_3}{\overset{CH_3}{C}}\text{—}CH_2^{\oplus}} \xrightarrow{\text{rearr.}} \underset{3^\circ\ \text{cation}}{CH_3\text{—}\underset{\oplus}{\overset{CH_3}{C}}\text{—}CH_2\text{—}CH_3}$$

$$\xrightarrow{H_2O} CH_3\text{—}\underset{OH}{\overset{CH_3}{C}}\text{—}CH_2CH_3$$

$$\xrightarrow{-H^+} CH_3\text{—}\overset{CH_3}{C}{=}CH\text{—}CH_3$$

$$\xrightarrow{-H^+} CH_2{=}\overset{CH_3}{C}\text{—}CH_2CH_3$$

29. (a) $CH_3CH_2CH_2CH_2NH_2$ *n*-Butylamine (b) $HCONHCH_3$ *N*-Methylformamide (c) *m*-$CH_3OC_6H_4NH_2$ *m*-Anisidine

(See the labeled spectra on page 715 of this Study Guide.)

30. (a) $C_6H_5\underset{NH_2}{C}HCH_3$ α-Phenylethylamine (b) $C_6H_5CH_2CH_2NH_2$ β-Phenylethylamine (c) *p*-$CH_3C_6H_4NH_2$ *p*-Toluidine

(See the labeled spectra on page 716 of this Study Guide.)

31. (a) Cyclohexylamine (b) 4-Methylpiperidine (c) 4-Ethylpyridine

(See the labeled spectra on page 717 of this Study Guide.)

32. p-$CH_3CH_2OC_6H_4NH_2$ — p-Phenetidine (p-Ethoxyaniline) — AA

$C_6H_5CH_2NHCH_2CH_3$ — Ethylbenzylamine — BB

$(CH_3)_2N\text{-}C_6H_4\text{-}C(=O)\text{-}C_6H_4\text{-}N(CH_3)_2$ — Michler's ketone — CC

(See the labeled spectra on pages 718 and 719 of this Study Guide.)

24

Phenols

24.1 *Inter*molecular hydrogen bonding in the *meta* and *para* isomers is diminished or eliminated by dilution with $CHCl_3$; *intra*molecular hydrogen bonding in the *ortho* isomer (page 891) is unaffected.

24.2 Intramolecular hydrogen bonding:

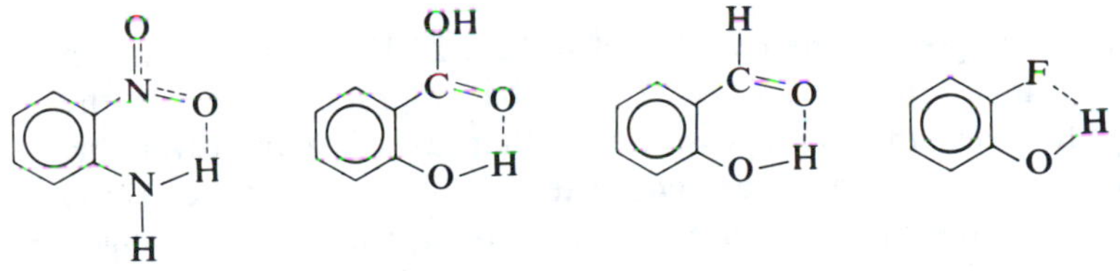

The geometry is wrong for the nitrile: the $—C{\equiv}N$ group, with digonal (*sp*) carbon, is linear, putting nitrogen too far away from —OH. The $—CH_3$ group in *o*-cresol does not form hydrogen bonds.

24.3 See Figure 24.8, page 426 of this Study Guide.

24.4 Friedel–Crafts alkylation: benzene, propylene (from cracking), HF.

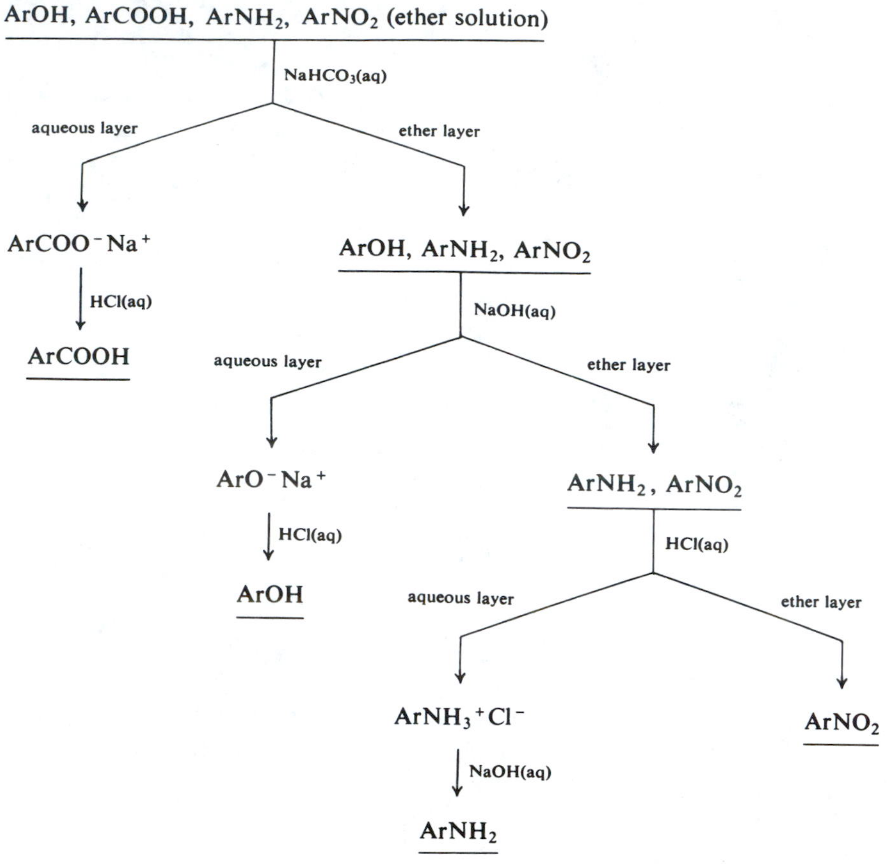

Figure 24.8 Separation of the mixture of Problem 24.3.

24.5 The conclusion is exactly analogous to the one drawn in Sec. 22.17 for the Hofmann rearrangement. The substituent effect (page 896) shows that migration occurs in the rate-determining step. There are, then, two possibilities: that (2) is fast and reversible followed by a slow (3); or that (2) and (3) are concerted. But the reverse of (2) is combination of the electron-deficient particle with water; if this happened, unrearranged hydroperoxide would contain oxygen-18—contrary to fact. We are left with the concerted reaction (2,3).

(Bassey, M.; Bunton, C. A.; Davies, A. G.; Lewis, T. A.; Llewellyn, D. R. "Organic Peroxides. Part V. Isotopic Tracer Studies on the Formation and Decomposition of Organic Peroxides"; *J. Chem. Soc.* **1955**, 2471.)

24.6 (a) *H migration*

$$\text{Ar—CH(H)—O—OH}_2^+ \xrightarrow{-H_2O} \text{Ar—CH}\text{=}\overset{+}{\text{O}}\text{—H} \xrightarrow{-H^+} \text{Ar—CH}\text{=O}$$

(Ar = *p*-Tolyl) *p*-Methylbenzaldehyde (*61%*)

Aryl migration

$$\mathrm{H{-}\overset{H}{\underset{Ar}{C}}{-}O{-}OH_2^+ \xrightarrow{-H_2O} H{-}\overset{H}{C}{=}\overset{+}{O}{-}Ar \xrightarrow[-H^+]{H_2O} H{-}\overset{H}{\underset{OH}{C}}{-}O{-}Ar \longrightarrow H{-}\overset{H}{C}{=}O + HOAr}$$

Hemiacetal; Formaldehyde; *p*-Cresol (*38%*)

(b) H migrates faster than *p*-tolyl.

24.7

$$\mathrm{R{-}\overset{H}{\underset{H}{C}}{-}O{-}OH_2^+ \xrightarrow{-H_2O} R{-}\overset{H}{C}{=}\overset{+}{O}H \xrightarrow{-H^+} RCHO}$$ **H migration**

$$\mathrm{H{-}\overset{H}{\underset{R}{C}}{-}O{-}OH_2^+ \xrightarrow{-H_2O} H{-}\overset{H}{C}{=}\overset{+}{O}R \xrightarrow{H_2O} HCHO + ROH}$$ *R migration*

$$\mathrm{R{-}\overset{R}{\underset{H}{C}}{-}O{-}OH_2^+ \xrightarrow{-H_2O} R{-}\overset{R}{C}{=}\overset{+}{O}H \xrightarrow{-H^+} R_2C{=}O}$$ **H migration**

$$\mathrm{R{-}\overset{H}{\underset{R}{C}}{-}O{-}OH_2^+ \xrightarrow{-H_2O} R{-}\overset{H}{C}{=}\overset{+}{O}R \xrightarrow{H_2O} RCHO + ROH}$$ *R migration*

Since alcohols (ROH) are *not* obtained, H must migrate much faster than R.

24.8 (a) 1,2-shift of R from boron to oxygen. In view of what is observed for other 1,2-shifts (Sec. 22.16), we expect retention of configuration in the migrating group, R.

(1) $\mathrm{R_3B + OOH^- \longrightarrow R_3\bar{B}{-}O{-}OH}$

(2) $\mathrm{R_2\overset{R}{B}{-}O{-}OH \longrightarrow R_2B{-}O{-}R + OH^-}$

then (1), (2), (1), (2) for other R's

(b) The overall *syn*-addition could *a priori* be the result of either of two combinations: *syn*-addition in the hydroboration, and retention in the oxidation; or *anti*-addition in the hydroboration, and inversion in the oxidation. If our expectation in (a) is correct, hydroboration must involve *syn*-addition—as it has actually been found to do.

24.9 Because of acid-strengthening nitro groups, the acidity of these phenols is high enough (K_a values 10^{-4} and "very large") to permit reaction with weakly basic HCO_3^-.

24.10 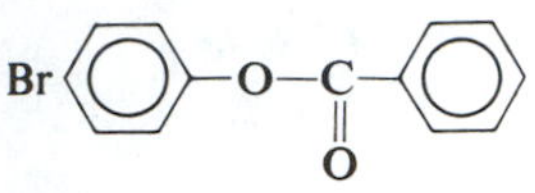 Br–C_6H_4–O–C(=O)–C_6H_5 $\qquad$ –O–C(=O)–R *activates*; –C(=O)–OR *deactivates*

Compare Problem 5(c), page 547.

24.11 The volatility of the *ortho* isomer is greater due to intramolecular hydrogen bonding. (Compare Problem 24.2, page 892.)

24.12 (a) This is *nitrodesulfonation:* the $—SO_3H$ is displaced by electrophilic reagents, in this case by $^+NO_2$.

$$ArSO_3^- + NO_2^+ \longrightarrow \overset{\oplus}{Ar}(SO_3^-)(NO_2) \longrightarrow ArNO_2 + SO_3$$

(b) There is less destructive oxidation by the nitrating agent.

24.13 Sulfonation is reversible. The *ortho* isomer is formed more rapidly; the *para* isomer is more stable. At 15–20 °C, there is rate control of product composition; at 100 °C, there is equilibrium control. (See Sec. 11.23.)

24.14 p-$O_2NC_6H_4OH$ $\xleftarrow{HNO_3}$ p-$ON{-}C_6H_4OH$ (N=O) $\xleftarrow[7\text{–}8\ °C]{NaNO_2,\ H_2SO_4}$ C_6H_5OH

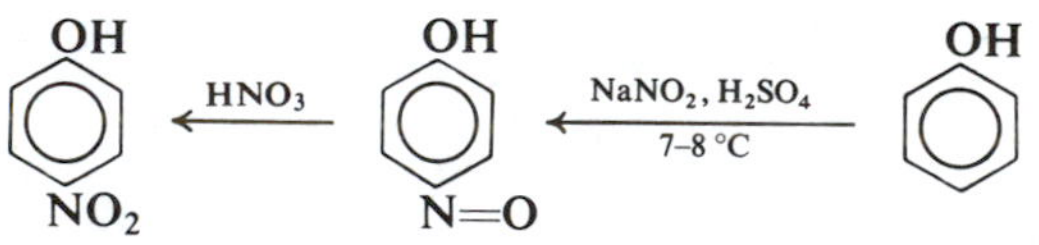

24.15 (a) Removal of the proton is easier from $CHCl_3$, which is more acidic than $CHCl_2$; it is impossible for CCl_4, which has no protons. Confirms step (1).

(b) Removal of the proton from $CHCl_3$ occurs to form an anion that can acquire D from D_2O to yield $CDCl_3$. Confirms step (1), and its reversibility.

(c) Step (2)—and with it step (1)—is reversed by excess Cl^-. Confirms step (2), and its reversibility.

(d) Step (2)—and with it step (1)—is reversed, this time by I^-. Confirms steps (1) and (2), and their reversibility.

(e) Nucleophilic Cl_3C^- adds to the carbonyl group. Confirms step (1).

24.16 (a) The anions, $CH_3OSO_3^-$ and SO_4^{2-}, are very weakly basic and hence are very good leaving groups.

(b) Sulfonic acid esters, such as the tosylates (Sec. 6.14).

24.17 Phenoxide ion ($C_6H_5O^-$) is more basic than ethylene oxide, and would gain the proton from any added acid.

24.18 (a) $-OCH_3 > -CH_3$

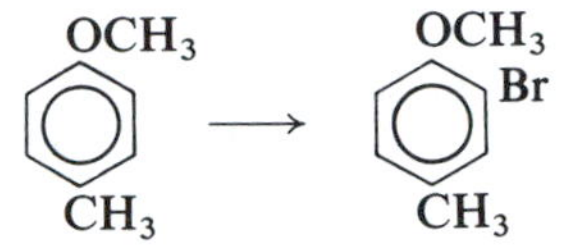

p-$CH_3C_6H_4OCH_3$ $\longrightarrow$ 2-bromo-4-methylanisole (OCH_3, Br ortho; CH_3 para)

(b) $-OCH_3 > -NO_2$

m-nitroanisole $\longrightarrow$ (OCH_3; NO_2, NO_2) + (O_2N, OCH_3, NO_2)

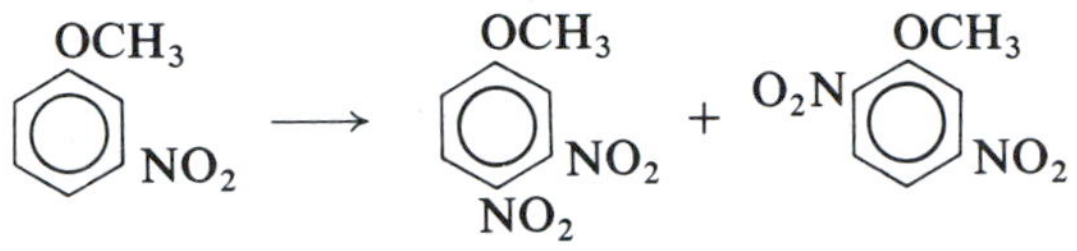

(c) $-OR > -R$

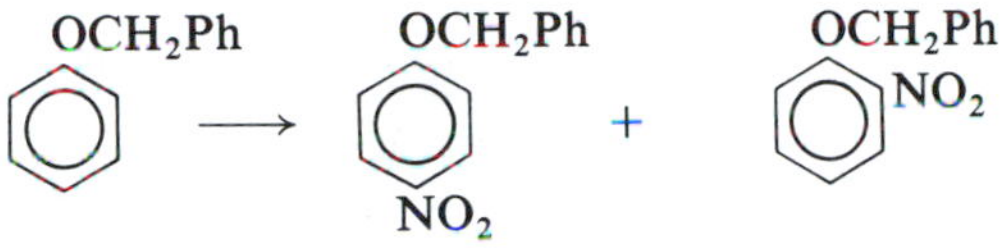

$C_6H_5OCH_2Ph$ $\longrightarrow$ p-$O_2NC_6H_4OCH_2Ph$ + o-$O_2NC_6H_4OCH_2Ph$

24.19 $PhOCH_2C_6H_4NO_2$ $\longleftarrow$ PhONa + $BrCH_2C_6H_4NO_2$ $\xleftarrow{Br_2,\ heat}$ $CH_3C_6H_4NO_2$ $\xleftarrow{HNO_3,\ H_2SO_4}$ toluene

PhONa $\uparrow$ NaOH phenol

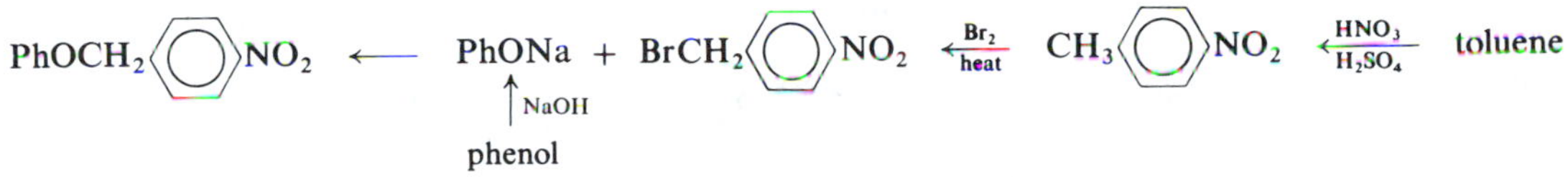

24.20 ArONa, $ClCH_2COONa$, warm, acidify. Neutralization equivalents (Sec. 19.21) of the aryloxyacetic acids would be useful in identifying phenols.

1. (a) 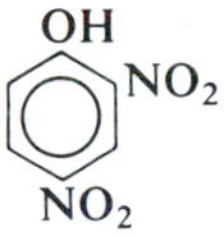 OH, NO_2, NO_2 (2,4-dinitrophenol) (b) 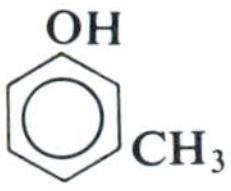 OH, CH_3 (m-cresol) (c)  OH, OH (hydroquinone) (d) 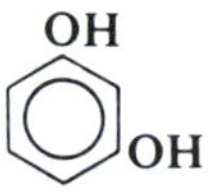 OH, OH (resorcinol)

(e) 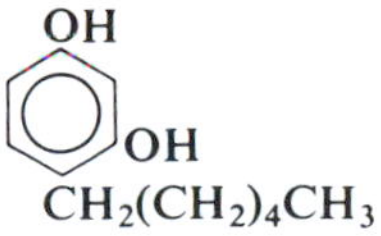 OH, OH, $CH_2(CH_2)_4CH_3$ (f) 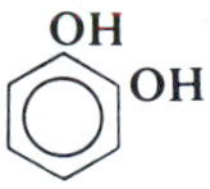 OH, OH (catechol) (g) 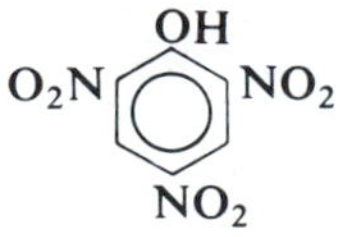 O_2N, OH, NO_2, NO_2 (picric acid)

(h)  $C_6H_5OCCH_3$ (with =O) (i) 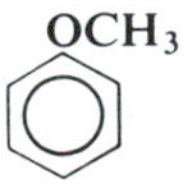 $C_6H_5OCH_3$ (j)  OH, COOH (k) 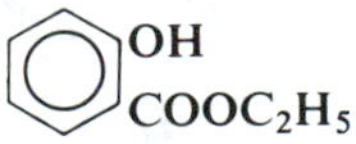 OH, $COOC_2H_5$

2. (a) $C_6H_5NH_2 \xrightarrow{HONO} C_6H_5N_2^+ \xrightarrow[\text{warm}]{H_2O,\ H^+} C_6H_5OH$

(b) $C_6H_6 \xrightarrow{SO_3,\ H_2SO_4} C_6H_5SO_3H \xrightarrow{NaOH} C_6H_5SO_3^-Na^+ \xrightarrow[\text{strong heat}]{NaOH} C_6H_5O^-Na^+ \xrightarrow{H^+} C_6H_5OH$

(c) $C_6H_5Cl \xrightarrow[\text{heat, press.}]{NaOH} C_6H_5ONa \xrightarrow{H^+} C_6H_5OH$

(d) $C_6H_5C(CH_3)_2H \xrightarrow[\text{heat}]{\text{air}} C_6H_5C(CH_3)_2OOH \xrightarrow{H^+} C_6H_5OH + (CH_3)_2C{=}O$

3. (a) $o\text{-}C_6H_4(OH)_2 \xleftarrow[\text{heat}]{HBr} o\text{-}C_6H_4(OH)(OCH_3)$

(b) $o\text{-}C_6H_4(OH)_2 \xleftarrow{H^+} o\text{-}C_6H_4(ONa)_2 \xleftarrow[\text{strong heat}]{NaOH} o\text{-}C_6H_4(ONa)(SO_3Na) \xleftarrow{NaOH} o\text{-}C_6H_4(OH)(SO_3H) \xleftarrow[\text{low temp.}]{H_2SO_4} C_6H_5OH$

(c) $m\text{-}C_6H_4(OH)_2 \xleftarrow{H^+} m\text{-}C_6H_4(ONa)_2 \xleftarrow[\text{strong heat}]{NaOH} m\text{-}C_6H_4(SO_3Na)_2 \xleftarrow{NaOH} m\text{-}C_6H_4(SO_3H)_2 \xleftarrow[\text{heat}]{2SO_3,\ H_2SO_4} C_6H_6$

(d) $2,4,6\text{-}(O_2N)_3C_6H_2OH \xleftarrow{HNO_3,\ H_2SO_4} 2,4\text{-}(NO_2)_2C_6H_3OH \xleftarrow{H^+} 2,4\text{-}(NO_2)_2C_6H_3ONa \xleftarrow{NaOH} 2,4\text{-}(NO_2)_2C_6H_3Cl \xleftarrow{HNO_3,\ H_2SO_4} C_6H_5Cl$ (from C_6H_6)

(e) $o\text{-}C_6H_4(OCH_3)_2 \xleftarrow{2Me_2SO_4,\ NaOH} o\text{-}C_6H_4(OH)_2$

4. (a) $o\text{-}CH_3C_6H_4OH \xleftarrow{H_2O,\ H^+} o\text{-}CH_3C_6H_4N_2^+ \xleftarrow{HONO} o\text{-}CH_3C_6H_4NH_2$ (page 871)

(b) $m\text{-}CH_3C_6H_4OH \xleftarrow{\text{as in (a)}} m\text{-}CH_3C_6H_4NH_2 \xleftarrow{Fe,\ H^+} m\text{-}CH_3C_6H_4NO_2$ (Problem 23.12, page 872)

(c) $p\text{-}CH_3C_6H_4OH \xleftarrow{H_2O,\ H^+} p\text{-}CH_3C_6H_4N_2^+ \xleftarrow{HONO} p\text{-}CH_3C_6H_4NH_2$ (page 871)

(d) OH, I $\xleftarrow{H_2O, H^+}$ N_2^+, I $\xleftarrow{HONO}$ NH_2, I $\xleftarrow[H^+]{Fe}$ NO_2, I $\xleftarrow{KI}$ NO_2, N_2^+ $\xleftarrow{HONO}$ NO_2, NH_2 (page 861)

(e) OH, Br $\xleftarrow{H_2O, H^+}$ N_2^+, Br $\xleftarrow{HONO}$ NH_2, Br $\xleftarrow{Fe, H^+}$ NO_2, Br $\xleftarrow{Br_2, Fe}$ NO_2 (from C_6H_6)

(f) OH, Br $\xleftarrow{H_2O, H^+}$ N_2^+, Br $\xleftarrow{HONO}$ NH_2, Br $\xleftarrow[H^+]{Fe}$ NO_2, Br $\xleftarrow[H_2SO_4]{HNO_3}$ Br (from C_6H_6)

(g) OH, Br, CH_3 $\xleftarrow[warm]{H_2O, H^+}$ N_2^+, Br, CH_3 $\xleftarrow{HONO}$ NH_2, Br, CH_3 $\xleftarrow[H^+]{Fe}$ NO_2, Br, CH_3 $\xleftarrow[Fe]{Br_2}$ NO_2, CH_3 (from $PhCH_3$)

(h) OH, Br, CH_3 $\xleftarrow[warm]{H_2O, H^+}$ N_2^+, Br, CH_3 $\xleftarrow{HONO}$ NH_2, Br, CH_3 (page 872)

(i) OH, Br, CH_3 $\xleftarrow[warm]{H_2O, H^+}$ N_2^+, Br, CH_3 $\xleftarrow{HONO}$ $\xleftarrow[H^+]{Fe}$ NO_2, Br, CH_3 $\xleftarrow{CuBr}$ $\xleftarrow{HONO}$ NO_2, NH_2, CH_3

NO_2, NH_2, CH_3 $\xleftarrow[H^+]{H_2O}$ NO_2, NHAc, CH_3 $\xleftarrow{HNO_3, H_2SO_4}$ NHAc, CH_3 (page 871)

(j) OH, CH_3, Br $\xleftarrow[warm]{H_2O, H^+}$ N_2^+, CH_3, Br $\xleftarrow{HONO}$ $\xleftarrow[H^+]{Fe}$ NO_2, CH_3, Br $\xleftarrow[Fe]{Br_2}$ NO_2, CH_3 (from $PhCH_3$)

(k) OH, NO_2, NO_2 $\xleftarrow{H^+}$ ONa, NO_2, NO_2 $\xleftarrow[warm]{NaOH}$ Cl, NO_2, NO_2 (as in 3d, above)

(l) OH, $CH(CH_3)_2$ $\xleftarrow{H_2O, H^+}$ N_2^+, $CH(CH_3)_2$ $\xleftarrow{HONO}$ NH_2, $CH(CH_3)_2$ $\xleftarrow[H^+]{Fe}$ NO_2, $CH(CH_3)_2$ $\xleftarrow[H_2SO_4]{HNO_3}$ $CH(CH_3)_2$ (Problem 18d, page 5

(m) OH, Br, Br, $CH(CH_3)_2$ $\xleftarrow{2Br_2(aq)}$ OH, $CH(CH_3)_2$ $\longleftarrow$ as in (l)

(n) 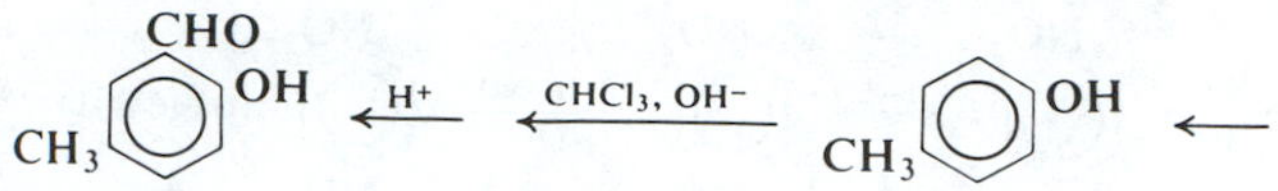 CHO, OH, CH_3 (ring) $\xleftarrow{H^+}$ $\xleftarrow{CHCl_3,\ OH^-}$ OH, CH_3 (ring) $\longleftarrow$ as in (c)

(o) 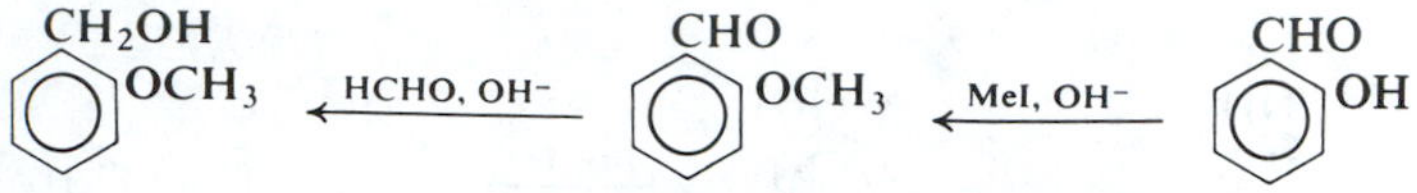 CH_2OH, OCH_3 (ring) $\xleftarrow{HCHO,\ OH^-}$ CHO, OCH_3 (ring) $\xleftarrow{MeI,\ OH^-}$ CHO, OH (ring) (page 908)

5. (a)  ONa, CH_3 (ring) (b) no reaction (c) no reaction (d)  OCH_3, CH_3 (ring)

(e)  OCH_2Ph, CH_3 (ring) (f) no reaction (g) 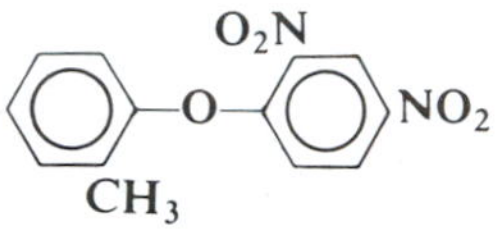 CH_3-ring–O–ring(O_2N)–NO_2 (h) 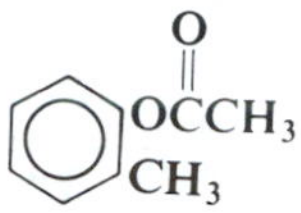 $OCCH_3$ (C=O), CH_3 (ring) (see Sec. 19.16)

(i) 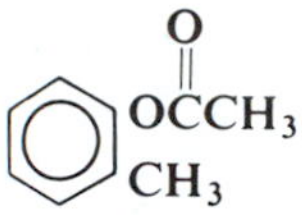 $OCCH_3$ (C=O), CH_3 (ring) (j) 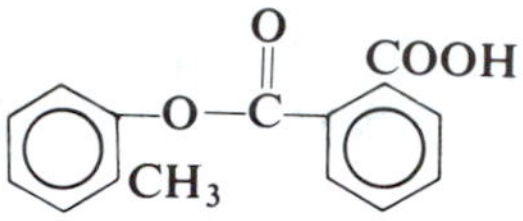 CH_3-ring–O–C(=O)–ring($COOH$) (k) 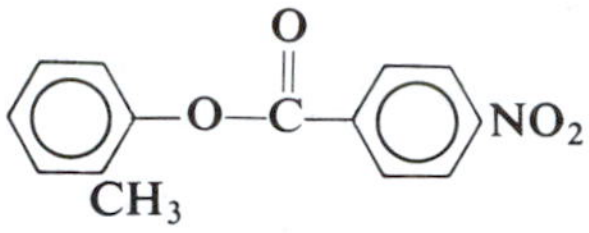 CH_3-ring–O–C(=O)–ring–NO_2

(l)  OSO_2Ph, CH_3 (ring) (m) 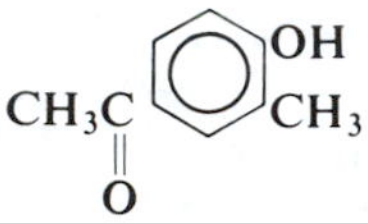 $CH_3C(=O)$, OH, CH_3 (ring) (n) no reaction (o) color produced

(p) 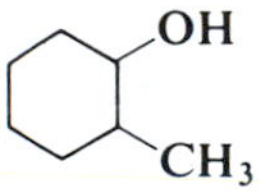 OH, CH_3 (cyclohexane) (q)  O_2N, OH, CH_3 (ring) (r) 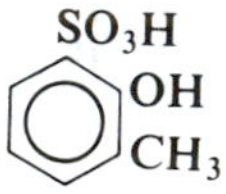 SO_3H, OH, CH_3 (ring) (s)  HO_3S, OH, CH_3 (ring)

(t) 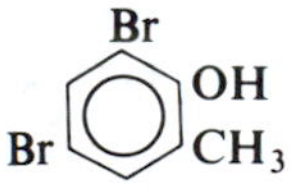 Br, Br, OH, CH_3 (ring) (u) 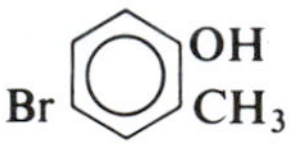 Br, OH, CH_3 (ring) (v)  O=N, OH, CH_3 (ring) (w)  O_2N, OH, CH_3 (ring)

(x) 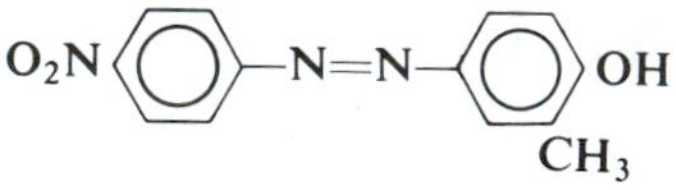 O_2N-ring–N=N–ring(CH_3)–OH (y)  $COOH$, OH, CH_3 (ring) (z) 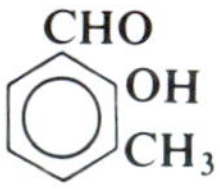 CHO, OH, CH_3 (ring)

6. (c) PhOH + MeBr (p) 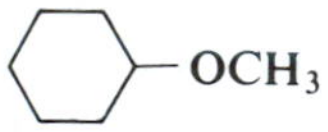 cyclohexyl–OCH_3 (r)  OCH_3, SO_3H (ring) and HO_3S, OCH_3 (ring)

(s) same as (r) (t)  OCH_3, Br (ring) and Br, OCH_3 (ring) (u) same as (t)

All others: no reaction

7. (c) $PhCH_2Br$ (h) $PhCH_2OCCH_3$ (with C=O) (i) same as (h) (j) $C_6H_4(OCH_2Ph)(COOH)$

(k) $PhCH_2O{-}C(=O){-}C_6H_4{-}NO_2$ (l) $PhCH_2O{-}S(=O)_2{-}Ph$ (n) $PhCH_2Cl$

All others: no reaction

8. (a) $PhSO_3H > PhCOOH > PhOH > PhCH_2OH$

(b) $H_2SO_4 > H_2CO_3 > PhOH > H_2O$

(c) *m*-NO_2 > *m*-Br > unsubstd. > *m*-CH_3

(d) 2,4,6- > 2,4- > *p*-

9. Activating effect: $—OH > —OMe > —CH_3 > —H > —Cl > —NO_2$

(a) $PhOH > PhOCH_3 > C_6H_6 > PhCl > PhNO_2$
(b) *m*-HOC_6H_4OMe > *m*-MeC_6H_4OMe > *o*-MeC_6H_4OMe > C_6H_5OMe
(c) *p*-$C_6H_4(OH)_2$ > *p*-$MeOC_6H_4OH$ > *p*-$C_6H_4(OMe)_2$

10. (a) NaOH(aq)

(b) NaOH(aq) to detect phenol. Then CrO_3/H_2SO_4 to distinguish alcohol from ether.

(c) $NaHCO_3$(aq) to detect the acid. Then cold NaOH(aq) to distinguish phenol from ester.

(d) Dilute HCl.

(e) The two acids are soluble in $NaHCO_3$(aq), but only salicylic acid gives a color with $FeCl_3$. Of the remaining compounds, only ethyl salicylate (free phenolic —OH) is soluble in cold NaOH(aq).

(f) $NaHCO_3$(aq) to detect the acid (CO_2 evolved; dissolution). Then NaOH(aq) dissolves the phenol. Then dilute HCl dissolves the amine. The neutral dinitrobenzene is unaffected.

11. (a)

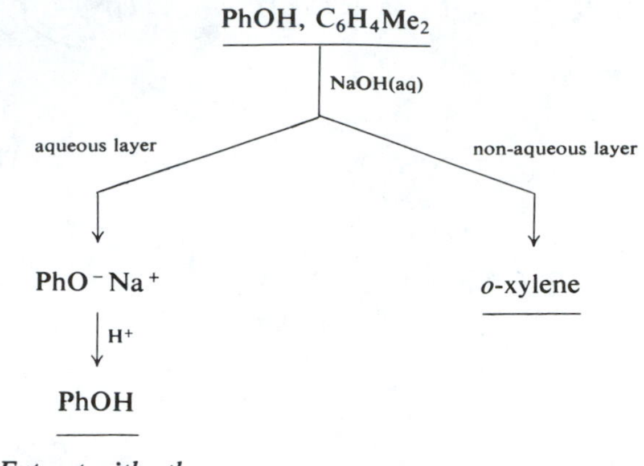

Extract with ether

(c)

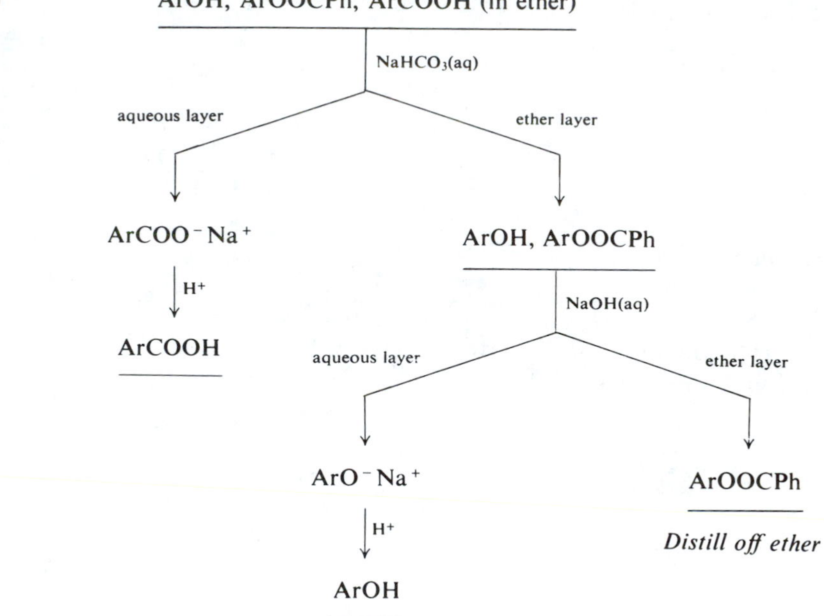

Distill off ether

(d)

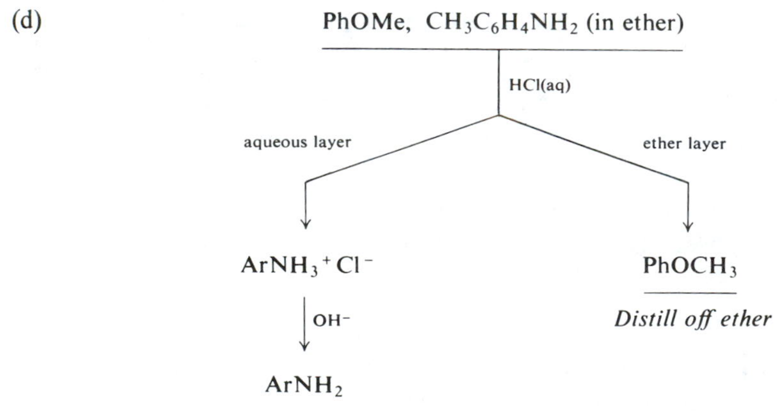

Distill off ether

(f)

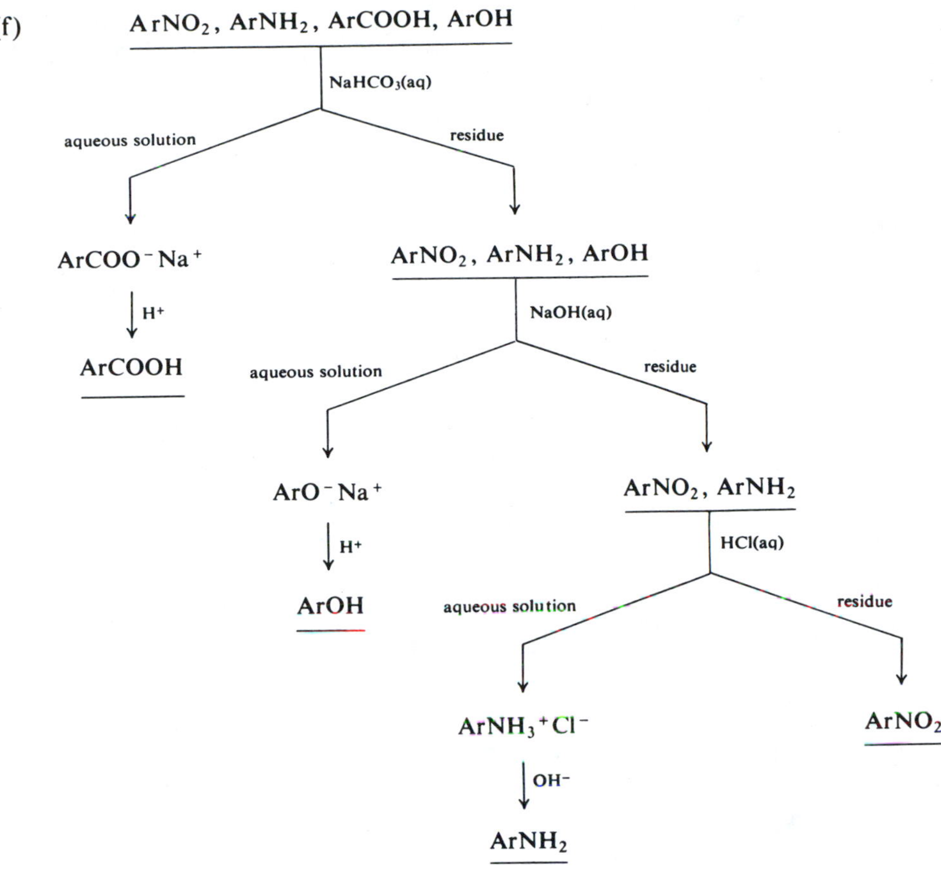

12. Aspirin (o-$OCOCH_3$, $COOH$) $\xleftarrow{Ac_2O,\ H^+}$ Salicylic acid (from phenol) (OH, COOH) $\xrightarrow[\text{warm}]{MeOH,\ H^+}$ Methyl salicylate (OH, $COOCH_3$)

13. (a) 2,4-diaminophenol (OH, NH_2, NH_2) $\xleftarrow{Sn,\ H^+}$ 2,4-dinitrophenol (OH, NO_2, NO_2) $\longleftarrow$ as in 4(k) and 3(d), above

(b) (OCH_3, OCH_3, NH_2) $\xleftarrow{Fe,\ H^+}$ (OCH_3, OCH_3, NO_2) $\xleftarrow{HNO_3,\ H_2SO_4}$ (OCH_3, OCH_3) $\xleftarrow{2Me_2SO_4,\ OH^-}$ (OH, OH)

(c) (OH, NO_2, OH) $\xleftarrow[\text{heat}]{H_2O,\ H^+}$ (HO_3S, OH, NO_2, OH, SO_3H) $\xleftarrow{HNO_3,\ H_2SO_4}$ (HO_3S, OH, OH, SO_3H) $\xleftarrow{2H_2SO_4}$ (OH, OH)

(d) 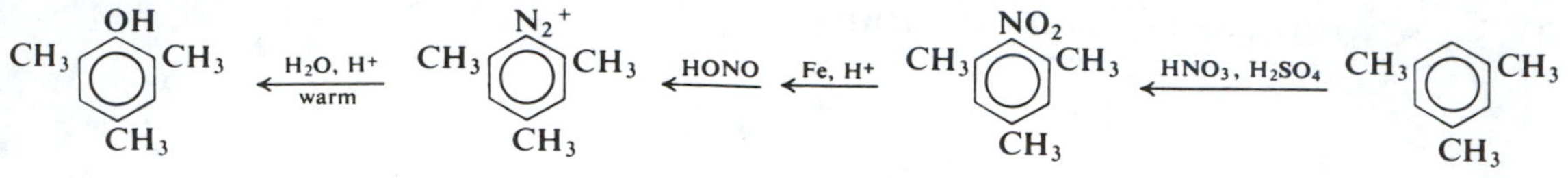 2,4,6-trimethylphenol (CH_3, CH_3, CH_3, OH) $\xleftarrow[\text{warm}]{H_2O,\ H^+}$ N_2^+ (2,4,6-CH_3) $\xleftarrow{HONO}$ $\xleftarrow{Fe,\ H^+}$ NO_2 (2,4,6-CH_3) $\xleftarrow{HNO_3,\ H_2SO_4}$ Mesitylene (CH_3, CH_3, CH_3)

Mesitylene

(e) 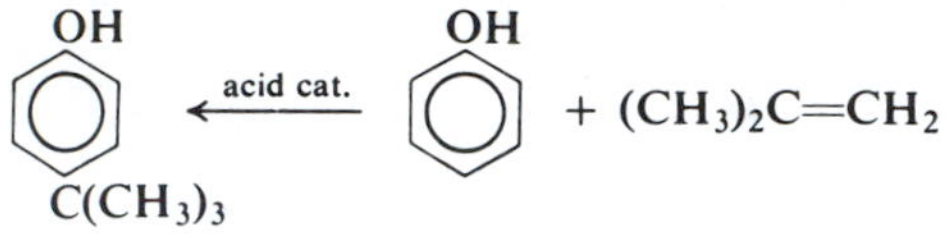 p-$C(CH_3)_3$-phenol (OH, $C(CH_3)_3$) $\xleftarrow{\text{acid cat.}}$ phenol (OH) + $(CH_3)_2C{=}CH_2$

(f) 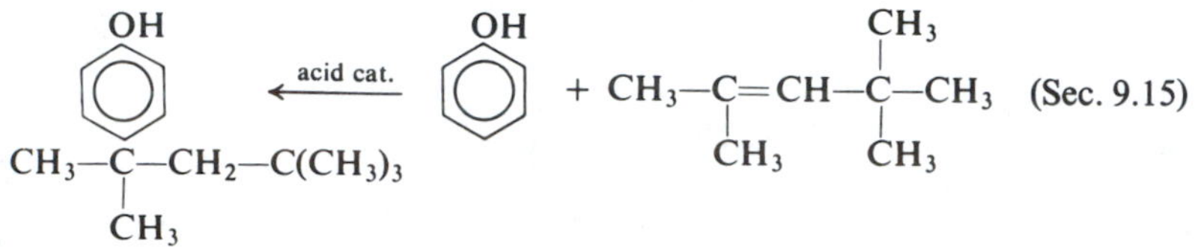 OH, $CH_3-\underset{CH_3}{C}-CH_2-C(CH_3)_3$ $\xleftarrow{\text{acid cat.}}$ phenol (OH) + $CH_3-\underset{CH_3}{C}{=}CH-\overset{CH_3}{\underset{CH_3}{C}}-CH_3$ (Sec. 9.15)

(g) $BrCH_2CH_2OPh \longleftarrow BrCH_2CH_2Br + NaOPh \xleftarrow{NaOH(aq)} PhOH$

(h) $CH_2{=}CH{-}O{-}Ph \xleftarrow[\text{heat}]{KOH} BrCH_2CH_2OPh \longleftarrow$ (g)

(i) $CH_2{=}CH{-}O{-}Ph \xrightarrow{H^+} \underset{H}{CH_2}{-}CH{=}\overset{+}{O}{-}Ph \xrightarrow[-H^+]{H_2O} \left[CH_3{-}\underset{OH}{CH}{-}O{-}Ph\right] \longrightarrow CH_3CHO + PhOH$

A hemiacetal

(j) 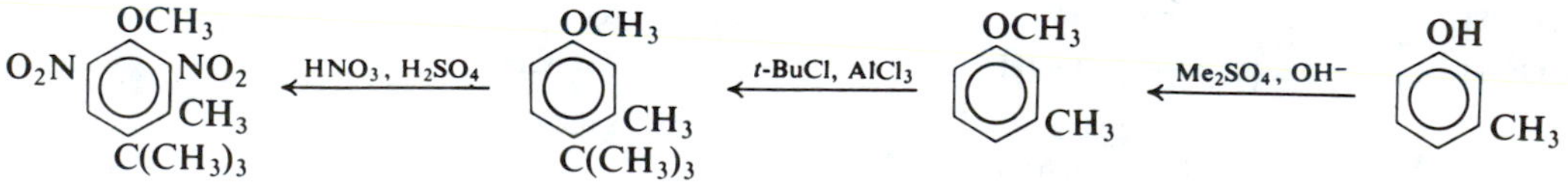 O_2N, NO_2, OCH_3, CH_3, $C(CH_3)_3$ benzene $\xleftarrow{HNO_3,\ H_2SO_4}$ OCH_3, CH_3, $C(CH_3)_3$ benzene $\xleftarrow{t\text{-BuCl, AlCl}_3}$ OCH_3, CH_3 benzene $\xleftarrow{Me_2SO_4,\ OH^-}$ OH, CH_3 benzene

(k) 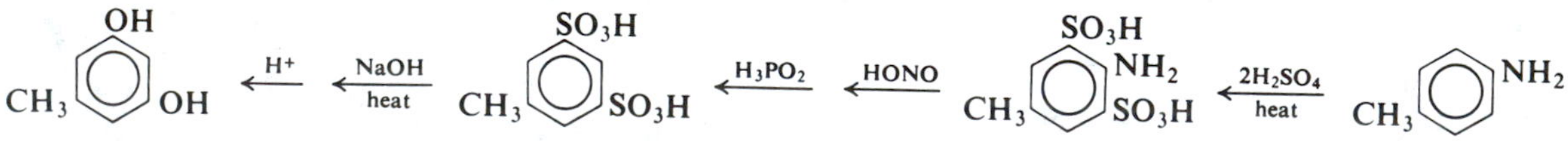 CH_3, OH, OH benzene $\xleftarrow{H^+}$ $\xleftarrow[\text{heat}]{NaOH}$ CH_3, SO_3H, SO_3H benzene $\xleftarrow{H_3PO_2}$ $\xleftarrow{HONO}$ CH_3, SO_3H, NH_2, SO_3H benzene $\xleftarrow[\text{heat}]{2H_2SO_4}$ CH_3, NH_2 benzene

(page 871)

14. (a) 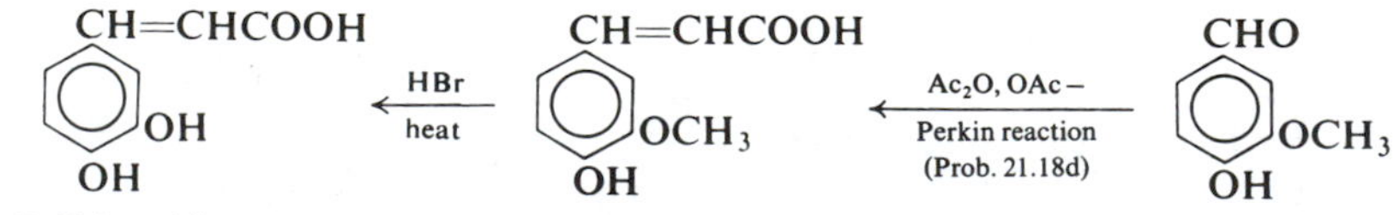 Caffeic acid ($CH{=}CHCOOH$, OH, OH) $\xleftarrow[\text{heat}]{HBr}$ $CH{=}CHCOOH$, OCH_3, OH benzene $\xleftarrow[\text{Perkin reaction (Prob. 21.18d)}]{Ac_2O,\ OAc^-}$ Vanillin (CHO, OCH_3, OH)

Caffeic acid

Vanillin

(b) Tyramine (p-$HOC_6H_4CH_2CH_2NH_2$) $\xleftarrow{HI}$ p-$CH_3OC_6H_4CH_2CH_2NH_2$ $\xleftarrow{H_2,\ Ni}$ p-$CH_3OC_6H_4CH{=}CHNO_2$ $\xleftarrow[\text{(Prob. 21.18a)}]{CH_3NO_2,\ OH^-}$ p-$CH_3OC_6H_4CHO$ $\xleftarrow{H_2SO_4,\ K_2Cr_2O_7}$ p-$CH_3OC_6H_4CH{=}CHCH_3$ (Anethole)

(c) Noradrenaline ($3,4$-$(HO)_2C_6H_3CHOHCH_2NH_2$) $\xleftarrow{H_2,\ Ni}$ $3,4$-$(HO)_2C_6H_3CHOHC{\equiv}N$ $\xleftarrow{CN^-,\ H^+}$ $3,4$-$(HO)_2C_6H_3CHO$ $\xleftarrow[\text{heat}]{HBr}$ Vanillin (4-HO-3-$CH_3OC_6H_3CHO$)

15. (a) First, a hydroperoxide is formed, probably by a mechanism like this:

$$Ar_3C{-}OH \underset{}{\overset{H^+}{\rightleftharpoons}} Ar_3C{-}OH_2^+ \xrightarrow{-H_2O} Ar_3C^{\oplus} \xrightarrow{H_2O_2} Ar_3C{-}\overset{H}{\overset{|}{\overset{\oplus}{O}}}{-}OH \xrightarrow{-H^+} \underset{\text{A hydroperoxide}}{Ar_3C{-}O{-}OH}$$

This then undergoes rearrangement as discussed in Sec. 24.5.

$$\underset{\textit{Hydroperoxide}}{Ar_3C{-}O{-}OH} \overset{H^+}{\rightleftharpoons} Ar_2\overset{Ar}{C}{-}O{-}OH_2^+ \longrightarrow Ar_2C{=}\overset{+}{O}{-}Ar \xrightarrow[-H^+]{H_2O} \underset{\textit{Hemiketal}}{Ar_2\overset{OH}{C}{-}OAr} \longrightarrow \underset{\textit{Ketone}}{Ar_2C{=}O} + \underset{\textit{Phenol}}{ArOH}$$

(b) The relative migratory aptitude (Sec. 24.6) of the aryl groups involved is

$$p\text{-}CH_3OC_6H_4{-} > Ph{-} > p\text{-}ClC_6H_4{-}$$

The expected major products are

$$C_6H_5{-}\underset{O}{\underset{\|}{C}}{-}C_6H_5 + p\text{-}CH_3OC_6H_4OH \quad \text{and} \quad p\text{-}ClC_6H_4{-}\underset{O}{\underset{\|}{C}}{-}C_6H_5 + C_6H_5OH$$

16.

Citral (CHO) $\xrightarrow[(1)]{H^+}$ $\overset{\oplus}{C}HOH$ $\xrightarrow[(2)]{\text{olivetol}}$ (H, OH, HO, HO, C_5H_{11}) $\xrightarrow[(3)]{H^+,\ -H_2O}$ (H, HO, $C^{\oplus}$, HO, C_5H_{11}) $\xrightarrow{(4)}$ (HO, $\oplus$, HO, C_5H_{11}) $\xrightarrow{(5)}$ (HO, $\oplus$OH, C_5H_{11}) $\xrightarrow[(6)]{-H^+}$ I (HO, O, C_5H_{11})

(1) Protonation of an aldehyde. (2) Electrophilic substitution, with a protonated aldehyde as the electrophile. (3) Protonation of an —OH and loss of H_2O to form an allylic cation. (4) Electrophilic addition of an allylic cation to a double bond to form a 3° cation and generate a second ring. (5) Combination of a 3° cation with an HO— of a phenol to form a protonated ether and generate a third ring. (6) Loss of a proton to yield I.

(Taylor, E. C.; Lenard, K.; Shvo, Y. "Active Constituents of Hashish. Synthesis of *dl*-Δ^6-3,4-*trans*-Tetrahydrocannabinol"; *J. Am. Chem. Soc.* **1966**, *88*, 367.)

17. Strong acid converts phloroglucinol (R = H) and its ethers (R = Me or Et) reversibly into protonated species, hybrids of this structure

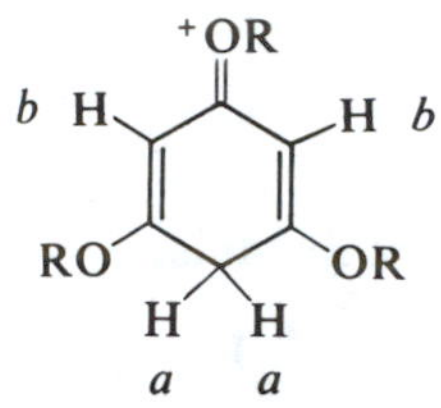

$^+$OR; *b* H, H *b*; RO, OR; H H; *a a*

R = H, Me, or Et

and corresponding structures with the charge on the other oxygens. These species are the familiar benzenonium ions (*sigma* complexes) that are intermediates in electrophilic aromatic substitution—this time with two protons at the site of attack (see Sec. 15.8; Problem 15.10, page 535; and Problem 4, Chapter 17).

The spectrum is due to protons *a* (δ 4.15) and *b* (δ 6.12), allylic and vinylic protons deshielded by the positive charge. On dilution with water, a proton is lost, and the original compounds are regenerated.

In D_2SO_4, all ring protons are gradually replaced by deuterons in electrophilic aromatic substitutions via similar benzenonium ions. From 1,3,5-$C_6H_3(OCH_3)_3$ in D_2SO_4, we would expect to isolate 1,3,5-$C_6D_3(OCH_3)_3$.

18. Migration of ring carbon occurs.

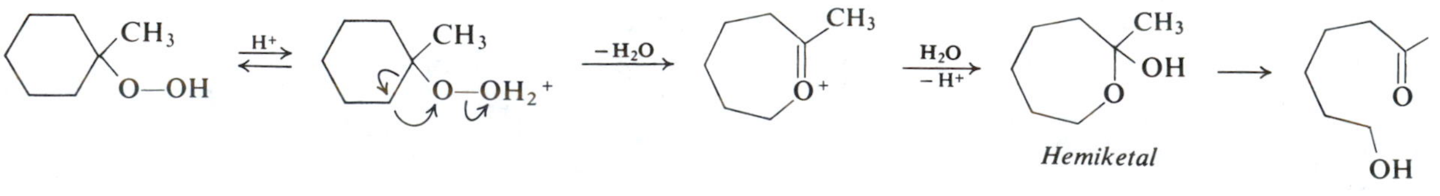

CH_3, O—OH $\xrightleftharpoons{H^+}$ CH_3, O—OH_2^+ $\xrightarrow{-H_2O}$ CH_3, O^+ $\xrightarrow[-H^+]{H_2O}$ CH_3, OH, O → O, OH

Hemiketal

or

$HOCH_2CH_2CH_2CH_2CH_2\overset{\displaystyle O}{\overset{\|}{C}}CH_3$

Alcohol–ketone

7-Hydroxy-2-heptanone

19. (a) $O_2N\text{-}C_6H_4\text{-}OH \xrightarrow{OH^-} O_2N\text{-}C_6H_4\text{-}O^- \xrightarrow{EtBr} O_2N\text{-}C_6H_4\text{-}OEt$ (A)

$O_2N\text{-}C_6H_4\text{-}OEt$ (A) $\xrightarrow{Sn,\ HCl} H_2N\text{-}C_6H_4\text{-}OEt$ (B) $\xrightarrow{HONO} {}^+N_2\text{-}C_6H_4\text{-}OEt \xrightarrow{PhOH} HO\text{-}C_6H_4\text{-}N{=}N\text{-}C_6H_4\text{-}OEt$ (C)

$HO\text{-}C_6H_4\text{-}N{=}N\text{-}C_6H_4\text{-}OEt$ (C) $\xrightarrow{OH^-} {}^-O\text{-}C_6H_4\text{-}N{=}N\text{-}C_6H_4\text{-}OEt \xrightarrow{Et_2SO_4} EtO\text{-}C_6H_4\text{-}N{=}N\text{-}C_6H_4\text{-}OEt$ (D)

$EtO\text{-}C_6H_4\text{-}N{=}N\text{-}C_6H_4\text{-}OEt$ (D) $\xrightarrow{SnCl_2} 2EtO\text{-}C_6H_4\text{-}NH_2$ (E) $\xrightarrow{CH_3COCl} EtO\text{-}C_6H_4\text{-}NH\text{-}C(=O)\text{-}CH_3$

Phenacetin

(b) o-$HO\text{-}C_6H_4\text{-}CH_2CH_2OH \xrightarrow{HBr}$ o-$HO\text{-}C_6H_4\text{-}CH_2CH_2Br$ (F) $\xrightarrow{OH^-}$ o-${}^-O\text{-}C_6H_4\text{-}CH_2CH_2Br \xrightarrow{-Br^-}$ Coumarane

F

Coumarane
No phenolic —OH;
NaOH-insoluble

(c) $C_6H_5OH \xrightarrow{OH^-} C_6H_5O^- \xrightarrow{ClCH_2COO^-} C_6H_5\text{-}O\text{-}CH_2\text{-}COO^- \xrightarrow{H^+} C_6H_5\text{-}O\text{-}CH_2\text{-}COOH$ (G)

$C_6H_5\text{-}O\text{-}CH_2\text{-}COOH$ (G) $\xrightarrow{SOCl_2} C_6H_5\text{-}O\text{-}CH_2\text{-}C(=O)Cl$ (H) $\xrightarrow{AlCl_3}$ 3-Cumaranone

(d) p-Cymene $\xrightarrow{H_2SO_4}$ [SO_3H ortho to CH_3] + [SO_3H ortho to isopropyl] $\xrightarrow[heat]{KOH} \xrightarrow{H^+}$ [OH ortho to CH_3] + [OH ortho to isopropyl]

p-Cymene

I and J
Assignment uncertain

Carvacrol and thymol
Assignment uncertain

I $\xrightarrow[oxidn.]{HNO_3}$ [CH_3, SO_3H, COOH benzene] *or* [CH_3, SO_3H, COOH benzene]

K
Assignment uncertain

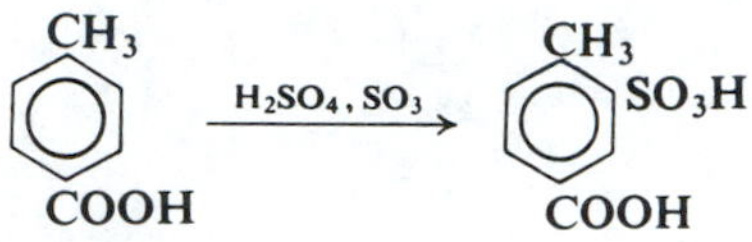

$$p\text{-}CH_3C_6H_4COOH \xrightarrow{H_2SO_4,\ SO_3} \text{K (}CH_3, SO_3H, COOH\text{)}$$

K

Therefore:

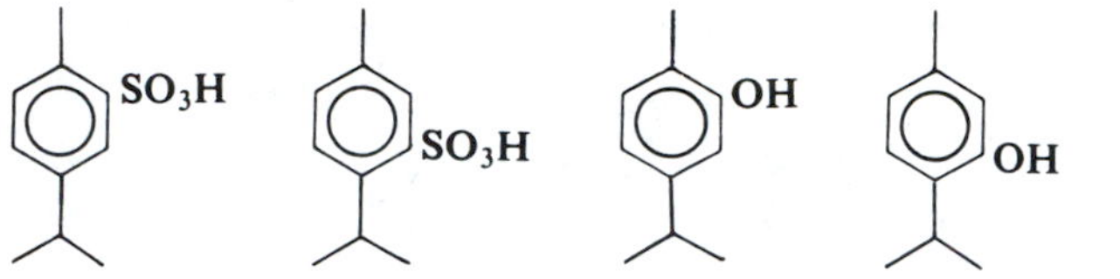

I J Carvacrol Thymol

(e) $CH_3OC_6H_4CH{=}CHCH_3$ (Anethole) $\xrightarrow{HBr}$ $CH_3OC_6H_4CHBrCH_2CH_3$ (L) $\xrightarrow{Mg}$ $CH_3OC_6H_4CH(Et)-CH(Et)C_6H_4OCH_3$ (M) $\xrightarrow{HBr,\ heat}$ $HOC_6H_4CH(Et)-CH(Et)C_6H_4OH$ (Hexestrol)

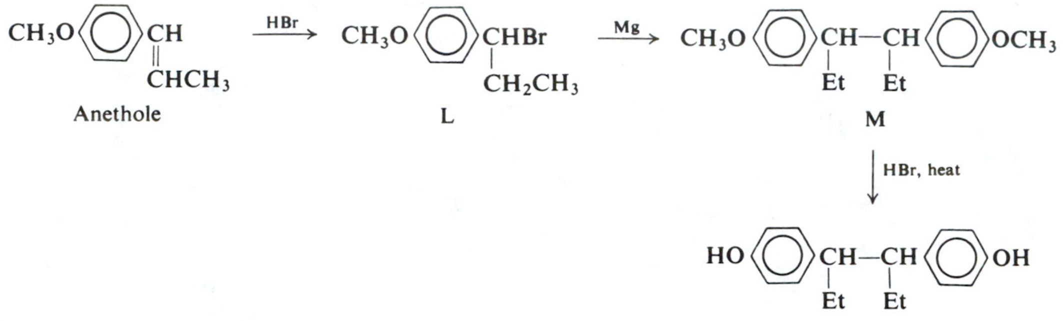

Anethole L M Hexestrol

20. (a) $C_6H_4(OH)_2 \xrightarrow{OH^-} C_6H_4(O^-)(OH)$ (III) $\xrightarrow{(ClCH_2CH_2)_2O} C_6H_4(OCH_2CH_2OCH_2CH_2Cl)(OH)$ (IV)

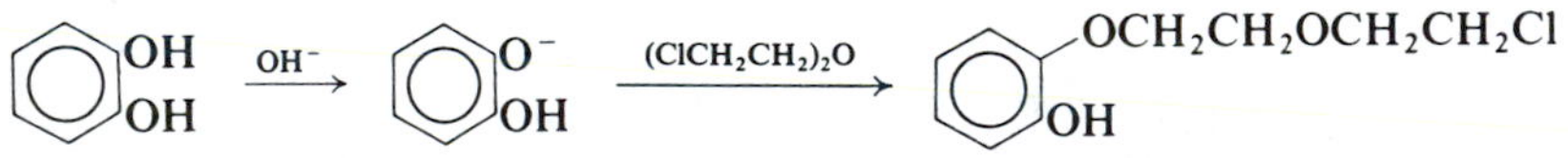

III IV

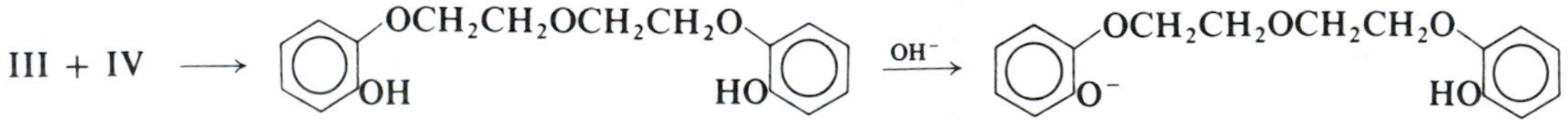

III + IV $\longrightarrow$ $HOC_6H_4OCH_2CH_2OCH_2CH_2OC_6H_4OH$ $\xrightarrow{OH^-}$ $^-OC_6H_4OCH_2CH_2OCH_2CH_2OC_6H_4OH$ (V)

V

V $\xrightarrow{(ClCH_2CH_2)_2O}$ $HOC_6H_4OCH_2CH_2OCH_2CH_2OC_6H_4OCH_2CH_2OCH_2CH_2Cl$ (VI)

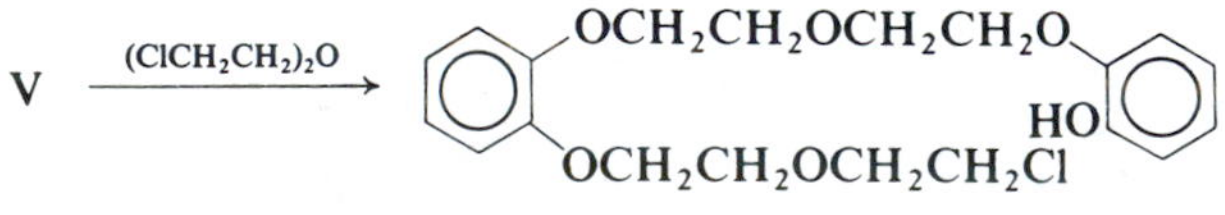

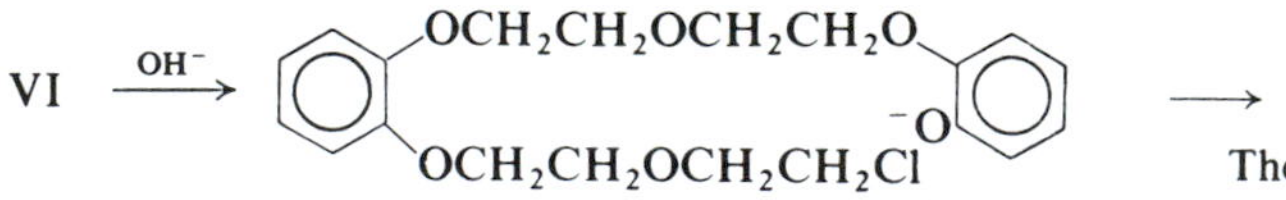

VI

VI $\xrightarrow{OH^-}$ $^-OC_6H_4OCH_2CH_2OCH_2CH_2OC_6H_4OCH_2CH_2OCH_2CH_2Cl$ (VII) $\longrightarrow$ II

VII

The crown ether

(b) A phenol would, of course, be soluble in aqueous NaOH.

(c) A water-soluble complex between Na^+ and the crown ether forms.

(d) Ordinarily, we would expect that collision—and reaction—between the two ends of a molecule like VII would be highly unlikely. Instead, we would expect VII to react with a new molecule of the chloroether to lengthen the chain.

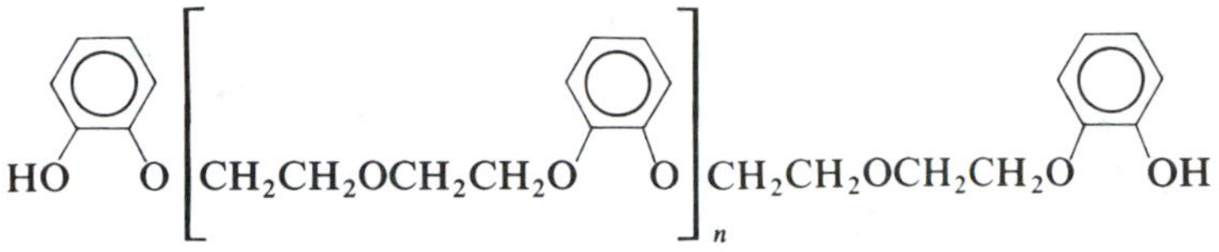

HO O [$CH_2CH_2OCH_2CH_2O$ O]$_n$ $CH_2CH_2OCH_2CH_2O$ OH

(e) The Na^+ or K^+ ion acts as a template for the formation of the crown ether. As the ether linkages are formed, the molecule wraps itself *about* the cation; the two ends of VII are held in just the right position for ring-closing to occur.

This is, actually, a symphoric effect; the reacting atoms are brought together and held in the proper spatial relationship by the metal cation. The metal cation thus exerts its effect in fundamentally the same way as a transition metal or an enzyme (see page 1046).

21. (a) There are three possible shifts to the electron-deficient oxygen.

(i) Migration of hydrogen:

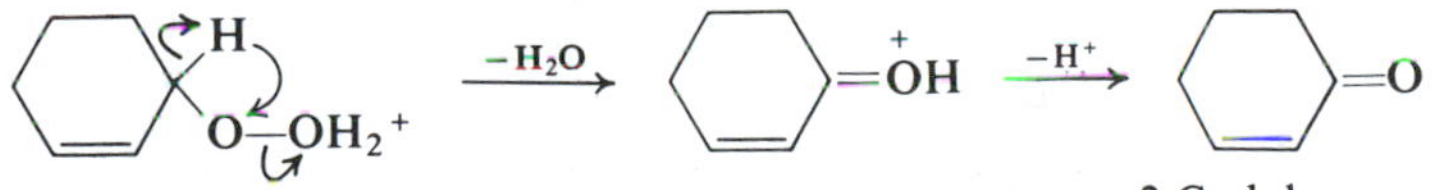

H, O—OH_2^+ $\xrightarrow{-H_2O}$ =$\overset{+}{O}H$ $\xrightarrow{-H^+}$ =O

2-Cyclohexenone

(ii) Migration of an alkyl group, —CH_2:

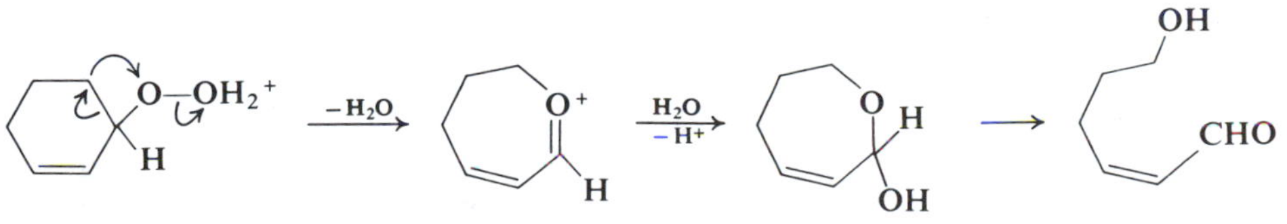

O—OH_2^+, H $\xrightarrow{-H_2O}$ O^+, H $\xrightarrow[-H^+]{H_2O}$ O, H, OH $\longrightarrow$ OH, CHO

Hemiacetal

or

$HOCH_2CH_2CH_2CH{=}CHCHO$

Alcohol–aldehyde

6-Hydroxy-2-hexenal

(iii) Migration of a vinyl group, —CH=CH:

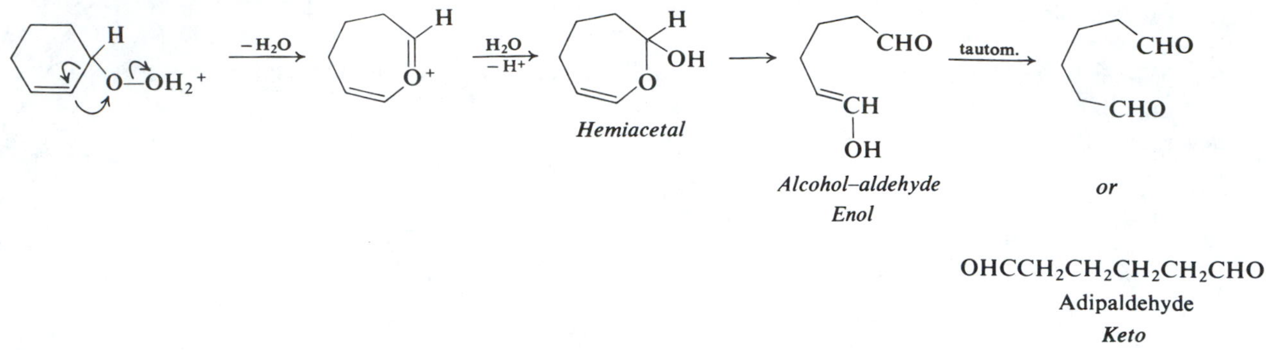
H
O—OH2+
– H2O
H
O+
H2O
– H+
H
OH
O
Hemiacetal
CHO
CH
OH
Alcohol–aldehyde
Enol
tautom.
CHO
CHO
or
OHCCH2CH2CH2CH2CHO
Adipaldehyde
Keto

Of these only the last gives adipaldehyde. Furthermore, we realize that an intramolecular (acid-catalyzed) aldol condensation of adipaldehyde can give cyclopentene-1-carboxaldehyde. (In an acid-catalyzed aldol condensation, we remember (Problem 16, page 818), acid performs two functions: it catalyzes the enolization of one molecule (the nucleophile); and it protonates the other carbonyl group and makes it more electrophilic.)

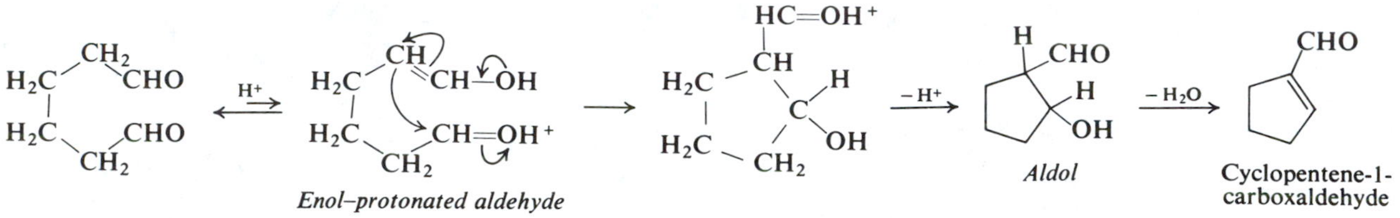
H2C CH2 CHO
H2C CH2 CHO
H+
H2C CH CH—OH
H2C CH2 CH=OH+
Enol–protonated aldehyde
HC=OH+
CH
H2C
H2C
C
H
OH
CH2
– H+
H
CHO
H
OH
Aldol
– H2O
CHO
Cyclopentene-1-
carboxaldehyde

(b) Evidently a vinyl group migrates faster than hydrogen or alkyl. We can see why this might be so: like aryl, vinyl has extra π electrons available for bonding. Migration of vinyl need not involve a transition state containing pentavalent carbon (IV, page 897), but instead can proceed via an epoxide intermediate.

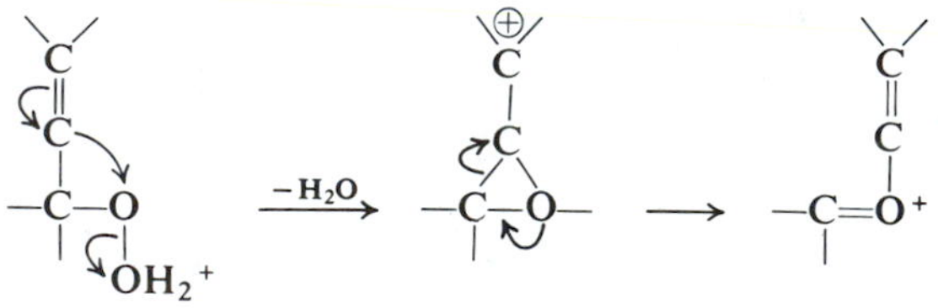
– H2O
OH2+
C=O+

22. 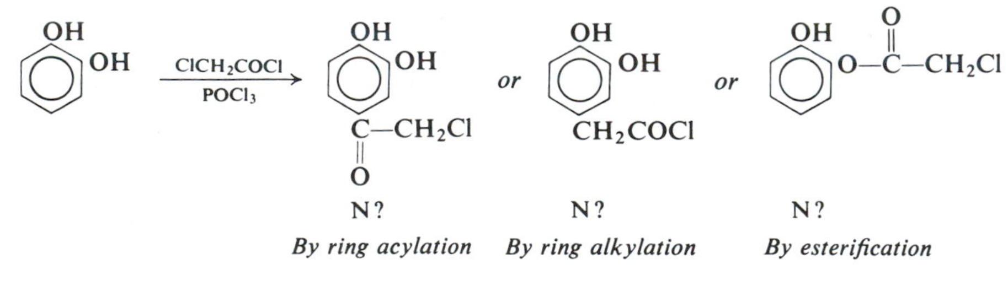
OH
OH
ClCH2COCl
POCl3
OH
OH
C—CH2Cl
O
or
OH
OH
CH2COCl
or
OH
O
O—C—CH2Cl
N?
N?
N?
By ring acylation
By ring alkylation
By esterification

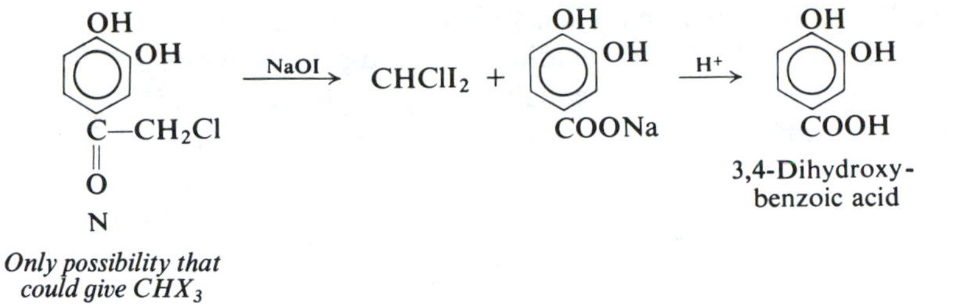

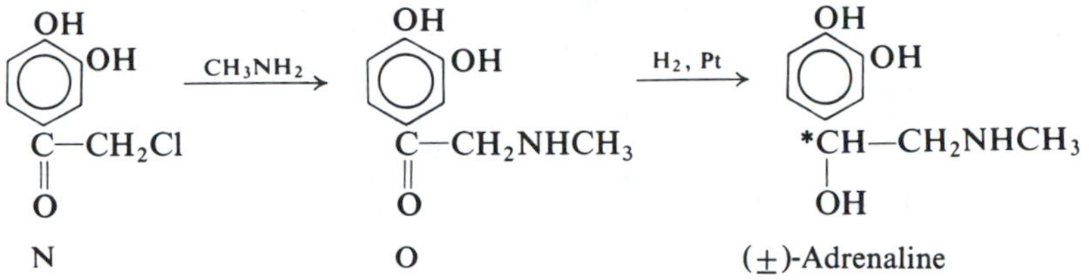

OH OH C—CH_2Cl ‖ O N —NaOI→ $CHCl I_2$ + OH OH COONa —H^+→ OH OH COOH

3,4-Dihydroxy-benzoic acid

Only possibility that could give CHX_3

OH OH C—CH_2Cl ‖ O N —CH_3NH_2→ OH OH C—CH_2NHCH_3 ‖ O O —H_2, Pt→ OH OH *CH—CH_2NHCH_3 | OH

(±)-Adrenaline

23. (a) Phellandral is an aldehyde (Tollens' test), oxidizable to an acid that contains one reducible unsaturation, undoubtedly a carbon–carbon double bond. With these facts in mind, let us do some chemical arithmetic on the saturated acid.

$C_{10}H_{18}O_2$	C_9H_{19} if R were open-chain
$-COOH$	$-C_9H_{17}$ R
C_9H_{17} *means* R	2H missing *means* 1 ring

We now know that phellandral contains one ring, one double bond, and a —CHO group. Synthesis by an unambiguous route gives us its structure.

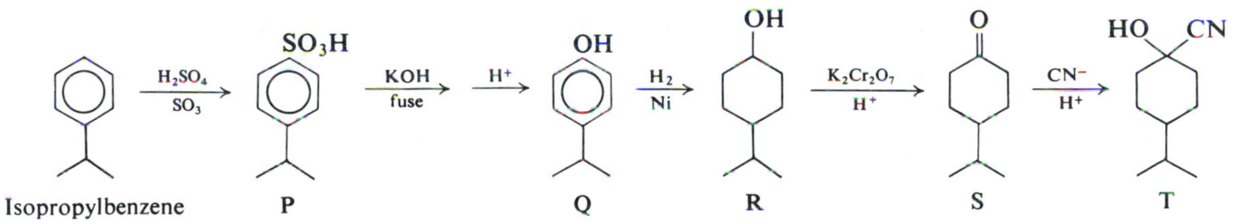

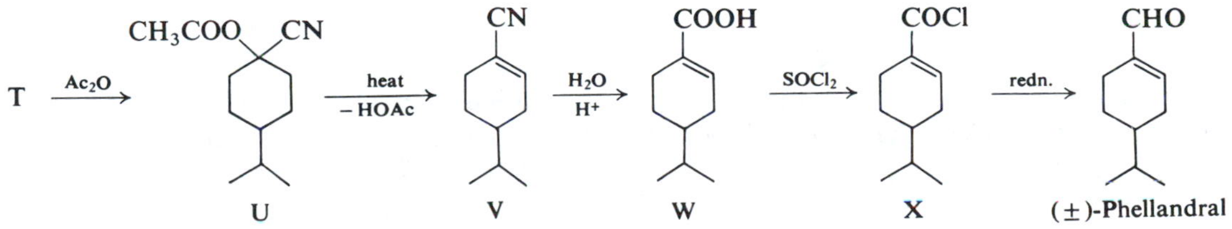

Isopropylbenzene —H_2SO_4, SO_3→ P (SO_3H) —KOH fuse→ —H^+→ Q (OH) —H_2 Ni→ R (OH) —$K_2Cr_2O_7$ H^+→ S (O) —CN^- H^+→ T (HO CN)

T —Ac_2O→ U (CH_3COO CN) —heat − HOAc→ V (CN) —H_2O H^+→ W (COOH) —$SOCl_2$→ X (COCl) —redn.→ (±)-Phellandral (CHO)

(b) Synthetic phellandral is the racemic modification:

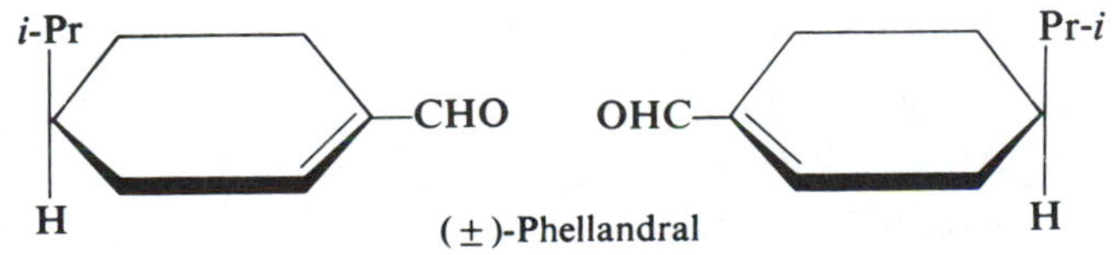

i-Pr, H, —CHO OHC—, Pr-*i*, H

(±)-Phellandral

The chirality that persists into phellandral is generated when U is converted into V: the double bond can form in either direction from the carbon holding —CN, to give either of a pair of enantiomers. Elimination in either direction is equally likely, and the racemic modification is obtained. From this stage onward, there are no stereochemical changes.

i-Pr —CN *i*-Pr —CN

Enantiomers
Formed in equal amounts

(c) Compare Problem 13.12(e) and 13.12(f), page 471.

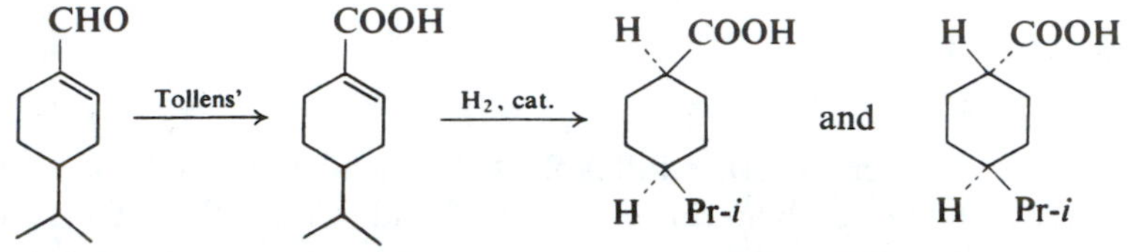

CHO —Tollens'→ COOH —H_2, cat.→ H COOH H Pr-*i* and H COOH H Pr-*i*

(−)-Phellandral (−)-Phellandric acid Dihydrophellandric acid
Diastereomers: each is achiral

24. Y is *m*-cresol, the only cresol that can give a tribromo compound.

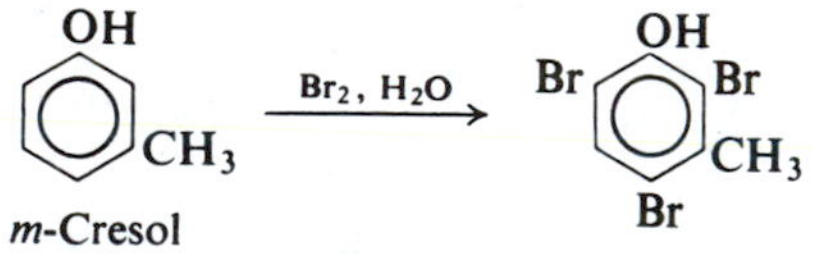

OH CH_3 —Br_2, H_2O→ OH Br Br CH_3 Br

m-Cresol

25. (a) Z and AA are neutral: oxygen is —OH, —OR, or —C=O. They are unsaturated. Vigorous oxidation to anisic acid shows the presence (and retention) of an ether linkage, but loss of two carbons from a side chain. At this point, we can say that the following structural elements are present:

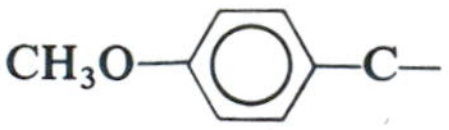

CH_3O—C_6H_4—C— + two more C's on side chain

Chemical arithmetic

$C_{10}H_{12}O$	Z and AA
$-C_7H_7O$	$CH_3OC_6H_4$
C_3H_5	side chain

shows the side chain to be C_3H_5; the chemical tests have shown that this is unsaturated, not a cyclopropane ring.

Possible structures are, then:

cis- or *trans*-CH_3O⟨◯⟩$CH{=}CHCH_3$ CH_3O⟨◯⟩$CH_2CH{=}CH_2$ CH_3O⟨◯⟩$C(CH_3){=}CH_2$

(b) Hydrogenation to the same compound, $C_{10}H_{14}O$, shows that the carbon skeleton of the unsaturated side chain is the same in Z as in AA. This rules out the branched chain structure, and we are now down to three possibilities:

cis- or *trans*-CH_3O⟨◯⟩$CH{=}CHCH_3$ CH_3O⟨◯⟩$CH_2CH{=}CH_2$

(c) Ozonolysis; or isomerization of one into the other by strong heating with KOH.

(d) This synthesis involves the coupling of a Grignard reagent with an allyl halide.

CH_3O⟨◯⟩Br $\xrightarrow{Mg}$ CH_3O⟨◯⟩MgBr + $BrCH_2CH{=}CH_2$ ⟶ CH_3O⟨◯⟩$CH_2CH{=}CH_2$ + $MgBr_2$

p-Bromoanisole   Allyl bromide   *p*-Allylanisole Z

(e) Heating with strong base converts the allylbenzene Z into the more stable AA, with the double bond conjugated with the ring. Of the two geometric isomers, we would expect to get the more stable *trans* isomer.

p-$CH_3OC_6H_4CH_2CH{=}CH_2$ $\underset{+H^+}{\overset{-H^+}{\rightleftharpoons}}$ *p*-$CH_3OC_6H_4\overbrace{CH{\cdots}CH{\cdots}CH_2}^{\ominus}$ $\underset{-H^+}{\overset{+H^+}{\rightleftharpoons}}$ *p*-$CH_3OC_6H_4(H)C{=}C(H)CH_3$ (*cis*-*p*-Propenylanisole)

$\downarrow\uparrow$ $+H^+$ / $-H^+$

p-$CH_3OC_6H_4(H)C{=}C(CH_3)H$

trans-*p*-Propenylanisole

AA

Most stable isomer

(f) CH_3O⟨◯⟩$CH{=}CHCH_3$ (*trans*) $\xleftarrow{Na,\ NH_3}$ CH_3O⟨◯⟩$C{\equiv}CCH_3$ $\xleftarrow{CH_3I}$ $\xleftarrow{NaNH_2}$ CH_3O⟨◯⟩$C{\equiv}CH$

CH_3O⟨◯⟩$C{\equiv}CH$ $\xleftarrow{NaNH_2}$ $\xleftarrow{KOH(alc)}$ CH_3O⟨◯⟩$CH(Br){-}CH_2(Br)$ $\xleftarrow{Br_2}$ CH_3O⟨◯⟩$CH{=}CH_2$

CH_3O⟨◯⟩$CH{=}CH_2$ $\underset{-H_2O}{\xleftarrow{H^+}}$ CH_3O⟨◯⟩$CH(OH)CH_3$ $\xleftarrow{CH_3CHO}$ CH_3O⟨◯⟩MgBr $\xleftarrow{Mg}$ CH_3O⟨◯⟩Br

26. BB is evidently a phenol; this accounts for (at least) one oxygen. It also appears to be an ester (accounting for the other two oxygens), slowly undergoing hydrolysis to an alcohol that gives a positive haloform test, that is, hydrolysis to $RCH(OH)CH_3$.

$$\begin{array}{ll} C_{10}H_{12}O_3 & \text{BB} \\ -OH & \text{phenolic} \\ \hline C_{10}H_{11}O_2 & \end{array} \qquad \begin{array}{ll} C_{10}H_{11}O_2 & \\ -C_6H_4 & \text{aromatic ring} \\ \hline C_4H_7O_2 & \end{array} \qquad \begin{array}{ll} C_4H_7O_2 & \\ -CO_2 & \text{ester} \\ \hline C_3H_7 & \end{array}$$

A residue of C_3H_7 shows that the alcohol is *iso*-PrOH.

$$\underset{\text{BB}}{C_6H_4\begin{cases}OH \\ COOPr\text{-}i\end{cases}} \xrightarrow[\text{heat}]{OH^-} \underset{\substack{\downarrow OI^- \\ CHI_3}}{i\text{-PrOH}} + C_6H_4\begin{cases}O^- \\ COO^-\end{cases} \xrightarrow{H^+} \underset{\text{CC}}{C_6H_4\begin{cases}OH \\ COOH\end{cases}}$$

The only hydroxybenzoic acid that is steam-volatile is the *ortho* isomer, salicylic acid; CC must be this, and BB must be isopropyl salicylate.

27. Its solubility behavior shows that chavibetol is a phenol.

(a) Methylation with $Me_2SO_4/NaOH$ introduces one methyl,

$$\begin{array}{ll} C_{11}H_{14}O_2 & \text{DD} \\ -C_{10}H_{12}O_2 & \text{chavibetol} \\ \hline CH_2 & \end{array}$$

indicating that there is one phenolic group in chavibetol.

(b) Cleavage with hot HI to give CH_3I shows that chavibetol is a methyl ether, accounting for the second oxygen. Glancing ahead to the conversion into vanillin, we tentatively

$$\begin{array}{ll} C_{10}H_{12}O_2 & \text{chavibetol} \\ -OH & \text{phenolic} \\ \hline C_{10}H_{11}O & \end{array} \qquad \begin{array}{ll} C_{10}H_{11}O & \\ -OCH_3 & \text{ether} \\ \hline C_9H_8 & \end{array} \qquad \begin{array}{ll} C_9H_8 & \\ -C_6H_3 & \text{trisubstituted aromatic} \\ \hline C_3H_5 & \end{array}$$

assume that, like vanillin, chavibetol is a trisubstituted aromatic, and that the ether oxygen is attached to the ring.

The residue of C_3H_5 from our chemical arithmetic corresponds to an unsaturated side chain (compare Problem 25, above).

(c) Treatment of chavibetol with strong, hot base brings about isomerization to EE; we can rationalize this conversion as

$$\underset{\substack{\text{Chavibetol} \\ C_{10}H_{12}O_2}}{\begin{matrix}HO \\ CH_3O\end{matrix}\!\!>\!C_6H_3{-}CH_2{-}CH{=}CH_2} \xrightarrow[\text{heat}]{OH^-} \underset{\substack{\text{EE} \\ C_{10}H_{12}O_2}}{\begin{matrix}HO \\ CH_3O\end{matrix}\!\!>\!C_6H_3CH{=}CHCH_3}$$

The same isomerization undoubtedly occurs when DD is similarly treated.

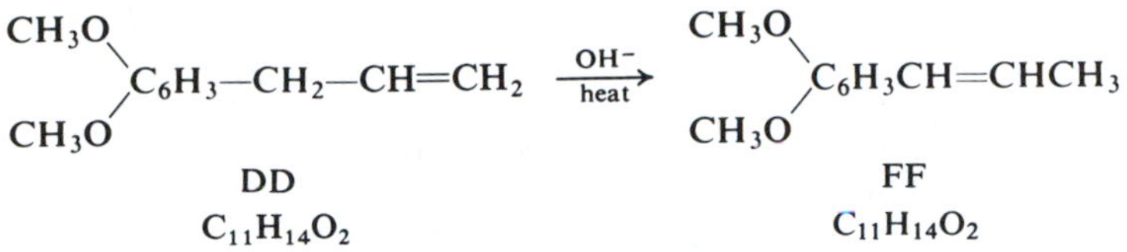

$$(CH_3O)_2C_6H_3{-}CH_2{-}CH{=}CH_2 \xrightarrow[\text{heat}]{OH^-} (CH_3O)_2C_6H_3CH{=}CHCH_3$$

DD $C_{11}H_{14}O_2$ FF $C_{11}H_{14}O_2$

Now we have to work out the orientation in the compounds. This problem is simplified by the ozonolysis.

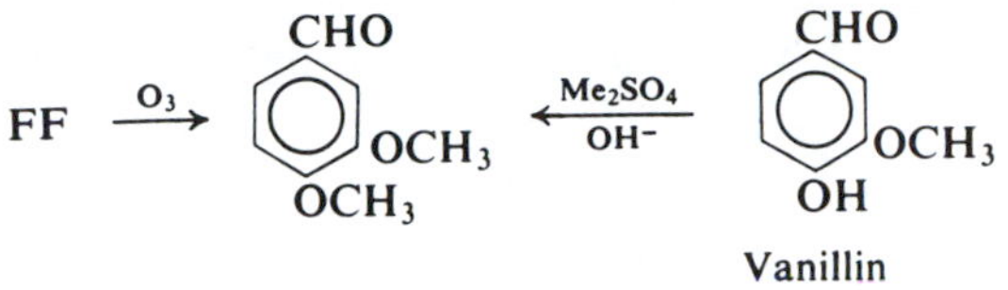

FF $\xrightarrow{O_3}$ 3,4-$(OCH_3)_2C_6H_3CHO$ $\xleftarrow[OH^-]{Me_2SO_4}$ 4-OH-3-$OCH_3C_6H_3CHO$ (Vanillin)

The result from FF is easy to interpret:

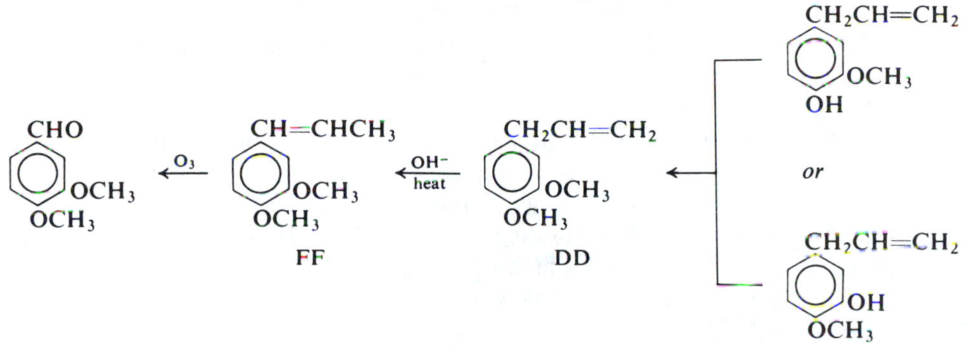

CHO / OCH₃ / OCH₃ benzene $\xleftarrow{O_3}$ FF (CH=CHCH₃, OCH₃, OCH₃) $\xleftarrow[\text{heat}]{OH^-}$ DD (CH₂CH=CH₂, OCH₃, OCH₃) ← [CH₂CH=CH₂, OCH₃, OH] *or* [CH₂CH=CH₂, OH, OCH₃] *One of these is chavibetol*

Which of the two possibilities is chavibetol? We know that EE has essentially the same orientation of groups (1-carbon, 3-oxygen, 4-oxygen) as FF; hence the only possible isomer of vanillin that can be formed by cleavage of EE is the one in which the —OH and —OCH_3 groups are reversed.

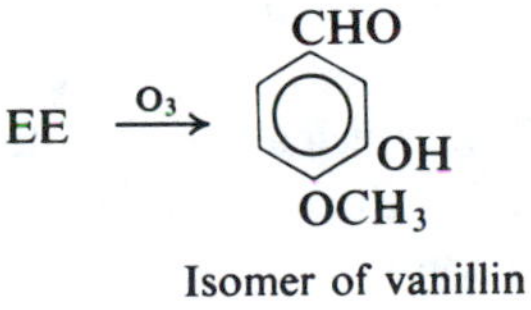

EE $\xrightarrow{O_3}$ CHO / OH / OCH₃ benzene

Isomer of vanillin

Finally, then, we arrive at the structure of chavibetol:

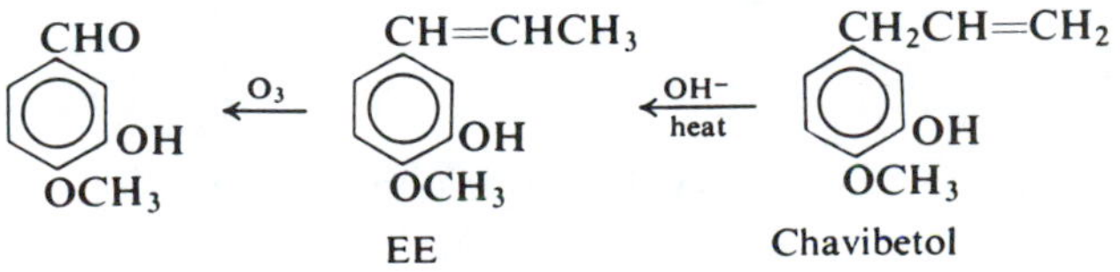

CHO / OH / OCH₃ benzene $\xleftarrow{O_3}$ EE (CH=CHCH₃, OH, OCH₃) $\xleftarrow[\text{heat}]{OH^-}$ Chavibetol (CH₂CH=CH₂, OH, OCH₃)

28. (a) Piperine is an amide of piperidine (Sec. 30.12) and an acid called piperic acid, RCOOH in which R = $C_{11}H_9O_2$.

$$\text{R—C(=O)—N}\bigcirc \xrightarrow{H_2O,\ OH^-} \text{H—N}\bigcirc + \text{RCOOH}$$

Piperidine (R = $C_{11}H_9O_2$)

(b) The m.w. of piperic acid ($C_{12}H_{10}O_4$) is 218; its N.E. is 215 ± 6. Clearly, piperic acid is a monocarboxylic acid.

From the low hydrogen content of R, it is apparent that it contains rings and/or unsaturated bonds. Bromination (without substitution) gives $C_{12}H_{10}O_4Br_4$; we conclude that there are probably two active double bonds present, requiring four of the twelve carbons.

$C_{12}H_{10}O_4$	piperic acid	$C_8H_6O_4$	
$-C_4H_4$	two double bonds	$-$COOH	carboxylic acid
$C_8H_6O_4$		$C_7H_5O_2$	

The residue points toward an aromatic ring (six C), leaving us with one as-yet-unaccounted-for carbon, and two oxygens. These cannot be in another —COOH, since piperic acid has only one such group (remember, m.w. = N.E.), which we have already taken into account.

Side-stepping this last problem for the moment, let us now take a look at the unsaturation. Oxidative cleavage of piperic acid gives a number of products, two of which account in a clear-cut way for all the carbons: tartaric acid, HOOCCHOHCHOHCOOH, and an acid, $C_8H_6O_4$, found to be different from any of the phthalic acids, $C_6H_4(COOH)_2$, and given the name of piperonylic acid by the original investigators.

We can relate piperic and piperonylic acid in the following way:

$$C_7H_5O_2\text{—COOH} + \text{HOOC—CH(OH)—CH(OH)—COOH} \xleftarrow{\text{oxidn.}} C_7H_5O_2\text{—CH=CH—CH=CH—COOH}$$

Piperonylic acid — Tartaric acid — Piperic acid *Partial*

We can delay no longer our study of the "missing" carbon and two oxygens. Phenols and alcohols are impossible—oxidation would ruin them. A one-carbon carbonyl group would be an aldehyde; oxidized to —COOH it would lead to a phthalic acid instead of piperonylic acid. We are left with ether linkages as the answer for the oxygens. But how does one get two ether linkages from only *one* carbon?

Cleavage of the ether gives us our clue: formaldehyde and *two* phenolic groups are generated.

$$\text{HCHO} + (\text{HO})_2C_6H_3\text{COOH} \xleftarrow[200\,°C]{\text{HCl}} \text{piperonylic acid}$$

Furthermore, chemical arithmetic shows that only one H_2O is lost in the formation of these two ether linkages.

$$\begin{array}{ll} C_7H_6O_4 & \text{3,4-dihydroxybenzoic acid} \\ +HCHO & \text{formaldehyde} \\ \hline C_8H_8O_5 & \end{array} \qquad \begin{array}{ll} C_8H_8O_5 & \\ -C_8H_6O_4 & \text{piperonylic acid} \\ \hline H_2O & \end{array}$$

The double ether is an acetal, and a cyclic one. Piperonylic acid can have only one structure:

$$H_2C{=}O + HO(C_6H_3)(OH)COOH \xrightarrow{-H_2O} \text{(}H_2C\text{-dioxy)}C_6H_3COOH$$

Piperonylic acid

From this structure, the way is clear to piperic acid and piperine:

$$\text{(}H_2C\text{-dioxy)}C_6H_3CH{=}CH{-}CH{=}CH{-}COOH \qquad \text{(}H_2C\text{-dioxy)}C_6H_3CH{=}CH{-}CH{=}CH{-}C({=}O){-}NC_5H_{10}$$

Piperic acid Piperine

(c) The synthetic sequence confirms our structure in every respect.

$$HO(C_6H_4)OH \xrightarrow[OH^-]{CHCl_3} HO(C_6H_3)(OH)CHO \xrightarrow{2NaOH} H_2CI_2 \;\; {}^-O(C_6H_3)(O^-)CHO \longrightarrow \text{(}H_2C\text{-dioxy)}C_6H_3CHO$$

Catechol GG HH Piperonal

$$\text{piperonal} \xrightarrow[OH^-]{CH_3CHO} \text{(}H_2C\text{-dioxy)}C_6H_3CH{=}CHCHO \xrightarrow[NaOAc]{Ac_2O} \text{(}H_2C\text{-dioxy)}C_6H_3CH{=}CHCH{=}CHCOOH$$

II Piperic acid

$$\text{piperic acid} \xrightarrow{SOCl_2} \text{(}H_2C\text{-dioxy)}C_6H_3CH{=}CHCH{=}CHCOCl \xrightarrow{\text{piperidine}} \text{(}H_2C\text{-dioxy)}C_6H_3CH{=}CHCH{=}CHC({=}O){-}NC_5H_{10}$$

JJ Piperine

29. Its solubility behavior shows that hordinene is a phenol and an amine. The Hinsberg test indicates that the amino grouping is tertiary; while the benzenesulfonyl chloride may react at the phenolic group, it leaves the amino group unchanged, and still basic.

Again, oxidative degradation gives much information but, because of the sensitivity of phenols, hordinene must be methylated first.

$$HOArR \xrightarrow[OH^-]{Me_2SO_4} CH_3OArR \xrightarrow{KMnO_4} CH_3O(C_6H_4)COOH$$

Hordinene LL Anisic acid

Now we know that hordinene has a phenolic —OH *para* to a single side chain that carries the amino group. Since the amino is tertiary (requiring a minimum of three carbons), the side chain can contain no more than two carbons:

$$\begin{array}{ll} C_{10}H_{15}ON & \text{hordinene} \\ -\,HOC_6H_4 & \\ \hline C_4H_{10}N & \end{array} \qquad \begin{array}{ll} C_4H_{10}N & \\ -\,(CH_3)_2N & \text{smallest } 3^\circ \text{ amine} \\ \hline C_2H_4 & \text{left for side chain} \end{array}$$

This is confirmed by the formation of *p*-methoxystyrene from LL with, presumably, elimination of Me_2NH.

$$Me_2NH + CH_3O\text{-}C_6H_4\text{-}CH{=}CH_2 \xleftarrow{\text{heat}} CH_3O\text{-}C_6H_4\text{-}C_2H_4NMe_2$$

p-Methoxystyrene LL

(a) So far, two structures are consistent with the data:

$$HO\text{-}C_6H_4\text{-}CH_2CH_2\text{-}N(CH_3)\text{-}CH_3 \quad or \quad HO\text{-}C_6H_4\text{-}CH(CH_3)\text{-}N(CH_3)\text{-}CH_3$$

(b) Either structure could be made by reductive amination (Sec. 22.11).

$$HO\text{-}C_6H_4\text{-}CH_2C(H){=}O + HN(CH_3)_2 \xrightarrow{H_2,\ Ni} HO\text{-}C_6H_4\text{-}CH_2CH_2\text{-}N(CH_3)\text{-}CH_3$$

$$HO\text{-}C_6H_4\text{-}C(CH_3){=}O + HN(CH_3)_2 \xrightarrow{H_2,\ Ni} HO\text{-}C_6H_4\text{-}CH(CH_3)\text{-}N(CH_3)\text{-}CH_3$$

The first of these syntheses actually gives hordinene, and proves the structure to be

$$HO\text{-}C_6H_4\text{-}CH_2CH_2\text{-}N(CH_3)\text{-}CH_3$$

Hordinene

30.

$$\underset{p\text{-Toluic acid}}{p\text{-}CH_3C_6H_4COOH} \xrightarrow[SO_3]{H_2SO_4} \underset{\text{MM}}{CH_3C_6H_3(SO_3H)COOH} \xrightarrow[\text{fuse}]{KOH} \xrightarrow{H^+} \underset{\text{NN}}{CH_3C_6H_3(OH)COOH} \xrightarrow[EtOH]{Na} \underset{\text{OO}}{\text{4-methyl-3-hydroxycyclohexanecarboxylic acid (OH, COOH)}} \xrightarrow{HBr} \underset{\text{PP}}{\text{3-bromo-4-methylcyclohexanecarboxylic acid (Br, COOH)}}$$

$$\text{PP} \xrightarrow[\text{heat}]{\text{base}} \underset{\text{QQ}}{\text{4-methylcyclohex-3-enecarboxylic acid (COOH)}} \xrightarrow[H^+]{EtOH} \underset{\text{RR}}{\text{(COOEt)}} \xrightarrow{CH_3MgI} \xrightarrow{H_2O} \underset{\alpha\text{-Terpineol}}{\text{(C(CH}_3)_2\text{OH)}}$$

31. Coniferyl alcohol is soluble in NaOH but not in $NaHCO_3$, and hence must be a phenol. Treatment with benzoyl chloride (to give SS) increases the carbon number by 14, indicating that two benzoyl groups have been added, and hence that two —OH's have been esterified. Reaction with cold HBr to replace —OH by —Br shows that one of the —OH's is alcoholic. Cleavage by HI to give CH_3I reveals the presence of a methyl ether grouping.

Chemical arithmetic suggests the presence of an unsaturated side chain:

$C_{10}H_{12}O_3$	coniferyl alcohol	$C_{10}H_{11}O_2$		$C_{10}H_{10}O$		C_9H_7	
$-OH$	phenolic	$-OH$	alcoholic	$-OCH_3$	ether	$-C_6H_3$	aromatic ring
$C_{10}H_{11}O_2$		$C_{10}H_{10}O$		C_9H_7		C_3H_4	

In agreement, SS (with phenolic and alcoholic functions protected) was found to decolorize both $KMnO_4$ and Br_2/CCl_4: evidence for unsaturation. Structures possible at this point are:

$C_6H_3(OH)(OCH_3)CH{=}CHCH_2OH$ $\qquad$ $C_6H_3(OH)(OCH_3)C(CH_2OH){=}CH_2$

Finally, ozonolysis gives results that leave no doubts about the structures of coniferyl alcohol, SS, and TT.

Vanillin (HO, CH_3O on ring, CHO) $\xleftarrow{O_3}$ Coniferyl alcohol (HO, CH_3O on ring, $CH{=}CH{-}CH_2OH$)

Coniferyl alcohol $\xrightarrow{PhCOCl}$ SS: $PhC(=O)O$, CH_3O on ring, $CH{=}CH{-}CH_2OC(=O)Ph$

Coniferyl alcohol $\xrightarrow{MeI,\ OH^-}$ TT: CH_3O, CH_3O on ring, $CH{=}CH{-}CH_2OH$

Vanillin

Coniferyl alcohol

SS

TT

32. (a) UU is an acetal (strictly, a ketal) and lactone. It is formed by intramolecular (nucleophilic) trapping of the benzenonium ion that ordinarily leads to electrophilic aromatic substitution. (This seems to be the first such case to be discovered.)

UU $\leftarrow$ [benzenonium ion: $(CH_3)_2C(COO^-)-O^+{=}$ cyclohexadiene bearing CH_3, Br] $\leftarrow$ $(CH_3)_2C(COO^-)-O-C_6H_4-CH_3$ + Br—Br

UU

A benzenonium ion
(A *sigma* complex)

(b)

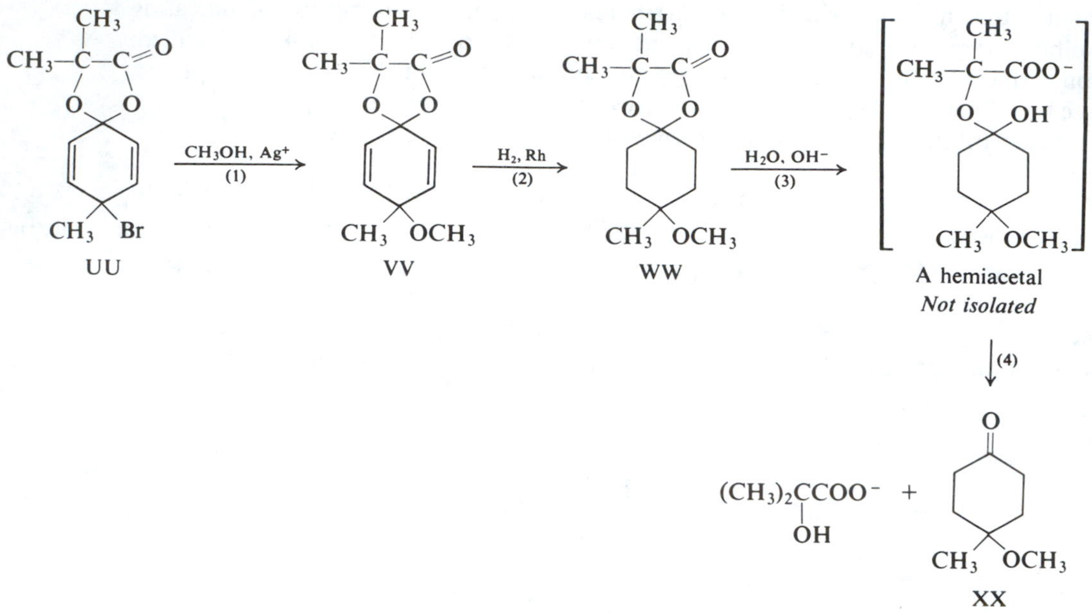

UU $\xrightarrow[(1)]{CH_3OH,\ Ag^+}$ VV $\xrightarrow[(2)]{H_2,\ Rh}$ WW $\xrightarrow[(3)]{H_2O,\ OH^-}$ [A hemiacetal, *Not isolated*] $\xrightarrow{(4)}$ $(CH_3)_2C(OH)COO^-$ + XX

(1) Nucleophilic substitution with (Lewis) acidic catalysis by Ag^+, which helps pull Br^- away to permit entry of $—OCH_3$. (2) Reduction of two double bonds. (3) Alkaline hydrolysis of a lactone (an ester) to yield an alcohol (a hemiacetal) and a carboxylate anion. (4) Like other hemiacetals, this one is unstable and yields the carbonyl compound and an alcohol (a hydroxy acid).

(c) This approach, with modifications, permits synthesis of a variety of non-benzenoid cyclic compounds from benzenoid starting materials. It is an alternative to reduction of aromatic rings by Li or Na in NH_3, which may affect other reducible functional groups in the molecule as well.

For example, cleavage of the ether linkage ($ROCH_3$) in **XX** and dehydration of the resulting alcohol would give rise to VIII, and thus complete the transformation of a phenol into a polyfunctional aliphatic system. With a carbonyl group in the molecule—the most important functional group in organic chemistry—dozens of synthetic routes are open.

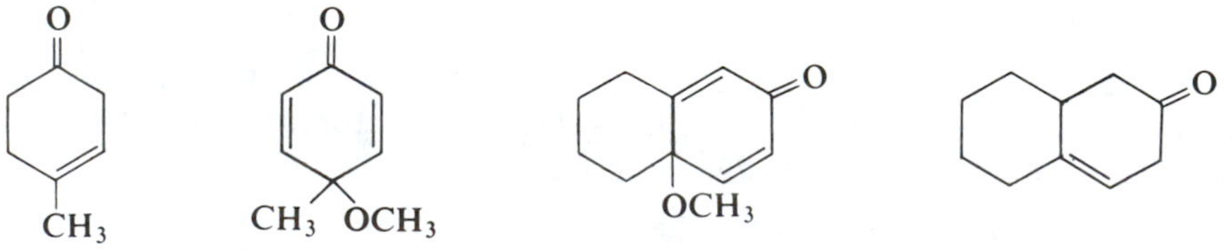

VIII IX X XI

Groups other than $—OCH_3$ can be introduced in step (1): acetate, for example. The ketal–lactone can be cleaved before hydrogenation, to produce structures like IX. The process can be extended to polycyclic aromatic systems (Sec. 14.12), with formation of compounds like X and XI.

(Corey, E. J.; Barcza, S.; Klotmann, G. "A New Method for the Directed Conversion of the Phenoxy Grouping into a Variety of Cyclic Polyfunctional Systems"; *J. Am. Chem. Soc.* **1969**, *91*, 4782.)

33.

O–CH₂ / O (methylenedioxy) ring with CHO

Piperonal (page 919)

AAA

(and HH in Problem 28, above)

OH, OCH_3, CHO

Vanillin (page 894)

BBB

OH, OCH_3, $CH_2CH{=}CH_2$

Eugenol (page 894)

CCC

OH, $CH(CH_3)_2$, CH_3

Thymol (page 894)

DDD

OH, OCH_3, $CH{=}CHCH_3$

Isoeugenol (page 894)

EEE

O–CH_2–O ring with $CH_2CH{=}CH_2$

Safrole (page 894)

FFF

First, check your answers against those above. Whether you were right or wrong, try to fit the correct structures to the spectra. Next, turn to the labeled spectra on pages 720–722 of this Study Guide, and check your peak assignments against the ones shown there. Finally, turn to the analysis of these spectra outlined on page 723 of this Study Guide.

25

Carbanions II

Malonic Ester and Acetoacetic Ester Syntheses

25.1 We use the general procedure as given on pages 924–926.

Acid wanted:	*Requires:*	*Alkylate malonic ester with:*
(a) $CH_3CH_2CH_2—CH_2COOH$	R = *n*-Pr—	*n*-PrBr
$CH_3CH(CH_3)—CH_2COOH$	R = *iso*-Pr—	*iso*-PrBr
$CH_3CH_2—CH(CH_3)COOH$	R = Et— R′ = Me—	EtBr MeBr

One cannot make trisubstituted acetic acids by the malonic ester synthesis, because only *two* alkyl groups can be introduced.

(b) $(CH_3)_2CHCH_2—CH(NH_2)COOH \xleftarrow{NH_3} (CH_3)_2CHCH_2—CH(Br)COOH \xleftarrow[P]{Br_2} (CH_3)_2CHCH_2—CH_2COOH$

Leucine

$\uparrow$ Malonic ester synthesis with R = *iso*-Bu—

(Bromination and ammonolysis are best carried out on the free acid rather than on the ester.)

(c) $CH_3CH_2CH(CH_3)—CH(NH_2)COOH \longleftarrow$ as in (b), from $CH_3CH_2CH(CH_3)—CH_2COOH$

Isoleucine

$\uparrow$ Malonic ester synthesis with R = *sec*-Bu—

25.2 (a) In the synthesis of adipic acid, the excess of sodiomalonic ester leads to displacement of both Br's from ethylene bromide.

$$(EtOOC)_2CH^- + BrCH_2CH_2Br + {}^-CH(COOEt)_2$$

$$\downarrow$$

$$(EtOOC)_2CH{-}CH_2CH_2{-}CH(COOEt)_2 \xrightarrow[\text{heat}]{OH^-} \xrightarrow[-CO_2]{H^+,\ \text{heat}} HOOCCH_2{-}CH_2CH_2{-}CH_2COOH$$

Adipic acid

In the synthesis of the cyclic acid, the constant excess of organic halide ensures monosubstitution:

$$(EtOOC)_2CH^- + BrCH_2CH_2Br \longrightarrow (EtOOC)_2CH{-}CH_2CH_2Br + Br^-$$

Addition then of a second mole of ethoxide generates a carbanion which, by intramolecular nucleophilic attack, closes the ring.

$$(EtOOC)_2CH{-}CH_2CH_2Br \xrightarrow{OEt^-} (EtOOC)_2\overset{\ominus}{C}H{-}CH_2{-}CH_2{-}Br \longrightarrow (EtOOC)_2C\langle^{CH_2}_{CH_2} + Br^-$$

$$\downarrow OH^-,\ \text{heat}$$

$$\downarrow H^+,\ \text{heat},\ -CO_2$$

$$HOOC{-}HC\langle^{CH_2}_{CH_2}$$

Cyclopropanecarboxylic acid

(b) As for cyclopropanecarboxylic acid in (a), except use $BrCH_2CH_2CH_2CH_2Br$.

$$H_2C\langle^{CH_2}_{H_2C-CH_2}\rangle CH{-}COOH$$

Cyclopentanecarboxylic acid

25.3 (a) The Knoevenagel reaction is an aldol-like condensation (Sec. 21.9).

$$CH_2(COOEt)_2 + PhCHO \xrightarrow{\text{base}} \left[PhCH(OH){-}CH(COOEt)_2\right] \xrightarrow{-H_2O} PhCH{=}C(COOEt)_2$$

(b) $PhCH{=}C(COOEt) \xrightarrow[\text{heat}]{OH^-,\ H_2O} PhCH{=}C(COO^-)_2 \xrightarrow{H^+} PhCH{=}C(COOH)_2 \xrightarrow[-CO_2]{\text{heat}} PhCH{=}CHCOOH$

Cinnamic acid

(c) By the Perkin condensation (Problem 21.18d, page 811).

25.4 (a) The Cope reaction is an aldol-like condensation (Sec. 21.9).

$$\text{C}_6\text{H}_{10}{=}O + CH_2(CN){-}COOEt \xrightarrow{\text{base}} \left[\text{C}_6\text{H}_{10}(OH){-}CH(CN){-}COOEt\right] \xrightarrow{-H_2O} \text{C}_6\text{H}_{10}{=}C(CN){-}COOEt$$

(b) $\text{C}_6\text{H}_{10}{=}C(CN){-}COOEt \xrightarrow[\text{heat}]{H_2O,\ H^+} \text{C}_6\text{H}_{10}{=}C(COOH){-}COOH \xrightarrow{-CO_2} \text{C}_6\text{H}_{10}{=}CH{-}COOH$

Cyclohexylideneacetic acid

25.5
$$\underset{\text{A}}{\underset{|}{CH_3CHCH_2COOEt}\atop CH(COOEt)_2}$$

A is formed as described in Sec. 27.7.

25.6 Nucleophilic substitution with competing elimination (Sec. 8.25). Yield of substitution product: 1° > 2° ≫ 3°. Aryl halides cannot be used (Sec. 26.4).

25.7 (a) $[CH_3COCHCOOEt]^-Na^+ + BrCH_2COOEt \longrightarrow CH_3C(=O)\text{—}CH(CH_2COOEt)\text{—}COOEt + NaBr$

$$CH_3C(=O)\text{—}CH(CH_2COOEt)\text{—}COOEt \longrightarrow CH_3C(=O)\text{—}CH(CH_2COOH)\text{—}COOH \xrightarrow{-CO_2} \underset{\text{A }\gamma\text{-keto acid}}{CH_3\overset{\gamma}{C}(=O)\text{—}\overset{\beta}{C}H_2\text{—}\overset{\alpha}{C}H_2COOH}$$

We notice that, of the two carboxyl groups, it is the one to which the keto group is *beta* that is lost.

Use of the acidic bromoacetic acid would simply convert the sodio compound back into acetoacetic ester.

(b) $[CH_3COCHCOOEt]^-Na^+ + Cl\text{—}C(=O)\text{—}Ph \longrightarrow CH_3C(=O)\text{—}CH(C(=O)Ph)\text{—}COOEt + NaCl$

$$CH_3C(=O)\text{—}CH(C(=O)Ph)\text{—}COOEt \longrightarrow CH_3C(=O)\text{—}CH(C(=O)Ph)\text{—}COOH \xrightarrow{-CO_2} \underset{\text{A }\beta\text{-diketone}}{CH_3\overset{\beta}{C}(=O)\text{—}\overset{\alpha}{C}H_2\text{—}C(=O)Ph}$$

$$[CH_3COCHCOOEt]^-Na^+ + ClCH_2C(=O)CH_3 \longrightarrow CH_3C(=O)\text{—}CH(CH_2C(=O)CH_3)\text{—}COOEt + NaCl$$

$$CH_3C(=O)\text{—}CH(CH_2C(=O)CH_3)\text{—}COOEt \longrightarrow CH_3C(=O)\text{—}CH(CH_2C(=O)CH_3)\text{—}COOH \xrightarrow{-CO_2} \underset{\text{A }\gamma\text{-diketone}}{CH_3\overset{\gamma}{C}(=O)\text{—}\overset{\beta}{C}H_2\text{—}\overset{\alpha}{C}H_2\text{—}C(=O)CH_3}$$

25.8 We follow the general procedure on pages 927–929.

	Ketone wanted:	*Requires:*	*Alkylate acetoacetic ester with:*
(a)	$CH_3CH_2CH_2{-}CH_2\overset{O}{\overset{\|\|}{C}}CH_3$	R = *n*-Pr—	*n*-PrBr
(b)	$CH_3\overset{CH_3}{\overset{\|}{C}}H{-}CH_2\overset{O}{\overset{\|\|}{C}}CH_3$	R = *iso*-Pr—	*iso*-PrBr
(c)	$CH_3CH_2{-}\overset{CH_3}{\overset{\|}{C}}H\overset{O}{\overset{\|\|}{C}}CH_3$	R = Et— R′ = Me—	EtBr MeBr

(d) One cannot make trisubstituted acetones by the acetoacetic ester synthesis, because only *two* alkyl groups can be introduced.

(e) $CH_3\overset{O}{\overset{\|\|}{C}}{-}CH_2\overset{O}{\overset{\|\|}{C}}CH_3 \longleftarrow$ acetoacetic ester synthesis with R = $CH_3\overset{O}{\overset{\|\|}{C}}{-}$ (RX = CH_3COCl)

(f) $CH_3\overset{O}{\overset{\|\|}{C}}CH_2{-}CH_2\overset{O}{\overset{\|\|}{C}}CH_3 \longleftarrow$ acetoacetic ester synthesis with R = $CH_3\overset{O}{\overset{\|\|}{C}}CH_2{-}$ (RX = CH_3COCH_2Cl)

Or, more conveniently, proceed as follows:

$$2Na^{+}\,{}^{-}[CH_3CO\,CHCOOEt] + I_2 \longrightarrow 2NaI + CH_3\overset{O}{\overset{\|\|}{C}}\underset{}{\overset{EtOOC}{\overset{\|}{C}}}H{-}\overset{COOEt}{\overset{\|}{C}}H\overset{O}{\overset{\|\|}{C}}CH_3 \xrightarrow{OH^-} \xrightarrow{H^+} \xrightarrow{-CO_2} CH_3\overset{O}{\overset{\|\|}{C}}CH_2{-}CH_2\overset{O}{\overset{\|\|}{C}}CH_3$$

(g) $Ph\overset{O}{\overset{\|\|}{C}}CH_2{-}CH_2\overset{O}{\overset{\|\|}{C}}CH_3 \longleftarrow$ acetoacetic ester synthesis with R = $Ph\overset{O}{\overset{\|\|}{C}}CH_2{-}$ (RX = $Ph\overset{O}{\overset{\|\|}{C}}CH_2Br$)

Phenacyl bromide

25.9 There is a crossed Claisen condensation to give B, followed by the hydrolysis and decarboxylation of a β-keto acid. (Here, as in Problem 25.7(a), we see that, of two carboxyl groups, it is the one to which the keto group is *beta* that is lost.)

$$EtOOC{-}\overset{O}{\overset{\|\|}{C}}{-}OEt + \overset{CH_3}{\overset{\|}{C}}H_2COOEt \xrightarrow{-EtOH} \underset{\textbf{B}}{EtOOC{-}\overset{O}{\overset{\|\|}{C}}{-}\overset{CH_3}{\overset{\|}{C}}HCOOEt}$$

$$\textbf{B} \xrightarrow[H_2O]{H^+} HOOC{-}\overset{O}{\overset{\|\|}{C}}{-}\overset{CH_3}{\overset{\|}{C}}HCOOH \xrightarrow{-CO_2} HOOC{-}\overset{O}{\overset{\|\|}{C}}{-}CH_2CH_3$$

An α-keto acid

25.10

(a) $HOOCC(=O)—CH_2CH(CH_3)_2 \xleftarrow[-CO_2]{H^+} EtOOCC(=O)—CH(COOEt)CH(CH_3)_2 \xleftarrow{OEt^-} EtOOC—C(=O)—OEt + EtOOCCH_2CH(CH_3)_2$

α-Ketoisocaproic acid; Ethyl oxalate; Ethyl isovalerate

(b) $HOOCC(=O)—CH_2Ph \xleftarrow[-CO_2]{H^+} HOOCC(=O)—CH(COOEt)Ph \xleftarrow{OEt^-} EtOOC—C(=O)—OEt + EtOOCCH_2Ph$

α-Keto-β-phenylpropionic acid; Ethyl oxalate; Ethyl phenylacetate

(c) $HOOCC(=O)—CH_2CH_2COOH \xleftarrow[-CO_2]{H^+} EtOOCC(=O)—CH(COOEt)CH_2COOEt$

α-Ketoglutaric acid

$\uparrow OEt^-$

$EtOOC—C(=O)—OEt + EtOOCCH_2CH_2COOEt$

Ethyl oxalate; Ethyl succinate

(d) $HOOCCH(NH_2)—CH_2CH(CH_3)_2 \xleftarrow[Ni]{NH_3,\ H_2} HOOCC(=O)—CH_2CH(CH_3)_2 \longleftarrow$ as in (a)

Leucine

(e) $HOOCCH(NH_2)—CH_2CH_2COOH \xleftarrow[Ni]{NH_3,\ H_2} HOOCC(=O)—CH_2CH_2COOH \longleftarrow$ as in (c)

Glutamic acid

25.11 (a) The charged end loses CO_2. The resulting anion (like I, page 930) is stabilized through accommodation of the negative charge by oxygen.

$$HO—C(=O)—CH_2—COO^- \longrightarrow HO—C(—O^-)=CH_2 + CO_2$$

(b) A doubly charged anion, $^-OOC—CH_2^-$, would be less stable, and the transition state leading to it would be reached with difficulty.

25.12 In this case decarboxylation gives 2,4,6-$(NO_2)_3C_6H_2^-$, which is stabilized relative to most aryl anions by the three powerfully electron-withdrawing groups.

25.13 Intermediates that can react with Br_2 and I_2 are involved. Both the proposed intermediates in decarboxylation, the carbanion I and the enol (page 931), are known to react with halogens. (Secs. 21.3 and 21.4.)

25.14 $HO{-}\underset{\underset{O}{\|}}{C}{-}CH_2{-}COOH \rightleftarrows HO{=}\underset{\underset{\oplus OH}{\vdots}}{C}{-}CH_2{-}COO^- \longrightarrow CO_2 + HO{-}\underset{\underset{OH}{|}}{C}{=}CH_2$ (*Enol*) $\longrightarrow HO{-}\underset{\underset{O}{\|}}{C}{-}CH_3$

25.15 $PhC{\equiv}C{-}CO_2^- \longleftarrow CO_2 + PhC{\equiv}C{:}^{\ominus} \xrightarrow{H_2O} PhC{\equiv}CH$

Decarboxylation gives $PhC{\equiv}C^-$, which, like anions from other terminal acetylenes, is relatively stable (Sec. 12.11).

25.16 We follow the general procedure on page 932.

	Acid wanted:	*Requires:*	*Starting materials:*
(a)	$CH_3CH_2{-}CH(H){-}COOH$	$R = H{-}$ $R' = CH_3CH_2{-}$	CH_3COOH $R'X = EtBr$
(b)	$CH_3{-}CH(CH_3){-}COOH$	$R = H{-}$ $R' = CH_3{-}$ $R'' = CH_3{-}$	CH_3COOH $R'X = MeBr$ $R''X = MeBr$
(c)	$CH_3{-}CH(CH_3){-}COOH$	$R = CH_3{-}$ $R' = CH_3{-}$	CH_3CH_2COOH $R'X = MeBr$
(d)	$PhCH_2{-}CH(H){-}COOH$	$R = H{-}$ $R' = PhCH_2{-}$	CH_3COOH $R'X = PhCH_2Cl$

(Meyers, A. I., *et al.* "Oxazolines. IX. Synthesis of Homologated Acetic Acids and Esters"; *J. Org. Chem.* **1974**, *39*, 2778.)

25.17 The oxazoline ring protects the carboxyl group, so that Grignard reactions can be carried out on other parts of the molecule.

(a) HO–(cyclohexyl)–COOH + HO–CH₂–C(CH₃)₂–NH₂ ⟶ HO–(cyclohexyl)–(4,4-dimethyloxazolinyl) **C**

C $\xrightarrow{CrO_3}$ O=(cyclohexylidene)–(4,4-dimethyloxazolinyl) **D**

D $\xrightarrow{PhMgBr}$ Ph(HO)C(cyclohexyl)–(4,4-dimethyloxazolinyl) $\xrightarrow[H^+]{EtOH}$ Ph(HO)C(cyclohexyl)–COOEt $\xrightarrow[-H_2O]{H^+}$ Ph–(cyclohexenyl)–COOEt **E**

(b) $Ph\text{-}\!\!\vdots\!\!\text{-}C(=O)CH_2CH_2COOH \xleftarrow{AlCl_3} PhH + $ Succinic anhydride (cyclic H_2C–$C(=O)$–O–$C(=O)$–CH_2) *Friedel–Crafts acylation*

A γ-keto acid

Succinic anhydride

(c) $Ph(Et)C{=}CHCH_2COOH \xleftarrow{-H_2O} Et\text{–}C(Ph)(OH)\text{–}CH_2CH_2COOH \longleftarrow Et\text{–}C(Ph)(OH)\text{–}CH_2CH_2\text{–}$(4,4-dimethyloxazolinyl, with CH_3, CH_3)

↑ EtMgBr

as in (b) $\longrightarrow PhC(=O)CH_2CH_2COOH + HOCH_2C(CH_3)_2NH_2 \longrightarrow PhC(=O)CH_2CH_2\text{–}$(4,4-dimethyloxazolinyl, with CH_3, CH_3)

(d) $EtCH(OH)C_6H_4COOEt \xleftarrow[H^+]{EtOH} EtCH(OH)C_6H_4\text{–}$(4,4-dimethyloxazolinyl) $\xleftarrow{EtCHO} BrMgC_6H_4\text{–}$(4,4-dimethyloxazolinyl)

↑ Mg

$BrC_6H_4COOH + HOCH_2C(CH_3)_2NH_2 \longrightarrow BrC_6H_4\text{–}$(4,4-dimethyloxazolinyl, with CH_3, CH_3)

(from p-$BrC_6H_4CH_3$)

(e) $PhCH(OH)C_6H_4COOEt \longleftarrow$ as in (d), except use PhCHO in place of EtCHO

(Meyers, A. I., *et al.* "Oxazolines. XI. Synthesis of Functionalized Aromatic and Aliphatic Acids. A Useful Protecting Group for Carboxylic Acids against Grignard and Hydride Reagents"; *J. Org. Chem.* **1974**, *39*, 2787.)

25.18 An acid–base complex forms; within this, a proton is transferred via a cyclic transition state.

$R_2\overset{\ominus}{B}(R)\text{–}\overset{\oplus}{O}{=}C(R')\text{–}OH \longrightarrow R_2B\text{–}O\text{–}C(R'){=}O \quad R\text{–}H$

25.19 (a) $CH_3CH_2CH_2CH_2 \vdots CH_2C(=O)CH_3 \xleftarrow{\text{base}}$ [$n\text{-Bu—B}$(9-BBN) $\xleftarrow{\text{9-BBN}} CH_3CH_2CH{=}CH_2$; $BrCH_2COCH_3 \xleftarrow{CuBr_2} CH_3COCH_3$]

(b) $CH_3CH(CH_3)CH_2 \vdots CH_2COOH \xleftarrow[H^+]{H_2O} \xleftarrow{\text{base}}$ [$iso\text{-Bu—B}$(9-BBN) $\xleftarrow{\text{9-BBN}} CH_3C(CH_3){=}CH_2$; $BrCH_2COOEt$ (from CH_3COOH)]

(c) $CH_3CH_2CH(CH_3) \vdots CH_2C(=O)CH_3 \longleftarrow$ as in (a), except start with $CH_3CH{=}CHCH_3$

(d) cyclohexyl $\vdots CH_2C(=O)CH_3 \longleftarrow$ as in (a), except start with cyclohexene

(e) 2-methylcyclopentyl $\vdots CH_2COOEt \longleftarrow$ as in (b), except start with 1-methylcyclopentene (CH_3)

(f) $CH_3CH(CH_3)CH_2 \vdots CH_2C(=O)Ph \xleftarrow{\text{base}}$ [$iso\text{-Bu—B}$(9-BBN) $\xleftarrow{\text{9-BBN}} CH_3C(CH_3){=}CH_2$; $BrCH_2COPh \xleftarrow{CuBr_2} CH_3COPh$]

(g) cyclopentyl $\vdots CH_2C(=O){-}C(CH_3)_2{-}CH_3 \xleftarrow{\text{base}}$ [cyclopentyl—B(9-BBN) $\xleftarrow{\text{9-BBN}}$ cyclopentene ; $BrCH_2COC(CH_3)_3 \xleftarrow{CuBr_2} CH_3COC(CH_3)_3$ *tert*-Butyl methyl ketone]

(Brown, H. C.; *Boranes in Organic Chemistry*; Cornell University Press: Ithaca, NY, 1972; pp 372–391; or Brown, H. C.; Nambu, H.; Rogic, M. M.; *J. Am. Chem. Soc.* **1969**, *91*, 6852, 6855.)

25.20 We follow the general procedure as illustrated on page 937.

To make:	*Alkylate or acylate enamine of:*	*With:*
(a) 2-benzylcyclohexanone (cyclohexanone $\vdots CH_2Ph$)	cyclohexanone	$PhCH_2Br$

To make:	*Alkylate or acylate enamine of*:	*With*:
(b) $CH_2{=}CHCH_2{-}C(CH_3)_2{-}CHO$	$H{-}C(CH_3)_2{-}CHO$	$CH_2{=}CHCH_2Br$
(c) 2-(CH(CH_3)COOEt)cyclohexanone	cyclohexanone	$BrCH(CH_3)COOEt$
(d) 2-(CH_2COCH_3)cyclohexanone	cyclohexanone	$BrCH_2COCH_3$
(e) 2-(2,4-dinitrophenyl)cyclohexanone (O_2N, NO_2)	cyclohexanone	1-bromo-2,4-dinitrobenzene (O_2N, Br, NO_2)
(f) $CH_3CO{-}C(CH_3)_2{-}CHO$	$H{-}C(CH_3)_2{-}CHO$	$CH_3{-}CO{-}Cl$

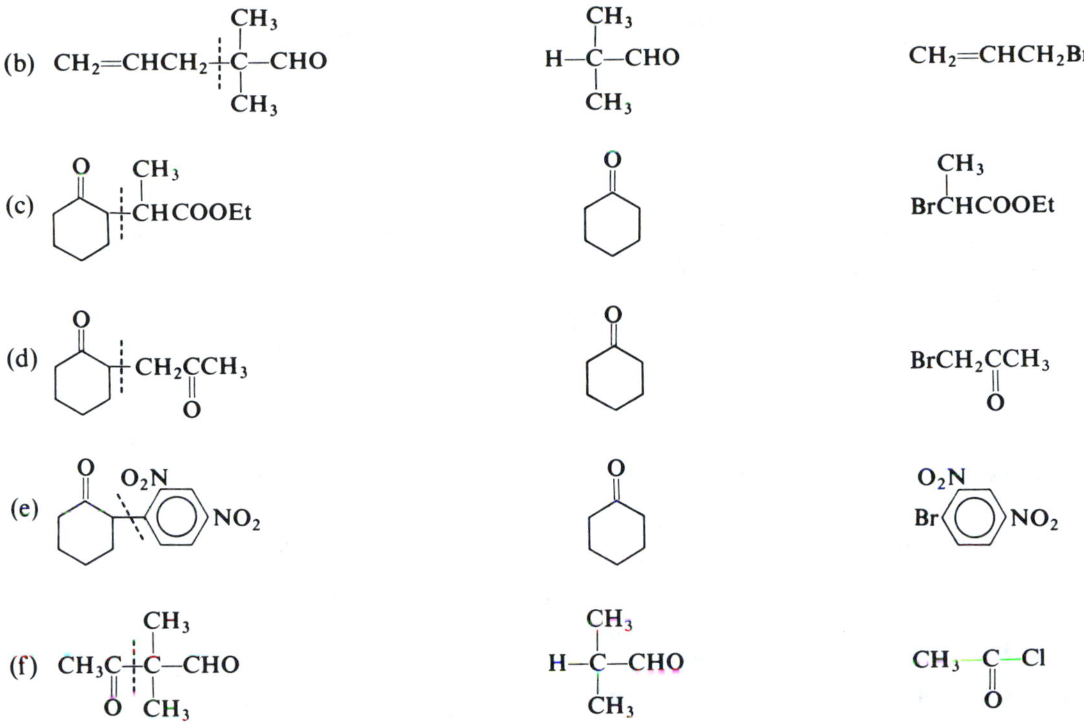

(House, H. O. *Modern Synthetic Reactions*; W. A. Benjamin: New York, 2nd ed., 1972; pp 766–772.)

25.21 (a) In an aldol-like condensation, nucleophilic carbon of the enamine attacks carbonyl carbon of the aldehyde.

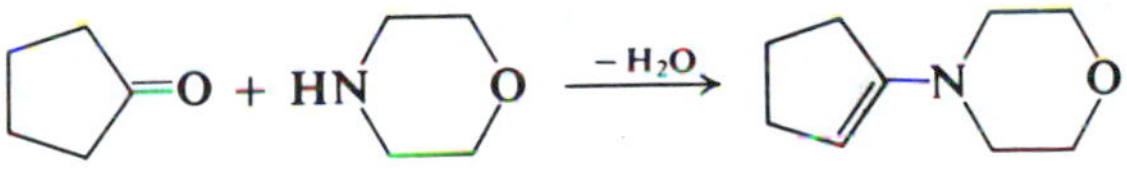

cyclopentanone + HN(morpholine) $\xrightarrow{-H_2O}$ 1-morpholinocyclopentene

F

enamine F + Ph—CH=O ⟶ iminium (⊕N=) with Ph—CH—O⁻ $\xrightarrow{H_2O}$ iminium with Ph—CH—OH $\xrightarrow[H^+]{H_2O}$ 2-(PhCH=)cyclopentanone

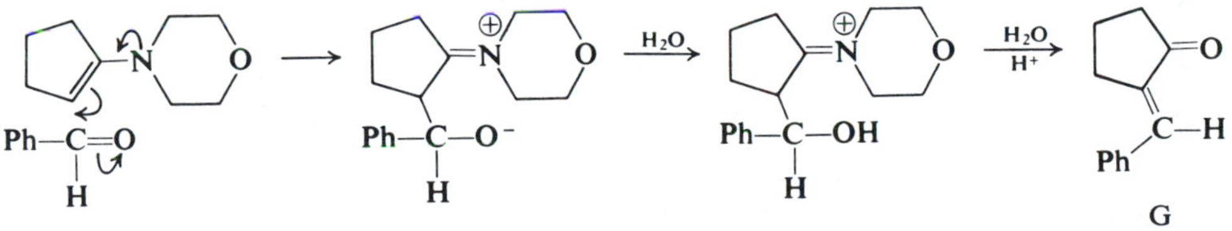

G

(b) Imine H is acidic enough to decompose the Grignard reagent and form the magnesium salt; this contains nucleophilic carbon, and is alkylated by benzyl chloride.

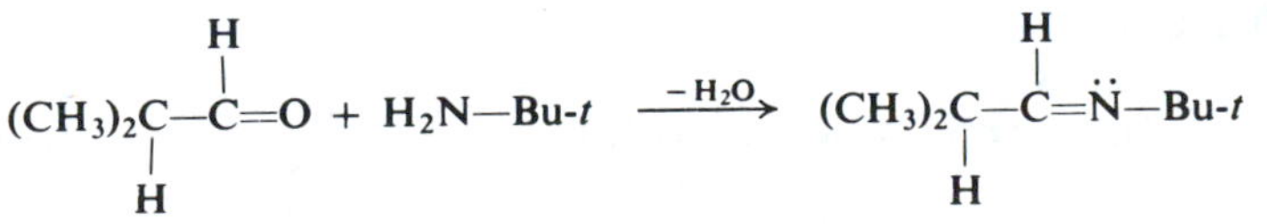

$$(CH_3)_2CH{-}CH{=}O + H_2N{-}Bu\text{-}t \xrightarrow{-H_2O} (CH_3)_2CH{-}CH{=}\ddot{N}{-}Bu\text{-}t$$

H

An imine

$$(CH_3)_2\underset{\underset{\displaystyle C_2H_5\text{—}MgBr}{H}}{C}\text{—}\overset{H}{C}{=}\ddot{N}\text{—Bu-}t + C_2H_5MgBr \longrightarrow \underset{I}{(CH_3)_2C{=}\overset{H}{C}\text{—}\overset{\ominus}{\underset{MgBr^+}{\ddot{N}}}\text{—Bu-}t} + \underset{J}{C_2H_6}$$

$$\underset{PhCH_2\text{—}Cl}{(CH_3)_2C{=}\overset{H}{C}\text{—}\overset{\ominus}{\ddot{N}}\text{—Bu-}t} \xrightarrow{-Cl^-} PhCH_2\text{—}\overset{CH_3}{\underset{CH_3}{C}}\text{—}CH{=}\ddot{N}\text{—Bu-}t \xrightarrow[H^+]{H_2O} \underset{K}{PhCH_2\text{—}\overset{CH_3}{\underset{CH_3}{C}}\text{—}CHO} + t\text{-}BuNH_2$$

1. We follow the general procedure outlined on pages 924–926.

	Acid wanted:	*Requires:*	*Alkylate*
(a)	$CH_3CH_2CH_2CH_2\text{—}CH_2COOH$	R = *n*-Bu—	*n*-BuBr
(b)	$CH_3\text{—}\overset{CH_3}{C}HCOOH$	R = CH_3—	CH_3Br
		R′ = CH_3—	CH_3Br
(c)	$CH_3\overset{CH_3}{C}H\text{—}CH_2COOH$	R = *iso*-Pr—	*iso*-PrBr
(d)	$CH_3\overset{CH_3}{C}H\text{—}\underset{CH_3}{C}HCOOH$	R = *iso*-Pr—	*iso*-PrBr
		R′ = Me—	MeBr
(e)	$C_2H_5\text{—}\underset{C_2H_5}{C}HCOOH$	R = Et—	EtBr
		R′ = Et—	EtBr
(f)	$PhCH_2\text{—}\underset{CH_2Ph}{C}HCOOH$	R = $PhCH_2$—	$PhCH_2Br$
		R′ = $PhCH_2$—	$PhCH_2Br$
(g)	$HOOC\underset{H_3C}{C}H\text{—}\underset{CH_3}{C}HCOOH$	R = $CH_3\underset{COOEt}{C}H$—	$CH_3CHBrCOOEt$
		R′ = Me—	MeBr

Or, more conveniently, oxidize two moles of the sodiomethylmalonic ester with I_2. (Compare Problem 25.8(f), pages 929–930.)

$$2Na^{+\,-}[CH_3C(COOEt)_2] + I_2 \longrightarrow 2NaI + (EtOOC)_2\underset{H_3C}{C}\text{—}\underset{CH_3}{C}(COOEt)_2 \xrightarrow{OH^-} \xrightarrow{H^+} \xrightarrow{-CO_2} HOOC\underset{H_3C}{C}H\text{—}\underset{CH_3}{C}HCOOH$$

(h) $HOOCCH_2 \vdots CH_2 \vdots CH_2COOH$

Add one mole of CH_2Br_2 to two moles of sodiomalonic ester; then hydrolyze and decarboxylate as usual. (See Problem 25.2, page 927.)

(i) cyclobutane–CHCOOH (ring drawn as a square, with dashed bonds)

Add one mole of sodiomalonic ester to excess $BrCH_2CH_2CH_2Br$; then add one mole NaOEt; finally, hydrolyze and decarboxylate as usual. (See Problem 25.2, page 927.)

2. We follow the general procedure as outlined on pages 927–929.

	Ketone wanted:	*Requires:*	*Alkylate acetoacetic ester with:*
(a)	CH_3—CH_2COCH_3	R = Me—	MeBr
(b)	C_2H_5—$CH(C_2H_5)COCH_3$	R = Et—; R′ = Et—	EtBr; EtBr
(c)	$CH_3CH_2CH_2$—$CH(C_2H_5)COCH_3$	R = *n*-Pr—; R′ = Et—	*n*-PrBr; EtBr
(d)	$C_2H_5CH(CH_3)CH_2$—CH_2COCH_3	R = $EtCH(Me)CH_2$—	$EtCH(Me)CH_2Br$ (from active amyl alcohol)
(e)	$(CH_3)_2CHCH_2CH_2$—$CH(CH_3)COCH_3$	R = *iso*-Pe—; R′ = Me—	*iso*-PeBr; MeBr
(f)	$HOOCCH(CH_3)$—CH_2COCH_3	R = $HOOCCH(CH_3)$—	$CH_3CH(Br)COOEt$

(g) $HOOCCH_2$—$CH_2CH(OH)CH_3$ $\xleftarrow{H_2,\ Ni}$ $HOOCCH_2$—$CH_2C(=O)CH_3$ ⟵ a.a.e. synthesis with RX = $ClCH_2COOEt$

(h) $CH_3CH_2CH_2$—$CH(CH_3)CH(OH)CH_3$ $\xleftarrow{LiAlH_4}$ $CH_3CH_2CH_2$—$CH(CH_3)C(=O)CH_3$ ⟵ a.a.e. synthesis with RX = *n*-PrBr and R′X = MeBr

(i) $CH_3CH_2CH(CH_3)CH_2$—$CH_2CH(CH_3)CH_3$ $\xleftarrow{H_2,\ Ni}$ $CH_3CH_2CH(CH_3)CH_2$—$CH{=}C(CH_3)CH_3$ (and 1-alkene) $\xleftarrow{KOH(alc)}$ $CH_3CH_2CH(CH_3)CH_2$—$CH_2C(CH_3)(Br)CH_3$

↑ PBr_3

(d) ⟶ $CH_3CH_2CH(CH_3)CH_2$—$CH_2C(=O)CH_3$ $\xrightarrow{MeMgBr}$ $CH_3CH_2CH(CH_3)CH_2$—$CH_2C(CH_3)(OH)CH_3$

(j) $CH_3CH_2CH_2CH(CH_3)$—CH_2COOH $\xleftarrow[\text{Prob. 11}]{\text{conc. OH}^-}$ a.a.e. alkylated by 2-PeBr

(k) $CH_3CH(CH_3)—CH_2COOH$ $\xleftarrow[\text{Prob. 11}]{\text{conc. }OH^-}$ a.a.e. alkylated by *iso*-PrBr

(l) $HOOCCH(CH_3)—CH_2COOH$ $\xleftarrow[\text{Prob. 11}]{\text{conc. }OH^-}$ a.a.e. alkylated by $CH_3CHBrCOOEt$

(m) $CH_3CH(OH)CH_2—CH_2CH(OH)CH_3$ $\xleftarrow{H_2,\ Ni}$ $CH_3C(=O)CH_2 \vdots CH_2C(=O)CH_3$ $\longleftarrow$ Problem 25.8(f)

3. 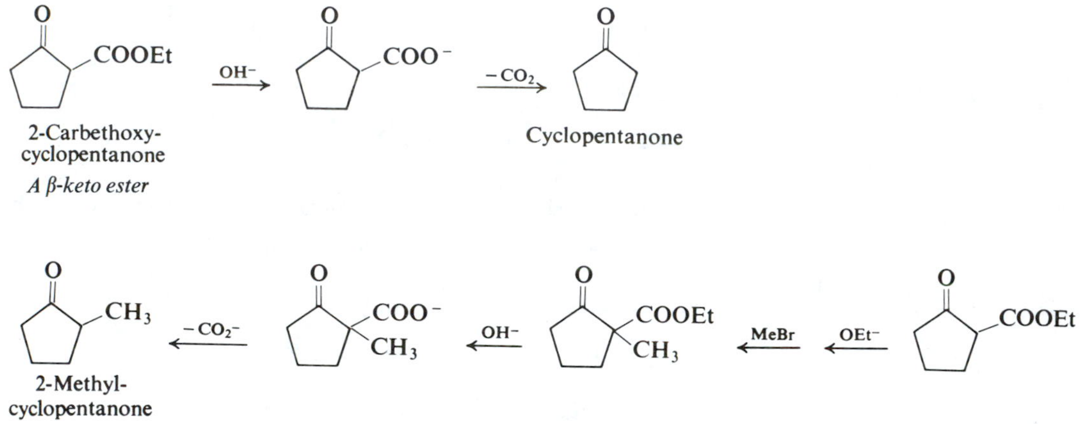 2-Carbethoxycyclopentanone (*A β-keto ester*) $\xrightarrow{OH^-}$ 2-carboxylatocyclopentanone (COO^-) $\xrightarrow{-CO_2}$ Cyclopentanone

2-Carbethoxycyclopentanone (COOEt) $\xrightarrow{OEt^-}$ $\xrightarrow{MeBr}$ 2-carbethoxy-2-methylcyclopentanone (COOEt, CH_3) $\xrightarrow{OH^-}$ 2-carboxylato-2-methylcyclopentanone (COO^-, CH_3) $\xrightarrow{-CO_2^-}$ 2-Methylcyclopentanone

4. (a)

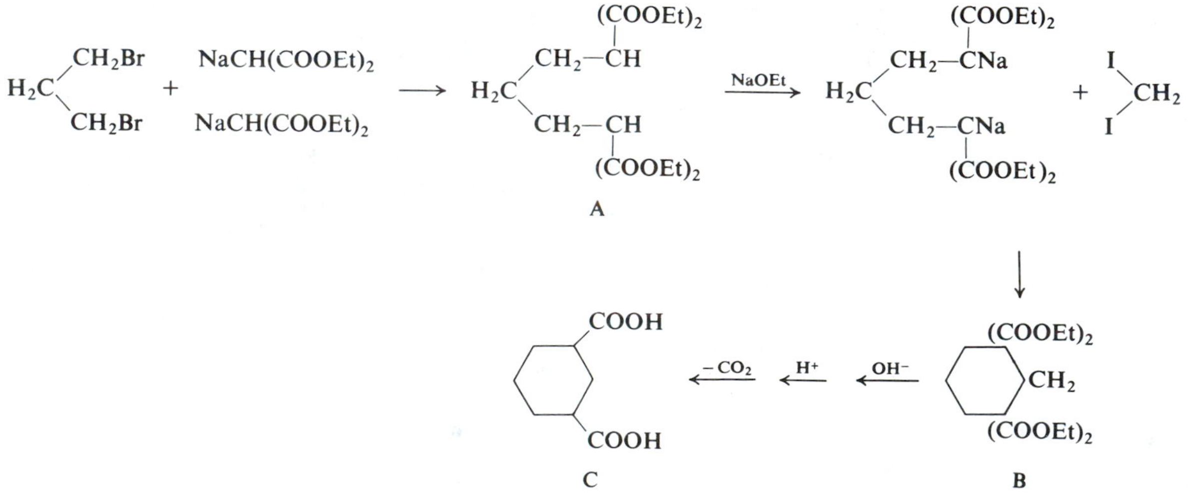

$H_2C(CH_2Br)_2$ + $2\ NaCH(COOEt)_2$ $\longrightarrow$ $H_2C[CH_2—CH(COOEt)_2]_2$ (A) $\xrightarrow{NaOEt}$ $H_2C[CH_2—CNa(COOEt)_2]_2$ + CH_2I_2

$\downarrow$

(B) cyclohexane-1,1,3,3-tetracarboxylic ester [ring with $(COOEt)_2$, CH_2, $(COOEt)_2$] $\xrightarrow{OH^-}$ $\xrightarrow{H^+}$ $\xrightarrow{-CO_2}$ (C) cyclohexane-1,3-dicarboxylic acid (COOH, COOH)

(b)

$$\begin{matrix}H_2C{-}Br\\ |\\ H_2C{-}Br\end{matrix} + \begin{matrix}NaCH(COOEt)_2\\ NaCH(COOEt)_2\end{matrix} \longrightarrow \underset{\mathbf{D}}{\begin{matrix}(COOEt)_2\\ |\\ CH\\ /\\ H_2C\\ |\\ H_2C\\ \backslash\\ CH\\ |\\ (COOEt)_2\end{matrix}} \xrightarrow{NaOEt} \begin{matrix}(COOEt)_2\\ |\\ CNa\\ /\\ H_2C\\ |\\ H_2C\\ \backslash\\ CNa\\ |\\ (COOEt)_2\end{matrix} + \begin{matrix}Br\\ \backslash\\ CH_2\\ |\\ CH_2\\ /\\ Br\end{matrix}$$

$$\downarrow$$

$$\underset{\mathbf{F}}{\text{HOOC-cyclohexane-COOH (1,4)}} \xleftarrow{-CO_2} \xleftarrow{H^+} \xleftarrow{OH^-} \underset{\mathbf{E}}{(COOEt)_2\text{-cyclohexane-}(COOEt)_2\ (1,1,4,4)}$$

(c)

$$\begin{matrix}EtOOC\\ |\\ HC^{\ominus}\\ |\\ EtOOC\end{matrix} + \begin{matrix}COOEt\\ |\\ {}^{\ominus}CH\\ |\\ COOEt\end{matrix} + I_2 \longrightarrow \underset{\mathbf{G}}{\begin{matrix}EtOOC & COOEt\\ | & |\\ HC & -CH\\ | & |\\ EtOOC & COOEt\end{matrix}} + 2I^-$$

$$G \xrightarrow[heat]{OH^-} \xrightarrow{H^+} \begin{matrix}HOOC & COOH\\ | & |\\ HC & -CH\\ | & |\\ HOOC & COOH\end{matrix} \xrightarrow[-2CO_2]{heat} \underset{\substack{\mathbf{H}\\ \text{Succinic acid}}}{HOOCCH_2{-}CH_2COOH}$$

(d)

$$D \xrightarrow{2OEt^-} \begin{matrix}COOEt\\ |\\ \overset{\ominus}{C}{-}COOEt\\ /\\ H_2C\\ |\\ H_2C\\ \backslash\\ \underset{\ominus}{C}{-}COOEt\\ |\\ COOEt\end{matrix} \xrightarrow{I_2} \underset{\mathbf{I}}{\begin{matrix}COOEt\\ |\\ H_2C{-}C{-}COOEt\\ |\quad\ |\\ H_2C{-}C{-}COOEt\\ |\\ COOEt\end{matrix}} + 2I^-$$

$$I \xrightarrow[heat]{OH^-} \xrightarrow{H^+} \begin{matrix}COOH\\ |\\ H_2C{-}C{-}COOH\\ |\quad\ |\\ H_2C{-}C{-}COOH\\ |\\ COOH\end{matrix} \xrightarrow[-2CO_2]{heat} \underset{\mathbf{J}}{\begin{matrix}H\\ |\\ H_2C{-}C{-}COOH\\ |\quad\ |\\ H_2C{-}C{-}COOH\\ |\\ H\end{matrix}}$$

(e)

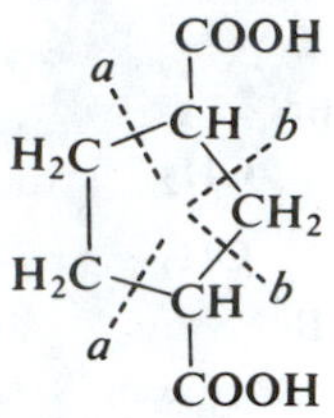

1,3-Cyclopentanedicarboxylic acid

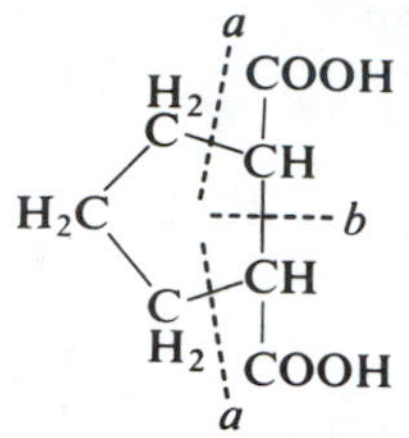

1,2-Cyclopentanedicarboxylic acid

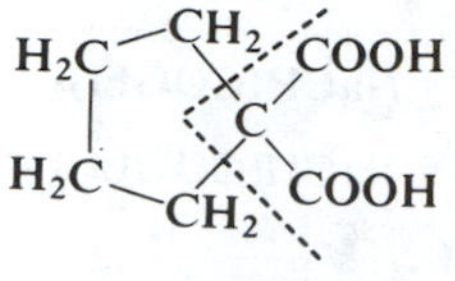

1,1-Cyclopentanedicarboxylic acid

Cyclopentane-1,3-dicarboxylic acid. Add one mole of $BrCH_2CH_2Br$ to an excess of sodiomalonic ester (as in Problem 25.2). Then add 2NaOEt, followed by CH_2I_2. Hydrolyze and decarboxylate.

Cyclopentane-1,2-dicarboxylic acid. As for the 1,3-compound, except that we use $BrCH_2CH_2CH_2Br$ instead of $BrCH_2CH_2Br$, and I_2 instead of CH_2I_2. (Compare Problem 4(c), above.)

Cyclopentane-1,1-dicarboxylic acid. As in Problem 1(i), above, except use $BrCH_2CH_2CH_2CH_2Br$ instead of $BrCH_2CH_2CH_2Br$, and do not decarboxylate after hydrolysis.

5. $2CH_2{=}CHCH_2Br + Mg + BrCH_2CH{=}CH_2 \longrightarrow CH_2{=}CHCH_2{-}CH_2CH{=}CH_2 + MgBr_2$ (K)

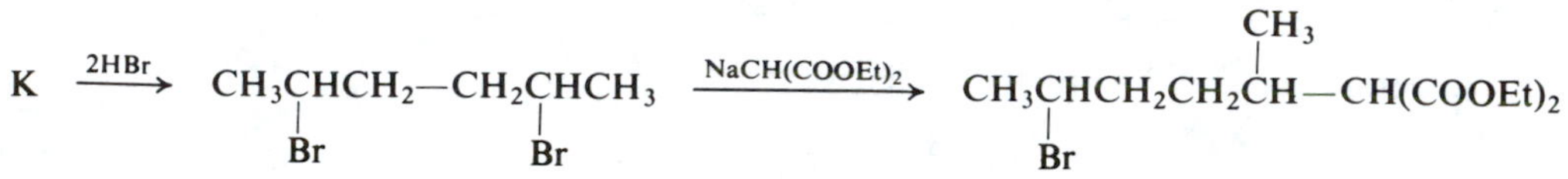

$K \xrightarrow{2HBr} CH_3CH(Br)CH_2{-}CH_2CH(Br)CH_3$ (L) $\xrightarrow{NaCH(COOEt)_2} CH_3CH(Br)CH_2CH_2CH(CH_3){-}CH(COOEt)_2$ (M)

M $\xrightarrow{OEt^-}$ [carbanion $^-C(COOEt)_2$ attacks CH–Br intramolecularly, displacing Br^-] $\longrightarrow$ 1,1-bis(COOEt)-2,5-dimethylcyclopentane (N)

N $\xrightarrow{OH^-}\xrightarrow{H^+}$ 2,5-dimethylcyclopentane-1,1-dicarboxylic acid $\xrightarrow[-CO_2]{heat}$ 2,5-dimethylcyclopentanecarboxylic acid (O)

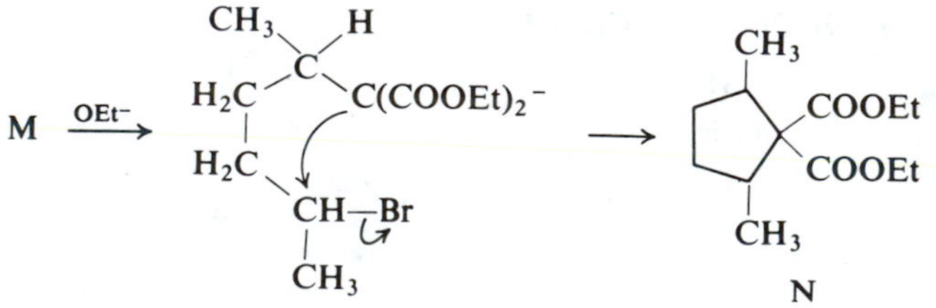

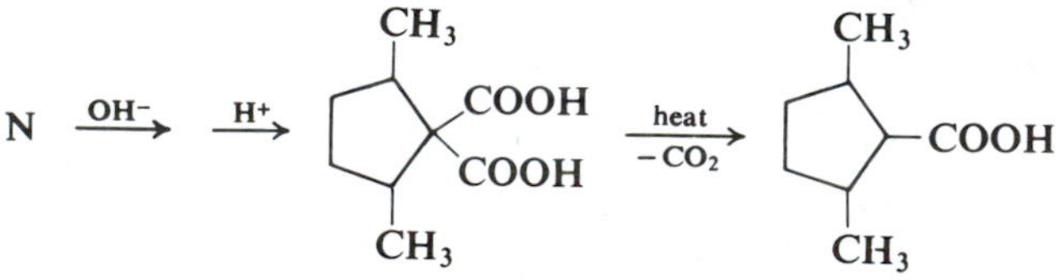

6. (1) $Cl_3C{-}COO^- \xrightarrow{heat} CO_2 + Cl_3C{:}^-$

Trichloroacetate

(2) $Cl_3C{:}^- \rightleftharpoons {:}CCl_2 + Cl^-$

Dichlorocarbene

(3) 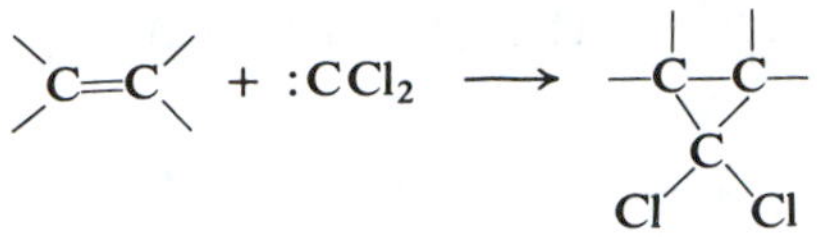 $>C{=}C< + :CCl_2 \longrightarrow$ (dichlorocyclopropane: $-\overset{|}{C}-\overset{|}{C}-$ bridged by CCl_2)

Decarboxylation (1) of Cl_3CCOO^- is comparatively easy because of the stability of the carbanion being formed. (Compare Problem 25.12, page 930, and Problem 25.15, page 931.) Once Cl_3C^- is formed, (2) and (3) follow, as on page 477.

7. (a) Much as in Problem 25.2(a), except that we use two moles of acetoacetic ester instead of malonic ester.

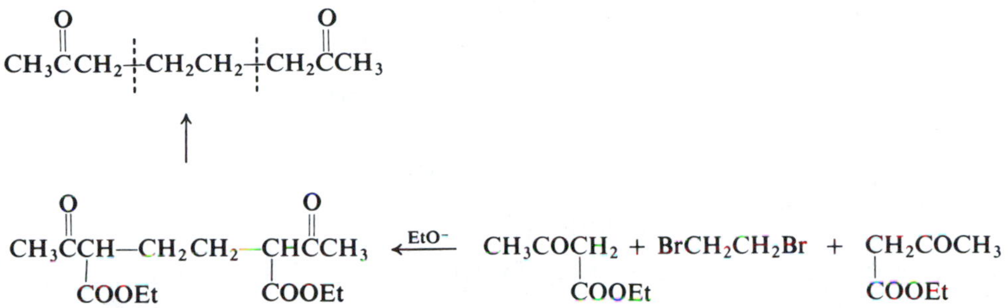

$$CH_3\overset{O}{\overset{\|}{C}}CH_2 \,\vdots\, CH_2CH_2 \,\vdots\, CH_2\overset{O}{\overset{\|}{C}}CH_3$$

$$\uparrow$$

$$CH_3\overset{O}{\overset{\|}{C}}\underset{COOEt}{\underset{|}{CH}}-CH_2CH_2-\underset{COOEt}{\underset{|}{CH}}\overset{O}{\overset{\|}{C}}CH_3 \xleftarrow{EtO^-} CH_3CO\underset{COOEt}{\underset{|}{CH_2}} + BrCH_2CH_2Br + \underset{COOEt}{\underset{|}{CH_2}}COCH_3$$

(b) An intramolecular aldol condensation takes place. Of the two possibilities, the reaction takes place in the way that gives a five-membered ring, not a seven-membered ring.

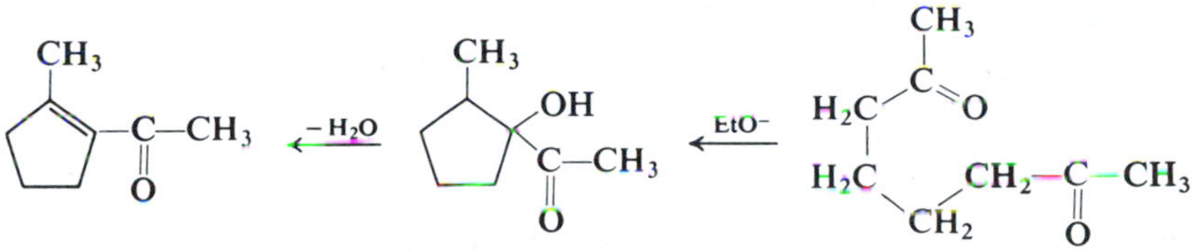

1-acetyl-2-methylcyclopentene $\xleftarrow{-H_2O}$ 1-acetyl-2-methylcyclopentanol (OH, $\overset{O}{\overset{\|}{C}}-CH_3$) $\xleftarrow{EtO^-}$ $CH_3\overset{O}{\overset{\|}{C}}CH_2CH_2CH_2CH_2\overset{O}{\overset{\|}{C}}CH_3$

(c) As in (a), except use CH_2I_2 instead of CH_2BrCH_2Br.

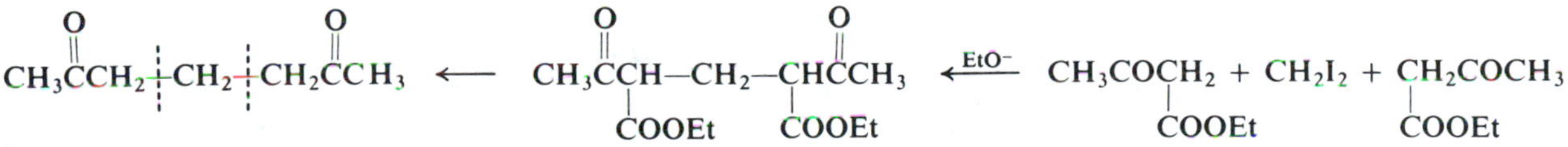

$$CH_3\overset{O}{\overset{\|}{C}}CH_2 \,\vdots\, CH_2 \,\vdots\, CH_2\overset{O}{\overset{\|}{C}}CH_3 \longleftarrow CH_3\overset{O}{\overset{\|}{C}}\underset{COOEt}{\underset{|}{CH}}-CH_2-\underset{COOEt}{\underset{|}{CH}}\overset{O}{\overset{\|}{C}}CH_3 \xleftarrow{EtO^-} CH_3CO\underset{COOEt}{\underset{|}{CH_2}} + CH_2I_2 + \underset{COOEt}{\underset{|}{CH_2}}COCH_3$$

(d) Probably an intramolecular aldol condensation, and in the way that gives a six-membered ring, not a four-membered ring.

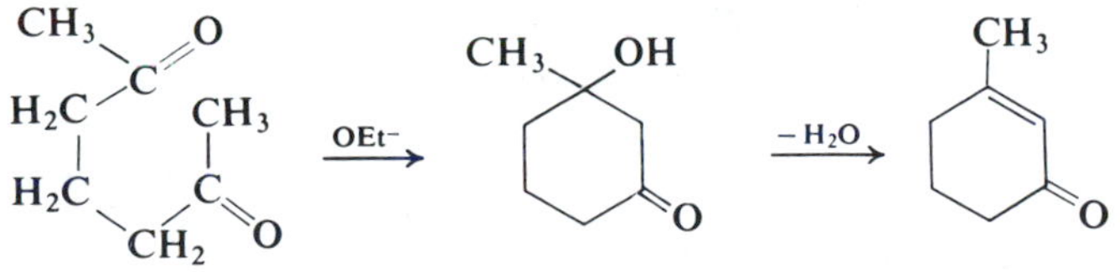

$CH_3\overset{O}{\overset{\|}{C}}CH_2CH_2CH_2\overset{O}{\overset{\|}{C}}CH_3 \xrightarrow{OEt^-}$ 3-hydroxy-3-methylcyclohexanone (CH_3, OH; =O) $\xrightarrow{-H_2O}$ 3-methylcyclohex-2-enone (CH_3; =O)

8. (a) In compound I, page 816, we see two —COOEt's *beta* to C=O and therefore susceptible to easy decarboxylation.

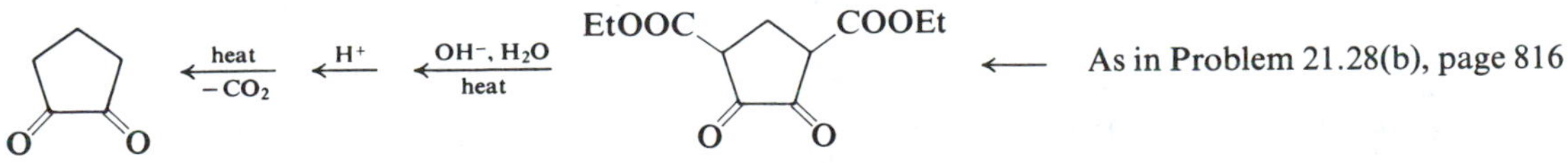

cyclopentane-1,2-dione $\xleftarrow[-CO_2]{heat}$ $\xleftarrow{H^+}$ $\xleftarrow[heat]{OH^-, H_2O}$ EtOOC, COOEt-substituted cyclopentane-1,2-dione $\longleftarrow$ As in Problem 21.28(b), page 816

(b) We follow the general procedure for preparation of α-keto acids, as illustrated in Problem 25.9 (page 930).

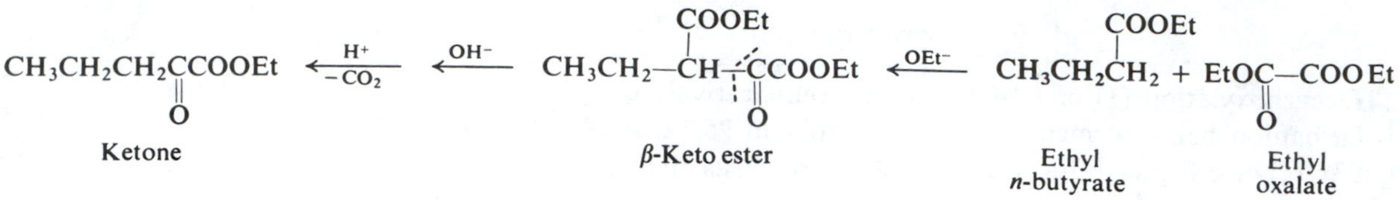

$$\underset{\text{Ketone}}{CH_3CH_2CH_2\underset{\underset{O}{\|}}{C}COOEt} \xleftarrow[-CO_2]{H^+} \xleftarrow{OH^-} \underset{\beta\text{-Keto ester}}{CH_3CH_2-\overset{\overset{COOEt}{|}}{CH}-\underset{\underset{O}{\|}}{C}COOEt} \xleftarrow{OEt^-} \underset{\substack{\text{Ethyl}\\ n\text{-butyrate}}}{CH_3CH_2\overset{\overset{COOEt}{|}}{CH_2}} + \underset{\substack{\text{Ethyl}\\ \text{oxalate}}}{EtO\underset{\underset{O}{\|}}{C}-COOEt}$$

9. These are all prepared by reaction of urea with the appropriate substituted malonic ester:

$$\underset{\text{Urea}}{O{=}C(NH_2)_2} + \underset{\text{Disubstituted malonic ester}}{(C_2H_5O-CO)_2CRR'} \xrightarrow[EtOH]{OEt^-} \underset{\text{Disubstituted barbituric acid}}{\text{O=C}_2\text{(N}_3\text{H-C}_4\text{(=O))(N}_1\text{H-C}_6\text{(=O))}{>}\text{C}_5\text{RR}'}$$

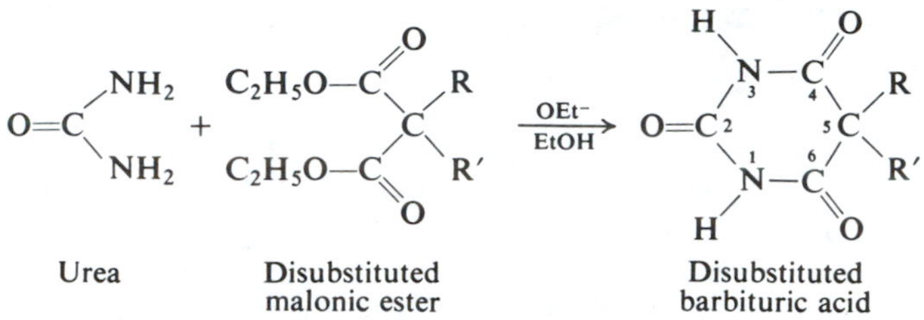

The malonic esters are prepared through successive alkylation with:

(a) C_2H_5Br, twice.

(b) $CH_3CH_2CH_2CHBrCH_3$; $CH_2{=}CHCH_2Br$.

(c) $(CH_3)_2CHCH_2CH_2Br$; C_2H_5Br.

10. (a) Removal of a C–5 proton from barbituric acid gives a resonance-stabilized aromatic anion (six π electrons: two from C–5 and two from each of the two nitrogens); Veronal has no C–5 protons, and cannot form such an anion.

(b) Veronal is an imide, with two acyl groups for each N—H group.

11. (a) This is a *retro*-Claisen condensation: a series of steps that are essentially the reverse of those by which acetoacetic ester is made. Here, the equilibria are shifted in the direction that produces resonance-stabilized, unreactive acetate ions. This reaction is called the *acid cleavage* of acetoacetic ester, in contrast to the more common *ketonic cleavage* used to produce ketones.

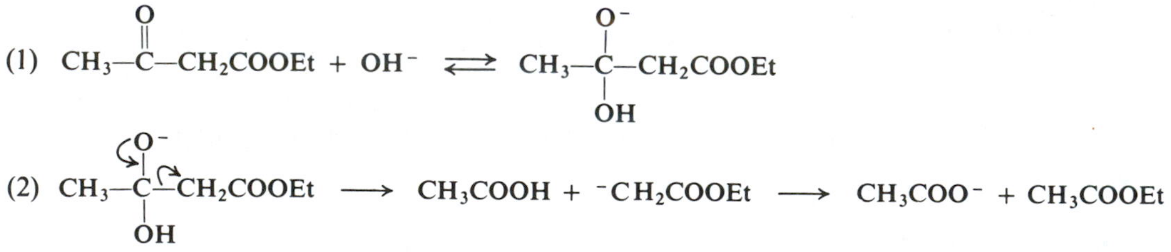

(1) $$CH_3-\overset{\overset{O}{\|}}{C}-CH_2COOEt + OH^- \rightleftarrows CH_3-\overset{\overset{O^-}{|}}{\underset{\underset{OH}{|}}{C}}-CH_2COOEt$$

(2) $$CH_3-\overset{\overset{O^-}{|}}{\underset{\underset{OH}{|}}{C}}-CH_2COOEt \longrightarrow CH_3COOH + {}^-CH_2COOEt \longrightarrow CH_3COO^- + CH_3COOEt$$

(3) $$CH_3COOEt + OH^- \longrightarrow CH_3COO^- + EtOH$$

(b) Alkylate acetoacetic ester as usual, then heat with concentrated base to cause *retro*-Claisen condensation (acid cleavage).

$$CH_3\underset{\underset{O}{\|}}{C}{-}CH_2COOEt \longrightarrow CH_3\underset{\underset{O}{\|}}{C}{-}\underset{\underset{R'}{|}}{\overset{\overset{R}{|}}{C}}{-}COOEt \xrightarrow{\text{conc. } OH^-} CH_3COO^- + RR'CHCOO^-$$

(c)

$$CH_3COCHCOOEt^- + CH_3CH_2CH_2Br \longrightarrow CH_3CH_2CH_2{-}\overset{\overset{COOEt}{|}}{C}H\underset{\underset{O}{\|}}{C}CH_3$$

$$CH_3CH_2CH_2\overset{\overset{COOEt}{|}}{C}H\underset{\underset{O}{\|}}{C}CH_3 \xrightarrow{OH^-}\xrightarrow{H^+} CH_3CH_2CH_2CH_2COCH_3 \qquad \textit{Ketonic cleavage}$$

$$\xrightarrow{OH^-}\xrightarrow{H^+} \underset{n\text{-Valeric acid}}{CH_3CH_2CH_2CH_2COOH} + \underset{\text{Acetic acid}}{CH_3COOH} \qquad \textit{Acid cleavage: side reaction}$$

The crude ketone will contain water, EtOH, acetic acid, and *n*-valeric acid. Remove the acids by washing with aqueous NaOH. Then distill, collecting water, EtOH, and the ketone. Remove the EtOH and some water by shaking with concentrated aqueous $CaCl_2$. Separate the ketone from the aqueous layer, dry over anhydrous $MgSO_4$, and distill.

12. (a) In Problem 21.27 (page 816), we saw that ketones undergo a crossed Claisen condensation with esters to give *β*-diketones. Like the ordinary Claisen condensation (Problem 11, above) this reaction, too, can be reversed, to give the ketone and (instead of the original ester) a carboxylate anion. To use the example in Problem 21.27:

$$CH_3\underset{\underset{O}{\|}}{C}CH_2\underset{\underset{O}{\|}}{C}CH_3 + OH^- \rightleftarrows CH_3{-}\underset{\underset{OH}{|}}{\overset{\overset{O^-}{|}}{C}}{-}CH_2\underset{\underset{O}{\|}}{C}CH_3$$

$$CH_3{-}\underset{\underset{OH}{|}}{\overset{\overset{O^-}{|}}{C}}{-}CH_2\underset{\underset{O}{\|}}{C}CH_3 \longrightarrow CH_3COOH + {}^-CH_2COCH_3 \longrightarrow CH_3COO^- + CH_3\underset{\underset{O}{\|}}{C}CH_3$$

If this *retro*-Claisen is applied to *cyclic β*-diketones, useful products can be obtained with carboxylate and keto groups in the same molecule.

$$\text{2-acylcyclohexanone (ring C=O, ring–C(=O)–R)} + OH^- \rightleftarrows \text{1-hydroxy-1-oxido-2-acylcyclohexane } ({}^-O,\ OH;\ \text{ring–}\underset{\underset{O}{\|}}{C}{-}CH_3)$$

$$\text{(}{}^-O\text{, OH adduct; ring–}\underset{\underset{O}{\|}}{C}{-}R) \longrightarrow \text{open chain: } HOOC\text{–}(H_2C\text{–}H_2C\text{–}CH_2\text{–}CH_2)\text{–}CHCOR^- \longrightarrow {}^-OOC\text{–}(H_2C\text{–}H_2C\text{–}CH_2\text{–}CH_2)\text{–}CH_2COR \equiv R\underset{\underset{O}{\|}}{C}(CH_2)_5COO^-$$

(b) Ketone (cyclohexanone) $\xrightarrow[\text{TsOH}]{\text{pyrroline}}$ Enamine $\xrightarrow{\text{PhCOCl}}$ $\xrightarrow[\text{H}^+]{\text{H}_2\text{O}}$ β-Diketone (2-benzoylcyclohexanone) $\xrightarrow{\text{KOH}}$ $\xrightarrow{\text{H}^+}$ $HOOC(CH_2)_5\overset{O}{\overset{\|}{C}}Ph$ $\xrightarrow[\text{base}]{N_2H_4}$ $HOOC(CH_2)_6Ph$

Ketone   Enamine   β-Diketone

(c) As in (b), except use $ClCO(CH_2)_7COOEt$ instead of PhCOCl. (Azelaic acid, $HOOC(CH_2)_7COOH$, is available from the oxidative cleavage of oleic acid, from fats (page 713).)

13. $n\text{-}C_6H_{13}CHO + BrCH_2COOEt \xrightarrow{\text{Zn}} \xrightarrow{H_2O} n\text{-}C_6H_{13}CH(OH)CH_2COOEt$ (P)

$n\text{-}C_6H_{13}CH(OH)CH_2COOEt$ (P) $\xrightarrow{CrO_3}$ $n\text{-}C_6H_{13}C(=O)CH_2COOEt$ (Q)

A β-keto ester

$n\text{-}C_6H_{13}C(=O)CH_2COOEt$ (Q) $\xrightarrow{\text{NaOEt}} \xrightarrow{PhCH_2Cl}$ $n\text{-}C_6H_{13}C(=O){-}CH(CH_2Ph)COOEt$ (R)

$n\text{-}C_6H_{13}C(=O){-}CH(CH_2Ph)COOEt$ (R) $\longrightarrow$ $n\text{-}C_6H_{13}C(=O){-}CH(CH_2Ph)COO^-$ $\xrightarrow{-CO_2}$ $CH_3(CH_2)_5C(=O)CH_2CH_2Ph$ (S)

1-Phenyl-3-nonanone

14. 1,5-Cyclooctadiene $\xrightarrow{B_2H_6}$ BH (1,4-) + BH (1,5-) $\xrightarrow{\text{heat}}$ BH (1,5-)

1,4-   1,5-   1,5-

A mixture T   *Pure* U

$\downarrow H_2O_2, OH^-$

HO, HO: 1,4-diol   HO, OH: 1,5-diol

Hydroboration of the double bonds takes place in the two possible ways to give T, a mixture of 1,4- and 1,5-isomers. Heating causes equilibration, presumably through reversible hydroboration, to yield almost entirely the more stable 1,5-isomer, U, the one containing two six-membered rings. U is 9-BBN (page 934).

The oxidation takes place with retention of configuration (Problem 24.8, page 898). Formation of the *cis*-diols shows, then, that bridging in both 1,4- and 1,5-boranes is *cis*, which is what we would expect on steric grounds.

H_2B —(B_2H_6)→ ; HB 1,4- $\xrightarrow[OH^-]{H_2O_2}$ *cis*-1,4-diol ; HB 1,5- $\xrightarrow[OH^-]{H_2O_2}$ *cis*-1,5-diol

(Knight, E. F.; Brown, H. C. "Cyclic Hydroboration of 1,5-Cyclooctadiene. A Simple Synthesis of 9-Borabicyclo[3.3.1]nonane, an Unusually Stable Dialkylborane"; *J. Am. Chem. Soc.* **1968**, *90*, 5280.)

15. (a)

$$ArCHO + OH^- \rightleftharpoons Ar{-}\overset{H}{\underset{OH}{C}}{-}O^- + H_2O$$

$$Ar{-}\overset{H}{\underset{OH}{C}}{-}O^- + OH^- \rightleftharpoons Ar{-}\overset{H}{\underset{O^-}{C}}{-}O^- + H_2O$$

$$Ar{-}\overset{H}{\underset{O^-}{C}}{-}O^- \longrightarrow Ar^- + HCOO^-$$

$$Ar^- + H_2O \longrightarrow ArH + OH^-$$

(b) The two *o*-Cl's stabilize Ar^- and the transition state leading to its formation.

(Bunnett, J. F.; Miles, J. H.; Nahabedian, K. V. "Kinetics and Mechanism of the Alkali Cleavage of 2,6-Dihalobenzaldehydes"; *J. Am. Chem. Soc.* **1961**, *83*, 2512.)

16. Chemical arithmetic,

$$2C_4H_6O_2 - C_8H_{10}O_3 = H_2O$$

indicates combination of two moles of lactone with loss of one H_2O, and strongly suggests an aldol-like condensation followed by dehydration:

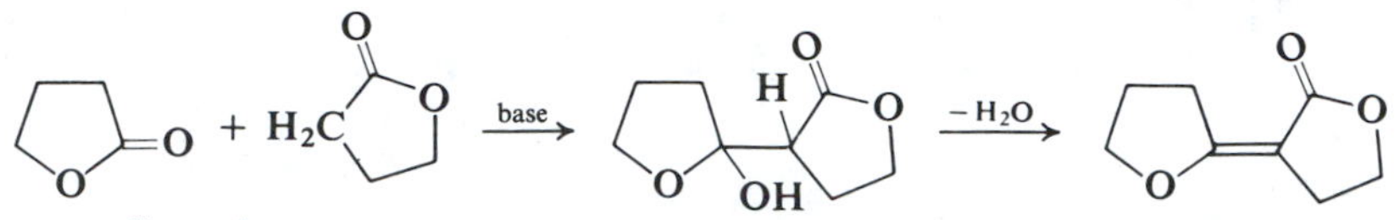

γ-Butyrolactone
Two moles

base

$-H_2O$

V

V is a lactone (an ester) and a vinyl ether. Both these functional groupings are cleaved by concentrated HCl: the vinyl ether to give a ketone and alcohol (see the answer to Problem 17, Chapter 18); the lactone to give an alcohol and carboxylic acid. At this point we have a

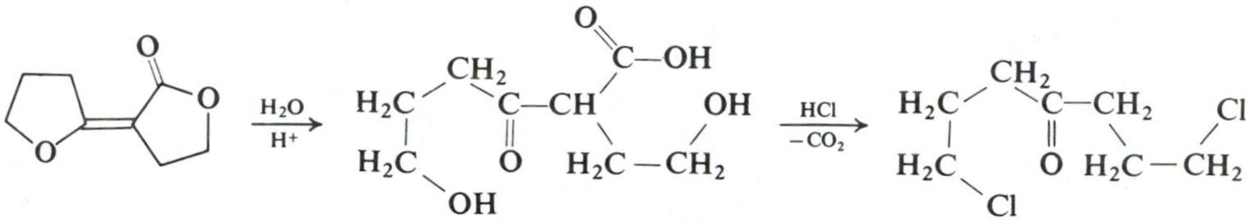

Vinyl ether–ester
V

$\xrightarrow[H^+]{H_2O}$

Ketone–alcohol *Acid–alcohol*
A dihydroxy β-keto acid

$\xrightarrow[-CO_2]{HCl}$

Dichloroketone
W

dihydroxy β-keto acid; decarboxylation and replacement of the —OH's by —Cl's give W.

In the presence of acid, the ketone W undergoes intramolecular alkylation at each of the α-carbons, to give dicyclopropyl ketone.

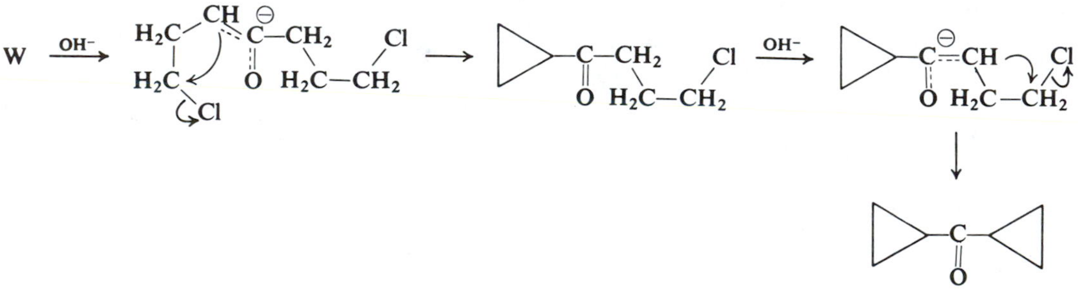

$W \xrightarrow{OH^-}$ → $\xrightarrow{OH^-}$ →

Dicyclopropyl ketone

17. (a) $RX + NaCH(COOEt)CCH_3(=O) \longrightarrow RCH(COOEt)CCH_3(=O) \xrightarrow{OH^-} RCH(COO^-)CCH_3(=O) \xrightarrow[warm]{-CO_2} RCH_2CCH_3(=O)$

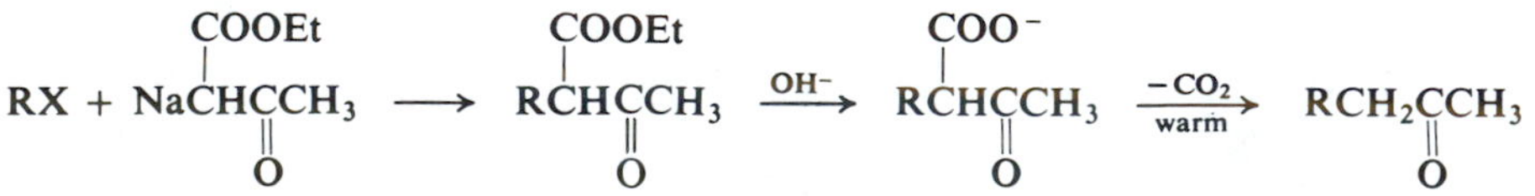

X Y

$RCH_2C(CH_3)=O + {}^-:C{\equiv}CH \xrightarrow{+H^+} RCH_2C(CH_3)(OH)-C{\equiv}CH \xrightarrow{redn.} RCH_2C(CH_3)(OH)-CH{=}CH_2$

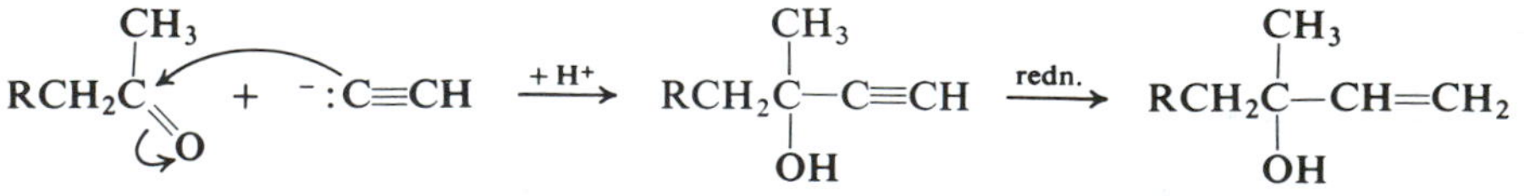

Y Z AA
Nerolidol

(b) 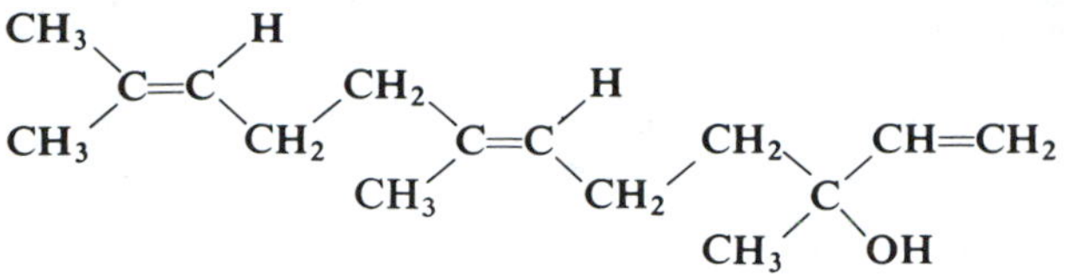 CH_3, CH_3 — C=C — H; CH_2, CH_2 — C=C — H, CH_3; CH_2, CH_2 — C — $CH=CH_2$, CH_3, OH

Nerolidol

18. (a) In the first step, C_2H_5OH is lost ($C_{12}H_{22}O_4 - C_{10}H_{16}O_3 = C_2H_5OH$) suggesting an intramolecular Claisen condensation (Dieckmann condensation, Problem 21.25, page 815). This could take place in either of two ways:

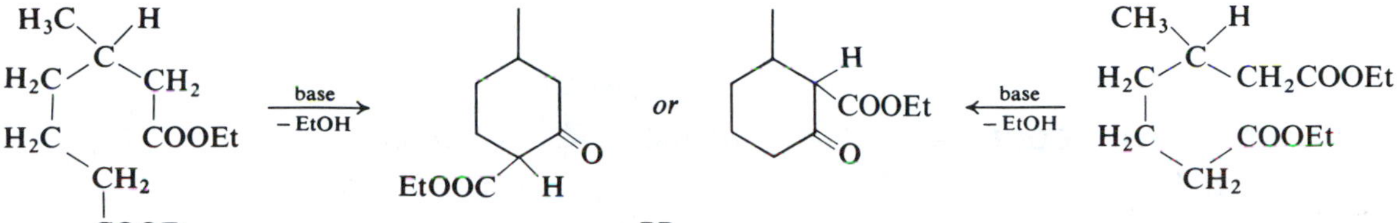

H_3C, H, C, H_2C, CH_2, H_2C, COOEt, CH_2, COOEt —base, − EtOH→ [ring] O, EtOOC, H *or* [ring] H, COOEt, O ←base, − EtOH— CH_3, H, C, H_2C, CH_2COOEt, H_2C, COOEt, CH_2

BB

Possible structures

Depending upon which is the correct structure for BB, menthone could have either of two structures:

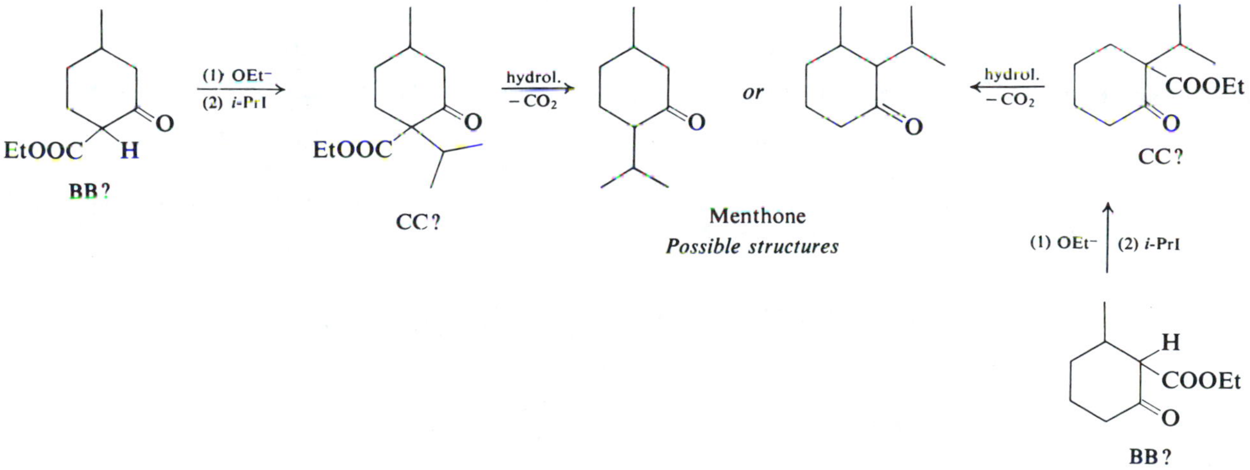

O, EtOOC, H BB? (1) OEt^- (2) *i*-PrI O, EtOOC CC? hydrol. − CO_2 O *or* O hydrol. − CO_2 COOEt, O CC?

Menthone

Possible structures

(1) OEt^- (2) *i*-PrI ↑

[ring] H, COOEt, O

BB?

(b) Of the two structures possible at this point, the more likely is the one that follows the isoprene rule.

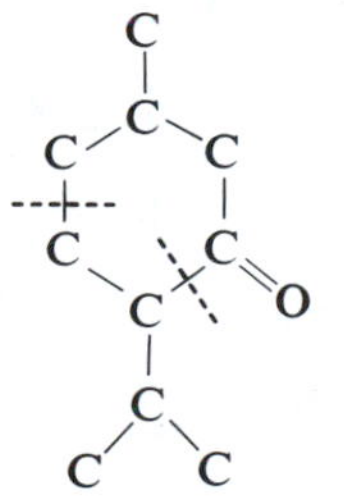

C, C, C, C, C, C, C, O, C, C, C, C

More likely: two isoprene units

(c) The reduction confirms the structure indicated in (b).

$$\text{4-Isopropyl-1-methylcyclohexane} \xleftarrow[\text{reduction}]{\text{vigorous}} \text{Menthone}$$

4-Isopropyl-1-methylcyclohexane

Menthone

19.

$$[CH_3COCHCOOEt]^-Na^+ \xrightarrow{CH_3I} \underset{\textbf{DD}}{CH_3COCH(CH_3)COOEt} \xrightarrow{NaOEt} \xrightarrow{CH_3I} \underset{\textbf{EE}}{CH_3COC(CH_3)_2COOEt}$$

$$\underset{\textbf{EE}}{CH_3-C(=O)-C(CH_3)_2-COOEt} + BrCH_2COOEt \xrightarrow{Zn} \xrightarrow{H_2O} \underset{\textbf{FF}}{EtOOCCH_2-C(H_3C)(HO)-C(CH_3)_2-COOEt}$$

$$\textbf{FF} \xrightarrow{PCl_5} EtOOCCH_2-C(H_3C)(Cl)-C(CH_3)_2-COOEt \xrightarrow{CN^-} \underset{\textbf{GG}}{EtOOCCH_2-C(H_3C)(NC)-C(CH_3)_2-COOEt} \xrightarrow{H_2O,\ H^+} \underset{\text{Camphoronic acid}}{HOOCCH_2-C(H_3C)(HOOC)-C(CH_3)_2-COOH}$$

20.

$$CH_3COCH(COOEt)^-Na^+ \xrightarrow{ClCH_2COOEt} \underset{\textbf{HH}}{CH_3COCH(COOEt)-CH_2COOEt} \xrightarrow{CH_3MgI} \xrightarrow{H_2O} \underset{\textbf{II}}{CH_3-C(OH)(H_3C)-CH(COOEt)-CH_2COOEt}$$

(*Ketone reacts faster than ester*)

$$\textbf{II} \xrightarrow[\text{heat}]{OH^-} \xrightarrow{H^+} \underset{\substack{\textbf{JJ} \\ \textit{A } \gamma\textit{-hydroxy acid}}}{CH_3-C(OH)(H_3C)-CH(COOH)-CH_2COOH} \xrightarrow{-H_2O} \underset{\substack{\text{Terebic acid} \\ \textit{A } \gamma\textit{-lactone}}}{\text{Terebic acid}}$$

Terebic acid

A γ-lactone

$$\underset{\textbf{HH}}{CH_3COCH(COOEt)CH_2COOEt} \xrightarrow{OEt^-} \underset{\textit{Most acidic H removed}}{CH_3CO\overset{\ominus}{C}(COOEt)CH_2COOEt} \xrightarrow{ClCH_2COOEt} \underset{\textbf{KK}}{CH_3COC(CH_2COOEt)(COOEt)CH_2COOEt}$$

$$\underset{\substack{\textbf{KK} \\ \textit{A } \beta\textit{-keto ester}}}{CH_3COC(CH_2COOEt)(COOEt)CH_2COOEt} \xrightarrow{OH^-} \xrightarrow[-CO_2]{H^+} \underset{\textbf{LL}}{CH_3COCH(CH_2COOH)CH_2COOH} \xrightarrow{2EtOH,\ H^+} \underset{\textbf{MM}}{CH_3COCH(CH_2COOEt)CH_2COOEt}$$

$$\text{MM} \xrightarrow{CH_3MgI} \xrightarrow{H_2O} CH_3-\overset{\overset{HO}{|}}{\underset{\underset{H_3C}{|}}{C}}-\overset{\overset{CH_2COOEt}{|}}{\underset{\underset{H}{|}}{C}}CH_2COOEt \xrightarrow{OH^-} \xrightarrow{H^+} CH_3-\overset{\overset{HO}{|}}{\underset{\underset{H_3C}{|}}{C}}-\overset{\overset{CH_2COOH}{|}}{\underset{\underset{H}{|}}{C}}CH_2COOH \xrightarrow{-H_2O}$$

NN

OO
A γ-hydroxy acid

Terpenylic acid (lactone ring with CH_3, CH_3, CH_2COOH, H)
A γ-lactone

21. As usual, the organolithium compound adds to the carbonyl group of the aldehyde to give a 2° alcohol.

(a) CH_3–(4,4-dimethyloxazoline) I (R = H) $\xrightarrow{n\text{-BuLi}}$ $Li^{+}\,{}^{-}CH_2$–(4,4-dimethyloxazoline) $\xrightarrow{CH_3(CH_2)_5CHO}$ $n\text{-}C_6H_{13}-\overset{\overset{H}{|}}{\underset{\underset{OLi}{|}}{C}}-CH_2$–(4,4-dimethyloxazoline)

PP

$$\text{PP} \xrightarrow[H_2SO_4]{EtOH} n\text{-}C_6H_{13}\underset{\underset{OH}{|}}{C}H-CH_2COOEt$$

QQ
A β-hydroxy ester

(b) As in (a), except use $(n\text{-Pr})_2C{=}O$ as the carbonyl compound.

$$CH_3CH_2CH_2-\overset{\overset{CH_3CH_2CH_2}{|}}{\underset{\underset{OH}{|}}{C}}\,\vdots\,CH_2COOEt$$

(c) As in (a), except use PhCHO and I (R = Et).

$$Ph-\underset{\underset{OH}{|}}{C}H\,\vdots\,\overset{\overset{C_2H_5}{|}}{C}HCOOEt$$

26

Aryl Halides

Nucleophilic Aromatic Substitution

26.1 (a) See Problem 15.11 (page 544). (b) See Sec. 15.19.

26.2 There is a stronger Ar—O bond in phenols, due to partial double-bond character and/or sp^2 hybridization of aromatic carbon.

26.3 (a)

$$O{=}N\text{-}C_6H_4\text{-}NMe_2 \xrightarrow{OH^-} {}^{\ominus}\!:\ddot{O}\text{—}\ddot{N}{=}C_6H_4(OH)(NMe_2)\ (\text{I}) \xrightarrow{H_2O} Me_2NH + O{=}N\text{-}C_6H_4\text{-}OH \xrightarrow{KOH} O{=}N\text{-}C_6H_4\text{-}O^-K^+$$

Potassium *p*-nitrosophenoxide

(b) Nucleophilic aromatic substitution.

(c) Electron withdrawal—or, more properly, electron acceptance—by the nitroso group, which stabilizes the transition state leading to intermediate I.

(d)

$$C_6H_5NO_2 \xrightarrow[H^+]{Fe} C_6H_5NH_2 \xrightarrow{2EtI} C_6H_5NEt_2 \xrightarrow{HONO} p\text{-}O{=}N\text{-}C_6H_4\text{-}NEt_2 \xrightarrow{OH^-} Et_2NH + p\text{-}O{=}N\text{-}C_6H_4\text{-}O^-$$

$$EtOH \xrightarrow{P,\ I_2} EtI$$

26.4 (a) Nucleophilic aromatic substitution of —OH for —OCH_3, with activation by the two —NO_2 groups *ortho* and *para* to —OCH_3.

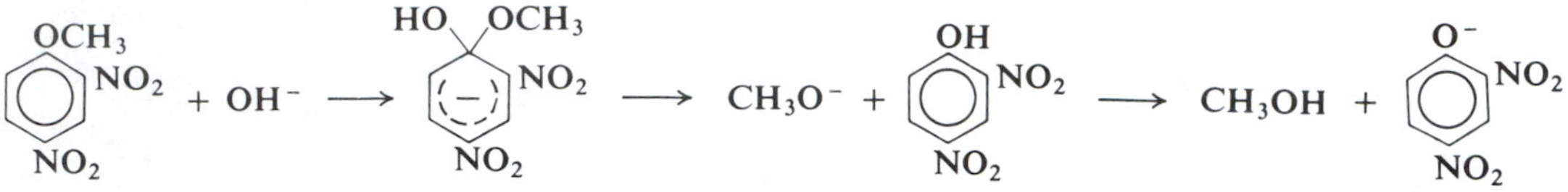

(b) The hydrolysis product, *p*-nitroaniline, is activated toward nucleophilic aromatic substitution by the *p*-nitro group. In the alkaline medium, some of the product suffers displacement of —NH_2 by —OH to give *p*-nitrophenol; the yield is lowered, and the phenol and its oxidation products make the product harder to purify.

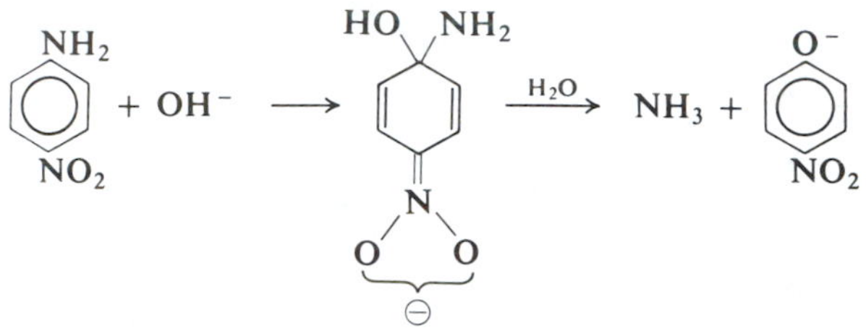

(c) Nucleophilic displacement of —Cl by —SO_3Na. Here the reagent, SO_3^{2-}, is a nucleophile, with an unshared pair of electrons on sulfur. In ordinary sulfonation, the reagent, SO_3, is an electrophile, with electron-deficient sulfur.

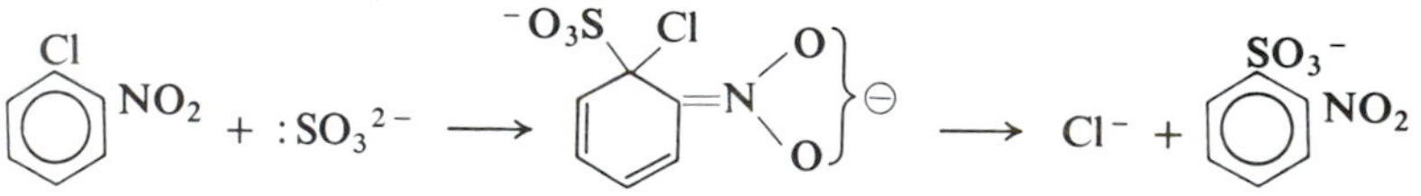

(d) It is not a general method, since it requires the ring to be activated by substituents like —NO_2 *ortho* and/or *para* to the point of attack. Furthermore, this kind of substitution cannot be used to displace H^-, but only weakly basic leaving groups like Cl^-. It could not be used to prepare benzenesulfonic acid.

(e) As in (c), there is nucleophilic aromatic substitution with SO_3^{2-} as nucleophile and, this time, NO_2^- as leaving group. In the *ortho* or *para* isomer, each NO_2 activates the other, and a water-soluble benzenesulfonate salt is obtained:

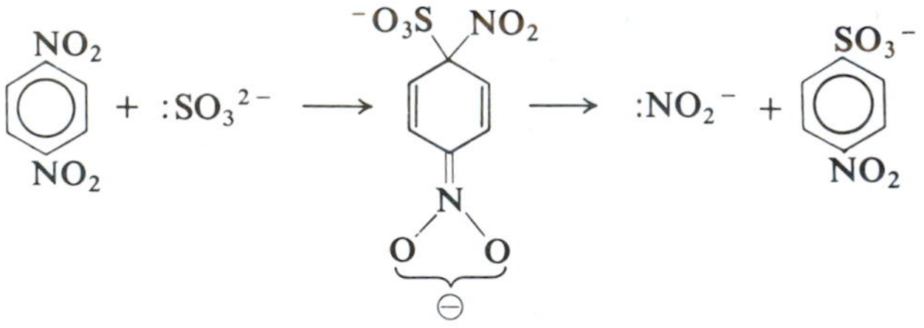

The *meta* isomer does not react, and remains insoluble.

26.5 The —NO can help disperse either kind of charge, negative or positive, developing in the transition state leading to the intermediate in aromatic substitution.

Z:⁻ + X–C₆H₄–N̈=Ö: ⟶ (X)(Z)C₆H₄=N–Ö:⁻ **Nucleophilic**

Y⁺ + C₆H₅–N̈=Ö: ⟶ (H)(Y)C₆H₅=N⁺=Ö: **Electrophilic**

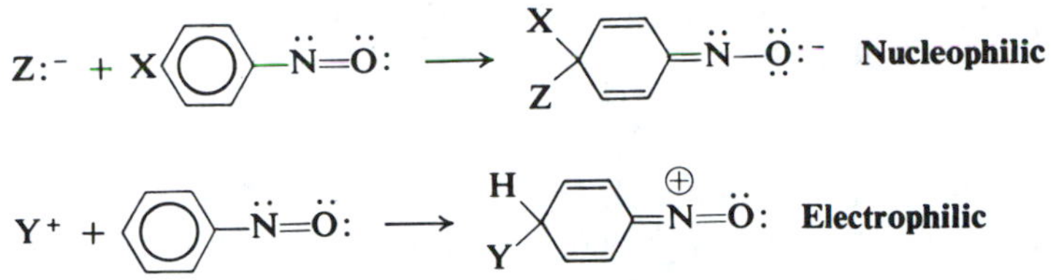

26.6 Chemical arithmetic indicates the addition of sodium ethoxide,

$C_9H_{10}O_8N_3Na$	product
$-C_7H_5O_7N_3$	starting material
C_2H_5ONa	equivalent to sodium ethoxide

or, in the second case, of sodium methoxide. The product II is the same from both reac-

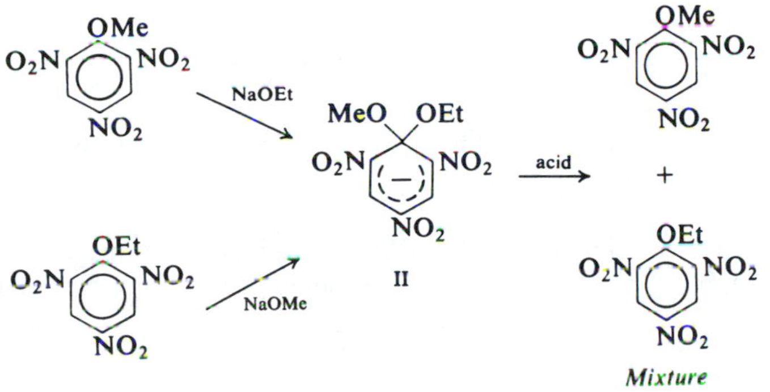

2,4,6-trinitroanisole (OMe, O_2N, NO_2, NO_2) —NaOEt→ II (MeO, OEt; O_2N, NO_2, NO_2; (−)) ←NaOMe— 2,4,6-trinitrophenetole (OEt, O_2N, NO_2, NO_2)

II —acid→ OMe, O_2N, NO_2, NO_2 + OEt, O_2N, NO_2, NO_2

Mixture

tions, and is a stable example of the intermediate in nucleophilic aromatic substitution by the bimolecular mechanism.

26.7 The phenyl anion, $C_6H_5^-$, is a stronger base than the 2-F-3-$CH_3OC_6H_3^-$ anion, which contains two electron-withdrawing, base-weakening substituents, —F and —OCH_3. Conversely, *o*-fluoroanisole is a stronger acid than benzene.

26.8 The organolithium compound, like a Grignard reagent, reacts:

(a) with CO_2 to give a carboxylic acid

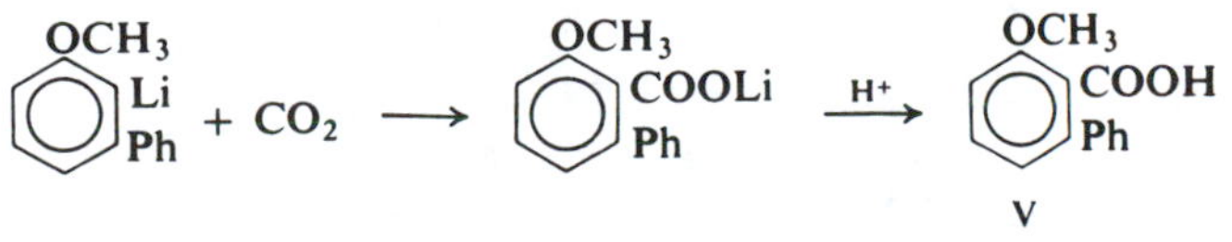

OCH_3, Li, Ph (benzene) + CO_2 ⟶ OCH_3, COOLi, Ph (benzene) —H^+→ OCH_3, COOH, Ph (benzene)

V

(b) with a ketone to give a tertiary alcohol

$$o\text{-}CH_3O\text{-}C_6H_3(Li)(Ph) + Ph{-}\underset{\underset{O}{\|}}{C}{-}Ph \longrightarrow o\text{-}CH_3O\text{-}C_6H_3(C(OLi)Ph_2)(Ph) \xrightarrow{H^+} \underset{\textbf{VI}}{o\text{-}CH_3O\text{-}C_6H_3(C(OH)Ph_2)(Ph)}$$

(c) Magnesium reacts at the —Br bond to give the Grignard reagent, which is analogous to the organolithium product of reaction (5), page 966, and reacts as in (6) to give the benzyne.

$$o\text{-}BrC_6H_4F + Mg \longrightarrow o\text{-}\overset{\delta-}{C_6H_4}(\overset{\delta+}{MgBr})F \longrightarrow MgBrF + \underset{\text{Benzyne}}{C_6H_4}$$

26.9 (a) $AgNO_3$

(b) $AgNO_3$

(c) Br_2/CCl_4, or $KMnO_4$

(d) iodoform test; or CrO_3/H_2SO_4

(e) iodoform test

26.10 (a) Oxidation to acids; then determination of their m.p.s (but both are very high), or their neutralization equivalents (201 *vs.* 123).

(b) Ozonolysis and identification of fragments (CH_3CHO *vs.* HCHO), or isomerization of one isomer (the allylic compound) into the other (the propenyl isomer) by hot KOH (see Problem 25, Chapter 24).

1. (a) PhMgBr (h) *o*- and *p*-$BrC_6H_4NO_2$ (i) *o*- and *p*-$BrC_6H_4SO_3H$ (j) *o*- and *p*-BrC_6H_4Cl (m) *o*- and *p*-$BrC_6H_4C_2H_5$

No reaction: b, c, d, e, f, g, k, l, n, o.

2. (a) *n*-BuMgBr

(b) *n*-BuOH

(c) $CH_3CH_2CH{=}CH_2$

(d) *n*-BuC≡CH

(e) *n*-Bu—O—Et

(f) *n*-$BuNH_2$

(g) *n*-BuCN

(l) Ph—Bu-*n* and Ph—Bu-*sec*

No reaction: h, i, j, k, m, n, o.

3. (b) 2,4-$(NO_2)_2C_6H_3OH$ (e) 2,4-$(NO_2)_2C_6H_3OEt$ (f) 2,4-$(NO_2)_2C_6H_3NH_2$ (g) 2,4-$(NO_2)_2C_6H_3CN$

4. (a) Mg, anhydrous Et_2O; H_2O (b) HNO_3, H_2SO_4, warm

(c) Cl_2, Fe (d) SO_3, H_2SO_4

(e) 1,2,4-$C_6H_3Br_3$ $\xleftarrow{Br_2,\ Fe,\ heat}$ p-BrC_6H_4Br + o-BrC_6H_4Br $\xleftarrow{Br_2,\ Fe}$ C_6H_5Br

(f) CH_3Cl, $AlCl_3$

(g) C_6H_5⋮CH_2OH $\xleftarrow{HCHO}$ C_6H_5MgBr $\xleftarrow{Mg}$ C_6H_5Br

(h) C_6H_5⋮$CH(OH)CH_3$ $\xleftarrow{CH_3CHO}$ C_6H_5MgBr (as in g)

(i) C_6H_5⋮$C(CH_3)_2OH$ $\xleftarrow{CH_3COCH_3}$ C_6H_5MgBr (as in g)

(j) 2,4-$(O_2N)_2C_6H_3OH$ $\xleftarrow{H^+}$ $\xleftarrow[warm]{OH^-}$ 2,4-$(O_2N)_2C_6H_3Br$ $\xleftarrow[twice]{HNO_3,\ H_2SO_4}$ C_6H_5Br

(See Problem 3b, above.)

(k) C_6H_5—$CH_2CH{=}CH_2$ ⟵ C_6H_5MgBr (as in g) + $BrCH_2CH{=}CH_2$

(l) C_6H_5COOH $\xleftarrow{H^+}$ $\xleftarrow{CO_2}$ C_6H_5MgBr (as in g)

(m) $NaNH_2/NH_3$(liq)

5. (a) C_6H_6 (b) C_6H_6 (c) C_6H_6 (d) $PhCH_2CH{=}CH_2$

(e) $PhCH_2OH$ (f) $PhCHOHCH_3$ (g) PhCHOHPh (h) *p*-$CH_3C_6H_4CHOHPh$

(i) Ph—C(CH_3)(OH)—CH_3 (j) 1-phenylcyclohexanol (OH, Ph on same carbon) (k) 1-phenyl-3,3-dimethylcyclohexanol (OH, Ph; CH_3, CH_3) (l) Ph—C(Ph)(OH)—CH_3

(m) Ph_3COH (n) Ph—C(Ph)(OH)—CH(CH_3)—C_2H_5 (o) C_6H_6 + HC≡CMgBr

Optically active

Racemic modification: f, h, k. (Optically inactive.)
Optically active single compound: n.
All others: optically inactive single compounds.

6. (a) 2,4,6-triNO_2 > 2,4-diNO_2 > *o*-NO_2 > *m*-NO_2, unsubstd.

(b) $PhCH_3$ > PhH > PhCl > $PhNO_2$

(c) $CH_3CH(Br)CH{=}CH_2$ (Allylic) > $CH_2(Br)CH_2CH{=}CH_2$ (1° alkyl) > $CH_3CH_2CH{=}CH(Br)$ (Vinylic)

(d) PhMe > *p*-BrC_6H_4Me > PhBr > *p*-BrC_6H_4Br

(e) $PhCH_2Cl$ (Benzylic) > EtCl (1° alkyl) > PhCl (Aryl)

(f) $PhCH(Br)CH_3$ (Benzylic) > $PhCH_2CH_2Br$ (1° alkyl) > $PhCH{=}CH(Br)$ (Vinylic)

7. Aqueous $NaHCO_3$, to minimize nucleophilic substitution by OH^- that would convert part of the product into 2,4-dinitrophenol (compare Problem 26.4b).

8. We must not forget the preference for the reaction of one kind of halogen over another: (a) allyl over vinyl; (b) benzyl over aryl; (e) *para* over *meta*; (f) ArBr over ArCl; (i) benzylic over aryl.

(a) $CH_2{=}C(Br){-}CH_2OH$ (b) *p*-$BrC_6H_4CH_2NH_2$ (c) *p*-ClC_6H_4COOH (d) *m*-$BrC_6H_4CH(Br){-}CH_2Br$

(e) 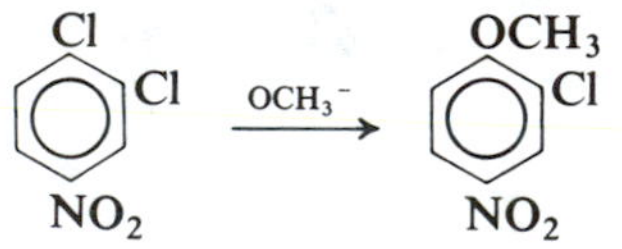 2,4-dichloronitrobenzene (Cl, Cl, NO_2 on ring) $\xrightarrow{OCH_3^-}$ (OCH_3, Cl, NO_2 on ring) (f) *p*-ClC_6H_4MgBr (g) no reaction

(h) *p*-$BrC_6H_4CH_2Br$ (i) 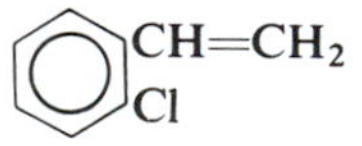 ring bearing $CH{=}CH_2$ and Cl (j) *p*-$BrC_6H_4CH_2Cl$

(k) 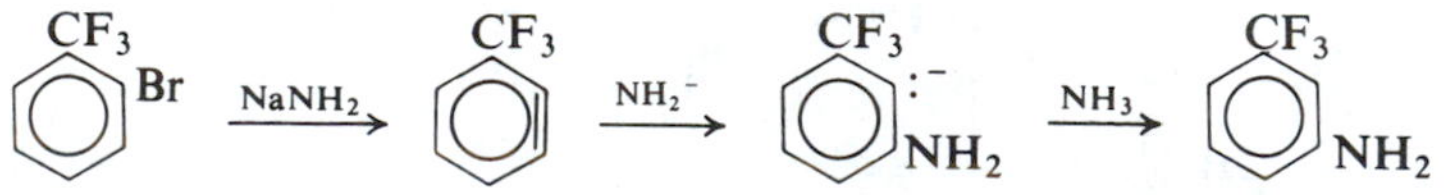 CF_3 Br $\xrightarrow{NaNH_2}$ CF_3 $\xrightarrow{NH_2^-}$ CF_3 :$^-$ NH_2 $\xrightarrow{NH_3}$ CF_3 NH_2

Aryne — *More stable anion* (See Sec. 26.14)

(l) 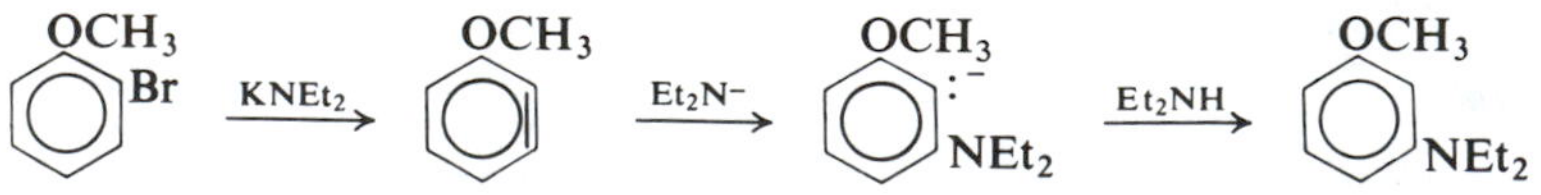 OCH_3 Br $\xrightarrow{KNEt_2}$ OCH_3 $\xrightarrow{Et_2N^-}$ OCH_3 :$^-$ NEt_2 $\xrightarrow{Et_2NH}$ OCH_3 NEt_2

Aryne — *More stable anion* (See Sec. 26.14)

9. (a) $C_6H_4(NO_2)Cl$ (ring with NO_2, Cl) $\xleftarrow{Cl_2,\ Fe}$ $C_6H_5NO_2$ (ring with NO_2) $\xleftarrow{HNO_3,\ H_2SO_4}$ C_6H_6

(b) ring with NO_2, Cl $\xleftarrow{HNO_3,\ H_2SO_4}$ ring with Cl $\xleftarrow{Cl_2,\ Fe}$ C_6H_6

(c) ring with COOH, Br $\xleftarrow[heat]{Br_2,\ Fe}$ ring with COOH $\xleftarrow{KMnO_4}$ ring with CH_3

(d) ring with COOH, Br $\xleftarrow{KMnO_4}$ ring with CH_3, Br $\xleftarrow{Br_2,\ Fe}$ ring with CH_3

(e) ring with CCl_3, Cl $\xleftarrow{Cl_2,\ Fe}$ ring with CCl_3 $\xleftarrow{3Cl_2,\ heat}$ ring with CH_3

(f) ring with NO_2, Br, Br $\xleftarrow{Br_2,\ Fe}$ ring with NO_2, Br $\xleftarrow{HNO_3,\ H_2SO_4}$ ring with Br $\xleftarrow{Br_2,\ Fe}$ C_6H_6

(g) ring with $CHCl_2$, Br $\xleftarrow{2Cl_2,\ heat}$ ring with CH_3, Br $\xleftarrow{Br_2,\ Fe}$ ring with CH_3

(h) ring with NH_2, NO_2, NO_2 $\xleftarrow[warm]{NH_3}$ ring with Cl, NO_2, NO_2 $\xleftarrow[twice]{HNO_3,\ H_2SO_4}$ ring with Cl $\xleftarrow{Cl_2,\ Fe}$ C_6H_6

(i) ring with $CH{=}CH_2$, Br $\xleftarrow{KOH(alc)}$ ring with $CHClCH_3$, Br $\xleftarrow[heat]{Cl_2}$ ring with CH_2CH_3, Br $\xleftarrow[Fe]{Br_2}$ ring with Et (page 578)

(j) ring with COOH, Br, Br $\xleftarrow{KMnO_4}$ ring with CH_3, Br, Br $\xleftarrow[twice]{Br_2,\ Fe}$ ring with CH_3

(k) ring with CH_3, I $\xleftarrow{KI}$ ring with CH_3, N_2^+ $\xleftarrow{HONO}$ ring with CH_3, NH_2 $\xleftarrow{Fe,\ H^+}$ ring with CH_3, NO_2 (Problem 23.12, page 872)

(l) p-Bromobenzenesulfonic acid (SO_3H, Br) $\xleftarrow{SO_3, H_2SO_4}$ Bromobenzene $\xleftarrow{Br_2, Fe}$ C_6H_6

(m) p-Chlorobenzyl alcohol (CH_2OH, Cl) $\xleftarrow{NaOH(aq)}$ p-Chlorobenzyl chloride (CH_2Cl, Cl) $\xleftarrow{Cl_2, heat}$ p-Chlorotoluene (CH_3, Cl) $\xleftarrow{Cl_2, Fe}$ Toluene (CH_3)

(n) $H_3C\diagdown CH(CH_3)$ (CH_3) $\xleftarrow[Ni]{H_2}$ $H_3C-C(=CH_2)$ (CH_3) $\xleftarrow[heat]{H^+}$ $CH_3-C(OH)-CH_3$ (CH_3) $\xleftarrow{CH_3COCH_3}$ MgBr (CH_3) $\xleftarrow{Mg}$ Br (CH_3) (as in d)

10. See Sec. 30.10.

11. Hydrogen exchange, we have seen, takes place via the carbanion (pages 964–965). Its rate depends upon how fast the carbanion is formed. This, in turn, depends upon accommodation of the negative charge developing in the transition state leading to the formation of the carbanion. Through its inductive effect, fluorine helps to accommodate this negative charge. The inductive effect depends on the distance from the negative carbon, and is *much* stronger from the *ortho* position than from the *meta* or *para* positions. (See the reference given in the answer to Problem 17, below.)

12. The $—N_2^+$ group very powerfully activates the ring toward nucleophilic aromatic substitution. Chloride ion displaces much —Br before reduction occurs.

13. (a) 28, N_2; 44, CO_2; 76, benzyne; 152, biphenylene, a dimer of benzyne which gradually forms.

Benzyne + Benzyne ⟶ Biphenylene

Benzyne
Two moles

Biphenylene

(b) NH_2, $COOH$

o-Aminobenzoic acid
(Anthranilic acid)

(The dimerization of benzyne in part (a) is a [2 + 2] thermal cycloaddition, and is *symmetry-forbidden* (Sec. 28.9). Presumably, either the high energy of the particles involved here makes the difficult reaction possible, or reaction proceeds by a non-concerted, stepwise mechanism.)

14. Some chemical arithmetic first:

C_6H_5Cl	chlorobenzene	$C_{25}H_{20}ClK$	reactants
$+C_{19}H_{15}K$	Ph_3CK	$-C_{25}H_{20}$	products
$C_{25}H_{20}ClK$		KCl is lost; suggests product is Ph_4C	

On the assumption that the product is tetraphenylmethane, we can write the following series of reactions:

Chlorobenzene $\xrightarrow[NH_3]{NH_2^-}$ Benzyne $\xrightarrow{Ph_3C:^-}$ $C_6H_5^- CPh_3$ $\xrightarrow{NH_3}$ Ph_4C

Chlorobenzene Benzyne Tetraphenylmethane

KNH_2 is needed to produce benzyne; the resonance-stabilized Ph_3C^- anion is too weakly basic to produce benzyne, and it is too weakly nucleophilic to bring about bimolecular displacement.

But a benzylic carbanion, like an aromatic amine (Problem 23.11, page 866, and Problem 23.16, page 875), contains two sites of nucleophilic reactivity. Here, they are the $-CH_2:^-$, and the π electrons of the ring. We should consider the possibility that—for steric reasons—benzyne might react not at side-chain carbon to give Ph_4C but at the ring to give X, a structure reminiscent of the dimer of triphenylmethyl (I, page 572).

X

15. (a) At 340 °C reaction proceeds, at least partly, via aryne XI, which adds water in two different ways to give both *p*- and *m*-cresol.

p-$CH_3C_6H_4I$ $\xrightarrow[340\,°C]{NaOH}$ XI $\xrightarrow{H_2O}$ *p*-$CH_3C_6H_4OH$ and *m*-$CH_3C_6H_4OH$

At 250 °C, reaction proceeds by bimolecular displacement to give only *p*-cresol.

p-$CH_3C_6H_4I$ + OH^- $\xrightarrow{250\,°C}$ [carbanion intermediate with I and OH] $\longrightarrow$ *p*-$CH_3C_6H_4OH$ + I^-

(b) Aryne XII is formed (compare Problem 13, above) and adds *tert*-BuOH in the two possible ways.

O_2N-aminobenzoic acid (COOH, NH_2) $\xrightarrow{HONO}$ O_2N-(COO^-, N_2^+) $\xrightarrow{heat}$ $CO_2 + N_2$ + O_2N-benzyne (XII) $\xrightarrow{t\text{-BuOH}}$ O_2N-C_6H_4-OBu-*t* and O_2N-C_6H_4-OBu-*t*

(c) Aryne XIII is formed and adds not only NH_3 (as $^-NH_2$) but CH_3CN (as $^-CH_2CN$). Here the possible products are *ortho* and *meta* isomers and, as on page 965, the *meta* isomer

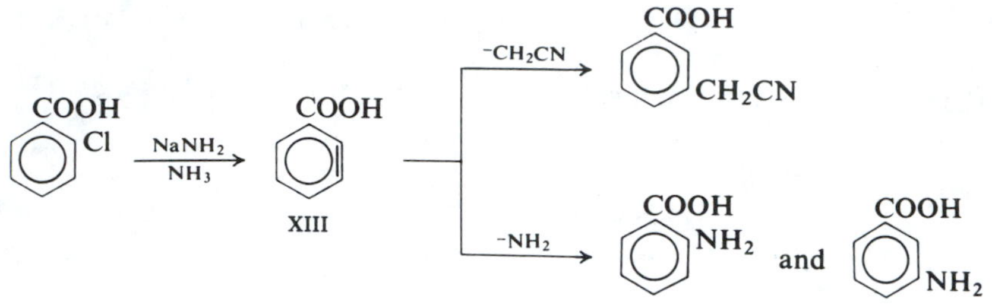

COOH Cl $\xrightarrow[NH_3]{NaNH_2}$ COOH XIII → ($^-CH_2CN$) COOH CH_2CN; ($^-NH_2$) COOH NH_2 and COOH NH_2

predominates: generation of the negative charge on the carbon next to the electron-withdrawing substituent is greatly preferred. (In parts (a) and (b) above, the choice is between *meta* and *para* isomers, and little preference for one over the other is shown.)

16. Both II and III give the same aryne, XIV, which by intramolecular reaction gives XV.

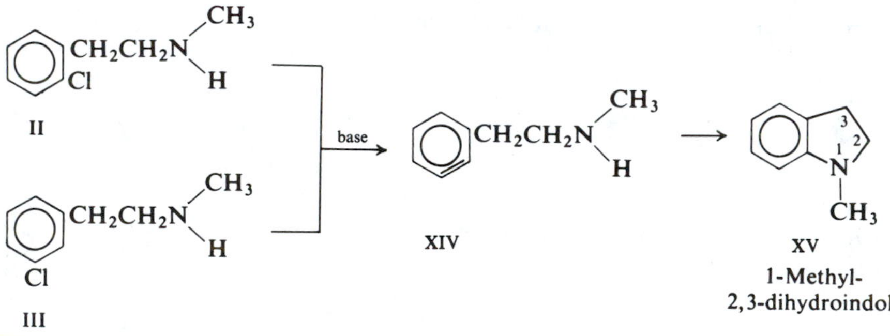

II: $CH_2CH_2N(CH_3)H$, Cl

III: $CH_2CH_2N(CH_3)H$, Cl

base → XIV: $CH_2CH_2N(CH_3)H$ → XV (N, CH_3)

XV
1-Methyl-
2,3-dihydroindole

17. (a) A benzyne mechanism will account for *some* of the facts.

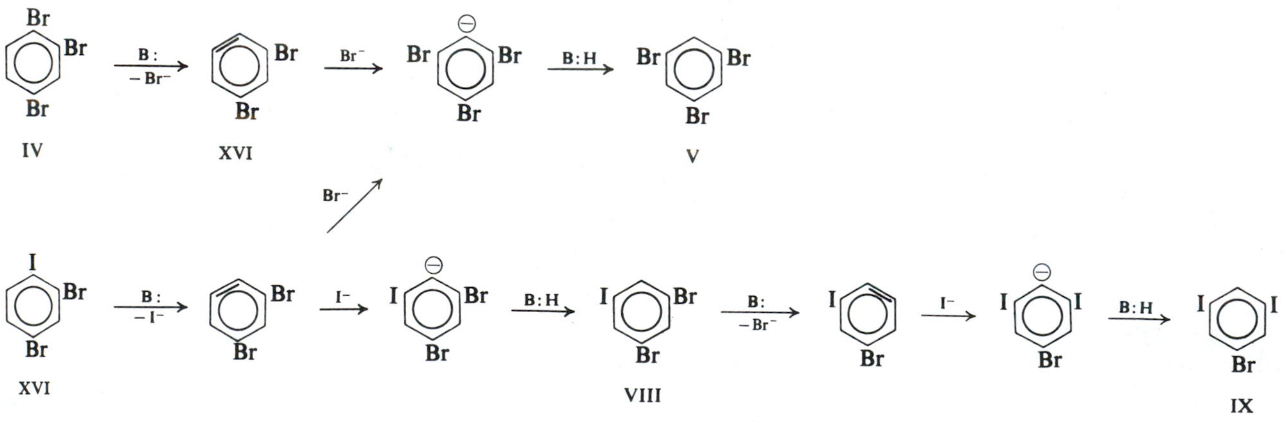

IV (Br, Br, Br) $\xrightarrow[-Br^-]{B:}$ XVI (Br, Br) $\xrightarrow{Br^-}$ (Br, Br, Br, ⊖) $\xrightarrow{B:H}$ V (Br, Br, Br)

Br^- ↗

XVI (I, Br, Br) $\xrightarrow[-I^-]{B:}$ (Br, Br) $\xrightarrow{I^-}$ (I, Br, Br, ⊖) $\xrightarrow{B:H}$ VIII (I, Br, Br) $\xrightarrow[-Br^-]{B:}$ (I, Br) $\xrightarrow{I^-}$ (I, I, Br, ⊖) $\xrightarrow{B:H}$ IX (I, I, Br)

(b) Benzynes are known to react with halide ion. If benzyne XVI is an intermediate, as postulated in (a), it should react with the added iodide to give VIII—contrary to fact.

IV $\xrightarrow{B:}$ XVI $\xrightarrow{I^-}$ (carbanion) $\xrightarrow{B:H}$ VIII

Is not obtained under these conditions

(c) The halogen dance is, basically, a two-step.

(i) $Ar{-}H + B: \rightleftharpoons Ar^{\ominus} + B{:}H$

(ii) $Ar^{\ominus} + Ar'{-}X \rightleftharpoons Ar{-}X + Ar'^{\ominus}$

Step (i) is the familiar abstraction of a proton by base to give a carbanion. Step (ii) is nucleophilic displacement *on halogen*, with an aryl carbanion as nucleophile, and another aryl carbanion as leaving group.

(ii) $Ar^{\ominus} + X{-}Ar' \longrightarrow \left[\overset{\delta-}{Ar}\text{---}X\text{---}\overset{\delta-}{Ar'}\right] \longrightarrow Ar{-}X + Ar'^{\ominus}$

To account for all the facts, we need just these two steps plus one guiding rule: *the only carbanions stable enough to be involved are those with the negative charge ortho to halogen.* As exemplified by the data of Problem 11, above, the inductive effect of halogen is *much* stronger from the *ortho* position than from the more distant *meta* and *para* positions.

As suggested in the problem, let us start with V and the base, in the presence of VI, and see how IV can be formed. The key to this, we find, is the transformation of the tribromo carbanion XVII into the isomeric carbanion XVIII, with Br's distributed as in IV.

(1) V + B: $\rightleftharpoons$ XVII + B:H

(2) XVII + VI $\rightleftharpoons$ VI + XVIII

(3) XVIII + B:H $\rightleftharpoons$ IV + B:

By the exact reverse of the above steps, we can account for the conversion of IV into V in the presence of VI.

(4) IV + B: $\rightleftharpoons$ XVIII + B:H

(5) 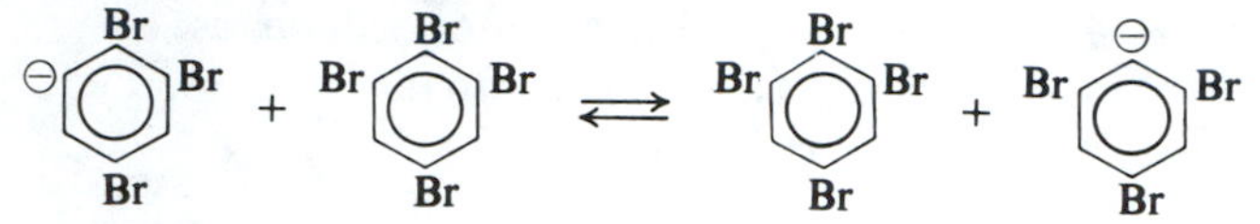 XVIII + VI ⇄ VI + XVII

(6) 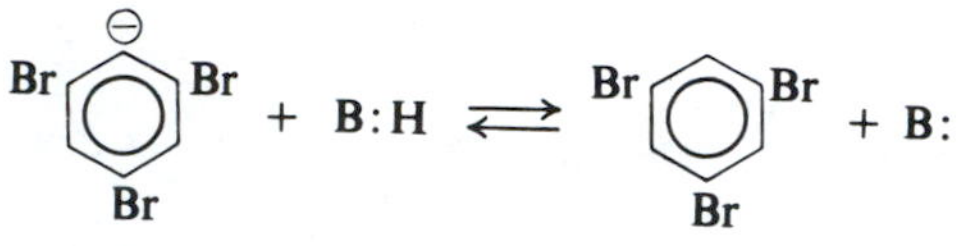 XVII + B:H ⇄ V + B:

We begin to see the special role played by the tetrabromo compound VI: it contains Br's distributed *both* as in IV and as in V, and, by loss of one Br or another, can be converted into either IV or V.

Next, let us start with *only* IV and base, and see how all the products are formed.

(4) 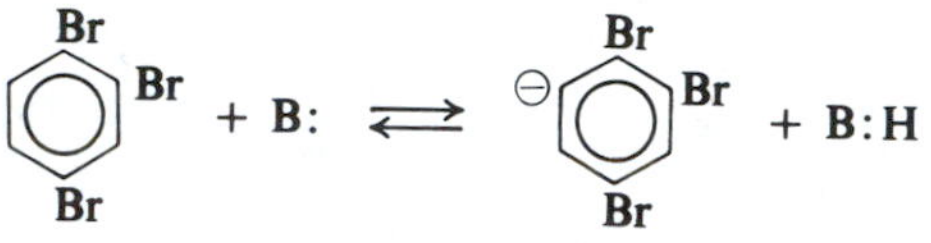 IV + B: ⇄ XVIII + B:H

(7) 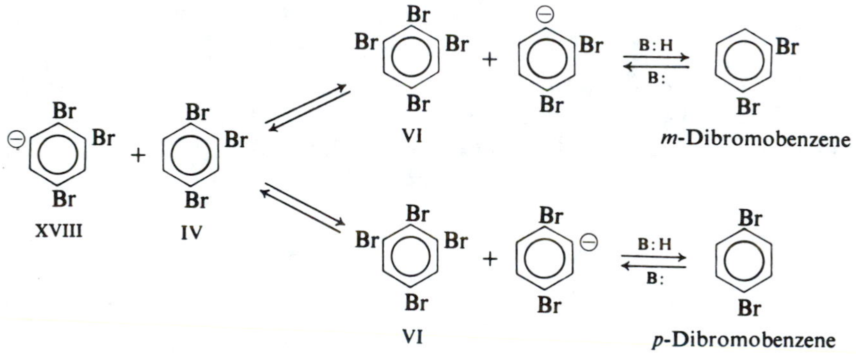 XVIII + IV ⇄ VI + (anion) $\underset{B:}{\overset{B:H}{\rightleftarrows}}$ *m*-Dibromobenzene

XVIII + IV ⇄ VI + (anion) $\underset{B:}{\overset{B:H}{\rightleftarrows}}$ *p*-Dibromobenzene

Sequence (4),(7) yields *m*- and *p*-dibromobenzenes and generates the tetrabromo compound VI. With VI now present, sequence (5),(6) can take place to yield V.

The reverse of (7),(4) converts *m*- and *p*-dibromobenzenes back into IV, but the Br that is gained need not be the same as the one originally lost from IV; the net result, after many cycles, is complete scrambling of a Br label in IV. (Write a set of equations to prove to yourself that this is so.)

The reactions of the dibromoiodo compound VII are analogous to (4) and (7). Tribromo and bromodiiodo compounds are obtained by a kind of "scrambling", this time not with isotopes but with the different halogens, Br and I. (Write equations to show this, too.)

Finally, the hardest part: why must VI be added to bring about isomerization of V but not the isomerization of IV? We write equations analogous to (4) and (7) starting with V instead of IV.

(1) 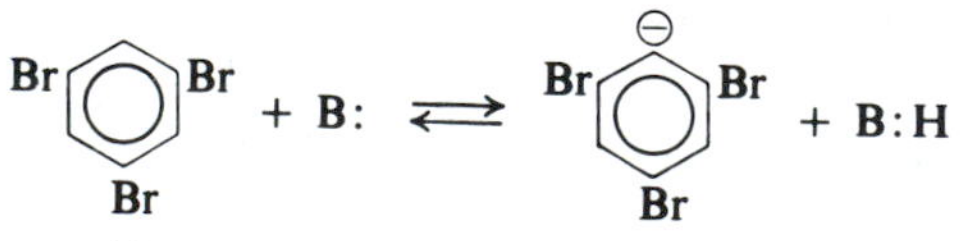 V + B: ⇄ XVII + B:H

(8) 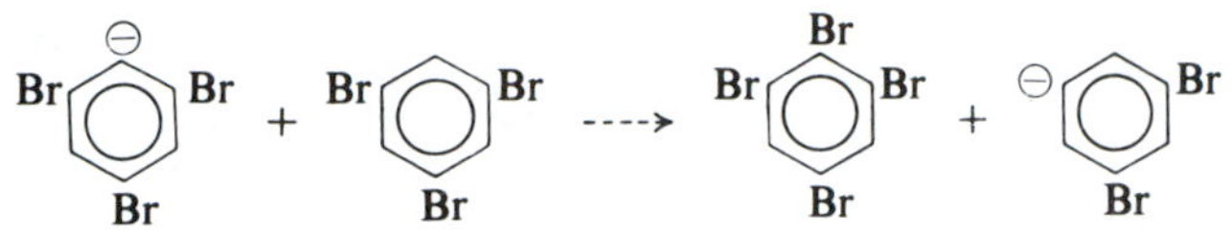 XVII + V ----> VI + XIX *Does not happen*

⊖ *is not ortho to X*

For V to generate the key tetrabromo compound VI, it must abstract Br from another molecule of V; but this it cannot do, because the reaction would require formation of carbanion XIX, in which the negative charge is not *ortho* to halogen.

Why is it *not* necessary to add V to bring about isomerization of IV? The answer, as Bunnett points out, "is already before our eyes": in sequence (4),(7) VI is *generated* from the reaction of base with IV alone.

(Bunnett, J. F. "The Base-Catalyzed Halogen Dance, and Other Reactions of Aryl Halides"; *Acc. Chem. Res.* **1972**, *5*, 139.)

27

α, β-Unsaturated Carbonyl Compounds

Conjugate Addition

27.1 (a) $CH_3CH_2CH_2COOH$

(b) $CH_3CH{=}CHCOO^- + EtOH$

(c) $PhCH{=}CHCOO^- + CHI_3$

(d) $CH_3CH{=}CHCH{=}NNHPh$

(e) $CH_3CH{=}CHCOO^- + Ag$

(f) $PhCHO + PhCOCHO$

(g) $CH_3CH_2CH_2CH_2OH$

(h) *meso*-HOOCCHBrCHBrCOOH

(i) racemic HOOCCHOHCHOHCOOH

27.2 A, $PhCH_2CH_2CHO$ B, $PhCH_2CH_2CH_2OH$ C, $PhCH{=}CHCH_2OH$

27.3 (a) $CH_3CH{=}CHCHO \xleftarrow[-H_2O]{H^+} CH_3\underset{\displaystyle OH}{\underset{|}{C}}H{-}CHCHO \xleftarrow{OH^-} CH_3CHO \xleftarrow[H^+,\,Hg^{2+}]{H_2O} HC{\equiv}CH$

(b) $PhCH{=}CHCHO \xleftarrow{-H_2O} \left[Ph\underset{\displaystyle OH}{\underset{|}{C}}H{-}CH_2CHO\right] \xleftarrow{OH^-} PhCHO + CH_3\ HO$

(c) $PhCH{=}CHCOOH \xleftarrow[-H_2O]{OAc^-} PhCHO + (CH_3CO)_2O$ (Perkin condensation, Problem 21.18(d))

(d) $(CH_3)_2CHCH{=}CHCOOH \xleftarrow{KOH(alc)} (CH_3)_2CHCH_2\underset{\displaystyle Br}{\underset{|}{C}}HCOOH \longleftarrow$ as in Problem 25.1(b)

27.4 (a) $\underset{\text{Acrylonitrile}}{CH_2{=}CHCN} \xleftarrow[-H_2O]{H^+} \underset{\displaystyle OH}{CH_2}CH_2CN \xleftarrow{CN^-} \underset{\displaystyle OH}{CH_2}{-}\underset{\displaystyle Cl}{CH_2} \xleftarrow{HOCl} CH_2{=}CH_2$

(b) $\underset{\text{Methyl acrylate}}{CH_2{=}CHCOOMe} \xleftarrow[-H_2O]{H^+} \underset{\displaystyle OH}{CH_2}CH_2COOMe \xleftarrow[H^+]{MeOH} \underset{\displaystyle OH}{CH_2}CH_2CN \longleftarrow$ as in (a)

(c) $\underset{\text{Methyl methacrylate}}{CH_2{=}\overset{\displaystyle CH_3}{C}COOMe} \xleftarrow[-H_2O]{H^+} CH_3\overset{\displaystyle CH_3}{\underset{\displaystyle OH}{C}}COOMe \xleftarrow[H^+]{MeOH} CH_3\overset{\displaystyle CH_3}{\underset{\displaystyle OH}{C}}CN \xleftarrow[H^+]{CN} CH_3\underset{\displaystyle \parallel \atop O}{C}CH_3$

(d) $\underset{\text{Orlon}}{\sim CH_2\underset{\displaystyle CN}{CH}\sim} \qquad \underset{\text{Acryloid}}{\sim CH_2\underset{\displaystyle COOMe}{CH}\sim} \qquad \underset{\text{Lucite, Plexiglas}}{\sim CH_2\overset{\displaystyle CH_3}{\underset{\displaystyle COOMe}{C}}\sim}$

27.5 (a) $\underset{\text{Glycerol}}{\underset{\displaystyle OH}{CH_2}\underset{\displaystyle OH}{CH}{-}\underset{\displaystyle OH}{CH_2}} \xrightarrow{-H_2O} \underset{\displaystyle OH}{CH_2}CH{=}\underset{\displaystyle OH}{CH} \xrightarrow{\text{tautom.}} \underset{\displaystyle OH}{CH_2}CH_2CHO \xrightarrow{-H_2O} \underset{\text{Acrolein}}{CH_2{=}CHCHO}$

(b) $\underset{\text{Acrolein}}{CH_2{=}CHCHO} \xrightarrow{Ag(NH_3)_2^+} \underset{\text{Acrylic acid}}{CH_2{=}CHCOOH}$

27.6 (a) $HOOC{-}CH{=}CH{-}CH{=}CH{-}COOH \xleftarrow{KOH(alc)} HOOC\underset{\displaystyle Br}{C}HCH_2CH_2\underset{\displaystyle Br}{C}HCOOH$

$\uparrow 2Br_2,\ P$

$\underset{\text{Adipic acid}}{HOOCCH_2CH_2CH_2CH_2COOH}$

(b) $CH_3{-}\underset{\displaystyle \parallel \atop O}{C}{-}CH{=}CH_2 \xleftarrow[-H_2O]{H^+} \underset{\text{Aldol}}{CH_3{-}\underset{\displaystyle \parallel \atop O}{C}{-}CH_2 \vdots \underset{\displaystyle OH}{CH_2}} \xleftarrow{OH^-} CH_3{-}\underset{\displaystyle \parallel \atop O}{C}{-}CH_3 + O{=}CH_2$

(c) $\underset{\textit{Keto}}{CH_3{-}\underset{\displaystyle \parallel \atop O}{C}{-}CH{=}CH_2} \xleftarrow{\text{tautom.}} \underset{\textit{Enol}}{CH_2{=}\underset{\displaystyle OH}{C}{-}CH{=}CH_2} \xleftarrow[H^+]{H_2O,\ Hg^{2+}} \underset{\text{Vinylacetylene}}{HC{\equiv}C{-}CH{=}CH_2}$

27.7 All are less stable than I.

27.8 IV is an amide, formed by nucleophilic addition and subsequent ring closure through amide formation.

$$\underset{\text{IV}}{\text{(ring: H, CH}_3\text{ on C; CH}_2\text{; C=O; N–C}_6\text{H}_5\text{; N–H)}} \xleftarrow{-H_2O} \text{CH}_3\text{CH(NH–NH–C}_6\text{H}_5\text{)CH}_2\text{COOH} \longleftarrow CH_3CH{=}CHCOOH + C_6H_5NHNH_2$$

27.9 Two successive nucleophilic additions.

$$CH_2{=}CHCN + NH_3 \longrightarrow \underset{NH_2}{CH_2CH_2CN}$$

$$CH_2{=}CHCN + H_2NCH_2CH_2CN \longrightarrow H{-}N(CH_2CH_2CN)_2$$

27.10 Two successive nucleophilic additions.

$$CH_2{=}CHCOOEt + CH_3NH_2 \longrightarrow \underset{CH_3NH}{CH_2CH_2COOEt}$$

$$CH_2{=}CHCOOEt + \underset{CH_3NH}{CH_2CH_2COOEt} \longrightarrow CH_3{-}N(CH_2CH_2COOEt)_2$$

27.11 (a)

$$\begin{matrix} CH_3CH{=}CHCOOEt \\ H_2C{-}COOEt \\ | \\ COOEt \end{matrix} \longrightarrow \underset{\text{A}}{\begin{matrix} CH_3CHCH_2COOEt \\ | \\ H{-}C{-}COOEt \\ | \\ COOEt \end{matrix}} \longrightarrow \longrightarrow \longrightarrow \underset{\text{B}}{\begin{matrix} CH_3CHCH_2COOH \\ | \\ CH_2COOH \end{matrix}}$$

(b)

$$\begin{matrix} CH_2{=}CHCOOEt \\ H_2C{-}COCH_3 \\ | \\ COOEt \end{matrix} \longrightarrow \underset{\text{C}}{\begin{matrix} CH_2CH_2COOEt \\ | \\ H{-}C{-}COCH_3 \\ | \\ COOEt \end{matrix}} \xrightarrow{\text{ketonic cleavage}} \underset{\text{D}}{\begin{matrix} CH_2CH_2COOH \\ | \\ CH_2COCH_3 \end{matrix}}$$

(c)

$$\begin{matrix} CH_2{=}CHCOCH_3 \\ H_2C{-}COOEt \\ | \\ COOEt \end{matrix} \longrightarrow \underset{\text{E}}{\begin{matrix} CH_2CH_2COCH_3 \\ | \\ H{-}C{-}COOEt \\ | \\ COOEt \end{matrix}}$$

(d)

$$\begin{matrix} PhCH{=}CHCOPh \\ HCH_2COPh \end{matrix} \longrightarrow \underset{\text{F}}{\begin{matrix} PhCHCH_2COPh \\ | \\ CH_2COPh \end{matrix}}$$

(e) 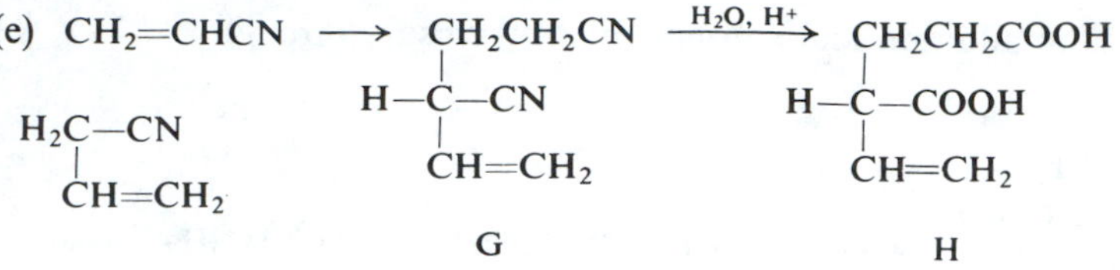 $CH_2{=}CHCN \longrightarrow$ CH_2CH_2CN / $H{-}C{-}CN$ / $CH{=}CH_2$ $\xrightarrow{H_2O,\ H^+}$ CH_2CH_2COOH / $H{-}C{-}COOH$ / $CH{=}CH_2$

$H_2C{-}CN$ / $CH{=}CH_2$

G H

(f) 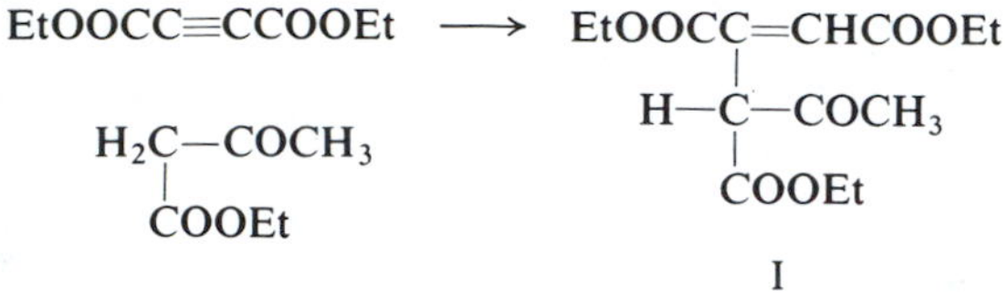 $EtOOCC{\equiv}CCOOEt \longrightarrow$ $EtOOCC{=}CHCOOEt$ / $H{-}C{-}COCH_3$ / $COOEt$

$H_2C{-}COCH_3$ / $COOEt$

I

(g) $EtOOCC{=}CHCOOEt$ / $H{-}C{-}COCH_3$ / $COOEt$ $\xrightarrow{\text{strong } OH^-}$ $HOOCC{=}CHCOOH$ / CH_2COOH *Cleavage to acids*

I J

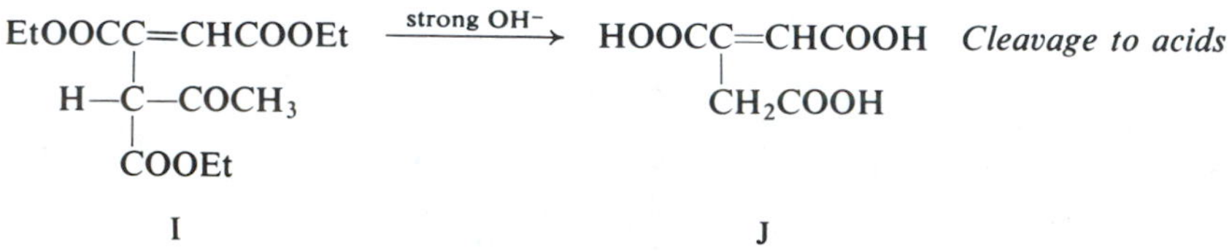

27.12 (a)

$C_7H_{12}O_4$ malonic ester
$+CH_2O$ formaldehyde
$C_8H_{14}O_5$

$C_8H_{14}O_5$
$-C_8H_{12}O_4$ product K
H_2O lost: suggests aldol condensation

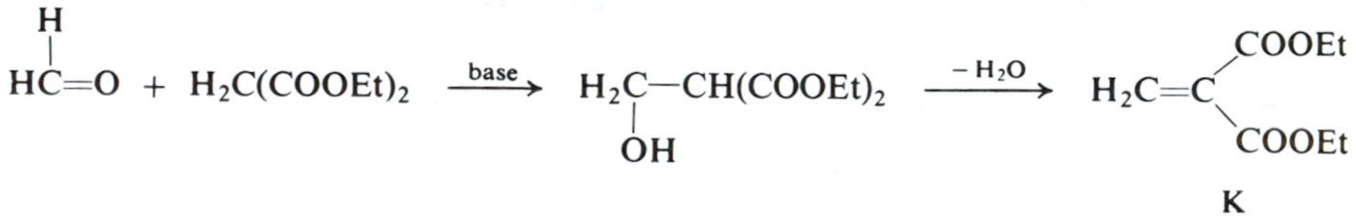

$$HC(H){=}O + H_2C(COOEt)_2 \xrightarrow{\text{base}} H_2C(OH){-}CH(COOEt)_2 \xrightarrow{-H_2O} H_2C{=}C(COOEt)_2$$

K

(b) 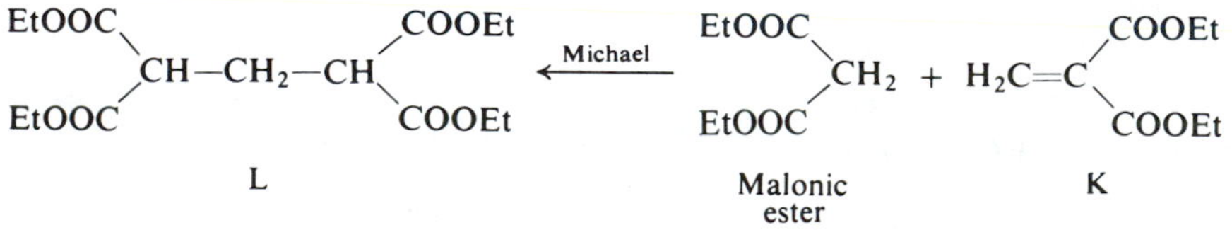 $(EtOOC)_2CH{-}CH_2{-}CH(COOEt)_2 \xleftarrow{\text{Michael}} (EtOOC)_2CH_2 + H_2C{=}C(COOEt)_2$

L Malonic ester K

(c) 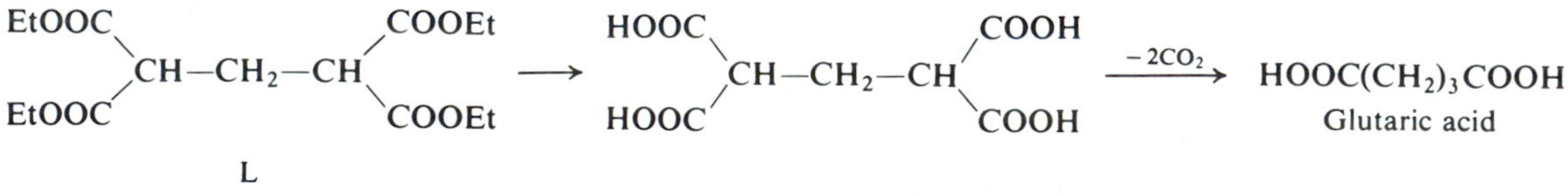 $(EtOOC)_2CH{-}CH_2{-}CH(COOEt)_2 \longrightarrow (HOOC)_2CH{-}CH_2{-}CH(COOH)_2 \xrightarrow{-2CO_2} HOOC(CH_2)_3COOH$

L Glutaric acid

27.13 $\Delta^{1,9}$-Octalone $\xleftarrow[-H_2O]{\text{aldol}}$ 2-($CH_2CH_2C(=O)CH_3$)cyclohexanone $\xleftarrow{\text{Michael}}$ Cyclohexanone + MVK ($CH_3C(=O)CH{=}CH_2$)

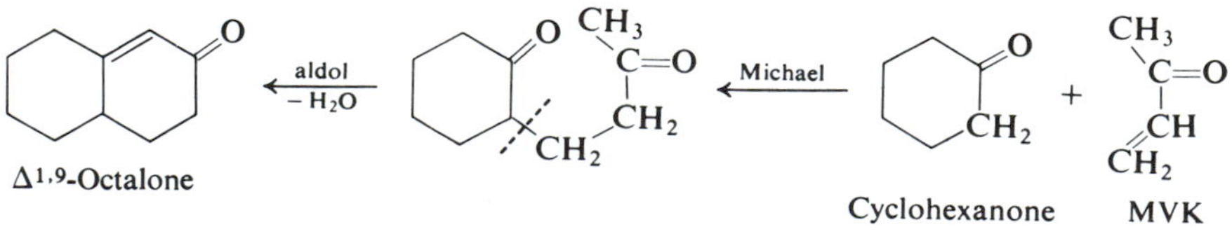

27.14 (a)

$$\mathbf{M}\ \xleftarrow[-OEt^-]{\text{Claisen}}\ (CH_3)_2C(CH_2COCH_3)-CH(COOEt)_2\ \xleftarrow{\text{Michael}}\ (CH_3)_2C{=}CH-CO-CH_3 + H_2C(COOEt)_2$$

M: 6-membered ring: H_2C–C(=O)–CH_2–C=O–CH(COOEt)–$C(CH_3)_2$

Mesityl oxide Malonic ester

(b)

$$\mathbf{M}\ \xrightarrow{OH^-,\ \text{heat}}\ \xrightarrow{H^+}\ \xrightarrow{-CO_2}\ \text{5,5-dimethyl-1,3-cyclohexanedione}$$

M

27.15

$$\underset{\mathbf{N}}{PhCH(C_5H_5)-CH_2-\overset{O}{\overset{\|}{C}}-C_6H_4Br}\ \xleftarrow{\text{Michael}}\ PhCH{=}CH-\overset{O}{\overset{\|}{C}}-C_6H_4Br + C_5H_6$$

1,3-Cyclopentadiene
Acidic (*Sec. 14.10*)

27.16 (a)

$$PhCH{=}CH-\overset{Me}{\overset{|}{C}}{=}O + Me_2NH \rightleftarrows PhCH{=}CH-\overset{Me}{\overset{|}{C}}{=}\overset{\oplus}{N}Me_2 + OH$$

An iminium ion

$$PhCH{=}CH-\overset{Me}{\overset{|}{C}}{=}\overset{\oplus}{N}Me_2 + CH(COOEt)_2^- \longrightarrow \underset{\mathbf{V}}{PhCH(CH(COOEt)_2)-CH{=}\overset{Me}{\overset{|}{C}}-NMe_2} \xrightarrow{H_2O} PhCH(CH(COOEt)_2)-CH_2-\overset{Me}{\overset{|}{C}}{=}O + Me_2NH$$

(b) When the anion from ethyl malonate adds to the iminium ion, there is formed, not a carbanion like I on page 977, but the much more stable *neutral molecule*, V.

(c) A tertiary amine has no proton to lose, and hence cannot form an iminium ion.

27.17

$$\text{Diels–Alder adduct } (C_6H_5)_2 \xleftarrow{\text{Diels–Alder}} C_6H_5CH{=}CH-CH{=}CHC_6H_5 + \text{maleic anhydride}$$

1,4-Diphenyl-1,3-butadiene Maleic anhydride

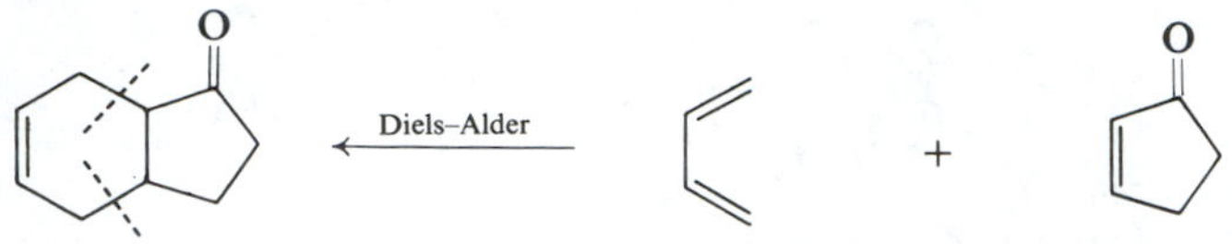

Diels–Alder

1,3-Butadiene + 2-Cyclopentenone

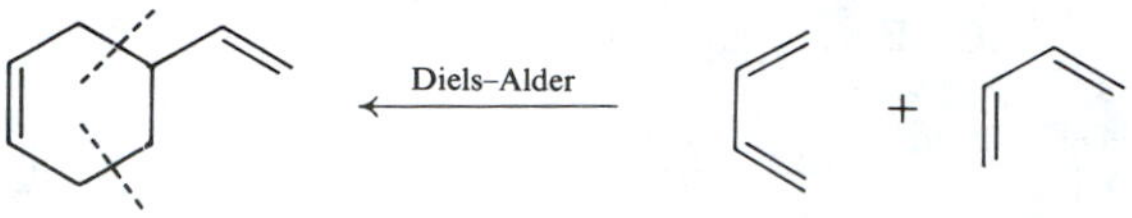

Diels–Alder

1,3-Butadiene *Two moles*

27.18 (a) Me, EtO, O, O

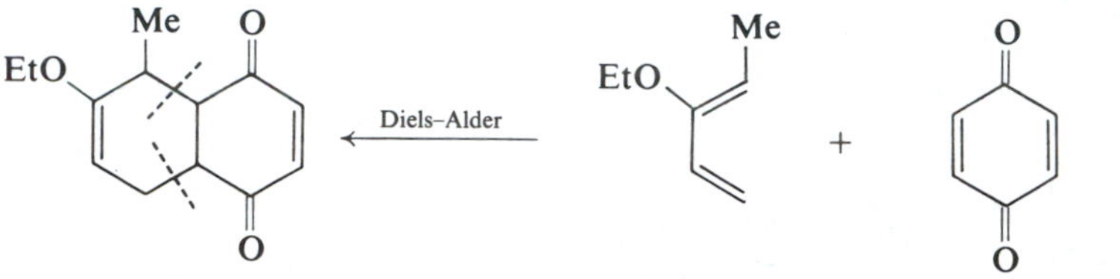

I ← Diels–Alder — 3-Ethoxy-1,3-pentadiene + *p*-Benzoquinone

(b) O, CH_3, CH_3O, O

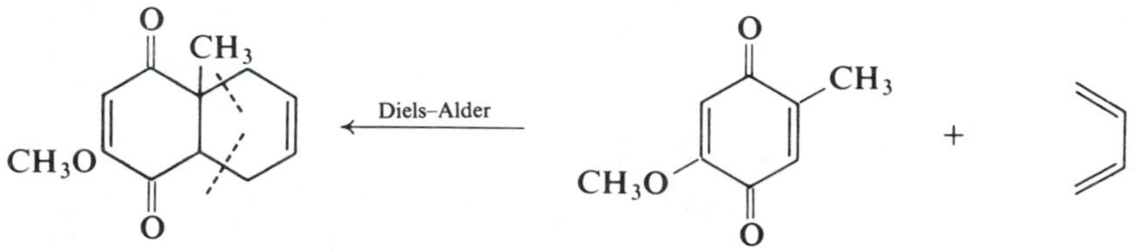

II ← Diels–Alder — 5-Methoxy-2-methyl-1,4-benzoquinone + 1,3-Butadiene

27.19 $\xrightarrow[\text{1,4-addn.}]{\text{HCl}}$ $\xrightarrow{\text{tautom.}}$

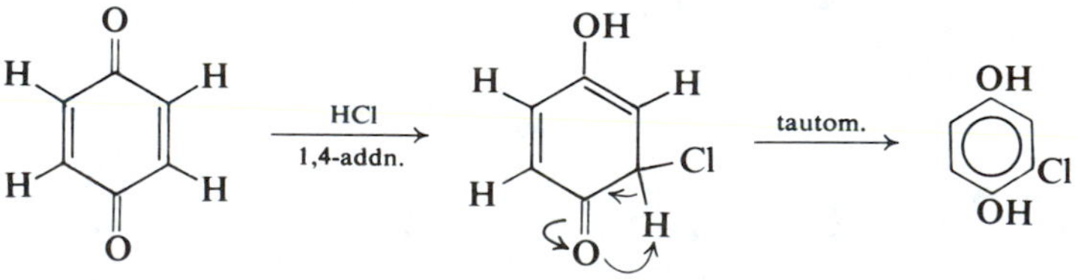

27.20 (a) Ease of oxidation.

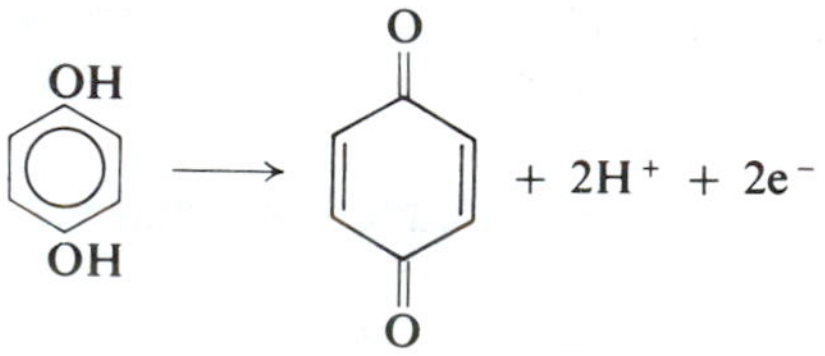

$$\longrightarrow \quad + 2H^+ + 2e^-$$

(b) Ease of reduction.

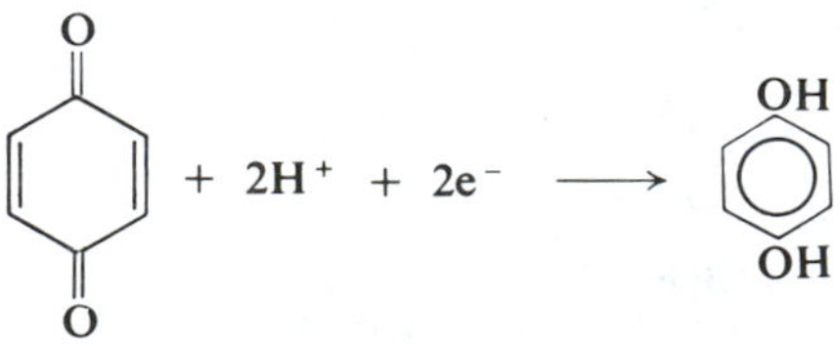

$$+ 2H^+ + 2e^- \longrightarrow$$

27.21 Phenol $\xrightarrow{HONO}$ *p*-Nitrosophenol (OH, N=O) $\xrightarrow{tautom.}$ *p*-Benzoquinone monoxime (O, N—OH)

1. Acrolein: $CH_2{=}CHCHO \xleftarrow[-H_2O]{H^+} \xleftarrow{OH^-} HCHO + CH_3CHO$ (Or Problem 27.5, page 974.)

Crotonaldehyde (2-butenal): page 800.

Cinnamaldehyde (3-phenyl-2-propenal): page 801.

Mesityl oxide (4-methyl-3-pentan-2-one): page 800.

Benzalacetone (4-phenyl-3-buten-2-one): page 801.

Dibenzalacetone: $PhCH{=}CHC({=}O)CH{=}CHPh \xleftarrow[-2H_2O]{aldol} PhCHO + CH_3C({=}O)CH_3 + OHCPh$

Benzalacetophenone: page 801.

Dypnone: $PhC(CH_3){=}CHCOPh \xleftarrow[-H_2O]{aldol} PhC(CH_3){=}O + CH_3COPh$

Acrylic acid: from acrylonitrile, Problem 27.4(a), page 973.

Crotonic acid: *trans*-$CH_3CH{=}CHCOOH \xleftarrow{Ag(NH_3)_2^+}$ *trans*-$CH_3CH{=}CHCHO$ (page 800)

Isocrotonic acid: *cis*-$CH_3CH{=}CHCOOH \xleftarrow{H_2,\ Pd} CH_3C{\equiv}CCOOH \xleftarrow{CO_2} CH_3C{\equiv}CMgBr \xleftarrow{MeMgBr} CH_3C{\equiv}CH$

Methacrylic acid: page 673.

Sorbic acid: $CH_3CH{=}CHCH{=}CHCOOH \xleftarrow{Ag(NH_3)_2^+} CH_3CH{=}CHCH{=}CHCHO$

$\uparrow$ OH^-, $-H_2O$

$CH_3CHO + CH_3CH{=}CHCHO$ (page 800)

Cinnamic acid: Problem 21.18(d) (Perkin condensation), page 811.

Maleic acid: *cis*-$HOOCCH{=}CHCOOH \xleftarrow{H_2O}$ maleic anhydride (HC=CH, C=O, O, C=O ring) $\xleftarrow{250\text{–}300\ °C}$ *cis*- and *trans*-$HOOCCH{=}CHCOOH$

$\uparrow$ KOH(alc)

$HOOCCH_2CH_2COOH \xrightarrow[P]{Br_2} HOOCCH_2CH(Br)COOH$

(Problem 5a, page 747)

(Actually made via maleic anhydride by catalytic oxidation of benzene.)

Fumaric acid: *trans*-$HOOCCH{=}CHCOOH$ $\xleftarrow[\text{light}]{\text{trace HBr}}$ *cis*-$HOOCCH{=}CHCOOH$
(See Problem 12, page 583) Maleic acid (above)

Maleic anhydride: $HC{=}CH$, $C(=O)$–O–$C(=O)$ ring $\xleftarrow{\text{heat}}$ maleic or fumaric acid (see above)

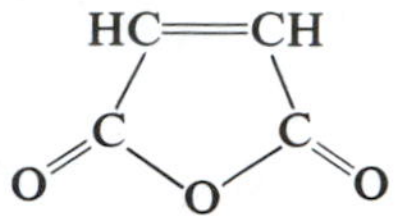

Methyl acrylate: Problem 27.4(b), page 973.

Methyl methacrylate: Problem 27.4(c), page 973.

Ethyl cinnamate: $PhCH{=}CHCOOEt \xleftarrow{OEt^-} PhCHO + CH_3COOEt$

Acrylonitrile: Problem 27.4(a), page 973.

2. (a) $PhCH_2CH_2\underset{\underset{O}{\|}}{C}CH_3 + PhCH_2CH_2\underset{\underset{OH}{|}}{C}HCH_3$

(b) $PhCH{=}CH\underset{\underset{OH}{|}}{C}HCH_3$

(c) $PhCH{=}CHCOONa + CHI_3$

(d) $PhCHO + CH_3\underset{\underset{O}{\|}}{C}CHO$

(e) $Ph\underset{\underset{Br}{|}}{CH}{-}\underset{\underset{Br}{|}}{CH}{-}\underset{\underset{O}{\|}}{C}CH_3$

(f) $Ph\underset{\underset{Cl}{|}}{C}HCH_2\underset{\underset{O}{\|}}{C}CH_3$

(g) $Ph\underset{\underset{Br}{|}}{C}HCH_2\underset{\underset{O}{\|}}{C}CH_3$

(h) $Ph\underset{\underset{OH}{|}}{C}HCH_2\underset{\underset{O}{\|}}{C}CH_3$

(i) $Ph\underset{\underset{OCH_3}{|}}{C}HCH_2\underset{\underset{O}{\|}}{C}CH_3$

(j) $Ph\underset{\underset{CN}{|}}{C}HCH_2\underset{\underset{O}{\|}}{C}CH_3$

(k) $Ph\underset{\underset{CH_3-NH}{|}}{C}HCH_2\underset{\underset{O}{\|}}{C}CH_3$

(l) $Ph\underset{\underset{PhNH}{|}}{C}HCH_2\underset{\underset{O}{\|}}{C}CH_3$

(m) $Ph\underset{\underset{NH_2}{|}}{C}HCH_2\underset{\underset{O}{\|}}{C}CH_3$

(n) $Ph\underset{\underset{NHOH}{|}}{C}HCH_2\underset{\underset{O}{\|}}{C}CH_3$

(o) $PhCH{=}CH\underset{\underset{O}{\|}}{C}CH{=}CHPh$

(p) $CH_3\underset{\underset{O}{\|}}{C}CH_2\underset{\underset{Ph}{|}}{C}HCH(COOEt)_2$

(q) $CH_3\underset{\underset{O}{\|}}{C}CH_2\underset{\underset{Ph}{|}}{CH}{-}\underset{\underset{CN}{|}}{C}HCOOEt$

(r) $CH_3\underset{\underset{O}{\|}}{C}CH_2\underset{\underset{Ph}{|}}{CH}{-}\underset{\underset{CH_3}{|}}{C}(COOEt)_2$

(s) $CH_3\underset{\underset{O}{\|}}{C}CH_2\underset{\underset{Ph}{|}}{CH}{-}\underset{\underset{COCH_3}{|}}{C}HCOOEt$

(t) 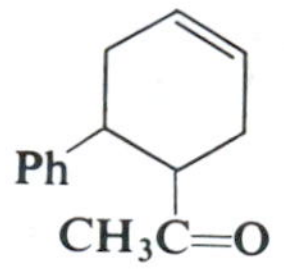 Ph; $CH_3C{=}O$

(u) 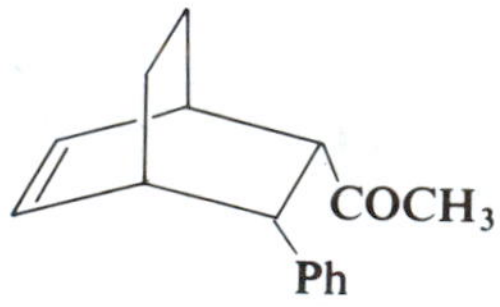 $COCH_3$; Ph

(v) 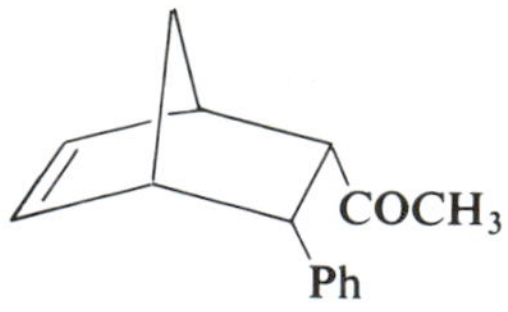 $COCH_3$; Ph

3. The C—C bond formed in each Michael addition is indicated by the broken line. The —COO— lost in subsequent decarboxylations (see Problem 4, below) is indicated by boldface type in parts (a)–(i).

(a) $Ph-C(=O)-CH_2CH(Ph)-CH(COOEt)-CN$

(b) $EtOOCCH_2CH(Ph)-CH(COOEt)-CN$

(c) $EtOOCCH_2CH(COOEt)-CH(COOEt)-COOEt$

(d) $EtOOCCH{=}C(COOEt)-CH(COOEt)-COOEt$

(e) $CH_3-C(=O)-CH_2C(CH_3)_2-CH(COOEt)-COOEt$

(f) $CH_3-C(=O)-CH_2C(CH_3)_2-CH(COOEt)-C(=O)-CH_3$

(g) $EtOOCCH_2CH(CH_3)-C(CH_3)(COOEt)-COOEt$

(h) $EtOOC-CH(COOEt)-CH_2-CH(COOEt)-COOEt$

(i) $CH_3-C(=O)-CH(COOEt)-CH(CH_3)-CH(COOEt)-C(=O)-CH_3$

(j) $MeOOCCH_2CH_2-CH_2NO_2$

(k) $EtOOCCH_2CH(CH_3)-CH(NO_2)-CH(CH_3)CH_2COOEt$

(l) $NCCH_2CH_2-C(CH_2CH_2CN)(NO_2)-CH_2CH_2CN$

(m) $NCCH_2CH_2-CCl_3$

4. (a) $Ph-C(=O)-CH_2CH(Ph)-CH_2COOH$

(b) $HOOCCH_2CH(Ph)-CH_2COOH$

(c) $HOOCCH_2CH(COOH)-CH_2COOH$

(d) $HOOCCH{=}C(COOH)-CH_2COOH$

(e) $CH_3-C(=O)-CH_2C(CH_3)_2-CH_2COOH$

(f) $CH_3-C(=O)-CH_2C(CH_3)_2-CH_2-C(=O)-CH_3$

(g) $HOOCCH_2CH(CH_3)-CH(CH_3)COOH$

(h) $HOOCCH_2-CH_2-CH_2COOH$

(i) 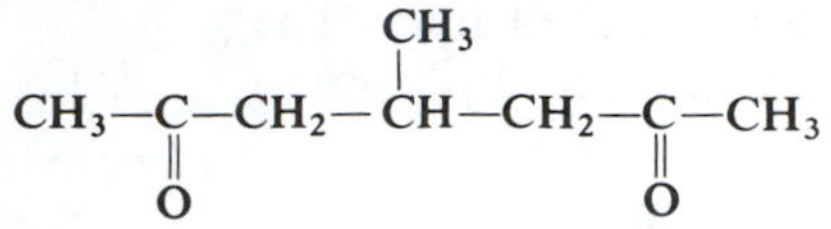 $CH_3-\underset{\underset{O}{\|}}{C}-CH_2-\overset{\overset{CH_3}{|}}{C}H-CH_2-\underset{\underset{O}{\|}}{C}-CH_3$

5.

$$(EtOOC)_2CH_2 + PhCH{=}CH\underset{\underset{O}{\|}}{C}CH{=}CHPh + H_2C(COOEt)_2 \xrightarrow[\text{Michael}]{\text{double}} EtOOC-\underset{EtOOC}{\overset{H}{C}}-\underset{H}{\overset{Ph}{C}}-CH_2\underset{\underset{O}{\|}}{C}CH_2-\underset{H}{\overset{Ph}{C}}-\underset{COOEt}{\overset{H}{C}}-COOEt$$

A

$$PhCH{=}CH\underset{\underset{O}{\|}}{C}CH{=}CHPh + H_2C(COOEt)_2 \xrightarrow{\text{Michael}} PhCH{=}CH\underset{\underset{O}{\|}}{C}-CH_2-\underset{H}{\overset{Ph}{C}}-\underset{COOEt}{\overset{H}{C}}-COOEt$$

B

B (O=C with HC=PhCH and CH_2–CHPh–CH($COOEt$)$_2$ chain) $\xrightarrow{\text{internal Michael}}$ C (ring: C=O, H_2C, CH_2, PhCH, CHPh, C(EtOOC)(COOEt))

B C

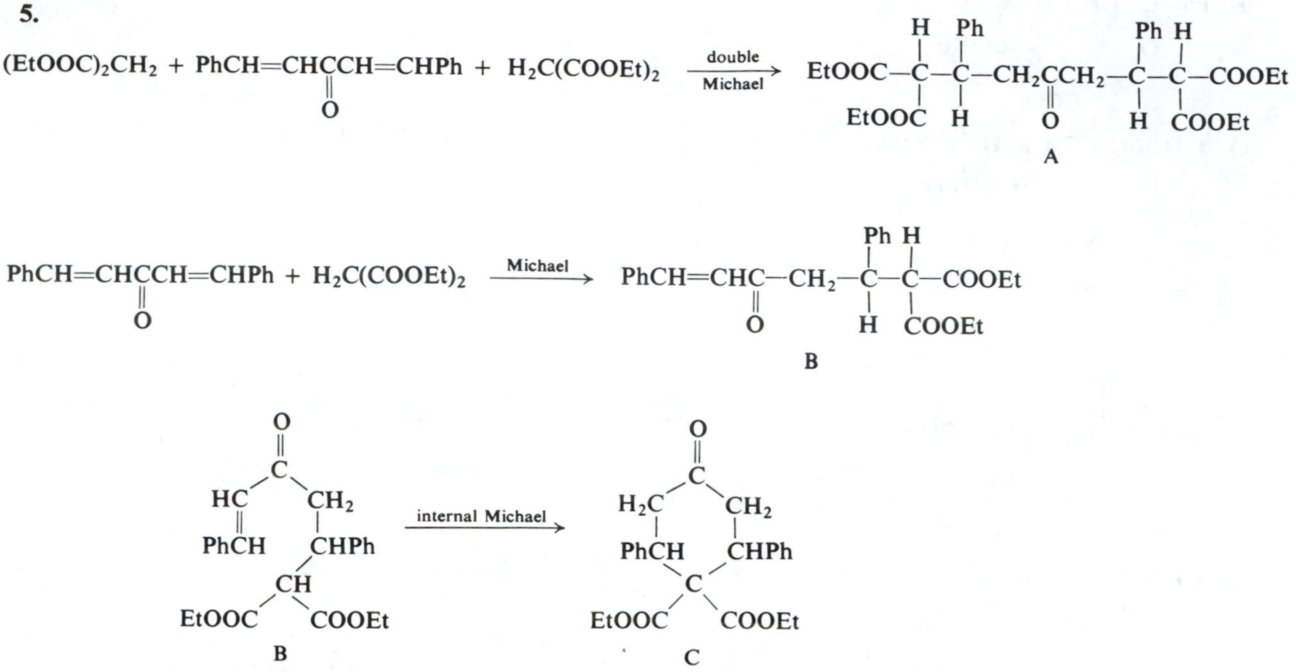

6. $$\text{Spermine} \xleftarrow[\text{Ni}]{H_2} NC-CH_2CH_2-\overset{H}{N}(CH_2)_4\overset{H}{N}-CH_2CH_2CN \xleftarrow[\text{addn.}]{\text{nucleo.}} 2CH_2{=}CHCN + NH_2(CH_2)_4NH_2$$

7. (a) CH_3 (b) (c) (d) CH_3

(e) CHO, CH_3 (f) Ph Ph (g) Ph, NO_2 (h)

(i) (j) (k)

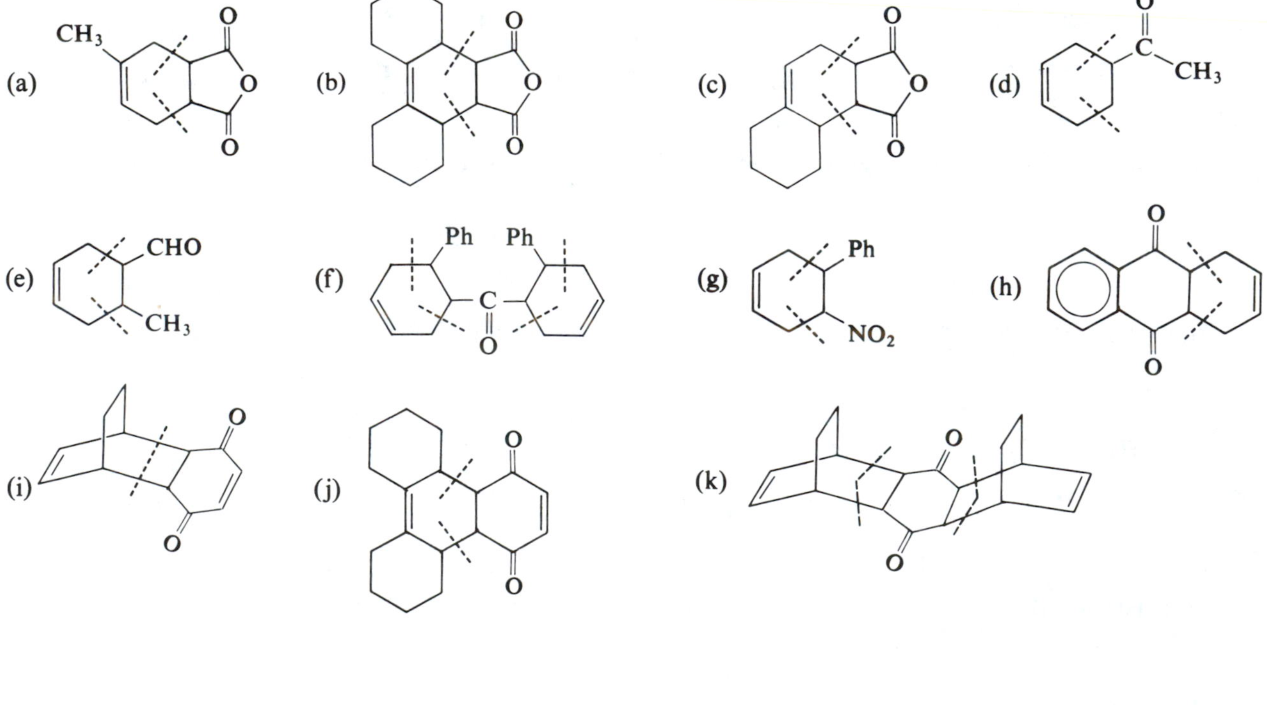

(l)

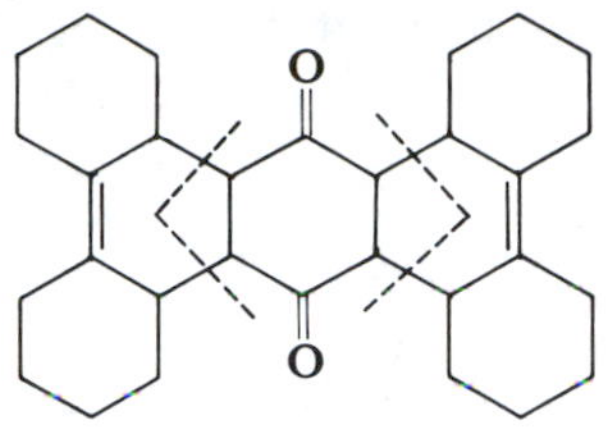

(m) 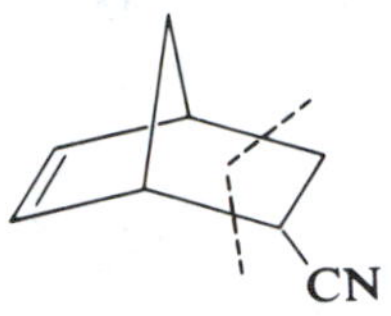 CN

(n) 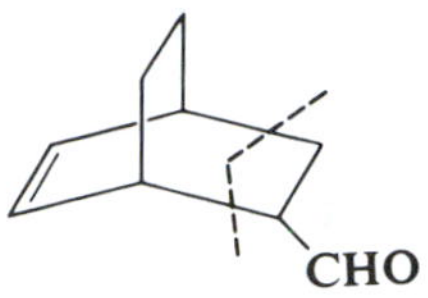 CHO

8. (a) 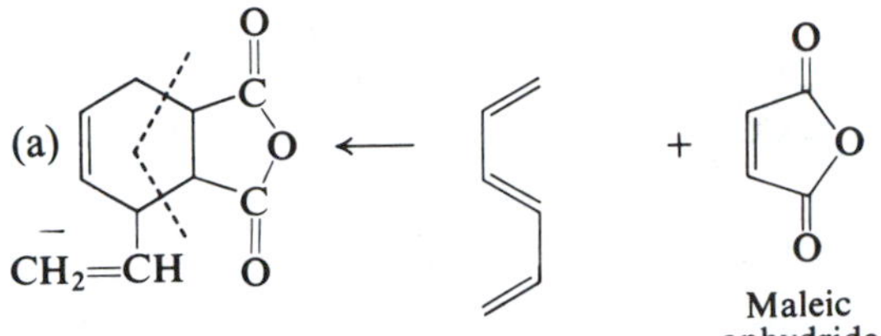 $CH_2{=}CH$ ⟵ 1,3,5-Hexatriene + Maleic anhydride

(b) 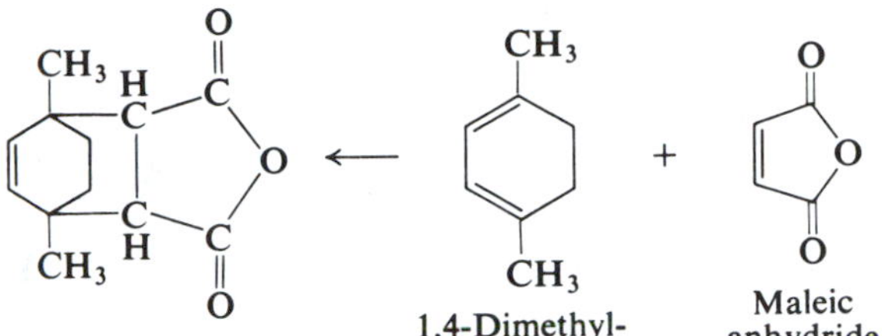 CH_3 H C O ⟵ CH_3 / CH_3 1,4-Dimethyl-1,3-cyclohexadiene + Maleic anhydride

(c) 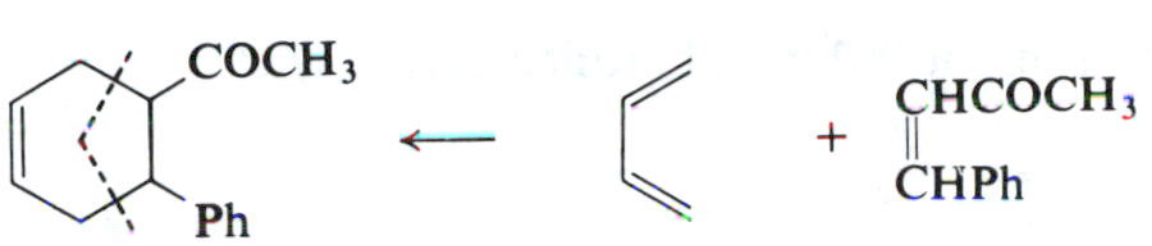 $COCH_3$ Ph ⟵ 1,3-Butadiene + $CHCOCH_3$ ‖ CHPh Benzalacetone

(d) 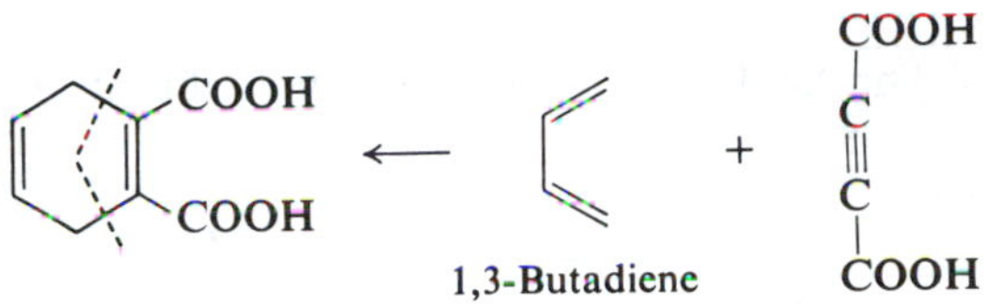 COOH COOH ⟵ 1,3-Butadiene + COOH–C≡C–COOH Acetylene-dicarboxylic acid

(e) 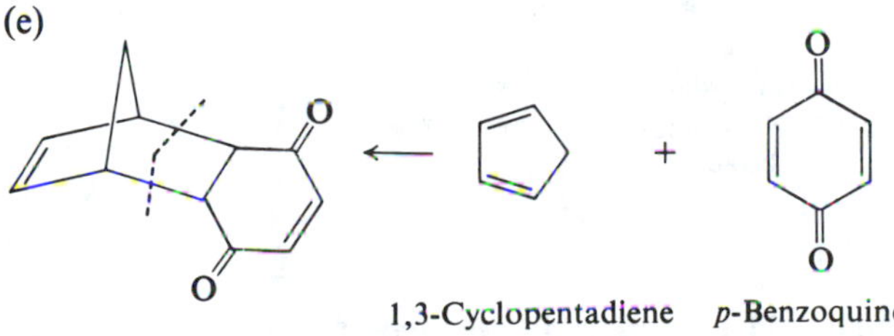 ⟵ 1,3-Cyclopentadiene + *p*-Benzoquinone

(f) 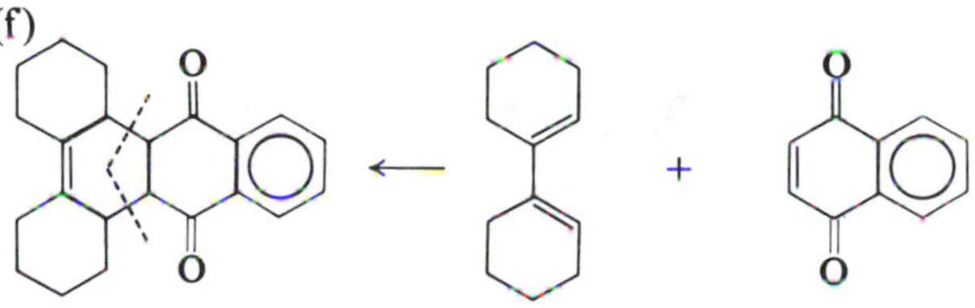 ⟵ 1,1′-Bicyclohexenyl (I, page 986) + 1,4-Naphthoquinone (II, page 986)

(g) 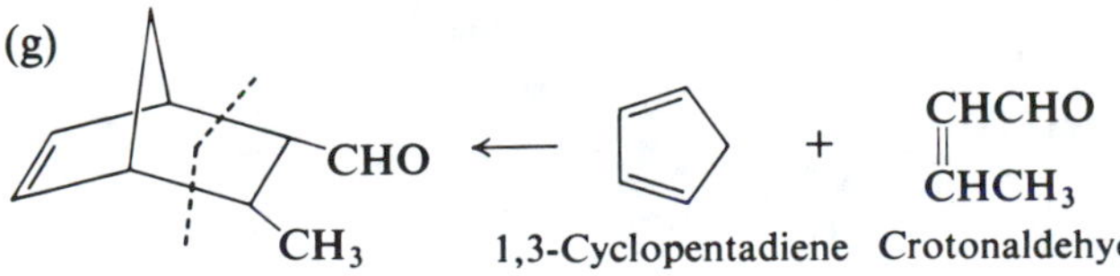 CHO CH_3 ⟵ 1,3-Cyclopentadiene + CHCHO ‖ $CHCH_3$ Crotonaldehyde

(h) 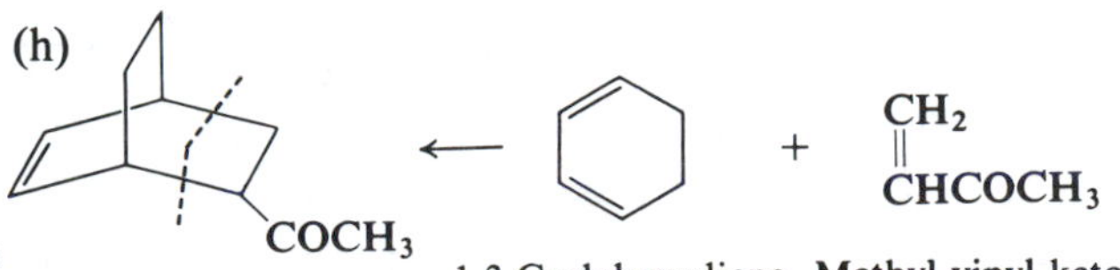 $COCH_3$ ⟵ 1,3-Cyclohexadiene + CH_2 ‖ $CHCOCH_3$ Methyl vinyl ketone (MVK)

(i) 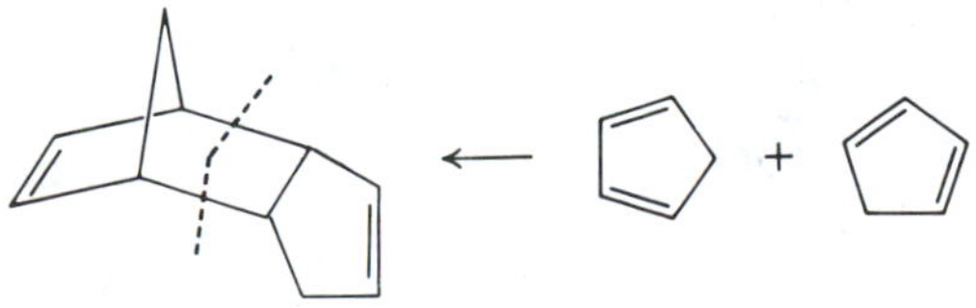 ⟵ 1,3-Cyclopentadiene + 1,3-Cyclopentadiene

1,3-Cyclopentadiene *Two moles*

9. If we work through this problem, neglecting for the moment stereoisomerism, we arrive at:

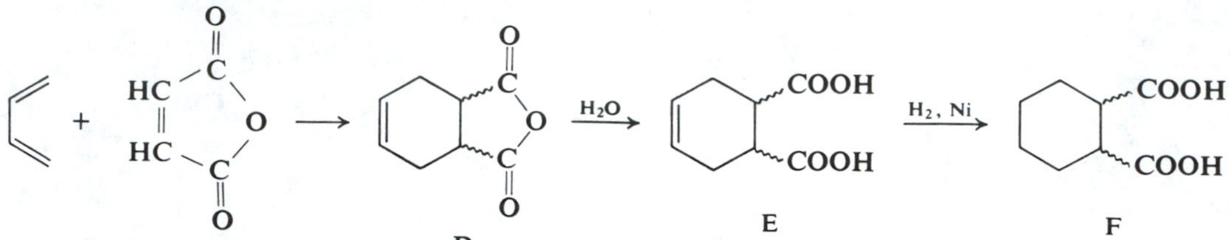

Maleic anhydride

D

E

F

1,2-Cyclohexane-dicarboxylic acid

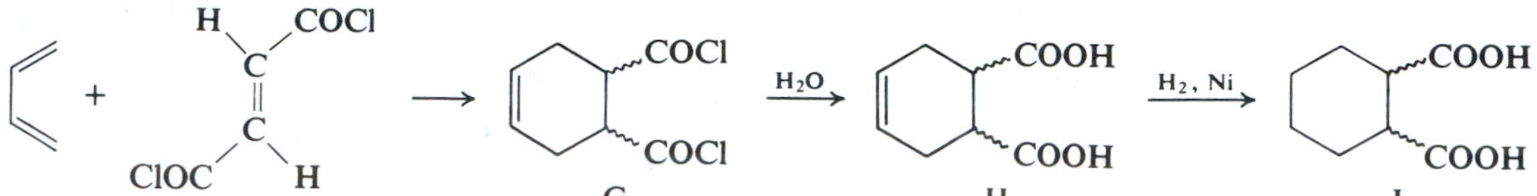

Fumaryl chloride

G

H

I

1,2-Cyclohexane-dicarboxylic acid

The final products F and I must be, then, the stereoisomeric 1,2-cyclohexanedicarboxylic acids:

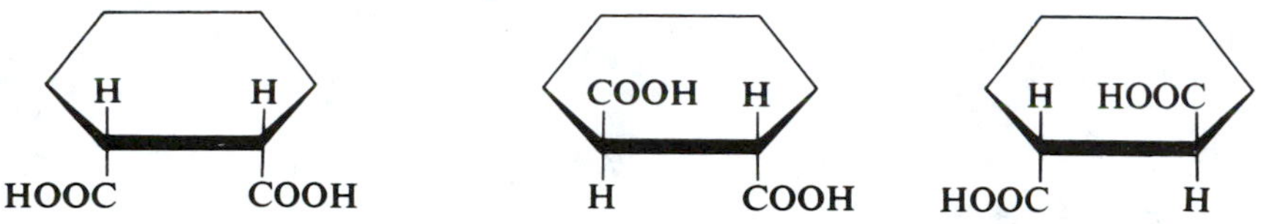

F

Meso: non-resolvable

cis-1,2-Cyclohexane-dicarboxylic acid

I

Racemic: resolvable

trans-1,2-Cyclohexane-dicarboxylic acid

Working backwards from these, we see that **the Diels–Alder reaction involves *syn*-addition.**

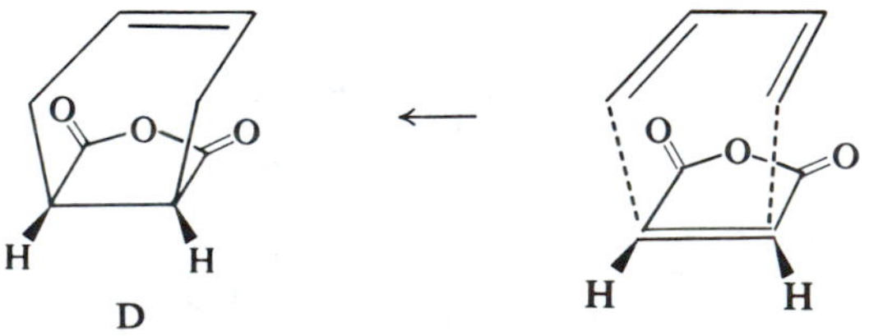

D

cis product

***Syn*-addition**

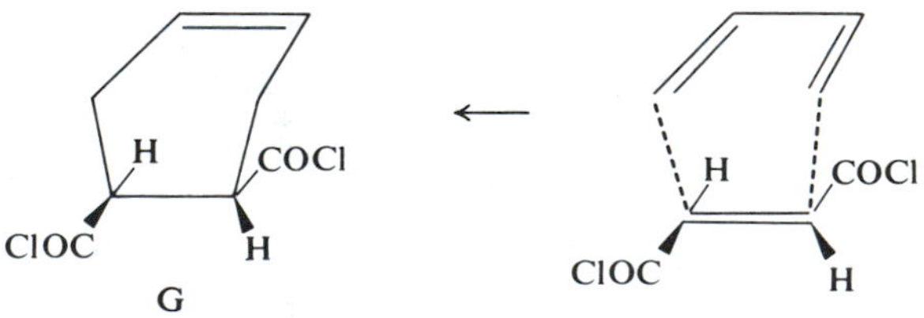

G

trans product

***Syn*-addition**

10. (a) 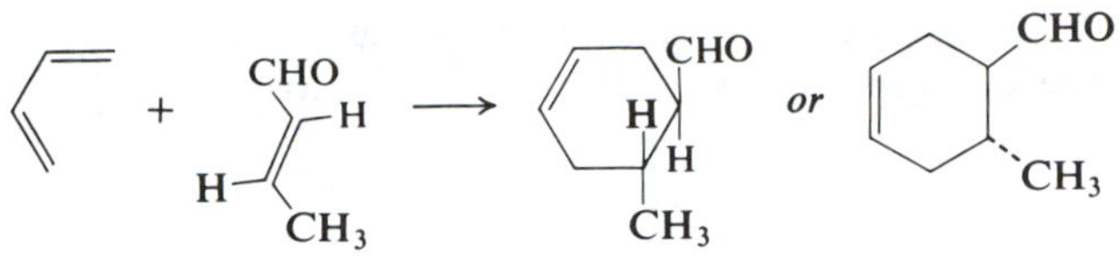 [1,3-butadiene] + [trans-crotonaldehyde: CHO, H; H, CH_3] ⟶ [cyclohexene with CHO, H, H, CH_3] *or* [cyclohexene with CHO and CH_3] + enantiomer

(b) 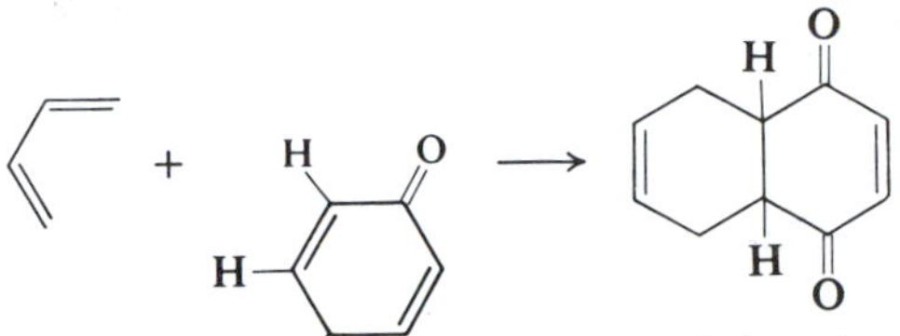 [1,3-butadiene] + [p-benzoquinone: H, H, O, O] ⟶ [bicyclic dione with H, H, O, O]

cis (*meso*)

(c) 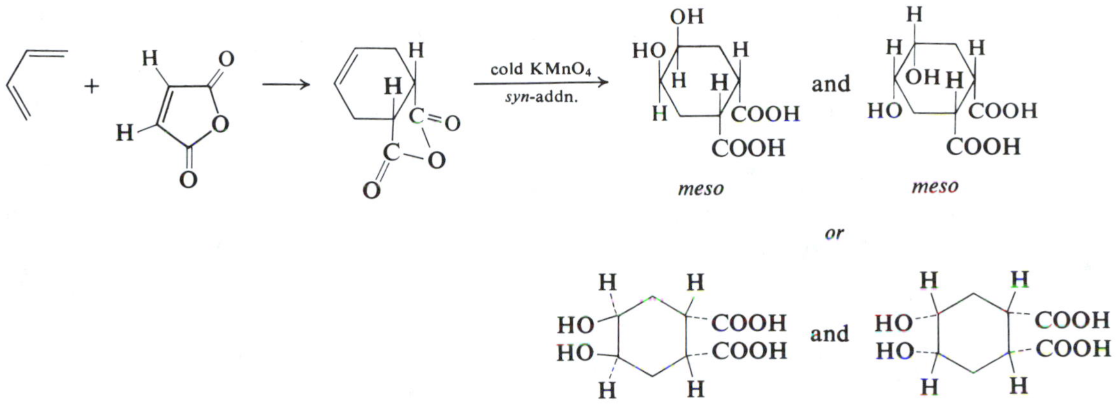 [1,3-butadiene] + [maleic anhydride: H, H, O, O, O] ⟶ [cyclohexene anhydride: H, H, C=O, C—O, O] $\xrightarrow[\textit{syn}\text{-addn.}]{\text{cold } KMnO_4}$ [OH, HO, H, H, H, H, COOH, COOH] and [H, H, H, OH H, HO, COOH, COOH]

meso *meso*

or

[H, H, HO, HO, COOH, COOH, H, H] and [H, H, HO, HO, COOH, COOH, H, H]

(d) 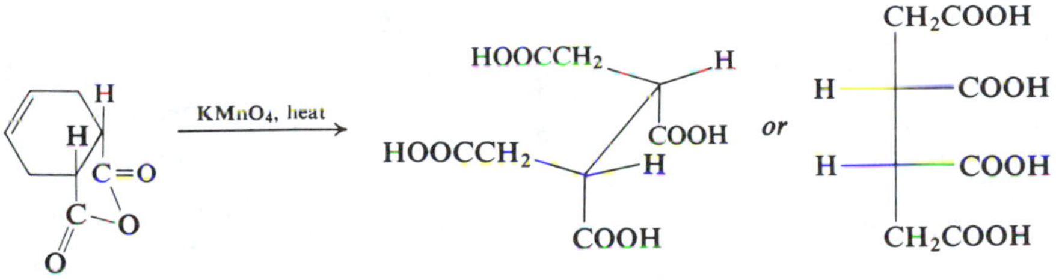 [cyclohexene anhydride: H, H, C=O, C—O, O] $\xrightarrow{KMnO_4,\ \text{heat}}$ [$HOOCCH_2$, H, COOH; $HOOCCH_2$, H, COOH] *or* [CH_2COOH; H—COOH; H—COOH; CH_2COOH]

(from (c), above)

meso

11. (a) There is a familiar 1,2-shift of methyl to give a more stable carbocation, followed by elimination of a proton.

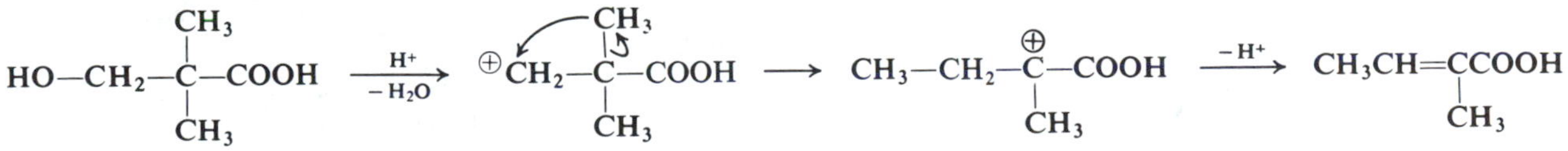

$$HO—CH_2—\underset{CH_3}{\overset{CH_3}{C}}—COOH \xrightarrow[-H_2O]{H^+} {}^{\oplus}CH_2—\underset{CH_3}{\overset{CH_3}{C}}—COOH \longrightarrow CH_3—CH_2—\underset{CH_3}{\overset{\oplus}{C}}—COOH \xrightarrow{-H^+} CH_3CH=\underset{CH_3}{C}COOH$$

(b) As we saw in Problem 21.16 (page 810), the γ-hydrogens of α,β-unsaturated carbonyl compounds are acidic; the carbanion formed by abstraction of such a proton is involved in a Claisen condensation.

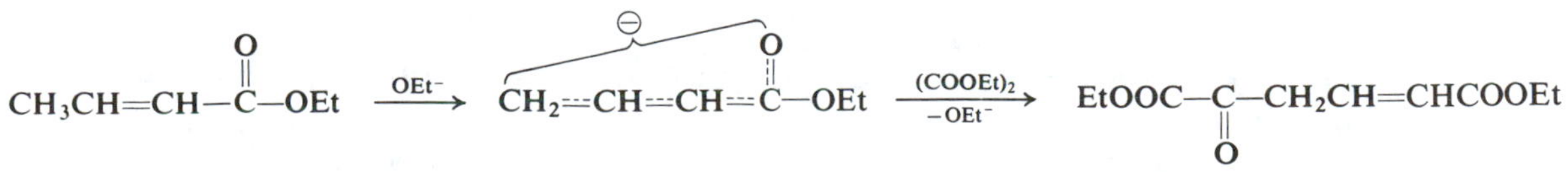

$$CH_3CH=CH—\overset{O}{\overset{\|}{C}}—OEt \xrightarrow{OEt^-} \overset{\ominus}{CH_2\text{:::}CH\text{:::}CH\text{:::}\overset{O}{\overset{\|}{C}}}—OEt \xrightarrow[-OEt^-]{(COOEt)_2} EtOOC—\underset{O}{\underset{\|}{C}}—CH_2CH=CHCOOEt$$

(c) The electron-withdrawing phosphonium group activates the double bond toward nucleophilic attack by the phenoxide ion; the addition product XIII then undergoes an intramolecular Wittig reaction.

$o\text{-}C_6H_4(CHO)O^- + CH_2{=}CH{-}\overset{\oplus}{P}Ph_3Br^- \xrightarrow{H^+}$ $o\text{-}C_6H_4(CH{=}O)(O{-}CH_2{-}CH_2{-}\overset{\oplus}{P}Ph_3\ Br^-)$ (XIII) $\xrightarrow{\text{Wittig}}$ 2H-chromene $+ Ph_3PO$

(d) Benzyne is formed, and undergoes a Diels–Alder reaction with furan.

Furan (*Chap. 30*) + Benzyne $\xrightarrow{\text{Diels–Alder}}$ oxygen-bridged adduct

Benzyne ← (Li) ← F,Br-benzene

12. (a) $HC{\equiv}C{-}CHO \longleftarrow HC{\equiv}C{-}CH(OEt)_2 \xleftarrow{\text{base}} HCBr{-}CHBr{-}CH(OEt)_2 \xleftarrow{Br_2} CH_2{=}CH{-}CH(OEt)_2$

(L of Problem 13a, below)
Protect —CHO

(b) $HOOCCH_2{-}CH(Ph)CH_2COOH \xleftarrow{-CO_2} \xleftarrow{H^+} \xleftarrow{OH^-} (EtOOC)_2CH{-}CH(Ph)CH_2COOEt$

↑ Michael addn.

$(EtOOC)_2CH_2 + PhCH{=}CHCOOEt$
Ethyl cinnamate
(Problem 1, above)

Or via $(EtOOC)_2CH_2 + PhCHO + H_2C(COOEt)_2$ (Knoevenagel followed by Michael)

(c) $HOOC{-}CH(Ph){-}CH_2COOH \xleftarrow[H^+]{H_2O} N{\equiv}C{-}CH(Ph){-}CH_2COOEt \xleftarrow{CN^-} PhCH{=}CHCOOEt$
Ethyl cinnamate
(Problem 1, above)

(d) $$CH_3\underset{\underset{O}{\|}}{C}CH_2-\overset{\overset{Ph}{|}}{C}H-CH_2\underset{\underset{O}{\|}}{C}CH_3 \xleftarrow[-CO_2]{H^+} \xleftarrow{OH^-} CH_3\underset{\underset{O}{\|}}{C}-\underset{\underset{COOEt}{|}}{C}H-\overset{\overset{Ph}{|}}{C}H-CH_2\underset{\underset{O}{\|}}{C}CH_3$$

$$\uparrow \text{Michael}$$

$$PhCHO + CH_3COCH_3 \xrightarrow[OH^-]{-H_2O} PhCH{=}CH\underset{\underset{O}{\|}}{C}CH_3 + CH_3COCH_2COOEt$$

13. (a) $$\underset{OH}{\underset{|}{CH_2}}-\underset{OH}{\underset{|}{CH}}-\underset{OH}{\underset{|}{CH_2}} \xrightarrow[\text{(Prob. 27.5)}]{H_2SO_4,\ \text{heat}} \underset{\text{J, acrolein}}{CH_2{=}CH-CHO}$$

$$\underset{\mathbf{J}}{CH_2{=}CH-CHO} \xrightarrow{HCl} \left[\underset{Cl}{\underset{|}{CH_2}}-CH_2-CHO\right] \xrightarrow[H^+]{EtOH} \underset{\mathbf{K}}{\underset{Cl}{\underset{|}{CH_2}}-CH_2-\overset{\overset{H}{|}}{\underset{\underset{OEt}{|}}{C}}-OEt}$$

Aldehyde protected by acetal formation

$$\underset{\mathbf{K}}{\underset{Cl}{\underset{|}{CH_2}}-CH_2-CH(OEt)_2} \xrightarrow[-HCl]{OH^-} \underset{\mathbf{L}}{CH_2{=}CH-CH(OEt)_2} \xrightarrow{KMnO_4} \underset{\mathbf{M}}{\underset{OH}{\underset{|}{CH_2}}-\underset{OH}{\underset{|}{CH}}-CH(OEt)_2}$$

$$\underset{\mathbf{M}}{\underset{OH}{\underset{|}{CH_2}}-\underset{OH}{\underset{|}{CH}}-CH(OEt)_2} \xrightarrow{H^+} \underset{\text{N, glyceraldehyde}}{\underset{OH}{\underset{|}{CH_2}}-\underset{OH}{\underset{|}{CH}}-CHO}$$

Cleavage of acetal to aldehyde

(b) $$EtOOC-C{\equiv}C-COOEt + H_2\underset{\underset{COOEt}{|}}{C}-COOEt \xrightarrow[\text{addn.}]{\text{Michael}} EtOOC-\underset{\underset{\underset{\underset{COOEt}{|}}{HC-COOEt}}{|}}{C}{=}CH-COOEt$$

O (*Problem 27.11f*)

$$\mathbf{O} \xrightarrow[\text{heat}]{OH^-} {}^-OOC-\underset{\underset{\underset{\underset{COO^-}{|}}{H-C-COO^-}}{|}}{C}{=}CH-COO^- \xrightarrow{H^+} \xrightarrow[-CO_2]{\text{heat}} HOOC-\underset{\underset{CH_2COOH}{|}}{C}{=}CH-COOH$$

P, aconitic acid

(c) $$EtOOC-CH{=}CH-COOEt + H_2\underset{\underset{COOEt}{|}}{C}-COOEt \xrightarrow[\text{addn.}]{\text{Michael}} EtOOC-\underset{\underset{\underset{\underset{COOEt}{|}}{H-C-COOEt}}{|}}{CH}-CH_2-COOEt$$

Q

$$\mathbf{Q} \xrightarrow[\text{heat}]{OH^-} {}^-OOC-\underset{\underset{\underset{\underset{COO^-}{|}}{H-C-COO^-}}{|}}{CH}-CH_2-COO^- \xrightarrow{H^+} \xrightarrow[-CO_2]{\text{heat}} HOOC-\underset{\underset{CH_2COOH}{|}}{CH}-CH_2-COOH$$

R, tricarballylic acid

(d) PhC(=O)–C(=O)Ph + $PhCH_2$–C(=O)–CH_2Ph —base, double aldol→ S, tetraphenylcyclopentadienone ("Tetracyclone")

Ring forms by interaction of two bifunctional molecules

S + Maleic anhydride —Diels–Alder→ T —heat, $-CO, -H_2$→ U Tetraphenylphthalic anhydride

Aromatization through loss of $CO + H_2$

(e) S + PhC≡CH —Diels–Alder→ V —heat, $-CO$→ W Pentaphenylbenzene

Aromatization through loss of CO

(f) CH_3–C(=O)–CH_3 + BrMgC≡COEt ⟶ CH_3–C(CH_3)(OMgBr)–C≡COEt —H^+→ CH_3–C(CH_3)(OH)–C≡COEt X

X —Pd/$CaCO_3$, H_2→ CH_3–C(CH_3)(OH)–CH=CHOEt Y

$$\underset{\substack{\text{Y}\\ \textit{A vinyl ether}}}{CH_3-\overset{\overset{CH_3}{|}}{\underset{\underset{OH}{|}}{C}}-CH{=}CHOEt} \xrightarrow[H^+]{H_2O} CH_3-\overset{\overset{CH_3}{|}}{\underset{\underset{OH}{|}}{C}}-CH_2-\overset{\overset{H}{|}}{C}{=}O$$

Easy cleavage of a vinyl ether (*Problem 17, page 704*)

$$CH_3-\overset{\overset{CH_3}{|}}{\underset{\underset{OH}{|}}{C}}-CH_2-\overset{\overset{H}{|}}{C}{=}O \xrightarrow[-H_2O]{H^+} \underset{\text{Z}}{CH_3-\overset{\overset{CH_3}{|}}{C}{=}CH-\overset{\overset{H}{|}}{C}{=}O}$$

β-Methylcrotonaldehyde
(*Has isoprene skeleton: useful in synthesis*)

(g) $$CH_3-\overset{\overset{CH_3}{|}}{C}{=}CHCOOEt + H_2\underset{\underset{COOEt}{|}}{C}-CN \xrightarrow[\text{addn.}]{\text{Michael}} \underset{\text{AA}}{CH_3-\overset{\overset{CH_3}{|}}{\underset{\underset{H-\underset{\underset{COOEt}{|}}{C}-CN}{|}}{C}}-CH_2COOEt} \xrightarrow[\text{heat}]{OH^-} CH_3-\overset{\overset{CH_3}{|}}{\underset{\underset{H-\underset{\underset{COO^-}{|}}{C}-COO^-}{|}}{C}}-CH_2COO^-$$

$$\xrightarrow{H^+} \xrightarrow[-CO_2]{\text{heat}} \underset{\text{BB}}{CH_3-\overset{\overset{CH_3}{|}}{\underset{\underset{CH_2COOH}{|}}{C}}-CH_2COOH}$$

(h) $$CH_3-\overset{\overset{CH_3}{|}}{C}{=}CHCOCH_3 + H_2\underset{\underset{COOEt}{|}}{C}-COOEt \xrightarrow[\text{addn.}]{\text{Michael}} \underset{\text{CC}}{CH_3-\overset{\overset{CH_3}{|}}{\underset{\underset{H-\underset{\underset{COOEt}{|}}{C}-COOEt}{|}}{C}}-CH_2COCH_3}$$

$$\text{CC} \xrightarrow{OBr^-} CH_3-\overset{\overset{CH_3}{|}}{\underset{\underset{H-\underset{\underset{COOEt}{|}}{C}-COOEt}{|}}{C}}-CH_2COO^- + CHBr_3 \xrightarrow[\text{heat}]{OH^-} CH_3-\overset{\overset{CH_3}{|}}{\underset{\underset{H-\underset{\underset{COO^-}{|}}{C}-COO^-}{|}}{C}}-CH_2COO^- \xrightarrow{H^+} \xrightarrow[-CO_2]{\text{heat}} \text{BB (see above)}$$

(i) 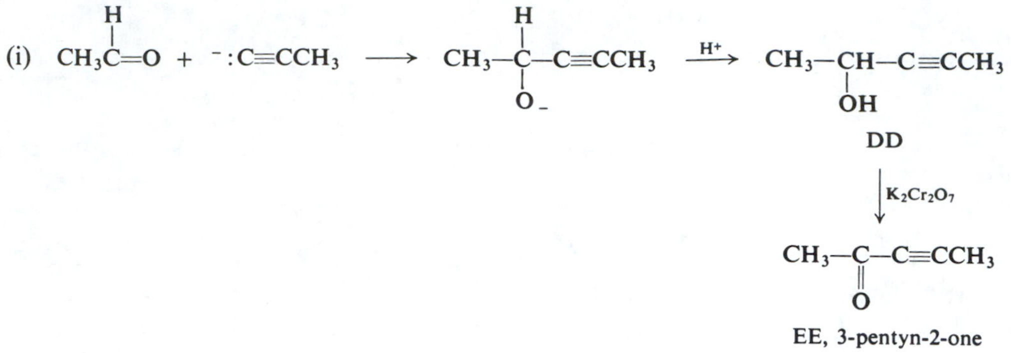 $CH_3CH{=}O + {}^{-}{:}C{\equiv}CCH_3 \longrightarrow CH_3{-}CH(O^-){-}C{\equiv}CCH_3 \xrightarrow{H^+} CH_3{-}CH(OH){-}C{\equiv}CCH_3$ (DD)

$$\text{DD} \xrightarrow{K_2Cr_2O_7} CH_3{-}\underset{\|}{C}(O){-}C{\equiv}CCH_3$$

EE, 3-pentyn-2-one

(j) 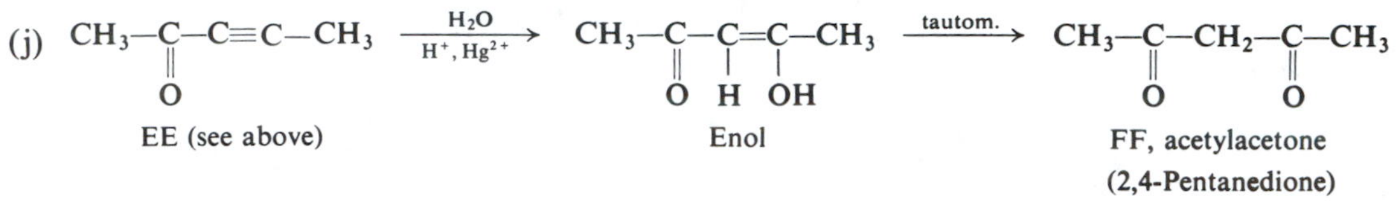 $CH_3{-}C({=}O){-}C{\equiv}C{-}CH_3 \xrightarrow[H^+,\ Hg^{2+}]{H_2O} CH_3{-}C({=}O){-}CH{=}C(OH){-}CH_3 \xrightarrow{\text{tautom.}} CH_3{-}C({=}O){-}CH_2{-}C({=}O){-}CH_3$

EE (see above) → Enol → FF, acetylacetone (2,4-Pentanedione)

Hydration of alkyne (*Sec. 12.10*)

(k) 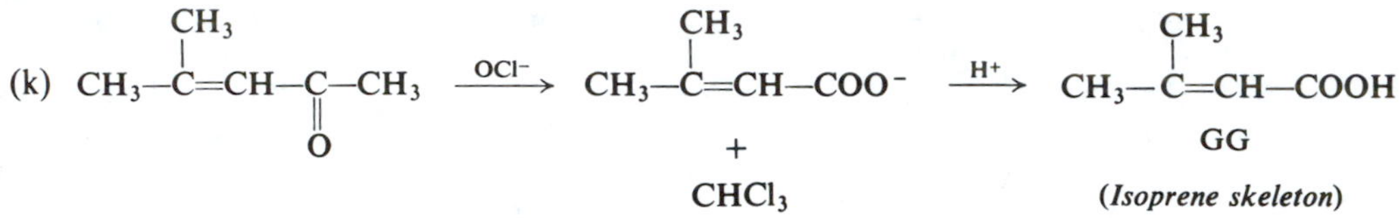 $CH_3{-}C(CH_3){=}CH{-}C({=}O){-}CH_3 \xrightarrow{OCl^-} CH_3{-}C(CH_3){=}CH{-}COO^- + CHCl_3 \xrightarrow{H^+} CH_3{-}C(CH_3){=}CH{-}COOH$

GG (*Isoprene skeleton*)

(l) 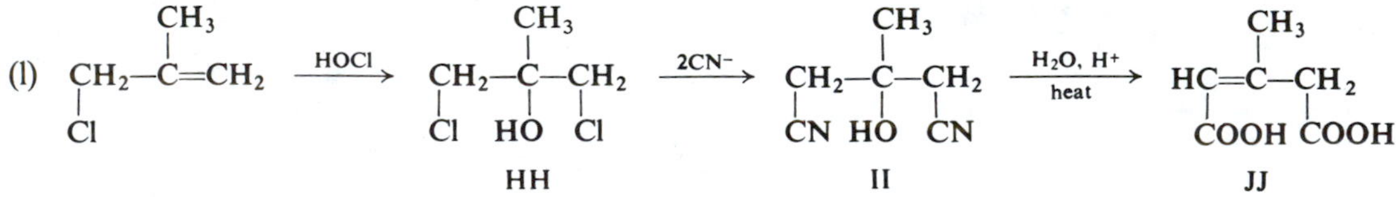 $ClCH_2{-}C(CH_3){=}CH_2 \xrightarrow{HOCl} ClCH_2{-}C(CH_3)(HO){-}CH_2Cl \xrightarrow{2CN^-} NCCH_2{-}C(CH_3)(HO){-}CH_2CN \xrightarrow[\text{heat}]{H_2O,\ H^+} HC(COOH){=}C(CH_3){-}CH_2COOH$

HH → II → JJ

(m) 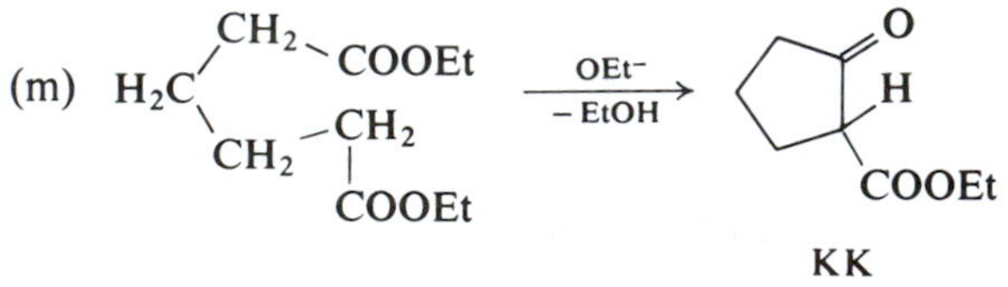 $H_2C(CH_2COOEt)(CH_2CH_2COOEt) \xrightarrow[-\,EtOH]{OEt^-}$ 2-carbethoxycyclopentanone (KK)

Intramolecular Claisen (Dieckmann) condensation (*Problem 21.25, page 815*)

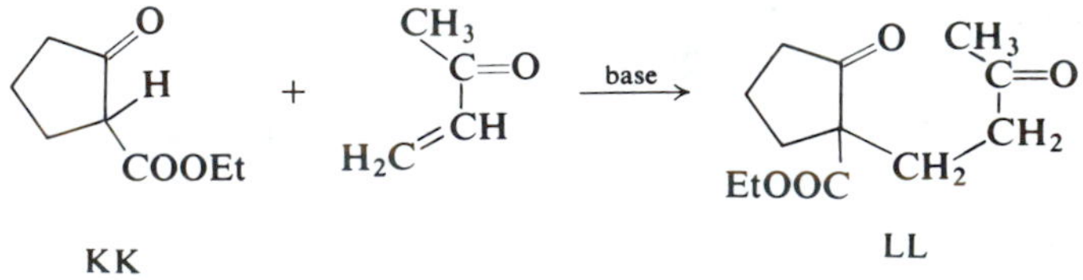

KK + $CH_3{-}C({=}O){-}CH{=}CH_2 \xrightarrow{\text{base}}$ 2-carbethoxy-2-(3-oxobutyl)cyclopentanone (LL)

Michael addition

LL $\xrightarrow[-\,H_2O]{\text{base}}$ MM (bicyclic enone with COOH at ring fusion)

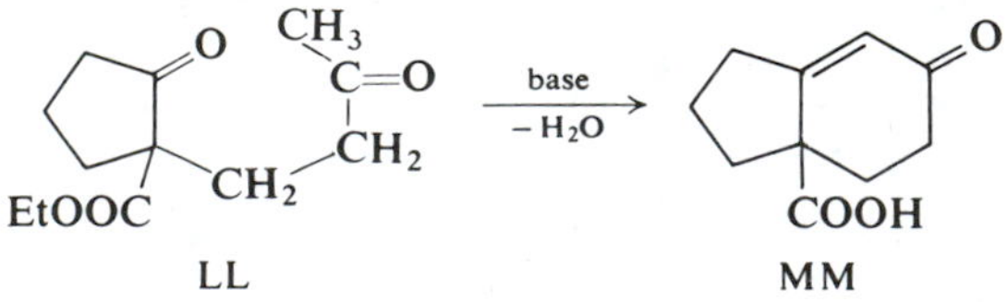

Intramolecular aldol condensation

(n) Hexachlorocyclopentadiene + $2CH_3O^-$ ⟶ NN, a ketal

Allylic Cl's react; vinylic do not

NN + $CH_2{=}CH_2$ —Diels–Alder⟶ OO —redn.⟶ PP —H^+⟶ QQ, 7-ketonorbornene

Cleavage of ketal unmasks ketone

(o) $CH_3CONHCH(COOEt)_2$ —$CH_2{=}CHCHO$ / Michael⟶ $CH_3CONHC(COOEt)_2{-}CH_2CH_2CHO$ (RR) —CN^- / H^+⟶ $CH_3CONHC(COOEt)_2CH_2CH_2CH(OH)CN$ (SS)

SS —acid / heat⟶ $CH_3CONHC(COOEt)_2CH_2CH{=}CHCN$ (TT) —H_2 / cat.⟶ $CH_3CONHC(COOEt)_2CH_2CH_2CH_2CH_2NH_2$ (UU)

UU —Ac_2O⟶ $CH_3CONHC(COOEt)_2(CH_2)_4NHAc$ (VV) —OH^- / heat⟶ —H^+⟶ —$-CO_2$⟶ $H_2NCH(COO^-)(CH_2)_4NH_3^+$ (WW, (±)-Lysine)

(p) $HCH(COOEt)_2$ —$CH_2{=}CHCN$ / Michael⟶ $HC(COOEt)_2{-}CH_2CH_2CN$ (XX) —H_2 / cat.⟶ [$HC(COOEt)_2{-}CH_2CH_2CH_2NH_2$] (YY) —$-EtOH$⟶ ZZ (3-carbethoxy-2-piperidone, EtOOC, NH, O)

An amide

ZZ $\xrightarrow[\alpha\text{-substn.}]{SO_2Cl_2}$ AAA (3-chloro-3-carbethoxy-2-piperidone) $\xrightarrow[-EtOH,\ -CO_2]{HCl,\ heat}$ $^{+}H_3NCH_2CH_2CH_2CH(Cl)COO^{-}$ (BBB) $\xrightarrow[-HCl]{base}$ CCC (±)-Proline

AAA BBB CCC (±)-Proline

14. (a) $EtOOC{-}CH(COCH_3){-}CH(CH_3){-}CH(COCH_3){-}COOEt$ $\xleftarrow{Michael}$ $EtOOC{-}CH_2{-}COCH_3 + CH_3CH{=}C(COCH_3){-}COOEt$

III
A δ-diketone

α,β-Unsaturated ketone

↑ aldol, $-H_2O$

$CH_3CHO + H_2C(COCH_3){-}COOEt$

IV (A β-hydroxy ketone) $\xleftarrow{aldol}$ III

IV
A β-hydroxy ketone

III

(b) IV is the correct structure.

IV

The NMR spectrum is too complex for a molecule as symmetrical as III, from which we would expect only six signals.

$\overset{c}{CH_3}\overset{f}{CH_2}OC(O){-}\overset{d}{CH}(COCH_3^{\,e}){-}\overset{b}{CH}(CH_3^{\,a}){-}\overset{d}{CH}(COCH_3^{\,e}){-}C(O)O\overset{f}{CH_2}\overset{c}{CH_3}$

III

a doublet, 3H
b multiplet, 1H
c triplet, 6H
d doublet, 2H
e singlet, 6H
f quartet, 4H

(Binns, T. D.; Brettle, R. "Anodic Oxidation. Part I. The Electrolysis of Ethyl Sodioacetoacetate in Ethanolic Solutions"; *J. Chem. Soc.* (*C*) **1966**, 336.)

15. Like other nucleophilic reagents, the Grignard reagent can undergo conjugate addition to α,β-unsaturated carbonyl compounds.

$$\text{PhCH=CH–C(=O)–CH}_3 \xrightarrow{\text{EtMgBr}} \text{PhCH(Et)–CH}{\cdots}\text{C(}{\cdots}\text{O}^{\ominus}\text{)–CH}_3\ \text{MgBr}^+ \xrightarrow{\text{H}^+} \text{PhCH(Et)–CH}_2\text{–C(=O)–CH}_3 \xrightarrow{\text{I}_2,\ \text{OH}^-} \text{CHI}_3$$

16.

$$^-\text{OOC–CH=CH–COO}^- \xrightarrow{\text{Br}_2} {}^-\text{OOC–CH(Br)–C}^+\text{H(COO}^-) \longrightarrow \text{V}$$

V (β-lactone: ring of H(COO⁻)C, O, C=O, CHBr)

This experiment gives evidence that the addition of bromine is two-step and electrophilic: the intermediate carbocation (or bromonium ion) is trapped—just as it was (Sec. 9.13) by Cl^-, Br^-, I^-, $NO_3{}^-$, etc.—by carboxylate ion within the same molecule.

(Tarbell, D. S.; Bartlett, P. D. *J. Am. Chem. Soc.* **1937**, *59*, 407.)

17.

butadiene + HC≡C–COOH $\xrightarrow{\text{Diels–Alder}}$ DDD (1,4-cyclohexadiene-1-carboxylic acid) $\xrightarrow{\text{LiAlH}_4}$ EEE (CH_2OH)

EEE + Cl–C(=O)–OMe ⟶ FFF (CH_2–O–C(=O)–OMe) $\xrightarrow[-\text{CO}_2,\ -\text{MeOH}]{\text{heat}}$ Toluene (CH_3) + GGG (=CH_2)

GGG + $C(CN)_2{=}C(CN)_2$ $\xrightarrow{\text{Diels–Alder}}$ HHH (H_2C=, $(CN)_2$, $(CN)_2$)

The structure assigned to GGG is indicated most strongly by its easy conversion into its isomer toluene; taken together with the fact that GGG is *not* 1,3,5-cycloheptatriene, this conversion severely limits the number of possible structures. The ultraviolet spectrum (Sec. 17.8) indicates strong conjugation. (Actually, absorption is at precisely the wavelength predicted for this structure.) Conjugation is also shown by the occurrence of the Diels–Alder reaction. The infrared spectrum (Sec. 17.5) shows trigonal C—H stretch at 3020 cm^{-1}, conjugated C=C stretch at 1595, and the expected (see page 593) C—H out-of-plane bend, at 864 and 692.

Finally, the structure is consistent with the method of synthesis.

The conversion of EEE into GGG is worth looking at. EEE is an allylic ester, and is believed to undergo an *allylic rearrangement* to XV before elimination. Such a rearrangement *could* proceed by ionization to an allylic cation and CH_3OCOO^- followed by recombination at a different carbon of the hybrid cation; here, it seems more likely to involve concerted bond-breaking and bond-making as shown.

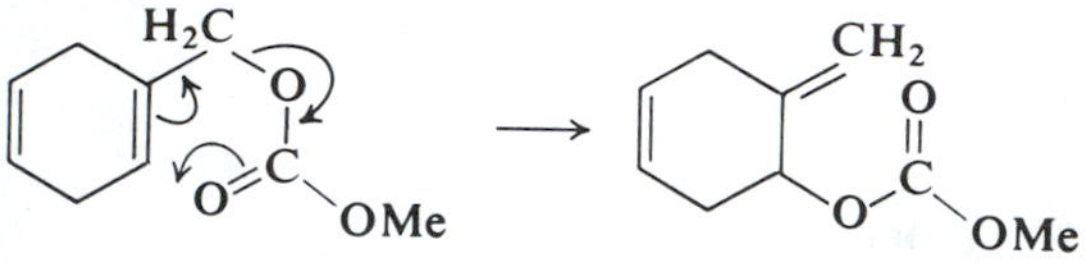

FFF → XIV *An allylic rearrangement*

Formation of GGG involves pyrolysis of an ester, a well-known and—as demonstrated here—mild method of elimination. From the stereochemistry of ester pyrolyses (*syn*), it seems likely that here, too, a concerted reaction is involved.

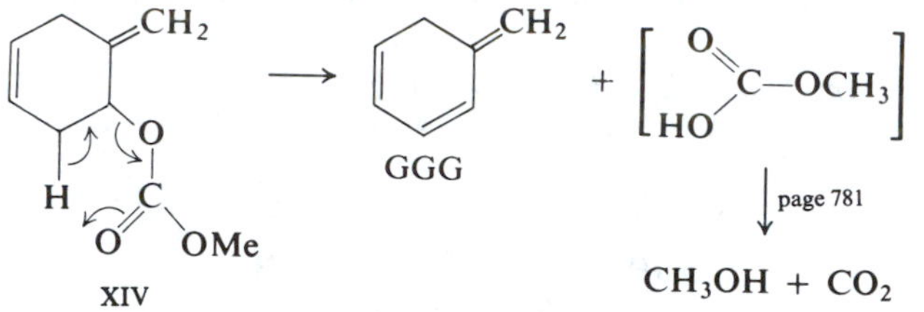

XIV → GGG + [HO–C(=O)–OCH_3] → (page 781) $CH_3OH + CO_2$

The most interesting thing about all this is that GGG is isolated at all. It is less stable by some 30 kcal than its aromatic isomer toluene, and we might expect it to rearrange practically instantaneously into the more stable compound. (In fact, such a prediction was made, in 1943.) Actually, the conversion, although easy, is far from instantaneous; it takes hours at room temperature. Evidently, this highly exothermic reaction has a sizable energy of activation, surprising for a simple transfer of hydrogen. The explanation for this did not emerge until the years following 1965, with the concept of *orbital symmetry* (Chapter 28). As we shall see (Problem 15b, Chapter 28), the easy route from GGG to toluene is "forbidden", and reaction is forced to follow a more difficult path.

(Bailey, W. J.; Baylouny, R. A. "Cyclic Dienes. XXVI. 5-Methylene-1,3-cyclohexadiene, an Alicyclic Isomer of Toluene"; *J. Org. Chem.* **1962**, *27*, 3476.)

18. The enamine III contains nucleophilic carbon, and undergoes a Michael addition to the α,β-unsaturated ester.

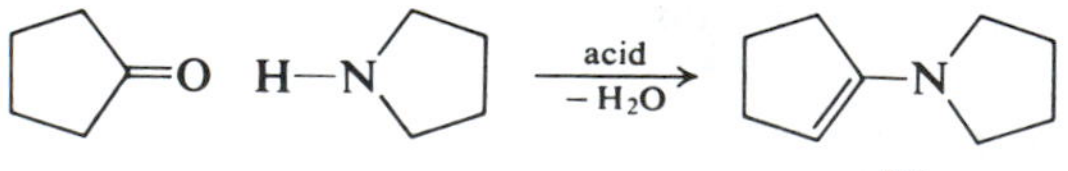

Cyclopentanone + H–N(pyrrolidine) —acid, $-H_2O$→ III

An enamine (*Sec. 25.8*)

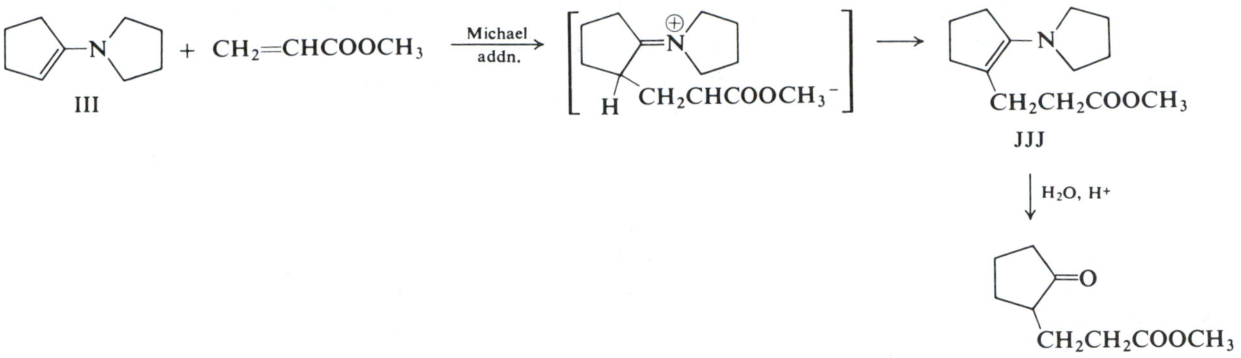

III + $CH_2{=}CHCOOCH_3$ —Michael addn.→ [$CH_2CHCOOCH_3^-$] → JJJ $CH_2CH_2COOCH_3$ —H_2O, H^+→ KKK $CH_2CH_2COOCH_3$

19. (a)

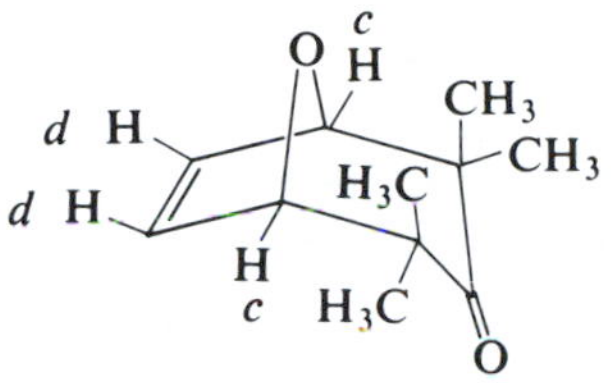

VIII

CH_3's *a* and *b*

The four methyls are not equivalent: two are closer to the oxygen bridge than the other two.

(b)

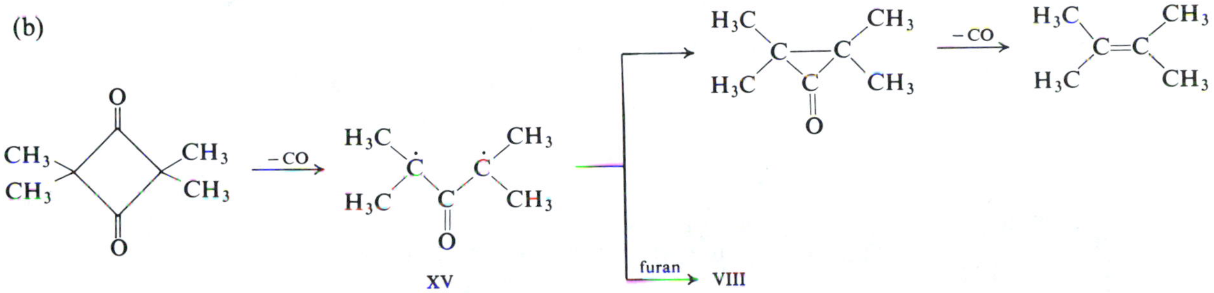

—CO→ XV → (—CO→) ; furan → VIII

Loss of one CO gives XV (which may be singlet or triplet, Sec. 13.16). The terminal carbons of XV each contribute an electron to form a bond and generate cyclopropanone, which loses a second CO to give the alkene. The intermediacy of XV is shown by the fact that it can be trapped by furan to give the 1,4-addition product VIII.

20. $C_6H_5CH{=}O + PhNH_3^+ \rightleftarrows C_6H_5CH{=}\overset{\oplus}{N}HPh + H_2O$

An iminium ion

$C_6H_5CH{=}\overset{\oplus}{N}HPh + NH_2NHCONH_2 \longrightarrow C_6H_5CH(\overset{\oplus}{N}H_2NHCONH_2)NHPh \longrightarrow$ semicarbazone

The anilinium ion reacts with benzaldehyde to form an *iminium ion*, which is the compound that semicarbazide actually adds to. After the addition, the —NHPh group is hydrolyzed away, and the semicarbazone forms.

Addition takes place much faster to the iminium ion than to the carbonyl group because there is no developing negative charge to be accommodated by the molecule (see Sec. 18.7). This same kind of activation could be accomplished by protonation of the carbonyl group (see page 671), but this would require an acidity so high that much of the semicarbazide would exist in the unreactive protonated form (Sec. 18.11).

Here, as in Problem 27.16 above, we see catalysis of nucleophilic addition brought about by amines, not just because of their basicity, but because of their ability to form intermediate iminium ions.

21. We have already seen (Problem 13, Chapter 26) that benzyne can be formed by loss of N_2 and CO_2 from an *ortho* diazonium carboxylate (I, page 969). We know (Sec. 14.10) that the cyclopentadienyl anion is aromatic. We consider it likely, therefore, that loss of N_2 and CO_2 from the salt of IX gives the aryne XVI related to this non-benzenoid aromatic compound.

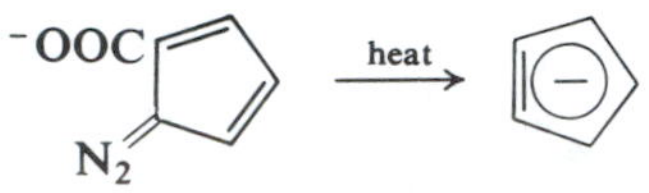

$^-$OOC, N_2 — heat →

IX XVI

Dehydrocyclo-
pentadienyl anion
An aryne

Aryne XVI is trapped by X to give the Diels–Alder adduct XVII which, by loss of CO, generates the new, central benzene ring of XI.

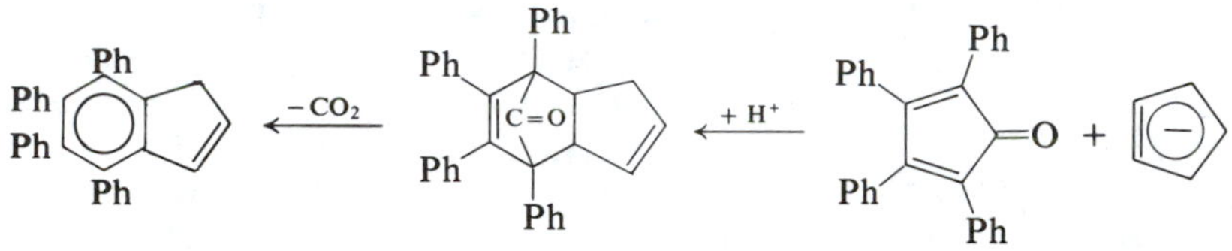

Ph Ph Ph Ph ← $-CO_2$ — Ph Ph C=O Ph Ph ← $+H^+$ — Ph Ph =O Ph Ph +

XI XVII X XVI

These findings provide one more piece of evidence for the aromaticity of the cyclopentadienyl anion and for the validity of the Hückel $4n + 2$ rule (Secs. 14.10 and 28.6).

(Martin, J. C.; Bloch, D. R. "The Dehydrocyclopentadienyl Anion. A New Aryne"; *J. Am. Chem. Soc.* **1971**, *93*, 451.)

22. We see here a series of nucleophilic reactions: addition of the Michael kind, then addition to a carbonyl group, and finally intramolecular nucleophilic substitution.

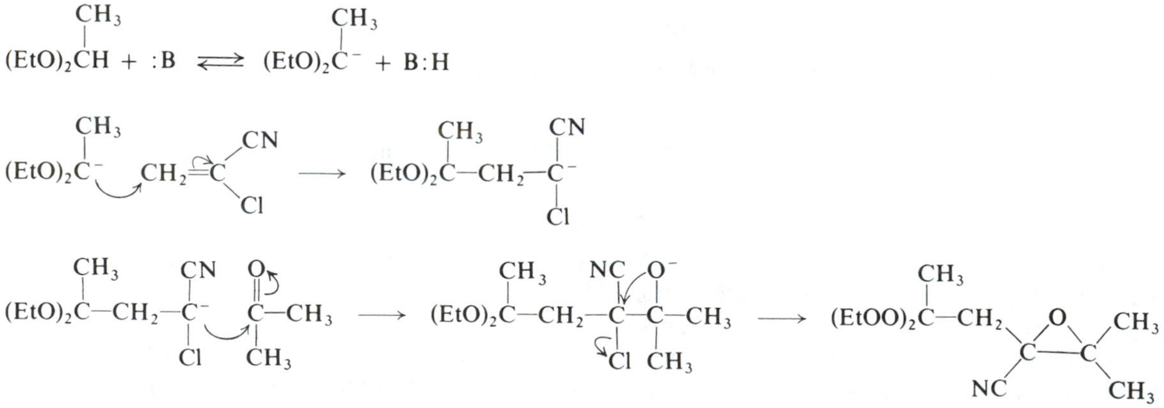

$$(EtO)_2CH(CH_3) + :B \rightleftarrows (EtO)_2C^-(CH_3) + B{:}H$$

$$(EtO)_2C^-(CH_3) \quad CH_2{=}C(CN)Cl \longrightarrow (EtO)_2C(CH_3){-}CH_2{-}C^-(CN)(Cl)$$

$$(EtO)_2C(CH_3){-}CH_2{-}C^-(CN)(Cl) \quad O{=}C(CH_3){-}CH_3 \longrightarrow (EtO)_2C(CH_3){-}CH_2{-}C(NC)(Cl){-}C(O^-)(CH_3){-}CH_3 \longrightarrow (EtOO)_2C(CH_3){-}CH_2{-}\underset{NC}{C}\overset{O}{—}C(CH_3)CH_3$$

XII

28

Molecular Orbitals. Orbital Symmetry

28.1 The first-formed monocation is slowly converted into the—evidently more stable—dication. The stability of the latter species, despite its double positive charge, is consistent with aromaticity; it contains 2 π electrons ($4n + 2$, $n = 0$).

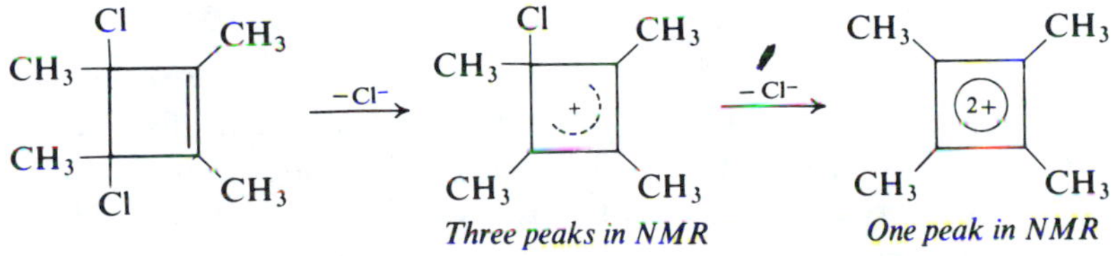

Three peaks in NMR

One peak in NMR

(Olah, G. A.; Bollinger, J. M.; White, A. W. "Stable Carbonium Ions. LXXXIX. The Tetramethylcyclobutenium Dication as an Aromatic 2 π-Electron System"; *J. Am. Chem. Soc.* **1969**, *91*, 3667.)

28.2 (a) The facts indicate that compound I has a dipolar structure; it is basic because acceptance of a proton eliminates the separation of charge and thus stabilizes the molecule.

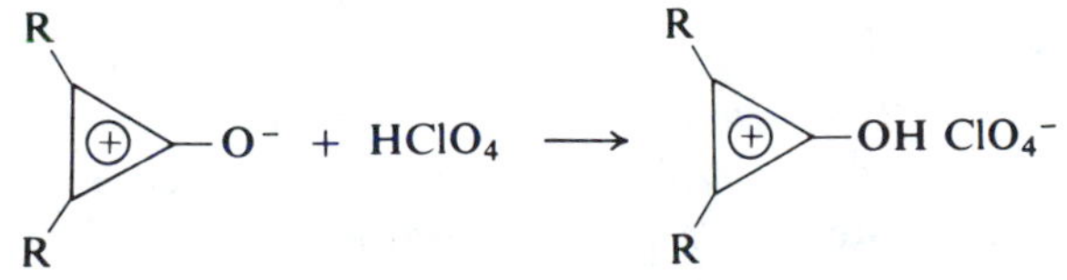

All this stems from the stability of the positively charged ring. This, in turn, we attribute to its aromaticity. "Carbonyl" carbon shares only one pair of electrons with oxygen; the empty p orbital of this carbon contributes no π electrons, and the total for the cyclic system is the Hückel number 2 ($4n + 2$, with $n = 0$).

(b) The product is the aromatic 1,2,3-triphenylcyclopropenyl cation (see the answer to Problem 14.6, page 507).

Ph Ph Ph Ph Ph

$\xrightarrow{\text{PhMgBr}}$ Ph OMgBr $\xrightarrow{H^+}$ Ph OH $\xrightarrow{H^+}$ Ph OH_2^+ $\xrightarrow{-H_2O}$ Ph

Ph Ph Ph Ph Ph

(c) Decarboxylation of trichloroacetate is easy because of the relative stability of the Cl_3C^- anion being formed.

$$Cl_3CCOO^- \xrightarrow{\text{heat}} CO_2 + Cl_3C{:}^- \xrightarrow{-Cl^-} {:}CCl_2$$

(For a general discussion of aromaticity, see Breslow, R. "Aromatic Character"; *Chem. Eng. News* **June 28, 1965**, 90.)

28.3 (a) The ring closure involves four double bonds of the polyene, and hence 8 π electrons; this corresponds to $4n$ ($n = 2$). Table 28.1 predicts that for such a system a thermal electrocyclic reaction will take place with conrotatory motion.

I or III ⟶ *trans* product

II ⟶ *cis* product

(b) Photochemical ring closure in this same system will take place with disrotatory motion.

I or III ⟶ *cis* product

II ⟶ *trans* product

Let us look at the reverse process, ring-opening. In principle, if not practically, the *trans*-dimethylcyclooctatriene is related thermally to *two* polyenes, I and III, *both* by symmetry-allowed conrotatory motion. The *cis*-dimethylcyclooctatriene is related photochemically to *two* polyenes, I and III.

There will always be two possible conrotatory modes,

↷ ↷ and ↶ ↶ **Conrotatory**

and two possible disrotatory modes.

↶ ↷ and ↷ ↶ **Disrotatory**

These may lead to the same product or to different products, depending on the particular case. Where two products are possible, both may or may not be formed.

In Figure 28.12 (page 1005), for example, the *cis*-dimethylcyclobutene can give by conrotatory motion only one product, the *cis,trans*-hexadiene. The *trans*-dimethylcyclobutene could give either the *trans,trans* diene (with H's turning toward each other in the transition state) or the *cis,cis* diene (with CH_3's turning toward each other). Only the *trans,trans* diene is obtained, undoubtedly for steric reasons.

In the reverse reaction, ring closure (Figure 28.15, page 1008), the *cis,trans*-hexadiene can give only the *cis*-dimethylcyclobutene. The *trans,trans*-hexadiene yields one enantiomer of the dimethylcyclobutene by one conrotatory mode, and the other enantiomer by the other mode; here the two modes are equally likely, and the racemic product is obtained.

We shall encounter other reactions where this factor enters in, and must carefully consider each case individually.

28.4 (a) ψ_1 (Figure 28.7, page 999). The allyl cation has 2 π electrons.

(b) $4n + 2$ thermal: expect disrotatory motion.

(c) In the allyl anion, there are 4 π electrons. $4n$ thermal: expect conrotatory motion.

(d) In the pentadienyl cation, there are 4 π electrons. $4n$ thermal: expect conrotatory motion.

28.5 (a) In the polyene (the tropylidene) three double bonds are involved, or 6 π electrons. $4n + 2$ thermal: disrotatory ring-opening.

This interconversion takes place so fast that only at $-120\,°C$ does the NMR spectrum of the equilibrium mixture show completely separate signals for the two compounds. (Ciganek, E. "The Direct Observation of a Norcaradiene–Cycloheptatriene Equilibrium"; *J. Am. Chem. Soc.* **1965**, *87*, 1149.)

We notice that, of the two possible disrotatory modes, only one can occur: the angular H's must turn *away from* each other as the ring opens. The alternative mode would generate two *trans* double bonds, geometrically impossible in a ring this size. Photochemical ring-opening cannot occur, since this would require conrotatory motion, with the formation of one *trans* double bond in the seven-membered ring, also geometrically impossible.

(b) In the polyene (the reactant) three double bonds are involved, or 6 π electrons. $4n + 2$ thermal: disrotatory ring-closing.

(c) Three double bonds or 6 π electrons ($4n + 2$) are involved throughout.

First step. Thermal disrotatory ring-closing.

Second step. Photochemical conrotatory ring-opening.

Third step. Thermal disrotatory ring-closing.

(d) In the first reaction, only two double bonds of the polyene are involved; in the second reaction, all three bonds are involved.

First step. 4 π electrons ($4n$): thermal conrotatory ring-opening.

Second step. 6 π electrons ($4n + 2$): thermal disrotatory ring closure.

(e) 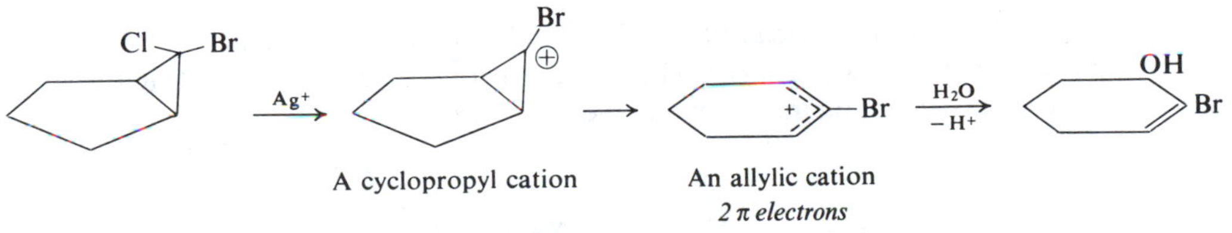

This is an example of the conversion of a cyclopropyl cation into an allylic cation (Problem 28.4, page 1011): 2 π electrons ($4n + 2$), thermal disrotatory ring-opening. The allylic cation then combines with water.

Actually, ring-opening is stereospecific in a further way, indicating that ring-opening is concerted with loss of halide ion. Calculations lead to another rule, by which one can predict *which* of the two possible disrotatory modes will occur: the substituents on the same side of the cyclopropyl ring as the leaving group will rotate inward (toward each other), not outward. In the present case, the substituents on one side of the cyclopropyl ring are carbons of the five-membered ring; these must rotate toward each other since they are tied together in a smallish (5 ⟶ 6) ring. As a result, the leaving group is the *endo* halogen, as shown.

(For a discussion of this point, see one of the following: Woodward, R. B.; Hoffmann, R. *The Conservation of Orbital Symmetry*; Academic Press: New York, 1970; pp 46–48; DePuy, C. H. "The Chemistry of Cyclopropanols"; *Acc. Chem. Res.* **1968**, *1*, 33.)

(f) The protonated ketone is equivalent to the pentadienyl cation VIII (4 π electrons, 4*n*), which undergoes thermal conrotatory ring-closing to the cyclopentenyl cation IX. (See Problem 28.4(d), page 1011.)

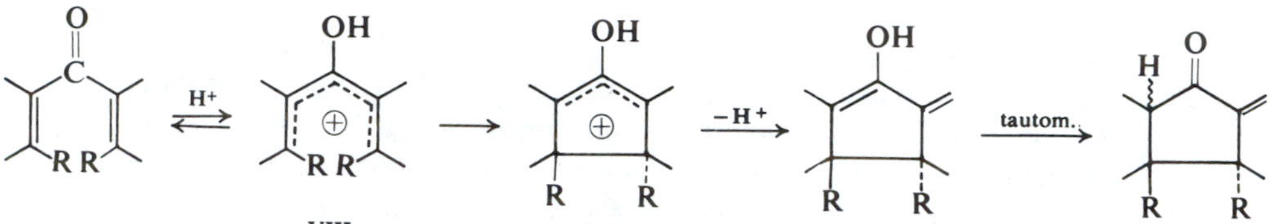

VIII
A pentadienyl cation

IX
A cyclopentenyl cation

28.6 Interconversion takes place via the cyclobutene. Two double bonds of the polyene are involved, or 4 π electrons. 4*n* thermal: conrotatory ring-closing and ring-opening.

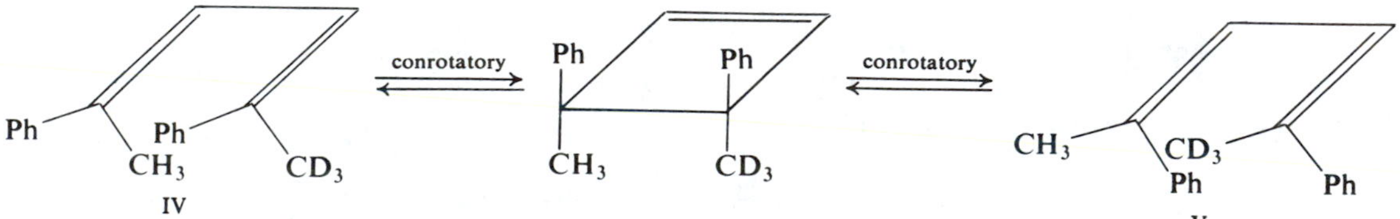

IV

V

Here we see the two possible conrotatory modes of ring-opening (see the answer to Problem 28.3, page 1011), one giving IV and the other V.

28.7 (a) This is a [4 + 2] thermal cycloaddition, the familiar Diels–Alder reaction. The symmetry-allowed reaction is *supra,supra*, which is also geometrically easy. Putting the diene in the necessary *s-cis* configuration, we arrive at our product:

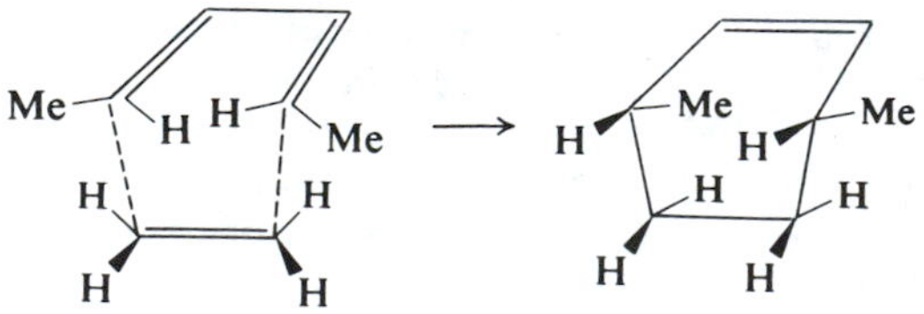

We predict, then, formation of *cis*-3,6-dimethylcyclohexene.

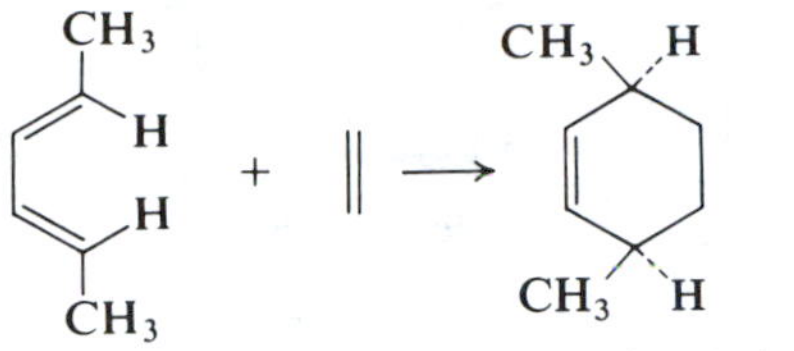

trans,trans-2,4-Hexadiene + Ethylene → *cis*-3,6-Dimethylcyclohexene

(b) Another [4 + 2] thermal cycloaddition: *supra,supra*. With the diene in the *s-cis* configuration, we bring up the dienophile. With regard to the dienophile, reaction leads to *syn*-addition: in the product, the two angular H's are *cis* to each other. The reaction is also *endo*, with the anhydride bridge near the developing double bond (beneath it, in the representation above): in the product, the methyl group is *cis* to the anhydride bridge.

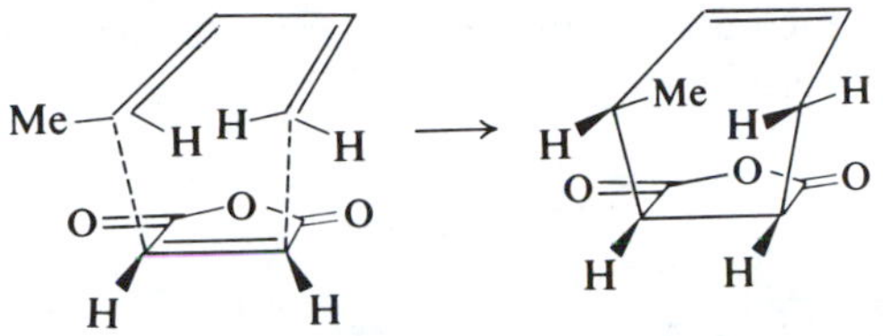

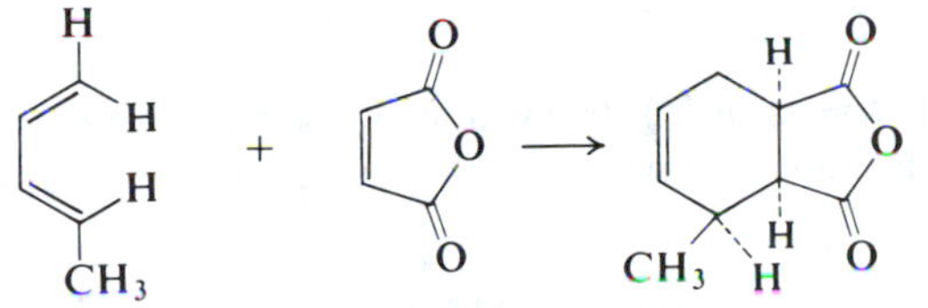

trans-1,3-Pentadiene + Maleic anhydride →

(c) As in part (b), [4 + 2] thermal *supra,supra* cycloaddition. Once more we remember: *s-cis*, *syn*, and *endo*.

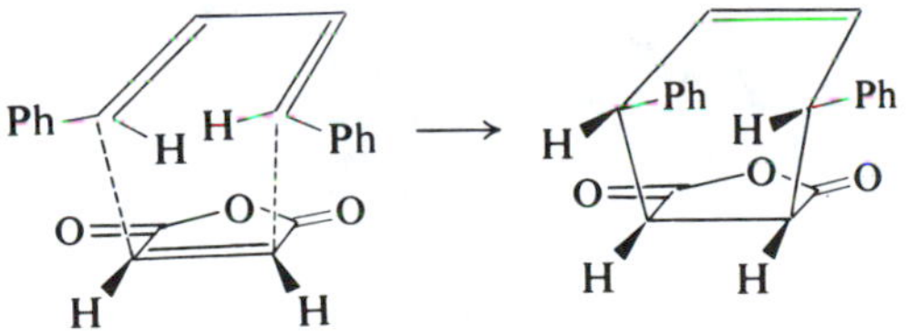

In the product the Ph's are *cis* to each other (*syn*-addition) and *cis* to the anhydride bridge (*endo* reaction).

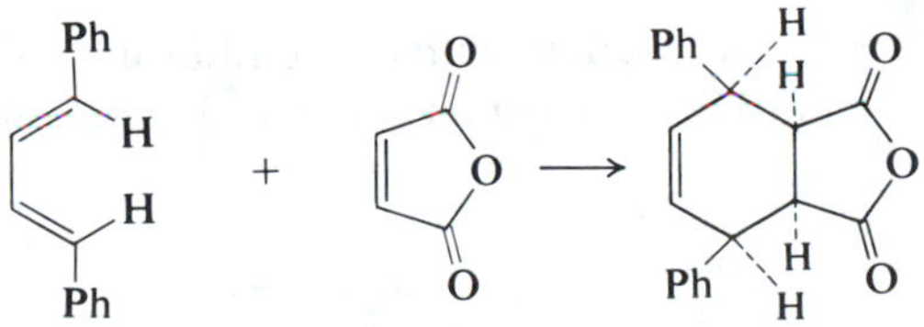

trans,trans-1,4-Diphenyl-1,3-butadiene + Maleic anhydride →

(d) [2 + 2] photochemical cycloaddition: *supra,supra*. This amounts to *syn*-addition to each alkene, so that the two methyls of each alkene unit remain *cis* to each other. But

cis-2-Butene *Two moles* → B + A ← *cis*-2-Butene *Two moles*

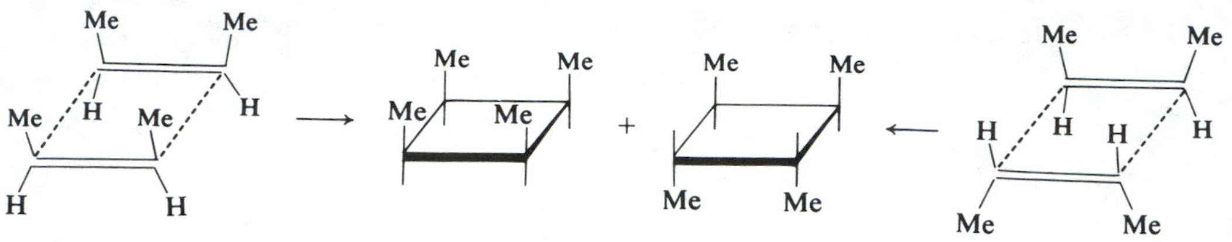

there are two ways in which the alkene units can come together, and two products; one set of methyls can be *cis* or *trans* to the other set.

(e) As in (d). Here the methyls of each alkene unit remain *trans* to each other. Again there are two combinations, and two products.

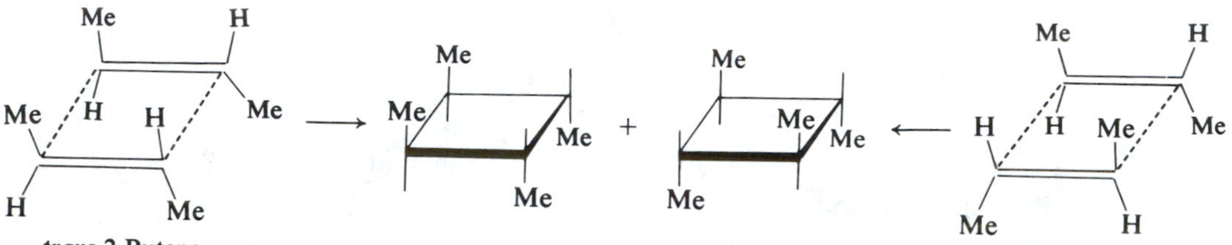

trans-2-Butene *Two moles* → A + C ← *trans*-2-Butene *Two moles*

(f) *cis*-Alkene can add to *cis*-alkene as in (d), *trans* to *trans* as in (e), and, besides, *cis* to *trans*. There are four products, all the possible 1,2,3,4-tetramethylcyclobutanes.

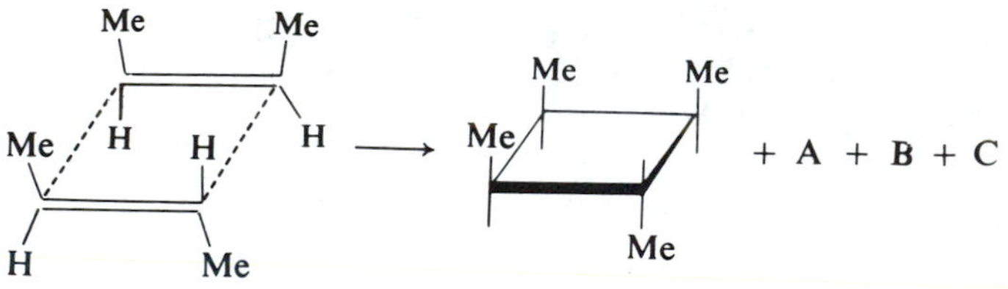

cis-2-Butene and *trans*-2-Butene → D + A + B + C

28.8 (a) Dicyclopentadiene is formed by a Diels–Alder reaction in which one molecule of cyclopentadiene acts as diene, and the other molecule acts as dienophile. Regeneration of cyclopentadiene is a *retro*-Diels–Alder reaction, with the equilibrium being shifted to replace the more volatile component, which is being removed by distillation.

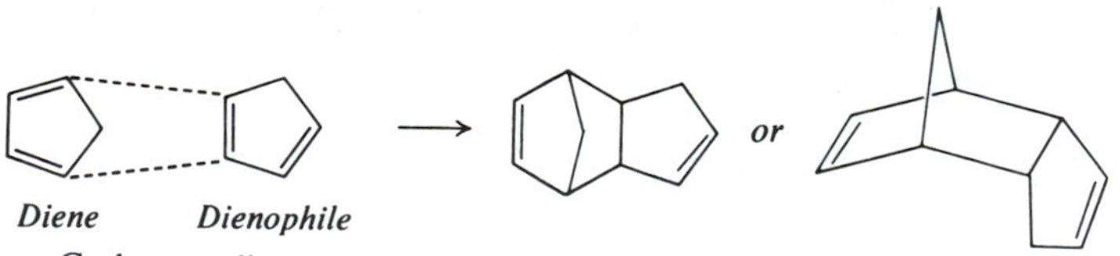

Diene *Dienophile*
Cyclopentadiene
Two moles

→ *or*

I
Dicyclopentadiene

(b) Reaction is *endo*, with the ring of the dienophile close to the developing double bond in the diene moiety.

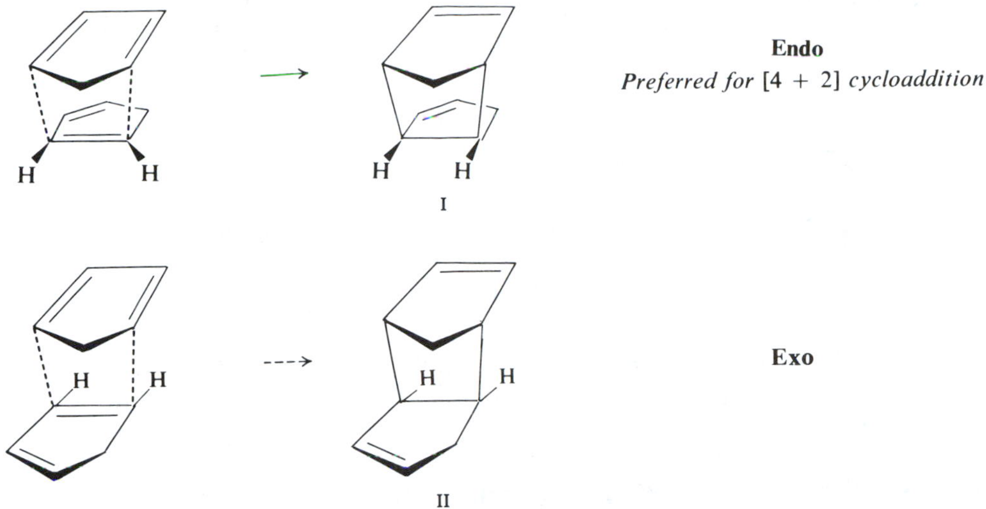

28.9 (a) Only two double bonds of the triene are involved. Reaction is thus [4 + 2] cycloaddition, which is *supra,supra* and easy, instead of [6 + 2] cycloaddition, which would have to be *supra,antara* and hence geometrically difficult.

(b) An intramolecular [2 + 2] photochemical cycloaddition. The *supra,supra* stereochemistry gives the all-*cis* configuration in the cyclobutane ring.

(c) Reaction is a [6 + 4] thermal cycloaddition: predicted to be *supra,supra*—and hence geometrically easy—on the basis that $i + j$ equals 10 and hence is $4n + 2$.

This was the first example of a [6 + 4] cycloaddition to be reported. (Cookson, R. C.; Drake, B. V.; Hudec, J.; Morrison, A. "The Adduct of Tropone and Cyclopentadiene: A New Type of Cyclic Reaction"; *Chem. Commun.* **1966**, 15.)

A closer look shows us that, in contrast to [4 + 2] cycloaddition, this reaction takes place in the *exo* sense.

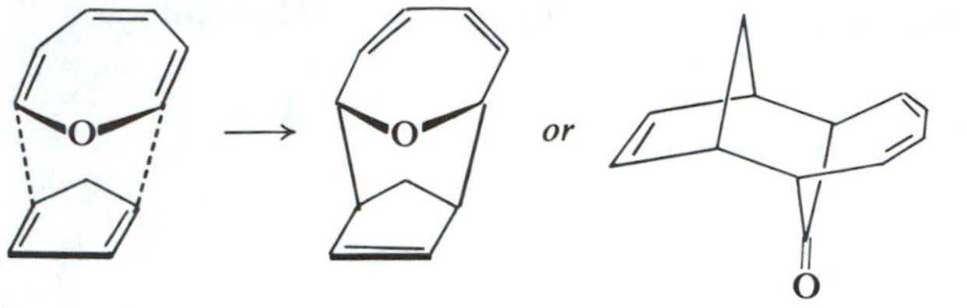

or **Exo**
Preferred for [6 + 4] *cycloaddition*

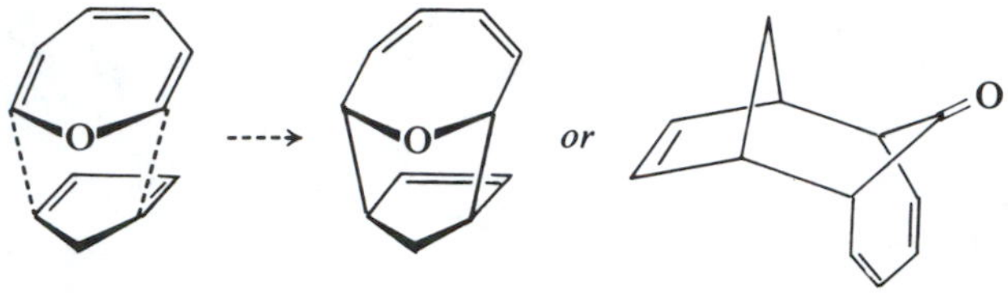

or **Endo**

We shall see later (Problem 13, page 1029) that both the *exo* preference in [6 + 4] cycloaddition and the *endo* preference in [4 + 2] cycloaddition are accounted for by orbital symmetry theory.

(d) Reaction is [8 + 2] thermal cycloaddition. As in (c), $i + j$ equals 10, and reaction was predicted to be *supra,supra* and hence geometrically easy—as it has turned out to be.

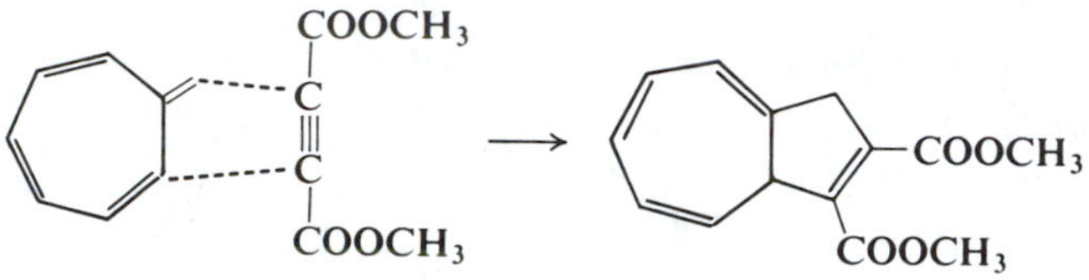

$COOCH_3$ $COOCH_3$ → $COOCH_3$ $COOCH_3$

(e) Reaction is [14 + 2] cycloaddition. Since $i + j$ equals 16 ($4n$), reaction is predicted to be *supra,antara*. This is in fact the case: in the representation shown on page 1018, attachment on the left-hand side is from beneath the polyene (the angular H is up, toward us), and on the right-hand side attachment is from above (the angular H is down, away from us).

28.10 (a) The reaction is a [1,5]-H shift, and is predicted to be *supra*. This is actually the case, with transfer taking place to either face of the trigonal carbon.

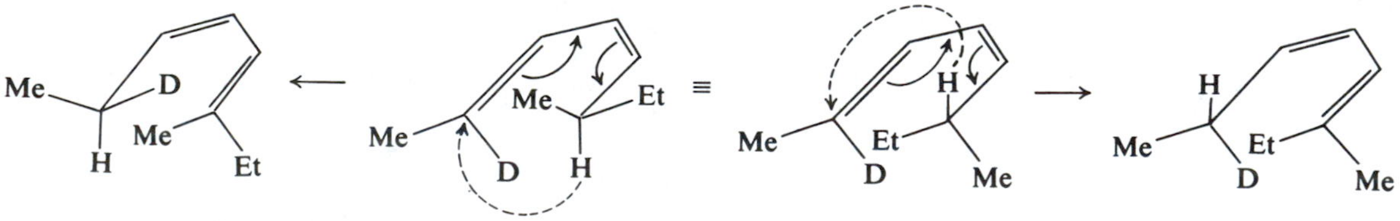

(b) These results illustrate the preference for [1,5] shifts of hydrogen over [1,3] or [1,7] shifts. A series of [1,5]-D shifts will distribute deuterium only among the 3, 4, 7, and 8 positions, as is observed.

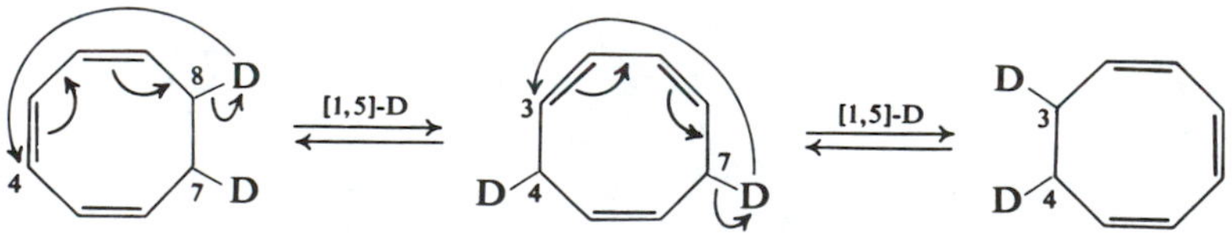

[1,5]-D [1,5]-D

A series of [1,3]-D shifts (or of [1,7]-D shifts), on the other hand, would scramble the label among all positions—contrary to fact.

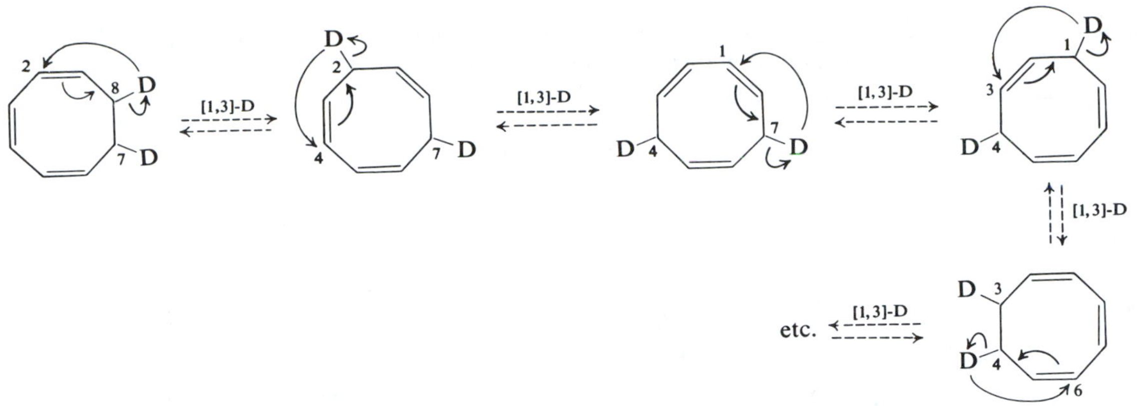

(c) A [1,3]-C shift (*supra*) occurs with *inversion* in the migrating group, as predicted. (If, instead, the front leg has simply *stepped* from C-1 to C-3 with retention, the $-CH_3$ would have swung into the *endo* position, over the ring—contrary to fact.)

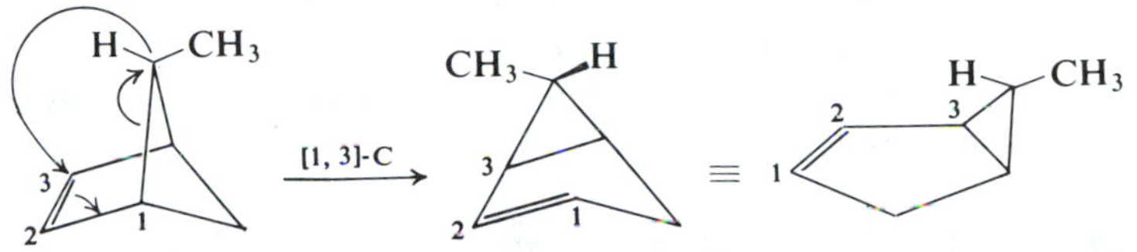

1. (a) In chemical properties tropolone resembles phenols. It is not adequately represented by structure I.

(b) To be a phenol, tropolone must have an aromatic ring. Aromaticity is also indicated by other properties; flatness, bond lengths, heat of combustion. The dipole moment suggests a dipolar structure.

Tropolone is a hybrid of seven dipolar structures like these:

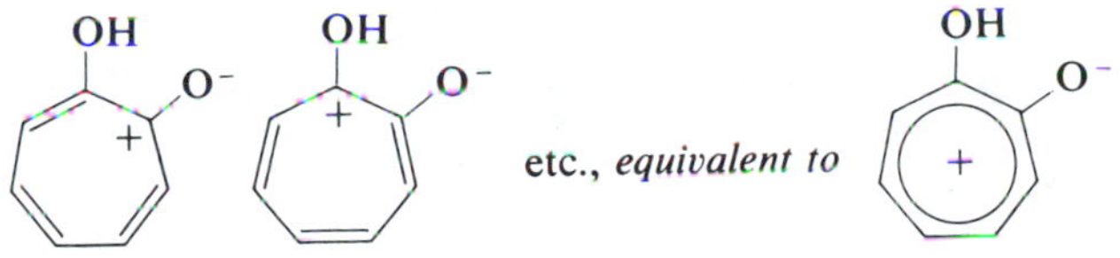

etc., *equivalent to*

Tropolone
Six π electrons
Aromatic

The ring is that of the cycloheptatrienyl cation, which has 6 π electrons and is aromatic (see Secs. 14.10 and 28.6).

(c) The dipole moment of 5-bromotropolone is smaller than that of tropolone, showing that the strong dipole of tropolone is in the opposite direction from that of C—Br, that is, *toward* "carbonyl" oxygen. This is consistent with the dipolar structure. (The dipole moment for

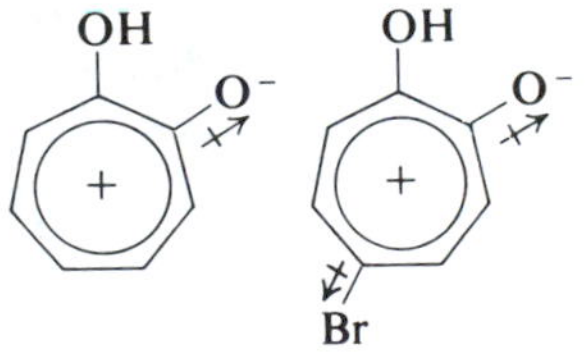

Dipole $\mu = 3.71$ D

Net dipole $\mu = 2.07$ D

bromobenzene (Table 26.2, page 951) is 1.71 D. If the C—Br moment of 5-bromotropolone were roughly the same, we would expect the compound to have $\mu = 3.71 - 1.71 = 2.00$ D, rather close to the observed value of 2.07 D.)

(d) There is intramolecular hydrogen bonding in tropolone.

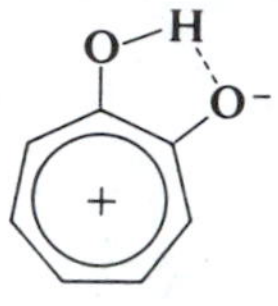

Tropolone

Hückel proposed his $4n + 2$ rule in 1931 (Sec. 28.6). In 1945, M. J. S. Dewar (page 414) postulated tropolone as the unit responsible for the unusual—that is, aromatic—properties of several natural products. Tropolone itself was synthesized in 1951 by W. von E. Doering and L. H. Knox, and it was this confirmation of Dewar's proposal—coupled with the recognition in 1952 of the aromaticity of ferrocene (page 505)—that set off a great wave of experimental work on non-benzenoid aromaticity.

(Doering, W. von E.; Knox, L. H. "Tropolone"; *J. Am. Chem. Soc.* **1951**, *73*, 828; Dauben, H. J., Jr.; Ringold, H. J. "Synthesis of Tropone"; *J. Am. Chem. Soc.* **1951**, *73*, 876; Doering, W. von E.; Detert, F. L. "Cycloheptatrienylium Oxides"; *J. Am. Chem. Soc.* **1951**, *73*, 876. For a review, see Ginsberg, D., Ed.; *Non-Benzenoid Aromatic Compounds*; Interscience: New York, 1959.)

2. So that you can keep track of carbons in the following transformations, the numbers used in the starting material are retained throughout (even though the numbers thus assigned in a product may not conform to accepted usage). In many cases, only one of two (equivalent) possibilities is shown: for example, in the product of the first step in (a), the *trans* double bond is shown between C-6 and C-7 and not between C-1 and C-8.

(a) **First step.** The cyclobutene ring opens. Two double bonds in the polyene are involved, or 4 π electrons. $4n$ thermal: conrotatory ring-opening.

Second step. A [1,5]-H shift, thermal: *supra*.

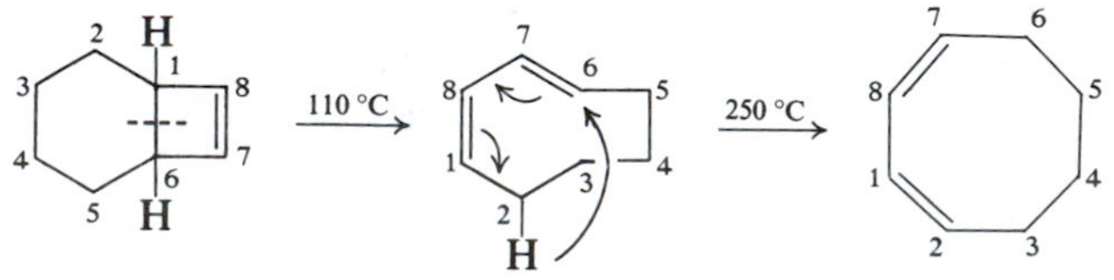

110 °C → 250 °C →

(b) **First step.** The cyclobutene ring opens. Two double bonds of the polyene are involved, or 4 π electrons. $4n$ thermal: conrotatory ring-opening.

heat →

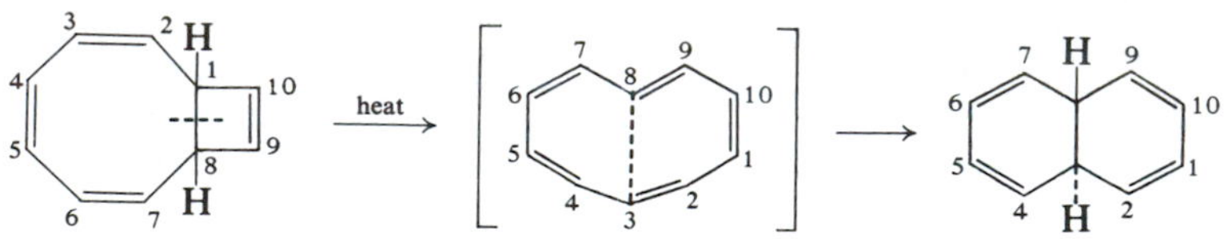

Second step. Three double bonds of the polyene are involved, or 6 π electrons. $4n + 2$ thermal: disrotatory ring-closing.

(c) For some molecules there is closure of a cyclobutene ring. Two double bonds of the polyene are involved, or 4 π electrons. $4n$ photochemical: disrotatory ring-closing.

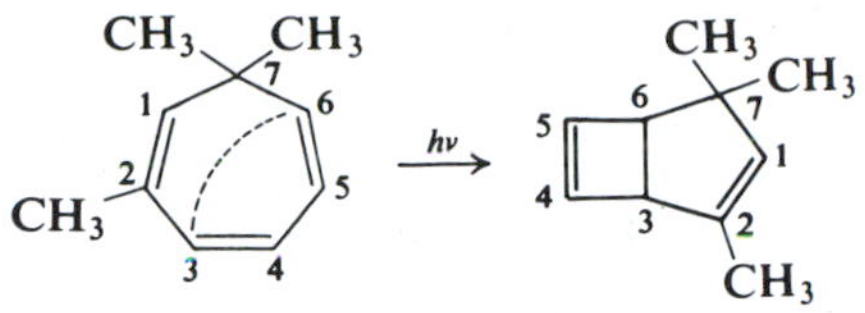

The rest of the molecules undergo the following.

First step. A [1,7]-C shift, photochemical: *supra*.

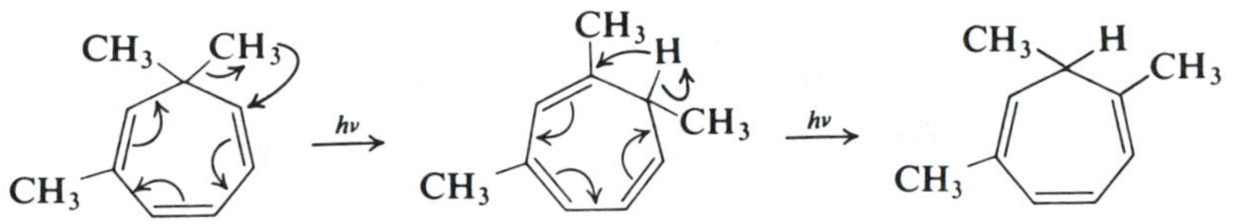

Second step. A [1,7]-H shift, photochemical: *supra*.

(d) **First step.** An intramolecular [4 + 4] cycloaddition, $i + j$ equals 8 ($4n$), photochemical: *supra,supra*.

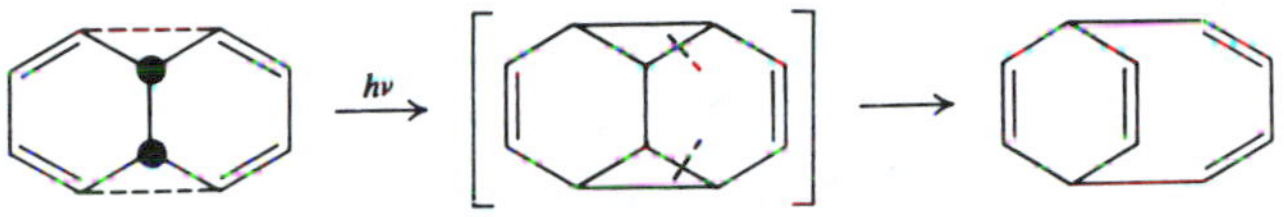

Second step. A [4 + 2] cycloreversion, $i + j$ equals 6 ($4n + 2$), presumably thermal: *supra,supra*.

(e) An allylic cation (2 π electrons) undergoes a [4 + 2] cycloaddition (thermal, *supra,supra*) to give cation XIII, which then loses a proton from either of two positions.

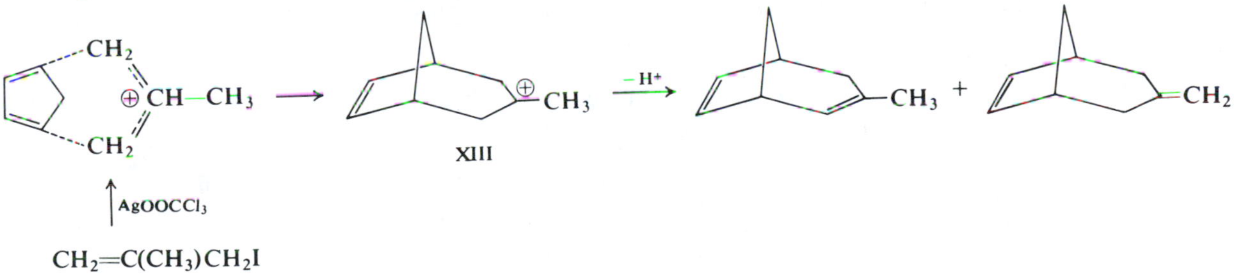

XIII

$CH_2{=}C(CH_3)CH_2I$

(f) The bridge walks around the ring in a series of *supra* [1,5]-C shifts (with, presumably, retention at the migrating carbon).

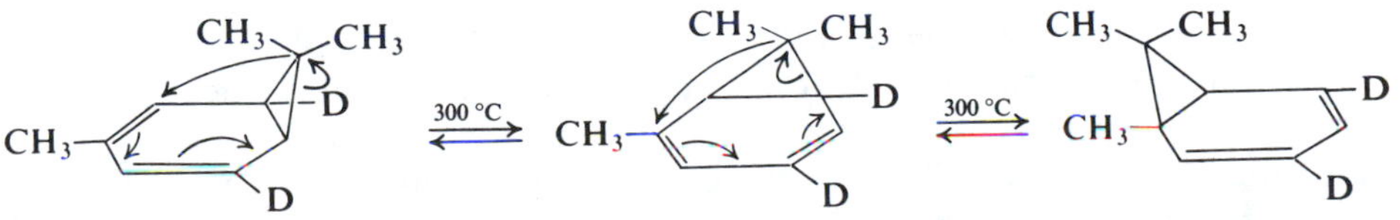

(g) An intramolecular thermal [4 + 2] cycloaddition (a Diels–Alder reaction): *supra,supra*.

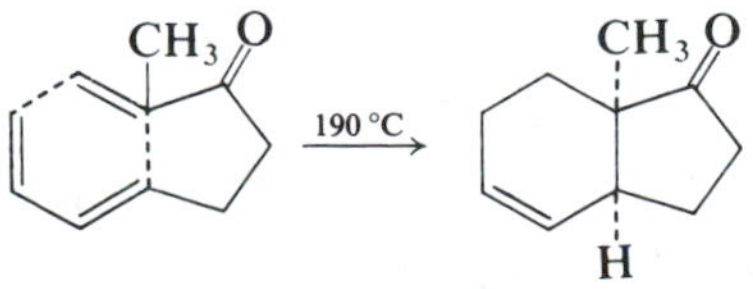

3. (a) **First step.** Ring closure involving four double bonds of the polyene, or 8 π electrons ($4n$): thermal, conrotatory. One R swings up, the other R down, so that they are *trans* in A.

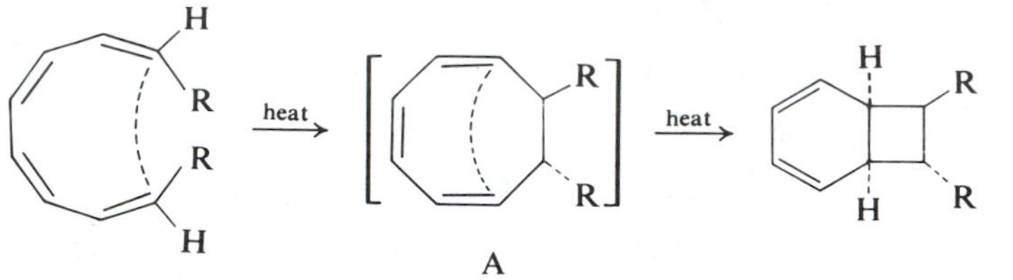

A

trans-7,8-Dialkyl-*cis,cis,cis*-cycloocta-1,3,5-triene

Second step. Ring closure involving three double bonds of the polyene, or 6 π electrons ($4n + 2$): thermal, disrotatory. The R's retain, of course, their *trans* relationship; the disrotatory motion gives a *cis* fusion of the rings.

(b) **First step.** There is only one possible [1,5]-H shift (which, as we have seen, is preferred over a [1,3]-H shift).

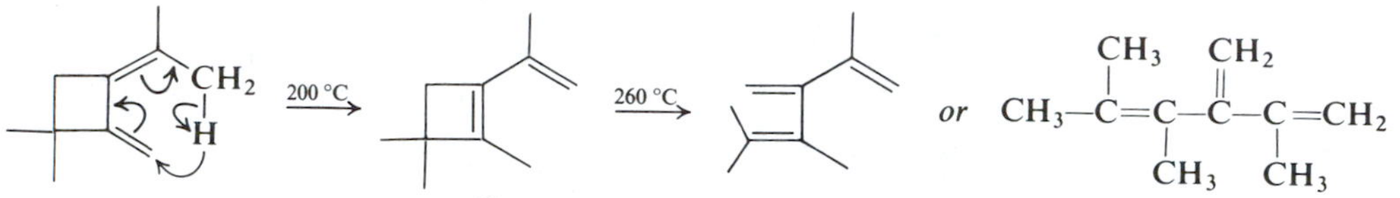

B

1-Isopropenyl-2,3,3-trimethylcyclobutene

C

2,4,5-Trimethyl-3-methylene-1,4-hexadiene

Second step. There is only one possible electrocyclic ring-opening. It involves two double bonds of the polyene, or 4 π electrons ($4n$); being thermal, it is presumably conrotatory.

(c) **First step.** The cyclopropane ring opens. Four double bonds in the polyene (D) are involved, or 8 π electrons. $4n$ thermal: conrotatory.

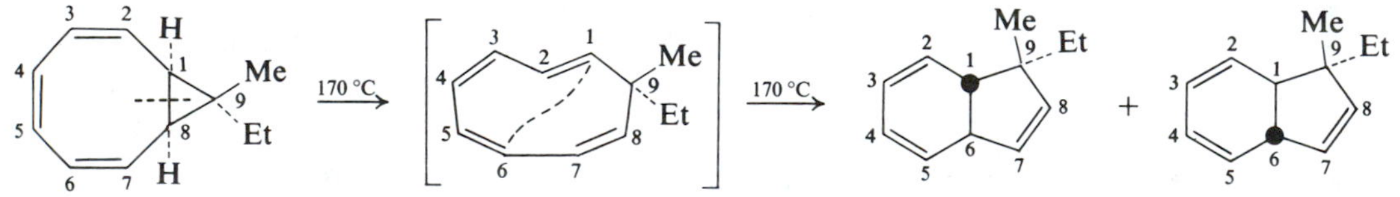

D

Second step. A ring closure involving three double bonds, or 6 π electrons. $4n + 2$ thermal: disrotatory. This leads to *trans* fusion of the rings: as we draw our formulas, one angular H comes toward us, and the other goes away from us.

But there are *two* disrotatory modes possible (see the answer to Problem 28.3). The angular H on C–1 can come toward us or go away from us; that is, it can be on the same side of the ring as Me, or on the same side as Et. (The angular H on C–6, of course, does the opposite in each case.) The two modes give different (diastereomeric) products.

(d) **First step.** Closure of the cyclobutene ring. Two double bonds of the polyene are involved, or 4 π electrons. $4n$ photochemical: disrotatory.

E F G

cis,cis-Cyclo-nona-1,3-diene

cis-Bicyclo[5.2.0]-nona-8-ene

cis,trans-Cyclo-nona-1,3-diene

trans-Bicyclo[5.2.0]-nona-8-ene

Second step. As in the first step, but thermal: conrotatory.

Third step. Closure of the cyclobutene ring. As in the first step, 4*n* photochemical: disrotatory.

(e) The H shifts are *supra* for geometric reasons. They are [1,5] if thermal, [1,7] if photochemical.

Starting material

H

I

Starting material

I

H

4. Two double bonds of the polyene are involved, or 4 π electrons. 4*n* thermal: conrotatory. Conrotatory motion must convert a *cis*-3,4-disubstituted cyclobutene into a *cis,trans* diene. This is easy for the first reactant, and is what actually happens. The second reactant, however, yields a seven-membered ring, which is too small to accommodate a *trans* double bond. The second reaction takes the disrotatory path, probably by a high-energy nonconcerted (stepwise) mechanism.

5. **First step.** Closure of the cyclobutane ring. Three double bonds of the polyene are involved or 6 π electrons. $4n + 2$ thermal: disrotatory to give K with *cis* fusion.

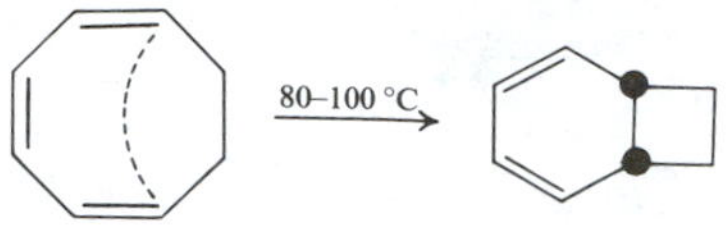

80–100 °C

K

cis,cis,cis-Cyclo-octa-1,3,5-triene

cis-Bicyclo[4.2.0]-octa-2,4-diene

Second step. K undergoes a [4 + 2] cycloaddition (Diels–Alder reaction) to give L: *supra,supra* and *endo*.

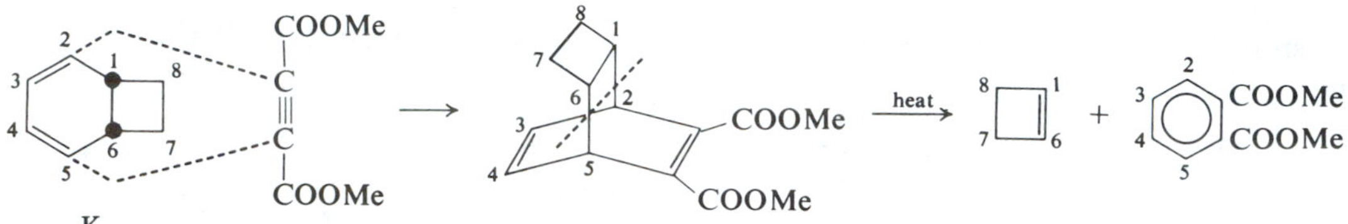

COOMe COOMe K L heat Cyclobutene Dimethyl phthalate

Third step. L undergoes a [4 + 2] cycloreversion (*retro*-Diels–Alder reaction) but with breaking of different bonds than those formed in the second step.

6. (a) The rearrangement of a carbocation, we have said (Sec. 5.22), involves a 1,2-shift, and takes place via a transition state, XIV, in which the migrating alkyl group (or hydrogen) is bonded to both migration source and migration terminus.

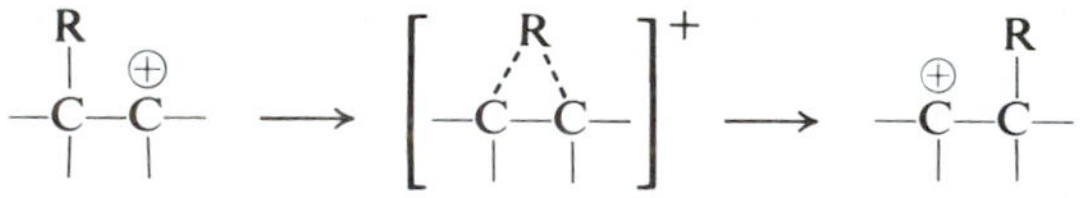

XIV

Using the approach of Sec. 28.10, we treat XIV as arising from overlap between an orbital of a free radical (the migrating group R) and a π framework. The π framework here is not an allylic free radical, but a vinyl radical–cation: ethylene minus one π electron. HOMO is π in Figure 28.5 (page 997).

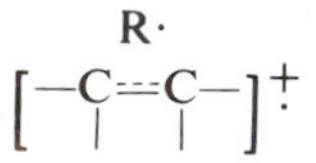

XIV HOMO π **1,2-shift in carbocation**

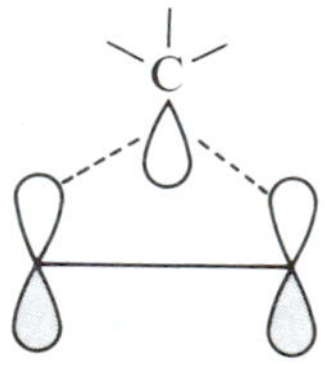

Suprafacial

Retention in migrating group

On this basis, we see, the symmetry-allowed process is suprafacial and hence feasible geometrically. It should proceed with retention in the migrating group, the stereochemistry actually observed for 1,2-shifts to electron-deficient atoms (Sec. 22.16).

Rearrangement of carbocations by [1,2] *suprafacial sigmatropic shifts proceeds with retention of configuration in the migrating group.*

(b) By the approach of part (a), we consider the π framework to be a diene radical–cation: a diene minus one π electron. HOMO is ψ_2 in Figure 28.6 (page 998). For a [1,4] suprafacial sigmatropic shift in an allylic cation, we predict *inversion of configuration* in the migrating group.

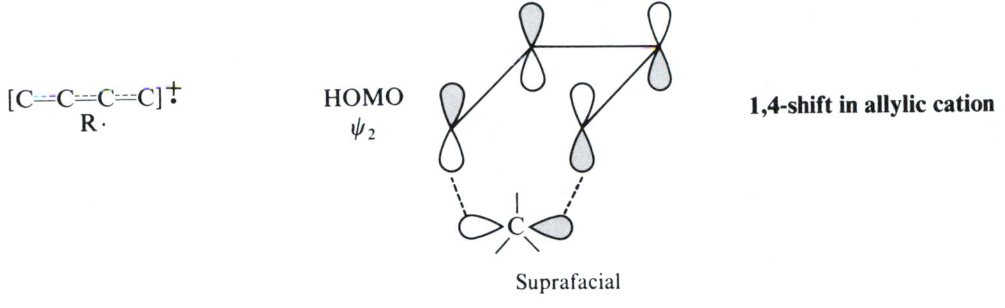

Suprafacial

Inversion in migrating group

Although such [1,4] shifts are not common, they do occur, and have been found to proceed with the predicted inversion of configuration. (See, for example, Hart, H.; Rodgers, T. R.; Griffiths, J. "Stereochemistry of the Rapid Equilibration of Protonated Bicyclo-[3.1.0]hexenones"; *J. Am. Chem. Soc.* **1969**, *91*, 754.)

7. For geometric reasons, such a reaction would have to be *supra,supra*, and that is symmetry-forbidden. For H_2, HOMO is σ and LUMO is σ^* (Figure 28.4, page 996). For ethylene, LUMO is π^* and HOMO is π (Figure 28.5, page 997).

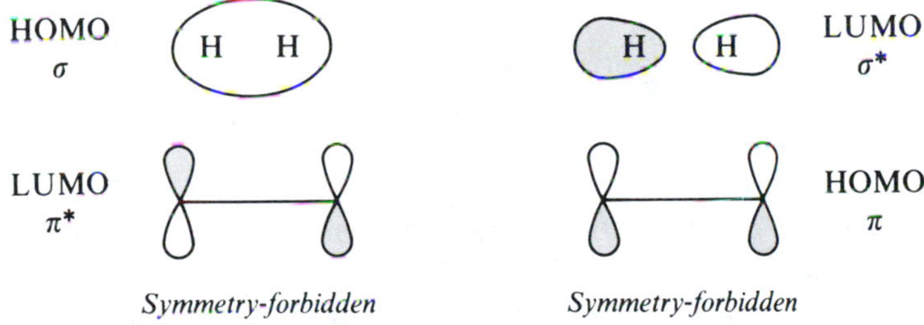

Symmetry-forbidden *Symmetry-forbidden*

8. An intramolecular Diels–Alder reaction converts II into the intermediate XV. This undergoes either of two (almost) equivalent *retro*-Diels–Alder reactions: one, to regenerate II, and the other—by breaking of bonds as shown—to form III.

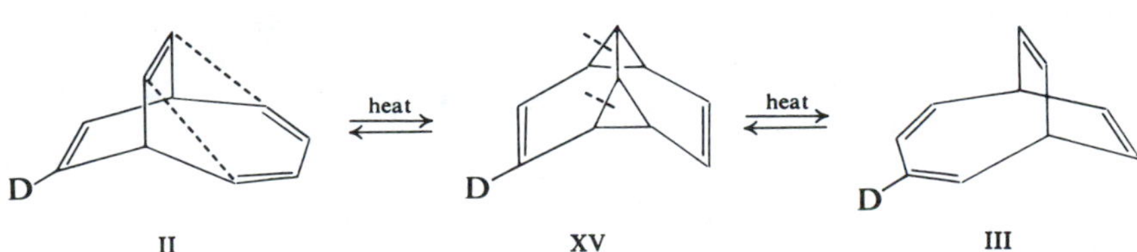

9. (a) The diazonium salt IV forms benzyne (Problem 13, Chapter 26),

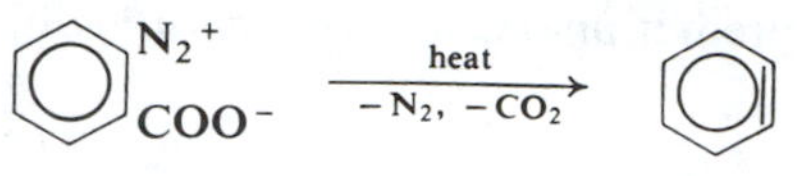

N_2^+ COO^- heat $-N_2, -CO_2$

IV Benzyne

which undergoes a [4 + 2] *supra,supra* cycloaddition with the diene to give V. (See the answer to Problem 11(d), Chapter 27.)

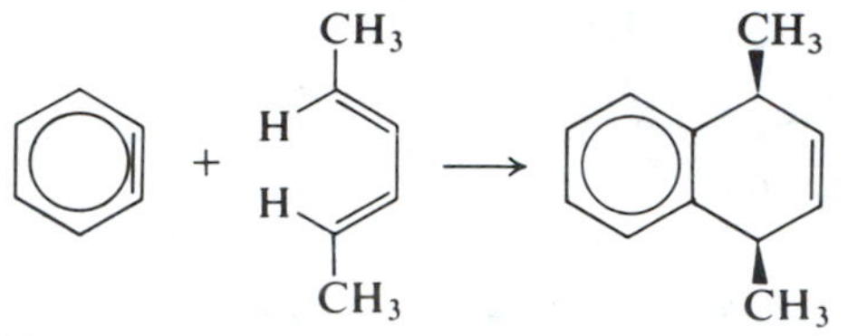

CH_3 H H CH_3 + → CH_3 CH_3

Benzyne *trans,trans*-2,4-Hexadiene *cis*-V

(b) A [2 + 2] thermal cycloaddition is symmetry-forbidden. Reaction is non-stereospecific because it is not concerted: it probably takes place stepwise via diradicals. For example:

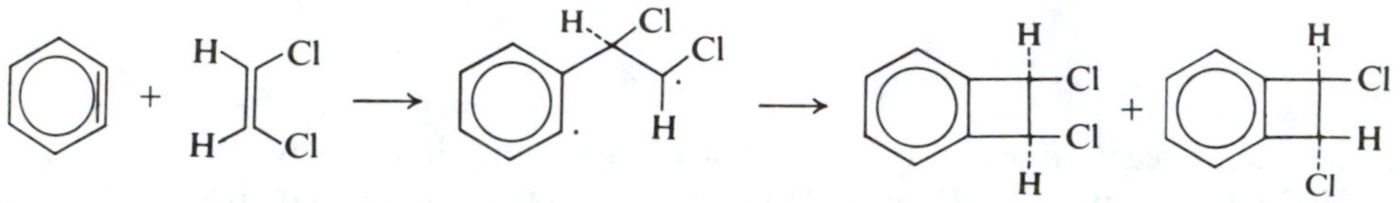

H Cl H Cl + → H Cl Cl H → H Cl Cl H + H Cl H Cl

Benzyne *cis*-1,2-Dichloroethene *A diradical* *cis*-VI *trans*-VI

(Jones, M., Jr.; Levin, R. H. "The Stereochemistry of the 2 + 2 and 2 + 4 Cycloadditions of Benzyne"; *J. Am. Chem. Soc.* **1969**, *91*, 6411.)

10. In each case the cyclobutane ring opens to give a non-aromatic intermediate. Two double bonds of the polyene are involved, or 4 π electrons. $4n$ thermal: conrotatory. This intermediate then undergoes a Diels–Alder reaction with maleic anhydride.

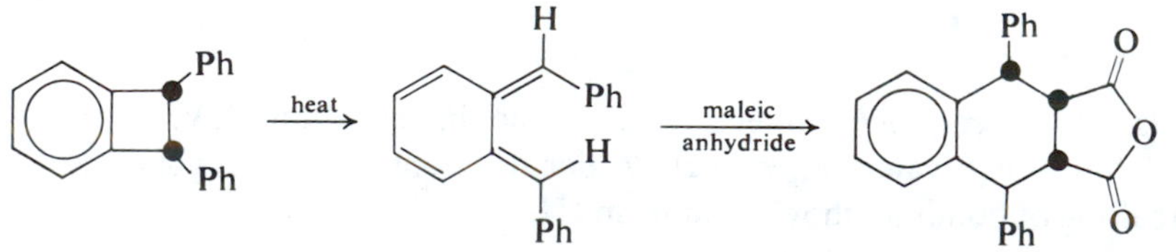

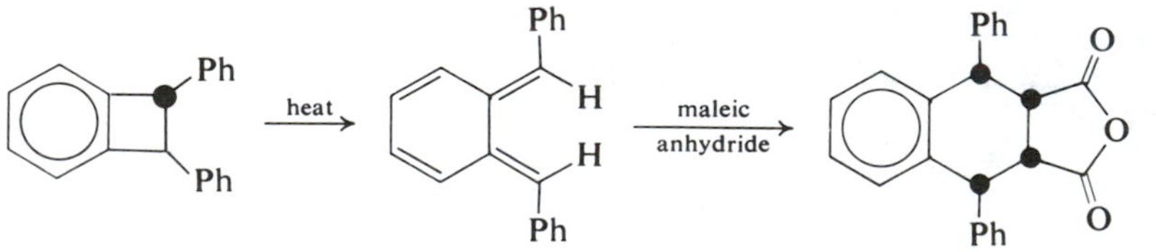

Ph Ph heat → H Ph H Ph maleic anhydride → Ph O O O Ph

Ph Ph heat → Ph H H Ph maleic anhydride → Ph O O O Ph

11. (a) There are two successive nucleophilic substitutions: the nucleophile is the cyclopentadienyl anion in the first, and the substituted cyclopentadienyl anion in the second.

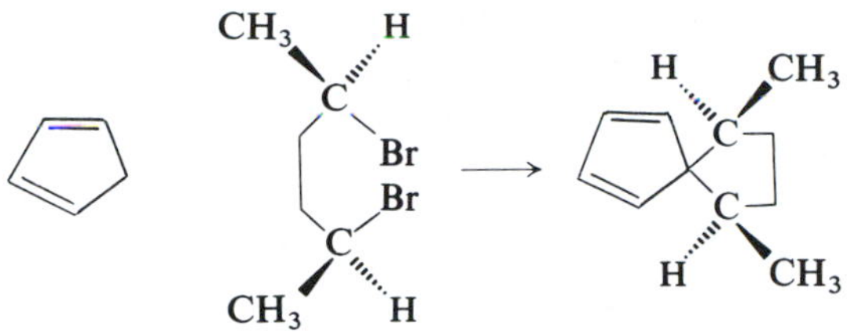

meso Dibromide → *cis*-VII (page 1024)

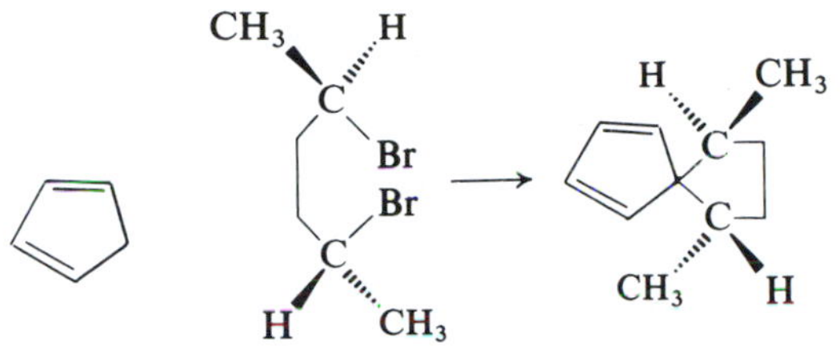

racemic Dibromide → *trans*-VII

(b) *cis*-VII contains four non-equivalent olefinic protons. *trans*-VII contains two pairs of equivalent protons. (Remember, the planes of the two rings are perpendicular to each other. *Use models.*)

(Kloosterziel, H., *et al.* "Stereospecificity and Stereochemistry of a Thermal Sigmatropic [1,5]-Shift of an Alkyl Group"; *Chem. Commun.* **1970**, 1168.)

12. (a) M and N are position isomers, both from *syn exo* addition of B_2D_6. Similarly, O and P are position isomers.

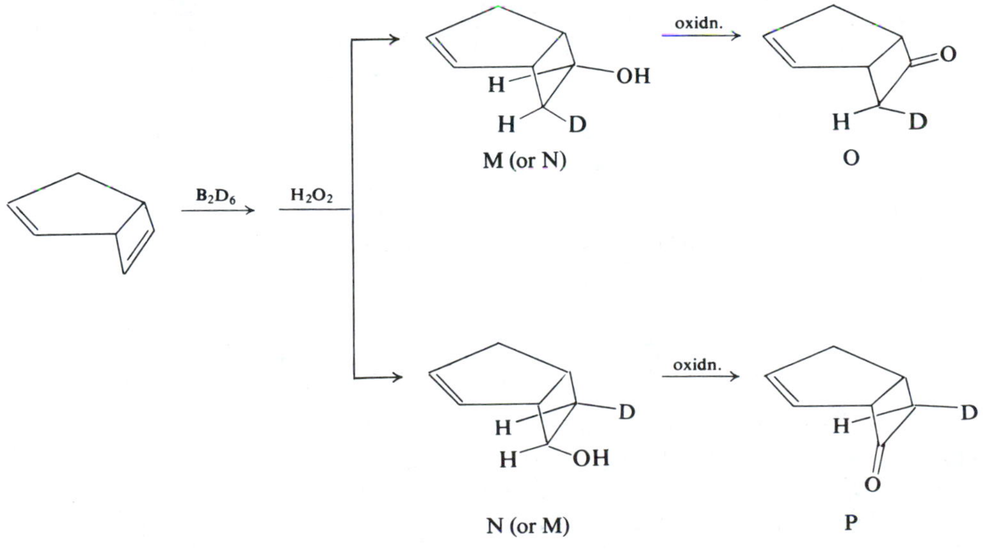

B_2D_6 → H_2O_2 → M (or N) —oxidn.→ O

→ N (or M) —oxidn.→ P

(We know that hydroboration–oxidation is *syn* (Problem 11, page 490). We have not learned, however, that hydroboration is stereospecifically *exo* in reactions like these—although this is the case. Let us consider the possibility that, instead of being position isomers, M and N were stereoisomers resulting from *exo* and *endo* hydroboration. In that case, O and P would also be stereoisomers; but they would differ only as to whether —H or —D were *exo* or *endo*, and would hardly have been separable.)

Reduction of O converts the carbonyl group into a new chiral center, and gives both possible configurations about that center:

O —redn.→ Ac_2O → Q (V on page 1023) + R

(b) V undergoes a *retro*-Diels–Alder reaction:

VI (page 1023) —heat→ Cyclopentadiene + Vinyl acetate

(Berson, J. A.; Nelson, G. L. "Inversion of Configuration in the Migrating Group of a Thermal 1,3-Sigmatropic Rearrangement"; *J. Am. Chem. Soc.* **1967**, *89*, 5503; Berson, J. A. "The Stereochemistry of Sigmatropic Rearrangements. Test of the Predictive Power of Orbital Symmetry Rules"; *Acc. Chem. Res.* **1968**, *1*, 152.)

13. (a) Let us make a diagram similar to Figure 28.20 (page 1015) for the *endo* dimerization of butadiene.

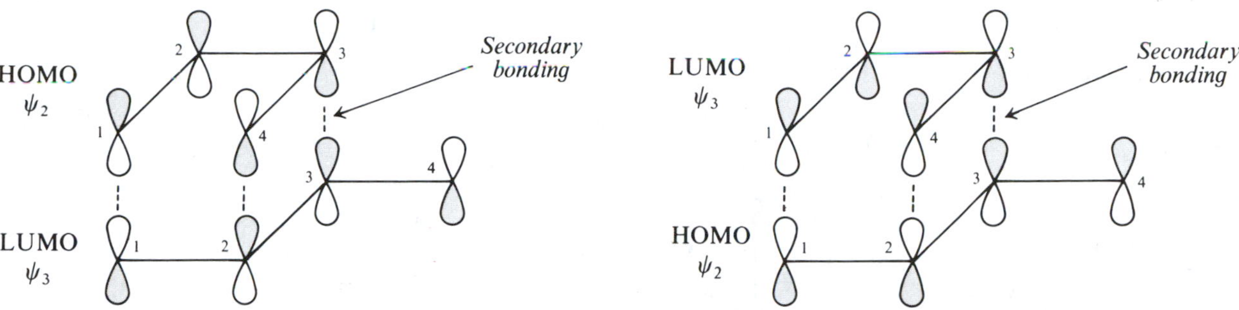

We see, of course, the overlap between lobes on C–1 and C–4 of the diene and lobes on C–1 and C–2 of the ene. In addition, we see that a lobe on C–3 of the diene is brought close to a lobe *of the same phase* on C–3 of the ene—carbons that are not even bonded to each other in the product. Weak (temporary) bonding between these atoms helps to stabilize the transition state. Such secondary bonding could not occur if reaction were *exo*, since the atoms concerned would be far apart.

(b) Let us make a diagram similar to the one in (a) for the *endo* [6 + 4] cycloaddition of a triene with a diene.

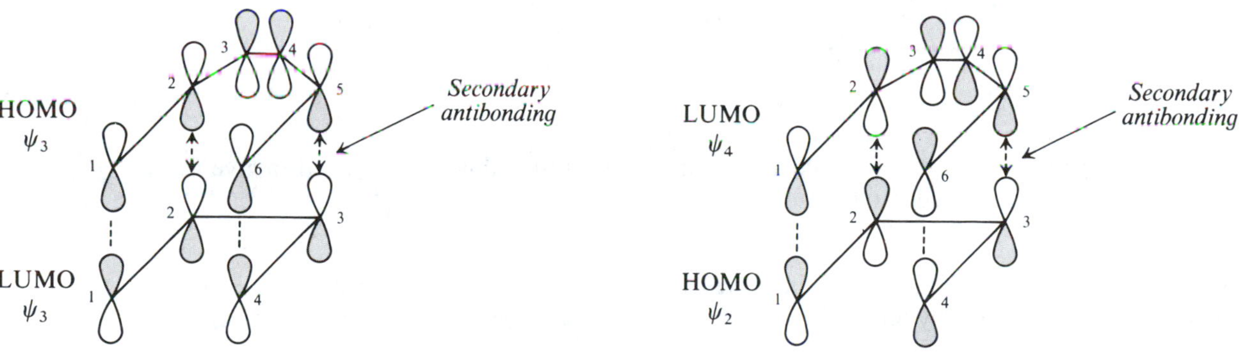

We see the overlap that leads to bond formation: between lobes on C–1 and C–6 of the triene and C–1 and C–4 of the diene. But juxtaposed lobes on C–2 and C–5 of the triene and C–2 and C–3 of the diene are *of opposite phase*: instead of giving secondary bonding in the transition state, such interactions would give antibonding. As a result, [6 + 4] cycloadditions take place in the *exo* manner.

This was predicted by Woodward and Hoffmann in 1965, and when, shortly afterward, the first [6 + 4] cycloaddition was recognized, the prediction was found to be correct (see Problem 28.9(c), page 1018).

14.

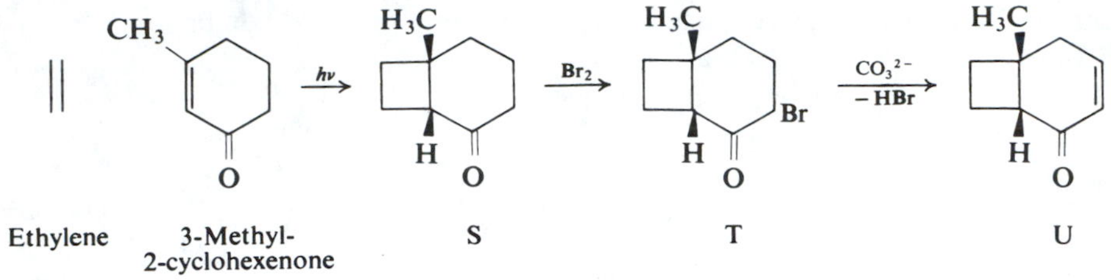

Ethylene 3-Methyl-2-cyclohexenone S T U

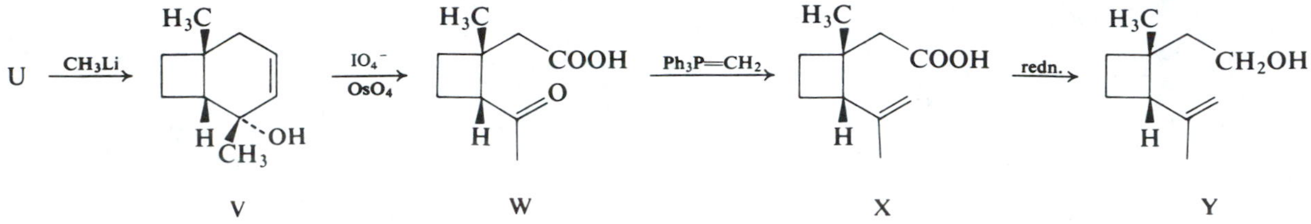

$h\nu$; Br_2; CO_3^{2-}, $-HBr$

U $\xrightarrow{CH_3Li}$ V $\xrightarrow{IO_4^-,\ OsO_4}$ W $\xrightarrow{Ph_3P{=}CH_2}$ X $\xrightarrow{redn.}$ Y

The first step is a [2 + 2] photochemical *supra,supra* cycloaddition. This is the key step in the synthesis. It generates the needed cyclobutane ring, and with the proper stereochemical relationship between —CH_3 and —H. At the same time, it provides the first of the carbonyl groups on whose properties depend most of the straightforward chemistry that follows: *alpha*-bromination to give T; addition of CH_3Li to give V, which opens the way to W, with a new carbonyl group; the Wittig reaction to give X with the terminal olefinic group. Several steps deserve a closer look.

Conversion of U into V. CH_3Li attacks U from the most open side (by path *a*), away from the fold in the molecule.

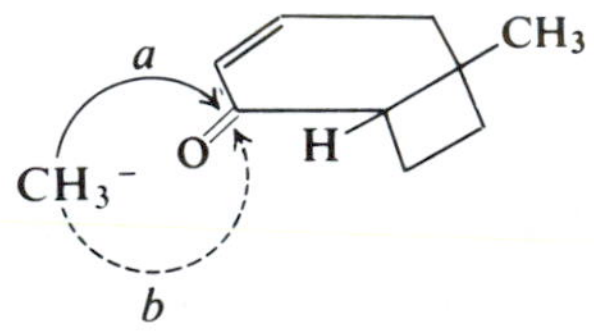

Conversion of V into W. The double bond of V is presumably hydroxylated to give a triol, which is then cleaved at two places to give W.

(b) The molecular formula for Z indicates intramolecular solvomercuration. This is geometrically possible only for the isomer in which the double bond and —CH_2OH are *cis*.

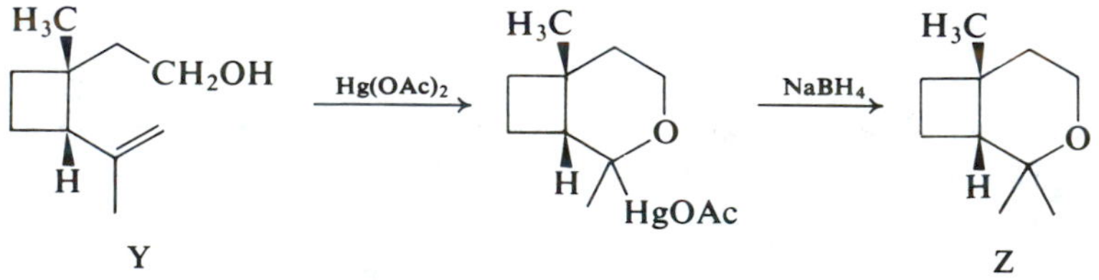

Y $\xrightarrow{Hg(OAc)_2}$ (HgOAc) $\xrightarrow{NaBH_4}$ Z

(Zurflüh, R.; Dunham, L. L.; Spain, V. L.; Siddall, J. B. "Synthetic Studies on Insect Hormones. IX. Stereoselective Total Synthesis of a Racemic Boll Weevil Pheromone"; *J. Am. Chem. Soc.* **1970**, *92*, 425.)

15. (a) Conversion of VII into benzene requires opening of a cyclobutene ring, and involves 4 π electrons of the "polyene" (benzene). The symmetry-allowed process is conrotatory,

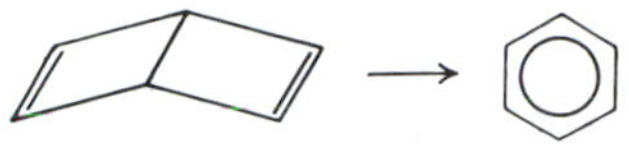

VII
Bicyclo[2,2,0]-
hexa-2,5-diene
"Dewar benzene"

Benzene

which would yield the impossibly strained *cis,cis,trans*-cyclohexa-1,3,5-triene. Reaction must go with forbidden (that is, difficult) disrotatory motion—or, perhaps, by a mechanism that is not concerted at all, but stepwise.

(b) Conversion of VIII into toluene requires a [1,3]-H shift (or a [1,7]-H shift). The symmetry-allowed *antara* shift is geometrically impossible. Reaction must go either by the

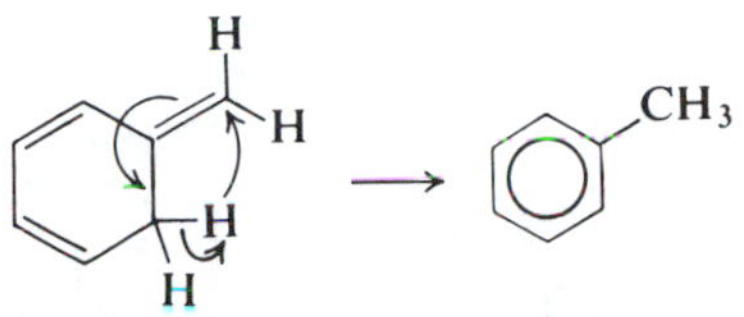

5-Methylene-
1,3-cyclohexadiene

Toluene

forbidden *supra* path, or by a non-concerted mechanism, also difficult.

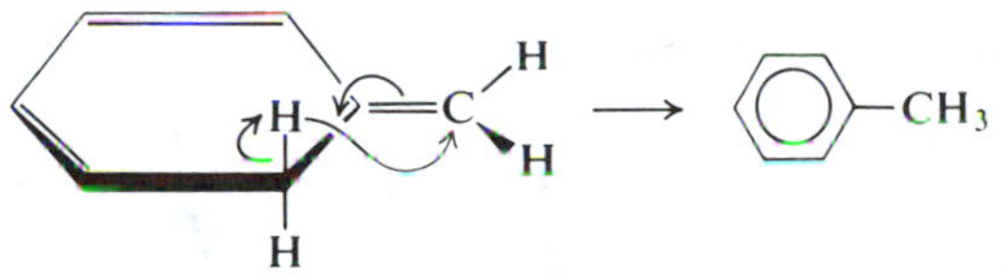

Antara
Symmetry-allowed, geometrically impossible

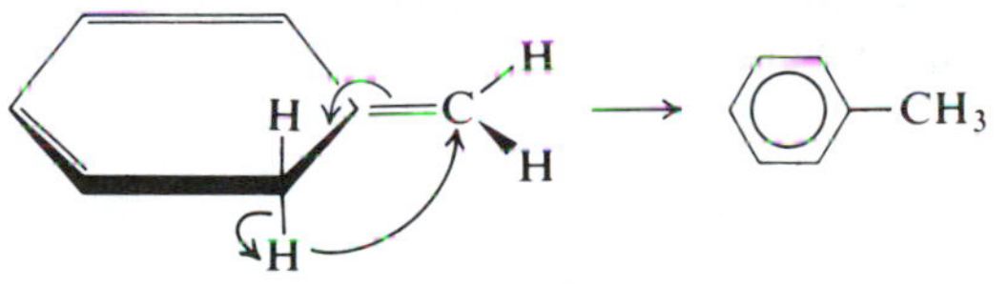

Supra
Geometrically easy, symmetry-forbidden

16. (a) **First step.** An electrocyclic opening of ring B. In the polyene three double bonds are involved, or 6 π electrons. $4n + 2$ photochemical: conrotatory opening.

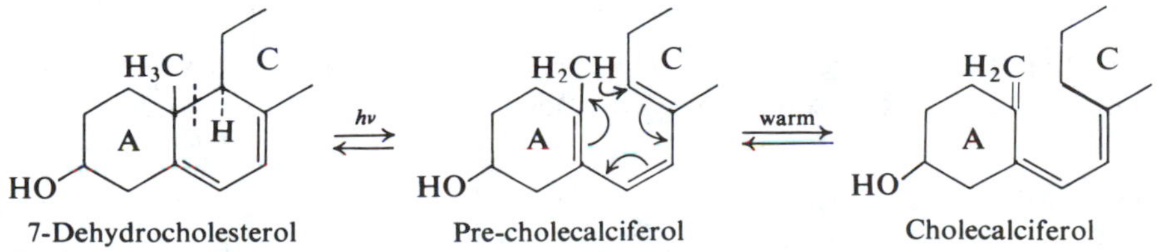

7-Dehydrocholesterol

Pre-cholecalciferol

Cholecalciferol

(A *thermal* ring-opening cannot take place: the symmetry-allowed disrotatory motion would generate a *trans* double bond in either ring A or ring C, geometrically impossible in a six-membered ring.)

Second step. A [1,7]-H shift which, since it is thermal, is presumably *antara*.

(b) These have structures analogous to those of pre-cholecalciferol and cholecalciferol, but with a nine-carbon unsaturated side chain:

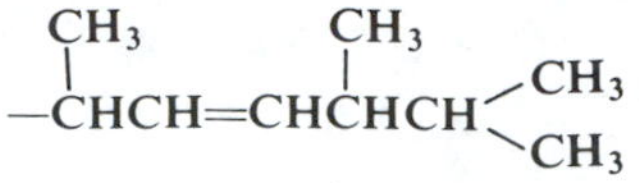

$$-\underset{}{\overset{CH_3}{\overset{|}{C}}}HCH{=}CH\overset{CH_3}{\overset{|}{C}}HCH\begin{matrix}CH_3\\CH_3\end{matrix}$$

(c) Electrocyclic closure of ring B, involving three double bonds of the polyene, or 6 π electrons. $4n + 2$ thermal: disrotatory.

As discussed in the answer to Problem 3(c), above, there are two possible disrotatory modes,

↷ ↶ and ↶ ↷

and here, as in that earlier problem, they yield different stereoisomers: IX and X. In ergosterol, the C–1 methyl is β and the C–9 H is α (see page 1135). In IX and X, both substituents are α in one and both are β in the other.

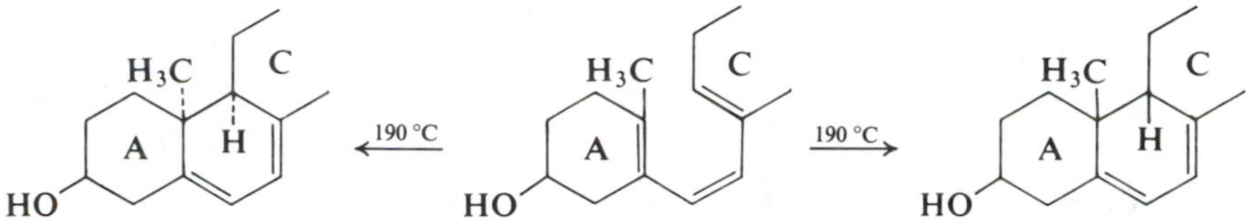

IX (or X) ← 190 °C — Pre-ergocalciferol — 190 °C → X (or IX)

(d) Electrocyclic opening of ring B, involving 6 π electrons. $4n + 2$ photochemical: conrotatory, as in the first step of part (b).

Consider the ring-closing process. There are *two* conrotatory modes,

↷ ↷ and ↶ ↶

and in this case they yield different stereoisomers. Pre-ergocalciferol is thus related by conrotatory motion to two compounds: ergosterol and XI. In ergosterol the C–1 methyl is β, and the C–9 H is α. In XI the C–1 methyl is α, and the C–9 H is β.

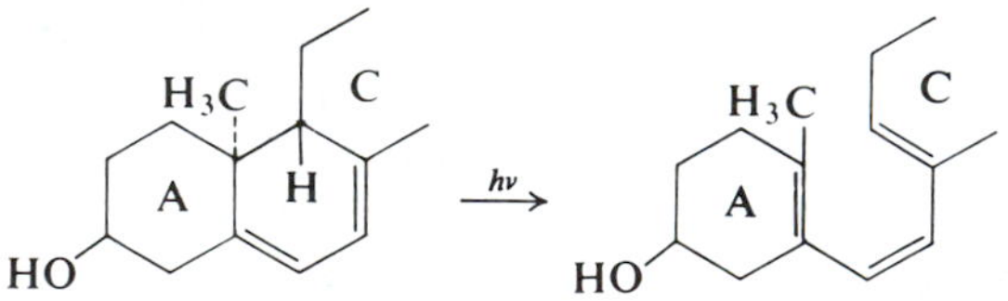

XI —$h\nu$→ Pre-ergocalciferol

17. (a) *trans*-XII undergoes ring-opening to give a cyclodecapentaene. This involves three double bonds of the polyene, or 6 π electrons. $4n + 2$ photochemical: conrotatory ring-opening.

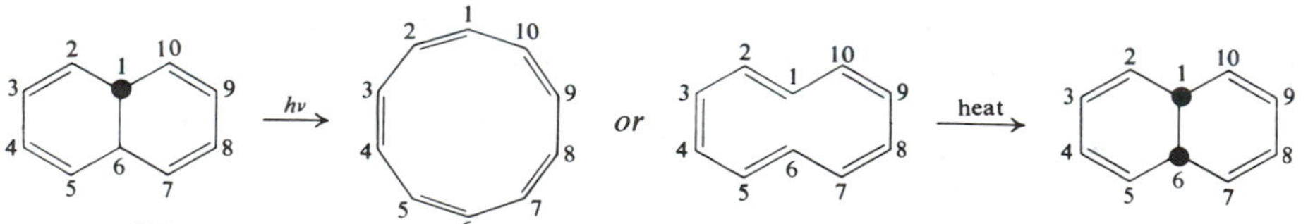

trans-XII —$h\nu$→ XVIa *or* XVIb —heat→ *cis*-XII

XVIa XVIb
Cyclodeca-1,3,5,7,9-pentaene

There are two conrotatory modes (see the answer to Problem 28.3, above) to give two possible products: in XVIa the three new double bonds are all *cis*; in XVIb they are *trans,cis,trans*.

XVI is evidently thermally unstable, and at room temperature undergoes a ring closure involving three double bonds, or 6 π electrons. $4n + 2$ thermal: disrotatory motion to give *cis*-XII. XVI is thus related to both stereoisomers of XII: to one photochemically, and to the other, thermally.

At -190 °C, XVI is stable, and is not converted into *cis*-XII. It can, however, be reduced to cyclodecane.

When *trans*-XII is photolyzed at -190 °C, XVI is formed. Warming to room temperature converts it into *cis*-XII, which remains unchanged when the temperature is lowered again.

(b) Cyclodecapentaene is a cyclic, completely conjugated polyene with five double bonds, or 10 π electrons; that is, it is [10]*annulene* (Problem 9, Chapter 14). But 10 is a Hückel number ($4n + 2$, with $n = 2$), and—*geometry permitting*—should give rise to aromaticity. (As we saw in Problem 13, page 516, the cyclooctatetraenyl dianion, with 10 π electrons, *is* aromatic.)

The question is, *does* the geometry permit aromaticity? And the answer seems clearly to be: *no*. There is little doubt that cyclodecapentaene is the intermediate in these reactions; there is equally little doubt that it is highly unstable and hence not aromatic.

The next question is: *why* is it not aromatic? Aromaticity requires delocalization of π electrons, and this in turn requires that the molecule be flat. If XVIa were flat, it would be a regular decagon, with bond angles of 144°, 24° more than the normal trigonal angle of 120°. The angle strain accompanying this deviation is evidently more than the molecule can afford to accept for aromaticity. (The aromatic cyclooctatetraenyl dianion has bond angles of 135°, 15° bigger than normal trigonal angles: this much strain the molecule evidently *can* accept—along with a double negative charge.)

XVIb could be flat with no C—C—C angle strain, but here another problem arises: the hydrogens on C-1 and C-6 would point toward each other and be hopelessly crowded. (Not until the [18]annulene (page 1000) is the ring big enough to accommodate the inward-pointing hydrogens.)

The isolation of a stable compound to establish aromaticity can be difficult, but is straightforward. To interpret *failure to isolate* a supposed intermediate as evidence of non-aromaticity is much trickier. Besides isolation of the reduction product, cyclodecane, the strength of the argument here depends on the design of the experiments, and rests ultimately on the strength of the orbital symmetry rules.

(van Tamelen, E. E.; Burkoth, T. L.; Greeley, R. H. "The *trans*-9,10-Dihydronaphthalene–Cyclodecapentaene Valence Bond Isomer System"; *J. Am. Chem. Soc.* **1971**, *93*, 6120.)

29

Symphoria

Neighboring Group Effects. Catalysis by Transition Metal Complexes

29.1

Br, H, CH_3, CH_3, H, $\overset{\frown}{O}H_2^+$ —H_2O→ ⊕Br, H, CH_3, CH_3, H, Br^-, c, d

Protonated *erythro*-3-bromo-2-butanol *Optically active*

Bromonium ion

c → Br, H, CH_3, CH_3, H, Br ≡ CH_3; H—Br; H—Br; CH_3

d → H, CH_3, Br, Br, CH_3, H ≡ CH_3; Br—H; Br—H; CH_3

meso-2,3-Dibromobutane *Optically inactive*

Attacks *c* and *d* both yield the same product: *meso*-2,3-dibromobutane.

(Winstein, S.; Lucas, H. J. "Retention of Configuration in the Reaction of the 3-Bromo-2-butanols with Hydrogen Bromide"; *J. Am. Chem. Soc.* **1939**, *61*, 1576; "The Loss of Optical Activity in the Reaction of the Optically Active *erythro*- and *threo*-3-Bromo-2-butanols with Hydrogen Bromide"; *J. Am. Chem. Soc.* **1939**, *61*, 2845.)

29.2 There are two successive nucleophilic substitutions: first, intramolecular attack by —COO^- to give an α-lactone, and then attack on this lactone by hydroxide ion. Each proceeds with inversion, to give net retention.

(Cowdrey, W. A.; Hughes, E. D.; Ingold, C. K. *J. Chem. Soc.* **1937**, 1208.)

29.3

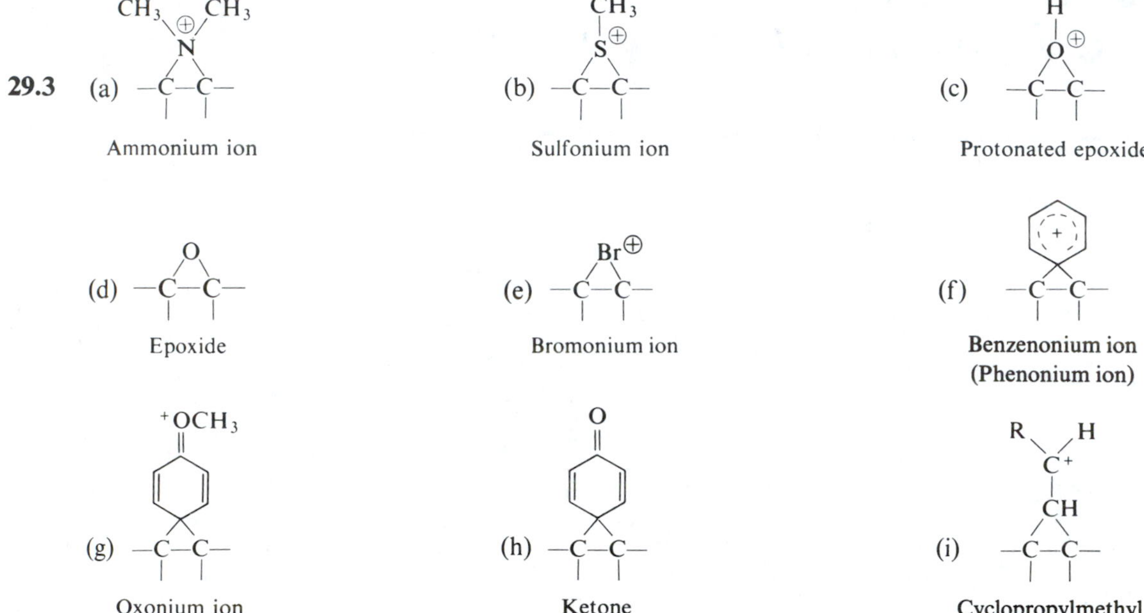

29.4 In solvolysis with S_N1 character (Sec. 7.9), halogen at the 2-position could exert two opposing effects. By its electron-withdrawing inductive effect, it could tend to slow down reaction by intensifying the positive charge developing on the carbon. Through anchimeric assistance, it could tend to speed up reaction. Only in the *trans* isomers can —X and —OBs take up the diaxial configuration required for anchimeric assistance.

The data indicate no anchimeric assistance by *cis*-Cl or *trans*-Cl, or by *cis*-Br; there is only the strong deactivation expected from the inductive effect. The *trans*-Br compound is much more reactive than the *cis*-Br compound, indicating anchimeric assistance—although not strong enough to offset the inductive effect completely. Finally, *trans*-I provides powerful anchimeric assistance, more than offsetting any inductive effect, and giving a rate 18 000 times as fast as for the unsubstituted brosylate.

Clearly, the ability of halogens to give nucleophilic assistance falls in the order

$$I \gg Br > Cl$$

The —I group is the most nucleophilic: it is the least electronegative and hence the least reluctant to hold the positive charge that develops on it. (Another factor contributing to its nucleophilicity may be its size: it is the "softest" and most easily deformable in the transition state; there are empty orbitals of not-too-high energy available for bonding.)

1. (a) Nucleophilic substitution in which both nucleophile and substrate are parts of the same molecule.

(b)

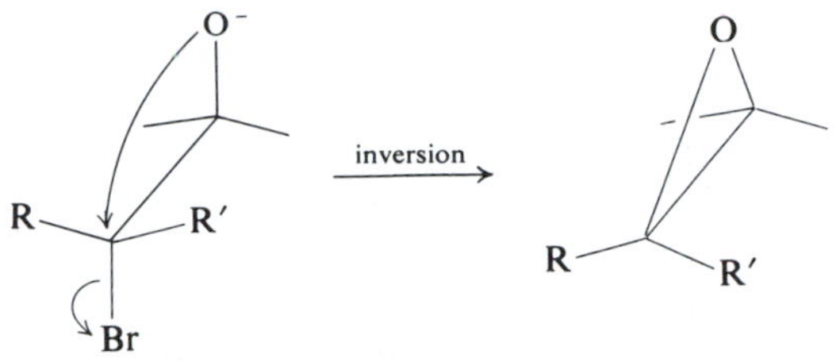

O^-, R, R′, Br —inversion→ O, R, R′

(c) In the *trans* isomer, it is possible for both $—O^-$ and —Cl to be axial and thus permit back-side (*anti*-periplanar) displacement.

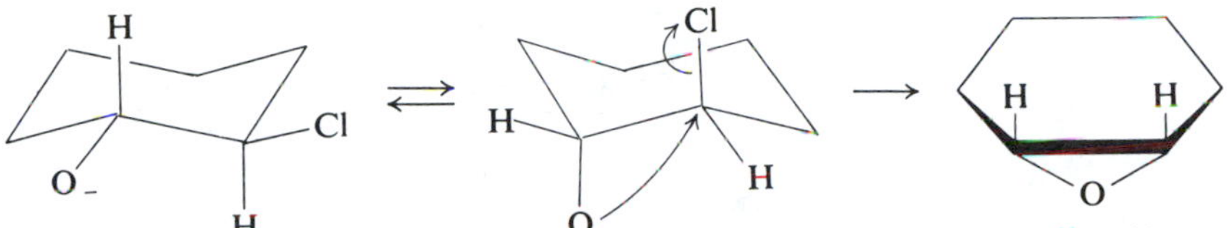

trans Isomer ⇄ *anti-periplanar displacement* → *cis*-Epoxide

This is not possible for the *cis* isomer, and reaction must follow another course.

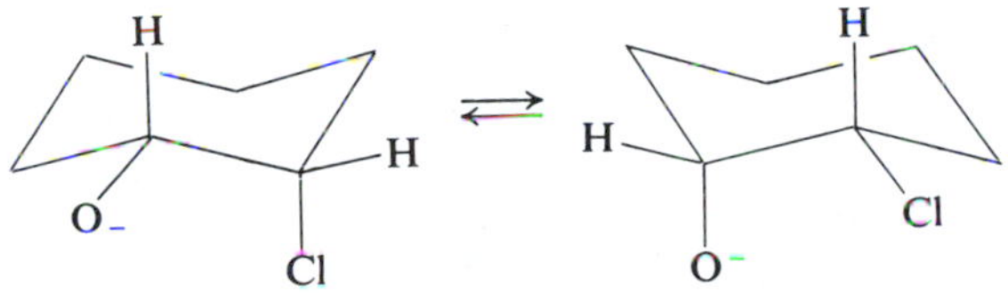

cis Isomer

anti-periplanar (back-side) displacement of —Cl *by* $—O^-$ *not possible*

2. (a) Attack by the strong nucleophile OH^- is at the less hindered position.

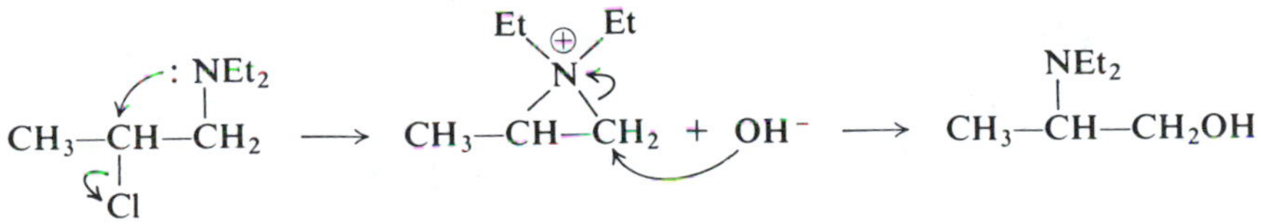

$$CH_3—CH(Cl)—CH_2NEt_2 \longrightarrow CH_3—CH—CH_2\ (\text{bridged } N^{\oplus}Et_2) + OH^- \longrightarrow CH_3—CH(NEt_2)—CH_2OH$$

(b) Formation of the bridged sulfonium ion may be either S_N1-like (two steps) or S_N2-like (one step).

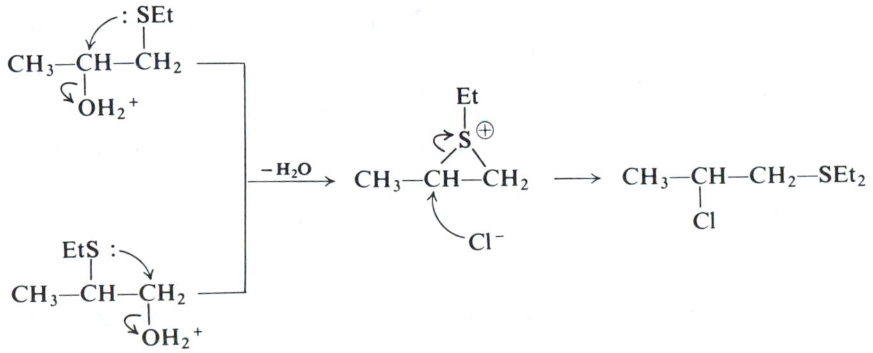

$$CH_3—CH(OH_2^+)—CH_2SEt \quad \text{or} \quad CH_3—CH(SEt)—CH_2OH_2^+ \xrightarrow{-H_2O} CH_3—CH—CH_2\ (\text{bridged } S^{\oplus}Et) + Cl^- \longrightarrow CH_3—CH(Cl)—CH_2—SEt$$

Cleavage involves attack by the weak nucleophile Cl^-, and takes place either by an S_N1 reaction (two steps) or, like the ring-opening of protonated epoxides (Sec. 13.24), by a single step that has considerable S_N1-character; in either case, considerable positive charge develops in the transition state, and reaction occurs at the carbon that can best accommodate that positive charge, the secondary carbon.

(c) In each case there is nucleophilic attack by OH^- at the least hindered position: at $—CH_2Br$ in I to give III directly:

$$\underset{\text{I}}{CH_3—\overset{}{CH}—CH—CH_2—Br}\ (\text{epoxide O bridging C-2, C-3})\quad OH^- \xrightarrow{-Br^-} \underset{\text{III}}{CH_3—CH—CH—CH_2OH}\ (\text{epoxide O bridging C-2, C-3})$$

at C–1 of II to form IV, which reacts further intramolecularly to yield III. (See the answer to Problem 13.21, page 482.)

$$\underset{\text{II}}{CH_3—\underset{Br}{CH}—CH—CH_2}\ (\text{epoxide O bridging C-3, C-4})\quad OH^- \longrightarrow \underset{\text{IV}}{CH_3—\overset{Br}{CH}—\underset{O^-}{CH}—CH_2OH} \xrightarrow{-Br^-} \underset{\text{III}}{CH_3—CH—CH—CH_2OH}\ (\text{epoxide O bridging C-2, C-3})$$

3. (a)

$$\underset{^{82}Br}{CH_2}—CH{=}CH_2 \xrightarrow{Br_2} \begin{cases} \underset{\text{V}}{\underset{^{82}Br}{CH_2}—CH—CH_2}\ (\text{bridged by } Br^{\oplus}) \xrightarrow{H_2O} \begin{cases} \xrightarrow{-H^+} \underset{^{82}Br}{CH_2}—\underset{Br}{CH}—\underset{OH}{CH_2} \\ \xrightarrow{-H^+} \underset{^{82}Br}{CH_2}—\underset{OH}{CH}—\underset{Br}{CH_2} \end{cases} \\ \underset{\text{VI}}{CH_2—CH—\underset{Br}{CH_2}}\ (\text{bridged by } ^{82}Br^{\oplus}) \xrightarrow{H_2O} \begin{cases} \xrightarrow{-H^+} \underset{^{82}Br}{CH_2}—\underset{OH}{CH}—\underset{Br}{CH_2} \\ \xrightarrow{-H^+} \underset{OH}{CH_2}—\underset{^{82}Br}{CH}—\underset{Br}{CH_2} \end{cases} \end{cases}$$

Addition of bromine gives not only the expected bromonium ion V, but also bromonium ion VI, formed through bridging by the ^{82}Br already in the molecule. VI is exactly equivalent to V, except for the isotopic label, and undergoes ring-opening in the same way to yield bromohydrins. (We cannot tell how much of the secondary alcohol is formed from each bromonium ion, but presumably it is formed, as the minor product, from both of them.)

We cannot say from this evidence just how VI arises. It could be formed from V through intramolecular attack by $—^{82}Br$, in either an S_N2 or S_N1 manner. It could be that VI is initially formed (in part) via a short-lived open cation, with ring-closing involving either of the adjacent bromines.

(b) The more electronegative —Cl has less tendency to act as an internal nucleophile and accept a positive charge.

4. *trans*-1,2-Dibromocyclohexane.

Both *cis* and *trans* alcohols give the same product because they form the same cyclic bromonium ion upon loss of water from an initial oxonium ion: the *trans* alcohol probably by an S_N2-like reaction (reaction (1), page 1033); the *cis* alcohol necessarily via an open carbocation. Back-side attack by Br^- on the bridged ion gives the *trans* dibromide.

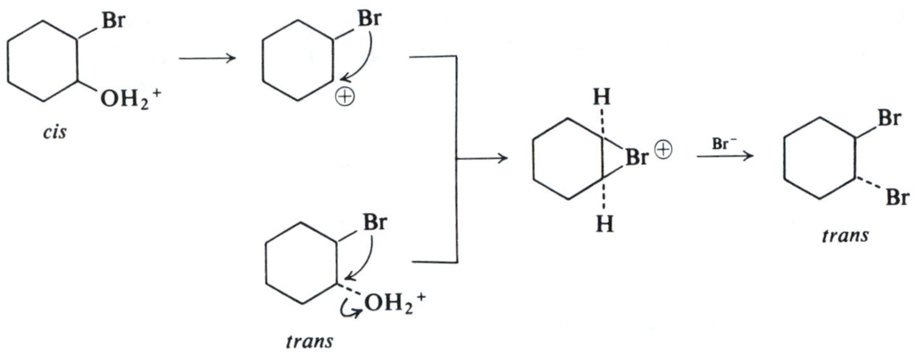

Br OH$_2^+$ *cis* → Br ⊕ ; Br OH$_2^+$ *trans* → H Br⊕ H $\xrightarrow{Br^-}$ Br Br *trans*

5. H R H R′ $\xrightarrow{Cl_2}$ ⊕Cl H R H R′ +enantiomer $\xrightarrow{H_2O}$ Cl H R H R′ OH +enantiomer $\xrightarrow{OH^-}$ Cl H R H R′ O⁻ +enantiomer ↓

H R H R′ $\xrightarrow{PhCO_2OH}$ H R H R′ O +enantiomer

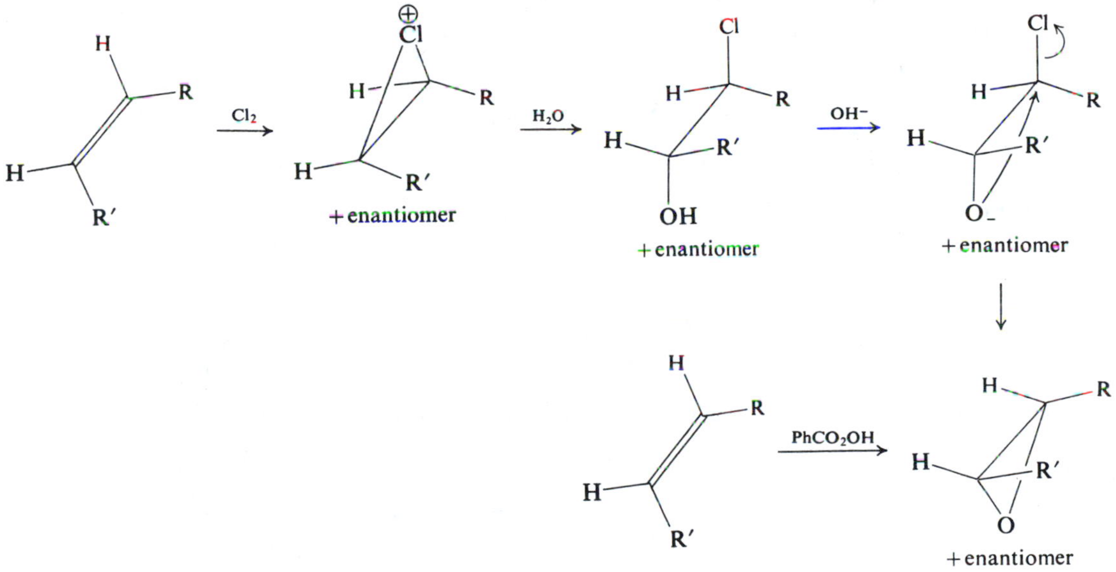

R = —$(CH_2)_7COOH$, R′ = $CH_3(CH_2)_7$—

6. The oxo process amounts to overall *syn*-addition of —CHO and —H.

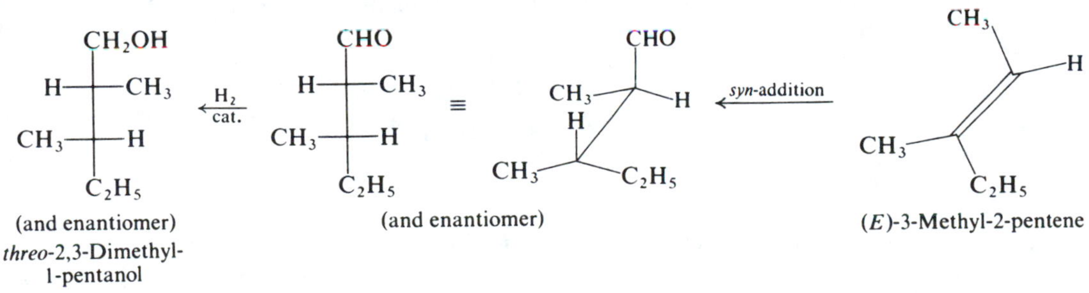

CH_2OH / H—CH_3 / CH_3—H / C_2H_5 (and enantiomer) *threo*-2,3-Dimethyl-1-pentanol $\xleftarrow[\text{cat.}]{H_2}$ CHO / H—CH_3 / CH_3—H / C_2H_5 (and enantiomer) ≡ CHO CH_3 H H CH_3 C_2H_5 $\xleftarrow{\textit{syn}\text{-addition}}$ CH_3 H CH_3 C_2H_5 (*E*)-3-Methyl-2-pentene

The situation is exactly analogous to that in homogeneous hydrogenation, which also involves *syn*-addition (Sec. 29.6), and the reasoning follows the same lines. The observed stereochemistry is the net result of *syn*-addition in step (3), and migration with *retention* in step (4). (See pages 1052–1053.) In step (3) the metal and hydrogen attach themselves simultaneously to the doubly bonded carbons. For geometric reasons—the metal and hydrogen are bonded to each other in the reactant—this addition must take place to the same face of the double bond. In step (4) the alkyl group migrates from the metal to a carbon monoxide ligand. Again for geometric reasons, CO attaches itself to the *front side* of alkyl carbon—the same side that was attached to the metal in the reactant. Back-side attack is just not possible, since the metal holds both CO carbon and alkyl carbon in the transition state.

7. (a) $n\text{-}C_{10}H_{21}Br \xrightarrow{HC\equiv CNa} \underset{A}{n\text{-}C_{10}H_{21}C\equiv CH} \xrightarrow{n\text{-BuLi}} \underset{B}{n\text{-}C_{10}H_{21}C\equiv CLi}$

$(CH_3)_2CHCH_2CH_2CH{=}CH_2 \xrightarrow[\text{peroxides}]{HBr} \underset{C}{i\text{-HexCH}_2Br}$

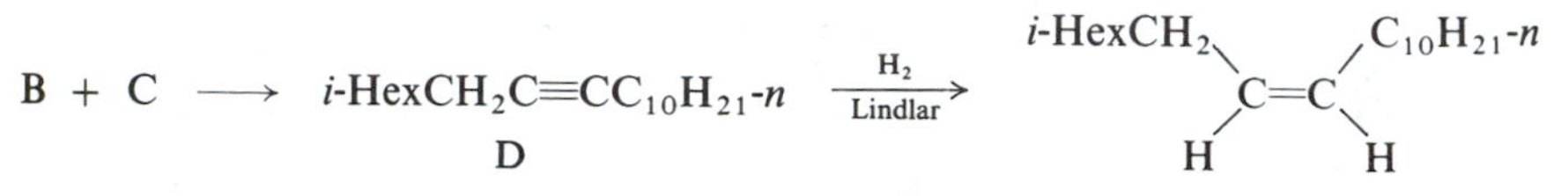

$B + C \longrightarrow \underset{D}{i\text{-HexCH}_2C\equiv CC_{10}H_{21}\text{-}n} \xrightarrow[\text{Lindlar}]{H_2}$ E: *i*-HexCH$_2$(H)C=C(H)C$_{10}$H$_{21}$-*n* (cis)

$E \xrightarrow{ArCO_2OH}$ *i*-HexCH$_2$—CH—CH—C$_{10}$H$_{21}$-*n* (cis epoxide, O bridging the two carbons)

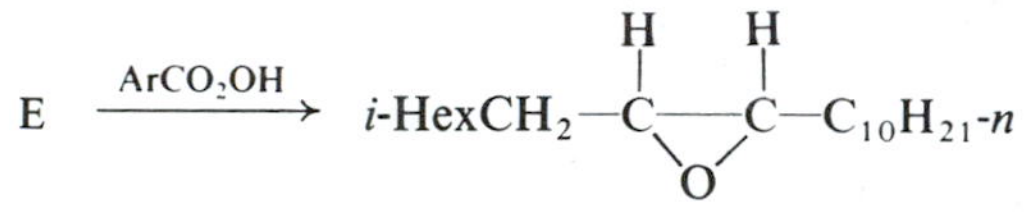

Racemic

Sex attractant of gypsy moth

Disparlure

(b) The molecule is chiral, as shown by the non-superimposability of mirror images:

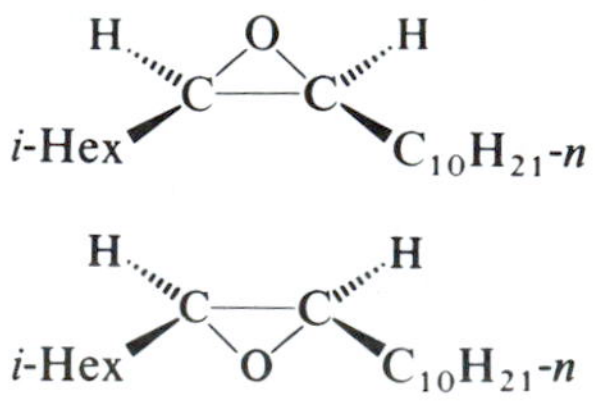

[Two mirror-image cis epoxides: H, *i*-Hex on one C; H, C$_{10}$H$_{21}$-*n* on the other; O above the ring in one, below in the other]

The natural material consists of (an excess of) one of these enantiomers. The synthesis in (a) gives the racemic modification; under achiral conditions attachment of O to either face of the alkene is equally likely.

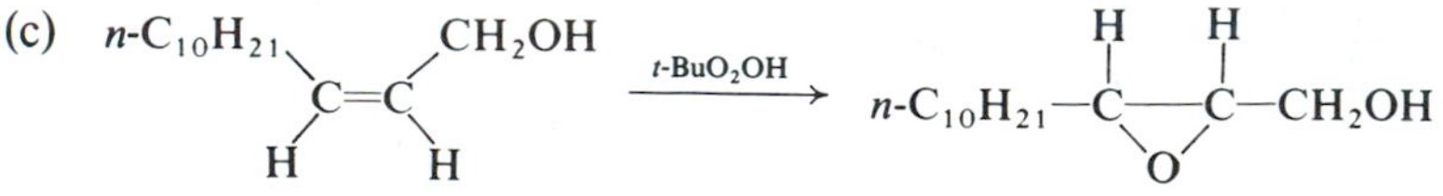

(c) *n*-C$_{10}$H$_{21}$(H)C=C(H)CH$_2$OH (cis) $\xrightarrow{t\text{-BuO}_2\text{OH}}$ *n*-C$_{10}$H$_{21}$—CH—CH—CH$_2$OH (epoxide, O bridging) **F**

(d) To provide a chiral environment in which formation of one enantiomeric epoxide is favored over the other.

(e) The response of the female gypsy moth, like other biological processes, is stereospecific (see Sec. 4.11).

8. This work was reported in one of a series of papers by Donald J. Cram (a recent Nobel Laureate) at the University of California at Los Angeles.

(a) Solvolysis, we see, is completely stereospecific and proceeds, it at first appears, with *retention* of configuration: racemic *erythro* tosylate gives only racemic *erythro* acetate,

```
    CH3            CH3                           CH3            CH3
 H ──┼── Ph    Ph ──┼── H      HOAc         H ──┼── Ph     Ph ──┼── H
 H ──┼── OTs  TsO ──┼── H    ────────→      H ──┼── OAc   AcO ──┼── H
    CH3            CH3         KOAc            CH3            CH3
```

Erythro
Racemic

Erythro
Racemic

and racemic *threo* tosylate gives only racemic *threo* acetate.

```
    CH3            CH3                           CH3            CH3
 H ──┼── Ph    Ph ──┼── H      HOAc         H ──┼── Ph     Ph ──┼── H
TsO ──┼── H     H ──┼── OTs  ────────→    AcO ──┼── H       H ──┼── OAc
    CH3            CH3         KOAc            CH3            CH3
```

Threo
Racemic

Threo
Racemic

When, however, optically active *threo* tosylate is used, it is found to yield optically *inactive* product, racemic *threo* acetate. We see here the same pattern as in Sec. 29.2: retention at both carbons in half the molecules of the product, inversion at both carbons in the other half.

```
    CH3                        CH3            CH3
 H ──┼── Ph     HOAc       H ──┼── Ph     Ph ──┼── H
TsO ──┼── H   ────────→   AcO ──┼── H       H ──┼── OAc
    CH3         KOAc           CH3            CH3
```

(+)-*Threo*
Optically active

Threo
Racemic

Cram interpreted these results in the following way. The neighboring phenyl group, with its π electrons, helps to push out (1) the tosylate anion. There is formed an intermediate

(1) 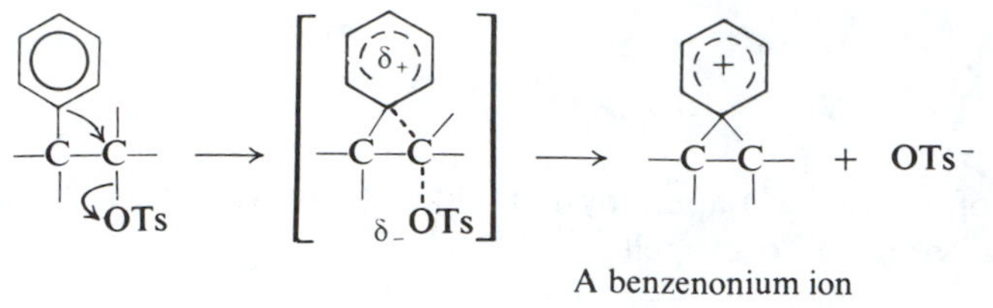 $-\overset{|}{C}-\overset{|}{C}-$ (Ph, OTs) $\longrightarrow$ [bridged, $\delta+$, $\delta-$OTs] $\longrightarrow$ bridged $-\overset{|}{C}-\overset{|}{C}-$ (+) $+ \ OTs^-$

A benzenonium ion

(2) 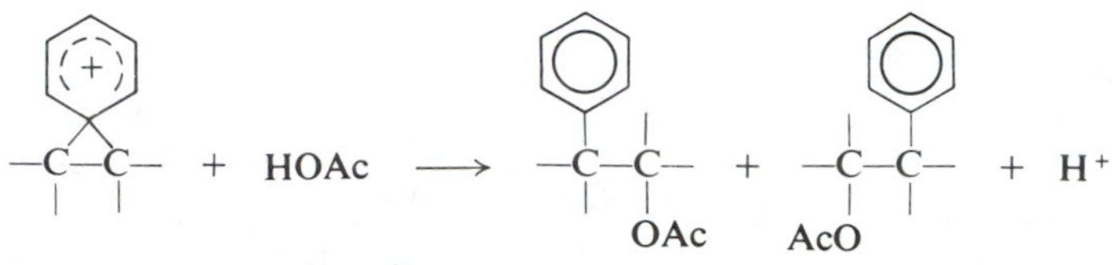 bridged ion (+) $+$ HOAc $\longrightarrow$ $-\overset{|}{C}-\overset{|}{C}-$ (OAc) $+$ $-\overset{|}{C}-\overset{|}{C}-$ (AcO) $+ \ H^+$

bridged ion. This ion is of the kind we have encountered (Sec. 15.8) as the intermediate in electrophilic aromatic substitution: a *benzenonium ion.* The ring is bonded to both carbons by full-fledged σ bonds, to give a symmetrical structure. The ion owes its stability to the fact that the positive charge is distributed about the ring, being strongest at the positions *ortho* and *para* to the point of attachment.

In step (2) the benzenonium ion is attacked by the acetic acid at either of the two equivalent carbons to yield the product.

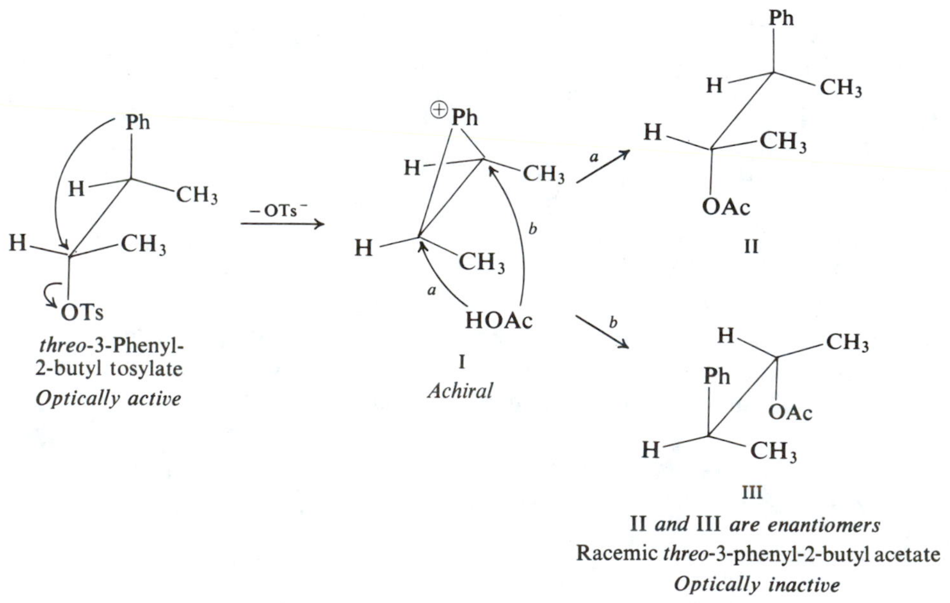

Ph, H, CH_3, H, CH_3, OTs $\xrightarrow{-OTs^-}$ $\oplus$Ph, H, CH_3, H, CH_3, HOAc (a, b) → a: II; b: III

threo-3-Phenyl-2-butyl tosylate
Optically active

I
Achiral

II: Ph, H, CH_3, H, CH_3, OAc

III: H, CH_3, Ph, OAc, H, CH_3

II *and* III *are enantiomers*
Racemic *threo*-3-phenyl-2-butyl acetate
Optically inactive

Bridged ion I is achiral, and must give optically inactive product: specifically, because attacks *a* and *b* give enantiomers, and in equal amounts.

(b)

erythro-3-Phenyl-2-butyl tosylate *Optically active* $\xrightarrow{-OTs^-}$ IV *Chiral* → V, VI

V *and* VI *are the same* *erythro*-3-Phenyl-2-butyl acetate *Optically active*

Bridged ion IV is chiral. Attack by either path *c* or path *d* gives the same, optically active compound.

(Cram, D. J. "Studies in Stereochemistry. V. Phenonium Sulfonate Ion-pairs as Intermediates in the Intramolecular Rearrangements and Solvolysis Reactions that Occur in the 3-Phenyl-2-butanol System"; *J. Am. Chem. Soc.* **1952**, *74*, 104.)

9. (a) Formation of II could arise from straightforward substitution. But the abnormally fast rate makes us curious, and isolation of the cyclopropyl compound III gives us the clue. Reaction occurs with anchimeric assistance by the π electrons of the double bond, to give the cyclopropylcarbinyl cation VIII. This cation can undergo elimination to yield III or substitution, with ring opening, to yield II.

$(CH_3)_2C{=}CH{-}CH_2{-}CH_2{-}OTs$ (I) $\longrightarrow$ $[(CH_3)_2C^{\delta+}\cdots CH{-}CH_2{-}CH_2\cdots OTs^{\delta-}]$ $\xrightarrow{-OTs}$ $(CH_3)_2C^{\oplus}{-}CH(CH_2)_2$ (VIII)

VIII $\xrightarrow{-H^+}$ $CH_2{=}C(CH_3){-}CH(CH_2)_2$ (III)

VIII + HOAc $\longrightarrow$ $(CH_3)_2C{=}CH_2{-}CH_2{-}CH_2{-}OAc$ (II)

This interpretation is confirmed by the behavior of the labeled compound Ia, which yields the mixture of isomers, IIa and IIb, that is predicted from attack by paths *a* and *b* on the cation VIIIa.

$$(CH_3)_2C{=}CH{-}CH_2{-}CD_2{-}OTs \;(\text{Ia}) \longrightarrow \text{VIIIa} \xrightarrow{HOAc} \begin{cases} \xrightarrow{a} (CH_3)_2C{=}CH{-}CH_2{-}CD_2{-}OAc \;(\text{IIa}) \\ \xrightarrow{b} (CH_3)_2C{=}CH{-}CD_2{-}CH_2{-}OAc \;(\text{IIb}) \end{cases}$$

(Rogan, J. B. "The Acetolysis of 4-Methyl-3-Penten-1-yl *p*-Toluenesulfonate"; *J. Org. Chem.* **1962**, *27*, 3910. The labeling experiment was not actually carried out on this reaction, but complete scrambling was obtained in the closely related substitution of chloride for chloride (Hart, H., *et al.* "Alkylation of Phenol with a Homoallylic Halide"; *J. Am. Chem. Soc.* **1963**, *85*, 3269).)

10. (a) IV is a lactone and hence an ester. Like any ester, it reacts with OCH_3^- to undergo nucleophilic acyl substitution: transesterification (Sec. 20.20). The product is a new ester and a new alkoxide (or, in this case, a phenoxide); these are part of the same molecule, IX.

IV + OCH_3^- ⟶ IX ≡ IX $\xrightarrow{-Br^-}$ V

(IV: Ph, CH_2CH_2Br; IX: O^-, $COOCH_3$, Ph, CH_2CH_2Br; V: Ph, $COOCH_3$)

IX undergoes intramolecular nucleophilic substitution—a Williamson synthesis, actually—to give the cyclic ether V.

An analogous reaction with R_2NH gives X, which is both an amide and a phenoxide. As before, there is an internal Williamson reaction to yield the ether, VI.

IV + HNR_2 $\xrightarrow{-H^+}$ X ≡ X $\xrightarrow{-Br^-}$ VI

(X: O^-, $CONR_2$, Ph, CH_2CH_2Br; VI: Ph, $CONR_2$)

(b) Nucleophilic acyl substitution, it has been proposed (Sec. 20.4), involves attachment of the nucleophile to acyl carbon to give a tetrahedral intermediate which then, in a second step, ejects the leaving group. We have seen (Sec. 20.17) evidence of isotopic exchange which supports that proposal. In the reaction of IV with HNR_2, for example, the intermediate would be XI. Expulsion of phenoxide would yield the amide X.

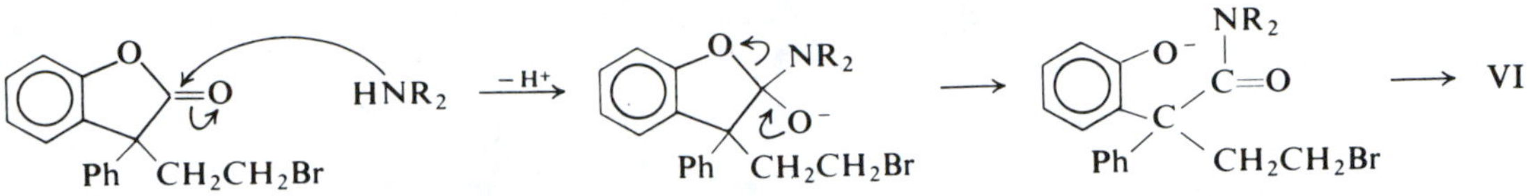

IV —HNR₂, $-H^+$→ XI → X → VI

But in intermediate XI the acyl group is temporarily converted into —O⁻. This group is capable of nucleophilic attack on the carbon carrying —Br to yield VII. Compound VII is thus the product of internal trapping of the tetrahedral intermediate, and its isolation is strong evidence for the existence of this intermediate.

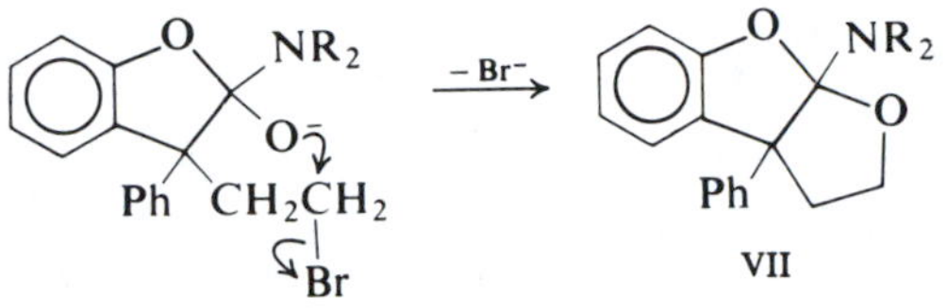

XI —$-Br^-$→ VII

30

Heterocyclic Compounds

30.1

$$\mathrm{CH_3COCH_2COOEt} \xrightarrow{\text{NaOEt}} \underset{\mathbf{A}}{[\mathrm{CH_3COCHCOOEt}]^-} \xrightarrow{\mathrm{I_2}} \underset{\substack{\mathbf{B}\\ \textit{A double }\beta\textit{-keto ester}}}{\mathrm{CH_3CO\overset{\overset{\displaystyle EtOOC}{|}}{C}H{-}\overset{\overset{\displaystyle COOEt}{|}}{C}HCOCH_3}} \xrightarrow[\mathrm{H_2O}]{\mathrm{H^+}} \mathrm{CH_3CO\overset{\overset{\displaystyle HOOC}{|}}{C}H{-}\overset{\overset{\displaystyle COOH}{|}}{C}HCOCH_3}$$

$$\xrightarrow{-\,2\mathrm{CO_2}} \underset{\text{2,5-Hexanedione}}{\mathrm{CH_3\overset{O}{\overset{\|}{C}}CH_2CH_2\overset{O}{\overset{\|}{C}}CH_3}}$$

30.2

$$\underset{\text{2,5-Diphenylfuran}}{\text{Ph–(furan)–Ph}} \xleftarrow{\mathrm{P_2O_5}} \underset{\text{A 1,4-diketone}}{\mathrm{PhC(=O)CH_2{-}CH_2C(=O)Ph}} \xleftarrow[\text{Prob. 30.1}]{\text{as in}} \mathrm{PhCOCH_2COOEt} \xleftarrow{\mathrm{OEt^-}} \mathrm{PhCOOEt} + \mathrm{CH_3COOEt}$$

30.3 Ring-opening involves electrophilic attack by H^+ (see the answer to Problem 4(b), below), which is slowed down by the electron-withdrawing —COOH group.

30.4 In Sec. 31.7 we see that, in acid or base, phenols react with formaldehyde by what is both carbonyl addition and electrophilic aromatic substitution. A hydroxymethyl group is introduced into the ring, and, on further reaction, rings become connected by $—CH_2—$ linkages.

Like phenol, furan contains an activated ring, and it undergoes an analogous reaction.

$$\text{(EtOOC, CH}_3\text{-furan)–CH}_2\text{–(furan-CH}_3\text{, COOEt, CH}_3) \xleftarrow[-\,\mathrm{H_2O}]{\mathrm{H^+}} \text{EtOOC, CH}_3, \mathrm{CH_3}\text{-furan} + \mathrm{HOCH_2}\text{-furan(CH}_3\text{, COOEt, CH}_3) \xleftarrow[\mathrm{H^+}]{\mathrm{HCHO}} \text{CH}_3\text{, COOEt, CH}_3\text{-furan}$$

30.5 Like an aryl aldehyde, furfural has no α-hydrogens, and undergoes the Cannizzaro reaction (Sec. 18.13).

2 Furfural (2-furyl–CHO) $\xrightarrow{\text{conc. OH}^-}$ Sodium furoate (2-furyl–COO^-Na^+) + Furfuryl alcohol (2-furyl–CH_2OH)

30.6 Pyridine–N^+–SO_3^- + ArH ⟶ $\overset{\oplus}{Ar}(H)(SO_3^-)$ + pyridine ⟶ $ArSO_3^-$ + pyridine–N^+–H

30.7 (a) Overlap of sextets of pyrrole and benzene rings, much as in naphthalene (Sec. 14.12).

(b) Attack at the 3-position gives the most stable cation: aromaticity of the six-membered ring is preserved, and the positive charge is accommodated by nitrogen. (See Secs. 15.21 and 30.4.)

Indole (positions 1–7, N–H at 1) $\xrightarrow{Y^+}$ 3-H,3-Y indoleninium cation ($\overset{\oplus}{N}$H) $\xrightarrow{-H^+}$ 3-Y-indolenine (N)

30.8 THF $\xrightarrow{HCl}$ $Cl(CH_2)_4Cl$ $\xrightarrow{2CN^-}$ $NC(CH_2)_4CN$

→ $\xrightarrow[H^+]{H_2O}$ $HOOC(CH_2)_4COOH$ Adipic acid

→ $\xrightarrow[Ni]{H_2}$ $H_2NCH_2(CH_2)_4CH_2NH_2$ Hexamethylenediamine

30.9 (a) Pyrrolidinium ($\overset{\oplus}{N}H_2$) Cl^- (b) no reaction (c) Pyrrolidine N–C(=O)CH_3 (d) Pyrrolidine N–SO_2–Ph (O–S–O, Ph) (e) Pyrrolidine N–Me

(f) N-Me pyrrolidine ⟶ $\overset{\oplus}{N}$(Me)(Me) pyrrolidinium OH^- ⟶ $Me_2NCH_2CH_2CH{=}CH_2$

Hofmann elimination (*Sec. 23.6*)

30.10 Hygrine is basic, a 3° amine. It contains a carbonyl group, and is a methyl ketone (Sec. 18.21). Vigorous oxidation removes one more carbon (and two more hydrogens) than the haloform reaction: a tentative conclusion is that there is a $—CH_2COCH_3$ group attached to a resistant nucleus.

$$\underset{\text{Hygrine}}{RCH_2COCH_3} \xrightarrow{\text{oxidn.}} \underset{\text{Hygrinic acid}}{RCOOH}$$

Synthesis of hygrinic acid gives us the structures.

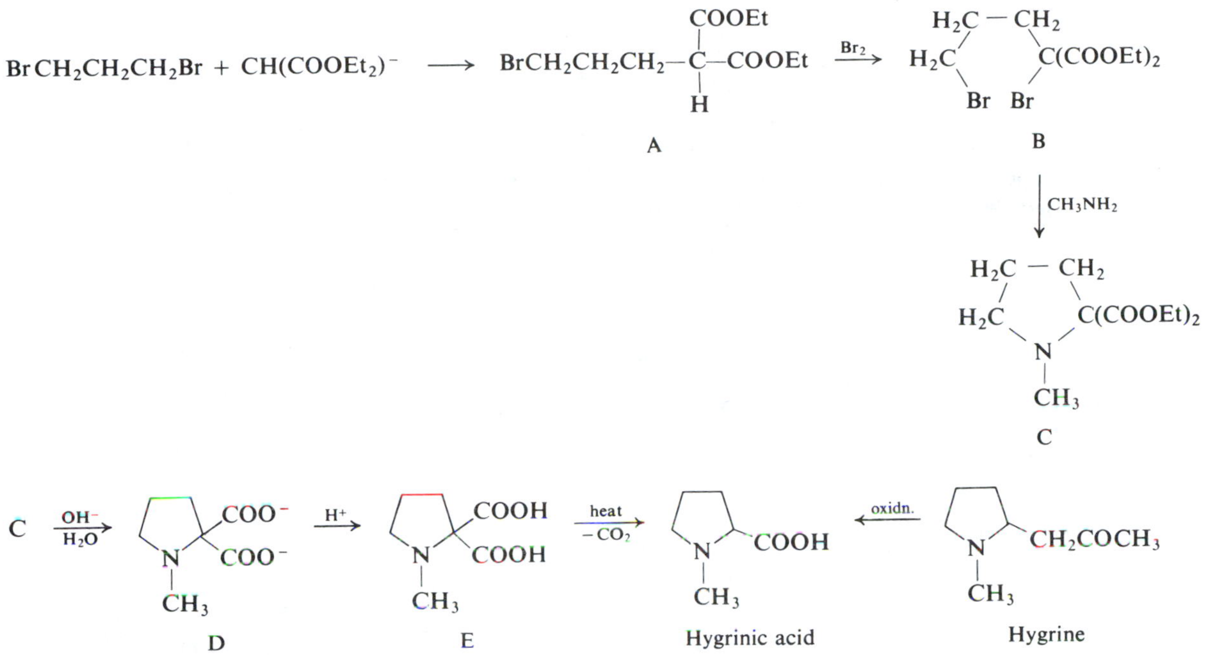

$$BrCH_2CH_2CH_2Br + CH(COOEt_2)^- \longrightarrow \underset{\mathbf{A}}{BrCH_2CH_2CH_2—CH(COOEt)—COOEt} \xrightarrow{Br_2} \mathbf{B}$$

B: $H_2C—CH_2$, H_2C, $C(COOEt)_2$, Br Br

$\xrightarrow{CH_3NH_2}$ C: $H_2C—CH_2$, H_2C, $C(COOEt)_2$, N, CH_3

C $\xrightarrow[H_2O]{OH^-}$ D (N-CH_3 pyrrolidine with COO^-, COO^-) $\xrightarrow{H^+}$ E (N-CH_3 pyrrolidine with COOH, COOH) $\xrightarrow[-CO_2]{\text{heat}}$ Hygrinic acid (N-CH_3 pyrrolidine with COOH) $\xleftarrow{\text{oxidn.}}$ Hygrine (N-CH_3 pyrrolidine with CH_2COCH_3)

30.11 The orientation ("para") and reactivity are controlled by the activating $—NH_2$ group (as in aniline).

30.12 3-Aminopyridine (NH_2) $\xleftarrow[OH^-]{OBr^-}$ ($CONH_2$) $\xleftarrow{NH_3}$ (COCl) $\xleftarrow{SOCl_2}$ (COOH) $\xleftarrow{KMnO_4}$ (CH_3) β-Picoline

30.13 (a) The nitrogen-containing ring is deactivated, as in pyridine; electrophilic substitution thus occurs at an α-position in the less deactivated ring. (Compare Sec. 15.21 and Problem 15.13.)

(b) The benzenoid ring is more easily attacked by oxidizing agents than the deactivated nitrogen-containing ring.

30.14 Basicity falls off with increasing *s*-character of the orbital holding the unshared pair.

$$\underset{(sp^3)}{\text{amine}} > \underset{(sp^2)}{\text{imine}} > \underset{(sp)}{\text{nitrile}}$$

30.15 Pyridine is the base needed for the dehydrohalogenation. It does not cleave the ester as alcoholic KOH would.

30.16 Electrophilic substitution proceeds via intermediate I in which the positive charge is carried by nitrogen, and all atoms have complete octets.

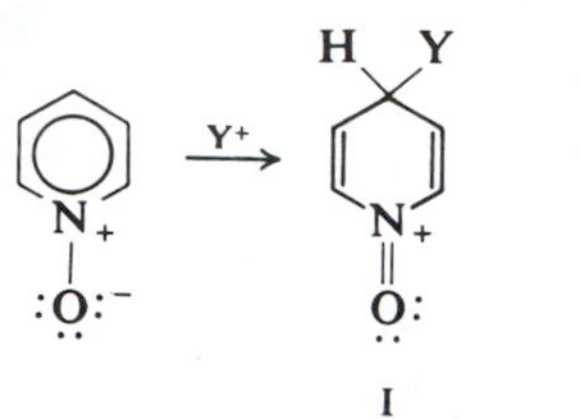

Electrophilic substitution

30.17 Nucleophilic substitution proceeds via intermediate II in which the negative charge is carried by the electronegative element oxygen.

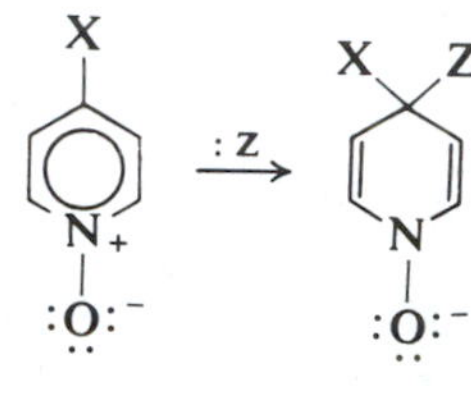

Nucleophilic substitution

30.18 Piperidine, a 2° amine, would itself be acylated.

1. (a) Br, N — 3-Bromopyridine

(b) SO_3H, N — 3-Pyridinesulfonic acid

(c) no reaction

(d) NO_2, N — 3-Nitropyridine

(e) NH_2, N — 2-Aminopyridine (*Sec. 30.10*)

(f) Ph, N — 2-Phenylpyridine (*Sec. 30.10*)

(g) N^+–H Cl^- — Pyridinium chloride

(h) no reaction

(i) no reaction

(j) no reaction

(k) N^+–Et Br^- — *N*-Ethylpyridinium bromide

(l) N^+–CH_2Ph Cl^- — *N*-Benzylpyridinium chloride

(m) NO_2, N^+–O^- — 4-Nitropyridine *N*-oxide (*Problem 30.16*)

(n) N–H — Piperidine (*Sec. 30.12*)

2. (a) 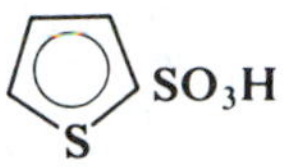 2-Thiophenesulfonic acid (thiophene ring, SO_3H at C-2)

(b) 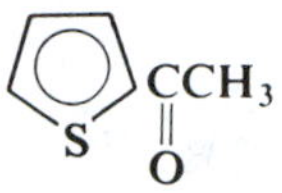 2-Acetylthiophene (thiophene ring, $\overset{O}{\overset{\|}{C}}CH_3$ at C-2)

(c) 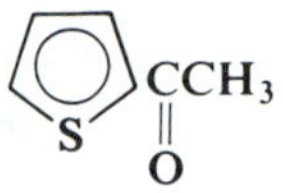 2-Acetylthiophene (thiophene ring, $\overset{O}{\overset{\|}{C}}CH_3$ at C-2)

(d)  2-Nitrothiophene (thiophene ring, NO_2 at C-2)

(e)  2-Aminothiophene (thiophene ring, NH_2 at C-2)

(f)  2-Bromothiophene (thiophene ring, Br at C-2)

(g)  2-Thiophenecarboxylic acid (thiophene ring, COOH at C-2)

(h) 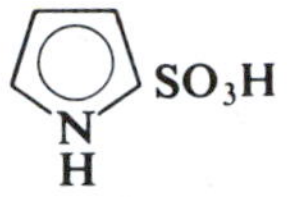 2-Pyrrolesulfonic acid (pyrrole ring with N–H, SO_3H at C-2)

(i) 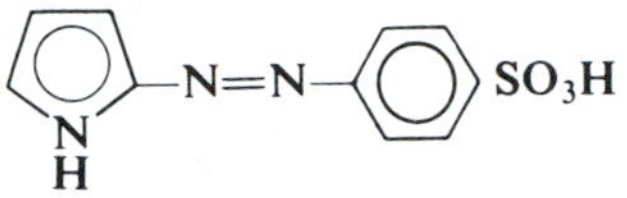 *p*-(2-Pyrrylazo)benzenesulfonic acid (pyrrole ring with N–H, –N=N–C_6H_4–SO_3H at C-2)

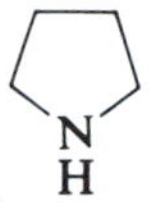

(j) Pyrrolidine (saturated five-membered ring with N–H)

(k) Furfurylideneacetone (furan ring, $CH{=}CH\overset{O}{\overset{\|}{C}}CH_3$ at C-2)

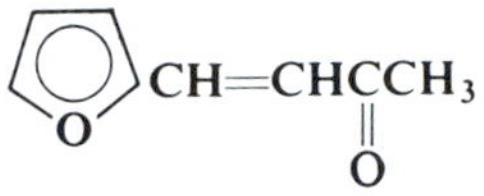

3. (a) 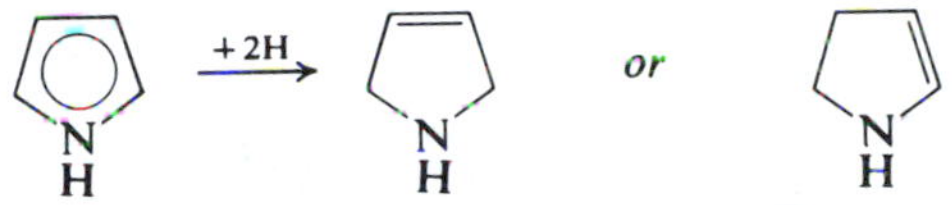 Pyrrole $\xrightarrow{+2H}$ Δ^3-Pyrroline (2,5-Dihydropyrrole) *or* Δ^2-Pyrroline (2,3-Dihydropyrrole)

(b) 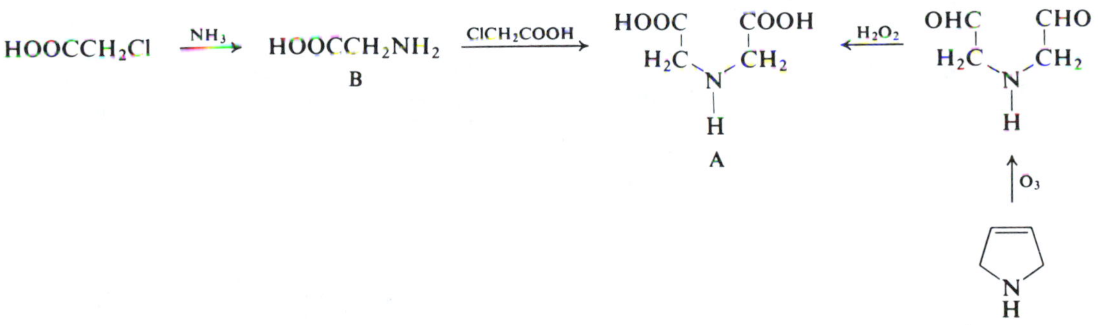 $HOOCCH_2Cl \xrightarrow{NH_3} HOOCCH_2NH_2$ (B) $\xrightarrow{ClCH_2COOH}$ $HOOCCH_2$–NH–CH_2COOH (A)

A $\xleftarrow{H_2O_2}$ $OHCCH_2$–NH–CH_2CHO $\xleftarrow{O_3}$ Δ^3-Pyrroline (2,5-Dihydropyrrole)

4. (a) $2CHI_3 + HOOCCH_2CH_2COOH$ (Succinic acid) $\xleftarrow{NaOI}$ $CH_3\overset{O}{\overset{\|}{C}}CH_2CH_2\overset{O}{\overset{\|}{C}}CH_3$

C

2,5-Hexanedione

(b) 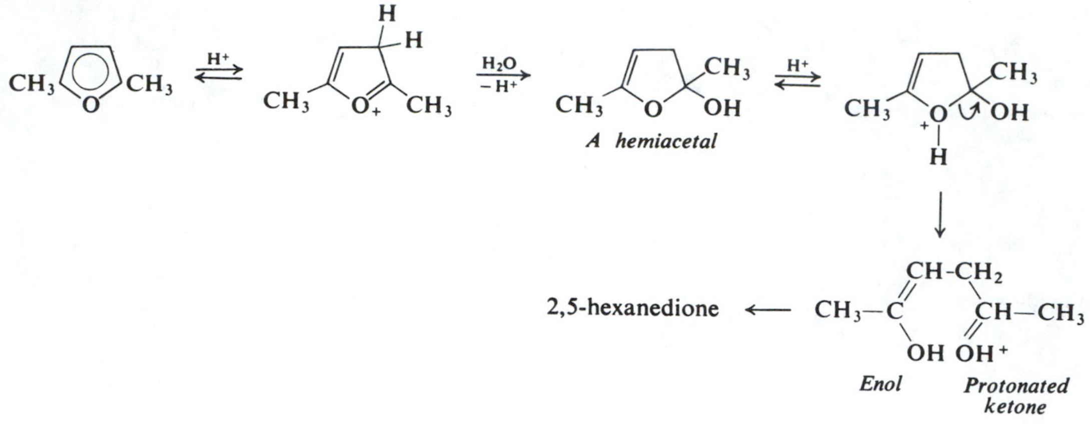 CH_3–furan–CH_3 $\underset{}{\overset{H^+}{\rightleftarrows}}$ (protonated ring, O_+) $\xrightarrow[-H^+]{H_2O}$ (hemiacetal) $\overset{H^+}{\rightleftarrows}$ (O-protonated hemiacetal)

A hemiacetal

↓

2,5-hexanedione ⟵ $CH_3-C(OH)=CH-CH_2-CH=\overset{+}{O}H-CH_3$

Enol *Protonated ketone*

5. First, some arithmetic:

$(C_4H_5N)_4$	4 pyrrole		$(C_5H_7ON)_4$	
$+(HCHO)_4$	4 HCHO		$-(C_5H_5N)_4$	product
$(C_5H_7ON)_4$			$4H_2O$ lost	

Like phenol (or furan, Problem 30.4, page 1063), pyrrole contains an activated ring, and we suspect that a reaction analogous to that described in Sec. 31.7 has taken place here:

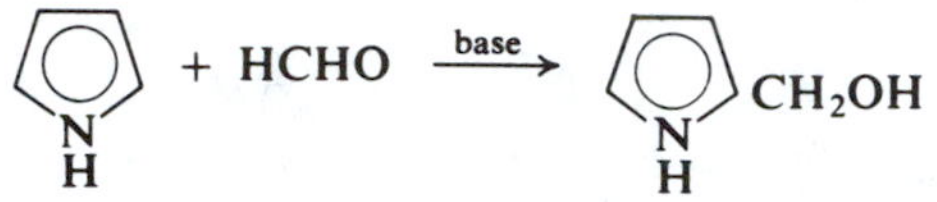

pyrrole + HCHO $\xrightarrow{\text{base}}$ 2-pyrrolyl–CH_2OH

Four of these units are joined together, with loss of $4H_2O$. The structure must therefore be *cyclic*: four units, four new linkages.

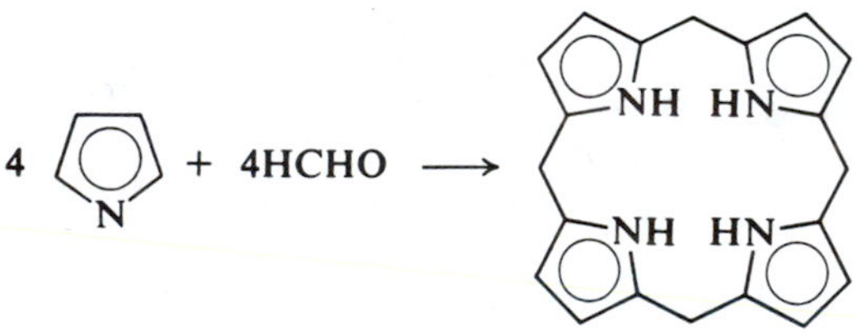

4 pyrrole + 4HCHO ⟶ Porphin

Compare ring structure of chlorophyll (*p. 1059*) *and* heme (*p. 1228*)

6. 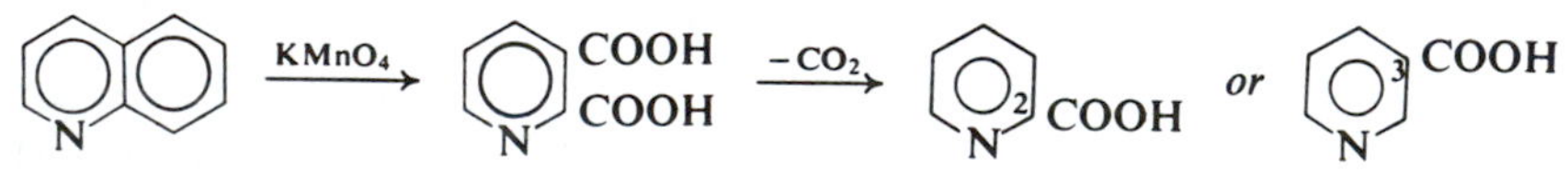 Quinoline $\xrightarrow{KMnO_4}$ 2,3-pyridinedicarboxylic acid $\xrightarrow{-CO_2}$ 2-pyridinecarboxylic acid *or* 3-pyridinecarboxylic acid

E *must be one of these*

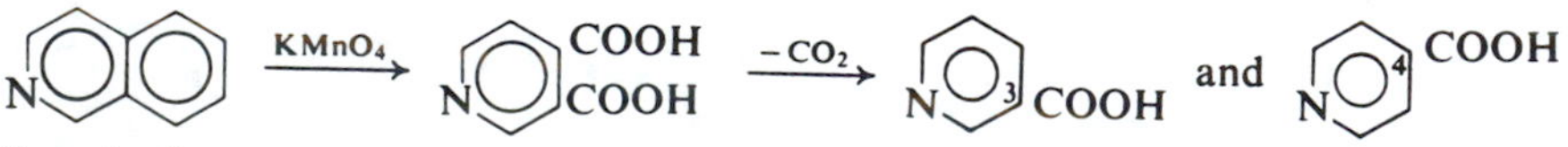

Isoquinoline $\xrightarrow{KMnO_4}$ 3,4-pyridinedicarboxylic acid $\xrightarrow{-CO_2}$ 3-pyridinecarboxylic acid and 4-pyridinecarboxylic acid

This must be E *This must be* F

The only acid left must be D:

2-pyridinecarboxylic acid

D

7. (a) 3-Pyridine-CN $\xleftarrow{CuCN}$ 3-Pyridine-N_2^+ $\xleftarrow{HONO}$ 3-Pyridine-NH_2 (Problem 30.12)

(b) 2-Methylpiperidine (N–H, CH_3) $\xleftarrow[Ni]{3H_2}$ 2-Methylpyridine (CH_3) $\xleftarrow{CH_3Li}$ Pyridine

(c) O_2N-(5)furan(2)-COOEt $\xleftarrow[H_2SO_4]{HNO_3}$ Furan-COOEt $\xleftarrow[H^+]{EtOH}$ Furan-COOH $\xleftarrow{Ag_2O}$ Furan-CHO

(d) Furan-CH=CHCOOH $\xleftarrow[\text{(Problem 21.18d, p. 811)}]{\text{Perkin reaction}}$ Furan-CHO + Ac_2O + NaOAc

(e) $H_2C—CH_2$ / H_2C $CH—CH_2$ / Cl Cl Cl $\xleftarrow{\text{conc. HCl}}$ Tetrahydrofuran-CH_2OH $\xleftarrow[Ni]{3H_2}$ Furan-CHO

(f) Indole-3-CHO (N–H) $\xleftarrow{CHCl_3, OH^-}$ Indole (N–H)

8. 3-Pyridine-COOH $\xrightarrow{SOCl_2}$ 3-Pyridine-COCl $\xrightarrow{EtO(CH_2)_3CdCl}$ 3-Pyridine-C(=O)-CH_2-CH_2-CH_2-OEt (G) $\xrightarrow[Ni]{NH_3, H_2}$ 3-Pyridine-CH(NH_2)-CH_2-CH_2-CH_2-OEt (H)

H $\xrightarrow{HBr}$ [3-Pyridine-CH(H_2N)-CH_2-CH_2-CH_2-Br] $\longrightarrow$ 3-Pyridyl-pyrrolidine (N–H, H, *) (I, *2° amine*) $\xrightarrow[\text{alkylation}]{CH_3I, OH^-}$ 3-Pyridyl-N-methylpyrrolidine (CH_3, H, *) ($\pm$)-Nicotine

J and K are diastereomeric salts.

9. (a) Benzene(CHO)(OH) + $CH_3—\overset{\displaystyle O}{\overset{\|}{C}}—Ph$ $\xrightarrow{\text{aldol}}$ Benzene(OH)-CH=CH-C(=O)-Ph (L)

(b) 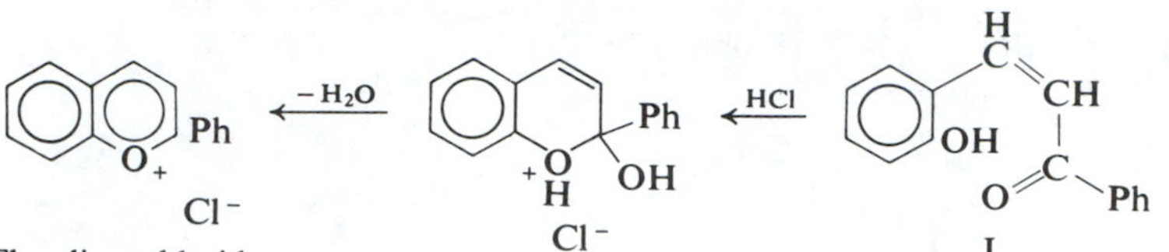 Flavylium chloride $\xleftarrow{-H_2O}$ ... $\xleftarrow{HCl}$ L

(c) Oxygen contributes a pair of electrons to complete the aromatic sextet, thus giving a system analogous to that of naphthalene.

10. (a) "Pyrrole" nitrogen—the ring N with the —H attached—contributes two π electrons, and the other atoms (including the "pyridine" N) contribute one each, to give the aromatic sextet (Sec. 30.2).

(b) 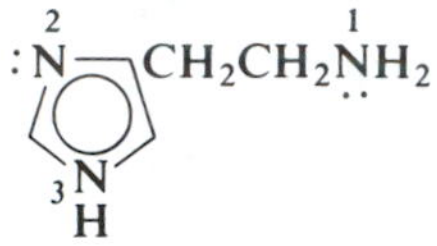 Histamine

aliphatic $—NH_2$ > "pyridine" N > "pyrrole" N

The unshared pair is in, respectively, an sp^3 orbital, an sp^2 orbital, and the π cloud. (See Secs. 30.2 and 30.11.)

(c) 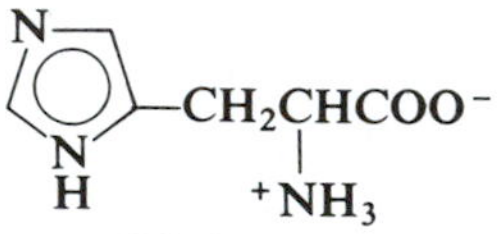 $—CH_2CHCOO^-$, $^+NH_3$

Histidine

The proton goes to the most basic nitrogen.

11. Tropinic acid, $C_8H_{13}O_4N$, has a m.w. of 187. Since its N.E. is 94 $\pm$ 1 (= m.w./2), it must be a diacid. It is a 3° amine, contains no easily oxidizable functions, and is saturated. Chemical arithmetic,

$C_6H_{15}N$	sat'd open-chain amine	$C_8H_{15}O_4N$	
$+C_2O_4$	two —COO— groups	$-C_8H_{13}O_4N$	tropinic acid
$C_8H_{15}O_4N$		2H missing: *means* one ring	

indicates that tropinic acid contains *one ring.*

Let us turn to the exhaustive methylation data:

$C_9H_{16}O_4NI$	M
$-C_8H_{13}O_4N$	tropinic acid
CH_3I *means* one Me introduced	

$$\text{M} \xrightarrow{-H_2O} \text{N} \quad \textit{with no loss of N}$$

Tropinic acid must be a 3° amine. The nitrogen must be part of the ring, since otherwise it would have been lost as $MeNR_2$.

A repetition of the procedure does split out Me_3N, and leaves an unsaturated molecule,

P, which undoubtedly contains the two —COOH's. Only one carbon is lost ($C_8 \longrightarrow C_7$), clearly indicating a $-\overset{|}{N}-CH_3$ unit in tropinic acid.

$$\begin{array}{ll} C_7H_{12}O_4 & \text{sat'd open-chain diacid} \\ \underline{-\,C_7H_8O_4} & \text{P} \\ \text{4H missing: } \textit{means} \text{ two double bonds} & \end{array}$$

P is a doubly unsaturated diacid; hydrogenation to heptanedioic acid shows that the carbon chain is unbranched.

P thus can be one of several acids,

$HOOC-CH=CH-CH=CH-CH_2-COOH$ $HOOC-CH=CH-CH_2-CH=CH-COOH$

$HOOC-CH=C=CH-CH_2-CH_2-COOH$ $HOOC-CH_2-CH=C=CH-CH_2-COOH$

although the allenic compounds are unlikely, and we are discounting the likelihood of an acetylenic structure. (From what size ring could an alkyne structure, or an allenic structure, arise through exhaustive methylation?)

(a) At this point, depending upon which structure for P and which point of cleavage of the ring we postulate, possible structures for tropinic acid include these:

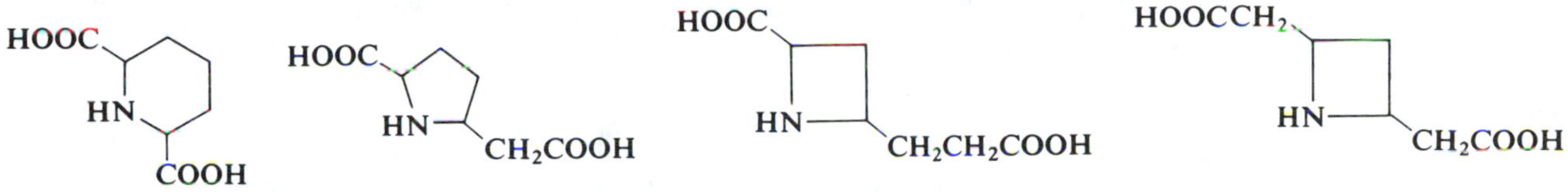

HOOC, HN, COOH (piperidine ring); HOOC, HN, CH_2COOH (pyrrolidine ring); HOOC, HN, CH_2CH_2COOH (azetidine ring); $HOOCCH_2$, HN, CH_2COOH (azetidine ring)

(Starting with each of these, give possible structures for N and P.)

(b) Tropinone (N–CH_3, C=O) $\xrightarrow{\text{oxidn.}}$ Tropinic acid (N–CH_3, $-CH_2COOH$, $-COOH$)

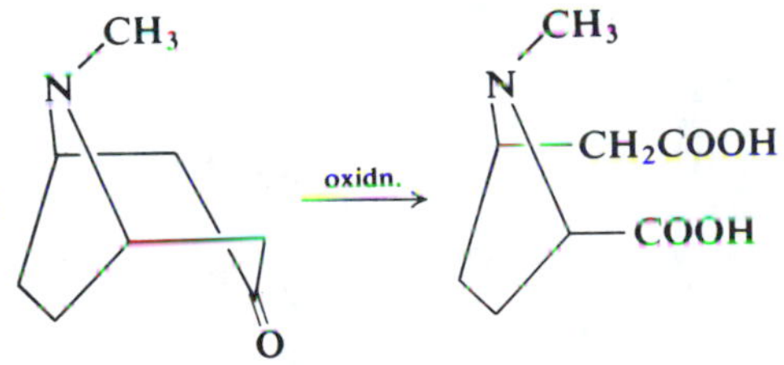

Tropinone   Tropinic acid

The various other structures are:

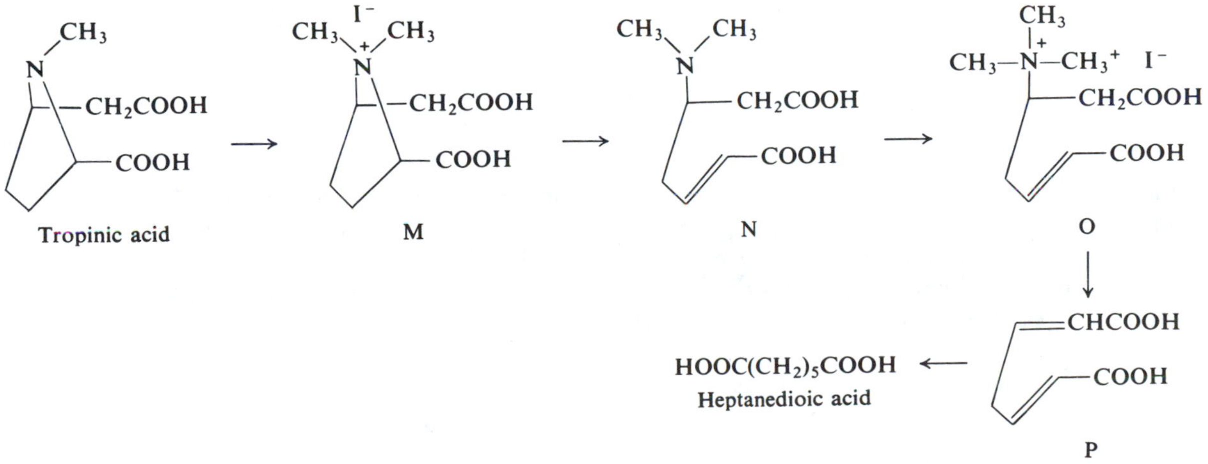

Tropinic acid (N–CH_3; $-CH_2COOH$; $-COOH$) ⟶ M (CH_3, CH_3, $\overset{+}{N}$, I^-; $-CH_2COOH$; $-COOH$) ⟶ N (CH_3, CH_3, N; $-CH_2COOH$; $-COOH$) ⟶ O ($CH_3-\overset{+}{N}-CH_3$, CH_3, I^-; $-CH_2COOH$; $-COOH$)

O ⟶ P ($=CHCOOH$; $-COOH$)

$HOOC(CH_2)_5COOH$ ⟵ P

Heptanedioic acid

12. NCH$_3$ =O $\xrightarrow{LiAlH_4}$ NCH$_3$ —OH $\xrightarrow{-H_2O}$ NCH$_3$ $\xrightarrow[\text{methyln.}]{\text{exhaust.}}$ N(CH$_3$)$_2$ $\xrightarrow[\text{methyln.}]{\text{exhaust.}}$ Tropilidene 1,3,5-Cycloheptatriene

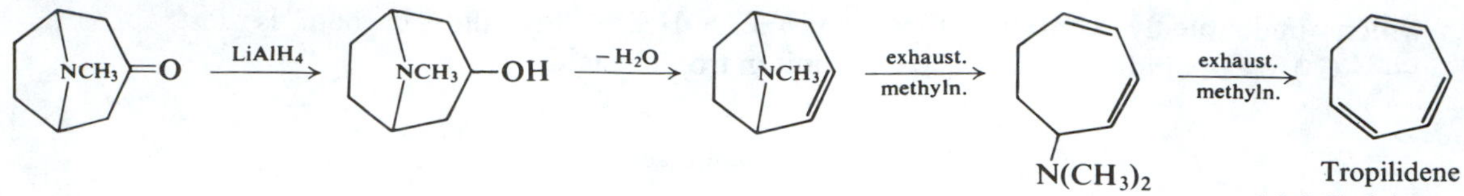

13. Tropine (N–CH$_3$; HO, H) $\xrightarrow{\text{heat}}$ Pseudotropine (N–CH$_3$; H, OH)

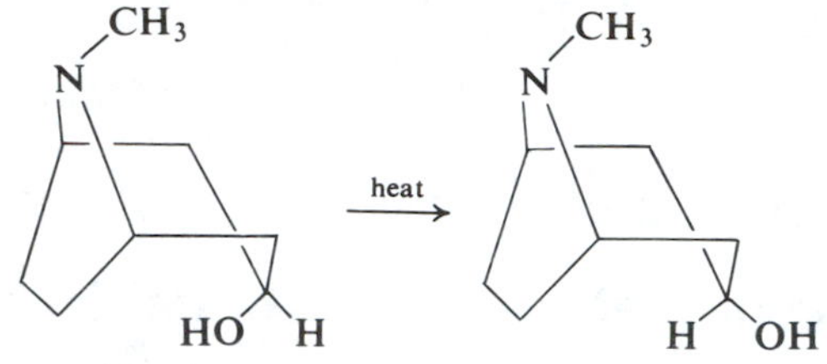

Tropine Pseudotropine

Pseudotropine is the more stable, and hence is probably the one with the equatorial —OH.

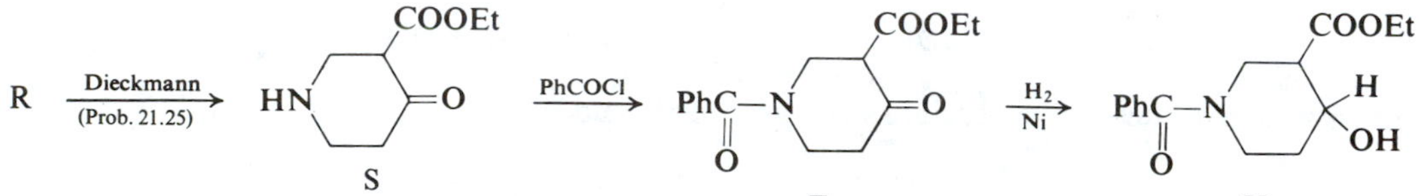

14. (a) $CH_2{=}CHCOOEt \xrightarrow{NH_3}$ CH_2CH_2COOEt (NH$_2$) Q $\xrightarrow{CH_2{=}CHCOOEt}$ HN(CH$_2$CH$_2$COOEt)$_2$ R

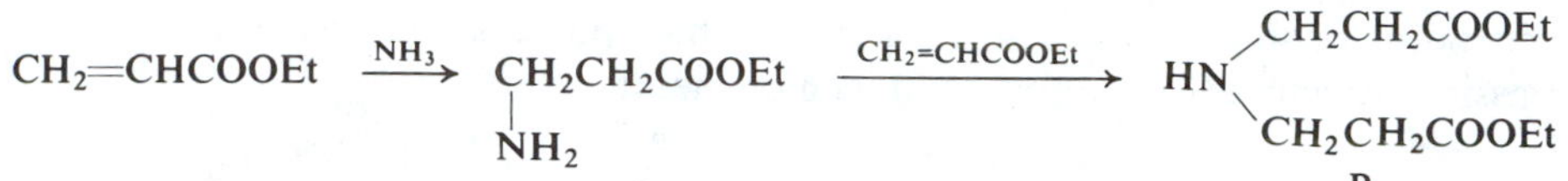

R $\xrightarrow[\text{(Prob. 21.25)}]{\text{Dieckmann}}$ HN ring (COOEt, =O) S $\xrightarrow{PhCOCl}$ PhC(=O)—N ring (COOEt, =O) T $\xrightarrow[Ni]{H_2}$ PhC(=O)—N ring (COOEt, H, OH) U

U $\xrightarrow[H_2O]{H^+}$ HN ring (COOH, H, OH) $\xrightarrow{-H_2O}$ HN ring (COOH) V Guvacine $\xrightarrow{MeI}$ CH$_3$N ring (COOH) Arecaidine

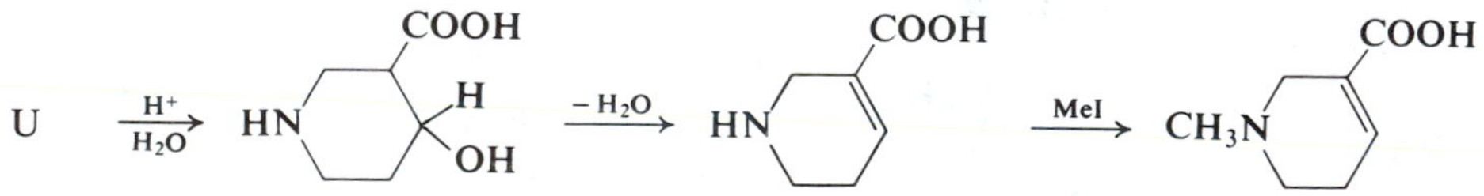

V

Guvacine

Arecaidine

(b) Guvacine $\xrightarrow{-4H}$ N COOH

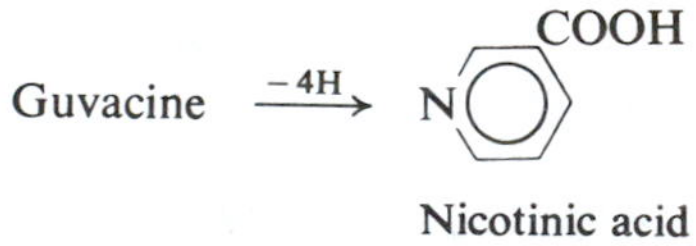

Nicotinic acid

15. Phenol (Sec. 31.7), furan (Problem 30.4, page 1063), and pyrrole (Problem 5, page 1074) contain rings reactive enough to react with formaldehyde: first, with the introduction of —CH$_2$OH; and then, with the connecting of rings by —CH$_2$— linkages.

Here we have another reactive aromatic ring, and here we have, if not formaldehyde, another carbonyl compound. Chemical arithmetic,

$C_8H_8S_2$	2 thiophene	$C_{14}H_{20}OS_2$	
$+C_6H_{12}O$	3-hexanone	$-C_{14}H_{18}S_2$	W
$C_{14}H_{20}OS_2$		H_2O missing	

indicates that two moles of thiophene combine with one of ketone and lose one H_2O. We postulate, therefore:

Thiophene + Et—C(=O)—Pr $\xrightarrow{H^+}$ 2-thienyl—C(Et)(OH)—Pr $\xrightarrow[H^+]{\text{thiophene}}$ (2-thienyl)—C(Et)(Pr)—(2-thienyl)
W

Next, Friedel–Crafts acylation at the usual 2-position, followed by Wolff–Kishner reduction (Sec. 18.9):

W $\xrightarrow[H^+]{Ac_2O}$ (2-thienyl)—C(Et)(Pr)—(5-acetyl-2-thienyl) (X, CCH_3, O) $\xrightarrow{N_2H_4,\ OH^-}$ (2-thienyl)—C(Et)(Pr)—(5-ethyl-2-thienyl) (Y)

We have not previously encountered an amide as an acylating agent (although, of course, an amide acylates water, in hydrolysis), but the hint plus an atom count leaves us with only a decision about the position of formylation, and we pick the only 2-position left open. Oxidation with Tollens' reagent gives AA; the acidic handle permits resolution (see Problem 20.8, page 766).

Y $\xrightarrow{\text{PhN(Me)CHO}}$ OHC—(thienyl)—C(Et)(Pr)—(thienyl)—Et (Z) $\xrightarrow{Ag_2O}$ HOOC—(thienyl)—C(Et)(Pr)—(thienyl)—Et (AA)

AA
Resolved into enantiomers

Following decarboxylation there is, finally, hydrogenation, which removes sulfur, adds hydrogen, and clearly opens the rings. We are left with a saturated open-chain alkane (C_nH_{2n+2}).

(+)-AA $\xrightarrow{-CO_2}$ (2-thienyl)—C(Et)(Pr)—(5-ethyl-2-thienyl) ((+)-BB) $\xrightarrow[Ni]{H_2}$ $CH_3CH_2CH_2CH_2CH_2CH_2—C(CH_2CH_3)(CH_2CH_2CH_3)—CH_2CH_2CH_2CH_3$

(+)-BB

CC
Ethyl-*n*-propyl-*n*-butyl-*n*-hexylmethane
Optically inactive

CC is a single enantiomer of a chiral compound and, by all rights, should be optically active. But, as we learned earlier (Sec. 4.13), chirality does not always lead to *measurable* optical activity. This particular compound, it has been estimated, should have the undetectable specific rotation of only 0.00001°.

(Wynberg, H., *et al.* "The Optical Activity of Butylethylhexylpropylmethane"; *J. Am. Chem. Soc.* **1965**, *87*, 2635.)

In the last step of this synthesis, the heterocyclic rings are opened and even the hetero-atom, sulfur, is removed. Yet the presence of these rings was vital to each preceding step, providing the regioselectivity necessary for the building of the molecule that was wanted. (For an overview of the potential of the heterocyclic field, see Katritzky, A. R. "Heterocycles"; *Chem. Eng. News* **April 13, 1970**, 80.)

16. (a) 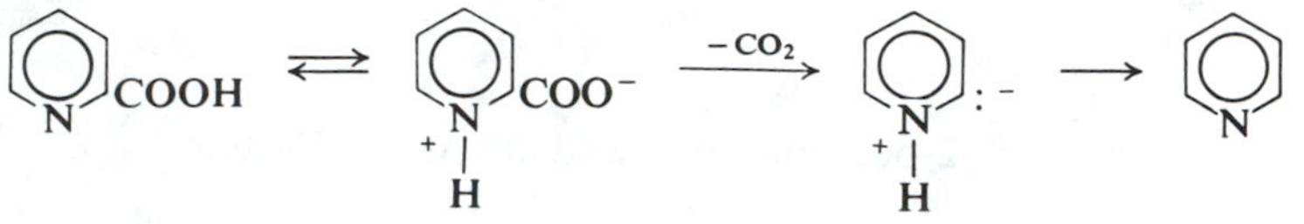 [2-pyridinecarboxylic acid] ⇄ V $\xrightarrow{-CO_2}$ VI → pyridine

V

VI

The dipolar ion (V) loses CO_2 to form a carbanion (VI) stabilized by the positive charge on nitrogen. Either acid or base decreases the concentration of dipolar ion, and slows down reaction. The *N*-methyl derivative necessarily exists entirely as dipolar ion, and hence reacts faster.

Like other carbanions, VI can add to a carbonyl group, and in the presence of the ketone is (partially) trapped.

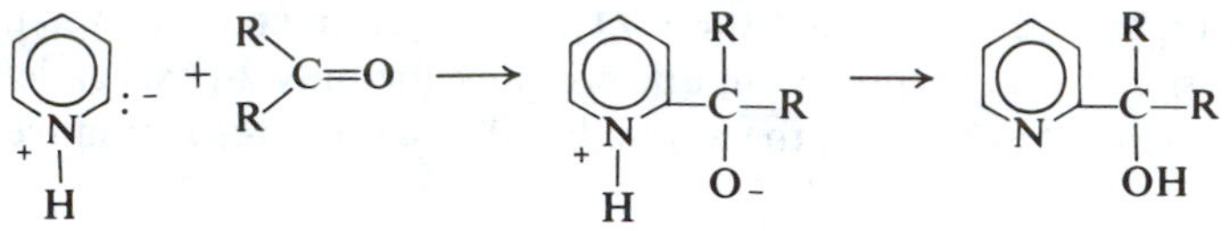

VI + $R_2C{=}O$ → → II

VI

II

(b) Carbanion VI and its isomers are stabilized by the inductive effect of the positive charge on nitrogen, which becomes weaker with distance. The 2- and 4-pyridineacetic acids give benzylic-like carbanions (VII) stabilized by resonance involving structures like VIII. Such resonance is not possible for the "*meta*" 3-pyridineacetic acid.

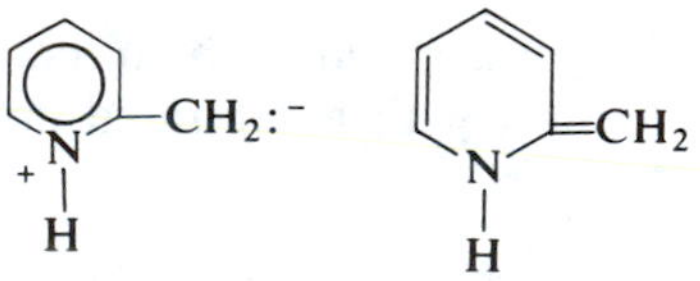

VII VIII

31

Macromolecules. Polymers and Polymerization

31.1 (a) An amide (see Sec. 31.7). $HOOC(CH_2)_4COOH + H_2N(CH_2)_6NH_2$
Adipic acid   Hexamethylenediamine

(b) An amide. $HOOC(CH_2)_5NH_2$ (The monomer is actually caprolactam, Problem 12, page 1098.)
ε-Aminocaproic acid

(c) An ether. $H_2C\!-\!CH_2$ (bridged by O), ethylene oxide

(d) A chloroalkene. $CH_2{=}C(Cl){-}CH{=}CH_2$, chloroprene

(e) A chloroalkane. $CH_2{=}C(Cl){-}Cl$, vinylidene chloride

31.2 (a) An amide. $^{+}H_3NCHRCOO^{-}$, an amino acid

(b) An ester. H_3PO_4 +
Phosphoric acid

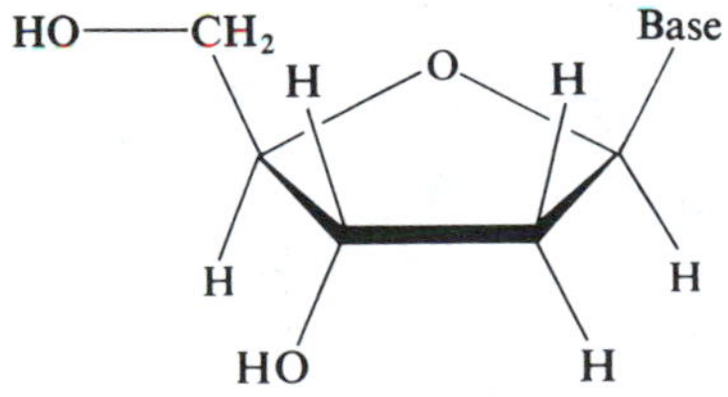

HO—CH₂ furanose ring with Base (2-deoxy: H, H at C-2; HO at C-3) *or* HO—CH₂ furanose ring with Base (HO at C-3, OH at C-2)

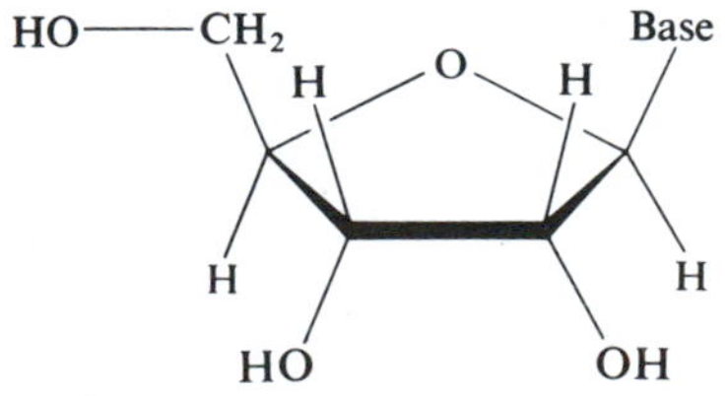

A sugar

(c) An acetal.

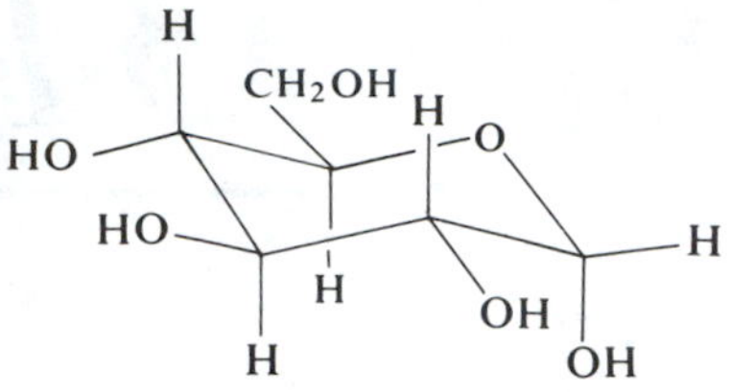

α-D-Glucopyranose

(d) An acetal.

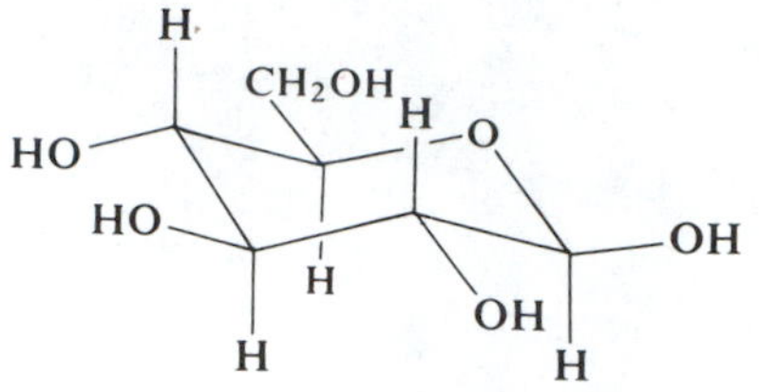

β-D-Glucopyranose

31.3 Rad· + $CH_2{=}CH{-}CH{=}CH_2$ ⟶ Rad—CH_2—CH⋯CH⋯CH_2 (I)

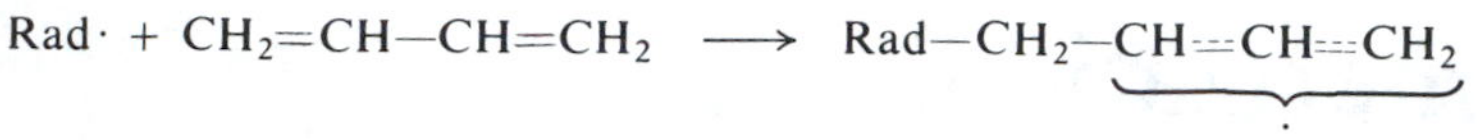

I + $CH_2{=}CH{-}CH{=}CH_2$ —

1,4-addn. ⟶ Rad—CH_2—CH═CH—CH_2—CH_2—CH⋯CH⋯CH_2 ⟶ *etc.*

1,2-addn. ⟶ Rad—CH_2—CH($CH{=}CH_2$)—CH_2—CH⋯CH⋯CH_2 ⟶ *etc.*

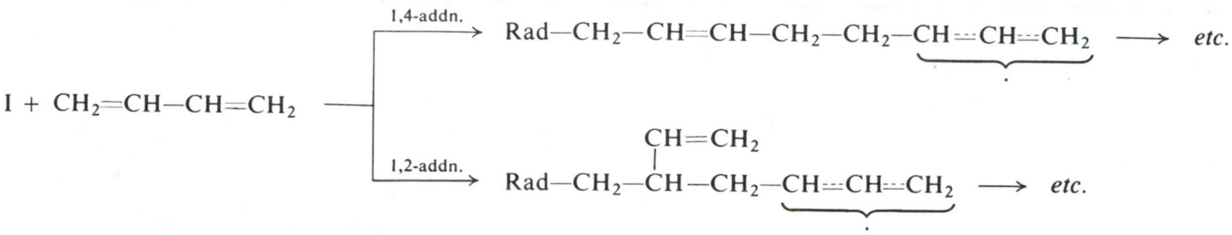

31.4 Combination of two growing free radicals.

31.5 The effectiveness of a chain-transfer agent depends upon the ease of abstraction of an atom from it; this depends on, among other things, the stability of the radical being formed.

(a) H: aryl, 1°, 1° benzylic, 2° benzylic, 3° benzylic.

(b) H: 2°, allylic.

(c) X: C—Br is weaker than C—Cl.

31.6 (a) The chain-transfer agent can be another molecule of polymer. For example:

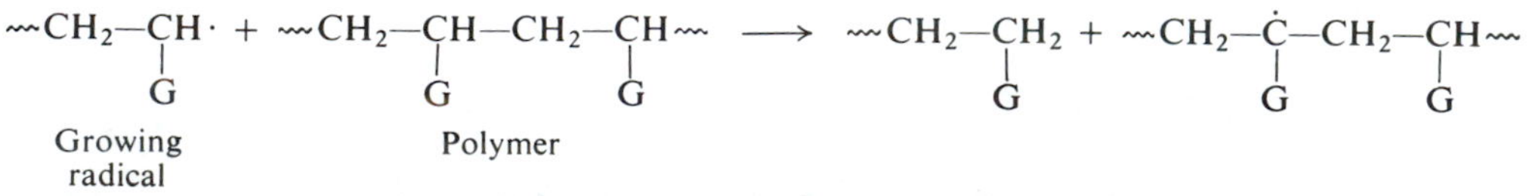

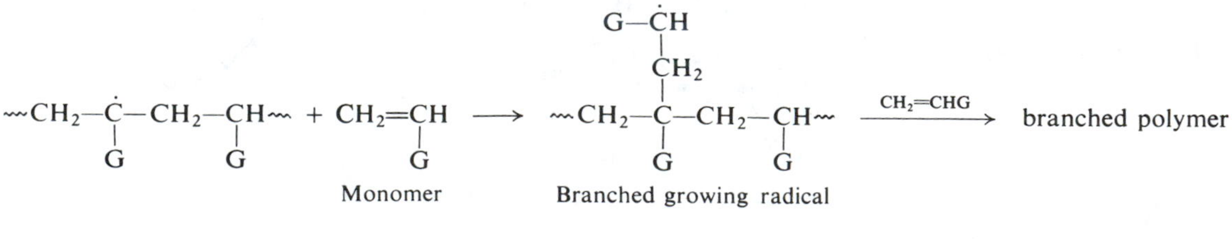

~CH_2—CH(G)· + ~CH_2—CH(G)—CH_2—CH(G)~ ⟶ ~CH_2—CH_2(G) + ~CH_2—Ċ(G)—CH_2—CH(G)~

Growing radical; Polymer

~CH_2—Ċ(G)—CH_2—CH(G)~ + $CH_2{=}CH$(G) ⟶ ~CH_2—C(CH_2—ĊH—G)(G)—CH_2—CH(G)~ $\xrightarrow{CH_2=CHG}$ branched polymer

Monomer; Branched growing radical

(b) The chain-transfer agent can be the growing polymer molecule itself: the growing end abstracts ("bites") hydrogen from a position four or five carbons back along the chain. For example:

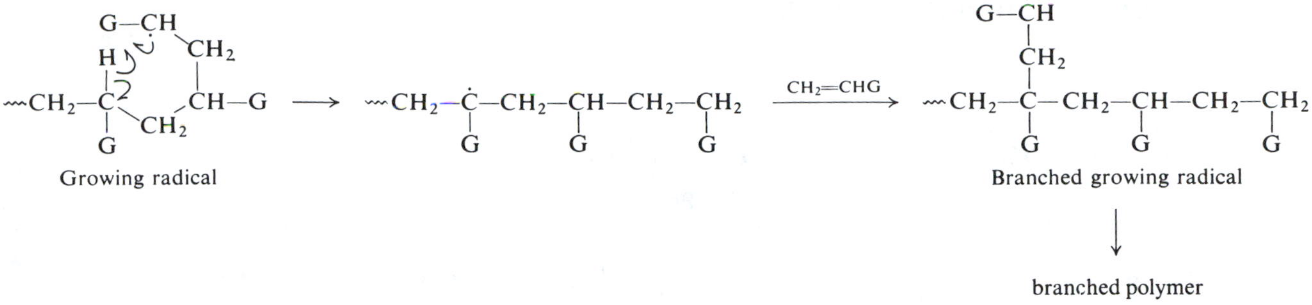

$$\text{~~CH}_2\text{—C(H)(G)—[cyclic: CH}_2\text{—CH(G)—CH}_2\text{—}\dot{\text{C}}\text{H—G]} \longrightarrow \text{~~CH}_2\text{—}\dot{\text{C}}\text{(G)—CH}_2\text{—CH(G)—CH}_2\text{—CH}_2\text{(G)} \xrightarrow{CH_2=CHG} \text{~~CH}_2\text{—C(G)(CH}_2\text{—}\dot{\text{C}}\text{H—G)—CH}_2\text{—CH(G)—CH}_2\text{—CH}_2\text{(G)}$$

Growing radical — Branched growing radical

↓

branched polymer

31.7 (a) The polybutadiene still contains double bonds and hence easily abstracted allylic hydrogens. The allylic free radicals thus formed add to styrene to start a reaction sequence that grafts polystyrene onto the polybutadiene chain. The process resembles the branching in Problem 31.6(a).

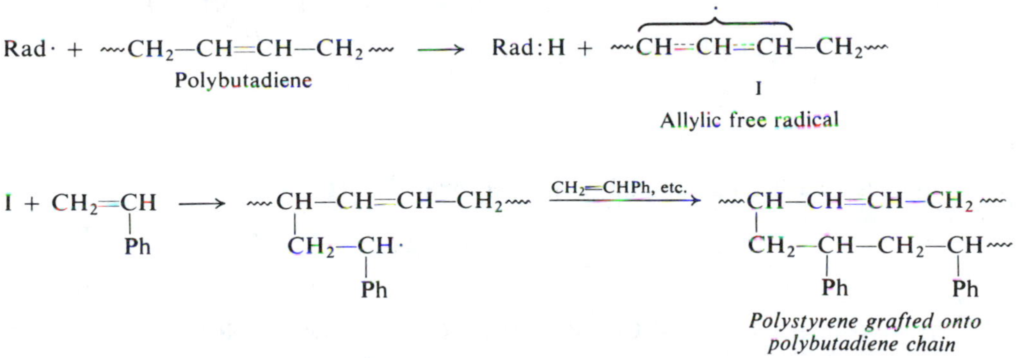

$$\text{Rad}\cdot + \text{~~CH}_2\text{—CH=CH—CH}_2\text{~~} \longrightarrow \text{Rad:H} + \text{~~}\overbrace{\text{CH}\text{=-CH=-CH}}^{\cdot}\text{—CH}_2\text{~~}$$

Polybutadiene; I, Allylic free radical

$$\text{I} + \text{CH}_2\text{=CH(Ph)} \longrightarrow \text{~~CH(CH}_2\text{—}\dot{\text{C}}\text{H—Ph)—CH=CH—CH}_2\text{~~} \xrightarrow{CH_2=CHPh,\ etc.} \text{~~CH(CH}_2\text{—CH(Ph)—CH}_2\text{—CH(Ph)~~)—CH=CH—CH}_2\text{~~}$$

Polystyrene grafted onto polybutadiene chain

(b) Similar to part (a), with hydrogen being abstracted from the carbon carrying —Cl.

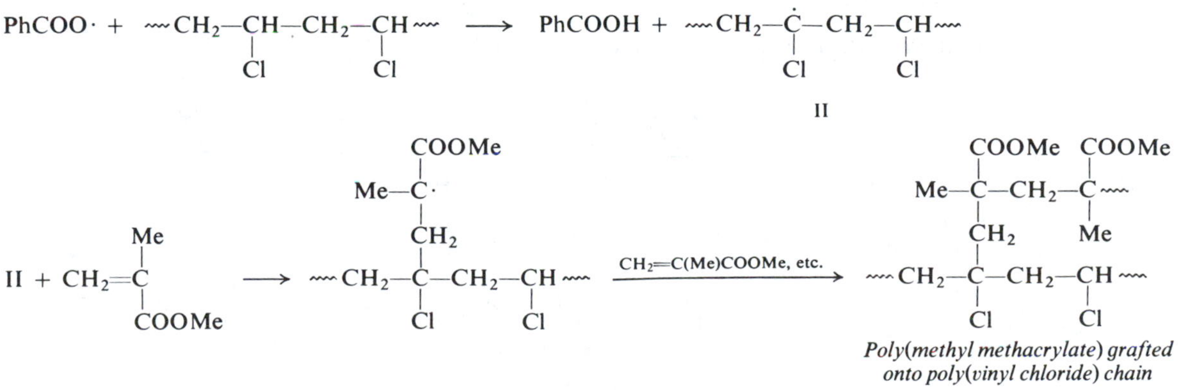

$$\text{PhCOO}\cdot + \text{~~CH}_2\text{—CH(Cl)—CH}_2\text{—CH(Cl)~~} \longrightarrow \text{PhCOOH} + \text{~~CH}_2\text{—}\dot{\text{C}}\text{(Cl)—CH}_2\text{—CH(Cl)~~}$$

II

$$\text{II} + \text{CH}_2\text{=C(Me)COOMe} \longrightarrow \text{~~CH}_2\text{—C(Cl)(CH}_2\text{—}\dot{\text{C}}\text{(Me)COOMe)—CH}_2\text{—CH(Cl)~~} \xrightarrow{CH_2=C(Me)COOMe,\ etc.} \text{~~CH}_2\text{—C(Cl)(CH}_2\text{—C(Me)(COOMe)—CH}_2\text{—C(Me)(COOMe)~~)—CH}_2\text{—CH(Cl)~~}$$

Poly(methyl methacrylate) grafted onto poly(vinyl chloride) chain

31.8 (a) Chain transfer is involved.

$$CH_3O(CH_2CH_2O)_nCH_2CH_2O^- + CH_3OH \longrightarrow CH_3O(CH_2CH_2O)_nCH_2CH_2OH + CH_3O^-$$

(b) $CH_3OCH_2CH_2OH$, 2-methoxyethanol.

31.9 (a) $\sim CH_2{-}CH(Ph)\!:^- + H_2O \longrightarrow \sim CH_2{-}CH_2(Ph) + OH^-$

(b) $\sim CH_2{-}CH(Ph)\!:^- + CO_2 \longrightarrow \sim CH_2{-}CH(Ph){-}COO^- \xrightarrow{H_2O} \sim CH_2CH(Ph)COOH + OH^-$

(c) $\sim CH_2{-}CH(Ph)\!:^- + H_2C{-}CH_2$ (epoxide, O) $\longrightarrow \sim CH_2CH(Ph)CH_2CH_2O^- \xrightarrow{H_2O} \sim CH_2CH(Ph)CH_2CH_2OH + OH^-$

(d) $\sim CH_2{-}CH(Ph)CH_2CH_2O^-$ (from part c) $+ n\text{-}H_2C{-}CH_2$ (epoxide, O) $\longrightarrow \sim CH_2CH(Ph)(CH_2CH_2O)_nCH_2CH_2O^-$

$\downarrow H_2O$

$\sim CH_2CH(Ph)(CH_2CH_2O)_nCH_2CH_2OH + OH^-$

31.10 (a) A primary —OH is esterified more rapidly than a secondary —OH.

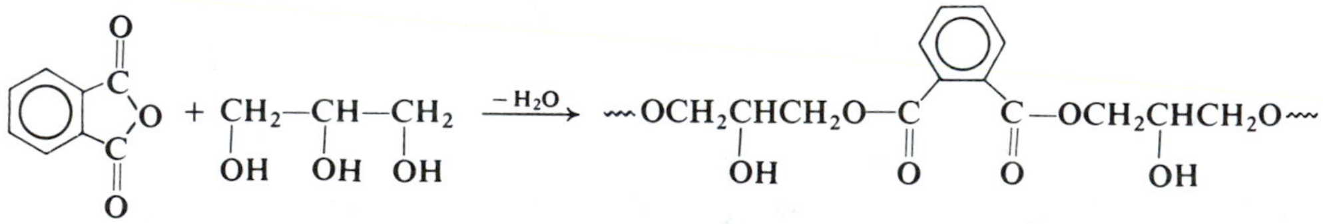

Phthalic anhydride $+ CH_2(OH){-}CH(OH){-}CH_2(OH) \xrightarrow{-H_2O} \sim OCH_2CH(OH)CH_2O{-}C(=O){-}C_6H_4{-}C(=O){-}OCH_2CH(OH)CH_2O\sim$

(b) Cross-linking results from esterification of the secondary —OH groups.

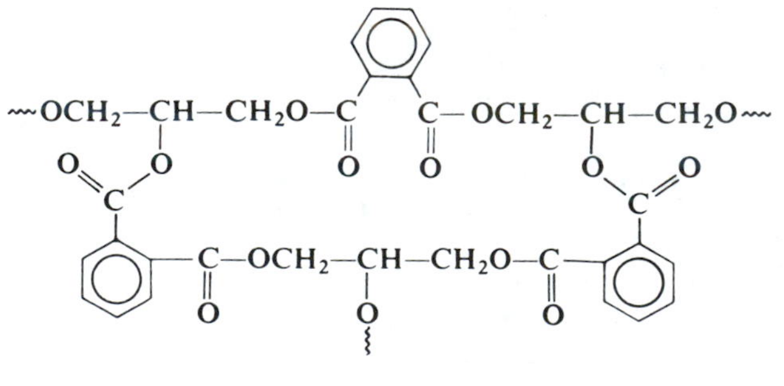

$\sim OCH_2{-}CH{-}CH_2O{-}C(=O){-}C_6H_4{-}C(=O){-}OCH_2{-}CH{-}CH_2O\sim$, with each CH bearing $O{-}C(=O){-}C_6H_4{-}C(=O){-}OCH_2{-}CH(O\sim){-}CH_2O{-}C(=O){-}C_6H_4{-}C(=O){-}O$ cross-links

31.11 $HOCH_2CH_2OH$ + O=C=N–(C$_6$H$_3$)(CH$_3$)–N=C=O + $HOCH_2CH_2OH$

Diol *Diisocyanate* *Diol*

↓

~OCH$_2$CH$_2$O–C(=O)–N(H)–(C$_6$H$_3$)(CH$_3$)–N(H)–C(=O)–OCH$_2$CH$_2$O~

Polyurethane

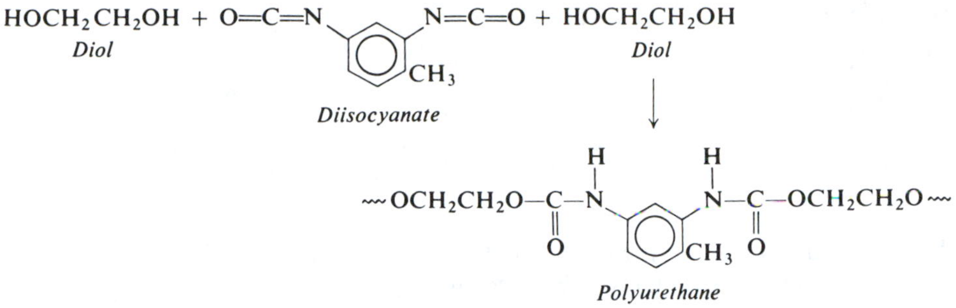

31.12 (a) Hydrolysis of an amide: hot aqueous acid or base.

(b) Hydrolysis of an ester: hot aqueous acid or base.

(c) Hydrolysis of an acetal: aqueous acid.

(d) Hydrolysis of an acetal: aqueous acid.

(Or, in each case, an enzyme-catalyzed hydrolysis.)

31.13 (a) Transesterification.

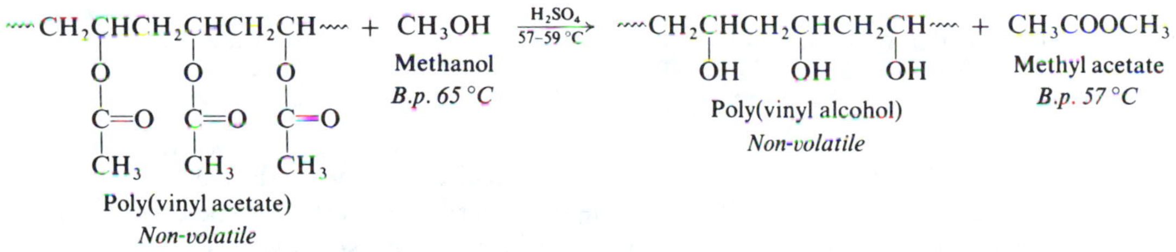

~CH$_2$CH(OCOCH$_3$)CH$_2$CH(OCOCH$_3$)CH$_2$CH(OCOCH$_3$)~ + CH_3OH $\xrightarrow[57-59\,°C]{H_2SO_4}$ ~CH$_2$CH(OH)CH$_2$CH(OH)CH$_2$CH(OH)~ + CH_3COOCH_3

Poly(vinyl acetate) *Non-volatile*; Methanol *B.p. 65 °C*; Poly(vinyl alcohol) *Non-volatile*; Methyl acetate *B.p. 57 °C*

The hypothetical monomer of the new polymer is vinyl alcohol, which exists in the keto form, acetaldehyde.

(b) Formation of a cyclic acetal.

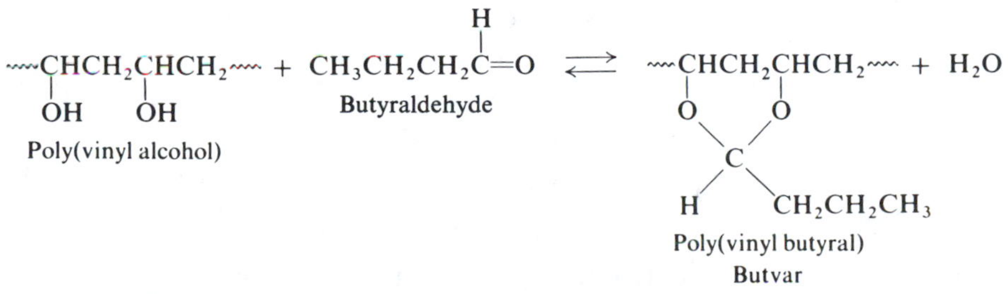

~CH(OH)CH$_2$CH(OH)CH$_2$~ + $CH_3CH_2CH_2CH{=}O$ ⇌ ~CHCH$_2$CHCH$_2$~ (O–C(H)(CH$_2$CH$_2$CH$_3$)–O ring) + H_2O

Poly(vinyl alcohol); Butyraldehyde; Poly(vinyl butyral) Butvar

31.14

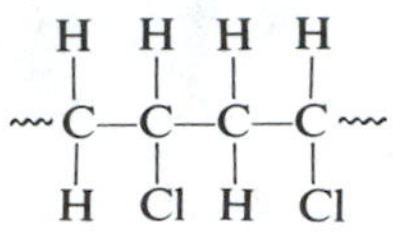

$$\begin{array}{c} \text{H}\quad\text{H}\quad\text{H}\quad\text{H} \\ | \quad\ \ | \quad\ \ | \quad\ \ | \\ \sim\text{C}-\text{C}-\text{C}-\text{C}\sim \\ | \quad\ \ | \quad\ \ | \quad\ \ | \\ \text{H}\quad\text{Cl}\quad\text{H}\quad\text{Cl} \end{array} \qquad\qquad \begin{array}{c} \text{H}\quad\text{Cl}\quad\text{H}\quad\text{Cl} \\ | \quad\ \ | \quad\ \ | \quad\ \ | \\ \sim\text{C}-\text{C}-\text{C}-\text{C}\sim \\ | \quad\ \ | \quad\ \ | \quad\ \ | \\ \text{H}\quad\text{Cl}\quad\text{H}\quad\text{Cl} \end{array}$$

Poly(vinyl chloride) PVC — Poly(vinylidene chloride) Saran

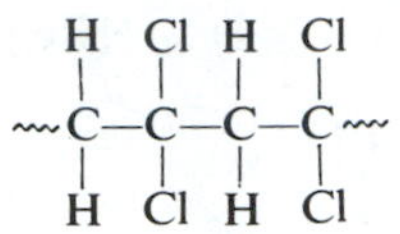

Poly(vinyl chloride) can show the same stereoisomerism as polypropylene (Figure 31.1, page 1089, with —Cl in place of $—CH_3$). Formed by a free-radical process, it is atactic; the molecules fit together poorly.

Poly(vinylidene chloride) has two identical substituents on each carbon, and the chains fit together well.

31.15 (a) The chains are irregularly substituted, and fit together poorly; the intermolecular forces are weak. (The diene provides double bonds, and hence allylic hydrogen, in the polymer; vulcanization causes cross-linking as with rubber, page 420.)

(b) Abstraction of —H from the polymer generates free radicals, which combine. The cross-links here are carbon–carbon bonds.

1. The more stable particle is formed in each step, with the same orientation.

2. There is acid-catalyzed polymerization of the alkene that is easily formed from a 2° or 3° alcohol.

3. Each carbon of polyisobutylene carries two identical substituents (two —H's or two $—CH_3$'s); this symmetry prevents the existence of stereoisomeric polymers. On the other hand, 1-butene should form exactly the same kind of isomeric polymers as propylene (Figure 31.1, page 1089, with $—C_2H_5$ in place of $—CH_3$).

4. Nucleophilic carbonyl addition; anionic chain-reaction polymerization.

(1) $$CH_3O^- + \begin{array}{l}\text{H}\diagdown \\ \quad\ \text{C=O} \\ \text{H}\diagup\end{array} \longrightarrow CH_3O-\begin{array}{c}\text{H}\\|\\\text{C}\\|\\\text{H}\end{array}-O^-$$

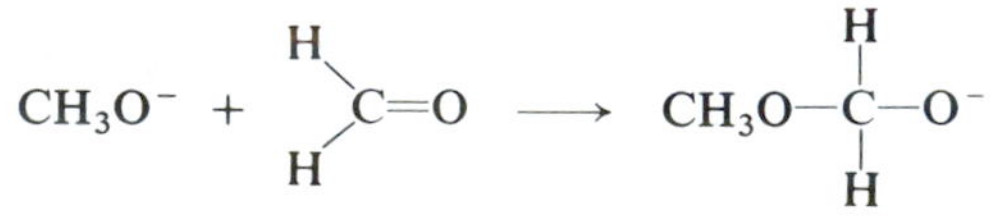

(2) $$CH_3O-\begin{array}{c}\text{H}\\|\\\text{C}\\|\\\text{H}\end{array}-O^- + \begin{array}{l}\text{H}\diagdown \\ \quad\ \text{C=O} \\ \text{H}\diagup\end{array} \longrightarrow CH_3O-\begin{array}{c}\text{H}\\|\\\text{C}\\|\\\text{H}\end{array}-O-\begin{array}{c}\text{H}\\|\\\text{C}\\|\\\text{H}\end{array}-O^-$$

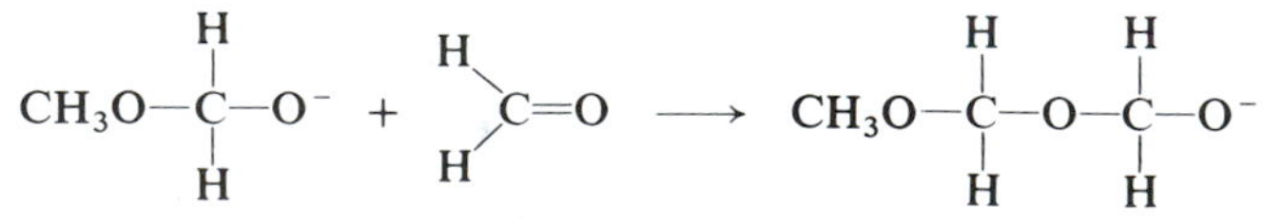

then steps like (2), *to give* $CH_3O(CH_2—O)_nCH_2OH$

5. Polyurethanes contain ester and amide linkages. Hydrolysis—with water in sealed containers at 200 °C—gives diols, diamines, and CO_2. For example:

$$\sim OCH_2CH_2O-\overset{\overset{\displaystyle O}{\|}}{C}-\overset{H}{\overset{|}{N}}-C_6H_3(CH_3)-\overset{H}{\overset{|}{N}}-\overset{\overset{\displaystyle O}{\|}}{C}-OCH_2CH_2O\sim \xrightarrow[\text{heat}]{H_2O} HOCH_2CH_2OH + CO_2 + H_2N-C_6H_3(CH_3)-NH_2$$

Polyurethane — Ethylene glycol — 2,4-Diaminotoluene

6.

$$C_6H_5CH_2OH \xrightarrow[-H_2O]{H^+} C_6H_5CH_2^{\oplus} \xrightarrow{PhCH_2OH} C_6H_5CH_2C_6H_4CH_2OH$$

Benzyl cation

$$C_6H_5CH_2C_6H_4CH_2OH \xrightarrow[-H_2O]{H^+} C_6H_5CH_2C_6H_4CH_2^{\oplus} \xrightarrow{\text{many repetitions}} \sim C_6H_4CH_2C_6H_4CH_2C_6H_4CH_2\sim$$

Polymer

7. Let us examine the mechanism for this reaction given on page 355.

(1) $\text{peroxide} \longrightarrow \text{Rad}\cdot$

(2) $\text{Rad}\cdot + Cl{:}CCl_3 \longrightarrow \text{Rad}{:}Cl + \cdot CCl_3$

(3) $\cdot CCl_3 + RCH{=}CH_2 \longrightarrow R\dot{C}H-CH_2-CCl_3$

(4) $R\dot{C}H-CH_2-CCl_3 + Cl{:}CCl_3 \longrightarrow RCH(Cl)-CH_2-CCl_3 + \cdot CCl_3$

then (3), (4), (3), (4), etc.

(a) The free radicals produced in step (3) are shown as abstracting an atom from CCl_4 in step (4). They can, instead, *add to the alkene* (5) to form a new, bigger radical which then attacks (6) CCl_4 to yield the 2:1 adduct.

(5) $RCH{=}CH_2 + R\dot{C}H-CH_2-CCl_3 \longrightarrow R\dot{C}H-CH_2-CH(R)-CH_2-CCl_3$

(6) $R\dot{C}H-CH_2-CH(R)-CH_2-CCl_3 + Cl{:}CCl_4 \longrightarrow RCH(Cl)-CH_2-CH(R)-CH_2-CCl_3 + \cdot CCl_3$

(b) Two reagents are competing for the organic free radical: CX_4 in reaction (4), and $RCH{=}CH_2$ in reaction (5). CBr_4 is more reactive than CCl_4—it contains weaker C—X bonds—and, other things being equal, abstraction (4) is favored.

$$RC\dot{H}CH_2CX_3 \xrightarrow[(4)]{CX_4} RCH(X)CH_2CX_3 \quad \textbf{Abstraction}$$

$$RC\dot{H}CH_2CX_3 \xrightarrow[(5)]{RCH=CH_2} RC\dot{H}CH_2CH(R)CH_2CX_3 \quad \textbf{Addition}$$

(c) Again there is competition. Styrene is a more reactive alkene than 1-octene and, other things being equal, addition (5) is favored—in this case, to such an extent that only polymerization is observed. (There is an additional factor. The intermediate benzylic radical formed from styrene is relatively unreactive and therefore selective (Secs. 2.24 and 3.28). In choosing between abstraction and the easier addition, it tends to choose addition.)

In these examples of free-radical addition we see a type intermediate between the simple 1:1 addition at one end of the reaction spectrum, and straightforward polymerization at the other end.

8. (a) An anionic chain-reaction polymerization to give I, a polymer with terminal —OH groups; then a step-reaction polymerization with the diisocyanate to give the block copolymer.

$$\underset{\text{(epoxide, O bridging)}}{CH_2{-}CH_2} + \underset{\textit{Limited amount}}{OH^-} \longrightarrow \underset{\text{I}}{HOCH_2CH_2(OCH_2CH_2)_nOCH_2CH_2OH}$$

$$\text{I} + OCN{-}C_6H_3(NCO){-}CH_3 \longrightarrow \text{polymer}$$

(b) A step-reaction polymerization (esterification) to give the unsaturated linear polyester II; then chain-reaction vinyl copolymerization with styrene to provide cross-links.

$$HOCH_2CH_2OH + \underset{\text{(maleic anhydride: OC–O–CO ring)}}{CH{=}CH} \longrightarrow \underset{\text{II}}{\sim OCH_2CH_2O{-}CO{-}CH{=}CH{-}CO{-}OCH_2CH_2O\sim}$$

$$\text{II} + C_6H_5CH{=}CH_2 \xrightarrow[\text{initiator}]{\text{free-radical}} \text{polymer}$$

(c) An anionic chain-reaction polymerization. (Compare the reaction of ethylene oxide with methylamine, Problem 13.25(c), page 485.)

$$\underset{\text{(epoxide, O bridging)}}{CH_3CH{-}CH_2} + \underset{\textit{Limited amount}}{H_2NCH_2CH_2NH_2} \longrightarrow \text{polymer}$$

(d) An anionic chain-reaction polymerization to give living polystyrene, which is then killed with ethylene oxide to give a block copolymer.

$$PhCH{=}CH_2 \xrightarrow[\text{naphthalene}]{\text{Na}} {}^{-}\underset{Ph}{CH}CH_2(\underset{Ph}{CH}CH_2)_n(CH_2\underset{Ph}{CH})_nCH_2\underset{Ph}{CH}{}^{-}$$

III
Living polymer

$$\text{III} + \underset{\diagdown O \diagup}{CH_2{-}CH_2} \longrightarrow \text{polymer}$$

(e) A free-radical chain-reaction vinyl copolymerization to give the saturated linear polymer IV. Abstraction of hydrogen from IV by free radicals from benzoyl peroxide generates free-radical sites on IV; at these places, poly(methyl methacrylate) branches grow by free-radical chain-reaction polymerization.

$$CH_2{=}\underset{Cl}{CH} + CH_2{=}\underset{OAc}{CH} \xrightarrow[\text{initiator}]{\text{free-radical}} \sim CH_2\underset{OAc}{CH}CH_2\underset{Cl}{CH}CH_2\underset{OAc}{CH}CH_2\underset{Cl}{CH}CH_2\underset{Cl}{CH}CH_2\underset{OAc}{CH}\sim$$

IV

$$\text{IV} + CH_2{=}\overset{CH_3}{\underset{COOCH_3}{C}} \xrightarrow[\text{peroxide}]{\text{benzoyl}} \text{polymer}$$

9. The product is a polyester, formed by an anionic chain-reaction polymerization; each step involves nucleophilic substitution at the acyl group of the cyclic ester.

$$\text{Base:} \curvearrowright \underset{O}{\overset{CH_2-CH_2}{C{=}\!\!-\!\!O}} \longrightarrow \text{Base}{-}\underset{\|\atop O}{C}{-}CH_2CH_2{-}O^-$$

$$\text{Base}{-}\underset{\|\atop O}{C}{-}CH_2CH_2{-}O^- \curvearrowright \underset{O}{\overset{CH_2-CH_2}{C{-}O}} \longrightarrow \text{Base}{-}\underset{\|\atop O}{C}{-}CH_2CH_2{-}O{-}\underset{\|\atop O}{C}{-}CH_2CH_2{-}O^-, \textit{etc.}$$

10. $NH_2CH_2\underset{Ph}{CH}\sim CH_2\underset{Ph}{CH}{}^- + NH_3 \longrightarrow NH_2CH_2\underset{Ph}{CH}\sim CH_2\underset{Ph}{CH_2}$

11. Cleavage by HIO_4 indicates occasional vicinal diol groupings, —CH(OH)CH(OH)—, and hence vicinal diacetate groups, —CH(OAc)CH(OAc)—. These show that *some* head-to-head polymerization has taken place along with the predominant head-to-tail orientation.

12. (a) $\sim\!\!\text{N(H)(CH}_2)_5\text{C(=O)—N(H)(CH}_2)_5\text{C(=O)}\!\!\sim$

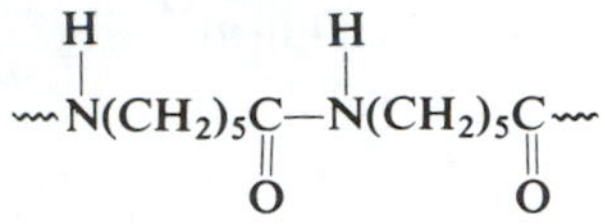

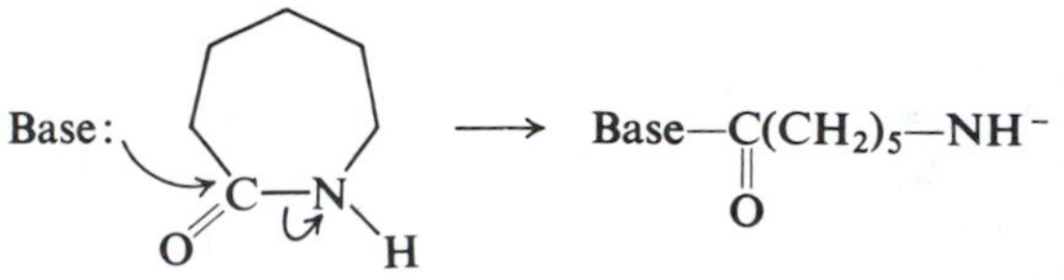

Nylon-6

(b) Base: + Caprolactam $\longrightarrow$ Base—$\text{C(=O)(CH}_2)_5$—NH^-

Caprolactam

Base—$\text{C(=O)(CH}_2)_5$—NH^- + Caprolactam $\longrightarrow$ Base—$\text{C(=O)(CH}_2)_5$—N(H)—$\text{C(=O)(CH}_2)_5\text{NH}^-$, *etc.*

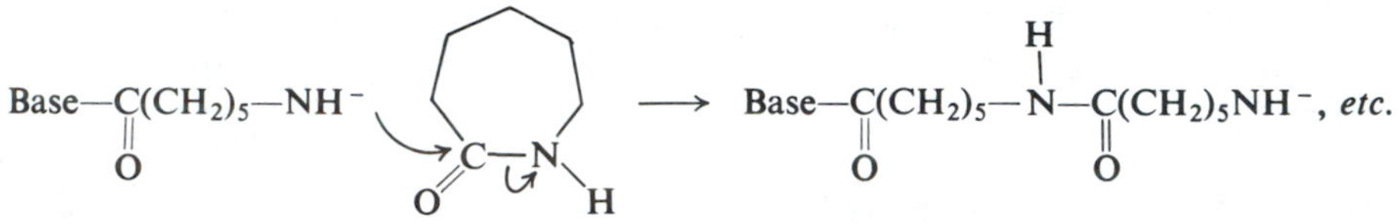

Caprolactam

The reaction is anionic chain-reaction polymerization, involving nucleophilic substitution at the acyl group of the cyclic amide. The base could be OH^- itself or the anion formed by abstraction of the —NH proton from a molecule of lactam.

13. Caprolactam $\xleftarrow{\text{acid}}$ Cyclohexanonoxime (=NOH) $\xleftarrow{NH_2OH}$ Cyclohexanone

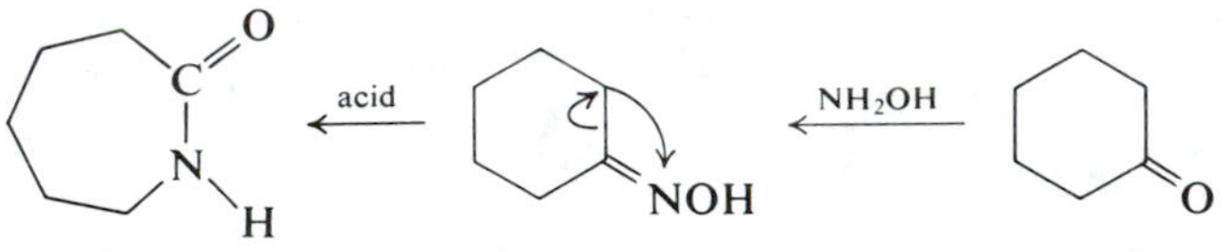

Caprolactam Cyclohexanonoxime Cyclohexanone

14. The *para* isomer gives straight, symmetrical chains that fit together well.

15. These compounds are ionic—or essentially so—due to the stability of benzylic anions.

16. (a) They are diastereomers: one is *meso* and the other is racemic.

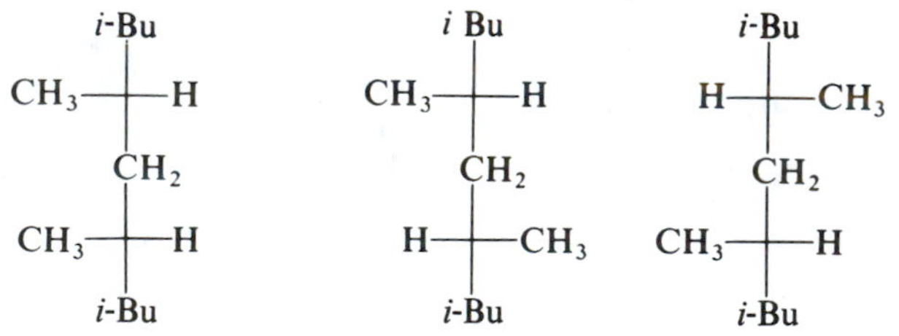

i-Bu	*i* Bu	*i*-Bu
CH_3—C—H	CH_3—C—H	H—C—CH_3
CH_2	CH_2	CH_2
CH_3—C—H	H—C—CH_3	CH_3—C—H
i-Bu	*i*-Bu	*i*-Bu
R,S	*R,R*	*S,S*

Meso A   Enantiomers B

(b) A is *meso*; B is chiral, and was obtained in racemic form in part (a).

(c) If we make drawings of A and B in the style of Figure 31.1 (page 1089), the answer leaps to our eye: A resembles the isotactic polymer and B resembles the syndiotactic polymer.

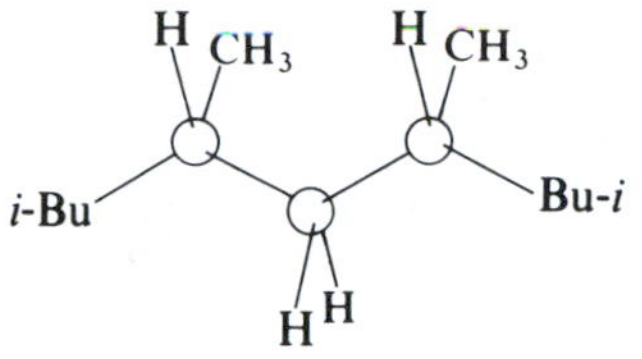

A

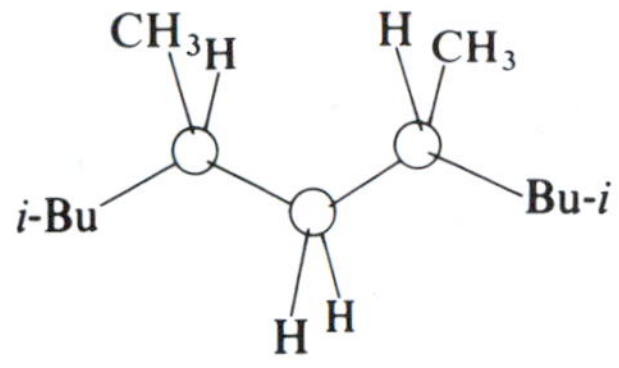

B (R,R enantiomer)

17. First step. Excess 1,2-ethanediol ensures terminal —OH's in C.

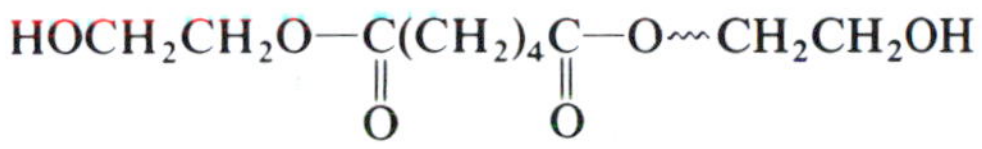

$$HOCH_2CH_2O{-}\underset{\displaystyle O}{\underset{\|}{C}}(CH_2)_4\underset{\displaystyle O}{\underset{\|}{C}}{-}O{\sim}CH_2CH_2OH$$

C

Second step. Excess diisocyanate ensures terminal —NCO's in D:

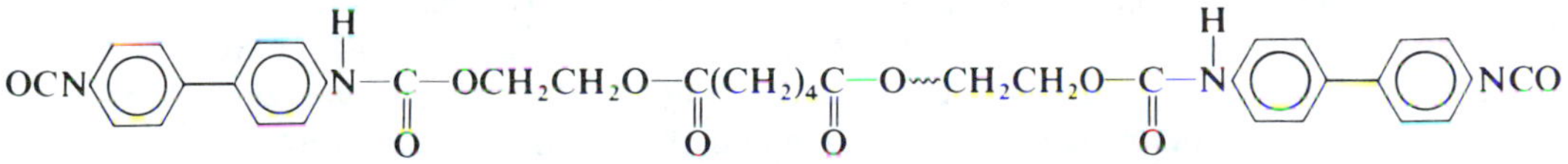

D

It also provides cross-linking through reaction at urethane (amide) —NH.

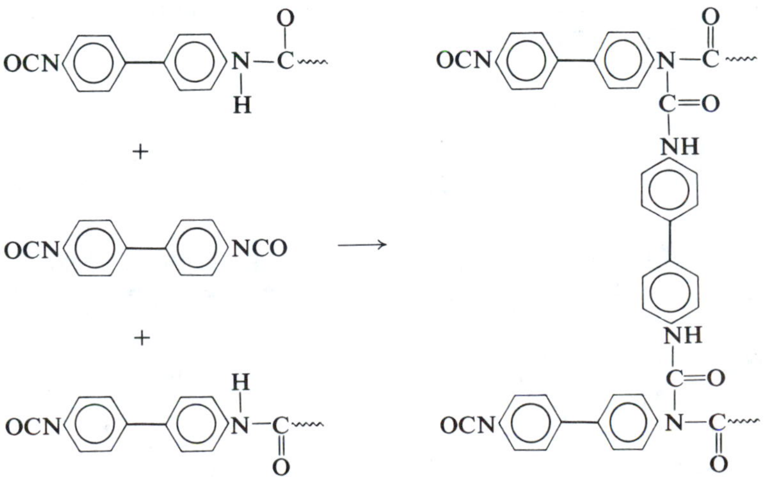

Third step. Water reacts with some of the —NCO groups in D to release CO_2 and provide —NH_2 groups.

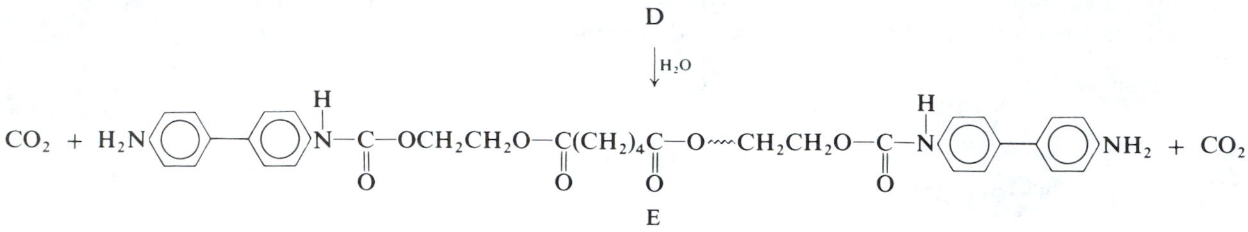

D

$\downarrow H_2O$

CO_2 + H_2N–C₆H₄–C₆H₄–NH–C(=O)–OCH_2CH_2O–C(=O)$(CH_2)_4$C(=O)–O〰CH_2CH_2O–C(=O)–NH–C₆H₄–C₆H₄–NH_2 + CO_2

E

The —NH_2 groups react with other free —NCO groups to give ureas (amides) with further cross-linking. The CO_2 is dispersed in the polymer to give a foam.

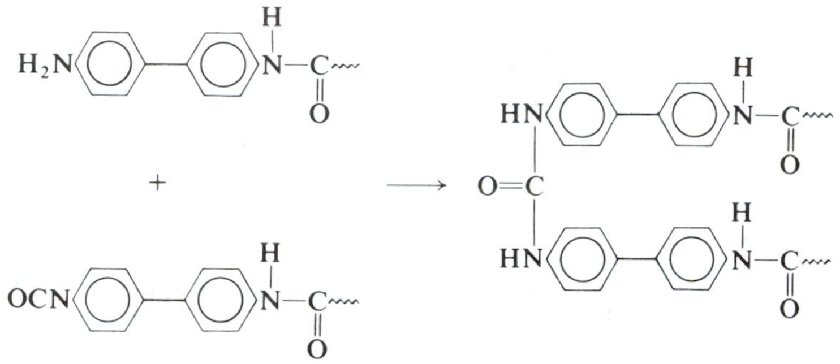

H_2N–C₆H₄–C₆H₄–NH–C(=O)〰 + OCN–C₆H₄–C₆H₄–NH–C(=O)〰 ⟶ O=C(–HN–C₆H₄–C₆H₄–NH–C(=O)〰)₂

18. The monomer (which can lose a hydrogen atom to give an allylic free radical) serves as a chain-transfer agent, and limits the size of the polymer. Abstraction of deuterium is more difficult, and the labeled monomer is a poorer chain-transfer agent. (Here again we see a free radical undergoing the competing reactions, addition and abstraction—this time at two sites in the *same* molecule.)

19. (a) The two reactive groups of the epichlorohydrin are (i) C—Cl, which reacts with phenoxide in a Williamson synthesis, and (ii) the epoxide group, which reacts with phenoxide as on page 911. The epoxide group is the less reactive; the excess of epichlorohydrin assures unreacted epoxide groups at both ends. The cement is:

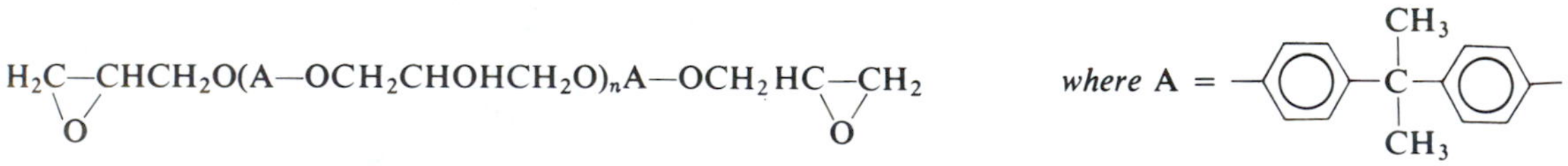

H_2C—$CHCH_2O(A$—$OCH_2CHOHCH_2O)_nA$—OCH_2HC—CH_2 (each end an epoxide ring, O bridging the two carbons)

where A = —C₆H₄—C(CH_3)₂—C₆H₄—

(b) During hardening, the amine reacts—at three sites in the molecule—with the epoxide groups in the cement, to form cross-links and thus generate a space-network polymer.

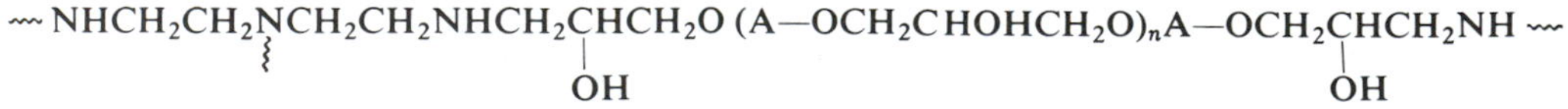

〰 $NHCH_2CH_2NCH_2CH_2NHCH_2CHCH_2O\,(A$—$OCH_2CHOHCH_2O)_nA$—$OCH_2CHCH_2NH$ 〰 (the N bearing a further chain ⌇; each CH bearing OH)

(c) Use excess phenol, acetone, and acid or base (compare Sec. 31.7).

20. Oxygen abstracts hydrogen atoms to form allylic free radicals, R·. These combine with oxygen to produce —O—O— cross-links; the gain in weight is due to the gain of oxygen. The mechanism seems to be:

$$R\cdot + O_2 \longrightarrow RCOO\cdot$$

$$2RCOO\cdot \longrightarrow ROOR + O_2$$

(Oxygen serves the same purpose here that sulfur does in the vulcanization of rubber, Sec. 11.24.)

21. In F the methylene protons (—CH_2—) are equivalent, and give one signal, a singlet (2H per monomer unit). F must therefore be *syndiotactic*.

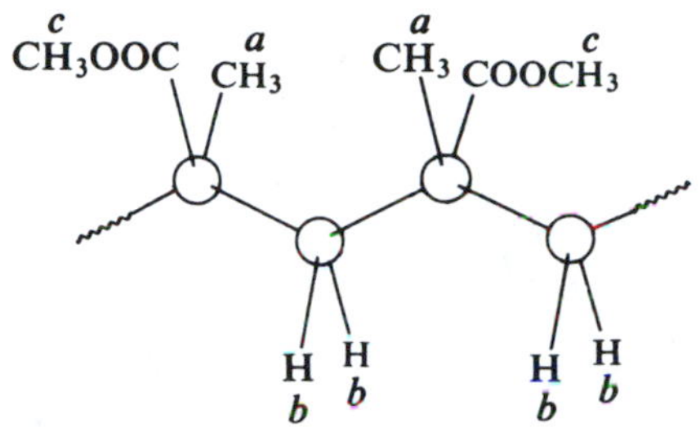

F
Syndiotactic

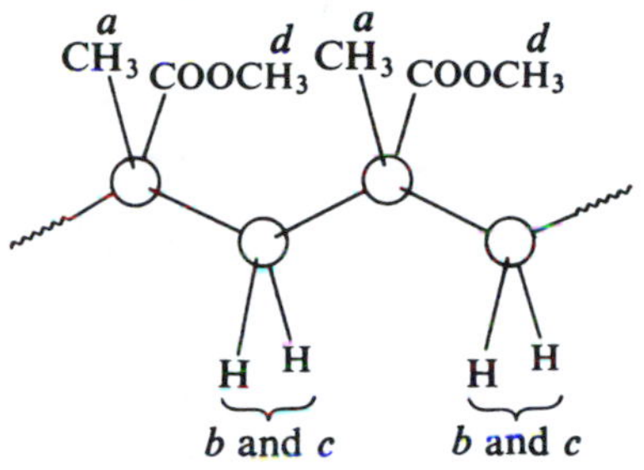

G
Isotactic

In G the methylene protons are not equivalent, and give two 1H signals, each split into a doublet by the other. G must therefore be *isotactic*. The methylene protons here are diastereotopic (Sec. 32.5); in the representation we have shown, for example, one H is near two —CH_3's and the other H is near two —$COOCH_3$'s.

32

Stereochemistry III. Enantiotopic and Diastereotopic Ligands and Faces

32.1 (a)

H
H—H
H—H
H—H
H

None

(b)

CH_3
H_R—H_S
H_S—H_R
CH_3

Two equivalent pairs

(c)

CH_3
CH_3—H
CH_3

None

(d)

Cl
H_S—H_R
H_S—H_R
CH_3

Two pairs

(e)

Cl
${}_SCH_3$—${}_RCH_3$
H

One pair

(f)

Cl
H_S—H_R
${}_SCH_3$—${}_RCH_3$
H

Two pairs

(g)

H
Cl_R—Cl_S
CH_3

One pair

(h)

CH_3
H_R—H_S
OH

One pair

(i)

OH
H_S—H_R
H_S—H_R
CH_3

Two pairs

(j)

OH
${}_SCH_3$—${}_RCH_3$
H

One pair

(k)

${}_SCH_3$
H_S—OH_S
H_R—OH_R
${}_RCH_3$

Three pairs

(l)

CH_3
HO—H
H—OH
CH_3

None

32.2 (a) H, H / C=C / H, CH_3 — *One pair*

(b) H, H / C=C / CH_3, CH_3 — *None*

(c) CH_3, CH_3 / C=C / H, CH_3 — *One pair*

(d) H, H / C=C / H, Cl — *One pair*

(e) CH_3; H—C—Cl; H_S—C—H_R; CH_3 — *One pair*

(f) CH_3; H—C—Cl; H_S—C—H_R; H_S—C—H_R; CH_3 — *Two pairs*

32.3 In each case, the specification refers to the face turned upward towards us.

(a) C_2H_5, H / C=O — *Re*; H, C_2H_5 / C=O — *Si*; *Enantiotopic*

(b) CH_3, CH_3 / C=O — *None*

(c) C_2H_5; CH_3—C—H; C(=O)CH_3 (O left, CH_3 right) — *Re*; C_2H_5; CH_3—C—H; C(=O)CH_3 (CH_3 left, O right) — *Si*; *Diastereotopic*

(d) In this example, C–2 has enantiotopic faces. To decide the relative priority of the three ligands attached to this trigonal carbon, we must apply Sequence Rule 3 (page 143): where there is a double bond, both atoms are considered to be duplicated. In this case, we do not duplicate the trigonal carbon itself, since *Re* and *Si* are concerned with only three ligands. But we do consider that

=C(H)(H) equals —C(C)(H)—H

and therefore has priority over —CH_3.

CH_3, Br / C=C / H, H — *Re*; Br, CH_3 / C=C / H, H — *Si*

Br > —C(H)(H)—C > —C(H)(H)—H

Enantiotopic

(e) $CH_3(CH_3)C{=}CH_2$

None

(f) $CH_3(CH_2Br)C{=}CH_2$ — *Re*; $BrCH_2(CH_3)C{=}CH_2$ — *Si*

$-CH_2Br > -CH_2-C > -CH_3$

Enantiotopic

(g) Here, as we see, *both* trigonal carbons of the double bond have heterotopic faces; we simply treat each of them in turn as though it were the only heterotopic face in the molecule.

$H(Cl)C{=}C(H)CH_3$ — *Re*, *Re*; $Cl(H)C{=}C(CH_3)H$ — *Si*, *Si*

$-C-Cl > -C-H > -H$

Two pairs of enantiotopic faces

(h) C_6H_5, CH_3, H; C=C, H, H, H — *Re*; C_6H_5, CH_3, H; C=C, H, H, H — *Si*

Diastereotopic

1\. (a) CH_3; H_R—H; H—Cl; H_S—H; CH_3 — *Enantiotopic ligands*

CH_3; H—H_S; H—Cl; H—H_R; CH_3 — *Enantiotopic ligands*

Diastereotopic ligands

$_R CH_3$; CH_2; H—Cl; CH_2; $_S CH_3$ — *Enantiotopic ligands*

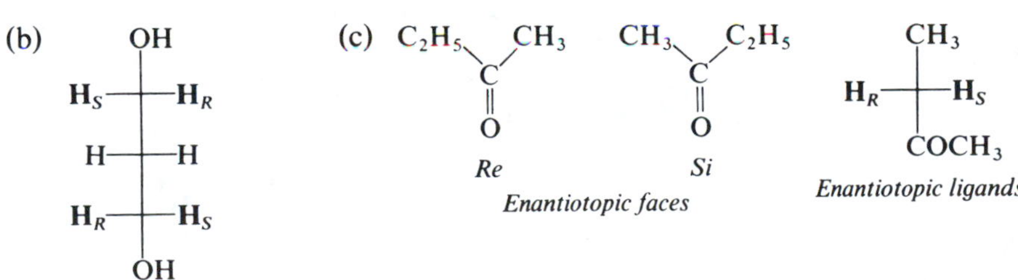

Two equivalent pairs of enantiotopic ligands

Enantiotopic faces

Enantiotopic ligands

(d)

CH_3 / H—C—OH / H_R—C—H_S / H—C—OH / CH_3

Diastereotopic ligands

(See answer to Problem 8(g), page 59 of this Study Guide)

CH_3 / H_S—C—OH / CH_2 / H_R—C—OH / CH_3

Enantiotopic ligands

CH_3 / H—C—OH_S / CH_2 / H—C—OH_R / CH_3

Enantiotopic ligands

$_SCH_3$ / H—C—OH / CH_2 / H—C—OH / $_RCH_3$

Enantiotopic ligands

(e)

CH_3 / HO—C—H / H—C—H / H—C—OH / CH_3

None

(f)

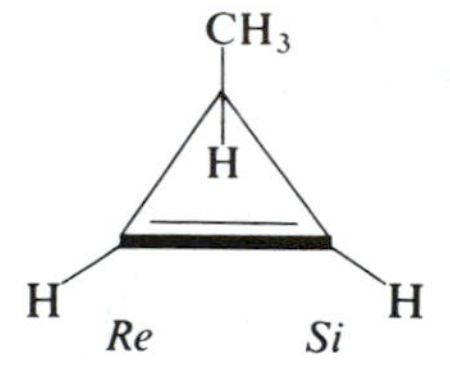

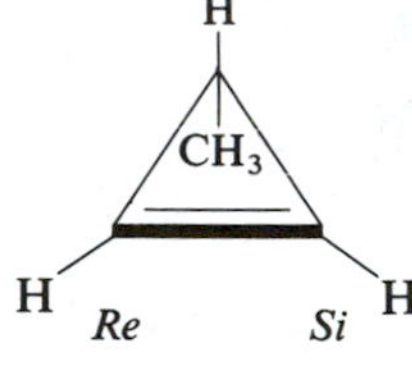

Diastereotopic faces

(g)

CH_3 / H_R H H_S / H H

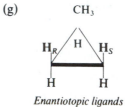

Enantiotopic ligands

CH_3 / H H H / H_S H_R

Enantiotopic ligands

Diastereotopic ligands

(h)

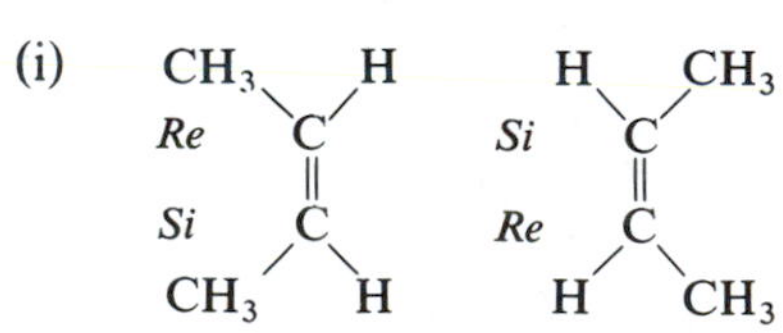

Enantiotopic ligands

CH_3 / H / H H / *Re* *Si*

Enantiotopic faces

H / CH_3 / H H / *Re* *Si*

Enantiotopic faces

Diastereotopic faces

(i)

CH_3 H / *Re* C / *Si* C / CH_3 H

H CH_3 / *Si* C / *Re* C / H CH_3

Two equivalent pairs of enantiotopic faces

(j)

CH_3 H / C=C / CH_3 CH_3

Diastereotopic ligands

CH_3 H / C=C / CH_3 CH_3 / *Re*

H CH_3 / C=C / CH_3 CH_3 / *Si*

Enantiotopic faces

2. (a) Alcohol I. (b) NADD.

$$\underset{\text{II}}{\text{H—}\overset{CH_3}{\underset{OH}{C}}\text{—D}} \xrightarrow{TsCl} \text{H—}\overset{CH_3}{\underset{OTs}{C}}\text{—D} \xrightarrow{OH^-} \underset{\text{I}}{\text{D—}\overset{CH_3}{\underset{OH}{C}}\text{—H}} \xrightarrow{NAD^+} \text{NADD} + CH_3CHO + H^+$$

Inversion in the hydrolysis of the tosylate gives ethanol I, which, of course, transfers D to NAD^+ to give NADD.

3. (a) (*S*)-Alanine would be formed from amide V. No bond to the chiral carbon is broken in this step, and reaction would take place with retention.

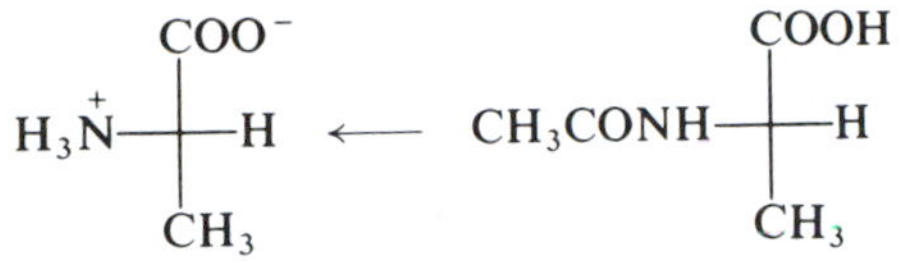

COO⁻ / $H_3\overset{+}{N}$—C—H / CH_3 ← COOH / CH_3CONH—C—H / CH_3

(*S*)-Alanine V

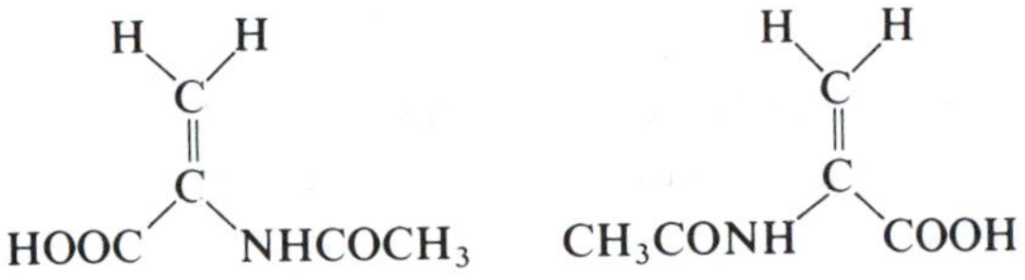

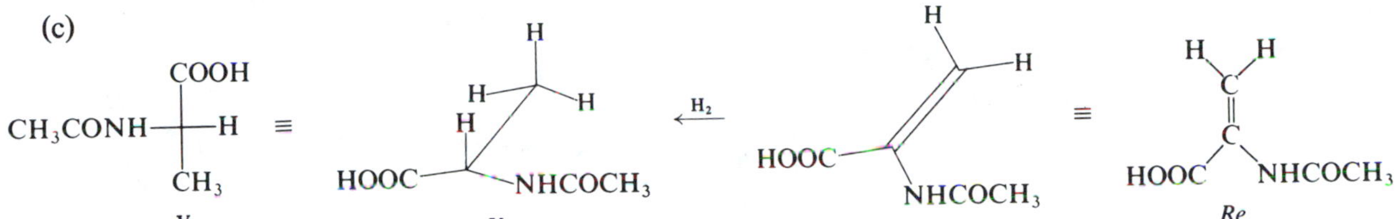

(b) H H C=C HOOC NHCOCH$_3$ *Re*   H H C=C CH_3CONH COOH *Si*

(c) COOH / CH_3CONH—C—H / CH_3 V ≡ H, H, H, H, HOOC, NHCOCH$_3$ V ← H_2 — H, H, HOOC, NHCOCH$_3$ 1-Acetamido-propenoic acid ≡ H H C=C HOOC NHCOCH$_3$ *Re*

4. We can use the same site as in Figure 32.3 (page 1112). We place the ethanol molecule on the bonding site (*a*) the only way it can fit: with H_S in the little hole, and H_R pointing upward. Only this upward-pointing H can be transferred to NAD^+.

5.

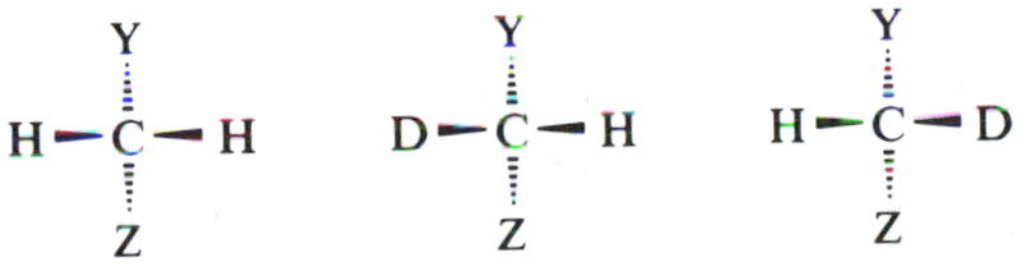

Y / H—C—H / Z NADH   Y / D—C—H / Z NADD(i)   Y / H—C—D / Z NADD(iv)

NADD(v) is a mixture: 44% NADD(i) and 56% NADD(iv). Not surprisingly, the chemical reduction of NAD^+ is (nearly) non-stereospecific.

Clearly, hydride transfer is stereospecific not only with regard to the ethanol/acetaldehyde (Sec. 32.4) but with regard to the NAD^+/NADH as well. NADH contains two hydrogens, chemically equivalent but stereochemically non-equivalent: enantiotopic or diastereotopic. (Actually, because of the chirality of the rest of the NADH molecule, they are diastereotopic.) NAD^+ contains only one hydrogen at this position; NAD^+ is evidently unsaturated, and contains a pair of enantiotopic or diastereotopic faces.

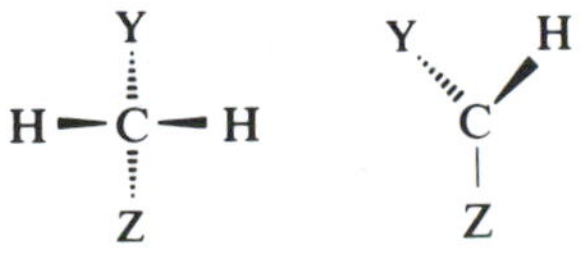

Y / H—C—H / Z NADH   Y H C / Z NAD^+

(i) When reduced (enzymatically) by CH_3CD_2OH, all the D is transferred to just one of the faces of NAD^+; the NADD(i) formed contains D in only one of the two stereochemical locations.

ethanol—D + NAD$^+$ ⟶ NADD(i)

NAD$^+$ NADD(i)

(ii) Now, when NADD(i) is used to reduce CH_3CHO, the reaction is the reverse of the one by which the NADD(i) was formed—catalyzed by the same enzyme and passing through the same transition state—and hydrogen is transferred back from the same stereochemical location as that at which it had been received; only D is transferred, and the NAD^+(ii) formed is left without any D.

acetaldehyde + NADD(i) ⟶ NAD$^+$(ii)

NADD(i) NAD$^+$(ii)
Contains no D

(iii) Next, the NADD(i) is used to reduce a different compound—a different aldehyde (Sec. 34.3)—glucose. This reaction is also completely stereospecific. It is catalyzed by a different enzyme and, as it happens, has exactly the opposite stereospecificity to that of the acetaldehyde reaction. Now only the protium in NADD(i) is transferred, and all the deuterium is left in the NAD^+(iii).

NADD(i) + glucose ⟶ NAD$^+$(iii)

NADD(i) NAD$^+$(iii)
Contains one D per molecule

(iv) When NAD(iii) is reduced by ordinary CH_3CH_2OH, it receives, of course, ordinary protium. Reaction (iv) has exactly the same stereochemistry as reaction (i), and the hydrogen becomes attached to the same face of NAD^+. But this time the D was already in the molecule, and it is H that is being attached. The product NADD(iv) is the stereoisomer of NADD(i).

ethanol—H + NAD$^+$(iii) ⟶ NADD(iv)

NAD$^+$(iii) NADD(iv)

(v) We begin again with ordinary NAD^+, and reduce it chemically in the presence of D_2O. Again the NADD formed contains one D per molecule. But this time the reaction involves ordinary chemical reagents, and is almost completely non-stereospecific. D becomes attached to either face of NAD^+, and the NADD(v) obtained is a mixture of NADD(i) and NADD(iv).

Y H C Z $\xrightarrow[Na_2S_2O_3]{D_2O}$ D—C—H (Y, Z) + H—C—D (Y, Z)

NAD^+ NADD(i) (56%) NADD(iv) (44%)

NADD(v)

(vi) When NADD(v) is re-oxidized by CH_3CH_2OH, hydrogen from only one stereochemical location is, of course, transferred: some of this (56%) is deuterium; the rest is protium. There is left in the NAD^+(vi) the deuterium (44%) that occupied the opposite location. NAD^+(vi) is thus a 56:44 mixture of NAD^+(ii) and NAD^+(iii).

acetaldehyde D—C—H (Y, Z) ⟶ Y H C Z

NADD(i) (56%) NAD^+(ii) (56%)

acetaldehyde H—C—D (Y, Z) ⟶ Y D C Z

NADD(iv) (44%) NAD^+(iii) (44%)

For a general discussion, see Levy, H. R.; Talalay, P.; Vennesland, B. "The Steric Course of Enzymatic Reaction at Meso Carbon Atoms: Application of Hydrogen Isotopes"; *Progress in Stereochemistry*; Butterworths: Washington, 1962; Vol. 3, pp 299–344. See also Loewus, F. A.; Westheimer, F. H.; Vennesland, B. "Enzymatic Synthesis of Enantiomorphs of Ethanol-1-*d*"; *J. Am Chem. Soc.* **1953**, *75*, 5018; Westheimer, F. H. *Adv. Phys. Org. Chem.* **1985**, *21*, 1.

6. (a) Krebs considered the citric acid molecule to be symmetrical and to contain two exactly equivalent terminal —COOH groups. In the conversion into α-ketoglutaric acid, he expected "scrambling" of the isotopic label between the two —COOH groups of the product.

This line of reasoning was accepted by the two sets of workers who carried out the labeling experiments (in 1941 and 1942), and they both concluded that citric acid could not be an intermediate in the formation of α-ketoglutaric acid.

(b) The line of reasoning is not valid. We recognize the two —COOH groups as enantiotopic and stereochemically non-equivalent; an enzyme could distinguish between them and give the observed labeling results.

(That an enzyme could differentiate between such groups was pointed out by A. G. Ogston in 1948; his ideas eventually led to our present understanding of stereochemical discrimination.)

(c) Citric acid *can* be an intermediate in the process—and, indeed, it is generally accepted that it *is*.

7. Yes: a, b, d, e, f. No: c.

(a) The carbonyl carbon has non-equivalent faces which, because of the chirality at C–2, are diastereotopic; attachment of hydride could take place preferentially to either of these faces. (Compare the reduction of acetaldehyde in Figure 32.2, page 1106.)

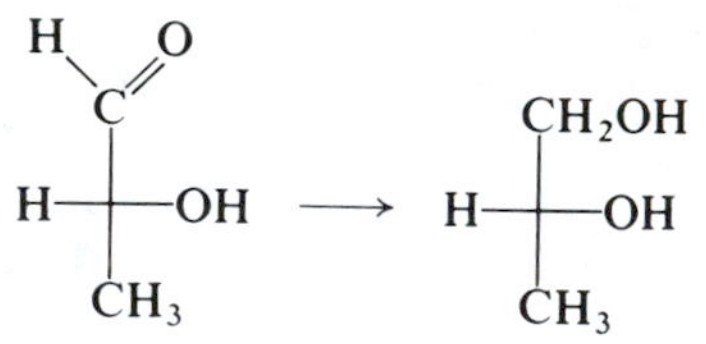

$$\text{H}\!-\!\overset{\displaystyle \overset{\text{H}\diagdown\ \diagup\!\!\text{O}}{\text{C}}}{\underset{\displaystyle \text{CH}_3}{\text{C}}}\!-\!\text{OH} \longrightarrow \text{H}\!-\!\overset{\displaystyle \text{CH}_2\text{OH}}{\underset{\displaystyle \text{CH}_3}{\text{C}}}\!-\!\text{OH}$$

Reaction at diastereotopic faces

(Deuterium studies have shown that, with alcohol dehydrogenase as the enzyme, hydride actually adds only to the *Re* face.)

(b) The —CH_2OH groups at the two ends of the molecule are enantiotopic; depending upon which of these reacts, one or the other of a pair of enantiomers will be formed.

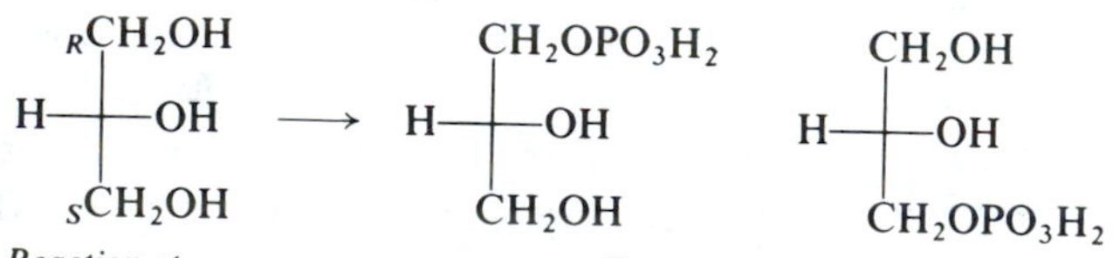

$$\text{H}\!-\!\overset{\displaystyle {}_R\text{CH}_2\text{OH}}{\underset{\displaystyle {}_S\text{CH}_2\text{OH}}{\text{C}}}\!-\!\text{OH} \longrightarrow \text{H}\!-\!\overset{\displaystyle \text{CH}_2\text{OPO}_3\text{H}_2}{\underset{\displaystyle \text{CH}_2\text{OH}}{\text{C}}}\!-\!\text{OH} \qquad \text{H}\!-\!\overset{\displaystyle \text{CH}_2\text{OH}}{\underset{\displaystyle \text{CH}_2\text{OPO}_3\text{H}_2}{\text{C}}}\!-\!\text{OH}$$

Reaction at enantiotopic ligands *Enantiomers*

(Work with ^{14}C-labeling has shown that glycerol kinase catalyzes phosphorylation only at the pro-*R* group.)

(c) The two —CH_2OH groups are stereochemically equivalent; whichever end of the molecule reacts, the product is the same.

CH_2OH | H—C—OH | HO—C—H | CH_2OH → $CH_2OPO_3H_2$ | H—C—OH | HO—C—H | CH_2OH CH_2OH | H—C—OH | HO—C—H | $CH_2OPO_3H_2$

Identical

(d) As in (b), the $-CH_2OH$ groups at the two ends of the molecule are enantiotopic; depending upon which of these reacts, one or the other of a pair of enantiomers will be formed.

${}_RCH_2OH$ | H—C—OH | H—C—OH | ${}_SCH_2OH$ → $CH_2OPO_3H_2$ | H—C—OH | H—C—OH | CH_2OH CH_2OH | H—C—OH | H—C—OH | $CH_2OPO_3H_2$

Reaction at enantiotopic ligands   *Enantiomers*

(This reaction has actually been found to be stereospecific in a biological system.)

(e) Fumaric acid has two sets of enantiotopic faces; attachment of —OH could take place to either face (at either carbon) to give either of a pair of enantiomers.

H, COOH on *Re* C; *Re* C bearing HOOC, H   HOOC, H on *Si* C; *Si* C bearing H, COOH → COOH | H—C—OH | CH_2OH COOH | HO—C—H | CH_2OH

(*R*) *From attack at Re face*   (*S*) *From attack at Si face*

Enantiomers

But —H, too, is being added to the double bond; this could be attached to the same face as the —OH (*syn*-addition) or to the opposite face (*anti*-addition).

(In the presence of the enzyme fumarase, attachment of —OH actually takes place only to the *Si* face, to give only (*S*)-malic acid. Deuterium studies have shown that —H becomes attached only to the *Re* face, and hence that addition is *anti*. Thus, of four possible stereoisomers from the addition of —D and —OH, only *one* is actually obtained, the 2*S*,3*R*. There are, however, two equally likely combinations that give this stereochemistry: —OH can add to either doubly bonded carbon, and —H to the other one.)

(f) Succinic acid contains two equivalent pairs of enantiotopic hydrogens. Whether the elimination is *anti* or *syn*, two hydrogens from different carbons must be lost *in certain pairs*, as determined by their stereochemistry: if *anti*, two H_R or two H_S; if *syn*, one H_R and one H_S. (Actually, this elimination has been found to be *anti*.)

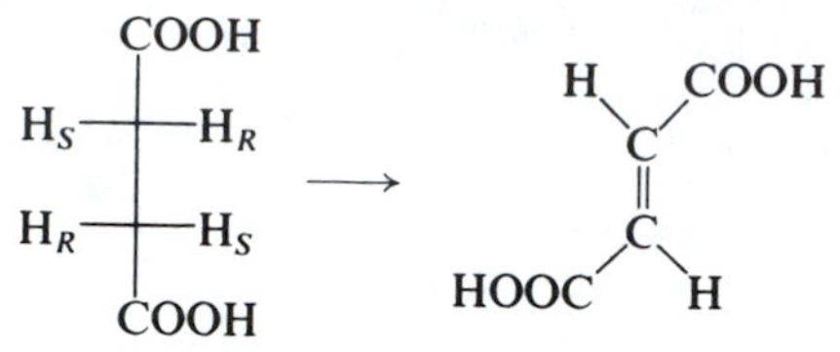

COOH
H_S—+—H_R
H_R—+—H_S
COOH

→

H COOH
C
||
C
HOOC H

Two equivalent pairs of enantiotopic hydrogens

33

Lipids

Fats and Steroids

33.1 Decarboxylation. This finding supports the theory that the hydrocarbons in oil shales were formed by such algae.

$$\underset{\text{[18-}^{14}\text{C] Stearic acid}}{{}^{14}CH_3(CH_2)_{15}CH_2COOH} \longrightarrow \underset{\text{[1-}^{14}\text{C] }n\text{-Heptadecane}}{{}^{14}CH_3(CH_2)_{15}CH_3} + CO_2$$

(Han, J.; Chau, H. W.-S.; Calvin, M. "A Biosynthesis of Alkanes in *Nostoc muscorum*"; *J. Am. Chem. Soc.* **1969**, *91*, 5156.)

33.2 (a)

```
  C           C           C           C
  |           |           |           |
C—C—C—C ¦ C—C—C—C ¦ C—C—C—C ¦ C—C—C—COOH
        ¦         ¦         ¦
              Isoprene units
```

(b) The configurations at C–7 and C–11 are identical in phytol and the carboxylic acid. In phytol, C–3 is not a chiral center and, significantly, *both* configurations at this center were found in the carboxylic acid. These facts strongly indicate that these acids came from chlorophyll and hence from green plants.

(Cox, R. E.; Maxwell, J. R.; Eglinston, G.; Pillinger, C. T. "The Geological Fate of Chlorophyll: the Absolute Stereochemistries of a Series of Acyclic Isoprenoid Acids in a 50 Million Year Old Lacustrine Sediment"; *Chem. Commun.* **1970**, 1639.)

Actually, acetate is the building block here, too, since the isoprene units come from isopentenyl pyrophosphate (page 1136), which comes ultimately from acetate. (See Problem 14, in this chapter, and its answer.)

33.3 Tung oil contains a high proportion of eleostearic acid (Table 33.1, page 1121). Abstraction of hydrogen atoms gives allylic free radicals with delocalization over three double bonds.

33.4 Alkoxide is a poor leaving group.

33.5 Unsaturation preserves the semiliquidity of the membranes in the colder part of the body.

33.6 Biological oxidation of fatty acids removes two carbons at a time, starting at the carboxyl end: "*beta*-oxidation".

33.7 (a)

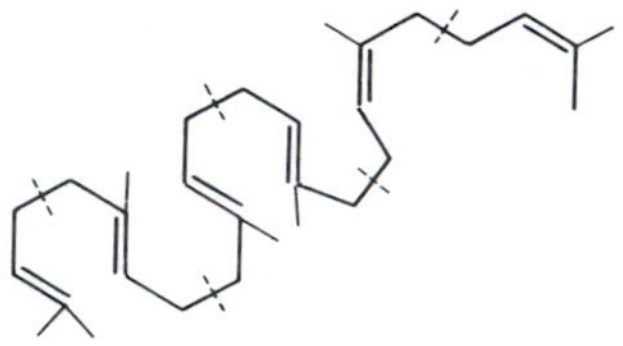

Squalene

(b) At the center, the molecule is probably made by the head-to-head combination of two identical C_{15} units.

(c) Four ring closures, and a methyl migration. Loss of three methyl groups.

33.8 (a) Loss of OPP^- to form the dimethylallyl cation, which attacks the terminal unsaturated carbon of isopentenyl pyrophosphate to form a 3° cation, which in turn loses a proton to yield geranyl pyrophosphate.

$$CH_3{-}\overset{\displaystyle CH_3}{\overset{|}{C}}{=}CH{-}CH_2{-}OPP \longrightarrow CH_3{-}\underbrace{\overset{\displaystyle CH_3}{\overset{|}{C}}{\cdots}CH{\cdots}CH_2}_{\oplus} + OPP^-$$

Dimethylallyl pyrophosphate

$$CH_3{-}\underbrace{\overset{\displaystyle CH_3}{\overset{|}{C}}{\cdots}CH{\cdots}CH_2}_{\oplus} + CH_2{=}\overset{\displaystyle CH_3}{\overset{|}{C}}{-}CH_2{-}CH_2{-}OPP \longrightarrow CH_3{-}\overset{\displaystyle CH_3}{\overset{|}{C}}{=}CH{-}CH_2{-}CH_2{-}\underset{\oplus}{\overset{\displaystyle CH_3}{\overset{|}{C}}}{-}CH_2{-}CH_2{-}OPP$$

Isopentenyl pyrophosphate

$$\downarrow -H^+$$

$$CH_3{-}\overset{\displaystyle CH_3}{\overset{|}{C}}{=}CH{-}CH_2{-}CH_2{-}\overset{\displaystyle CH_3}{\overset{|}{C}}{=}CH{-}CH_2{-}OPP$$

Geranyl pyrophosphate

(b)

$$(CH_3)_2C{=}CHCH_2CH_2C(CH_3){=}CHCH_2CH_2C(CH_3){=}CHCH_2OPP$$

Farnesyl pyrophosphate

(c) Two farnesyl units, head-to-head, form the squalene skeleton. (See Problem 33.7, above.)

(d) Continuation of the sequence started in (a) and (b).

We shall catch other glimpses of the organic chemistry underlying the biogenesis of these complex and important compounds. It begins with two-carbon acetate units, which combine to form four-carbon acetoacetate (Sec. 33.10), and then six-carbon mevalonate; this loses CO_2 (Problem 14, in this chapter) to give five-carbon isopentenyl pyrophosphate with the characteristic isoprene skeleton.

(Clayton, R. B. "Biosynthesis of Sterols, Steroids, and Terpenoids. Part I"; *Q. Rev., Chem. Soc.* **1965**, *19*, 168, especially pp 169–170.)

1.

$$CH_3(CH_2)_{22}COOH \xleftarrow[Ni]{H_2} CH_3(CH_2)_7CH{=}CH(CH_2)_{13}COOH \xrightarrow{oxidn.} CH_3(CH_2)_7COOH + HOOC(CH_2)_{13}COOH$$

Tetracosanoic acid — *cis*- or *trans*-15-Tetracosenoic acid, Nervonic acid (*Actually trans*) — Nonanoic acid, N.E. = m.w. = 158 — Pentadecanedioic acid, N.E. = m.w./2 = 136

2. Since the acid units (and the alcohol unit, glycerol) remain the same, the only change can be in the *distribution* of acyl groups among the glyceride molecules. Something like this, say:

$$\begin{matrix} CH_2O{-}COR \\ | \\ CHO{-}COR \\ | \\ CH_2O{-}COR \end{matrix} + \begin{matrix} CH_2O{-}COR' \\ | \\ CHO{-}COR' \\ | \\ CH_2O{-}COR' \end{matrix} + \begin{matrix} CH_2O{-}COR'' \\ | \\ CHO{-}COR'' \\ | \\ CH_2O{-}COR'' \end{matrix} \overset{base}{\rightleftharpoons} 3 \begin{matrix} CH_2O{-}COR \\ | \\ CHO{-}COR' \\ | \\ CH_2O{-}COR'' \end{matrix}$$

This would involve transesterification (Sec. 20.20), for which sodium methoxide would be an excellent catalyst:

$$-\overset{|}{\underset{|}{HCO}}{-}COR + OCH_3^- \rightleftharpoons -\overset{|}{\underset{|}{HCO}}^- + CH_3O{-}COR$$

$$-\overset{|}{\underset{|}{HCO}}^- + -\overset{|}{\underset{|}{HCO}}{-}COR' \rightleftharpoons -\overset{|}{\underset{|}{HCO}}{-}COR' + -\overset{|}{\underset{|}{HCO}}^- \quad etc.$$

3.

$$R{-}H + O_2 \longrightarrow H{-}O{-}O\cdot + R\cdot$$

$$R\cdot + O_2 \longrightarrow R{-}O{-}O\cdot$$

$$R{-}O{-}O\cdot + R{-}H \longrightarrow R{-}O{-}O{-}H + R\cdot \quad etc.$$

In the case of methyl oleate:

$$-\overset{11}{C}H_2{-}\overset{10}{C}H{=}\overset{9}{C}H{-}\overset{8}{C}H_2{-} \longrightarrow -\overset{11}{C}H_2{-}\overset{10}{C}H{\cdots}\overset{9}{C}H{\cdots}\overset{8}{C}H{-}\ (\text{radical}) \longrightarrow -\overset{11}{C}H_2{-}\overset{10}{C}H{=}\overset{9}{C}H{-}\underset{OOH}{\overset{8}{C}H}{-} + -\overset{11}{C}H_2{-}\underset{OOH}{\overset{10}{C}H}{-}\overset{9}{C}H{=}\overset{8}{C}H_2{-}$$

$$-\overset{11}{C}H_2{-}\overset{10}{C}H{=}\overset{9}{C}H{-}\overset{8}{C}H_2{-} \longrightarrow -\overset{11}{C}H{\cdots}\overset{10}{C}H{\cdots}\overset{9}{C}H{-}\overset{8}{C}H_2{-}\ (\text{radical}) \longrightarrow -\underset{OOH}{\overset{11}{C}H}{-}\overset{10}{C}H{=}\overset{9}{C}H{-}\overset{8}{C}H_2{-} + -\overset{11}{C}H{=}\overset{10}{C}H{-}\underset{OOH}{\overset{9}{C}H}{-}\overset{8}{C}H_2{-}$$

4. The weakly basic 2,4-dinitrophenoxide ion is a good leaving group.

5. $$n\text{-}C_{15}H_{31}COO^- + n\text{-}C_{15}H_{31}CH_2OH \xleftarrow[H_2O]{OH^-} n\text{-}C_{15}H_{31}\text{—}\overset{\overset{\displaystyle O}{\|}}{C}\text{—}O\text{—}CH_2C_{15}H_{31}\text{-}n$$

B: 1-Hexadecanol

Spermaceti: *n*-Hexadecyl hexadecanoate, N.E. = 480

$$n\text{-}C_{15}H_{31}COO^- \xrightarrow{H^+} n\text{-}C_{15}H_{31}COOH$$

A: Hexadecanoic acid (Palmitic acid), M.p. 63 °C; N.E. = m.w. = 256

6. $$n\text{-}C_6H_{13}Cl \xrightarrow{NaC{\equiv}CH} \underset{C}{n\text{-}C_6H_{13}C{\equiv}CH} \xrightarrow{Na,\ NH_3} n\text{-}C_6H_{13}C{\equiv}C^-Na^+ \xrightarrow{I(CH_2)_9Cl} \underset{D}{n\text{-}C_6H_{13}C{\equiv}C(CH_2)_9Cl}$$

$$D \xrightarrow{KCN} \underset{E}{n\text{-}C_6H_{13}C{\equiv}C(CH_2)_9CN} \xrightarrow[heat]{OH^-} \xrightarrow{H^+} \underset{F}{n\text{-}C_6H_{13}C{\equiv}C(CH_2)_9COOH}$$

$$F \xrightarrow{Pd,\ H_2} (n\text{-}C_6H_{13})HC{=}CH((CH_2)_9COOH)$$

Vaccenic acid (*cis*)

7. $$n\text{-}C_{13}H_{27}CH_2Br + Na[CH(COOEt)_2] \longrightarrow \underset{G}{n\text{-}C_{13}H_{27}CH_2CH(COOEt)_2} \xrightarrow{1KOH} \underset{H}{n\text{-}C_{13}H_{27}CH_2CH(COOH)(COOEt)}$$

$$H \xrightarrow{DHP} \underset{I}{n\text{-}C_{13}H_{27}CH_2CH(COOTHP)(COOEt)}$$

$$cis\text{-}n\text{-}C_6H_{13}CH{=}CH(CH_2)_7COOH \xrightarrow{SOCl_2} \underset{J}{cis\text{-}n\text{-}C_6H_{13}CH{=}CH(CH_2)_7COCl}$$

$$\text{I} \xrightarrow{\text{Na}} \xrightarrow{\text{J}} cis\text{-}n\text{-}C_{13}H_{27}CH_2-\underset{\text{THPOOC}}{\overset{\text{COOEt}}{\underset{|}{\overset{|}{C}}}}-\underset{\text{O}}{\underset{\|}{C}}(CH_2)_7CH{=}CHC_6H_{13}\text{-}n$$

K

$$\Big\downarrow \text{dil. } H^+$$

$$cis\text{-}n\text{-}C_{13}H_{27}CH_2-\underset{\text{H}}{\overset{\text{COOEt}}{\underset{|}{\overset{|}{C}}}}-\underset{\text{O}}{\underset{\|}{C}}(CH_2)_7CH{=}CHC_6H_{13}\text{-}n + CO_2$$

L

$$\Big\downarrow NaBH_4$$

$$cis\text{-}n\text{-}C_{13}H_{27}CH_2-\underset{\text{H}}{\overset{\text{COOEt}}{\underset{|}{\overset{|}{C}}}}-\underset{\text{OH}}{\underset{|}{C}H}(CH_2)_7CH{=}CHC_6H_{13}\text{-}n$$

M

$$\text{heat}\Big\downarrow OH^-$$

$$\Big\downarrow H^+$$

$$cis\text{-}n\text{-}C_{13}H_{27}CH_2-\overset{\text{COOH}}{\overset{|}{C}H}-CHOH(CH_2)_7CH{=}CHC_6H_{13}\text{-}n$$

(±)-Corynomycolenic acid

8. $$n\text{-}C_8H_{17}-\underset{\text{OH}}{\underset{|}{C}H}-CH_3 \xrightarrow{PBr_3} n\text{-}C_8H_{17}-\underset{\text{Br}}{\underset{|}{C}H}-CH_3 + Na[CH(COOEt)_2] \longrightarrow \begin{matrix} n\text{-}C_8H_{17} \\ CH_3 \end{matrix}\!\!>CH-CH(COOEt)_2$$

N

$$\begin{matrix} n\text{-}C_8H_{17} \\ CH_3 \end{matrix}\!\!>CHCH(COOEt)_2 \xrightarrow[\text{heat}]{OH^-} \xrightarrow{H^+} \xrightarrow{\text{heat}} \begin{matrix} n\text{-}C_8H_{17} \\ CH_3 \end{matrix}\!\!>CHCH_2COOH \xrightarrow{SOCl_2} \begin{matrix} n\text{-}C_8H_{17} \\ CH_3 \end{matrix}\!\!>CHCH_2COCl$$

O P

$$\text{P} \xrightarrow{\text{EtOH}} \begin{matrix} n\text{-}C_8H_{17} \\ CH_3 \end{matrix}\!\!>CHCH_2COOEt \xrightarrow{LiAlH_4} \begin{matrix} n\text{-}C_8H_{17} \\ CH_3 \end{matrix}\!\!>CHCH_2CH_2OH \xrightarrow{PBr_3} \begin{matrix} n\text{-}C_8H_{17} \\ CH_3 \end{matrix}\!\!>CHCH_2CH_2Br$$

Q R S

$$\text{S} \xrightarrow{\text{Mg}} \begin{matrix} n\text{-}C_8H_{17} \\ CH_3 \end{matrix}\!\!>CHCH_2CH_2MgBr \xrightarrow{CdCl_2} \left[\begin{matrix} n\text{-}C_8H_{17} \\ CH_3 \end{matrix}\!\!>CHCH_2CH_2\right]_2 Cd \xrightarrow{EtOOC(CH_2)_5COCl} \begin{matrix} n\text{-}C_8H_{17} \\ CH_3 \end{matrix}\!\!>CHCH_2CH_2-\underset{\underset{O}{\|}}{C}-(CH_2)_5COOEt$$

T

$$\text{T} \xrightarrow[\text{HCl}]{\text{Zn(Hg)}} \begin{matrix} n\text{-}C_8H_{17} \\ CH_3 \end{matrix}\!\!>CHCH_2CH_2CH_2-(CH_2)_5COOEt \xrightarrow[\text{heat}]{OH^-} \xrightarrow{H^+} \begin{matrix} n\text{-}C_8H_{17} \\ CH_3 \end{matrix}\!\!>CHCH_2CH_2CH_2(CH_2)_5COOH$$

U

Tuberculostearic acid
10-Methyloctadecanoic acid

9. $C_{26}H_{53}COOH$ sat'd open-chain
$-C_{26}H_{51}COOH$

2H missing *means* one double bond (or one ring)

$$-\overset{|}{C}\,\vdots\!=\!\overset{\overset{CH_3}{|}}{C}-COOH \xrightarrow{O_3} \xrightarrow[H_2O]{Zn} O=\overset{\overset{CH_3}{|}}{C}-COOH$$

C_{27}-Phthienoic acid

$$-\overset{\overset{H}{|}}{C}\,\vdots\!=\!\overset{\overset{CH_3}{|}}{C}-COOH \xrightarrow{KMnO_4} -COOH$$

C_{27}-Phthienoic acid ($C_{27}H_{52}O_2$) V ($C_{24}H_{48}O_2$)

$$-\underset{\underset{H}{|}}{\overset{|}{C}}-COOH \xrightarrow[H^+]{MeOH} -\underset{\underset{H}{|}}{\overset{|}{C}}-COOMe \xrightarrow{2PhMgBr} \xrightarrow{H_2O} -\underset{\underset{H}{|}}{\overset{|}{C}}-\underset{\underset{OH}{|}}{\overset{\overset{Ph}{|}}{C}}-Ph \xrightarrow[-H_2O]{H^+} -\overset{|}{C}=\overset{\overset{Ph}{|}}{C}-Ph$$

V ($C_{24}H_{48}O_2$) W ($C_{36}H_{58}O$) X ($C_{36}H_{56}$)

$$\xrightarrow{CrO_3} -\overset{|}{C}=O + O=\overset{\overset{Ph}{|}}{C}-Ph$$

Y, a ketone ($C_{23}H_{46}O$)

$$C_{21}H_{43}-\overset{\overset{CH_3}{|}}{C}=O \xrightarrow[NaOH]{I_2} C_{21}H_{43}COONa + CHI_3$$

Y
A methyl ketone

At this point, then:

$$\underset{\mathbf{Y}}{C_{21}H_{43}-\overset{\overset{\large CH_3}{|}}{C}=O} \qquad \underset{\mathbf{V}}{C_{21}H_{43}-\overset{\overset{\large CH_3}{|}}{\underset{\underset{\large H}{|}}{C}}-COOH} \qquad \underset{C_{27}\text{-Phthienoic acid}}{C_{21}H_{43}-\overset{\overset{\large CH_3}{|}}{\underset{\underset{\large H}{|}}{C}}-CH=\overset{\overset{\large CH_3}{|}}{C}-COOH}$$

Next:

$$\underset{\substack{\mathbf{V}\\(C_{24}H_{48}O_2)}}{-\underset{\underset{\large H}{|}}{\overset{|}{C}}-\underset{\underset{\large H}{|}}{\overset{\overset{\large CH_3}{|}}{C}}-COOH} \xrightarrow[P]{Br_2} \underset{\mathbf{Z}}{-\underset{\underset{\large H}{|}}{\overset{|}{C}}-\underset{\underset{\large Br}{|}}{\overset{\overset{\large CH_3}{|}}{C}}-COOH} \xrightarrow[-HBr]{KOH(alc)} \underset{\substack{\mathbf{AA}\\(C_{24}H_{46}O_2)}}{-\overset{|}{C}=\overset{\overset{\large CH_3}{|}}{C}-COOH}$$

$$\underset{\substack{\mathbf{AA}\\(C_{24}H_{46}O_2)}}{CH_3(CH_2)_{17}\overset{\overset{\large CH_3}{|}}{C}H-\!\!\!\!\!\vdots\;CH=\overset{\overset{\large CH_3}{|}}{C}-COOH} \xrightarrow{KMnO_4} \underset{\substack{\textit{Identified}\\(C_{20}H_{40}O)}}{CH_3(CH_2)_{17}\overset{\overset{\large CH_3}{|}}{C}=O} + \text{other products}$$

Hence:

$$\underset{\mathbf{Y}}{CH_3(CH_2)_{17}\overset{\overset{\large CH_3}{|}}{C}HCH_2\overset{\overset{\large CH_3}{|}}{C}=O} \qquad \underset{\mathbf{V}}{CH_3(CH_2)_{17}\overset{\overset{\large CH_3}{|}}{C}HCH_2\overset{\overset{\large CH_3}{|}}{C}HCOOH} \qquad \underset{C_{27}\text{-Phthienoic acid}}{CH_3(CH_2)_{17}\overset{\overset{\large CH_3}{|}}{C}HCH_2\overset{\overset{\large CH_3}{|}}{C}HCH=\overset{\overset{\large CH_3}{|}}{C}-COOH}$$

10. $\overset{a}{CH_3}(\overset{b}{CH_2})_{15}\overset{c}{CH_2}\overset{d}{COOH}$ — **CC**, Octadecanoic acid, Stearic acid

$$\overset{a}{CH_3}(\overset{c}{CH_2})_{14}\underset{\underset{\substack{\large H\\ d}}{|}}{\overset{\overset{\substack{b\\ \large CH_3}}{|}}{C}}-\overset{e}{COOH}$$

DD — 2-Methyl heptadecanoic acid

11. $C_2H_5-\overset{\overset{\large CH_3}{|}}{C}=O + [(MeO)_2P(O)CHCOOMe]^-Na^+ \longrightarrow \underset{\mathbf{EE}}{\left[C_2H_5-\underset{\underset{\large Na^+\,^-O}{|}}{\overset{\overset{\large CH_3}{|}}{C}}-\underset{\underset{\large P(O)(OMe)_2}{|}}{C}HCOOMe\right]} \longrightarrow \underset{\mathbf{FF}}{\overset{CH_3}{\underset{C_2H_5}{\diagdown}}C=C\overset{H}{\underset{COOMe}{\diagup}}}$

$$\text{FF} \xrightarrow{\text{LiAlH}_4} \underset{\textbf{GG}}{\overset{CH_3\quad\quad H}{\underset{C_2H_5\quad\quad CH_2OH}{C{=}C}}} \xrightarrow{\text{PBr}_3} \underset{\textbf{HH}}{\overset{CH_3\quad\quad H}{\underset{C_2H_5\quad\quad CH_2Br}{C{=}C}}}$$

$$\downarrow [C_2H_5COCHCOOEt]^-Na^+$$

$$\underset{\textbf{II}}{\overset{CH_3\quad\quad H\quad\quad COOEt}{\underset{C_2H_5\quad\quad CH_2CHCOC_2H_5}{C{=}C}}}$$

$$\text{II} \xrightarrow[H_2O]{OH^-} \xrightarrow{H^+} \xrightarrow{\text{heat}} \underset{\textbf{JJ}}{\overset{CH_3\quad\quad H}{\underset{C_2H_5\quad\quad CH_2CH_2COC_2H_5}{C{=}C}}}$$

$$\downarrow [(MeO)_2P(O)CHCOOMe]^-Na^+$$

$$\left[\overset{CH_3\quad\quad H}{\underset{C_2H_5\quad\quad CH_2CH_2\overset{C_2H_5}{\underset{Na^{+}\,{}^{-}O}{C}}-\underset{P(O)(OMe)_2}{C}HCOOMe}{C{=}C}}\right]$$

KK

$$\downarrow$$

$$\underset{\textbf{LL}}{\overset{CH_3\quad\quad H\quad C_2H_5\quad\quad COOMe}{\underset{C_2H_5\quad\quad CH_2{-}CH_2\quad\quad H}{C{=}C\qquad\qquad C{=}C}}}$$

$$\text{LL} \xrightarrow{\text{LiAlH}_4} \underset{\textbf{MM}}{\overset{CH_3\quad\quad H\quad C_2H_5\quad\quad CH_2OH}{\underset{C_2H_5\quad\quad CH_2{-}CH_2\quad\quad H}{C{=}C\qquad\qquad C{=}C}}} \xrightarrow{\text{PBr}_3} \underset{\textbf{NN}}{\overset{CH_3\quad\quad H\quad C_2H_5\quad\quad CH_2Br}{\underset{C_2H_5\quad\quad CH_2{-}CH_2\quad\quad H}{C{=}C\qquad\qquad C{=}C}}}$$

$$\text{NN} \xrightarrow{[CH_3COCHCOOEt]^-Na^+} \underset{\textbf{OO}}{\overset{CH_3\quad\quad H\quad C_2H_5\quad\quad CH_2CHCOCH_3}{\underset{C_2H_5\quad\quad CH_2{-}CH_2\quad\quad H\quad COOEt}{C{=}C\qquad\qquad C{=}C}}}$$

$$\downarrow OH^-;\ H^+;\ \text{heat}$$

$$\underset{\textbf{PP}}{\overset{CH_3\quad\quad H\quad C_2H_5\quad\quad CH_2CH_2COCH_3}{\underset{C_2H_5\quad\quad CH_2{-}CH_2\quad\quad H}{C{=}C\qquad\qquad C{=}C}}}$$

$$\downarrow [(MeO)_2P(O)CHCOOMe]^-Na^+$$

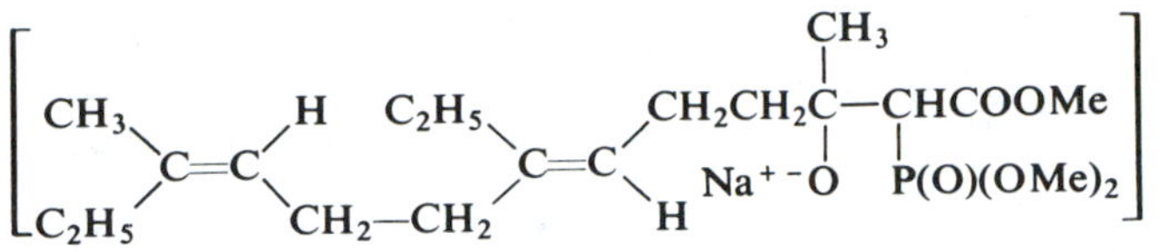

$$\left[(CH_3)(C_2H_5)C{=}CH{-}CH_2{-}CH_2{-}C(C_2H_5){=}CH{-}CH_2CH_2C(CH_3)(O^-\,Na^+){-}CH(COOMe)P(O)(OMe)_2 \right]$$

QQ

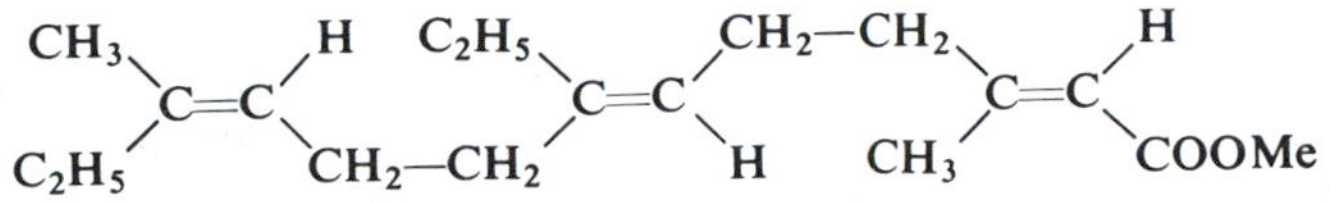

$\downarrow$

$$(CH_3)(C_2H_5)C{=}CH{-}CH_2{-}CH_2{-}C(C_2H_5){=}CH{-}CH_2{-}CH_2{-}C(CH_3){=}CH{-}COOMe$$

RR

$\downarrow$ $ArCO_2OH$

$$(CH_3)(C_2H_5)\overset{O}{C{-}C}H{-}CH_2{-}CH_2{-}C(C_2H_5){=}CH{-}CH_2{-}CH_2{-}C(CH_3){=}CH{-}COOCH_3$$

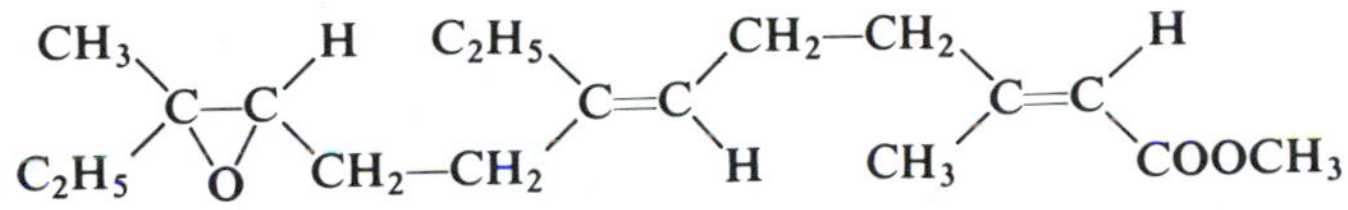

SS, the juvenile hormone

(Dahm, K. H.; Trost, B. M.; Röller, H. "The Juvenile Hormone. V. Synthesis of the Racemic Juvenile Hormone"; *J. Am. Chem. Soc.* **1967**, *89*, 5292; Trost, B. M. "The Juvenile Hormone of *Hyalophora cecropia*"; *Acc. Chem. Res.* **1970**, *3*, 120.)

In the preparation of FF, LL, and RR, we encounter a modification of the Wittig reaction (Sec. 21.10), in which a *phosphonate* is used in place of the usual ylide. A base

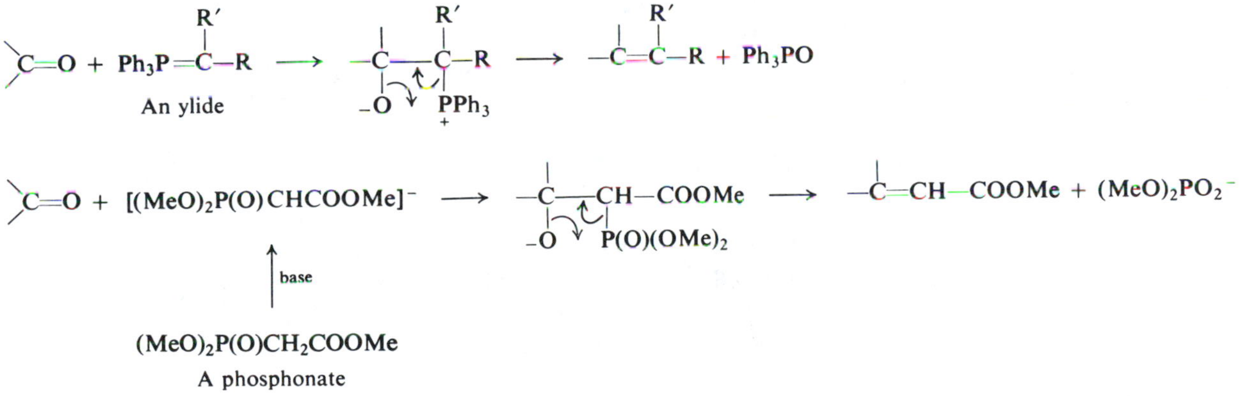

$$>C{=}O + Ph_3P{=}C(R'){-}R \longrightarrow {-}\underset{-O}{C}{-}\underset{\overset{+}{P}Ph_3}{C}(R'){-}R \longrightarrow {-}C{=}C(R'){-}R + Ph_3PO$$

An ylide

$$>C{=}O + [(MeO)_2P(O)CHCOOMe]^- \longrightarrow {-}\underset{-O}{C}{-}\underset{P(O)(OMe)_2}{CH}{-}COOMe \longrightarrow {-}C{=}CH{-}COOMe + (MeO)_2PO_2^-$$

$\uparrow$ base

$(MeO)_2P(O)CH_2COOMe$

A phosphonate

(NaH, in this particular example) converts the phosphonate into an anion which adds, in a Reformatsky-like reaction, to the carbonyl group. Elimination of $(MeO)_2PO_2^-$, the anion of dimethyl phosphate, occurs readily to generate the double bond. (Trippett, S. "The Wittig Reaction"; *Q. Rev., Chem. Soc.* **1963**, *17*, 406, especially pp 431–434.)

12. CO_2 becomes the —COOH of malonyl CoA in reaction (1), page 1133; this is the carbon lost in reaction (4).

13. (a) An aldol-like condensation between the ester and the keto group of oxaloacetate.

(b) An aldol-like condensation between the ester and the keto group of acetoacetyl CoA. Then reduction of an ester to a primary alcohol by hydride transfer.

14. $$\begin{array}{c} ^{-}OOC \\ -\overset{|}{\underset{|}{C}}-\overset{|}{\underset{|}{C}}- \\ \quad OPO_3H_2 \end{array} \longrightarrow -\overset{|}{C}=\overset{|}{C}- + CO_2 + H_2PO_4^-$$

The dihydrogen phosphate ion, $H_2PO_4^-$, is a much better leaving group than the more strongly basic OH^-. Loss of CO_2 leads, not to the unstable carbanion, but—with $H_2PO_4^-$ carrying away the electrons—to a stable, neutral alkene.

Carbanion chemistry—or something resembling it—runs throughout these biosyntheses. Acetate units are combined into acetoacetate by what amounts to a malonic ester synthesis (Sec. 33.10). In an aldol-like condensation, another acetate adds to acetoacetate to give 3-methyl-3-hydroxyglutarate, a six-carbon precursor of mevalonic acid. Here, we see the loss of CO_2 that finally gives the isoprenoid skeleton—and the way is clear to geraniol, farnesol, squalene, lanosterol, and cholesterol; or, in a different organism, to rubber. (See the reference given in the answer to Problem 33.8.)

FARNESOL

(*To the tune of:* "Jingle Bells")

Take an acetate,
Condense it with a mate,
Pretty soon you have
Acetoacetate.
Let 'em have a ball,
You get geraniol.
Add another isoprene
And you've got farnesol.

Farnesol, farnesol, good old farnesol,
First it goes to squalene, then you get cholesterol.
Farnesol, farnesol, good old farnesol.
First it goes to squalene, then you get cholesterol.

Now squalene makes a roll,
Becomes lanosterol.
The extra methyls do
Come off as CO_2.
Then comes zymosterol,
And then desmosterol,
If you don't take Triparanol,
You get cholesterol.

Farnesol, farnesol, *etc.*

David Kritchevsky
The Wistar Institute
Philadelphia, Pennsylvania

15. These are not terribly difficult syntheses, if we realize that most of the molecule is simply going along for the ride.

(a) The key step is anti-Markovnikov introduction of —OH.

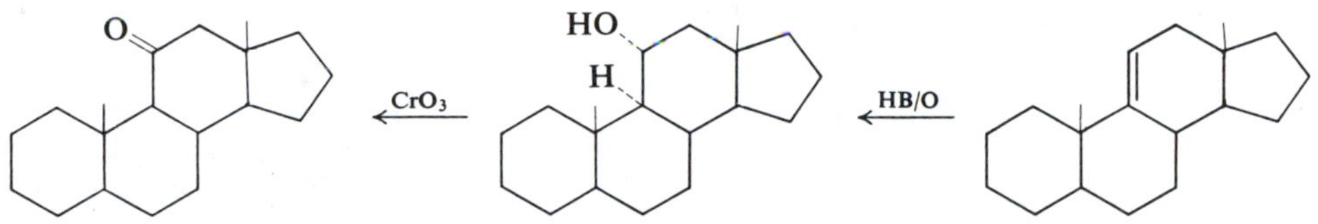

Androstan-11-one $\xleftarrow{CrO_3}$ Androstan-11 α-ol $\xleftarrow{HB/O}$ Androst-9(11)-ene

(b) Again, anti-Markovnikov introduction of —OH is needed.

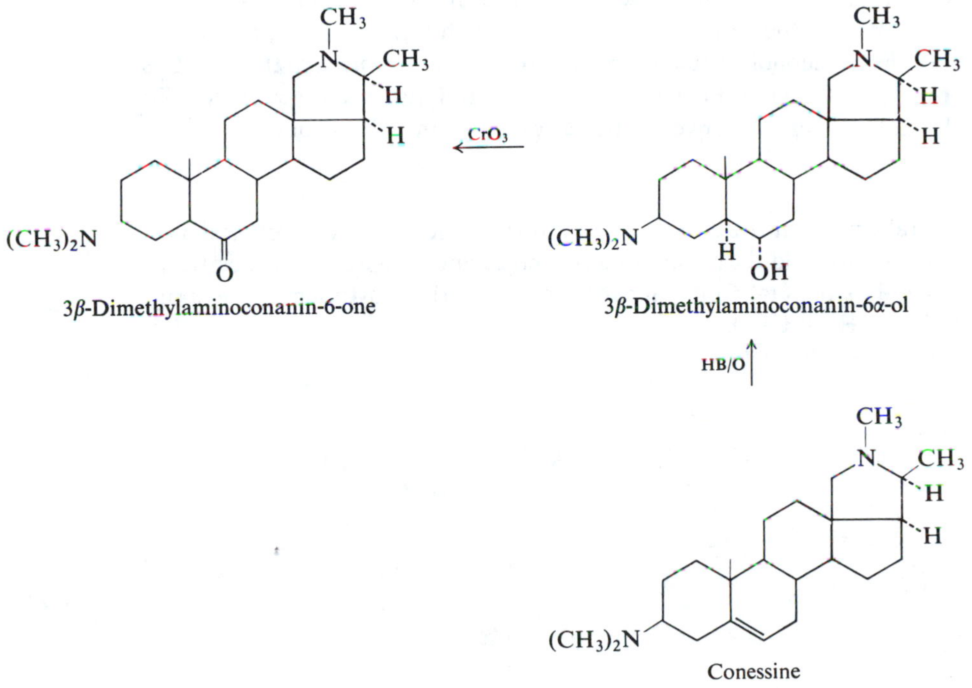

3β-Dimethylaminoconanin-6-one $\xleftarrow{CrO_3}$ 3β-Dimethylaminoconanin-6α-ol

$\uparrow$ HB/O

Conessine (3β-Dimethylaminocon-5-enine)

(c) The transformation at C–3 is done last:

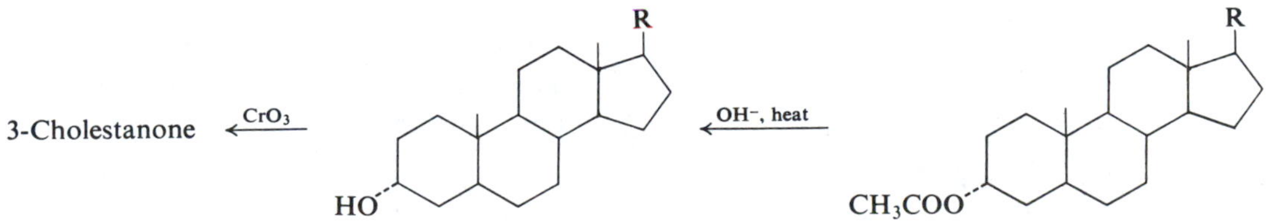

3-Cholestanone $\xleftarrow{CrO_3}$ Cholestane-3α-ol $\xleftarrow{OH^-,\ heat}$ Acetate ester of cholestane-3α-ol

First, to get the acetate of cholestane-3α-ol, we build the saturated eight-carbon chain R:

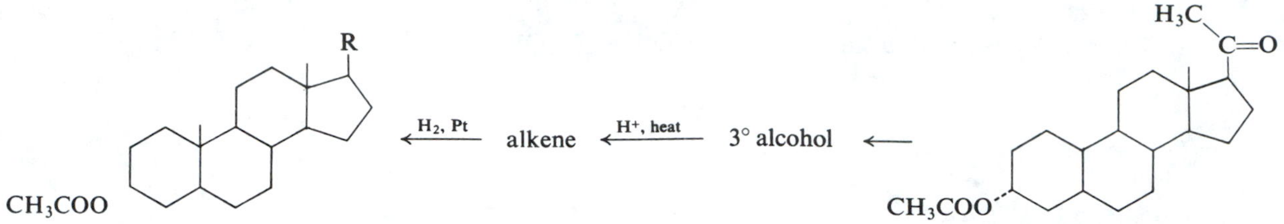

CH_3COO — Acetate ester of cholestane-3α-ol $\xleftarrow{H_2, Pt}$ alkene $\xleftarrow{H^+, heat}$ 3° alcohol $\longleftarrow$ Acetate ester of 5α-pregnane-3α-ol-20-one (CH_3COO, H_3C–C=O) + $BrMgCH_2CH_2CH_2CH(CH_3)_2$ Isohexylmagnesium bromide

Alternative sequences would not work. We cannot introduce the C–3 keto group before we build the chain, since that keto group would react with the Grignard reagent, too. A free —OH at C–3 would decompose the Grignard reagent—or, if things got that far, undergo dehydration along with the 3° alcohol. These problems are solved by *protecting* the —OH as an ester. Esters do react with Grignard reagents, but not so rapidly as ketones. Once the side chain has been built, the ester can conveniently be converted into the alcohol.

16. (a) *syn*-Hydration can take place in either of two ways: from "beneath" to yield cholestane-3β,6α-diol or from "above" to yield the stereoisomeric cholestane-3β,6β-diol. Attack from "beneath" is less hindered than attack from "above", because of the substituents projecting "upward", particularly —CH_3 at C–10.

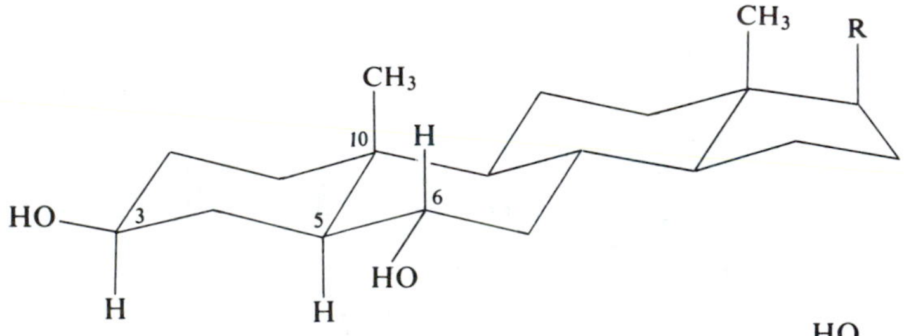

Cholestane-3β, 6α-diol
Greatly predominates

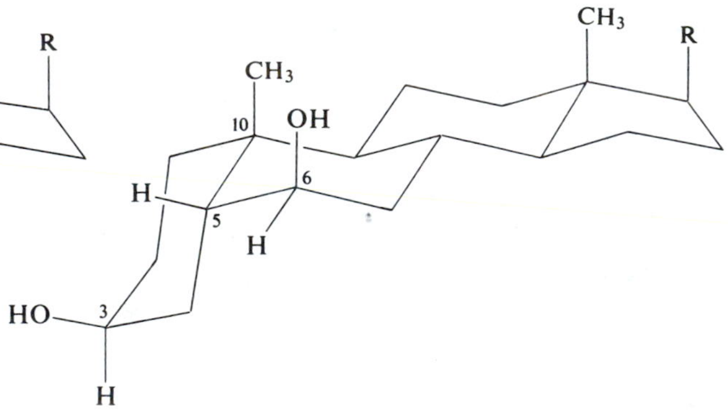

Cholestane-3β, 6β-diol

(b) *syn*-Hydration from beneath gives an α-OH at C–11, and an α-H at C–9.

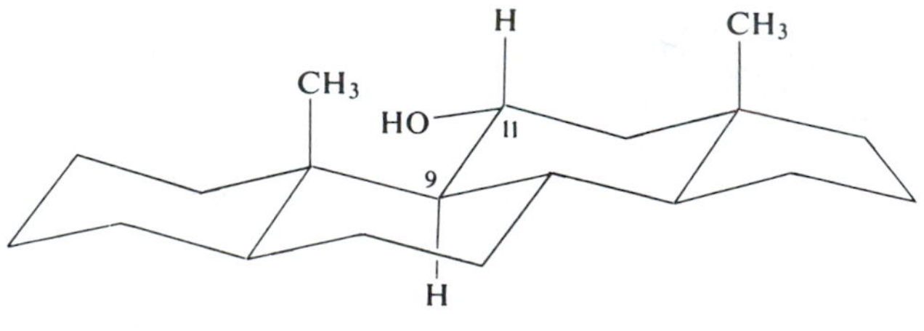

17. (a)

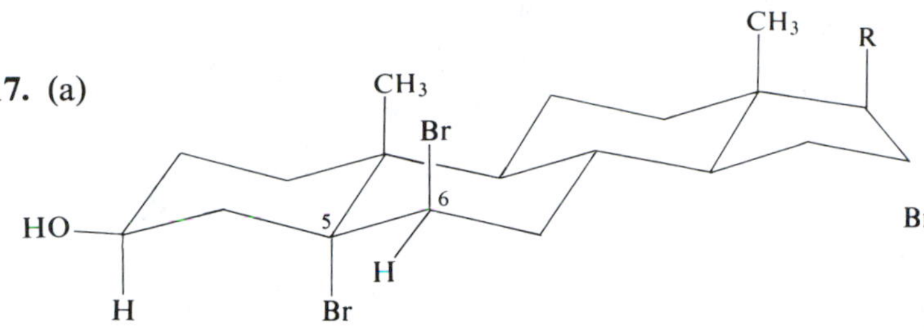

CH3 R CH3 Br 6 HO 5 H H Br

5α, 6β-dibromo compound
Greatly predominates

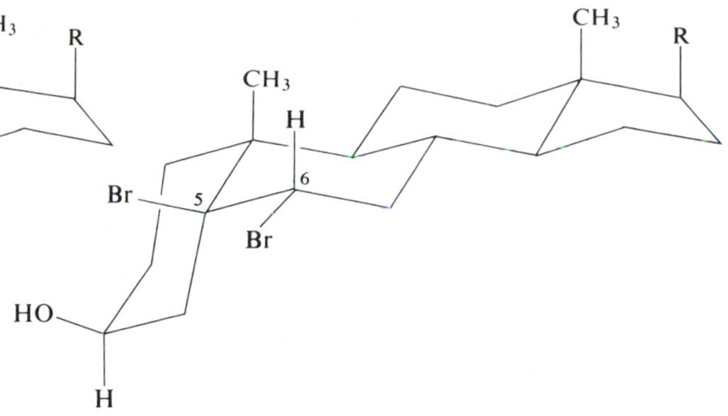

CH3 R CH3 H 6 Br 5 Br HO H

5β, 6α-dibromo compound

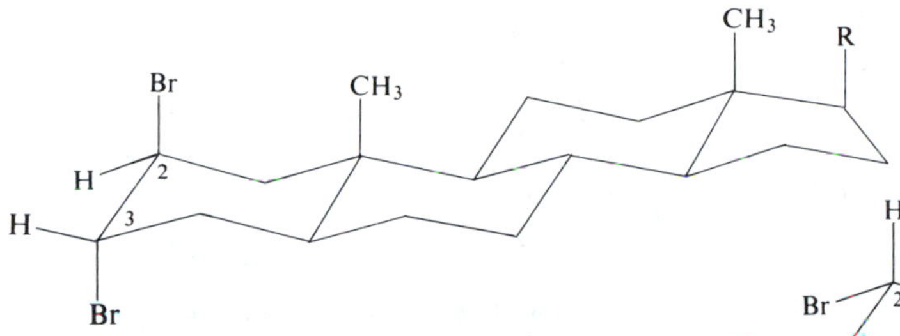

CH3 R Br CH3 H 2 H 3 Br

2β, 3α-dibromo compound
Greatly predominates

CH3 R H CH3 Br 2 3 Br H

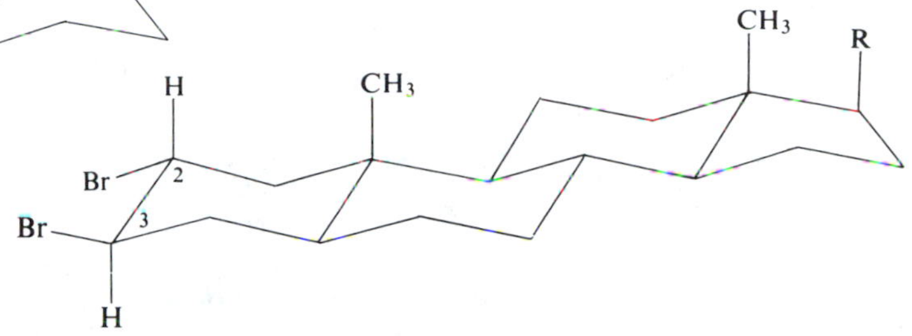

2α, 3β-dibromo compound

(b) There is preferred formation of the bromonium ion by the less hindered attack from beneath. The bromonium ion is then opened via an *anti* transition state to yield a diaxial dibromide.

18.

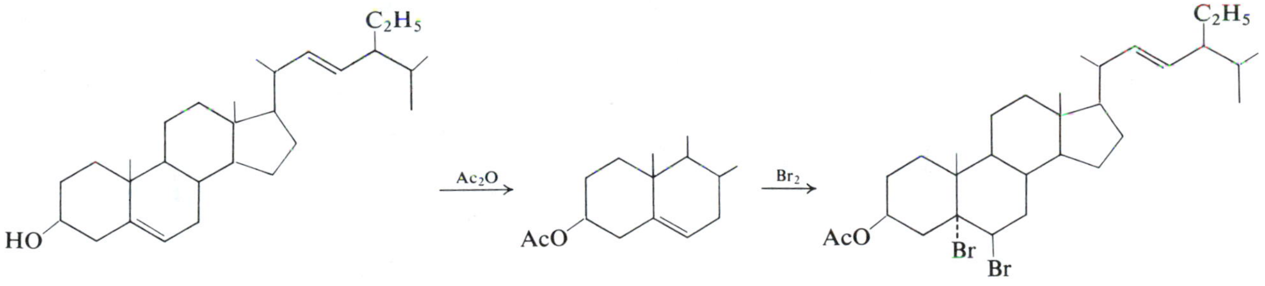

C_2H_5 HO Stigmasterol $\xrightarrow{Ac_2O}$ AcO TT $\xrightarrow{Br_2}$ C_2H_5 AcO Br Br UU

UU $\xrightarrow{O_3}$ $\xrightarrow{Ag_2O}$ COOH AcO Br Br VV $\xrightarrow{Zn}$ COOH AcO WW $\xrightarrow[H^+]{EtOH}$ COOEt XX

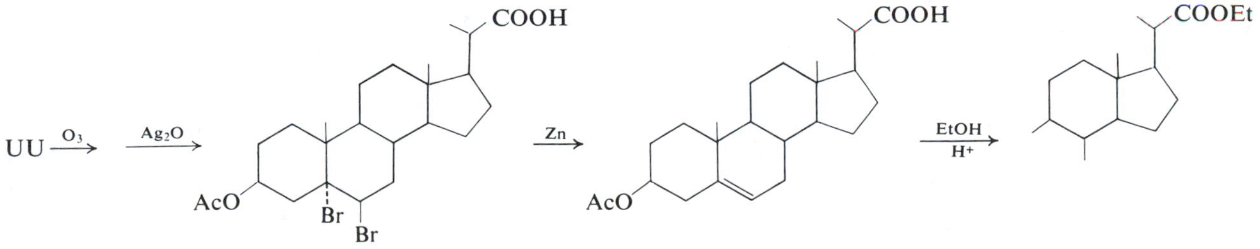

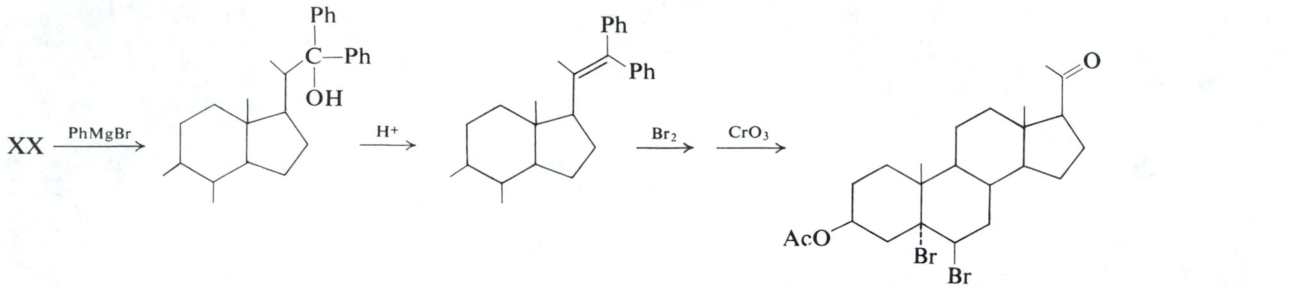

XX $\xrightarrow{PhMgBr}$ YY $\xrightarrow{H^+}$ ZZ $\xrightarrow{Br_2}$ $\xrightarrow{CrO_3}$ AAA

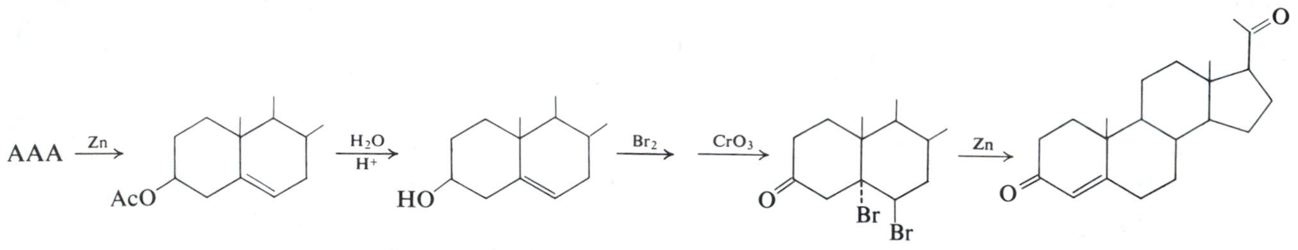

AAA $\xrightarrow{Zn}$ BBB $\xrightarrow[H^+]{H_2O}$ CCC (Pregnenolone) $\xrightarrow{Br_2}$ $\xrightarrow{CrO_3}$ DDD $\xrightarrow{Zn}$ Progesterone

(b) The ultraviolet spectrum shows that progesterone has a double bond conjugated with the C–3 keto group. In the last step of the synthesis, the double bond shifts to yield the more stable product.

34

Carbohydrates I. Monosaccharides

34.1 (a) In each case, you would observe a change in the molecular formula which, interpreted in light of the reagent used and supported by the properties of the product, would tell you what had happened. For example, in the second reaction,

$$C_6H_{12}O_6 \xrightarrow{Br_2(aq)} C_6H_{12}O_7$$

addition of one O with no loss of C or H indicates oxidation of an aldehyde to a (mono)-carboxylic acid:

$$RCHO \longrightarrow RCOOH$$

This would be supported by the acidic properties of the product, and the fact that N.E. = m.w.

(b) Starting at the top, each reaction shows that glucose contains:

(i) —C(=O)— (ii) —CHO (iii) —CHO and —CH_2OH (iv) five —OH

(v) C—C—C—C—C—C (vi) C—C—C—C—C—CHO

34.2 Formulas I–VIII, page 1156.

34.3 (a) Three chiral centers:

```
      CH2OH
      |
      C=O
      |
  HO--+--H
      |
   H--+--OH
      |
   H--+--OH
      |
      CH2OH
```

Fructose

(b) There should be eight 2-ketohexoses ($2^3 = 8$): four pairs of enantiomers.

(c)

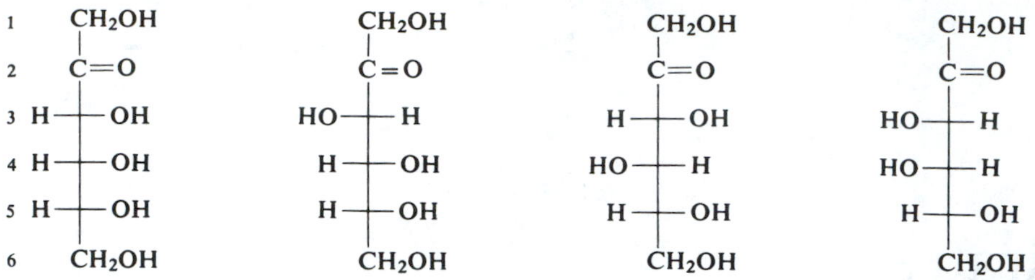

	Psicose	Fructose	Sorbose	Tagatose
1	CH_2OH	CH_2OH	CH_2OH	CH_2OH
2	C=O	C=O	C=O	C=O
3	H—OH	HO—H	H—OH	HO—H
4	H—OH	H—OH	HO—H	HO—H
5	H—OH	H—OH	H—OH	H—OH
6	CH_2OH	CH_2OH	CH_2OH	CH_2OH

34.4 Remembering our approach to Problem 18.19 (page 699), we write:

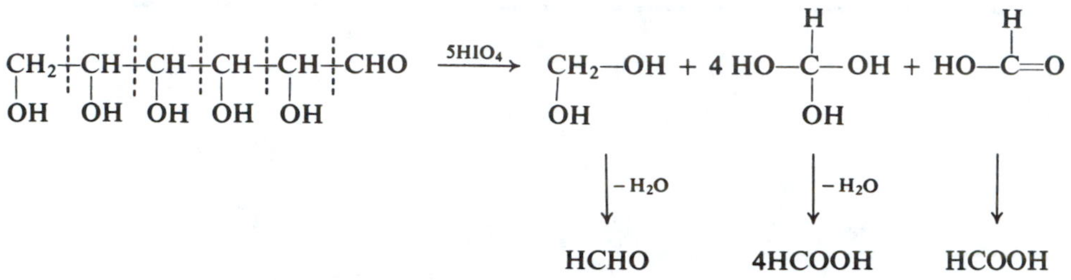

$$\underset{OH}{CH_2} \vdots \underset{OH}{CH} \vdots \underset{OH}{CH} \vdots \underset{OH}{CH} \vdots \underset{OH}{CH} \vdots CHO \xrightarrow{5HIO_4} \underset{OH}{CH_2{-}OH} + 4\ HO{-}\overset{H}{\underset{OH}{C}}{-}OH + HO{-}\overset{H}{C}{=}O$$

$$\downarrow -H_2O \qquad \downarrow -H_2O \qquad \downarrow$$

$$HCHO \qquad 4HCOOH \qquad HCOOH$$

34.5 Remembering our approach to Problem 18.20 (page 699), we write:

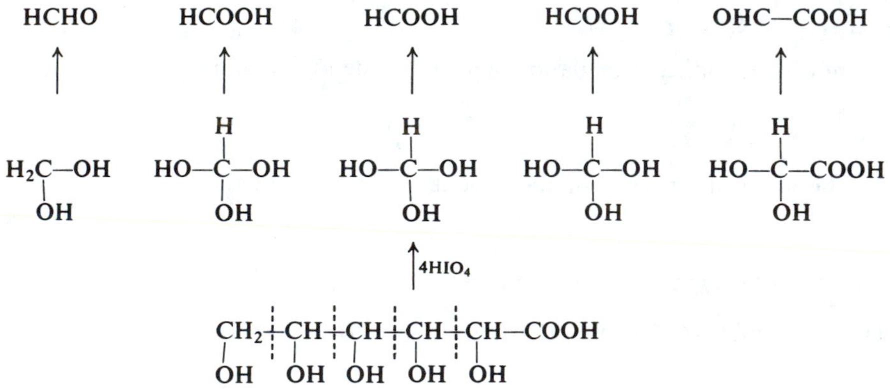

$$HCHO \qquad HCOOH \qquad HCOOH \qquad HCOOH \qquad OHC{-}COOH$$

$$\uparrow \qquad \uparrow \qquad \uparrow \qquad \uparrow \qquad \uparrow$$

$$\underset{OH}{H_2C{-}OH} \qquad HO{-}\overset{H}{\underset{OH}{C}}{-}OH \qquad HO{-}\overset{H}{\underset{OH}{C}}{-}OH \qquad HO{-}\overset{H}{\underset{OH}{C}}{-}OH \qquad HO{-}\overset{H}{\underset{OH}{C}}{-}COOH$$

$$\uparrow 4HIO_4$$

$$\underset{OH}{CH_2} \vdots \underset{OH}{CH} \vdots \underset{OH}{CH} \vdots \underset{OH}{CH} \vdots \underset{OH}{CH}{-}COOH$$

A

Gluconic acid

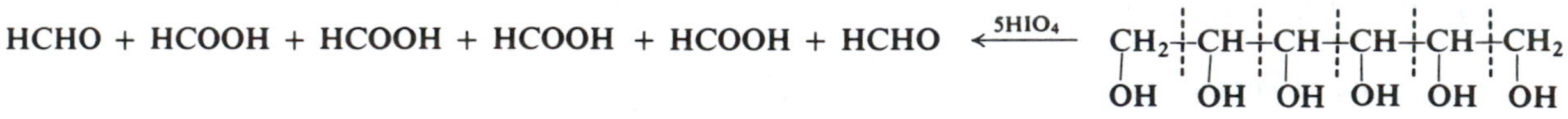

$$HCHO + HCOOH + HCOOH + HCOOH + HCOOH + HCHO \xleftarrow{5HIO_4} \underset{OH}{CH_2} \vdots \underset{OH}{CH} \vdots \underset{OH}{CH} \vdots \underset{OH}{CH} \vdots \underset{OH}{CH} \vdots \underset{OH}{CH_2}$$

B

Glucitol

$$HOOC{-}CHO + HCOOH + HCOOH + OHC{-}COOH \xleftarrow{3HIO_4} HOOC{-}\underset{OH}{CH} \vdots \underset{OH}{CH} \vdots \underset{OH}{CH} \vdots \underset{OH}{CH}{-}COOH$$

C

Glucaric acid

$$HCOOH + HCOOH + HCOOH + HCOOH + OHC{-}COOH \xleftarrow{4HIO_4} O{=}\underset{H}{C}{-}\underset{OH}{CH}{-}\underset{OH}{CH}{-}\underset{OH}{CH}{-}\underset{OH}{CH}{-}COOH$$

D

Glucuronic acid

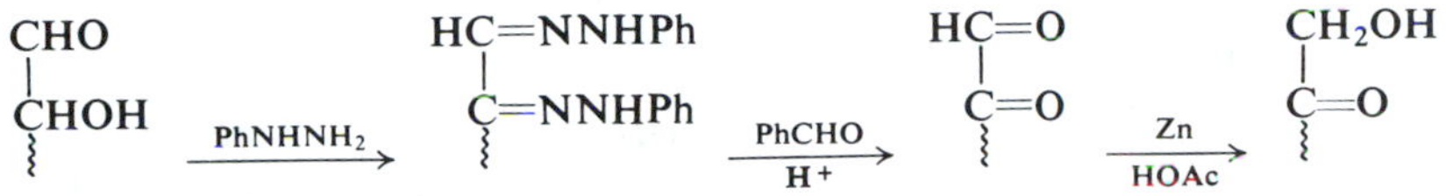

34.6

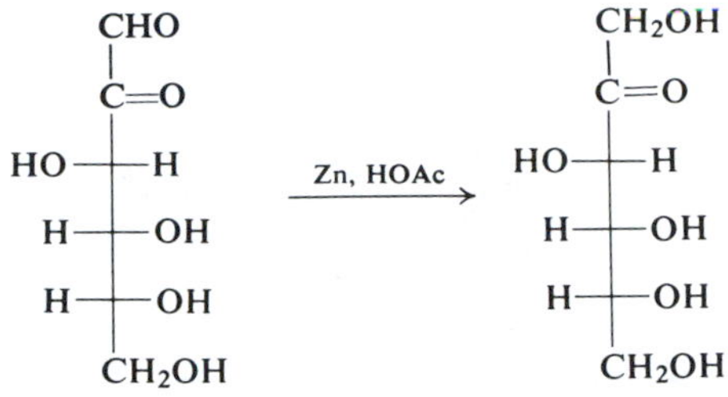

Glucosone: CHO, C=O, HO—H, H—OH, H—OH, CH_2OH $\xrightarrow{Zn,\ HOAc}$ Fructose: CH_2OH, C=O, HO—H, H—OH, H—OH, CH_2OH

Aldose (CHO, CHOH) $\xrightarrow{PhNHNH_2}$ Osazone (HC=NNHPh, C=NNHPh) $\xrightarrow[H^+]{PhCHO}$ Osone (HC=O, C=O) $\xrightarrow[HOAc]{Zn}$ 2-Ketose (CH_2OH, C=O)

34.7 All three sugars have identical configurations at C–3, C–4, and C–5.

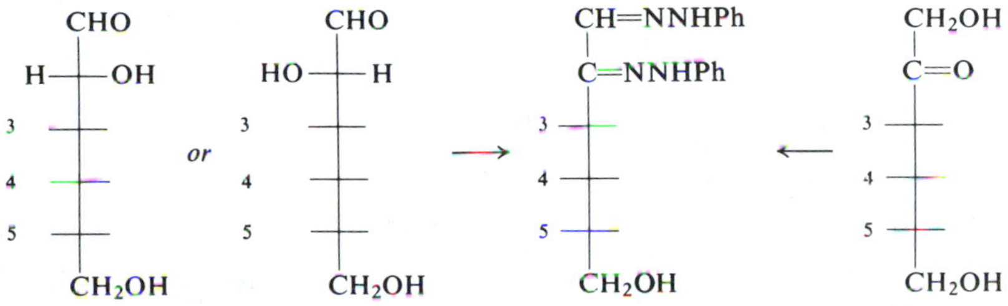

Glucose or mannose (CHO, H—OH, 3, 4, 5, CH_2OH) *or* Mannose or glucose (CHO, HO—H, 3, 4, 5, CH_2OH) ⟶ Osazone (CH=NNHPh, C=NNHPh, 3, 4, 5, CH_2OH) ⟵ Fructose (CH_2OH, C=O, 3, 4, 5, CH_2OH)

34.8 Addition of the lactone to $NaBH_4$—even to a limited quantity—means a temporary excess of reducing agent, which reduces the lactone all the way to the alcohol. For example:

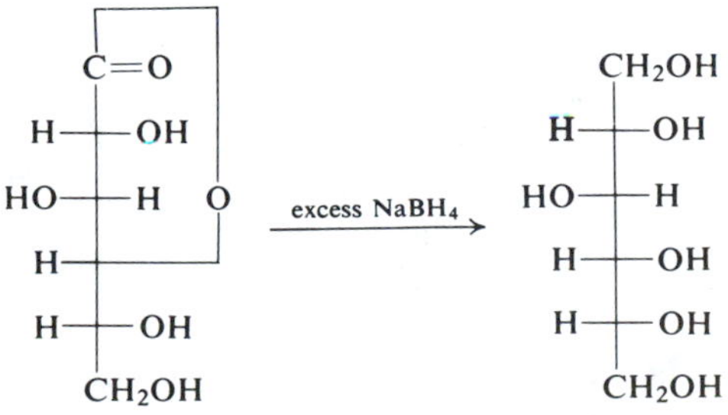

Gluconolactone: C=O, H—OH, HO—H, H—(O, ring), H—OH, CH_2OH $\xrightarrow{\text{excess } NaBH_4}$ Glucitol: CH_2OH, H—OH, HO—H, H—OH, H—OH, CH_2OH

34.9

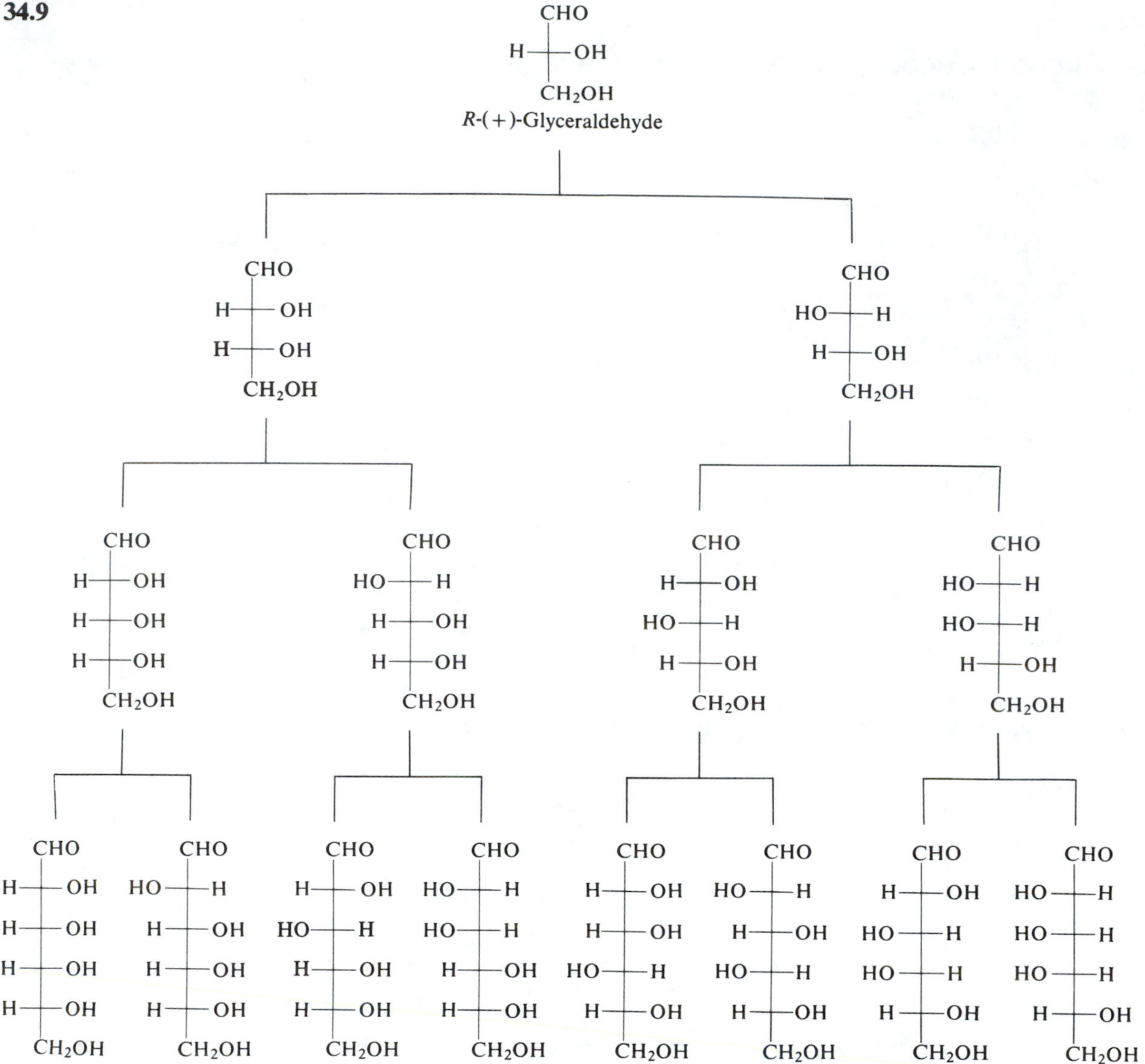

The lowest chiral center has the —OH on the right, since this configuration is undisturbed in building up from *R*-(+)-glyceraldehyde.

34.10 (a)

CHO / H—OH / H—OH / CH_2OH ⟶ COOH / H—OH / H—OH / COOH

(*R*,*S*)-Tartaric acid
(Mesotartaric acid)
Inactive

CHO / HO—H / H—OH / CH_2OH ⟶ COOH / HO—H / H—OH / COOH

(*S*,*S*)-Tartaric acid

Active

(b) Simply carry out the oxidation, and see whether the product is optically active or inactive.

34.11 (a) Because they give the same osazone, (+)-galactose and (+)-talose must differ only in the configuration about C–2; they are a pair of epimers. Thus, they can be I and II, or VII and VIII, but not VI.

$$\begin{array}{ccc} & CHO & \\ H & - & OH \\ H & - & OH \\ H & - & OH \\ H & - & OH \\ & CH_2OH & \\ & \text{I} & \end{array} \qquad \begin{array}{ccc} & CHO & \\ HO & - & H \\ H & - & OH \\ H & - & OH \\ H & - & OH \\ & CH_2OH & \\ & \text{II} & \end{array} \qquad \begin{array}{ccc} & CHO & \\ H & - & OH \\ HO & - & H \\ HO & - & H \\ H & - & OH \\ & CH_2OH & \\ & \text{VII} & \end{array} \qquad \begin{array}{ccc} & CHO & \\ HO & - & H \\ HO & - & H \\ HO & - & H \\ H & - & OH \\ & CH_2OH & \\ & \text{VIII} & \end{array}$$

Of these, I and II would be degraded to ribose:

$$\text{I or II} \longrightarrow \begin{array}{ccc} & CHO & \\ H & - & OH \\ H & - & OH \\ H & - & OH \\ & CH_2OH & \\ & \text{Ribose} & \end{array}$$

VII and VIII would be degraded to lyxose, and hence must be the structures of galactose and talose.

$$\text{VII or VIII} \longrightarrow \begin{array}{ccc} & CHO & \\ HO & - & H \\ HO & - & H \\ H & - & OH \\ & CH_2OH & \\ & \text{Lyxose} & \end{array}$$

Since (−)-lyxose belongs to the same family as (+)-glucose—the *R*-(+)-glyceraldehyde family—so must (+)-galactose and (+)-talose.

This leaves only the problem of deciding which configuration, VII or VIII, represents galactose and which represents talose. On oxidation with nitric acid, VII would give an optically inactive (*meso*) aldaric acid, and VIII would give an optically active aldaric acid.

$$\begin{array}{ccc} & CHO & \\ H & - & OH \\ HO & - & H \\ HO & - & H \\ H & - & OH \\ & CH_2OH & \\ & \text{VII} & \\ & \text{(+)-Galactose} & \end{array} \xrightarrow{HNO_3} \begin{array}{ccc} & COOH & \\ H & - & OH \\ HO & - & H \\ HO & - & H \\ H & - & OH \\ & COOH & \\ & \textit{Meso} & \\ & \textbf{Inactive} & \end{array} \qquad \begin{array}{ccc} & CHO & \\ HO & - & H \\ HO & - & H \\ HO & - & H \\ H & - & OH \\ & CH_2OH & \\ & \text{VIII} & \\ & \text{(+)-Talose} & \end{array} \xrightarrow{HNO_3} \begin{array}{ccc} & COOH & \\ HO & - & H \\ HO & - & H \\ HO & - & H \\ H & - & OH \\ & COOH & \\ & & \\ & \textbf{Active} & \end{array}$$

(+)-Galactose must therefore be VII and (+)-talose must be VIII.

(b) (+)-Allose and (+)-altrose are epimers, differing only in configuration about C–2, the new chiral center generated in the Kiliani–Fischer synthesis. Since (−)-ribose belongs

$$\begin{array}{c} \text{CHO} \\ \text{H}-\!\!\!-\text{OH} \\ \text{H}-\!\!\!-\text{OH} \\ \text{H}-\!\!\!-\text{OH} \\ \text{CH}_2\text{OH} \\ (-)\text{-Ribose} \end{array} \longrightarrow \begin{array}{c} \text{CHO} \\ \text{H}-\!\!\!-\text{OH} \\ \text{H}-\!\!\!-\text{OH} \\ \text{H}-\!\!\!-\text{OH} \\ \text{H}-\!\!\!-\text{OH} \\ \text{CH}_2\text{OH} \\ \text{I} \end{array} \quad \text{and} \quad \begin{array}{c} \text{CHO} \\ \text{HO}-\!\!\!-\text{H} \\ \text{H}-\!\!\!-\text{OH} \\ \text{H}-\!\!\!-\text{OH} \\ \text{H}-\!\!\!-\text{OH} \\ \text{CH}_2\text{OH} \\ \text{II} \end{array}$$

to the same family as (+)-glucose, so do (+)-allose and (+)-altrose. They must be I and II, the only enantiomeric pair left.

Aldose I would give an optically inactive (*meso*) aldaric acid on oxidation, and an optically inactive alditol on reduction. The corresponding products from II would be optically active. (+)-Allose must therefore be I, and (+)-altrose must be II.

$$\begin{array}{c} \text{CHO} \\ \text{H}-\!\!\!-\text{OH} \\ \text{H}-\!\!\!-\text{OH} \\ \text{H}-\!\!\!-\text{OH} \\ \text{H}-\!\!\!-\text{OH} \\ \text{CH}_2\text{OH} \\ \text{I} \\ (+)\text{-Allose} \end{array} \xrightarrow{\text{redn.}} \begin{array}{c} \text{CH}_2\text{OH} \\ \text{H}-\!\!\!-\text{OH} \\ \text{H}-\!\!\!-\text{OH} \\ \text{H}-\!\!\!-\text{OH} \\ \text{H}-\!\!\!-\text{OH} \\ \text{CH}_2\text{OH} \\ \textit{Meso} \\ \textbf{Inactive} \end{array} \qquad \begin{array}{c} \text{CHO} \\ \text{HO}-\!\!\!-\text{H} \\ \text{H}-\!\!\!-\text{OH} \\ \text{H}-\!\!\!-\text{OH} \\ \text{H}-\!\!\!-\text{OH} \\ \text{CH}_2\text{OH} \\ \text{II} \\ (+)\text{-Altrose} \end{array} \xrightarrow{\text{redn.}} \begin{array}{c} \text{CH}_2\text{OH} \\ \text{HO}-\!\!\!-\text{H} \\ \text{H}-\!\!\!-\text{OH} \\ \text{H}-\!\!\!-\text{OH} \\ \text{H}-\!\!\!-\text{OH} \\ \text{CH}_2\text{OH} \\ \\ \textbf{Active} \end{array}$$

(c) This leaves only structure VI for idose. Since (−)-idose and (−)-gulose give the same osazone, they are epimers. (−)-Gulose (structure V, page 1160) belongs to the

$$\begin{array}{c} \text{CHO} \\ \text{HO}-\!\!\!-\text{H} \\ \text{H}-\!\!\!-\text{OH} \\ \text{HO}-\!\!\!-\text{H} \\ \text{H}-\!\!\!-\text{OH} \\ \text{CH}_2\text{OH} \\ \text{VI} \\ (-)\text{-Idose} \end{array} \longrightarrow \begin{array}{c} \text{HC=NNHPh} \\ \text{C=NNHPh} \\ \text{H}-\!\!\!-\text{OH} \\ \text{HO}-\!\!\!-\text{H} \\ \text{H}-\!\!\!-\text{OH} \\ \text{CH}_2\text{OH} \\ \\ \text{Osazone} \end{array} \longleftarrow \begin{array}{c} \text{CHO} \\ \text{H}-\!\!\!-\text{OH} \\ \text{H}-\!\!\!-\text{OH} \\ \text{HO}-\!\!\!-\text{H} \\ \text{H}-\!\!\!-\text{OH} \\ \text{CH}_2\text{OH} \\ \\ (-)\text{-Gulose} \end{array}$$

same family as (+)-glucose, and so, therefore, does (−)-idose.

34.12 See Figure 34.17, page 609 of this Study Guide.

Derived in this systematic way, the aldoses fall into the order given by: **All Altr**uists **Gl**adly **Ma**ke **Gu**m **In Ga**llon **Ta**nks.

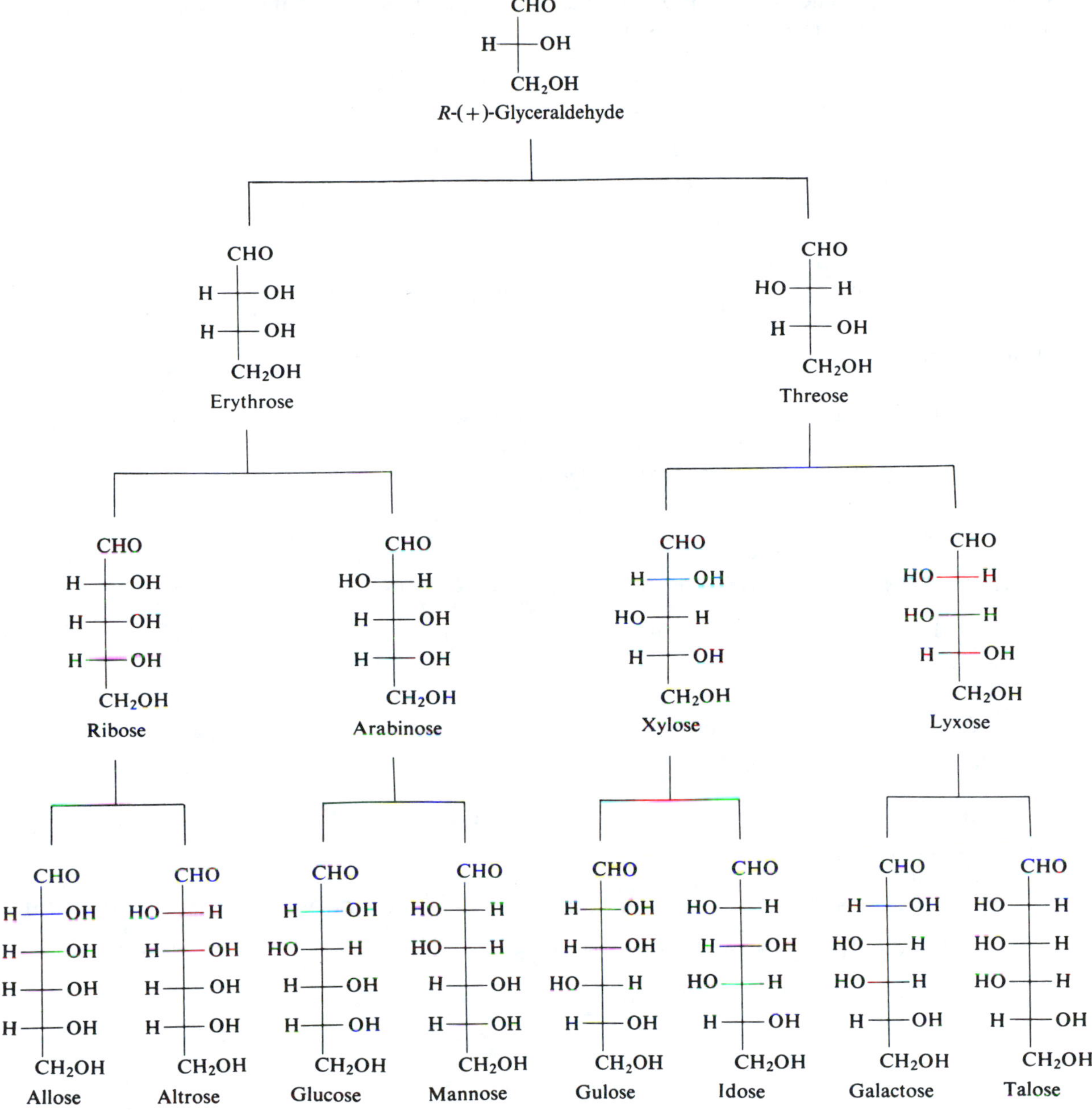

Figure 34.17 Names of descendants of *R*-(+)-glyceraldehyde (Problem 34.12).

34.13 Since (−)-fructose gives the same osazone as (+)-glucose, it must have this configuration:

CH_2OH
C=O
HO—H
H—OH
H—OH
CH_2OH

(−)-Fructose

34.14 These are the enantiomers of III (page 1159) and IV (page 1160), and of the fructose structure in the answer to Problem 34.13 (above).

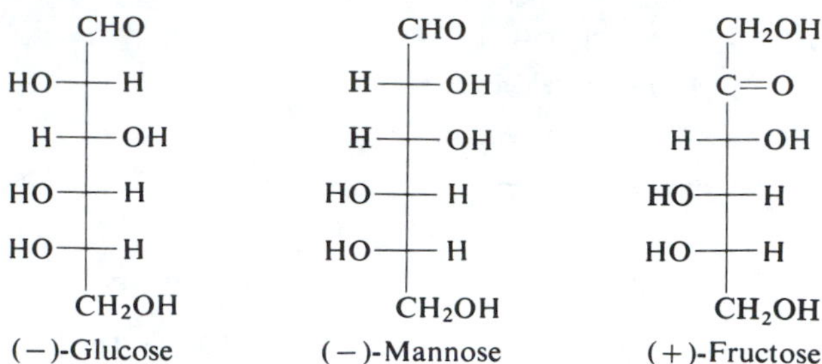

34.15 (a) See the answer to Problem 34.3(c).

(b) D-Psicose: gives the same osazone as D-allose or D-altrose
D-Sorbose: gives the same osazone as D-gulose or D-idose
D-Tagatose: gives the same osazone as D-galactose or D-talose

34.16

(a) CHO
H—C—OH
CH_2OH
R

(b) COOH
H—C—OH
CH_2OH
R

(c) COOH
H—C—OH
CH_2Br
S

(d) COOH
H—C—OH
CH_3
R

34.17

COOH
HO—C—H
CH_3
L-(+)-Lactic acid ⟶ COOEt
HO—C—H
CH_3
A ⟶ CH_2OH
HO—C—H
CH_3
B ⟶ CH_2Br
HO—C—H
CH_3
C

C ⟶ CH_2CN
HO—C—H
CH_3
D ⟶ CH_2COOH
HO—C—H
CH_3
E ⟶ CH_2COOMe
HO—C—H
CH_3
F ⟶ CH_2CH_2OH
HO—C—H
CH_3
G

G ⟶ CH_2CH_2I
HO—C—H
CH_3
H ⟶ C_2H_5
HO—C—H
CH_3
S-(+)-2-Butanol

34.18

(a) COOH
HO—C—H
H—C—OH
COOH
(−)-Tartaric acid
(*S*,*S*)

(b) COOH
H—C—OH
HO—C—H
COOH
(+)-Tartaric acid
(*R*,*R*)

(c) COOH
H—C—OH
H—C—OH
COOH
Mesotartaric acid
(*R*,*S*)

34.19 (a) This ratio is determined in the initial step: formation of the cyanohydrins. A chiral center is already present in the reactants; the transition states, like the products, are diastereomeric and of different stabilities.

(b)

L-(−)-Glyceraldehyde: CHO; HO—H; CH_2OH $\xrightarrow{HCN}$

Upper path: CN; HO—H; HO—H; CH_2OH $\xrightarrow{Ba(OH)_2}$ COOH; HO—H; HO—H; CH_2OH $\xrightarrow{HNO_3}$ COOH; HO—H; HO—H; COOH

meso-Tartaric acid

Lower path: CN; H—OH; HO—H; CH_2OH $\xrightarrow{Ba(OH)_2}$ COOH; H—OH; HO—H; CH_2OH $\xrightarrow{HNO_3}$ COOH; H—OH; HO—H; COOH

L-(+)-Tartaric acid

The ratio *meso*:active will again be 1:3. In cyanohydrin formation, the two transition states are enantiomers of the transition states in the reaction of D-(+)-glyceraldehyde. The energy difference between transition states is the same as in the other case, and so, also, is the ratio between the products.

(c) Inactive, because the enantiomeric cyanohydrins will be formed in equal amounts. The isomer favored in the L-series will be exactly balanced by its enantiomer, favored in the D-series.

34.20

D-(+)-Glucose: CHO; H—OH; HO—H; H—OH; H—OH; CH_2OH → (+)-Glucaric acid: COOH; H—OH; HO—H; H—OH; H—OH; COOH

Upper path: A (Lactone): COOH; H—OH; —H; O H—OH; H—OH; C=O → C (Gluconic acid): COOH; H—OH; HO—H; H—OH; H—OH; CH_2OH → D (Gluconolactone): C=O; H—OH; HO—H O; H—; H—OH; CH_2OH → D-(+)-Glucose: CHO; H—OH; HO—H; H—OH; H—OH; CH_2OH

A
Lactone

C
Gluconic acid

D
Gluconolactone

D-(+)-Glucose

Lower path: B (Lactone): C=O; H—OH; HO—H O; H—; H—OH; COOH → E (Gulonic acid): CH_2OH; H—OH; HO—H; H—OH; H—OH; COOH → F (Gulonolactone): CH_2OH; H—OH; —H; O H—OH; H—OH; C=O → L-(+)-Gulose: CH_2OH; H—OH; HO—H; H—OH; H—OH; CHO

B
Lactone

E
Gulonic acid

F
Gulonolactone

L-(+)-Gulose

(+)-Gulose is a member of the L-family since, with —CHO at the top, the —OH on the bottom chiral carbon is on the left. It is the enantiomer of V on page 1160.

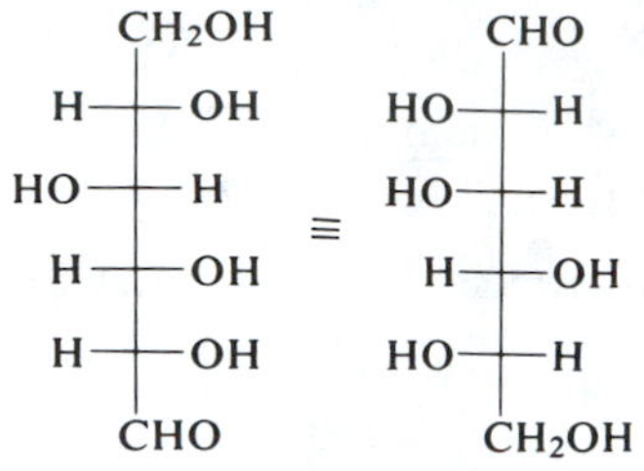
CH2OH
H—OH
HO—H
H—OH
H—OH
CHO
≡
CHO
HO—H
HO—H
H—OH
HO—H
CH2OH

L-(+)-Gulose

34.21 (a)

$$N_\alpha \times 112° + N_\beta \times 19° = 52.7°$$

and

$$N_\alpha + N_\beta = 1$$

where N is the mole fraction of each anomer.

Solving, we find

$$N_\alpha = 36.2\% \qquad N_\beta = 63.8\%$$

(b) The β-form predominates because in it the anomeric —OH group is equatorial, not axial as it is in the α-form.

34.22

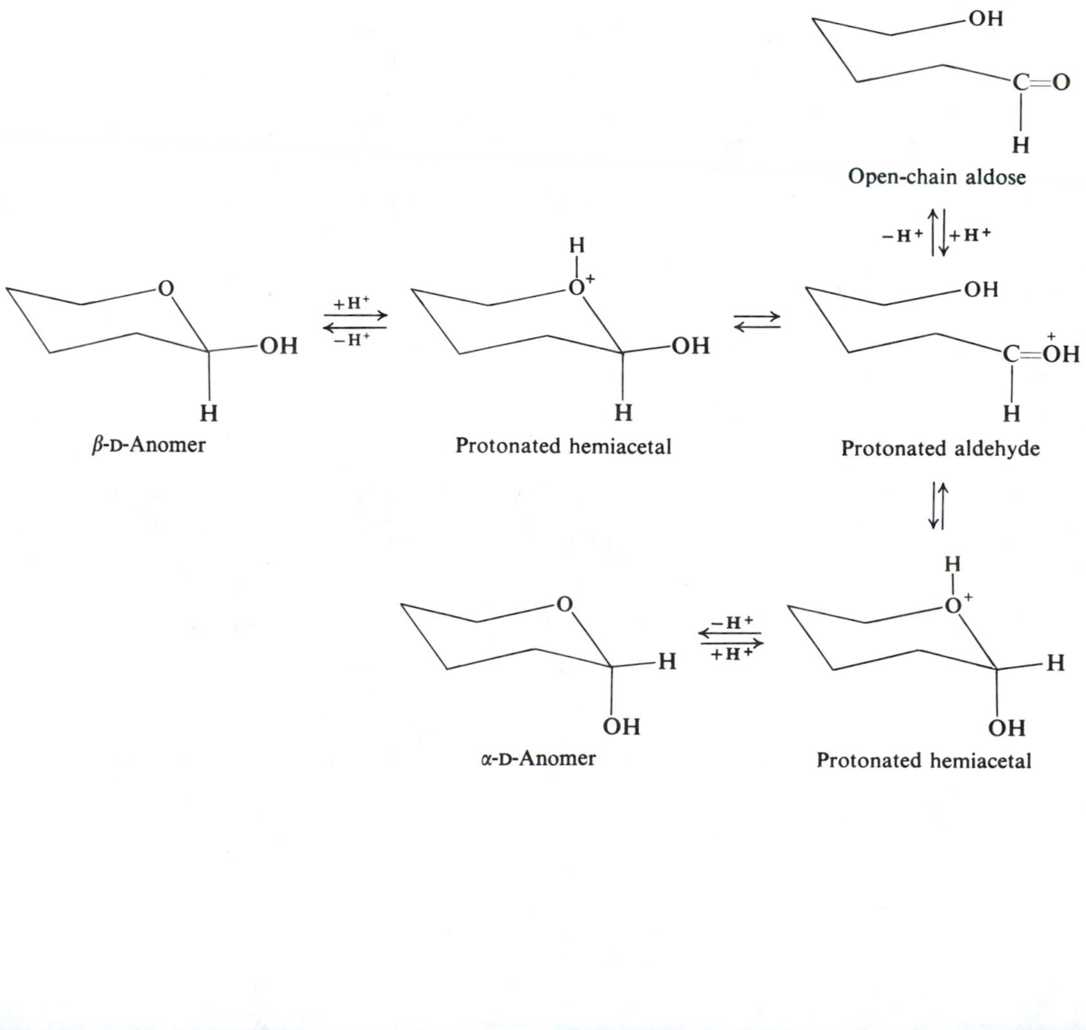
Open-chain aldose
−H⁺ +H⁺
β-D-Anomer
+H⁺
−H⁺
Protonated hemiacetal
Protonated aldehyde
α-D-Anomer
−H⁺
+H⁺
Protonated hemiacetal

34.23 (+)-Glucose $\xrightarrow{Ac_2O}$ β-Pentaacetate and α-Pentaacetate

β-Pentaacetate

α-Pentaacetate

Acetylation of D-glucose does not give, as we might naively have assumed, an open-chain aldehyde with acetate groups at C–2 through C–6. Instead, the products are cyclic; acetylation occurs, not at C–5, but at C–1, and to give α- and β-forms.

34.24 (a) Acetal $\xrightarrow[H^+]{H_2O}$ H—C=O / COOH (Glyoxylic acid) + CH_3OH + COOH / H—C—OH / CH_2OH (D-(−)-Glyceric acid)

Acetal

Glyoxylic acid

D-(−)-Glyceric acid

These products are common to *all* methyl D-aldohexopyranosides.

(b) (+)-Glucose gave the same (−)-glyceric acid as that obtained by oxidation of (+)-glyceraldehyde.

34.25 $\xrightarrow{2HIO_4}$ ... + HCHO

The easily detected difference: production of HCHO.

34.26 (a) Trimethoxyglutaric acid $\xleftarrow{HNO_3}$ 2,3,4-Tri-*O*-methylpentose $\xleftarrow[H^+]{H_2O}$ Methyl 2,3,4-tri-*O*-methylpentoside $\xleftarrow[OH^-]{Me_2SO_4}$ Methyl pentoside

Trimethoxyglutaric acid

2,3,4-Tri-*O*-methylpentose

Methyl 2,3,4-tri-*O*-methylpentoside

Methyl pentoside

The —OH on C–5 was not methylated because it was not available: it must have been tied up in the ring; and the ring must therefore have been six-membered (C–1 through C–5, plus an oxygen).

(b) 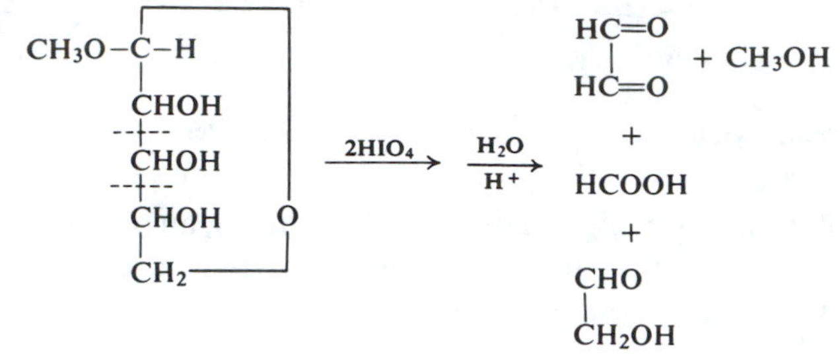 CH_3O–C–H / CHOH / CHOH / CHOH / CH_2 (ring O) $\xrightarrow{2HIO_4}$ $\xrightarrow[H^+]{H_2O}$ HC=O / HC=O + CH_3OH + HCOOH + CHO / CH_2OH

34.27 (a) 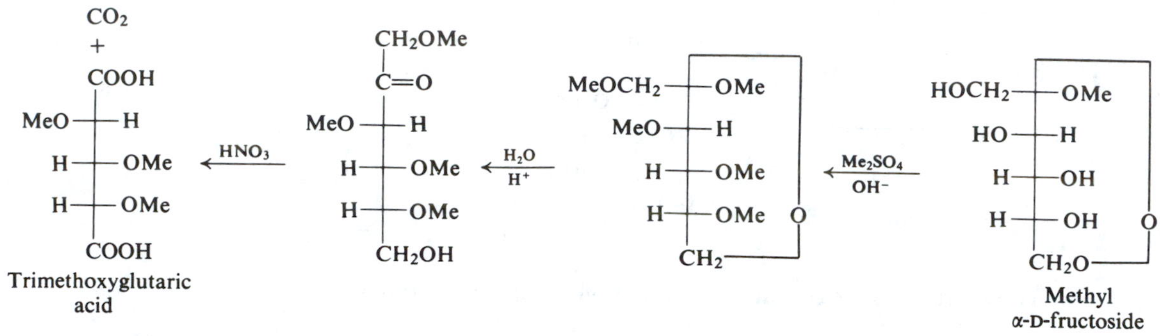 CO_2 + COOH / MeO—H / H—OMe / H—OMe / COOH (Trimethoxyglutaric acid) $\xleftarrow{HNO_3}$ CH_2OMe / C=O / MeO—H / H—OMe / H—OMe / CH_2OH $\xleftarrow[H^+]{H_2O}$ $MeOCH_2$—OMe / MeO—H / H—OMe / H—OMe / CH_2 (ring O) $\xleftarrow[OH^-]{Me_2SO_4}$ $HOCH_2$—OMe / HO—H / H—OH / H—OH / CH_2O (ring) (Methyl α-D-fructoside)

The —OH on C–6 can be oxidized because it was not methylated: it must have been tied up in the ring; and the ring must therefore have been six-membered (C–2 through C–6, plus one oxygen).

(b) 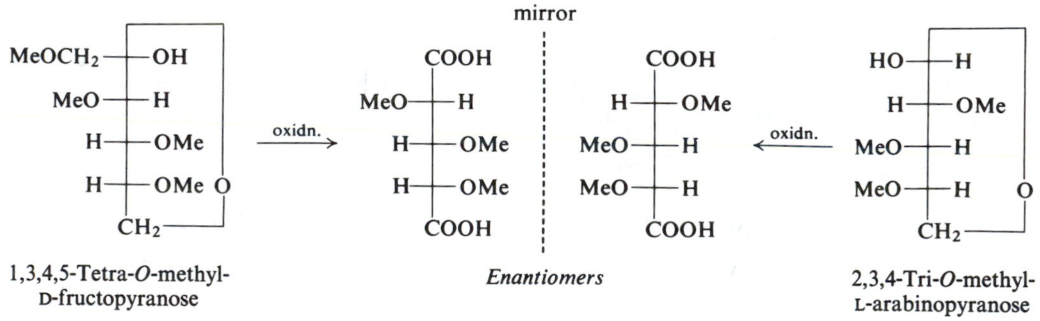 $MeOCH_2$—OH / MeO—H / H—OMe / H—OMe / CH_2 (ring O) (1,3,4,5-Tetra-*O*-methyl-D-fructopyranose) $\xrightarrow{oxidn.}$ COOH / MeO—H / H—OMe / H—OMe / COOH | mirror | COOH / H—OMe / MeO—H / MeO—H / COOH (*Enantiomers*) $\xleftarrow{oxidn.}$ HO—H / H—OMe / MeO—H / MeO—H / CH_2 (ring O) (2,3,4-Tri-*O*-methyl-L-arabinopyranose)

34.28 (a) 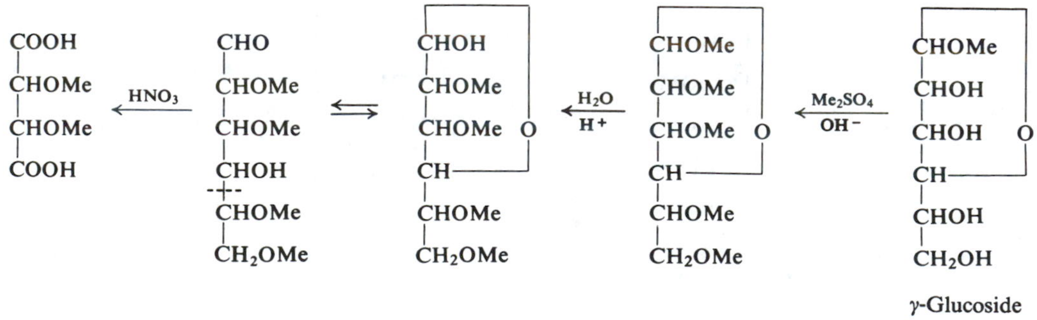 COOH / CHOMe / CHOMe / COOH $\xleftarrow{HNO_3}$ CHO / CHOMe / CHOMe / CHOH / CHOMe / CH_2OMe $\rightleftharpoons$ CHOH / CHOMe / CHOMe / CH / CHOMe / CH_2OMe (ring O) $\xleftarrow[H^+]{H_2O}$ CHOMe / CHOMe / CHOMe / CH / CHOMe / CH_2OMe (ring O) $\xleftarrow[OH^-]{Me_2SO_4}$ CHOMe / CHOH / CHOH / CH / CHOH / CH_2OH (ring O) (γ-Glucoside)

The formation of a dimethoxysuccinic acid (without accompanying glutaric acid) indicates a "free"—OH on C–4. This means a five-membered ring in the γ-glucoside.

(b) 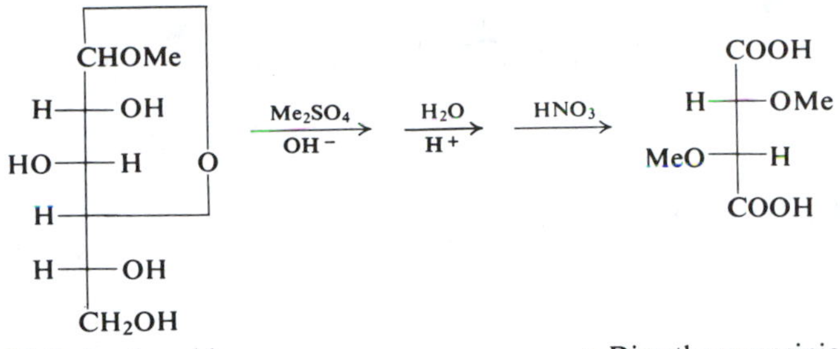 CHOMe, H—OH, HO—H, H—, H—OH, CH_2OH (ring O) $\xrightarrow[OH^-]{Me_2SO_4}$ $\xrightarrow[H^+]{H_2O}$ $\xrightarrow{HNO_3}$ COOH, H—OMe, MeO—H, COOH

Methyl γ-glucoside

L-Dimethoxysuccinic acid
Optically active

(c) Following the argument of (a), the γ-fructoside also contains a five-membered ring, the dimethoxysuccinic acid coming from C–2, C–3, C–4, and C–5.

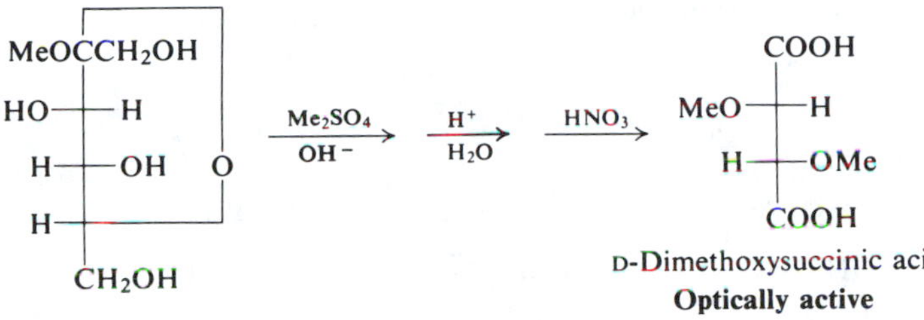

$MeOCCH_2OH$, HO—H, H—OH, H—, CH_2OH (ring O) $\xrightarrow[OH^-]{Me_2SO_4}$ $\xrightarrow[H_2O]{H^+}$ $\xrightarrow{HNO_3}$ COOH, MeO—H, H—OMe, COOH

D-Dimethoxysuccinic acid
Optically active

This acid, too, is optically active; it is the enantiomer of the acid from the γ-glucoside.

34.29 In each case, the prediction given below has been confirmed by experiment.

(a) 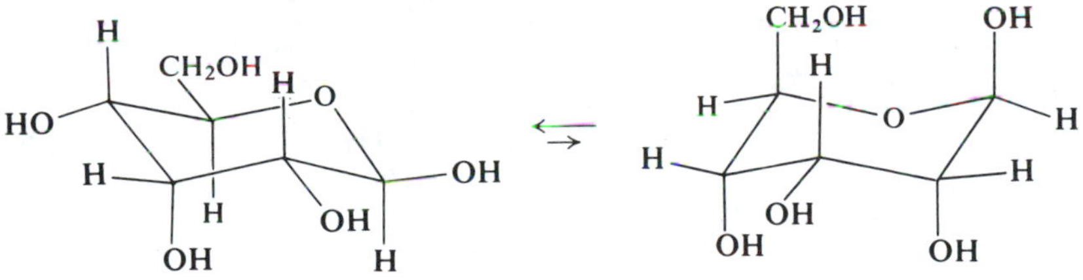 [chair conformations] ⇄

More stable

β-D-Allopyranose

(b) 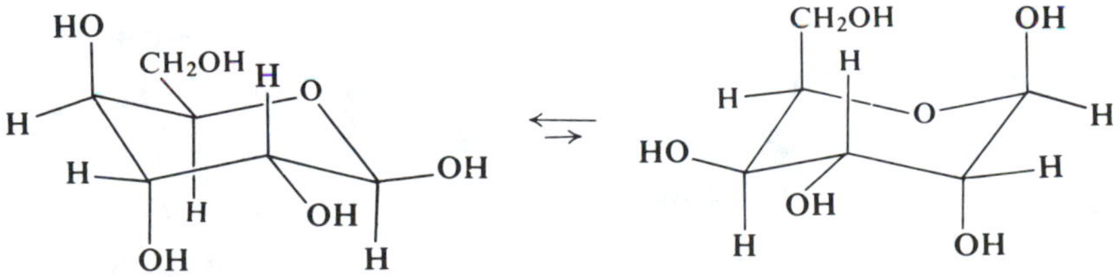 [chair conformations] ⇄

More stable

β-D-Gulopyranose

(c)

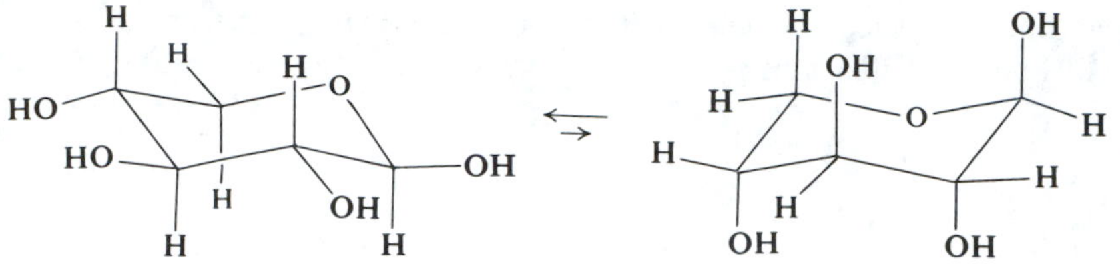

More stable

β-D-Xylopyranose

(d)

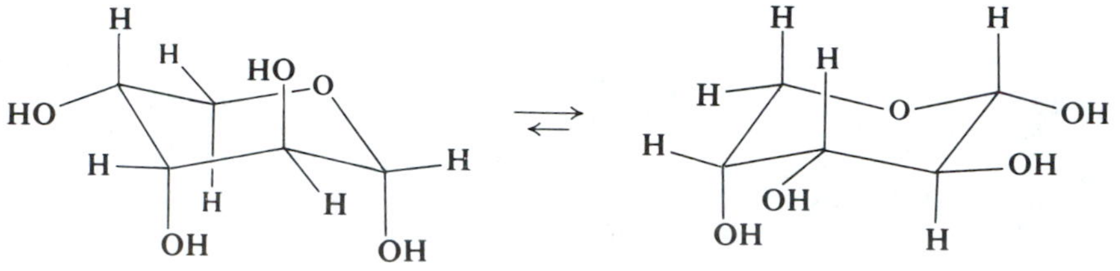

More stable

α-D-Arabinopyranose

(e)

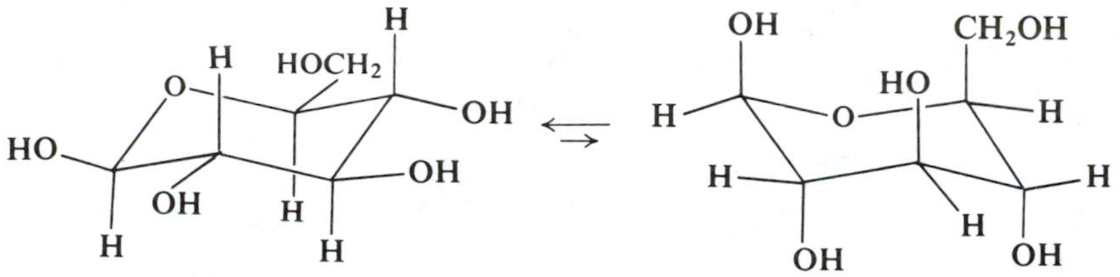

More stable

β-L-Glucopyranose

(f)

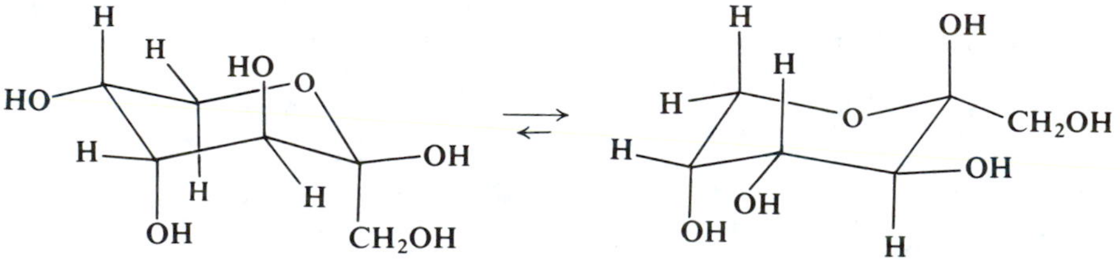

More stable

β-D-Fructopyranose

1. (a)

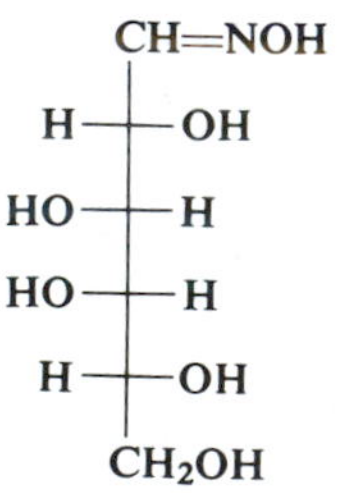

CH=NOH
H—OH
HO—H
HO—H
H—OH
CH_2OH

D-Galactosoxime

(b)

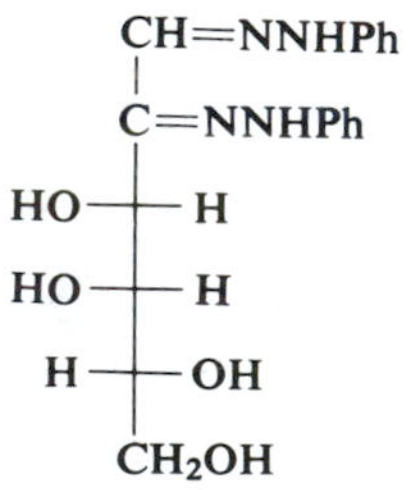

CH=NNHPh
C=NNHPh
HO—H
HO—H
H—OH
CH_2OH

D-Galactosazone

(c)

COOH
H—OH
HO—H
HO—H
H—OH
CH_2OH

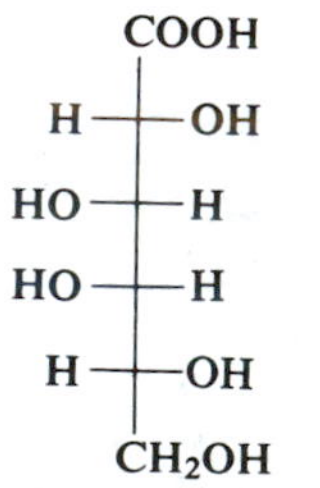

D-Galactonic acid
(or lactone)

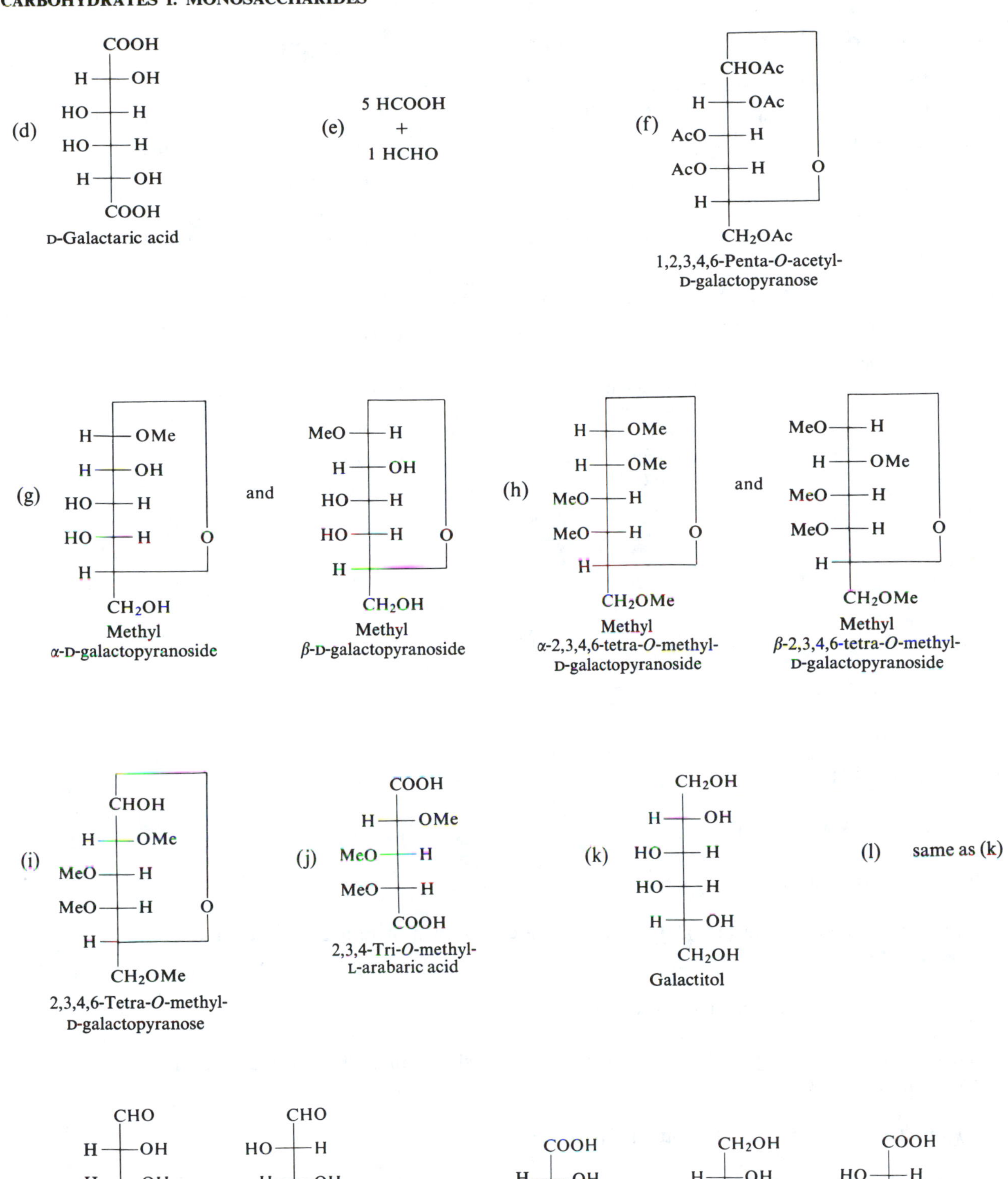
(d)
D-Galactaric acid
(e)
5 HCOOH
+
1 HCHO
(f)
1,2,3,4,6-Penta-O-acetyl-D-galactopyranose
(g)
and
Methyl α-D-galactopyranoside
Methyl β-D-galactopyranoside
(h)
and
Methyl α-2,3,4,6-tetra-O-methyl-D-galactopyranoside
Methyl β-2,3,4,6-tetra-O-methyl-D-galactopyranoside
(i)
2,3,4,6-Tetra-O-methyl-D-galactopyranose
(j)
2,3,4-Tri-O-methyl-L-arabaric acid
(k)
Galactitol
(l) same as (k)
(m)
Two epimeric aldoheptoses
(n)
D-Galactonic acid
L-Galactonic acid

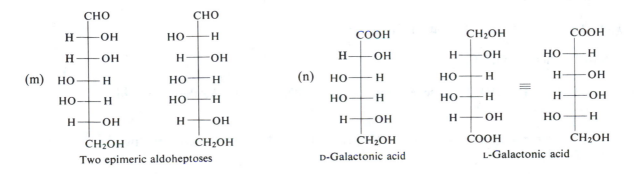

(o)

CHO
HO—H
HO—H
HO—H
H—OH
CH_2OH
D-Talose

and D-galactose

(p)

CHO
C=O
HO—H
HO—H
H—OH
CH_2OH
D-Galactosone

(q)

CH_2OH
C=O
HO—H
HO—H
H—OH
CH_2OH
D-Tagatose

(r)

CHO
HO—H
HO—H
H—OH
CH_2OH
D-Lyxose

(s) No reaction

(t) Product on page 1174 from α-anomer; β-anomer gives opposite configuration at C–1

(u)

CHO
COOH
\+
COOH
H—OH
CH_2OH
D-Glyceric acid

2. (a) MeOH, HCl.

(b) Product (a), Me_2SO_4, OH^-.

(c) Product (b), dil. HCl.

(d) See Figure 34.5 (page 1155). The starting material shown there is D-glucose, and the product is D-mannose.

(e) See the answer to Problem 34.20, above.

(f) See Figure 34.4 (page 1154). The starting material shown there is D-glucose, and the product is D-arabinose.

(g) Product (f) treated as at the bottom of page 1161.

(h) $NaBH_4$; then Ac_2O, H^+.

(i) $3PhNHNH_2$; PhCHO, H^+ (osone formation); Zn, HOAc. (Compare the answer to Problem 34.6, above.)

(j) Kiliani–Fischer synthesis (Sec. 34.8): CN^-, H^+; H_2O, H^+, warm; separate the two lactones; Na(Hg), CO_2.

3. (a) $2ClCH_2CHO$ + $BrMgC{\equiv}CMgBr$ $\longrightarrow$

CH_2Cl
H—OH
C
|||
C
H—OH
CH_2Cl
Meso
A

$\xrightarrow{KOH}$

CH_2—O (epoxide)
H—
C
|||
C
H—
CH_2—O (epoxide)
B

$\xrightarrow[H_2O]{OH^-}$

CH_2OH
H—OH
C
|||
C
H—OH
CH_2OH
C

$\xrightarrow[Pd]{H_2}$

CH_2OH
H—OH
CH
||
CH
H—OH
CH_2OH
cis
D

$\downarrow KMnO_4$

CH_2OH		CH_2OH
H—OH		H—OH
H—OH	+	HO—H
H—OH		HO—H
H—OH		H—OH
CH_2OH		CH_2OH
D-Allitol		D-Galactitol
Meso		*Meso*

E and E′

D $\xrightarrow{HCO_2H}$

CH_2OH		CH_2OH		CH_2OH
H—OH		H—OH		HO—H
HO—H	+	H—OH	≡	H—OH
H—OH		HO—H		HO—H
H—OH		H—OH		HO—H
CH_2OH		CH_2OH		CH_2OH

Glucitol (or gulitol)
Racemic
F

C $\xrightarrow[NH_3]{Na}$

CH_2OH	CH_2OH
H—OH	H—OH
H(C=C)H (*trans*)	H(C=C)H (*trans*)
H—OH	H—OH
CH_2OH	CH_2OH

Two diastereomers
G

$\xrightarrow{KMnO_4}$

CH_2OH	CH_2OH		CH_2OH
H—OH	H—OH		HO—H
HO—H	H—OH	≡	H—OH
H—OH	HO—H		HO—H
H—OH	H—OH		HO—H
CH_2OH	CH_2OH		CH_2OH

Glucitol (or gulitol)
Racemic
H

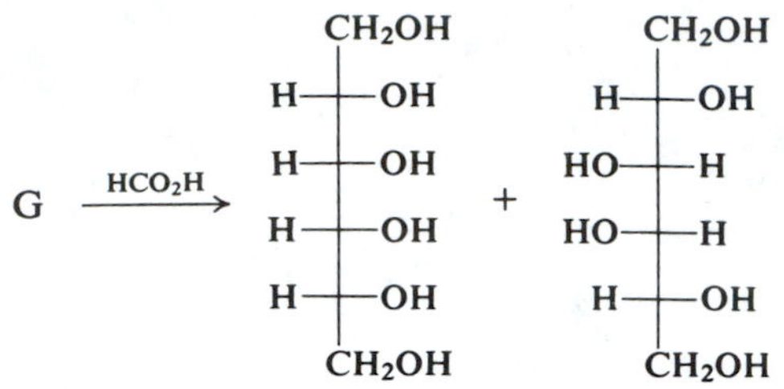

G $\xrightarrow{HCO_2H}$

CH_2OH
H—OH
H—OH
H—OH
H—OH
CH_2OH
D-Allitol
Meso

+

CH_2OH
H—OH
HO—H
HO—H
H—OH
CH_2OH
D-Galactitol
Meso

I and I′

(b)

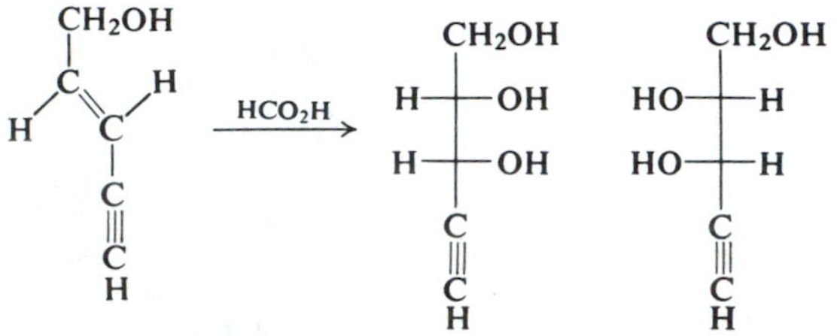

CH_2OH — C(H)=C(H) — C≡C—H

$\xrightarrow{HCO_2H}$

CH_2OH
H—OH
H—OH
C≡CH

CH_2OH
HO—H
HO—H
C≡CH

Racemic
J

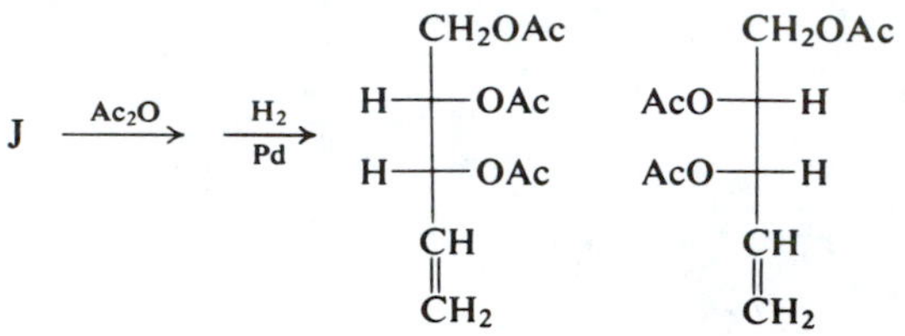

J $\xrightarrow{Ac_2O}$ $\xrightarrow[Pd]{H_2}$

CH_2OAc
H—OAc
H—OAc
CH=CH_2

CH_2OAc
AcO—H
AcO—H
CH=CH_2

Racemic
K

↓ HOBr

CH_2OAc
H—OAc
H—OAc
H—OH
CH_2Br

CH_2OAc
H—OAc
H—OAc
HO—H
CH_2Br

CH_2OAc
AcO—H
AcO—H
H—OH
CH_2Br

CH_2OAc
AcO—H
AcO—H
HO—H
CH_2Br

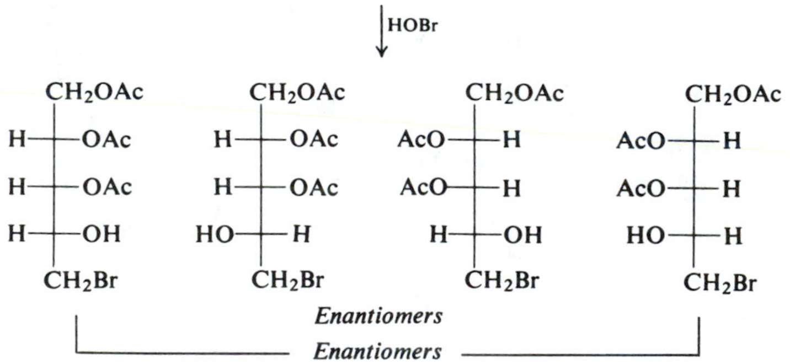

Enantiomers
Enantiomers
L and M

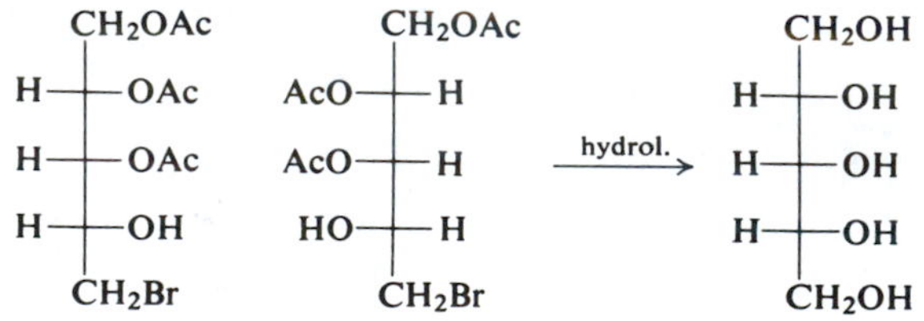

CH_2OAc
H—OAc
H—OAc
H—OH
CH_2Br

CH_2OAc
AcO—H
AcO—H
HO—H
CH_2Br

L

$\xrightarrow{\text{hydrol.}}$

CH_2OH
H—OH
H—OH
H—OH
CH_2OH
Meso
N
Ribitol

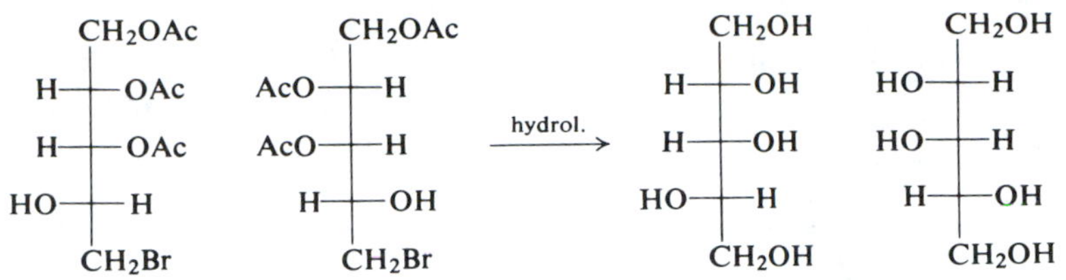

$$\begin{array}{c} CH_2OAc \\ H-C-OAc \\ H-C-OAc \\ HO-C-H \\ CH_2Br \end{array} \quad \begin{array}{c} CH_2OAc \\ AcO-C-H \\ AcO-C-H \\ H-C-OH \\ CH_2Br \end{array} \xrightarrow{\text{hydrol.}} \begin{array}{c} CH_2OH \\ H-C-OH \\ H-C-OH \\ HO-C-H \\ CH_2OH \end{array} \quad \begin{array}{c} CH_2OH \\ HO-C-H \\ HO-C-H \\ H-C-OH \\ CH_2OH \end{array}$$

M

Racemic
O
Arabitol
(Lyxitol)

(c) There are two possible routes to each.

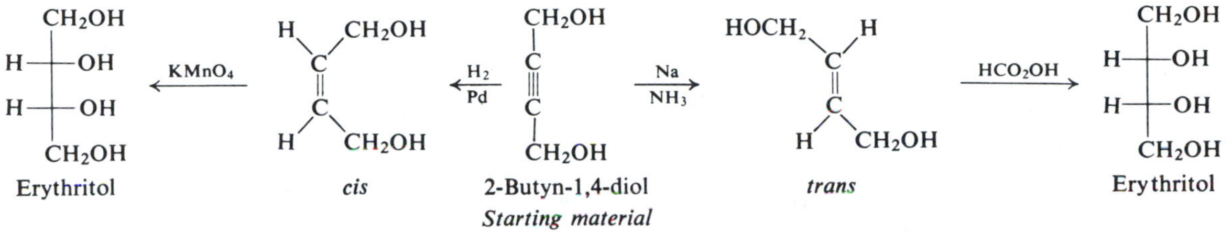

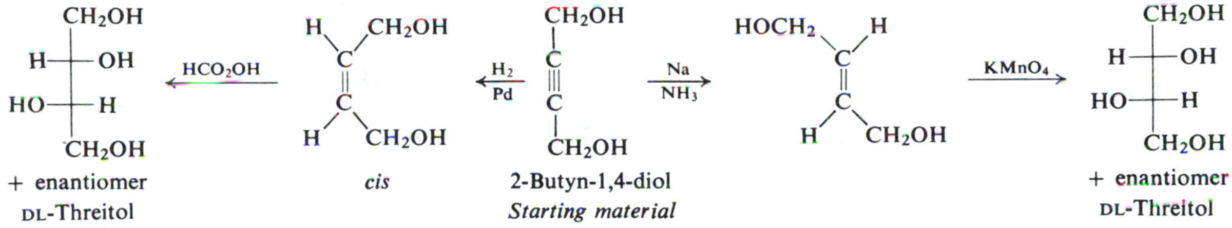

$$\text{Erythritol} \xleftarrow{KMnO_4} \textit{cis}\ HOCH_2CH{=}CHCH_2OH \xleftarrow[Pd]{H_2} \underset{\substack{\text{2-Butyn-1,4-diol}\\ \textit{Starting material}}}{HOCH_2C{\equiv}CCH_2OH} \xrightarrow[NH_3]{Na} \textit{trans}\ HOCH_2CH{=}CHCH_2OH \xrightarrow{HCO_2OH} \text{Erythritol}$$

Erythritol: $\begin{array}{c} CH_2OH \\ H-C-OH \\ H-C-OH \\ CH_2OH \end{array}$

$$\text{DL-Threitol} \xleftarrow{HCO_2OH} \textit{cis}\ HOCH_2CH{=}CHCH_2OH \xleftarrow[Pd]{H_2} \underset{\substack{\text{2-Butyn-1,4-diol}\\ \textit{Starting material}}}{HOCH_2C{\equiv}CCH_2OH} \xrightarrow[NH_3]{Na} \textit{trans}\ HOCH_2CH{=}CHCH_2OH \xrightarrow{KMnO_4} \text{DL-Threitol}$$

$\begin{array}{c} CH_2OH \\ H-C-OH \\ HO-C-H \\ CH_2OH \end{array}$ + enantiomer
DL-Threitol

(d) Nucleophilic carbonyl addition.

4. α- (chair structure: COOH, H, HO, HO, OH, H, OR axial) *or* β- (chair structure: COOH, H, HO, HO, OH, H, OR equatorial)

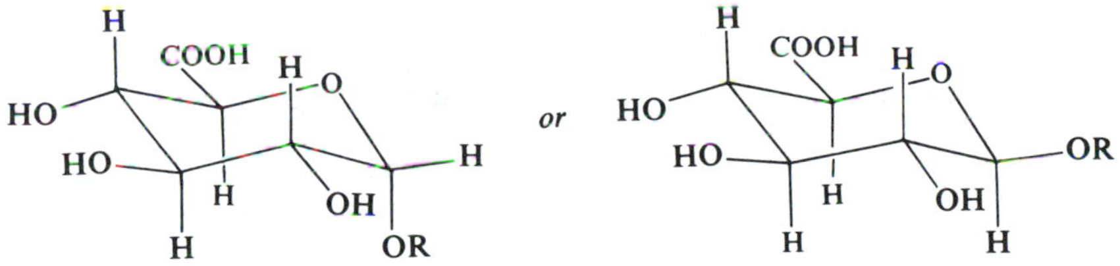

P

5. The rate-determining step involves OH^- before reaction with Cu^{2+}: probably abstraction of a proton leading to the formation of the enediol (page 1150). (Compare Section 21.3 and the answer to Problem 23, Chapter 21.)

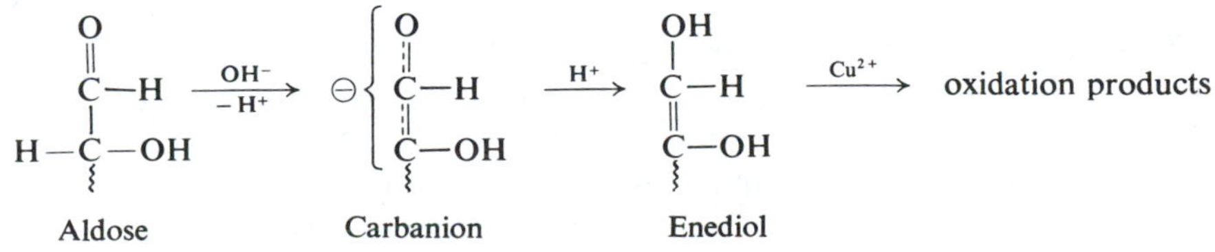

$$\underset{\text{Aldose}}{\begin{array}{c} O \\ \| \\ C-H \\ | \\ H-C-OH \end{array}} \xrightarrow[-H^+]{OH^-} \ominus\left\{\underset{\text{Carbanion}}{\begin{array}{c} O \\ \vdots \\ C-H \\ \vdots \\ C-OH \end{array}}\right. \xrightarrow{H^+} \underset{\text{Enediol}}{\begin{array}{c} OH \\ | \\ C-H \\ \| \\ C-OH \end{array}} \xrightarrow{Cu^{2+}} \text{oxidation products}$$

6. (a)

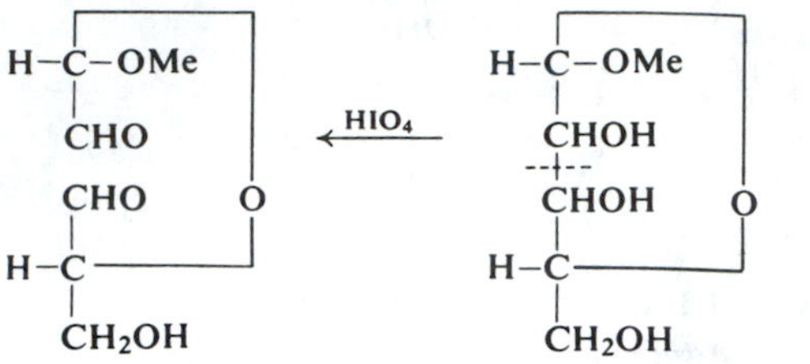

H–C–OMe / CHO / CHO / H–C / CH_2OH (ring O) ← HIO_4 — H–C–OMe / CHOH / CHOH / H–C / CH_2OH (ring O)

Q

Five-membered ring

(b) The configuration is known for C–1 and C–4.

(c)

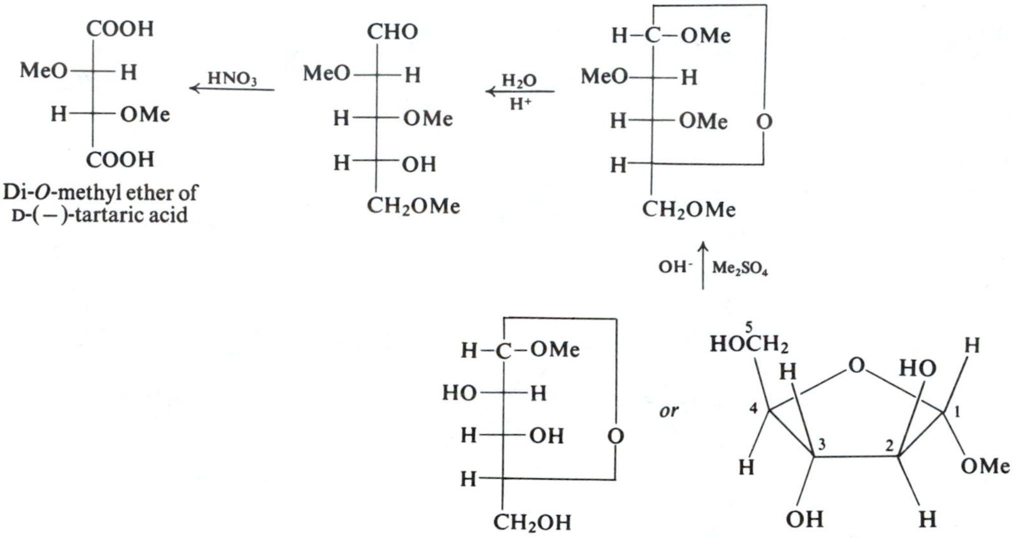

COOH / MeO–H / H–OMe / COOH

Di-*O*-methyl ether of D-(−)-tartaric acid

← HNO_3 — CHO / MeO–H / H–OMe / H–OH / CH_2OMe ← H_2O, H^+ — H–C–OMe / MeO–H / H–OMe / H / CH_2OMe (ring O)

↑ OH^- Me_2SO_4

H–C–OMe / HO–H / H–OH / H / CH_2OH (ring O) *or* [furanose ring: 5 $HOCH_2$, H, O, HO, H, 4, 1, 3, 2, H, OMe, OH, H]

Methyl α-D-arabinofuranoside

Q

7. Salicin is a non-reducing sugar and evidently a β-glucoside (hydrolyzed by emulsin). From

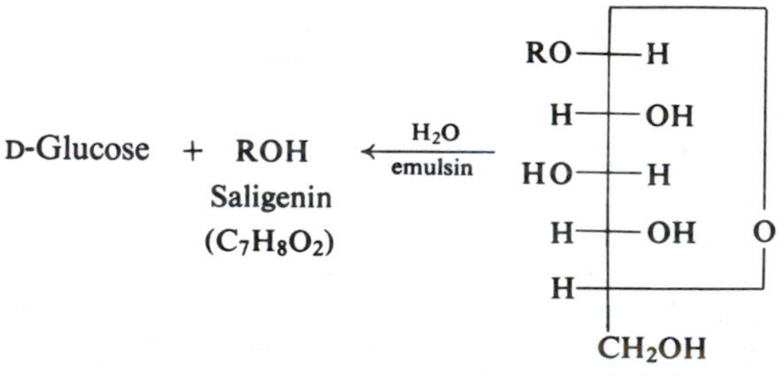

D-Glucose + ROH ← H_2O, emulsin — RO–H / H–OH / HO–H / H–OH / H / CH_2OH (ring O)

Saligenin ($C_7H_8O_2$)

Salicin

(*Pyranose ring assumed*)

its formula ($C_7H_8O_2$) saligenin, ROH, appears to be aromatic.

The HNO_3 treatment evidently does not affect the glucose moiety—subsequent hydrolysis gives D-glucose unchanged—and hence it must be a group in R of saligenin that is oxidized. The isolation of salicylaldehyde shows that this can only involve oxidation of —CH_2OH to —CHO. We now know the structure of a saligenin and hence of salicin.

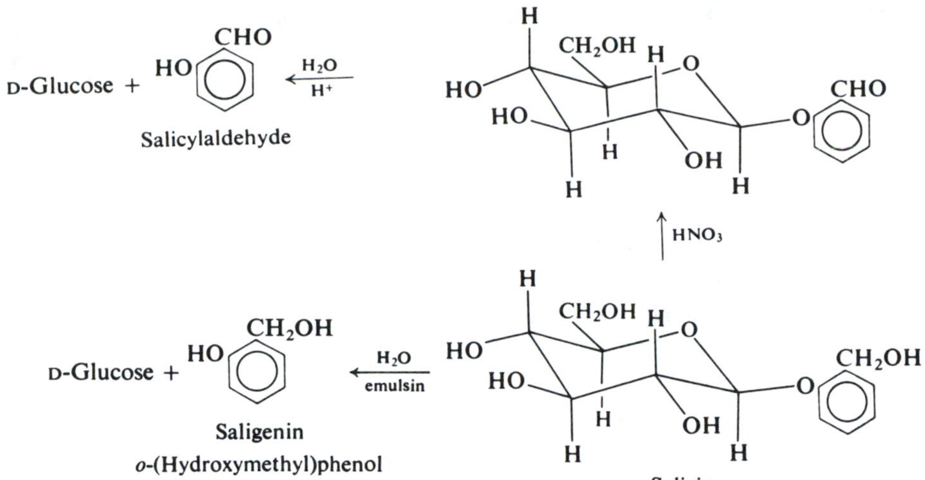

D-Glucose + Salicylaldehyde $\xleftarrow[H^+]{H_2O}$ (o-formylphenyl β-D-glucopyranoside)

↑ HNO_3

D-Glucose + Saligenin $\xleftarrow[\text{emulsin}]{H_2O}$ Salicin

Saligenin
o-(Hydroxymethyl)phenol

Salicin
o-(Hydroxymethyl)phenyl
β-D-glucopyranoside

Methylation and hydrolysis confirms the assumption that salicin is a pyranoside, not a furanoside. (The fifth methyl of pentamethylsalicin is in $-CH_2OCH_3$ attached to the aromatic ring.)

8. Let us examine R and S first. Chemical arithmetic,

$C_6H_{14}O_6$	sat'd open-chain
$-\ C_6H_{12}O_6$	R or S
2H	missing *means* one ring

indicates—in the absence of unsaturation—the presence of one ring. The acetylation data indicate six —OH's in R and S, and the HIO_4 results narrow the possibilities down to stereoisomers of 1,2,3,4,5,6-hexahydroxycyclohexane (*inositol*).

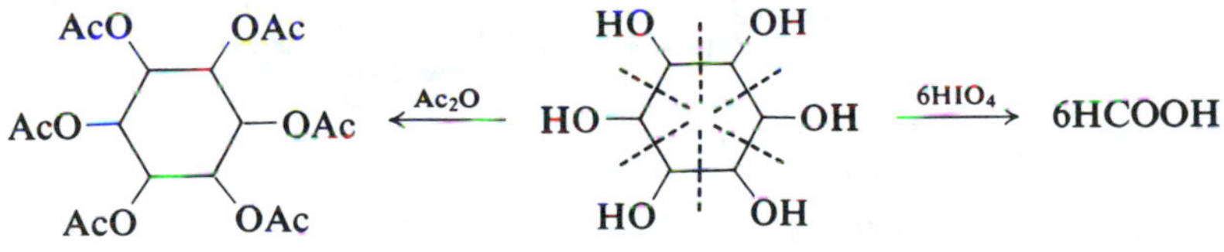

Hexaacetates $\xleftarrow{Ac_2O}$ R and S Inositols $\xrightarrow{6HIO_4}$ 6HCOOH

Hexaacetates
$C_{18}H_{24}O_{12}$

R and S
Inositols
$C_6H_{12}O_6$

Bio-inonose is, then, a pentahydroxycyclohexanone. It reduces Benedict's solution because it is an α-hydroxy ketone (Sec. 34.6). It does not reduce bromine water because it is not an aldehyde or hemiacetal.

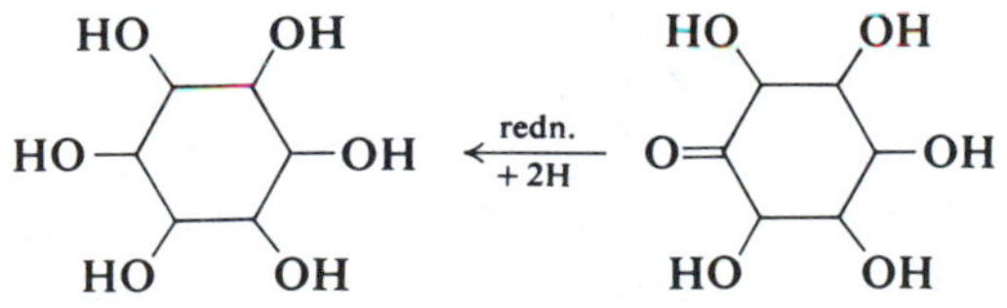

R and S Inositols $\xleftarrow[+2H]{\text{redn.}}$ Bio-inonose

R and S
Inositols

Bio-inonose (*partial structure*)
Positive Benedict's

But which of the many stereoisomeric possibilities is bio-inonose? Vigorous oxidation gives the answer.

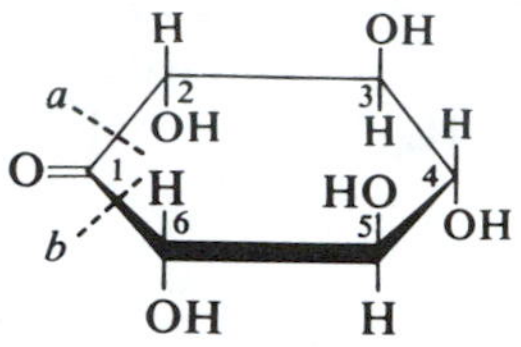

Bio-inonose
Meso

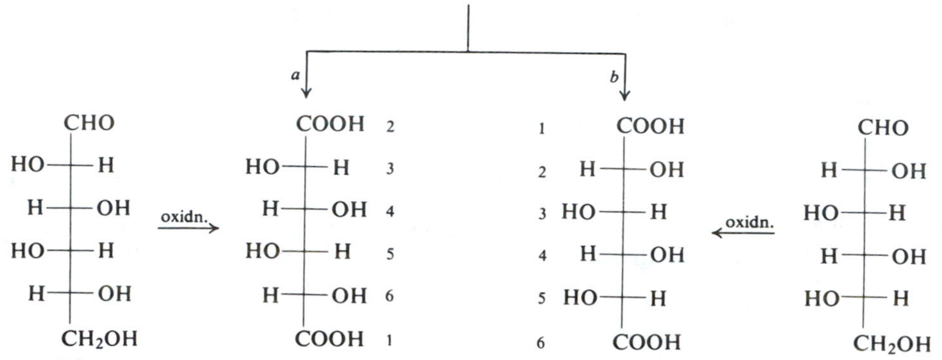

CHO	COOH 2	1 COOH	CHO
HO—H	HO—H 3	2 H—OH	H—OH
H—OH	H—OH 4	3 HO—H	HO—H
HO—H	HO—H 5	4 H—OH	H—OH
H—OH	H—OH 6	5 HO—H	HO—H
CH_2OH	COOH 1	6 COOH	CH_2OH
D-Idose	D-Idaric acid	L-Idaric acid	L-Idose

Reduction of bio-inonose gives both possible configurations about the new chiral center, and the diastereomers R and S are formed.

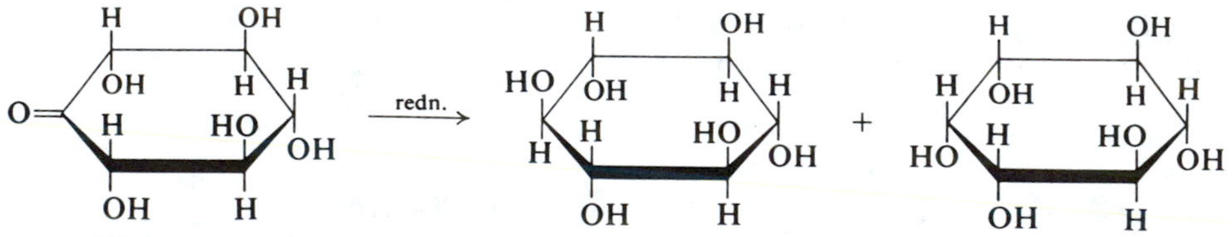

Bio-inonose —redn.→ Scyllitol (*Meso*) + Mesoinositol (*Meso*)

R and S

It is not hard to figure out the more stable configuration of each compound.

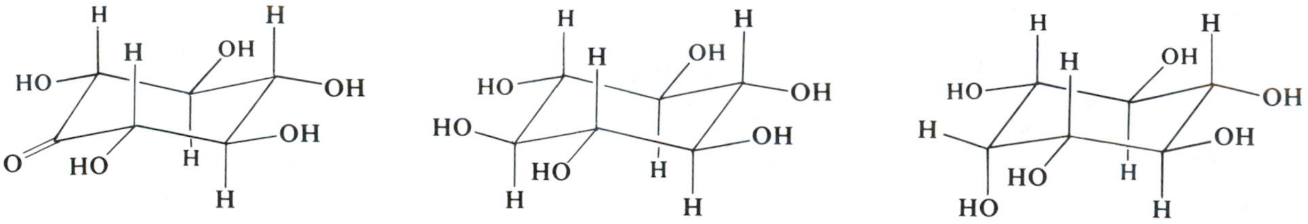

Bio-inonose
Meso

Scyllitol
Meso

Mesoinositol
Meso

R and S

9. (a)

$$\overset{6}{C}H_2OH-\overset{5}{C}HOH-\overset{4}{C}HOH-\overset{3}{C}HOH-\overset{2}{C}HOH\,\vdots\,\overset{1}{C}HO \xrightarrow{\text{Ruff}} \overset{6}{C}H_2OH-\overset{5}{C}HOH-\overset{4}{C}HOH-\overset{3}{C}HOH-\overset{2}{C}HO + \underset{\text{C-1}}{CO_2}$$

Glucose → Arabinose

$$\overset{6}{C}H_2OH-\overset{5}{C}HOH-\overset{4}{C}HOH-\overset{3}{C}HOH\,\vdots\,\overset{2}{C}HO \xrightarrow{\text{Ruff}} \underset{\text{C-2}}{CO_2}$$

Arabinose

$$\overset{6}{C}H_2OH\,\vdots\,\overset{5}{C}HOH\,\vdots\,\overset{4}{C}HOH\,\vdots\,\overset{3}{C}HOH\,\vdots\,\overset{2}{C}HOH\,\vdots\,\overset{1}{C}HO \xrightarrow{HIO_4} \underset{\text{C-6}}{HCHO} + 5HCOOH$$

Glucose

$$\text{Glucose} \xrightarrow[HCl]{CH_3OH} \overset{6}{C}H_2OH-\overset{5}{C}H-\overset{4}{C}HOH\,\vdots\,\overset{3}{C}HOH\,\vdots\,\overset{2}{C}HOH-\overset{1}{C}H(OCH_3)\ (\text{ring: C-5–O–C-1}) \xrightarrow{HIO_4} \underset{\text{C-3}}{HCOOH}$$

(*see page 1174*)

$$\overset{6}{C}H_2OH\overset{5}{C}HOH\overset{4}{C}HOH\,\vdots\,\overset{3}{C}HOH\overset{2}{C}HOH\overset{1}{C}HO \xrightarrow{L.\ casei} \overset{6}{C}H_3\overset{5}{C}HOH\,\vdots\,\overset{4}{C}OOH \;+\; \overset{3}{H}OOC\,\vdots\,\overset{2}{C}HOH\overset{1}{C}H_3$$

Glucose

$$\overset{6}{C}H_3\overset{5}{C}HOH\,\vdots\,\overset{4}{C}OOH \xrightarrow{KMnO_4} \overset{6}{C}H_3\,\vdots\,\overset{5}{C}HO + \underset{\text{C-4}}{CO_2}$$

$$\overset{6}{C}H_3\,\vdots\,\overset{5}{C}HO \xrightarrow{NaOI} \underset{\text{C-6}}{CHI_3} + \underset{\text{C-5}}{HCOOH}$$

$$HOO\overset{3}{C}\,\vdots\,\overset{2}{C}HOH\overset{1}{C}H_3 \xrightarrow{KMnO_4} OH\overset{2}{C}\,\vdots\,\overset{1}{C}H_3 + \underset{\text{C-3}}{CO_2}$$

$$OH\overset{2}{C}\,\vdots\,\overset{1}{C}H_3 \xrightarrow{NaOI} \underset{\text{C-2}}{HCOOH} + \underset{\text{C-1}}{CHI_3}$$

(b)

$$\overset{5}{C}H_2OH\,\vdots\,\overset{4}{C}HOH\,\vdots\,\overset{3}{C}HOH\,\vdots\,\overset{2}{\underset{\|}{C}}(=O)-\overset{1}{C}H_2OH \xrightarrow{HIO_4} \underset{\text{C-5}}{HCHO} + \underset{\text{C-4 and C-3}}{2HCOOH} + \underset{\text{C-2 and C-1}}{HOO\overset{2}{C}-\overset{1}{C}H_2OH}$$

Ribulose

$$\text{Ribulose} \xrightarrow[Pt]{H_2} \overset{5}{C}H_2OH\,\vdots\,\overset{4}{C}HOH\,\vdots\,\overset{3}{C}HOH\,\vdots\,\overset{2}{C}HOH\,\vdots\,\overset{1}{C}H_2OH \xrightarrow{HIO_4} \underset{\text{C-5 and C-1}}{2HCHO} + \underset{\text{C-4, C-3, and C-2}}{3HCOOH}$$

$$\text{Ribulose} \xrightarrow{PhNHNH_2} \overset{5}{CH_2OH} \vdots \overset{4}{CHOH} \vdots \overset{3}{CHOH}-\underset{\underset{NNHPh}{\|}}{\overset{2}{C}}-\overset{1}{CH}=NNPh \xrightarrow{HIO_4}$$

$$\underset{\text{C-5}}{HCHO} + \underset{\text{C-4}}{HCOOH} + \underset{\text{C-3, C-2, and C-1}}{OHC-\underset{\underset{NNHPh}{\|}}{C}-CH=NNHPh}$$

A problem in algebra: can you determine the activity of *each individual position* of glucose and of ribulose?

10. (a) Sugar T is $C_5H_{10}O_5$ and is oxidized by Br_2(aq): it is an aldopentose. It is levorotatory and gives an optically inactive aldaric acid: it must be either D-(−)-ribose (IX, page 1160) or L-(−)-xylose (enantiomer of XI, page 1161).

D-(−)-Ribose T?	$\xrightarrow{HNO_3}$	*Meso* aldaric acid $C_5H_8O_7$ **Inactive**
CHO		COOH
H—C—OH		H—C—OH
H—C—OH		H—C—OH
H—C—OH		H—C—OH
CH_2OH		COOH

L-(−)-Xylose T?	$\xrightarrow{HNO_3}$	*Meso* aldaric acid $C_5H_8O_7$ **Inactive**
CHO		COOH
HO—C—H		HO—C—H
H—C—OH		H—C—OH
HO—C—H		HO—C—H
CH_2OH		COOH

The epimer of D-(−)-ribose is D-(−)-arabinose (X, page 1160); the epimer of L-(−)-xylose is L-(+)-lyxose (enantiomer of XII, page 1161). The epimer of T is (−)-U. From its sign of rotation alone, we conclude that (−)-U must be D-(−)-arabinose. This is confirmed by its degradation and oxidation of the resulting tetrose to inactive (*meso*) tartaric acid. (Similar treatment of L-(+)-lyxose would have given active L-(+)-tartaric acid.)

Mesotartaric acid	$\xleftarrow{HNO_3}$		$\xleftarrow{\text{degradn.}}$	D-(−)-Arabinose U		D-(−)-Ribose T
				CHO		CHO
COOH		CHO		HO—C—H		H—C—OH
H—C—OH		H—C—OH		H—C—OH		H—C—OH
H—C—OH		H—C—OH		H—C—OH		H—C—OH
COOH		CH_2OH		CH_2OH		CH_2OH

T, then, is D-(−)-ribose.

(b) Of the possible phosphates of D-(−)-ribose, only the 3-phosphate can give the *meso* compound V.

T phosphate	$\xrightarrow[Pt]{H_2}$	V *Meso* **Inactive**	$\xrightarrow{H_2O}$	W *Meso* **Inactive**	$\xrightarrow{Ac_2O}$	X *Meso* **Inactive**
CHO		CH_2OH		CH_2OH		CH_2OAc
H—C—OH		H—C—OH		H—C—OH		H—C—OAc
H—C—OPO_3H_2		H—C—OPO_3H_2		H—C—OH		H—C—OAc
H—C—OH		H—C—OH		H—C—OH		H—C—OAc
CH_2OH		CH_2OH		CH_2OH		CH_2OAc

(c) Phosphate is located, then, at the 3-position of ribose. At which point is adenosine attached? Adenylic acid does not reduce Tollens' reagent or Benedict's solution. *Conclusion*: that C–1 of ribose is tied up as in a glycoside. If the nitrogen of adenosine is joined to a carbon atom in T, it must be to C–1.

Finally, what is the ring size of the ribose unit? This is revealed by the usual sequence of methylation, hydrolysis, and vigorous oxidation.

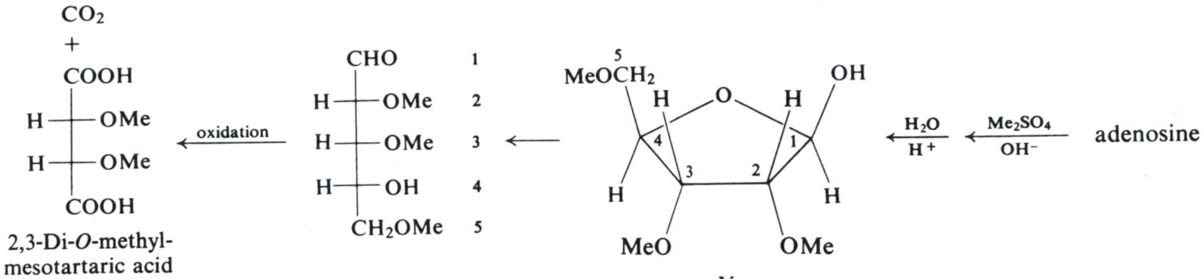

CO_2 + COOH, H—OMe, H—OMe, COOH
2,3-Di-*O*-methyl-mesotartaric acid

$\xleftarrow{\text{oxidation}}$ CHO (1), H—OMe (2), H—OMe (3), H—OH (4), CH_2OMe (5) ⟵ Y $C_8H_{16}O_5$ $\xleftarrow[H^+]{H_2O}$ $\xleftarrow[OH^-]{Me_2SO_4}$ adenosine

Taking the evidence of synthesis, we arrive at the structure of adenylic acid (compare Figure 36.10, page 1241).

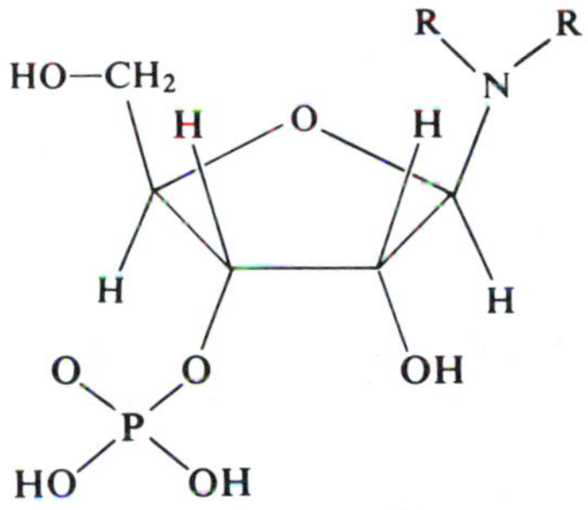

Adenylic acid

11. (a) Chemical arithmetic,

$$\begin{array}{ll} C_6H_{12}O_6 & \text{glucose} \\ + C_3H_6O & \text{1 acetone} \\ \hline C_9H_{18}O_7 & \end{array} \qquad \begin{array}{ll} C_9H_{18}O_7 & \\ - C_9H_{16}O_6 & \text{AA} \\ \hline H_2O & \text{lost} \end{array} \qquad \begin{array}{ll} C_6H_{12}O_6 & \text{glucose} \\ + C_6H_{12}O_2 & \text{2 acetone} \\ \hline C_{12}H_{24}O_8 & \end{array} \qquad \begin{array}{ll} C_{12}H_{24}O_8 & \\ - C_{12}H_{20}O_6 & \text{Z} \\ \hline 2H_2O & \text{lost} \end{array}$$

shows that glucose combines with one (AA) or two (Z) molecules of acetone to split out, respectively, one or two molecules of water. The products are resistant to alkali and hydrolyzed by acid, properties we expect of acetals. Each acetone reacts with two —OH's of glucose to give a cyclic ketal, called an *O-isopropylidene derivative*:

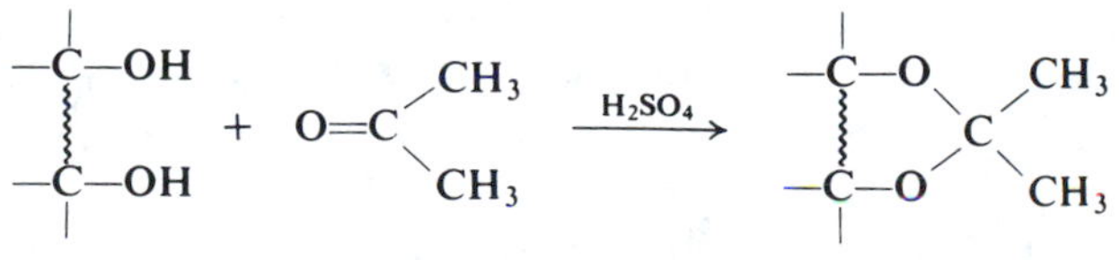

$$\text{Glucose} + \text{Acetone } (O{=}C(CH_3)_2) \xrightarrow{H_2SO_4} \text{O-Isopropylideneglucose}$$

Glucose Acetone *O*-Isopropylideneglucose
A cyclic ketal

Hydrolysis of Z gives AA as an intermediate. Since aqueous acid is an unfavorable medium for re-formation of acetals, it seems highly likely that the ketal bridge AA is the same as one of the ketal bridges in Z.

Now, to which pairs of carbons in glucose are the *O*-isopropylidene groups attached?

(b) Neither Z nor AA is a reducing sugar. *Conclusion*: in both compounds an *O*-isopropylidene group is attached to C–1.

(c) Z is methylated at one position. Acid cleaves the ketals, but not the ether: the methoxyl group remains. This *O*-methyl ether gives an osazone. Evidently methylation did not occur at either C–1 or C–2. *Conclusion*: in Z, an *O*-isopropylidene group is attached to C–1 and C–2.

CH=NNHPh
C=NNHPh
DD ← $PhNHNH_2$ —

CHOH
CHOH O
CC ← H_2O, H^+ — BB ← Me_2SO_4, OH^- —

1 CH—O
C$(CH_3)_2$
2 CH—O O

One *O*-isopropylidene grouping in Z

(d) The evidence is as in (c). *Conclusion*: in AA, the only *O*-isopropylidene group is attached to C–1 and C–2.

CH=NNHPh
C=NNHPh
GG ← $PhNHNH_2$ —

CHOH
CHOH O
FF ← H_2O, H^+ — EE ← Me_2SO_4, OH^- —

1 CH—O
C$(CH_3)_2$
2 CH—O O

The *O*-isopropylidene grouping in AA

(e) Methylation shows that C–4 was tied up in the acetal ring of FF, and hence of EE and AA, too. *Conclusion*: these compounds contain furanose rings. We now know the structure of AA.

CHOH
CHOMe
CHOMe O
CHO
CHOMe
CH_2OMe

2,3,5,6-Tetra-*O*-methyl-D-glucofuranose

← Me_2SO_4, OH^- —

CHOH
CHOH
CHOMe O
CH
CHOMe
CH_2OMe

FF

← H_2O, H^+ —

CH—O
C$(CH_3)_2$
CH—O
CH—OMe O
CH
CH—OMe
CH_2—OMe

EE

← Me_2SO_4, OH^- —

CH—O
C$(CH_3)_2$
CH—O
CH—OH O
CH
CH—OH
CH_2—OH

AA

1,2-*O*-Isopropylidene-D-glucofuranose

(f) HH is a six-carbon dicarboxylic acid. One carbon is in the *O*-methyl group, leaving five carbons for the chain; the carboxyl groups are C–1 and C–5 of the *O*-methyl-D-glucose, CC. *Conclusion*: C_5-OH is not methylated, and hence must have been tied up by an *O*-isopropylidene group.

In CC, methoxyl is not at C–1 or C–2 (*O*-isopropylidene bridge), not at C–4 (furanose ring), and, now, evidently not at C–5, leaving only C–3 or C–6.

(g) Formation of the δ-lactone II shows that C–5 of this compound, *and hence C–4 of CC*, carries a free —OH. *Conclusion*: the methoxyl group of CC is at C–3.

This leaves only C–5 and C–6 for the second *O*-isopropylidene bridge in Z. We now know the structure of Z.

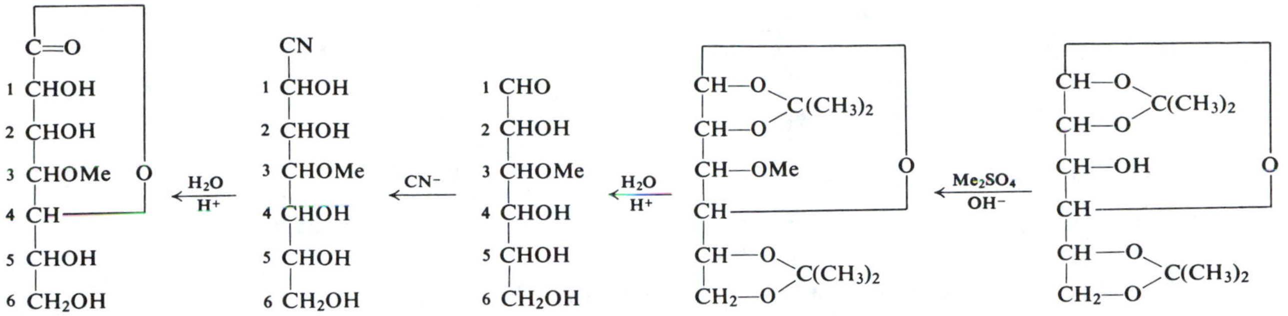

C=O
1 CHOH
2 CHOH
3 CHOMe O
4 CH
5 CHOH
6 CH_2OH
and epimer
II
A δ-lactone

$\xleftarrow[H^+]{H_2O}$

CN
1 CHOH
2 CHOH
3 CHOMe
4 CHOH
5 CHOH
6 CH_2OH

$\xleftarrow{CN^-}$

1 CHO
2 CHOH
3 CHOMe
4 CHOH
5 CHOH
6 CH_2OH
CC
(As open-chain)

$\xleftarrow[H^+]{H_2O}$

CH—O
CH—O $C(CH_3)_2$
CH—OMe O
CH
CH—O
CH_2—O $C(CH_3)_2$
BB

$\xleftarrow[OH^-]{Me_2SO_4}$

CH—O
CH—O $C(CH_3)_2$
CH—OH O
CH
CH—O
CH_2—O $C(CH_3)_2$
Z
1,2:5,6-Di-*O*-isopropylidene-D-glucofuranose

12. Carbon–carbon bonds are formed; with the reactants aldehydes and ketones, and the reagent base, we think of the aldol condensation. Either D-glyceraldehyde (an aldose) or dihydroxyacetone (a ketose) gives the same products; Sec. 34.6 reminds us that base can cause isomerization—among other things, between an aldose and a ketose.

We postulate an equilibrium between the two three-carbon reactants:

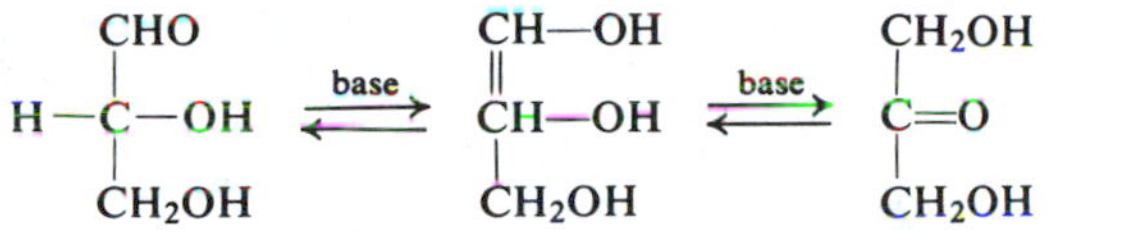

CHO, H—C—OH, CH_2OH $\underset{}{\overset{base}{\rightleftharpoons}}$ CH—OH, ‖ CH—OH, CH_2OH $\overset{base}{\rightleftharpoons}$ CH_2OH, C=O, CH_2OH

D-Glyceraldehyde *An aldose* — Enediol — Dihydroxyacetone *A ketose*

Aldol and crossed aldol condensations yield the final products.

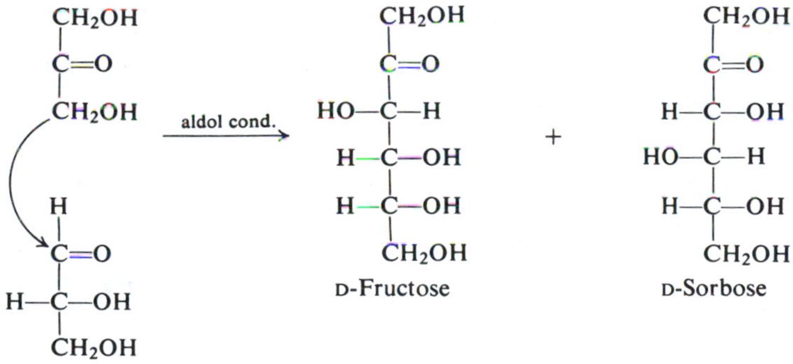

CH_2OH—C(=O)—CH_2OH + H—C(=O)—(H—C—OH)—CH_2OH $\xrightarrow{\text{aldol cond.}}$

CH_2OH, C=O, HO—C—H, H—C—OH, H—C—OH, CH_2OH
D-Fructose

\+

CH_2OH, C=O, H—C—OH, HO—C—H, H—C—OH, CH_2OH
D-Sorbose

CH_2OH—C(=O)—CH_2OH + $HOCH_2$—C(=O)—CH_2OH $\xrightarrow{\text{aldol cond.}}$

CH_2OH, C=O, H—C—OH, $HOCH_2$—C—OH, CH_2OH

\+

CH_2OH, C=O, HO—C—H, HO—C—CH_2OH, CH_2OH

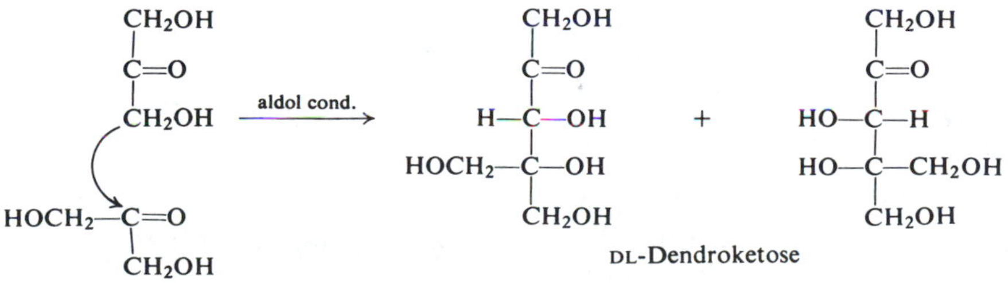

DL-Dendroketose

13. Hydrolysis is S_N1-like, with the separation of an especially stable cation: an oxonium ion, not a carbocation.

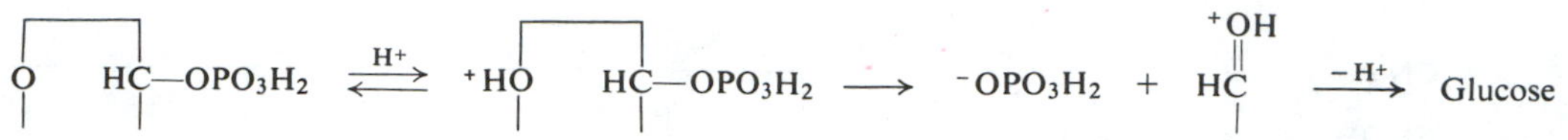

Glucose-1-phosphate $\underset{}{\overset{H^+}{\rightleftharpoons}}$ Protonated acetal $\longrightarrow$ Phosphate ($^{-}OPO_3H_2$) + Protonated aldehyde (Oxonium ion) $\xrightarrow{-H^+}$ Glucose

14. (a) The downfield signal is due to the proton on C–1, which is deshielded by *two* oxygens.

(b) Our interpretation (based on Chapter 17) would be: KK has an equatorial —H on C–1 (signal farther downfield); JJ has *anti* (axial,axial) protons on C–1 and C–2 (larger *J* value). Only in the *β*-anomer are the protons on C–1 and C–2 *trans*, and capable of being axial,axial. We are led to these structures:

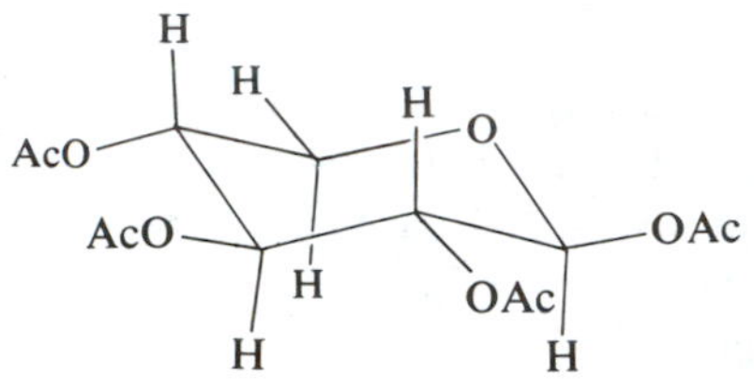

D-Tetra-*O*-acetylxylopyranose

JJ
β-Anomer
Axial —H on C–1
Anti protons on C–1, C–2

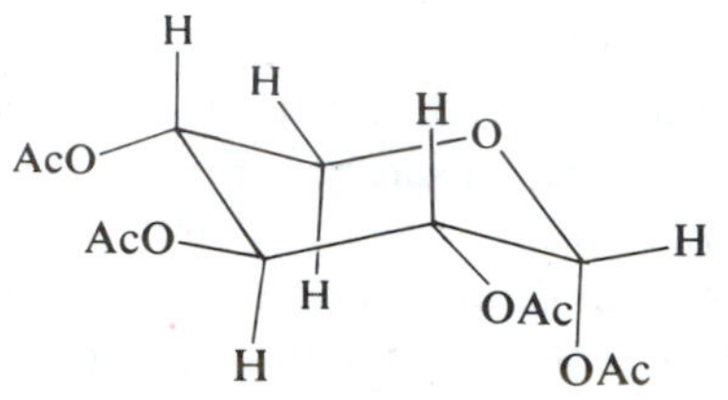

KK
α-Anomer
Equatorial —H on C–1
Gauche protons on C–1, C–2

Each conformation, we find, is the one we would expect to be the more stable for the particular anomer. In the *β*-anomer, the anomeric effect (page 1180) is weaker than the steric effect: four equatorial —OAc's. (In the *α*-anomer, the anomeric effect simply reinforces the steric effect.)

(c) Our interpretation: MM has an equatorial —H on C–1 (signal farther downfield); LL has *anti* (axial,axial) protons on C–1 and C–2 (larger *J* value). Again, only in the *β*-anomer can the protons on C–1 and C–2 be axial,axial.

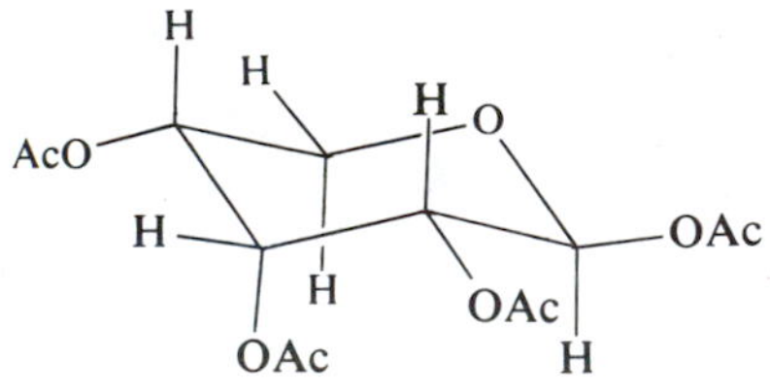

D-Tetra-*O*-acetylribopyranose

LL
β-Anomer
Axial —H on C–1
Anti protons on C–1, C–2

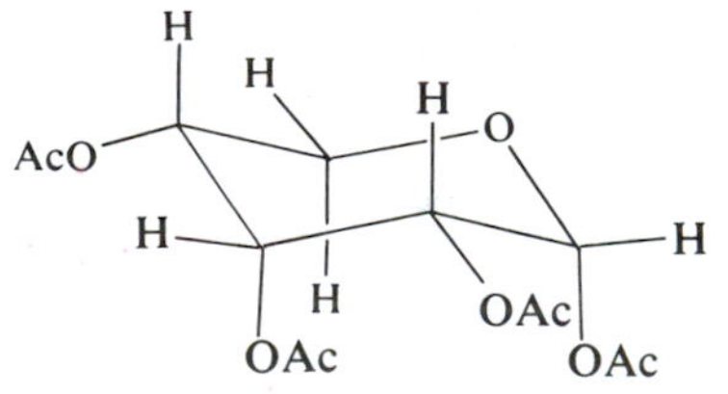

MM
α-Anomer
Anomeric effect
(Sec. 34.20)

In the *β*-anomer, the anomeric effect is weaker than the steric effect: three out of four —OAc's equatorial. In the *α*-anomer, where there are two equatorial —OAc's in either conformation, the anomeric effect is the deciding factor.

(d) This time, let us begin by drawing what we would expect to be the more stable conformation of each compound.

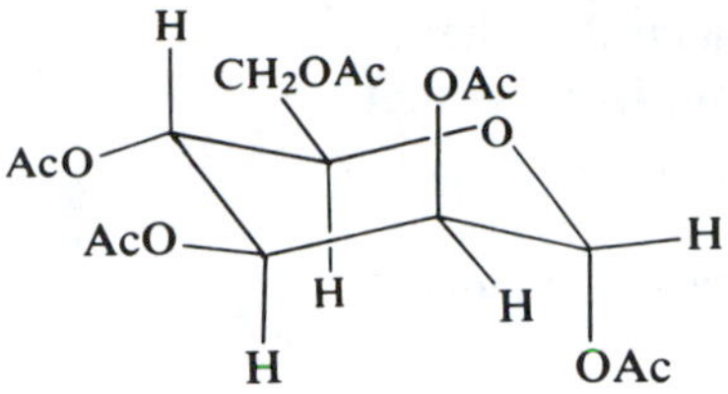

D-Penta-*O*-acetylmannopyranose

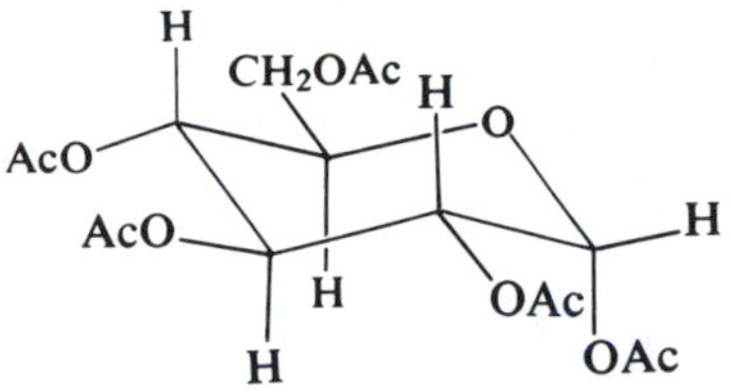

NN
α-Anomer
Equatorial —H on C–1
Gauche protons on C–1, C–2

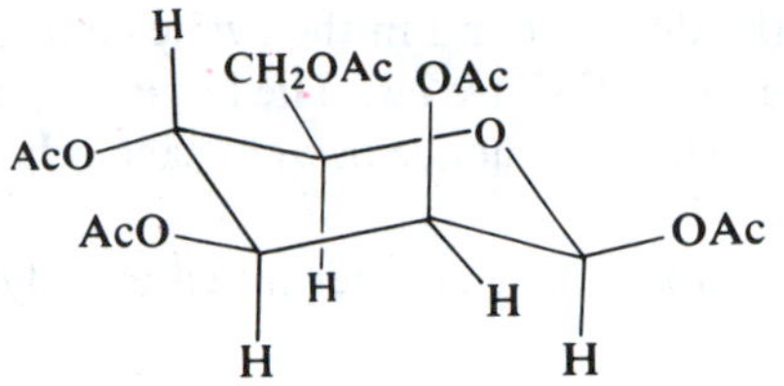

OO
β-Anomer
Axial —H on C–1
Gauche protons on C–1, C–2

D-Penta-*O*-acetylglucopyranose

QQ
α-Anomer
Equatorial —H on C–1
Gauche protons on C–1, C–2

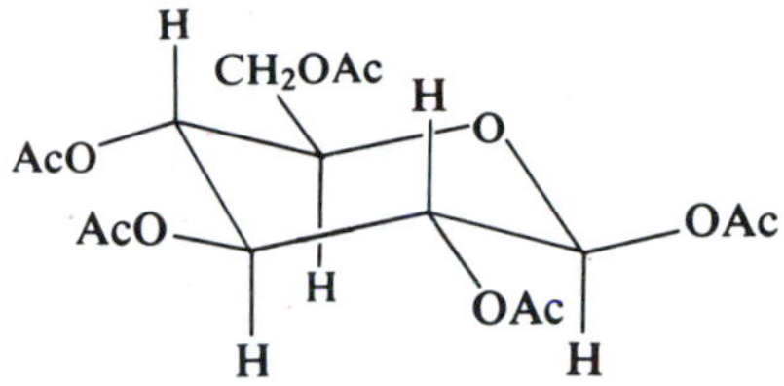

PP
β-Anomer
Axial —H on C–1
Anti protons on C–1, C–2

We can readily assign an α-structure (equatorial —H) to NN (δ 5.97) and QQ (δ 5.99), and a β-structure (axial —H) to OO (δ 5.68) and PP (δ 5.54). Only one of the four anomers, the β-glucose derivative, has *anti* protons on C–1 and C–2; only PP has a large value for *J*, the coupling constant. Therefore PP is the β-glucose derivative; its anomer, QQ, must be the α-glucose derivative. Of necessity, then, NN is the α-mannose derivative, and OO is the β-mannose derivative.

15. (a)

(i)

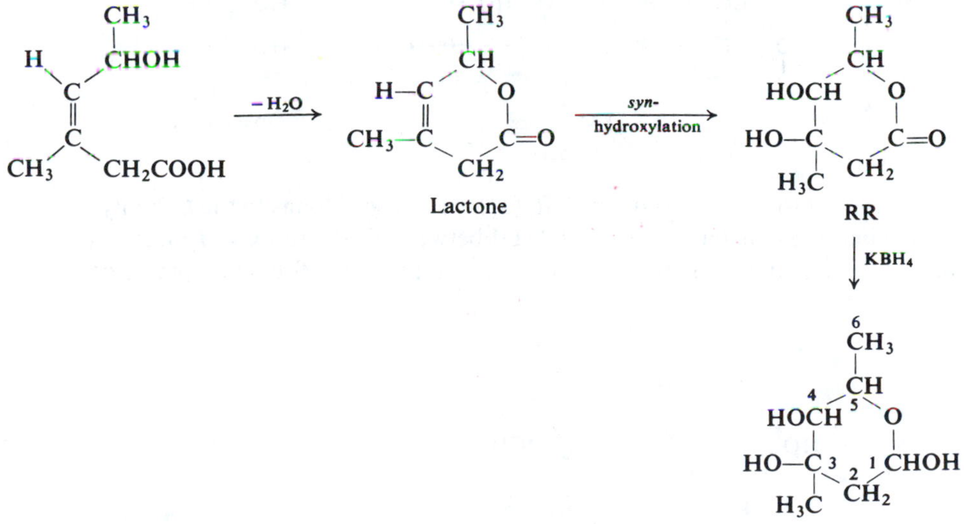

$\xrightarrow{-H_2O}$ Lactone $\xrightarrow{syn\text{-hydroxylation}}$ RR $\xrightarrow{KBH_4}$ Mycarose

Mycarose
Stereochemistry disregarded

(Only the *Z* isomer of the starting material can form a lactone.)

(b) *syn*-Hydroxylation (i) of the double bond in the cyclic compound gives a *cis*-diol; hence in RR and mycarose C_3-OH and C_4-OH are *cis*. The large coupling constant (ii) shows that C_4-H and C_5-H are *anti* (axial,axial) and hence *trans* to each other.

(c) The absolute configuration about C–5 is determined exactly as was done for D-glucose (Sec. 34.17).

(iii) 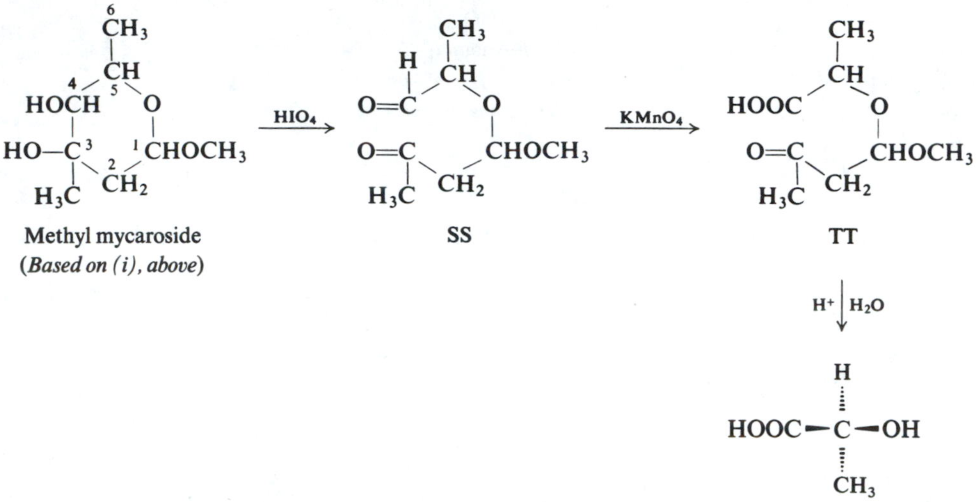 Methyl mycaroside (*Based on (i), above*) $\xrightarrow{HIO_4}$ SS $\xrightarrow{KMnO_4}$ TT $\xrightarrow{H^+,\ H_2O}$ L-Lactic acid ((*S*)-Lactic acid)

Methyl mycaroside
(*Based on (i), above*)

SS

TT

L-Lactic acid
((*S*)-Lactic acid)

(d) Taking the information gathered so far, we can draw:

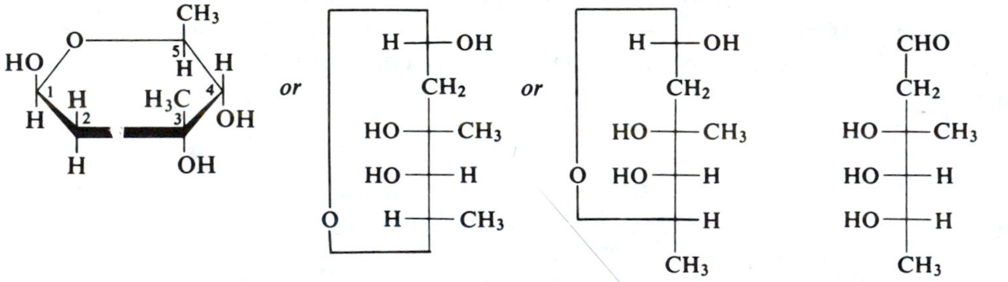

L-(−)-Mycarose
(β-anomer)

Open-chain

Since —OH on the lowest chiral carbon is on the left, (−)-mycarose belongs to the L-family.

The size of the coupling constant ($J = 9.5$–9.7 Hz) between C_4-H and C_5-H indicates that these protons are axial,axial. Starting from there, we arrive at the following preferred conformation:

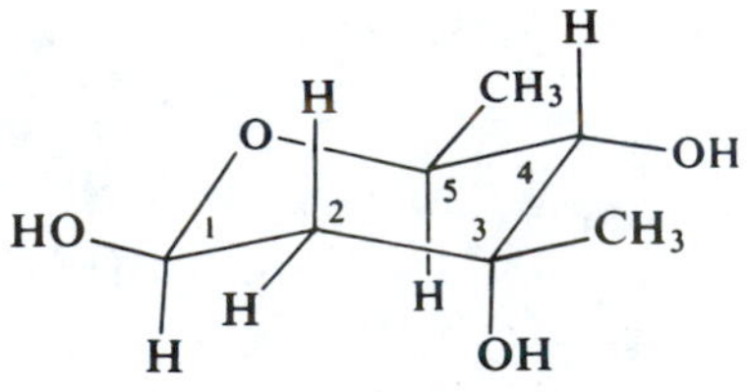

L-(−)-Mycarose

Disregarding C–1 for the moment, we see that in this conformation three of four large groups are equatorial; this is the conformation we would have predicted to be the more stable.

(e) The downfield peak is, of course, due to C_1-H. In the α-anomer C_1-H is *gauche* to both H's on C–2. The signal is split into a triplet, with a low *J* value, as observed.

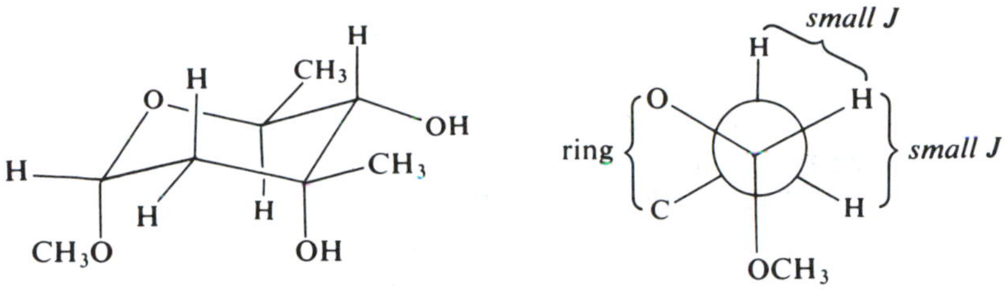

α-Anomer

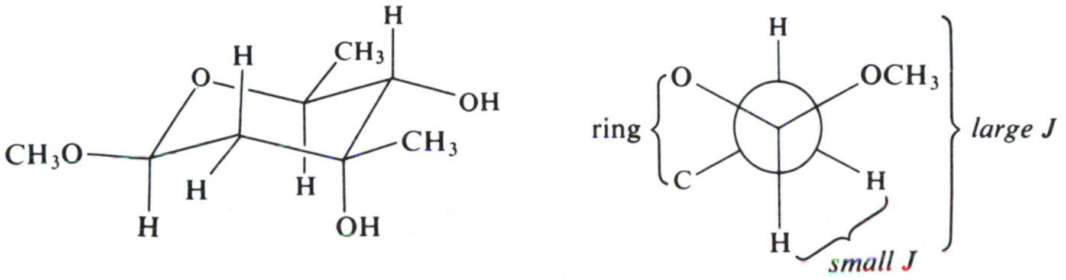

β-Anomer

In the β-anomer, C_1-H is *gauche* to one C_2-H and *anti* to the other C_2-H. We expect two signals for C_1-H, each a doublet. There will be two *J* values: one small, comparable to that (2.4 Hz) for the α-anomer; and the other larger.

(f) Here we see exactly the kind of NMR spectrum just predicted. Evidently free mycarose exists preferentially as the β-anomer, as shown in (d) above.

16. (a) The anomeric effect (page 1180) stabilizes the α-anomer relative to the β-anomer.

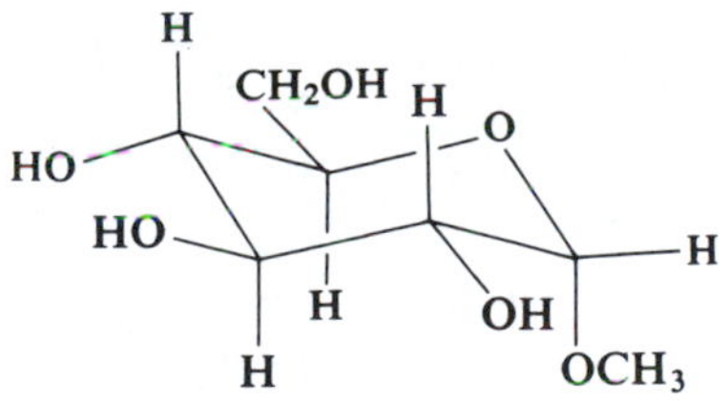

Methyl α-D-glucoside
Favored anomer

(*Note*: Here we are not just discussing which conformation of a particular compound is more stable. We are discussing which *compound*—the α- or β-anomer—is the more stable. Our interpretation is this: that the α-anomer is more stable than the β-anomer because the preferred conformation of the α-anomer is more stable than the preferred conformation of the β-anomer.)

(b) This is evidently an anomeric-like effect. Repulsion between the dipoles associated with the ring oxygens and the C—Cl bonds favors the diaxial conformation.

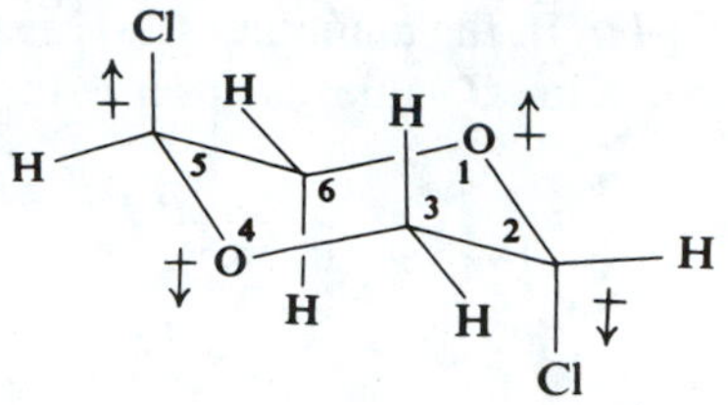

trans-2,5-Dichloro-1,4-dioxane
Favored conformation

35

Carbohydrates II. Disaccharides and Polysaccharides

35.1

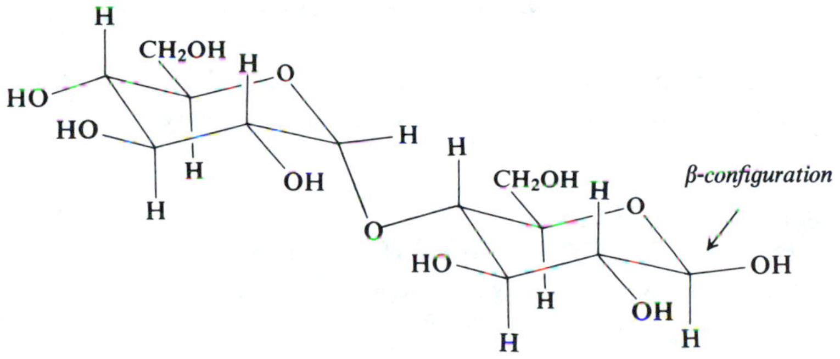

β-configuration

β-Maltose

α- and β-maltose differ in configuration only at C–1 of the reducible glucose unit. (If the two glucose moieties were *joined* by a β-linkage, the compound would not be maltose, but cellobiose.)

35.2

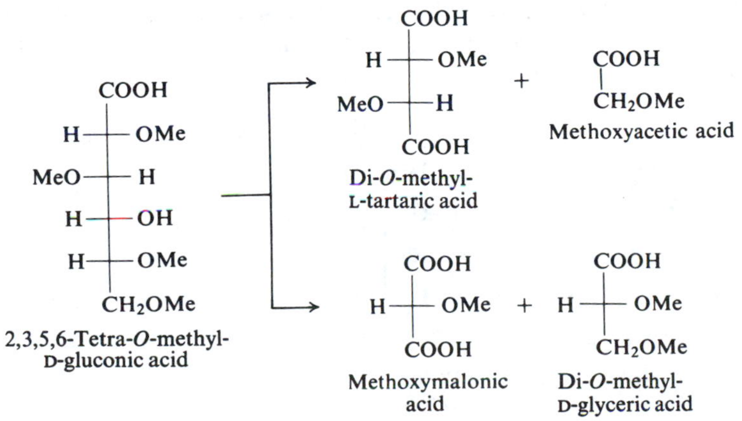

2,3,5,6-Tetra-*O*-methyl-D-gluconic acid

Di-*O*-methyl-L-tartaric acid + Methoxyacetic acid

Methoxymalonic acid + Di-*O*-methyl-D-glyceric acid

35.3

(β-anomer) → 2,3,4,6-Tetra-*O*-methyl-D-glucose + 2,3,6-Tri-*O*-methyl-D-glucose

Besides, of course, knowing that maltose is a glucosyl glucose, we would know that the non-reducing glucose unit contains a pyranose ring. But we would be lacking knowledge about the reducing unit: its ring size, and the point of attachment of the glycosidic linkage. It could be either a 4-*O*-substituted-D-glucopyranose *or* a 5-*O*-substituted-D-glucofuranose.

Oxidation before methylation and hydrolysis opens the hemiacetal ring. The only unmethylated position is then the point of attachment; with that known, we can deduce the ring size.

35.4 See Figure 35.5, page 638 of this Study Guide.

After loss of two carbons, the C_3-CHO has no —OH *alpha* to it (that is, no free —OH on C_4), so that osazone formation is no longer possible. This shows that the two rings are joined at C–4 of the reducing moiety of maltose.

35.5 There is less chance of breaking the glycosidic linkage, with formation of monosaccharides.

35.6 Everything is as in Figure 35.1 (page 1187), except that there is a β-glycosidic linkage in the first three formulas. The final products are identical to those from (+)-maltose.

35.7 (a) Similar to Figure 35.1 (page 1187), except that one starts with lactose: there is a β-glycosidic linkage, and C–4 in the non-reducing moiety has the opposite configuration. The final methylated monosaccharides are:

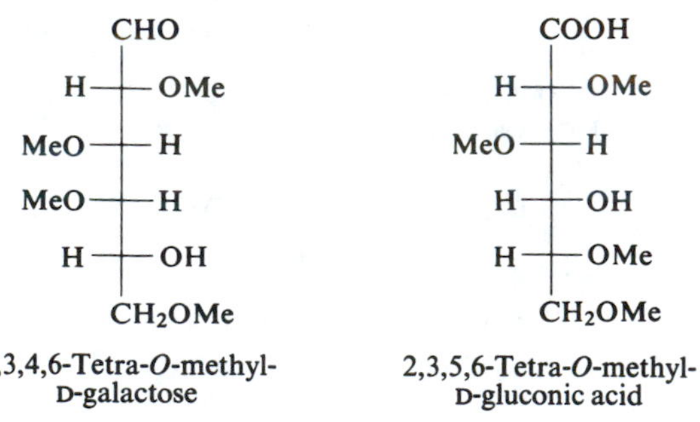

2,3,4,6-Tetra-*O*-methyl-D-galactose   2,3,5,6-Tetra-*O*-methyl-D-gluconic acid

(b)

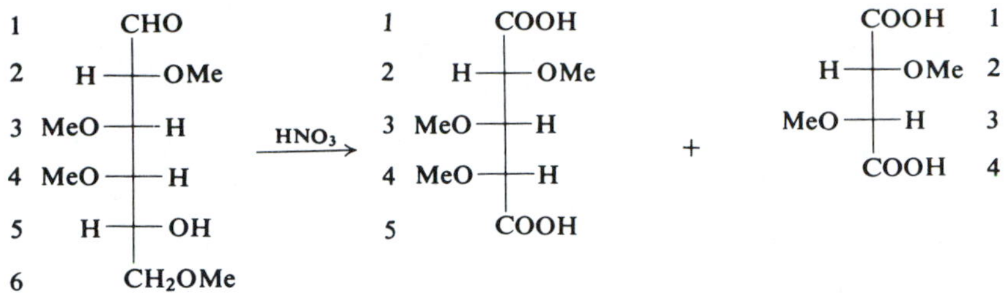

	2,3,4,6-Tetra-*O*-methyl-D-galactose		2,3,4-Tri-*O*-methyl-L-arabinaric (or L-lyxaric) acid		Di-*O*-methyl-L-tartaric acid	
1	CHO		1 COOH		COOH	1
2	H—OMe	$\xrightarrow{HNO_3}$	2 H—OMe	+	H—OMe	2
3	MeO—H		3 MeO—H		MeO—H	3
4	MeO—H		4 MeO—H		COOH	4
5	H—OH		5 COOH			
6	CH_2OMe					

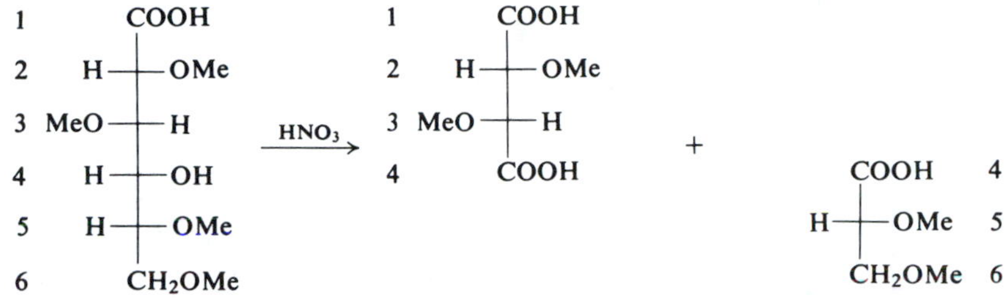

	2,3,5,6-Tetra-*O*-methyl-D-gluconic acid		Di-*O*-methyl-L-tartaric acid		Di-*O*-methyl-D-glyceric acid	
1	COOH		1 COOH			
2	H—OMe	$\xrightarrow{HNO_3}$	2 H—OMe	+		
3	MeO—H		3 MeO—H			
4	H—OH		4 COOH		COOH	4
5	H—OMe				H—OMe	5
6	CH_2OMe				CH_2OMe	6

35.8

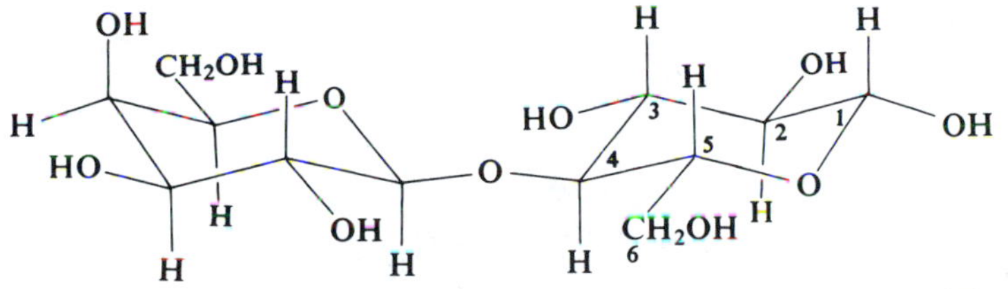

Lactose
(*β*-anomer)

$\downarrow$ −2C

$\downarrow$ hydrol.

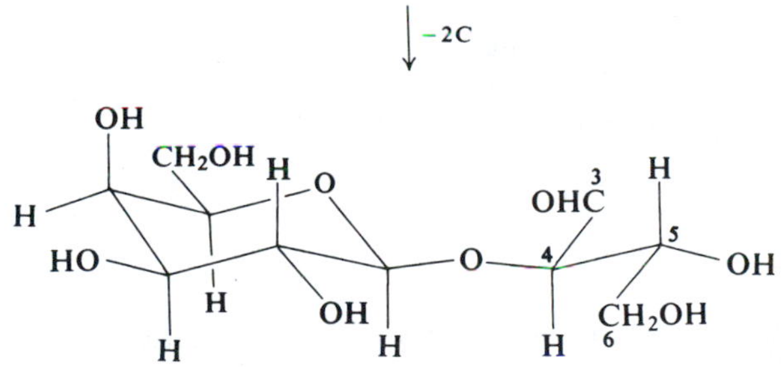

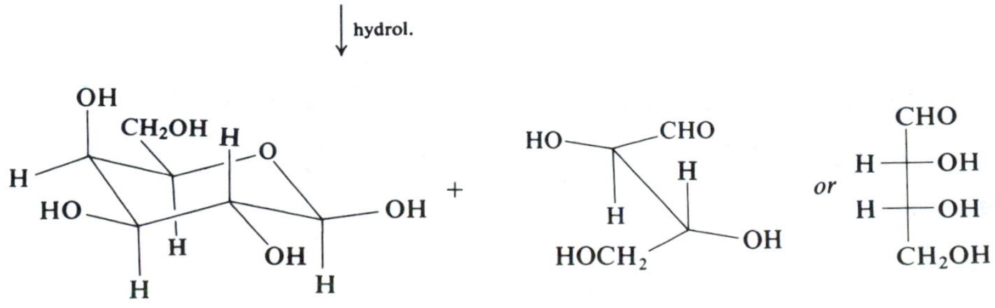

D-Galactose + D-Erythrose

or

CHO
H—OH
H—OH
CH_2OH

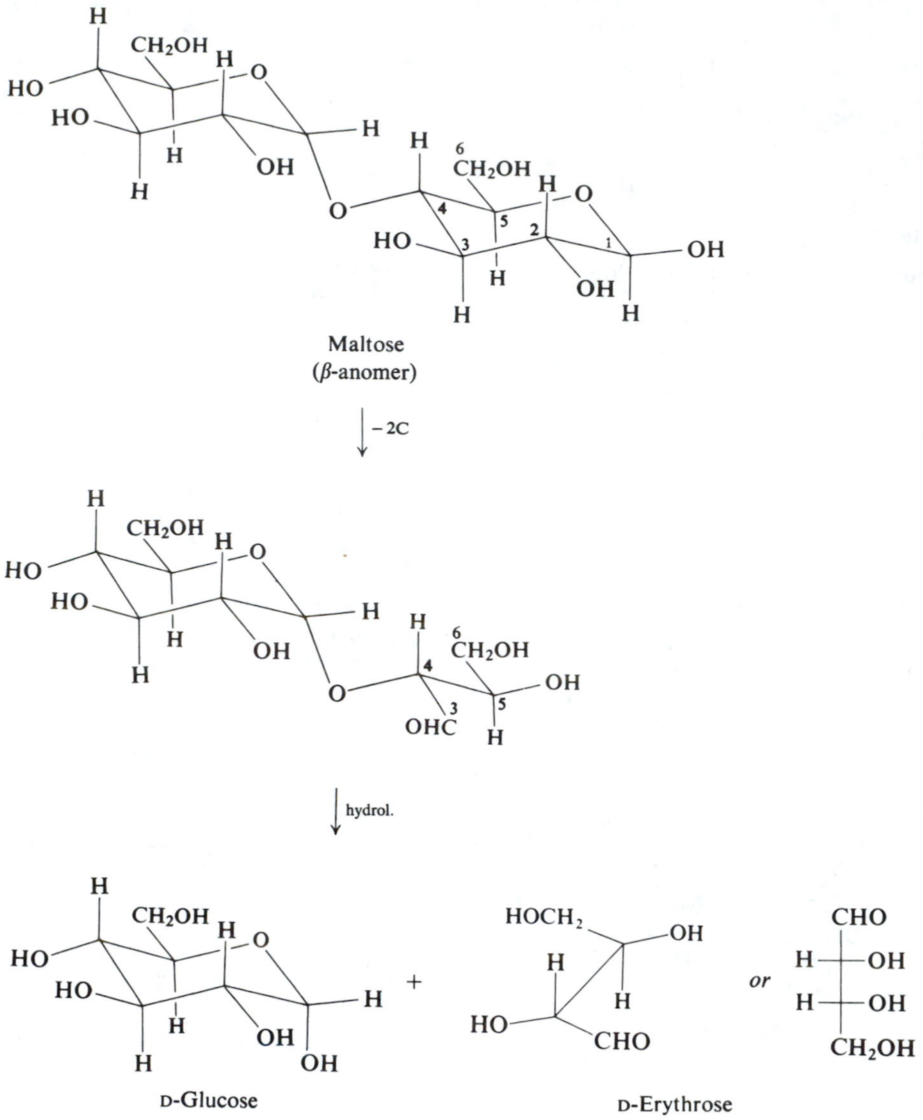

Figure 35.5 Reactions of Problem 35.4.

35.9

$$(-92.4° + 52.7°)/2 = -19.9°$$

Divided by 2 since g/mL is based on the combined weight of the *two* (isomeric) compounds.

35.10 There would be two glycosidic linkages, and hence loss of two moles of water:

$$C_6H_{12}O_6 + C_6H_{12}O_6 - 2H_2O = C_{12}H_{20}O_{10}$$

It would be non-reducing.

35.11 The glucose unit probably has the α-configuration, and hence sucrose is an α-glucoside.

35.12 (a) The methylated monosaccharides from sucrose are:

2,3,4,6-Tetra-*O*-methyl D-glucose	1,3,4,6-Tetra-*O*-methyl-D-fructose
CHO	CH_2OMe
H—OMe	C=O
MeO—H	MeO—H
H—OMe	H—OMe
H—OH	H—OH
CH_2OMe	CH_2OMe

(b)

	2,3,4,6-Tetra-*O*-methyl-D-glucose			2,3,4-Tri-*O*-methyl-xylaric acid			Di-*O*-methyl L-tartaric acid	
1	CHO		1	COOH			COOH	1
2	H—OMe		2	H—OMe			H—OMe	2
3	MeO—H	$\xrightarrow{HNO_3}$	3	MeO—H	+		MeO—H	3
4	H—OMe		4	H—OMe			COOH	4
5	H—OH		5	COOH				
6	CH_2OMe							

	1,3,4,6-Tetra-*O*-methyl-D-fructose		Di-*O*-methyl-D-tartaric acid	
1	CH_2OMe			
2	C=O	$\xrightarrow{HNO_3}$	COOH	2
3	MeO—H		MeO—H	3
4	H—OMe		H—OMe	4
5	H—OH		COOH	5
6	CH_2OMe			

35.13

One break	⟶	2000 avg. length	1/4000 *or* 0.025% links broken
Three breaks	⟶	1000 avg. length	3/4000 *or* 0.075% links broken
Nine breaks	⟶	400 avg. length	9/4000 *or* 0.225% links broken

35.14 (a) The largest groups attached to each ring are the other glucose units of the chain. One or the other of these would have to be axial in a chair conformation.

(b) In a twist-boat conformation, these (as well as the $—CH_2OH$ at C–5) would be in quasi-equatorial positions: at, say, the positions marked 1, 4, and 5 below. (Compare Sec. 13.11.)

35.15

2,3,4-Tri-*O*-methyl-D-glucose
Chief product ← *Attachment at C–1 and C–6* **Chain-forming unit**

2,3,4,6-Tetra-*O*-methyl-D-glucose ← *Attachment at C–1* **Chain-terminating unit**

2,4-Di-*O*-methyl-D-glucose ← *Attachment at C–1, C–3, C–6* **Chain-linking unit**

Dextran is a poly-α-D-glucopyranoside.

Chain-forming unit: attachment at C–1 and C–6.
Chain-terminating unit: attachment at C–1.
Chain-linking unit: attachment at C–1, C–3, and C–6.

Thus, much of the dextran is made up of simple α-glucoside chains with units joined to each other by linkages between C–1 and C–6; but every once in a while two of these chains are joined together by a linkage involving C–3. Schematically, something like this:

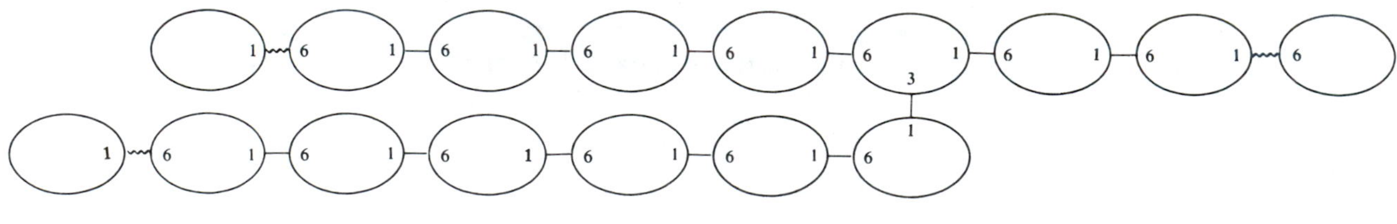

A dextran

35.16

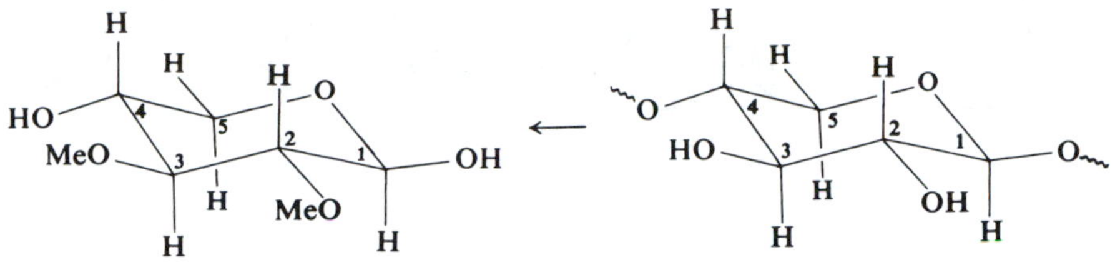

2,3-Di-*O*-methyl-D-xylose ← *Attachment at C–1 and C–4* **Chain-forming unit**

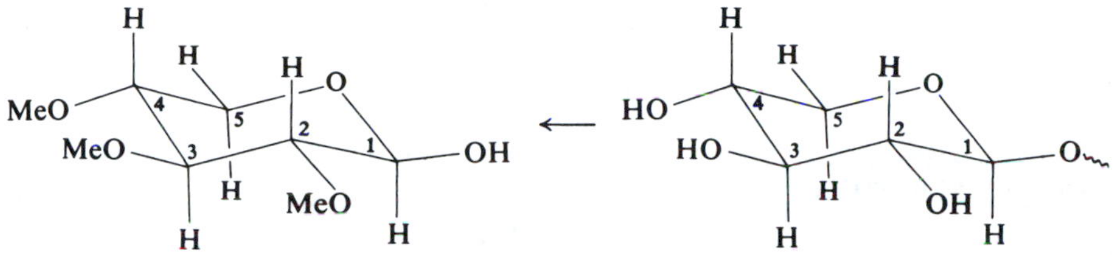

2,3,4-Tri-*O*-methyl-D-xylose ← *Attachment at C–1* **Chain-terminating unit**

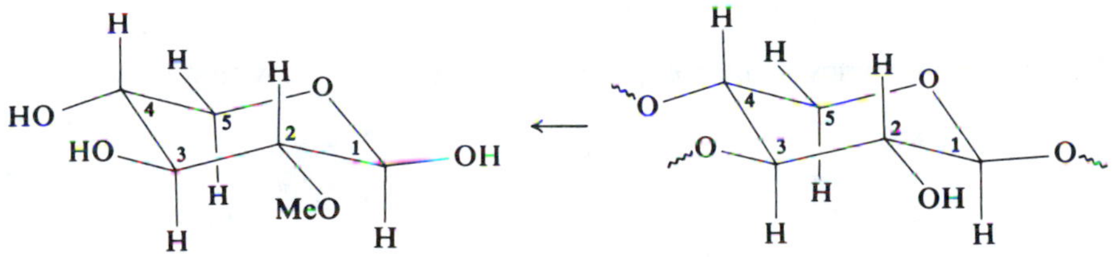

2-*O*-Methyl-D-xylose ← *Attachment at C–1, C–3, and C–4* **Chain-linking unit**

The xylan is a poly-β-D-xylopyranoside.

Chain-forming unit: attachment at C–1 and C–4.
Chain-terminating unit: attachment at C–1.
Chain-linking unit: attachment at C–1, C–3, and C–4.

Much of the xylan is made up of simple β-xyloside chains, with units joined to each other by linkages between C–1 and C–4; every once in a while two chains are joined together by a linkage involving C–3.

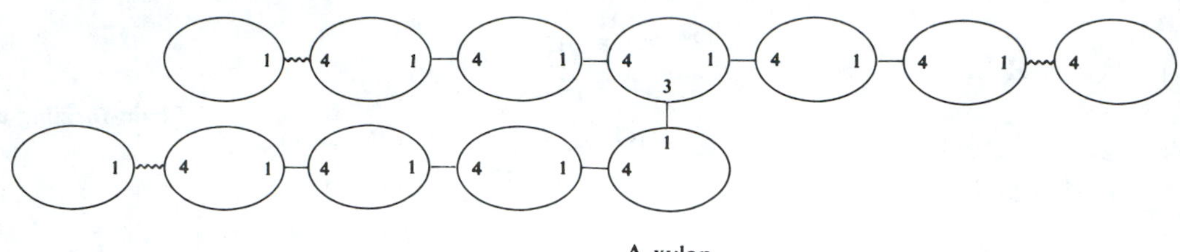

A xylan

35.17 (a) Non-reducing. There is no terminal glucose unit, and hence no "free" carbonyl group.

(b) Only D-(+)-glucose.

(c) Only 2,3,6-tri-*O*-methyl-D-glucose.

(d) No formaldehyde or formic acid.

(e) M.w. = 162.1*n*, where 162.1 is the value for a glucose residue and *n* is a whole number.

35.18 The polar $—SO_3^-Na^+$ group protrudes. The non-polar benzene ring is held within the lipophilic cavity.

35.19 Each cyclodextrin is precipitated by the compound of a size to fit its cavity. In order of increasing size, these are: α-cyclodextrin by cyclohexane, β-cyclodextrin by fluorobenzene, and γ-cyclodextrin by anthracene.

35.20 With regard to fitting into a cyclodextrin pail, we recognize that the isomeric isopropyltoluenes are of different sizes: the *para* is the smallest, and the *meta* is the biggest. We might expect to separate them by the successive use of cyclodextrins of different sizes: the smallest to complex the *para* isomer, a bigger one for the *ortho*, and a still bigger one for the *meta*.

1.

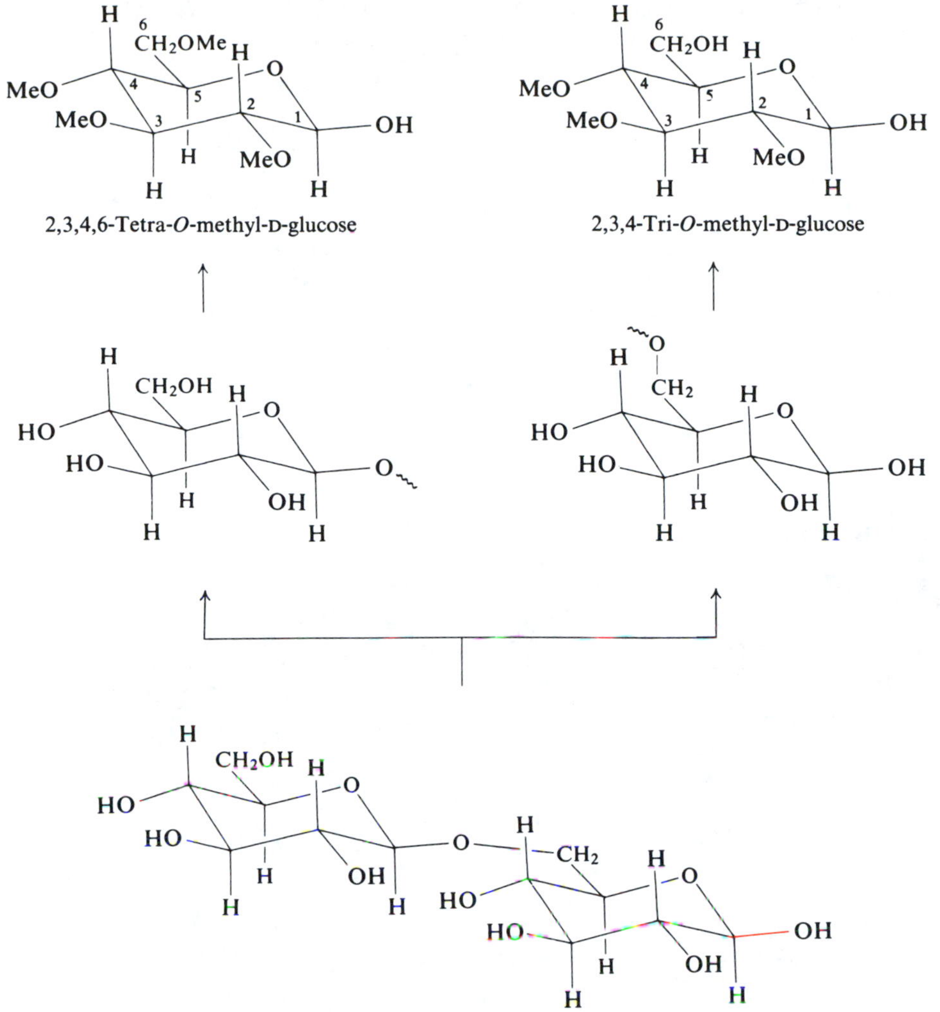

2,3,4,6-Tetra-*O*-methyl-D-glucose

2,3,4-Tri-*O*-methyl-D-glucose

6-*O*-(β-D-Glucopyranosyl)-D-glucopyranose
(β-anomer)
(+)-Gentiobiose

2. (a) Trehalose is a non-reducing sugar: the glucose units must be joined by a C(1)—C(1) linkage. It is hydrolyzed by maltase only: the glycosidic linkage is *alpha* to each ring. The —OH on C–5 is not methylated: the rings are pyranose.

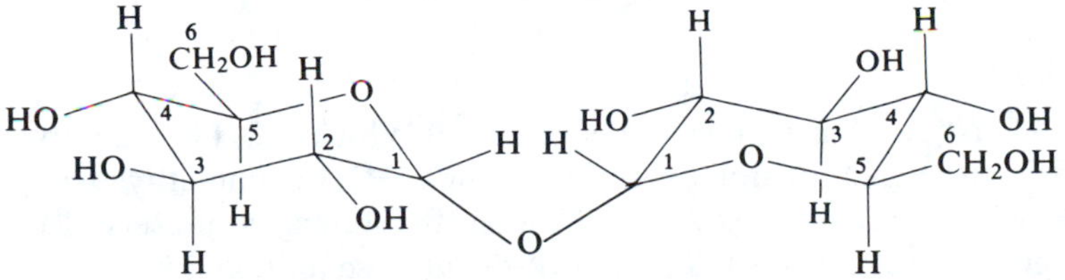

α-D-Glucopyranosyl-α-D-glucopyranoside
Trehalose

(b) Isotrehalose is hydrolyzed by either emulsin or maltose: the glycoside linkage is *alpha* to one ring and *beta* to the other. In neotrehalose, the linkage is *beta* to each ring.

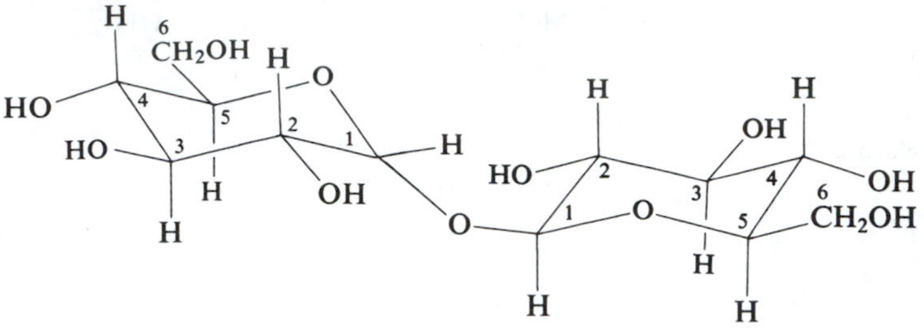

α-D-Glucopyranosyl-β-D-glucopyranoside
Isotrehalose

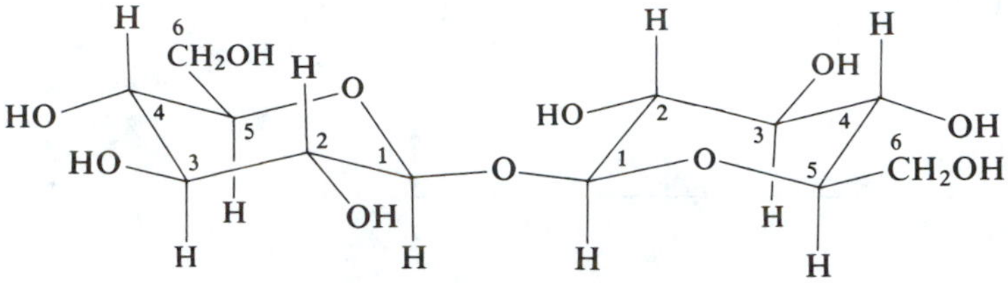

β-D-Glucopyranosyl-β-D-glucopyranoside
Neotrehalose

3.

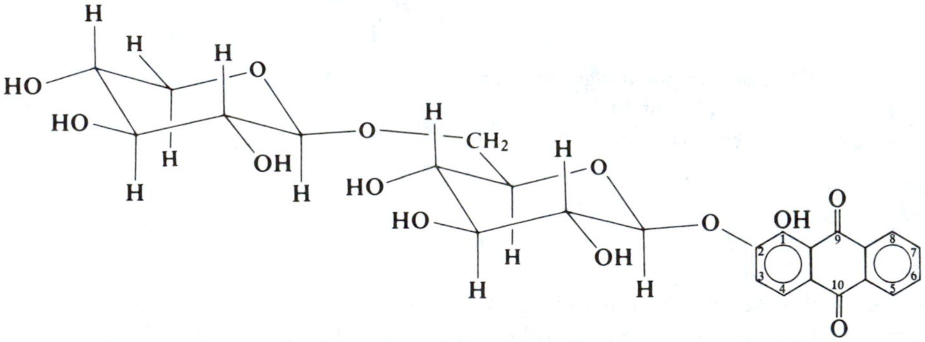

Ruberythric acid
2-(1-Hydroxy-9,10-anthraquinonyl)-β-6-*O*-(β-D-xylopyranosyl)-D-glucopyranoside

From the data provided two questions remain unanswered. (*i*) Which —OH of alizarin is involved? Methylate ruberythric acid, hydrolyze, and see which —OH is unmethylated. (Glycoside linkages are cleaved under conditions where phenolic ether linkage is preserved.) (*ii*) Are the glycoside linkages (between alizarin and C–1 of the glucose unit, and between C–6 of the glucose unit and C–1 of the xylose unit) *alpha* or *beta*? Study the enzymatic hydrolysis of ruberythric acid and of primeverose.

4.

Sucrose

Glucose Fructose

Galactose

Melibiose

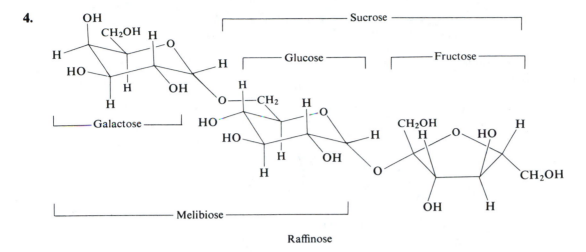

Raffinose

6-*O*-(α-D-Galactopyranosyl)-D-glucopyranose
(β-anomer)
Melibiose

5. (a) See Figure 35.6, page 646 of this Study Guide. In melezitose, the α-D-glucopyranosyl unit (maltase test) is attached at C–3 of the fructose unit of sucrose (no methylation of C_3-OH of fructose).

(b)

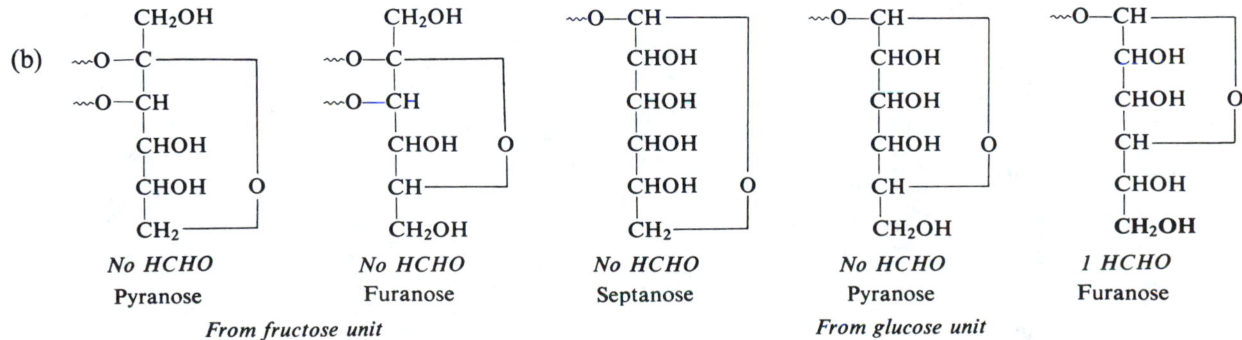

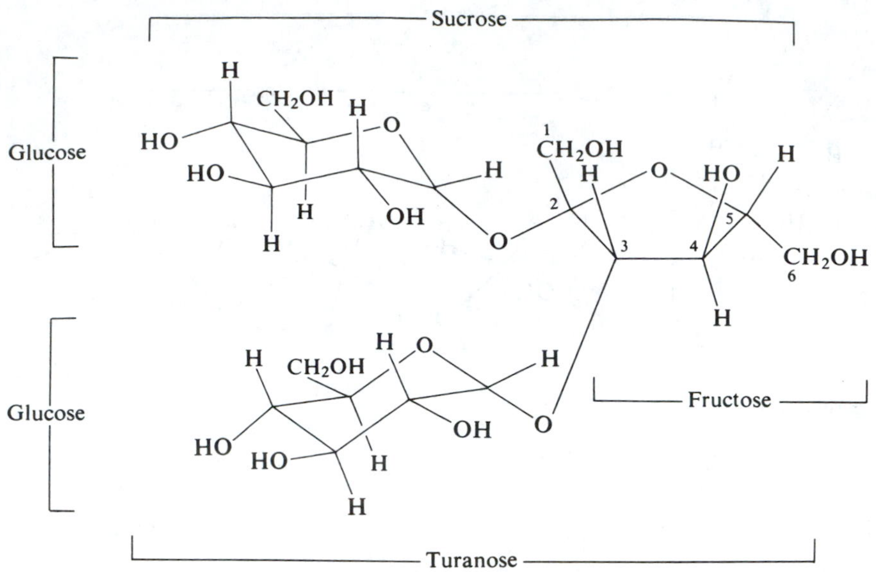

Sucrose

Glucose

Glucose

Fructose

Turanose

Melezitose

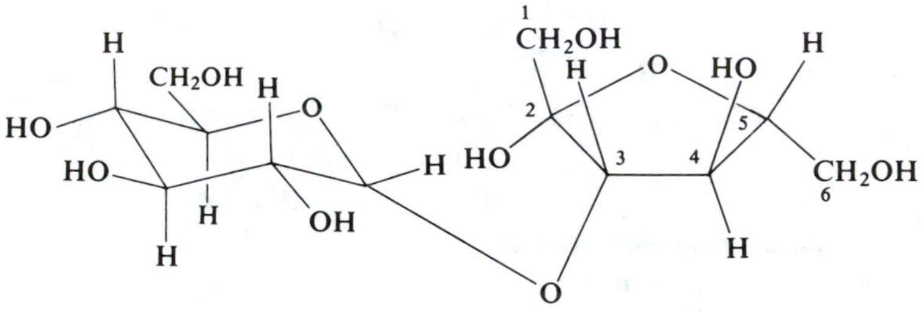

3-*O*-(α-D-Glucopyranosyl)-D-fructofuranose
Turanose

Figure 35.6 Melezitose and turanose (Problem 5a).

(c) 2 Septanose $\xrightarrow{6HIO_4}$ 4 HCOOH

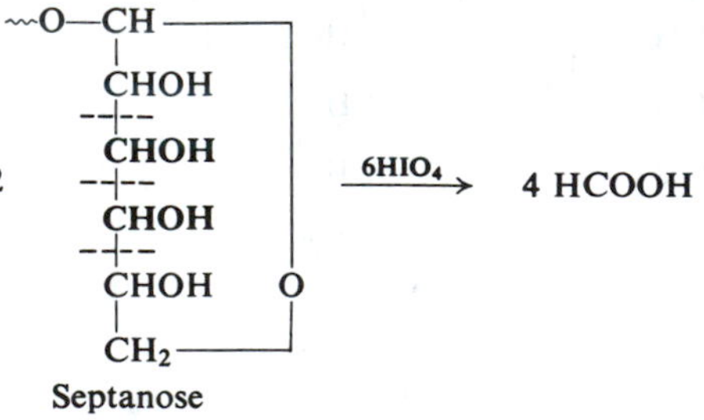

(d) 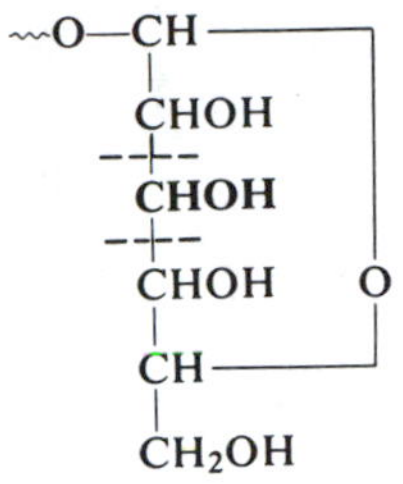 ~O—CH—CHOH—CHOH—CHOH—CH(—O—)—CH_2OH $\xrightarrow{2HIO_4}$ 1 HCOOH

Pyranose

$3HIO_4$ (septanose) + $2HIO_4$ (pyranose) = $5HIO_4$

2HCOOH (septanose) + $1HIO_4$ (pyranose) = 3HCOOH

(e) $2HIO_4$ (pyranose) × 2 = $4HIO_4$

1HCOOH (pyranose) × 2 = 2HCOOH

(f) Both glucose rings must be pyranose, since actually only $4HIO_4$ was consumed and 2HCOOH was formed.

(g) 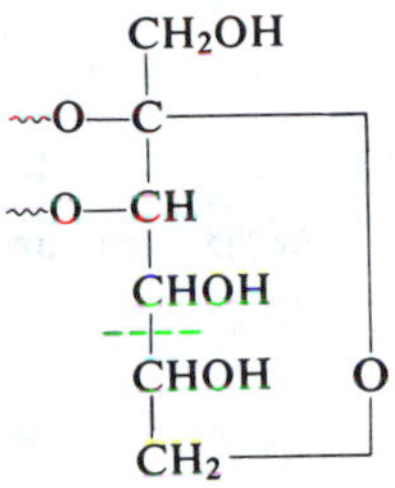 Pyranose: CH_2OH, ~O—C, ~O—CH, CHOH, CHOH, CH_2 (ring O) $\xrightarrow{1HIO_4}$ *no* HCOOH

Furanose: CH_2OH, ~O—CH, ~O—CH, CHOH, CH (ring O), CH_2OH $\xrightarrow{HIO_4}$ *no reaction*

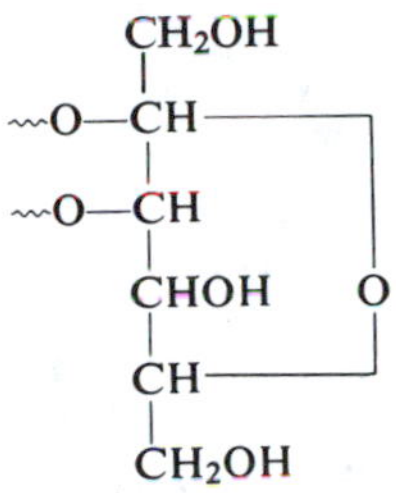

(h) The fructose unit must have a furanose ring: all the HIO_4 was used up by the glucose units ($4HIO_4$), and hence the fructose ring must have consumed no HIO_4.

(i) Perfectly consistent, as examination of the structure given in (a) will show.

6. (a)

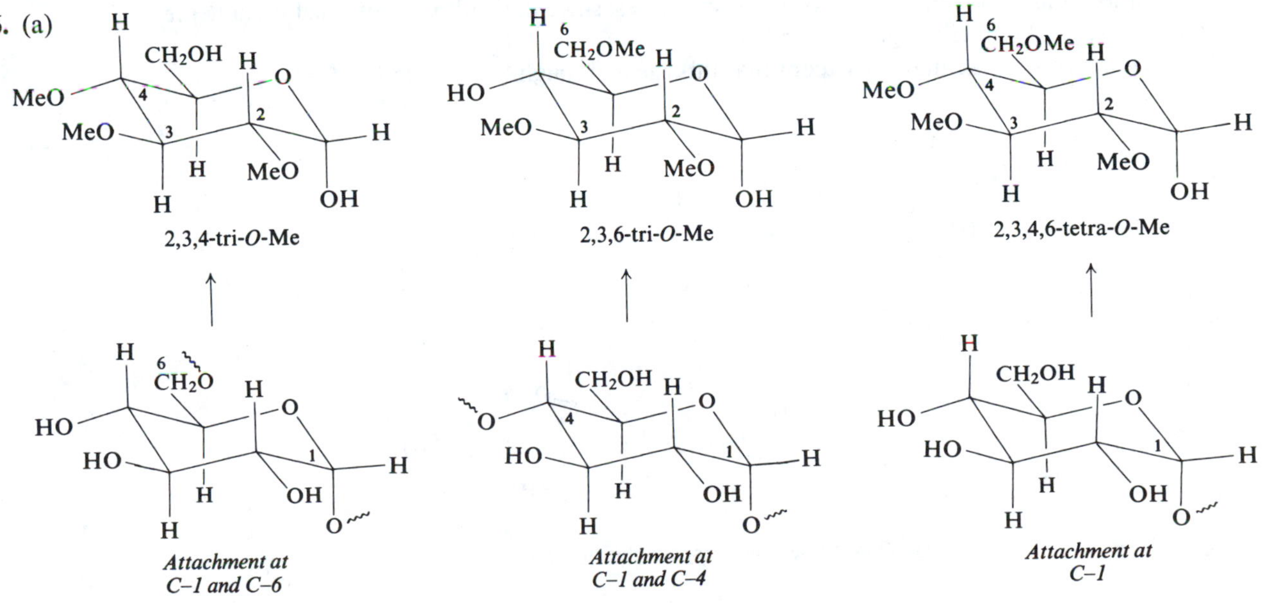

2,3,4-tri-*O*-Me ↑ *Attachment at C-1 and C-6*

2,3,6-tri-*O*-Me ↑ *Attachment at C-1 and C-4*

2,3,4,6-tetra-*O*-Me ↑ *Attachment at C-1*

These three units could be arranged in two different ways:

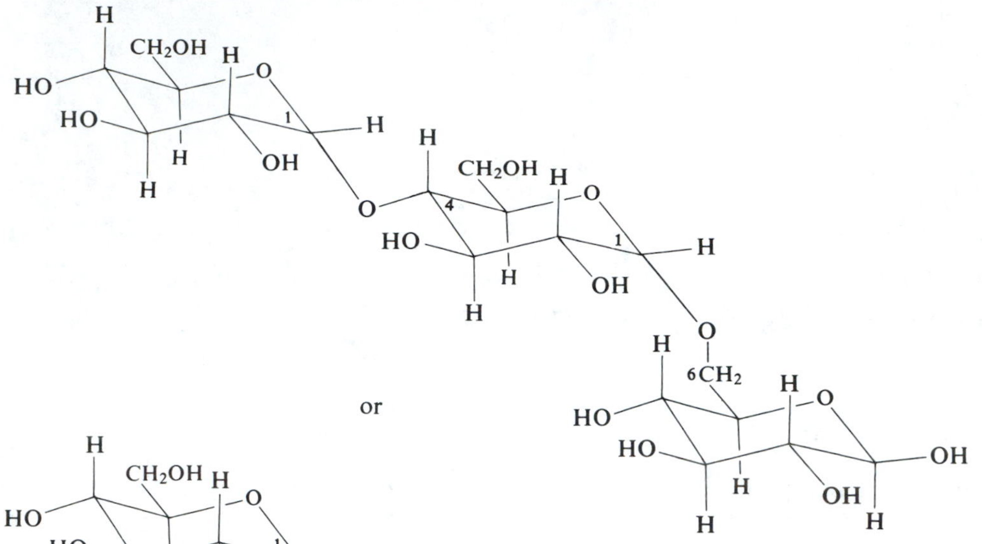

or

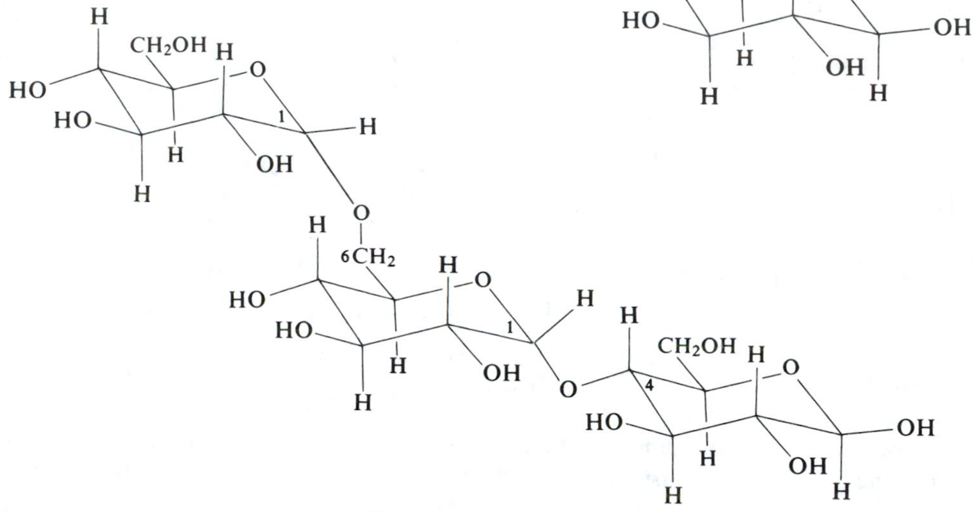

Actual structure

(*Chair conformations assumed*)
Panose

(b) Panose must have the second of these structures, since the maltose unit must include the oxidizable (and reducible) glucose unit.

Isomaltose is like maltose except that linkage is through C–6 instead of C–4.

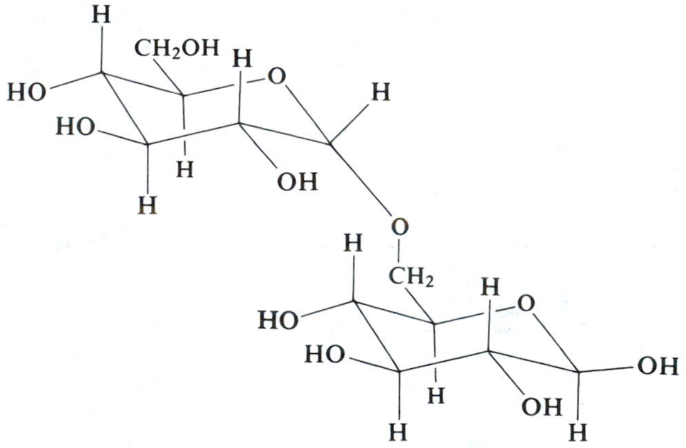

6-*O*-(α-D-Glucopyranosyl)-D-glucopyranose
Isomaltose

(c) This confirms the branched structure of amylopectin (page 1197).

Isomaltose

Panose unit

Maltose

Amylopectin

7. (a) The araban is a poly-L-arabinofuranoside.

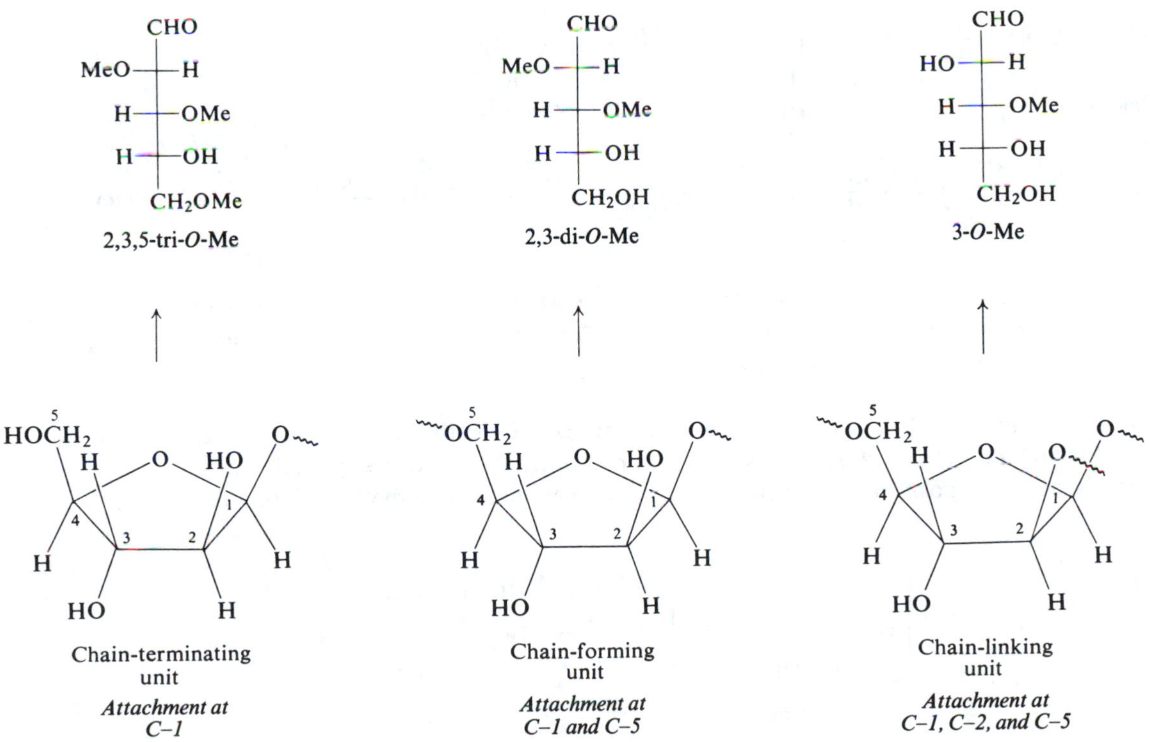

Chain-terminating unit
Attachment at C-1

Chain-forming unit
Attachment at C-1 and C-5

Chain-linking unit
Attachment at C-1, C-2, and C-5

It evidently has many two unit branches:

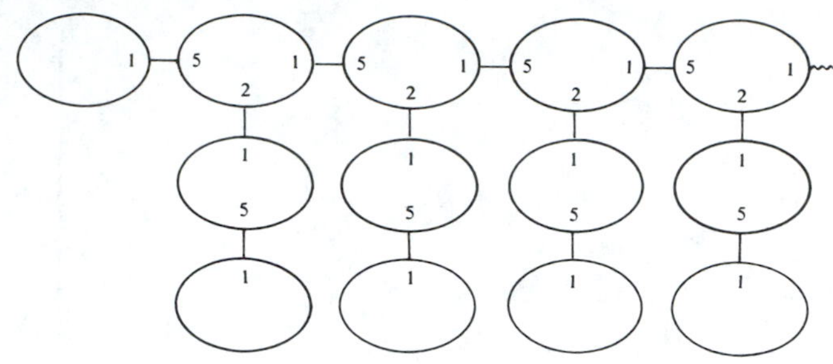

An araban

(b) The mannan is a poly-D-mannopyranoside. (See Figure 35.7, page 651 of this Study Guide.)

It has one-unit branches at (on the average) every other chain-forming unit. If we assume a maximum of regularity in its biosynthesis, we might arrive at something like this: three different units alternate regularly along the chain; those linked through C–1 and C–6 generally (but not quite always—witness the small amount of the 2,3,4-tri-*O*-methyl compound) branch at C–2.

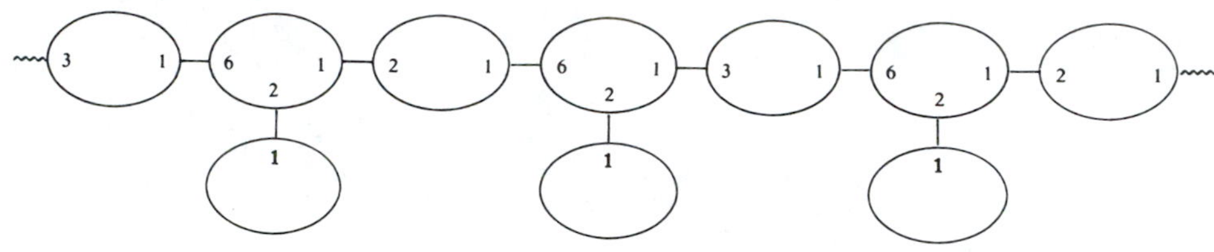

A mannan

8. $HOOC(CH_2)_4COOH \xleftarrow[H^+]{H_2O} N{\equiv}C(CH_2)_4C{\equiv}N \xleftarrow{CN^-} Cl(CH_2)_4Cl$

F — Adipic acid; E — $C_6H_8N_2$; D — $C_4H_8Cl_2$

$Cl(CH_2)_4Cl \xleftarrow{HCl}$ C $\xleftarrow[cat.]{H_2}$ B $\xleftarrow[-CO_2]{heat}$ A (2-furyl–COOH) $\xleftarrow{KMnO_4}$ (2-furyl–CHO)

D; C — C_4H_8O, Tetrahydrofuran, THF; B — C_4H_4O, Furan (Sec. 30.2); A — $C_5H_4O_3$, Furoic acid; $C_5H_4O_2$, Furfural

All this does not reveal the point of attachment of —CHO to the furan ring. Since furfural is formed here from a straight-chain pentose unit, it would seem likely that attachment is at the 2-position. (It actually *is.*) We can picture something like this happening:

An ether (furan ring: HC—CH, HC, C—CHO, O) $\xleftarrow{-H_2O}$ [HC—CH, HC(HO) , C(OH)—CHO] A di(enol) $\xleftarrow{-H_2O}$ HOCH—CHOH, H_2C(HO), CH(OH)—CHO A pentose

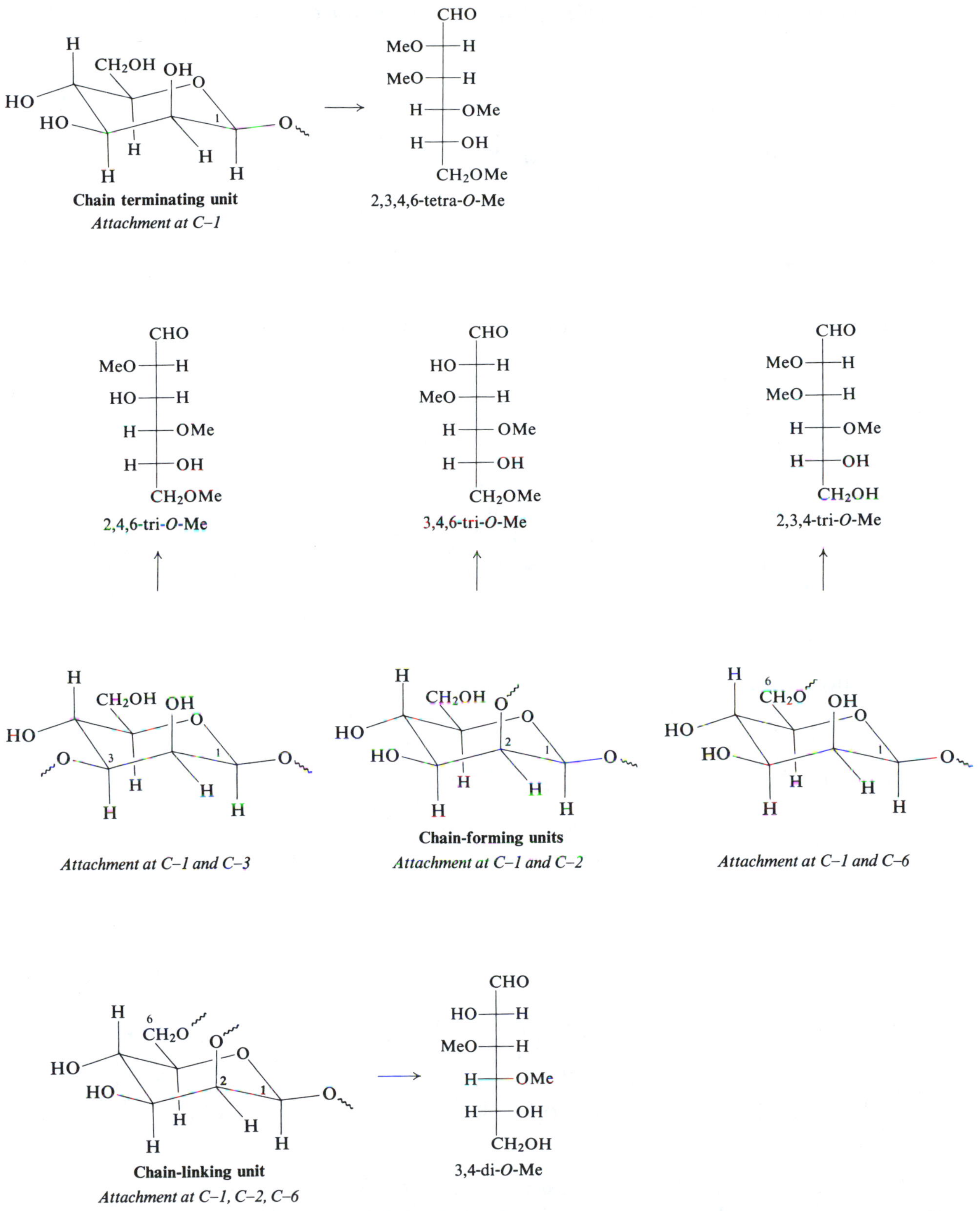

Chain terminating unit
Attachment at C–1

2,3,4,6-tetra-*O*-Me

2,4,6-tri-*O*-Me

3,4,6-tri-*O*-Me

2,3,4-tri-*O*-Me

Attachment at C–1 and C–3

Chain-forming units
Attachment at C–1 and C–2

Attachment at C–1 and C–6

Chain-linking unit
Attachment at C–1, C–2, C–6

3,4-di-*O*-Me

Figure 35.7 A mannan: methylation, hydrolysis (Problem 7b).

9. (a) Alginic acid is a poly(β-D-mannuronic acid), with units attached at C–1 and C–4.

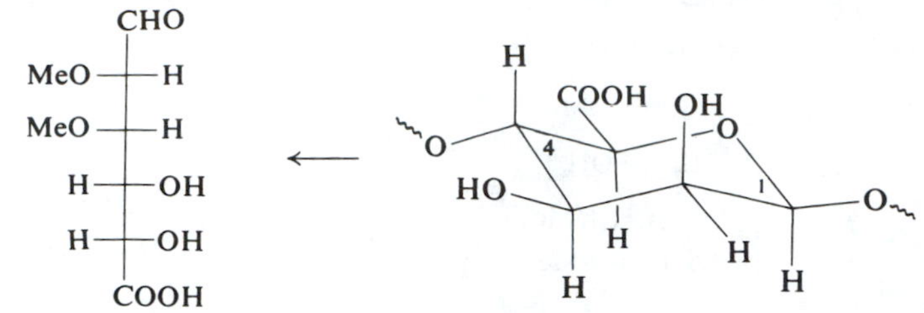

2,3-Di-*O*-methyl-D-mannuronic acid

Attachment at C–1 and C–4

(b) Pectic acid is a poly(α-D-galacturonic acid), with units attached at C–1 and C–4.

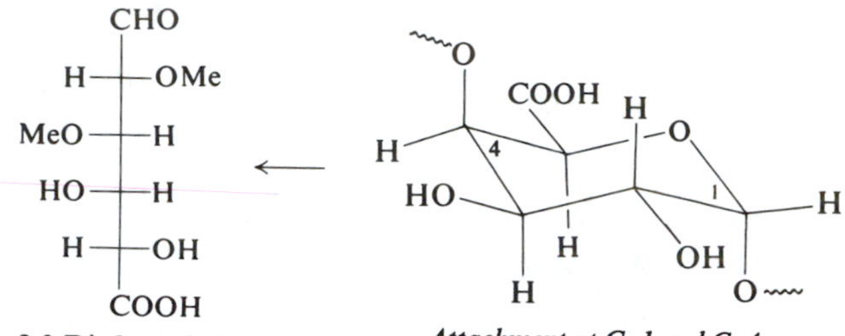

2,3-Di-*O*-methyl-D-galacturonic acid

Attachment at C–1 and C–4

(c) Agar is a polygalactopyranoside.

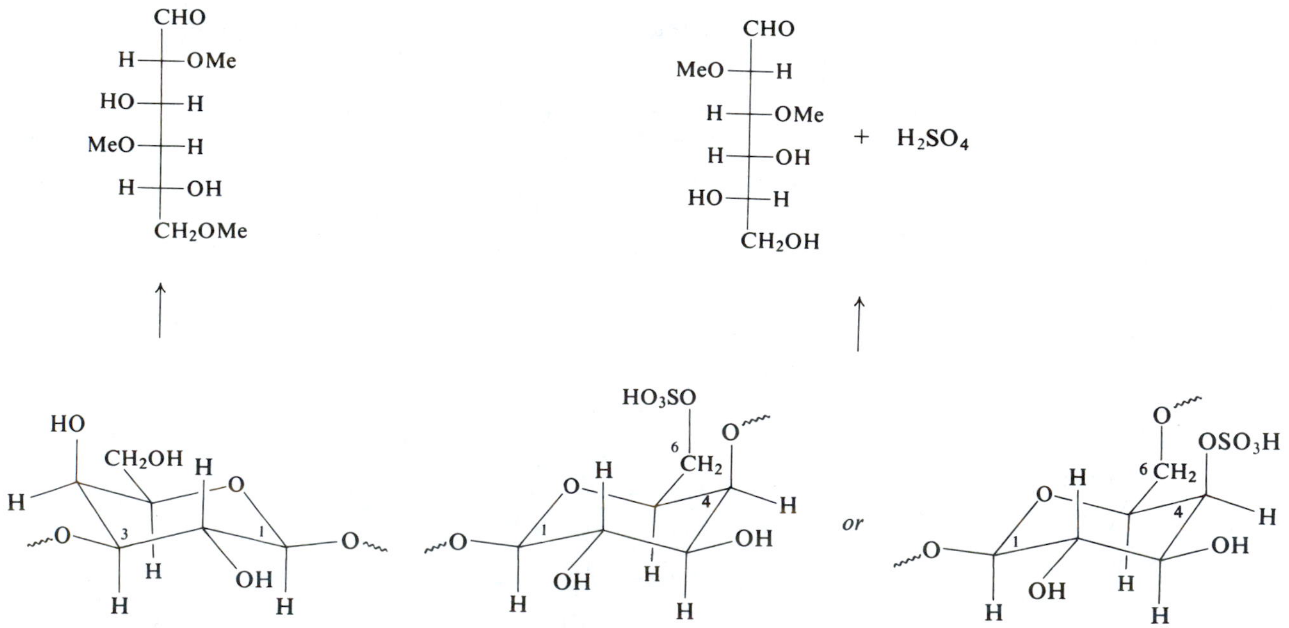

$+ \ H_2SO_4$

or

Most of the units are D-galactose, attached at C–1 and C–3. About every tenth unit is L-galactose esterified by sulfuric acid. This unit is either attached through C–1 and C–4 with sulfate at C–6, or attached through C–1 and C–6 with sulfate at C–4. (Or, conceivably, it has a furanose ring, with attachment at C–6 and sulfate at C–5, or vice versa.)

10. This capsule polysaccharide is a poly(cellobiuronic acid) with pyranose rings, and chain-forming attachments at C–1 and C–4 of the D-glucose units and at C–1 and C–3 of the D-glucuronic acid units. (See Figure 35.8, page 655 of this Study Guide.)

11.

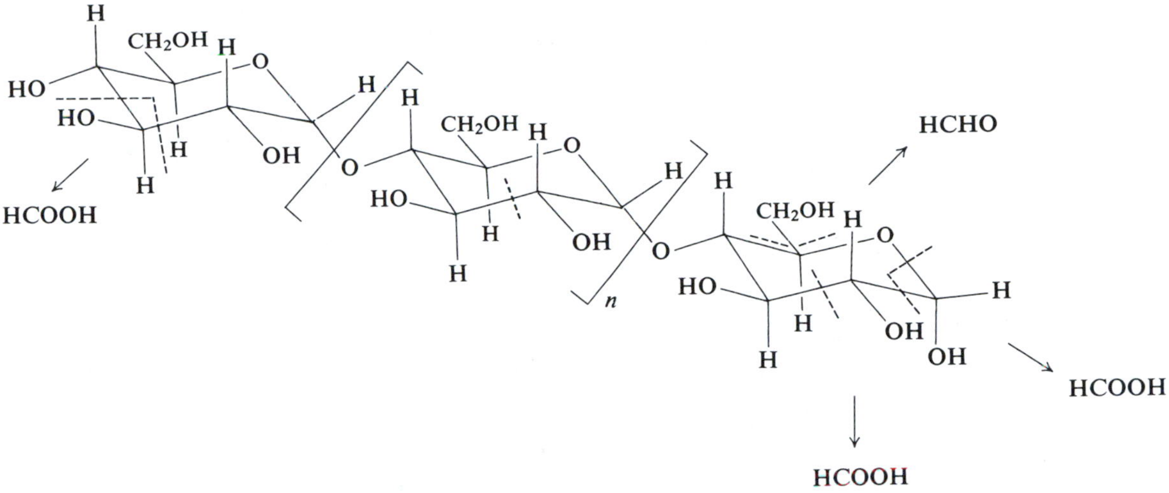

HCOOH HCHO HCOOH HCOOH

Amylose

$\downarrow HIO_4$

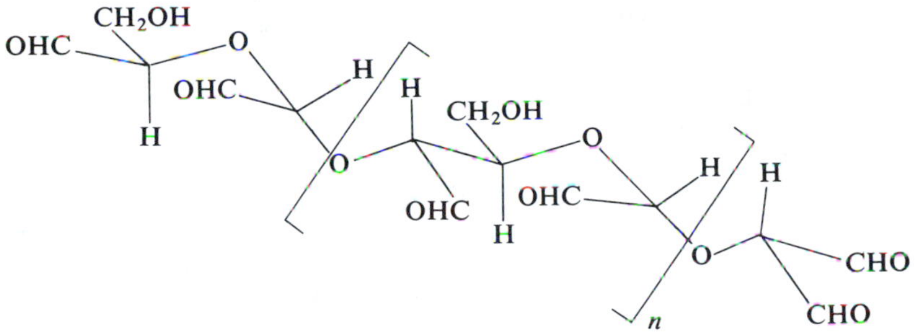

G

$\downarrow Br_2(aq)$

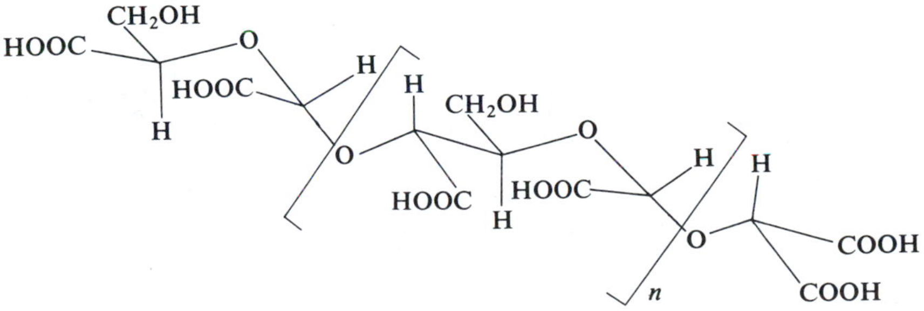

H

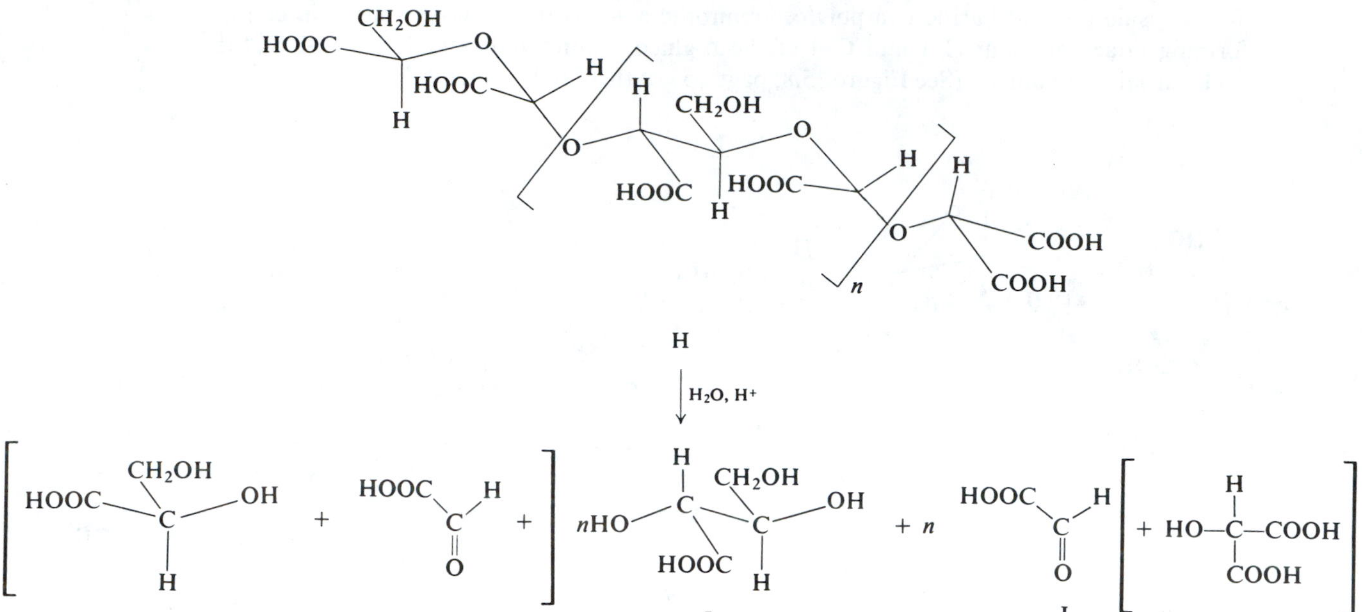

HOOC, CH_2OH, O, H, HOOC, H, H, CH_2OH, O, O, H, H, HOOC, HOOC, H, O, COOH, COOH, n

H

$\downarrow$ H_2O, H^+

[HOOC—C(H)(CH_2OH)—OH + HOOC—C(H)=O] Negligible product

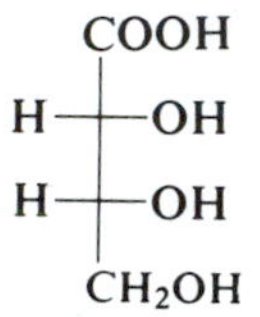

nHO—C(H)(HOOC)—C(H)(CH_2OH)—OH + n HOOC—C(H)=O [+ HO—C(H)(COOH)—COOH] Negligible product

I $C_4H_8O_5$ J $C_2H_2O_3$

COOH
H—OH
H—OH
CH_2OH
D-Erythronic acid
I

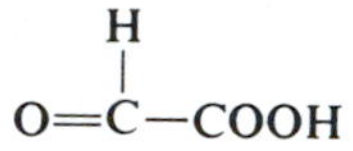

O=C(H)—COOH
Glyoxylic acid
J

12. (a) Protonated *tert*-butyl hypochlorite acts as the source of positive chlorine, which becomes attached to the ring in typical electrophilic aromatic substitution (Sec. 15.11).

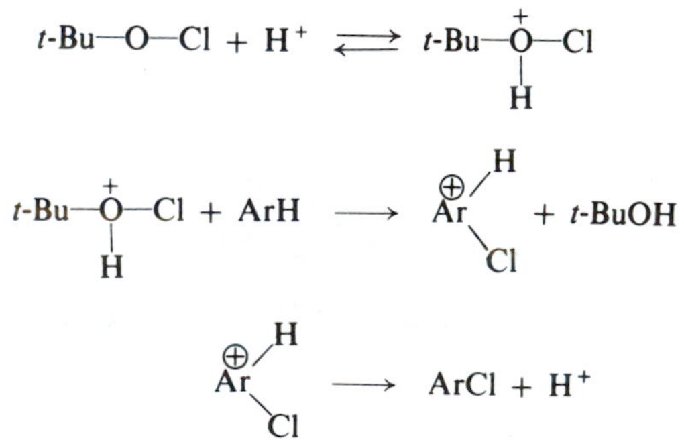

$$t\text{-Bu—O—Cl} + H^+ \rightleftharpoons t\text{-Bu—}\overset{+}{O}(H)\text{—Cl}$$

$$t\text{-Bu—}\overset{+}{O}(H)\text{—Cl} + ArH \longrightarrow \overset{\oplus}{Ar}(H)(Cl) + t\text{-BuOH}$$

$$\overset{\oplus}{Ar}(H)(Cl) \longrightarrow ArCl + H^+$$

(b) We can account for the regioselectivity on purely steric grounds. The phenyl group of anisole is held in the lipophilic cavity of the cyclodextrin, with the polar $—OCH_3$ protruding.

The *ortho* positions are thus shielded from attack, while the *para* position is exposed at the other end of the cavity.

To account for the acceleration of rate at the *para* position, we must speculate further. Just as HOCl converts *t*-BuOH into *t*-BuOCl, so it converts an —OH of the cyclodextrin into an —OCl. This group, lying on the rim of the cyclodextrin pail, is in just the right position to transfer positive chlorine to the nearby *para* position of anisole. The enhanced rate is thus a neighboring group effect.

(c) Chlorination is brought about by —OCl attached to a C–3 position. This means that the *para* position of anisole is near the "upper" rim of the pail, where the C–3 hydroxyls are (Figure 35.4, page 1199), and that the —OCH_3 group must protrude through the "bottom".

(Breslow, R. "Biomimetic Chemistry"; *Chem. Soc. Rev.* **1972**, *1*, 573.)

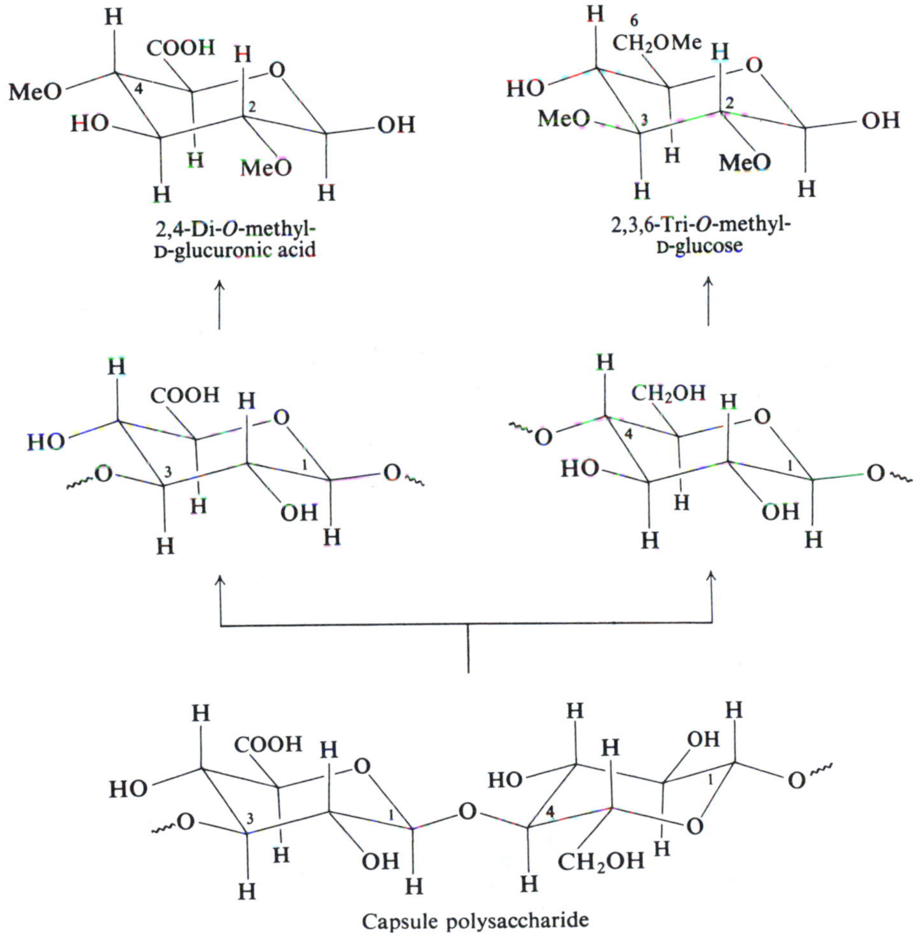

2,4-Di-*O*-methyl-D-glucuronic acid

2,3,6-Tri-*O*-methyl-D-glucose

Capsule polysaccharide

371 **Figure 35.8** Capsule polysaccharide: methylation, hydrolysis (Problem 10).

13. (a) As the structural formula given on page 1204 shows, we get three molecules of HCOOH (and one molecule of HCHO) per molecule of amylose:

1HCOOH from the terminal non-reducing unit;
no small cleavage products from non-terminal units;
2HCOOH + 1HCHO from the terminal reducing unit.

(b)

moles HCOOH/3	=	moles amylose
wt. amylose/moles amylose	=	m.w. amylose
m.w. amylose/wt. (of 162) per glucose unit	=	glucose units per molecule of amylose

(c) 980 glucose units per molecule. (What is the approximate m.w. of this sample of amylose?)

36

Proteins and Nucleic Acids

Molecular Biology

36.1 $—NH_2 \quad > \quad —COO^-$

$$K_b \sim 10^{-4} \qquad K_b = \frac{K_w}{K_a} = \frac{10^{-14}}{\sim 10^{-3}} = \sim 10^{-11}$$

Proton goes to $—NH_2$: $H_2NCHRCOO^- + H^+ \longrightarrow {}^+H_3NCHRCOO^-$

36.2 $—COOH \quad > \quad —NH_3{}^+$

$$K_a \sim 10^{-3} \qquad K_a = \frac{K_w}{K_b} = \frac{10^{-14}}{\sim 10^{-4}} = \sim 10^{-10}$$

Proton leaves $—COOH$: $+H_3NCHRCOOH \longrightarrow H^+ + {}^+H_3NCHRCOO^-$

36.3 K_b for aromatic amines is quite low, about 10^{-10}. (The $—COO^-$ group, with K_b 10^{-9}, is actually a slightly stronger base than aromatic $—NH_2$.) As a result, $—NH_2$ is not neutralized by $—COOH$. But $—SO_3H$ is strongly acidic and can neutralize an aromatic $—NH_2$ group.

36.4 (a) $^+H_3N(CH_2)_4\underset{\displaystyle NH_2}{\underset{|}{C}}HCOO^-$ or $H_2N(CH_2)_4\underset{\displaystyle {}^+NH_3}{\underset{|}{C}}HCOO^-$

Actual structure
Lysine

The $—COO^-$ group is electron-withdrawing and hence base-weakening. The more distant ε-amino group is affected less, is more basic, and gets the proton.

(b) 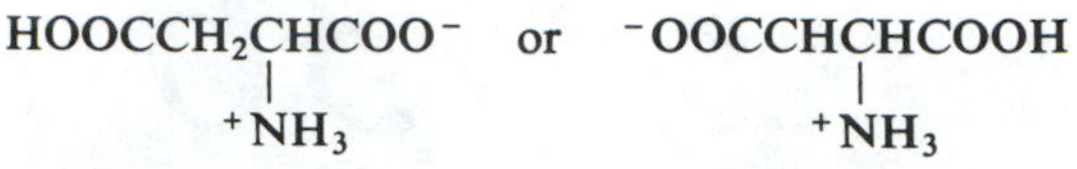 $HOOCCH_2\underset{\displaystyle {}^+NH_3}{\underset{|}{C}}HCOO^-$ or $^-OOCCH\underset{\displaystyle {}^+NH_3}{\underset{|}{C}}HCOOH$

Actual structure

Aspartic acid

The inductive effect of —NH_2 is electron-withdrawing and hence acid-strengthening. The nearer —COOH is affected more, is more acidic, and loses the proton.

(c) $H_2N\text{—}\underset{\displaystyle {}^+NH_2}{\underset{\|}{C}}\text{—}\overset{\displaystyle H}{\overset{|}{N}}(CH_2)_3\underset{\displaystyle NH_2}{\underset{|}{C}}HCOO^-$ $\qquad$ $^+H_3N\text{—}\underset{\displaystyle NH}{\underset{\|}{C}}\text{—}\overset{\displaystyle H}{\overset{|}{N}}(CH_2)_3\underset{\displaystyle NH_2}{\underset{|}{C}}HCOO^-$

Actual structure

Arginine

$H_2N\text{—}\underset{\displaystyle HN}{\underset{\|}{C}}\text{—}\underset{\displaystyle H}{\underset{|}{\overset{\displaystyle H}{\overset{+|}{N}}}}(CH_2)_3\underset{\displaystyle NH_2}{\underset{|}{C}}HCOO^-$ $\qquad$ $H_2N\text{—}\underset{\displaystyle NH}{\underset{\|}{C}}\text{—}\overset{\displaystyle H}{\overset{|}{N}}(CH_2)_3\underset{\displaystyle {}^+NH_3}{\underset{|}{C}}HCOO^-$

The =NH nitrogen is the most basic, since the acceptance of a proton there gives a resonance-stabilized cation (compare the guanidium ion, Problem 20.23, page 783).

(d) 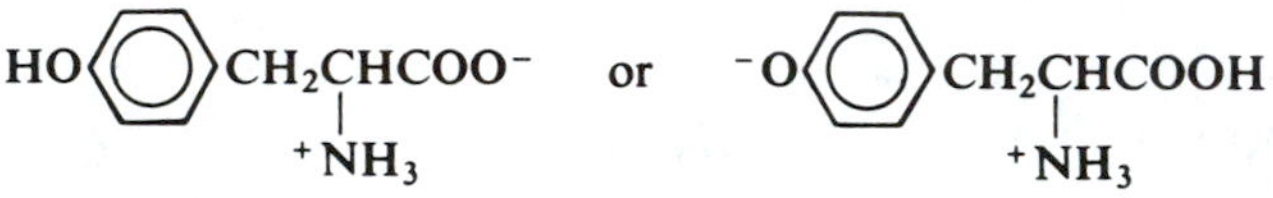 $HO\text{—}C_6H_4\text{—}CH_2\underset{\displaystyle {}^+NH_3}{\underset{|}{C}}HCOO^-$ or $^-O\text{—}C_6H_4\text{—}CH_2\underset{\displaystyle {}^+NH_3}{\underset{|}{C}}HCOOH$

Actual structure

Tyrosine

The —COOH group (K_a about 10^{-3}) is much more acidic than a phenolic —OH (K_a about 10^{-10}), and loses the proton.

36.5 (a) Esterify amino acids in strongly acidic solutions where the major component, $^+H_3NCHRCOOH$, contains —COOH.

(b) Acylate amino acids in strongly alkaline solutions where the major component, $H_2NCH_2COO^-$, contains —NH_2.

36.6 (a) On the acid side. The ionization of the extra —COOH ($K_a \sim 10^{-5}$) has to be suppressed by a considerable excess of H^+.

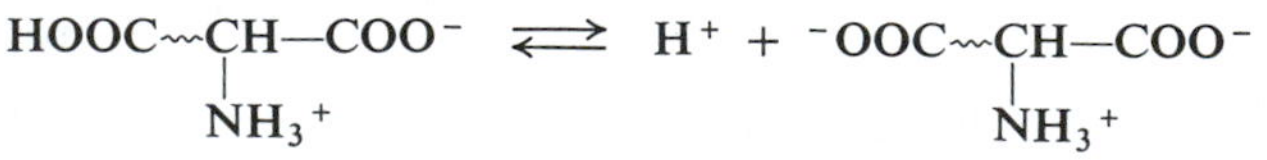

$$HOOC\sim\underset{\displaystyle NH_3^+}{\underset{|}{C}}H\text{—}COO^- \rightleftarrows H^+ + {}^-OOC\sim\underset{\displaystyle NH_3^+}{\underset{|}{C}}H\text{—}COO^-$$

Neutral species

(b) On the basic side. The ionization of the extra —NH_2 ($K_b \sim 10^{-4}$) to give OH^- has to be suppressed by a considerable excess of OH^-.

$$^{+}H_3N\text{\~}\underset{\substack{|\\ NH_2}}{CH}\text{—}COO^- + H_2O \rightleftarrows OH^- + {}^{+}H_3N\text{\~}\underset{\substack{|\\ NH_3^+}}{CH}\text{—}COO^-$$

Neutral species

(c) More acidic and more basic than for glycine.

36.7 In alkali, sulfanilic acid, p-$^{+}H_3NC_6H_4SO_3^-$, gives up a proton to OH^- to form a water-soluble sulfonate, p-$H_2NC_6H_4SO_3^-$. But sulfonic acids are very strong acids, and sulfonate anions are correspondingly weak bases; in acid solution —SO_3^- does not accept a proton, and sulfanilic acid remains unchanged.

36.8 Use differential migration at different pH values; or use differences in solubility at different pH values; or precipitate acidic amino acids as salts of certain bases, and basic amino acids as salts of certain acids.

36.9 (–)-Cysteine and (–)-cystine.

In all the amino acids, H has the lowest priority and NH_3^+ has the highest. In most amino acids, C=O has priority over R because of the O once removed from the chiral center. But in cysteine and cystine, R contains S once removed from the chiral center, and this gives R priority over C=O. The reversal of priority of these groups results in a change in specification from *S* to *R*.

36.10

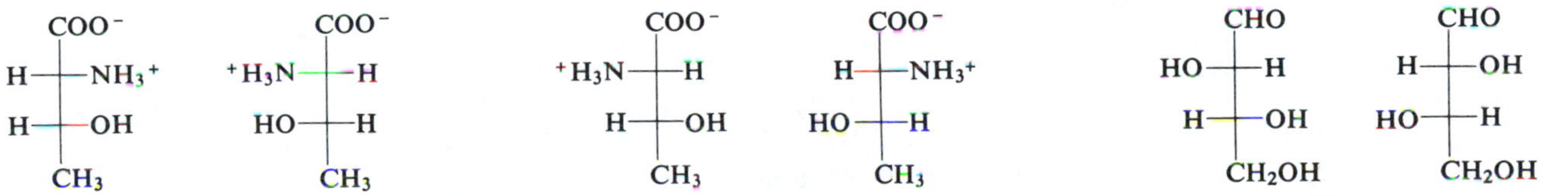

COO^-	COO^-	COO^-	COO^-	CHO	CHO
H—NH_3^+	$^{+}H_3N$—H	$^{+}H_3N$—H	H—NH_3^+	HO—H	H—OH
H—OH	HO—H	H—OH	HO—H	H—OH	HO—H
CH_3	CH_3	CH_3	CH_3	CH_2OH	CH_2OH
		L-Threonine (L *based on C-2*)		Threose (Sec. 34.12)	

36.11 Cys—Cys: one *meso,* one pair of enantiomers. Hyl: two pairs of enantiomers. Hyp: two pairs of enantiomers. Ile: two pairs of enantiomers.

36.12 **Direct ammonolysis** (as on page 1213).

$$\underset{\substack{|\\ ^{+}NH_3}}{CH_2}COO^- \longleftarrow CH_3COOH$$

Glycine ← Acetic acid

$$CH_3\underset{\substack{|\\ ^{+}NH_3}}{CH}COO^- \longleftarrow CH_3CH_2COOH$$

Alanine ← Propionic acid

$$(CH_3)_2CH\underset{\substack{|\\ ^{+}NH_3}}{CH}COO^- \longleftarrow (CH_3)_2CHCH_2COOH$$

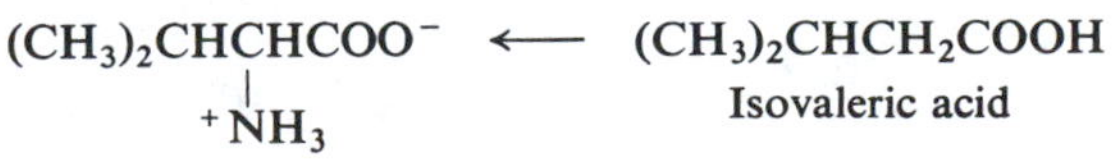

Valine ← Isovaleric acid

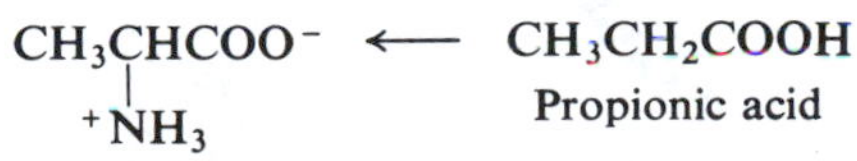

$(CH_3)_2CHCH_2CH(NH_3^+)COO^-$ ⟵ $(CH_3)_2CHCH_2CH_2COOH$

Leucine — Isocaproic acid

$HOOCCH_2CH(NH_3^+)COO^-$ ⟵ $HOOCCH_2CH_2COOH$

Aspartic acid — Succinic acid

Gabriel synthesis (as on page 1213).

$CH_2(NH_3^+)COO^-$ ⟵ $ClCH_2COOEt$

Glycine — Ethyl chloroacetate

$(CH_3)_2CHCH_2CH(NH_3^+)COO^-$ ⟵ $(CH_3)_2CHCH_2CH(Cl)COOEt$

Leucine — Ethyl α-chloroisocaproic acid

Malonic ester synthesis (as on page 1213).

$(CH_3)_2CH{-}CH(NH_3^+)COO^-$ ⟵ $(CH_3)_2CHBr$

Valine — Isopropyl bromide

$CH_3CH_2CH(CH_3){-}CH(NH_3^+)COO^-$ ⟵ $CH_3CH_2CH(CH_3)Br$

Isoleucine — *sec*-Butyl bromide

Phthalimidomalonic ester method (as on page 1214).

$HOCH_2{-}CH(NH_3^+)COO^-$ ⟵ HCHO, OH^- (aldol condensation)

Serine

$HOOCCH_2CH_2{-}CH(NH_3^+)COO^-$ ⟵ $EtOOCCH_2CH_2Br$

Glutamic acid — Ethyl β-bromopropionate

$HOOCCH_2{-}CH(NH_3^+)COO^-$ ⟵ $EtOOCCH_2Br$

Aspartic acid — Ethyl bromoacetate

36.13 The Strecker synthesis converts an aldehyde into an α-amino nitrile:

$$\underset{\text{Alanine}}{CH_3\underset{\underset{^+NH_3}{|}}{C}HCOO^-} \xleftarrow{H_2O,\ H^+} CH_3\underset{\underset{NH_2}{|}}{C}H{-}CN \xleftarrow[NH_4^+]{CN^-} CH_3CHO \qquad \underset{\text{Glycine}}{\underset{\underset{^+NH_3}{|}}{C}H_2COO^-} \longleftarrow \text{HCHO (from MeOH)}$$

$$\underset{\text{Leucine}}{(CH_3)_2CHCH_2\underset{\underset{^+NH_3}{|}}{C}HCOO^-} \longleftarrow (CH_3)_2CHCH_2CHO \longleftarrow \underset{\text{Isopentyl alcohol}}{(CH_3)_2CHCH_2CH_2OH}$$

$$\underset{\text{Isoleucine}}{CH_3CH_2CH(CH_3)\underset{\underset{^+NH_3}{|}}{C}HCOO^-} \longleftarrow CH_3CH_2CH(CH_3)CHO \longleftarrow \underset{\text{2-Methyl-1-butanol}}{CH_3CH_2CH(CH_3)CH_2OH}$$

$$\underset{\text{Valine}}{(CH_3)_2CH\underset{\underset{^+NH_3}{|}}{C}HCOO^-} \longleftarrow (CH_3)_2CHCHO \longleftarrow \underset{\text{Isobutyl alcohol}}{(CH_3)_2CHCH_2OH}$$

$$\underset{\text{Serine}}{HOCH_2\underset{\underset{^+NH_3}{|}}{C}HCOO^-} \xleftarrow{\text{acid}} EtOCH_2\underset{\underset{^+NH_3}{|}}{C}HCOO^- \xleftarrow[H^+]{H_2O} EtOCH_2\underset{\underset{NH_2}{|}}{C}HCN \xleftarrow[NH_4^+]{CN^-} EtOCH_2CHO$$

36.14 (a) $$\underset{\text{Ethyl isovalerate}}{(CH_3)_2CHCH_2COOEt} + EtOOC{-}COOEt \xrightarrow{OEt^-} \underset{\mathbf{A}}{(CH_3)_2CH\overset{\overset{COOEt}{|}}{C}H{-}\underset{\underset{O}{\|}}{C}{-}COOEt}$$

$$\mathbf{A} \xrightarrow[\text{heat}]{H^+} (CH_3)_2CH\overset{\overset{COOH}{|}}{C}H{-}\underset{\underset{O}{\|}}{C}{-}COOH \xrightarrow{-CO_2} \underset{\mathbf{B}}{(CH_3)_2CHCH_2{-}\underset{\underset{O}{\|}}{C}{-}COOH} \xrightarrow[Pd]{NH_3,\ H_2} \underset{\text{Leucine}}{(CH_3)_2CHCH_2{-}\underset{\underset{^+NH_3}{|}}{C}H{-}COO^-}$$

(b) $$\underset{\text{Alanine}}{CH_3\underset{\underset{^+NH_3}{|}}{C}HCOO^-} \longleftarrow \underset{\text{Ethyl acetate}}{CH_3COOEt} \text{ in place of ethyl isovalerate}$$

$$\underset{\text{Glutamic acid}}{HOOCCH_2CH_2\underset{\underset{^+NH_3}{|}}{C}HCOO^-} \longleftarrow \underset{\text{Ethyl succinate}}{EtOOCCH_2CH_2COOEt} \text{ in place of ethyl isovalerate}$$

36.15 (a) $H_2NCH_2COO^-Na^+$ (b) $Cl^-\,{}^+H_3NCH_2COOH$

(c) $Ph\underset{\underset{O}{\|}}{C}{-}NHCH_2COOH$ (d) $CH_3\underset{\underset{O}{\|}}{C}{-}NHCH_2COOH$

(e) $HOCH_2COOH + N_2$ (f) $HSO_4^-\,{}^+H_3NCH_2COOEt$

(g) $PhCH_2O\underset{\underset{O}{\|}}{C}{-}NHCH_2COOH$ (Cbz-Gly)

36.16 (a) $PhC(=O)NHCH_2COCl$ (b) $PhC(=O)NHCH_2CONH_2$

(c) $PhC(=O)NHCH_2C(=O)NHCH(CH_3)COOH$ (Bz-Gly-Ala)

(d) $PhC(=O)NHCH_2COOEt$

(e) 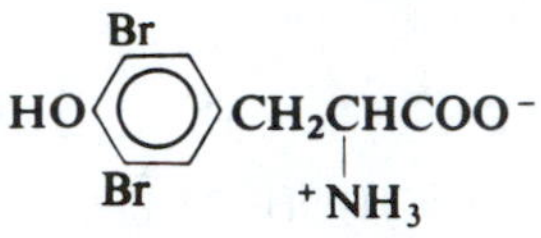 $HO\text{-}C_6H_2Br_2\text{-}CH_2CH(^+NH_3)COO^-$ (Br at 3,5 positions)

3,5-Dibromotyrosine

(f) $^-OOCCH_2CH(^+NH_3)COO^-$

(g) 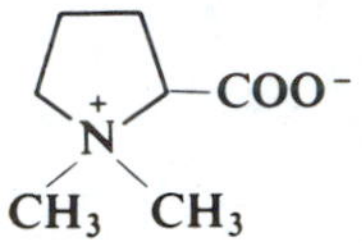 Pyrrolidine ring with $-COO^-$ on C-2 and $\overset{+}{N}(CH_3)_2$

N,*N*-Dimethylproline

(h)  $MeO\text{-}C_6H_4\text{-}CH_2CH(^+NH_3)COO^-$

(i) $Na^+\ ^-OOCCH_2CH_2CH(^+NH_3)COO^-$

Monosodium glutamate

(j) $EtOOCCH_2CH_2CH(^+NH_3HSO_4^-)COOEt$

36.17 (a) 0.001 mol or 22.4 mL N_2 from a monoamine (Leu).

(b) 0.002 mol or 44.8 mL N_2 from a diamine (Lys).

(c) No N_2 from a 2° amine.

36.18

$$2.01 \times \frac{748}{760} \times \frac{273}{293} = \text{mL } N_2 \text{ at S.T.P.}$$

$$\frac{\text{mL } N_2}{22.4 \text{ mL/mmol}} = \text{millimoles } N_2 = \text{millimoles} -NH_2 = \text{millimoles amino acid if one} -NH_2$$

if monoamine: 0.0821 mmol amino acid

$$\frac{9.36 \text{ mg}}{0.0821 \text{ mmol}} = \text{mol. wt.} = 114 = \text{minimum mol. wt.}$$

The amino acid could be valine, $C_5H_{11}O_2N$, m.w. 117. It could *not* be proline (m.w. 115), since proline has no $-NH_2$ present.

36.19 (a)  $-C(=\ddot{O}:)-\ddot{N}(H)-$ ⟷ $-C(-\ddot{O}:^\ominus)=\overset{\oplus}{N}(H)-$

(b) Overlap of *p* orbital of N with *π* orbital of $>C=O$.

36.20 (a)

$$H-C(=O)-N(CH_3)-CH_3 \quad \longleftrightarrow \quad H-C(-O^-)=\overset{\oplus}{N}(CH_3)-CH_3$$

Partial double-bond character of the carbon–nitrogen bond causes hindered rotation, resulting in diastereotopic —CH_3 groups that produce separate NMR signals. These signals coalesce at higher temperatures as the —CH_3's become equivalent through rapid rotation. (Compare Problem 19, Chapter 20.)

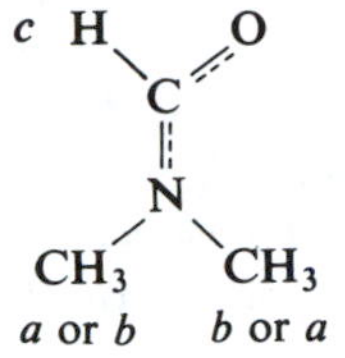

c H, O on C; C⋯N; N bearing CH_3 (*a* or *b*) and CH_3 (*b* or *a*)

(b) This confirms the partial double-bond character of the carbon–nitrogen bond in the peptide linkage.

36.21

$$\text{Ala:} \quad \frac{0.89\text{ g}}{89\text{ g/mol}} = 0.01\text{ mol in 100 g salmine}$$

$$\text{Arg:} \quad \frac{86.04\text{ g}}{174\text{ g/mol}} = 0.50\text{ mol}$$

$$\text{Gly:} \quad \frac{3.01\text{g}}{75\text{ g/mol}} = 0.04\text{ mol}$$

$$\text{Ile:} \quad \frac{1.28\text{ g}}{131\text{ g/mol}} = 0.01\text{ mol}$$

$$\text{Pro:} \quad \frac{6.90\text{ g}}{115\text{ g/mol}} = 0.06\text{ mol}$$

$$\text{Ser:} \quad \frac{7.29\text{ g}}{105\text{ g/mol}} = 0.07\text{ mol}$$

$$\text{Val:} \quad \frac{3.68\text{ g}}{117\text{ g/mol}} = 0.03\text{ mol}$$

The empirical formula is thus $AlaArg_{50}Gly_4IlePro_6Ser_7Val_3$. The weights of the amino acids add up to more than 100 g because water is taken up in the hydrolysis of the peptide links.

36.22 Since there is 0.01 mol Ala in 100 g of salmine (Problem 36.21, above) there must be 1 mol Ala in 10 000 g of salmine. This means that there is only one Ala per molecule, and the molecular formula is the same as the empirical formula.

36.23 If we assume only one Tyr per molecule,

$$0.29\%\text{ Tyr} = \frac{\text{m.w. Tyr}}{\text{m.w. protein}} \times 100 = \frac{204}{\text{m.w.}} \times 100$$

$$\text{m.w.} = 70\,300$$

This is the *minimum* m.w.; if there are two Tyr per molecule, the m.w. is 2 × 70 300, etc.

36.24 (a) If we assume only one Fe per molecule,

$$0.335\%\ \text{Fe} = \frac{\text{at.wt. Fe}}{\text{m.w. protein}} \times 100 = \frac{55.8}{\text{m.w.}} \times 100$$

$$\text{m.w.} = 16\,700$$

This is the *minimum* m.w.

(b)
$$\frac{67\,000}{16\,700} = 4\text{Fe per molecule}$$

36.25

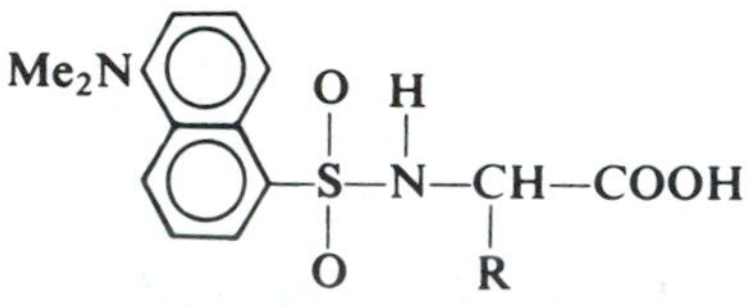

$$\text{Me}_2\text{N}\text{-C}_{10}\text{H}_6\text{-}\overset{\displaystyle\text{O}}{\underset{\displaystyle\text{O}}{\overset{|}{\underset{|}{\text{S}}}}}\text{—}\overset{\displaystyle\text{H}}{\overset{|}{\text{N}}}\text{—}\underset{\displaystyle\text{R}}{\underset{|}{\text{CH}}}\text{—COOH}$$

Dansyl derivative

A sulfonamide: not easily hydrolyzed

36.26 After the *C*-terminal residue has been removed from a peptide, the *next* residue becomes the *C*-terminal residue; this is, of course, also subject to removal by the enzyme. And so on with the next and the next. There will thus be obtained not just one amino acid cut from the chain, but several. The original *C*-terminal residue will be the amino acid that is liberated fastest at the beginning of the treatment.

36.27 (a)

$$\sim\text{NH}\underset{\displaystyle\text{R}'}{\underset{|}{\text{CH}}}\overset{\displaystyle\text{O}}{\overset{\|}{\text{C}}}\text{NH}\underset{\displaystyle\text{R}}{\underset{|}{\text{CH}}}\text{COOH} \xrightarrow{\text{LiBH}_4} \sim\text{NH}\underset{\displaystyle\text{R}'}{\underset{|}{\text{CH}}}\overset{\displaystyle\text{O}}{\overset{\|}{\text{C}}}\text{NH}\underset{\displaystyle\text{R}}{\underset{|}{\text{CH}}}\text{CH}_2\text{OH} \xrightarrow{\text{H}_2\text{O, H}^+} \text{NH}_2\underset{\displaystyle\text{R}}{\underset{|}{\text{CH}}}\text{CH}_2\text{OH} + {}^+\text{NH}_3\underset{\displaystyle\text{R}'}{\underset{|}{\text{CH}}}\text{COO}^-\text{, etc.}$$

An amino alcohol *C-terminal residue* — Amino acids

(b) Hydrazine, like the closely related ammonia, is a base and nucleophile, and cleaves the amide linkages to give acyl hydrazides—but not from the *C*-terminal residue.

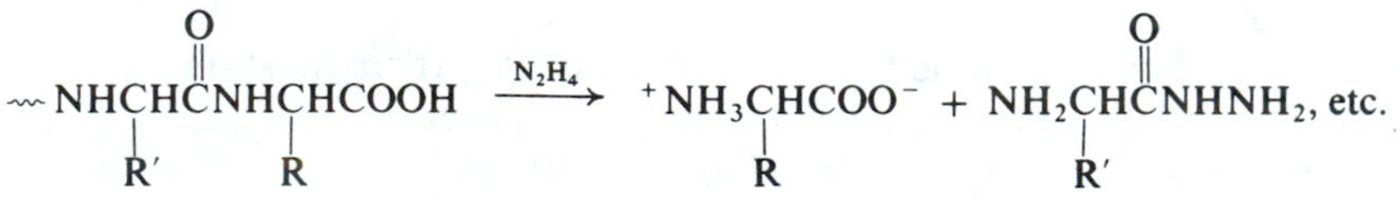

$$\sim\text{NH}\underset{\displaystyle\text{R}'}{\underset{|}{\text{CH}}}\overset{\displaystyle\text{O}}{\overset{\|}{\text{C}}}\text{NH}\underset{\displaystyle\text{R}}{\underset{|}{\text{CH}}}\text{COOH} \xrightarrow{\text{N}_2\text{H}_4} {}^+\text{NH}_3\underset{\displaystyle\text{R}}{\underset{|}{\text{CH}}}\text{COO}^- + \text{NH}_2\underset{\displaystyle\text{R}'}{\underset{|}{\text{CH}}}\overset{\displaystyle\text{O}}{\overset{\|}{\text{C}}}\text{NHNH}_2\text{, etc.}$$

An amino acid *C-terminal residue* — Acyl hydrazides

36.28 (a)

```
        Val-Asp
    Phe-Val
            Asp-Glu
                Glu-His
-----------------------
    Phe-Val-Asp-Glu-His
```

(b)

Cys-Gly-Ser
His-Leu-Cys
Ser-His-Leu

His-Leu-Cys-Gly-Ser-His-Leu

(c)

Val-Cys-Gly
Tyr-Leu-Val
Gly-Glu-Arg
Glu-Arg-Gly
Gly-Phe-Phe

Tyr-Leu-Val-Cys-Gly-Glu-Arg-Gly-Phe-Phe

(One must be careful of placement of Gly units.)

36.29 (a) Cbz-Gly-Ala + $SOCl_2$; then Phe; then H_2, Pd.

(b) $PhCH_2OCOCl$ + Ala; then $SOCl_2$; then Gly; finally H_2, Pd.

36.30

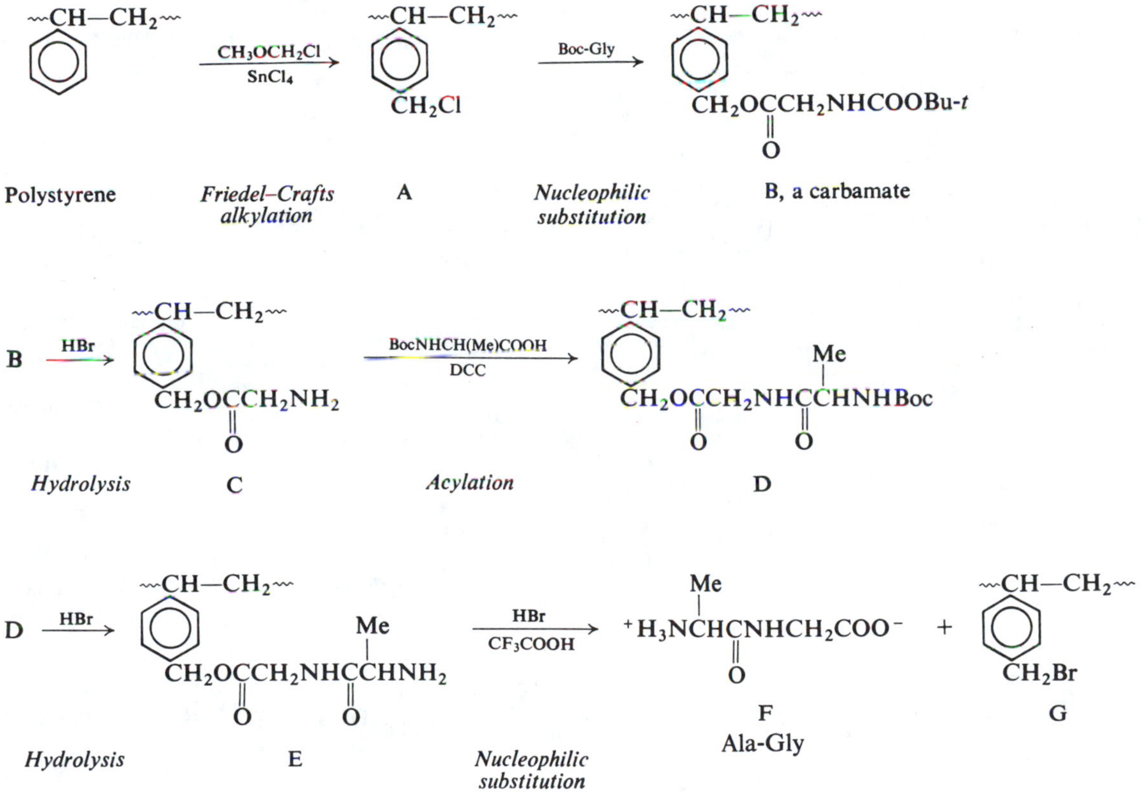

Polystyrene (~CH—CH₂~ on benzene) $\xrightarrow[SnCl_4]{CH_3OCH_2Cl}$ A (~CH—CH₂~ benzene with CH_2Cl) $\xrightarrow{Boc\text{-}Gly}$ B, a carbamate (~CH—CH₂~ benzene with $CH_2OC(=O)CH_2NHCOOBu\text{-}t$)

Polystyrene — *Friedel–Crafts alkylation* — A — *Nucleophilic substitution* — B, a carbamate

B $\xrightarrow{HBr}$ C (~CH—CH₂~ benzene with $CH_2OC(=O)CH_2NH_2$) $\xrightarrow[DCC]{BocNHCH(Me)COOH}$ D (~CH—CH₂~ benzene with $CH_2OC(=O)CH_2NHC(=O)CH(Me)NHBoc$)

Hydrolysis — C — *Acylation* — D

D $\xrightarrow{HBr}$ E (~CH—CH₂~ benzene with $CH_2OC(=O)CH_2NHC(=O)CH(Me)NH_2$) $\xrightarrow[CF_3COOH]{HBr}$ $^+H_3NCH(Me)C(=O)NHCH_2COO^-$ + G (~CH—CH₂~ benzene with CH_2Br)

Hydrolysis — E — *Nucleophilic substitution* — F Ala-Gly — G

(Henahan, John F. "R. Bruce Merrifield. Designer of Protein-Making Machine"; *Chem. Eng. News* **August 2, 1971**, 22–25.)

1. (a) $PhCH_2CH(NH_3^+)COO^-$ $\xleftarrow[NH_3]{excess}$ $PhCH_2CH(Br)COOH$ $\xleftarrow[P]{Br_2}$ $PhCH_2CH_2COOH$ $\xleftarrow[ester]{via\ malonic}$ $PhCH_2Cl$

(b) $PhCH_2CH(^+NH_3)COO^- \xleftarrow{\text{hydrol.}}$ N-phthalimido–$CH(COOEt)CH_2Ph$ $\xleftarrow{\text{K phthalimide}} PhCH_2CH(Br)COOEt$

$PhCH_2CH(Br)COOEt \xleftarrow{H^+,\ EtOH} PhCH_2CH(Br)COOH$ (As in (a))

(c) $PhCH_2{-}CH(^+NH_3)COO^- \xleftarrow[NH_3]{\text{excess}} PhCH_2{-}CH(Br)COOH \xleftarrow{-CO_2} PhCH_2{-}C(Br)(COOH){-}COOH$

$PhCH_2{-}C(Br)(COOH){-}COOH \xleftarrow{Br_2} PhCH_2{-}CH(COOH){-}COOH \xleftarrow[\text{ester}]{\text{via malonic}} PhCH_2Cl$

(d) $PhCH_2{-}CH(^+NH_3)COO \xleftarrow{\text{hydrol.}}$ N-phthalimido–$CH(COOH){-}CH_2Ph$ $\xleftarrow[-CO_2]{H^+,\ \text{heat}} \xleftarrow{PhCH_2Cl}$ N-phthalimido–$CH(COOEt)_2$

Phthalimidomalonic ester

$\uparrow$ K phthalimide

$BrCH(COOEt)_2$

Bromomalonic ester

(e) $PhCH_2CH(^+NH_3){-}COO^- \xleftarrow[H^+]{H_2O} PhCH_2CH(NH_2){-}CN \xleftarrow[NH_4^+]{CN^-} PhCH_2CHO \xleftarrow{LiAlH(OBu\text{-}t)_3} PhCH_2COCl$

$PhCH_2COCl \xleftarrow{SOCl_2} PhOH_2COOH$ (page 722)

(f) $PhCH_2CH(^+NH_3)COO^- \xleftarrow[Pd]{NH_3,\ H_2} PhCH_2C(=O)COOH \xleftarrow[-CO_2]{H^+,\ \text{heat}} PhCH(COOEt)C(=O)COOEt \xleftarrow[NaOEt]{(COOEt)_2} PhCH_2COOEt$

2. (a) Phthalimide N^-K^+ + $BrCH(COOEt)_2$ ⟶ Phthalimido–N–$CH(COOEt)_2$ (A) $\xrightarrow[\text{NaOEt}]{Br(CH_2)_3Br}$ Phthalimido–N–$C(COOEt)_2$–$CH_2CH_2CH_2Br$ (B)

B $\xrightarrow{\text{KOAc}}$ Phthalimido–N–$C(COOEt)_2$–$CH_2CH_2CH_2OCCH_3$ (with C=O) (C) $\xrightarrow[\text{heat}]{OH^-}$ $\xrightarrow[-CO_2]{H^+}$ ^+H_3N–$CH(COO^-)$–$CH_2CH_2CH_2OH$ (D)

D $\xrightarrow{\text{HCl}}$ $[\,^+H_3N$–$CH(COO^-)$–$CH_2CH_2CH_2Cl\,]$ (E) ⟶ Proline (pyrrolidine ring with $\overset{+}{N}H_2$, COO^-)

(b) We can see a good example of the "backwards" procedure of working out a synthesis in the structure of lysine; here the various synthetic units are easily recognized, once we commit ourselves to using the phthalimidomalonic ester procedure.

NH_3 ⇢ ^+H_3N ⋮ $CH_2CH_2CH_2CH_2$ ⋮ $CH(NH_2)COO^-$ ⇠ phthalimidomalonic ester

$BrCH_2CH_2CH_2CH_2Br$ ⇢ (the $CH_2CH_2CH_2CH_2$ unit)

Lysine

(from phthalimidomalonic ester) Phthalimido–N–$\ddot{C}^-Na^+(COOEt)_2$ $\xrightarrow{Br(CH_2)_4Br}$ Phthalimido–N–$C(COOEt)_2$–$(CH_2)_4Br$ $\xrightarrow[NH_3]{\text{excess}}$ Phthalimido–N–$C(COOEt)_2$–$(CH_2)_4$–NH_2 $\xrightarrow{\text{base}}$ $^+H_3N(CH_2)_4CH(NH_2)COO^-$

Lysine

3. (a)

$$\begin{array}{c} {}^{+}NH_3 \\ H_2C \quad COO^- \\ {}^-OOC \quad CH_2 \\ H_3N_+ \end{array} \xrightarrow[-2H_2O]{\text{heat}} \begin{array}{c} H \\ N \\ H_2C \quad C{=}O \\ O{=}C \quad CH_2 \\ N \\ H \end{array}$$

Diketopiperazine
A cyclic diamide

(b)

$$\underset{{}^{+}NH_3}{CH_3\overset{}{C}HCH_2COO^-} \xrightarrow[-NH_3]{\text{heat}} CH_3CH{=}CHCOOH$$

α,β-Unsaturated acid

(c)

$$\begin{array}{c} CH_2 \\ H_2C \quad COO^- \\ CH_3{-}CH \\ NH_3^+ \end{array} \xrightarrow[-H_2O]{\text{heat}} \begin{array}{c} H_2C{-}CH_2 \\ \quad C{=}O \\ CH_3{-}C \quad NH \\ H \end{array}$$

A γ-lactam
A cyclic amide

(d)

$$\begin{array}{c} CH_2 \\ H_2C \quad COO^- \\ H_2C \quad NH_3^+ \\ CH_2 \end{array} \xrightarrow[-H_2O]{\text{heat}} \begin{array}{c} CH_2 \\ H_2C \quad C{=}O \\ H_2C \quad NH \\ CH_2 \end{array}$$

A δ-lactam
A cyclic amide

4. Chemical arithmetic,

$C_2H_3O_2Cl$ $ClCH_2COOH$	$C_5H_{12}O_2NCl$	$C_2H_5O_2N$ glycine	$C_5H_{14}O_2NI_3$
+ C_3H_9N Me_3N	− $C_5H_{11}O_2N$ betaine	+ $C_3H_9I_3$ 3MeI	− $C_5H_{11}O_2N$ betaine
$C_5H_{12}O_2NCl$	HCl lost	$C_5H_{14}O_2NI_3$	3HI lost

leads us to the same product for both reactions:

$$Me_3N + ClCH_2COOH \xrightarrow{-HCl} Me_3\overset{+}{N}CH_2COO^-$$

Betaine

$$3MeI + {}^{+}H_3NCH_2COO^- \xrightarrow{-3HI} Me_3\overset{+}{N}CH_2COO^-$$

Betaine

The properties are those expected of a dipolar compound: a crystalline, water-soluble high-melting solid. Reaction with acid involves the —COO^- group.

$$\underset{\text{Betaine}}{Me_3\overset{+}{N}CH_2COO^-} + H^+ + Cl^- \longrightarrow \underset{Cl^-}{Me_3\overset{+}{N}CH_2COOH}$$

5. (a) Boc-Gly-Ala benzyl ester + HBr; then Phe benzyl ester + DCC; then HBr.
(b) Ala + (t-BuOCO)$_2$O; then Gly benzyl ester + DCC; then HBr.

6. As the water content of the solvent is lowered, the lipophilic parts no longer hide themselves. This breaks up soap micelles, and changes the characteristic shape—and with it characteristic properties—of a globular protein.

7.

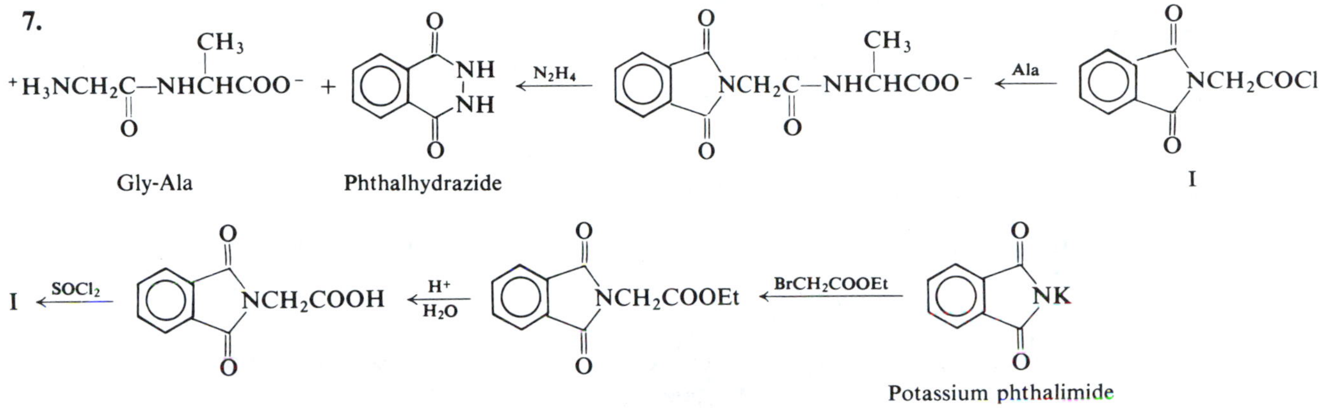

$$^{+}H_3NCH_2C(=O)\text{—}NHCH(CH_3)COO^- + \text{Phthalhydrazide} \xleftarrow{N_2H_4} C_6H_4(CO)_2NCH_2C(=O)\text{—}NHCH(CH_3)COO^- \xleftarrow{\text{Ala}} C_6H_4(CO)_2NCH_2COCl$$

Gly-Ala Phthalhydrazide I

$$\text{I} \xleftarrow{SOCl_2} C_6H_4(CO)_2NCH_2COOH \xleftarrow[H_2O]{H^+} C_6H_4(CO)_2NCH_2COOEt \xleftarrow{BrCH_2COOEt} C_6H_4(CO)_2NK$$

Potassium phthalimide

For Ala-Gly: start with $BrCH(CH_3)COOEt$, and then use $^{+}H_3NCH_2COO^-$

8. If one Fe atom per molecule:

$$0.43\%\ \text{Fe} = \frac{55.8}{\text{m.w.}} \times 100$$

$$\text{m.w.} = 13\,000$$

If one S atom per molecule:

$$1.48\%\ \text{S} = \frac{32.1}{\text{m.w.}} \times 100$$

$$\text{m.w.} = 2170$$

The minimum m.w. is 13 000: 1 Fe and 6 S.

9. (a)

$$\frac{1.31\ \text{mg NH}_3}{17\ \text{mg/mmol}} = 0.077\ \text{mmol NH}_3$$

$$\frac{100\ \text{mg protein}}{42\,020\ \text{mg/mmol}} = 0.00238\ \text{mmol protein}$$

$$\frac{0.077\ \text{mmol NH}_3}{0.00238\ \text{mmol protein}} = \text{approx. } 32\ \text{—CONH}_2 \text{ per molecule}$$

(b)

$$0.00238 \text{ mmol protein requires} \frac{17 \text{ mg } H_2O}{18 \text{ mg/mmol}} = 0.944 \text{ mmol } H_2O$$

$$\frac{0.944 \text{ mmol } H_2O}{0.00238 \text{ mmol protein}} = \text{approx. 397 amide linkages (peptide + 32 —}CONH_2\text{)}$$

(c) 397 − 32 = 365 peptide links + 4 *N*-terminal groups = 369 amino acid residues

10. (a)

Leu (m.w.)	131
Orn	132
Phe	165
Pro	115
Val	117
	660

A m.w. of 1300 means that there are two of each residue, and the molecular formula is $Leu_2Orn_2Phe_2Pro_2Val_2$.

(b) The DNP derivative is simply the one expected from the side chain of ornithine. Hence there is no *C*-terminal *or N*-terminal group: the polypeptide must be *cyclic*.

(c)

```
Leu-Phe
    Phe-Pro
    Phe-Pro-Val
            Val-Orn-Leu
                Orn-Leu
            Val-Orn
        Pro-Val-Orn
-----------------------
Leu-Phe-Pro-Val-Orn-Leu
```

From the molecular formula, we see that Orn, Phe, Pro, Val are still to be accounted for. These must be Phe-Pro-Val-Orn. The Orn combines with Leu at the other end to complete the ring:

```
Val-Orn-Leu-Phe
-Pro-        -Pro-
Val-Orn-Leu-Phe
```

Gramicidin S

Cyclic decapeptide

11. (a) In chain B, six amino acids occur only once: Pro, Arg, Lys, Asp, Ser, Thr. Let us look first for tripeptides and dipeptides that contain these amino acids.

Thr Pro Lys	**Arg**	**Ser**	**Asp**
Thr-Pro	Arg-Gly	Ser-His-Leu	Phe-Val-Asp
Pro-Lys-Ala	Gly-Glu-Arg		Val-Asp-Glu
Thr-Pro-Lys-Ala	Gly-Glu-Arg-Gly		Phe-Val-Asp-Glu

Four amino acids occur only twice: Ala, Cys, His, Tyr. Let us look for these next.

His

Ser-His-Leu Glu-His-Leu His-Leu-Cys (Two of these overlap)

Cys

His-Leu-Cys Leu-Val-Cys
Val-Cys-Gly

Leu-Val-Cys-Gly

Tyr

Tyr-Leu-Val-

Ala

Val-Glu-Ala Thr-Pro-Lys-Ala (derived above)

One Tyr is missing.

Gly occurs three times, and we have located all three:

Gly-Glu-Arg-Gly Leu-Cys-Gly Leu-Val-Cys-Gly

Two of these three peptides must overlap (four Gly appear here), but which two is uncertain at this point.

Val occurs three times:

Leu-Val-Cys-Gly Leu-Val-Glu
Val-Glu-Ala Phe-Val-Asp-Glu Tyr-Leu-Val

Leu-Val-Glu-Ala

There is overlapping of Tyr-Leu-Val with one of the first two peptides.

Glu occurs three times:

Glu-His-Leu Gly-Glu-Arg Leu-Val-Glu
Val-Asp-Glu Gly-Glu-Arg-Gly Val-Glu-Ala

Val-Asp-Glu-His-Leu Gly-Glu-Arg-Gly Leu-Val-Glu-Ala
Phe-Val-Asp-Glu

Phe-Val-Asp-Glu-His-Leu

There is no overlapping.

Phe occurs three times; two of these are missing from the tripeptides.

Now for **Leu**, which occurs four times. We find seven tripeptides containing Leu; there must be much overlapping.

Glu-His-Leu His-Leu-Cys Leu-Cys-Gly Leu-Val-Cys
Leu-Val-Glu Ser-His-Leu Tyr-Leu-Val

Certainly Leu-Cys-Gly, Leu-Val-Cys, and Leu-Val-Glu have no overlapping, and the fourth Leu must be found in Glu-His-Leu or Ser-His-Leu. This means, then, that we can propose the following combinations.

Leu-Cys-Gly Leu-Val-Cys or Leu-Val-Glu
His-Leu-Cys Tyr-Leu-Val Tyr-Leu-Val

His-Leu-Cys-Gly Tyr-Leu-Val-Cys or Tyr-Leu-Val-Glu

Phe-Val-Asp-Glu-His-Leu (derived above, under Glu)
or
Ser-His-Leu

We have located three Leu; one is still missing.

At this point we have derived the following:

One hexapeptide:	Phe-Val-Asp-Glu-His-Leu
Five tetrapeptides:	Thr-Pro-Lys-Ala Gly-Glu-Arg-Gly Leu-Val-Cys-Gly Leu-Val-Glu-Ala His-Leu-Cys-Gly
Two tripeptides:	Ser-His-Leu Tyr-Leu-Val
Missing:	2Phe, 1Tyr, 1Leu

(b)

```
Tyr-Leu-Val-Cys
    Leu-Val-Cys-Gly
-------------------
Tyr-Leu-Val-Cys-Gly
```

```
Ser-His-Leu-Val
    His-Leu-Val-Glu
-------------------
Ser-His-Leu-Val-Glu
        Leu-Val-Glu-Ala
-----------------------
Ser-His-Leu-Val-Glu-Ala
```

```
                Glu-His-Leu
                    His-Leu-Cys-Gly
                -------------------
                Glu-His-Leu-Cys-Gly
Phe-Val-Asp-Glu-His-Leu
-----------------------------------
Phe-Val-Asp-Glu-His-Leu-Cys-Gly
```

This takes care of most of the overlapping except for Gly.

Now we have the following pieces put together:

One octapeptide:	Phe-Val-Asp-Glu-His-Leu-Cys-Gly
One hexapeptide:	Ser-His-Leu-Val-Glu-Ala
One pentapeptide:	Tyr-Leu-Val-Cys-Gly
Two tetrapeptides:	Thr-Pro-Lys-Ala Gly-Glu-Arg-Gly
Missing:	2Phe, 1Tyr, 1Leu

(c)

```
                                            Val-Glu-Ala-Leu
                                Ser-His-Leu-Val-Glu-Ala
                                ---------------------------
                                Ser-His-Leu-Val-Glu-Ala-Leu
                    His-Leu-Cys-Gly-Ser-His-Leu
                    ---------------------------------------
                    His-Leu-Cys-Gly-Ser-His-Leu-Val-Glu-Ala-Leu
Phe-Val-Asp-Glu-His-Leu-Cys-Gly
-----------------------------------------------------------
Phe-Val-Asp-Glu-His-Leu-Cys-Gly-Ser-His-Leu-Val-Glu-Ala-Leu
```

This takes care of 15 residues out of 30. Ten more are found in

Tyr-Leu-Val-Cys-Gly-Glu-Arg-Gly-Phe-Phe

and the remaining five in

Tyr-Thr-Pro-Lys-Ala

Since Phe-Val is the *N*-terminal end (DNP data) and Ala is the *C*-terminal end (carboxypeptidase), the three pieces can be put together in only one way, and the structure of the B chain of beef insulin must be:

Phe-Val-Asp-Glu-His-Leu-Cys-Gly-Ser-His-Leu-Val-Glu-Ala-Leu-Tyr-Leu-Val-Cys-Gly-Glu-Arg-Gly-
Phe-Phe-Tyr-Thr-Pro-Lys-Ala

B-chain

(d) In chain A, three amino acids appear once: Gly, Ala, Ile. They are the first ones to look for among the peptides.

Gly Ile Gly-Ile-Val-Glu-Glu

Ala Cys-Cys-Ala

Five amino acids appear twice: Val, Leu, Asp, Ser, Tyr.

Val

Ser-Val-Cys Gly-Ile-Val-Glu-Glu (*N*-terminal residue)

Leu

Glu-Leu-Glu

Leu-Tyr-Glu
Ser-Leu-Tyr
———
Ser-Leu-Tyr-Glu

Tyr

Leu-Tyr-Glu
Ser-Leu-Tyr-
———
Ser-Leu-Tyr-Glu

Tyr-Cys
Glu-Asp-Tyr
———
Glu-Asp-Tyr-Cys

Asp

Glu-Asp-Tyr Cys-Asp (*C*-terminal residue)

Ser

Ser-Leu-Tyr-Glu Ser-Val-Cys

Two amino acids appear four times: **Glu, Cys**

Gly-Ile-Val-Glu-Glu (*N*-terminal residue)
Glu-Glu-Cys
Glu-Cys-Cys
———
Gly-Ile-Val-Glu-Glu-Cys-Cys
Cys-Cys-Ala
———
Gly-Ile-Val-Glu-Glu-Cys-Cys-Ala

Glu-Leu-Glu
Leu-Tyr-Glu
———
Leu-Tyr-Glu-Leu-Glu
Ser-Leu-Tyr-Glu
———
Ser-Leu-Tyr-Glu-Leu-Glu
Glu-Asp-Tyr-Cys
———
Ser-Leu-Tyr-Glu-Leu-Glu-Asp-Tyr-Cys

Two other fragments are found, one of which is the *C*-terminal residue.

Ser-Val-Cys Cys-Asp (*C*-terminal residue)

At this point, there are no amino acids missing, and we have put together the following pieces:

One nonapeptide: Ser-Leu-Tyr-Glu-Leu-Glu-Asp-Tyr-Cys

One octapeptide: Gly-Ile-Val-Glu-Glu-Cys-Cys-Ala (*N*-terminal)

One tripeptide: Ser-Val-Cys

One dipeptide: Cys-Asp (*C*-terminal)

We still do not know which of the non-terminal fragments comes first, since the *C*-terminal Cys of either could overlap the *N*-terminal Cys of the dipeptide.

(e) The only way to get Ser-Val-Cys and Ser-Leu together is to have a center portion that runs . . . Ser-Val-Cys-Ser-Leu-Tyr- . . .. The complete A chain of beef insulin is thus:

Gly-Ile-Val-Glu-Glu-Cys-Cys-Ala-Ser-Val-Cys-Ser-Leu-Tyr-Glu-Leu-Glu-Asp-Tyr-Cys-Asp

A-chain

(f)

```
Chain A:                    S———————————S
                            |           |          NH2         NH2
                            |           |          |           |
Gly-Ile-Val-Glu-Glu-Cy-Cy-Ala-Ser-Val-Cy-Ser-Leu-Tyr-Glu-Leu-Glu-Asp-Tyr-Cy-Asn
                |    |                                                  |
               NH2   S                                                  S
Chain B:             |                                                  |
                     S                                                  |
  NH2 NH2            |                                                  S
   |   |             |                                                  |
Phe-Val-Asp-Glu-His-Leu-Cy-Gly-Ser-His-Leu-Val-Glu-Ala-Leu-Tyr-Leu-Val-Cy-Gly-Glu-Arg-Gly-Phe—┐
                                                                        └Phe-Tyr-Thr-Pro-Lys-Ala
```

Beef insulin

(g) The other could have been

```
         H
         |
DNP-N(CH2)4CHCOO⁻
           |
          +NH3
```

from the ε-amino group of Lys. But if Lys had been *N*-terminal, we would have obtained a double DNP derivative of it, and no DNP-Phe—contrary to fact.

12. That the helixes are single-stranded.

13. Two different RNA molecules would be generated, and these would direct the synthesis of entirely different sets of amino acids.

17.32 (a) isobutylbenzene (b) *tert*-butylbenzene (c) *p*-isopropyltoluene

(See the labeled spectra on page 693 of this Study Guide.)

17.33 (a) 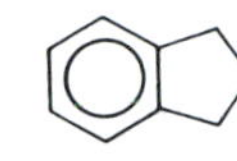
Indane

(b) 2,2,4-trimethylpentane (c) *p-tert*-butyltoluene

(See the labeled spectra on page 694 of this Study Guide.)

17.34 (a) 1,2,4-trimethylbenzene (b) mesitylene (c) isopropylbenzene

(See the labeled spectra on page 695 of this Study Guide.)

17.35

(a) $CH_3\overset{CH_3}{\overset{|}{C}}HCH_2CH_2OH$
Isopentyl alcohol

(b) $CH_3CH_2\overset{CH_2CH_3}{\overset{|}{C}}HCH_2OH$
2-Ethyl-1-butanol

(c) $CH_3\overset{CH_3}{\overset{|}{C}}HCH_2CHOHCH_3$
4-Methyl-2-pentanol

(See the labeled spectra on page 696 of this Study Guide.)

17.36 K, *p*-methoxybenzyl alcohol, *p*-$CH_3OC_6H_4CH_2OH$.

(See the labeled spectra on page 697 of this Study Guide.)

17.37 $CH_3-\underset{H_3C}{\overset{CH_3}{C}}-\underset{OH}{CH}-CH_3$

3,3-Dimethyl-2-butanol
L

(See the labeled spectra on page 698 of this Study Guide.)

20.30 (a) $EtOOC(CH_2)_4COOEt$
Ethyl adipate

(b) $Et-\underset{COOEt}{\overset{COOEt}{C}}-Ph$
Ethyl ethylphenylmalonate

(c) $CH_3-\underset{O}{\overset{}{\underset{\|}{C}}}-\overset{H}{N}-\underset{COOEt}{\overset{COOEt}{C}}-H$
Ethyl acetamidomalonate

(See the labeled spectra on page 713 of this Study Guide.)

Analysis of Spectra

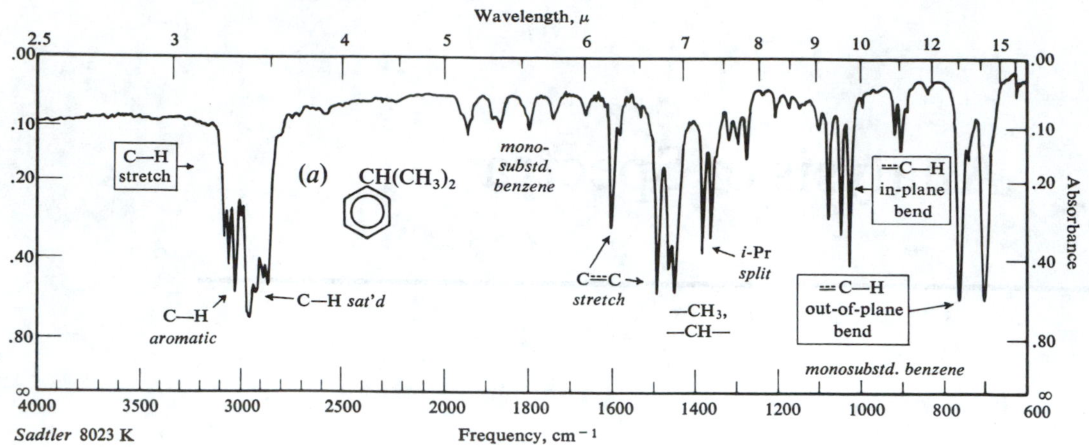

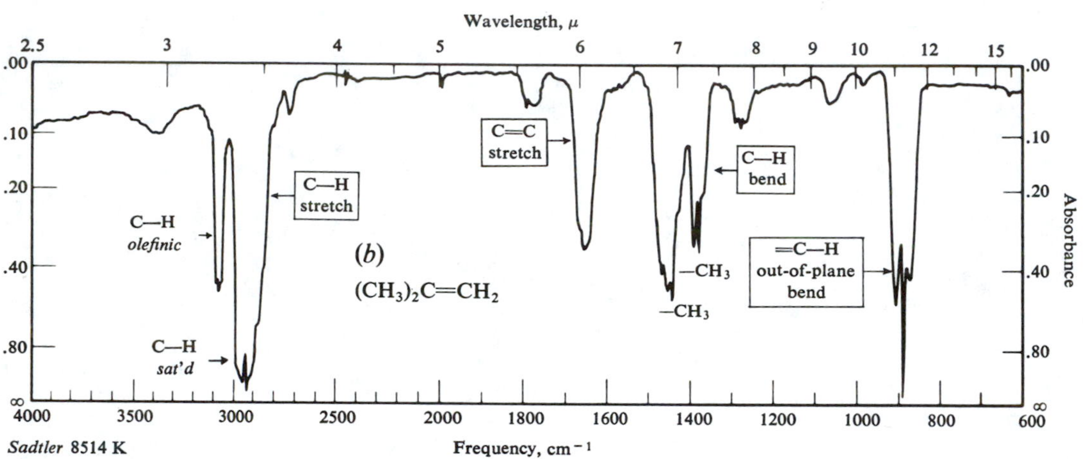

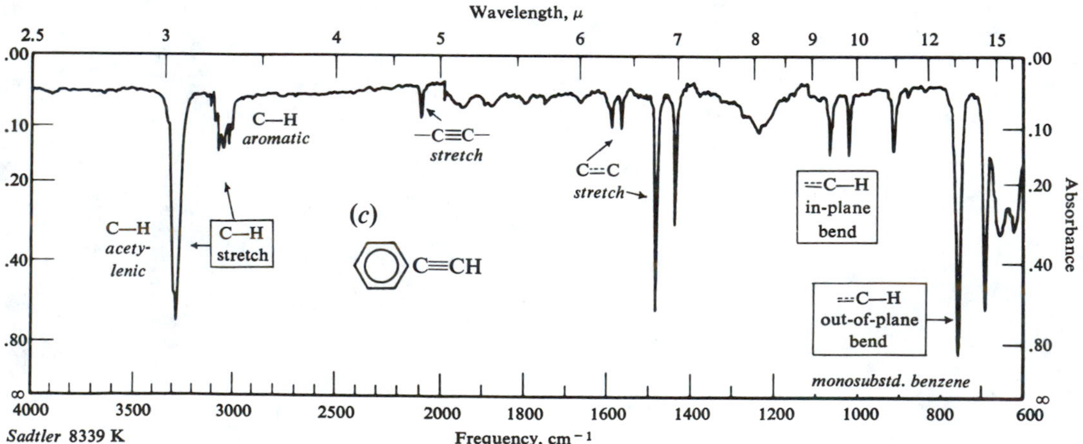

Figure 17.4 Infrared spectra for Problem 17.3, p. 594.

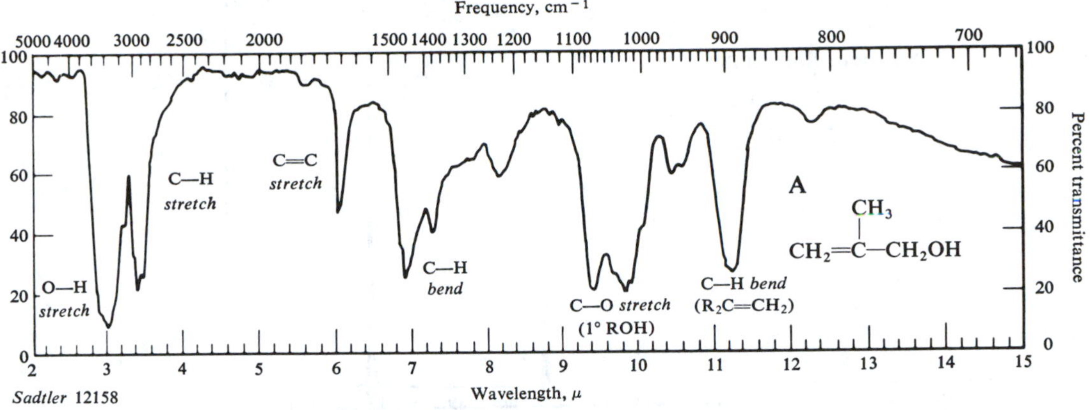

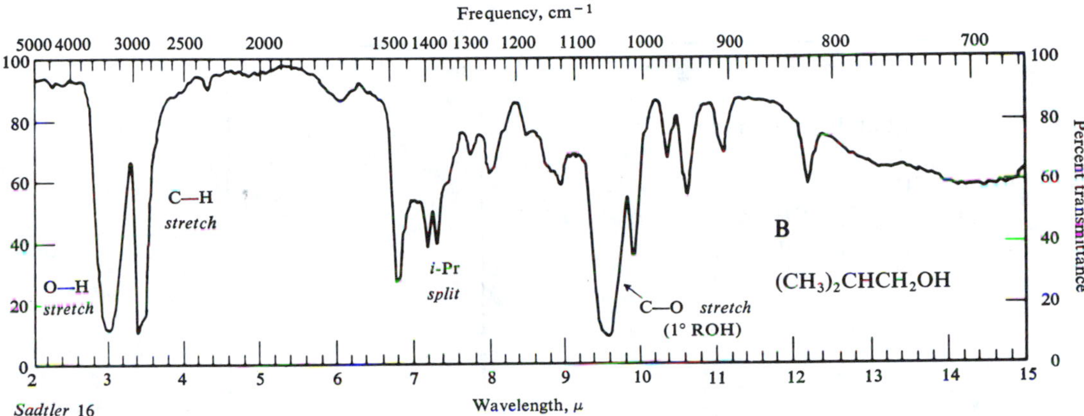

Figure 17.6 Infrared spectra for Problem 17.4, p. 594.

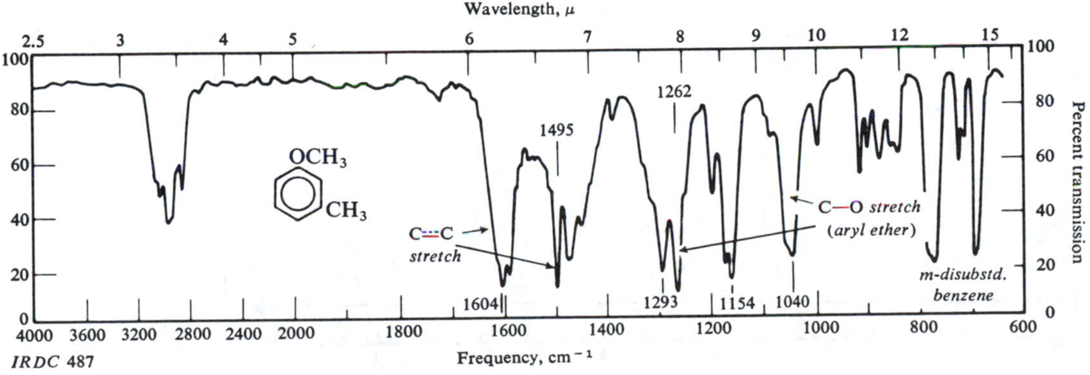

Figure 17.7 Infrared spectrum for Problem 17.5, p. 596.

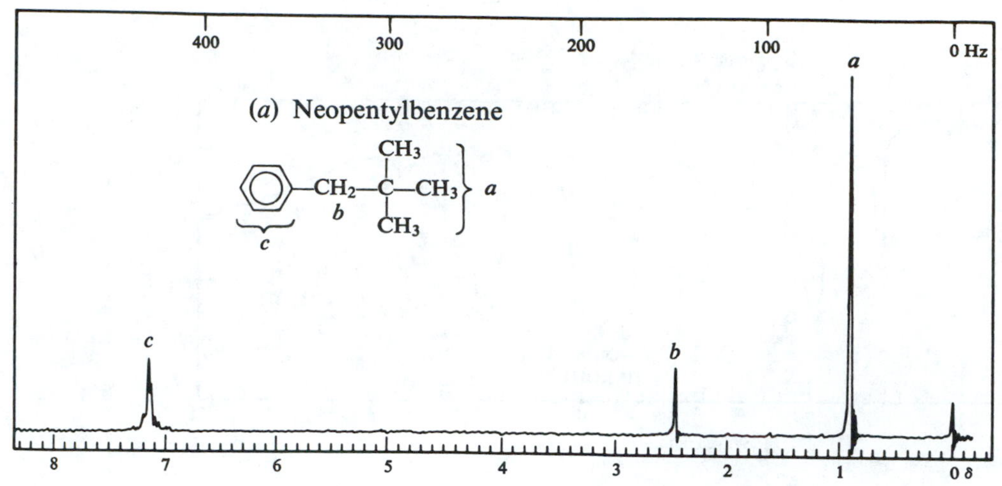

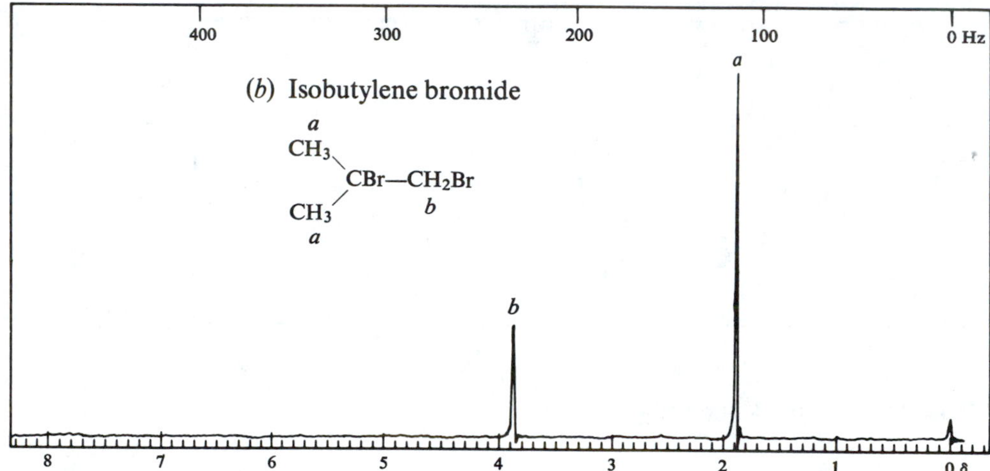

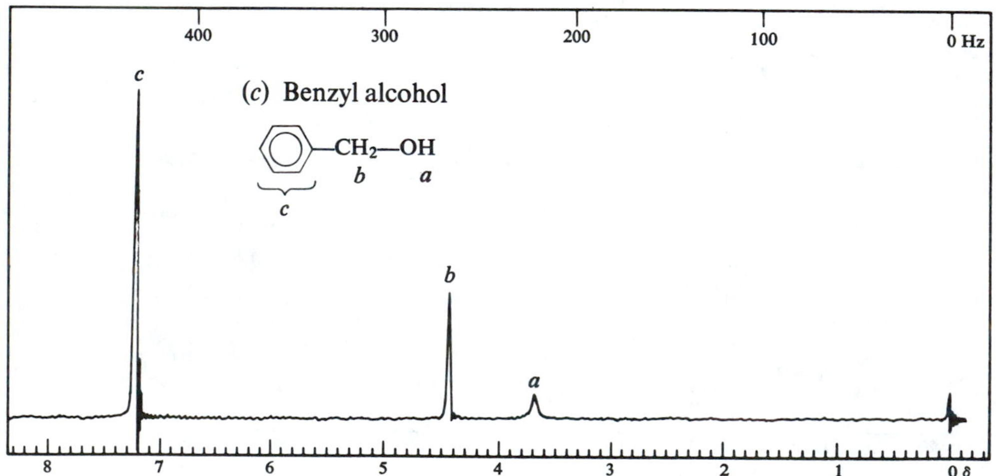

Figure 17.12 Proton NMR spectra for Problem 17.12, p. 610.

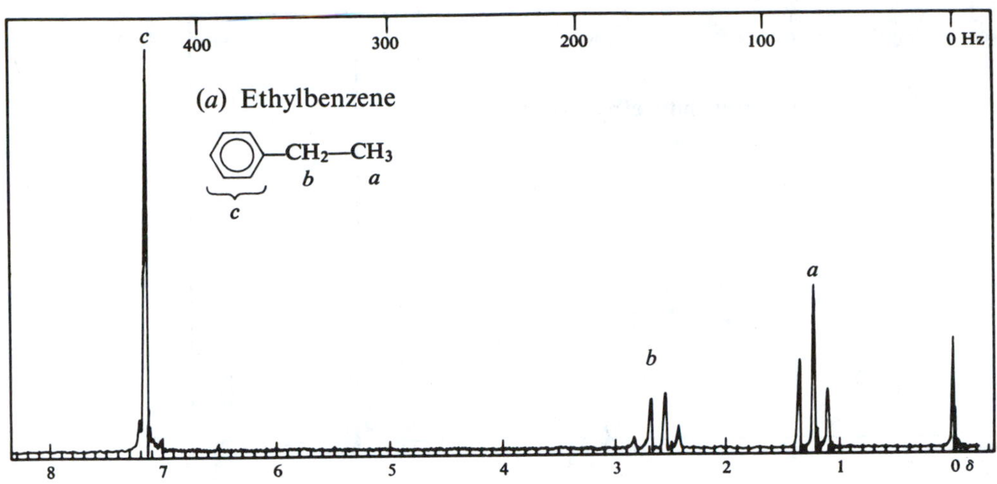

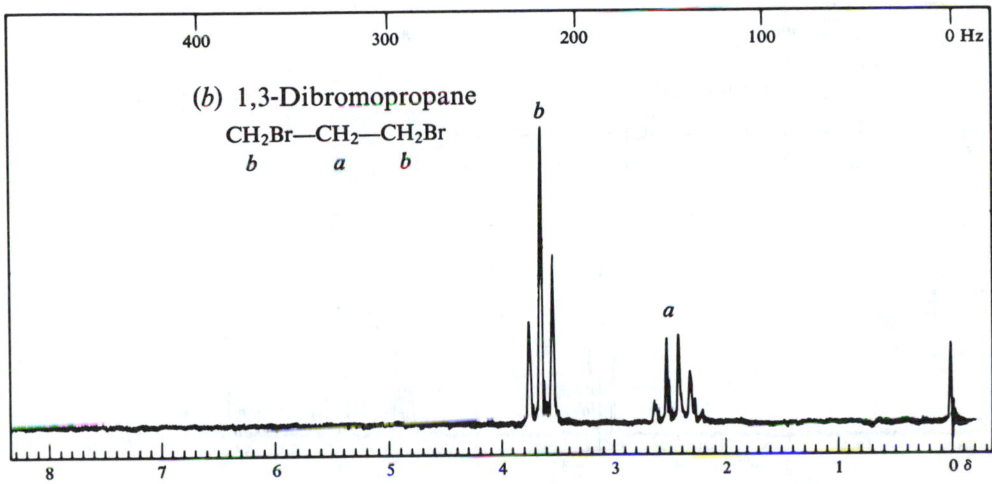

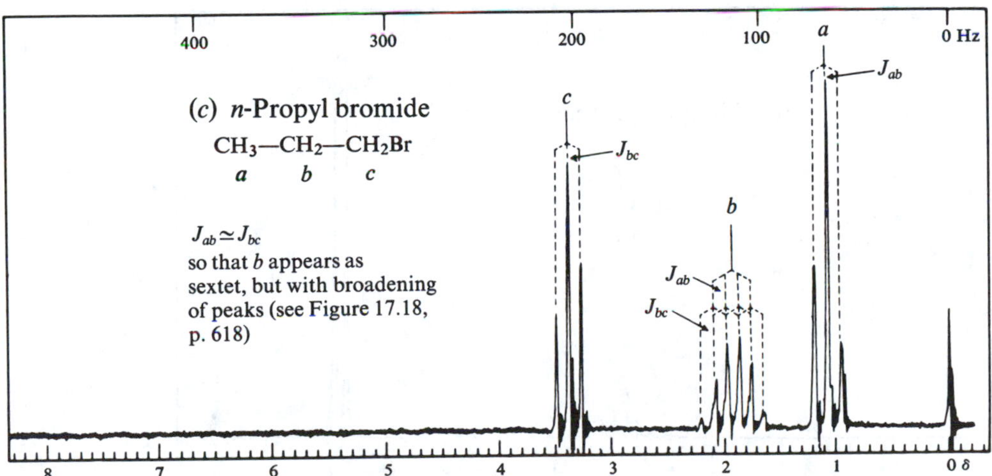

Figure 17.21 Proton NMR spectra for Problem 17.15, p. 620.

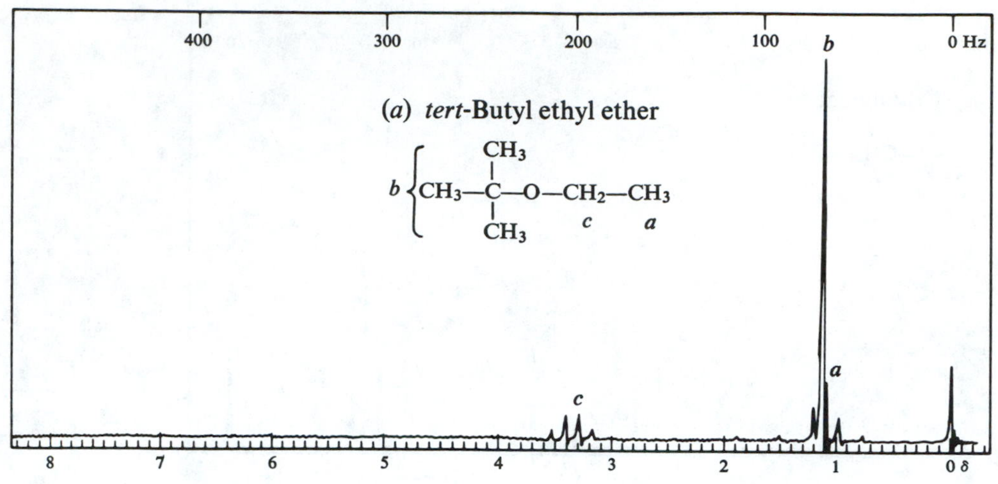

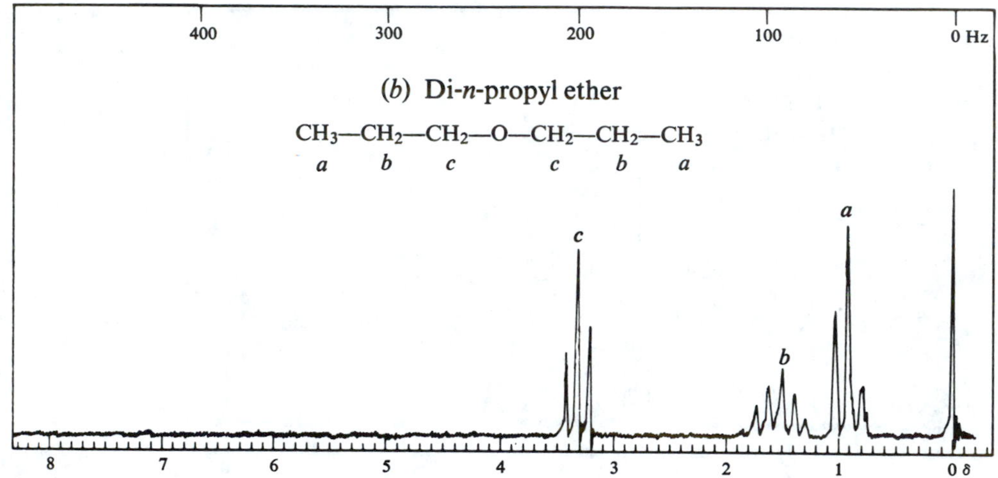

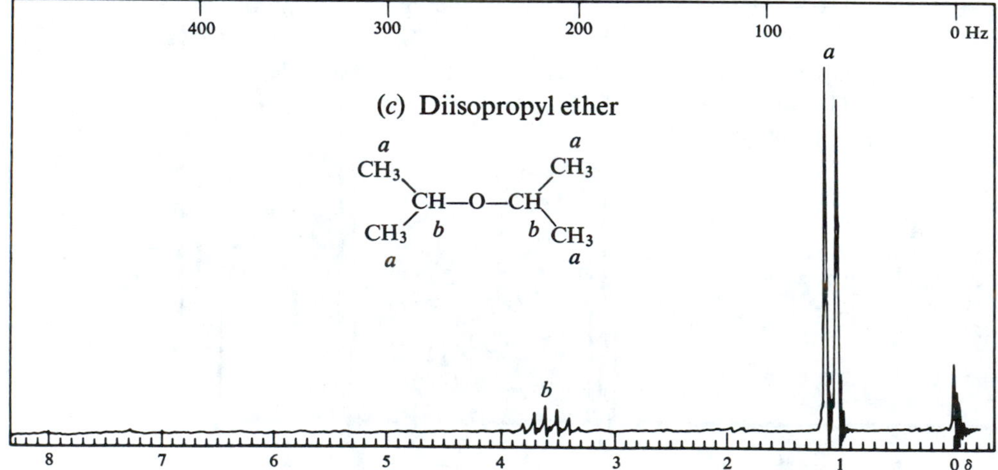

Figure 17.22 Proton NMR spectra for Problem 17.16, p. 620.

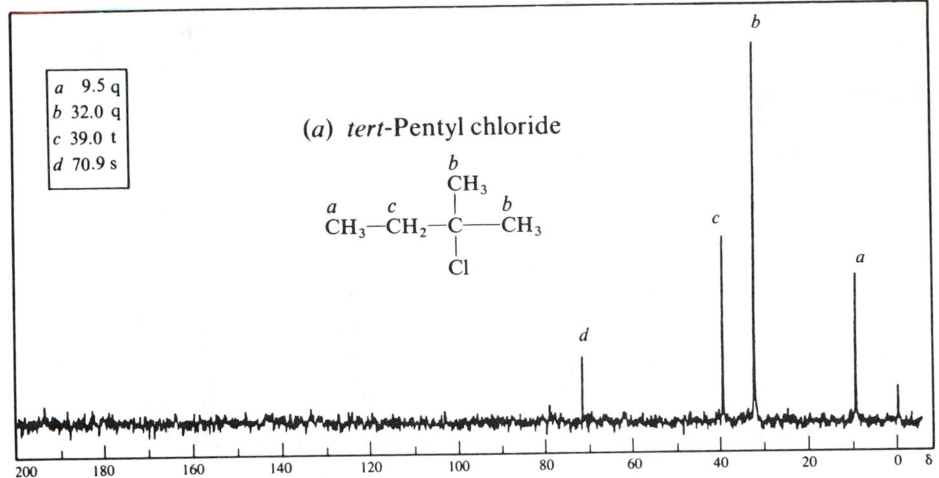

Sadtler 1014C © Sadtler Research Laboratories, Division of Bio-Rad, Inc., (1976).

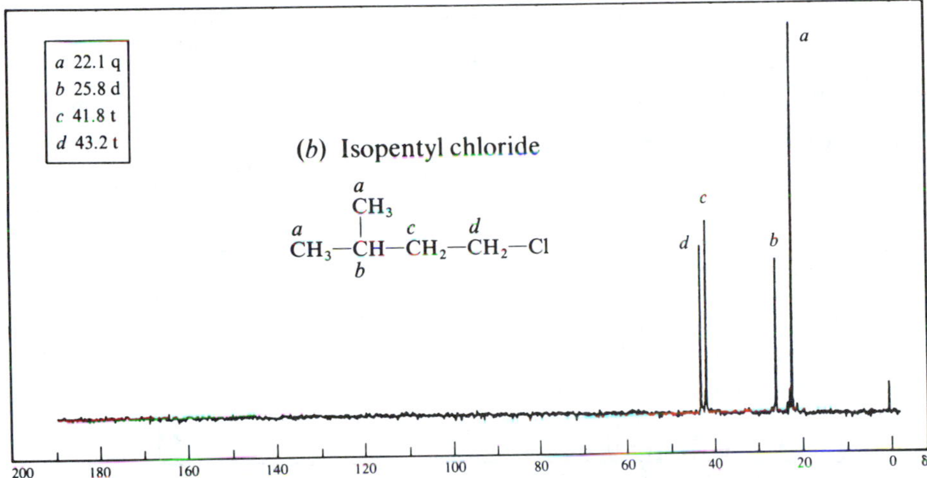

Sadtler 6450C © Sadtler Research Laboratories, Division of Bio-Rad, Inc., (1979).

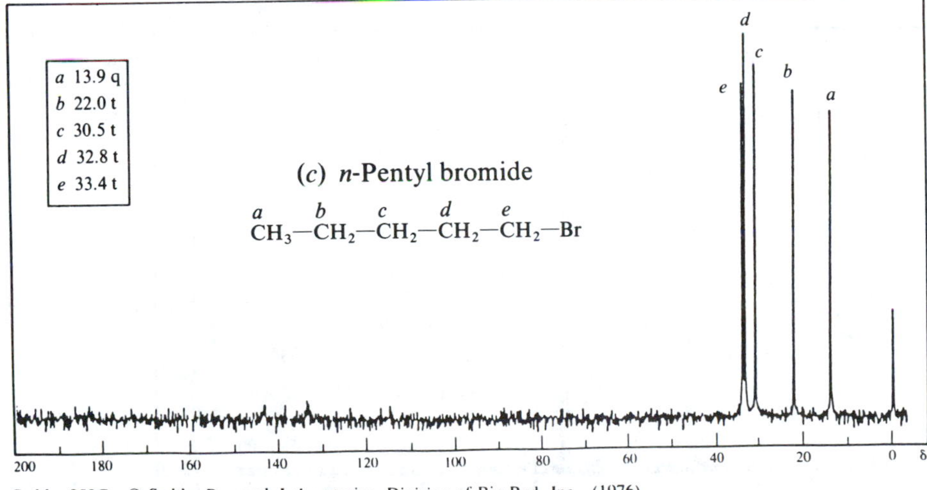

Sadtler 252C © Sadtler Research Laboratories, Division of Bio-Rad, Inc., (1976).

Figure 17.26 CMR spectra for Problem 17.25, p. 631.

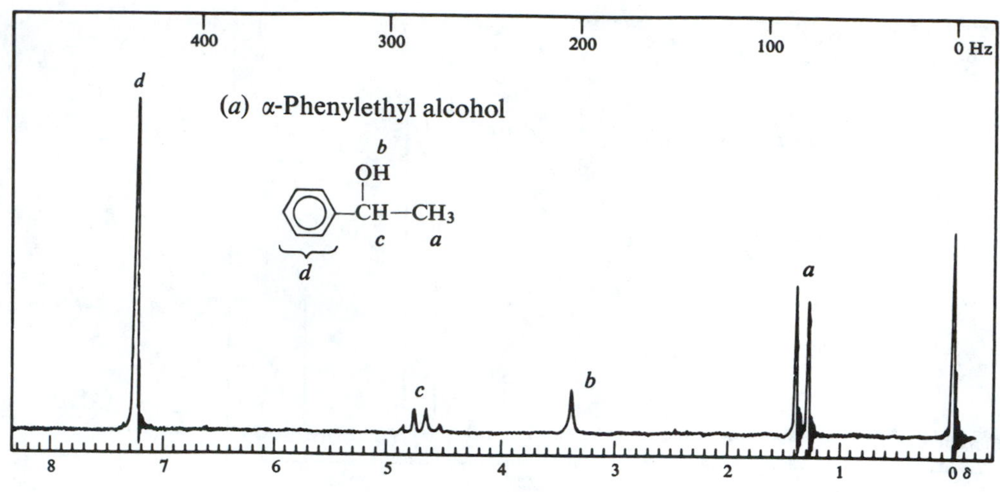

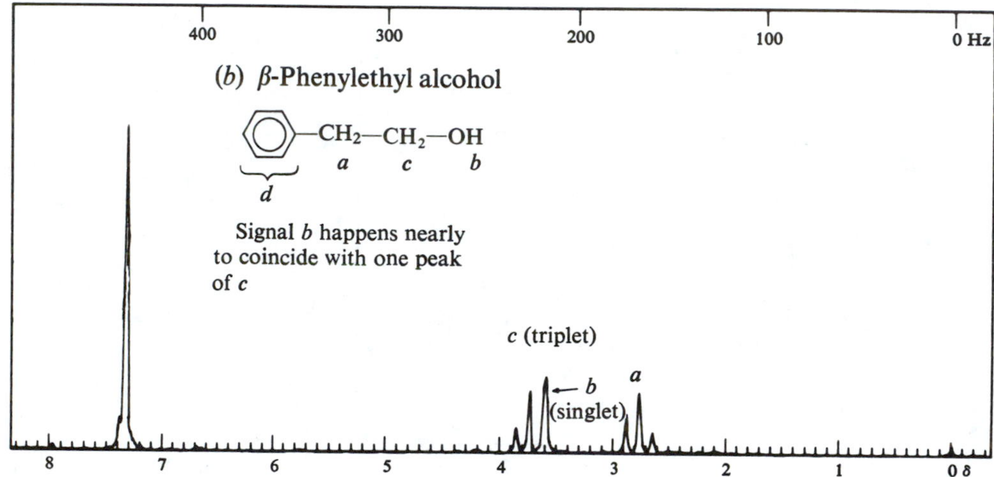

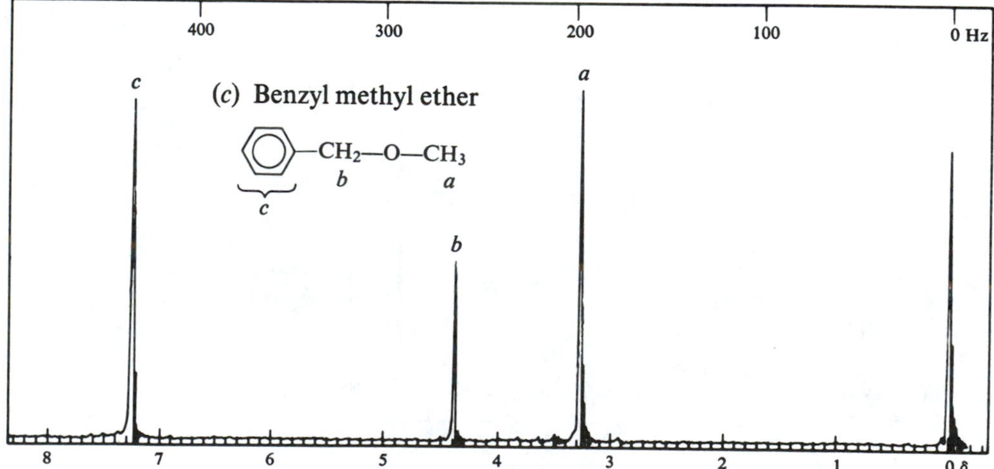

Figure 17.30 Proton NMR spectra for Problem 17.29, p. 640.

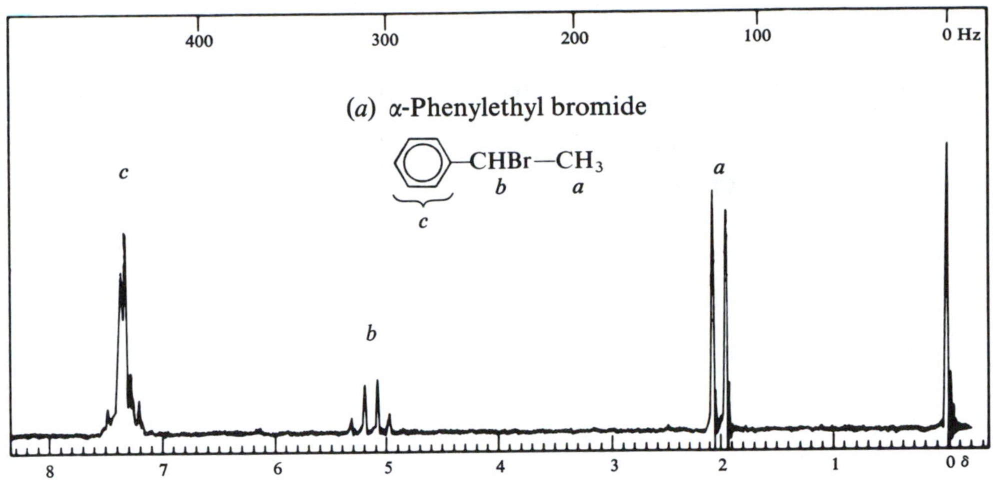

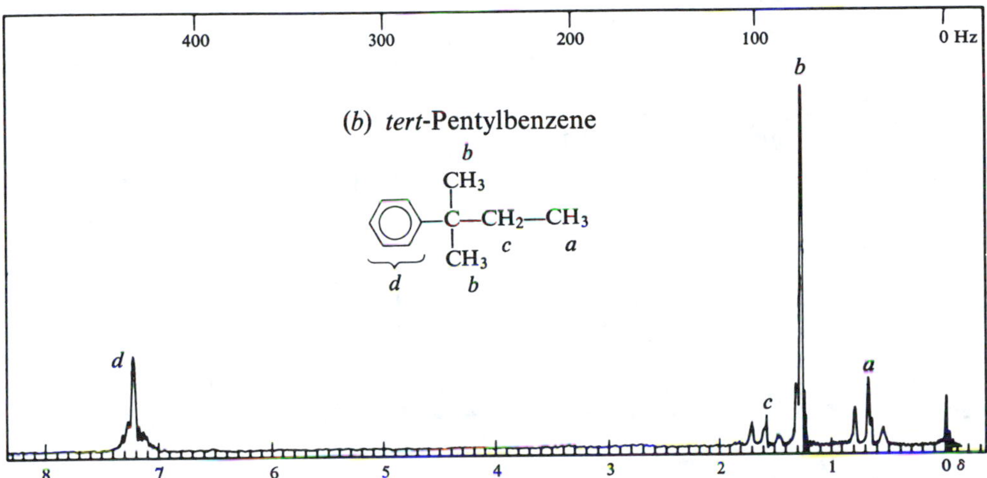

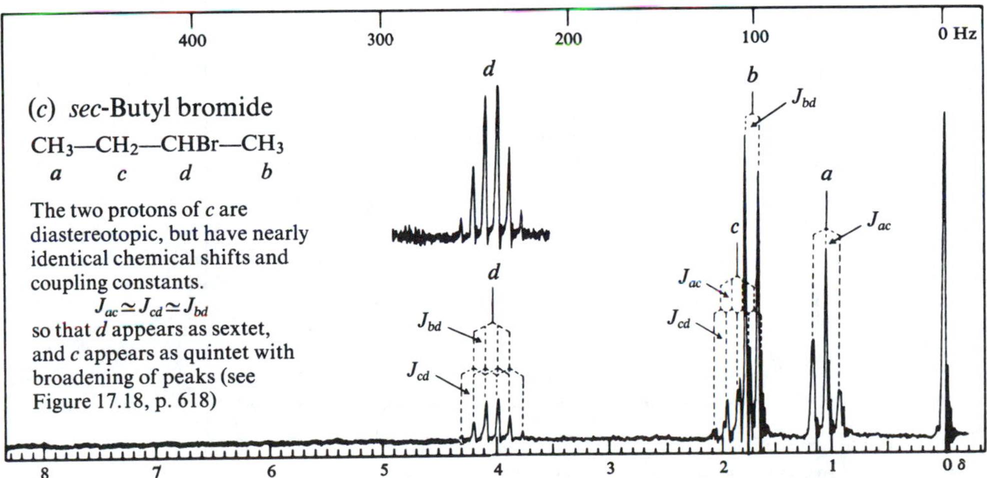

Figure 17.31 Proton NMR spectra for Problem 21, p. 646.

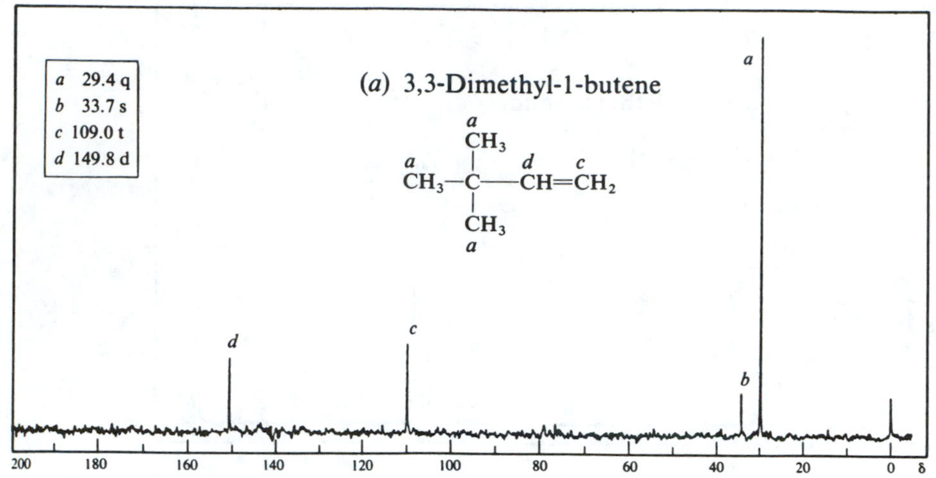

Sadtler 232C © Sadtler Research Laboratories, Division of Bio-Rad, Inc., (1976).

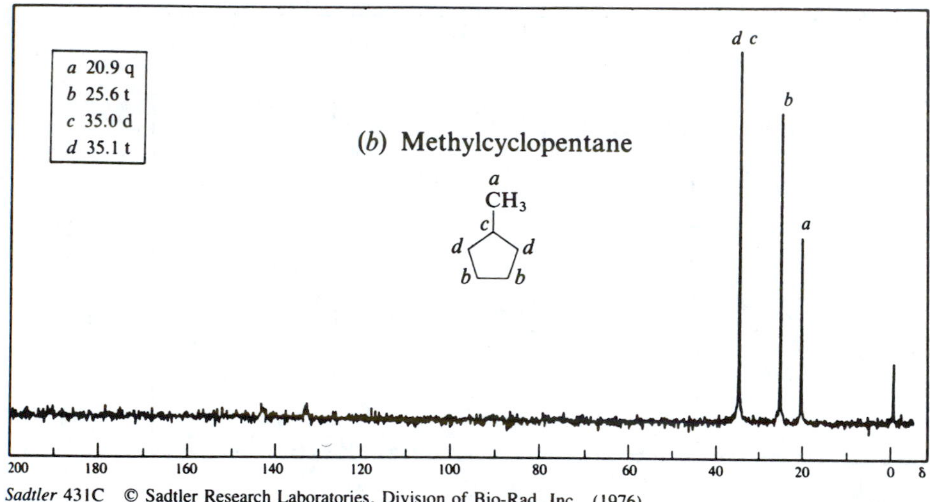

Sadtler 431C © Sadtler Research Laboratories, Division of Bio-Rad, Inc., (1976).

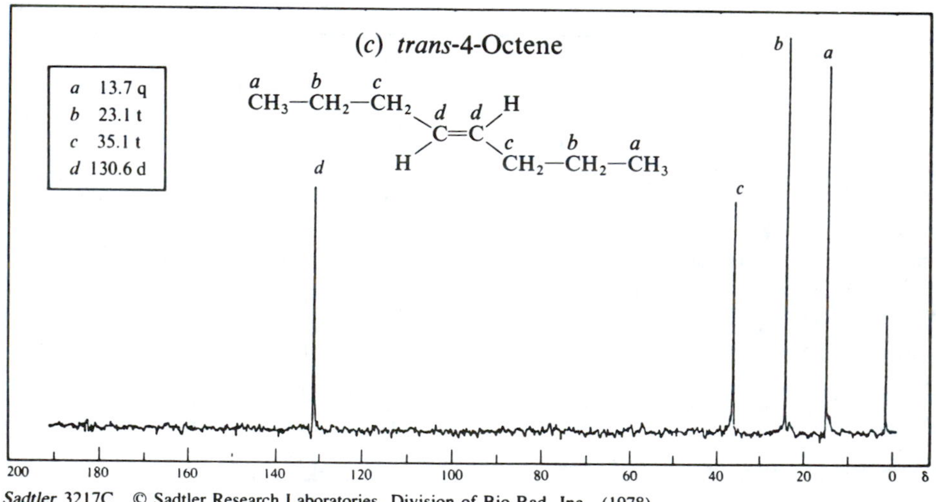

Sadtler 3217C © Sadtler Research Laboratories, Division of Bio-Rad, Inc., (1978).

Figure 17.32 CMR spectra for Problem 22, p. 646.

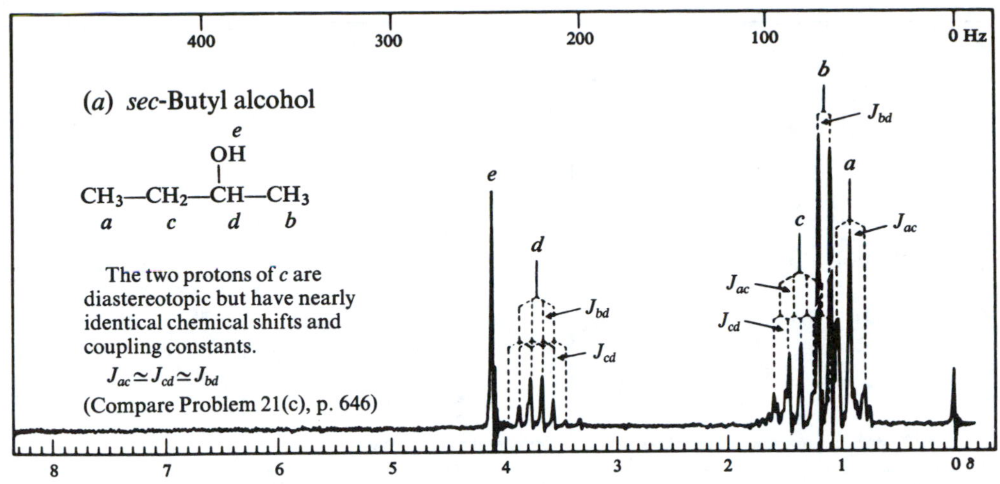

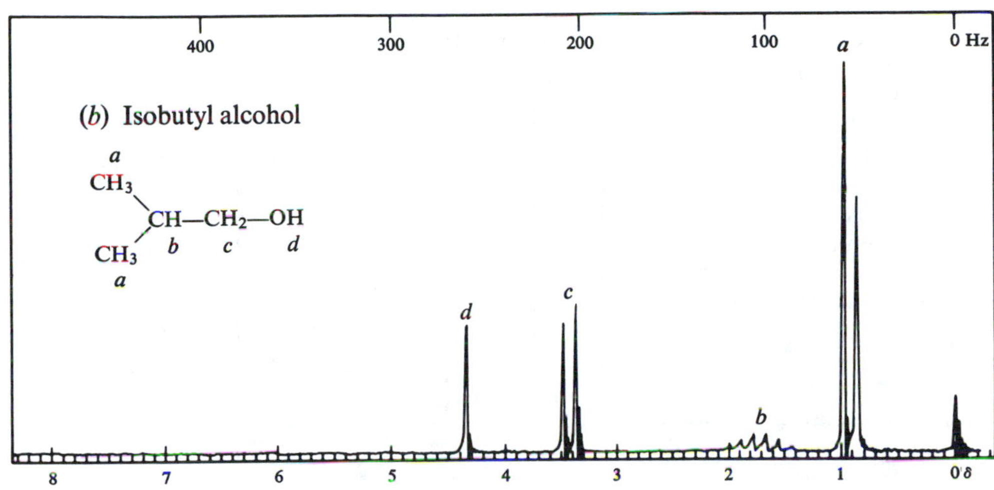

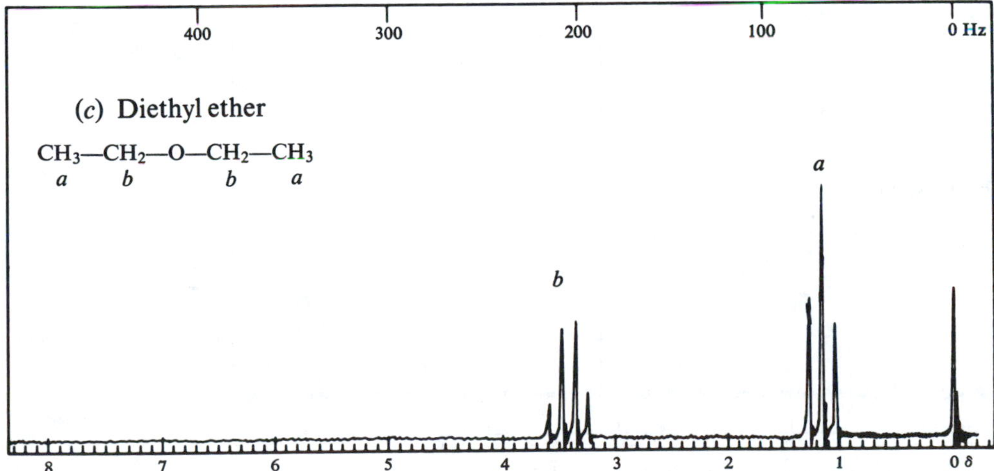

Figure 17.33 Proton NMR spectra for Problem 23, p. 646.

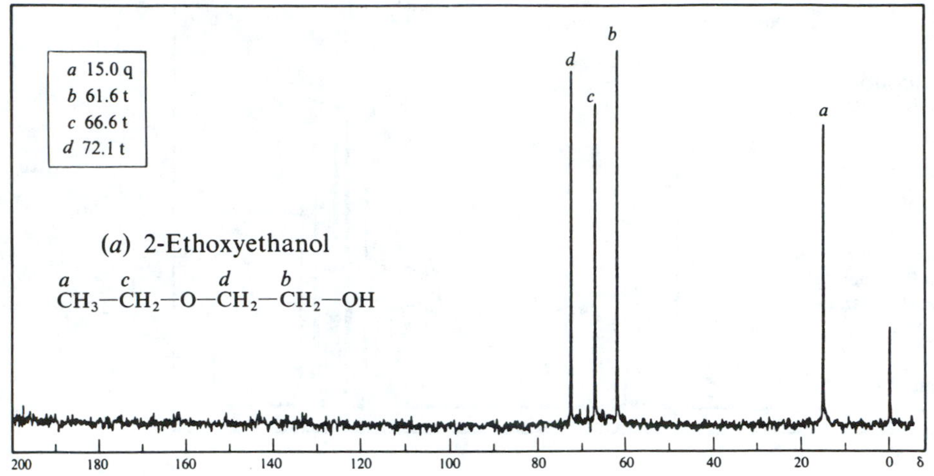

Sadtler 692C © Sadtler Research Laboratories, Division of Bio-Rad, Inc., (1976).

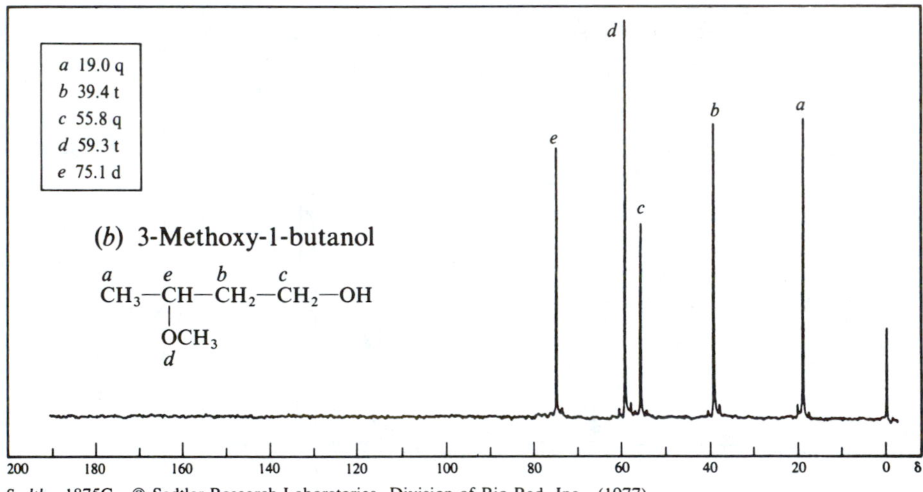

Sadtler 1875C © Sadtler Research Laboratories, Division of Bio-Rad, Inc., (1977).

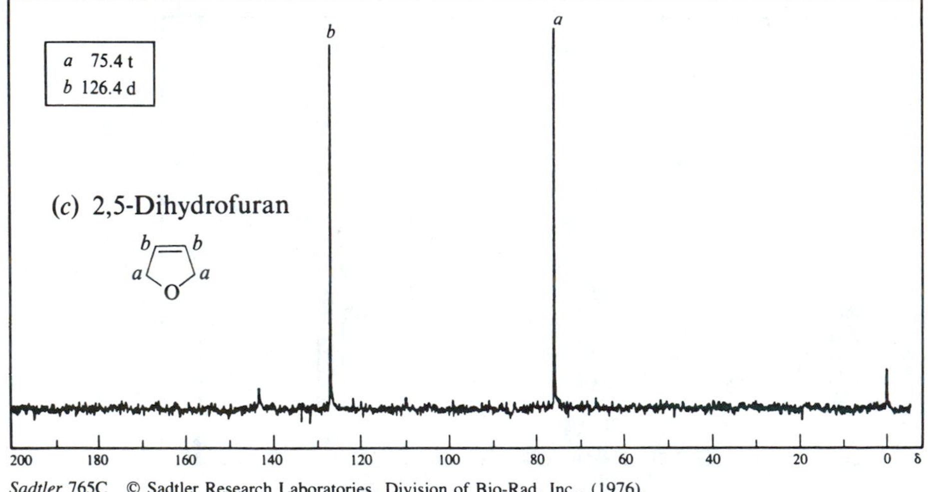

Sadtler 765C © Sadtler Research Laboratories, Division of Bio-Rad, Inc., (1976).

Figure 17.34 CMR spectra for Problem 24, p. 646.

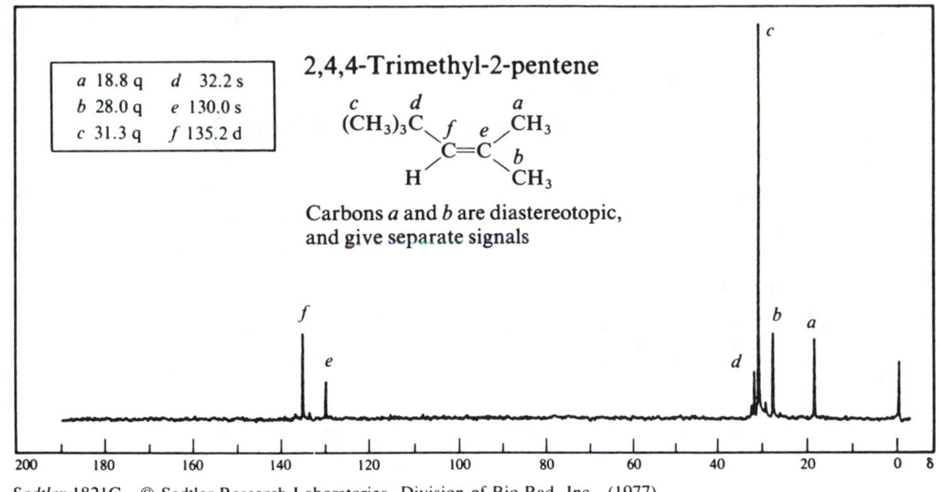

Figure 17.35 CMR spectrum for Problem 25, p. 646.

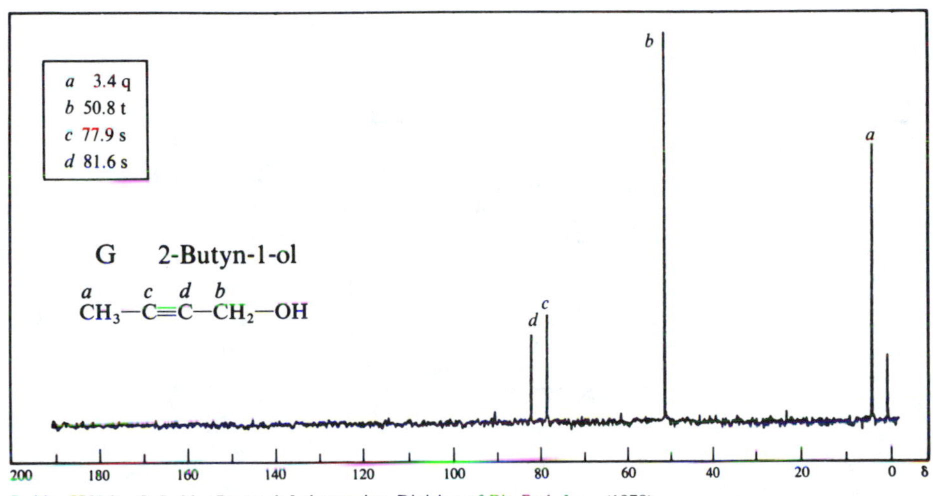

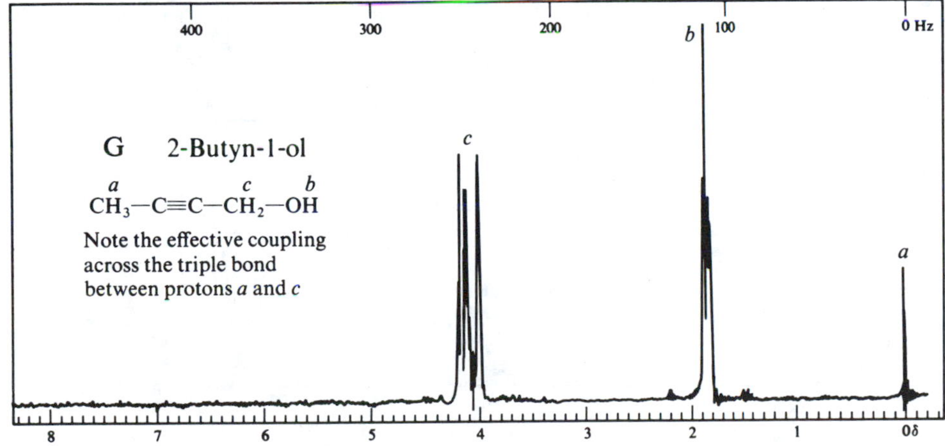

Figure 17.36 CMR and proton NMR spectra for Problem 26, p. 646.

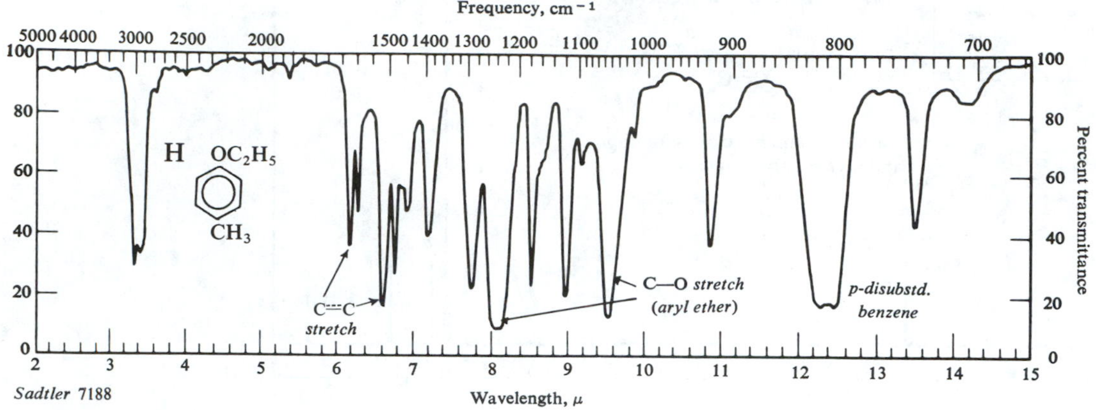

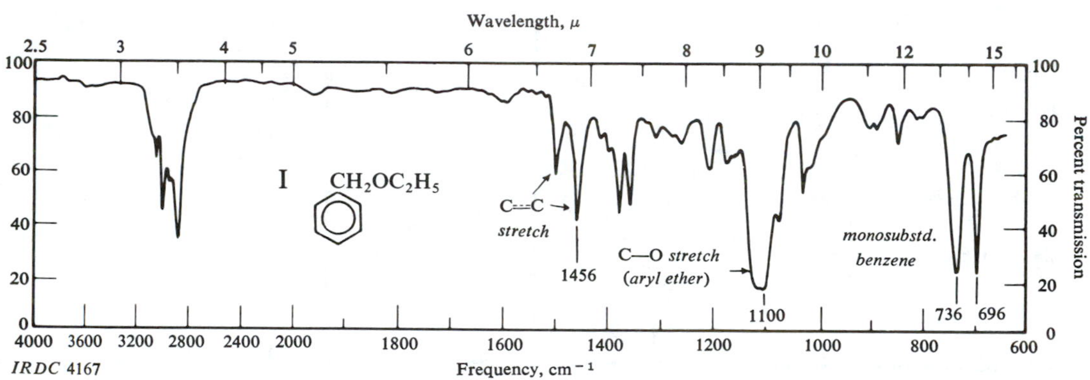

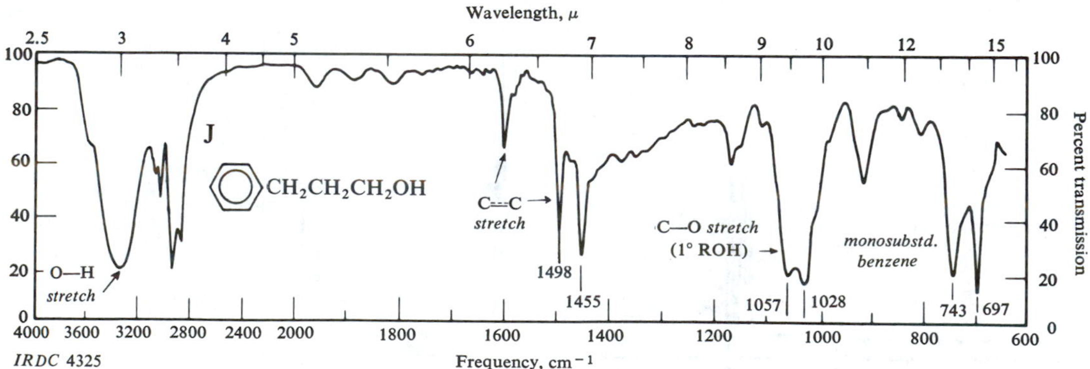

Figure 17.37 Infrared spectra for Problem 27, p. 646.

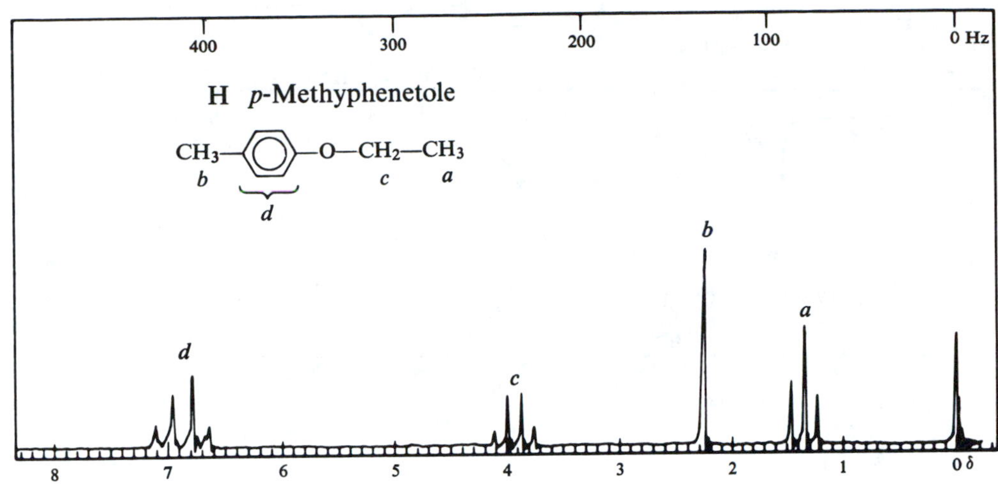

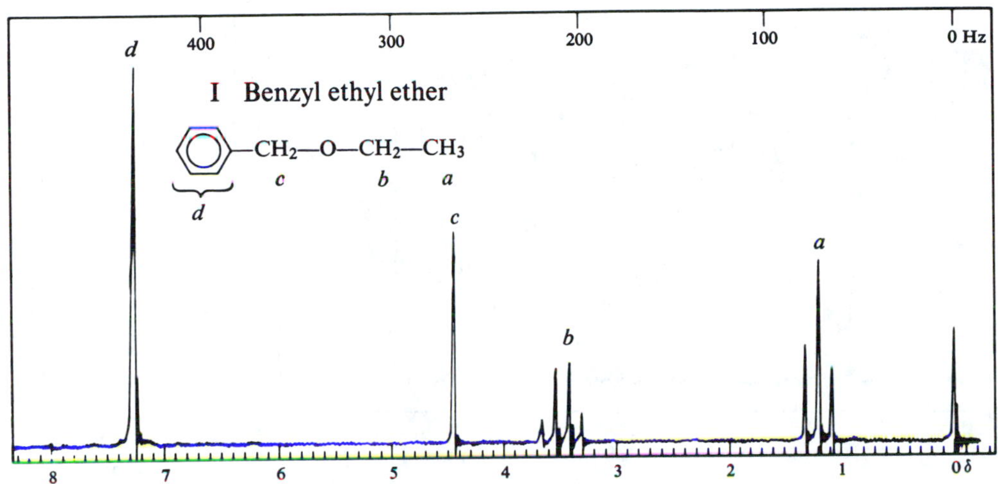

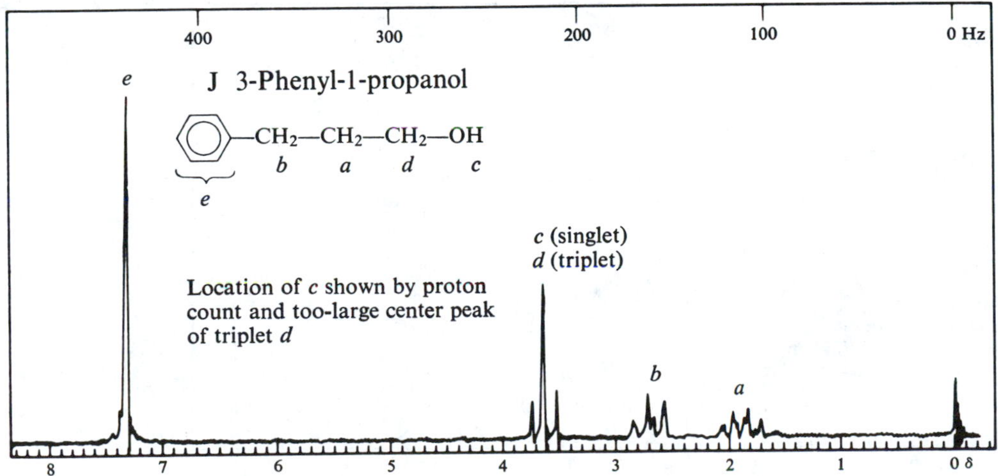

Figure 17.38 Proton NMR spectra for Problem 27, p. 646.

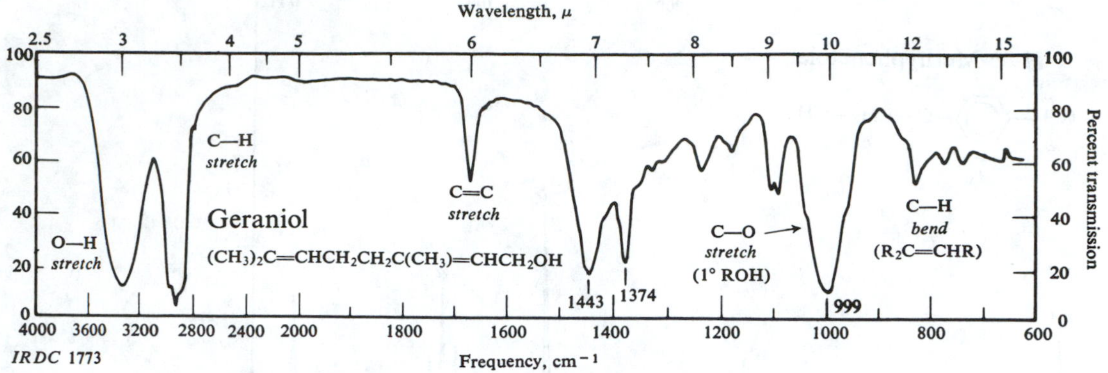

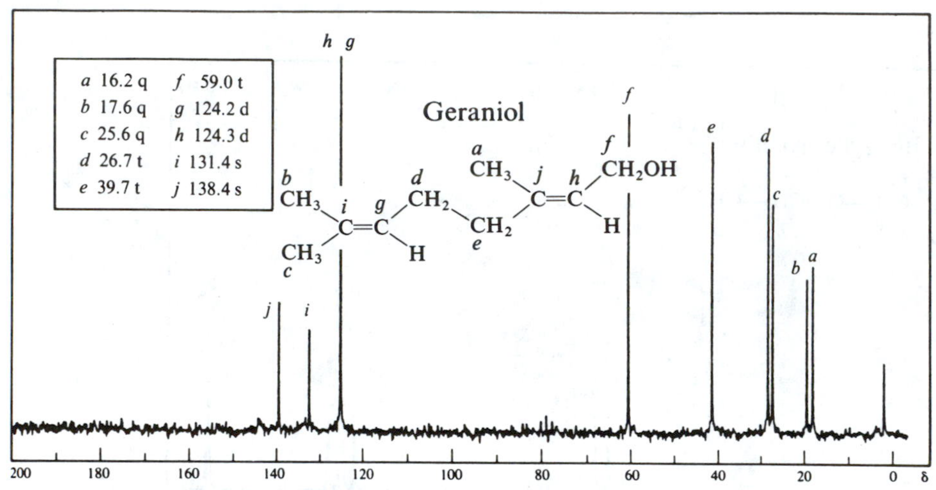

Sadtler 1249C © Sadtler Research Laboratories, Division of Bio-Rad, Inc., (1976).

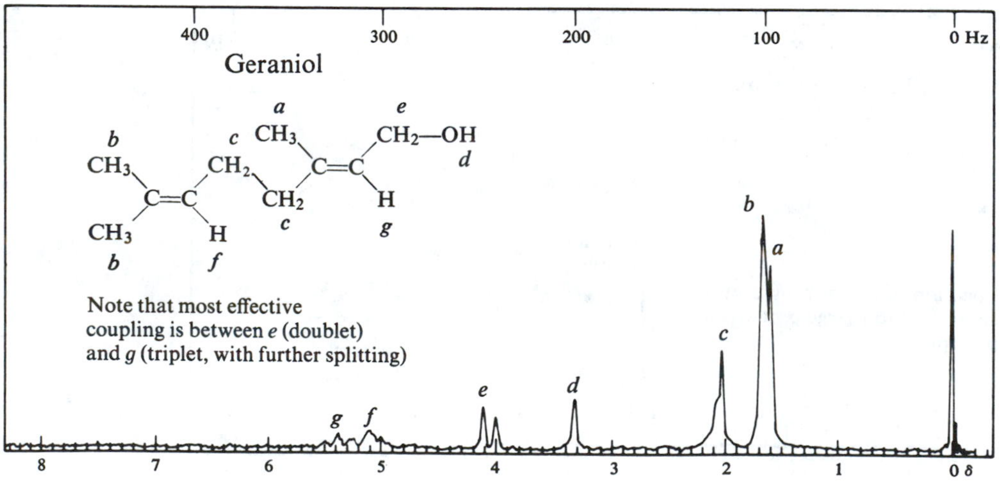

Geraniol

Note that most effective coupling is between *e* (doublet) and *g* (triplet, with further splitting)

400 300 200 100 0 Hz

8 7 6 5 4 3 2 1 0 δ

Figure 17.39 Infrared, CMR, and proton NMR spectra for Problem 28, p. 647.

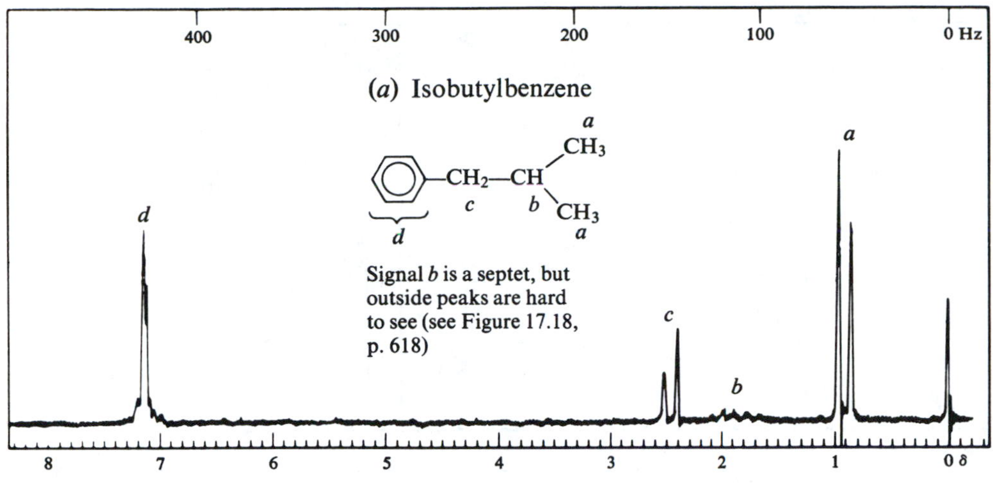

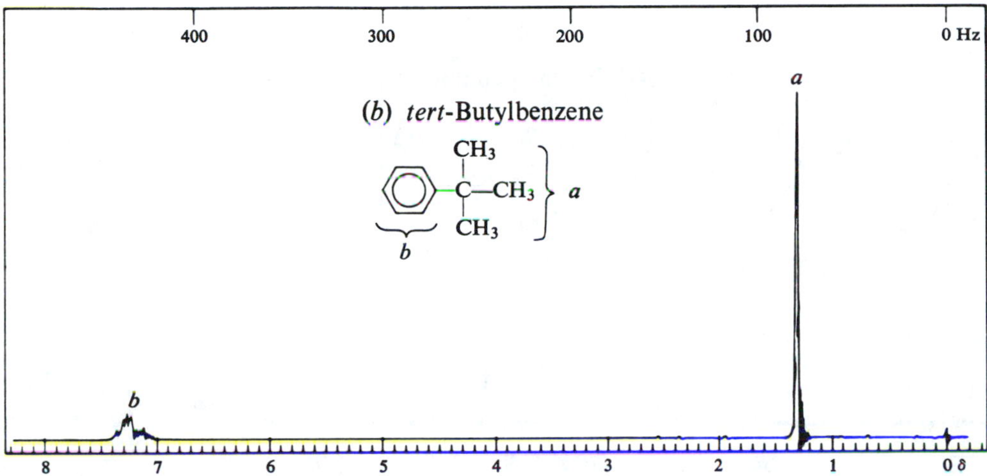

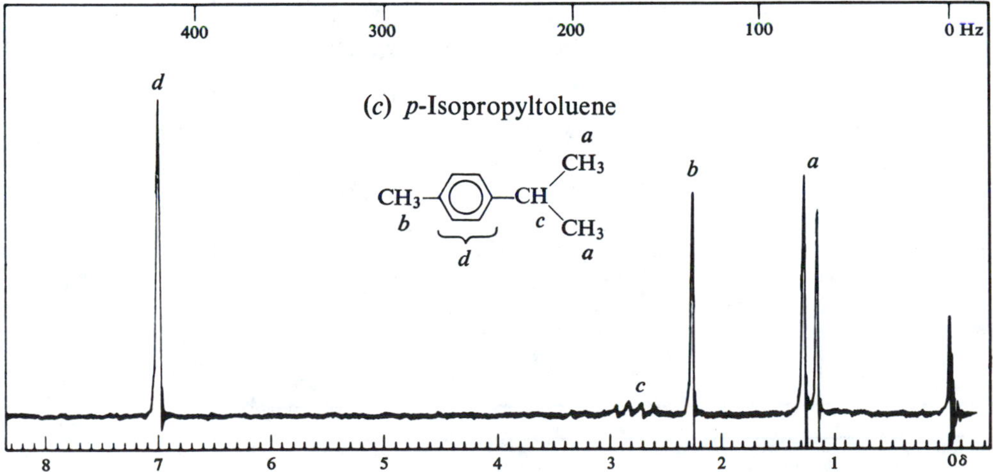

Figure 17.40 Proton NMR spectra for Problem 17.32, p. 283 of this Study Guide.

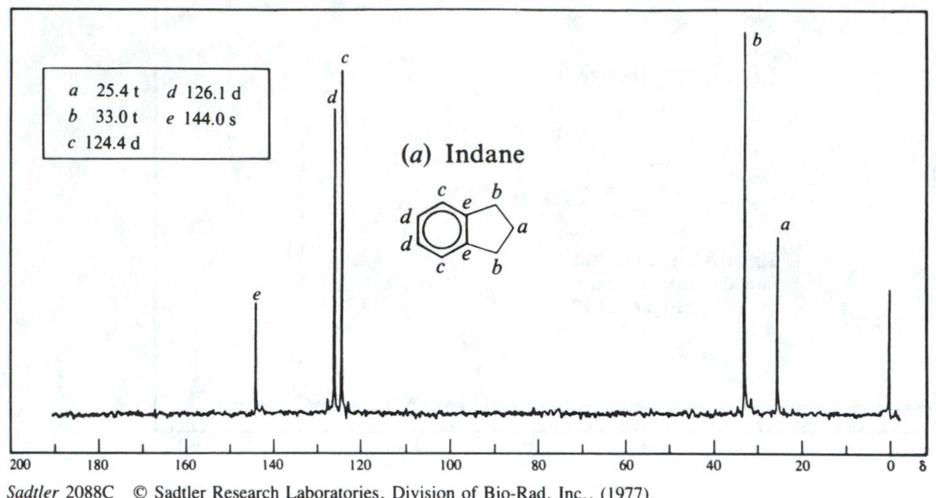

Sadtler 2088C © Sadtler Research Laboratories, Division of Bio-Rad, Inc., (1977).

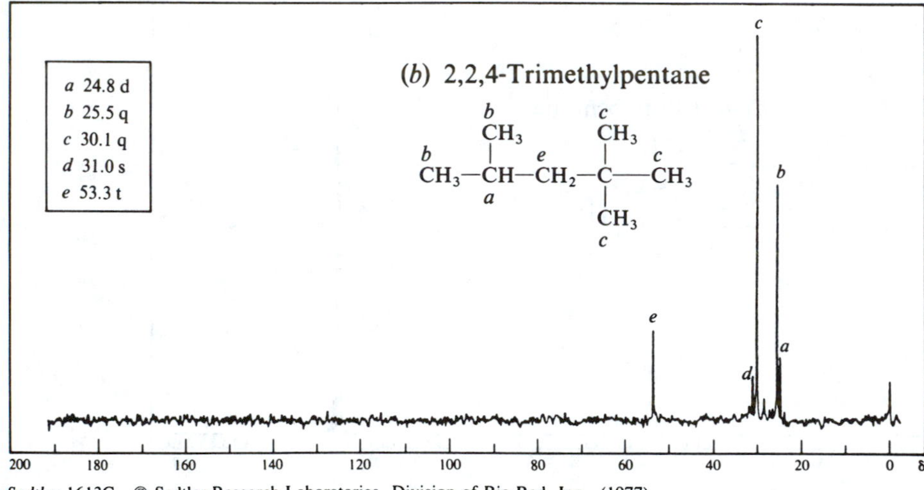

Sadtler 1613C © Sadtler Research Laboratories, Division of Bio-Rad, Inc., (1977).

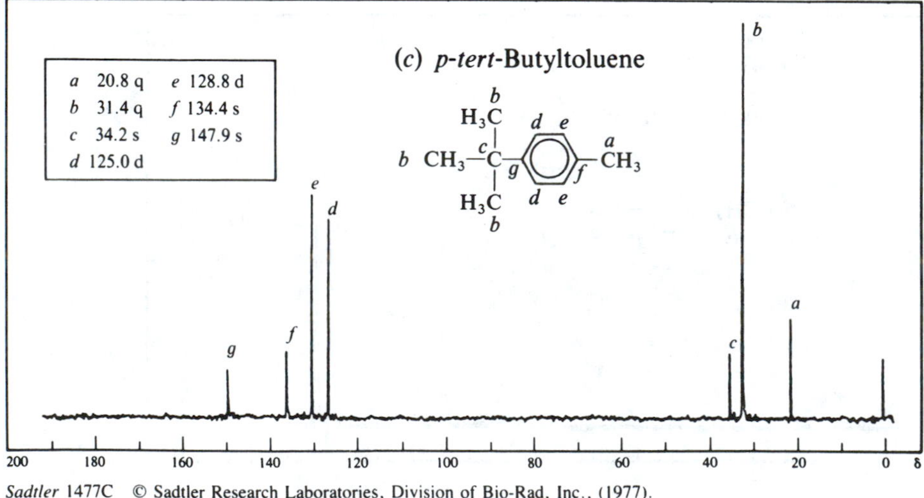

Sadtler 1477C © Sadtler Research Laboratories, Division of Bio-Rad, Inc., (1977).

Figure 17.41 CMR spectra for Problem 17.33, p. 283 of this Study Guide.

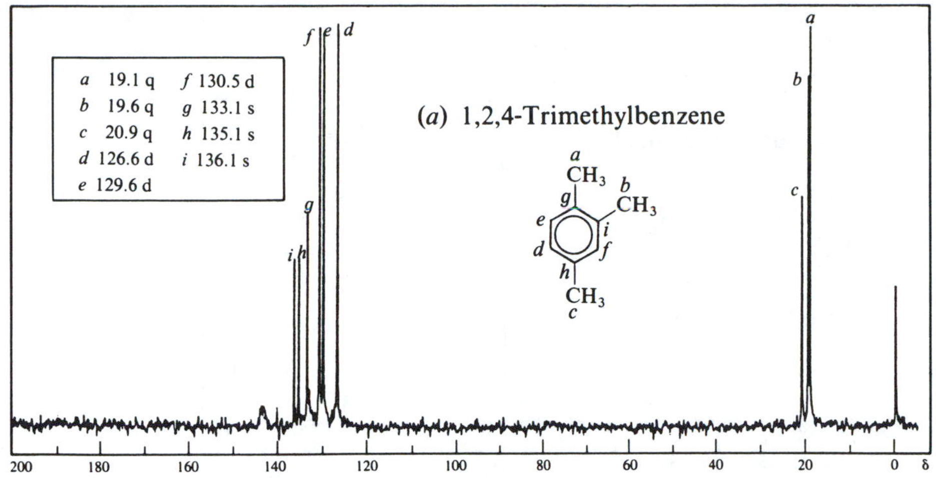

Sadtler 170C © Sadtler Research Laboratories, Division of Bio-Rad, Inc., (1976).

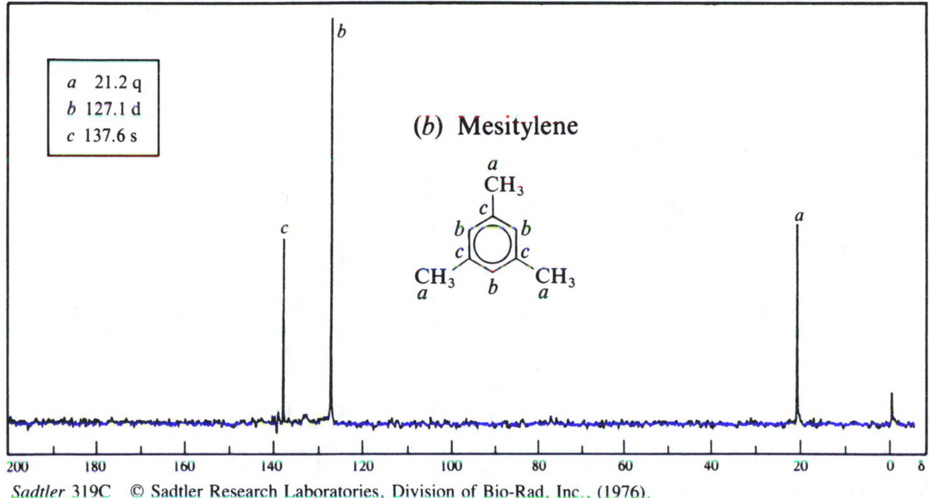

Sadtler 319C © Sadtler Research Laboratories, Division of Bio-Rad, Inc., (1976).

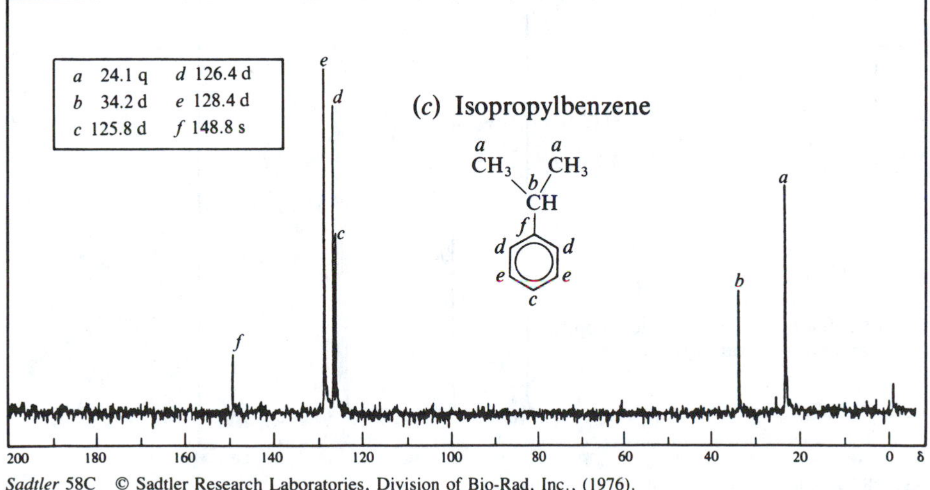

Sadtler 58C © Sadtler Research Laboratories, Division of Bio-Rad, Inc., (1976).

Figure 17.42 CMR spectra for Problem 17.34, p. 283 of this Study Guide.

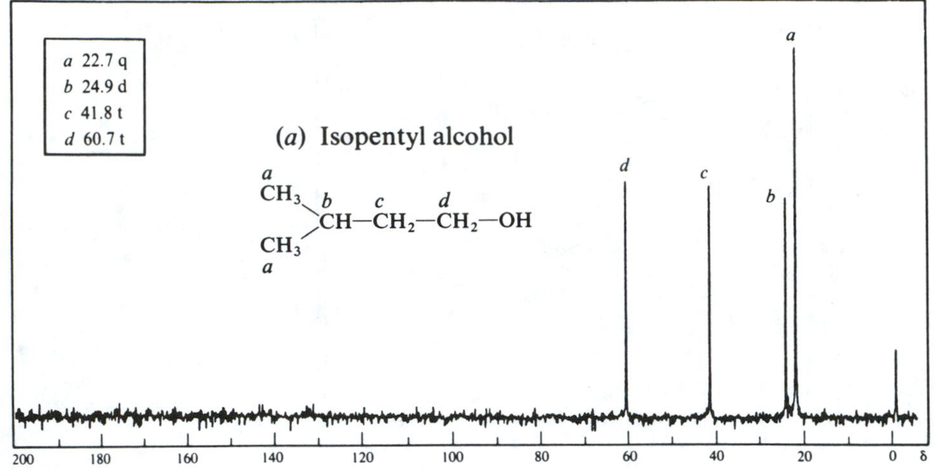

Sadtler 135C © Sadtler Research Laboratories, Division of Bio-Rad, Inc., (1976).

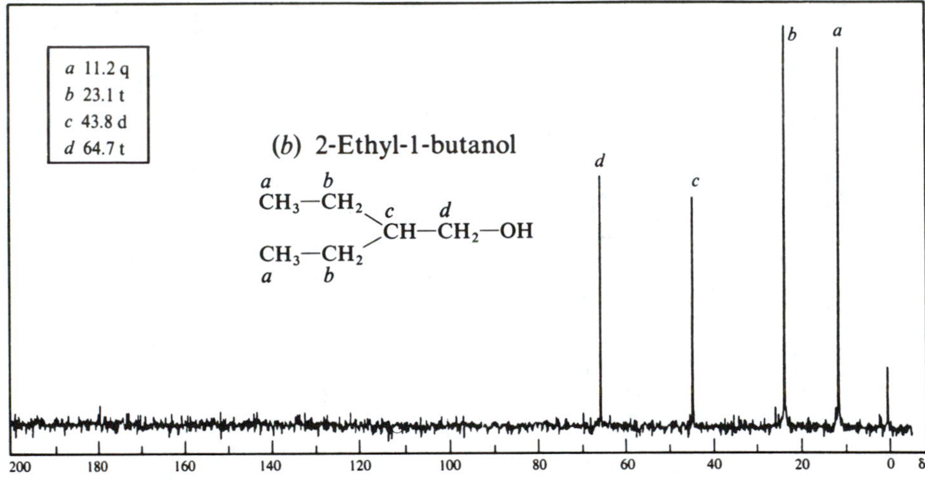

Sadtler 30C © Sadtler Research Laboratories, Division of Bio-Rad, Inc., (1976).

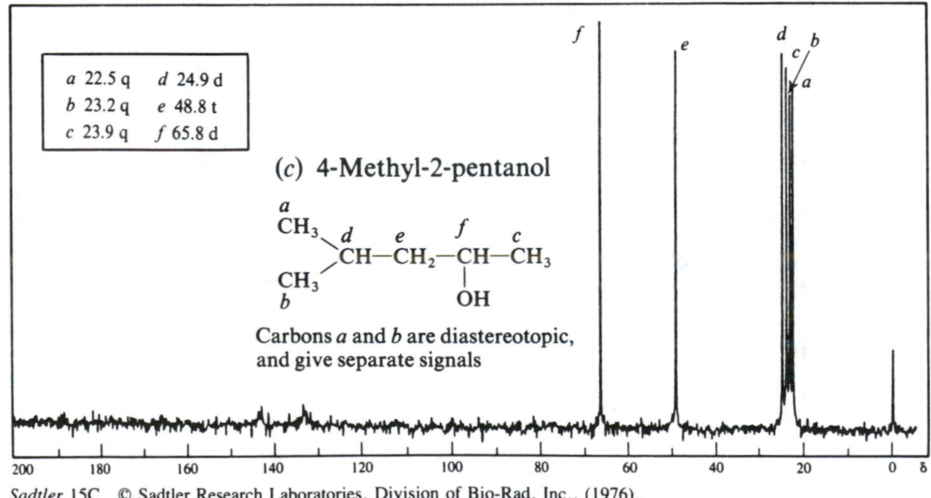

Sadtler 15C © Sadtler Research Laboratories, Division of Bio-Rad, Inc., (1976).

Figure 17.43 CMR spectra for Problem 17.35, p. 283 of this Study Guide.

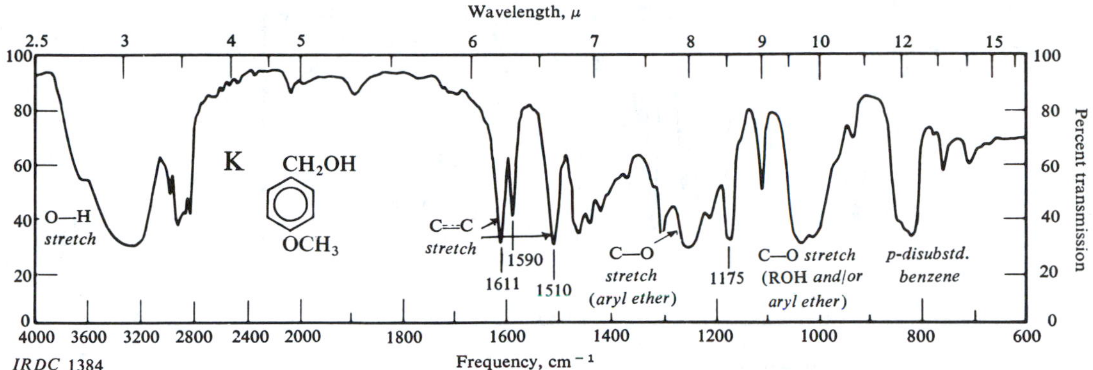

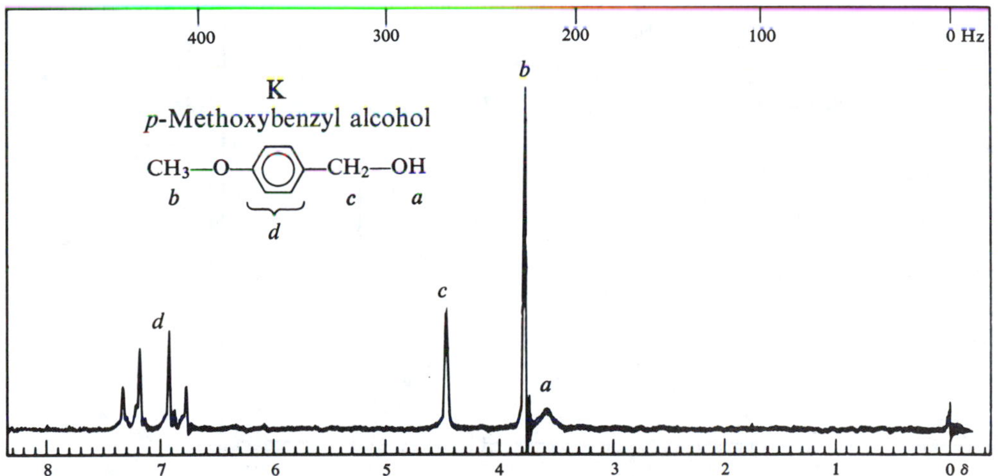

Figure 17.44 Infrared and proton NMR spectra for Problem 17.36, p. 283 of this Study Guide.

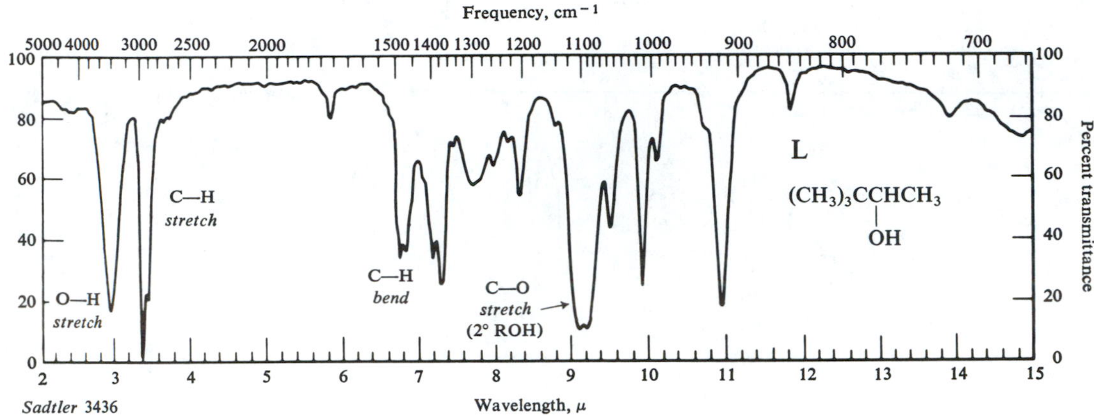

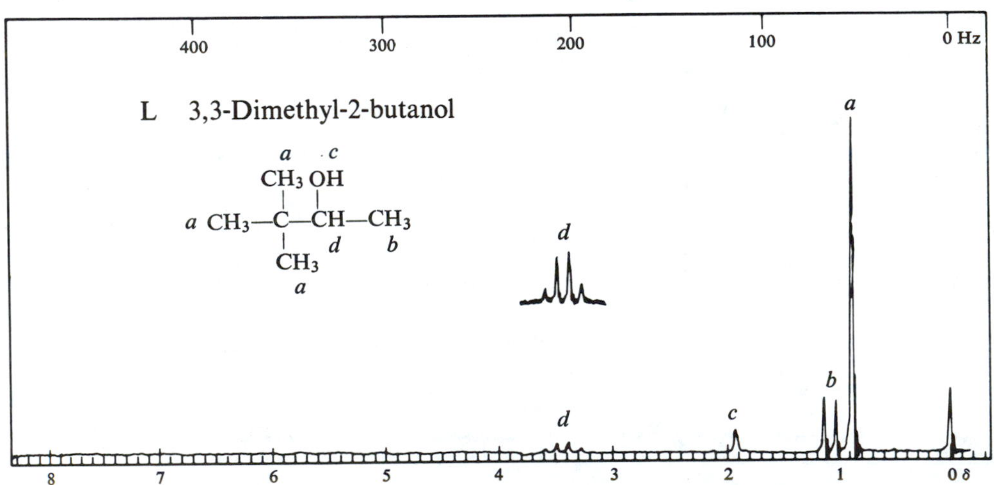

Figure 17.45 Infrared and proton NMR spectra for Problem 17.37, p. 283 of this Study Guide.

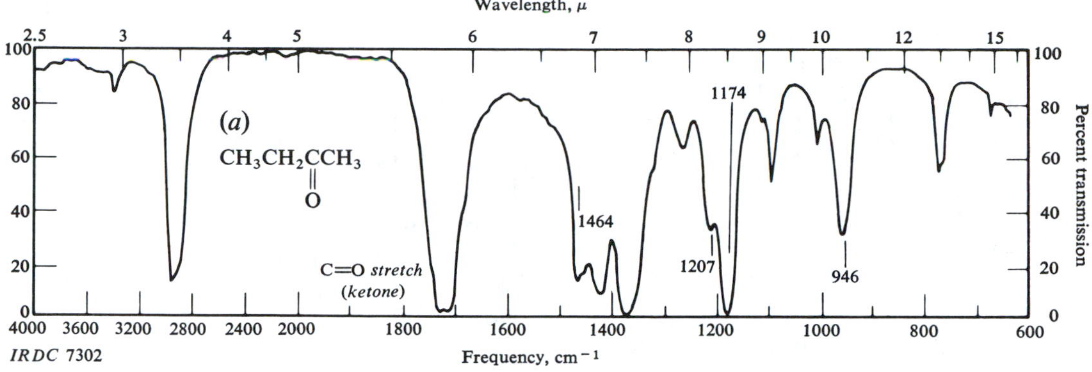

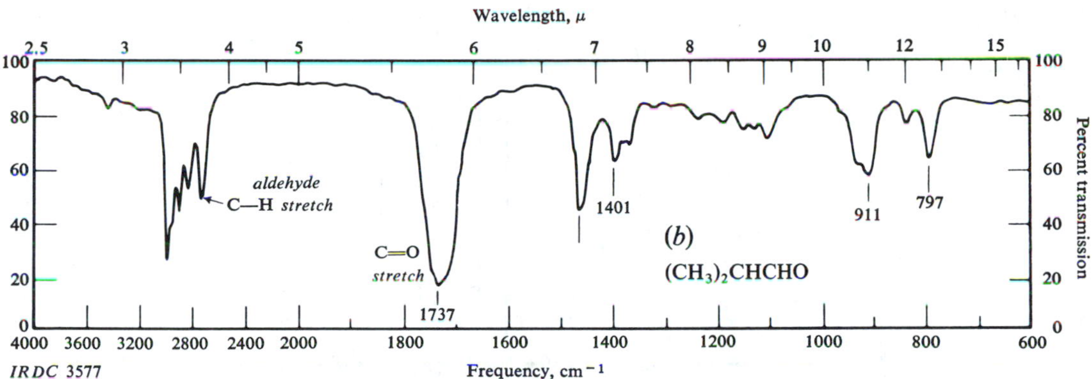

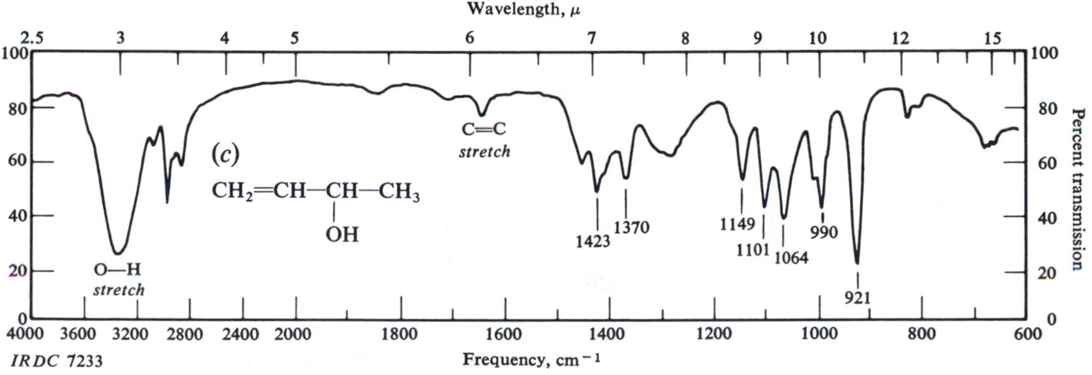

Figure 18.4 Infrared spectra for Problem 30, p. 706.

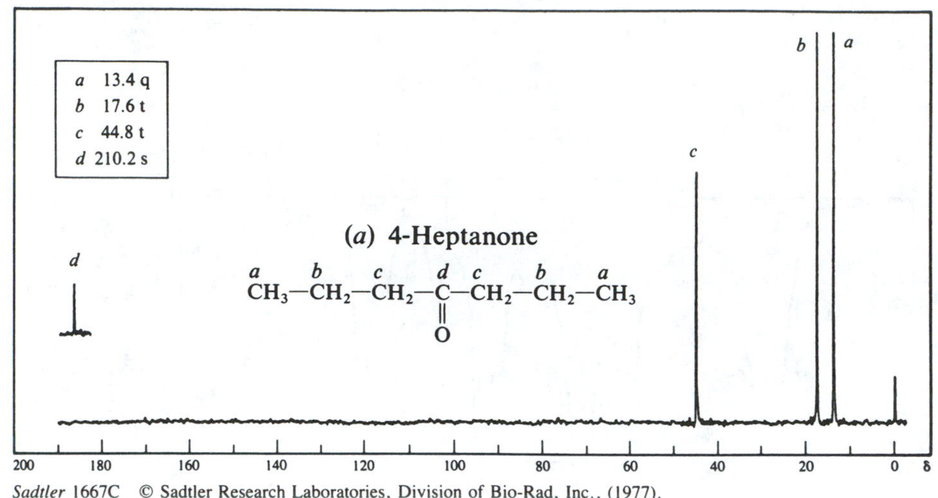

Sadtler 1667C © Sadtler Research Laboratories, Division of Bio-Rad, Inc., (1977).

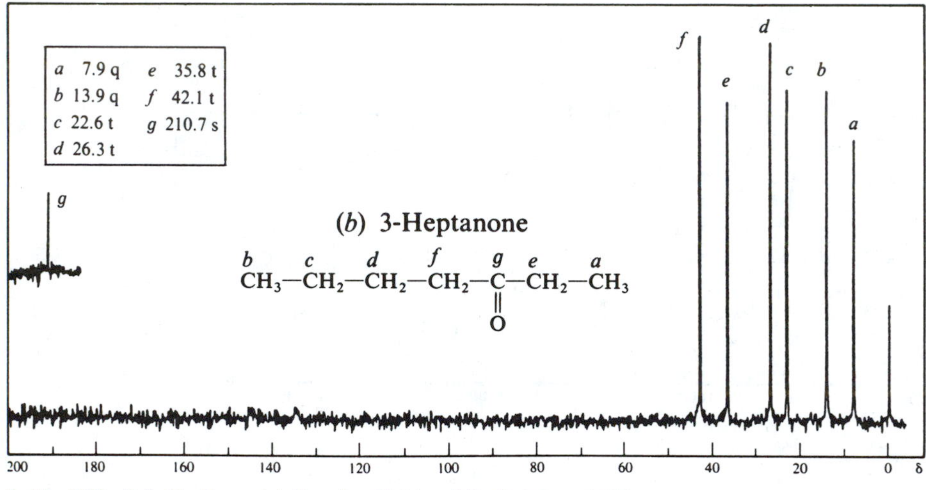

Sadtler 470C © Sadtler Research Labortories, Division of Bio-Rad, Inc., (1976).

a 13.9 q	*e* 31.7 t
b 22.7 t	*f* 43.7 t
c 23.8 t	*g* 208.4 s
d 29.6 q	

(*c*) 2-Heptanone

$\overset{a}{CH_3}-\overset{b}{CH_2}-\overset{e}{CH_2}-\overset{c}{CH_2}-\overset{f}{CH_2}-\overset{g}{\underset{\|}{C}}-\overset{d}{CH_3}$
O

200 180 160 140 120 100 80 60 40 20 0 δ

Sadtler 420C © Sadtler Research Laboratories, Division of Bio-Rad, Inc., (1976).

Figure 18.5 CMR spectra for Problem 31, p. 706.

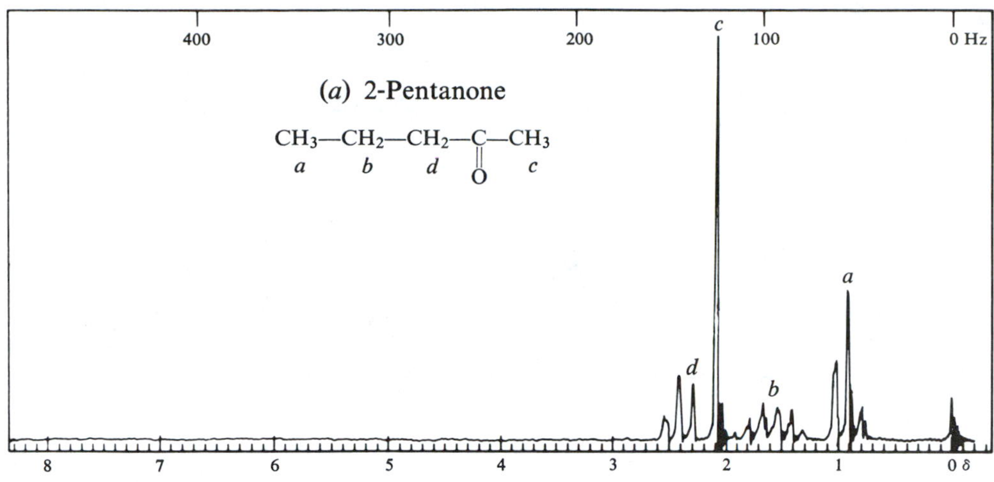

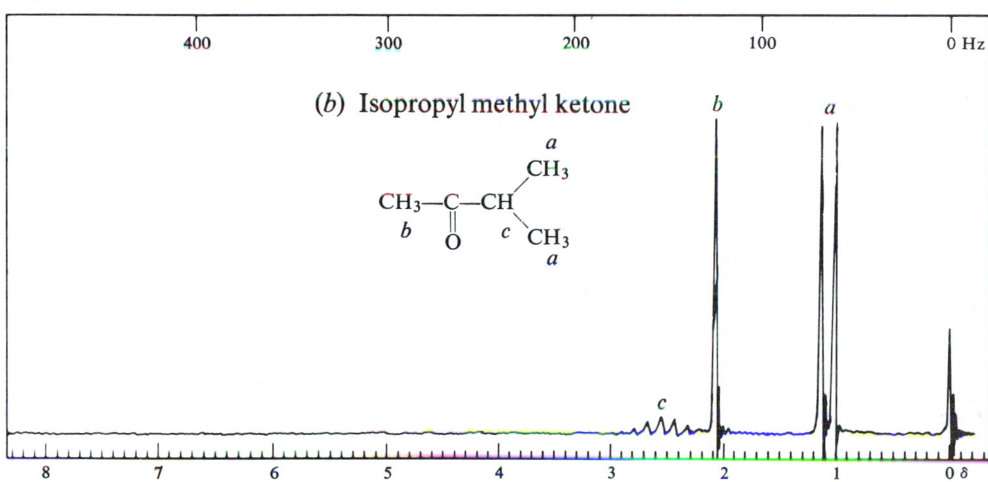

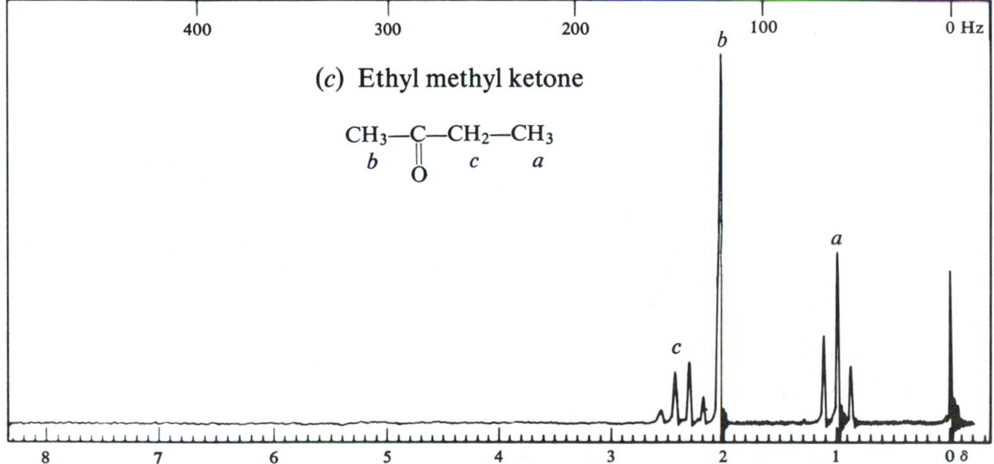

Figure 18.6 Proton NMR spectra for Problem 32, p. 706.

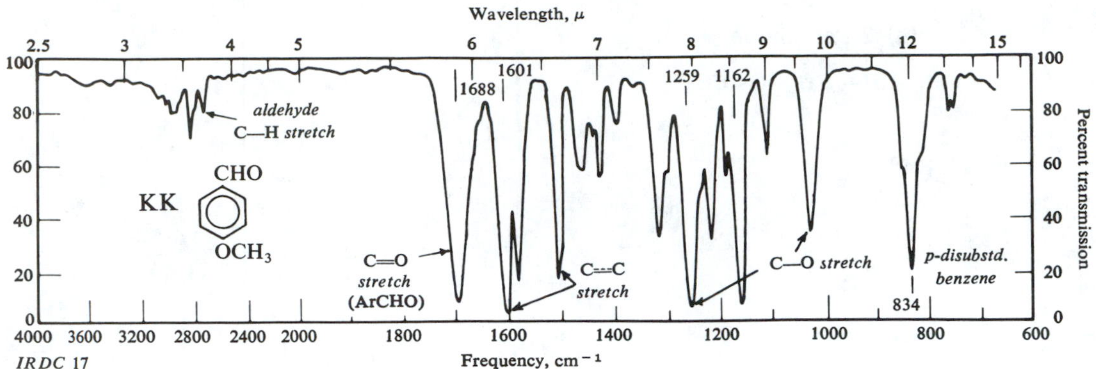

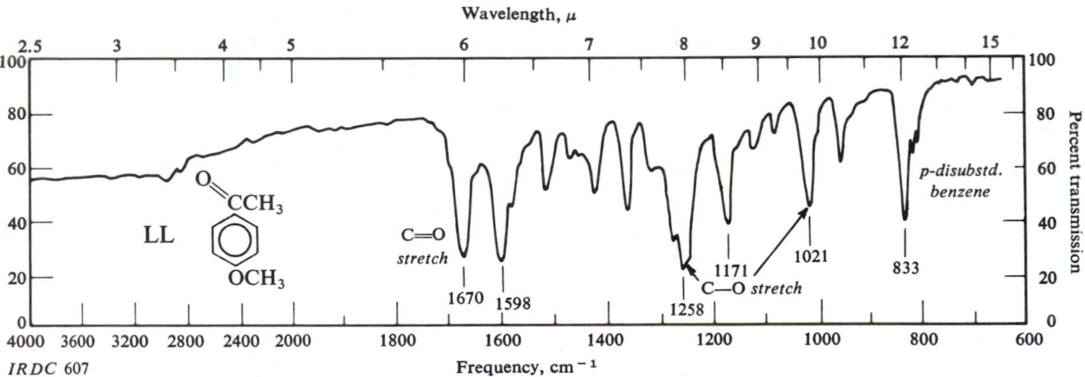

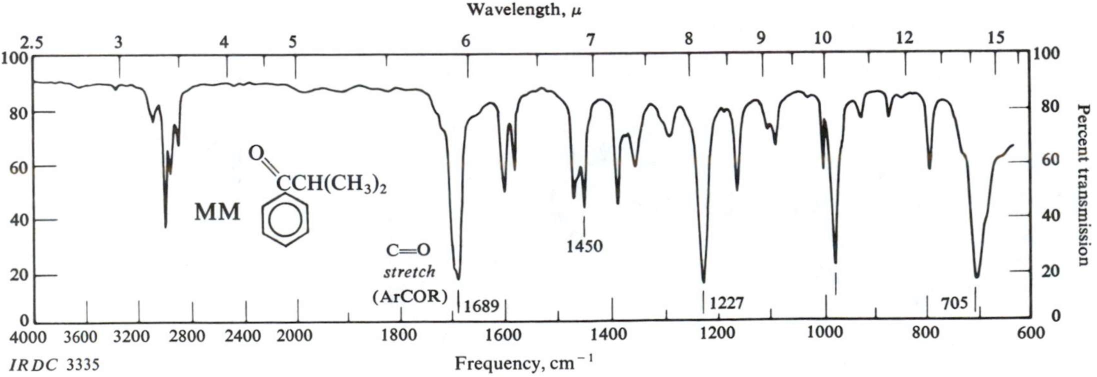

Figure 18.7 Infrared spectra for Problem 33, p. 706.

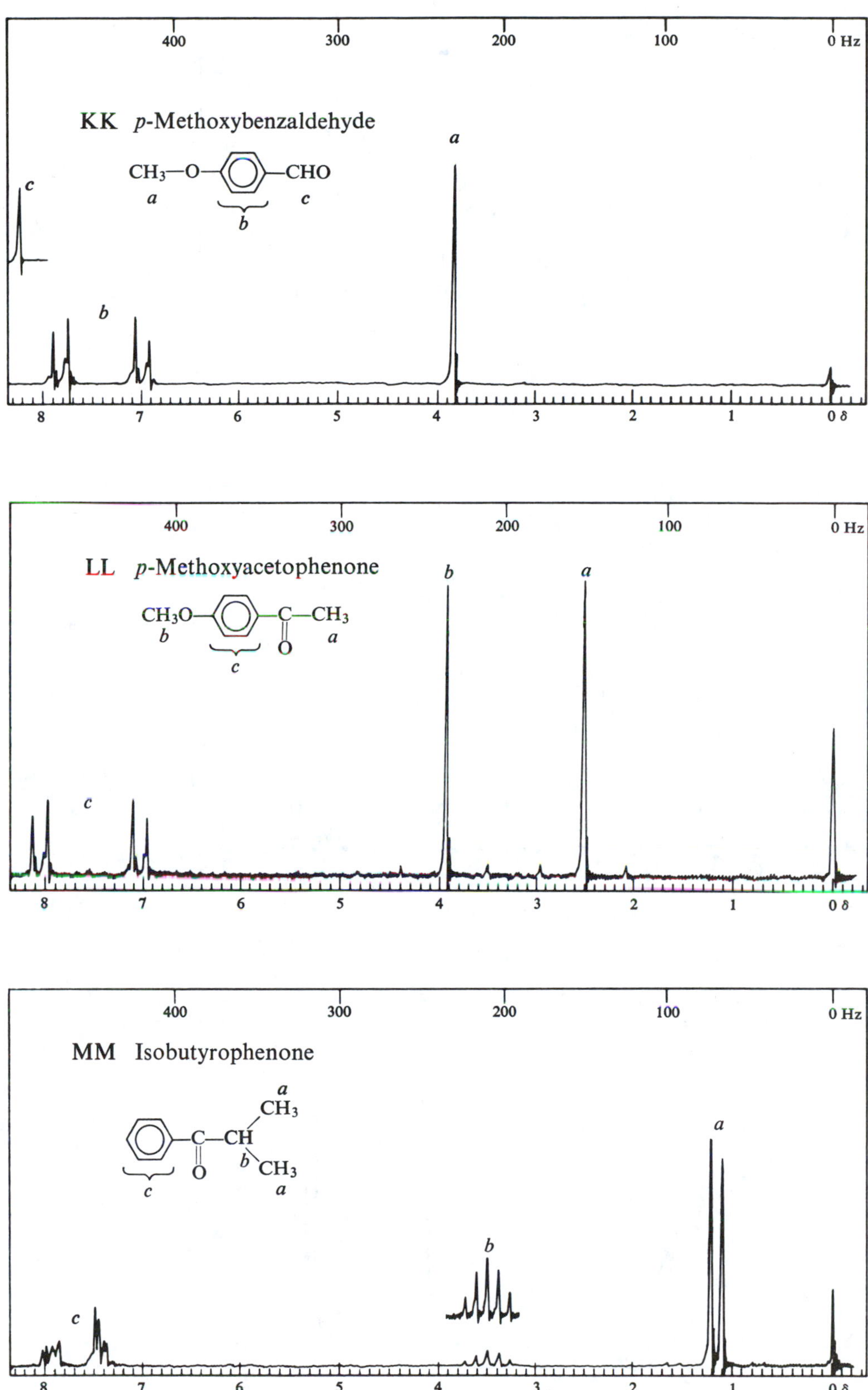

Figure 18.8 Proton NMR spectra for Problem 33, p. 706.

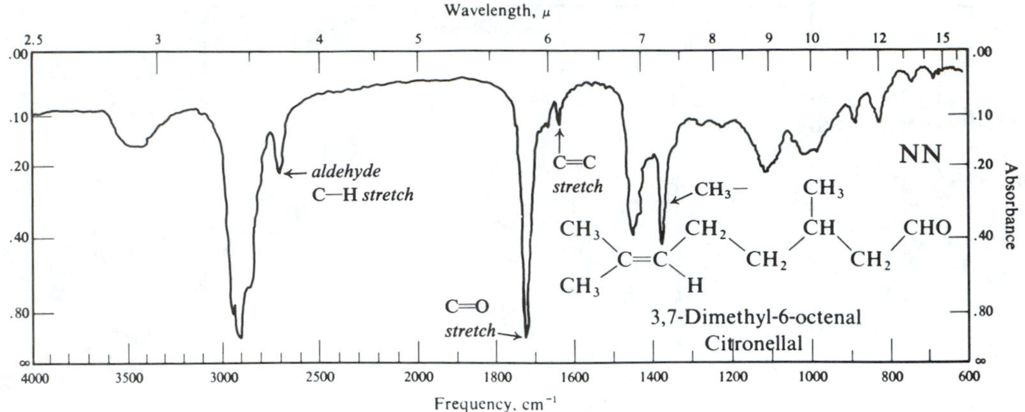

Sadtler 15514K © Sadtler Research Laboratories, Division of Bio-Rad, Inc., (1969).

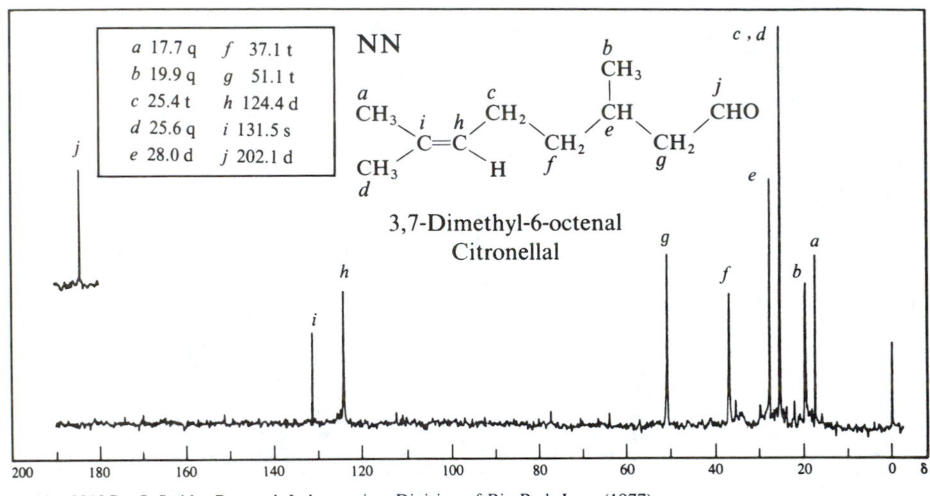

Sadtler 2918C © Sadtler Research Laboratories, Division of Bio-Rad, Inc., (1977).

NN

3,7-Dimethyl-6-octenal

Citronellal

400 300 200 100 0 Hz

8 7 6 5 4 3 2 1 0δ

Sadtler 10936M © Sadtler Research Laboratories, Division of Bio-Rad, Inc., (1971).

Figure 18.9 Infrared, CMR, and proton NMR spectra for Problem 34, p. 706.

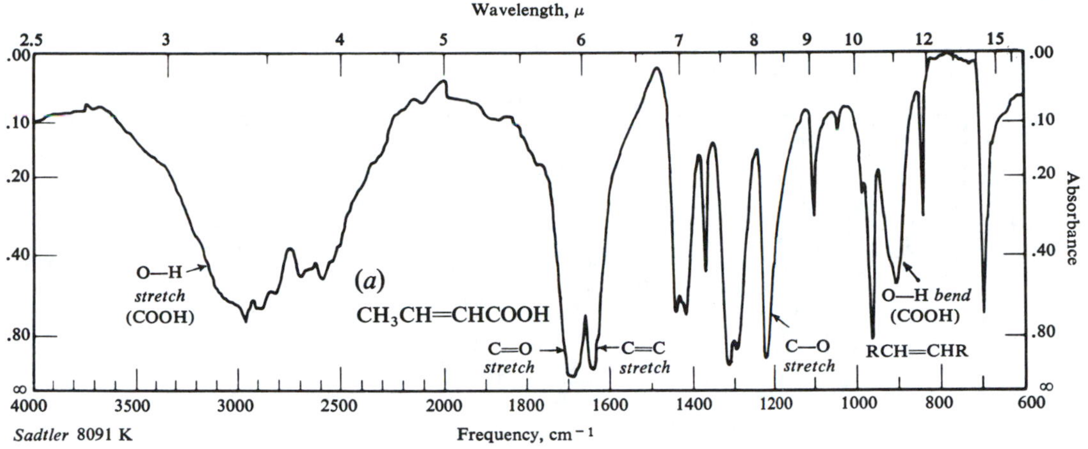

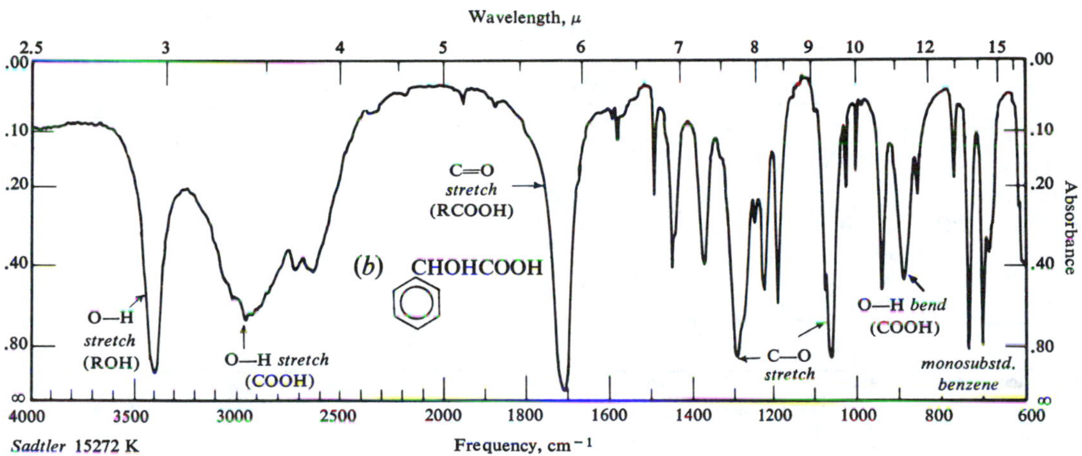

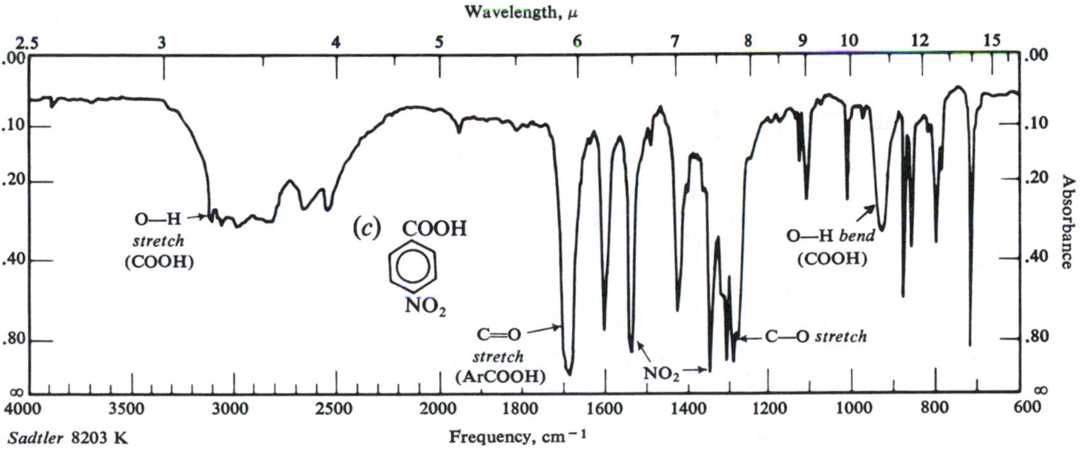

Figure 19.7 Infrared spectra for Problem 27, p. 751.

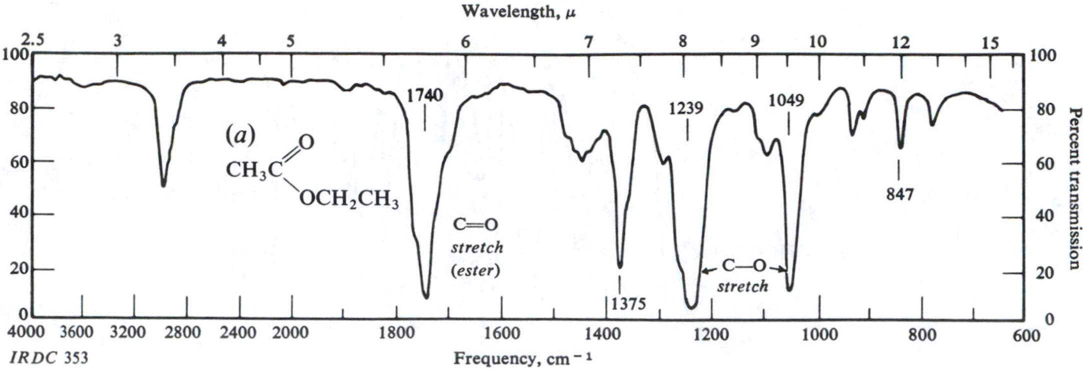

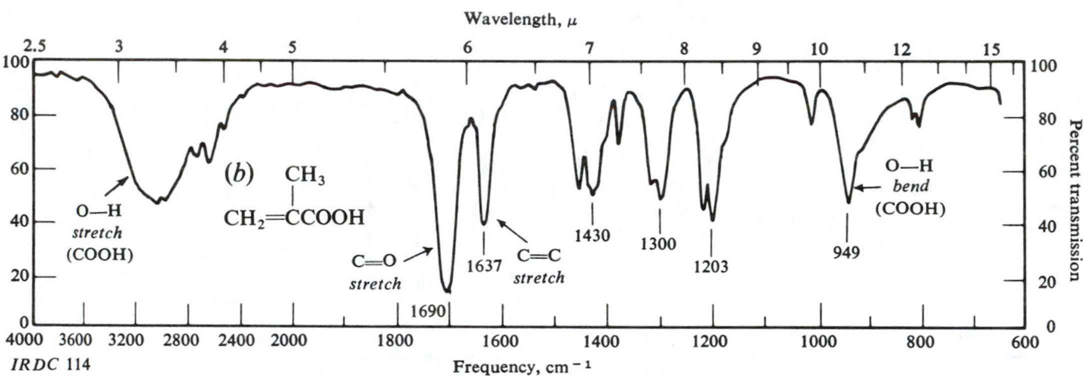

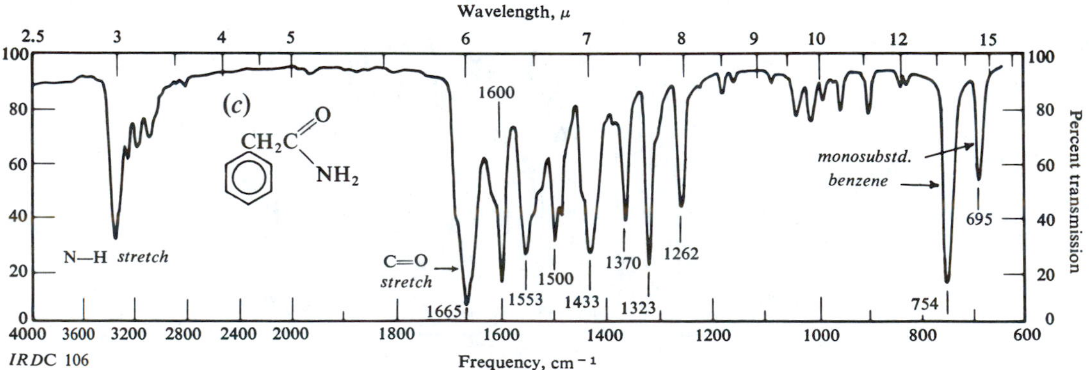

Figure 20.3 Infrared spectra for Problem 20, p. 789.

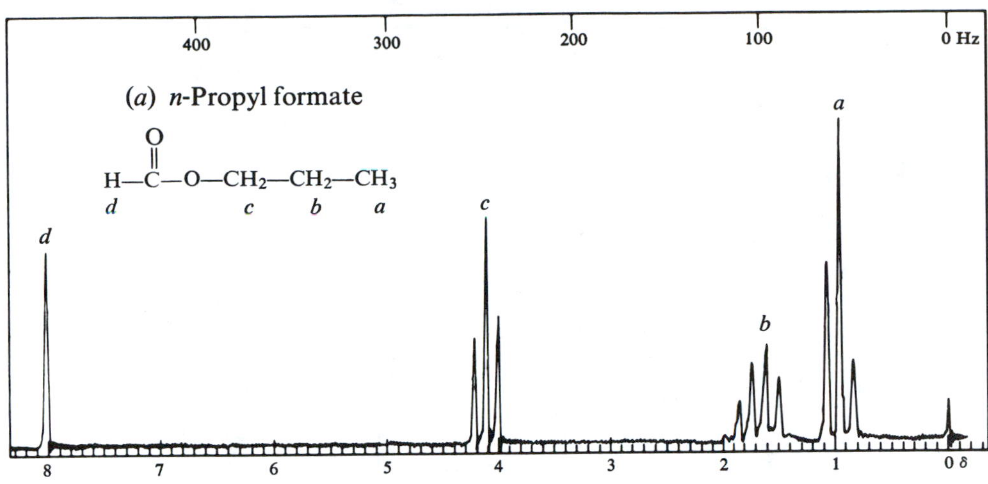

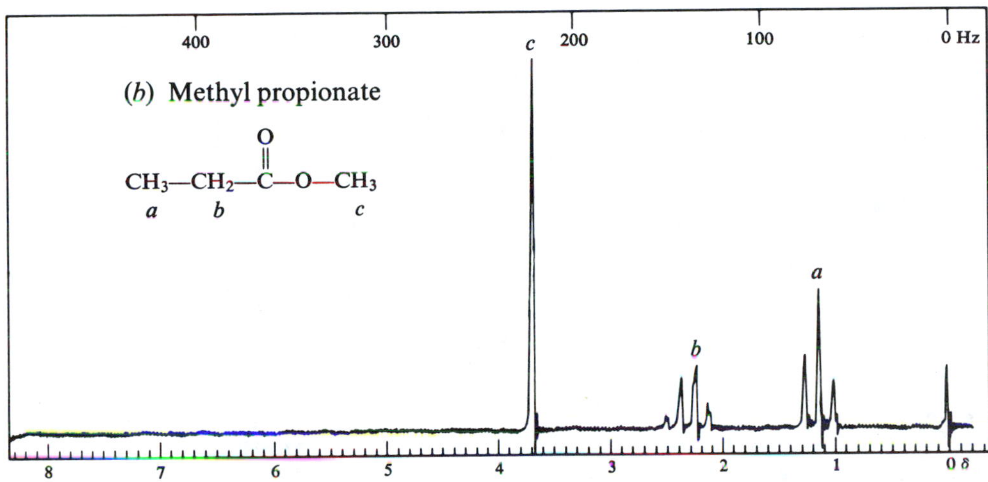

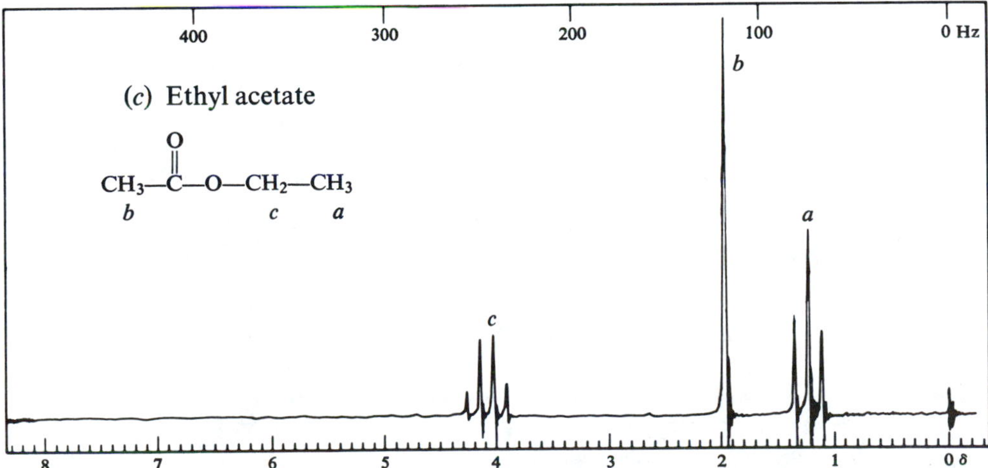

Figure 20.4 Proton NMR spectra for Problem 21, p. 789.

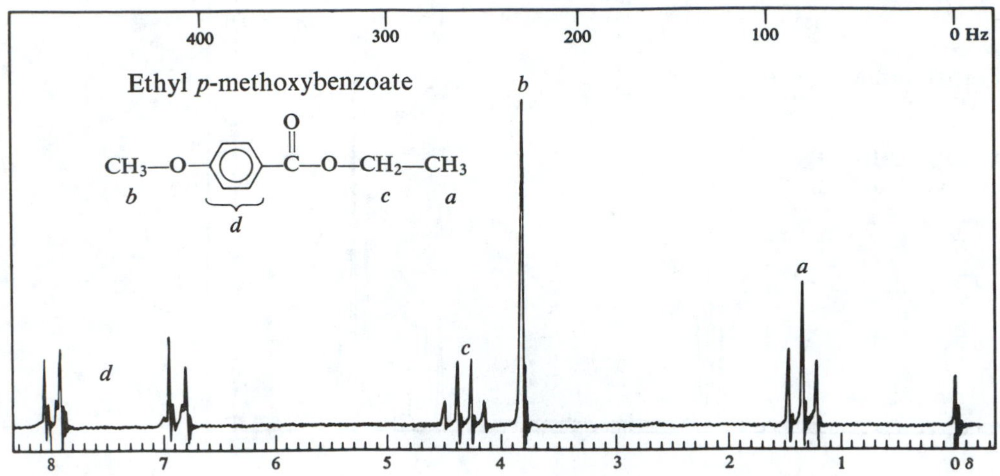

Figure 20.5 Proton NMR spectrum for Problem 22, p. 789.

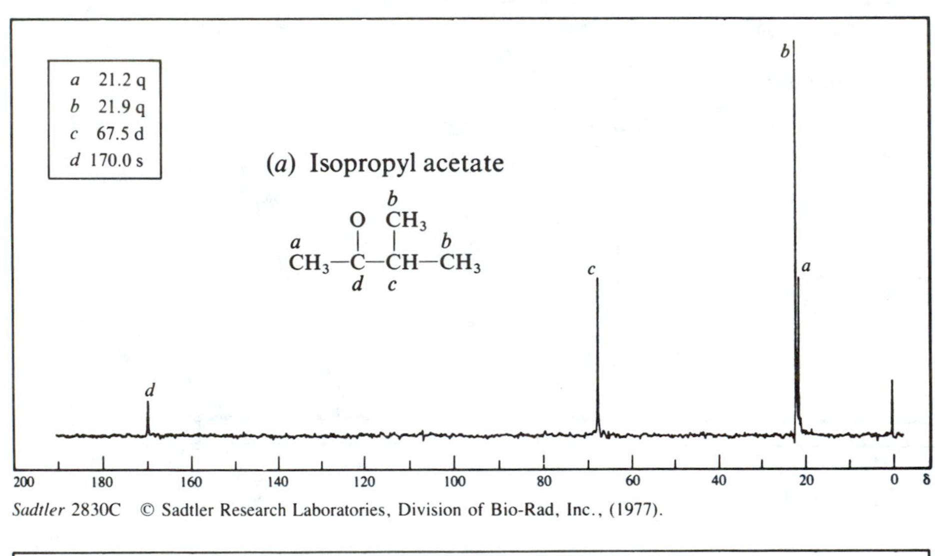

Sadtler 2830C © Sadtler Research Laboratories, Division of Bio-Rad, Inc., (1977).

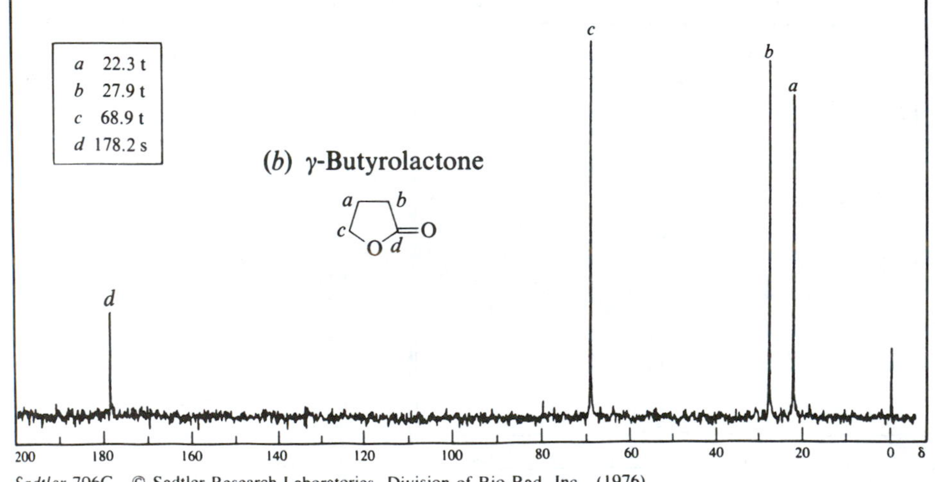

Sadtler 706C © Sadtler Research Laboratories, Division of Bio-Rad, Inc., (1976).

Figure 20.6 CMR spectra for Problem 23, p. 789.

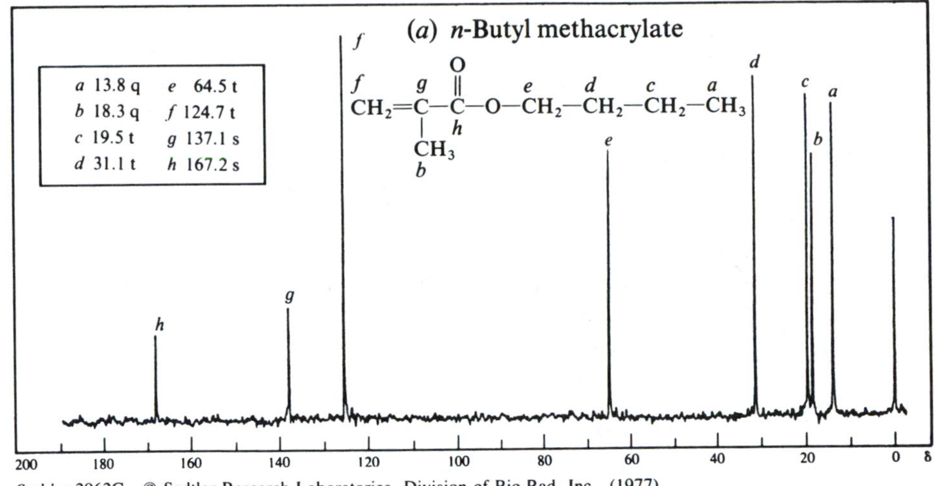

Sadtler 2062C © Sadtler Research Laboratories, Division of Bio-Rad, Inc., (1977).

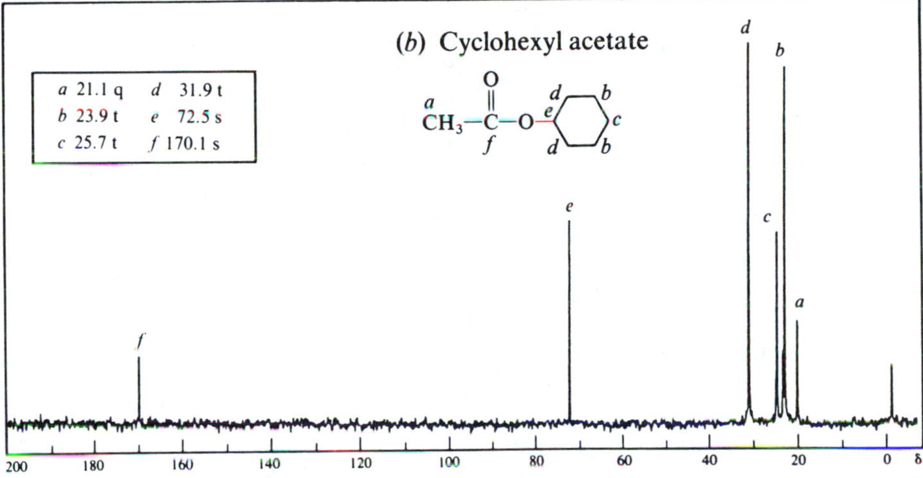

Sadtler 1149C © Sadtler Research Laboratories, Division of Bio-Rad, Inc., (1976).

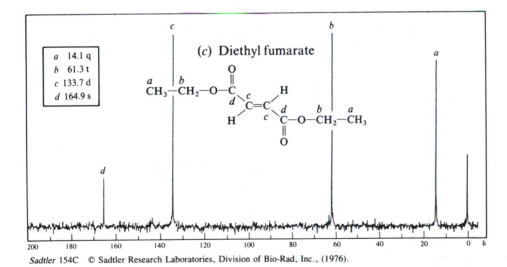

(c) Diethyl fumarate

a	14.1 q
b	61.3 t
c	133.7 d
d	164.9 s

$CH_3-CH_2-O-C(=O)-CH=CH-C(=O)-O-CH_2-CH_3$

200 180 160 140 120 100 80 60 40 20 0 δ

Sadtler 154C © Sadtler Research Laboratories, Division of Bio-Rad, Inc., (1976).

Figure 20.7 CMR spectra for Problem 24, p. 789.

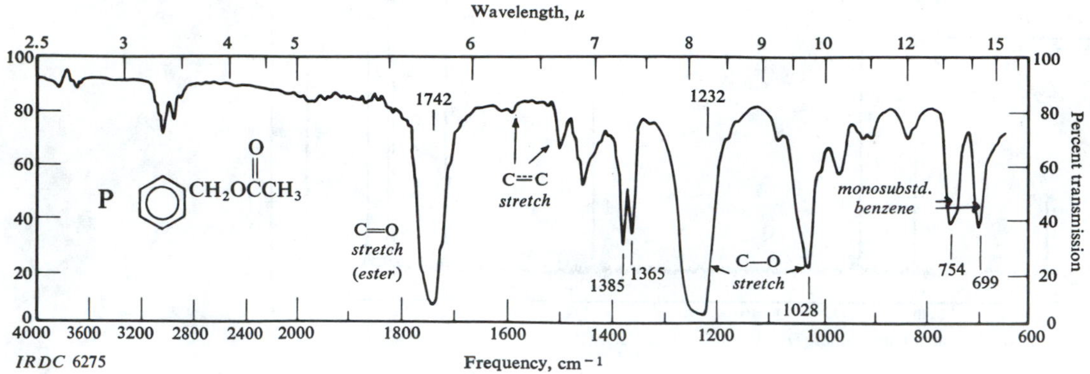

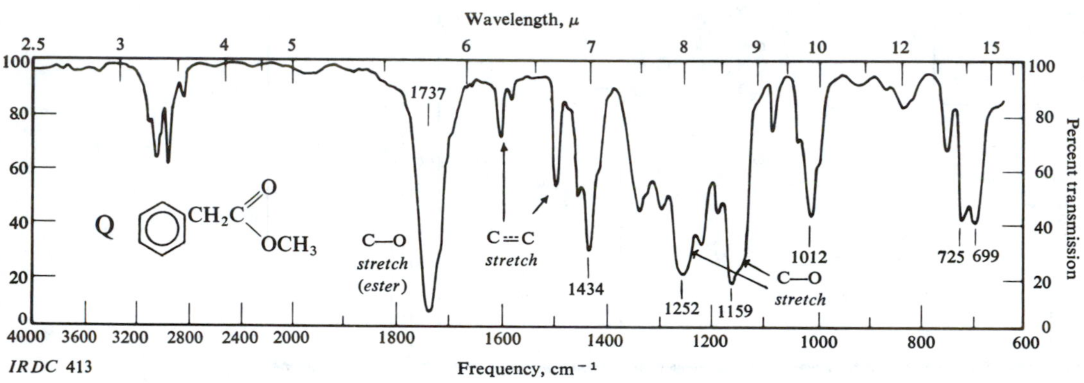

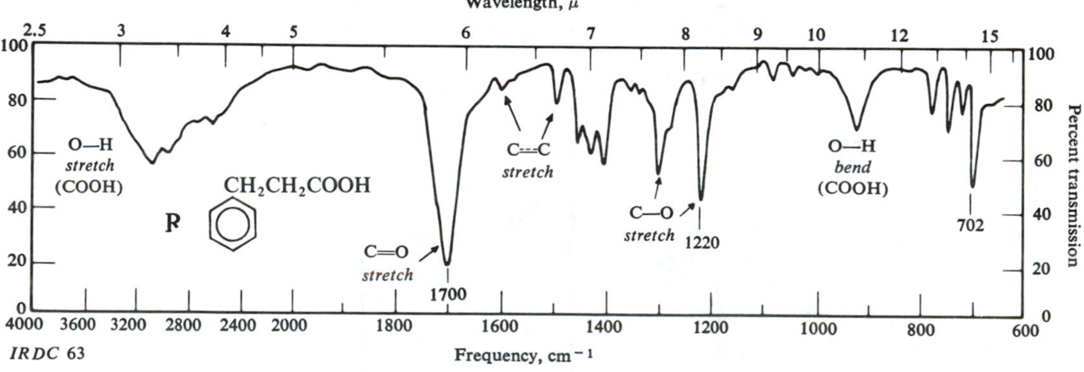

Figure 20.8 Infrared spectra for Problem 25, p. 789.

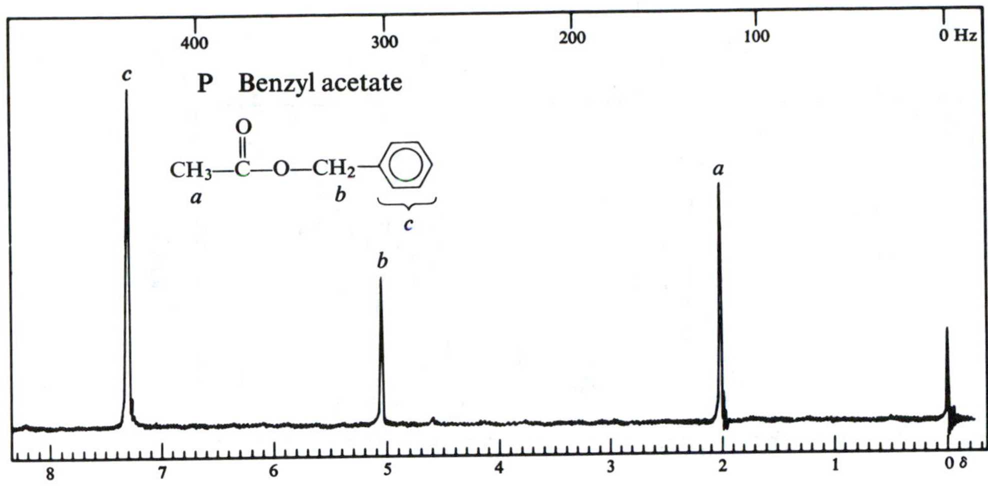

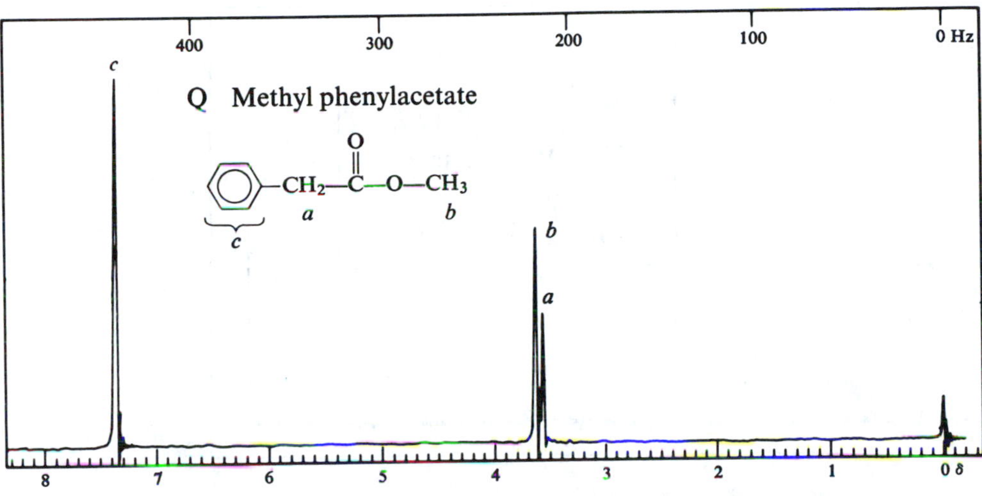

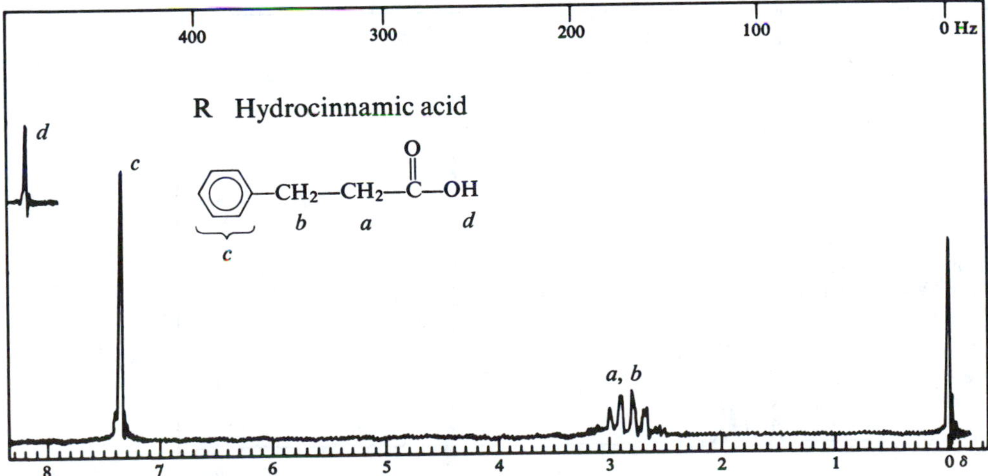

Figure 20.9 Proton NMR spectra for Problem 25, p. 789.

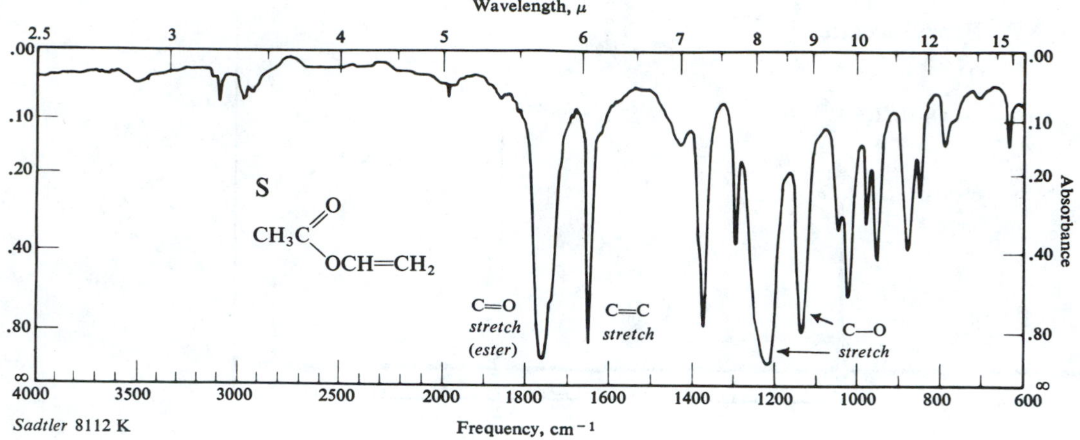

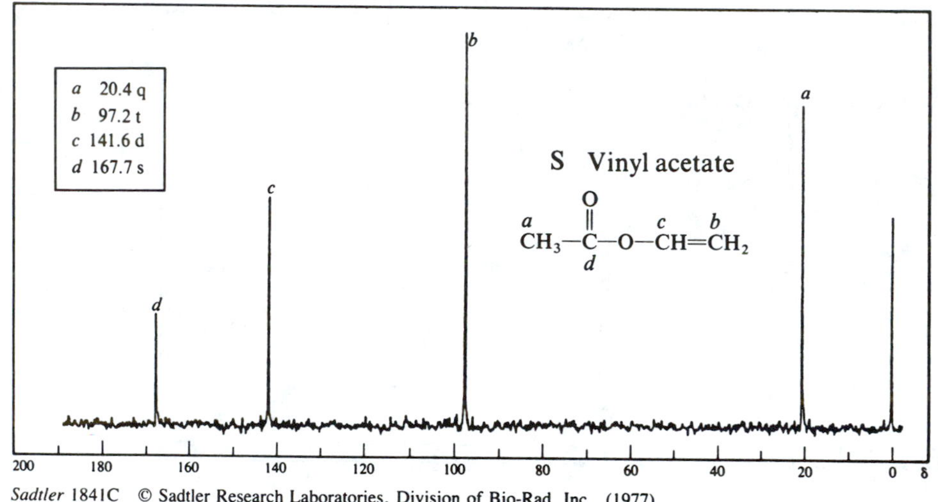

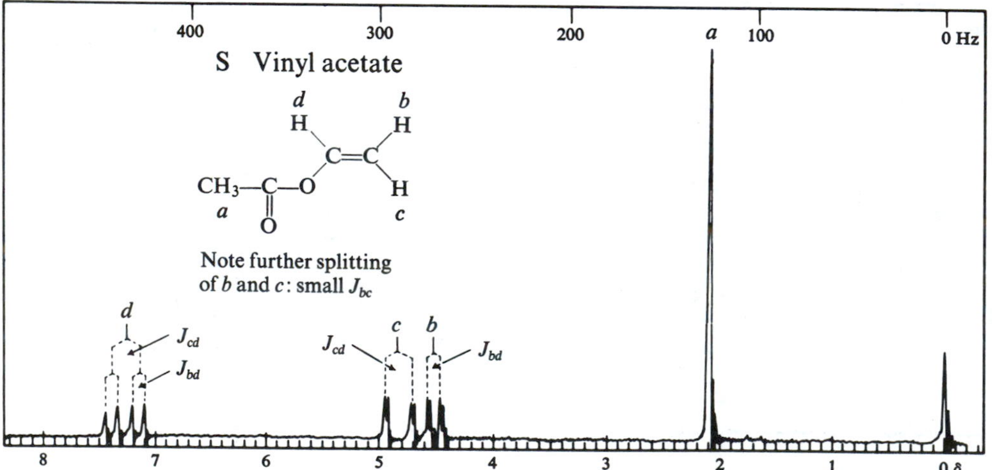

Figure 20.10 Infrared, CMR, and proton NMR spectra for Problem 26, p. 789.

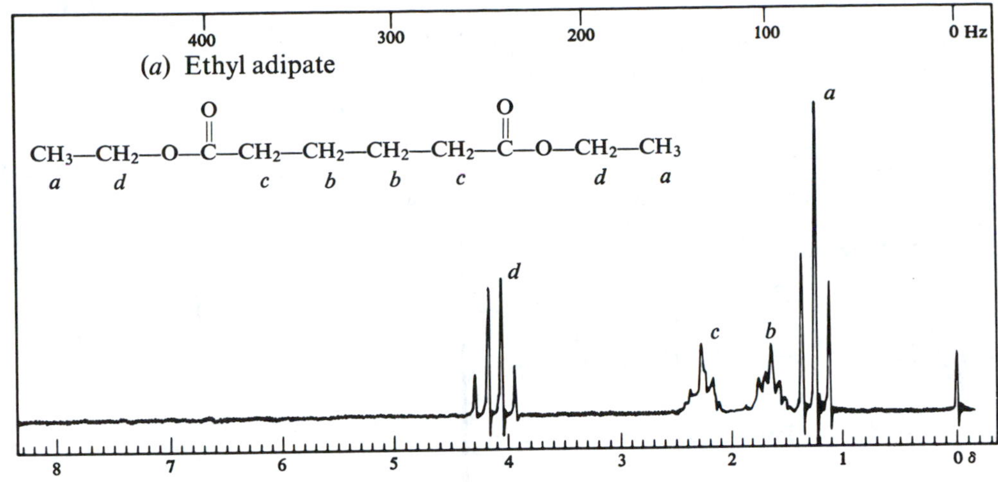

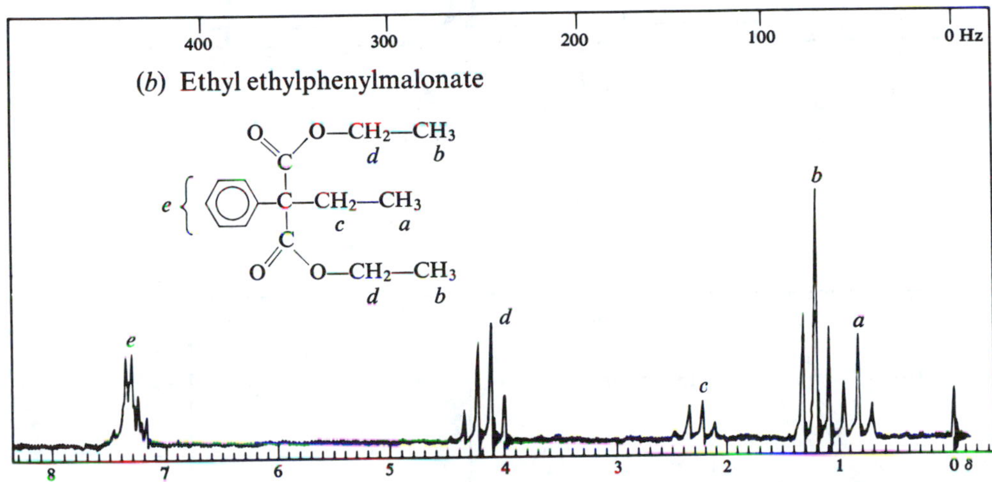

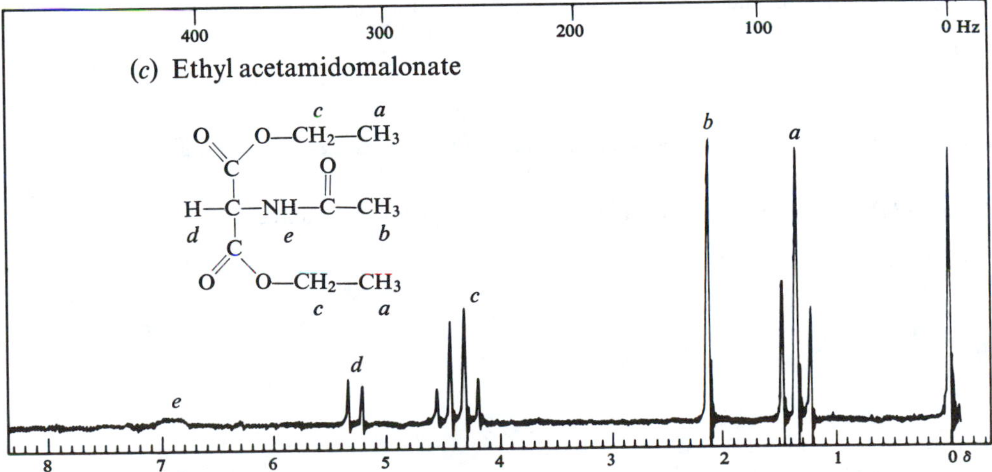

Figure 20.11 Proton NMR spectra for Problem 20.30, p. 361 of this Study Guide.

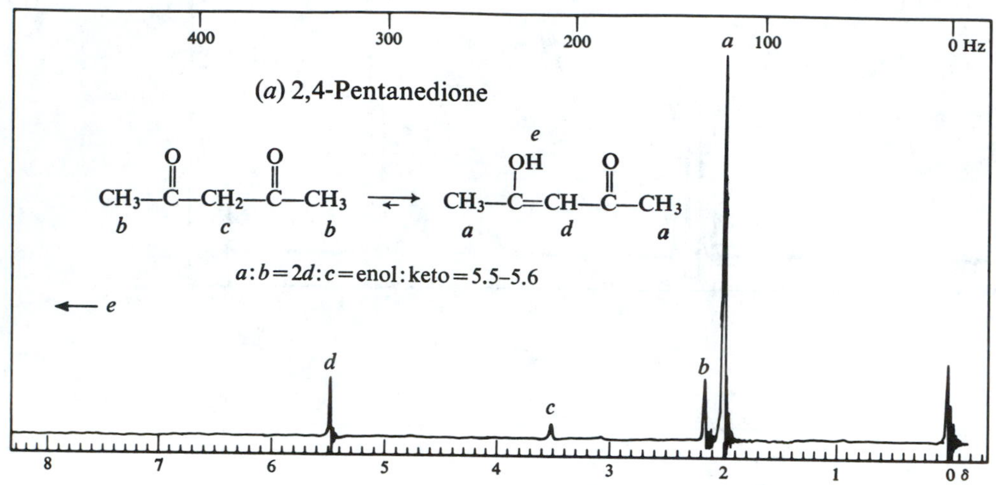

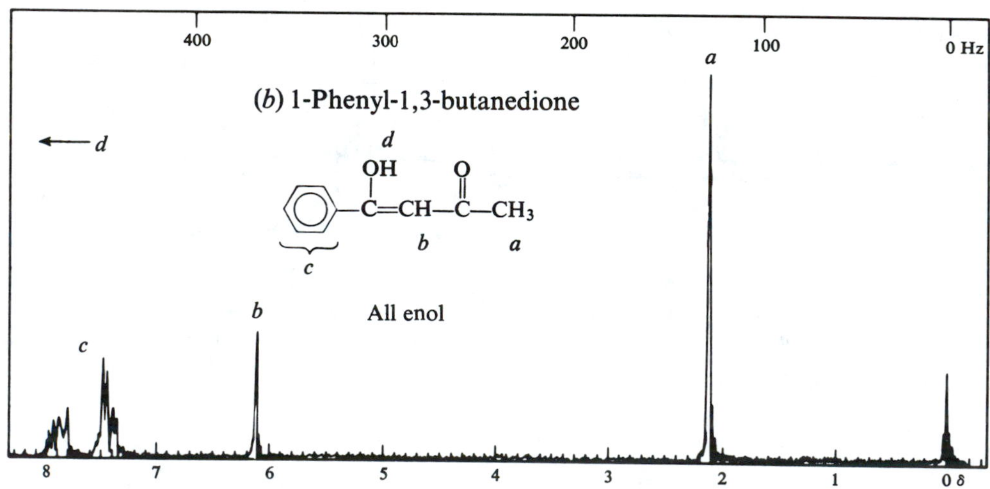

Figure 21.1 Proton NMR spectra of (*a*) 2,4-pentanedione and (*b*) 1-phenyl-1,3-butanedione.

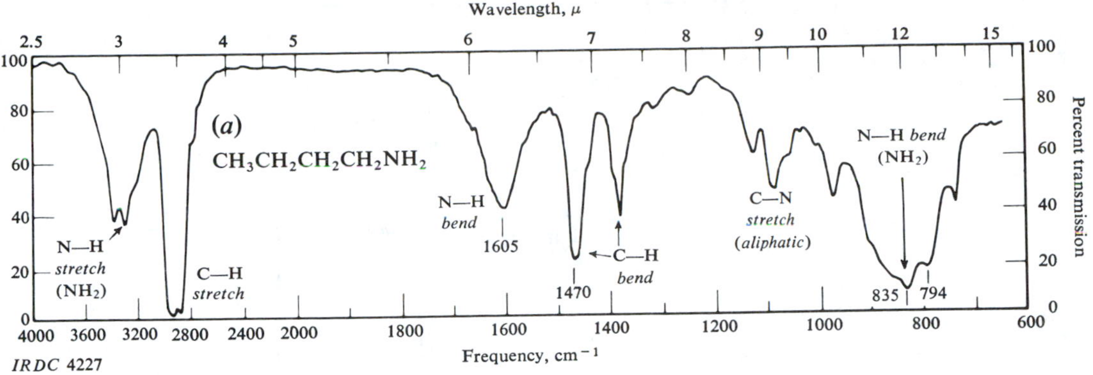

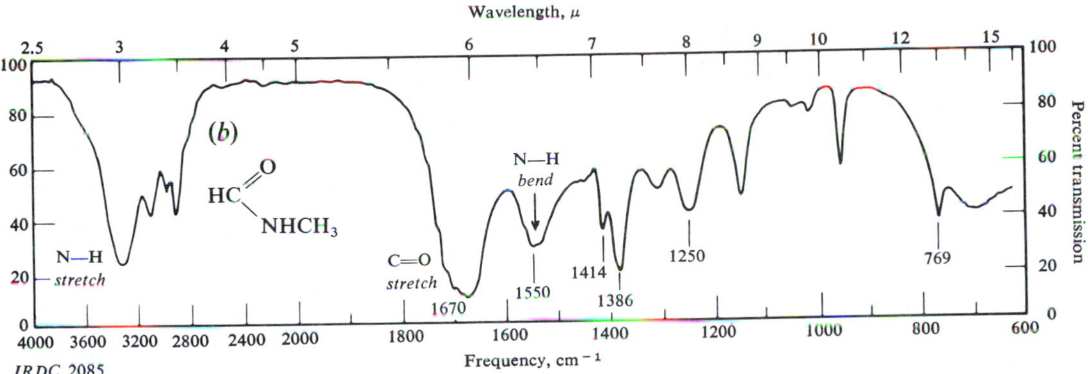

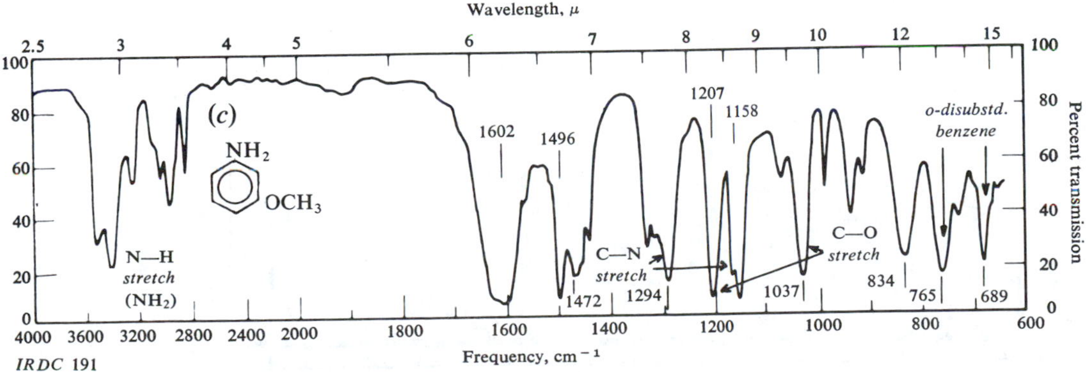

Figure 23.5 Infrared spectra for Problem 29, p. 883.

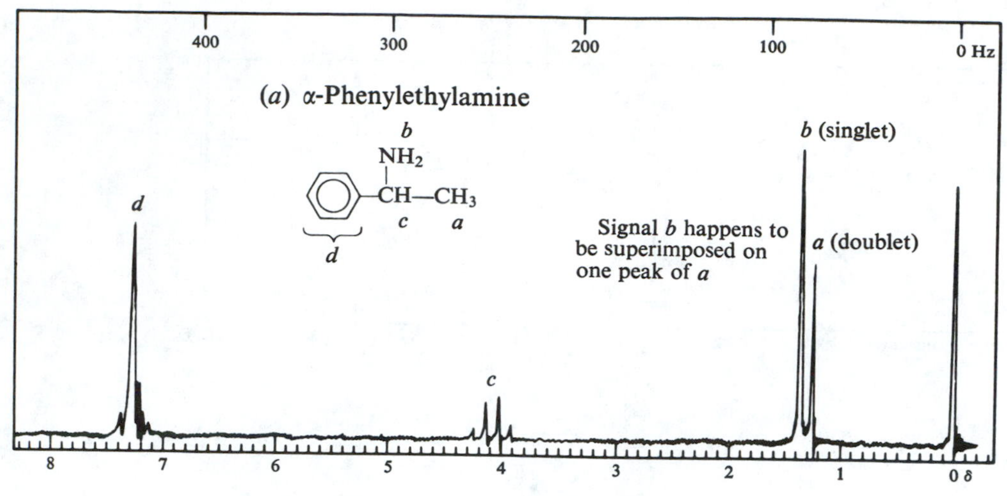

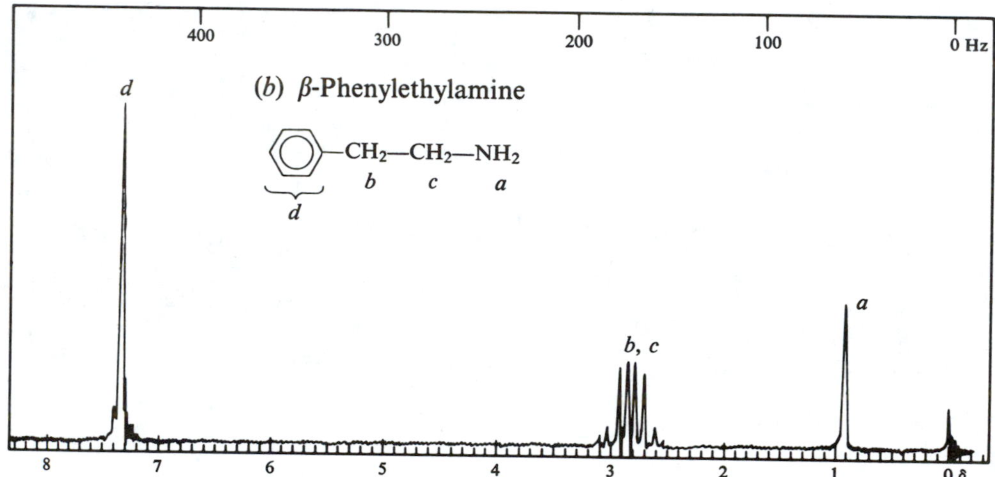

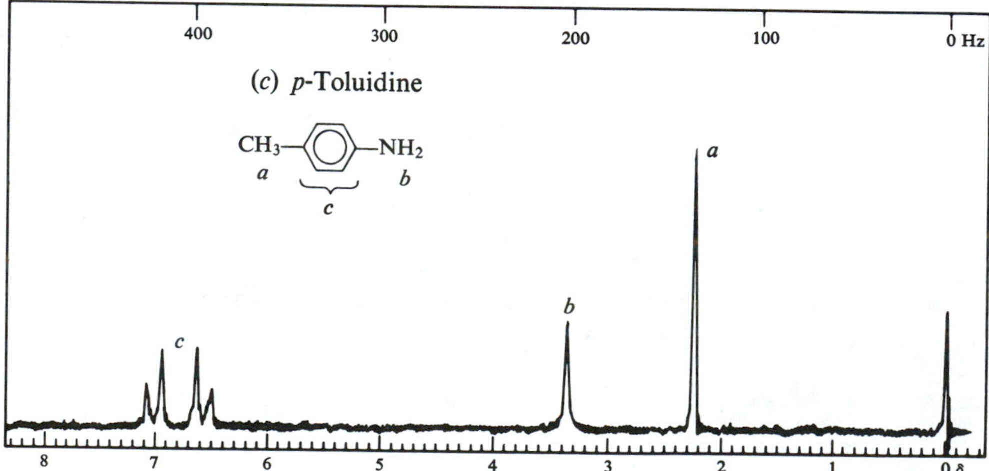

Figure 23.6 Proton NMR spectra for Problem 30, p. 883.

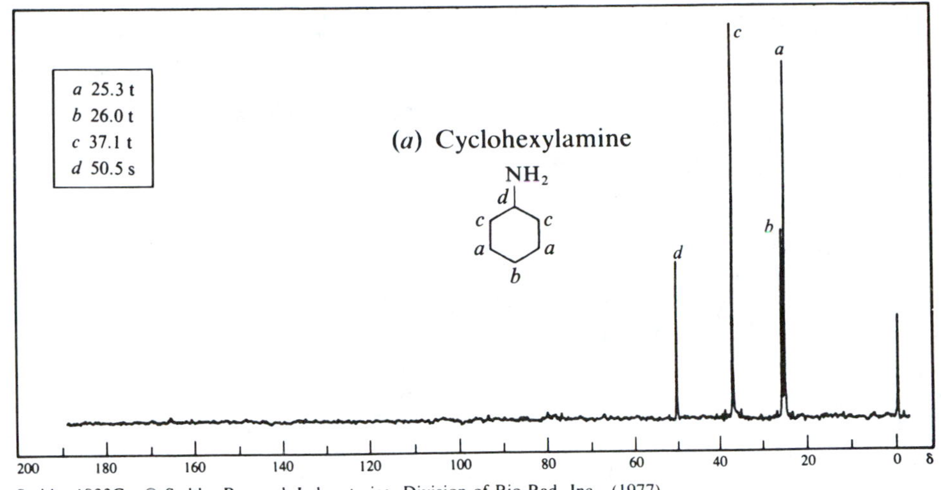

Sadtler 1833C © Sadtler Research Laboratories, Division of Bio-Rad, Inc., (1977).

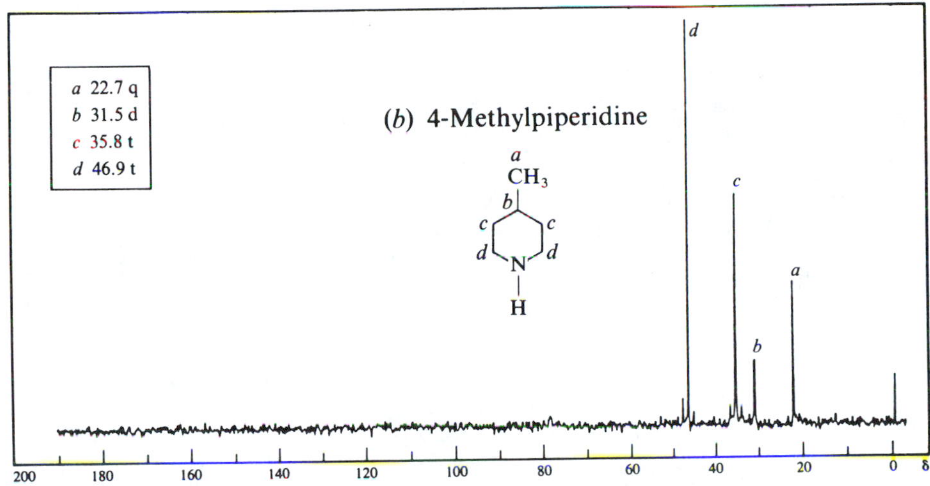

Sadtler 3072C © Sadtler Research Laboratories, Division of Bio-Rad, Inc., (1977).

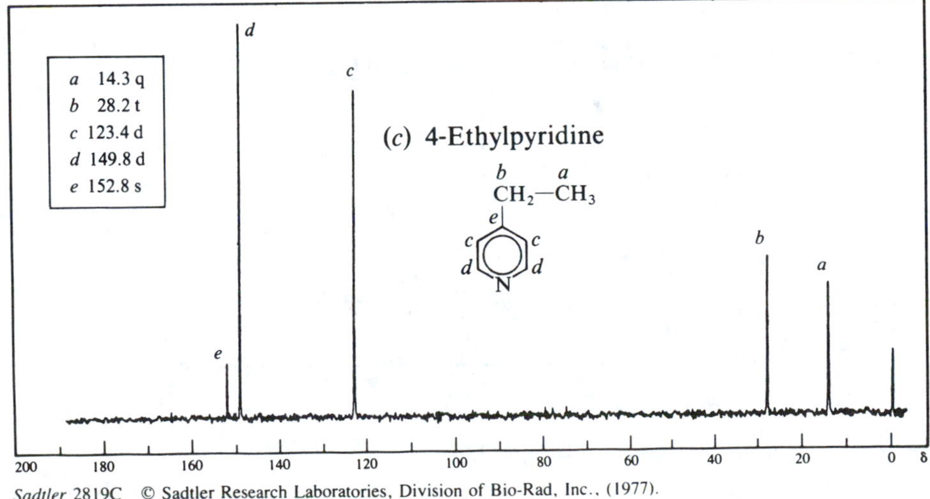

Sadtler 2819C © Sadtler Research Laboratories, Division of Bio-Rad, Inc., (1977).

Figure 23.7 CMR spectra for Problem 31, p. 883.

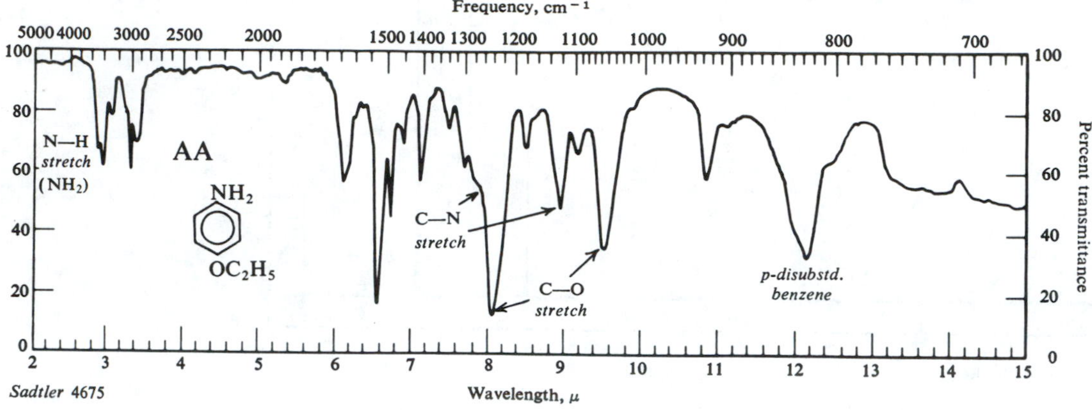

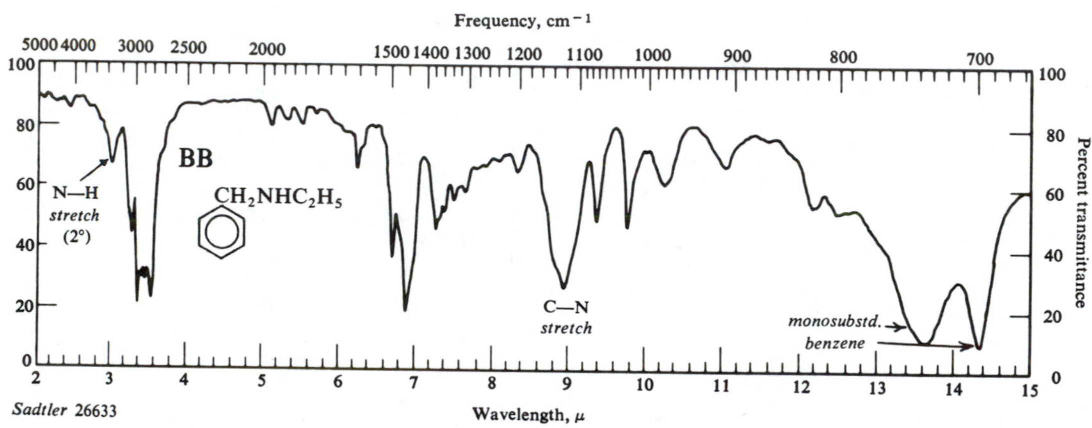

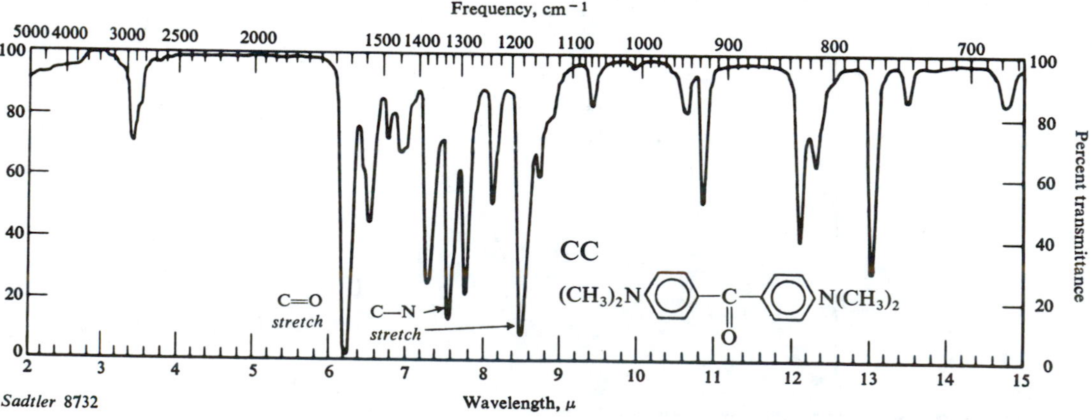

Figure 23.8 Infrared spectra for Problem 32, p. 883.

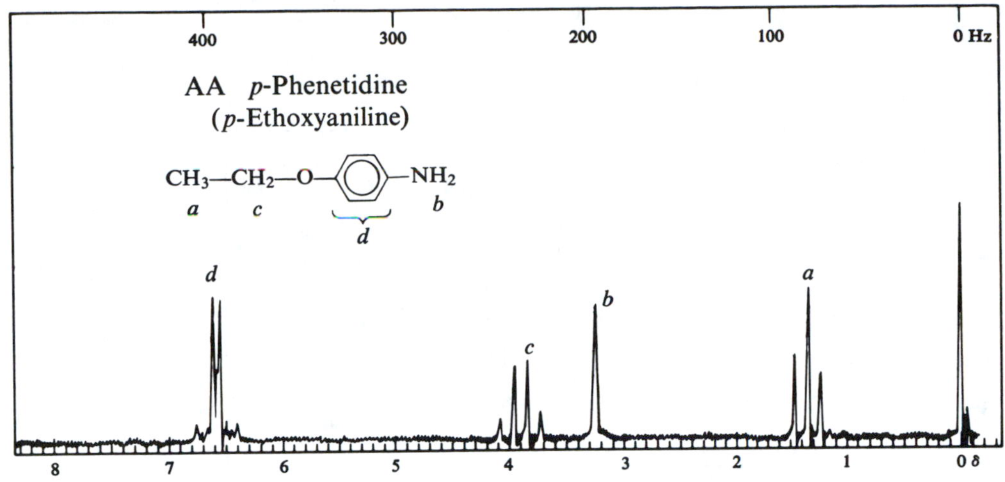

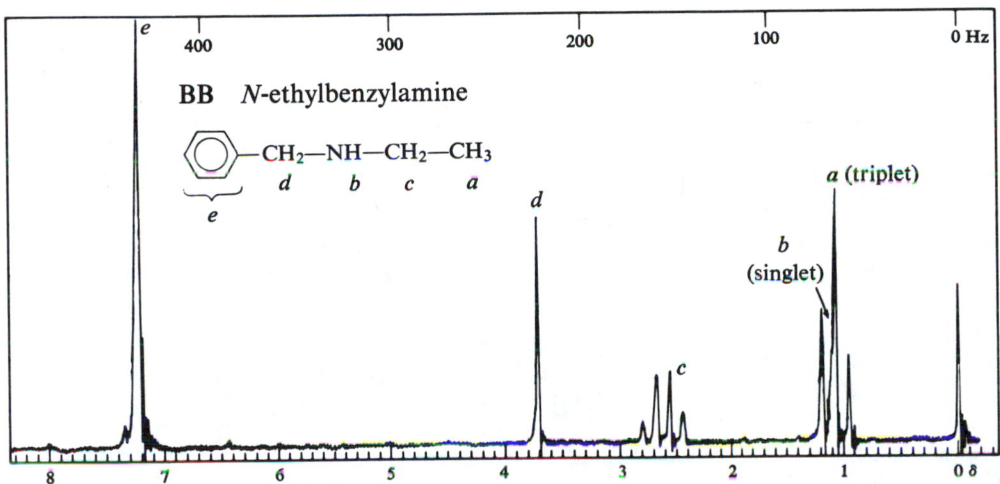

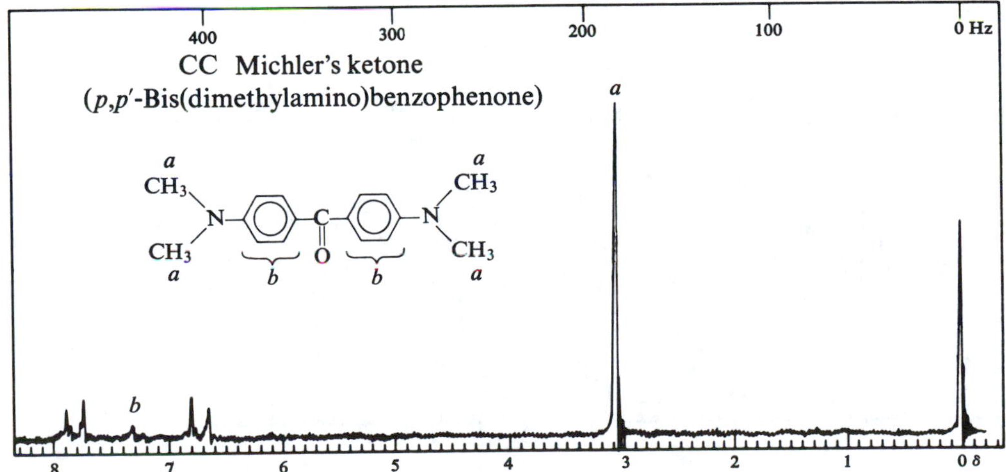

Figure 23.9 Proton NMR spectra for Problem 32, p. 883.

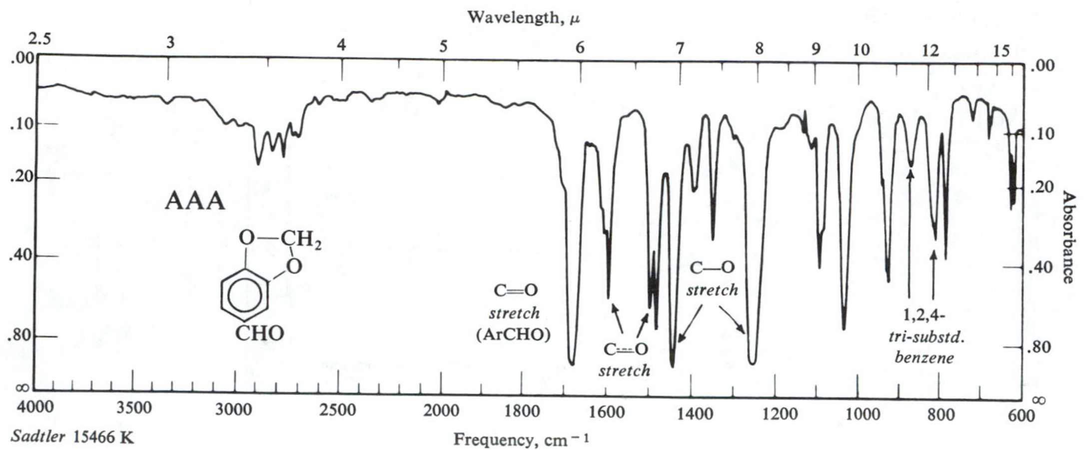

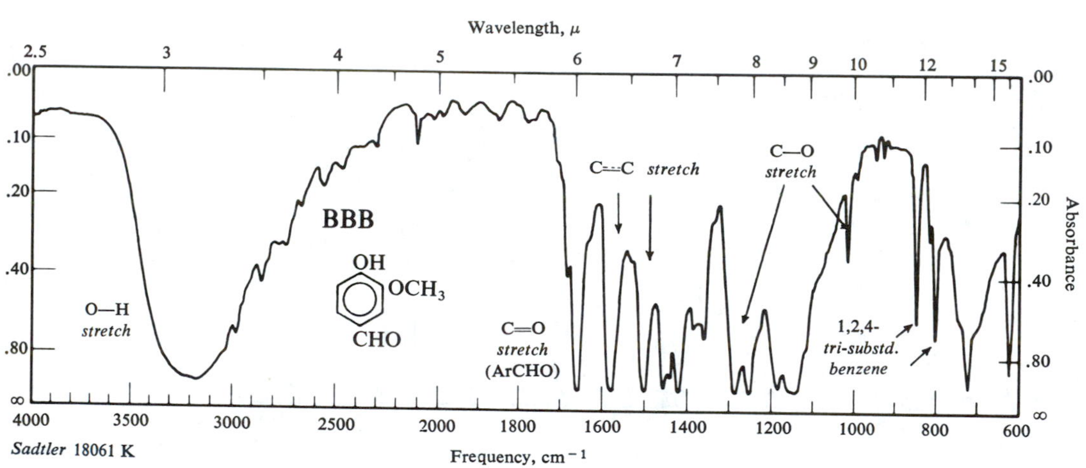

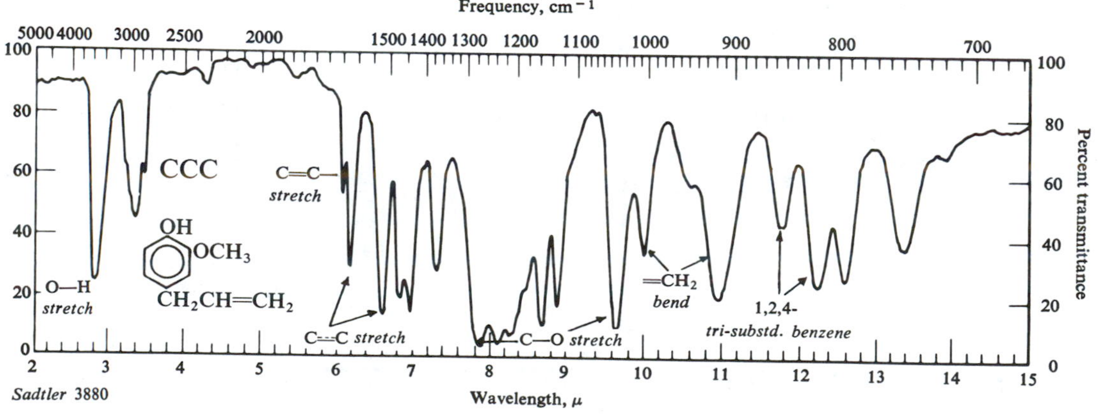

Figure 24.4 Infrared spectra for Problem 33, p. 919.

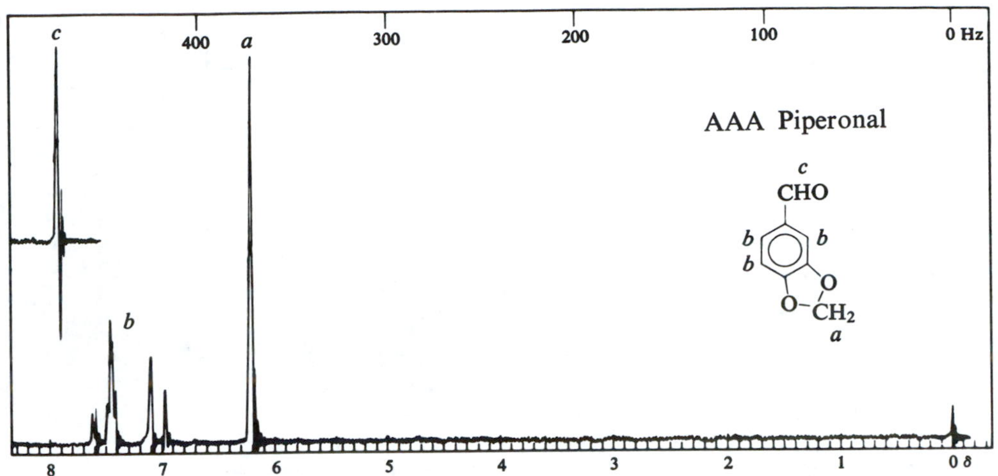

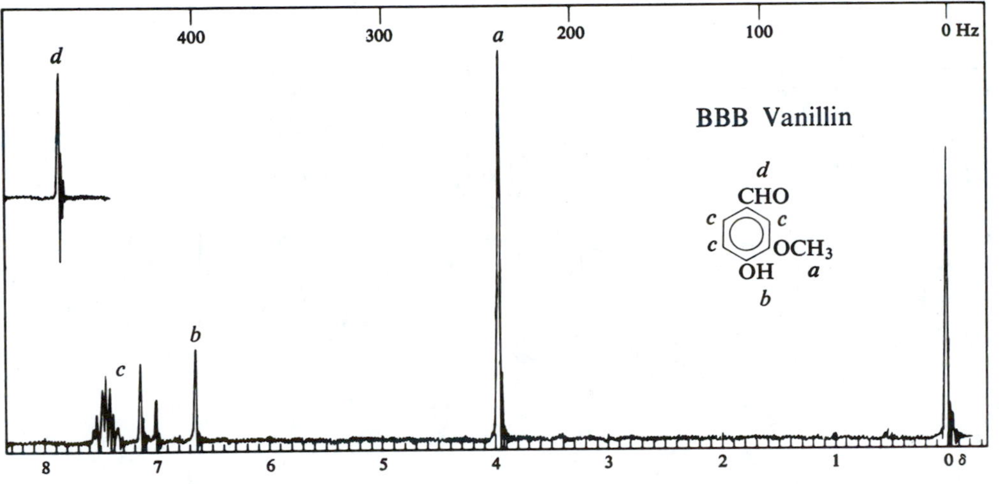

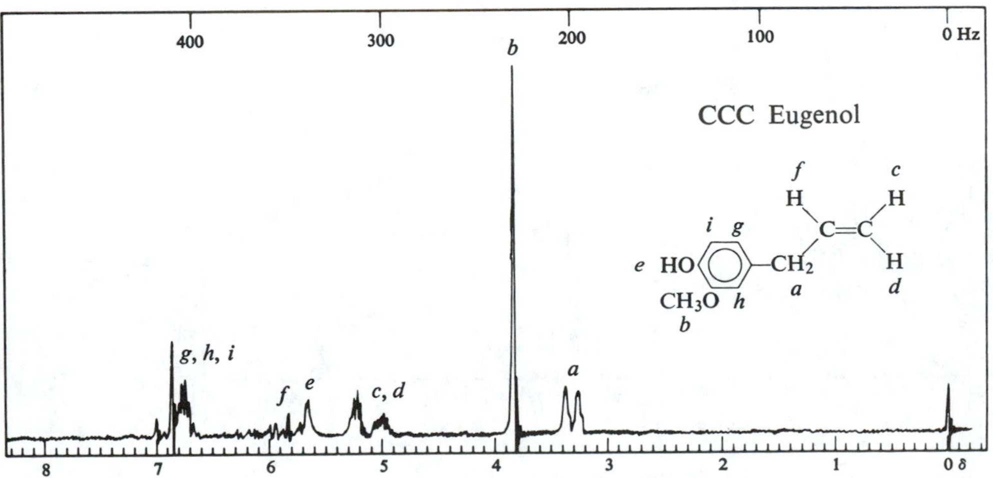

Figure 24.5 Proton NMR spectra for Problem 33, p. 919.

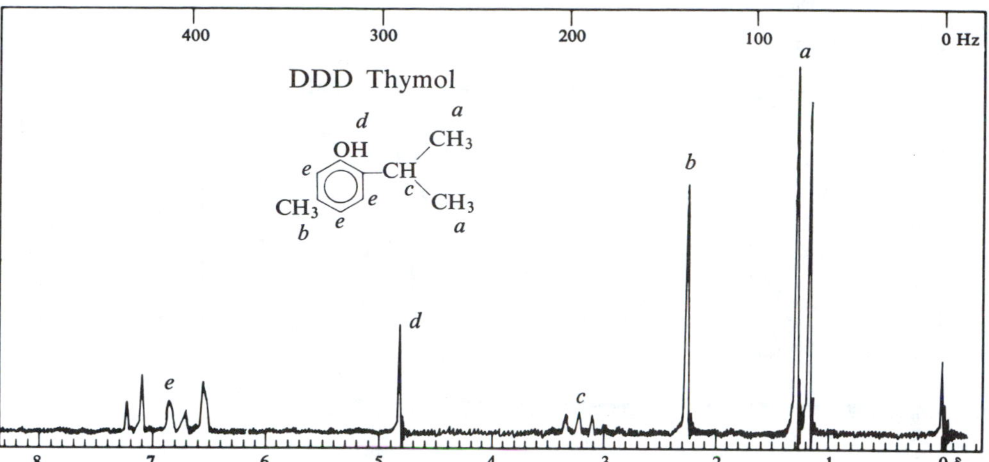

Figure 24.6 Proton NMR spectrum for Problem 33, p. 919.

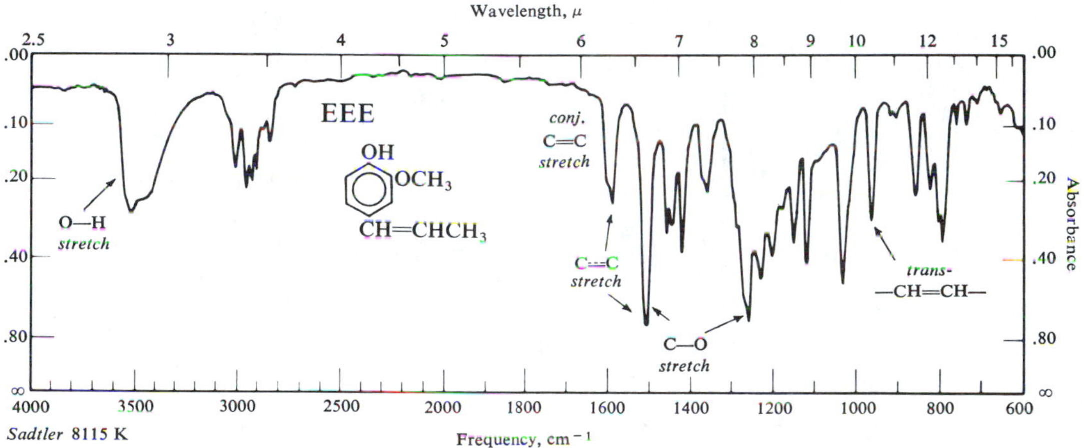

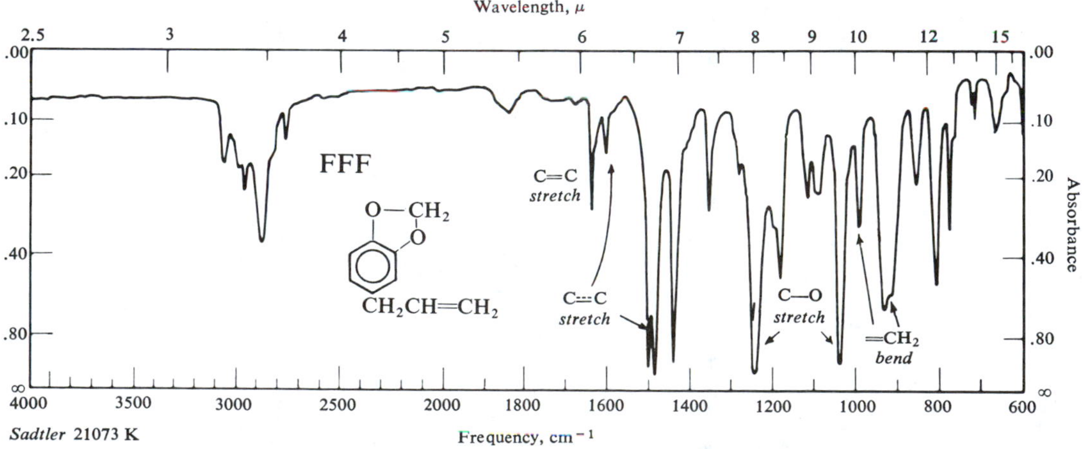

Figure 24.7 Infrared spectra for Problem 33, p. 919.

33. (Chapter 24, page 919.)

AAA and BBB show the C=O stretching band at 1700 cm^{-1}, and must be piperonal and vanillin, the only carbonyl compounds of the set; this is confirmed by the far downfield —CHO proton absorption in their NMR spectra. Of the two, BBB shows O—H stretching at 3200 cm^{-1}, and hence is vanillin; AAA is piperonal. These assignments are confirmed by the NMR spectra: proton counting (relative to the —CHO) reveals $—OCH_3$ (plus —OH) in contrast to $—OCH_2O—$; the two oxygens of $—OCH_2O—$ cause a much stronger downfield shift than the single oxygen of $—OCH_3$.

Of the remaining, CCC and EEE show O—H stretching, and must belong to the group of unassigned phenols: eugenol, isoeugenol, and thymol. Of the two, CCC shows the C=C stretching at 1650 cm^{-1} expected of an unconjugated C=C, and hence is eugenol; this is confirmed by the C—H out-of-plane bending bands at about 915 and 1000 cm^{-1}, characteristic of a terminal $=CH_2$. Compound EEE is, then, either isoeugenol or thymol. In contrast to thymol, isoeugenol has an unsaturated side chain, but C=C stretching in the conjugated system might well be hidden by the aromatic stretching band at 1600 cm^{-1}.

The NMR spectrum of DDD shows that it can be, of all seven possibilities, only thymol. The large doublet at δ 1.25 is too far upfield to be due to any allylic, vinylic, or alkoxy protons, and can be due only to the six β-protons of the isopropyl side chain of thymol. Inspection shows the entire spectrum to fit this structure neatly.

With thymol eliminated, EEE must now be isoeugenol. The shape of the 1600 cm^{-1} band hints at a hidden C=C absorption, and we see a band (about 965 cm^{-1}) where we would expect C—H bending for a *trans* —CH=CH—.

Finally, FFF shows no O—H band, and so must be, of the remaining possibilities, either safrole or anethole. The C=C stretching band at 1650 cm^{-1} indicates that the side chain is unconjugated, and hence that FFF is safrole; this is confirmed by the C—H bending bands (about 920 and 1000 cm^{-1}) expected of a terminal $=CH_2$.